Grid Computing
Making the Global Infrastructure a Reality

WILEY SERIES IN COMMUNICATIONS NETWORKING & DISTRIBUTED SYSTEMS.

Series Editor: David Hutchison, *Lancaster University*
Series Advisers: Harmen van As, *TU Vienna*
                          Serge Fdida, *University of Paris*
                          Joe Sventek, *Agilent Laboratories, Edinburgh*

The 'Wiley Series in Communications Networking & Distributed Systems' is a series of expert-level, technically detailed books covering cutting-edge research and brand new developments in networking, middleware and software technologies for communications and distributed systems. The books will provide timely, accurate and reliable information about the state-of-the-art to researchers and development engineers in the Telecommunications and Computing sectors.

Other titles in the series:

Grid Computing
Making the Global Infrastructure a Reality

Edited by

FRAN BERMAN
University of California, San Diego & San Diego Supercomputer Center, USA

GEOFFREY FOX
Community Grids Lab, Indiana University, USA

TONY HEY
Director e-Science Core Programme & University of Southampton, UK

WILEY

Other Wiley Editorial Offices

John Wiley & Sons Inc., 111 River Street, Hoboken, NJ 07030, USA

Jossey-Bass, 989 Market Street, San Francisco, CA 94103-1741, USA

Wiley-VCH Verlag GmbH, Boschstr. 12, D-69469 Weinheim, Germany

John Wiley & Sons Australia Ltd, 33 Park Road, Milton, Queensland 4064, Australia

John Wiley & Sons (Asia) Pte Ltd, 2 Clementi Loop #02-01, Jin Xing Distripark, Singapore 129809

John Wiley & Sons Canada Ltd, 22 Worcester Road, Etobicoke, Ontario, Canada M9W 1L1

Wiley also publishes its books in a variety of electronic formats. Some content that appears
in print may not be available in electronic books.

Library of Congress Cataloging-in-Publication Data

Grid computing : making the global infrastructure a reality / edited by Fran Berman,
  Geoffrey Fox, Tony Hey.
     p. cm. – (Wiley series in communications networking & distributed systems)
  Includes bibliographical references and index.
  ISBN 0-470-85319-0 (alk. paper)
     1. Computational grids (Computer systems) I. Berman, Fran. II. Fox, Geoffrey. III.
  Hey, Anthony J. G. IV. Series.

  QA76.9.C58G755 2003
  004′.36 – dc21

                                                                         2002192438

British Library Cataloguing in Publication Data

A catalogue record for this book is available from the British Library

ISBN 0-470-85319-0

Typeset in 10/12pt Times by Laserwords Private Limited, Chennai, India
Printed and bound in Great Britain by Antony Rowe Ltd, Chippenham, Wiltshire
This book is printed on acid-free paper responsibly manufactured from sustainable forestry
in which at least two trees are planted for each one used for paper production.

Contents

Table of Contents

List of Contributors

Sudesh Agrawal
Innovative Computing Lab.
University of Tennessee
1122 Volunteer Blvd.
Knoxville
Tennessee 37996-3450
United States

Walt Akers
Thomas Jefferson National
    Accelerator Facility
Newport News
Virginia 23606
United States

Gabrielle Allen
Max-Planck-Institut für
    Gravitationsphysik
Albert-Einstein-Institut
Am Mühlenberg 1
Golm D-14476
Germany

Rachana Ananthakrishnan
Department of Computer Science
Indiana University
Bloomington
Indiana 47405-7104
United States

Malcolm Atkinson
National e-Science Centre
15 South College Street
Edinburgh EH8 9AA

Scotland
United Kingdom

Mark A. Baker
Institute of Cosmology and Gravitation
University of Portsmouth
Mercantile House
Hampshire Terrace
Portsmouth, PO1 2EG
United Kingdom

Kim Baldridge
San Diego Supercomputer Center and
    University of California
San Diego
9500 Gilman Drive
La Jolla
California 92093-0114
United States

Chaitan Baru
San Diego Supercomputer Center and
    University of California
San Diego
9500 Gilman Drive
La Jolla
California 92093-0114
United States

Fran Berman
San Diego Supercomputer Center, and
    Department of Computer Science and
    Engineering
University of California

San Diego
9500 Gilman Drive
La Jolla
California 92093-0114
United States

John R. Boisseau
Texas Advanced Computing Center
The University of Texas at Austin
10100 Burnet Road
Austin
Texas 78758
United States

Philip E. Bourne
San Diego Supercomputer Center and
    Department of Pharmacology
University of California
San Diego
9500 Gilman Drive
La Jolla
California 92093-0537
United States

Mark Bradley
Department of Chemistry
University of Southampton
Southampton, SO17 1BJ
United Kingdom

Michael Brady
Medical Vision Laboratory
Department of Engineering Science
Oxford University
Parks Road
Oxford, OX1 3PJ
United Kingdom

John Brevik
Mathematics Department
Wheaton College
Norton

Massachusetts 02766
United States
                                          .

Todd Bryan
Computer Science Department
University of California
Santa Barbara
California 93106
United States

Julian J. Bunn
California Institute of Technology
Pasadena
California 91125
United States

Henri Casanova
Department of Computer Science and
    Engineering and San Diego
    Supercomputer Center
University of California
San Diego
9500 Gilman Drive
La Jolla
California 92093-0114
United States

Charles E. Catlett
Argonne National Laboratory
9700 Cass Avenue
Building 221
Argonne
Illinois 60439
United States

Jie Chen
Thomas Jefferson National
    Accelerator Facility
Newport News
Virginia 23606
United States

Ying Chen
Thomas Jefferson National
    Accelerator Facility
Newport News
Virginia 23606
United States

Andrew A. Chien
Entropia Inc.
10145 Pacific Heights
Suite 800
San Diego
California 92121
United States
and
Department of Computer Science and
    Engineering
University of California
San Diego
9500 Gilman Drive
La Jolla
California 92093
United States

David C. De Roure
Department of Electronics &
    Computer Science
University of Southampton
Southampton, SO17 1BJ
United Kingdom

Jack Dongarra
University of Tennessee
and
Oak Ridge National Laboratory
Tennessee
United States

Kattamuri Ekanadham
T. J. Watson Research Center
Yorktown Heights
New York 10598
United States

Jonathan W. Essex
Department of Chemistry
University of Southampton
Southampton, SO17 1BJ
United Kingdom

Adam J. Ferrari
Endeca Technologies Inc.
55 Cambridge Parkway
Cambridge
Massachusetts 02142
United States

Ian Foster
Mathematics and Computer
    Science Division
Argonne National Laboratory
9700 S. Cass Ave.
Argonne
Illinois 60439
United States
and
Department of Computer Science
The University of Chicago
Chicago
Illinois 60657
United States

Geoffrey Fox
Community Grids Lab
Indiana University
501 N. Morton Street
Suite 224
Bloomington
Indiana 47404
United States
and
Department of Computer Science
215 Lindley Hall
150 S. Woodlawn Ave.
Bloomington
Indiana 47405-7104
United States

and
School of Informatics and
   Physics Department
Indiana University
Indiana 47405
United States

Jeremy G. Frey
Department of Chemistry
University of Southampton
Southampton, SO17 1BJ
United Kingdom

Dennis Gannon
Department of Computer Science
Indiana University
215 Lindley Hall
150 S. Woodlawn Ave.
Bloomington
Indiana 47405-7104
United States

David Gavaghan
Computing Laboratory
Wolfson Building
Parks Road
Oxford
United Kingdom

Jarek Gawor
Argonne National Laboratory
9700 S. Cass Ave.
Argonne
Illinois 60439
United States

Jonathan Geisler
Argonne National Laboratory
9700 S. Cass Ave.
Argonne
Illinois 60439
United States

Tom Goodale
Max-Planck-Institut für
   Gravitationsphysik
Albert-Einstein-Institut
Am Mühlenberg 1
Golm D-14476
Germany

Madhusudhan Govindaraju
Department of Computer Science
Indiana University
Bloomington
Indiana 47405-7104
United States

Andrew S. Grimshaw
Department of Computer Science
University of Virginia
151 Engineer's Way
Charlottesville
Virginia 22904-4740
United States
and
CTO & Founder
Avaki Corporation
Burlington
Massachusetts
United States

Leanne P. Guy
IT Division – Database Group
CERN
1211 Geneva
Switzerland

Tomasz Haupt
Engineering Research Center
Mississippi State University
Starkville
Mississippi 39762
United States

Tony Hey
EPSRC
Polaris House
North Star Avenue
Swindon, SN2 1ET
United Kingdom
and
Department of Electronics and
    Computer Science
University of Southampton
Southampton, SO17 1BJ
United Kingdom

Ralph Highnam
Mirada Solutions Limited
Oxford Centre for Innovation
Mill Street
Oxford, OX2 0JX
United Kingdom

Wolfgang Hoschek
CERN IT Division
European Organization for
    Nuclear Research
1211 Geneva 23
Switzerland

Marty A. Humphrey
University of Virginia
Charlottesville
Virginia
United States

Michael B. Hursthouse
Department of Chemistry
University of Southampton
Southampton, SO17 1BJ
United Kingdom

Keith Jackson
Lawrence Berkeley National Laboratory
1 Cyclotron Road

Berkeley
California 94720
United States

Joefon Jann
T. J. Watson Research Center
Yorktown Heights
New York 10598
United States

Nicholas Jennings
Department of Electronics and
    Computer Science
University of Southampton
Southampton, SO17 1BJ
United Kingdom

William E. Johnston
Lawrence Berkeley National Laboratory
Berkeley
California 94720
United States
and
NASA Ames Research Center
Moffett Field
California 94035
United States

John F. Karpovich
Avaki Corporation
300 Preston Avenue, 5th Floor
Charlottesville
Virginia 22902
United States

Carl Kesselman
Information Sciences Institute
The University of Southern California
Marina del Rey
California 90292
United States

Sung-Hoon Ko
Community Grids Lab
Indiana University
Bloomington
Indiana 47404
United States

Sriram Krishnan
Department of Computer Science
150 S. Woodlawn Ave.
Bloomington
Indiana 47405
United States
and
Argonne National Laboratory
9700 S. Cass Ave.
Argonne
Illinois 60439
United States

Peter Z. Kunszt
IT Division – Database Group
CERN
1211 Geneva
Switzerland

Gregor von Laszewski
Argonne National Laboratory
9700 S. Cass Ave.
Argonne
Illinois 60439
United States

Craig Lee
Computer Systems Research
   Department
The Aerospace Corporation
P.O. Box 92957
Los Angeles
California 90009
United States

Sangmi Lee
Community Grids Lab
Indiana University
Bloomington
Indiana 47404
United States
and
Department of Computer Science
Florida State University
Tallahassee
Florida 32306
United States

Michael J. Lewis
Department of Computer Science
State University of New York at
   Binghamton
PO Box 6000
Binghamton
New York, 13902-6000
United States

Susan M. Lewis
Department of Mathematics
University of Southampton
Southampton, SO17 1BJ
United Kingdom

Miron Livny
Department of Computer Sciences
University of Wisconsin-Madison
1210 W. Dayton Street
Madison
Wisconsin 53705
United States

Michael M. Luck
Department of Mathematics
University of Southampton
Southampton, SO17 1BJ
United Kingdom

Vijay Mann
The Applied Software
  Systems Laboratory
Department of Electrical and Computer
  Engineering
Rutgers, The State University of
  New Jersey
94 Brett Road
Piscataway
New Jersey 08854
United States

Satoshi Matsuoka
Tokyo Institute of Technology
Global Scientific Information and
  Computing Center
2-12-1 Ookayama
Meguro-ku
Tokyo 152-8550
Japan

Luc Moreau
Department of Mathematics
University of Southampton
Southampton, SO17 1BJ
United Kingdom

Reagan Moore
San Diego Supercomputer Center and
  University of California
San Diego
9500 Gilman Drive
La Jolla
California 92093-0114
United States

Mark M. Morgan
Avaki Corporation
300 Preston Avenue, 5th Floor
Charlottesville
Virginia 22902
United States

Hidemoto Nakada
National Institute of Advanced Industrial
  Science and Technology
Grid Technology Research Center
Tsukuba Central 2
1-1-1 Umezono
Tsukuba
Ibaraki 305-8568
Japan
and
Tokyo Institute of Technology
Global Scientific Information and
  Computing Center
2-12-1 Ookayama
Meguro-ku
Tokyo 152-8550
Japan

Anand Natrajan
Avaki Corporation
300 Preston Avenue, 5th Floor
Charlottesville
Virginia 22902
United States

Harvey B. Newman
Department of Physics
California Institute of Technology
1201 E. California Blvd.
Pasadena
California 91125
United States

Anh Nguyen-Tuong
Avaki Corporation
300 Preston Avenue, 5th Floor
Charlottesville
Virginia 22902
United States

Jeffrey M. Nick
IBM Corporation

Poughkeepsie
New York 12601
United States

Bill Nickless
Argonne National Laboratory
9700 S. Cass Ave.
Argonne
Illinois 60439
United States

Jason Novotny
1 Cyclotron Road
Lawrence Berkeley National Laboratory
Berkeley
California 94704
United States

Shrideep Pallickara
Community Grids Lab
Indiana University
Bloomington
Indiana 47404
United States

Miguel Mulet Parada
Mirada Solutions Limited
Oxford Centre for Innovation
Mill Street
Oxford
United Kingdom

Manish Parashar
The Applied Software
  Systems Laboratory
Department of Electrical and Computer
  Engineering
Rutgers, The State University of
  New Jersey
94 Brett Road
Piscataway
New Jersey 08854
United States

Pratap Pattnaik
T. J. Watson Research Center
Yorktown Heights
New York 10598
United States

Marlon Pierce
Community Grids Lab
Indiana University
Bloomington
Indiana 47404
United States

James S. Plank
Computer Science Department
University of Tennessee
Knoxville
Tennessee 37996
United States

Xiaohong Qiu
EECS Department
Syracuse University
Syracuse
New York 13244
United States
and
Community Grids Lab
Indiana University
Bloomington
Indiana 47404
United States

Lavanya Ramakrishnan
Department of Computer Science
Indiana University
Bloomington
Indiana 47405-7104
United States

Xi Rao
Community Grids Lab

Indiana University
Bloomington
Indiana 47404
United States
and
Department of Computer Science
Indiana University
Bloomington
Indiana 47405
United States

Michael Russell
Max-Planck-Institut für
    Gravitationsphysik
Albert-Einstein-Institut
Am Mühlenberg 1
Golm D-14476
Germany

Edward Seidel
Max-Planck-Institut für
    Gravitationsphysik
Albert-Einstein-Institut
Am Mühlenberg 1
Golm D-14476
Germany

Satoshi Sekiguchi
National Institute of Advanced Industrial
    Science and Technology
Grid Technology Research Center
Tsukuba Central 2
1-1-1 Umezono
Tsukuba
Ibaraki 305-8568
Japan

Keith Seymour
University of Tennessee
306 Claxton Complex
1122 Volunteer Blvd.
Knoxville

Tennessee 37996
United States

Nigel R. Shadbolt
Department of Electronics and
    Computer Science
University of Southampton
Southampton, SO17 1BJ
United Kingdom

John Shalf
Lawrence Berkeley National Laboratory
Berkeley
California
United States

Andrew Simpson
Computing Laboratory
Wolfson Building
Parks Road
Oxford
United Kingdom

Aleksander Slominski
Department of Computer Science
Indiana University
Bloomington
Indiana 47405-7104
United States

Warren Smith
Argonne National Laboratory
9700 S. Cass Ave.
Argonne
Illinois 60439
United States

Larry Smarr
Cal-(IT)^2
University of California, San Diego
EBU1, 7th Floor

9500 Gilman Drive
La Jolla
California 92093-0405
United States

David Snelling
Fujitsu Laboratories of Europe
Hayes Park Central
Hayes End Road
Hayes
Middlesex, UB4 8FE
United Kingdom

Mike Surridge
Operations Director
IT Innovation Centre
2 Venture Road
Chilworth Science Park
Southampton, SO16 7NP
United Kingdom

Domenico Talia
DEIS
Università della Calabria
Via P. Bucci
cubo 41 c
87036 Rende, CS
Italy

Yoshio Tanaka
National Institute of Advanced Industrial
    Science and Technology
Grid Technology Research Center
Tsukuba Central 2
1-1-2 Umezono
Tsukuba
Ibaraki 305-8568
Japan

Todd Tannenbaum
Department of Computer Sciences
University of Wisconsin-Madison

1210 W. Dayton Street
Madison
Wisconsin 53706-1685
United States

Douglas Thain
Department of Computer Sciences
University of Wisconsin-Madison
1210 W. Dayton Street
Madison
Wisconsin 53705
United States

Mary P. Thomas
Texas Advanced Computing Center
The University of Texas at Austin
10100 Burnet Road
Austin
Texas 78758
United States

Anne Trefethen
EPSRC
Polaris House
North Star Avenue
Swindon, SN2 1ET
United Kingdom

Steven Tuecke
Mathematics and Computer
    Science Division
Argonne National Laboratory
9700 S. Cass Ave.
Argonne
Illinois 60439
United States

Ahmet Uyar
EECS Department
Syracuse University
Syracuse
New York 13244

United States
and
Community Grids Lab
Indiana University
Bloomington
Indiana 47404
United States

Sathish Vadhiyar
University of Tennessee
Tennessee
United States

Minjun Wang
EECS Department
Syracuse University
Syracuse
New York 13244
United States
and
Community Grids Lab
Indiana University
Bloomington
Indiana 47404
United States

Paul Watson
School of Computing Science
University of Newcastle
Newcastle upon Tyne, NE1 7RU
United Kingdom

William A. Watson III
Thomas Jefferson National
   Accelerator Facility
Newport News
Virginia 23606
United States

Alan H. Welsh
Department of Mathematics
University of Southampton
Southampton, SO17 1BJ
United Kingdom

Roy Williams
Caltech Center for Advanced
   Computing Research
California Institute of Technology
California
United States

Rich Wolski
Computer Science Department
University of California
Santa Barbara
California 93106
United States

Wenjun Wu
Community Grids Lab
Indiana University
Bloomington
Indiana 47404
United States

About the Editors

Francine Berman is a pioneer in Grid Computing and a leader in the international effort to build a comprehensive modern infrastructure to support research in science and engineering. Dr Berman serves as director for the San Diego Supercomputer Center, director of the National Partnership for Advanced Computational Infrastructure, and holds the High Performance Computing Chair in the Computer Science and Engineering Department at U.C. San Diego. Professor Berman was elected  a Fellow of the ACM in 1993. Her research over the last two decades has focused on High Performance and Grid Computing, in particular, in the areas of programming environments, adaptive middleware, scheduling and performance prediction. Most recently, Dr Berman has led or co-led the AppLeS (Application-Level Scheduling) Project, the design and development of adaptive middleware for Grid environments, the large NSF "Virtual Instrument/MCell" ITR project, and other research projects in the area of Grid Computing.

Dr Berman currently serves as director of the San Diego Supercomputer Center (SDSC), a national facility and organized research unit of the University of California, San Diego, devoted to advancing science through the development and use of technology. She is director of the National Partnership for Advanced Computational Infrastructure, a consortium of over 40 research groups, institutions, and university partners with the goal of building a national infrastructure to improve and extend the reach of science and engineering. Dr Berman is one of the principal investigators of the NSF-supported TeraGrid (DTF) and Extensible Terascale Facility (ETF), the largest coordinated US Grid development and deployment project to date. Dr Berman serves on numerous technical, advisory, conference, and scientific committees providing expertise in the areas of parallel computing and Grid Computing.

Geoffrey C. Fox, well-known for his work in the development and application of parallel computers, is currently a Professor of Computer Science, Informatics and Physics at the Bloomington campus of Indiana University. Previously he served as director of the Computational and Information Science Laboratory at Florida State University. He was also the director of the Northeast Parallel Architectures Center at Syracuse University from 1990 to 2000. His research has led to two commercial spin-offs – WebWisdom.com and Anabas Inc.

Geoffrey Fox has worked with many distinguished scientists in his career; while professor of physics at CalTech, he worked with co-editor Tony Hey and Nobel Prize winner in physics, Richard Feynman. His current projects include developing the Online Knowledge Center for the Department of Defense High Performance Computing and Modernization Program, which is creating technology to allow users to more easily access and update information on high-performance computing.

Professor Fox is editor of the Wiley journal *Concurrency and Computation: Practice and Experience* which publishes many Grid-related papers. Fox's Grid research is discussed in Chapters 18 and 30 of this book.

Tony Hey is Professor of Computation at the University of Southampton and has been Head of the Department of Electronics and Computer Science and Dean of Engineering and Applied Science at Southampton. From March 31, 2001, he has been seconded to the EPSRC and DTI as director of the UK's Core e-Science Programme. He is a Fellow of the Royal Academy of Engineering, the British Computer Society and the Institution of Electrical Engineers. Professor Hey is European editor of the journal *Concurrency and Computation: Practice and Experience* and is on the organising committee of many international conferences.

Professor Hey has worked in the field of parallel and distributed computing since the early 1980s. He was instrumental in the development of the MPI message-passing standard and in the Genesis Distributed Memory Parallel Benchmark suite. In 1991, he founded the Southampton Parallel Applications Centre (now the IT Innovation Centre) that has played a leading technology transfer role in Europe and the United Kingdom in collaborative industrial projects. His personal research interests are concerned with performance engineering for Grid applications but he also retains an interest in experimental explorations of quantum computing and quantum information theory. He also edited *The Feynman Lectures on Computation* and is co-author of two best-selling popular science books: *The Quantum Universe* and *Einstein's Mirror*. As the director of the UK e-Science Programme, Tony Hey is currently excited by the vision of the increasingly global scientific collaborations being enabled by the development of the next-generation 'Grid' middleware. The successful development of the Grid will have profound implications for industry and he is much involved with industry in the move towards OpenSource/OpenStandard Grid software.

Foreword

The prospect of a global infrastructure for Grid computing becoming a reality is a most enticing one.

Grid computing has been identified as an important new technology by a remarkable breadth of scientific and engineering fields as well as by many commercial and industrial enterprises. This widespread adoption of the Grid computing paradigm has taken place very rapidly, even faster than was the case for the web. In only a decade since the formulation of the first concepts that led to the Grid, there are scores of Grid computing projects underway – or in the planning stages – in dozens of countries, and there are even a few production Grids for both research and commercial applications. What makes Grid computing such a compelling concept?

Grid computing enables or facilitates the conduct of virtual organizations – geographically and institutionally distributed projects – and such organizations have become essential for tackling many projects in commerce and research. With Grid computing, one can readily bring to bear the most appropriate and effective human, information, and computing resources for tackling highly complex and multidisciplinary projects.

In commerce, Grids will facilitate the integration of efforts across large enterprises as well as the contributions of contractors for projects of finite duration. New services may be provided in health care as well as medical research.

It is becoming apparent that the use of Grids will be an enabler for major advances and new ways of doing science. Grids have the potential to integrate as never before the triad of scientific methods – theory, experiment, and computation – and to do so on a global scale. Gaining access to the huge data collections that are being assembled and curated by many disciplines has become a major driver for the use of Grids. Unlike computing power, such data archives are not so readily replicated at each user site; hence they must be accessed remotely. Further, multidisciplinary investigations often require the simultaneous access of multiple data collections, each of which is likely to be in a different location. In some cases, analysis of the data acquired from those collections requires powerful computer systems that are in another location, and the visualization of the results of the analysis might require the use of a system at yet another site. Grids also provide a way to increase greatly the number of individuals who analyze observational data, to facilitate telecollaboration, and to provide broader access to unique experimental or computational facilities.

Grid computing is not yet mature. There are many open issues to be addressed and missing functionality to be developed, and more will emerge as uses of computing Grids proliferate. A number of chapters in this book describe the challenges that remain. However, there are grounds for optimism that Grid computing will evolve to be the highly useful technology that it promises to be. The commercial and research applications that are driving the Grid are also providing the intellectual and financial resources that will lead to more and more production applications of Grid computing. Another positive sign is the growing interest in the computer science community in research related to Grid computing. Unlike traditional scientific computing, creation and use of Grids involve a number of mainstream computer science topics and issues, such as database technology, digital libraries, cybersecurity, ontologies, semantic webs, and web services.

This book is unique in that it provides a broad, up-to-date picture of a large fraction of the Grid computing activities worldwide, both those that are developing Grid technologies and those that are applying those technologies. With such a broad spectrum of current Grid computing activities described in one place, it will be a valuable resource for anyone who wants be become involved in Grid computing technologies as a user or a developer. For those who simply want to learn what Grid computing is, there are overview chapters that provide highly accessible introductions to Grid computing and its applications. The history, promise, and challenges of the creation of a global Grid computing infrastructure are portrayed in the words of many of the pioneers in the field. As one who has followed the field since its beginnings, I found the contents of this tome to provide impressive testimony of how Grid computing has blossomed and how many exciting activities are underway.

Paul Messina
Argonne National Laboratory and
European Organization for Nuclear Research (CERN)

PART A

Overview and motivation

Overview of the book: Grid computing – making the global infrastructure a reality

Fran Berman,[1,2] Geoffrey Fox,[3] and Tony Hey[4,5]

[1]*San Diego Supercomputer Center, and Department of Computer Science and Engineering, University of California, San Diego, California, United States,* [2]*Indiana University, Bloomington, Indiana, United States,* [3]*EPSRC, Swindon, United Kingdom,* [4]*University of Southampton, Southampton, United Kingdom*

SUMMARY OF THE BOOK

This book, *Grid Computing: Making the Global Infrastructure a Reality*, [1] brings together many of the major projects that are driving and shaping an emerging global Grid. In the chapters of this book you will find the perspectives of a pioneering group of Grid developers, researchers and application scientists whose vision forms the present and provides a view into the future of Grid computing.

Many of the chapters in this book provide definitions and characterizations of the Grid – peruse these and you will form your own view. Common to all perspectives is the notion that the Grid supports the integration of resources (computers, networks, data archives, instruments etc.). To build an integrated system, individuals and communities

Grid Computing – Making the Global Infrastructure a Reality. Edited by F. Berman, A. Hey and G. Fox
© 2003 John Wiley & Sons, Ltd ISBN: 0-470-85319-0

are working in coordination to ensure that the large and complex system of Grid software will be robust, useful, and provide an interoperable collection of services that support large-scale distributed computing and data management.

Grids are intrinsically distributed and heterogeneous but must be viewed by the user (whether an individual or another computer) as a virtual environment with uniform access to resources. Much of Grid software technology addresses the issues of resource scheduling, quality of service, fault tolerance, decentralized control and security and so on, which enable the Grid to be perceived as a single virtual platform by the user. For example, Grid security technologies must ensure that a single sign-on will generate security credentials that can be used for the many different actions (potentially on multiple resources) that may be needed in a Grid session. For some researchers and developers, the Grid is viewed as the future of the Web or Internet computing. The Web largely consists of individuals talking independently to servers; the Grid provides a collection of servers and clients often working collectively together to solve a problem. Computer-to-computer traffic is characteristic of Grids and both exploits and drives the increasing network backbone bandwidth.

The term *Grid* was originally used in analogy to other infrastructure (such as electrical power) grids. We will see in this book that the analogy correctly characterizes some aspects of Grid Computing (ubiquity, for example), but not others (the performance variability of different resources that can support the same code, for example).

There are several other places to find out about Grids, and given the increasing popularity and importance of this technology, additional literature will be added at a rapid rate. Reference [2] is the seminal book on Grids edited by Foster and Kesselman; Reference [3] is a link to the Global Grid Forum, an important community consortium for Grid activities, standardization, and best practices efforts. Reference [4] is the home page of the Globus project, which has had and continues to have a major influence on Grid Computing. References [5–9] are a selection of Web information sources for the Grid. The homepage of our book in Reference [10] will supplement the book with references, summaries and notes.

This book is a collection of separate chapters written by pioneers and leaders in Grid Computing. Different perspectives will inevitably lead to some inconsistencies of views and notation but will provide a comprehensive look at the state of the art and best practices for a wide variety of areas within Grid Computing. Most chapters were written especially for this book in the summer of 2002. A few chapters are reprints of seminal papers of technical and historical significance.

The Grid has both a technology pull and an application push. Both these aspects and their integration (manifested in the activity of programming the Grid) are covered in this book, which is divided into four parts. Chapters 1 to 5 (Part A) provide a basic motivation and overview of where we are today and how we got there. Part B, Chapters 6 to 19, covers the core Grid architecture and technologies. Part C, Chapters 20 to 34, describes how one can program the Grid and discusses Grid computing environments; these are covered both from the view of the user at a Web portal and of a computer program interacting with other Grid resources. Part D, Chapters 35 to 43, covers Grid applications – both ongoing and emerging applications representing the broad scope of programs that can be executed on Grids.

In the following sections, we will discuss the chapters of this book in more detail so that readers can better plan their perusal of the material. Extensive external citations are not given here but will be found in the individual chapters and on our book's home page [10].

PART A: OVERVIEW

Chapters 1 and 2 describe the Grid in broad terms, providing examples and motivation; Chapter 1 also gives an overview of the first two sections, Chapters 1 to 19. Chapter 20 provides an introduction to Grid computing environments, Chapters 20 to 34. Chapter 35 provides an overview of Grid applications and especially an introduction to Chapters 36 to 43.

There are a few critical historical papers with Chapter 3 giving a broad-based review of the evolution of the Grid. Chapter 37 by Smarr and Catlett describes their vision of metacomputing – a concept that has grown into that of Grid but originally contained many of the critical ideas of today's Grid. Chapters 4, 6 and 8 cover the evolution of the Globus project [4]; starting in Chapter 4 with the wonderful 'I-WAY metacomputing experiment' in 1995 at the SC95 (Supercomputing) conference in San Diego [11]. Finally, Chapter 5 provides a wealth of practical detail on production Grids from two of the best-regarded Grid deployments – the Information Power Grid from NASA and the DOE Science Grid. This chapter contains discussion of Globus as well as Condor (Chapter 11) and the Storage Resource Broker from SDSC (Chapter 16).

PART B: ARCHITECTURE AND TECHNOLOGIES OF THE GRID

Part B focuses on Grid architecture and begins with a reprinting of the now classic paper *Anatomy of the Grid* in Chapter 6 stressing the role of Grids in forming virtual organizations. The *Physiology of the Grid* of Chapter 8 describes exciting recent developments of OGSA (Open Grid Services Architecture) from Globus concepts and commercial Web services ideas. OGSA is further discussed in Chapters 7, 9 and 15. Chapter 9 describes some of the first experiences in integrating Web services with Grids. Chapter 7 discusses the next steps for the UK e-Science Grid, which is being built on Globus and OGSA.

The term *e-Science* is intended to capture a vision of the future for scientific research based on distributed resources – especially data-gathering instruments – and research groups. This is highlighted from the application view perspective in Chapter 36, while technological implications of applications are covered in Chapters 7, 14, 15 and 16. Chapter 14 takes a database perspective, Chapter 15 focuses on extending OGSA to Data Grids and Chapter 16 describes the experiences of the San Diego Supercomputer Center in the United States with the development of metadata and data to knowledge systems. Chapter 17 extends the Data Grid to a Semantic or Knowledge Grid and presents the challenges and opportunities of this attractive vision.

Chapters 10, 11 and 12 describe three major long-term Grid projects. Chapter 10 is devoted to the Legion project from the University of Virginia, which is currently being commercialized by Avaki. The Legion approach pioneered a uniform object-based app-roach to Grid infrastructure and provides a basis for ideas now being implemented in Web services and OGSA. Peer-to-peer (P2P) 'megacomputing' environments such as Entropia, which leverage the availability of a large amount of computational power on Internet clients, share many issues with the Grid and are discussed in Chapter 12. The system management and fault-tolerance issues are at the heart of both P2P and Grid technologies. Such approaches enable large collections of laptops, PCs and servers to rival the largest supercomputers (for some applications) in the computing cycles they can bring to bear to a single problem. Condor, which is described in Chapter 11, manages many thousands (or more) of workstations at a time – typically within enterprises – so they can be used in a load-balanced fashion as a reliable pool of computing resources. This system illustrates many Grid features – single sign-on, fault tolerance and resource management.

Grid concepts and infrastructure have been developed in largely academic research environments with science and engineering applications, but Grid ideas are currently the focus of intense commercial interest with many major computing companies announcing major Grid initiatives. IBM has stressed the linkage of Grid and autonomic computing described in Chapter 13 – autonomic computer systems have a built-in self-optimization and robustness designed in analogy to the autonomic nervous system of the body.

The final two chapters of Part B (Chapters 18 and 19) describe further the relationship between P2P and Grid computing. The Grid and P2P networks both support communities sharing resources but typically with different models for searching for and managing services. The P2P chapters cover collaboration, aspects of universal access and the power of P2P ideas in managing distributed metadata. Chapter 19 has a detailed discussion of the use of P2P ideas to manage Web service–related metadata and in particular to be able to discover efficiently information about dynamic distributed services. Note that P2P includes both the totally distributed systems generalizing projects like Gnutella [12] and the Internet Computing model originally popularized by SETI@Home [13]. Chapters 18 and 19 fall into the first class, while the Entropia project of Chapter 12 is of the second class. Thus in Part B of the book we have linked all types of P2P networks, autonomic computing together with seamless access and enterprise resource management characteristic of the original Grid concept.

PART C: GRID COMPUTING ENVIRONMENTS

The next set of Chapters 20 to 34 covers Grid computing environments and forms Part C of the book. Programming computers typically involves manipulating variables and data-structures in a fine-grain function. For the Grid, one is usually concerned with macroscopic objects – programs, files, library routines and services – and the programming paradigm must address different issues. Chapters 20 and 21 provide a broad discussion of Grid programming with Chapter 20 focusing on the portal (user) view and Chapter 21 focusing on the program interfaces. Chapter 22 describes the technology supporting message-based Grid computing environments and its relation to well-known commercial middleware with this paradigm. Chapters 23, 24 and 25 cover the programmatic interfaces to the Grid with a new European Gridlab project and the Grid Forum standard Grid RPC (Remote Procedure Call) to support distributed computing as a service.

Chapters 26, 27 and 28 describe toolkits written in a variety of languages to link Globus-level Grid capabilities to Web portals. Chapters 28, 29, 30 and 31 give examples of portals and problem-solving environments. These give the Grid user Web-based access to Grid functions at both low and higher levels. The low level is similar to the UNIX shell on a single computer, that is, at the level of functionality of general use such as a copy between distributed file systems. Problem-solving environments (Chapters 28, 30 and 31) typically offer a higher-level domain-specific user interface with ability to control a complete application with choices of solution method, data sets and parameters. In understanding the relevance of portals, one needs to distinguish interactions between backend computers, middleware and users. Each chapter in this section discusses different aspects of and approaches to this complex environment.

Chapters 32, 33 and 34 provide three examples of the important and emerging area of Grid tools. Chapter 33 presents an innovative parameter-sweep system allowing the control of multiple job instances with the same program but with different datasets. Chapter 32 describes a tool for resource allocation and sharing based on an economical model rather than the simple queuing algorithms usually adopted. Chapter 34 describes a Web-based Grid file utility.

PART D: GRID APPLICATIONS

The last part of the book is devoted to Grid applications, which are initially overviewed in Chapter 35 and are key to the success of the Grid. Chapter 37 gives a historical perspective with the 1992 vision of this area.

In 2002, it is clear that data-intensive applications will be of increasing importance to scientists for the next decade. With the integration of new instruments, and the increased focus on experiments, analysis, modeling and simulation, Grid infrastructure is expected to enable the next generation of science and engineering advances. Basic research in high-energy physics, astronomy and other areas are pioneering the use of the Grid for coping with the so-called data deluge described in Chapter 36. The virtual observatory of Chapter 38 describes a new type of astronomy using the Grid to analyze the data from multiple instruments observing at different wavelengths. High-energy physics described in Chapter 39 is preparing for the wealth of data (100 petabytes by 2007) expected from the new Large Hadron Collider at CERN with a carefully designed distributed computer and data architecture.

In the future, it may well be that biological and chemistry applications will be the new 'killer apps' for the Grid with the growing number of high throughput commodity instruments coming on-line. Ultimately, for example, personal gene profiles will enable a whole new approach to personalized healthcare. Aspects of this life sciences vision are described in Chapters 40, 41 and 42. Chapter 40 describes a variety of biology applications involving both distributed gene sequencing and parameter-sweep style simulations. Chapter 41 describes a different important biological problem in the healthcare area – using the Grid to manage a federated database of distributed mammograms.

The importance of high-quality provenance metadata and its implications for Grid-enabled medicine are stressed. Chapter 42 describes the role of the Grid for combinatorial chemistry – with new instruments producing and analyzing compounds in parallel. Here the Grid will manage both individual laboratories and their world-wide or company-wide integration. Early applications are mainly restricted to the academic research community but the Grid will soon be valuable in enterprise computing (see Chapter 13) and in the broader community. Chapter 43 describes the Grid supporting the collaboration between teachers, students and the public – the community Grid for education and outreach.

In aggregate, the chapters in this book provide a comprehensive and compelling description of the present and future of Grid Computing and provide a basis for further development and discussion of the global computational and data management infrastructure critical for science and society in the twenty-first century.

REFERENCES

1. Berman, F., Fox, G. and Hey, T. (2003) *Grid Computing: Making the Global Infrastructure a Reality*. Chichester: John Wiley & Sons.
2. Foster, I. and Kesselman, C. (eds) (1999) *The Grid: Blueprint for a New Computing Infrastructure*. San Francisco, CA: Morgan Kaufmann.
3. Global Grid Forum Web Site, http://www.gridforum.org.
4. Globus Project Web Site, http://www.globus.org.
5. Online Grid Magazine Published by EarthWeb, Which is a Division of INT Media Group, http://www.gridcomputingplanet.com/.
6. The Grid Computing Information Center, http://www.gridcomputing.com/.
7. Informal Listing of Grid News, http://www.thegridreport.com/.
8. Online Grid Magazine Published by Tabor Griffin Communications, http://www.gridtoday.com.
9. Grid Research Integration Deployment and Support Center, http://www.grids-center.org/.
10. Web site associated with book, Grid Computing: Making the Global Infrastructure a Reality, http://www.grid2002.org.
11. The SCxy Series of Conferences, http://www.supercomp.org.
12. Gnutella Peer-to-Peer (File-Sharing) Client, http://www.gnutella.com/ or http://www.gnutellanews.com/.
13. SETI@Home Internet Computing, http://setiathome.ssl.berkeley.edu/.

The Grid: past, present, future

Fran Berman,[1] Geoffrey Fox,[2] and Tony Hey[3,4]

[1]San Diego Supercomputer Center, and Department of Computer Science and Engineering, University of California, San Diego, California, United States, [2]Indiana University, Bloomington, Indiana, United States, [3]EPSRC, Swindon, United Kingdom, [4]University of Southampton, Southampton, United Kingdom

1.1 THE GRID

The Grid is the computing and data management infrastructure that will provide the electronic underpinning for a global society in business, government, research, science and entertainment [1–5]. Grids, illustrated in Figure 1.1, integrate networking, communication, computation and information to provide a virtual platform for computation and data management in the same way that the Internet integrates resources to form a virtual platform for information. The Grid is transforming science, business, health and society. In this book we consider the Grid in depth, describing its immense promise, potential and complexity from the perspective of the community of individuals working hard to make the Grid vision a reality.

> *Grid infrastructure will provide us with the ability to dynamically link together resources as an ensemble to support the execution of large-scale, resource-intensive, and distributed applications.*

Grid Computing – Making the Global Infrastructure a Reality. Edited by F. Berman, A. Hey and G. Fox
© 2003 John Wiley & Sons, Ltd   ISBN: 0-470-85319-0

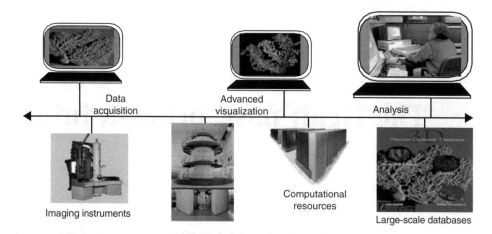

Figure 1.1 Grid resources linked together for neuroscientist Mark Ellisman's Telescience application (http://www.npaci.edu/Alpha/telescience.html).

Large-scale Grids are intrinsically distributed, heterogeneous and dynamic. They promise effectively infinite cycles and storage, as well as access to instruments, visualization devices and so on without regard to geographic location. Figure 1.2 shows a typical early successful application with information pipelined through distributed systems [6]. The reality is that to achieve this promise, complex systems of software and services must be developed, which allow access in a user-friendly way, which allow resources to be used together efficiently, and which enforce policies that allow communities of users to coordinate resources in a stable, performance-promoting fashion. Whether users access the Grid to use one resource (a single computer, data archive, etc.), or to use several resources in aggregate as a coordinated 'virtual computer', the Grid permits users to interface with the resources in a uniform way, providing a comprehensive and powerful platform for global computing and data management.

In the United Kingdom this vision of increasingly global collaborations for scientific research is encompassed by the term *e-Science* [7]. The UK e-Science Program is a major initiative developed to promote scientific and data-oriented Grid application development for both science and industry. The goals of the e-Science initiative are to assist in global efforts to develop a Grid e-Utility infrastructure for e-Science applications, which will support in silico experimentation with huge data collections, and assist the development of an integrated campus infrastructure for all scientific and engineering disciplines. e-Science merges a decade of simulation and compute-intensive application development with the immense focus on data required for the next level of advances in many scientific disciplines. The UK program includes a wide variety of projects including health and medicine, genomics and bioscience, particle physics and astronomy, environmental science, engineering design, chemistry and material science and social sciences. Most e-Science projects involve both academic and industry participation [7].

Box 1.1 Summary of Chapter 1

This chapter is designed to give a high-level motivation for the book. In Section 1.2, we highlight some historical and motivational building blocks of the Grid – described in more detail in Chapter 3. Section 1.3 describes the current community view of the Grid with its basic architecture. Section 1.4 contains four building blocks of the Grid. In particular, in Section 1.4.1 we review the evolution of the networking infrastructure including both the desktop and cross-continental links, which are expected to reach gigabit and terabit performance, respectively, over the next five years. Section 1.4.2 presents the corresponding computing backdrop with 1 to 40 teraflop performance today moving to petascale systems by the end of the decade. The U.S. National Science Foundation (NSF) TeraGrid project illustrates the state-of-the-art of current Grid technology. Section 1.4.3 summarizes many of the regional, national and international activities designing and deploying Grids. Standards, covered in Section 1.4.4 are a different but equally critical building block of the Grid. Section 1.5 covers the critical area of applications on the Grid covering life sciences, engineering and the physical sciences. We highlight new approaches to science including the importance of collaboration and the e-Science [7] concept driven partly by increased data. A short section on commercial applications includes the e-Enterprise/Utility [10] concept of computing power on demand. Applications are summarized in Section 1.5.7, which discusses the characteristic features of 'good Grid' applications like those illustrated in Figures 1.1 and 1.2. These show instruments linked to computing, data archiving and visualization facilities in a local Grid. Part D and Chapter 35 of the book describe these applications in more detail. Futures are covered in Section 1.6 with the intriguing concept of autonomic computing developed originally by IBM [10] covered in Section 1.6.1 and Chapter 13. Section 1.6.2 is a brief discussion of Grid programming covered in depth in Chapter 20 and Part C of the book. There are concluding remarks in Sections 1.6.3 to 1.6.5.

General references can be found in [1–3] and of course the chapters of this book [4] and its associated Web site [5]. The reader's guide to the book is given in the preceding preface. Further, Chapters 20 and 35 are guides to Parts C and D of the book while the later insert in this chapter (Box 1.2) has comments on Parts A and B of this book. Parts of this overview are based on presentations by Berman [11] and Hey, conferences [2, 12] and a collection of presentations from the Indiana University on networking [13–15].

In the next few years, the Grid will provide the fundamental infrastructure not only for e-Science but also for e-Business, e-Government, e-Science and e-Life. This emerging infrastructure will exploit the revolutions driven by Moore's law [8] for CPU's, disks and instruments as well as Gilder's law [9] for (optical) networks. In the remainder of this chapter, we provide an overview of this immensely important and exciting area and a backdrop for the more detailed chapters in the remainder of this book.

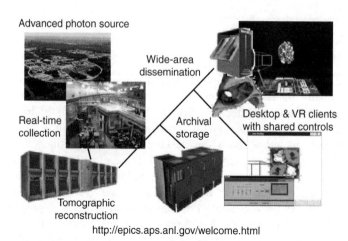

Advanced photon source

Wide-area dissemination

Real-time collection

Archival storage

Desktop & VR clients with shared controls

Tomographic reconstruction

http://epics.aps.anl.gov/welcome.html

Figure 1.2 Computational environment for analyzing real-time data taken at Argonne's advanced photon source was an early example of a data-intensive Grid application [6]. The picture shows data source at APS, network, computation, data archiving, and visualization. This figure was derived from work reported in "Real-Time Analysis, Visualization, and Steering of Microtomography Experiments at Photon Sources", Gregor von Laszewski, Mei-Hui Su, Joseph A. Insley, Ian Foster, John Bresnahan, Carl Kesselman, Marcus Thiebaux, Mark L. Rivers, Steve Wang, Brian Tieman, Ian McNulty, Ninth SIAM Conference on Parallel Processing for Scientific Computing, Apr. 1999.

1.2 BEGINNINGS OF THE GRID

It is instructive to start by understanding the influences that came together to ultimately influence the development of the Grid. Perhaps the best place to start is in the 1980s, a decade of intense research, development and deployment of hardware, software and applications for parallel computers. Parallel computing in the 1980s focused researchers' efforts on the development of algorithms, programs and architectures that supported simultaneity. As application developers began to develop large-scale codes that pushed against the resource limits of even the fastest parallel computers, some groups began looking at distribution beyond the boundaries of the machine as a way of achieving results for problems of larger and larger size.

During the 1980s and 1990s, software for parallel computers focused on providing powerful mechanisms for managing communication between processors, and development and execution environments for parallel machines. Parallel Virtual Machine (PVM), Message Passing Interface (MPI), High Performance Fortran (HPF), and OpenMP were developed to support communication for scalable applications [16]. Successful application paradigms were developed to leverage the immense potential of shared and distributed memory architectures. Initially it was thought that the Grid would be most useful in extending parallel computing paradigms from tightly coupled clusters to geographically distributed systems. However, in practice, the Grid has been utilized more as a platform for the integration of loosely coupled applications – some components of which might be running in parallel on a low-latency parallel machine – and for linking disparate resources (storage, computation, visualization, instruments). The fundamental Grid task of managing these heterogeneous components as we scale the size of distributed systems replaces that of the tight synchronization of the typically identical [in program but not data as in the SPMD (single program multiple data) model] parts of a domain-decomposed parallel application.

During the 1980s, researchers from multiple disciplines also began to come together to attack 'Grand Challenge' problems [17], that is, key problems in science and engineering for which large-scale computational infrastructure provided a fundamental tool to achieve new scientific discoveries. The Grand Challenge and multidisciplinary problem teams provided a model for collaboration that has had a tremendous impact on the way large-scale science is conducted to date. Today, interdisciplinary research has not only provided a model for collaboration but has also inspired whole disciplines (e.g. bioinformatics) that integrate formerly disparate areas of science.

The problems inherent in conducting multidisciplinary and often geographically dispersed collaborations provided researchers experience both with *coordination* and *distribution* – two fundamental concepts in Grid Computing. In the 1990s, the US Gigabit testbed program [18] included a focus on distributed metropolitan-area and wide-area applications. Each of the test beds – Aurora, Blanca, Casa, Nectar and Vistanet – was designed with dual goals: to investigate potential testbed network architectures and to explore their usefulness to end users. In this second goal, each testbed provided a venue for experimenting with distributed applications.

The first modern Grid is generally considered to be the information wide-area year (I-WAY), developed as an experimental demonstration project for SC95. In 1995, during the week-long Supercomputing conference, pioneering researchers came together to aggregate a national distributed testbed with over 17 sites networked together by the vBNS. Over 60 applications were developed for the conference and deployed on the I-WAY, as well as a rudimentary Grid software infrastructure (Chapter 4) to provide access, enforce security, coordinate resources and other activities. Developing infrastructure and applications for the I-WAY provided a seminal and powerful experience for the first generation of modern Grid researchers and projects. This was important as the development of Grid research requires a very different focus than distributed computing research. Whereas distributed computing research generally focuses on addressing the problems of geographical separation, Grid research focuses on addressing the problems of integration and management of software.

I-WAY opened the door for considerable activity in the development of Grid software. The Globus [3] (Chapters 6 and 8) and Legion [19–21] (Chapter 10) infrastructure projects explored approaches for providing basic system-level Grid infrastructure. The Condor project [22] (Chapter 11) experimented with high-throughput scheduling, while the AppLeS [23], APST (Chapter 33), Mars [24] and Prophet [25] projects experimented with high-performance scheduling. The Network Weather Service [26] project focused on resource monitoring and prediction, while the Storage Resource Broker (SRB) [27] (Chapter 16) focused on uniform access to heterogeneous data resources. The NetSolve [28] (Chapter 24) and Ninf [29] (Chapter 25) projects focused on remote computation via a client-server model. These, and many other projects, provided a foundation for today's Grid software and ideas.

In the late 1990s, Grid researchers came together in the Grid Forum, subsequently expanding to the Global Grid Forum (GGF) [2], where much of the early research is now evolving into the standards base for future Grids. Recently, the GGF has been instrumental in the development of the Open Grid Services Architecture (OGSA), which integrates Globus and Web services approaches (Chapters 7, 8, and 9). OGSA is being developed by both the United States and European initiatives aiming to define core services for a wide variety of areas including:

- *Systems Management and Automation*
- *Workload/Performance Management*
- *Security*
- *Availability/Service Management*
- *Logical Resource Management*
- *Clustering Services*
- *Connectivity Management*
- *Physical Resource Management.*

Today, the Grid has gone global, with many worldwide collaborations between the United States, European and Asia-Pacific researchers. Funding agencies, commercial vendors, academic researchers, and national centers and laboratories have come together to form a community of broad expertise with enormous commitment to building the Grid. Moreover, research in the related areas of networking, digital libraries, peer-to-peer computing, collaboratories and so on are providing additional ideas relevant to the Grid.

Although we tend to think of the Grid as a result of the influences of the last 20 years, some of the earliest roots of the Grid can be traced back to J.C.R. Licklider, many years before this. 'Lick' was one of the early computing and networking pioneers, who set the scene for the creation of the ARPANET, the precursor to today's Internet. Originally an experimental psychologist at MIT working on psychoacoustics, he was concerned with the amount of data he had to work with and the amount of time he required to organize and analyze his data. He developed a vision of networked computer systems that would be able to provide fast, automated support systems for human decision making [30]:

> *'If such a network as I envisage nebulously could be brought into operation, we could have at least four large computers, perhaps six or eight small computers, and a great assortment of disc files and magnetic tape units − not to mention remote consoles and teletype stations − all churning away'*

In the early 1960s, computers were expensive and people were cheap. Today, after thirty odd years of Moore's Law [8], the situation is reversed and individual laptops now have more power than Licklider could ever have imagined possible. Nonetheless, his insight that the deluge of scientific data would require the harnessing of computing resources distributed around the galaxy was correct. Thanks to the advances in networking and software technologies, we are now working to implement this vision.

In the next sections, we provide an overview of the present Grid Computing and its emerging vision for the future.

1.3 A COMMUNITY GRID MODEL

Over the last decade, the Grid community has begun to converge on a layered model that allows development of the complex system of services and software required to integrate Grid resources. This model, explored in detail in Part B of this book, provides a layered abstraction of the Grid. Figure 1.3 illustrates the *Community Grid Model* being developed in a loosely coordinated manner throughout academia and the commercial sector. We begin discussion by understanding each of the layers in the model.

The bottom horizontal layer of the Community Grid Model consists of the hardware *resources* that underlie the Grid. Such resources include computers, networks, data archives, instruments, visualization devices and so on. They are distributed, heterogeneous and have very different performance profiles (contrast performance as measured in FLOPS or memory bandwidth with performance as measured in bytes and data access time). Moreover, the resource pool represented by this layer is highly dynamic, both as a result of new resources being added to the mix and old resources being retired, and as a result of varying observable performance of the resources in the shared, multiuser environment of the Grid.

The next horizontal layer (*common infrastructure*) consists of the software services and systems which virtualize the Grid. Community efforts such as NSF's Middleware Initiative (NMI) [31], OGSA (Chapters 7 and 8), as well as emerging de facto standards such as Globus provide a commonly agreed upon layer in which the Grid's heterogeneous and dynamic resource pool can be accessed. The key concept at the common infrastructure layer is community agreement on software, which will represent the Grid as a unified virtual platform and provide the target for more focused software and applications.

The next horizontal layer (user and application-focused Grid *middleware, tools and services*) contains software packages built atop the common infrastructure. This software serves to enable applications to more productively use Grid resources by masking some of the complexity involved in system activities such as authentication, file transfer, and

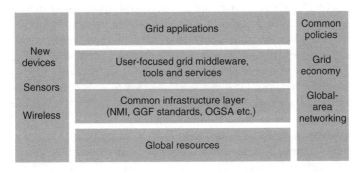

Figure 1.3 Layered architecture of the Community Grid Model.

so on. Portals, community codes, application scheduling software and so on reside in this layer and provide middleware that connects applications and users with the common Grid infrastructure.

The topmost horizontal layer (*Grid applications*) represents applications and users. The Grid will ultimately be only as successful as its user community and all of the other horizontal layers must ensure that the Grid presents a robust, stable, usable and useful computational and data management platform to the user. Note that in the broadest sense, even applications that use only a single resource on the Grid are Grid applications if they access the target resource through the uniform interfaces provided by the Grid infrastructure.

The vertical layers represent the next steps for the development of the Grid. The vertical layer on the left represents the *influence of new devices* – sensors, PDAs, and wireless. Over the next 10 years, these and other new devices will need to be integrated with the Grid and will exacerbate the challenges of managing heterogeneity and promoting performance. At the same time, the increasing globalization of the Grid will require serious consideration of *policies for sharing* and using resources, global-area networking and the development of *Grid economies* (the vertical layer on the right – see Chapter 32). As we link together national Grids to form a Global Grid, it will be increasingly important to develop Grid social and economic policies which ensure the stability of the system, promote the performance of the users and successfully integrate disparate political, technological and application cultures.

The Community Grid Model provides an abstraction of the large-scale and intense efforts of a community of Grid professionals, academics and industrial partners to build the Grid. In the next section, we consider the lowest horizontal layers (individual resources and common infrastructure) of the Community Grid Model.

1.4 BUILDING BLOCKS OF THE GRID

1.4.1 Networks

The heart of any Grid is its network – networks link together geographically distributed resources and allow them to be used collectively to support execution of a single application. If the networks provide 'big pipes', successful applications can use distributed resources in a more integrated and data-intensive fashion; if the networks provide 'small pipes', successful applications are likely to exhibit minimal communication and data transfer between program components and/or be able to tolerate high latency.

At present, Grids build on ubiquitous high-performance networks [13, 14] typified by the Internet2 Abilene network [15] in the United States shown in Figures 1.4 and 1.5. In 2002, such national networks exhibit roughly $10\,\text{Gb}\,\text{s}^{-1}$ backbone performance. Analogous efforts can be seen in the UK SuperJanet [40] backbone of Figure 1.6 and the intra-Europe GEANT network [41] of Figure 1.7. More globally, Grid efforts can leverage international networks that have been deployed (illustrated in Figure 1.8) including CA*net3 from Canarie in Canada [42] and the Asian network APAN [43], (shown in detail in Figure 1.9). Such national network backbone performance is typically complemented by

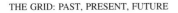

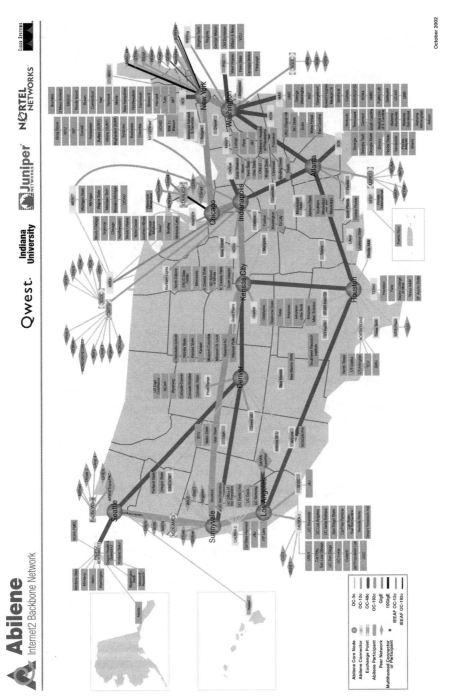

Figure 1.4 Sites on the Abilene Research Network.

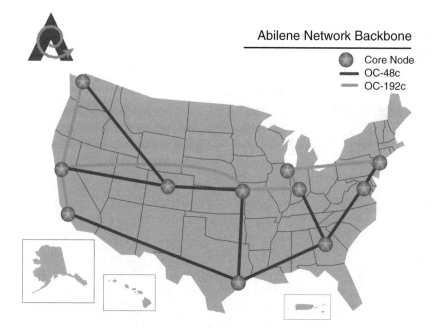

Figure 1.5 Backbone of Abilene Internet2 Network in USA.

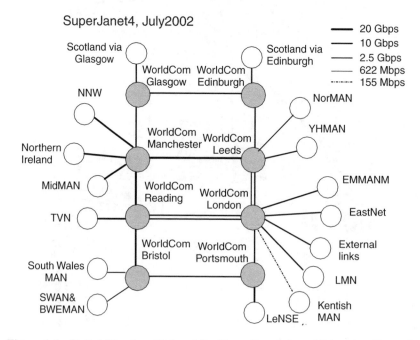

Figure 1.6 United Kingdom National Backbone Research and Education Network.

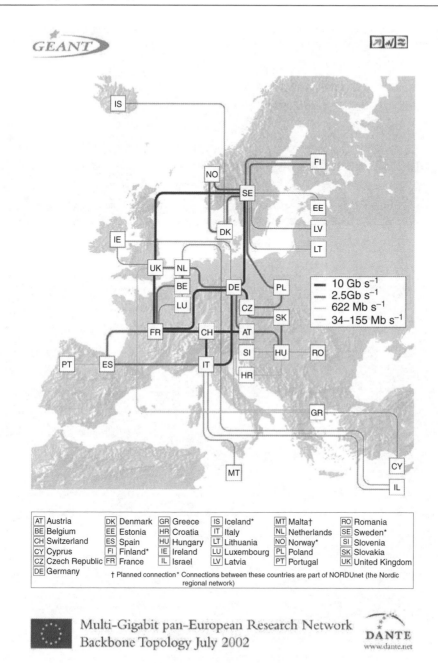

Figure 1.7   European Backbone Research Network GEANT showing countries and backbone speeds.

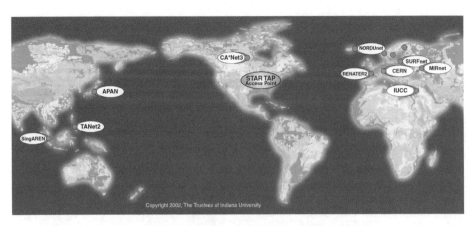

Figure 1.8 International Networks.

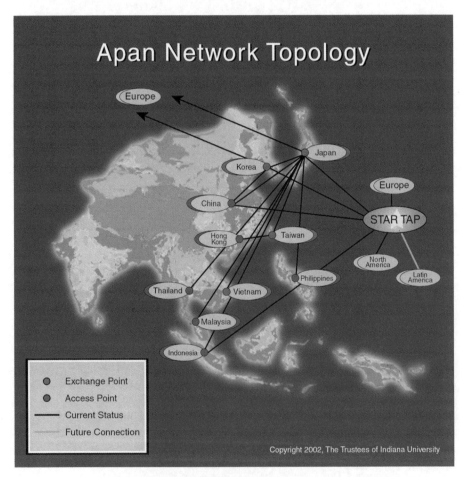

Figure 1.9 APAN Asian Network.

THE GRID: PAST, PRESENT, FUTURE   21

a $1\,\mathrm{Gb\,s^{-1}}$ institution-to-backbone link and by a 10 to $100\,\mathrm{Mb\,s^{-1}}$ desktop-to-institutional network link.

Although there are exceptions, one can capture a typical leading Grid research environment as a $10:1:0.1\,\mathrm{Gbs^{-1}}$ ratio representing *national: organization: desktop* links.

Today, new national networks are beginning to change this ratio. The GTRN or Global Terabit Research Network initiative shown in Figures 1.10 and 1.11 link national networks in Asia, the Americas and Europe with a performance similar to that of their backbones [44]. By 2006, GTRN aims at a $1000:1000:100:10:1$ gigabit performance ratio representing *international backbone: national: organization: optical desktop: Copper desktop* links. This implies a performance increase of over a factor of 2 per year in network performance, and clearly surpasses expected CPU performance and memory size increases of Moore's law [8] (with a prediction of a factor of two in chip density improvement every 18 months). This continued difference between network and CPU performance growth will continue to enhance the capability of distributed systems and lessen the gap between Grids and geographically centralized approaches. We should note that although network bandwidth will improve, we do not expect latencies to improve significantly. Further, as seen in the telecommunications industry in 2000–2002, in many ways network performance is increasing 'faster than demand' even though organizational issues lead to problems. A critical area of future work is network quality of service and here progress is less clear. Networking performance can be taken into account at the application level as in AppLeS and APST ([23] and Chapter 33), or by using the Network Weather Service [26] and NaradaBrokering (Chapter 22).

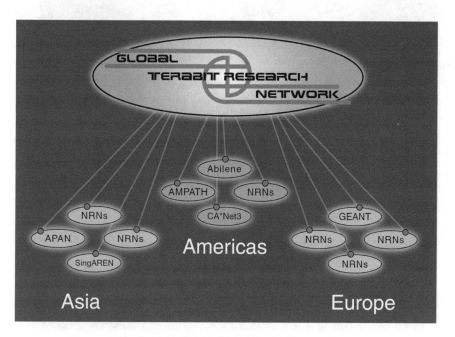

Figure 1.10 Logical GTRN Global Terabit Research Network.

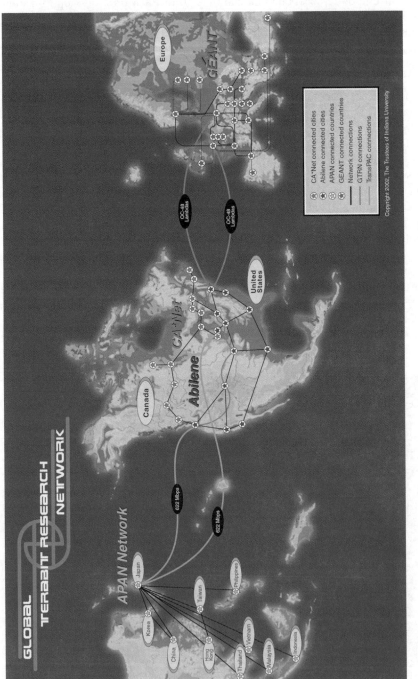

Figure 1.11 Physical GTRN Global Terabit Research Network.

High-capacity networking increases the capability of the Grid to support both parallel and distributed applications. In the future, wired networks will be further enhanced by continued improvement in wireless connectivity [45], which will drive integration of smaller and smaller devices into the Grid. The desktop connectivity described above will include the pervasive PDA (Personal Digital Assistant included in universal access discussion of Chapter 18) that will further promote the Grid as a platform for e-Science, e-Commerce and e-Education (Chapter 43).

1.4.2 Computational 'nodes' on the Grid

Networks connect resources on the Grid, the most prevalent of which are computers with their associated data storage. Although the computational resources can be of any level of power and capability, some of the most interesting Grids for scientists involve nodes that are themselves high-performance parallel machines or clusters. Such high-performance Grid 'nodes' provide major resources for simulation, analysis, data mining and other compute-intensive activities. The performance of the most high-performance nodes on the Grid is tracked by the Top500 site [46] (Figure 1.12). Extrapolations of this information indicate that we can expect a peak single machine performance of 1 petaflops/sec (10^{15} operations per second) by around 2010.

Contrast this prediction of power to the present situation for high-performance computing. In March 2002, Japan's announcement of the NEC Earth Simulator machine shown in Figure 1.13 [47], which reaches 40 teraflops s^{-1} with a good sustained to peak performance rating, garnered worldwide interest. The NEC machine has 640 eight-processor nodes and offers 10 terabytes of memory and 700 terabytes of disk space. It has already been used for large-scale climate modeling. The race continues with Fujitsu announcing

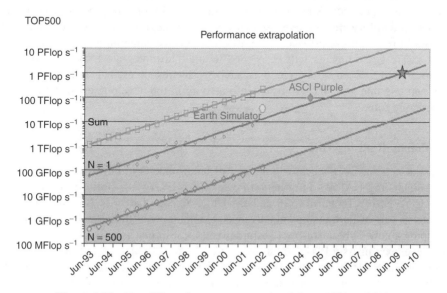

Figure 1.12 Top 500 performance extrapolated from 1993 to 2010.

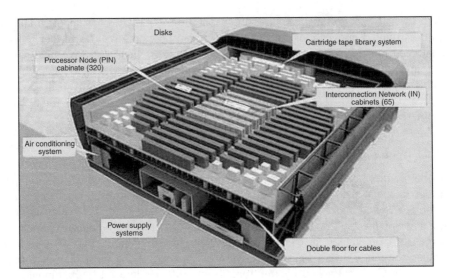

Figure 1.13   Japanese Earth Simulator 40 Teraflop Supercomputer.

in August 2002, the HPC2500 with up to 16 384 processors and 85 teraflops s^{-1} peak performance [48]. Until these heroic Japanese machines, DOE's ASCI program [49], shown in Figure 1.14, had led the pack with the ASCI White machine at Livermore National Laboratory peaking at 12 teraflops s^{-1}. Future ASCI machines will challenge for the Top 500 leadership position!

Such nodes will become part of future Grids. Similarly, large data archives will become of increasing importance. Since it is unlikely that it will be many years, if ever, before it becomes straightforward to move petabytes of data around global networks, data centers will install local high-performance computing systems for data mining and analysis. Complex software environments will be needed to smoothly integrate resources from PDAs (perhaps a source of sensor data) to terascale/petascale resources. This is an immense challenge, and one that is being met by intense activity in the development of Grid software infrastructure today.

1.4.3 Pulling it all together

The last decade has seen a growing number of large-scale Grid infrastructure deployment projects including NASA's Information Power Grid (IPG) [50], DoE's Science Grid [51] (Chapter 5), NSF's TeraGrid [52], and the UK e-Science Grid [7]. NSF has many Grid activities as part of Partnerships in Advanced Computational Infrastructure (PACI) and is developing a new Cyberinfrastructure Initiative [53]. Similar large-scale Grid projects are being developed in Asia [54] and all over Europe – for example, in the Netherlands [55], France [56], Italy [57], Ireland [58], Poland [59] and Scandinavia [60]. The DataTAG project [61] is focusing on providing a transatlantic lambda connection for HEP (High Energy Physics) Grids and we have already described the GTRN [14] effort. Some projects

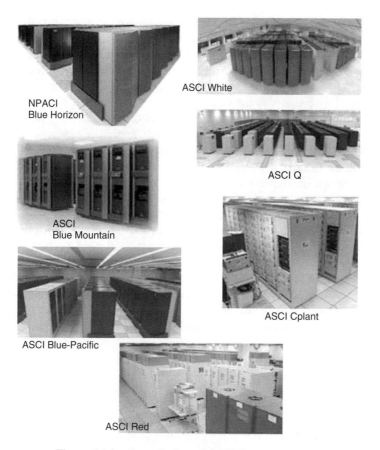

Figure 1.14 Constellation of ASCI Supercomputers.

are developing high-end, high-performance Grids with fast networks and powerful Grid nodes that will provide a foundation of experience for the Grids of the future. The European UNICORE system ([62] and Chapter 29) is being developed as a Grid computing environment to allow seamless access to several large German supercomputers. In the United States, the ASCI program and TeraGrid project are using Globus to develop Grids linking multi-teraflop computers together [63]. There are many support projects associated with all these activities including national and regional centers in the UK e-Science effort [64, 65], the European GRIDS activity [66] and the iVDGL (International Virtual Data Grid Laboratory) [67]. This latter project has identified a Grid Operation Center in analogy with the well-understood network operation center [68].

Much of the critical Grid software is built as part of infrastructure activities and there are important activities focused on software: the Grid Application Development System (GrADS) [69] is a large-scale effort focused on Grid program development and execution environment. Further, NSF's Middleware Initiative (NMI) is focusing on the development and documentation of ready-for-primetime Grid middleware. Europe has started several major software activities [62, 70–75]. Application Grid projects described in more detail in Section 1.5 include magnetic fusion [76], particle physics [68, 77, 78] (Chapter 39), astronomy [77, 79–81] (Chapter 38), earthquake engineering [82] and modeling [83], climate [84], bioinformatics [85, 86] (Chapters 40 and 41) and, more generally, industrial applications [87]. We finally note two useful Web resources [88, 89] that list, respectively, acronyms and major projects in the Grid area.

One of the most significant and coherent Grid efforts in Europe is the UK e-Science Program [7] discussed in Section 1.1. This is built around a coherent set of application Grids linked to a UK national Grid. The new 7 Teraflop (peak) HPC(X) machine from IBM will be located at Daresbury Laboratory and be linked to the UK e-Science Grid [90, 91] shown in Figure 1.15. In addition to the HPC(X) machine, the UK Grid will provide connections to the HPC Computer Services for Academic Research (CSAR) service in Manchester and high-performance clusters only accessible to UK university researchers via Grid digital certificates provided by the UK Grid Certification Authority. This is located at Rutherford Laboratory along with the UK Grid Support Centre and the Engineering Task Force. The UK e-Science Grid is intended to provide a model for a genuine production Grid that can be used by both academics for their research and industry for evaluation. The accompanying set of application projects are developing Grids that will connect and overlap with national Grid testing interoperability and security issues for different virtual communities of scientists. A striking feature of the UK e-Science

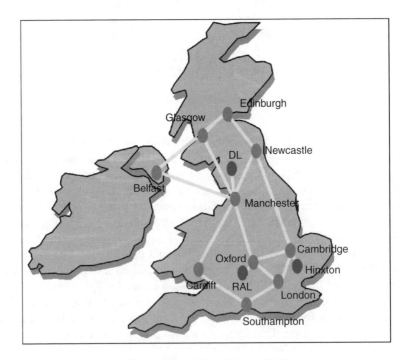

Figure 1.15   UK e-Science Grid.

initiative is the large-scale involvement of industry: over 50 companies are involved in the program, contributing over $30 M to supplement the $180 M funding provided by the UK Government.

The portfolio of the UK e-Science application projects is supported by the Core Program. This provides support for the application projects in the form of the Grid Support Centre and a supported set of Grid middleware. The initial starting point for the UK Grid was the software used by NASA for their IPG – Globus, Condor and SRB as described in Chapter 5. Each of the nodes in the UK e-Science Grid has $1.5 M budget for collaborative industrial Grid middleware projects. The requirements of the e-Science application projects in terms of computing resources, data resources, networking and remote use of facilities determine the services that will be required from the Grid middleware. The UK projects place more emphasis on data access and data federation (Chapters 14, 15 and 17) than traditional HPC applications, so the major focus of the UK Grid middleware efforts are concentrated in this area. Three of the UK e-Science centres – Edinburgh, Manchester and Newcastle – are working with the Globus team and with IBM US, IBM Hursley Laboratory in the United Kingdom, and Oracle UK in an exciting project on data access and integration (DAI). The project aims to deliver new data services within the Globus Open Grid Services framework.

Perhaps the most striking current example of a high-performance Grid is the new NSF TeraGrid shown in Figure 1.16, which links major subsystems at four different sites and will scale to the Pittsburgh Supercomputer Center and further sites in the next few years. The TeraGrid [52] is a high-performance Grid, which will connect the San Diego Supercomputer Center (SDSC), California Institute of Technology (Caltech), Argonne National Laboratory and the National Center for Supercomputing Applications (NCSA).

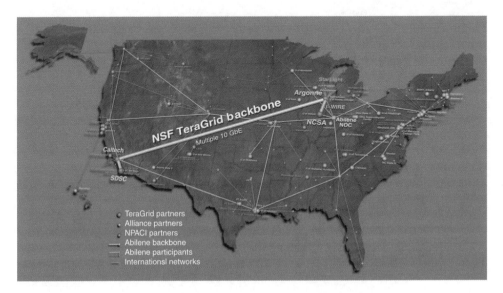

Figure 1.16   USA TeraGrid NSF HPCC system.

Once built, the TeraGrid will link the four in a Grid that will comprise in aggregate over 0.6 petabyte of on-line disk, over 13 teraflops compute performance, and will be linked together by a $40\,\text{Gb}\,\text{s}^{-1}$ network.

Each of the four TeraGrid sites specializes in different areas including visualization (Argonne), compute-intensive codes (NCSA), data-oriented computing (SDSC) and scientific collections (Caltech). An overview of the hardware configuration is shown in Figure 1.17. Each of the sites will deploy a cluster that provides users and application developers with an opportunity to experiment with distributed wide-area cluster computing as well as Grid computing. The Extensible Terascale Facility (ETF) adds the Pittsburgh Supercomputer Center to the original four TeraGrid sites. Beyond TeraGrid/ETF, it is the intention of NSF to scale to include additional sites and heterogeneous architectures as the foundation of a comprehensive 'cyberinfrastructure' for US Grid efforts [53]. With this as a goal, TeraGrid/ETF software and hardware is being designed to scale from the very beginning.

TeraGrid was designed to push the envelop on data capability, compute capability and network capability simultaneously, providing a platform for the community to experiment with data-intensive applications and more integrated compute-intensive applications. Key choices for the TeraGrid software environment include the identification of Linux as the operating system for each of the TeraGrid nodes, and the deployment of basic, core and advanced Globus and data services.

The goal is for the high-end Grid and cluster environment deployed on TeraGrid to resemble the low-end Grid and cluster environment used by scientists in their own laboratory settings. This will enable a more direct path between the development of test and prototype codes and the deployment of large-scale runs on high-end platforms.

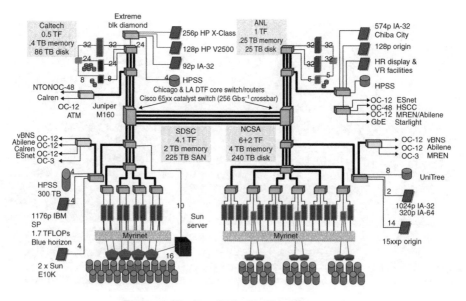

Figure 1.17 TeraGrid system components.

TeraGrid is being developed as a production Grid (analogous to the role that production supercomputers have played over the last two decades as the target of large-scale codes developed on laboratory workstations) and will involve considerable software and human infrastructure to provide access and support for users including portals, schedulers, operations, training, a distributed help-desk, and so on.

1.4.4 Common infrastructure: standards

For the foreseeable future, technology will continue to provide greater and greater potential capability and capacity and will need to be integrated within Grid technologies. To manage this ever-changing technological landscape, Grids utilize a common infrastructure to provide a virtual representation to software developers and users, while allowing the incorporation of new technologies. The development of key standards that allow the complexity of the Grid to be managed by software developers and users without heroic efforts is critical to the success of the Grid.

Both the Internet and the IETF [92], and the Web and the W3C consortium [93] have defined key standards such as TCP/IP, HTTP, SOAP, XML and now WSDL – the Web services definition language that underlines OGSA. Such standards have been critical for progress in these communities. The GGF [2] is now building key Grid-specific standards such as OGSA, the emerging *de facto* standard for Grid infrastructure. In addition, NMI [31] and the UK's Grid Core Program [7] are seeking to extend, standardize and make more robust key pieces of software for the Grid arsenal such as Globus [3] (Chapter 6), Condor [22] (Chapter 11), OGSA-DAI (Chapters 7 and 15) and the Network Weather Service [26]. In the last two decades, the development [16] of PVM [94] and MPI [95], which pre-dated the modern Grid vision, introduced parallel and distributed computing concepts to an entire community and provided the seeds for the community collaboration, which characterizes the Grid community today.

There are other important standards on which the Grid is being built. The last subsection stressed the key role of Linux as the standard for node operating systems [96]. Further within the commercial Web community, OASIS [97] is standardizing Web Services for Remote Portals (WSRP) – the portlet interface standard to define user-facing ports on Web services (Chapter 18). These standards support both commercial and noncommercial software and there is a growing trend in both arenas for open-source software. The Apache project [98] supplies key infrastructure such as servers [99] and tools to support such areas as WSDL-Java interfaces [100] and portals [101]. One expects these days that all software is either open source or provides open interfaces to proprietary implementations. Of course, the broad availability of modern languages like Java with good run-time and development environments has also greatly expedited the development of Grid and other software infrastructure.

Today, Grid projects seek to use common infrastructure and standards to promote interoperability and reusability, and to base their systems on a growing body of robust community software. Open source and standardization efforts are changing both the way software is written and the way systems are designed. This approach will be critical for the Grid as it evolves.

1.5 GRID APPLICATIONS AND APPLICATION MIDDLEWARE

The Grid will serve as an enabling technology for a broad set of applications in science, business, entertainment, health and other areas. However, the community faces a 'chicken and egg' problem common to the development of new technologies: applications are needed to drive the research and development of the new technologies, but applications are difficult to develop in the absence of stable and mature technologies. In the Grid community, Grid infrastructure efforts, application development efforts and middleware efforts have progressed together, often through the collaborations of multidisciplinary teams. In this section, we discuss some of the successful Grid application and application middleware efforts to date. As we continue to develop the software infrastructure that better realizes the potential of the Grid, and as common Grid infrastructure continues to evolve to provide a stable platform, the application and user community for the Grid will continue to expand.

1.5.1 Life science applications

One of the fastest-growing application areas in Grid Computing is the Life Sciences. Computational biology, bioinformatics, genomics, computational neuroscience and other areas are embracing Grid technology as a way to access, collect and mine data [e.g. the Protein Data Bank [102], the myGrid Project [103], the Biomedical Information Research Network (BIRN) [85]], accomplish large-scale simulation and analysis (e.g. MCell [104]), and to connect to remote instruments (e.g. in Telemicroscopy [105] and Chapter 33 [106]). The Biomedical Informatics Research Network links instruments and federated databases, illustrated in Figure 1.18. BIRN is a pioneering project that utilizes infrastructure to support cross-correlation studies of imaging and other data critical for neuroscience and biomedical advances.

The MCell collaboration between computational biologists and computer scientists to deploy large-scale Monte Carlo simulations using Grid technologies is a good example of a successful Grid-enabled life science application. Over the last decade, biologists have developed a community code called *MCell*, which is a general simulator for cellular microphysiology (the study of the physiological phenomena occurring at the microscopic level in living cells). MCell uses Monte Carlo diffusion and chemical reaction algorithms in 3D to simulate complex biochemical interactions of molecules inside and outside cells. MCell is one of the many scientific tools developed to assist in the quest to understand the form and function of cells, with specific focus on the nervous system. A local-area distributed MCell code is installed in laboratories around the world and is currently being used for several practical applications (e.g. the study of calcium dynamics in hepatocytes of the liver).

Grid technologies have enabled the deployment of large-scale MCell runs on a wide variety of target resources including clusters and supercomputers [107]. Computer scientists have worked with MCell biologists to develop the APST (AppLeS Parameter Sweep Template, described in Chapter 33) Grid middleware to efficiently deploy and schedule

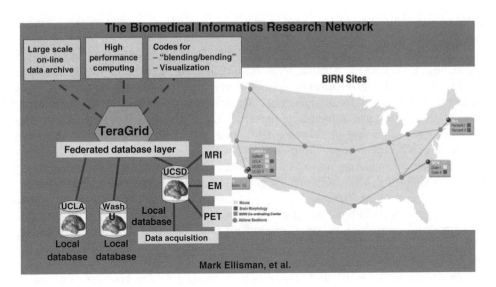

Figure 1.18 Biomedical Informatics Research Network – one of the most exciting new application models for the Grid.

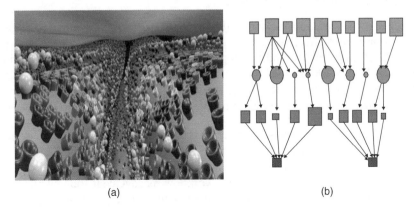

(a)   (b)

Figure 1.19 MCELL depiction of simulation of traversal of ligands in a cell (a) and program structure (b). On the right, we show linkage of shared input files, Monte Carlo " experiments" and shared output files.

large-scale runs in dynamic, distributed environments. APST has also been used by other distributed parameter sweep applications, forming part of the application-focused middleware layer of the Grid. Figure 1.19 shows MCell as seen by both disciplinary scientists and computer scientists. Figure 1.19a shows the traversal of ligands throughout the cell as simulated by MCell. Figure1.19b shows the program structure of MCell: the code comprises independent tasks that share common input files and output to common output files. Shared I/O requires data and output to be staged in order for the Grid to efficiently support application execution and this resulted in new computer science as well as computational science advances.

In the United Kingdom, the myGrid project [103] is a large consortium comprising the universities of Manchester, Southampton, Nottingham, Newcastle and Sheffield together with the European Bioinformatics Institute at Hinxton near Cambridge. In addition, GSK, AstraZeneca, IBM and SUN are industrial collaborators in the project. The goal of myGrid is to design, develop and demonstrate higher-level functionalities over the Grid to support scientists making use of complex distributed resources. The project is developing an e-Scientist's workbench providing support for the process of experimental investigation, evidence accumulation and result assimilation. A novel feature of the workbench will be provision for personalization facilities relating to resource selection, data management and process enactment. The myGrid design and development activity will be driven by applications in bioinformatics – one for the analysis of functional genomic data and another for supporting the annotation of a pattern database. The project intends to deliver Grid middleware services for automatic data annotation, workflow support and data access and integration. To support this last goal, the myGrid project will be a key application test case for the middleware being produced by the UK Core Programme project on OGSA – DAI [108].

1.5.2 Engineering-oriented applications

The Grid has provided an important platform for making resource-intensive engineering applications more cost-effective. One of the most comprehensive approaches to deploying production Grid infrastructure and developing large-scale engineering-oriented Grid applications is the NASA IPG [50] in the United States (Chapter 5). The NASA IPG vision provides a blueprint for revolutionizing the way in which NASA executes large-scale science and engineering problems via the development of

1. persistent Grid infrastructure supporting 'highly capable' computing and data management services that, on demand, will locate and co-schedule the multicenter resources needed to address large-scale and/or widely distributed problems,
2. ancillary services needed to support the workflow management frameworks that coordinate the processes of distributed science and engineering problems.

Figures 1.20 and 1.21 illustrate two applications of interest to NASA; in the first, we depict key aspects – airframe, wing, stabilizer, engine, landing gear and human factors – of the design of a complete aircraft. Each part could be the responsibility of a distinct, possibly geographically distributed, engineering team whose work is integrated together by a Grid realizing the concept of concurrent engineering. Figure 1.21 depicts possible Grid controlling satellites and the data streaming from them. Shown are a set of Web (OGSA) services for satellite control, data acquisition, analysis, visualization and linkage (assimilation) with simulations as well as two of the Web services broken up into multiple constituent services. Key standards for such a Grid are addressed by the new Space Link Extension international standard [109] in which part of the challenge is to merge a preGrid architecture with the still evolving Grid approach.

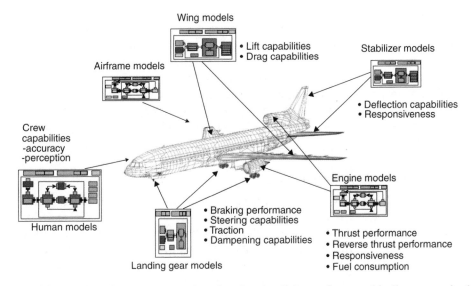

Figure 1.20   A Grid for aerospace engineering showing linkage of geographically separated sub-systems needed by an aircraft.

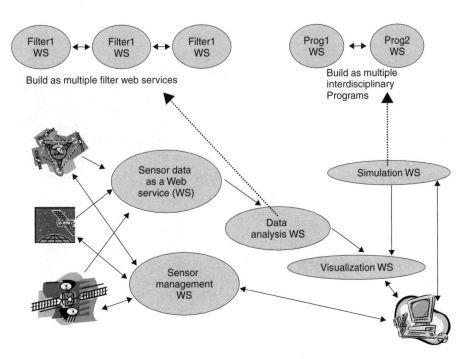

Figure 1.21   A possible Grid for satellite operation showing both spacecraft operation and data analysis. The system is built from Web services (WS) and we show how data analysis and simulation services are composed from smaller WS's.

In Europe, there are also interesting Grid Engineering applications being investigated. For example, the UK Grid Enabled Optimization and Design Search for Engineering (GEODISE) project [110] is looking at providing an engineering design knowledge repository for design in the aerospace area. Rolls Royce and BAESystems are industrial collaborators. Figure 1.22 shows this GEODISE engineering design Grid that will address, in particular the 'repeat engagement' challenge in which one wishes to build a semantic Grid (Chapter 17) to capture the knowledge of experienced designers. This of course is a research challenge and its success would open up many similar applications.

1.5.3 Data-oriented applications

As described in Chapter 36, data is emerging as the 'killer application' of the Grid. Over the next decade, data will come from everywhere – scientific instruments, experiments, sensors and sensornets, as well as a plethora of new devices. The Grid will be used to collect, store and analyze data and information, as well as to synthesize knowledge from data. Data-oriented applications described in Chapters 38 to 42 represent one of the most important application classes on the Grid and will be key to critical progress for both science and society. The importance of data for the Grid is also illustrated in several chapters: Chapters 7, 14 to 17 emphasize it in Part B of the book.

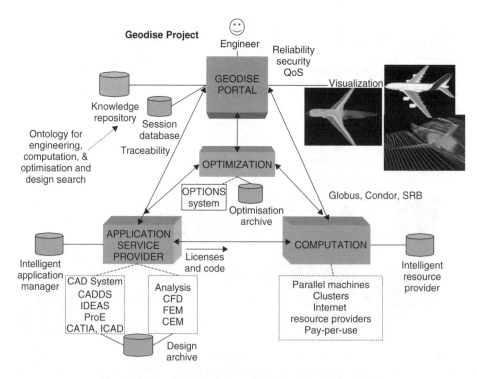

Figure 1.22   GEODISE aircraft engineering design Grid.

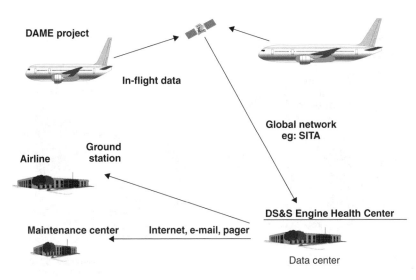

Figure 1.23 DAME Grid to manage data from aircraft engine sensors.

An example of a data-oriented application is Distributed Aircraft Maintenance Environment (DAME) [111], illustrated in Figure 1.23. DAME is an industrial application being developed in the United Kingdom in which Grid technology is used to handle the gigabytes of in-flight data gathered by operational aircraft engines and to integrate maintenance, manufacturer and analysis centers. The project aims to build a Grid-based distributed diagnostics system for aircraft engines and is motivated by the needs of Rolls Royce and its information system partner Data Systems and Solutions. The project will address performance issues such as large-scale data management with real-time demands. The main deliverables from the project will be a generic distributed diagnostics Grid application, an aero gas turbine application demonstrator for the maintenance of aircraft engines and techniques for distributed data mining and diagnostics. Distributed diagnostics is a generic problem that is fundamental in many fields such as medical, transport and manufacturing. DAME is an application currently being developed within the UK e-Science program.

1.5.4 Physical science applications

Physical science applications are another fast-growing class of Grid applications. Much has been written about the highly innovative and pioneering particle physics–dominated projects – the GriPhyN [77], Particle Physics Data Grid [78], and iVDGL [67] projects in the United States and the EU DataGrid [70], the UK GridPP [112] and the INFN (Italian National Institute for Research in Nuclear and Subnuclear Physics) Grid projects [57]. Figure 1.24 depicts the complex analysis of accelerator events being targeted to the Grid in these projects, which are described in more detail in Chapter 39. The pipelined structure of the solution allows the code to leverage the considerable potential of the Grid: In this case, the CERN linear accelerator will provide a deluge of data (perhaps $10\,\mathrm{Gb\,s}^{-1}$ of the

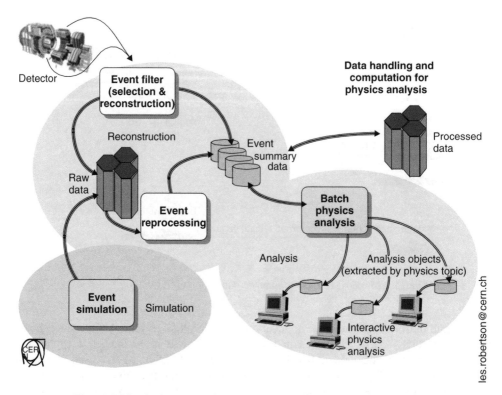

Figure 1.24 Architecture of particle physics analysis Grid (Chapter 39).

$100\,\text{Gb}\,\text{s}^{-1}$ GTRN network) while each physics event can be processed independently, resulting in trillion-way parallelism.

The astronomy community has also targeted the Grid as a means of successfully collecting, sharing and mining critical data about the universe. For example, the National Virtual Observatory Project in the United States [79] is using the Grid to federate sky surveys from several different telescopes, as discussed in Chapter 38. Using Grid technology, the sky surveys sort, index, store, filter, analyze and mine the data for important information about the night sky. High-performance Grids, such as TeraGrid, will enable NVO researchers to shorten the process of collecting, storing and analyzing sky survey data from 60 days to 5 days and will enable researchers to federate multiple sky surveys. Figures 1.25 and 1.26 show the NVO [113] and the potential impact of TeraGrid capabilities on this process.

In Europe, the EU AVO project and the UK AstroGrid project complement the US NVO effort in astronomy. The three projects are working together to agree to a common set of standards for the integration of astronomical data. Together, the NVO, AVO and AstroGrid efforts will provide scientists and educators with an unprecedented amount of accessible information about the heavens. Note that whereas in many other communities, data ownership is often an issue, astronomy data will be placed in the public domain in

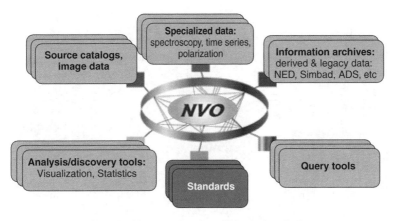

Figure 1.25 Architecture for the national virtual observatory.

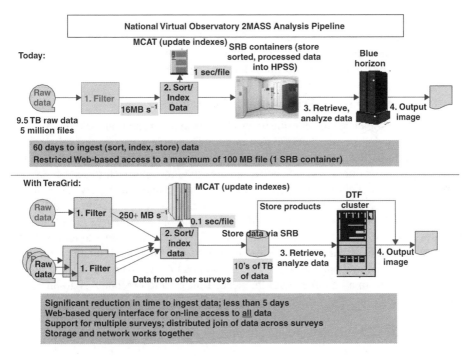

Figure 1.26 e-Science for the 2MASS astronomical data analyzed by the TeraGrid.

these Virtual Observatories after a set period of time in which the astronomers who took the original data have exclusive use of it. This points the way to a true 'democratization' of science and the emergence of a new mode of 'collection-based' research to be set alongside the traditional experimental, theoretical and computational modes.

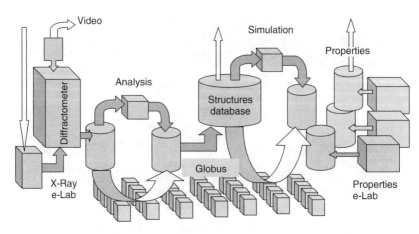

Figure 1.27   A combinatorial chemistry Grid (Chapter 42).

Finally, Figure 1.27 captures the combinatorial chemistry application of Chapter 42; experiments in this field create their deluge by parallel execution. Here we see 'experiment-on demand' with a smart laboratory (e-Lab) running miniGrid software and performing needed experiments in real time to fill in knowledge holes.

1.5.5 Trends in research: e-Science in a collaboratory

The interesting e-Science concept illustrates changes that information technology is bringing to the methodology of scientific research [114]. e-Science is a relatively new term that has become particularly popular after the launch of the major United Kingdom initiative described in Section 1.4.3. e-Science captures the new approach to science involving distributed global collaborations enabled by the Internet and using very large data collections, terascale computing resources and high-performance visualizations. e-Science is about global collaboration in key areas of science, and the next generation of infrastructure, namely the Grid, that will enable it. Figure 1.28 summarizes the e-Scientific method. Simplistically, we can characterize the last decade as focusing on simulation and its integration with science and engineering – this is computational science. e-Science builds on this adding data from all sources with the needed information technology to analyze and assimilate the data into the simulations.

Over the last half century, scientific practice has evolved to reflect the growing power of communication and the importance of collective wisdom in scientific discovery. Originally scientists collaborated by sailing ships and carrier pigeons. Now aircraft, phone, e-mail and the Web have greatly enhanced communication and hence the quality and real-time nature of scientific collaboration. The collaboration can be both 'real' or enabled electronically – as evidenced by Bill Wulf [115, 116] early influential work on the scientific collaboratory.

e-Science and hence the Grid is the infrastructure that enables collaborative science. The Grid can provide the basic building blocks to support real-time distance interaction, which has been exploited in distance education as described in Chapter 43. Particularly

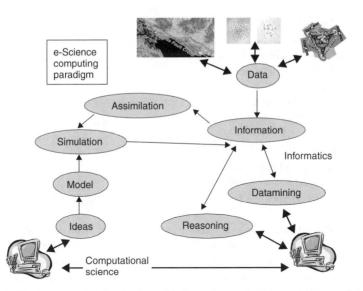

Figure 1.28 Computational science and information technology merge in e-Science.

important is the infrastructure to support shared resources – this includes many key services including security, scheduling and management, registration and search services
(Chapter 19) and the message-based interfaces of Web services (Chapter 18) to allow
powerful sharing (collaboration) mechanisms. All of the basic Grid services and infrastructure provide a critical venue for collaboration and will be highly important to the
community.

1.5.6 Commercial Applications

In the commercial world, Grid, Web and distributed computing, and information concepts
are being used in an innovative way in a wide variety of areas including inventory control,
enterprise computing, games and so on. The Butterfly Grid [117] and the Everquest multiplayer gaming environment [118] are current examples of gaming systems using Gridlike
environments. The success of SETI@home [36], a highly distributed data-mining application with the goal of identifying patterns of extraterrestrial intelligence from the massive
amounts of data received by the Arecibo radio telescope, has inspired both innovative
research and a cadre of companies to develop P2P technologies. Chapter 12 describes the
Entropia system, one of the intellectual leaders in this area of P2P or Megacomputing.
Another interesting application of this type, climateprediction.com [119], is being developed by the UK e-Science program. This will implement the ensemble (multiple initial
conditions and dynamical assumptions) method for climate prediction on a megacomputer.
    Enterprise computing areas where the Grid approach can be applied include [10]

- end-to-end automation,
- end-to-end security,

- virtual server hosting,
- disaster recovery,
- heterogeneous workload management,
- end-to-end systems management,
- scalable clustering,
- accessing the infrastructure,
- 'utility' computing,
- accessing new capability more quickly,
- better performance,
- reducing up-front investment,
- gaining expertise not available internally, and
- Web-based access (portal) for control (programming) of enterprise function.

Chapter 43 describes issues that arise in incorporating Web services into enterprise computing. In addition to these enterprise applications, the concept of 'e-Utility' has emerged to summarize 'X-on-demand': computing-on-demand, storage-on-demand, networking-on-demand, information-on-demand and so on. This generalizes the familiar concept of Application Service Providers (ASPs). Some clear examples today of computing-on-demand come from systems like Condor and Entropia (Chapters 11 and 12). The use of Grid technologies to support e-Utility can be merged with those of autonomic computing (Chapter 13 and Section 1.6.1) in a new generation of commercial systems. Other interesting commercial Grid activities include the Sun Grid Engine [32] and Platform Computing [34] implementing resource management and scheduling opportunities similar to those addressed by Condor in Chapter 11.

The growing partnership between the commercial sector and the academic community in the design and development of Grid technologies is likely to bear fruit in two important ways: as a vehicle for a new generation of scientific advances and as a vehicle for a new generation of successful commercial products.

1.5.7 Application Summary

Applications are key to the Grid, and the examples given above show that at this stage we have some clear successes and a general picture as to what works today. A major purpose of the broader Grid deployment activities described in the Section 1.4.3 is to encourage further application development. Ultimately, one would hope that the Grid will be the operating system of the Internet and will be viewed in this fashion. Today we must strive to improve the Grid software to make it possible that more than the 'marine corps' of application developers can use the Grid. We can identify three broad classes of applications that today are 'natural for Grids' [120].

- *Minimal communication applications*: These include the so-called 'embarrassingly parallel' applications in which one divides a problem into many essentially independent pieces. The successes of Entropia (Chapter 12), SETI@Home and other megacomputing projects are largely from this category.

- *Staged/linked applications (do Part A then do Part B)*: These include remote instrument applications in which one gets input from the instrument at Site A, compute/analyze data at Site B and visualizes at Site C. We can coordinate resources including computers, data archives, visualization and multiple remote instruments.

- *Access to resources (get something from/do something at Site A)*: This includes portals, access mechanisms and environments described in Part C of the book.

Chapters 35 to 42 describe many of the early successes as do several of the chapters in Part C that describe Grid environments used to develop problem-solving environments and portals. One influential project was the numerical relativity simulations of colliding black holes where Grids have provided the largest simulations. The Cactus Grid software was developed for this (Chapter 23) and an early prototype is described in Chapter 37. An interesting example is Synthetic Forces (SF) Express [121], which can be considered as an example of Grid technology applied to military simulations. This large-scale distributed interactive battle simulation decomposed terrain (Saudi Arabia, Kuwait, Iraq) contiguously among supercomputers and performed a simulation of 100 000 vehicles in early 1998 with vehicle (tanks, trucks and planes) location and state updated several times a second. Note that the military simulation community has developed its own sophisticated distributed object technology High-Level Architecture (HLA) [122] and the next step should involve integrating this with the more pervasive Grid architecture.

Next-generation Grid applications will include the following:

- *Adaptive applications (run where you can find resources satisfying criteria X),*
- *Real-time and on-demand applications (do something right now),*
- *Coordinated applications (dynamic programming, branch and bound)* and
- *Poly-applications (choice of resources for different components).*

Note that we still cannot 'throw any application at the Grid' and have resource management software determine where and how it will run.

There are many more Grid applications that are being developed or are possible. Major areas of current emphasis are health and medicine (brain atlas, medical imaging [123] as in Chapter 41, telemedicine, molecular informatics), engineering, particle physics (Chapter 39), astronomy (Chapter 38), chemistry and materials (Chapter 42 and [124]), environmental science (with megacomputing in [119]), biosciences and genomics (see Chapter 40 and [125, 126]), education (Chapter 43) and finally digital libraries (see Chapter 36).

Grid applications will affect everybody – scientists, consumers, educators and the general public. They will require a software environment that will support unprecedented diversity, globalization, integration, scale and use. This is both the challenge and the promise of the Grid.

1.6 FUTURES – GRIDS ON THE HORIZON

In many ways, the research, development and deployment of large-scale Grids are just beginning. Both the major application drivers and Grid technology itself will greatly change over the next decade. The future will expand existing technologies and integrate new technologies. In the future, more resources will be linked by more and better networks. At the end of the decade, sensors, PDAs, health monitors and other devices will be linked to the Grid. Petabyte data resources and petaflop computational resources will join low-level sensors and sensornets to constitute Grids of unprecedented heterogeneity and performance variation. Over the next decade, Grid software will become more sophisticated, supporting unprecedented diversity, scale, globalization and adaptation. Applications will use Grids in sophisticated ways, adapting to dynamic configurations of resources and performance variations to achieve goals of Autonomic computing.

Accomplishing these technical and disciplinary achievements will require an immense research, development and deployment effort from the community. Technical requirements will need to be supported by the human drivers for Grid research, development and education. Resources must be made available to design, build and maintain Grids that are of high capacity (rich in resources), of high capability (rich in options), persistent (promoting stable infrastructure and a knowledgeable workforce), evolutionary (able to adapt to new technologies and uses), usable (accessible, robust and easy-to-use), scalable (growth must be a part of the design), and able to support/promote new applications.

Today, many groups are looking beyond the challenges of developing today's Grids to the research and development challenges of the future. In this section, we describe some key areas that will provide the building blocks for the Grids of tomorrow.

1.6.1 Adaptative and autonomic computing

The Grid infrastructure and paradigm is often compared with the Electric Power Grid [127]. On the surface, the analogy holds up – the Grid provides a way to seamlessly virtualize resources so that they can provide access to effectively infinite computing cycles and data storage for the user who 'plugs in' to the Grid. The infrastructure managing which machines, networks, storage and other resources are used is largely hidden from the user in the same way as individuals generally do not know which power company, transformer, generator and so on are being used when they plug their electric appliance into a socket.

The analogy falls short when it comes to performance. Power is either there or not there. To the first order, the location of the plug should not make electrical devices plugged into it run better. However, on the Grid, the choice of the machine, the network and other component impacts greatly the performance of the program. This variation in performance can be leveraged by systems that allow programs to adapt to the dynamic performance that can be delivered by Grid resources. Adaptive computing is an important area of Grid middleware that will require considerable research over the next decade.

Early work in the academic community (e.g. the AppLeS project on adaptive scheduling [23], the GrADS project on adaptive program development and execution environments [69], Adaptive Middleware projects [128], the SRB [27], Condor [22] and others) have provided fundamental building blocks, but there is an immense amount of work that remains to be done. Current efforts in the commercial sector by IBM on 'Autonomic Computing', as discussed in Chapter 13, provide an exciting current focus likely to have a strong impact on the Grid. Through 'Project Eliza', IBM is exploring the concepts of software which is self-optimizing, self-configuring, self-healing and self-protecting to ensure that software systems are flexible and can adapt to change [129]. Moore's law [8] has of course a profound impact on computing. It describes the technology improvement that governs increasing CPU performance, memory size and disk storage. Further, it also underlies the improvement in sensor technology that drives the data deluge underlying much of e-Science. However, technology progress may provide increased capability at the cost of increased complexity. There are orders of magnitude more servers, sensors and clients to worry about. Such issues are explored in depth in Chapter 13 and we expect this to be an important aspect of Grid developments in the future. Both the nodes of the Grid and their organization must be made robust – internally fault-tolerant, as well as resilient to changes and errors in their environment. Ultimately, the Grid will need self-optimizing, self-configuring, self-healing and self-protecting components with a flexible architecture that can adapt to change.

1.6.2 Grid programming environments

A curious observation about computational environments is that as the environment becomes more complex, fewer robust and usable tools seem to be available for managing the complexity and achieving program performance. The 1980s saw more maturity in the development of parallel architecture models than effective parallel software, and currently, efforts to develop viable programming environments for the Grid are limited to just a few forward-looking groups.

In order for the Grid to be fully usable and useful, this state of affairs will need to change. It will be critical for developers and users to be able to debug programs on the Grid, monitor the performance levels of their programs on Grid resources and ensure that the appropriate libraries and environments are available on deployed resources. Part C of the book, which discusses about Grid computing environments, and is summarized in Chapter 20, describes this area. To achieve the full vision of the Grid, we will need compilers that can interact with resource discovery and resource selection systems to best target their programs and run-time environments that allow the migration of programs during execution to take advantage of more optimal resources. Robust, useful and usable programming environments will require coordinated research in many areas as well as test beds to test program development and run-time ideas. The GrADS project [69] provides a first example of an integrated approach to the design, development and prototyping of a Grid programming environment.

A key part of the user experience in computational environments is the way in which the user interacts with the system. There has been considerable progress in the important area of portals but the increasing complexity of Grid resources and the sophisticated manner in which applications will use the Grid will mandate new ways to access the Grid. 'Programming the Grid' really consists of two activities: preparation of the individual application nuggets associated with a single resource and integrating the nuggets to form a complete Grid program. An application nugget can be many things – the Structured Query Language (SQL) interface to a database, a parallel image processing algorithm and a finite element solver. Integrating nuggets to form a complete system may involve the dynamic integration of all the Grid and Portal system services.

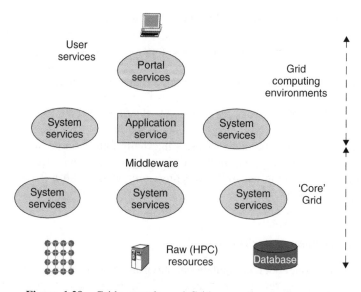

Figure 1.29   Grids, portals, and Grid computing environments.

An important area of research will target the development of appropriate models for interaction between users and applications and the Grid. Figure 1.29 illustrates the interrelation of Grid components involved in developing portals and Grid Computing Environments (GCEs): The horizontal direction corresponds to application and/or resource functionality (parallel simulation, sensor data gather, optimization, database etc.). The vertical direction corresponds to system functionality from scheduling to composition to portal rendering. Note that system state is determined by its environment, by user request and by the running application in some dynamic fashion. Currently, there is no 'consensus complete model' from user to resource and correspondingly no clear distinction between GCE and 'core' Grid capabilities (shown at the top and bottom of Figure 1.29, respectively). The matrix of capabilities sketched in Figure 1.29 and elaborated in Chapter 20 is very rich and we can expect different approaches to have value for different applications.

1.6.3 New Technologies

At the beginning of the twenty-first century, we are witnessing an immense explosion in telecommunications. The ubiquitous cell phones and PDAs of today are just the beginning of a deeper paradigm shift predicated upon the increasing availability of comprehensive information about the world around us.

Over the next decade, it will become increasingly important to application developers to integrate new devices and new information sources with the Grid. Sensors and sensor-nets embedded in bridges, roads, clothing and so on will provide an immense source of data. Real-time analysis of information will play an even greater role in health, safety, economic stability and other societal challenges. The integration of new devices will provide software and application challenges for the Grid community but will create a whole new level of potential for scientific advances.

1.6.4 Grid Policy and Grid Economies

Large-scale entities, such as the Science Collaboratories of Section 1.5.5, require organization in order to accomplish their goals. Complex systems from the Internet to the human cardiovascular system are organized hierarchically to manage/coordinate the interaction of entities via organizational structures that ensure system stability. The Grid will similarly require policies, organizational structure and an economy in order to maintain stability and promote individual and group performance. An important activity over the next decade will be the research, development and testing required to identify useful Grid policies, economies and 'social structures', which ensure the stability and efficiency of the Grid.

The Grid provides an interesting venue for policy. Grid resources may lie in different administrative domains and are governed by different local and national policies; however, the process of building and using the Grid is predicated on shared resources, agreement and coordination. Global collaboration heightens the need for community and culturally sensitive trust, policy, negotiation and payment services.

Most important, the Grid provides an exercise in cooperation: resource usage and administration must bridge technological, political and social boundaries, and Grid policies will need to provide an incentive to the individual (users and applications) to contribute to the success (stability) of the group.

1.6.5 A final note

The Grid vision is absolutely critical to future advances of science and society, but vision alone will not build the Grid. The promise and potential of the Grid must drive agendas for research, development and deployment over the next decade. In this book, we have asked a community of researchers, Grid developers, commercial partners and professionals to describe the current state of Grid middleware and their vision for the efforts that must drive future agendas. Building the Grid is one of the most challenging and exciting efforts in the science and technology community today, and more so because it must be done cooperatively and as a community effort. We hope that this book provides you, the reader, an insider's view of the challenges and issues involved in building the Grid and a sense of excitement about its potential and promise.

Box 1.2   Summary of Parts A and B of book (Chapters 1 to 19)

The initial chapter gives an overview of the whole book. Chapter 20 summarizes Part C and Chapter 35 summarizes Part D of the book. Here we summarize Parts A and B. Part A of this book, Chapters 1 to 5, provides an overview and motivation for Grids. Further, Chapter 37 is an illuminating discussion on Metacomputing from 1992–a key early concept on which much of the Grid has been built. Chapter 2 is a short overview of the Grid reprinted from Physics Today. Chapter 3 gives a detailed recent history of the Grid, while Chapter 4 describes the software environment of the seminal I-WAY experiment at SC95. As discussed in the main text of Section 1.2, this conference project challenged participants – including, for instance, the Legion activity of Chapter 10–to demonstrate Gridlike applications on an OC-3 backbone. Globus [3] grew out of the software needed to support these 60 applications at 17 sites; the human intensive scheduling and security used by I-WAY showed the way to today's powerful approaches. Many of these applications employed visualization including Computers and Advanced Visualization Environments (CAVE) virtual reality stations as demonstrated in the early Metacomputing work of Chapter 37. Chapter 5 brings us to 2002 and describes the experience building Globus-based Grids for NASA and DoE.

Turning to Part B of the book, Chapters 6 to 9 provide an overview of the community Grid approach in which the components of the Globus toolkit are being reimplemented as generalized OGSA Web services. Chapter 11 also fits into this thrust as Condor can operate as a stand-alone Grid but can also be thought of as providing workload and scheduling services for a general (Globus) Grid. Chapter 10 describes the Legion and Avaki approach, which pioneered the object model for Grids and provides an end-to-end solution that is compatible with the architecture of Figure 1.3. We will need to see how Globus, Condor and Avaki look after reformulation as Web services and if interoperability is possible and useful. All these systems should support the autonomic principles defined and described in Chapter 13.

This book illustrates Industry interest with Chapters 8, 10, 12, and 13 highlighting Avaki, Entropia and IBM; other key participation from Sun Microsystems (from the Sun Grid engine scheduling system [32] to the JXTA peer-to-peer network) [33]) and Platform Computing [34] are discussed in Chapters 3 and 18.

Chapter 12 on Entropia illustrates *megacomputing* or the harnessing of unused time on Internet clients [35]. Entropia can be thought of as a specialized Grid supplying needed management and fault tolerance for a megacomputing Grid with disparate unmanaged Internet or more structured enterprise nodes providing computing-on demand. Although early efforts of this type were part of I-WAY (see Fafner in Chapter 3), these ideas were developed most intensively in projects like SETI@home [36], which uses millions of Internet clients to analyze data looking for extraterrestrial life and for the newer project examining the folding of proteins [37]. These are building distributed computing solutions for applications, which can be divided into a huge number of essentially independent computations, and a central server system doles out separate work chunks to each participating client. In the parallel computing community, these problems are called 'pleasingly or embarrassingly parallel'. Other projects of this type include United Devices [38] and Parabon computation [39]. As explained in Chapter 12, other applications for this type of system include financial modeling, bioinformatics, Web performance and the scheduling of different jobs to use idle time on a network of workstations. Here the work links to Condor of Chapter 11, which focuses on more managed environments.

Chapters 14 to 17 address critical but different features of the data Grid supporting both the deluge from sensors and the more structured database and XML metadata resources. This is imperative if the Grid is to be used for what appear as the most promising applications. These chapters cover both the lower-level integration of data into the Grid fabric and the critical idea of a Semantic Grid – can knowledge be created in an emergent fashion by the linking of metadata enriched but not intrinsically intelligent Grid components.

Peer-to-peer (P2P) networks are an example of an approach to Gridlike systems which although crude today appears to offer both the scaling and autonomic self-sufficiency needed for the next generation Grid systems. Chapter 18 explains how to integrate P2P and Grid architectures, while Chapter 19 uses discovery of Web services to illustrate how P2P technology can provide the federation of disparate dynamic data resources.

REFERENCES

1. Foster, I. and Kesselman, C. (eds) (1999) *The Grid: Blueprint for a New Computing Infrastructure*. San Francisco, CA: Morgan Kaufmann.
2. The Global Grid Forum Web Site, http://www.gridforum.org.
3. The Globus Project Web Site, http://www.globus.org.
4. Berman, F., Fox, G. and Hey, T. (2003) *Grid Computing: Making the Global Infrastructure a Reality*. Chichester: John Wiley & Sons.
5. Web Site associated with book, *Grid Computing: Making the Global Infrastructure a Reality*, http://www.grid2002.org.
6. von Laszewski, G., Su, M.-H., Foster, I. and Kesselman, C. (2002) Chapter 8, in Dongarra, J., Foster, I., Fox, G., Gropp, W., Kennedy, K., Torczon, L., and White, A. (eds) *The Sourcebook of Parallel Computing*, ISBN 1-55860-871-0, San Francisco: Morgan Kaufmann Publishers.
7. Taylor, J. M. and e-Science, http://www.e-science.clrc.ac.uk and http://www.escience-grid.org.uk/.
8. Moore's Law as Explained by Intel, http://www.intel.com/research/silicon/mooreslaw.htm.
9. Gilder, G. (ed.) (2002) Gilder's law on network performance. *Telecosm: The World After Bandwidth Abundance*. ISBN: 0743205472, Touchstone Books.
10. Wladawsky-Berger, I. (Kennedy Consulting Summit) (2001) November 29, 2001, http://www-1.ibm.com/servers/events/pdf/transcript.pdf and http://www-1.ibm.com/servers/events/gridcomputing.pdf.
11. Berman, F. Presentations 2001 and 2002, http://share.sdsc.edu/dscgi/ds.py/View/Collection-551.
12. US/UK Grid Workshop, San Francisco, August 4, 5, 2001, http://www.isi.edu/us-uk.gridworkshop/presentations.html.
13. McRobbie, M. and Wallace, S. (2002) Spring 2002, Arlington, VA, Internet2 Member Meeting, May 6–8, 2002, http://www.internet2.edu/activities/html/spring_02.html and http://www.indiana.edu/~ovpit/presentations/.
14. Global Terabit Research Network, http://www.gtrn.net.
15. Abilene Network and Control Center, http://www.abilene.iu.edu/.
16. Dongarra, J., Foster, I., Fox, G., Gropp, W., Kennedy, K., Torczon, L. and White, A. (eds) (2002) *The Sourcebook of Parallel Computing*, ISBN 1-55860-871-0, San Francisco: Morgan Kaufmann Publishers.
17. NSF Grand Challenge as part of 1997 HPCC Implementation Plan http://www.itrd.gov/pubs/imp97/; their 2002 view at

http://www.itrd.gov/pubs/blue02/national-grand.html, National Grand Challenge Applications, National Coordination Office for Information Technology Research and Development.

18. CNRI, Corporation for National Research Initiatives, Gigabit Testbed Initiative Final Report, December, 1996, http://www1.cnri.reston.va.us/gigafr/.

19. Grimshaw, A. S. and Wulf, W. A. (1997) The legion vision of a worldwide virtual computer. *Communications of the ACM*, **40**(1), 39–45.

20. Grimshaw, A. S., Ferrari, A. J., Lindahl, G. and Holcomb, K. (1998) Metasystems. *Communications of the ACM*, **41**(11), 46–55.

21. Legion Worldwide Virtual Computer Home Page, http://legion.virginia.edu/index.html.

22. The Condor Project, http://www.cs.wisc.edu/condor/.

23. Berman, F., Wolski, R., Figueira, S., Schopf, J. and Shao, G. (1996) Application level scheduling on distributed heterogeneous networks. Proceedings of Supercomputing '96, 1996.

24. Gehrinf, J. and Reinfeld, A. (1996) Mars – a framework for minimizing the job execution time in a metacomputing environment. Proceedings of Future general Computer Systems, 1996.

25. Weissman, J. B. (1999) Prophet: automated scheduling of SPMD programs in workstation networks. *Concurrency: Practice and Experience*, **11**(6), 301–321.

26. The Network Weather Service, http://nws.cs.ucsb.edu/.

27. SDSC Storage Resource Broker, http://www.npaci.edu/DICE/SRB/.

28. Netsolve RPC Based Networked Computing, http://icl.cs.utk.edu/netsolve/.

29. Ninf Global Computing Infrastructure, http://ninf.apgrid.org/.

30. Internet Pioneer *J.C.R. Licklider,* http://www.ibiblio.org/pioneers/licklider.

31. NSF Middleware Initiative, http://www.nsf-middleware.org/.

32. Sun Grid Engine, http://wwws.sun.com/software/gridware/.

33. JXTA Peer-to-Peer Technology, http://www.jxta.org.

34. Platform Grid Computing, http://www.platform.com/grid/index.asp.

35. Foster, I. (2000) Internet computing and the emerging grid. *Nature*, **7**, http://www.nature.com/nature/webmatters/grid/grid.html.

36. SETI@Home Internet Computing, http://setiathome.ssl.berkeley.edu/.

37. Folding@home Internet Protein Structure, http://www.stanford.edu/group/pandegroup/Cosm/.

38. United Devices Internet Computing, http://www.ud.com/home.htm.

39. Parabon Java Computing, http://www.parabon.com.

40. superJANET4 United Kingdom Network, http://www.superjanet4.net/.

41. GEANT European Network by DANTE, http://www.dante.net/geant/.

42. CANARIE Advanced Canadian Networks, http://www.canarie.ca/advnet/advnet.html.

43. Asia-Pacific Advanced Network Consortium (APAN), http://www.apan.net/.

44. McRobbie, M. A. (Indiana University & Internet2), Wallace, S. (Indiana University & Internet2), van Houweling, D. (Internet2), Boyles, H. (Internet2), Liello, F. (European NREN Consortium), and Davies, D. (DANTE), A Global Terabit Research Network, http://www.gtrn.net/global.pdf.

45. California Institute for Telecommunications and Information Technology, http://www.calit2.net/.

46. Top 500 Supercomputers, http://www.top500.org.

47. Earth Simulator in Japan, http://www.es.jamstec.go.jp/esc/eng/index.html.

48. Fujitsu PRIMEPOWER HPC2500, http://pr.fujitsu.com/en/news/2002/08/22.html.

49. ASCI DoE Advanced Simulation and Computing Program, http://www.lanl.gov/asci/, http://www.llnl.gov/asci/, and http://www.sandia.gov/ASCI/.

50. NASA Information Power Grid, http://www.ipg.nasa.gov/.

51. DoE Department of Energy Science Grid, http://www.doesciencegrid.org.

52. TeraGrid Project, http://www.teragrid.org/.

53. NSF Advisory Committee on Cyberinfrastructure, http://www.cise.nsf.gov/b_ribbon/.

54. Asian Grid Center, http://www.apgrid.org/.

55. DutchGrid: Distributed Computing in the Netherlands, http://www.dutchgrid.nl/.

56. INRIA French Grid, http://www-sop.inria.fr/aci/grid/public/acigrid.html.

57. INFN and CNR Grids in Italy,
    http://www.ercim.org/publication/Ercim_News/enw45/codenotti.html.
58. CosmoGrid – National Computational Grid for Ireland, http://www.grid-ireland.org/.
59. PIONIER: Polish Optical Internet – Advanced Applications, Services and Technologies for
    Information Society, http://www.kbn.gov.pl/en/pionier/.
60. NorduGrid Scandinavian Grid, http://www.nordugrid.org/.
61. Transatlantic Grid, http://datatag.web.cern.ch/datatag/.
62. GRIP Unicore (Chapter 29 and http://www.unicore.org/links.htm), Globus Interoperability
    Project, http://www.grid-interoperability.org/.
63. Rheinheimer, R., Humphries, S. L., Bivens, H. P. and Beiriger, J. I. (2002) The ASCI com-
    putational grid: initial deployment. *Concurrency and Computation: Practice and Experience*
    14, Grid Computing Environments Special Issue 13–14.
64. National United Kingdom e-Science Center, http://umbriel.dcs.gla.ac.uk/NeSC/.
65. UK Grid Support Center, http://www.grid-support.ac.uk/.
66. GRIDS Grid Research Integration Deployment and Support Center,
    http://www.grids-center.org/.
67. International Virtual Data Grid Laboratory, http://www.ivdgl.org/.
68. iVDGL Grid Operations Center, http://igoc.iu.edu/.
69. GrADS Grid Application Development Software Project, http://nhse2.cs.rice.edu/grads/.
70. European DataGrid at CERN Accelerator Center, http://eu-datagrid.web.cern.ch/eu-datagrid/.
71. EUROGRID Grid Infrastructure, http://www.eurogrid.org/.
72. European Grid Application Toolkit and Testbed, http://www.gridlab.org/.
73. European Cross Grid Infrastructure Project, http://www.crossgrid.org/.
74. DAMIEN Metacomputing Project on Distributed Applications and Middleware for Industrial
    use of European Networks, http://www.hlrs.de/organization/pds/projects/damien/.
75. Grid Resource Broker Project, http://sara.unile.it/grb/grb.html.
76. DoE National Magnetic Fusion Collaboratory, http://www.fusiongrid.org/.
77. Grid Physics (Particle Physics, Astronomy, Experimental Gravitational waves) Network Gri-
    Phyn, http://www.griphyn.org/.
78. Particle Physics Data Grid, http://www.ppdg.net/.
79. National Virtual (Astronomical) Observatory, http://www.us-vo.org/.
80. European Astrophysical Virtual Observatory, http://www.eso.org/avo/.
81. European Grid of Solar Observations EGSO,
    http://www.mssl.ucl.ac.uk/grid/egso/egso_top.html.
82. NEES Grid, National Virtual Collaboratory for Earthquake Engineering Research,
    http://www.neesgrid.org/.
83. Solid Earth Research Virtual Observatory, http://www.servogrid.org.
84. DoE Earth Systems (Climate) Grid, http://www.earthsystemgrid.org/.
85. Biomedical Informatics Research Network BIRN Grid, http://www.nbirn.net/.
86. North Carolina Bioinformatics Grid, http://www.ncbiogrid.org/.
87. European Grid Resources for Industrial Applications, http://www.gria.org/.
88. Grid Acronym Soup Resource, http://www.gridpp.ac.uk/docs/GAS.html.
89. List of Grid Projects, http://www.escience-grid.org.uk/docs/briefing/nigridp.htm.
90. UK e-Science Network, http://www.research-councils.ac.uk/escience/documents/gridteam.pdf.
91. HPC(x) Press Release July 15 2002,
    http://www.research-councils.ac.uk/press/20020715supercomp.shtml.
92. The Internet Engineering Task Force IETF, http://www.ietf.org/.
93. The World Wide Web Consortium, http://www.w3c.org.
94. Parallel Virtual Machine, http://www.csm.ornl.gov/pvm/pvm_home.html.
95. Parallel Computing Message Passing Interface, http://www.mpi-forum.org/.
96. Linux Online, http://www.linux.org/.
97. OASIS Standards Organization, http://www.oasis-open.org/.
98. Apache Software Foundation, http://www.apache.org/.
99. Apache HTTP Server Project, http://httpd.apache.org/.

100. Apache Axis SOAP and WSDL Support, http://xml.apache.org/axis/index.html.
101. Apache Jakarta Jetspeed Portal, http://jakarta.apache.org/jetspeed/site/index.html.
102. Protein Data Bank Worldwide Repository for the Processing and Distribution of 3-D Biological Macromolecular Structure Data, http://www.rcsb.org/pdb/.
103. MyGrid – Directly Supporting the e-Scientist, http://www.mygrid.info/.
104. Mcell: General Monte Carlo Simulator of Cellular Microphysiology, http://www.mcell.cnl.salk.edu/.
105. National Center for Microscopy and Imaging Research Web-Based Telemicroscopy, http://ncmir.ucsd.edu/CMDA/jsb99.html.
106. Telescience for Advanced Tomography Applications Portal, http://gridport.npaci.edu/Telescience/.
107. Casanova, H., Bartol, T., Stiles, J. and Berman, F. (2001) Distributing MCell simulations on the grid. *International Journal of High Performance Computing Applications*, **14**(3), 243–257.
108. Open Grid Services Architecture Database Access and Integration (OGSA-DAI) UK e-Science Project, http://umbriel.dcs.gla.ac.uk/NeSC/general/projects/OGSA_DAI/.
109. Space Link Extension Standard, http://www.ccsds.org/rpa121/sm_review/.
110. GEODISE Grid for Engineering Design Search and Optimization Involving Fluid Dynamics, http://www.geodise.org/.
111. Distributed Aircraft Maintenance Environment DAME, http://www.iri.leeds.ac.uk/Projects/IAProjects/karim1.htm.
112. The Grid for UK Particle Physics, http://www.gridpp.ac.uk/.
113. Djorgovski Caltech, S. G. New Astronomy With a Virtual Observatory, and other presentations, http://www.astro.caltech.edu/~george/vo.
114. Fox, G. (2002) e-Science meets computational science and information technology. *Computing in Science and Engineering*, **4**(4), 84–85, http://www.computer.org/cise/cs2002/c4toc.htm.
115. W. Wulf. (1989).The National Collaboratory – A White Paper in Towards a National Collaboratory, Unpublished report of a NSF workshop, Rockefeller University, New York, March 17–18, 1989.
116. Kouzes, R. T., Myers, J. D. and Wulf, W. A. (1996) Collaboratories: doing science on the Internet, IEEE Computer August 1996, IEEE Fifth Workshops on Enabling Technology: Infrastructure for Collaborative Enterprises (WET ICE '96), Stanford, CA, USA, June 19–21, 1996, http://www.emsl.pnl.gov:2080/docs/collab/presentations/papers/IEEECollaboratories.html.
117. Butterfly Grid for Multiplayer Games, http://www.butterfly.net/.
118. Everquest Multiplayer Gaming Environment, http://www.everquest.com.
119. Ensemble Climate Prediction, http://www.climateprediction.com.
120. Fox, G. (2002) Chapter 4, in Dongarra, J., Foster, I., Fox, G., Gropp, W., Kennedy, K., Torczon, L. and White, A. (eds) *The Sourcebook of Parallel Computing*, ISBN 1-55860-871-0, San Francisco: Morgan Kaufmann Publishers.
121. Synthetic Forces Express, http://www.cacr.caltech.edu/SFExpress/.
122. Defense Modeling and Simulation Office High Level Architecture, https://www.dmso.mil/public/transition/hla/.
123. MIAS Medical Image Grid, http://www.gridoutreach.org.uk/docs/pilots/mias.htm.
124. Condensed Matter Reality Grid, http://www.realitygrid.org/.
125. ProteomeGRID for Structure-Based Annotation of the Proteins in the Major Genomes (Proteomes), http://umbriel.dcs.gla.ac.uk/Nesc/action/projects/project_action.cfm?title =34.
126. BiosimGRID for Biomolecular Simulations, http://umbriel.dcs.gla.ac.uk/Nesc/action/projects/project_action.cfm?title =35.
127. Chetty, M. and Buyya, R. (2002) Weaving computational grids: how analogous are they with electrical grids?. *Computing in Science and Engineering*, **4**, 61–71.
128. Darema, F. (2000) Chair of NSF Sponsored Workshop on Dynamic Data Driven Application Systems, March 8–10, 2000, http://www.cise.nsf.gov/eia/dddas/dddas-workshop-report.htm.
129. Horn, P. IBM, Autonomic Computing: IBM's Perspective on the State of Information Technology, http://www.research.ibm.com/autonomic/manifesto/autonomic_computing.pdf

Reprint from *Physics Today* © 2002 American Institute of Physics.
Minor changes to the original have been made to conform with house style.

2

The Grid: A new infrastructure for 21st century science

Ian Foster

Argonne National Laboratory, Argonne, Illinois, United States

As computer networks become cheaper and more powerful, a new computing paradigm is poised to transform the practice of science and engineering.

Driven by increasingly complex problems and propelled by increasingly powerful technology, today's science is as much based on computation, data analysis, and collaboration as on the efforts of individual experimentalists and theorists. But even as computer power, data storage, and communication continue to improve exponentially, computational resources are failing to keep up with what scientists demand of them.

A personal computer in 2001 is as fast as a supercomputer of 1990. But 10 years ago, biologists were happy to compute a single molecular structure. Now, they want to calculate the structures of complex assemblies of macromolecules (see Figure 2.1) and screen thousands of drug candidates. Personal computers now ship with up to 100 gigabytes (GB) of storage – as much as an entire 1990 supercomputer center. But by 2006, several physics projects, CERN's Large Hadron Collider (LHC) among them, will produce multiple petabytes (10^{15} byte) of data per year. Some wide area networks now operate at 155 megabits per second (Mb s^{-1}), three orders of magnitude faster than the state-of-the-art 56 kilobits per second (Kb s^{-1}) that connected U.S. supercomputer centers in 1985. But

Grid Computing – Making the Global Infrastructure a Reality.   Edited by F. Berman, A. Hey and G. Fox
© 2003 John Wiley & Sons, Ltd   ISBN: 0-470-85319-0

Figure 2.1 Determining the structure of a complex molecule, such as the cholera toxin shown here, is the kind of computationally intense operation that Grids are intended to tackle. (Adapted from G. von Laszewski *et al.*, *Cluster Computing*, **3**(3), page 187, 2000).

to work with colleagues across the world on petabyte data sets, scientists now demand tens of gigabits per second ($Gb\,s^{-1}$).

What many term the 'Grid' offers a potential means of surmounting these obstacles to progress [1]. Built on the Internet and the World Wide Web, the Grid is a new class of infrastructure. By providing scalable, secure, high-performance mechanisms for discovering and negotiating access to remote resources, the Grid promises to make it possible for scientific collaborations to share resources on an unprecedented scale and for geographically distributed groups to work together in ways that were previously impossible [2–4].

The concept of sharing distributed resources is not new. In 1965, MIT's Fernando Corbató and the other designers of the Multics operating system envisioned a computer facility operating 'like a power company or water company' [5]. And in their 1968 article 'The Computer as a Communications Device,' J. C. R. Licklider and Robert W. Taylor anticipated Gridlike scenarios [6]. Since the late 1960s, much work has been devoted to developing distributed systems, but with mixed success.

Now, however, a combination of technology trends and research advances makes it feasible to realize the Grid vision – to put in place a new international scientific infrastructure with tools that, together, can meet the challenging demands of twenty-first century science. Indeed, major science communities now accept that Grid technology is important for their future. Numerous government-funded R&D projects are variously developing core technologies, deploying production Grids, and applying Grid technologies to challenging applications. (For a list of major Grid projects, see http://www.mcs.anl.gov/~foster/grid-projects.)

2.1 TECHNOLOGY TRENDS

A useful metric for the rate of technological change is the average period during which speed or capacity doubles or, more or less equivalently, halves in price. For storage, networks, and computing power, these periods are around 12, 9, and 18 months, respectively. The different time constants associated with these three exponentials have significant implications.

The annual doubling of data storage capacity, as measured in bits per unit area, has already reduced the cost of a terabyte (10^{12} bytes) disk farm to less than \$10 000. Anticipating that the trend will continue, the designers of major physics experiments are planning petabyte data archives. Scientists who create sequences of high-resolution simulations are also planning petabyte archives.

Such large data volumes demand more from our analysis capabilities. Dramatic improvements in microprocessor performance mean that the lowly desktop or laptop is now a powerful computational engine. Nevertheless, computer power is falling behind storage. By doubling 'only' every 18 months or so, computer power takes five years to increase by a single order of magnitude. Assembling the computational resources needed for large-scale analysis at a single location is becoming infeasible.

The solution to these problems lies in dramatic changes taking place in networking. Spurred by such innovations as doping, which boosts the performance of optoelectronic devices, and by the demands of the Internet economy [7], the performance of wide area networks doubles every nine months or so; every five years it increases by two orders of magnitude. The NSFnet network, which connects the National Science Foundation supercomputer centers in the U.S., exemplifies this trend. In 1985, NSFnet's backbone operated at a then-unprecedented $56\,Kb\,s^{-1}$. This year, the centers will be connected by the $40\,Gb\,s^{-1}$ TeraGrid network (http://www.teragrid.org/) – an improvement of six orders of magnitude in 17 years.

The doubling of network performance relative to computer speed every 18 months has already changed how we think about and undertake collaboration. If, as expected, networks outpace computers at this rate, communication becomes essentially free. To exploit this bandwidth bounty, we must imagine new ways of working that are communication intensive, such as pooling computational resources, streaming large amounts of data from databases or instruments to remote computers, linking sensors with each other and with computers and archives, and connecting people, computing, and storage in collaborative environments that avoid the need for costly travel [8].

If communication is unlimited and free, then we are not restricted to using local resources to solve problems. When running a colleague's simulation code, I do not need to install the code locally. Instead, I can run it remotely on my colleague's computer. When applying the code to datasets maintained at other locations, I do not need to get copies of those datasets myself (not so long ago, I would have requested tapes). Instead, I can have the remote code access those datasets directly. If I wish to repeat the analysis many hundreds of times on different datasets, I can call on the collective computing power of my research collaboration or buy the power from a provider. And when I obtain interesting results, my geographically dispersed colleagues and I can look at and discuss large output datasets by using sophisticated collaboration and visualization tools.

Although these scenarios vary considerably in their complexity, they share a common thread. In each case, I use remote resources to do things that I cannot do easily at home. High-speed networks are often necessary for such remote resource use, but they are far from sufficient. Remote resources are typically owned by others, exist within different administrative domains, run different software, and are subject to different security and access control policies.

Actually using remote resources involves several steps. First, I must discover that they exist. Next, I must negotiate access to them (to be practical, this step cannot involve using the telephone!). Then, I have to configure my hardware and software to use the resources effectively. And I must do all these things without compromising my own security or the security of the remote resources that I make use of, some of which I may have to pay for.

Implementing these steps requires uniform mechanisms for such critical tasks as creating and managing services on remote computers, supporting single sign-on to distributed resources, transferring large datasets at high speeds, forming large distributed virtual communities, and maintaining information about the existence, state, and usage policies of community resources.

Today's Internet and Web technologies address basic communication requirements, but not the tasks just outlined. Providing the infrastructure and tools that make large-scale, secure resource sharing possible and straightforward is the Grid's raison d'être.

2.2 INFRASTRUCTURE AND TOOLS

An infrastructure is a technology that we can take for granted when performing our activities. The road system enables us to travel by car; the international banking system allows us to transfer funds across borders; and the Internet allows us to communicate with virtually any electronic device.

To be useful, an infrastructure technology must be broadly deployed, which means, in turn, that it must be simple, extraordinarily valuable, or both. A good example is the set of protocols that must be implemented within a device to allow Internet access. The set is so small that people have constructed matchbox-sized Web servers. A Grid infrastructure needs to provide more functionality than the Internet on which it rests, but it must also remain simple. And of course, the need remains for supporting the resources that power the Grid, such as high-speed data movement, caching of large datasets, and on-demand access to computing.

Tools make use of infrastructure services. Internet and Web tools include browsers for accessing remote Web sites, e-mail programs for handling electronic messages, and search engines for locating Web pages. Grid tools are concerned with resource discovery, data management, scheduling of computation, security, and so forth.

But the Grid goes beyond sharing and distributing data and computing resources. For the scientist, the Grid offers new and more powerful ways of working, as the following examples illustrate:

- *Science portals*: We are accustomed to climbing a steep learning curve when installing and using a new software package. Science portals make advanced problem-solving methods easier to use by invoking sophisticated packages remotely from Web browsers or other simple, easily downloaded 'thin clients.' The packages themselves can also run remotely on suitable computers within a Grid. Such portals are currently being developed in biology, fusion, computational chemistry, and other disciplines.
- *Distributed computing*: High-speed workstations and networks can yoke together an organization's PCs to form a substantial computational resource. Entropia Inc's Fight-AIDSAtHome system harnesses more than 30 000 computers to analyze AIDS drug candidates. And in 2001, mathematicians across the U.S. and Italy pooled their computational resources to solve a particular instance, dubbed 'Nug30,' of an optimization problem. For a week, the collaboration brought an average of 630 – and a maximum of 1006 – computers to bear on Nug30, delivering a total of 42 000 CPU-days. Future

improvements in network performance and Grid technologies will increase the range of problems that aggregated computing resources can tackle.

- *Large-scale data analysis*: Many interesting scientific problems require the analysis of large amounts of data. For such problems, harnessing distributed computing and storage resources is clearly of great value. Furthermore, the natural parallelism inherent in many data analysis procedures makes it feasible to use distributed resources efficiently. For example, the analysis of the many petabytes of data to be produced by the LHC and other future high-energy physics experiments will require the marshalling of tens of thousands of processors and hundreds of terabytes of disk space for holding intermediate results. For various technical and political reasons, assembling these resources at a single location appears impractical. Yet the collective institutional and national resources of the hundreds of institutions participating in those experiments can provide these resources. These communities can, furthermore, share more than just computers and storage. They can also share analysis procedures and computational results.

- *Computer-in-the-loop instrumentation*: Scientific instruments such as telescopes, synchrotrons, and electron microscopes generate raw data streams that are archived for subsequent batch processing. But quasi-real-time analysis can greatly enhance an instrument's capabilities. For example, consider an astronomer studying solar flares with a radio telescope array. The deconvolution and analysis algorithms used to process the data and detect flares are computationally demanding. Running the algorithms continuously would be inefficient for studying flares that are brief and sporadic. But if the astronomer could call on substantial computing resources (and sophisticated software) in an on-demand fashion, he or she could use automated detection techniques to zoom in on solar flares as they occurred.

- *Collaborative work*: Researchers often want to aggregate not only data and computing power but also human expertise. Collaborative problem formulation, data analysis, and the like are important Grid applications. For example, an astrophysicist who has performed a large, multiterabyte simulation might want colleagues around the world to visualize the results in the same way and at the same time so that the group can discuss the results in real time.

Real Grid applications will frequently contain aspects of several of these – and other–scenarios. For example, our radio astronomer might also want to look for similar events in an international archive, discuss results with colleagues during a run, and invoke distributed computing runs to evaluate alternative algorithms.

2.3 GRID ARCHITECTURE

Close to a decade of focused R&D and experimentation has produced considerable consensus on the requirements and architecture of Grid technology (see Box 2.1 for the early history of the Grid). Standard protocols, which define the content and sequence of message exchanges used to request remote operations, have emerged as an important and essential means of achieving the interoperability that Grid systems depend on. Also essential are standard application programming interfaces (APIs), which define standard interfaces to code libraries and facilitate the construction of Grid components by allowing code components to be reused.

Box 2.1 Historical origins

Grid concepts date to the earliest days of computing, but the genesis of much current Grid R&D lies in the pioneering work conducted on early experimental high-speed networks, such as the gigabit test beds that were established in the U.S. in the early 1990s [9].

One of these test beds was the CASA network, which connected four laboratories in California and New Mexico. Using CASA, Caltech's Paul Messina and his colleagues developed and demonstrated applications that coupled massively parallel and vector supercomputers for computational chemistry, climate modeling, and other sciences. Another test bed, Blanca, connected sites in the Midwest. Charlie Catlett of the National Center for Supercomputing Applications and his colleagues used Blanca to build multimedia digital libraries and demonstrated the potential of remote visualization. Two other test beds investigated remote instrumentation. The gigabit test beds were also used for experiments with wide area communication libraries and high-bandwidth communication protocols. Similar test beds were created in Germany and elsewhere.

Within the U.S. at least, the event that moved Grid concepts out of the network laboratory and into the consciousness of ordinary scientists was the I-WAY experiment [10]. Led by Tom DeFanti of the University of Illinois at Chicago and Rick Stevens of Argonne National Laboratory, this ambitious effort linked 11 experimental networks to create, for a week in November 1995, a national high-speed network infrastructure that connected resources at 17 sites across the U.S. and Canada. Some 60 application demonstrations, spanning the gamut from distributed computing to virtual reality collaboration, showed the potential of high-speed networks. The I-WAY also saw the first attempt to construct a unified software infrastructure for such systems, the I-Soft system. Developed by the author and others, I-Soft provided unified scheduling, single sign-on, and other services that allowed the I-WAY to be treated, in some important respects, as an integrated infrastructure

As Figure 2.2 shows schematically, protocols and APIs can be categorized according to the role they play in a Grid system. At the lowest level, the fabric, we have the physical devices or resources that Grid users want to share and access, including computers, storage systems, catalogs, networks, and various forms of sensors.

Above the fabric are the connectivity and resource layers. The protocols in these layers must be implemented everywhere and, therefore, must be relatively small in number. The connectivity layer contains the core communication and authentication protocols required for Grid-specific network transactions. Communication protocols enable the exchange of data between resources, whereas authentication protocols build on communication services to provide cryptographically secure mechanisms for verifying the identity of users and resources.

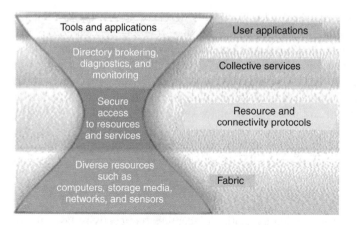

Figure 2.2 Grid architecture can be thought of a series of layers of different widths. At the center are the resource and connectivity layers, which contain a relatively small number of key protocols and application programming interfaces that must be implemented everywhere. The surrounding layers can, in principle, contain any number of components.

The resource layer contains protocols that exploit communication and authentication protocols to enable the secure initiation, monitoring, and control of resource-sharing operations. Running the same program on different computer systems depends on resource-layer protocols. The Globus Toolkit (which is described in Box 2.2) is a commonly used source of connectivity and resource protocols and APIs.

Box 2.2   The Globus Toolkit

The Globus Toolkit (http://www.globus.org/) is a community-based, open-architecture, open-source set of services and software libraries that supports Grids and Grid applications. The toolkit includes software for security, information infrastructure, resource management, data management, communication, fault detection, and portability. It is packaged as a set of components that can be used either independently or together to develop applications.

For each component, the toolkit both defines protocols and application programming interfaces (APIs) and provides open-source reference implementations in C and (for client-side APIs) Java. A tremendous variety of higher-level services, tools, and applications have been implemented in terms of these basic components. Some of these services and tools are distributed as part of the toolkit, while others are available from other sources. The NSF-funded GRIDS Center (http://www.grids-center.org/) maintains a repository of components.

Globus Project and Globus Toolkit are trademarks of the University of Chicago and University of Southern California.

The collective layer contains protocols, services, and APIs that implement interactions across collections of resources. Because they combine and exploit components from the relatively narrower resource and connectivity layers, the components of the collective layer can implement a wide variety of tasks without requiring new resource-layer components. Examples of collective services include directory and brokering services for resource discovery and allocation; monitoring and diagnostic services; data replication services; and membership and policy services for keeping track of who in a community is allowed to access resources.

At the top of any Grid system are the user applications, which are constructed in terms of, and call on, the components in any other layer. For example, a high-energy physics analysis application that needs to execute several thousands of independent tasks, each taking as input some set of files containing events, might proceed by

- *obtaining* necessary authentication credentials (connectivity layer protocols);
- *querying* an information system and replica catalog to determine availability of computers, storage systems, and networks, and the location of required input files (collective services);
- *submitting* requests to appropriate computers, storage systems, and networks to initiate computations, move data, and so forth (resource protocols); and
- *monitoring* the progress of the various computations and data transfers, notifying the user when all are completed, and detecting and responding to failure conditions (resource protocols).

Many of these functions can be carried out by tools that automate the more complex tasks. The University of Wisconsin's Condor-G system (http://www.cs.wisc.edu/condor) is an example of a powerful, full-featured task broker.

2.4 AUTHENTICATION, AUTHORIZATION, AND POLICY

Authentication, authorization, and policy are among the most challenging issues in Grids. Traditional security technologies are concerned primarily with securing the interactions between clients and servers. In such interactions, a client (that is, a user) and a server need to mutually authenticate (that is, verify) each other's identity, while the server needs to determine whether to authorize requests issued by the client. Sophisticated technologies have been developed for performing these basic operations and for guarding against and detecting various forms of attack. We use the technologies whenever we visit e-Commerce Web sites such as Amazon to buy products on-line.

In Grid environments, the situation is more complex. The distinction between client and server tends to disappear, because an individual resource can act as a server one moment (as it receives a request) and as a client at another (as it issues requests to other resources). For example, when I request that a simulation code be run on a colleague's computer, I am the client and the computer is a server. But a few moments later, that same code and computer act as a client, as they issue requests – on my behalf – to other computers to access input datasets and to run subsidiary computations. Managing that kind of transaction turns out to have a number of interesting requirements, such as

- *Single sign-on*: A single computation may entail access to many resources, but requiring a user to reauthenticate on each occasion (by, e.g., typing in a password) is impractical and generally unacceptable. Instead, a user should be able to authenticate once and then assign to the computation the right to operate on his or her behalf, typically for a specified period. This capability is achieved through the creation of a proxy credential. In Figure 2.3, the program run by the user (the user proxy) uses a proxy credential to authenticate at two different sites. These services handle requests to create new processes.
- *Mapping to local security mechanisms*: Different sites may use different local security solutions, such as Kerberos and Unix as depicted in Figure 2.3. A Grid security infrastructure needs to map to these local solutions at each site, so that local operations can proceed with appropriate privileges. In Figure 2.3, processes execute under a local

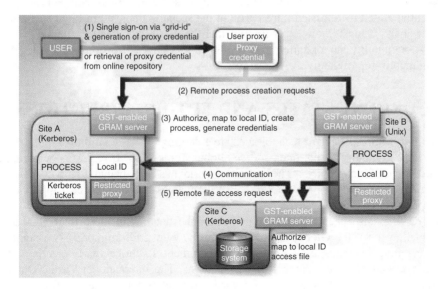

Figure 2.3 Smooth and efficient authentication and authorization of requests are essential for Grid operations. Here, a user calls on the computational resources of sites A and B, which then communicate with each other and read files located at a third site, C. Each step requires authorization and authentication, from the single sign-on (or retrieval of the proxy credential) that initiates the task to the remote file access request. Mediating these requests requires the Grid Security Infrastructure (GSI), which provides a single sign-on, run-anywhere authentication service, with support for delegation of credentials to subcomputations, local control over authorization, and mapping from global to local user identities. Also required is the Grid Resource Access and Management (GRAM) protocol and service, which provides remote resource allocation and process creation, monitoring, and management services.

ID and, at site A, are assigned a Kerberos 'ticket,' a credential used by the Kerberos authentication system to keep track of requests.

- *Delegation*: The creation of a proxy credential is a form of delegation, an operation of fundamental importance in Grid environments [11]. A computation that spans many resources creates subcomputations (subsidiary computations) that may themselves generate requests to other resources and services, perhaps creating additional subcomputations, and so on. In Figure 2.3, the two subcomputations created at sites A and B both communicate with each other and access files at site C. Authentication operations – and hence further delegated credentials – are involved at each stage, as resources determine whether to grant requests and computations determine whether resources are trustworthy. The further these delegated credentials are disseminated, the greater the risk that they will be acquired and misused by an adversary. These delegation operations and the credentials that enable them must be carefully managed.

- *Community authorization and policy*: In a large community, the policies that govern who can use which resources for what purpose cannot be based directly on individual identity. It is infeasible for each resource to keep track of community membership and privileges. Instead, resources (and users) need to be able to express policies in terms of other criteria, such as group membership, which can be identified with a cryptographic credential issued by a trusted third party. In the scenario depicted in Figure 2.3, the file server at site C must know explicitly whether the user is allowed to access a particular file. A community authorization system allows this policy decision to be delegated to a community representative.

2.5 CURRENT STATUS AND FUTURE DIRECTIONS

As the Grid matures, standard technologies are emerging for basic Grid operations. In particular, the community-based, open-source Globus Toolkit (see Box 2.2) is being applied by most major Grid projects. The business world has also begun to investigate Grid applications (see Box 2.3). By late 2001, 12 companies had announced support for the Globus Toolkit.

Progress has also been made on organizational fronts. With more than 1000 people on its mailing lists, the Global Grid Forum (http://www.gridforum.org/) is a significant force for setting standards and community development. Its thrice-yearly meetings attract hundreds of attendees from some 200 organizations. The International Virtual Data Grid Laboratory is being established as an international Grid system (Figure 2.4).

It is commonly observed that people overestimate the short-term impact of change but underestimate long-term effects [14]. It will surely take longer than some expect before Grid concepts and technologies transform the practice of science, engineering, and business, but the combination of exponential technology trends and R&D advances noted in this article are real and will ultimately have dramatic impacts.

In a future in which computing, storage, and software are no longer objects that we possess, but utilities to which we subscribe, the most successful scientific communities are likely to be those that succeed in assembling and making effective use of appropriate

Grid infrastructures and thus accelerating the development and adoption of new problem-solving methods within their discipline.

Box 2.3 Commercial Grids and the Open Grid Services Architecture

Grid concepts are becoming increasingly relevant to commercial information technology (IT). With the rise of e-Business and IT outsourcing, large-scale "enterprise" applications no longer run exclusively within the friendly confines of a central computing facility. Instead, they must operate on heterogeneous collections of resources that may span multiple administrative units within a company, as well as various external networks. Delivering high-quality service within dynamic virtual organizations is just as important in business as it is in science and engineering.

One consequence of this convergence is a growing interest in the integration of Grid technologies with previously distinct commercial technologies, which tend to be based on so-called Web services. Despite the name, Web services are not particularly concerned with Web sites, browsers, or protocols, but rather with standards for defining interfaces to, and communicating with, remote processes ("services"). Thus, for example, a distributed astronomical data system might be constructed as a set of Web services concerned variously with retrieving, processing, and visualizing data. By requiring input, such as a customer's address, in a certain format, Web services end up setting standards for remote services on the Web. Several major industrial distributed computing technologies, such as the Microsoft®. NET, IBM Corp's WebSphere, and Sun's Java™ 2 Enterprise Edition, are based on Web services [12].

To effect the integration of Grid technologies and Web services, the Globus Project and IBM's Open Service Architecture group have proposed the Open Grid Services Architecture [13]. In this blueprint, the two technologies are combined to define, among other things, standard behaviors and interfaces for what could be termed a Grid service: a Web service that can be created dynamically and that supports security, lifetime management, manageability, and other functions required in Grid scenarios. These features are being incorporated into the Globus Toolkit and will likely also appear in commercial products.

ACKNOWLEDGMENTS

I am grateful to David Abramson, Paul Avery, Fabrizio Gagliardi, Tony Hey, Satoshi Matsuoka, and Harvey Newman for their comments on an early draft of this article. My research is supported, in part, by grants from the U.S. Department of Energy, NSF, the Defense Advanced Research Projects Agency, NASA, and Microsoft.

Figure 2.4 The International Virtual Data Grid Laboratory (iVDGL) (http://www.ivdgl.org/) is being established to support both Grid research and production computing. The figure shows the approximate distribution of sites and networks planned for the initial rollout. (The actual sites could change by the time iVDGL becomes operational.) Major international projects, including EU DataGrid, Grid Physics Network, and Particle Physics Data Grid, are collaborating on the establishment of iVDGL.

REFERENCES

1. Foster, I. and Kesselman, C. (eds) (1999) *The Grid: Blueprint for a New Computing Infrastructure*. San Francisco, CA: Morgan Kaufmann Publishers.
2. Foster, I., Kesselman, C. and Tuecke, S. (2001) *International Journal of High Performance Computating Applications* **15**(3), 200, Also available at http://www.globus.org/research/papers/anatomy.pdf.
3. *National Collaboratories: Applying Information Technology for Scientific Research*. Washington, D.C.: National Academy Press, 1993, Also available at http://www.nap.edu/books/0309048486/html.
4. Teasley, S. and Wolinsky, S. (2001) *Science* **292**, 2254.
5. Vyssotsky, V. A., Corbató, F. J. and Graham, R. M. (1965) *Fall joint computer conference. AFIPS Conf. Proc.*, **27**, 203, Available at http://www.multicians.org/fjcc3.html.
6. Licklider, J. C. R. and Taylor, R. W. (1968) *Sci. Technol.*, Also available at http://memex.org/licklider.pdf.
7. Leiner, B. M. *et al.* (2000) *A Brief History of the Internet*. Reston, VA: Internet Society, Available at a http://www.isoc.org/internet-history/brief.html.
8. Kleinrock, L. (1992) *IEEE Communications Magazine*, **30**(4), 36.
9. Catlett, C. (1992) *IEEE Communications Magazine*, 42.
10. DeFanti, T. *et al.* (1996) *International Journal of Supercomputing Applications*, **10**(2), 123.
11. Gasser, M. and McDermott, E. (1990) *Proc. 1990 IEEE Symposium on Research in Security and Privacy*. Piscataway, NJ: IEEE Press, p. 20.

12. Graham, S. *et al.* (2001) *Building Web Services with Java: Making Sense of XML, SOAP, WSDL, and UDDI.* Indianapolis, IN: Sams Publishing.
13. Foster, I. *et al.* (2002) *The Physiology of the Grid: An Open Grid Services Architecture for Distributed Systems Integration.* Argonne, IL: Argonne National Laboratory, Available at http://www.globus.org/research/papers/ogsa.pdf.
14. Licklider, J. C. R. (1965) *Libraries of the Future.* Cambridge, MA: MIT Press.

The evolution of the Grid

David De Roure,[1] Mark A. Baker,[2] Nicholas R. Jennings,[1]
and Nigel R. Shadbolt[1]

[1]*University of Southampton, Southampton, United Kingdom,*
[2]*University of Portsmouth, Portsmouth, United Kingdom*

3.1 INTRODUCTION

The last decade has seen a substantial change in the way we perceive and use computing resources and services. A decade ago, it was normal to expect one's computing needs to be serviced by localised computing platforms and infrastructures. This situation has changed; the change has been caused by, among other factors, the take-up of commodity computer and network components, the result of faster and more capable hardware and increasingly sophisticated software. A consequence of these changes has been the capability for effective and efficient utilization of widely distributed resources to fulfil a range of application needs.

As soon as computers are interconnected and communicating, we have a distributed system, and the issues in designing, building and deploying distributed computer systems have now been explored over many years. An increasing number of research groups have been working in the field of wide-area distributed computing. These groups have implemented middleware, libraries and tools that allow the cooperative use of geographically distributed resources unified to act as a single powerful platform for the execution of

Grid Computing – Making the Global Infrastructure a Reality. Edited by F. Berman, A. Hey and G. Fox
© 2003 John Wiley & Sons, Ltd   ISBN: 0-470-85319-0

a range of parallel and distributed applications. This approach to computing has been known by several names, such as metacomputing, scalable computing, global computing, Internet computing and lately as Grid computing.

More recently there has been a shift in emphasis. In Reference [1], the 'Grid problem' is defined as 'Flexible, secure, coordinated resource sharing among dynamic collections of individuals, institutions, and resources'. This view emphasizes the importance of information aspects, essential for resource discovery and interoperability. Current Grid projects are beginning to take this further, from information to knowledge. These aspects of the Grid are related to the evolution of Web technologies and standards, such as XML to support machine-to-machine communication and the Resource Description Framework (RDF) to represent interchangeable metadata.

The next three sections identify three stages of Grid evolution: first-generation systems that were the forerunners of Grid computing as we recognise it today; second-generation systems with a focus on middleware to support large-scale data and computation; and current third-generation systems in which the emphasis shifts to distributed global collaboration, a service-oriented approach and information layer issues. Of course, the evolution is a continuous process and distinctions are not always clear-cut, but characterising the evolution helps identify issues and suggests the beginnings of a Grid roadmap. In Section 3.5 we draw parallels with the evolution of the World Wide Web and introduce the notion of the 'Semantic Grid' in which semantic Web technologies provide the infrastructure for Grid applications. A research agenda for future evolution is discussed in a companion paper (see Chapter 17).

3.2 THE EVOLUTION OF THE GRID: THE FIRST GENERATION

The early Grid efforts started as projects to link supercomputing sites; at this time this approach was known as metacomputing. The origin of the term is believed to have been the CASA project, one of several US Gigabit test beds deployed around 1989. Larry Smarr, the former NCSA Director, is generally accredited with popularising the term thereafter [2].

The early to mid 1990s mark the emergence of the early metacomputing or Grid environments. Typically, the objective of these early metacomputing projects was to provide computational resources to a range of high-performance applications. Two representative projects in the vanguard of this type of technology were FAFNER [3] and I-WAY [4]. These projects differ in many ways, but both had to overcome a number of similar hurdles, including communications, resource management, and the manipulation of remote data, to be able to work efficiently and effectively. The two projects also attempted to provide metacomputing resources from opposite ends of the computing spectrum. Whereas FAFNER was capable of running on any workstation with more than 4 Mb of memory, I-WAY was a means of unifying the resources of large US supercomputing centres.

3.2.1 FAFNER

The Rivest, Shamri and Adelman (RSA) public key encryption algorithm, invented by Rivest, Shamri and Adelman at MIT's Laboratory for Computer Science in 1976–1977

[5], is widely used; for example, in the Secure Sockets Layer (SSL). The security of RSA is based on the premise that it is very difficult to factor extremely large numbers, in particular, those with hundreds of digits. To keep abreast of the state of the art in factoring, RSA Data Security Inc. initiated the RSA Factoring Challenge in March 1991. The Factoring Challenge provides a test bed for factoring implementations and provides one of the largest collections of factoring results from many different experts worldwide.

Factoring is computationally very expensive. For this reason, parallel factoring algorithms have been developed so that factoring can be distributed. The algorithms used are trivially parallel and require no communications after the initial set-up. With this set-up, it is possible that many contributors can provide a small part of a larger factoring effort. Early efforts relied on electronic mail to distribute and receive factoring code and information. In 1995, a consortium led by Bellcore Labs., Syracuse University and Co-Operating Systems started a project, factoring via the Web, known as Factoring via Network-Enabled Recursion (FAFNER).

FAFNER was set up to factor RSA130 using a new numerical technique called the *Number Field Sieve* (NFS) factoring method using computational Web servers. The consortium produced a Web interface to NFS. A contributor then used a Web form to invoke server side Common Gateway Interface (CGI) scripts written in Perl. Contributors could, from one set of Web pages, access a wide range of support services for the sieving step of the factorisation: NFS software distribution, project documentation, anonymous user registration, dissemination of sieving tasks, collection of relations, relation archival services and real-time sieving status reports. The CGI scripts produced supported cluster management, directing individual sieving workstations through appropriate day/night sleep cycles to minimize the impact on their owners. Contributors downloaded and built a sieving software daemon. This then became their Web client that used HTTP protocol to GET values from and POST the resulting results back to a CGI script on the Web server.

Three factors combined to make this approach successful:

- The NFS implementation allowed even workstations with 4 Mb of memory to perform useful work using small bounds and a small sieve.
- FAFNER supported anonymous registration; users could contribute their hardware resources to the sieving effort without revealing their identity to anyone other than the local server administrator.
- A consortium of sites was recruited to run the CGI script package locally, forming a hierarchical network of RSA130 Web servers, which reduced the potential administration bottleneck and allowed sieving to proceed around the clock with minimal human intervention.

The FAFNER project won an award in TeraFlop challenge at Supercomputing 95 (SC95) in San Diego. It paved the way for a wave of Web-based metacomputing projects.

3.2.2 I-WAY

The information wide area year (I-WAY) was an experimental high-performance network linking many high-performance computers and advanced visualization environments

(CAVE). The I-WAY project was conceived in early 1995 with the idea not to build a network but to integrate existing high bandwidth networks. The virtual environments, datasets, and computers used resided at 17 different US sites and were connected by 10 networks of varying bandwidths and protocols, using different routing and switching technologies.

The network was based on Asynchronous Transfer Mode (ATM) technology, which at the time was an emerging standard. This network provided the wide-area backbone for various experimental activities at SC95, supporting both Transmission Control Protocol/Internet Protocol (TCP/IP) over ATM and direct ATM-oriented protocols.

To help standardize the I-WAY software interface and management, key sites installed point-of-presence (I-POP) servers to act as gateways to I-WAY. The I-POP servers were UNIX workstations configured uniformly and possessing a standard software environment called *I-Soft*. I-Soft attempted to overcome issues concerning heterogeneity, scalability, performance, and security. Each site participating in I-WAY ran an I-POP server. The I-POP server mechanisms allowed uniform I-WAY authentication, resource reservation, process creation, and communication functions. Each I-POP server was accessible via the Internet and operated within its site's firewall. It also had an ATM interface that allowed monitoring and potential management of the site's ATM switch.

The I-WAY project developed a resource scheduler known as the Computational Resource Broker (CRB). The CRB consisted of user-to-CRB and CRB-to-local-scheduler protocols. The actual CRB implementation was structured in terms of a single central scheduler and multiple local scheduler daemons – one per I-POP server. The central scheduler maintained queues of jobs and tables representing the state of local machines, allocating jobs to machines and maintaining state information on the Andrew File System (AFS) (a distributed file system that enables co-operating hosts to share resources across both local area and wide-area networks, based on the 'AFS' originally developed at Carnegie-Mellon University).

In I-POP, security was handled by using a telnet client modified to use Kerberos authentication and encryption. In addition, the CRB acted as an authentication proxy, performing subsequent authentication to I-WAY resources on a user's behalf. With regard to file systems, I-WAY used AFS to provide a shared repository for software and scheduler information. An AFS cell was set up and made accessible from only I-POPs. To move data between machines in which AFS was unavailable, a version of remote copy was adapted for I-WAY.

To support user-level tools, a low-level communications library, Nexus [6], was adapted to execute in the I-WAY environment. Nexus supported automatic configuration mechanisms that enabled it to choose the appropriate configuration depending on the technology being used, for example, communications via TCP/IP or AAL5 (the ATM adaptation layer for framed traffic) when using the Internet or ATM. The MPICH library (a portable implementation of the Message Passing Interface (MPI) standard) and CAVEcomm (networking for the CAVE virtual reality system) were also extended to use Nexus.

The I-WAY project was application driven and defined several types of applications:

- Supercomputing,
- Access to Remote Resources,

- Virtual Reality, and
- Video, Web, GII-Windows.

The I-WAY project was successfully demonstrated at SC'95 in San Diego. The I-POP servers were shown to simplify the configuration, usage and management of this type of wide-area computational test bed. I-Soft was a success in terms that most applications ran, most of the time. More importantly, the experiences and software developed as part of the I-WAY project have been fed into the Globus project (which we discuss in Section 3.2.2).

3.2.3 A summary of early experiences

Both FAFNER and I-WAY attempted to produce metacomputing environments by integrating resources from opposite ends of the computing spectrum. FAFNER was a ubiquitous system that worked on any platform with a Web server. Typically, its clients were low-end computers, whereas I-WAY unified the resources at multiple supercomputing centres.

The two projects also differed in the types of applications that could utilise their environments. FAFNER was tailored to a particular factoring application that was in itself trivially parallel and was not dependent on a fast interconnect. I-WAY, on the other hand, was designed to cope with a range of diverse high-performance applications that typically needed a fast interconnect and powerful resources. Both projects, in their way, lacked scalability. For example, FAFNER was dependent on a lot of human intervention to distribute and collect sieving results, and I-WAY was limited by the design of components that made up I-POP and I-Soft.

FAFNER lacked a number of features that would now be considered obvious. For example, every client had to compile, link, and run a FAFNER daemon in order to contribute to the factoring exercise. FAFNER was really a means of task-farming a large number of fine-grain computations. Individual computational tasks were unable to communicate with one another or with their parent Web-server. Likewise, I-WAY embodied a number of features that would today seem inappropriate. The installation of an I-POP platform made it easier to set up I-WAY services in a uniform manner, but it meant that each site needed to be specially set up to participate in I-WAY. In addition, the I-POP platform and server created one, of many, single points of failure in the design of the I-WAY. Even though this was not reported to be a problem, the failure of an I-POP would mean that a site would drop out of the I-WAY environment.

Notwithstanding the aforementioned features, both FAFNER and I-WAY were highly innovative and successful. Each project was in the vanguard of metacomputing and helped pave the way for many of the succeeding second-generation Grid projects. FAFNER was the forerunner of the likes of SETI@home [7] and Distributed.Net [8], and I-WAY for Globus [9] and Legion [10].

3.3 THE EVOLUTION OF THE GRID: THE SECOND GENERATION

The emphasis of the early efforts in Grid computing was in part driven by the need to link a number of US national supercomputing centres. The I-WAY project (see Section 3.2.2) successfully achieved this goal. Today the Grid infrastructure is capable of binding together more than just a few specialised supercomputing centres. A number of key enablers have helped make the Grid more ubiquitous, including the take-up of high bandwidth network technologies and adoption of standards, allowing the Grid to be viewed as a viable distributed infrastructure on a global scale that can support diverse applications requiring large-scale computation and data. This vision of the Grid was presented in Reference [11] and we regard this as the second generation, typified by many of today's Grid applications.

There are three main issues that had to be confronted:

- *Heterogeneity*: A Grid involves a multiplicity of resources that are heterogeneous in nature and might span numerous administrative domains across a potentially global expanse. As any cluster manager knows, their only truly homogeneous cluster is their first one!
- *Scalability*: A Grid might grow from few resources to millions. This raises the problem of potential performance degradation as the size of a Grid increases. Consequently, applications that require a large number of geographically located resources must be designed to be latency tolerant and exploit the locality of accessed resources. Furthermore, increasing scale also involves crossing an increasing number of organisational boundaries, which emphasises heterogeneity and the need to address authentication and trust issues. Larger scale applications may also result from the composition of other applications, which increases the 'intellectual complexity' of systems.
- *Adaptability*: In a Grid, a resource failure is the rule, not the exception. In fact, with so many resources in a Grid, the probability of some resource failing is naturally high. Resource managers or applications must tailor their behaviour dynamically so that they can extract the maximum performance from the available resources and services.

Middleware is generally considered to be the layer of software sandwiched between the operating system and applications, providing a variety of services required by an application to function correctly. Recently, middleware has re-emerged as a means of integrating software applications running in distributed heterogeneous environments. In a Grid, the middleware is used to hide the heterogeneous nature and provide users and applications with a homogeneous and seamless environment by providing a set of standardised interfaces to a variety of services.

Setting and using standards is also the key to tackling heterogeneity. Systems use varying standards and system application programming interfaces (APIs), resulting in the need for port services and applications to the plethora of computer systems used in a Grid environment. As a general principle, agreed interchange formats help reduce complexity, because n converters are needed to enable n components to interoperate via one standard, as opposed to n^2 converters for them to interoperate with each other.

In this section, we consider the second-generation requirements, followed by representatives of the key second-generation Grid technologies: core technologies, distributed object systems, Resource Brokers (RBs) and schedulers, complete integrated systems and peer-to-peer systems.

3.3.1 Requirements for the data and computation infrastructure

The data infrastructure can consist of all manner of networked resources ranging from computers and mass storage devices to databases and special scientific instruments. Additionally, there are computational resources, such as supercomputers and clusters. Traditionally, it is the huge scale of the data and computation, which characterises Grid applications.

The main design features required at the data and computational fabric of the Grid are the following:

- *Administrative hierarchy*: An administrative hierarchy is the way that each Grid environment divides itself to cope with a potentially global extent. The administrative hierarchy, for example, determines how administrative information flows through the Grid.
- *Communication services*: The communication needs of applications using a Grid environment are diverse, ranging from reliable point-to-point to unreliable multicast communication. The communications infrastructure needs to support protocols that are used for bulk-data transport, streaming data, group communications, and those used by distributed objects. The network services used also provide the Grid with important Quality of Service (QoS) parameters such as latency, bandwidth, reliability, fault tolerance, and jitter control.
- *Information services*: A Grid is a dynamic environment in which the location and type of services available are constantly changing. A major goal is to make all resources accessible to any process in the system, without regard to the relative location of the resource user. It is necessary to provide mechanisms to enable a rich environment in which information about the Grid is reliably and easily obtained by those services requesting the information. The Grid information (registration and directory) services provide the mechanisms for registering and obtaining information about the structure, resources, services, status and nature of the environment.
- *Naming services*: In a Grid, like in any other distributed system, names are used to refer to a wide variety of objects such as computers, services or data. The naming service provides a uniform namespace across the complete distributed environment. Typical naming services are provided by the international X.500 naming scheme or by the Domain Name System (DNS) used by the Internet.
- *Distributed file systems and caching*: Distributed applications, more often than not, require access to files distributed among many servers. A distributed file system is therefore a key component in a distributed system. From an application's point of view it is important that a distributed file system can provide a uniform global namespace, support a range of file I/O protocols, require little or no program modification, and provide means that enable performance optimisations to be implemented (such as the usage of caches).
- *Security and authorisation*: Any distributed system involves all four aspects of security: confidentiality, integrity, authentication and accountability. Security within a Grid environment is a complex issue requiring diverse resources autonomously administered to interact in a manner that does not impact the usability of the resources and that does not introduce security holes/lapses in individual systems or the environments as a whole. A security infrastructure is key to the success or failure of a Grid environment.

- *System status and fault tolerance*: To provide a reliable and robust environment it is important that a means of monitoring resources and applications is provided. To accomplish this, tools that monitor resources and applications need to be deployed.
- *Resource management and scheduling*: The management of processor time, memory, network, storage, and other components in a Grid are clearly important. The overall aim is the efficient and effective scheduling of the applications that need to utilise the available resources in the distributed environment. From a user's point of view, resource management and scheduling should be transparent and their interaction with it should be confined to application submission. It is important in a Grid that a resource management and scheduling service can interact with those that may be installed locally.
- *User and administrative GUI*: The interfaces to the services and resources available should be intuitive and easy to use as well as being heterogeneous in nature. Typically, user and administrative access to Grid applications and services are Web-based interfaces.

3.3.2 Second-generation core technologies

There are growing numbers of Grid-related projects, dealing with areas such as infrastructure, key services, collaborations, specific applications and domain portals. Here we identify some of the most significant to date.

3.3.2.1 Globus

Globus [9] provides a software infrastructure that enables applications to handle distributed heterogeneous computing resources as a single virtual machine. The Globus project is a US multi-institutional research effort that seeks to enable the construction of computational Grids. A computational Grid, in this context, is a hardware and software infrastructure that provides dependable, consistent, and pervasive access to high-end computational capabilities, despite the geographical distribution of both resources and users. A central element of the Globus system is the Globus Toolkit, which defines the basic services and capabilities required to construct a computational Grid. The toolkit consists of a set of components that implement basic services, such as security, resource location, resource management, and communications.

It is necessary for computational Grids to support a wide variety of applications and programming paradigms. Consequently, rather than providing a uniform programming model, such as the object-oriented model, the Globus Toolkit provides a bag of services that developers of specific tools or applications can use to meet their own particular needs. This methodology is only possible when the services are distinct and have well-defined interfaces (APIs) that can be incorporated into applications or tools in an incremental fashion.

Globus is constructed as a layered architecture in which high-level global services are built upon essential low-level core local services. The Globus Toolkit is modular, and an application can exploit Globus features, such as resource management or information infrastructure, without using the Globus communication libraries. The Globus Toolkit currently consists of the following (the precise set depends on the Globus version):

- An HTTP-based 'Globus Toolkit resource allocation manager' (GRAM) protocol is used for allocation of computational resources and for monitoring and control of computation on those resources.
- An extended version of the file transfer protocol, GridFTP, is used for data access; extensions include use of connectivity layer security protocols, partial file access, and management of parallelism for high-speed transfers.
- Authentication and related security services (GSI – Grid security infrastructure).
- Distributed access to structure and state information that is based on the lightweight directory access protocol (LDAP). This service is used to define a standard resource information protocol and associated information model.
- Remote access to data via sequential and parallel interfaces (GASS – global access to secondary storage) including an interface to GridFTP.
- The construction, caching and location of executables (GEM – Globus executable management).
- Resource reservation and allocation (GARA – Globus advanced reservation and allocation).

Globus has evolved from its original first-generation incarnation as I-WAY, through Version 1 (GT1) to Version 2 (GT2). The protocols and services that Globus provided have changed as it has evolved. The emphasis of Globus has moved away from supporting just high-performance applications towards more pervasive services that can support virtual organisations. The evolution of Globus is continuing with the introduction of the Open Grid Services Architecture (OGSA) [12], a Grid architecture based on Web services and Globus (see Section 3.4.1 for details).

3.3.2.2 Legion

Legion [10] is an object-based 'metasystem', developed at the University of Virginia. Legion provided the software infrastructure so that a system of heterogeneous, geographically distributed, high-performance machines could interact seamlessly. Legion attempted to provide users, at their workstations, with a single integrated infrastructure, regardless of scale, physical location, language and underlying operating system.

Legion differed from Globus in its approach to providing to a Grid environment: it encapsulated all its components as objects. This methodology has all the normal advantages of an object-oriented approach, such as data abstraction, encapsulation, inheritance and polymorphism.

Legion defined the APIs to a set of core objects that support the basic services needed by the metasystem. The Legion system had the following set of core object types:

- *Classes and metaclasses*: Classes can be considered as managers and policy makers. Metaclasses are classes of classes.
- *Host objects*: Host objects are abstractions of processing resources; they may represent a single processor or multiple hosts and processors.
- *Vault objects*: Vault objects represent persistent storage, but only for the purpose of maintaining the state of object persistent representation.

- *Implementation objects and caches*: Implementation objects hide details of storage object implementations and can be thought of as equivalent to an executable in UNIX.
- *Binding agents*: A binding agent maps object IDs to physical addressees.
- *Context objects and context spaces*: Context objects map context names to Legion object IDs, allowing users to name objects with arbitrary-length string names.

Legion was first released in November 1997. Since then the components that make up Legion have continued to evolve. In August 1998, Applied Metacomputing was established to exploit Legion commercially. In June 2001, Applied Metacomputing was relaunched as Avaki Corporation [13].

3.3.3 Distributed object systems

The Common Object Request Broker Architecture (CORBA) is an open distributed object-computing infrastructure being standardised by the Object Management Group (OMG) [14]. CORBA automates many common network programming tasks such as object registration, location, and activation; request de-multiplexing; framing and error-handling; parameter marshalling and de-marshalling; and operation dispatching. Although CORBA provides a rich set of services, it does not contain the Grid level allocation and scheduling services found in Globus (see Section 3.2.1), however, it is possible to integrate CORBA with the Grid.

The OMG has been quick to demonstrate the role of CORBA in the Grid infrastructure; for example, through the 'Software Services Grid Workshop' held in 2001. Apart from providing a well-established set of technologies that can be applied to e-Science, CORBA is also a candidate for a higher-level conceptual model. It is language-neutral and targeted to provide benefits on the enterprise scale, and is closely associated with the Unified Modelling Language (UML). One of the concerns about CORBA is reflected by the evidence of intranet rather than Internet deployment, indicating difficulty crossing organisational boundaries; for example, operation through firewalls. Furthermore, real-time and multimedia support were not part of the original design.

While CORBA provides a higher layer model and standards to deal with heterogeneity, Java provides a single implementation framework for realising distributed object systems. To a certain extent the Java Virtual Machine (JVM) with Java-based applications and services are overcoming the problems associated with heterogeneous systems, providing portable programs and a distributed object model through remote method invocation (RMI). Where legacy code needs to be integrated, it can be 'wrapped' by Java code.

However, the use of Java in itself has its drawbacks, the main one being computational speed. This and other problems associated with Java (e.g. numerics and concurrency) are being addressed by the likes of the Java Grande Forum (a 'Grande Application' is 'any application, scientific or industrial, that requires a large number of computing resources, such as those found on the Internet, to solve one or more problems') [15]. Java has also been chosen for UNICORE (see Section 3.6.3). Thus, what is lost in computational speed might be gained in terms of software development and maintenance times when taking a broader view of the engineering of Grid applications.

3.3.3.1 Jini and RMI

Jini [16] is designed to provide a software infrastructure that can form a distributed computing environment that offers network plug and play. A collection of Jini-enabled processes constitutes a Jini community – a collection of clients and services all communicating by the Jini protocols. In Jini, applications will normally be written in Java and communicated using the Java RMI mechanism. Even though Jini is written in pure Java, neither Jini clients nor services are constrained to be pure Java. They may include Java wrappers around non-Java code, or even be written in some other language altogether. This enables a Jini community to extend beyond the normal Java framework and link services and clients from a variety of sources.

More fundamentally, Jini is primarily concerned with communications between devices (not what devices do). The abstraction is the service and an interface that defines a service. The actual implementation of the service can be in hardware, software, or both. Services in a Jini community are mutually aware and the size of a community is generally considered that of a workgroup. A community's lookup service (LUS) can be exported to other communities, thus providing interaction between two or more isolated communities.

In Jini, a device or software service can be connected to a network and can announce its presence. Clients that wish to use such a service can then locate it and call it to perform tasks. Jini is built on RMI, which introduces some constraints. Furthermore, Jini is not a distributed operating system, as an operating system provides services such as file access, processor scheduling and user logins. The five key concepts of Jini are

- *Lookup*: to search for a service and to download the code needed to access it,
- *Discovery*: to spontaneously find a community and join,
- *Leasing*: time-bounded access to a service,
- *Remote events*: service A notifies service B of A's state change. Lookup can notify all services of a new service, and
- *Transactions*: used to ensure that a system's distributed state stays consistent.

3.3.3.2 The common component architecture forum

The Common Component Architecture Forum [17] is attempting to define a minimal set of standard features that a high-performance component framework would need to provide, or can expect, in order to be able to use components developed within different frameworks. Like CORBA, it supports component programming, but it is distinguished from other component programming approaches by the emphasis on supporting the abstractions necessary for high-performance programming. The core technologies described in the previous section, Globus or Legion, could be used to implement services within a component framework.

The idea of using component frameworks to deal with the complexity of developing interdisciplinary high-performance computing (HPC) applications is becoming increasingly popular. Such systems enable programmers to accelerate project development by introducing higher-level abstractions and allowing code reusability. They also provide clearly defined component interfaces, which facilitate the task of team interaction; such a standard will promote interoperability between components developed by different teams across different institutions. These potential benefits have encouraged research groups within a number of laboratories and universities to develop and experiment with prototype systems. There is a need for interoperability standards to avoid fragmentation.

3.3.4 Grid resource brokers and schedulers

3.3.4.1 Batch and scheduling systems

There are several systems available whose primary focus is batching and resource scheduling. It should be noted that all the packages listed here started life as systems for managing jobs or tasks on locally distributed computing platforms. A fuller list of the available software can be found elsewhere [18, 19].

- *Condor* [20] is a software package for executing batch jobs on a variety of UNIX platforms, in particular, those that would otherwise be idle. The major features of Condor are automatic resource location and job allocation, check pointing, and the migration of processes. These features are implemented without modification to the underlying UNIX kernel. However, it is necessary for a user to link their source code with Condor libraries. Condor monitors the activity on all the participating computing resources; those machines that are determined to be available are placed in a resource pool. Machines are then allocated from the pool for the execution of jobs. The pool is a dynamic entity – workstations enter when they become idle and leave when they get busy.
- The *portable batch system (PBS)* [21] is a batch queuing and workload management system (originally developed for NASA). It operates on a variety of UNIX platforms, from clusters to supercomputers. The PBS job scheduler allows sites to establish their own scheduling policies for running jobs in both time and space. PBS is adaptable to a wide variety of administrative policies and provides an extensible authentication and security model. PBS provides a GUI for job submission, tracking, and administrative purposes.
- The *sun Grid engine (SGE)* [22] is based on the software developed by Genias known as Codine/GRM. In the SGE, jobs wait in a holding area and queues located on servers provide the services for jobs. A user submits a job to the SGE, and declares a requirements profile for the job. When a queue is ready for a new job, the SGE determines suitable jobs for that queue and then dispatches the job with the highest priority or longest waiting time; it will try to start new jobs on the most suitable or least loaded queue.
- The *load sharing facility (LSF)* is a commercial system from Platform Computing Corp. [23]. LSF evolved from the Utopia system developed at the University of Toronto [24] and is currently the most widely used commercial job management system. LSF comprises distributed load sharing and batch queuing software that manages, monitors and analyses the resources and workloads on a network of heterogeneous computers, and has fault-tolerance capabilities.

3.3.4.2 Storage resource broker

The Storage Resource Broker (SRB) [25] has been developed at San Diego Supercomputer Centre (SDSC) to provide 'uniform access to distributed storage' across a range of storage devices via a well-defined API. The SRB supports file replication, and this can occur either off-line or on the fly. Interaction with the SRB is via a GUI. The SRB servers can be federated. The SRB is managed by an administrator, with authority to create user groups. A key feature of the SRB is that it supports metadata associated with a distributed file system, such as location, size and creation date information. It also supports the notion of application-level (or domain-dependent) metadata, specific to the content, which cannot be generalised across all data sets. In contrast with traditional network file systems, SRB is attractive for Grid applications in that it deals with large volumes of data, which can transcend individual storage devices, because it deals with metadata and takes advantage of file replication.

3.3.4.3 Nimrod/G resource broker and GRACE

Nimrod-G is a Grid broker that performs resource management and scheduling of param- eter sweep and task-farming applications [26, 27]. It consists of four components:

- A task-farming engine,
- A scheduler,
- A dispatcher, and
- Resource agents.

A Nimrod-G task-farming engine allows user-defined schedulers, customised applications or problem-solving environments (e.g. ActiveSheets [28]) to be 'plugged in', in place of default components. The dispatcher uses Globus for deploying Nimrod-G agents on remote resources in order to manage the execution of assigned jobs. The Nimrod-G scheduler has the ability to lease Grid resources and services depending on their capability, cost, and availability. The scheduler supports resource discovery, selection, scheduling, and the execution of user jobs on remote resources. The users can set the deadline by which time their results are needed and the Nimrod-G broker tries to find the best resources available in the Grid, uses them to meet the user's deadline and attempts to minimize the costs of the execution of the task.

Nimrod-G supports user-defined deadline and budget constraints for scheduling optimi- sations and manages the supply and demand of resources in the Grid using a set of resource trading services called *Grid Architecture for Computational Economy* (GRACE) [29]. There are four scheduling algorithms in Nimrod-G [28]:

- Cost optimisation uses the cheapest resources to ensure that the deadline can be met and that computational cost is minimized.
- Time optimisation uses all the affordable resources to process jobs in parallel as early as possible.
- Cost-time optimisation is similar to cost optimisation, but if there are multiple resources with the same cost, it applies time optimisation strategy while scheduling jobs on them.
- The conservative time strategy is similar to time optimisation, but it guarantees that each unprocessed job has a minimum budget-per-job.

The Nimrod-G broker with these scheduling strategies has been used in solving large- scale data-intensive computing applications such as the simulation of ionisation chamber calibration [27] and the molecular modelling for drug design [30].

3.3.5 Grid portals

A Web portal allows application scientists and researchers to access resources specific to a particular domain of interest via a Web interface. Unlike typical Web subject portals, a Grid portal may also provide access to Grid resources. For example, a Grid portal may authenticate users, permit them to access remote resources, help them make decisions about scheduling jobs, and allow users to access and manipulate resource information obtained and stored on a remote database. Grid portal access can also be personalised by the use of profiles, which are created and stored for each portal user. These attributes, and others, make Grid portals the appropriate means for Grid application users to access Grid resources.

3.3.5.1 The NPACI HotPage

The NPACI HotPage [31] is a user portal that has been designed to be a single point-of-access to computer-based resources, to simplify access to resources that are distributed across member organisations and allows them to be viewed either as an integrated Grid system or as individual machines.

The two key services provided by the HotPage are information and resource access and management services. The information services are designed to increase the effectiveness of users. It provides links to

- user documentation and navigation,
- news items of current interest,
- training and consulting information,
- data on platforms and software applications, and
- Information resources, such as user allocations and accounts.

The above are characteristic of Web portals. HotPage's interactive Web-based service also offers secure transactions for accessing resources and allows the user to perform tasks such as command execution, compilation, and running programs. Another key service offered by HotPage is that it provides status of resources and supports an easy mechanism for submitting jobs to resources. The status information includes

- CPU load/percent usage,
- processor node maps,
- queue usage summaries, and
- current queue information for all participating platforms.

3.3.5.2 The SDSC Grid port toolkit

The SDSC Grid port toolkit [32] is a reusable portal toolkit that uses HotPage infrastructure. The two key components of GridPort are the Web portal services and the application APIs. The Web portal module runs on a Web server and provides secure (authenticated) connectivity to the Grid. The application APIs provide a Web interface that helps end users develop customised portals (without having to know the underlying portal infrastructure). GridPort is designed to allow the execution of portal services and the client applications on separate Web servers. The GridPortal toolkit modules have been used to develop science portals for applications areas such as pharmacokinetic modelling, molecular modelling, cardiac physiology and tomography.

3.3.5.3 Grid portal development kit

The Grid Portal Collaboration is an alliance between NCSA, SDSC and NASA IPG [33]. The purpose of the Collaboration is to support a common set of components and utilities to make portal development easier and to allow various portals to interoperate by using the same core infrastructure (namely, the Grid Security Infrastructure (GSI) and Globus).

Example portal capabilities include the following:

- Either running simulations interactively or submitted to a batch queue.
- File transfer including file upload, file download, and third-party file transfers (migrating files between various storage systems).
- Querying databases for resource/job specific information.
- Maintaining user profiles that contain information about past jobs submitted, resources used, results information and user preferences.

The portal architecture is based on a three-tier model, in which a client browser securely communicates to a Web server over secure sockets (via https) connection. The Web server is capable of accessing various Grid services using the Globus infrastructure. The Globus Toolkit provides mechanisms for securely submitting jobs to a Globus gatekeeper, querying for hardware/software information using LDAP, and a secure PKI infrastructure using GSI.

The portals discussion in this subsection highlights the characteristics and capabilities that are required in Grid environments.

3.3.6 Integrated systems

As the second generation of Grid components emerged, a number of international groups started projects that integrated these components into coherent systems. These projects were dedicated to a number of exemplar high-performance wide-area applications. This section of the chapter discusses a representative set of these projects.

3.3.6.1 Cactus

Cactus [34] is an open-source problem-solving environment designed for scientists and engineers. Cactus has a modular structure that enables the execution of parallel applications across a range of architectures and collaborative code development between distributed groups. Cactus originated in the academic research community, where it was developed and used by a large international collaboration of physicists and computational scientists for black hole simulations.

Cactus provides a frontend to many core backend services, provided by, for example, Globus, the HDF5 parallel file I/O (HDF5 is a general purpose library and file format for storing scientific data), the PETSc scientific library (a suite of data structures and routines for parallel solution of scientific applications modelled by partial differential equations, using MPI) and advanced visualisation tools. The portal contains options to compile and deploy applications across distributed resources.

3.3.6.2 DataGrid

The European DataGrid project [35], led by CERN, is funded by the European Union with the aim of setting up a computational and data-intensive Grid of resources for the analysis of data coming from scientific exploration [36]. The primary driving application of the DataGrid project is the Large Hadron Collider (LHC), which will operate at CERN from about 2005 to 2015 and represents a leap forward in particle beam energy, density, and collision frequency. This leap is necessary in order to produce some examples of previously undiscovered particles, such as the Higgs boson or perhaps super-symmetric quarks and leptons.

The LHC will present a number of challenges in terms of computing. The project is designing and developing scalable software solutions and test beds in order to handle many Petabytes of distributed data, tens of thousands of computing resources (processors, disks, etc.), and thousands of simultaneous users from multiple research institutions. The main challenge facing the project is providing the means to share huge amounts of distributed data over the current network infrastructure. The DataGrid relies upon emerging Grid technologies that are expected to enable the deployment of a large-scale computational environment consisting of distributed collections of files, databases, computers, scientific instruments, and devices.

The objectives of the DataGrid project are

- to implement middleware for fabric and Grid management, including the evaluation, testing, and integration of existing middleware and research and development of new software as appropriate,
- to deploy a large-scale test bed, and
- to provide production quality demonstrations.

The DataGrid project is divided into twelve work packages distributed over four working groups: test bed and infrastructure, applications, computational and DataGrid middleware, management and dissemination. The work emphasizes on enabling the distributed processing of data-intensive applications in the area of high-energy physics, earth observation, and bioinformatics.

The DataGrid is built on top of Globus and includes the following components:

- *Job description language (JDL)*: a script to describe the job parameters.
- *User interface (UI)*: sends the job to the RB and receives the results.
- *Resource broker (RB)*: locates and selects the target Computing Element (CE).
- *Job submission service (JSS)*: submits the job to the target CE.
- *Logging and book keeping (L&B)*: records job status information.
- *Grid information service (GIS)*: Information Index about state of Grid fabric.
- *Replica catalogue*: list of data sets and their duplicates held on storage elements (SE).

The DataGrid test bed 1 is currently available and being used for high-energy physics experiments, and applications in Biology and Earth Observation. The final version of the DataGrid software is scheduled for release by the end of 2003.

3.3.6.3 UNICORE

UNIform Interface to COmputer REsources (UNICORE) [37] is a project funded by the German Ministry of Education and Research. The design goals of UNICORE include a uniform and easy to use GUI, an open architecture based on the concept of an abstract job, a consistent security architecture, minimal interference with local administrative procedures, exploitation of existing and emerging technologies through standard Java and Web technologies. UNICORE provides an interface for job preparation and secure submission to distributed supercomputer resources.

Distributed applications within UNICORE are defined as multi-part applications in which the different parts may run on different computer systems asynchronously or sequentially synchronized. A UNICORE job contains a multi-part application augmented by the information about destination systems, the resource requirements, and the dependencies between the different parts. From a structural viewpoint a UNICORE job is a recursive object containing job groups and tasks. UNICORE jobs and job groups carry the information of the destination system for the included tasks. A task is the unit that boils down to a batch job for the destination system.

The main UNICORE components are

- the job preparation agent (JPA),
- the job monitor controller (JMC),
- the UNICORE https server, also called the *Gateway*,
- the network job supervisor (NJS), and
- a Java applet-based GUI with an online help and assistance facility.

The UNICORE client enables the user to create, submit, and control jobs from any workstation or PC on the Internet. The client connects to a UNICORE gateway, which authenticates both the client and the user, before contacting the UNICORE servers, which in turn manage the submitted UNICORE jobs. Tasks destined for local hosts are executed via the native batch subsystem. Tasks to be run at a remote site are transferred to peer UNICORE gateways. All necessary data transfers and synchronisations are performed by the servers. These servers also retain status information and job output, passing it to the client upon user request.

The protocol between the components is defined in terms of Java objects. A low-level layer called the *UNICORE Protocol Layer* (UPL) handles authentication, SSL communication and transfer of data as in-lined byte-streams and a high-level layer (the Abstract Job Object or AJO class library) contains the classes to define UNICORE jobs, tasks and resource requests. Third-party components, such as Globus, can be integrated into the UNICORE framework to extend its functionality.

UNICORE is being extensively used and developed for the EuroGrid [38] project. EuroGrid is a project being funded by the European Commission. It aims to demonstrate the use of Grids in selected scientific and industrial communities, to address the specific requirements of these communities, and to highlight their use in the areas of biology, meteorology and computer-aided engineering (CAE). The objectives of the EuroGrid project include the support of the EuroGrid software infrastructure, the development of software components, and demonstrations of distributed simulation codes from different application areas (biomolecular simulations, weather prediction, coupled CAE simulations, structural analysis and real-time data processing).

3.3.6.4 WebFlow

WebFlow [39, 40] is a computational extension of the Web model that can act as a framework for wide-area distributed computing. The main design goal of WebFlow was to build a seamless framework for publishing and reusing computational modules on the Web, so that end users, via a Web browser, can engage in composing distributed applications using WebFlow modules as visual components and editors as visual authoring tools. WebFlow has a three-tier Java-based architecture that could be considered a visual dataflow system. The frontend uses applets for authoring, visualization, and control of the environment. WebFlow uses a servlet-based middleware layer to manage and interact with backend modules such as legacy codes for databases or high-performance simulations.

WebFlow is analogous to the Web; Web pages can be compared to WebFlow modules and hyperlinks that connect Web pages to intermodular dataflow channels. WebFlow content developers built and published modules by attaching them to Web servers. Application integrators used visual tools to link outputs of the source modules with inputs of the destination modules, thereby forming distributed computational graphs (or compute-Webs) and publishing them as composite WebFlow modules. A user activated these compute Webs by clicking suitable hyperlinks or customizing the computation either in terms of available parameters or by employing some high-level commodity tools for visual graph authoring. The backend of WebFlow was implemented using the Globus Toolkit, in particular, Meta Directory Service (MDS), GRAM, and GASS.

WebFlow was based on a mesh of Java-enhanced Web servers (Apache), running servlets that managed and coordinated distributed computation. This management infrastructure was implemented by three servlets: session manager, module manager, and connection manager. These servlets use URLs and can offer dynamic information about their services and current state. Each management servlet can communicate with others via sockets. The servlets are persistent and application independent.

Various implementations of WebFlow were developed; one version used CORBA, as the base distributed object model. WebFlow has evolved into the Gateway Computational Web portal [41] and the Mississippi Computational Web Portal [42].

3.3.7 Peer-to-Peer computing

One very plausible approach to address the concerns of scalability can be described as decentralisation, though this is not a simple solution. The traditional client-server model can be a performance bottleneck and a single point of failure, but is still prevalent because decentralisation brings its own challenges. Peer-to-Peer (P2P) computing [43] (as implemented, for example, by Napster [44], Gnutella [45], Freenet [46] and project JXTA [47]) and Internet computing (as implemented, for example, by the SETI@home [7], Parabon [48], and Entropia systems [49]) are examples of the more general computational structures that are taking advantage of globally distributed resources.

In P2P computing, machines share data and resources, such as spare computing cycles and storage capacity, via the Internet or private networks. Machines can also communicate directly and manage computing tasks without using central servers. This permits P2P computing to scale more effectively than traditional client-server systems that must expand a server's infrastructure in order to grow, and this 'clients as servers' decentralisation is attractive with respect to scalability and fault tolerance for the reasons discussed above. However, there are some obstacles to P2P computing:

- PCs and workstations used in complex P2P applications will require more computing power to carry the communications and security overhead that servers would other-wise handle.
- Security is an issue as P2P applications give computers access to other machines' resources (memory, hard drives, etc). Downloading files from other computers makes the systems vulnerable to viruses. For example, Gnutella users were exposed to the VBS_GNUTELWORM virus. Another issue is the ability to authenticate the identity of the machines with their peers.
- P2P systems have to cope with heterogeneous resources, such as computers using a variety of operating systems and networking components. Technologies such as Java and XML will help overcome some of these interoperability problems.
- One of the biggest challenges with P2P computing is enabling devices to find one another in a computing paradigm that lacks a central point of control. P2P computing needs the ability to find resources and services from a potentially huge number of globally based decentralised peers.

A number of P2P storage systems are being developed within the research community. In the Grid context, the following points raise important issues of security and anonymity:

- The Federated, Available, and Reliable Storage for an Incompletely Trusted Environment (FARSITE) is a serverless distributed file system [50].
- OceanStore, a global persistent data store that aims to provide a 'consistent, highly available, and durable storage utility atop an infrastructure composed of untrusted servers' and scale to very large numbers of users [51, 52].
- The self-certifying file system (SFS) is a network file system that aims to provide strong security over untrusted networks without significant performance costs [53].
- PAST is a 'large-scale peer-to-peer storage utility' that aims to provide high availability, scalability, and anonymity. Documents are immutable and the identities of content owners, readers, and storage providers are protected [54].

See Chapter 8 for further discussion of P2P in the Grid context.

3.3.7.1 JXTA

In April 2001, Sun Microsystems announced the creation of Project JXTA, an open source development community for P2P infrastructure and applications. JXTA describes a network system of five layers:

- Device, the bottom-most layer, means anything with an electronic pulse – desktop PCs, laptops, palmtops, set-tops, game consoles, cellular telephones, pagers, embedded controllers, and so on.
- Network-enabled operating environment consists of the JVM, which provides TCP/IP services and other resources to participate in a network.
- Platform is made up of functions that allow peers to identify each other (peering, and peer groups), know when a particular peer is available (presence), work out what resources (discovery) they have to offer, and communicate with them.
- Services provide the basic building blocks of applications, such as storage, resource aggregation, security, indexing, and searching.
- Application accommodates the likes of Napster, Groove, AIM, and SETI@home.

Much of what is in JXTA corresponds with the top three layers, leaving the JVM to handle basic device and network needs. It describes a suite of six protocols governing discovery, organization, monitoring, and intra-peer communication. A common XML-based messaging layer binds the protocols to the appropriate underlying network transports.

JXTA is a specification, rather than a software. The concepts and specification are separated from its reference implementation by a set of Java class libraries. Apart from the core JXTA components, the project provides rudimentary applications, including JXTA Shell, which provides a view of the JXTA environment via a UNIX-like command-line interface, a file-sharing/chat GUI interface and a content management system.

3.3.8 A summary of experiences of the second generation

In the second generation, the core software for the Grid has evolved from that provided by the early vanguard offerings, such as Globus (GT1) and Legion, which were dedicated to the provision of proprietary services to large and computationally intensive high-performance applications, through to the more generic and open deployment of Globus (GT2) and Avaki. Alongside this core software, the second generation also saw the development of a range of accompanying tools and utilities, which were developed to provide higher-level services to both users and applications, and spans resource schedulers and brokers as well as domain-specific users interfaces and portals. Peer-to-peer techniques have also emerged during this period.

3.4 THE EVOLUTION OF THE GRID: THE THIRD GENERATION

The second generation provided the interoperability that was essential to achieve large-scale computation. As further Grid solutions were explored, other aspects of the engineering of the Grid became apparent. In order to build new Grid applications it was desirable to be able to reuse existing components and information resources, and to assemble these components in a flexible manner. The solutions involved increasing adoption of a service-oriented model and increasing attention to metadata – these are two key characteristics of third-generation systems. In fact, the service-oriented approach itself has implications for the information fabric: the flexible assembly of Grid resources into a Grid application requires information about the functionality, availability and interfaces of the various components, and this information must have an agreed interpretation that can be processed by machine. For further discussion of the service-oriented approach, see the companion 'Semantic Grid' paper.

Whereas the Grid had traditionally been described in terms of large-scale data and computation, the shift in focus in the third generation was apparent from new descriptions. In particular, the terms 'distributed collaboration' and 'virtual organisation' were adopted in the 'anatomy' paper [1]. The third generation is a more holistic view of Grid computing and can be said to address the infrastructure for e-Science – a term that reminds us of the requirements (of doing new science, and of the e-Scientist) rather than the enabling technology. As Fox notes [55], the anticipated use of massively parallel computing facilities is only part of the picture that has emerged: there are also a lot more users, hence, loosely coupled distributed computing has not been dominated by the deployment of massively parallel machines.

There is a strong sense of automation in third-generation systems; for example, when humans can no longer deal with the scale and heterogeneity but delegate to processes to do so (e.g. through scripting), which leads to autonomy within the systems. This implies a need for coordination, which, in turn, needs to be specified programmatically at various levels – including process descriptions. Similarly, the increased likelihood of failure implies a need for automatic recovery: configuration and repair cannot remain manual tasks. These requirements resemble the self-organising and healing properties of biological systems, and have been termed *autonomic* after the autonomic nervous system. According to the definition in Reference [56], an autonomic system has the following eight properties:

1. Needs detailed knowledge of its components and status,
2. Must configure and reconfigure itself dynamically,
3. Seeks to optimise its behaviour to achieve its goal,
4. Is able to recover from malfunction,
5. Protect itself against attack,
6. Be aware of its environment,
7. Implement open standards, and
8. Make optimised use of resources.

The third-generation Grid systems now under development are beginning to exhibit many of these features.

In this section we consider first the service-oriented approach, looking at Web services and agent-based computing, and then the information layer issues.

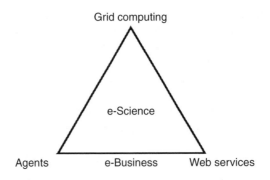

Figure 3.1   A view of e-Science technologies.

3.4.1 Service-oriented architectures

By 2001, a number of Grid architectures were apparent in a variety of projects. For example, the 'anatomy' paper proposed a layered model, and the Information Power Grid project [57] featured an extensive set of services, again arranged in layers. Around this time the Web services model was also gaining popularity, promising standards to support a service-oriented approach. In fact, one research community, agent-based computing, had already undertaken extensive work in this area: software agents can be seen as producers, consumers and indeed brokers of services. Altogether, it was apparent that the service-oriented paradigm provided the flexibility required for the third-generation Grid. Figure 3.1 depicts the three technologies.

3.4.1.1 Web services

The creation of Web services standards is an industry-led initiative, with some of the emerging standards in various states of progress through the World Wide Web Consortium (W3C). The established standards include the following:

- *SOAP (XML protocol)*: Simple object access protocol (SOAP) provides an envelope that encapsulates XML data for transfer through the Web infrastructure (e.g. over HTTP, through caches and proxies), with a convention for Remote Procedure Calls (RPCs) and a serialisation mechanism based on XML Schema datatypes. SOAP is being developed by W3C in cooperation with the Internet Engineering Task Force (IETF).
- *Web services description language (WSDL)*: Describes a service in XML, using an XML Schema; there is also a mapping to the RDF. In some ways WSDL is similar to an interface definition language IDL. WSDL is available as a W3C note [58].
- *Universal description discovery and integration (UDDI)*: This is a specification for distributed registries of Web services, similar to yellow and white pages services. UDDI supports 'publish, find and bind': a service provider describes and publishes the service details to the directory, service requestors make requests to the registry to find the providers of a service, the services 'bind' using the technical details provided by UDDI. It also builds on XML and SOAP [59].

The next Web service standards attracting interest are at the process level. For example, Web Services Flow Language (WSFL) [60] is an IBM proposal that defines workflows as combinations of Web services and enables workflows themselves to appear as services; XLANG [61] from Microsoft supports complex transactions that may involve multiple Web Services. A combined proposal called *Business Process Execution Language for Web Services* (BPEL4WS) is anticipated.

There are several other proposals for standards that address various aspects of Web services. In addition to the necessary machinery for Web services, there are important efforts that address the design of Web services systems. For example, the Web services Modelling Framework (WSMF) provides a conceptual model for developing and describing Web services based on the principles of maximal decoupling and a scalable mediation service [62].

Web services are closely aligned to the third-generation Grid requirements: they support a service-oriented approach and they adopt standards to facilitate the information aspects such as service description. In fact, WSDL describes how to interface with a service rather than the functionality of that service. Further work is required on service descriptions; one such activity is DAML-S [63].

3.4.1.2 The Open Grid Services Architecture (OGSA) framework

The OGSA Framework, the Globus-IBM vision for the convergence of Web services and Grid computing was presented at the Global Grid Forum (GGF) [82] meeting held in Toronto in February 2002. OGSA is described in the 'physiology' paper [12]. The GGF has set up an Open Grid Services working group to review and refine the Grid services architecture and documents that form the technical specification.

The OGSA supports the creation, maintenance, and application of ensembles of services maintained by Virtual Organizations (VOs). Here a service is defined as a network-enabled entity that provides some capability, such as computational resources, storage resources, networks, programs and databases. It tailors the Web services approach to meet some Grid-specific requirements. For example, these are the standard interfaces defined in OGSA:

- *Discovery*: Clients require mechanisms for discovering available services and for determining the characteristics of those services so that they can configure themselves and their requests to those services appropriately.
- *Dynamic service creation*: A standard interface (Factory) and semantics that any service creation service must provide.
- *Lifetime management*: In a system that incorporates transient and stateful service instances, mechanisms must be provided for reclaiming services and state associated with failed operations.
- *Notification*: A collection of dynamic, distributed services must be able to notify each other asynchronously of interesting changes to their state.
- *Manageability*: The operations relevant to the management and monitoring of large numbers of Grid service instances are provided.

- *Simple hosting environment*: A simple execution environment is a set of resources located within a single administrative domain and supporting native facilities for service management: for example, a J2EE application server, Microsoft. NET system, or Linux cluster.

The parts of Globus that are impacted most by the OGSA are

- The Grid resource allocation and management (GRAM) protocol.
- The information infrastructure, metadirectory service (MDS-2), used for information discovery, registration, data modelling, and a local registry.
- The Grid security infrastructure (GSI), which supports single sign-on, restricted delegation, and credential mapping.

It is expected that the future implementation of Globus Toolkit will be based on the OGSA architecture. Core services will implement the interfaces and behaviour described in the Grid service specification. Base services will use the Core services to implement both existing Globus capabilities, such as resource management, data transfer and information services, as well as new capabilities such as resource reservation and monitoring. A range of higher-level services will use the Core and Base services to provide data management, workload management and diagnostics services.

3.4.1.3 Agents

Web services provide a means of interoperability, the key to Grid computing, and OGSA is an important innovation that adapts Web services to the Grid and quite probably anticipates needs in other applications also. However, Web services do not provide a new solution to many of the challenges of large-scale distributed systems, nor do they yet provide new techniques for the engineering of these systems. Hence, it is valuable to look at other service-oriented models. The OGSA activity sits on one side of the triangle in Figure 3.1, and we suggest agent-based computing [64] as another important input to inform the service-oriented Grid vision.

The agent-based computing paradigm provides a perspective on software systems in which entities typically have the following properties, known as *weak agency* [65]. Note the close relationship between these characteristics and those of autonomic computing listed above.

1. *Autonomy*: Agents operate without intervention and have some control over their actions and internal state,
2. *Social ability*: Agents interact with other agents using an agent communication language,
3. *Reactivity*: Agents perceive and respond to their environment, and
4. *Pro-activeness*: Agents exhibit goal-directed behaviour.

Agent-based computing is particularly well suited to a dynamically changing environment, in which the autonomy of agents enables the computation to adapt to changing circumstances. This is an important property for the third-generation Grid. One of the techniques for achieving this is on-the-fly negotiation between agents, and there is a significant body of research in negotiation techniques [66]. Market-based approaches, in particular, provide an important approach to the computational economies required in Grid applications.

Hence, we can view the Grid as a number of interacting components, and the information that is conveyed in these interactions falls into a number of categories. One of those is the domain-specific content that is being processed. Additional types include

- information about components and their functionalities within the domain,
- information about communication with the components, and
- information about the overall workflow and individual flows within it.

These must be tied down in a standard way to promote interoperability between components, with agreed common vocabularies. Agent Communication Languages (ACLs) address exactly these issues. In particular, the Foundation for Intelligent Physical Agents (FIPA) activity [67] provides approaches to establishing a semantics for this information in an interoperable manner. FIPA produces software standards for heterogeneous and interacting agents and agent-based systems, including extensive specifications. In the FIPA abstract architecture:

- Agents communicate by exchanging messages that represent speech acts, and that are encoded in an agent communication language.
- Services provide support agents, including directory-services and message-transport-services.
- Services may be implemented either as agents or as software that is accessed via method invocation, using programming interfaces (e.g. in Java, C++ or IDL).

Again, we can identify agent-to-agent information exchange and directory entries as information formats that are required by the infrastructure 'machinery'.

3.4.2 Information aspects: relationship with the World Wide Web

In this section, we focus firstly on the Web. The Web's information handling capabilities are clearly an important component of the e-Science infrastructure, and the Web infrastructure is itself of interest as an example of a distributed system that has achieved global deployment. The second aspect addressed in this section is support for collaboration, something that is the key to the third-generation Grid. We show that the Web infrastructure itself lacks support for synchronous collaboration between users, and we discuss technologies that do provide such support.

It is interesting to consider the rapid uptake of the Web and how this might inform the design of the Grid, which has similar aspirations in terms of scale and deployment. One principle is clearly simplicity – there was little new in HTTP and HTML, and this facilitated massive deployment. We should, however, be aware of a dramatic contrast between Web and Grid: despite the large scale of the Internet, the number of hosts involved in a typical Web transaction is still small, significantly lower than that envisaged for many Grid applications.

3.4.2.1 The Web as a Grid information infrastructure

The Web was originally created for distribution of information in an e-Science context at CERN. So an obvious question to ask is, does this information distribution architecture described in Section 3.4.1 meet Grid requirements? A number of concerns arise.

- *Version control*: The popular publishing paradigm of the Web involves continually updating pages without version control. In itself, the Web infrastructure does not explicitly support versioning.
- *Quality of service*: Links are embedded, hardwired global references and they are fragile, are rendered useless by changing the server, location, name or content of the destination document. Expectations of link consistency are low and e-Science may demand a higher quality of service.
- *Provenance*: There is no standard mechanism to provide legally significant evidence that a document has been published on the Web at a particular time [68].
- *Digital rights management*: e-Science demands particular functionality with respect to management of the digital content, including, for example, copy protection and intellectual property management.
- *Curation*: Much of the Web infrastructure focuses on the machinery for delivery of information rather than the creation and management of content. Grid infrastructure designers need to address metadata support from the outset.

To address some of these issues we can look to work in other communities. For example, the multimedia industry also demands support for digital rights management. MPEG-21 aims to define 'a multimedia framework to enable transparent and augmented use of multimedia resources across a wide range of networks and devices used by different communities' [69], addressing the multimedia content delivery chain. Its elements include declaration, identification, content handling, intellectual property management and protection. Authoring is another major concern, especially collaborative authoring. The Web-based Distributed Authoring and Versioning (WebDAV) activity [70] is chartered 'to define the HTTP extensions necessary to enable distributed Web authoring tools to be broadly interoperable, while supporting user needs'.

   In summary, although the Web provides an effective layer for information transport, it does not provide a comprehensive information infrastructure for e-Science.

3.4.2.2 Expressing content and meta-content

The Web has become an infrastructure for distributed applications, in which information is exchanged between programs rather than being presented for a human reader. Such information exchange is facilitated by the Extensible Markup Language (XML) family of recommendations from W3C. XML is designed to mark up documents and has no fixed tag vocabulary; the tags are defined for each application using a Document Type Definition (DTD) or an XML Schema. A well-formed XML document is a labelled tree. Note that the DTD or Schema addresses syntactic conventions and does not address semantics. XML Schema are themselves valid XML expressions. Many new 'formats' are expressed in XML, such as synchronised multimedia integration language (SMIL).

RDF is a standard way of expressing metadata, specifically resources on the Web, though in fact it can be used to represent structured data in general. It is based on 'triples', where each triple expresses the fact that an object O has attribute A with value V, written A(O,V). An object can also be a value, enabling triples to be 'chained', and in fact, any RDF statement can itself be an object or attribute – this is called *reification* and permits nesting. RDF Schema are to RDF what XML Schema are to XML: they permit definition of a vocabulary. Essentially RDF schema provides a basic type system for RDF such as Class, subClassOf and subPropertyOf. RDF Schema are themselves valid RDF expressions.

XML and RDF (with XML and RDF schema) enable the standard expression of content and metacontent. Additionally, a set of tools has emerged to work with these formats, for example, parsers, and there is increasing support by other tools. Together this provides the infrastructure for the information aspects of the third-generation Grid.

W3C ran a 'Metadata Activity,' which addressed technologies including RDF, and this has been succeeded by the Semantic Web activity. The activity statement [71] describes the Semantic Web as follows:

'The Semantic Web is an extension of the current Web in which information is given well-defined meaning, better enabling computers and people to work in cooperation. It is the idea of having data on the Web defined and linked in a way that it can be used for more effective discovery, automation, integration, and reuse across various applications. The Web can reach its full potential if it becomes a place where data can be shared and processed by automated tools as well as by people.'

This vision is familiar – it shares much with the Grid vision. The Scientific American paper [72] provides motivation, with a scenario that uses agents. In a nutshell, the Semantic Web is intended to do for knowledge representation what the Web did for hypertext. We discuss this relationship further in Section 3.5.

The DARPA Agent Markup Language (DAML) Program [73], which began in 2000, brings Semantic Web technologies to bear on agent communication (as discussed in the previous section). DAML extends XML and RDF with ontologies, a powerful way of describing objects and their relationships. The Ontology Interchange Language (OIL) has been brought together with DAML to form DAML+OIL. W3C has created a Web Ontology Working Group, which is focusing on the development of a language based on DAML+OIL.

3.4.3 Live information systems

The third generation also emphasises distributed collaboration. One of the collaborative aspects builds on the idea of a 'collaboratory', defined in a 1993 US NSF study [74] as a 'centre without walls, in which the nation's researchers can perform their research without regard to geographical location – interacting with colleagues, accessing instrumentation, sharing data and computational resource, and accessing information in digital libraries.' This view accommodates 'information appliances' in the laboratory setting, which might, for example, include electronic logbooks and other portable devices.

3.4.3.1 Collaboration

As an information dissemination mechanism, the Web might have involved many users as 'sinks' of information published from major servers. However, in practice, part of the Web phenomenon has been widespread publishing by the users. This has had a powerful effect in creating online communities. However, the paradigm of interaction is essentially 'publishing things at each other', and is reinforced by e-mail and newsgroups, which also supports asynchronous collaboration.

Despite this, however, the underlying Internet infrastructure is entirely capable of supporting live (real-time) information services and synchronous collaboration. For example:

- Live data from experimental equipment,
- Live video feeds ('Webcams') via unicast or multicast (e.g. MBONE),
- Videoconferencing (e.g. H.323, coupled with T.120 to applications, SIP),
- Internet relay chat,
- Instant messaging systems,
- MUDs,
- Chat rooms, and
- Collaborative virtual environments.

All these have a role in supporting e-Science, directly supporting people, behind the scenes between processes in the infrastructure, or both. In particular, they support the extension of e-Science to new communities that transcend current organisational and geographical boundaries.

Although the histories of these technologies predate the Web, they can interoperate with the Web and build on the Web infrastructure technologies through adoption of appropriate standards. For example, messages that can be expressed in XML and URLs are routinely exchanged. In particular, the Web's metadata infrastructure has a role: data from experimental equipment can be expressed according to an ontology, enabling it to be processed by programs in the same way as static data such as library catalogues.

The application of computer systems to augment the capability of humans working in groups has a long history, with origins in the work of Doug Englebart [75]. In this context, however, the emphasis is on facilitating distributed collaboration, and we wish to embrace the increasingly 'smart' workplaces of the e-Scientist, including meeting rooms and laboratories. Amongst the considerable volume of work in the 'smart space' area, we note, in particular, the Smart Rooms work by Pentland [76] and Coen's work on the Intelligent Room [77]. This research area falls under the 'Advanced Collaborative Environments' Working group of the GGF (ACE Grid), which addresses both collaboration environments and ubiquitous computing.

3.4.3.2 Access Grid

The Access Grid [78] is a collection of resources that support human collaboration across the Grid, including large-scale distributed meetings and training. The resources include multimedia display and interaction, notably through room-based videoconferencing (group-to-group), and interfaces to Grid middleware and visualisation environments.

Access Grid nodes are dedicated facilities that explicitly contain the high, quality audio and video technology necessary to provide an effective user experience.

Current Access Grid infrastructure is based on IP multicast. The ISDN-based video-conferencing world (based on H.320) has evolved alongside this, and the shift now is to products supporting LAN-based videoconferencing (H.323). The T.120 protocol is used for multicast data transfer, such as remote camera control and application sharing. Meanwhile the IETF has developed Session Initiation Protocol (SIP), which is a signalling protocol for establishing real-time calls and conferences over the Internet. This resembles HTTP and uses Session Description Protocol (SDP) for media description.

During a meeting, there is live exchange of information, and this brings the information layer aspects to the fore. For example, events in one space can be communicated to other spaces to facilitate the meeting. At the simplest level, this might be slide transitions or remote camera control. These provide metadata that is generated automatically by software and devices, and can be used to enrich the conference and stored for later use. New forms of information may need to be exchanged to handle the large scale of meetings, such as distributed polling and voting.

Another source of live information is the notes taken by members of the meeting, or the annotations that they make on existing documents. Again, these can be shared and stored to enrich the meeting. A feature of current collaboration technologies is that sub-discussions can be created easily and without intruding – these also provide enriched content.

In videoconferences, the live video and audio feeds provide presence for remote participants – especially in the typical access Grid installation with three displays each with multiple views. It is also possible for remote participants to establish other forms of presence, such as the use of avatars in a collaborative virtual environment, and there may be awareness of remote participants in the physical meeting space.

The combination of Semantic Web technologies with live information flows is highly relevant to Grid computing and is an emerging area of activity [79]. Metadata streams may be generated by people, by equipment or by programs – for example, annotation, device settings, or data processed in real-time. Live metadata in combination with multimedia streams (such as multicast video) raises quality of service (QoS) demands on the network and raises questions about whether the metadata should be embedded (in which respect, the multimedia metadata standards are relevant). A scenario in which knowledge technologies are being applied to enhance collaboration is described in Reference [80].

3.5 SUMMARY AND DISCUSSION

In this chapter, we have identified the first three generations of the Grid:

- First-generation systems involved proprietary solutions for sharing high-performance computing resources;
- Second-generation systems introduced middleware to cope with scale and heterogeneity, with a focus on large-scale computational power and large volumes of data; and
- Third-generation systems are adopting a service-oriented approach, adopt a more holistic view of the e-Science infrastructure, are metadata-enabled and may exhibit autonomic features.

The evolution of the Grid has been a continuous process and these generations are not rigidly defined – they are perhaps best distinguished by philosophies rather than technologies. We suggest the book [11] marks the transition from first to second generation, and the 'anatomy' [1] and 'physiology' [12] papers mark the transition from a second to third-generation philosophy.

We have seen that in the third generation of the Grid, the early Semantic Web technologies provide the infrastructure for Grid applications. In this section, we explore further the relationship between the Web and the Grid in order to suggest future evolution.

3.5.1 Comparing the Web and the Grid

The state of play of the Grid today is reminiscent of the Web some years ago: there is limited deployment, largely driven by enthusiasts within the scientific community, with emerging standards and a degree of commercial uptake. The same might also be said of the Semantic Web. Meanwhile, the Web has seen a shift from machine-to human communications (HTML) to machine-to-machine (XML), and this is precisely the infrastructure needed for the Grid. Related to this, the Web services paradigm appears to provide an appropriate infrastructure for the Grid, though already Grid requirements are extending this model.

It is appealing to infer from these similarities that Grid deployment will follow the same exponential model as Web growth. However, a typical Grid application might involve large numbers of processes interacting in a coordinated fashion, while a typical Web transaction today still only involves a small number of hosts (e.g. server, cache, browser). Achieving the desired behaviour from a large-scale distributed system involves technical challenges that the Web itself has not had to address, though Web services take us into a similar world.

The Web provides an infrastructure for the Grid. Conversely, we can ask what the Grid offers to the Web. As a Web application, it raises certain challenges that motivate evolution of Web technologies – such as the enhancements to Web services in OGSA, which may well transcend Grid applications. It also provides a high-performance infrastructure for various aspects of Web applications, for example, in search, data mining, translation and multimedia information retrieval.

3.5.2 The Semantic Grid

The visions of the Grid and the Semantic Web have much in common but can perhaps be distinguished by a difference of emphasis: the Grid is traditionally focused on computation, while the ambitions of the Semantic Web take it towards inference, proof and trust. The Grid we are now building in this third generation is heading towards what we term the Semantic Grid: as the Semantic Web is to the Web, so the Semantic Grid is to the Grid. This is depicted in Figure 3.2.[1]

---

[1] The term was used by Erick Von Schweber in GGF2 and a comprehensive report on the Semantic Grid was written by the present authors for the UK e-Science Programme in July 2001 [81]. This particular representation of the Semantic Grid is due to Norman Paton of the University of Manchester, UK.

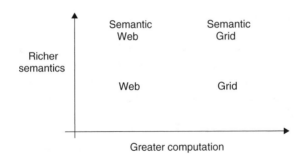

Figure 3.2 The Semantic Grid.

The Semantic Grid is achievable now in a simple but very powerful form – it is metadata-enabled and ontologically principled. Third-generation Grid is addressing the way that *information* is represented, stored, accessed, shared and maintained – information is understood as data equipped with meaning. We anticipate that the next generation will be concerned with the way that *knowledge* is acquired, used, retrieved, published and maintained to assist e-Scientists to achieve their particular goals and objectives – knowledge is understood as information applied to achieve a goal, solve a problem or enact a decision. The Semantic Grid involves all three conceptual layers of the Grid: knowledge, information and computation/data. These complementary layers will ultimately provide rich, seamless and pervasive access to globally distributed heterogeneous resources.

3.5.3 Research issues

The general view of the Grid is that of a three-layered system made up of computation/data, information and knowledge layers. Even though the computation/data layer of the Grid is the layer that is perhaps the most mature in terms of the time, experience and where most software is available and directly useable, it still lacks many essential aspects that will allow the provision of seamless, pervasive and secure use of system resources. A certain number of these aspects are now being addressed as the information and knowledge layers of the Grid evolve. The following generic areas are seen as ones that require further work:

- *Information services*: The mechanisms that are used to hold information about the resources in a Grid need to provide extendable, fast, reliable, secure, and scalable services.
- *Resource information*: All manner of Grid information will be necessary to enable the Grid to work correctly. This information will range from security data through to application requirements and from resource naming data through to user profiles. It is vital that all this information can be understood, interpreted and used, by all the services that require it.
- *Resource discovery*: Given a resource's unique name or characteristics there need to be mechanisms to locate the resource within the globally distributed system. Services

are resources. Some resources may persist, some may be transitory, and some may be created on demand.

- *Synchronisation and coordination*: How to orchestrate a complex sequence of computations over a variety of resources, given the inherent properties of both loosely- and tightly-coupled distributed systems. This may involve process description, and require an event-based infrastructure. It involves scheduling at various levels, including metascheduling and workflow.
- *Fault tolerance and dependability*: Environments need to cope with the failure of software and hardware components, as well as access issues – in general, accommodating the exception-handling that is necessary in such a dynamic, multi-user, multi-organisation system.
- *Security*: Authentication, authorisation, assurance, and accounting mechanisms need to be set in place, and these need to function in the context of increasing scale and automation. For example, a user may delegate privileges to processes acting on their behalf, which may in turn need to propagate some privileges further.
- *Concurrency and consistency*: The need to maintain an appropriate level of data consistency in the concurrent, heterogeneous environment. Weaker consistency may be sufficient for some applications.
- *Performance*: The need to be able to cope with non-local access to resources, through caching and duplication. Moving the code (or service) to the data (perhaps with scripts or mobile agents) is attractive and brings a set of challenges.
- *Heterogeneity*: The need to work with a multitude of hardware, software and information resources, and to do so across multiple organisations with different administrative structures.
- *Scalability*: Systems need to be able to scale up the number and size of services and applications, without scaling up the need for manual intervention. This requires automation, and ideally self-organisation.

At the information layer, although many of the technologies are available today (even if only in a limited form), a number of the topics still require further research. These include

- Issues relating to e-Science content types. Caching when new content is being produced. How will the Web infrastructure respond to the different access patterns resulting from automated access to information sources? These are issues that arise during the curation of e-Science content.
- Digital rights management in the e-Science context (as compared with multimedia and e-Commerce, for example).
- Provenance. Is provenance stored to facilitate reuse of information, repeat of experiments, or to provide evidence that certain information existed at a certain time?
- Creation and management of metadata, and provision of tools for metadata support.
- Service descriptions and tools for working with them. How best does one describe a service-based architecture?
- Workflow description and enaction, and tools for working with descriptions.
- Adaptation and personalisation. With the system 'metadata-enabled' throughout, how much knowledge can be acquired and how can it be used?

- Collaborative infrastructures for the larger community, including interaction between scientists, with e-Science content and visualisations, and linking smart laboratories and other spaces.
- Use of metadata in collaborative events, especially live metadata; establishing metadata schema to support collaboration in meetings and in laboratories.
- Capture and presentation of information using new forms of device; for example, for scientists working in the field.
- Representation of information about the underlying Grid fabric, as required by applications; for example, for resource scheduling and monitoring.

The Semantic Grid will be a place where data is equipped with a rich context and turned into information. This information will then be shared and processed by virtual organisations in order to achieve specific aims and objectives. Such actionable information constitutes knowledge. Hence, the knowledge layer is key to the next stage in the evolution of the Grid, to a fully fledged Semantic Grid. The research agenda to create the Semantic Grid is the subject of the companion paper 'The Semantic Grid: A Future e-Science Infrastructure'.

REFERENCES

1. Foster, I., Kesselman, C. and Tuecke, S. (2001) The anatomy of the Grid: enabling scalable virtual organizations. *International Journal of Supercomputer Applications and High Performance Computing*, **15**(3).
2. Catlett, C. and Smarr, L. (1992) Metacomputing. *Communications of the ACM*, **35**(6), 44–52.
3. FAFNER, http://www.npac.syr.edu/factoring.html.
4. Foster, I., Geisler, J., Nickless, W., Smith, W. and Tuecke, S. (1997) Software infrastructure for the I-WAY high performance distributed computing experiment. *Proc. 5th IEEE Symposium on High Performance Distributed Computing*, 1997, pp. 562–571.
5. Rivest, R. L., Shamir, A. and Adelman, L. (1977) *On Digital Signatures and Public Key Cryptosystems*, MIT Laboratory for Computer Science Technical Memorandum 82, April, 1977.
6. Foster, I., Kesselman, C. and Tuecke, S. (1996) The nexus approach to integrating multithreading and communication. *Journal of Parallel and Distributed Computing*, **37**, 70–82.
7. SETI@Home, http://setiathome.ssl.berkeley.edu/.
8. Distributed Net, http://www.distributed.net/.
9. Foster, I. and Kesselman, C. (1997) Globus: A metacomputing infrastructure toolkit. *International Journal of Supercomputer Applications*, **11**(2), 115–128, 1997.
10. Grimshaw, A. *et al.* (1997) The legion vision of a worldwide virtual computer. *Communications of the ACM*, **40**(1), 39–45.
11. Foster, I. and Kesselman, C. (eds) (1998) *The Grid: Blueprint for a New Computing Infrastructure*. San Francisco, CA: Morgan Kaufmann Publishers, ISBN 1-55860-475-8.
12. Foster, I., Kesselman, C., Nick, J. and Tuecke, S. (2002) *The Physiology of the Grid: Open Grid Services Architecture for Distributed Systems Integration*, Presented at GGF4, February, 2002, http://www.globus.og/research/papers/ogsa.pdf.
13. Avaki, http://www.avaki.com/.
14. OMG, http://www.omg.org.
15. Java Grande, http://www.javagrande.org/.
16. JINI, http://www.jini.org.
17. Armstrong, R. *et al.* (1999) Toward a common component architecture for high performance scientific computing. *Proceedings of the 8th High Performance Distributed Computing (HPDC '99)*, 1999.

18. Baker, M. A., Fox, G. C. and Yau, H. W. (1996) A review of cluster management software. *NHSE Review*, **1**(1), http://www.nhse.org/NHSEreview/CMS/.

19. Jones, J. P. (1996) *NAS Requirements Checklist for Job Queuing/Scheduling Software*. NAS Technical Report NAS-96-003, April, 1996, http://www.nas.nasa.gov/Research/Reports/Techreports/1996/nas-96-003-abstract.html.

20. Condor, http://www.cs.wisc.edu/condor/.

21. Portable Batch System, http://www.openpbs.org/.

22. Sun Grid Engine, http://www.sun.com/software/Gridware/.

23. Platform Computing, http://www.platform.com.

24. Zhou, S., Zheng, X., Wang, J. and Delisle, P. (1993) Utopia: A load sharing facility for large heterogeneous distributed computer systems. *Software Practice and Experience*, 1993, **23**, 1305–1336.

25. Rajasekar, A. K. and Moore, R. W. (2001) Data and metadata collections for scientific applications. *European High Performance Computing Conference*, Amsterdam, Holland, June 26, 2001.

26. Buyya, R., Abramson, D. and Giddy, J. (2000) Nimrod/G: An architecture for a resource management and scheduling system in a global computational Grid. *The 4th International Conference on High Performance Computing in Asia-Pacific Region (HPC Asia 2000)*, Beijing, China, 2000.

27. Abramson, D., Giddy, J. and Kotler, L. (2000) High performance parametric modeling with nimrod/G: killer application for the global Grid? *International Parallel and Distributed Processing Symposium (IPDPS)*. IEEE Computer Society Press.

28. Abramson, D., Roe, P., Kotler, L. and Mather, D. (2001) ActiveSheets: Super-computing with spreadsheets. *2001 High Performance Computing Symposium (HPC '01), Advanced Simulation Technologies Conference*, April, 2001.

29. Buyya, R., Giddy, J. and Abramson, D. (2000) An evaluation of economy-based resource trading and scheduling on computational power Grids for parameter sweep applications, *The Second Workshop on Active Middleware Services (AMS 2000), In Conjunction with HPDC 2001*. Pittsburgh, USA: Kluwer Academic Press.

30. Buyya, R. (2001) The virtual laboratory project: molecular modeling for drug design on Grid. *IEEE Distributed Systems Online*, **2**(5), http://www.buyya.com/vlab/.

31. HotPage, https://hotpage.npaci.edu/.

32. SDSC GridPort Toolkit, http://gridport.npaci.edu/.

33. NLANR Grid Portal Development Kit, http://dast.nlanr.net/Features/GridPortal/.

34. Allen, G., Dramlitsch, T., Foster, I., Karonis, N., Ripeanu, M., Seidel, Ed. and Toonen, B. (2001) *Supporting Efficient Execution in Heterogeneous Distributed Computing Environments with Cactus and Globus*, Winning Paper for Gordon Bell Prize (Special Category), Supercomputing 2001, August, 2001; Revised version.

35. The DataGrid Project, http://eu-datagrid.Web.cern.ch/.

36. Hoschek, W., Jaen-Martinez, J., Samar, A., Stockinger, H. and Stockinger, K. (2000) Data management in an international data Grid project. *Proceedings of the 1st IEEE/ACM International Workshop on Grid Computing (Grid 2000)*, Bangalore, India, December 17–20, 2000, Germany: Springer-Verlag.

37. Almond, J. and Snelling, D. (1999) UNICORE: Uniform access to supercomputing as an element of electronic commerce. *Future Generation Computer Systems*, **15**, 539–548.

38. EuroGrid, http://www.eurogrid.org/.

39. Akarsu, E., Fox, G. C., Furmanski, W. and Haupt, T. (1998) WebFlow: High-level programming environment and visual authoring toolkit for high performance distributed computing. *SC98: High Performance Networking and Computing*, Orlando, FL, 1998.

40. Haupt, T., Akarsu, E., Fox, G. and Furmanski, W. (1999) Web based metacomputing, Special Issue on Metacomputing, *Future Generation Computer Systems*. New York: North Holland.

41. Pierce, M. E., Youn, C. and Fox, G. C. The Gateway Computational Web Portal. Accepted for publication in Concurrency and Computation: Practice and Experience.

42. Haupt, T., Bangalore, P. and Henley, G. Mississippi Computational Web Portal. Accepted for publication in Concurrency and Computation: Practice and Experience. http://www.erc.msstate.edu/~haupt/DMEFS/welcome.html.

43. Clark, D. (2001) Face-to-face with peer-to-peer networking. *Computer*, **34**(1), 18–21.

44. Napster, http://www.napster.com/.

45. Gnutella, http://www.gnutella.co.uk/.

46. Clarke, I., Sandberg, O., Wiley, B. and Hong, T. W. (2001) Freenet: A distributed anonymous information storage and retrieval system. *ICSI Workshop on Design Issues in Anonymity and Unobservability*, 2001.

47. JXTA, http://www.jxta.org/.

48. Parabon, http://www.parabon.com/.

49. Entropia, http://entropia.com/.

50. Bolosky, W. J., Douceur, J. R., Ely, D. and Theimer, M. (2000) Feasibility of a serverless distributed file system deployed on an existing set of desktop PCs, *Proc. International Conference on Measurement and Modeling of Computer Systems (SIGMETRICS 2000)*. New York: ACM Press, pp. 34–43.

51. Kubiatowicz, J. *et al.* (2000) OceanStore: An architecture for global-scale persistent storage. *Proceedings of the Ninth International Conference on Architectural Support for Programming Languages and Operating Systems (ASPLOS 2000)*, November, 2000.

52. Zhuang, S. Q., Zhao, B. Y., Joseph, A. D., Katz, R. H. and Kubiatowicz, J. (2001) Bayeux: An architecture for scalable and fault-tolerant wide-area data dissemination. *Proceedings of the Eleventh International Workshop on Network and Operating System Support for Digital Audio and Video (NOSSDAV 2001)*, June, 2001.

53. Fu, Kevin, Frans Kaashoek, M. and Mazières, D. (2000) Fast and secure distributed read-only file system. *Proceedings of the 4th USENIX Symposium on Operating Systems Design and Implementation (OSDI 2000)*, San Diego, CA, October, 2000.

54. Druschel, P. and Rowstron, A. (2001) PAST: A large-scale, persistent peer-to-peer storage utility, *HotOS VIII*. Germany: Schoss Elmau.

55. Fox, G. C. (2002) From Computational Science to Internetics: Integration of Science with Computer Science, in Boisvert, R. F. and Houstis, E. (eds) *Computational Science, Mathematics and Software*. West Lafayette, Indiana: Purdue University Press, pp. 217–236, ISBN 1-55753-250-8.

56. IBM Autonomic Computing, http://www.research.ibm.com/autonomic/.

57. NASA Information Power Grid, http://www.ipg.nasa.gov/.

58. Web Services Description Language (WSDL) Version 1.1, W3C Note 15, March, 2001, http://www.w3.org/TR/wsdl.

59. http://www.uddi.org/.

60. Web Services Flow Language (WSFL) Version 1.0, http://www-4.ibm.com/software/solutions/Webservices/pdf/WSFL.pdf.

61. Web Services for Business Process Design, http://www.gotdotnet.com/team/xml_wsspecs/xlang-c/default.htm.

62. Fensel, D. and Bussler, C. *The Web Service Modeling Framework WSMF*, http://www.cs.vu.nl/~dieter/wese/.

63. DAML Services Coalition (Ankolenkar, A. *et al.*) (2002) DAML-S: Web service description for the semantic web. *The First International Semantic Web Conference (ISWC)*, June, 2002, pp. 348–363.

64. Jennings, N. R. (2001) An agent-based approach for building complex software systems. *Communications of the ACM*, **44**(4), 35–41.

65. Wooldridge, M. and Jennings, N. R. (1995) Intelligent agents: theory and practice. *Knowledge Engineering Review*, **10**(2), 115–152.

66. Jennings, N. R., Faratin, P., Lomuscio, A. R., Parsons, S., Sierra, C. and Wooldridge, M. (2001) Automated negotiation: prospects, methods and challenges. *Journal of Group Decision and Negotiation*, **10**(2), 199–215.

67. The Foundation for Physical Agents, http://www.fipa.org/.

68. Haber, S. and Stornetta, W. S. (1991) How to time-stamp a digital document. *Journal of Cryptography*, **3**(2), 99–111.
69. ISO/IEC JTC1/SC29/WG11, Coding of Moving Picture and Audio, MPEG-21 Overview, Document N4041.
70. Web-based Distributed Authoring and Versioning, http://www.Webdav.org/.
71. W3C Semantic Web Activity Statement, http://www.w3.org/2001/sw/Activity.
72. Berners-Lee, T., Hendler, J. and Lassila, O. (2001) The Semantic Web. *Scientific American*, **284**(5), 34–43.
73. Hendler, J. and McGuinness, D. (2000) The DARPA agent Markup language. *IEEE Intelligent Systems*, **15**(6), 72–73.
74. Cerf, V. G. *et al.* (1993) *National Collaboratories: Applying Information Technologies for Scientific Research*. Washington, D.C.: National Academy Press.
75. Englebart, D. (1962) *Augmenting Human Intellect: A Conceptual Framework*, AFOSR-3233, October, 1962,
http://sloan.stanford.edu/mousesite/EngelbartPapers/B5_F18_ConceptFrameworkInd.html.
76. Pentland, (1996) Smart rooms. *Scientific American*, **274**(4), 68–76.
77. Coen, M. A. (1998) A prototype intelligent environment, in Streitz, N., Konomi, S. and Burkhardt, H.-J.(eds) *Cooperative Buildings – Integrating Information, Organisation and Architecture*. LNCS 1370, Heidelberg: Springer-Verlag.
78. Access Grid, http://www.accessgrid.org.
79. Page, K. R., Cruickshank, D. and De Roure, D. (2001) It's about time: link streams as continuous metadata. *Proc. Twelfth ACM Conference on Hypertext and Hypermedia (Hypertext '01)*, 2001, pp. 93–102.
80. Buckingham Shum, S., De Roure, D., Eisenstadt, M., Shadbolt, N. and Tate, A. (2002) CoAKTinG: collaborative advanced knowledge technologies in the Grid. *Proceedings of the Second Workshop on Advanced Collaborative Environments, Eleventh IEEE Int. Symposium on High Performance Distributed Computing (HPDC-11)*, Edinburgh, Scotland July 24–26, 2002.
81. DeRoure, D., Jennings, N. R. and Shadbolt, N. R. (2001) 'Research Agenda for the Semantic Grid: A Future e-Science Infrastructure' UK e-Science Technical Report, UKeS-2002-01, National e-Science Centre, Edinburgh, UK.
82. Global Grid Forum, http://www.gridforum.org/.

Reprint from *Concurrency: Practice and Experience*, **10**(7), 1998, 567–581 © 1998 John Wiley & Sons, Ltd.
Minor changes to the original have been made to conform with house style.

4

Software infrastructure for the I-WAY high-performance distributed computing experiment

Ian Foster, Jonathan Geisler, Bill Nickless, Warren Smith, and Steven Tuecke

Argonne National Laboratory, Argonne, Illinois, United States

4.1 INTRODUCTION

Recent developments in high-performance networks, computers, information servers, and display technologies make it feasible to design *network-enabled tools* that incorporate remote compute and information resources into local computational environments and *collaborative environments* that link people, computers, and databases into collaborative sessions. The development of such tools and environments raises numerous technical problems, including the naming and location of remote computational, communication, and data resources; the integration of these resources into computations; the location, characterization, and selection of available network connections; the provision of security and reliability; and uniform, efficient access to data.

Grid Computing – Making the Global Infrastructure a Reality.   Edited by F. Berman, A. Hey and G. Fox
© 2003 John Wiley & Sons, Ltd   ISBN: 0-470-85319-0

Previous research and development efforts have produced a variety of candidate 'point solutions' [1]. For example, DCE, CORBA, Condor [2], Nimrod [3], and Prospero [4] address problems of locating and/or accessing distributed resources; file systems such as AFS [5], DFS, and Truffles [6] address problems of sharing distributed data; tools such as Nexus [7], MPI [8], PVM [9], and Isis [10] address problems of coupling distributed computational resources; and low-level network technologies such as Asynchronous Transfer Mode (ATM) promise gigabit/sec communication. However, little work has been done to integrate these solutions in a way that satisfies the scalability, performance, functionality, reliability, and security requirements of realistic high-performance distributed applications in large-scale internetworks.

It is in this context that the I-WAY project [11] was conceived in early 1995, with the goal of providing a large-scale testbed in which innovative high-performance and geographically distributed applications could be deployed. This application focus, argued the organizers, was essential if the research community was to discover the critical technical problems that must be addressed to ensure progress, and to gain insights into the suitability of different candidate solutions. In brief, the I-WAY was an ATM network connecting supercomputers, mass storage systems, and advanced visualization devices at 17 different sites within North America. It was deployed at the Supercomputing conference (SC'95) in San Diego in December 1995 and used by over 60 application groups for experiments in high-performance computing, collaborative design, and the coupling of remote supercomputers and databases into local environments.

A central part of the I-WAY experiment was the development of a management and application programming environment, called I-Soft. The I-Soft system was designed to run on dedicated I-WAY point of presence (I-POP) machines deployed at each participating site, and provided uniform authentication, resource reservation, process creation, and communication functions across I-WAY resources. In this article, we describe the techniques employed in I-Soft development, and we summarize the lessons learned during the deployment and evaluation process. The principal contributions are the design, prototyping, preliminary integration, and application-based evaluation of the following novel concepts and techniques:

1. *Point of presence* machines as a structuring and management technique for wide-area distributed computing.
2. A computational resource broker that uses *scheduler proxies* to provide a uniform scheduling environment that integrates diverse local schedulers.
3. The use of *authorization proxies* to construct a uniform authentication environment and define trust relationships across multiple administrative domains.
4. *Network-aware parallel programming tools* that use configuration information regarding topology, network interfaces, startup mechanisms, and node naming to provide a uniform view of heterogeneous systems and to optimize communication performance.

The rest of this article is as follows. In Section 4.2, we review the applications that motivated the development of the I-WAY and describe the I-WAY network. In Section 4.3, we introduce the I-WAY software architecture, and in Sections 4.4–4.8 we describe various components of this architecture and discuss lessons learned when these components were used in the I-WAY experiment. In Section 4.9, we discuss some related work. Finally, in Section 4.10, we present our conclusions and outline directions for future research.

4.2 THE I-WAY EXPERIMENT

For clarity, in this article we refer consistently to the I-WAY experiment in the past tense. However, we emphasize that many I-WAY components have remained in place after SC'95 and that follow-on systems are being designed and constructed.

4.2.1 Applications

A unique aspect of the I-WAY experiment was its application focus. Previous gigabit testbed experiments focused on network technologies and low-level protocol issues, using either synthetic network loads or specialized applications for experiments (e.g., see [12]). The I-WAY, in contrast, was driven primarily by the requirements of a large application suite. As a result of a competitive proposal process in early 1995, around 70 application groups were selected to run on the I-WAY (over 60 were demonstrated at SC'95). These applications fell into three general classes [11]:

1. Many applications coupled immersive virtual environments with remote supercomputers, data systems, and/or scientific instruments. The goal of these projects was typically to combine state-of-the-art interactive environments and backend supercomputing to couple users more tightly with computers, while at the same time achieving distance independence between resources, developers, and users.
2. Other applications coupled multiple, geographically distributed supercomputers in order to tackle problems that were too large for a single supercomputer or that benefited from executing different problem components on different computer architectures.
3. A third set of applications coupled multiple virtual environments so that users at different locations could interact with each other and with supercomputer simulations.

Applications in the first and second classes are prototypes for future 'network-enabled tools' that enhance local computational environments with remote compute and information resources; applications in the third class are prototypes of future collaborative environments.

4.2.2 The I-WAY network

The I-WAY network connected multiple high-end display devices (including immersive CAVE^{TM} and ImmersaDesk^{TM} virtual reality devices [13]); mass storage systems; specialized instruments (such as microscopes); and supercomputers of different architectures, including distributed memory multicomputers (IBM SP, Intel Paragon, Cray T3D, etc.), shared-memory multiprocessors (SGI Challenge, Convex Exemplar), and vector multiprocessors (Cray C90, Y-MP). These devices were located at 17 different sites across North America.

This heterogeneous collection of resources was connected by a network that was itself heterogeneous. Various applications used components of multiple networks (e.g., vBNS, AAI, ESnet, ATDnet, CalREN, NREN, MREN, MAGIC, and CASA) as well as additional connections provided by carriers; these networks used different switching technologies and were interconnected in a variety of ways. Most networks used ATM to provide OC-3 (155 Mb s^{-1}) or faster connections; one exception was CASA, which used HIPPI technology. For simplicity, the I-WAY standardized on the use of TCP/IP for application networking; in future experiments, alternative protocols will undoubtedly be explored. The need to configure both IP routing tables and ATM virtual circuits in this heterogeneous environment was a significant source of implementation complexity.

4.3 I-WAY INFRASTRUCTURE

We now describe the software (and hardware) infrastructure developed for I-WAY management and application programming.

4.3.1 Requirements

We believe that the routine realization of high-performance, geographically distributed applications requires a number of capabilities not supported by existing systems. We list first *user-oriented* requirements; while none has been fully addressed in the I-WAY software environment, all have shaped the solutions adopted.

1. *Resource naming and location*: The ability to name computational and information resources in a uniform, location-independent fashion and to locate resources in large internets based on user or application-specified criteria.
2. *Uniform programming environment*: The ability to construct parallel computations that refer to and access diverse remote resources in a manner that hides, to a large extent, issues of location, resource type, network connectivity, and latency.
3. *Autoconfiguration and resource characterization*: The ability to make sensible configuration choices automatically and, when necessary, to obtain information about resource characteristics that can be used to optimize configurations.
4. *Distributed data services*: The ability to access conceptually 'local' file systems in a uniform fashion, regardless of the physical location of a computation.
5. *Trust management*: Authentication, authorization, and accounting services that operate even when users do not have strong prior relationships with the sites controlling required resources.
6. *Confidentiality and integrity*: The ability for a computation to access, communicate, and process private data securely and reliably on remote sites.

Solutions to these problems must be scalable to large numbers of users and resources.
   The fact that resources and users exist at different sites and in different administrative domains introduces another set of *site-oriented* requirements. Different sites not only

provide different access mechanisms for their resources, but also have different policies governing their use. Because individual sites have ultimate responsibility for the secure and proper use of their resources, we cannot expect them to relinquish control to an external authority. Hence, the problem of developing management systems for I-WAY–like systems is above all one of defining protocols and interfaces that support a negotiation process between users (or brokers acting on their behalf) and the sites that control the resources that users want to access.

The I-WAY testbed provided a unique opportunity to deploy and study solutions to these problems in a controlled environment. Because the number of users (few hundred) and sites (around 20) were moderate, issues of scalability could, to a large extent, be ignored. However, the high profile of the project, its application focus, and the wide range of application requirements meant that issues of security, usability, and generality were of critical concern. Important secondary requirements were to minimize development and maintenance effort, for both the I-WAY development team and the participating sites and users.

4.3.2 Design overview

In principle, it would appear that the requirements just elucidated could be satisfied with purely software-based solutions. Indeed, other groups exploring the concept of a 'meta-computer' have proposed software-only solutions [14, 15]. A novel aspect of our approach was the deployment of a dedicated I-WAY point of presence, or I-POP, machine at each participating site. As we explain in detail in the next section, these machines provided a uniform environment for deployment of management software and also simplified validation of security solutions by serving as a 'neutral' zone under the joint control of I-WAY developers and local authorities.

Deployed on these I-POP machines was a software environment, I-Soft, providing a variety of services, including scheduling, security (authentication and auditing), parallel programming support (process creation and communication), and a distributed file system. These services allowed a user to log on to any I-POP and then schedule resources on heterogeneous collections of resources, initiate computations, and communicate between computers and with graphics devices – all without being aware of where these resources were located or how they were connected.

In the next four sections, we provide a detailed discussion of various aspects of the I-POP and I-Soft design, treating in turn the I-POPs, scheduler, security, parallel programming tools, and file systems. The discussion includes both descriptive material and a critical presentation of the lessons learned as a result of I-WAY deployment and demonstration at SC'95.

4.4 POINT OF PRESENCE MACHINES

We have explained why management systems for I-WAY–like systems need to interface to local management systems, rather than manage resources directly. One critical issue that arises in this context is the physical location of the software used to implement these interfaces. For a variety of reasons, it is desirable that this software execute behind site firewalls. Yet this location raises two difficult problems: sites may, justifiably, be reluctant to allow outside software to run on their systems; and system developers will be required to develop interfaces for many different architectures.

The use of I-POP machines resolve these two problems by providing a uniform, jointly administered physical location for interface code. The name is chosen by analogy with a comparable device in telephony. Typically, the telephone company is responsible for, and manages, the telephone network, while the customer owns the phones and in-house wiring. The interface between the two domains lies in a switchbox, which serves as the telephone company's 'point of presence' at the user site.

4.4.1 I-POP design

Figure 4.1 shows the architecture of an I-POP machine. It is a dedicated workstation, accessible via the Internet and operating inside a site's firewalls. It runs a standard set of software supplied by the I-Soft developers. An ATM interface allows it to monitor and, in principle, manage the site's ATM switch; it also allows the I-POP to use the ATM network for management traffic. Site-specific implementations of a simple management interface allow I-WAY management systems to communicate with other machines at the site to allocate resources to users, start processes on resources, and so forth. The Andrew distributed file system (AFS) [5] is used as a repository for system software and status information.

Development, maintenance, and auditing costs are significantly reduced if all I-POP computers are of the same type. In the I-WAY experiment, we standardized on Sun SPARCStations. A standard software configuration included SunOS 4.1.4 with latest

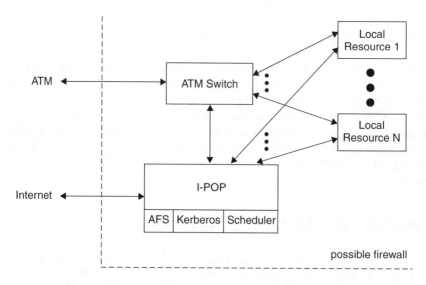

Figure 4.1 An I-WAY point of presence (I-POP) machine.

patches; a limited set of Unix utilities; the Cygnus release of Kerberos 4; AFS; the I-WAY scheduler; and various security tools such as Tripwire [16], TCP wrappers, and auditing software. This software was maintained at a central site (via AFS) and could be installed easily on each I-POP; furthermore, the use of Tripwire meant that it was straightforward to detect changes to the base configuration.

The I-POP represented a dedicated point of presence for the I-WAY at the user site. It was jointly managed: the local site could certify the I-POP's software configuration and could disconnect the I-POP to cut access to the I-WAY in the event of a security problem; similarly, the I-WAY security team could log accesses, check for modifications to its configuration, and so forth. The dedicated nature of the I-POP meant that its software configuration could be kept simple, facilitating certification and increasing trust.

4.4.2 I-POP discussion

Seventeen sites deployed I-POP machines. For the most part the effort required to install software, integrate a site into the I-WAY network, and maintain the site was small (in our opinion, significantly less than if I-POPs had not been used). The fact that all I-POPs shared a single AFS cell proved extremely useful as a means of maintaining a single, shared copy of I-Soft code and as a mechanism for distributing I-WAY scheduling information. The deployment of I-POPs was also found to provide a conceptual framework that simplified the task of explaining the I-WAY infrastructure, both to users and to site administrators.

While most I-POPs were configured with ATM cards, we never exploited this capability to monitor or control the ATM network. The principal reason was that at many sites, the ATM switch to which the I-POP was connected managed traffic for both I-WAY and non–I-WAY resources. Hence, there was a natural reluctance to allow I-POP software to control the ATM switches. These authentication, authorization, and policy issues will need to be addressed in future I-WAY–like systems.

We note that the concept of a point of presence machine as a locus for management software in a heterogeneous I-WAY–like system is a unique contribution of this work. The most closely related development is that of the ACTS ATM Internetwork (AAI) network testbed group: they deployed fast workstations at each site in a Gigabit testbed, to support network throughput experiments [12].

4.5 SCHEDULER

I-WAY–like systems require the ability to locate computational resources matching various criteria in a heterogeneous, geographically distributed pool. As noted above, political and technical constraints make it infeasible for this requirement to be satisfied by a single 'I-WAY scheduler' that replaces the schedulers that are already in place at various sites. Instead, we need to think in terms of a negotiation process by which requests (ideally, expressible in a fairly abstract form, for example, 'N Gigaflops,' or 'X nodes of type Y, with maximum latency Z') are handled by an independent entity, which then negotiates with the site schedulers that manage individual resources. We coin the term Computational Resource Broker (CRB) to denote this entity. In an Internet-scale distributed computing system, we can imagine a network of such brokers. In the I-WAY, one was sufficient.

4.5.1 Scheduler design

The practical realization of the CRB concept requires the development of fairly general user-to-CRB and CRB-to-resource scheduler protocols. Time constraints in the I-WAY project limited what we could achieve in each area. On the user-to-CRB side, we allowed users to request access only to predefined disjoint subsets of I-WAY computers called *virtual machines*; on the CRB-to-resource scheduler side, we required sites to turn over scheduling control of specified resources to the I-WAY scheduler, which would then use the resources to construct virtual machines. In effect, our simple CRB obtained access to a block of resources, which it then distributed to its users.

The scheduler that was defined to meet these requirements provided *management functions* that allowed administrators to configure dedicated resources into virtual machines, obtain status information, and so forth and *user functions* that allowed users to list available virtual machines and to determine status, list queued requests, or request time on a particular virtual machine.

The scheduler implementation was structured in terms of a single central scheduler and multiple local scheduler daemons. The *central scheduler* daemon maintained the queues and tables representing the state of the different virtual machines and was responsible for allocating time on these machines on a first-come, first-served basis. It also maintained state information on the AFS file system, so as to provide some fault tolerance in the case of daemon failures. The central scheduler communicated with *local scheduler* daemons, one per I-POP, to request that operations be performed on particular machines. Local schedulers performed site-dependent actions in response to three simple requests from the central scheduler.

- *Allocate resource*: This request enables a local scheduler to perform any site-specific initialization required to make a resource usable by a specified user, for example, by initializing switch configurations so that processors allocated to a user can communicate, and propagating configuration data.
- *Create process*: This request asks a local scheduler to create a process on a specified processor, as a specified user: it implements, in effect, a Unix remote shell, or rsh, command. This provides the basic functionality required to initiate remote computations; as we discuss below, it can be used directly by a user and is also used to implement other user-level functions such as ixterm (start an X-terminal process on a specified processor), ircp (start a copy process on a specified processor), and impirun (start an MPI program on a virtual machine).
- *Deallocate resource*: This request enables a local scheduler to perform any site-specific operations that may be required to terminate user access to a resource: for example, disabling access to a high-speed interconnect, killing processes, or deleting temporary files.

4.5.2 Scheduler discussion

The basic scheduler structure just described was deployed on a wide variety of systems (interfaces were developed for all I-WAY resources) and was used successfully at SC'95 to schedule a large number of users. Its major limitations related not to its basic structure but to the too-restrictive interfaces between user and scheduler and scheduler and local resources.

The concept of using fixed virtual machines as schedulable units was only moderately successful. Often, no existing virtual machine met user requirements, in which case new virtual machines had to be configured manually. This difficulty would have been avoided if even a very simple specification language that allowed requests of the form 'give me M nodes of type X and N nodes of type Y' had been supported. This feature could easily be integrated into the existing framework. The development of a more sophisticated resource description language and scheduling framework is a more difficult problem and will require further research.

A more fundamental limitation related to the often limited functionality provided by the non–I-WAY resource schedulers with which local I-WAY schedulers had to negotiate. Many were unable to inquire about completion time of scheduled jobs (and hence expected availability of resources) or to reserve computational resources for specified timeslots; several sites provided timeshared rather than dedicated access. In addition, at some sites, networking and security concerns required that processors intended for I-WAY use be specially configured. We compensated either by dedicating partitions to I-WAY users or by timesharing rather than scheduling. Neither solution was ideal. In particular, the use of dedicated partitions meant that frequent negotiations were required to adapt partition size to user requirements and that computational resources were often idle. The long-term solution probably is to develop more sophisticated schedulers for resources that are to be incorporated into I-WAY–like systems. However, applications also may need to become more flexible about what type and 'quality' of resources they can accept.

We note that while many researchers have addressed problems relating to scheduling computational resources in parallel computers or local area networks, few have addressed the distinctive problems that arise when resources are distributed across many sites. Legion [17] and Prospero [4] are two exceptions. In particular, Prospero's 'system manager' and 'node manager' processes have some similarities to our central and local managers. However, neither system supports interfaces to other schedulers: they require full control of scheduled resources.

4.6 SECURITY

Security is a major and multifaceted issue in I-WAY–like systems. Ease-of-use concerns demand a uniform authentication environment that allows a user to authenticate just once in order to obtain access to geographically distributed resources; performance concerns require that once a user is authenticated, the authorization overhead incurred when accessing a new resource be small. Both uniform authentication and low-cost authorization are complicated in scalable systems, because users will inevitably need to access resources located at sites with which they have no prior trust relationship.

4.6.1 Security design

When developing security structures for the I-WAY software environment, we focused on providing a uniform authentication environment. We did not address in any detail issues relating to authorization, accounting, or the privacy and integrity of user data. Our goal was to provide security at least as good as that existing at the I-WAY sites. Since all sites used clear-text password authentication, this constraint was not especially stringent. Unfortunately, we could not assume the existence of a distributed authentication system such as Kerberos (or DCE, which uses Kerberos) because no such system was available at all sites.

Our basic approach was to separate the authentication problem into two parts: authentication to the I-POP environment and authentication to the local sites. Authentication to I-POPs was handled by using a `telnet` client modified to use Kerberos authentication and encryption. This approach ensured that users could authenticate to I-POPs without passing passwords in clear text over the network. The scheduler software kept track of which user id was to be used at each site for a particular I-WAY user, and served as an 'authentication proxy,' performing subsequent authentication to other I-WAY resources on the user's behalf. This proxy service was invoked each time a user used the command language described above to allocate computational resources or to create processes.

The implementation of the authentication proxy mechanism was integrated with the site-dependent mechanisms used to implement the scheduler interface described above. In the I-WAY experiment, most sites implemented all three commands using a privileged (root) `rsh` from the local I-POP to an associated resource. This method was used because of time constraints and was acceptable only because the local site administered the local I-POP and the `rsh` request was sent to a local resource over a secure local network.

4.6.2 Security discussion

The authentication mechanism just described worked well in the sense that it allowed users to authenticate once (to an I-POP) and then access any I-WAY resource to which access was authorized. The 'authenticate-once' capability proved to be extremely useful and demonstrated the advantages of a common authentication and authorization environment.

One deficiency of the approach related to the degree of security provided. Root `rsh` is an unacceptable long-term solution even when the I-POP is totally trusted, because of the possibility of IP-spoofing attacks. We can protect against these attacks by using a remote shell function that uses authentication (e.g., one based on Kerberos [18] or PGP, either directly or via DCE). For similar reasons, communications between the scheduling daemons should also be authenticated.

A more fundamental limitation of the I-WAY authentication scheme as implemented was that each user had to have an account at each site to which access was required. Clearly, this is not a scalable solution. One alternative is to extend the mechanisms that map I-WAY user ids to local user ids, so that they can be used to map I-WAY user ids to preallocated 'I-WAY proxy' user ids at the different sites. The identity of the individual using different proxies at different times could be recorded for audit purposes.

However, this approach will work only if alternative mechanisms can be developed for the various functions provided by an 'account.' The formal application process that is typically associated with the creation of an account serves not only to authenticate the user but also to establish user obligations to the site (e.g., 'no commercial work' is a frequent requirement at academic sites) and to define the services provided by the site to the user (e.g., backup policies). Proxy accounts address only the authentication problem (if sites trust the I-WAY). Future approaches will probably require the development of formal representations of conditions of use, as well as mechanisms for representing transitive relationships. (For example, a site may agree to trust any user employed by an organization with which it has formalized a trust relationship. Similarly, an organization may agree on behalf of its employees to obligations associated with the use of certain resources.)

4.7 PARALLEL PROGRAMMING TOOLS

A user who has authenticated to an I-POP and acquired a set of computational resources then requires mechanisms for creating computations on these resources. At a minimum, these mechanisms must support the creation of processes on different processors and the communication of data between these processes. Because of the complexity and hetero-geneity of I-WAY–like environments, tools should ideally also relieve the programmer of the need to consider low-level details relating to network structure. For example, tools should handle conversions between different data representations automatically and be able to use different protocols when communicating within rather than between paral-lel computers. At the same time, the user should be able to obtain access to low-level information (at an appropriate level of abstraction) when it is required for optimization purposes.

4.7.1 Parallel tools design

The irsh and ixterm commands described above allow authenticated and authorized users to access and initiate computation on any I-WAY resource. Several users relied on these commands alone to initiate distributed computations that then communicated by using TCP/IP sockets. However, this low-level approach did not hide (or exploit) any details of the underlying network.

To support the needs of users desiring a higher-level programming model, we adapted the Nexus multithreaded communication library [7] to execute in an I-WAY environment. Nexus supports automatic configuration mechanisms that allow it to use information con-tained in resource databases to determine which startup mechanisms, network interfaces, and protocols to use in different situations. For example, in a virtual machine connecting IBM SP and SGI Challenge computers with both ATM and Internet networks, Nexus uses three different protocols (IBM proprietary MPL on the SP, shared memory on the Challenge, and TCP/IP or AAL5 between computers) and selects either ATM or Internet network interfaces, depending on network status. We modified the I-WAY scheduler to pro-duce appropriate resource database entries when a virtual machine was allocated to a user. Nexus could then use this information when creating a user computation. (Nexus support for multithreading should, in principle, also be useful – for latency hiding – although in practice it was not used for that purpose during the I-WAY experiment.)

Several other libraries, notably the CAVEcomm virtual reality library [19] and the MPICH implementation of MPI, were extended to use Nexus mechanisms [20]. Since MPICH is defined in terms of an 'abstract point-to-point communication device,' an implementation of this device in terms of Nexus mechanisms was not difficult. Other systems that use Nexus mechanisms include the parallel language CC++ and the parallel scripting language nPerl, used to write the I-WAY scheduler.

4.7.2 Parallel tools discussion

The I-WAY experiment demonstrated the advantages of the Nexus automatic configuration mechanisms. In many cases, user were able to develop applications with high-level tools such as MPI, CAVEcomm, and/or CC++, without any knowledge of low-level details relating to the compute and network resources included in a computation.

A significant difficulty revealed by the I-WAY experiment related to the mechanisms used to generate and maintain the configuration information used by Nexus. While resource database entries were generated automatically by the scheduler, the information contained in these entries (such as network interfaces) had to be provided manually by the I-Soft team. The discovery, entry, and maintenance of this information proved to be a significant source of overhead, particularly in an environment in which network status was changing rapidly. Clearly, this information should be discovered automatically whenever possible. Automatic discovery would make it possible, for example, for a parallel tool to use dedicated ATM links if these were available, but to fall back automatically to shared Internet if the ATM link was discovered to be unavailable. The development of such automatic discovery techniques remains a challenging research problem.

The Nexus communication library provides mechanisms for querying the resource database, which users could have used to discover some properties of the machines and networks on which they were executing. In practice, few I-WAY applications were configured to use this information; however, we believe that this situation simply reflects the immature state of practice in this area and that users will soon learn to write programs that exploit properties of network topology and so forth. Just what information users will find useful remains to be seen, but presumably enquiry functions that reveal the number of machines involved in a computation and the number of processors in each machine would definitely be required. One application that could certainly benefit from access to information about network topology is the I-WAY MPI implementation. Currently, this library implements collective operations using algorithms designed for multicomputer environments; presumably, communication costs can often be reduced by using communication structures that avoid intermachine communication.

4.8 FILE SYSTEMS

I-WAY–like systems introduce three related requirements with a file-system flavor. First, many users require access to various status data and utility programs at many different sites. Second, users running programs on remote computers must be able to access executables and configuration data at many different sites. Third, application programs must be able to read and write potentially large data sets. These three requirements have very different characteristics. The first requires support for multiple users, consistency across multiple sites, and reliability. The second requires somewhat higher performance (if executables are large) but does not require support for multiple users. The third requires, above all, high performance. We believe that these three requirements are best satisfied with different technologies.

The I-Soft system supported only the first of these requirements. An AFS cell (with three servers for reliability) was deployed and used as a shared repository for I-WAY software, and also to maintain scheduler status information. The AFS cell was accessible only from the I-POPs, since many I-WAY computers did not support AFS, and when they did, authentication problems made access difficult. The only assistance provided for the second and third requirements was a remote copy (`ircp`) command that supported the copying of data from one machine to another.

While the AFS system was extremely useful, the lack of distributed file system support on I-WAY nodes was a serious deficiency. Almost all users found that copying files and configuration data to remote sites was an annoyance, and some of the most ambitious I-WAY applications had severe problems postprocessing, transporting, and visualizing the large amounts of data generated at remote sites. Future I-WAY–like systems should support something like AFS on all nodes and if necessary provide specialized high-performance distributed data access mechanisms for performance-critical applications.

4.9 RELATED WORK

In preceding sections, we have referred to a number of systems that provide point solutions to problems addressed in I-Soft development. Here, we review systems that seek to provide an integrated treatment of distributed system issues, similar or broader in scope than I-Soft.

The Distributed Computing Environment (DCE) and Common Object Request Broker Architecture (CORBA) are two major industry-led attempts to provide a unifying framework for distributed computing. Both define (or will define in the near future) a standard directory service, remote procedure call (RPC), security service, and so forth; DCE also defines a Distributed File Service (DFS) derived from AFS. Issues such as fault tolerance and interoperability between languages and systems are addressed. In general, CORBA is distinguished from DCE by its higher-level, object-oriented architecture. Some DCE mechanisms (RPC, DFS) may well prove to be appropriate for implementing I-POP services; CORBA directory services may be useful for resource location. However, both DCE and CORBA appear to have significant deficiencies as a basis for application programming in I-WAY–like systems. In particular, the remote procedure call is not well suited to applications in which performance requirements demand asynchronous communication, multiple outstanding requests, and/or efficient collective operations.

The Legion project [17] is another project developing software technology to support computing in wide-area environments. Issues addressed by this wide-reaching effort include scheduling, file systems, security, fault tolerance, and network protocols. The

I-Soft effort is distinguished by its focus on high-performance systems and by its use of I-POP and proxy mechanisms to enhance interoperability with existing systems.

4.10 CONCLUSIONS

We have described the management and application programming environment developed for the I-WAY distributed computing experiment. This system incorporates a number of ideas that, we believe, may be useful in future research and development efforts. In particular, it uses point of presence machines as a means of simplifying system configuration and management, scheduler proxies for distributed scheduling, authentication proxies for distributed authentication, and network-aware tools that can exploit configuration information to optimize communication behavior. The I-Soft development also took preliminary steps toward integrating these diverse components, showing, for example, how a scheduler can provide network topology information to parallel programming tools.

The SC'95 event provided an opportunity for intense and comprehensive evaluation of the I-Soft and I-POP systems. I-Soft was a success in that most applications ran successfully at least some of the time; the network rather than the software proved to be the least reliable system component. Specific deficiencies and limitations revealed by this experience have been detailed in the text. More generally, we learned that system components that are typically developed in isolation must be more tightly integrated if performance, reliability, and usability goals are to be achieved. For example, resource location services in future I-WAY–like systems will need low-level information on network characteristics; schedulers will need to be able to schedule network bandwidth as well as computers; and parallel programming tools will need up-to-date information on network status.

We are now working to address some of the critical research issues identified in I-Soft development. The Globus project, involving Argonne, Caltech, the Aerospace Corporation, and Trusted Information Systems, is addressing issues of resource location (computational resource brokers), automatic configuration, scalable trust management, and high-performance distributed file systems. In addition, we and others are defining and constructing future I-WAY–like systems that will provide further opportunities to evaluate management and application programming systems such as I-Soft.

ACKNOWLEDGMENTS

The I-WAY was a multi-institutional, multi-individual effort. Tom DeFanti, Rick Stevens, Tim Kuhfuss, Maxine Brown, Linda Winkler, Mary Spada, and Remy Evard played major roles. We acknowledge in particular Carl Kesselman and Steve Schwab (I-Soft design), Gene Rackow (I-POP software), Judy Warren (AFS), Doru Marcusiu (I-Soft deployment), Bill Gropp and Ewing Lusk (MPI), and Gary Minden, Mike St Johns, and Ken Rowe (security architecture). This work was supported in part by the Mathematical, Information, and Computational Sciences Division subprogram of the Office of Computational and Technology Research, U.S. Department of Energy, under Contract W-31-109-Eng-38.

REFERENCES

1. Mullender, S. (ed.) (1989) *Distributed Systems*, ACM Press.
2. Litzkow, M., Livney, M. and Mutka, M. (1988) Condor – a hunter of idle workstations. *Proc. 8th Intl. Conf. on Distributed Computing Systems*, pp. 104–111, 1988.
3. Abramson, D., Sosic, R., Giddy, J. and Hall, B. (1995) Nimrod: A tool for performing parameterised simulations using distributed workstations. *Proc. 4th IEEE Symp. on High Performance Distributed Computing*, IEEE Press.
4. Clifford Neumann, B. and Rao, S. The Prospero resource manager: A scalable framework for processor allocation in distributed systems. *Concurrency: Practice and Experience*, June 1994.
5. Morris, J. H. *et al.* (1986) Andrew: A distributed personal computing environment. *CACM*, **29**(3).
6. Cook, J., Crocker, S. D. Jr., Page, T., Popek, G. and Reiher, P. (1993) Truffles: Secure file sharing with minimal system administrator intervention. *Proc. SANS-II, The World Conference on Tools and Techniques for System Administration, Networking, and Security*, 1993.
7. Foster, I., Kesselman, C. and Tuecke, S. (1996) The Nexus approach to integrating multithreading and communication. *Journal of Parallel and Distributed Computing*, to appear.
8. Gropp, W., Lusk, E. and Skjellum, A. (1995) *Using MPI: Portable Parallel Programming with the Message Passing Interface*, MIT Press.
9. Geist, A., Beguelin, A., Dongarra, J., Jiang, W., Manchek, B. and Sunderam, V. (1994) *PVM: Parallel Virtual Machine – A User's Guide and Tutorial for Network Parallel Computing*, MIT Press.
10. Birman, K. (1993) The process group approach to reliable distributed computing. *Communications of the ACM*, **36**(12), 37–53.
11. DeFanti, T., Foster, I., Papka, M., Stevens, R. and Kuhfuss, T. (1996) Overview of the I-WAY: Wide area visual supercomputing. *International Journal of Supercomputer Applications*, in press.
12. Ewy, B., Evans, J., Frost, V. and Minden, G. (1994) TCP and ATM in wide area networks. *Proc. 1994 IEEE Gigabit Networking Workshop*, IEEE.
13. Cruz-Neira, C., Sandin, D. J., DeFanti, T. A., Kenyon, R. V. and Hart, J. C. (1992) The CAVE: Audio visual experience automatic virtual environment. *Communications of the ACM*, **35**(6), 65–72.
14. Catlett, C. and Smarr, L. (1992) Metacomputing. *Communications of the ACM*, **35**(6), 44–52.
15. Grimshaw, A., Weissman, J., West, E. and Lyot, E. Jr. (1994) Metasystems: An approach combining parallel processing and heterogeneous distributed computing systems. *Journal of Parallel and Distributed Computing*, **21**(3), 257–270.
16. Kim, G. H. and Spafford, E. H. (1994) Writing, supporting, and evaluating Tripwire: A publically available security tool. *Proc. USENIX Unix Applications Development Symp.*, The USENIX Association, pp. 89–107.
17. Grimshaw, A., Wulf, W., French, J., Weaver, A. and Reynolds, P. Jr. Legion: The next logical step toward a nationwide virtual computer. Technical Report CS-94-21, Department of Computer Science, University of Virginia, 1994.
18. Steiner, J., Neumann, B. and Schiller, J. (1988) Kerberos: An authentication system for open network systems. Usenix Conference Proceedings, pp. 191–202.
19. Disz, T. L., Papka, M. E., Pellegrino, M. and Stevens, R. (1995) Sharing visualization experiences among remote virtual environments. *International Workshop on High Performance Computing for Computer Graphics and Visualization*, Springer-Verlag, pp. 217–237.
20. Foster, I., Geisler, J. and Tuecke, S. (1996) MPI on the I-WAY: A wide-area, multimethod implementation of the Message Passing Interface. *Proceedings of the 1996 MPI Developers Conference*, IEEE Computer Society Press.

5

Implementing production Grids

**William E. Johnston,[1] The NASA IPG Engineering Team,[2]
and The DOE Science Grid Team[3]**

[1]*Lawrence Berkeley National Laboratory, Berkeley, California, United States,* [2]*NASA
Ames Research Center and NASA Glenn Research Center,* [3]*Lawrence Berkeley National
Lab, Argonne National Lab, National Energy Research Scientific Computing Center,
Oak Ridge National Lab, and Pacific Northwest National Lab*

5.1 INTRODUCTION: LESSONS LEARNED FOR BUILDING LARGE-SCALE GRIDS

Over the past several years there have been a number of projects aimed at building
'production' Grids. These Grids are intended to provide identified user communities with
a rich, stable, and standard distributed computing environment. By 'standard' and 'Grids',
we specifically mean Grids based on the common practice and standards coming out of
the Global Grid Forum (GGF) (www.gridforum.org).

There are a number of projects around the world that are in various stages of putting
together production Grids that are intended to provide this sort of persistent cyber infrastructure for science. Among these are the UK e-Science program [1], the European
DataGrid [2], NASA's Information Power Grid [3], several Grids under the umbrella of

Grid Computing – Making the Global Infrastructure a Reality. Edited by F. Berman, A. Hey and G. Fox
© 2003 John Wiley & Sons, Ltd   ISBN: 0-470-85319-0

the DOE Science Grid [4], and (at a somewhat earlier stage of development) the Asia Pacific Grid [5].

In addition to these basic Grid infrastructure projects, there are a number of well-advanced projects aimed at providing the types of higher-level Grid services that will be used directly by the scientific community. These include, for example, Ninf (a network-based information library for global worldwide computing infrastructure [6, 7]) and GridLab [8].

This chapter, however, addresses the specific and actual experiences gained in building NASA's IPG and DOE's Science Grids, both of which are targeted at infrastructure for large-scale, collaborative science, and access to large-scale computing and storage facilities.

The IPG project at NASA Ames [3] has integrated the operation of Grids into the NASA Advanced Supercomputing (NAS) production supercomputing environment and the computing environments at several other NASA Centers, and, together with some NASA 'Grand Challenge' application projects, has been identifying and resolving issues that impede application use of Grids.

The DOE Science Grid [4] is implementing a prototype production environment at four DOE Labs and at the DOE Office of Science supercomputer center, NERSC [9]. It is addressing Grid issues for supporting large-scale, international, scientific collaborations.

This chapter only describes the experience gained from deploying a specific set of software: Globus [10], Condor [11], SRB/MCAT [12], PBSPro [13], and a PKI authentication substrate [14–16]. That is, these suites of software have provided the implementation of the Grid functions used in the IPG and DOE Science Grids.

The Globus package was chosen for several reasons:

- A clear, strong, and standards-based security model,
- Modular functions (not an all-or-nothing approach) providing all the Grid Common Services, except general events,
- A clear model for maintaining local control of resources that are incorporated into a Globus Grid,
- A general design approach that allows a decentralized control and deployment of the software,
- A demonstrated ability to accomplish large-scale Metacomputing (in particular, the SF-Express application in the Gusto test bed – see Reference [17]),
- Presence in supercomputing environments,
- A clear commitment to open source, and
- Today, one would also have to add 'market share'.

Initially, Legion [18] and UNICORE [19] were also considered as starting points, but both these failed to meet one or more of the selection criteria given above.

SRB and Condor were added because they provided specific, required functionality to the IPG Grid, and because we had the opportunity to promote their integration with Globus (which has happened over the course of the IPG project).

PBS was chosen because it was actively being developed in the NAS environment along with the IPG. Several functions were added to PBS over the course of the IPG project in order to support Grids.

Grid software beyond those provided by these suites are being defined by many organizations, most of which are involved in the GGF. Implementations are becoming available, and are being experimented within the Grids being described here (e.g. the Grid monitoring and event framework of the Grid Monitoring Architecture Working Group (WG) [20]), and some of these projects will be mentioned in this chapter. Nevertheless, the software of the prototype production Grids described in this chapter is provided primarily by the aforementioned packages, and these provide the context of this discussion.

This chapter recounts some of the lessons learned in the process of deploying these Grids and provides an outline of the steps that have proven useful/necessary in order to deploy these types of Grids. This reflects the work of a substantial number of people, representatives of whom are acknowledged below.

The lessons fall into four general areas – deploying operational infrastructure (what has to be managed operationally to make Grids work), establishing cross-site trust, dealing with Grid technology scaling issues, and listening to the users – and all of these will be discussed.

This chapter is addressed to those who are setting up science-oriented Grids, or who are considering doing so.

5.2 THE GRID CONTEXT

'Grids' [21, 22] are an approach for building dynamically constructed problem-solving environments using geographically and organizationally dispersed, high-performance computing and data handling resources. Grids also provide important infrastructure supporting multi-institutional collaboration.

The overall motivation for most current large-scale, multi-institutional Grid projects is to enable the resource and human interactions that facilitate large-scale science and engineering such as aerospace systems design, high-energy physics data analysis [23], climate research, large-scale remote instrument operation [9], collaborative astrophysics based on virtual observatories [24], and so on. In this context, Grids are providing significant new capabilities to scientists and engineers by facilitating routine construction of information- and collaboration-based problem-solving environments that are built on demand from large pools of resources.

Functionally, Grids are tools, middleware, and services for

- building the application frameworks that allow disciplined scientists to express and manage the simulation, analysis, and data management aspects of overall problem solving,
- providing a uniform and secure access to a wide variety of distributed computing and data resources,
- supporting construction, management, and use of widely distributed application systems,
- facilitating human collaboration through common security services, and resource and data sharing,
- providing support for remote access to, and operation of, scientific and engineering instrumentation systems, and
- managing and operating this computing and data infrastructure as a persistent service.

This is accomplished through two aspects: (1) a set of uniform software services that manage and provide access to heterogeneous, distributed resources and (2) a widely deployed infrastructure. The software architecture of a Grid is depicted in Figure 5.1.

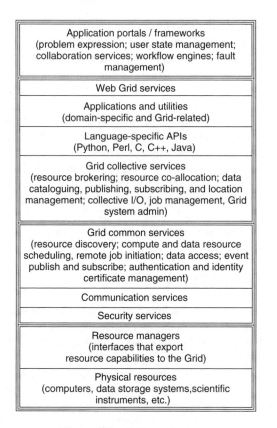

Figure 5.1   Grid architecture.

Grid software is not a single, monolithic package, but rather a collection of interoperating software packages. This is increasingly so as the Globus software is modularized and distributed as a collection of independent packages, and as other systems are integrated with basic Grid services.

In the opinion of the author, there is a set of basic functions that all Grids must have in order to be called a *Grid*: The Grid Common Services. These constitute the 'neck of the hourglass' of Grids, and include the Grid Information Service ('GIS' – the basic resource discovery mechanism) [25], the Grid Security Infrastructure ('GSI' – the tools and libraries that provide Grid security) [26], the Grid job initiator mechanism (e.g. Globus GRAM [27]), a Grid scheduling function, and a basic data management mechanism such as GridFTP [28]. It is almost certainly the case that to complete this set we need a Grid event mechanism. The Grid Forum's Grid Monitor Architecture (GMA) [29] addresses one approach to Grid events, and there are several prototype implementations of the GMA (e.g. References [30, 31]). A communications abstraction (e.g. Globus I/O [32]) that incorporates Grid security is also in this set.

At the resource management level – which is typically provided by the individual computing system, data system, instrument, and so on – important Grid functionality is provided as part of the resource capabilities. For example, job management systems (e.g. PBSPro [13], Maui [33], and under some circumstances the Condor Glide-in [34] – see Section 5.3.1.5) that support advance reservation of resource functions (e.g. CPU sets) are needed to support co-scheduling of administratively independent systems. This is because, in general, the Grid scheduler can request such service in a standard way but cannot provide these services unless they are supported on the resources.

Beyond this basic set of capabilities (provided by the Globus Toolkit [10] in this discussion) are associated client-side libraries and tools, and other high-level capabilities such as Condor-G [35] for job management, SRB/MCAT [12] for federating and cataloguing tertiary data storage systems, and the new Data Grid [10, 36] tools for Grid data management.

In this chapter, while we focus on the issues of building a Grid through deploying and managing the Grid Common Services (provided mostly by Globus), we also point out along the way other software suites that may be required for a functional Grid and some of the production issues of these other suites.

5.3 THE ANTICIPATED GRID USAGE MODEL WILL DETERMINE WHAT GETS DEPLOYED, AND WHEN

As noted, Grids are not built from a single piece of software but from suites of increasingly interoperable software. Having some idea of the primary, or at least initial uses of your Grid will help identify where you should focus your early deployment efforts. Considering the various models for computing and data management that might be used on your Grid is one way to select what software to install.

5.3.1 Grid computing models

There are a number of identifiable computing models in Grids that range from single resource to tightly coupled resources, and each requires some variations in Grid services. That is, while the basic Grid services provide all the support needed to execute a distributed program, things like coordinated execution of multiple programs [as in High Throughput Computing (HTC)] across multiple computing systems, or management of many thousands of parameter study or data analysis jobs, will require additional services.

5.3.1.1 Export existing services

Grids provide a uniform set of services to export the capabilities of existing computing facilities such as supercomputer centers to existing user communities, and this is accomplished by the Globus software. The primary advantage of this form of Grids is to provide a uniform view of several related computing systems, or to prepare for other types of uses. This sort of Grid also facilitates/encourages the incorporation of the supercomputers into user constructed systems.

By 'user constructed systems' we mean, for example, various sorts of portals or frameworks that run on user systems and provide for creating and managing related suites of Grid jobs. See, for example, The GridPort Toolkit [37], Cactus [38, 39], JiPANG (a Jini-based Portal Augmenting Grids) [40], GridRPC [41], and in the future, NetSolve [42].

User constructed systems may also involve data collections that are generated and maintained on the user systems and that are used as input, for example, supercomputer processes running on the Grid, or are added to by these processes. The primary issue here is that a Grid compatible data service such as GridFTP must be installed and maintained on the user system in order to accommodate this use. The deployment and operational implications of this are discussed in Section 5.7.11.

5.3.1.2 Loosely coupled processes

By loosely coupled processes we mean collections of logically related jobs that nevertheless do not have much in common once they are executing. That is, these jobs are given some input data that might, for example, be a small piece of a single large dataset, and they generate some output data that may have to be integrated with the output of other such jobs; however, their execution is largely independent of the other jobs in the collection.

Two common types of such jobs are data analysis, in which a large dataset is divided into units that can be analyzed independently, and parameter studies, where a design space of many parameters is explored, usually at low model resolution, across many different parameter values (e.g. References [43, 44]).

In the data analysis case, the output data must be collected and integrated into a single analysis, and this is sometimes done as part of the analysis job and sometimes by collecting the data at the submitting site where the integration is dealt with. In the case of parameter studies, the situation is similar. The results of each run are typically used to fill in some sort of parameter matrix.

In both cases, in addition to the basic Grid services, a job manager is required to track these (typically numerous) related jobs in order to ensure either that they have all run exactly once or that an accurate record is provided of those that ran and those that failed. (Whether the job manager can restart failed jobs typically depends on how the job is assigned work units or how it updates the results dataset at the end.)

The Condor-G job manager [35, 45] is a Grid task broker that provides this sort of service, as well as managing certain types of job dependencies.

Condor-G is a client-side service and must be installed on the submitting systems. A *Condor_manager* server is started by the user and then jobs are submitted to this user job manager. This manager deals with refreshing the proxy[1] that the Grid resource must have in order to run the user's jobs, but the user must supply new proxies to the

---

[1] A proxy certificate is the indirect representation of the user that is derived from the Grid identity credential. The proxy is used to represent the authenticated user in interactions with remote systems where the user does not have a direct presence. That is, the user authenticates to the Grid once, and this authenticated identity is carried forward as needed to obtain authorization to use remote resources. This is called *single sign-on*.

Condor manager (typically once every 12 h). The manager must stay alive while the jobs are running on the remote Grid resource in order to keep track of the jobs as they complete. There is also a Globus GASS server on the client side that manages the default data movement (binaries, stdin/out/err, etc.) for the job. Condor-G can recover from both server-side and client-side crashes, but not from long-term client-side outages. (That is, e.g. the client-side machine cannot be shutdown over the weekend while a lot of Grid jobs are being managed.)

This is also the job model being addressed by 'peer-to-peer' systems. Establishing the relationship between peer-to-peer and Grids is a new work area at the GGF (see Reference [46]).

5.3.1.3 Workflow managed processes

The general problem of workflow management is a long way from being solved in the Grid environment; however, it is quite common for existing application system frameworks to have *ad hoc* workflow management elements as part of the framework. (The 'framework' runs the gamut from a collection of shell scripts to elaborate Web portals.)

One thing that most workflow managers have in common is the need to manage events of all sorts. By 'event', we mean essentially any asynchronous message that is used for decision-making purposes. Typical Grid events include

- normal application occurrences that are used, for example, to trigger computational steering or semi-interactive graphical analysis,
- abnormal application occurrences, such as numerical convergence failure, that are used to trigger corrective action,
- messages that certain data files have been written and closed so that they may be used in some other processing step.

Events can also be generated by the Grid remote job management system signaling various sorts of things that might happen in the control scripts of the Grid jobs, and so on.

The Grid Forum, Grid Monitoring Architecture [29] defines an event model and management system that can provide this sort of functionality. Several prototype systems have been implemented and tested to the point where they could be useful prototypes in a Grid (see, e.g. References [30, 31]). The GMA involves a server in which the sources and sinks of events register, and these establish event channels directly between producer and consumer – that is, it provides the event publish/subscribe service. This server has to be managed as a persistent service; however, in the future, it may be possible to use the GIS/Monitoring and Discovery Service (MDS) for this purpose.

5.3.1.4 Distributed-pipelined/coupled processes

In application systems that involve multidisciplinary or other multicomponent simulations, it is very likely that the processes will need to be executed in a 'pipeline' fashion. That is, there will be a set of interdependent processes that communicate data back and forth throughout the entire execution of each process.

In this case, co-scheduling is likely to be essential, as is good network bandwidth between the computing systems involved.

Co-scheduling for the Grid involves scheduling multiple individual, potentially architecturally and administratively heterogeneous computing resources so that multiple processes are guaranteed to execute at the same time in order that they may communicate and coordinate with each other. This is quite different from co-scheduling within a 'single' resource, such as a cluster, or within a set of (typically administratively homogeneous) machines, all of which run one type of batch schedulers that can talk among themselves to co-schedule.

This coordinated scheduling is typically accomplished by fixed time or advance reservation scheduling in the underlying resources so that the Grid scheduling service can arrange for simultaneous execution of jobs on independent systems. There are currently a few batch scheduling systems that can provide for Grid co-scheduling, and this is typically accomplished by scheduling to a time of day. Both the PBSPro [13] and Maui Silver [33] schedulers provide time-of-day scheduling (see Section 5.7.7). Other schedulers are slated to provide this capability in the future.

The Globus job initiator can pass through the information requesting a time-of-day reservation; however, it does not currently include any automated mechanisms to establish communication among the processes once they are running. That must be handled in the higher-level framework that initiates the co-scheduled jobs.

In this Grid computing model, network performance will also probably be a critical issue. See Section 5.7.6.

5.3.1.5 Tightly coupled processes

MPI and Parallel Virtual Machine (PVM) support a distributed memory programming model.

MPICH-G2 (the Globus-enabled MPI) [47] provides for MPI style interprocess communication between Grid computing resources. It handles data conversion, communication establishment, and so on. Co-scheduling is essential for this to be a generally useful capability since different 'parts' of the same program are running on different systems.

PVM [48] is another distributed memory programming system that can be used in conjunction with Condor and Globus to provide Grid functionality for running tightly coupled processes.

In the case of MPICH-G2, it can use Globus directly to co-schedule (assuming the underlying computing resource supports the capability) and coordinates communication among a set of tightly coupled processes. The MPICH-G2 libraries must be installed and tested on the Grid compute resources in which they will be used. MPICH-G2 will use the manufacturer's MPI for local communication if one is available and currently will not operate correctly if other versions of MPICH are installed. (Note that there was a significant change in the MPICH implementation between Globus 1.1.3 and 1.1.4 in that the use of the Nexus communication libraries was replaced by the Globus I/O libraries, and there is no compatibility between programs using Globus 1.1.3 and below and 1.1.4 and above.) Note also that there are wide area network (WAN) version of MPI that are more mature than MPICH-G2 (e.g. PACX-MPI [49, 50]); however, to the author's knowledge, these implementations are not Grid services because they do not make use of the Common Grid Services. In particular, the MIPCH-G2 use of the Globus I/O library that, for example, automatically provides access to the Grid Security Services (GSS), since the I/O library incorporates GSI below the I/O interface.

In the case of PVM, one can use Condor to manage the communication and coordination. In Grids, this can be accomplished using the Personal Condor Glide-In [34]. This is essentially an approach that has Condor using the Globus job initiator (GRAM) to start the Condor job manager on a Grid system (a 'Glide-In'). Once the Condor Glide-In is started, then Condor can provide the communication management needed by PVM. PVM can also use Condor for co-scheduling (see the Condor User's Manual [51]), and then Condor, in turn, can use Globus job management. (The Condor Glide-In can provide co-scheduling within a Condor flock if it is running when the scheduling is needed. That is, it could drive a distributed simulation in which some of the computational resources are under the control of the user – for example, a local cluster – and some (the Glide-in) are scheduled by a batch queuing system. However, if the Glide-in is not the 'master' and co-scheduling is required, then the Glide-in itself must be co-scheduled using, e.g. PBS.) This, then, can provide a platform for running tightly coupled PVM jobs in Grid environments. (Note, however, that PVM does not use the 'has no' mechanism to make use of the GSS, and so its communication cannot be authenticated within the context of the GSI.)

This same Condor Glide-In approach will work for MPI jobs.

The Condor Glide-In is essentially self-installing: As part of the user initiating a Glide-In job, all the required supporting pieces of Condor are copied to the remote system and installed in user-space.

5.3.2 Grid data models

Many of the current production Grids are focused around communities whose interest in wide-area data management is at least as great as their interest in Grid-based computing. These include, for example, Particle Physics Data Grid (PPDG) [52], Grid Physics Network (GriPhyN) [23], and the European Union DataGrid [36].

Like computing, there are several styles of data management in Grids, and these styles result in different requirements for the software of a Grid.

5.3.2.1 Occasional access to multiple tertiary storage systems

Data mining, as, for example, in Reference [53], can require access to metadata and uniform access to multiple data archives.

SRB/MCAT provides capabilities that include uniform remote access to data and local caching of the data for fast and/or multiple accesses. Through its metadata catalogue, SRB provides the ability to federate multiple tertiary storage systems (which is how it is used in the data mining system described in Reference [53]). SRB provides a uniform interface by placing a server in front of (or as part of) the tertiary storage system. This server must directly access the tertiary storage system, so there are several variations depending on the particular storage system (e.g. HPSS, UniTree, DMF, etc.). The server should also have some local disk storage that it can manage for caching, and so on. Access control in SRB is treated as an attribute of the dataset, and the equivalent of a Globus mapfile is stored in the dataset metadata in MCAT. See below for the operational issues of MCAT.

GridFTP provides many of the same basic data access capabilities as SRB, however, for a single data source. GridFTP is intended to provide a standard, low-level Grid data access service so that higher-level services like SRB could be componentized. However, much of the emphasis in GridFTP has been WAN performance and the ability to manage huge files in the wide area for the reasons given in the next section. The capabilities of GridFTP (not all of which are available yet, and many of which are also found in SRB) are also described in the next section.

GridFTP provides uniform access to tertiary storage in the same way that SRB does, and so there are customized backends for different type of tertiary storage systems. Also like SRB, the GridFTP server usually has to be managed on the tertiary storage system, together with the configuration and access control information needed to support GSI. [Like most Grid services, the GridFTP control and data channels are separated, and the control channel is always secured using GSI (see Reference [54])].

The Globus Access to Secondary Storage service (GASS, [55]) provides a Unix I/O style access to remote files (by copying the entire file to the local system on file open, and back on close). Operations supported include read, write and append. GASS also provides for local caching of file so that they may be staged and accessed locally and reused during a job without recopying. That is, GASS provides a common view of a file cache within a single Globus job.

A typical configuration of GASS is to put a GASS server on or near a tertiary storage system. A second typical use is to locate a GASS server on a user system where files (such as simulation input files) are managed so that Grid jobs can access data directly on those systems.

The GASS server must be managed as a persistent service, together with the auxiliary information for GSI authentication (host and service certificates, Globus mapfile, etc.).

5.3.2.2 Distributed analysis of massive datasets followed by cataloguing and archiving

In many scientific disciplines, a large community of users requires remote access to large datasets. An effective technique for improving access speeds and reducing network loads can be to replicate frequently accessed datasets at locations chosen to be 'near' the eventual users. However, organizing such replication so that it is both reliable and efficient can be a challenging problem, for a variety of reasons. The datasets to be moved can be large, so issues of network performance and fault tolerance become important. The individual locations at which replicas may be placed can have different performance characteristics, in which case users (or higher-level tools) may want to be able to discover these characteristics and use this information to guide replica selection. In addition, different locations may have different access control policies that need to be respected.

From *A Replica Management Service for High-Performance Data Grids,* The Globus Project [56].

This quote characterizes the situation in a number of data-intensive science disciplines, including high-energy physics and astronomy. These disciplines are driving the development of data management tools for the Grid that provides naming and location transparency, and replica management for very large data sets. The Globus Data Grid tools include a replica catalogue [57], a replica manager [58], and a high-performance data movement tool (GridFTP, [28]). The Globus tools do not currently provide metadata catalogues. (Most of the aforementioned projects already maintain their own style of metadata catalogue.) The European Union DataGrid project provides a similar service for replica management that uses a different set of catalogue and replica management tools (GDMP [59]). It, however, also uses GridFTP as the low-level data service. The differences in the two approaches are currently being resolved in a joint US–EU Data Grid services committee.

Providing an operational replica service will involve maintaining both the replica manager service and the replica catalogue. In the long term, the replica catalogue will probably just be data elements in the GIS/MDS, but today it is likely to be a separate directory service. Both the replica manager and catalogue will be critical services in the science environments that rely on them for data management.

The data-intensive science applications noted above that are international in their scope have motivated the GridFTP emphasis on providing WAN high performance and the ability to manage huge files in the wide area. To accomplish this, GridFTP provides

- integrated GSI security and policy-based access control,
- third-party transfers (between GridFTP servers),
- wide-area network communication parameter optimization,
- partial file access,
- reliability/restart for large file transfers,
- integrated performance monitoring instrumentation,
- network parallel transfer streams,
- server-side data striping (cf. DPSS [60] and HPSS striped tapes),
- server-side computation,
- proxies (to address firewall and load-balancing).

Note that the operations groups that run tertiary storage systems typically have (an appropriately) conservative view of their stewardship of the archival data, and getting GridFTP (or SRB) integrated with the tertiary storage system will take a lot of careful planning and negotiating.

5.3.2.3 Large reference data sets

A common situation is that a whole set of simulations or data analysis programs will require the use of the same large reference dataset. The management of such datasets, the originals of which almost always live in a tertiary storage system, could be handled by one of the replica managers. However, another service that is needed in this situation is a network cache: a unit of storage that can be accessed and allocated as a Grid resource, and that is located 'close to' (in the network sense) the Grid computational resources that will run the codes that use the data. The Distributed Parallel Storage System (DPSS, [60]) can provide this functionality; however, it is not currently well integrated with Globus.

5.3.2.4 Grid metadata management

The Metadata Catalogue of SRB/MCAT provides a powerful mechanism for managing all types of descriptive information about data: data content information, fine-grained access control, physical storage device (which provides location independence for federating archives), and so on.

The flip side of this is that the service is fairly heavyweight to use (when its full capabilities are desired) and it requires considerable operational support. When the MCAT server is in a production environment in which a lot of people will manage lots of data via SRB/MCAT, it requires a platform that typically consists of an Oracle DBMS (Database Management System) running on a sizable multiprocessor Sun system. This is a common installation in the commercial environment; however, it is not typical in the science environment, and the cost and skills needed to support this in the scientific environment are nontrivial.

5.4 GRID SUPPORT FOR COLLABORATION

Currently, Grids support collaboration, in the form of Virtual Organizations (VO) (by which we mean human collaborators, together with the Grid environment that they share), in two very important ways.

The GSI provides a common authentication approach that is a basic and essential aspect of collaboration. It provides the authentication and communication mechanisms, and trust management (see Section 5.6.1) that allow groups of remote collaborators to interact with each other in a trusted fashion, and it is the basis of policy-based sharing of collaboration resources. GSI has the added advantage that it has been integrated with a number of tools that support collaboration, for example, secure remote login and remote shell – GSISSH [61, 62], and secure ftp – GSIFTP [62], and GridFTP [28].

The second important contribution of Grids is that of supporting collaborations that are VO and as such have to provide ways to preserve and share the organizational structure (e.g. the identities – as represented in X.509 certificates (see Section 5.6) – of all the participants and perhaps their roles) and share community information (e.g. the location and description of key data repositories, code repositories, etc.). For this to be effective over the long term, there must be a persistent publication service where this information may be deposited and accessed by both humans and systems. The GIS can provide this service.

A third Grid collaboration service is the Access Grid (AG) [63] – a group-to-group audio and videoconferencing facility that is based on Internet IP multicast, and it can be managed by an out-of-band floor control service. The AG is currently being integrated with the Globus directory and security services.

5.5 BUILDING AN INITIAL MULTISITE, COMPUTATIONAL AND DATA GRID

5.5.1 The Grid building team

Like networks, successful Grids involve almost as much sociology as technology, and therefore establishing good working relationships among all the people involved is essential.

The concept of an Engineering WG has proven successful as a mechanism for promoting cooperation and mutual technical support among those who will build and manage the Grid. The WG involves the Grid deployment teams at each site and meets weekly via teleconference. There should be a designated WG lead responsible for the agenda and managing the discussions. If at all possible, involve some Globus experts at least during the first several months while people are coming up to speed. There should also be a WG mail list that is archived and indexed by thread. Notes from the WG meetings should be mailed to the list. This, then, provides a living archive of technical issues and the state of your Grid.

Grid software involves not only root-owned processes on all the resources but also a trust model for authorizing users that is not typical. Local control of resources is maintained, but is managed a bit differently from current practice. It is therefore very important to set up liaisons with the system administrators for all systems that will provide computation and storage resources for your Grid. This is true whether or not these systems are within your organization.

5.5.2 Grid resources

As early as possible in the process, identify the computing and storage resources to be incorporated into your Grid. In doing this be sensitive to the fact that opening up systems to Grid users may turn lightly or moderately loaded systems into heavily loaded systems. Batch schedulers may have to be installed on systems that previously did not use them in order to manage the increased load.

When choosing a batch scheduler, carefully consider the issue of co-scheduling! Many potential Grid applications need this, for example, to use multiple Grid systems to run cross system MPI jobs or support pipelined applications as noted above, and only a few available schedulers currently provide the advance reservation mechanism that is used for Grid co-scheduling (e.g. PBSPro and Maui). If you plan to use some other scheduler, be very careful to critically investigate any claims of supporting co-scheduling to make sure that they actually apply to heterogeneous Grid systems. (Several schedulers support co-scheduling only among schedulers of the same type and/or within administratively homogeneous domains.) See the discussion of the PBS scheduler in Section 5.7.7.

5.5.3 Build the initial test bed

5.5.3.1 Grid information service

The Grid Information Service provides for locating resources based on the characteristics needed by a job (OS, CPU count, memory, etc.). The Globus MDS [25] provides this capability with two components. The Grid Resource Information Service (GRIS) runs on the Grid resources (computing and data systems) and handles the soft-state registration of the resource characteristics. The Grid Information Index Server (GIIS) is a user accessible directory server that supports searching for resource by characteristics. Other information may also be stored in the GIIS, and the GGF, Grid Information Services group is defining schema for various objects [64].

Plan for a GIIS at each distinct site with significant resources. This is important in order to avoid single points of failure, because if you depend on a GIIS at some other site and it becomes unavailable, you will not be able to examine your local resources. Depending upon the number of local resources, it may be necessary to set up several GIISs at a site in order to accommodate the search load.

The initial test bed GIS model can be independent GIISs at each site. In this model, either cross-site searches require explicit knowledge of each of the GIISs that have to be searched independently or all resources cross-register in each GIIS. (Where a resource register is a configuration parameter in the GRISs that run on each Grid resource.)

5.5.3.2 Build Globus on test systems

Use PKI authentication and initially use certificates from the Globus Certificate Authority ('CA') or any other CA that will issue you certificates for this test environment. (The OpenSSL CA [65] may be used for this testing.) Then validate access to, and operation of the, GIS/GIISs at all sites and test local and remote job submission using these certificates.

5.6 CROSS-SITE TRUST MANAGEMENT

One of the most important contributions of Grids to supporting large-scale collaboration is the uniform Grid entity naming and authentication mechanisms provided by the GSI.

However, for this mechanism to be useful, the collaborating sites/organizations must establish mutual trust in the authentication process. The software mechanism of PKI, X.509 identity certificates, and their use in the GSI through Transport Layer Security (TLS)/Secure Sockets Layer (SSL) [54], are understood and largely accepted. The real issue is that of establishing trust in the process that each 'CA' uses for issuing the identity certificates to users and other entities, such as host systems and services. This involves two steps. First is the 'physical' identification of the entities, verification of their association with the VO that is issuing identity certificates, and then the assignment of an appropriate name. The second is the process by which an X.509 certificate is issued. Both these steps are defined in the CA policy.

In the PKI authentication environment assumed here, the CA policies are encoded as formal documents associated with the operation of the CA that issues your Grid identity credentials. These documents are called *the Certificate Policy/Certification Practice Statement*, and we will use 'CP' to refer to them collectively. (See Reference [66].)

5.6.1 Trust

Trust is 'confidence in or reliance on some quality or attribute of a person or thing, or the truth of a statement'.[2] Cyberspace trust starts with clear, transparent, negotiated, and documented policies associated with identity. When a Grid identity token (X.509 certificate in the current context) is presented for remote authentication and is verified using the appropriate cryptographic techniques, then the relying party should have some level of confidence that the person or entity that initiated the transaction is the person or entity that it is expected to be.

The nature of the policy associated with identity certificates depends a great deal on the nature of your Grid community and/or the VO associated with your Grid. It is relatively easy to establish a policy for homogeneous communities, such as in a single organization, because an agreed upon trust model will probably already exist.

It is difficult to establish trust for large, heterogeneous VOs involving people from multiple, international institutions, because the shared trust models do not exist. The typical issues related to establishing trust may be summarized as follows:

- Across administratively similar systems
  - for example, within an organization
  - informal/existing trust model can be extended to Grid authentication and authorization
- Administratively diverse systems
  - for example, across many similar organizations (e.g. NASA Centers, DOE Labs)
  - formal/existing trust model can be extended to Grid authentication and authorization
- Administratively heterogeneous
  - for example, cross multiple organizational types (e.g. science labs and industry),
  - for example, international collaborations
  - formal/new trust model for Grid authentication and authorization will need to be developed.

The process of getting your CP (and therefore your user's certificates) accepted by other Grids (or even by multisite resources in your own Grid) involves identifying the people who can authorize remote users at all the sites/organizations that you will collaborate with and exchanging CPs with them. The CPs are evaluated by each party in order to ensure that the local policy for remote user access is met. If it is not, then a period of negotiation ensues. The sorts of issues that are considered are indicated in the European Union DataGrid Acceptance and Feature matrices [67].

Hopefully the sites of interest already have people who are (1) familiar with the PKI CP process and (2) focused on the scientific community of the institution rather than on the administrative community. (However, be careful that whomever you negotiate with actually has the authority to do so. Site security folks will almost always be involved at some point in the process, if that process is appropriately institutionalized.)

Cross-site trust may, or may not, be published. Frequently it is. See, for example, the European Union DataGrid list of acceptable CAs [68].

---

[2] Oxford English Dictionary, Second Edition (1989). Oxford University Press.

5.6.2 Establishing an operational CA[3]

Set up, or identify, a Certification Authority to issue Grid X.509 identity certificates to users and hosts. Both the IPG and DOE Science Grids use the Netscape CMS software [69] for their operational CA because it is a mature product that allows a very scalable usage model that matches well with the needs of science VO.

Make sure that you understand the issues associated with the CP of your CA. As noted, one thing governed by CP is the 'nature' of identity verification needed to issue a certificate, and this is a primary factor in determining who will be willing to accept your certificates as adequate authentication for resource access. Changing this aspect of your CP could well mean not just reissuing all certificates but requiring all users to reapply for certificates.

Do not try and invent your own CP. The GGF is working on a standard set of CPs that can be used as templates, and the DOE Science Grid has developed a CP that supports international collaborations, and that is contributing to the evolution of the GGF CP. (The SciGrid CP is at http://www.doegrids.org/ [66].)

Think carefully about the space of entities for which you will have to issue certificates. These typically include human users, hosts (systems), services (e.g. GridFTP), and possibly security domain gateways (e.g. the PKI to Kerberos gateway, KX509 [70]). Each of these must have a clear policy and procedure described in your CA's CP/CPS.

If you plan to interoperate with other CAs, then discussions about homogenizing the CPs and CPSs should begin as soon as possible, as this can be a lengthy process.

Establish and publish your Grid CP as soon as possible so that you will start to appreciate the issues involved.

5.6.2.1 Naming

One of the important issues in developing a CP is the naming of the principals (the 'subject,' i.e. the Grid entity identified by the certificate). While there is an almost universal tendency to try and pack a lot of information into the subject name (which is a multicomponent, X.500 style name), increasingly there is an understanding that the less information of any kind put into a certificate, the better. This simplifies certificate management and re-issuance when users forget pass phrases (which will happen with some frequency). More importantly, it emphasizes that *all trust is local* – that is, established by the resource owners and/or when joining a virtual community. The main reason for having a complicated subject name invariably turns out to be that people want to do some of the authorization on the basis of the components of the name (e.g. organization). However, this usually leads to two problems. One is that people belong to multiple organizations, and the other is that the authorization implied by the issuing of a certificate will almost certainly collide with some aspect the authorization actually required at any given resource.

The CA run by ESnet (the DOE Office of Science scientific networks organization [71]) for the DOE Science Grid, for example, will serve several dozen different VO, several

---

[3] Much of the work described in this section is that of Tony Genovese (tony@es.net) and Mike Helm (helm@es.net), ESnet, Lawrence Berkeley National Laboratory.

of which are international in their makeup. The certificates use what is essentially a flat namespace, with a 'reasonable' common name (e.g. a 'formal' human name) to which has been added a random string of alphanumeric digits to ensure name uniqueness.

However, if you do choose to use hierarchical institutional names in certificates, do not use colloquial names for institutions – consider their full organizational hierarchy in defining the naming hierarchy. Find out if anyone else in your institution, agency, university, and so on is working on PKI (most likely in the administrative or business units) and make sure that your names do not conflict with theirs, and if possible follow the same name hierarchy conventions.

It should be pointed out that CAs set up by the business units of your organization frequently do not have the right policies to accommodate Grid users. This is not surprising since they are typically aimed at the management of institutional financial transactions.

5.6.2.2 The certification authority model

There are several models for CAs; however, increasingly associated groups of collaborations/VO are opting to find a single CA provider. The primary reason for this is that it is a formal and expensive process to operate a CA in such a way that it will be trusted by others.

One such model has a central CA that has an overall CP and subordinate policies for a collection of VOs. The CA delegates to VOs (via Registration Agents) the responsibility of deciding who is a member of the particular VO and how the subscriber/user will be identified in order to be issued a VO certificate. Each VO has an appendix in the CP that describes VO specific issues. VO Registration Agents are responsible for applying the CP identity policy to their users and other entities. Once satisfied, the RA authorizes the CA to issue (generate and sign) a certificate for the subscriber.

This is the model of the DOE Science Grid CA, for example, and it is intended to provide a CA that is scalable to dozens of VO and thousands of users. This approach to scalability is the usual divide and conquer policy, together with a hierarchical organization that maintains the policy integrity. The architecture of the DOE Science Grid CA is indicated in Figure 5.2 and it has the following key features.

The Root CA (which is kept locked up and off-line) signs the certificates of the CA that issues user certificates. With the exception of the 'community' Registration Manager (RMs), all RMs are operated by the VOs that they represent. (The community RM addresses those 'miscellaneous' people who legitimately need DOE Grid certificates, but for some reason are not associated with a Virtual Organization.) The process of issuing a certificate to a user ('subscriber') is indicated in Figure 5.3.

ESnet [71] operates the CA infrastructure for DOE Science Grids; they do not interact with users. The VO RAs interface with certificate requestors. The overall policy oversight is provided by a Policy Management Authority, which is a committee that is chaired by ESnet and is composed of each RA and a few others.

This approach uses an existing organization (ESnet) that is set up to run a secure production infrastructure (its network management operation) to operate and protect the

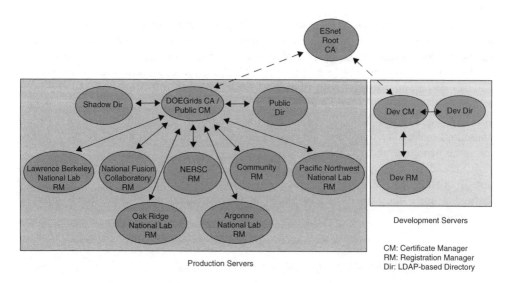

Figure 5.2 Software architecture for 5/15/02 deployment of the DOE Grids CA (Courtesy Tony Genovese (tony@es.net) and Mike Helm (helm@es.net), ESnet, Lawrence Berkeley National Laboratory).

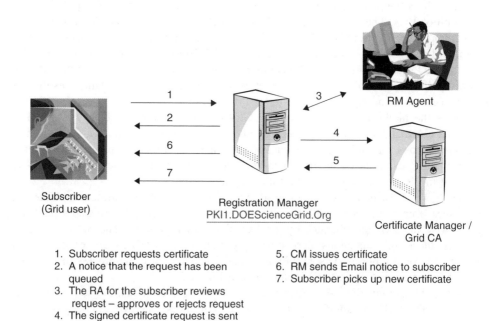

1. Subscriber requests certificate
2. A notice that the request has been queued
3. The RA for the subscriber reviews request – approves or rejects request
4. The signed certificate request is sent to CA
5. CM issues certificate
6. RM sends Email notice to subscriber
7. Subscriber picks up new certificate

Figure 5.3 Certificate issuing process (Courtesy Tony Genovese (tony@es.net) and Mike Helm (helm@es.net), ESnet, Lawrence Berkeley National Laboratory).

critical components of the CA. ESnet defers user contact to agents within the collaboration communities. In this case, the DOE Science Grid was fortunate in that ESnet personnel were also well versed in the issues of PKI and X.509 certificates, and so they were able to take a lead role in developing the Grid CA architecture and policy.

5.7 TRANSITION TO A PROTOTYPE PRODUCTION GRID

5.7.1 First steps

Issue host certificates for all the computing and data resources and establish procedures for installing them. Issue user certificates.

Count on revoking and re-issuing all the certificates at least once before going operational. This is inevitable if you have not previously operated a CA.

Using certificates issued by your CA, validate correct operation of the GSI [72], GSS libraries, GSISSH [62], and GSIFTP [73] and/or GridFTP [28] at all sites.

Start training a Grid application support team on this prototype.

5.7.2 Defining/understanding the extent of 'your' Grid

The 'boundaries' of a Grid are primarily determined by three factors:

- *Interoperability of the Grid software*: Many Grid sites run some variation of the Globus software, and there is fairly good interoperability between versions of Globus, so most Globus sites can potentially interoperate.
- *What CAs you trust*: This is explicitly configured in each Globus environment on a per CA basis.
  Your trusted CAs establish the maximum extent of your user population; however, there is no guarantee that every resource in what you think is 'your' Grid trusts the same set of CAs – that is, each resource potentially has a different space of users – this is a local decision. In fact, this will be the norm if the resources are involved in multiple VO as they frequently are, for example, in the high-energy physics experiment data analysis communities.
- *How you scope the searching of the GIS/GIISs or control the information that is published in them*: This depends on the model that you choose for structuring your directory services.

So, the apparent 'boundaries' of most Grids depend on who is answering the question.

5.7.3 The model for the Grid Information System

Directory servers above the local GIISs (resource information servers) are an important scaling mechanism for several reasons.

They expand the resource search space through automated cross-GIIS searches for resources, and therefore provide a potentially large collection of resources transparently to users. They also provide the potential for query optimization and query results caching. Furthermore, such directory services provide the possibility for hosting and/or defining VOs and for providing federated views of collections of data objects that reside in different storage systems.

There are currently two main approaches that are being used for building directory services above the local GIISs. One is a hierarchically structured set of directory servers and a managed namespace, al la X.500, and the other is 'index' servers that provide *ad hoc*, or 'VO' specific, views of a specific set of other servers, such as a collection of GIISs, data collections, and so on.

Both provide for 'scoping' your Grid in terms of the resource search space, and in both cases many Grids use o = grid as the top level.

5.7.3.1 An X.500 style hierarchical name component space directory structure

Using an X.500 Style hierarchical name component space directory structure has the advantage of organizationally meaningful names that represent a set of 'natural' boundaries for scoping searches, and it also means that you can potentially use commercial metadirectory servers for better scaling.

Attaching virtual organization roots, data namespaces, and so on to the hierarchy makes them automatically visible, searchable, and in some sense 'permanent' (because they are part of this managed namespace).

If you plan to use this approach, try very hard to involve someone who has some X.500 experience because the directory structures are notoriously hard to get right, a situation that is compounded if VOs are included in the namespace.

5.7.3.2 Index server directory structure

Using the Globus MDS [25] for the information directory hierarchy (see Reference [74]) has several advantages.

The MDS research and development work has added to the usual Lightweight Directory Access Protocol (LDAP)–based directory service capabilities several features that are important for Grids.

Soft-state registration provides for autoregistration and de-registration, and for registration access control. This is very powerful. It keeps the information up-to-date (via a keep-alive mechanism) and it provides for a self configuring and dynamic Grid: a new resource registering for the first time is essentially no different from an old resource that is reregistering after, for example, a system crash. The autoregistration mechanism also allows resources to participate in multiple information hierarchies, thereby easily accommodating membership in multiple VOs. The registration mechanism also provides a natural way to impose authorization on those who would register with your GIISs.

Every directory server from the GRIS on the resource, up to and including the root of the information hierarchy, is essentially the same, which simplifies the management of the servers.

Other characteristics of MDS include the following:

- Resources are typically named using the components of their Domain Name System (DNS) name, which has the advantage of using an established and managed namespace.
- One must use separate 'index' servers to define different relationships among GIISs, virtual organization, data collections, and so on; on the other hand, this allows you to establish 'arbitrary' relationships within the collection of indexed objects.
- Hierarchical GIISs (index nodes) are emerging as the preferred approach in the Grids community that uses the Globus software.

Apart from the fact that all the directory servers must be run as persistent services and their configuration maintained, the only real issue with this approach is that we do not have a lot of experience with scaling this to multiple hierarchies with thousands of resources.

5.7.4 Local authorization

As of yet, there is no standard authorization mechanism for Grids. Almost all current Grid software uses some form of access control lists ('ACL'), which is straightforward, but typically does not scale very well.

The Globus mapfile is an ACL that maps from Grid identities (the subject names in the identity certificates) to local user identification numbers (UIDs) on the systems where jobs are to be run. The Globus Gatekeeper [27] replaces the usual login authorization mechanism for Grid-based access and uses the mapfile to authorize access to resources after authentication. Therefore, managing the contents of the mapfile is the basic Globus user authorization mechanism for the local resource.

The mapfile mechanism is fine in that it provides a clear-cut way for locally controlling access to a system by Grid users. However, it is bad in that for a large number of resources, especially if they all have slightly different authorization policies, it can be difficult to manage.

The first step in the mapfile management process is usually to establish a connection between user account generation on individual platforms and requests for Globus access on those systems. That is, generating mapfile entries is done automatically when the Grid user goes through the account request process. If your Grid users are to be automatically given accounts on a lot of different systems with the same usage policy, it may make sense to centrally manage the mapfile and periodically distribute it to all systems. However, unless the systems are administratively homogeneous, a nonintrusive mechanism, such as e-mail to the responsible system admins to modify the mapfile, is best.

The Globus mapfile also allows a many-to-one mapping so that, for example, a whole group of Grid users can be mapped to a single account. Whether the individual identity is preserved for accounting purposes is typically dependent on whether the batch queuing system can pass the Grid identity (which is carried along with a job request, regardless of the mapfile mapping) back to the accounting system. PBSPro, for example, will provide this capability (see Section 5.7.7).

One way to address the issues of mapfile management and disaggregated accounting within an administrative realm is to use the Community Authorization Service (CAS), which is just now being tested. See the notes of Reference [75].

5.7.5 Site security issues

Incorporating any computing resource into a distributed application system via Grid services involves using a whole collection of IP communication ports that are otherwise not used. If your systems are behind a firewall, then these ports are almost certainly blocked, and you will have to negotiate with the site security folks to open the required ports.

Globus can be configured to use a restricted range of ports, but it still needs several tens, or so, in the mid-700s. (The number depending on the level of usage of the resources behind the firewall.) A Globus 'port catalogue' is available to tell what each Globus port is used for, and this lets you provide information that your site security folks will probably want to know. It will also let you estimate how many ports have to be opened (how many per process, per resource, etc.). Additionally, GIS/GIIS needs some ports open, and the CA typically uses a secure Web interface (port 443). The Globus port inventory is given in Reference [72]. The DOE Science Grid is in the process of defining Grid firewall policy document that we hope will serve the same role as the CA Certificate Practices Statement: It will lay out the conditions for establishing trust between the Grid administrators and the site security folks who are responsible for maintaining firewalls for site cyber protection.

It is important to develop tools/procedures to periodically check that the ports remain open. Unless you have a very clear understanding with the network security folks, the Grid ports will be closed by the first network engineer that looks at the router configuration files and has not been told why these nonstandard ports are open.

Alternate approaches to firewalls have various types of service proxies manage the intraservice component communication so that one, or no, new ports are used. One interesting version of this approach that was developed for Globus 1.1.2 by Yoshio Tanaka at the Electrotechnical Laboratory [ETL, which is now the National Institute of Advanced Industrial Science and Technology (AIST)] in Tsukuba, Japan, is documented in References [76, 77].

5.7.6 High performance communications issues

If you anticipate high data-rate distributed applications, whether for large-scale data movement or process-to-process communication, then enlist the help of a WAN networking specialist and check and refine the network bandwidth end-to-end using large packet size test data streams. (Lots of problems that can affect distributed application do not show up by pinging with the typical 32 byte packets.) Problems are likely between application host and site LAN/WAN gateways, WAN/WAN gateways, and along any path that traverses the commodity Internet.

Considerable experience exists in the DOE Science Grid in detecting and correcting these types of problems, both in the areas of diagnostics and tuning.

End-to-end monitoring libraries/toolkits (e.g. NetLogger [78] and pipechar [79]) are invaluable for application-level distributed debugging. NetLogger provides for detailed data path analysis, top-to-bottom (application to NIC) and end-to-end (across the entire network path) and is used extensively in the DOE Grid for this purpose. It is also being incorporated into some of the Globus tools. (For some dramatic examples of the use of NetLogger to debug performance problem in distributed applications, see References [80–83].)

If at all possible, provide network monitors capable of monitoring specific TCP flows and returning that information to the application for the purposes of performance debugging. (See, for example, Reference [84].)

In addition to identifying problems in network and system hardware and configurations, there are a whole set of issues relating to how current TCP algorithms work, and how they must be tuned in order to achieve high performance in high-speed, wide-area networks. Increasingly, techniques for automatic or semi-automatic setting of various TCP parameters based on monitored network characteristics are being used to relieve the user of having to deal with this complex area of network tuning that is critically important for high-performance distributed applications. See, for example, References [85–87].

5.7.7 Batch schedulers[4]

There are several functions that are important to Grids that Grid middleware cannot emulate: these must be provided by the resources themselves.

Some of the most important of these are the functions associated with job initiation and management on the remote computing resources. Development of the PBS batch scheduling system was an active part of the IPG project, and several important features were added in order to support Grids.

In addition to the scheduler providing a good interface for Globus GRAM/RLS (which PBS did), one of the things that we found was that people could become quite attached to the specific syntax of the scheduling system. In order to accommodate this, PBS was componentized and the user interfaces and client-side process manager functions were packaged separately and interfaced to Globus for job submission.

PBS was somewhat unique in this regard, and it enabled PBS-managed jobs to be run on Globus-managed systems, as well as the reverse. This lets users use the PBS frontend utilities (submit via PBS 'qsub' command-line and 'xpbs' GUI, monitor via PBS 'qstat', and control via PBS 'qdel', etc.) to run jobs on remote systems managed by Globus. At the time, and probably today, the PBS interface was a more friendly option than writing Globus RSL.

This approach is also supported in Condor-G, which, in effect, provides a Condor interface to Globus.

PBS can provide time-of-day-based advanced reservation. It actually creates a queue that 'owns' the reservation. As such, all the access control features (allowing/disallowing specific users/groups) can be used to control access to the reservation. It also allows one to submit a string of jobs to be run during the reservation. In fact, you can use the existing job-chaining features in PBS to do complex operations such as run X; if X fails, run Y; if X succeeds, run Z.

PBS passes the Grid user ID back to the accounting system. This is important for allowing, for example, the possibility of mapping all Grid users to a single account (and thereby not having to create actual user accounts for Grid user) but at the same time still maintaining individual accountability, typically for allocation management.

---

[4] Thanks to Bill Nitzberg (bill@computer.org), one of the PBS developers and area co-director for the GGF scheduling and resource management area, for contributing to this section.

Finally, PBS supports access-controlled, high-priority queues. This is of interest in scenarios in which you might have to 'commandeer' a lot of resources in a hurry to address a specific, potentially emergency situation. Let us say for example that we have a collection of Grid machines that have been designated for disaster response/management. For this to be accomplished transparently, we need both lots of Grid managed resources and ones that have high-priority queues that are accessible to a small number of preapproved people who can submit 'emergency' jobs. For immediate response, this means that they would need to be preauthorized for the use of these queues, and that PBS has to do per queue, UID-based access control. Further, these should be preemptive high-priority queues. That is, when a job shows up in the queue, it forces other, running, jobs to be checkpointed and rolled out, and/or killed, in order to make sure that the high-priority job runs.

PBS has full 'preemption' capabilities, and that, combined with the existing access control mechanisms, provides this sort of 'disaster response' scheduling capability.

There is a configurable 'preemption threshold' – if a queue's priority is higher than the preemption threshold, then any jobs ready to run in that queue will preempt all running work on the system with lower priority. This means you can actually have multiple levels of preemption. The preemption action can be configured to (1) try to checkpoint, (2) suspend, and/or (3) kill and requeue, in any order.

For access control, every queue in PBS has an ACL that can include and exclude specific users and groups. All the usual stuff is supported, for example, 'everyone except bill', 'groups foo and bar, but not joe', and so on.

5.7.8 Preparing for users

Try and find problems before your users do. Design test and validation suites that exercise your Grid in the same way that applications are likely to use your Grid.

As early as possible in the construction of your Grid, identify some test case distributed applications that require reasonable bandwidth and run them across as many widely separated systems in your Grid as possible, and then run these test cases every time something changes in your configuration.

Establish user help mechanisms, including a Grid user e-mail list and a trouble ticket system. Provide user-oriented Web pages with pointers to documentation, including a Globus 'Quick Start Guide' [88] that is modified to be specific to your Grid, and with examples that will work in your environment (starting with a Grid 'hello world' example).

5.7.9 Moving from test bed to prototype production Grid

At this point, Globus, the GIS/MDS, and the security infrastructure should all be operational on the test bed system(s). The Globus deployment team should be familiar with the install and operation issues and the system admins of the target resources should be engaged.

Deploy and build Globus on at least two production computing platforms at two different sites. Establish the relationship between Globus job submission and the local batch schedulers (one queue, several queues, a Globus queue, etc.).

Validate operation of this configuration.

5.7.10 Grid systems administration tools[5]

Grids present special challenges for system administration owing to the administratively heterogeneous nature of the underlying resources.

In the DOE Science Grid, we have built Grid monitoring tools from Grid services. We have developed pyGlobus modules for the NetSaint [89] system monitoring framework that test GSIFTP, MDS and the Globus gatekeeper. We have plans for, but have not yet implemented, a GUI tool that will use these modules to allow an admin to quickly test functionality of a particular host.

The harder issues in Grid Admin tools revolve around authorization and privilege management across site boundaries. So far we have concentrated only on tools for identifying problems. We still use e-mail to a privileged local user on the broken machine in order to fix things. In the long term, we have been thinking about a framework that will use a more autonomic model for continuous monitoring and restart of services.

In both Grids, tools and techniques are being developed for extending Trouble Ticket-based problem tracking systems to the Grid environment.

In the future, we will have to evolve a Grid account system that tracks Grid-user usage across a large number of machines and manages allocations in accordance with (probably varying) policy on the different systems. Some work by Jarosław Nabrzyski and his colleagues at the Poznan Supercomputing and Networking Center [90] in Poland is developing prototypes in this area. See Reference [91].

5.7.11 Data management and your Grid service model

Establish the model for moving data between *all* the systems involved in your Grid.

GridFTP servers should be deployed on the Grid computing platforms and on the Grid data storage platforms.

This presents special difficulties when data resides on user systems that are not usually Grid resources and raises the general issue of your Grid 'service model': what services are necessary to support in order to achieve a Grid that is useful for applications but are outside your core Grid resources (e.g. GridFTP on user data systems) and how you will support these services are issues that have to be recognized and addressed.

Determine if any user systems will manage user data that are to be used in Grid jobs. This is common in the scientific environment in which individual groups will manage their experiment data, for example, on their own systems. If user systems will manage data, then the GridFTP server should be installed on those systems so that data may be moved from user system to user job on the computing platform, and back.

Offering GridFTP on user systems may be essential; however, managing long-lived/root AG components on user systems may be 'tricky' and/or require you to provide some level of system admin on user systems.

Validate that all the data paths work correctly.

These issues are summarized in Figure 5.4.

---

[5] Thanks to Keith Jackson (krjackson@lbl.gov) and Stephen Chan (sychan@lbl.gov) for contributing to this section.

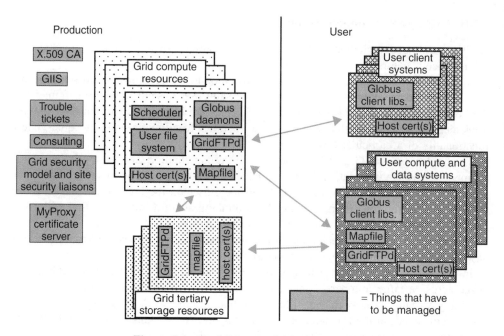

Figure 5.4   Establish your Grid service model.

5.7.12 Take good care of the users as early as possible

If at all possible, establish a Grid/Globus application specialist group. This group should be running sample jobs as soon as the test bed is stable, and certainly as soon as the prototype production system is operational. They should be able to assist generally with building Grid distributed applications, and specifically should serve as the interface between users and the Grid system administrators in order to solve Grid-related application problems.

Identify specific early users and have the Grid application specialists encourage/assist them in getting jobs running on the Grid.

One of the scaling/impediment-to-use issues currently is that extant Grid functions are relatively primitive (i.e. at a low level). This is being addressed by Grid middleware at various levels that provide aggregate functionality, more conveniently packaged functionality, toolkits for building Grid-based portals, and so on. Examples of such work in progress includes the Web Grid Services (e.g. the Open Grid Services Architecture OGSA [92] and the resulting Open Grid Services Interface [93] work at GGF), the Grid Web services test bed of the GGF (GCE) Grid Computing Environments WG [94], diverse interfaces to Grid functions (e.g. PyGlobus [95], CoG Kits [96–98]), and the Grid Portal Development Kit [99].

One approach that we have seen to be successful in the IPG and DOE Science Grid is to encourage applications that already have their own 'frameworks' to port those frameworks on the Grid. This is typically not too difficult because many of these frameworks already have some form of resource management built in, and this is easily replaced/augmented with Grid resource management functions. This hides some of the 'low-level functionality' problem. Examples of such frameworks deployed in IPG and/or DOE Science Grid are NPSS [100, 101] and Cactus [38]. Another example of this approach is Ninf [6].

Another useful tool for users is a Grid job-tracking and monitoring portal. IPG's LaunchPad [102] and NPACIs HotPage [103] are two examples.

5.7.12.1 MyProxy service

Consider providing a MyProxy service [104, 105] to simplify user management of certificates.

A frequent Grid user complaint relates to the constant management of GSI credentials and the frequent necessity of generating proxy credentials so that Grid work can proceed. A related, and functionally more serious, issue is that in order to minimize the risk of the relatively unprotected proxy credentials their lifetimes are kept relatively short (typically 12 h). This can create significant problems when, for example, large jobs take longer than that to execute on remote computing systems, or if the batch queues are long and the proxies expire before execution starts. In either case, the job is likely to fail.

The MyProxy service is designed to alleviate these problems, as well as to ease the problem of trying to move the user's permanent identity credential to all the systems from which the user will want to access the Grid.

The MyProxy service provides for creating and storing intermediate lifetime proxies that may be accessed by, for example, Web-based portals, job schedulers, and so forth, on behalf of the user. There are plans to extend the service so that it can manage the user's permanent identity credential as well.

MyProxy provides a set of client tools that lets the user create, store, and destroy proxies, and for programs acting on behalf of the user to obtain valid (short-term) proxies. The user can create a proxy with a lifetime of a week, or a few weeks, then store that proxy on a MyProxy server. The user and the MyProxy service establish a shared secret, and the user passes that secret to processes that need to obtain proxies on the user's behalf. In this way, a Grid service such as the Globus job initiator or the Condor job manager, can, after getting the user's access secret for MyProxy, contact the MyProxy service each time that they need a short-term proxy to perform a task. Now when a Grid job manager finds that a job's proxy is about to expire, it can ask the MyProxy service for a new proxy without user intervention.

The user still has to supply proxy-generating authority to MyProxy, but much less often than to a usual Grid task.

The security risks of this are analyzed in Reference [104] and are found to be not only acceptable compared to direct user management of the short-lived proxies but perhaps even less risky since the process is much less user error prone.

A key operational issue is that not only does the MyProxy server has to be managed as a persistent service but it also has to be a secure persistent service. This means that it should probably be in a controlled physical environment (e.g. a controlled access machine room), should be a strictly single purpose system, and should probably be behind a content filtering firewall.

5.8 CONCLUSIONS

We have presented the essence of experience gained in building two production Grids and provided some of the global context for this work.

As the reader might imagine, there were a lot of false starts, refinements to the approaches and to the software, and several substantial integration projects (SRB and Condor integrated with Globus) to get to where we are today.

However, the main point of this chapter is to try and make it substantially easier for others to get to the point where IPG and the DOE Science Grids are today. This is what is needed in order to move us toward the vision of a common cyber infrastructure for science.

The author would also like to remind the readers that this chapter primarily represents the actual experiences that resulted from specific architectural and software choices during the design and implementation of these two Grids. The choices made were dictated by the criteria laid out in Section 5.1.

There is a lot more Grid software available today than there was four years ago, and several of these packages are being integrated into IPG and the DOE Grids.

However, the foundation choices of Globus, SRB, and Condor would not be significantly different today than they were four years ago. Nonetheless, if the GGF is successful in its work – and we have every reason to believe that it will be – then in a few years we will see that the functions provided by these packages will be defined in terms of protocols and APIs, and there will be several robust implementations available for each of the basic components, especially the Grid common services.

The impact of the emerging Web Grid services work is not yet clear. It will probably have a substantial impact on building higher-level services; however, it is the opinion of the author that this will in no way obviate the need for the Grid common services. These are the foundation of Grids, and the focus of almost all the operational and persistent infrastructure aspects of Grids.

ACKNOWLEDGEMENTS

The experience represented in this chapter is the result of a lot of hard work by the NASA and DOE Science Grid Engineering Teams.

The principals in the NASA IPG team are Tony Lisotta, Chair, Warren Smith, George Myers, Judith Utley, and formerly Mary Hultquist, all of the NASA Ames Research Center, and Isaac Lopez, of the NASA Glenn Research Center. This project is lead by William Johnston, Tom Hinke, and Arsi Vaziri of NASA Ames Research Center.

The principals in the DOE Science Grid Engineering team are Keith Jackson, Chair, Lawrence Berkeley National Laboratory; Tony Genovese and Mike Helm, ESnet; Von Welch, Argonne National Laboratory; Steve Chan, NERSC; Kasidit Chanchio, Oak Ridge National Laboratory, and; Scott Studham, Pacific Northwest National Laboratory. This project is lead by William E. Johnston, Lawrence Berkeley National Laboratory; Ray Bair, Pacific Northwest National Laboratory; Ian Foster, Argonne National Laboratory; Al Geist, Oak Ridge National Laboratory, and William Kramer, LBNL/NERSC.

The IPG work is funded by NASA's Aero-Space Enterprise, Computing, Information, and Communication Technologies (CICT) Program (formerly the Information Technology), Computing, Networking, and Information Systems Project. Eugene Tu, Jerry Yan, and Cathy Schulbach are the NASA program managers.

The DOE Science Grid work is funded by the US Department of Energy, Office of Science, Office of Advanced Scientific Computing Research, Mathematical, Information, and Computational Sciences Division (http://www.sc.doe.gov/ascr/mics/) under contract DE-AC03-76SF00098 with the University of California. This program office is led by Walt Polansky. Mary Anne Scott is the Grids program manager and George Seweryniak is the ESnet program manager.

While not directly involved in funding the NASA or DOE Grids work, the author would also like to acknowledge the important support for Grids (e.g. the NSF Middleware Initiative [106]) provided by the National Science Foundation, in work funded by Alan Blatecky.

Without the support and commitment of program managers like these, and their counterparts in Europe and Asia Pacific, we would have little chance of realizing the vision of building a new and common cyber infrastructure for science.

Credit is also due to the intellectual leaders of the major software projects that formed the basis of IPG and the DOE Science Grid. The Globus team is led by Ian Foster, University of Chicago and Argonne National Laboratory, and by Carl Kesselman, University of Southern California, Information Sciences Institute. The SRB/MCAT team is led by Reagan Moore of the San Diego Supercomputer Center. The Condor project is led by Miron Livny at the University of Wisconsin, Madison.

Special thanks goes to Bill Feiereisen, who as NASA HPCC program manager conceived of the Information Power Grid and coined the term *Grid* in this context. While NAS Division Chief at NASA Ames, he provided the unfailing support for the project that was necessary for its success. Bill is currently head of the Computing Division at Los Alamos National Laboratory.

Important contributions are also being made by industry, and in the author's opinion among the most important of these is the support of Grid tools like the Globus Toolkit by computing systems manufacturers. This support – like the recently announced IBM support for Globus on the SP/AIX supercomputers [107] and Fujitsu's support for Globus on its VPP series supercomputers – will be very important for Grids as a sustainable infrastructure.

The author would also like to thank the several reviewers who took the time to read drafts of this chapter and to provide useful comments that improved the final version. One reviewer, in particular, made extensive and very useful comments, and that review is the basis of the Abstract.

NOTES AND REFERENCES

1. *UK eScience Program*, http://www.research-councils.ac.uk/escience/.

In November 2000 the Director General of Research Councils, Dr John Taylor, announced £98 M funding for a new UK e-Science programme. The allocations were £3 M to the ESRC, £7 M to the NERC, £8 M each to the BBSRC and the MRC, £17 M to EPSRC and £26 M to PPARC. In addition, £5 M was awarded to CLRC to 'Grid Enable' their experimental facilities and £9 M was allocated towards the purchase of a new Teraflop scale HPC system. A sum of £15 M was allocated to a Core e-Science Programme, a cross-Council activity to develop and broker generic technology solutions and generic middleware to enable e-Science and to form the basis for new commercial e-Business software. The £15 M funding from the OST for the Core e-Science Programme has been enhanced by an allocation of a further £20 M from the CII Directorate of the DTI which will be matched by a further £15 M from industry. The Core e-Science Programme will be managed by EPSRC on behalf of all the Research Councils.

The e-Science Programme will be overseen by a Steering Committee chaired by Professor David Wallace, Vice-Chancellor of Loughborough University. Professor Tony Hey, previously Dean of Engineering at the University of Southampton, has been seconded to EPSRC as Director of the e-Science Core Programme.

What is meant by e-Science? In the future, e-Science will refer to the large-scale science that will increasingly be carried out through distributed global collaborations enabled by the Internet. Typically, a feature of such collaborative scientific enterprises is that they will require access to very large data collections, very large-scale computing resources, and high performance visualization back to the individual user scientists.

The World Wide Web gave us access to information on Web pages written in html anywhere on the Internet. A much more powerful infrastructure is needed to support e-Science. Besides information stored in Web pages, scientists will need easy access to expensive remote facilities, to computing resources – either as dedicated Teraflop computers or cheap collections of PCs – and to information stored in dedicated databases.

The Grid is an architecture proposed to bring all these issues together and make a reality of such a vision for e-Science. Ian Foster and Carl Kesselman, inventors of the Globus approach to the Grid define the Grid as an enabler for Virtual Organizations: 'An infrastructure that enables flexible, secure, coordinated resource sharing among dynamic collections of individuals, institutions and resources.' It is important to recognize that resource in this context includes computational systems and data storage and specialized experimental facilities.

2. *EU DataGrid Project*, www.eu-datagrid.org/.

DataGrid is a project funded by European Union. The objective is to build the next-generation computing infrastructure providing intensive computation and analysis of shared large-scale databases, from hundreds of terabytes to petabytes, across widely distributed scientific communities.

3. *NASA's Information Power Grid*, http://www.ipg.nasa.gov.

The Information Power Grid (IPG) is NASA's high-performance computational Grid. Computational Grids are persistent networked environments that integrate geographically distributed supercomputers, large databases, and high-end instruments. These resources are managed by diverse organizations in widespread locations and shared by researchers from many different institutions. The IPG is a collaborative effort between NASA Ames, NASA Glenn, and NASA Langley Research Centers, and the NSF PACI programs at SDSC and NCSA.

4. *DOE Science Grid*, http://www.doesciencegrid.org.

The DOE Science Grid's major objective is to provide the advanced distributed computing infrastructure on the basis of Grid middleware and tools to enable the degree of scalability in scientific computing necessary for DOE to accomplish its missions in science.

5. *AP Grid*, http://www.apgrid.org/.

ApGrid is a partnership for Grid computing in the Asia Pacific region. ApGrid focuses on (1) sharing resources, (2) developing Grid technologies, (3) helping the use of our technologies to create new applications, (4) building on each other's work, and so on, and ApGrid is not restricted to just a few developed countries, neither to a specific network nor its related group of researchers.

6. *Ninf: A Network Based Information Library for Global World-Wide Computing Infrastructure*, http://ninf.apgrid.org/welcome.shtml.

Ninf is an ongoing global networkwide computing infrastructure project that allows users to access computational resources including hardware, software, and scientific data distributed across a wide area network with an easy-to-use interface. Ninf is intended not only to exploit high performance in network parallel computing but also to provide high-quality numerical computation services and access to scientific database published by other researchers. Computational resources are shared as Ninf remote libraries executable at a remote Ninf server. Users can build an application by calling the libraries with the Ninf Remote Procedure Call, which is designed to provide a programming interface similar to conventional function calls in existing languages and is tailored for scientific computation. In order to facilitate location transparency and network-wide parallelism, Ninf metaserver maintains global resource information regarding computational server and databases, allocating and scheduling coarse-grained computation to achieve good global load balancing. Ninf also interfaces with existing network service such as the WWW for easy accessibility.

7. *NetCFD: A Ninf CFD Component for Global Computing, and its Java Applet GUI*, Sato, M., Kusano, K., Nakada, H., Sekiguchi, S. and Matsuoka, S. (2000) *Proc. of HPC Asia 2000*, 2000, http://ninf.apgrid.org/papers/hpcasia00msato/HPCAsia2000-netCFD.pdf.

Ninf is a middleware for building a global computing system in wide area network environments. We designed and implemented a Ninf computational component, netCFD for CFD (Computational Fluid Dynamics). The Ninf Remote Procedure Call (RPC) provides an interface to a parallel CFD program running on any high-performance platforms. The netCFD turns high-performance platforms such as supercomputers and clusters into valuable components for use in global computing. Our experiment shows that the overhead of a remote netCFD computation for a typical application was about 10% compared to its conventional local execution. The netCFD applet GUI that is loaded in a Web browser allows a remote user to control and visualize the CFD computation results interactively.

8. *GridLab: A Grid Application Toolkit and Testbed*, http://www.gridlab.org/.

The GridLab project is currently running, and is being funded under the Fifth Call of the Information Society Technology (IST) Program.

The GridLab project will develop an easy-to-use, flexible, generic and modular Grid Application Toolkit (GAT), enabling today's applications to make innovative use of global computing resources. The project [has two thrusts]:

1. Co-development of Infrastructure and Applications: We [undertake] a balanced program with co-development of a range of Grid applications (based on Cactus, the leading, widely used Grid-enabled open source application framework and Triana, a dataflow framework used in gravitational wave research) alongside infrastructure development, working on transatlantic test beds of varied supercomputers and clusters. This practical approach ensures that the developed software truly enables easy and efficient use of Grid resources in a real environment. We [are] maintain[ing] and upgrad[ing] the test beds through deployment of new infrastructure and large-scale application technologies as they are developed. All deliverables will be immediately prototyped and continuously field-tested by several user communities. Our focus on specific application frameworks allows us immediately to create working Grid applications to gain experience for more generic components developed during the project.

2. Dynamic Grid Computing: We [are developing] capabilities for simulation and visualization codes to be self aware of the changing Grid environment and to be able to fully exploit dynamic resources for fundamentally new and innovative applications scenarios. For example, the applications themselves will possess the capability to migrate from site to site during the execution, both in whole or in part, to spawn related tasks, and to acquire/release additional resources demanded by both the changing availabilities of Grid resources, and the needs of the applications themselves.

This timely and exciting project will join together the following institutions and businesses: Poznan Supercomputing and Networking Center (PSNC), Poznan, Poland (Project Coordinator) Max-Planck Institut fuer Gravitationsphysik (AEI), Golm/Potsdam, Germany. Konrad-Zuse-Zentrum fuer Informationstechnik (ZIB), Berlin, Germany Masaryk University, Brno, Czech Republic MTA SZTAKI, Budapest, Hungary Vrije Universiteit (VU), Amsterdam, The Netherlands ISUFI/High Performance Computing Center (ISUFI/HPCC), Lecce, Italy Cardiff University, Cardiff, Wales National Technical University of Athens (NTUA), Athens, Greece University of Chicago, Chicago, USA Information Sciences Institute (ISI), Los Angeles, USA University of Wisconsin, Wisconsin, USA Sun Microsystems Gridware GmbH Compaq Computer EMEA.

9. *National Energy Research Scientific Computing Center*, www.nersc.gov.

NERSC is one of the largest unclassified scientific supercomputer centers in the US. Its mission is to accelerate the pace of scientific discovery in the DOE Office of Science community by providing high-performance computing, information, and communications services. NERSC is the principal provider of high-performance computing services to the Office of Science programs – Magnetic Fusion Energy, High-Energy and Nuclear Physics, Basic Energy Sciences, Biological and Environmental Research, and Advanced Scientific Computing Research.

10. *The Globus Project*, http://www.globus.org.

The Globus project is developing fundamental technologies needed to build computational Grids. Grids are persistent environments that enable software applications to integrate instruments, displays, computational and information resources that are managed by diverse organizations in widespread locations.

11. *The Condor Project*, http://www.cs.wisc.edu/condor/.

The goal of the Condor Project is to develop, implement, deploy, and evaluate mechanisms and policies that support High Throughput Computing (HTC) on large collections of distributively owned computing resources. Guided by both the technological and sociological challenges of such a computing environment, the Condor Team has been building software tools that enable scientists and engineers to increase their computing throughput.

12. *The Storage Resource Broker*, http://www.npaci.edu/DICE/SRB/.

The SDSC Storage Resource Broker (SRB) is a client-server middleware that provides a uniform interface for connecting to heterogeneous data resources over a network and accessing replicated data sets. SRB, in conjunction with the Metadata Catalog (MCAT), provides a way to access data sets and resources based on their attributes rather than their names or physical locations.

13. *The Portable Batch Scheduler*, http://www.pbspro.com/tech_overview.html.

The purpose of the PBS system is to provide additional controls over initiating or scheduling execution of batch jobs, and to allow routing of those jobs between different hosts [that run administratively coupled instances of PBS]. The batch system allows a site to define and implement policy as to what types of resources and how much of each resource can be used by different jobs. The batch system also provides a mechanism with which a user can insure a job will have access to the resources required to complete it.

14. *PKI Service – An ESnet White Paper*, Genovese, T. J. (2000) DOE Energy Sciences Network, September 15, 2000, http://envisage.es.net/Docs/old%20docs/WhitePaper-PKI.pdf.

This white paper will explore PKI technology of the ESnet community. The need in our community has been growing and expectations have been varied. With the deployment of large DOE computational Grids and the development of the Federal PKI Policy Authority's (FPKIPA) Federal Bridge Certificate Authority's (FBCA) CP and CPS, the importance for ESnet to deploy a PKI infrastructure has also grown.

15. *ESnet & DOE Science Grid PKI – Overview*, January 24, 2002, http://envisage.es.net/Docs/PKIwhitepaper.pdf.

ESnet is building a Public Key Infrastructure service to support the DOE Science Grid mission and other SciDAC projects. DOE scientists and engineers will be able to use the ESnet PKI service to participate in the growing national and international computational Grids. To build this service and to insure the widest possible acceptance of its certificates, we will be participating in two international forums. First, the GGF, which is working to establish international Grid standards/recommendations – specifically we are contributing to the Grid Certificate policy working group and the Grid Information Services Area. The Grid CP effort is focusing on the development of a common Certificate Policy that all Grid PKIs could use instead of custom individual CPs that hamper certificate validation. Second, we will be working with the European Data Grid CA operations group to insure that the EDG Test beds will accept our certificates. The project Website, Envisage.es.net will be used to track the progress of this project. The Website will contain all project documents and status reports. This paper will provide a project overview of the immediate requirements for the DOE Science Grid PKI support and cover the long-term project goals described in the ESnet PKI and Directory project document.

16. *ESnet's SciDAC PKI & Directory Project – Homepage*, Genovese, T. and Helm, M. DOE Energy Sciences Network, http://envisage.es.net/.

This is the ESnet PKI project site. ESnet is building a Public Key Infrastructure service to support the DOE Science Grid, SciDAC projects and other DOE research efforts. The main goal is to provide DOE scientists and engineers Identity and Service certificates that allow them to participate in the growing national and international computational Grids.

17. *Application Experiences with the Globus Toolkit*, Brunett, S. *et al.* (1998) *Proc. 7th IEEE Symp. on High Performance Distributed Computing*, IEEE Press, 1998, http://www.globus.org/research/papers.html#globus-apps.

'..., SF-Express, is a distributed interactive simulation (DIS) application that harnesses multiple supercomputers to meet the computational demands of large-scale network-based simulation environments. A large simulation may involve many tens of thousands of entities and requires thousands of processors. Globus services can be used to locate, assemble, and manage those resources. For example, in one experiment in March 1998, SF-Express was run on 1352 processors distributed over 13 supercomputers at nine sites.... This experiment involved over 100 000 entities, setting a new world record for simulation and meeting a performance goal that was not expected to be achieved until 2002.'

18. *Legion*, http://legion.virginia.edu/.

Legion, an object-based metasystems software project at the University of Virginia, is designed for a system of millions of hosts and trillions of objects tied together with high-speed links. Users working on their home machines see the illusion of a single computer, with access to all kinds of data and physical resources, such as digital libraries, physical simulations, cameras, linear accelerators, and video streams. Groups of users can construct shared virtual workspaces, to collaborate research and exchange information. This abstraction springs from Legion's transparent scheduling, data management, fault tolerance, site autonomy, and a wide range of security options.

Legion sits on top of the user's operating system, acting as liaison between its own host(s) and whatever other resources are required. The user is not bogged down with time-consuming negotiations with outside systems and system administrators, since Legion's scheduling and security policies act on his or her behalf. Conversely, it can protect its own resources against other Legion users, so that administrators can choose appropriate policies for who uses which resources under what circumstances. To allow users to take advantage of a wide range of possible resources, Legion offers a user-controlled naming system called *context space*, so that users can easily create and use objects in far-flung systems. Users can also run applications written in multiple languages, since Legion supports interoperability between objects written in multiple languages.

19. *UNICORE*, http://www.unicore.de/.

UNICORE lets the user prepare or modify structured jobs through a graphical user interface on a local Unix workstation or a Windows PC. Jobs can be submitted to any of the platforms of a UNICORE GRID and the user can monitor and control the submitted jobs through the job monitor part of the client.

A UNICORE job contains a number of interdependent tasks. The dependencies indicate temporal relations or data transfer. Currently, execution of scripts, compile, link, execute tasks and data transfer directives are supported. An execution system request associated with a job specifies where its tasks are to be run. Tasks can be grouped into subjobs, creating a hierarchical job structure and allowing different steps to execute on different systems within the UNICORE GRID.

20. *Grid Monitoring Architecture Working Group*, Global Grid Forum,
      http://www-didc.lbl.gov/GGF-PERF/GMA-WG/.

The Grid Monitoring Architecture working group is focused on producing a high-level architecture statement of the components and interfaces needed to promote interoperability between heterogeneous monitoring systems on the Grid. The main products of this work are the architecture document itself, and accompanying case studies that illustrate the concrete application of the architecture to monitoring problems.

21. *The Grid: Blueprint for a New Computing Infrastructure*, Foster, I. and Kesselman, C. (eds) (1998) San Francisco, CA: Morgan Kaufmann,
      http://www.mkp.com/books_catalog/1-55860-475-8.asp.

22. *The Anatomy of the Grid: Enabling Scalable Virtual Organizations*, Foster, I., Kesselman, C. and Tuecke, S. (2001) *International Journal of Supercomputer Applications*, **15**(3), pp. 200–222, http://www.globus.org/research/papers.html#anatomy.

Defines Grid computing and the associated research field, proposes a Grid architecture, and discusses the relationships between Grid technologies and other contemporary technologies.

23. *GriPhyN (Grid Physics Network)*, http://www.griphyn.org.

The GriPhyN collaboration is a team of experimental physicists and information technology (IT) researchers who are implementing the first petabyte-scale computational environments for data intensive science in the twenty-first century. Driving the project are unprecedented requirements for geographically dispersed extraction of complex scientific information from very large collections of measured data. To meet these requirements, which arise initially from the four physics experiments involved in this project but will also be fundamental to science and commerce in the twenty-first century, GriPhyN will deploy computational environments called *Petascale Virtual Data Grids* (PVDGs) that meet the data-intensive computational needs of a diverse community of thousands of scientists spread across the globe.

GriPhyN involves technology development and experimental deployment in four science projects. The CMS and ATLAS experiments at the Large Hadron Collider will search for the origins of mass and probe matter at the smallest length scales; LIGO (Laser Interferometer Gravitational-wave Observatory) will detect the gravitational waves of pulsars, supernovae and in-spiraling binary stars; and SDSS (Sloan Digital Sky Survey) will carry out an automated sky survey enabling systematic studies of stars, galaxies, nebulae, and large-scale structure.

24. *Virtual Observatories of the Future*, Caltech, http://www.astro.caltech.edu/nvoconf/.

Within the United States, there is now a major, community-driven push towards the National Virtual Observatory (NVO). The NVO will federate the existing and forthcoming digital sky archives, both ground-based and space-based.

25. *Grid Information Services/MDS*, Globus Project, http://www.globus.org/mds/.

Grid computing technologies enable widespread sharing and coordinated use of networked resources. Sharing relationships may be static and long-lived – for example, among the major resource centers of a company or university – or highly dynamic, for example, among the evolving membership of a scientific collaboration. In either case, the fact that users typically have little or no knowledge of the resources contributed by participants in the 'virtual organization' (VO) poses a significant obstacle to their use. For this reason, information services designed to support the initial discovery and ongoing monitoring of the existence and characteristics of resources, services, computations, and other entities are a vital part of a Grid system. ('Grid Information Services for Distributed Resource Sharing' – http://www.globus.org/research/papers/MDS-HPDC.pdf.)

The Monitoring and Discovery Service architecture addresses the unique requirements of Grid environments. Its architecture consists of two basic elements:

– A large, distributed collection of generic information providers provide access to information about individual entities, via local operations or gateways to other information sources (e.g. SNMP queries). Information is structured in terms of a standard data model, taken from LDAP: an entity is described by a set of 'objects' composed of typed attribute-value pairs.
– Higher-level services, collect, manage, index, and/or respond to information provided by one or more information providers. We distinguish in particular aggregate directory services, which facilitate resource discovery and monitoring for VOs by implementing both generic and specialized views and search methods for a collection of resources. Other higher-level services can use this information and/or information obtained directly from providers for the purposes of brokering, monitoring, troubleshooting, and so forth.

Interactions between higher-level services (or users) and providers are defined in terms of two basic protocols: a soft-state registration protocol for identifying entities participating in the information service and an enquiry protocol for retrieval of information about those entities, whether via query or subscription. In brief, a provider uses the registration protocol to notify higher-level services of its existence; a higher-level service uses the enquiry protocol to obtain information about the entities known to a provider, which it merges into its aggregate view. Integration with the Grid Security Infrastructure (GSI) provides for authentication and access control to information.

26. *Grid Security Infrastructure (GSI)*, Globus Project, 2002, http://www.globus.org/security/.

The primary elements of the GSI are identity certificates, mutual authentication, confidential communication, delegation, and single sign-on.

GSI is based on public key encryption, X.509 certificates, and the Secure Sockets Layer (SSL) communication protocol. Extensions to these standards have been added for single sign-on and delegation. The Globus Toolkit's implementation of the GSI adheres to the Generic Security Service API (GSS-API), which is a standard API for security systems promoted by the Internet Engineering Task Force (IETF).

27. *Globus Resource Allocation Manager (GRAM)*, Globus Project, 2002,
    http://www-fp.globus.org/gram/overview.html.

The Globus Resource Allocation Manager (GRAM) is the lowest level of Globus resource management architecture. GRAM allows you to run jobs remotely, providing an API for submitting, monitoring, and terminating your job.

To run a job remotely, a GRAM gatekeeper (server) must be running on a remote computer, listening at a port; and the application needs to be compiled on that remote machine. The execution begins when a GRAM user application runs on the local machine, sending a job request to the remote computer.

The request is sent to the gatekeeper of the remote computer. The gatekeeper handles the request and creates a job manager for the job. The job manager starts and monitors the remote program, communicating state changes back to the user on the local machine. When the remote application terminates, normally or by failing, the job manager terminates as well.

The executable, stdin and stdout, as well as the name and port of the remote computer, are specified as part of the job request. The job request is handled by the gatekeeper, which creates a job manager for the new job. The job manager handles the execution of the job, as well as any communication with the user.

28. *The GridFTP Protocol and Software*, Globus Project, 2002,
    http://www.globus.org/datagrid/gridftp.html.

GridFTP is a high-performance, secure, reliable data transfer protocol optimized for high-bandwidth wide-area networks. The GridFTP protocol is based on FTP, the highly popular Internet file transfer protocol. We have selected a set of protocol features and extensions defined already in IETF RFCs and added a few additional features to meet the requirement from current data Grid projects.

29. *A Grid Monitoring Architecture*, Tierney, B., Aydt, R., Gunter, D., Smith, W., Taylor, V., Wolski, R. and Swany, M. http://www-didc.lbl.gov/GGF-PERF/GMA-WG/.

The current GMA specification from the GGF Performance Working Group may be found in the documents section of the Working Group Web page.

30. *Distributed Monitoring Framework (DMF)*, Lawrence Berkeley National Laboratory, http://www-didc.lbl.gov/DMF/.

The goal of the Distributed Monitoring Framework is to improve end-to-end data throughput for data intensive applications in high-speed WAN environments and to provide the ability to do performance analysis and fault detection in a Grid computing environment. This monitoring framework will provide accurate, detailed, and adaptive monitoring of all distributed computing components, including the network. Analysis tools will be able to use this monitoring data for real-time analysis, anomaly identification, and response.

Many of the components of the DMF have already been prototyped or implemented by the DIDC Group. The NetLogger Toolkit includes application sensors, some system and network sensors, a powerful event visualization tool, and a simple event archive. The Network characterization Service has proven to be a very useful hop-by-hop network sensor. Our work on the Global Grid Forum Grid Monitoring Architecture (GMA) addressed the event management system. JAMM (Java Agents for

Monitoring Management) is preliminary work on sensor management. The Enable project produced a simple network tuning advice service.

31. *Information and Monitoring Services Architecture*, European Union DataGrid – WP3, http://hepunx.rl.ac.uk/edg/wp3/documentation/doc/arch/index.html.

The aim of this work package is to specify, develop, integrate and test tools and infrastructure to enable end-user and administrator access to status and error information in a Grid environment and to provide an environment in which application monitoring can be carried out. This will permit both job performance optimization as well as allow for problem tracing and is crucial to facilitating high-performance Grid computing.

32. *Globus I/O*, Globus Project, 2002, http://www-unix.globus.org/api/c-globus-2.0-beta1/globus_io/html/index.html.

The Globus_io library is motivated by the desire to provide a uniform I/O interface to stream and datagram style communications. The goals in doing this are 1) to provide a robust way to describe, apply, and query connection properties. These include the standard socket options (socket buffer sizes, etc.) as well as additional attributes. These include security attributes and, eventually, QoS attributes. 2) Support nonblocking I/O and handle asynchronous file and network events. 3) Provide a simple and portable way to implement communication protocols. Globus components such as GASS and GRAM can use this to redefine their control message protocol in terms of TCP messages.

33. *Maui Silver Metascheduler*, http://www.supercluster.org/documentation/silver/silveroverview.html.

Silver is an advance reservation metascheduler. Its design allows it to load balance workload across multiple systems in completely independent administrative domains. How much or how little a system participates in this load-sharing activity is completely up to the local administration. All workload is tracked and accounted for allowing 'allocation' exchanges to take place between the active sites.

34. *Personal Condor and Globus Glide-In*, Condor Project, http://www.cs.wisc.edu/condor/condorg/README.

A Personal Condor is a version of Condor running as a regular user, without any special privileges. The idea is that you can use your Personal Condor to run jobs on your local workstations and have Condor keep track of their progress, and then through 'flocking' access the resources of other Condor pools. Additionally, you can 'Glide-in' to Globus-Managed resources and create a virtual-condor pool by running the Condor daemons on the Globus resources, and then let your Personal Condor manage those resources.

35. *Condor-G*, Frey, J., Tannenbaum, T., Livny, M., Foster, I. and Tuecke, S. (2001) *Proceedings of the Tenth International Symposium on High Performance Distributed Computing (HPDC-10)*. IEEE Press, San Francisco, California, USA; http://www.globus.org/research/papers.html#Condor-G-HPDC.

In recent years, there has been a dramatic increase in the amount of available computing and storage resources. Yet few have been able to exploit these resources in an aggregated form. We present the Condor-G system, which leverages software from Globus and Condor to allow users to harness multidomain resources as if they all belong to one personal domain. We describe the structure of Condor-G and how it handles job management, resource selection, security, and fault tolerance.

36. *European Union DataGrid Project*, http://eu-datagrid.web.cern.ch/eu-datagrid/.

The DataGrid Project is a proposal made to the European Commission for shared cost research and technological development funding. The project has six main partners: CERN – The European Organization for Nuclear Research near Geneva, Swiss; CNRS – France – Le Comité National de la Recherche Scientifique; ESRIN – the European Space Agency's Centre in Frascati (near Rome), Italy; INFN – Italy – Istituto Nazionale di Fisica Nucleare; NIKHEF – The Dutch National Institute for Nuclear Physics and High Energy – Physics, Amsterdam, and; PPARC – United Kingdom – Particle Physics and Astronomy Research Council.

The objective of the project is to enable next-generation scientific exploration, which requires intensive computation and analysis of shared large-scale databases, from hundreds of terabytes to petabytes, across widely distributed scientific communities. We see these requirements emerging in many scientific disciplines, including physics, biology, and earth sciences. Such sharing is made complicated by the distributed nature of the resources to be used, the distributed nature of the communities, the size of the databases and the limited network bandwidth available. To address these problems we propose to build on emerging computational Grid technologies, such as that developed by the Globus Project.

37. *The GridPort Toolkit: A System for Building Grid Portals*, Thomas, M., Mock, S., Dahan, M., Mueller, K., Sutton, D. and Boisseau, J. R.
http://www.tacc.utexas.edu/~mthomas/pubs/GridPort_HPDC11.pdf.

Grid portals are emerging as convenient mechanisms for providing the scientific community with familiar and simplified interfaces to the Grid. Our experience in implementing Grid portals has led to the creation of GridPort: a unique, layered software system for building Grid Portals. This system has several unique features: the software is portable and runs on most Web servers; written in Perl/CGI, it is easy to support and modify; it is flexible and adaptable; it supports single login between multiple portals; and portals built with it may run across multiple sites and organizations. The feasibility of this portal system has been successfully demonstrated with the implementation of several application portals. In this paper we describe our experiences in building this system, including philosophy and design choices. We explain the toolkits we are building, and we demonstrate the benefits of this system with examples of several production portals. Finally, we discuss our experiences with Grid Web service architectures.

38. *Cactus*, http://www.cactuscode.org/.

Cactus is an open source problem-solving environment designed for scientists and engineers. Its modular structure easily enables parallel computation across different architectures and collaborative code development between different groups. Cactus originated in the academic research community, where it was developed and used over many years by a large international collaboration of physicists and computational scientists.

The name Cactus comes from the design of a central core (or 'flesh') that connects to application modules (or 'thorns') through an extensible interface. Thorns can implement custom-developed scientific or engineering applications, such as computational fluid dynamics. Other thorns from a standard computational toolkit provide a range of computational capabilities, such as parallel I/O, data distribution, or checkpointing. Cactus runs on many architectures. Applications, developed on standard workstations or laptops, can be seamlessly run on clusters or supercomputers. Cactus provides easy access to many cutting edge software technologies being developed in the academic research community, including the Globus Metacomputing Toolkit, HDF5 parallel file I/O, the PETSc scientific library, adaptive mesh refinement, Web interfaces, and advanced visualization tools.

39. *Supporting Efficient Execution in Heterogeneous Distributed Computing Environments with Cactus and Globus*, Allen, G., Dramlitsch, T., Foster, I., Karonis, N., Ripeanu, M., Seidel, E. and Toonen, B. (2001) *SC 2001*, 2001, http://www.globus.org/research/papers.html#sc01ewa.

Members of the Cactus and Globus projects have won one of this year's Gordon Bell Prizes in high-performance computing for the work described in their paper: Supporting Efficient Execution in Heterogeneous Distributed Computing Environments with Cactus and Globus. The international team composed of Thomas Dramlitsch, Gabrielle Allen and Ed Seidel, from the Max Planck Institute for Gravitational Physics, along with colleagues Matei Ripeanu, Ian Foster, Brian Toonen from the University of Chicago and Argonne National Laboratory, and Nicholas Karonis from Northern Illinois University. The special category award was presented during SC2001, a yearly conference showcasing high-performance computing and networking, this year held in Denver, Colorado.

The prize was awarded for the group's work on concurrently harnessing the power of multiple supercomputers to solve Grand Challenge problems in physics, which require substantially more resources than that can be provided by a single machine. The group enhanced the communication layer of Cactus, a generic programming framework designed for physicists and engineers, adding techniques capable of dynamically adapting the code to the available network bandwidth and latency between machines. The message-passing layer itself used MPICH-G2, a Grid-enabled implementation of the MPI protocol, which handles communications between machines separated by a wide area network. In addition, the Globus Toolkit was used to provide authentification and staging of simulations across multiple machines.

From 'Cactus, Globus and MPICH-G2 Win Top Supercomputing Award' at http://www.cactuscode.org/News/GordonBell2001.html

40. *A Jini-based Computing Portal System*, Suzumura, T., Matsuoka, S. and Nakada, H. (2001) *Proceeding of SC2001*, 2001, http://ninf.apgrid.org/papers/sc01suzumura/sc2001.pdf.

JiPANG (A Jini-based Portal Augmenting Grids) is a portal system and a toolkit that provides uniform access interface layer to a variety of Grid systems and is built on top of Jini distributed object technology. JiPANG performs uniform higher-level management of the computing services and resources being managed by individual Grid systems such as Ninf, NetSolve, Globus, and so on. In order to give the user a uniform interface to the Grids, JiPANG provides a set of simple Java APIs called *the JiPANG Toolkits*, and furthermore, allows the user to interact with Grid systems, again in a uniform way, using the JiPANG Browser application. With JiPANG, users need not install any client packages beforehand to interact with Grid systems, nor be concerned about updating to the latest version. Such uniform, transparent services available in a ubiquitous manner we believe is essential for the success of Grid as a viable computing platform for the next generation.

41. *GridRPC Tutorial*, Nakada, H., Matsuoka, S., Sato, M. and Sekiguchi, S. http://ninf.apgrid.org/papers/gridrpc_tutorial/gridrpc_tutorial_e.html.

GridRPC is a middleware that provides remote library access and task-parallel programming model on the Grid. Representative systems include Ninf, NetSolve, and so on. We employ Ninf to exemplify how to program Grid applications using Ninf, in particular, how to 'Gridify' a numerical library for remote RPC execution, how to perform parallel parameter sweep survey using multiple servers on the Grid.

42. *NetSolve*, http://icl.cs.utk.edu/netsolve/.

NetSolve is a client-server system that enables users to solve complex scientific problems remotely. The system allows users to access both hardware and software computational resources distributed across a network. NetSolve searches for computational resources on a network, chooses the best one available, and using retry for fault tolerance solves a problem, and returns the answers to the user. A load-balancing policy is used by the NetSolve system to ensure good performance by enabling the system to use the computational resources available as efficiently as possible. Our framework is based on the premise that distributed computations involve resources, processes, data, and users, and that secure yet flexible mechanisms for cooperation and communication between these entities is the key to metacomputing infrastructures.

43. *ILab: An Advanced User Interface Approach for Complex Parameter Study Process Specification on the Information Power Grid*, Yarrow, M., McCann, K. M., Biswas, R. and Wijngaart, R. F. V. D. (2000) *Grid 2000: First IEEE/ACM International Workshop*, Bangalore, India, 2000, http://www.ipg.nasa.gov/research/papers/nas-00-009.pdf.

The creation of parameter study suites has recently become a more challenging problem as the parameter studies have become multitiered and the computational environment has become a supercomputer Grid. The parameter spaces are vast, the individual problem sizes are getting larger, and researchers are seeking to combine several successive stages of parameterization and computation. Simultaneously, Grid-based computing offers immense resource opportunities but at the expense of great difficulty of use. We present ILab, an advanced graphical user interface approach to this problem. Our novel strategy stresses intuitive visual design tools for parameter study creation and complex process specification and also offers programming-free access to Grid-based supercomputer resources and process automation.

44. *US-CMS Testbed Production – Joint News Update with GriPhyN/iVDGL, Particle Physics Data Grid*, 2002, http://www.ppdg.net/ppdg_news.htm.

Members of the CMS experiment working in concert with PPDG, iVDGL, and GriPhyN have carried out the first production-quality simulated data generation on a data Grid comprising sites at Caltech, Fermilab, the University of California-San Diego, the University of Florida, and the University of Wisconsin-Madison. This is a combination of efforts supported by DOE SciDAC, HENP, MICS, and the NSF as well as the EU funded EU DataGrid project.

The deployed data Grid serves as an integration framework in which Grid middleware components are brought together to form the basis for distributed CMS Monte Carlo Production (CMS-MOP) and used to produce data for the global CMS physics program. The middleware components include Condor-G, DAGMAN, GDMP, and the Globus Toolkit packaged together in the first release of the Virtual Data Toolkit.

45. *Condor-G*, Condor Project, http://www.cs.wisc.edu/condor/condorg/.

Condor-G provides the Grid computing community with a powerful, full-featured task broker. Used as a frontend to a computational Grid, Condor-G can manage thousands of jobs destined to run at distributed sites. It provides job monitoring, logging, notification, policy enforcement, fault tolerance, credential management, and can handle complex job-interdependencies. Condor-G's flexible and intuitive commands are appropriate for use directly by end users, or for interfacing with higher-level task brokers and Web portals.

46. *Peer-to-Peer Area*, Global Grid Forum, http://www.gridforum.org/4_GP/P2P.htm.

This is a very new GGF activity, and initially the GGF Peer-to-Peer Area consists of the Working Groups of the previous Peer-to-Peer Working Group organization, which has merged with GGF. These WGs are as follows:

- NAT/Firewall
- Taxonomy
- Peer-to-Peer Security
- File Services
- Trusted Library

47. *MPICH-G2*, http://www.hpclab.niu.edu/mpi/.

MPICH-G2 is a Grid-enabled implementation of the MPI v1.1 standard. That is, using Globus services (e.g. job startup, security), MPICH-G2 allows you to couple multiple machines, potentially of different architectures, to run MPI applications. MPICH-G2 automatically converts data in messages sent between machines of different architectures and supports multiprotocol communication by automatically selecting TCP for intermachine messaging and (where available) vendor-supplied MPI for intramachine messaging.

48. *PVM*, http://www.csm.ornl.gov/pvm/pvm_home.html.

PVM (Parallel Virtual Machine) is a software package that permits a heterogeneous collection of Unix and/or Windows computers hooked together by a network to be used as a single large parallel computer. Thus, large computational problems can be solved more cost effectively by using the aggregate power and memory of many computers. The software is very portable. The source, which is available free through netlib, has been compiled on everything from laptops to CRAYs.

49. *PACX-MPI: Extending MPI for Distributed Computing*, Gabriel, E., Mueller, M. and Resch, M. High Performance Computing Center in Stuttgart, http://www.hlrs.de/organization/pds/projects/pacx-mpi/.

Simulation using several MPPs requires a communication interface, which enables both efficient message passing between the nodes inside each MPP and between the machines itself. At the same time the data exchange should rely on a standard interface.

PACX-MPI (PArallel Computer eXtension) was initially developed to connect a Cray-YMP to an Intel Paragon. Currently it has been extended to couple two and more MPPs to form a cluster of high-performance computers for Metacomputing.

50. *Trans-Atlantic Metacomputing*, http://www.hoise.com/articles/AE-PR-11-97-7.html.

Stuttgart, 15 November 97: An international team of computer experts combined the capacity of machines at three large supercomputer centers that exceeded three Tflop/s. The metacomputing effort used for the simulations linked 3 of the top 10 largest supercomputers in the world. Involved were HLRS, the High-Performance Computing-Center at the University of Stuttgart, Germany, Sandia National Laboratories, SNL, Albuquerque, NM, Pittsburgh Supercomputing Center, Pittsburgh, PA. They demonstrated a trans-Atlantic metacomputing and meta-visualization environment. The demonstration is a component of this official G7 Information Society pilot programme.

51. *Condor User's Manual*, Condor Project, http://www.cs.wisc.edu/condor.

The goal of the Condor Project is to develop, implement, deploy, and evaluate mechanisms and policies that support High Throughput Computing (HTC) on large collections of distributively owned computing resources. Guided by both the technological and sociological challenges of such a computing environment, the Condor Team has been building software tools that enable scientists and engineers to increase their computing throughput.

52. *Particle Physics Data Grid (PPDG)*, http://www.ppdg.net/.

The Particle Physics Data Grid collaboration was formed in 1999 because its members were keenly aware of the need for Data Grid services to enable the worldwide distributed computing model of current and future high-energy and nuclear physics experiments. Initially funded from the NGI initiative and later from the DOE MICS and HENP programs, it has provided an opportunity for early development of the Data Grid architecture as well as for evaluating some prototype Grid middleware.

PPDG involves work with and by four major high-energy physics experiments that are concerned with developing and testing Grid technology.

53. *Data Mining on NASA's Information Power Grid*, Hinke, T. and Novotny, J. (2000) *Ninth IEEE International Symposium on High Performance Distributed Computing*, 2000, http://www.ipg.nasa.gov/engineering/presentations/PDF_presentations/21-Hinke.pdf.

This paper describes the development of a data mining system that is to operate on NASA's Information Power Grid (IPG). Mining agents will be staged to one or more processors on the IPG. There they will grow using just-in-time acquisition of new operations. They will mine data delivered using just-in-time delivery. Some initial experimental results are presented.

54. *Overview of the Grid Security Infrastructure (GSI)*, Globus Project, 2002, http://www-fp.globus.org/security/overview.html.

The GSI uses public key cryptography (also known as asymmetric cryptography) as the basis for its functionality. Many of the terms and concepts used in this description of the GSI come from its use of public key cryptography. The PKI context is described here.

55. *Global Access to Secondary Storage (GASS)*, Globus Project, 2002, http://www-fp.globus.org/gass/.

GASS provides a Unix I/O style access to remote files. Operations supported include remote read, remote write, and append (achieved by copying the entire file to the local system on file open, and back on close). GASS also provides for local caching of file so that they may be reused during a job without recopying.
A typical use of GASS is to put a GASS server on or near a tertiary storage system that can access files by filename. This allows remote, filelike access by replicating the file locally. A second typical use is to locate a GASS server on a user system in which files (such as simulation input files) are managed so that Grid jobs can read data from those systems.

56. *A Replica Management Service for High-Performance Data Grids*, Globus Project, 2002, http://www-fp.globus.org/datagrid/replica-management.html.

Replica management is an important issue for a number of scientific applications. Consider a data set that contains one petabyte (one thousand million megabytes) of experimental results for a particle physics application. While the complete data set may exist in one or possibly several physical locations, it is likely that few universities, research laboratories, or individual researchers will have sufficient storage to hold a complete copy. Instead, they will store copies of the most relevant portions of the data set on local storage for faster access. Replica Management is the process of keeping track of where portions of the data set can be found.

57. *The Globus Replica Catalog*, Globus Project, 2002, http://www-fp.globus.org/datagrid/replica-catalog.html.

The Globus Replica Catalog supports replica management by providing mappings between logical names for files and one or more copies of the files on physical storage systems.

58. *Globus Replica Management*, Globus Project, 2002, http://www-fp.globus.org/datagrid/replica-management.html.

Replica management is an important issue for a number of scientific applications. Consider a data set that contains one petabyte (one thousand million megabytes) of experimental results for a particle physics application. While the complete data set may exist in one or possibly several physical locations, it is likely that few universities, research laboratories, or individual researchers will have sufficient storage to hold a complete copy. Instead, they will store copies of the most relevant portions of the data set on local storage for faster access. Replica Management is the process of keeping track of where portions of the data set can be found.

Globus Replica Management integrates the Globus Replica Catalog (for keeping track of replicated files) and GridFTP (for moving data) and provides replica management capabilities for data Grids.

59. *Grid Data Management Pilot (GDMP): A Tool for Wide Area Replication*, Samar, A. and Stockinger, H. (2001) *IASTED International Conference on Applied Informatics (AI2001)*, Innsbruck, Austria, February, 2001,
http://web.datagrid.cnr.it/pls/portal30/GRID.RPT_DATAGRID_PAPERS.show.

The stringent requirements of data consistency, security, and high-speed transfer of huge amounts of data imposed by the physics community need to be satisfied by an asynchronous replication mechanism. A pilot project called the *Grid Data Management Pilot* (GDMP) has been initiated which is responsible for asynchronously replicating large object-oriented data stores over the wide-area network to globally distributed sites.

The GDMP software consists of several modules that closely work together but are easily replaceable. In this section, we describe the modules and the software architecture of GDMP. The core modules are Control Communication, Request Manager, Security, Database Manager and the Data Mover. An application that is visible as a command-line tool uses one or several of these modules.

60. *Distributed-Parallel Storage System (DPSS)*, http://www-didc.lbl.gov/DPSS/.

The DPSS is a data block server that provides high-performance data handling and architecture for building high-performance storage systems from low-cost commodity hardware components. This technology has been quite successful in providing an economical, high-performance, widely distributed, and highly scalable architecture for caching large amounts of data that can potentially be used by many different users. Current performance results are 980 Mbps across a LAN and 570 Mbps across a WAN.

61. *Using GSI Enabled SSH*, NCSA,
http://www.ncsa.uiuc.edu/UserInfo/Alliance/GridSecurity/GSI/Tools/GSSH.html.

SSH is a well-known program for doing secure logon to remote hosts over an open network. GSI-enabled SSH (GSSH) is a modification of SSH version 1.2.27 that allows SSH to use Alliance certificates and designated proxy certificates for authentication.

62. *GSI-Enabled OpenSSH*, NCSA, http://www.ncsa.uiuc.edu/Divisions/ACES/GSI/openssh/.

NCSA maintains a patch to OpenSSH that adds support for GSI authentication.

63. *Access Grid*, http://www-fp.mcs.anl.gov/fl/accessgrid/default.htm.

The Access Grid (AG) is the ensemble of resources that can be used to support human interaction across the Grid. It consists of multimedia display, presentation and interactions environments, interfaces to Grid middleware, and interfaces to visualization environments. The Access Grid will support large-scale distributed meetings, collaborative work sessions, seminars, lectures, tutorials, and training. The Access Grid design point is group-to-group communication (thus differentiating it from desktop-to-desktop-based tools that focus on individual communication). The Access Grid environment must enable both formal and informal group interactions. Large-format displays integrated with intelligent or active meeting rooms are a central feature of the Access Grid nodes. Access Grid nodes are 'designed spaces' that explicitly contain the high-end audio and visual technology needed to provide a high-quality compelling user experience.

The Access Grid complements the computational Grid; indeed the Access Grid node concept is specifically targeted to provide 'group' access to the Grid. This access maybe for remote visualization or for interactive applications, or for utilizing the high-bandwidth environment for virtual meetings and events.

Access Grid Nodes (global AG sites) provide a research environment for the development of distributed data and visualization corridors and for studying issues relating to collaborative work in distributed environments.

64. *Grid Information Services*, Global Grid Forum, http://www.gridforum.org/1_GIS/gis.htm.

The Grid Information Services Area (GIS) is concerned with services that either provide information or consume information pertaining to the Grid.

65. *OpenSSL Certification Authority*, http://www.openssl.org/docs/apps/ca.html#.

The *ca* command is a minimal CA application. It can be used to sign certificate requests in a variety of forms and generate CRLs; it also maintains a text database of issued certificates and their status.

66. *DOE Science Grid PKI Certificate Policy and Certification Practice Statement*, http://www.doegrids.org/.

This document represents the policy for the DOE Science Grid Certification Authority operated by ESnet. It addresses Certificate Policy (CP) and Certification Practice Statement (CPS). The CP is a named set of rules that indicates the applicability of a certificate to a particular community and/or class of application with common security requirements. For example, a particular certificate policy might indicate applicability of a type of certificate to the authentication of electronic data interchange transactions for the trading of goods within a given price range. The CPS is a statement of the practices, which a certification authority employs in issuing certificates.

67. *Certification Authorities Acceptance and Feature Matrices*, European Union DataGrid, 2002, http://www.cs.tcd.ie/coghlan/cps-matrix/.

The Acceptance and Feature matrices are key aspects of establishing cross-site trust.

68. *Certification Authorities,* European Union DataGrid, 2002, http://marianne.in2p3.fr/datagrid/ca/ca-table-ca.html.

The current list of EU DataGrid recognized CAs and their certificates.

69. *Netscape Certificate Management System*, Netscape, http://wp.netscape.com/cms.

Use Netscape Certificate Management System, the highly scalable and flexible security solution, to issue and manage digital certificates for your extranet and e-Commerce applications.

70. *KX.509/KCA*, NSF Middleware Initiative, 2002, http://www.nsf-middleware.org/documentation/KX509KCA/.

KX.509, from the University of Michigan, is a Kerberized client-side program that acquires an X.509 certificate using existing Kerberos tickets. The certificate and private key generated by KX.509 are normally stored in the same cache alongside the Kerberos credentials. This enables systems that already have a mechanism for removing unused Kerberos credentials to also automatically remove the X.509 credentials. There is then a PKCS11 library that can be loaded by Netscape and Internet Explorer to access and use these credentials for https Web activity.

The Globus Toolkit normally uses X.509 credentials. KX.509 allows a user to authenticate a host that is running Globus software using Kerberos tickets instead of requiring X.509 certificates to be installed.

To use Globus utilities on a (local or remote) machine, a user is required to authenticate the machine using appropriate X.509 certificates. These long-term certificates are used to create a short-term proxy, which is used for authentication and Globus utilities. The proxy will expire after a preset amount of time, after which a new one must be generated from the long-term X.509 certificates again.

KX.509 can be used in place of permanent, long-term certificates. It does this by creating X.509 credentials (certificate and private key) using your existing Kerberos ticket. These credentials are then used to generate the Globus proxy certificate.

71. *Esnet*, http://www.es.net/.

The Energy Sciences Network, or ESnet, is a high-speed network serving thousands of Department of Energy scientists and collaborators worldwide. A pioneer in providing high-bandwidth, reliable connections, ESnet enables researchers at national laboratories, universities, and other institutions to communicate with each other using the collaborative capabilities needed to address some of the world's most important scientific challenges. Managed and operated by the ESnet staff at Lawrence Berkeley National Laboratory, ESnet provides direct connections to all major DOE sites with high-performance speeds, as well as fast interconnections to more than 100 other networks. Funded principally by DOE's Office of Science, ESnet services allow scientists to make effective use of unique DOE research facilities and computing resources, independent of time and geographic location. ESnet is funded by the DOE Office of Science to provide network and collaboration services in support of the agency's research missions.

72. *Using Globus/GSI with a Firewall*, Globus Project,
    http://www.globus.org/Security/v1.1/firewalls.html.

Describes the network traffic generated by Globus and GSI applications.

73. *GSI-Enabled FTP*, Globus Project, http://www.globus.org/security/v1.1/index.htm#gsiftp.

GSIFTP is a standard Unix FTP program modified to use the GSI libraries for authentication. It replaces the normal password authentication with Globus certificate authentication.

74. *Creating a Hierarchical GIIS*, Globus Project, http://www.globus.org/mds/.

This document describes by example how to create a hierarchical GIIS for use by MDS 2.1. The following topics are covered in this document: configuration files used in creating a hierarchical GIIS architecture; renaming of distinguished names (DNs) in the GIIS hierarchy; timing issues and registration control; additional information on site policy, configuration files, and command syntax.

75. *Community Authorization Service (CAS)*, Globus Project, 2002,
    http://www.globus.org/security/CAS/.

CAS allows resource providers to specify coarse-grained access control policies in terms of communities as a whole, delegating fine-grained access control policy management to the community itself. Resource providers maintain ultimate authority over their resources but are spared day-to-day policy administration tasks (e.g. adding and deleting users, modifying user privileges). Briefly, the process is as follows: 1) A CAS server is initiated for a community – a community representative acquires a GSI credential to represent that community as a whole and then runs a CAS server using that community identity. 2) Resource providers grant privileges to the community. Each resource provider verifies that the holder of the community credential represents that community and that the community's policies are compatible with the resource provider's own policies. Once a trust relationship has been established, the resource provider then grants rights to the community identity, using normal local mechanisms (e.g. Gridmap files and disk quotas, filesystem permissions, etc.). 3) Community representatives use the CAS to manage the community's trust relationships (e.g. to enroll users and resource providers into the community according to the community's standards) and grant fine-grained access control to resources. The CAS server is also used to manage its own access control policies; for example, community members who have the appropriate privileges may authorize additional community members to manage groups, grant permissions on some or all of the community's resources, and so on. 4) When a user wants to access resources served by the CAS, that user makes a request to the CAS server. If the CAS server's database indicates that the user has the appropriate privileges, the CAS issues the user a GSI restricted proxy credential with an embedded policy giving the user the right to perform the requested actions. 5) The user then uses the credentials from the CAS to connect to the resource with any normal Globus tool (e.g. GridFTP). The resource then applies its local policy to determine the amount of access granted to the community and further restricts that access based on the policy in the CAS credentials. This serves to limit the user's privileges to the intersection of those granted by the CAS to the user and those granted by the resource provider to the community.

76. *Resource Manager for Globus-based Wide-area Cluster Computing*, Tanaka, Y., Sato, M., Hirano, M., Nakada, H. and Sekiguchi, S. (1999) *1st IEEE International Workshop on Cluster Computing (IWCC '99)*, 1999, http://ninf.apgrid.org/papers/iwcc99tanaka/IWCC99.pdf.

77. *Performance Evaluation of a Firewall-compliant Globus-based Wide-area Cluster System*, Tanaka, Y., Sato, M., Hirano, M., Nakada, H. and Sekiguchi, S. (2000) *9th IEEE International Symposium on High Performance Distributed Computing*, 2000, http://ninf.apgrid.org/papers/hpdc00tanaka/HPDC00.pdf.

78. *NetLogger: A Toolkit for Distributed System Performance Analysis*, Gunter, D., Tierney, B., Crowley, B., Holding, M. and Lee, J. (2000) *IEEE Mascots 2000: Eighth International Symposium on Modeling, Analysis and Simulation of Computer and Telecommunication Systems*, 2000, http://www-didc.lbl.gov/papers/NetLogger.Mascots.paper.ieee.pdf.

Diagnosis and debugging of performance problems on complex distributed systems requires end-to-end performance information at both the application and system level. We describe a methodology, called *NetLogger*, that enables real-time diagnosis of performance problems in such systems. The methodology includes tools for generating precision event logs, an interface to a system event-monitoring framework and tools for visualizing the log data and real-time state of the distributed system. Low overhead is an important requirement for such tools, thereby we evaluate efficiency of the monitoring itself. The approach is novel in that it combines network, host, and application-level monitoring, providing a complete view of the entire system.

79. *Pipechar Network Characterization Service*, Jin, G. http://www-didc.lbl.gov/NCS/.

Tools based on hop-by-hop network analysis are increasingly critical to network troubleshooting on the rapidly growing Internet. Network characterization service (NCS) provides ability to diagnose and troubleshoot networks hop-by-hop in an easy and timely fashion. Using NCS makes applications capable to fully utilize the high-speed networks, for example, saturating 1Gbps local network from a single x86 platform. This page contains rich information about NCS and network measurement algorithms. Tutorials for using NCS to analyze and troubleshoot network problems are presented below.

80. *High-Speed Distributed Data Handling for On-Line Instrumentation Systems*, Johnston, W., Greiman, W., Hoo, G., Lee, J., Tierney, B., Tull, C. and Olson, D. (1997) *ACM/IEEE SC97: High Performance Networking and Computing*, 1997, http://www-itg.lbl.gov/~johnston/papers.html.

81. *The NetLogger Methodology for High Performance Distributed Systems Performance Analysis*, Tierney, B., Johnston, W., Crowley, B., Hoo, G., Brooks, C. and Gunter, D. (1998) *Proc. 7th IEEE Symp. on High Performance Distributed Computing*, 1998, http://www-didc.lbl.gov/NetLogger/.
82. *A Network-Aware Distributed Storage Cache for Data Intensive Environments*, Tierney, B., Lee, J., Crowley, B., Holding, M., Hylton, J. and Drake, F. (1999) *Proc. 8th IEEE Symp. on High Performance Distributed Computing*, 1999, http://www-didc.lbl.gov/papers/dpss.hpdc99.pdf.
83. *Using NetLogger for Distributed Systems Performance Analysis of the BaBar Data Analysis System*, Tierney, B., Gunter, D., Becla, J., Jacobsen, B. and Quarrie, D. (2000) *Proceedings of Computers in High Energy Physics 2000 (CHEP 2000)*, 2000, http://www-didc.lbl.gov/papers/chep.2K.Netlogger.pdf.
84. *Dynamic Monitoring of High-Performance Distributed Applications*, Gunter, D., Tierney, B., Jackson, K., Lee, J. and Stoufer, M. (2002) *11th IEEE Symposium on High Performance Distributed Computing, HPDC-11*, 2002, http://www-didc.lbl.gov/papers/HPDC02-HP-monitoring.pdf.
85. *Applied Techniques for High Bandwidth Data Transfers across Wide Area Networks*, Lee, J., Gunter, D., Tierney, B., Allock, W., Bester, J., Bresnahan, J. and Tuecke, S. (2001) *Proceedings of Computers in High Energy Physics 2001 (CHEP 2001)*, Beijing, China, 2001, http://www-didc.lbl.gov/papers/dpss_and_gridftp.pdf.
86. *TCP Tuning Guide for Distributed Applications on Wide Area Networks*, Tierney, B. In Usenix; login Journal, 2001, http://www-didc.lbl.gov/papers/usenix-login.pdf. Also see http://www-didc.lbl.gov/tcp-wan.html.
87. *A TCP Tuning Daemon*, Dunigan, T., Mathis, M. and Tierney, B. (2002) *Proceeding of IEEE Supercomputing 2002*, Baltimore, MD, 2002, http://www-didc.lbl.gov/publications.html.
88. *Globus Quick Start Guide*, Globus Project, 2002, http://www.globus.org/toolkit/documentation/QuickStart.pdf.

This document is intended for people who use Globus-enabled applications. It includes information on how to set up one's environment to use the Globus Toolkit and applications based on Globus software.

89. *NetSaint*, http://www.netsaint.org/.

NetSaint is a program that will monitor hosts and services on your network. It has the ability to e-mail or page you when a problem arises and when it gets resolved. ... It can run either as a normal process or as a daemon, intermittently running checks on various services that you specify. The actual service checks are performed by external 'plug-ins', which return service information to NetSaint. Several CGI programs are included with NetSaint in order to allow you to view the current service status, history, and so on, via a Web browser.

90. *Poznan Supercomputing and Networking Center*, http://www.man.poznan.pl/.
91. *Virtual User Account System*, Meyer, N. and Wolniewicz, P. http://www.man.poznan.pl/metacomputing/cluster/.

This system allows the running of jobs into a distributed cluster (between supercomputing centres) without additional administration overhead. A user does not have to have accounts on all supercomputers in the cluster; he only submits a job into the LSF on a local machine. The job is calculated on a virtual user account, but all billings and results are generated for the real user.

92. *The Physiology of the Grid: An Open Grid Services Architecture for Distributed Systems Integration*, Foster, I., Kesselman, C., Nick, J. and Tuecke, S. http://www.globus.org/research/papers.html#OGSA.

[We] define ... how a Grid functions and how Grid technologies can be implemented and applied. ... we focus here on the nature of the services that respond to protocol messages. We view a Grid as an extensible set of Grid services that may be aggregated in various ways to meet the needs of VOs, which themselves can be defined in part by the services that they operate and share. We then define the behaviors that such Grid services should possess in order to support distributed systems integration. By stressing functionality (i.e. 'physiology'), this view of Grids complements the previous protocol-oriented ('anatomical') description.

Second, we explain how Grid technologies can be aligned with Web services technologies ... to capitalize on desirable Web services properties, such as service description and discovery; automatic generation of client and server code from service descriptions; binding of service descriptions to interoperable network protocols; compatibility with emerging higher-level open standards, services, and tools; and broad commercial support. We call this alignment and augmentation of Grid and Web services technologies an *Open Grid Services Architecture* (OGSA), with the term *architecture* denoting here a well-defined set of basic interfaces from which can be constructed interesting systems, and *open* being used to communicate extensibility, vendor neutrality, and commitment to a community standardization process. This architecture uses the Web Services Description Language (WSDL) to achieve self-describing, discoverable services and interoperable protocols, with extensions to support multiple coordinated interfaces and to change management. OGSA leverages experience gained with the Globus Toolkit to define conventions and WSDL interfaces for a Grid service, a (potentially transient) stateful service instance supporting reliable and secure invocation (when required), lifetime management, notification, policy management, credential management, and virtualization. OGSA also defines interfaces for the discovery of Grid service instances and for the creation of transient Grid service instances. The result is a standards-based distributed service system (we avoid the term *distributed object system* owing to its overloaded meaning) that supports the creation of the sophisticated distributed services required in modern enterprise and interorganizational computing environments.

Third, we focus our discussion on commercial applications rather than the scientific and technical applications .... We believe that the same principles and mechanisms apply in both environments. However, in commercial settings we need, in particular, seamless integration with existing resources and applications, and with tools for workload, resource, security, network QoS, and availability management. OGSA's support for the discovery of service properties facilitates the mapping or adaptation of higher-level Grid service functions to such native platform facilities. OGSA's service orientation also allows us to virtualize resources at multiple levels, so that the same abstractions and mechanisms can be used both within distributed Grids supporting collaboration across organizational domains and within hosting environments spanning multiple tiers within a single IT domain. A common infrastructure means that differences (e.g. relating to visibility and accessibility) derive from policy controls associated with resource ownership, privacy, and security, rather than interaction mechanisms. Hence, as today's enterprise systems are transformed from separate computing resource islands to integrated, multitiered distributed systems, service components can be integrated dynamically and flexibly, both within and across various organizational boundaries.

93. *Open Grid Service Interface Working Group*, Global Grid Forum. http://www.gridforum.org/ogsi-wg/.

The purpose of the OGSI Working Group is to review and refine the Grid Service Specification and other documents that derive from this specification, including OGSA-infrastructure-related technical specifications and supporting informational documents.

94. *Grid Computing Environments Research Group*, Global Grid Forum, http://www.computingportals.org/.

Our working group is aimed at contributing to the coherence and interoperability of frameworks, portals, Problem Solving Environments, and other Grid-based computing environments and Grid services. We do this by choosing 'best practices' projects to derive standards, protocols, APIs and SDKs that are required to integrate technology implementations and solutions.

95. *Python Globus(pyGlobus)*, Jackson, K. Lawrence Berkeley National Laboratory, http://www-itg.lbl.gov/gtg/projects/pyGlobus/index.html.

– Provide a clean object-oriented interface to the Globus toolkit.
– Provide similar performance to using the underlying C code as much as possible.
– Minimize the number of changes necessary when aspects of Globus change.
– Where possible, make Globus as natural to use from Python as possible.
– For example, the gassFile module allows the manipulation of remote GASS files as Python file objects.

96. *CoG Kits: Enabling Middleware for Designing Science Appl*, Jackson, K. and Laszewski, G. V. (2001) Lawrence Berkeley National Laboratory and Argonne National Laboratory, March, 2001, Submitted SciDAC proposal.
97. *CoG Kits: A Bridge Between Commodity Distributed Computing and High-Performance Grids*, Laszewski, G. V., Foster, I. and Gawor, J. (2000) *ACM 2000 Java Grande Conference*, San Francisco, 2000, http://www.globus.org/cog/documentation/papers/index.html.
98. *A Java Commodity Grid Kit*, Laszewski, G. V., Foster, I., Gawor, J. and Lane, P. (2001) *Concurrency: Experience and Practice*, 2001, http://www.globus.org/cog/documentation/papers/index.html.
99. *The Grid Portal Development Kit*, Novotny, J. (2000) *Concurrency – Practice and Experience*, 2000. http://www.doesciencegrid.org/Grid/projects/GPDK/gpdkpaper.pdf.
100. *NPSS on NASA's IPG: Using CORBA and Globus to Coordinate Multidisciplinary Aeroscience Applications*, Lopez, G. *et al.* 2000, http://www.ipg.nasa.gov/research/papers/NPSS_CAS_paper.html.

[The] NASA Glenn Research Center is developing an environment for the analysis and design of aircraft engines called the *Numerical Propulsion System Simulation* (NPSS). The vision for NPSS is to create a 'numerical test cell' enabling full engine simulations overnight on cost-effective computing platforms. To this end, NPSS integrates multiple disciplines such as aerodynamics, structures, and heat transfer and supports 'numerical zooming' from zero-dimensional to one-, two-, and three-dimensional component engine codes. To facilitate the timely and cost-effective capture of complex physical processes, NPSS uses object-oriented technologies such as C++ objects to encapsulate individual engine components and Common Object Request Broker Architecture (CORBA) Object Request Brokers (ORBs) for object communication and deployment across heterogeneous computing platforms.

IPG implements a range of Grid services such as resource discovery, scheduling, security, instrumentation, and data access, many of which are provided by the Globus Toolkit. IPG facilities have the potential to benefit NPSS considerably. For example, NPSS should in principle be able to use Grid services to discover dynamically and then co-schedule the resources required for a particular engine simulation, rather than relying on manual placement of ORBs as at present. Grid services can also be used to initiate simulation components on massively parallel computers (MPPs) and to address intersite security issues that currently hinder the coupling of components across multiple sites.

... This project involves, first, development of the basic techniques required to achieve coexistence of commodity object technologies and Grid technologies, and second, the evaluation of these techniques in the context of NPSS-oriented challenge problems.

The work on basic techniques seeks to understand how 'commodity' technologies (CORBA, DCOM, Excel, etc.) can be used in concert with specialized Grid technologies (for security, MPP scheduling, etc.). In principle, this coordinated use should be straightforward because of the Globus and IPG philosophy of providing low-level Grid mechanisms that can be used to implement a wide variety of application-level programming models. (Globus technologies have previously been used to implement Grid-enabled message-passing libraries, collaborative environments, and parameter study tools, among others.) Results obtained to date are encouraging: a CORBA to Globus resource manager gateway has been successfully demonstrated that allows the use of CORBA remote procedure calls (RPCs) to control submission and execution of programs on workstations and MPPs; a gateway has been implemented from the CORBA Trader service to the Grid information service; and a preliminary integration of CORBA and Grid security mechanisms has been completed.

101. *A CORBA-based Development Environment for Wrapping and Coupling Legacy Codes*, Follen, G., Kim, C., Lopez, I., Sang, J. and Townsend, S. (2001) *Tenth IEEE International Symposium on High Performance Distributed Computing*, San Francisco, 2001, http://cnis.grc.nasa.gov/papers/hpdc-10_corbawrapping.pdf, http://cnis.grc.nasa.gov/papers/hpdc-10_corbawrapping.pdf.

102. *LaunchPad*, http://www.ipg.nasa.gov/launchpad/launchpad.

The IPG Launch Pad provides access to compute and other resources at participating NASA related sites. The initial release provides the ability to

- submit Jobs to 'batch' compute engines
- execute commands on compute resources
- transfer files between two systems
- obtain status on systems and jobs
- modify the user's environment.

Launch Pad was developed using the Grid Portal Development Kit created by Jason Novotny.

103. *HotPage*, https://hotpage.npaci.edu/.

HotPage [is] the NPACI Grid Computing Portal. HotPage enables researchers to find information about each of the resources in the NPACI computational Grid, including technical documentation, operational status, load and current usage, and queued jobs.
New tools allow you to

- obtain a portal account on-line
- personalize your view of the status bar
- access and manipulate your files and data once you are logged in, and
- submit, monitor, and delete jobs on HPC resources.

104. *An Online Credential Repository for the Grid: MyProxy*, Novotny, J., Tuecke, S. and Welch, V. (2001) *Tenth IEEE International Symposium on High Performance Distributed Computing*, San Francisco, 2001, http://www.globus.org/research/papers.html#MyProxy.

Grid Portals, based on standard Web technologies, are increasingly used to provide user interfaces for Computational and Data Grids. However, such Grid Portals do not integrate cleanly with existing Grid security systems such as the Grid Security Infrastructure (GSI), owing to lack of delegation capabilities in Web security mechanisms. We solve this problem using an on-line credentials repository system, called *MyProxy*. MyProxy allows Grid Portals to use the GSI to interact with Grid resources in a standard, secure manner. We examine the requirements of Grid Portals, give an overview of the GSI, and demonstrate how MyProxy enables them to function together. The architecture and security of the MyProxy system are described in detail.

105. *MyProxy*, NCSA, http://www.ncsa.uiuc.edu/Divisions/ACES/MyProxy/.

MyProxy provides a server with client-side utilities to store and retrieve medium-term lifetime (of the order of a week) delegated X.509 credentials via the Grid Security Infrastructure (GSI).

The myproxy-init program delegates a proxy credential to the myproxy server, which stores the proxy to disk. The myproxy-get-delegation program retrieves stored proxy credentials from the myproxy-server. The myproxy-destroy program removes credentials stored on a myproxy-server.

106. *NSF Middleware Initiative (NMI)*, http://www.nsf-middleware.org/.

A new package of software and other tools will make it easier for US scientists, engineers and educators to collaborate across the Internet and use the Grid, a group of high-speed successor technologies and capabilities to the Internet that link high-performance networks and computers nationwide and around the world.

The package of 'middleware', or software and services that link two or more otherwise unconnected applications across the Internet, was developed under the auspices of the National Science Foundation's (NSF) Middleware Initiative (NMI). NSF launched the initiative in September 2001 by committing $12 million over three years to create and deploy advanced network services that simplify access to diverse Internet information and services.

NMI Release 1.0 (NMI-R1) represents the first bundling of such Grid software as the Globus Toolkit, Condor-G and the Network Weather Service, along with security tools and best practices for enterprise computing such as eduPerson and Shibboleth. By wrapping them in a single package, NMI project leaders intend to ease the use and deployment of such middleware, making distributed, collaborative environments such as Grid computing and desktop video-conferencing more accessible.

107. *IBM and Department of Energy Supercomputing Center to Make DOE Grid Computing a Reality – DOE Science Grid to Transform Far-Flung Supercomputers into a Utility-like Service*, http://www.nersc.gov/news/IBMgrids032202.html.

ARMONK, NY and BERKELEY, CA, March 22, 2002 – IBM and the US Department of Energy's (DOE) National Energy Research Scientific Computing Center (NERSC) today announced a collaboration to begin deploying the first systems on a nationwide computing Grid, which will empower researchers to tackle scientific challenges beyond the capability of existing computers.

Beginning with two IBM supercomputers and a massive IBM storage repository, the DOE Science Grid will ultimately grow into a system capable of processing more than five trillion calculations per second and storing information equivalent to 200 times the number of books in the Library of Congress. The collaboration will make the largest unclassified supercomputer and largest data storage system within DOE available via the Science Grid by December 2002 – two years sooner than expected.

PART B

Grid architecture and technologies

Reprint from *International Journal of High Performance Computing Applications* © 2001 Sage Publications, Inc. (USA). Minor changes to the original have been made to conform with house style.

6

The anatomy of the Grid

Enabling Scalable Virtual Organizations*

Ian Foster,[1,2] Carl Kesselman,[3] and Steven Tuecke[1]

[1]*Mathematics and Computer Science Division, Argonne National Laboratory, Argonne, Illinois, United States,* [2]*Department of Computer Science, The University of Chicago, Chicago, Illinois, United States,* [3]*Information Sciences Institute, The University of Southern California, California, United States*

6.1 INTRODUCTION

The term 'the Grid' was coined in the mid-1990s to denote a proposed distributed computing infrastructure for advanced science and engineering [1]. Considerable progress has since been made on the construction of such an infrastructure (e.g., [2–5]), but the term 'Grid' has also been conflated, at least in popular perception, to embrace everything from advanced networking to artificial intelligence. One might wonder whether the term has any real substance and meaning. Is there really a distinct 'Grid problem' and hence a need for new 'Grid technologies'? If so, what is the nature of these technologies, and what is their domain of applicability? While numerous groups have interest in Grid concepts

---

* To appear: Intl J. Supercomputer Applications, 2001.

Grid Computing – Making the Global Infrastructure a Reality. Edited by F. Berman, A. Hey and G. Fox
© 2003 John Wiley & Sons, Ltd   ISBN: 0-470-85319-0

and share, to a significant extent, a common vision of Grid architecture, we do not see consensus on the answers to these questions.

Our purpose in this article is to argue that the Grid concept is indeed motivated by a real and specific problem and that there is an emerging, well-defined Grid technology base that addresses significant aspects of this problem. In the process, we develop a detailed architecture and roadmap for current and future Grid technologies. Furthermore, we assert that while Grid technologies are currently distinct from other major technology trends, such as Internet, enterprise, distributed, and peer-to-peer computing, these other trends can benefit significantly from growing into the problem space addressed by Grid technologies.

The real and specific problem that underlies the Grid concept is *coordinated resource sharing and problem solving in dynamic, multi-institutional virtual organizations*. The sharing that we are concerned with is not primarily file exchange but rather direct access to computers, software, data, and other resources, as is required by a range of collaborative problem-solving and resource-brokering strategies emerging in industry, science, and engineering. This sharing is, necessarily, highly controlled, with resource providers and consumers defining clearly and carefully just what is shared, who is allowed to share, and the conditions under which sharing occurs. A set of individuals and/or institutions defined by such sharing rules form what we call a *virtual organization* (VO).

The following are examples of VOs: the application service providers, storage service providers, cycle providers, and consultants engaged by a car manufacturer to perform scenario evaluation during planning for a new factory; members of an industrial consortium bidding on a new aircraft; a crisis management team and the databases and simulation systems that they use to plan a response to an emergency situation; and members of a large, international, multiyear high-energy physics collaboration. Each of these examples represents an approach to computing and problem solving based on collaboration in computation- and data-rich environments.

As these examples show, VOs vary tremendously in their purpose, scope, size, duration, structure, community, and sociology. Nevertheless, careful study of underlying technology requirements leads us to identify a broad set of common concerns and requirements. In particular, we see a need for highly flexible sharing relationships, ranging from client-server to peer-to-peer; for sophisticated and precise levels of control over how shared resources are used, including fine-grained and multistakeholder access control, delegation, and application of local and global policies; for sharing of varied resources, ranging from programs, files, and data to computers, sensors, and networks; and for diverse usage modes, ranging from single user to multiuser and from performance sensitive to cost-sensitive and hence embracing issues of quality of service, scheduling, co-allocation, and accounting.

Current distributed computing technologies do not address the concerns and requirements just listed. For example, current Internet technologies address communication and information exchange among computers but do not provide integrated approaches to the coordinated use of resources at multiple sites for computation. Business-to-business exchanges [6] focus on information sharing (often via centralized servers). So do virtual enterprise technologies, although here sharing may eventually extend to applications and physical devices (e.g., [7]). Enterprise distributed computing technologies such as CORBA and Enterprise Java enable resource sharing within a single organization. The Open Group's Distributed Computing Environment (DCE) supports secure resource sharing across sites, but most VOs would find it too burdensome and inflexible. Storage service providers (SSPs) and application service providers (ASPs) allow organizations to outsource storage and computing requirements to other parties, but only in constrained ways: for example, SSP resources are typically linked to a customer via a virtual private network (VPN). Emerging 'distributed computing' companies seek to harness idle computers on an international scale [31] but, to date, support only highly centralized access to those resources. In summary, current technology either does not accommodate the range of resource types or does not provide the flexibility and control on sharing relationships needed to establish VOs.

It is here that Grid technologies enter the picture. Over the past five years, research and development efforts within the Grid community have produced protocols, services, and tools that address precisely the challenges that arise when we seek to build scalable VOs. These technologies include security solutions that support management of credentials and policies when computations span multiple institutions; resource management protocols and services that support secure remote access to computing and data resources and the co-allocation of multiple resources; information query protocols and services that provide configuration and status information about resources, organizations, and services; and data management services that locate and transport datasets between storage systems and applications.

Because of their focus on dynamic, cross-organizational sharing, Grid technologies complement rather than compete with existing distributed computing technologies. For example, enterprise distributed computing systems can use Grid technologies to achieve resource sharing across institutional boundaries; in the ASP/SSP space, Grid technologies can be used to establish dynamic markets for computing and storage resources, hence overcoming the limitations of current static configurations. We discuss the relationship between Grids and these technologies in more detail below.

In the rest of this article, we expand upon each of these points in turn. Our objectives are to (1) clarify the nature of VOs and Grid computing for those unfamiliar with the area; (2) contribute to the emergence of Grid computing as a discipline by establishing a standard vocabulary and defining an overall architectural framework; and (3) define clearly how Grid technologies relate to other technologies, explaining both why emerging technologies do not yet solve the Grid computing problem and how these technologies can benefit from Grid technologies.

It is our belief that VOs have the potential to change dramatically the way we use computers to solve problems, much as the Web has changed how we exchange information. As the examples presented here illustrate, the need to engage in collaborative processes is fundamental to many diverse disciplines and activities: it is not limited to science, engineering, and business activities. It is because of this broad applicability of VO concepts that Grid technology is important.

6.2 THE EMERGENCE OF VIRTUAL ORGANIZATIONS

Consider the following four scenarios:

1. A company needing to reach a decision on the placement of a new factory invokes a sophisticated financial forecasting model from an ASP, providing it with access to appropriate proprietary historical data from a corporate database on storage systems operated by an SSP. During the decision-making meeting, what-if scenarios are run collaboratively and interactively, even though the division heads participating in the decision are located in different cities. The ASP itself contracts with a cycle provider for additional 'oomph' during particularly demanding scenarios, requiring of course that cycles meet desired security and performance requirements.
2. An industrial consortium formed to develop a feasibility study for a next-generation supersonic aircraft undertakes a highly accurate multidisciplinary simulation of the entire aircraft. This simulation integrates proprietary software components developed by different participants, with each component operating on that participant's computers and having access to appropriate design databases and other data made available to the consortium by its members.
3. A crisis management team responds to a chemical spill by using local weather and soil models to estimate the spread of the spill, determining the impact based on population location as well as geographic features such as rivers and water supplies, creating a short-term mitigation plan (perhaps based on chemical reaction models), and tasking emergency response personnel by planning and coordinating evacuation, notifying hospitals, and so forth.
4. Thousands of physicists at hundreds of laboratories and universities worldwide come together to design, create, operate, and analyze the products of a major detector at CERN, the European high energy physics laboratory. During the analysis phase, they pool their computing, storage, and networking resources to create a 'Data Grid' capable of analyzing petabytes of data [8–10].

These four examples differ in many respects: the number and type of participants, the types of activities, the duration and scale of the interaction, and the resources being shared. But they also have much in common, as discussed in the following (see also Figure 6.1).

In each case, a number of mutually distrustful participants with varying degrees of prior relationship (perhaps none at all) want to share resources in order to perform some task. Furthermore, sharing is about more than simply document exchange (as in 'virtual enterprises' [11]): it can involve direct access to remote software, computers, data, sensors, and other resources. For example, members of a consortium may provide access to specialized software and data and/or pool their computational resources.

Resource sharing is conditional: each resource owner makes resources available, subject to constraints on when, where, and what can be done. For example, a participant in VO P of Figure 6.1 might allow VO partners to invoke their simulation service only for 'simple' problems. Resource consumers may also place constraints on properties of the resources they are prepared to work with. For example, a participant in VO Q might accept only pooled computational resources certified as 'secure.' The implementation of

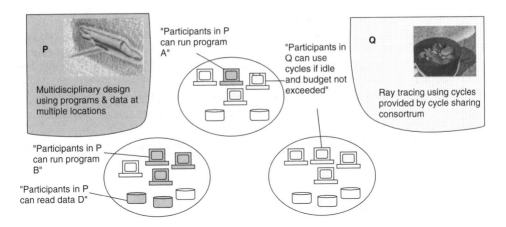

Figure 6.1 An actual organization can participate in one or more VOs by sharing some or all of its resources. We show three actual organizations (the ovals), and two VOs: P, which links participants in an aerospace design consortium, and Q, which links colleagues who have agreed to share spare computing cycles, for example, to run ray tracing computations. The organization on the left participates in P, the one to the right participates in Q, and the third is a member of both P and Q. The policies governing access to resources (summarized in quotes) vary according to the actual organizations, resources, and VOs involved.

such constraints requires mechanisms for expressing policies, for establishing the identity of a consumer or resource (authentication), and for determining whether an operation is consistent with applicable sharing relationships (authorization).

Sharing relationships can vary dynamically over time, in terms of the resources involved, the nature of the access permitted, and the participants to whom access is permitted. And these relationships do not necessarily involve an explicitly named set of individuals, but rather may be defined implicitly by the policies that govern access to resources. For example, an organization might enable access by anyone who can demonstrate that he or she is a 'customer' or a 'student.'

The dynamic nature of sharing relationships means that we require mechanisms for discovering and characterizing the nature of the relationships that exist at a particular point in time. For example, a new participant joining VO Q must be able to determine what resources it is able to access, the 'quality' of these resources, and the policies that govern access.

Sharing relationships are often not simply client-server, but peer to peer: providers can be consumers, and sharing relationships can exist among any subset of participants. Sharing relationships may be combined to coordinate use across many resources, each owned by different organizations. For example, in VO Q, a computation started on one pooled computational resource may subsequently access data or initiate subcomputations elsewhere. The ability to delegate authority in controlled ways becomes important in such situations, as do mechanisms for coordinating operations across multiple resources (e.g., coscheduling).

The same resource may be used in different ways, depending on the restrictions placed on the sharing and the goal of the sharing. For example, a computer may be used only to run a specific piece of software in one sharing arrangement, while it may provide generic compute cycles in another. Because of the lack of *a priori* knowledge about how a resource may be used, performance metrics, expectations, and limitations (i.e., quality of service) may be part of the conditions placed on resource sharing or usage.

These characteristics and requirements define what we term a *virtual organization*, a concept that we believe is becoming fundamental to much of modern computing. VOs enable disparate groups of organizations and/or individuals to share resources in a controlled fashion, so that members may collaborate to achieve a shared goal.

6.3 THE NATURE OF GRID ARCHITECTURE

The establishment, management, and exploitation of dynamic, cross-organizational VO sharing relationships require new technology. We structure our discussion of this technology in terms of a *Grid architecture* that identifies fundamental system components, specifies the purpose and function of these components, and indicates how these components interact with one another.

In defining a Grid architecture, we start from the perspective that effective VO operation requires that we be able to establish sharing relationships among *any* potential participants. Interoperability is thus the central issue to be addressed. In a networked environment, interoperability means common protocols. Hence, our Grid architecture is first and foremost a *protocol* architecture, with protocols defining the basic mechanisms by which VO users and resources negotiate, establish, manage, and exploit sharing relationships. A standards-based open architecture facilitates extensibility, interoperability, portability, and code sharing; standard protocols make it easy to define standard services that provide enhanced capabilities. We can also construct application programming interfaces and software development kits (see Appendix for definitions) to provide the programming abstractions required to create a usable Grid. Together, this technology and architecture constitute what is often termed middleware ('the services needed to support a common set of applications in a distributed network environment' [12]), although we avoid that term here because of its vagueness. We discuss each of these points in the following.

Why is interoperability such a fundamental concern? At issue is our need to ensure that sharing relationships can be initiated among arbitrary parties, accommodating new participants dynamically, across different platforms, languages, and programming environments. In this context, mechanisms serve little purpose if they are not defined and implemented so as to be interoperable across organizational boundaries, operational policies, and resource types. Without interoperability, VO applications and participants are forced to enter into bilateral sharing arrangements, as there is no assurance that the mechanisms used between any two parties will extend to any other parties. Without such assurance, dynamic VO formation is all but impossible, and the types of VOs that can be formed are severely limited. Just as the Web revolutionized information sharing by providing a universal protocol and syntax (HTTP and HTML) for information exchange, so we require standard protocols and syntaxes for general resource sharing.

Why are protocols critical to interoperability? A protocol definition specifies how distributed system elements interact with one another in order to achieve a specified behavior, and the structure of the information exchanged during this interaction. This focus on externals (interactions) rather than internals (software, resource characteristics) has important pragmatic benefits. VOs tend to be fluid; hence, the mechanisms used to discover resources, establish identity, determine authorization, and initiate sharing must be flexible and lightweight, so that resource-sharing arrangements can be established and changed quickly. Because VOs complement rather than replace existing institutions, sharing mechanisms cannot require substantial changes to local policies and must allow individual institutions to maintain ultimate control over their own resources. Since protocols govern the interaction between components, and not the implementation of the components, local control is preserved.

Why are services important? A service (see Appendix) is defined solely by the protocol that it speaks and the behaviors that it implements. The definition of standard services – for access to computation, access to data, resource discovery, coscheduling, data replication, and so forth – allows us to enhance the services offered to VO participants and also to abstract away resource-specific details that would otherwise hinder the development of VO applications.

Why do we also consider application programming interfaces (APIs) and software development kits (SDKs)? There is, of course, more to VOs than interoperability, protocols, and services. Developers must be able to develop sophisticated applications in complex and dynamic execution environments. Users must be able to operate these applications. Application robustness, correctness, development costs, and maintenance costs are all important concerns. Standard abstractions, APIs, and SDKs can accelerate code development, enable code sharing, and enhance application portability. APIs and SDKs are an adjunct to, not an alternative to, protocols. Without standard protocols, interoperability can be achieved at the API level only by using a single implementation everywhere – infeasible in many interesting VOs – or by having every implementation know the details of every other implementation. (The Jini approach [13] of downloading protocol code to a remote site does not circumvent this requirement.)

In summary, our approach to Grid architecture emphasizes the identification and definition of protocols and services, first, and APIs and SDKs, second.

6.4 GRID ARCHITECTURE DESCRIPTION

Our goal in describing our Grid architecture is not to provide a complete enumeration of all required protocols (and services, APIs, and SDKs) but rather to identify requirements for general classes of component. The result is an extensible, open architectural structure within which can be placed solutions to key VO requirements. Our architecture and the subsequent discussion organize components into layers, as shown in Figure 6.2. Components within each layer share common characteristics but can build on capabilities and behaviors provided by any lower layer.

In specifying the various layers of the Grid architecture, we follow the principles of the 'hourglass model' [14]. The narrow neck of the hourglass defines a small set of core

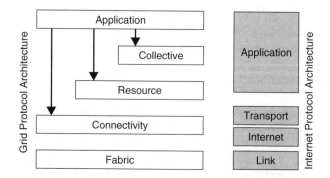

Figure 6.2 The layered Grid architecture and its relationship to the Internet protocol architecture. Because the Internet protocol architecture extends from network to application, there is a mapping from Grid layers into Internet layers.

abstractions and protocols (e.g., TCP and HTTP in the Internet), onto which many different high-level behaviors can be mapped (the top of the hourglass), and which themselves can be mapped onto many different underlying technologies (the base of the hourglass). By definition, the number of protocols defined at the neck must be small. In our architecture, the neck of the hourglass consists of *Resource* and *Connectivity* protocols, which facilitate the sharing of individual resources. Protocols at these layers are designed so that they can be implemented on top of a diverse range of resource types, defined at the *Fabric* layer, and can in turn be used to construct a wide range of global services and application-specific behaviors at the *Collective* layer – so called because they involve the coordinated ('collective') use of multiple resources.

Our architectural description is high level and places few constraints on design and implementation. To make this abstract discussion more concrete, we also list, for illustrative purposes, the protocols defined within the Globus Toolkit [15] and used within such Grid projects as the NSF's National Technology Grid [5], NASA's Information Power Grid [4], DOE's DISCOM [2], GriPhyN (www.griphyn.org), NEESgrid (www.neesgrid.org), Particle Physics Data Grid (www.ppdg.net), and the European Data Grid (www.eu-datagrid.org). More details will be provided in a subsequent paper.

6.4.1 Fabric: Interfaces to local control

The Grid *Fabric* layer provides the resources to which shared access is mediated by Grid protocols: for example, computational resources, storage systems, catalogs, network resources, and sensors. A 'resource' may be a logical entity, such as a distributed file system, computer cluster, or distributed computer pool; in such cases, a resource implementation may involve internal protocols (e.g., the NFS storage access protocol or a cluster resource management system's process management protocol), but these are not the concern of Grid architecture.

Fabric components implement the local, resource-specific operations that occur on specific resources (whether physical or logical) as a result of sharing operations at higher levels. There is thus a tight and subtle interdependence between the functions implemented at the Fabric level, on the one hand, and the sharing operations supported, on the other. Richer Fabric functionality enables more sophisticated sharing operations; at the same time, if we place few demands on Fabric elements, then deployment of Grid infrastructure is simplified. For example, resource-level support for advance reservations makes it possible for higher-level services to aggregate (coschedule) resources in interesting ways that would otherwise be impossible to achieve.

However, as in practice few resources support advance reservation 'out of the box,' a requirement for advance reservation increases the cost of incorporating new resources into a Grid.

Experience suggests that at a minimum, resources should implement *enquiry* mechanisms that permit discovery of their structure, state, and capabilities (e.g., whether they support advance reservation), on the one hand, and *resource management* mechanisms that provide some control of delivered quality of service, on the other. The following brief and partial list provides a resource-specific characterization of capabilities.

- *Computational resources*: Mechanisms are required for starting programs and for monitoring and controlling the execution of the resulting processes. Management mechanisms that allow control over the resources allocated to processes are useful, as are advance reservation mechanisms. Enquiry functions are needed for determining hardware and software characteristics as well as relevant state information such as current load and queue state in the case of scheduler-managed resources.
- *Storage resources*: Mechanisms are required for putting and getting files. Third-party and high-performance (e.g., striped) transfers are useful [16]. So are mechanisms for reading and writing subsets of a file and/or executing remote data selection or reduction functions [17]. Management mechanisms that allow control over the resources allocated to data transfers (space, disk bandwidth, network bandwidth, CPU) are useful, as are advance reservation mechanisms. Enquiry functions are needed for determining hardware and software characteristics as well as relevant load information such as available space and bandwidth utilization.
- *Network resources*: Management mechanisms that provide control over the resources allocated to network transfers (e.g., prioritization, reservation) can be useful. Enquiry functions should be provided to determine network characteristics and load.
- *Code repositories*: This specialized form of storage resource requires mechanisms for managing versioned source and object code: for example, a control system such as CVS.
- *Catalogs*: This specialized form of storage resource requires mechanisms for implementing catalog query and update operations: for example, a relational database [18].

Globus Toolkit: The Globus Toolkit has been designed to use (primarily) existing fabric components, including vendor-supplied protocols and interfaces. However, if a vendor does not provide the necessary Fabric-level behavior, the Globus Toolkit includes the missing functionality. For example, enquiry software is provided for discovering structure and state information for various common resource types, such as computers (e.g., OS version, hardware configuration, load [19], scheduler queue status), storage systems (e.g., available space), and networks (e.g., current and predicted future load [20, 21], and for packaging this information in a form that facilitates the implementation of higher-level protocols, specifically at the Resource layer. Resource management, on the other hand, is generally assumed to be the domain of local resource managers. One exception is the General-purpose Architecture for Reservation and Allocation (GARA) [22], which provides a 'slot manager' that can be used to implement advance reservation for resources that do not support this capability. Others have developed enhancements to the Portable Batch System (PBS) [23] and Condor [24, 25] that support advance reservation capabilities.

6.4.2 Connectivity: Communicating easily and securely

The *Connectivity* layer defines core communication and authentication protocols required for Grid-specific network transactions. Communication protocols enable the exchange of data between Fabric layer resources. Authentication protocols build on communication services to provide cryptographically secure mechanisms for verifying the identity of users and resources.

Communication requirements include transport, routing, and naming. While alternatives certainly exist, we assume here that these protocols are drawn from the TCP/IP protocol stack: specifically, the Internet (IP and ICMP), transport (TCP, UDP), and application (DNS, OSPF, RSVP, etc.) layers of the Internet layered protocol architecture [26]. This is not to say that in the future, Grid communications will not demand new protocols that take into account particular types of network dynamics.

With respect to security aspects of the Connectivity layer, we observe that the complexity of the security problem makes it important that any solutions be based on existing standards whenever possible. As with communication, many of the security standards developed within the context of the Internet protocol suite are applicable.

Authentication solutions for VO environments should have the following characteristics [27]:

- *Single sign-on*: Users must be able to 'log on' (authenticate) just once and then have access to multiple Grid resources defined in the Fabric layer, without further user intervention.
- *Delegation*: [28–30]. A user must be able to endow a program with the ability to run on that user's behalf, so that the program is able to access the resources on which the user is authorized. The program should (optionally) also be able to conditionally delegate a subset of its rights to another program (sometimes referred to as restricted delegation).
- *Integration with various local security solutions*: Each site or resource provider may employ any of a variety of local security solutions, including Kerberos and Unix security. Grid security solutions must be able to interoperate with these various local solutions. They cannot, realistically, require wholesale replacement of local security solutions but rather must allow mapping into the local environment.
- *User-based trust relationships*: In order for a user to use resources from multiple providers together, the security system must not require each of the resource providers to cooperate or interact with each other in configuring the security environment. For

example, if a user has the right to use sites A and B, the user should be able to use sites A and B together without requiring that A's and B's security administrators interact.

Grid security solutions should also provide flexible support for communication protection (e.g., control over the degree of protection, independent data unit protection for unreliable protocols, support for reliable transport protocols other than TCP) and enable stakeholder control over authorization decisions, including the ability to restrict the delegation of rights in various ways.

Globus Toolkit: The Internet protocols listed above are used for communication. The public-key based Grid Security Infrastructure (GSI) protocols [27, 28] are used for authentication, communication protection, and authorization. GSI builds on and extends the Transport Layer Security (TLS) protocols [31] to address most of the issues listed above: in particular, single sign-on, delegation, integration with various local security solutions (including Kerberos [32]), and user-based trust relationships. X.509-format identity certificates are used. Stakeholder control of authorization is supported via an authorization toolkit that allows resource owners to integrate local policies via a Generic Authorization and Access (GAA) control interface. Rich support for restricted delegation is not provided in the current toolkit release (v1.1.4) but has been demonstrated in prototypes.

6.4.3 Resource: Sharing single resources

The *Resource* layer builds on Connectivity layer communication and authentication protocols to define protocols (and APIs and SDKs) for the secure negotiation, initiation, monitoring, control, accounting, and payment of sharing operations on individual resources. Resource layer implementations of these protocols call Fabric layer functions to access and control local resources. Resource layer protocols are concerned entirely with individual resources and hence ignore issues of global state and atomic actions across distributed collections; such issues are the concern of the Collective layer discussed next.

Two primary classes of Resource layer protocols can be distinguished:

- *Information protocols* are used to obtain information about the structure and state of a resource, for example, its configuration, current load, and usage policy (e.g., cost).
- *Management protocols* are used to negotiate access to a shared resource, specifying, for example, resource requirements (including advanced reservation and quality of service) and the operation(s) to be performed, such as process creation, or data access. Since management protocols are responsible for instantiating sharing relationships, they must serve as a 'policy application point,' ensuring that the requested protocol operations are consistent with the policy under which the resource is to be shared. Issues that must be considered include accounting and payment. A protocol may also support monitoring the status of an operation and controlling (e.g., terminating) the operation.

While many such protocols can be imagined, the Resource (and Connectivity) protocol layers form the neck of our hourglass model and as such should be limited to a small and focused set. These protocols must be chosen so as to capture the fundamental mechanisms of sharing across many different resource types (e.g., different local resource management systems), while not overly constraining the types or performance of higher-level protocols that may be developed.

The list of desirable Fabric functionality provided in Section 6.4.1 summarizes the major features required in Resource layer protocols. To this list we add the need for 'exactly once' semantics for many operations, with reliable error reporting indicating when operations fail.

Globus Toolkit: A small and mostly standards-based set of protocols is adopted. In particular:

- A Grid Resource Information Protocol (GRIP, currently based on the Lightweight Directory Access Protocol: LDAP) is used to define a standard resource information protocol and associated information model. An associated soft-state resource registration protocol, the Grid Resource Registration Protocol (GRRP), is used to register resources with Grid Index Information Servers, discussed in the next section [33].
- The HTTP-based Grid Resource Access and Management (GRAM) protocol is used for allocation of computational resources and for monitoring and control of computation on those resources [34].
- An extended version of the File Transfer Protocol, GridFTP, is a management protocol for data access; extensions include use of Connectivity layer security protocols, partial file access, and management of parallelism for high-speed transfers [35]. FTP is adopted as a base data transfer protocol because of its support for third-party transfers and because its separate control and data channels facilitate the implementation of sophisticated servers.
- LDAP is also used as a catalog access protocol.

The Globus Toolkit defines client-side C and Java APIs and SDKs for each of these protocols. Server-side SDKs and servers are also provided for each protocol, to facilitate the integration of various resources (computational, storage, network) into the Grid. For example, the Grid Resource Information Service (GRIS) implements server-side LDAP functionality, with callouts allowing for publication of arbitrary resource information [33]. An important server-side element of the overall toolkit is the 'gatekeeper,' which provides what is in essence a GSI-authenticated 'inetd' that speaks the GRAM protocol and can be used to dispatch various local operations. The Generic Security Services (GSS) API [36] is used to acquire, forward, and verify authentication credentials and to provide transport layer integrity and privacy within these SDKs and servers, enabling substitution of alternative security services at the Connectivity layer.

6.4.4 Collective: Coordinating multiple resources

While the Resource layer is focused on interactions with a single resource, the next layer in the architecture contains protocols and services (and APIs and SDKs) that are not associated with any one specific resource but rather are global in nature and capture interactions across collections of resources. For this reason, we refer to the next layer of the architecture as the *Collective* layer. Because Collective components build on the narrow Resource and Connectivity layer 'neck' in the protocol hourglass, they can implement a wide variety of sharing behaviors without placing new requirements on the resources being shared. For example:

- *Directory services* allow VO participants to discover the existence and/or properties of VO resources. A directory service may allow its users to query for resources by name and/or by attributes such as type, availability, or load [33]. Resource-level GRRP and GRIP protocols are used to construct directories.
- *coallocation-allocation, scheduling, and brokering services* allow VO participants to request the allocation of one or more resources for a specific purpose and the scheduling of tasks on the appropriate resources. Examples include AppLeS [37, 38], Condor-G [39], Nimrod-G [40], and the DRM broker [2].
- *Monitoring and diagnostics services* support the monitoring of VO resources for failure, adversarial attack ('intrusion detection'), overload, and so forth.
- *Data replication services* support the management of VO storage (and perhaps also network and computing) resources to maximize data access performance with respect to metrics such as response time, reliability, and cost [9, 35].
- *Grid-enabled programming systems* enable familiar programming models to be used in Grid environments, using various Grid services to address resource discovery, security, resource allocation, and other concerns. Examples include Grid-enabled implementations of the Message Passing Interface [41, 42] and manager-worker frameworks [43, 44].
- *Workload management systems and collaboration frameworks* – also known as problem solving environments (PSEs) – provide for the description, use, and management of multistep, asynchronous, multicomponent workflows.
- *Software discovery services* discover and select the best software implementation and execution platform based on the parameters of the problem being solved [45]. Examples include NetSolve [46] and Ninf [47].
- *Community authorization servers* enforce community policies governing resource access, generating capabilities that community members can use to access community resources. These servers provide a global policy enforcement service by building on resource information, and resource management protocols (in the Resource layer) and security protocols in the Connectivity layer. Akenti [48] addresses some of these issues.
- *Community accounting and payment services* gather resource usage information for the purpose of accounting, payment, and/or limiting of resource usage by community members.
- *Collaboratory services* support the coordinated exchange of information within potentially large user communities, whether synchronously or asynchronously. Examples are CAVERNsoft [49, 50], Access Grid [51], and commodity groupware systems.

These examples illustrate the wide variety of Collective layer protocols and services that are encountered in practice. Notice that while Resource layer protocols must be general in nature and are widely deployed, Collective layer protocols span the spectrum from general purpose to highly application or domain specific, with the latter existing perhaps only within specific VOs.

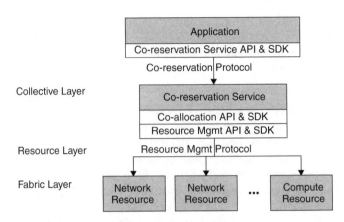

Figure 6.3 Collective and Resource layer protocols, services, APIs, and SDKs can be combined in a variety of ways to deliver functionality to applications.

Collective functions can be implemented as persistent services, with associated protocols, or as SDKs (with associated APIs) designed to be linked with applications. In both cases, their implementation can build on Resource layer (or other Collective layer) protocols and APIs. For example, Figure 6.3 shows a Collective co-allocation API and SDK (the middle tier) that uses a Resource layer management protocol to manipulate underlying resources. Above this, we define a co-reservation service protocol and implement a co-reservation service that speaks this protocol, calling the co-allocation API to implement co-allocation operations and perhaps providing additional functionality, such as authorization, fault tolerance, and logging. An application might then use the co-reservation service protocol to request end-to-end network reservations.

Collective components may be tailored to the requirements of a specific user community, VO, or application domain, for example, an SDK that implements an application-specific coherency protocol, or a co-reservation service for a specific set of network resources. Other Collective components can be more general purpose, for example, a replication service that manages an international collection of storage systems for multiple communities, or a directory service designed to enable the discovery of VOs. In general, the larger the target user community, the more important it is that a Collective component's protocol(s) and API(s) be standards based.

Globus Toolkit: In addition to the example services listed earlier in this section, many of which build on Globus Connectivity and Resource protocols, we mention the Meta Directory Service, which introduces Grid Information Index Servers (GIISs) to support arbitrary views on resource subsets, with the LDAP information protocol used to access resource-specific GRISs to obtain resource state and GRRP used for resource registration. Also, replica catalog and replica management services are used to support the management of dataset replicas in a Grid environment [35]. An on-line credential repository service ('MyProxy') provides secure storage for proxy credentials [52]. The DUROC co-allocation library provides an SDK and API for resource co-allocation [53].

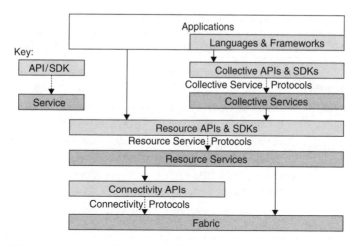

Figure 6.4 APIs are implemented by software development kits (SDKs), which in turn use Grid protocols to interact with network services that provide capabilities to the end user. Higher-level SDKs can provide functionality that is not directly mapped to a specific protocol but may combine protocol operations with calls to additional APIs as well as implement local functionality. Solid lines represent a direct call, dashed lines protocol interactions.

6.4.5 Applications

The final layer in our Grid architecture comprises the user applications that operate within a VO environment. Figure 6.4 illustrates an application programmer's view of Grid architecture. Applications are constructed in terms of, and by calling upon, services defined at any layer. At each layer, we have well-defined protocols that provide access to some useful service: resource management, data access, resource discovery, and so forth. At each layer, APIs may also be defined whose implementation (ideally provided by third-party SDKs) exchange protocol messages with the appropriate service(s) to perform desired actions.

We emphasize that what we label 'applications' and show in a single layer in Figure 6.4 may in practice call upon sophisticated frameworks and libraries (e.g., the Common Component Architecture [54], SCIRun [45], CORBA [55, 56], Cactus [57], workflow systems [58]) and feature much internal structure that would, if captured in our figure, expand it out to many times its current size. These frameworks may themselves define protocols, services, and/or APIs (e.g., the Simple Workflow Access Protocol [58]). However, these issues are beyond the scope of this article, which addresses only the most fundamental protocols and services required in a Grid.

6.5 GRID ARCHITECTURE IN PRACTICE

We use two examples to illustrate how Grid architecture functions in practice. Table 6.1 shows the services that might be used to implement the multidisciplinary simulation and cycle sharing (ray tracing) applications introduced in Figure 6.1. The basic Fabric elements

Table 6.1 The Grid services used to construct the two example applications of Figure 6.1

	Multidisciplinary Simulation	Ray Tracing
Collective (application-specific)	Solver coupler, distributed data archiver	Checkpointing, job management, failover, staging
Collective (generic)	Resource discovery, resource brokering, system monitoring, community authorization, certificate revocation	
Resource	Access to computation; access to data; access to information about system structure, state, performance.	
Connectivity	Communication (IP), service discovery (DNS), authentication, authorization, delegation	
Fabric	Storage systems, computers, networks, code repositories, catalogs	

are the same in each case: computers, storage systems, and networks. Furthermore, each resource speaks standard Connectivity protocols for communication and security and Resource protocols for enquiry, allocation, and management. Above this, each application uses a mix of generic and more application-specific Collective services.

In the case of the ray tracing application, we assume that this is based on a high-throughput computing system [25, 39]. In order to manage the execution of large numbers of largely independent tasks in a VO environment, this system must keep track of the set of active and pending tasks, locate appropriate resources for each task, stage executables to those resources, detect and respond to various types of failure, and so forth. An implementation in the context of our Grid architecture uses both domain-specific Collective services (dynamic checkpoint, task pool management, failover) and more generic Collective services (brokering, data replication for executables and common input files), as well as standard Resource and Connectivity protocols. Condor-G represents a first step toward this goal [39].

In the case of the multidisciplinary simulation application, the problems are quite different at the highest level. Some application framework (e.g., CORBA, CCA) may be used to construct the application from its various components. We also require mechanisms for discovering appropriate computational resources, for reserving time on those resources, for staging executables (perhaps), for providing access to remote storage, and so forth. Again, a number of domain-specific Collective services will be used (e.g., solver coupler, distributed data archiver), but the basic underpinnings are the same as in the ray tracing example.

6.6 'ON THE GRID': THE NEED FOR INTERGRID PROTOCOLS

Our Grid architecture establishes requirements for the protocols and APIs that enable sharing of resources, services, and code. It does not otherwise constrain the technologies that might be used to implement these protocols and APIs. In fact, it is quite feasible to define multiple instantiations of key Grid architecture elements. For example, we can construct both Kerberos-and PKI-based protocols at the Connectivity layer – and access these security mechanisms via the same API, thanks to GSS-API (see Appendix). However, Grids constructed with these different protocols are not interoperable and cannot share essential services – at least not without gateways. For this reason, the long-term success of Grid computing requires that we select and achieve widespread deployment of one set of protocols at the Connectivity and Resource layers – and, to a lesser extent, at the Collective layer. Much as the core Internet protocols enable different computer networks to interoperate and exchange information, these *Intergrid protocols* (as we might call them) enable different organizations to interoperate and exchange or share resources. Resources that speak these protocols can be said to be 'on the Grid.' Standard APIs are also highly useful if Grid code is to be shared. The identification of these Intergrid protocols and APIs is beyond the scope of this article, although the Globus Toolkit represents an approach that has had some success to date.

6.7 RELATIONSHIPS WITH OTHER TECHNOLOGIES

The concept of controlled, dynamic sharing within VOs is so fundamental that we might assume that Gridlike technologies must surely already be widely deployed. In practice, however, while the need for these technologies is indeed widespread, in a wide variety of different areas we find only primitive and inadequate solutions to VO problems. In brief, current distributed computing approaches do not provide a general resource-sharing framework that addresses VO requirements. Grid technologies distinguish themselves by providing this generic approach to resource sharing. This situation points to numerous opportunities for the application of Grid technologies.

6.7.1 World Wide Web

The ubiquity of Web technologies (i.e., IETF and W3C standard protocols – TCP/IP, HTTP, SOAP, etc. – and languages, such as HTML and XML) makes them attractive as a platform for constructing VO systems and applications. However, while these technologies do an excellent job of supporting the browser-client-to-Web-server interactions that are the foundation of today's Web, they lack features required for the richer interaction models that occur in VOs. For example, today's Web browsers typically use TLS for authentication but do not support single sign-on or delegation.

Clear steps can be taken to integrate Grid and Web technologies. For example, the single sign-on capabilities provided in the GSI extensions to TLS would, if integrated into Web browsers, allow for single sign-on to multiple Web servers. GSI delegation capabilities would permit a browser client to delegate capabilities to a Web server so that the server could act on the client's behalf. These capabilities, in turn, make it much easier to use Web technologies to build 'VO portals' that provide thin client interfaces to sophisticated VO applications. WebOS addresses some of these issues [59].

6.7.2 Application and storage service providers

Application service providers, storage service providers, and similar hosting companies typically offer to outsource specific business and engineering applications (in the case of ASPs) and storage capabilities (in the case of SSPs). A customer negotiates a service-level agreement that defines access to a specific combination of hardware and software. Security tends to be handled by using VPN technology to extend the customer's intranet to encompass resources operated by the ASP or SSP on the customer's behalf. Other SSPs offer file-sharing services, in which case access is provided via HTTP, FTP, or WebDAV with user ids, passwords, and access control lists controlling access.

From a VO perspective, these are low-level building-block technologies. VPNs and static configurations make many VO sharing modalities hard to achieve. For example, the use of VPNs means that it is typically impossible for an ASP application to access data located on storage managed by a separate SSP. Similarly, dynamic reconfiguration of resources within a single ASP or SSP is challenging and, in fact, is rarely attempted. The load sharing across providers that occurs on a routine basis in the electric power industry is unheard of in the hosting industry. A basic problem is that a VPN is not a VO: it cannot extend dynamically to encompass other resources and does not provide the remote resource provider with any control of when and whether to share its resources.

The integration of Grid technologies into ASPs and SSPs can enable a much richer range of possibilities. For example, standard Grid services and protocols can be used to achieve a decoupling of the hardware and software. A customer could negotiate an SLA for particular hardware resources and then use Grid resource protocols to dynamically provision that hardware to run customer-specific applications. Flexible delegation and access control mechanisms would allow a customer to grant an application running on an ASP computer direct, efficient, and securely access to data on SSP storage – and/or to couple resources from multiple ASPs and SSPs with their own resources, when required for more complex problems. A single sign-on security infrastructure able to span multiple security domains dynamically is, realistically, required to support such scenarios. Grid resource management and accounting/payment protocols that allow for dynamic provisioning and reservation of capabilities (e.g., amount of storage, transfer bandwidth) are also critical.

6.7.3 Enterprise computing systems

Enterprise development technologies such as CORBA, Enterprise Java Beans, Java 2 Enterprise Edition, and DCOM are all systems designed to enable the construction of distributed applications. They provide standard resource interfaces, remote invocation mechanisms, and trading services for discovery and hence make it easy to share resources within a single organization. However, these mechanisms address none of the specific VO requirements listed above. Sharing arrangements are typically relatively static and restricted to occur within a single organization. The primary form of interaction is client-server, rather than the coordinated use of multiple resources.

These observations suggest that there should be a role for Grid technologies within enterprise computing. For example, in the case of CORBA, we could construct an object request broker (ORB) that uses GSI mechanisms to address cross-organizational security issues. We could implement a Portable Object Adaptor that speaks the Grid resource management protocol to access resources spread across a VO. We could construct Grid-enabled Naming and Trading services that use Grid information service protocols to query information sources distributed across large VOs. In each case, the use of Grid protocols provides enhanced capability (e.g., interdomain security) and enables interoperability with other (non-CORBA) clients. Similar observations can be made about Java and Jini. For example, Jini's protocols and implementation are geared toward a small collection of devices. A 'Grid Jini' that employed Grid protocols and services would allow the use of Jini abstractions in a large-scale, multienterprise environment.

6.7.4 Internet and peer-to-peer computing

Peer-to-peer computing (as implemented, for example, in the Napster, Gnutella, and Freenet [60] file sharing systems) and Internet computing (as implemented, for example, by the SETI@home, Parabon, and Entropia systems) is an example of the more general ('beyond client-server') sharing modalities and computational structures that we referred to in our characterization of VOs. As such, they have much in common with Grid technologies.

In practice, we find that the technical focus of work in these domains has not overlapped significantly to date. One reason is that peer-to-peer and Internet computing developers have so far focused entirely on vertically integrated ('stovepipe') solutions, rather than seeking to define common protocols that would allow for shared infrastructure and interoperability. (This is, of course, a common characteristic of new market niches, in which participants still hope for a monopoly.) Another is that the forms of sharing targeted by various applications are quite limited, for example, file sharing with no access control, and computational sharing with a centralized server.

As these applications become more sophisticated and the need for interoperability becomes clearer, we will see a strong convergence of interests between peer-to-peer, Internet, and Grid computing [61]. For example, single sign-on, delegation, and authorization technologies become important when computational and data-sharing services must interoperate, and the policies that govern access to individual resources become more complex.

6.8 OTHER PERSPECTIVES ON GRIDS

The perspective on Grids and VOs presented in this article is of course not the only view that can be taken. We summarize here – and critique – some alternative perspectives (given in italics).

The Grid is a next-generation Internet: 'The Grid' is not an alternative to 'the Internet': it is rather a set of additional protocols and services that build on Internet protocols and services to support the creation and use of computation- and data-enriched environments. Any resource that is 'on the Grid' is also, by definition, 'on the Net.'

The Grid is a source of free cycles: Grid computing does not imply unrestricted access to resources. Grid computing is about controlled sharing. Resource owners will typically want to enforce policies that constrain access according to group membership, ability to pay, and so forth. Hence, accounting is important, and a Grid architecture must incorporate resource and collective protocols for exchanging usage and cost information, as well as for exploiting this information when deciding whether to enable sharing.

The Grid requires a distributed operating system: In this view (e.g., see [62]), Grid software should define the operating system services to be installed on every participating system, with these services providing for the Grid what an operating system provides for a single computer: namely, transparency with respect to location, naming, security, and so forth. Put another way, this perspective views the role of Grid software as defining a virtual machine. However, we feel that this perspective is inconsistent with our primary goals of broad deployment and interoperability. We argue that the appropriate model is rather the Internet Protocol suite, which provides largely orthogonal services that address the unique concerns that arise in networked environments. The tremendous physical and administrative heterogeneities encountered in Grid environments means that the traditional transparencies are unobtainable; on the other hand, it does appear feasible to obtain agreement on standard protocols. The architecture proposed here is deliberately open rather than prescriptive: it defines a compact and minimal set of protocols that a resource must speak to be 'on the Grid'; beyond this, it seeks only to provide a framework within which many behaviors can be specified.

The Grid requires new programming models: Programming in Grid environments introduces challenges that are not encountered in sequential (or parallel) computers, such as multiple administrative domains, new failure modes, and large variations in performance. However, we argue that these are incidental, not central, issues and that the basic programming problem is not fundamentally different. As in other contexts, abstraction and encapsulation can reduce complexity and improve reliability. But, as in other contexts, it is desirable to allow a wide variety of higher-level abstractions to be constructed, rather than enforcing a particular approach. So, for example, a developer who believes that a universal distributed shared-memory model can simplify Grid application development should implement this model in terms of Grid protocols, extending or replacing those protocols only if they prove inadequate for this purpose. Similarly, a developer who believes that all Grid resources should be presented to users as objects needs simply to implement an object-oriented API in terms of Grid protocols.

The Grid makes high-performance computers superfluous: The hundreds, thousands, or even millions of processors that may be accessible within a VO represent a significant source of computational power, if they can be harnessed in a useful fashion. This does not imply, however, that traditional high-performance computers are obsolete. Many problems require tightly coupled computers, with low latencies and high communication bandwidths; Grid computing may well increase, rather than reduce, demand for such systems by making access easier.

6.9 SUMMARY

We have provided in this article a concise statement of the 'Grid problem,' which we define as controlled and coordinated resource sharing and resource use in dynamic, scalable virtual organizations. We have also presented both requirements and a framework for a Grid architecture, identifying the principal functions required to enable sharing within VOs and defining key relationships among these different functions. Finally, we have discussed in some detail how Grid technologies relate to other important technologies.

We hope that the vocabulary and structure introduced in this document will prove useful to the emerging Grid community, by improving understanding of our problem and providing a common language for describing solutions. We also hope that our analysis will help establish connections among Grid developers and proponents of related technologies.

The discussion in this paper also raises a number of important questions. What are appropriate choices for the Intergrid protocols that will enable interoperability among Grid systems? What services should be present in a persistent fashion (rather than being duplicated by each application) to create usable Grids? And what are the key APIs and SDKs that must be delivered to users in order to accelerate development and deployment of Grid applications? We have our own opinions on these questions, but the answers clearly require further research.

ACKNOWLEDGMENTS

We are grateful to numerous colleagues for discussions on the topics covered here, in particular, Bill Allcock, Randy Butler, Ann Chervenak, Karl Czajkowski, Steve Fitzgerald, Bill Johnston, Miron Livny, Joe Mambretti, Reagan Moore, Harvey Newman, Laura Pearlman, Rick Stevens, Gregor von Laszewski, Rich Wellner, and Mike Wilde, and participants in the workshop on Clusters and Computational Grids for Scientific Computing (Lyon, September 2000) and the 4th Grid Forum meeting (Boston, October 2000), at which early versions of these ideas were presented.

This work was supported in part by the Mathematical, Information, and Computational Sciences Division subprogram of the Office of Advanced Scientific Computing Research, United States Department of Energy, under Contract W-31-109-Eng-38; by the Defense Advanced Research Projects Agency under contract N66001-96-C-8523; by the National Science Foundation; and by the NASA Information Power Grid program.

APPENDIX: DEFINITIONS

We define here four terms that are fundamental to the discussion in this article but are frequently misunderstood and misused, namely, protocol, service, SDK, and API.

Protocol: A *protocol* is a set of rules that end points in a telecommunication system use when exchanging information. For example:

- The Internet Protocol (IP) defines an unreliable packet transfer protocol.
- The Transmission Control Protocol (TCP) builds on IP to define a reliable data delivery protocol.
- The Transport Layer Security (TLS) protocol [31] defines a protocol to provide privacy and data integrity between two communicating applications. It is layered on top of a reliable transport protocol such as TCP.
- The Lightweight Directory Access Protocol (LDAP) builds on TCP to define a query-response protocol for querying the state of a remote database.

An important property of protocols is that they admit to multiple implementations: two end points need only implement the same protocol to be able to communicate. Standard protocols are thus fundamental to achieving *interoperability* in a distributed computing environment.

A protocol definition also says little about the behavior of an entity that speaks the protocol. For example, the FTP protocol definition indicates the format of the messages used to negotiate a file transfer but does not make clear how the receiving entity should manage its files.

As the above examples indicate, protocols may be defined in terms of other protocols.

Service: A *service* is a network-enabled entity that provides a specific capability, for example, the ability to move files, create processes, or verify access rights. A service is defined in terms of the protocol one uses to interact with it and the behavior expected in response to various protocol message exchanges (i.e., 'service = protocol + behavior.'). A service definition may permit a variety of implementations. For example:

- An FTP server speaks the File Transfer Protocol and supports remote read and write access to a collection of files. One FTP server implementation may simply write to and read from the server's local disk, while another may write to and read from a mass storage system, automatically compressing and uncompressing files in the process. From a Fabric-level perspective, the behaviors of these two servers in response to a store request (or retrieve request) are very different. From the perspective of a client of this service, however, the behaviors are indistinguishable; storing a file and then retrieving the same file will yield the same results regardless of which server implementation is used.
- An LDAP server speaks the LDAP protocol and supports response to queries. One LDAP server implementation may respond to queries using a database of information, while another may respond to queries by dynamically making SNMP calls to generate the necessary information on the fly.

A service may or may not be persistent (i.e., always available), be able to detect and/or recover from certain errors, run with privileges, and/or have a distributed implementation for enhanced scalability. If variants are possible, then *discovery* mechanisms that allow a client to determine the properties of a particular instantiation of a service are important.

Note also that one can define different services that speak the same protocol. For example, in the Globus Toolkit, both the replica catalog [35] and information service [33] use LDAP.

API: An Application Program Interface (API) defines a standard interface (e.g., set of subroutine calls, or objects and method invocations in the case of an object-oriented API) for invoking a specified set of functionality. For example:

- The generic security service (GSS) API [36] defines standard functions for verifying identify of communicating parties, encrypting messages, and so forth.
- The message-passing interface API [63] defines standard interfaces, in several languages, to functions used to transfer data among processes in a parallel computing system.

An API may define multiple language bindings or use an interface definition language. The language may be a conventional programming language such as C or Java, or it may be a shell interface. In the latter case, the API refers to particular a definition of command line arguments to the program, the input and output of the program, and the exit status of the program. An API normally will specify a standard behavior but can admit to multiple implementations.

It is important to understand the relationship between APIs and protocols. A protocol definition says nothing about the APIs that might be called from within a program to generate protocol messages. A single protocol may have many APIs; a single API may have multiple implementations that target different protocols. In brief, standard APIs enable *portability*; standard protocols enable *interoperability*. For example, both public key and Kerberos bindings have been defined for the GSS-API [36]. Hence, a program that uses GSS-API calls for authentication operations can operate in either a public key *or* a Kerberos environment without change. On the other hand, if we want a program to operate in a public key *and* a Kerberos environment at the same time, then we need a standard protocol that supports interoperability of these two environments. See Figure 6.5.

SDK: The term software development kit (SDK) denotes a set of code designed to be linked with, and invoked from within, an application program to provide specified functionality. An SDK typically implements an API. If an API admits to multiple implementations, then there will be multiple SDKs for that API. Some SDKs provide access to services via a particular protocol. For example:

- The OpenLDAP release includes an LDAP client SDK, which contains a library of functions that can be used from a C or C + + application to perform queries to an LDAP service.

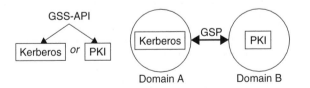

Figure 6.5 On the left, an API is used to develop applications that can target either Kerberos or PKI security mechanisms. On the right, protocols (the Grid security protocols provided by the Globus Toolkit) are used to enable interoperability between Kerberos and PKI domains.

- JNDI is a Java SDK that contains functions that can be used to perform queries to an LDAP service.
- Different SDKs implement GSS-API using the TLS and Kerberos protocols, respectively.

There may be multiple SDKs, for example, from multiple vendors, that implement a particular protocol. Further, for client-server oriented protocols, there may be separate client SDKs for use by applications that want to access a service and server SDKs for use by service implementers that want to implement particular, customized service behaviors.

An SDK need not speak any protocol. For example, an SDK that provides numerical functions may act entirely locally and not need to speak to any services to perform its operations.

REFERENCES

1. Foster, I. and Kesselman, C. (eds) (1999) *The Grid: Blueprint for a New Computing Infrastructure*. Morgan Kaufmann.
2. Beiriger, J., Johnson, W., Bivens, H., Humphreys, S. and Rhea, R. (2000) Constructing the ASCI Grid. *Proceedings of the 9th IEEE Symposium on High Performance Distributed Computing*, IEEE Press.
3. Brunett, S. *et al.* (1998) Application experiences with the Globus Toolkit. *Proceedings of the 7th IEEE Symposium on High Performance Distributed Computing*, IEEE Press, pp. 81–89.
4. Johnston, W. E., Gannon, D. and Nitzberg, B. (1999) Grids as production computing environments: The engineering aspects of NASA's Information Power Grid. *Proceedings of the 8th IEEE Symposium on High Performance Distributed Computing*, IEEE Press.
5. Stevens, R., Woodward, P., DeFanti, T. and Catlett, C. (1997) From the I-WAY to the National Technology Grid. *Communications of the ACM*, **40**(11), 50–61.
6. Sculley, A. and Woods, W. (2000) *B2B Exchanges: The Killer Application in the Business-to-Business Internet Revolution*. ISI Publications.
7. Barry, J. *et al.* (1998) NIIIP-SMART: An investigation of distributed object approaches to support MES development and deployment in a virtual enterprise. *2nd International Enterprise Distributed Computing Workshop*, IEEE Press.
8. Chervenak, A., Foster, I., Kesselman, C., Salisbury, C. and Tuecke, S. (2001) The Data Grid: Towards an architecture for the distributed management and analysis of large scientific data sets. *Journal of Network and Computer Applications*.
9. Hoschek, W., Jaen-Martinez, J., Samar, A., Stockinger, H. and Stockinger, K. (2000) Data management in an international Data Grid project. *Proceedings of the 1st IEEE/ACM International Workshop on Grid Computing*, Springer Verlag Press.
10. Moore, R., Baru, C., Marciano, R., Rajasekar, A. and Wan, M. (1999) Data-intensive computing, in Foster, I. and Kesselman, C. (eds) *The Grid: Blueprint for a New Computing Infrastructure*. Morgan Kaufmann, pp. 105–129.
11. Camarinha-Matos, L. M., Afsarmanesh, H., Garita, C. and Lima, C. Towards an architecture for virtual enterprises. *Journal of Intelligent Manufacturing*.
12. Aiken, R. *et al.* (2000) Network Policy and Services: A Report of a Workshop on Middleware, RFC 2768, IETF, http://www.ietf.org/rfc/rfc2768.txt.
13. Arnold, K., O'Sullivan, B., Scheifler, R. W., Waldo, J. and Wollrath, A. (1999) *The Jini Specification*. Addison-Wesley, See also www.sun.com/jini.
14. (1994) *Realizing the Information Future: The Internet and Beyond*. National Academy Press, http://www.nap.edu/readingroom/books/rtif/.

15. Foster, I. and Kesselman, C. (1998) The Globus Project: A status report. *Proceedings of the Heterogeneous Computing Workshop*, IEEE Press, pp. 4–18.
16. Tierney, B., Johnston, W., Lee, J. and Hoo, G. (1996) Performance analysis in high-speed wide area IP over ATM networks: Top-to-bottom end-to-end monitoring. *IEEE Networking*.
17. Beynon, M., Ferreira, R., Kurc, T., Sussman, A. and Saltz, J. (2000) DataCutter: middleware for filtering very large scientific datasets on archival storage systems. *Proceedings of the 8th Goddard Conference on Mass Storage Systems and Technologies/17th IEEE Symposium on Mass Storage Systems*, pp. 119–133.
18. Baru, C., Moore, R., Rajasekar, A. and Wan, M. (1998) The SDSC storage resource broker. *Proceedings of the CASCON'98 Conference*, 1998.
19. Dinda, P. and O'Hallaron, D. (1999) An evaluation of linear models for host load prediction. *Proceedings of the 8th IEEE Symposium on High-Performance Distributed Computing*, IEEE Press.
20. Lowekamp, B., Miller, N., Sutherland, D., Gross, T., Steenkiste, P. and Subhlok, J. (1998) A resource query interface for network-aware applications. *Proceedings of the 7th IEEE Symposium on High-Performance Distributed Computing*, IEEE Press.
21. Wolski, R. (1997) Forecasting network performance to support dynamic scheduling using the Network Weather Service. *Proceedings of the 6th IEEE Symposium on High Performance Distributed Computing*, Portland, Oregon, 1997.
22. Foster, I., Roy, A. and Sander, V. (2000) A quality of service architecture that combines resource reservation and application adaptation. *Proceedings of the 8th International Workshop on Quality of Service*, 2000.
23. Papakhian, M. (1998) Comparing job-management systems: The user's perspective. *IEEE Computational Science & Engineering*, See also http://pbs.mrj.com.
24. Litzkow, M., Livny, M. and Mutka, M. (1988) Condor – a hunter of idle workstations. *Proceedings of the 8th International Conference on Distributed Computing Systems*, pp. 104–111.
25. Livny, M. (1999) High-throughput resource management, in Foster, I. and Kesselman, C. (eds) *The Grid: Blueprint for a New Computing Infrastructure*. Morgan Kaufmann, pp. 311–337.
26. Baker, F. (1995) Requirements for IP Version 4 Routers, RFC 1812, IETF, http://www.ietf.org/rfc/rfc1812.txt.
27. Butler, R., Engert, D., Foster, I., Kesselman, C., Tuecke, S., Volmer, J. and Welch, V. (2000) Design and deployment of a national-scale authentication infrastructure. *IEEE Computer*, **33**(12), 60–66.
28. Foster, I., Kesselman, C., Tsudik, G. and Tuecke, S. (1998) A security architecture for computational Grids. *ACM Conference on Computers and Security*, pp. 83–91.
29. Gasser, M. and McDermott, E. (1990) An architecture for practical delegation in a distributed system. *Proceedings of the 1990 IEEE Symposium on Research in Security and Privacy*, IEEE Press, pp. 20–30.
30. Howell, J. and Kotz, D. (2000) End-to-end authorization. *Proceedings of the 2000 Symposium on Operating Systems Design and Implementation*, USENIX Association.
31. Dierks, T. and Allen, C. (1999) The TLS Protocol Version 1.0, RFC 2246, IETF, http://www.ietf.org/rfc/rfc2246.txt.
32. Steiner, J., Neuman, B. C. and Schiller, J. (1988) Kerberos: An authentication system for open network systems. *Proceedings of the Usenix Conference*, pp. 191–202.
33. Czajkowski, K., Fitzgerald, S., Foster, I. and Kesselman, C. (2001) Grid Information Services for Distributed Resource Sharing, 2001.
34. Czajkowski, K., Foster, I., Karonis, N., Kesselman, C., Martin, S., Smith, W. and Tuecke, S. (1998) A resource management architecture for metacomputing systems. *The 4th Workshop on Job Scheduling Strategies for Parallel Processing*, pp. 62–82.
35. Allcock, B. *et al.* (2001) Secure, efficient data transport and replica management for high-performance data-intensive computing. *Mass Storage Conference*, 2001.
36. Linn, J. (2000) Generic Security Service Application Program Interface Version 2, Update 1, RFC 2743, IETF, http://www.ietf.org/rfc/rfc2743.

37. Berman, F. (1999) High-performance schedulers, in Foster, I. and Kesselman, C. (eds) *The Grid: Blueprint for a New Computing Infrastructure*. Morgan Kaufmann, pp. 279–309.
38. Berman, F., Wolski, R., Figueira, S., Schopf, J. and Shao, G. (1996) Application-level scheduling on distributed heterogeneous networks. *Proceedings of the Supercomputing '96*, 1996.
39. Frey, J., Foster, I., Livny, M., Tannenbaum, T. and Tuecke, S. (2001) *Condor-G: A Computation Management Agent for Multi-Institutional Grids*. University of Wisconsin Madison.
40. Abramson, D., Sosic, R., Giddy, J. and Hall, B. (1995) Nimrod: A tool for performing parameterized simulations using distributed workstations. *Proceedings of the 4th IEEE Symposium on High Performance Distributed Computing*, 1995.
41. Foster, I. and Karonis, N. (1998) A Grid-enabled MPI: Message passing in heterogeneous distributed computing systems. *Proceedings of the SC'98*, 1998.
42. Gabriel, E., Resch, M., Beisel, T. and Keller, R. (1998) Distributed computing in a heterogeneous computing environment. *Proceedings of the EuroPVM/MPI'98*, 1998.
43. Casanova, H., Obertelli, G., Berman, F. and Wolski, R. (2000) The AppLeS parameter sweep template: User-level middleware for the Grid. *Proceedings of the SC'2000*, 2000.
44. Goux, J.-P., Kulkarni, S., Linderoth, J. and Yoder, M. (2000) An enabling framework for master-worker applications on the computational Grid. *Proceedings of the 9th IEEE Symposium on High Performance Distributed Computing*, IEEE Press.
45. Casanova, H., Dongarra, J., Johnson, C. and Miller, M. (1999) Application-specific tools, in Foster, I. and Kesselman, C. (eds) *The Grid: Blueprint for a New Computing Infrastructure*. Morgan Kaufmann, pp. 159–180.
46. Casanova, H. and Dongarra, J. (1997) NetSolve: A network server for solving computational science problems. *International Journal of Supercomputer Applications and High Performance Computing*, **11**(3), 212–223.
47. Nakada, H., Sato, M. and Sekiguchi, S. (1999) Design and implementations of Ninf: Towards a global computing infrastructure, *Future Generation Computing Systems*.
48. Thompson, M., Johnston, W., Mudumbai, S., Hoo, G., Jackson, K. and Essiari, A. (1999) Certificate-based access control for widely distributed resources. *Proceedings of the 8th Usenix Security Symposium*, 1999.
49. DeFanti, T. and Stevens, R. (1999) Teleimmersion, in Foster, I. and Kesselman, C. (eds) *The Grid: Blueprint for a New Computing Infrastructure*. Morgan Kaufmann, pp. 131–155.
50. Leigh, J., Johnson, A. and DeFanti, T. A. CAVERN: A distributed architecture for supporting scalable persistence and interoperability in collaborative virtual environments. *Virtual Reality: Research, Development and Applications*, **2**(2), 217–237.
51. Childers, L., Disz, T., Olson, R., Papka, M. E., Stevens, R. and Udeshi, T. (2000) Access Grid: Immersive group-to-group collaborative visualization. *Proceedings of the 4th International Immersive Projection Technology Workshop*, 2000.
52. Novotny, J., Tuecke, S. and Welch, V. (2001) Initial Experiences with an Online Certificate Repository for the Grid: MyProxy.
53. Czajkowski, K., Foster, I. and Kesselman, C. (1999) Coallocation services for computational Grids. *Proceedings of the 8th IEEE Symposium on High Performance Distributed Computing*, IEEE Press.
54. Armstrong, R., Gannon, D., Geist, A., Keahey, K., Kohn, S., McInnes, L. and Parker, S. (1999) Toward a common component architecture for high performance scientific computing. *Proceedings of the 8th IEEE Symposium on High Performance Distributed Computing*, 1999.
55. Gannon, D. and Grimshaw, A. (1999) Object-based approaches, in Foster, I. and Kesselman, C. (eds) *The Grid: Blueprint for a New Computing Infrastructure*. Morgan Kaufmann, pp. 205–236.
56. Lopez, I. *et al.* (2000) NPSS on NASA's IPG: Using CORBA and Globus to coordinate multidisciplinary aeroscience applications. *Proceedings of the NASA HPCC/CAS Workshop*, NASA Ames Research Center.
57. Benger, W., Foster, I., Novotny, J., Seidel, E., Shalf, J., Smith, W. and Walker, P. (1999) Numerical relativity in a distributed environment. *Proceedings of the 9th SIAM Conference on Parallel Processing for Scientific Computing*, 1999.

58. Bolcer, G. A. and Kaiser, G. (1999) SWAP: Leveraging the web to manage workflow. *IEEE Internet Computing*, 85–88.
59. Vahdat, A., Belani, E., Eastham, P., Yoshikawa, C., Anderson, T., Culler, D. and Dahlin, M. 1998 WebOS: Operating system services for wide area applications. *7th Symposium on High Performance Distributed Computing*, July 1998.
60. Clarke, I., Sandberg, O., Wiley, B. and Hong, T. W. (1999) Freenet: A distributed anonymous information storage and retrieval system. *ICSI Workshop on Design Issues in Anonymity and Unobservability*, 1999.
61. Foster, I. (2000) Internet computing and the emerging Grid, *Nature Web Matters*. http://www.nature.com/nature/webmatters/grid/grid.html.
62. Grimshaw, A. and Wulf, W. (1996) Legion – a view from 50,000 feet. *Proceedings of the 5th IEEE Symposium on High Performance Distributed Computing*, IEEE Press, pp. 89–99.
63. Gropp, W., Lusk, E. and Skjellum, A. (1994) *Using MPI: Portable Parallel Programming with the Message Passing Interface*. MIT Press.

Rationale for choosing the Open Grid Services Architecture

Malcolm Atkinson

National e-Science Centre, Edinburgh, Scotland, United Kingdom

7.1 INTRODUCTION

This chapter presents aspects of the UK e-Science communities' plans for generic Grid middleware. In particular, it derives from the discussions of the UK Architecture Task Force [1].

The UK e-Science Core Programme will focus on architecture and middleware development in order to contribute significantly to the emerging Open Grid Services Architecture (OGSA) [2]. This architecture views Grid technology as a generic integration mechanism assembled from Grid Services (GS), which are an extension of Web Services (WS) to comply with additional Grid requirements. The principal extensions from WS to GS are the management of state, identification, sessions and life cycles and the introduction of a notification mechanism in conjunction with Grid service data elements (SDE).

The UK e-Science Programme has many pilot projects that require integration technology and has an opportunity through its Core Programme to lead these projects towards adopting OGSA as a common framework. That framework must be suitable, for example, it must support adequate Grid service interoperability and portability. It must also be

Grid Computing – Making the Global Infrastructure a Reality. Edited by F. Berman, A. Hey and G. Fox
© 2003 John Wiley & Sons, Ltd   ISBN: 0-470-85319-0

populated with services that support commonly required functions, such as authorisation, accounting and data transformation.

To obtain effective synergy with the international community that is developing Grid standards and to best serve the United Kingdom's community of scientists, it is necessary to focus the United Kingdom's middleware development resources on a family of GS for which the United Kingdom is primarily responsible and to deliver their reference implementations. The UK e-Science and computing science community is well placed to contribute substantially to *structured* data integration services [3–15]. Richer information models should be introduced at the earliest opportunity to progressively approach the goal of a semantic Grid (see Chapter 17). The UK e-Science community also recognises an urgent need for accounting mechanisms and has the expertise to develop them in conjunction with international efforts.

This chapter develops the rationale for working with OGSA and a plan for developing commonly required middleware complementary to the planned baseline Globus Toolkit 3 provision. It takes the development of services for accessing and integrating structured data via the Grid as an example and shows how this will map to GS.

7.2 THE SIGNIFICANCE OF DATA FOR e-SCIENCE

The fundamental goal of the e-Science programme is to enable scientists to perform their science more effectively. The methods and principles of e-Science should become so pervasive that scientists can use them naturally whenever they are appropriate just as they use mathematics today. The goal is to arrive at the state where we just say 'science'. Just as there are branches of mathematics that support different scientific domains, so will there be differentiated branches of computation. We are in a pioneering phase, in which the methods and principles must be elucidated and made accessible and in which the differentiation of domain requirements must be explored. We are confident that, as with mathematics, these results will have far wider application than the scientific testing ground where we are developing them.

The transition that we are catalysing is driven by technology and is largely manifest in the tsunami of data (see Chapter 36). Detectors and instruments benefit from Moore's law, so that in astronomy for instance, the available data is doubling every year [16, 17]. Robotics and nanoengineering accelerates and multiplies the output from laboratories. For example, the available genetic sequence data is doubling every nine months [16]. The volume of data we can store at a given cost doubles each year. The rate at which we can move data is doubling every nine months. Mobile sensors, satellites, ocean-exploring robots, clouds of disposable micro-sensors, personal-health sensors, combined with digital radio communication are rapidly extending the sources of data.

These changes warrant a change in scientific behaviour. The norm should be to collect, annotate, curate and share data. This is already a trend in subjects such as large-scale physics, astronomy, functional genomics and earth sciences. But perhaps it is not yet as prevalent as it should be. For example, the output of many confocal microscopes, the raw data from many micro-arrays and the streams of data from automated pathology labs and digital medical scanners, do not yet appear as a matter of course for scientific use and analysis. It is reasonable to assume that if the benefits of data mining and correlating data from multiple sources become widely recognised, more data will be available in shared, often public, repositories.

This wealth of data has enormous potential. Frequently, data contains information relevant to many more topics than the specific science, engineering or medicine that motivated its original collection and determined its structure. If we are able to compose and study these large collections of data for correlations and anomalies, they may yield an era of rapid scientific, technological and medical progress. But discovering the valuable knowledge from the mountains of data is well beyond unaided human capacity. Sophisticated computational approaches must be developed. Their application will require the skills of scientists, engineers, computer scientists, statisticians and many other experts. Our challenge is to enable both the development of the sophisticated computation and the collaboration of all of those who should steer it. The whole process must be attainable by the majority of scientists, sustainable within a typical economy and trustable by scientists, politicians and the general public. Developing the computational approaches and the practices that exploit them will surely be one of the major differentiated domains of e-Science support.

The challenge of making good use of growing volumes of diverse data is not exclusive to science and medicine. In government, business, administration, health care, the arts and humanities, we may expect to see similar challenges and similar advantages in mastering those challenges. Basing decisions, judgements and understanding on reliable tests against trustworthy data must benefit industrial, commercial, scientific and social goals. It requires an infrastructure to support the sharing, integration, federation and analysis of data.

7.3 BUILDING ON AN OGSA PLATFORM

The OGSA emerged contemporaneously with the UK e-Science review of architecture and was a major and welcome influence. OGSA is the product of combining the flexible, dynamically bound integration architecture of WS with the scalable distributed architecture of the Grid. As both are still evolving rapidly, discussion must be hedged with the caveat that significant changes to OGSA's definition will have occurred by the time this chapter is read.

OGSA is well described in other chapters of this book (see Chapter 8) and has been the subject of several reviews, for example, References [18, 19]. It is considered as the basis for a data Grid (see Chapter 15) and is expected to emerge as a substantial advance over the existing Globus Toolkit (GT2) and as the basis for a widely adopted Grid standard.

7.3.1 Web services

Web Services are an emerging integration architecture designed to allow independently operated information systems to intercommunicate. Their definition is the subject of W3C-standards processes in which major companies, for example, IBM, Oracle, Microsoft and SUN, are participating actively. WS are described well in Reference [20], which offers the following definition:

'A Web service is a *platform and implementation independent* software component that can be

- *described* using a service description language,
- *published* to a registry of services,
- *discovered* through a standard mechanism (at run time or design time),
- *invoked* through a declared Application Programming Interface (API), usually over a network,
- *composed* with other services.'

WS are of interest to the e-Science community on two counts:

1. Their function of interconnecting information systems is similar to the Grid's intended function. Such interconnection is a common requirement as scientific systems are often composed using many existing components and systems.
2. The support of companies for Web services standards will deliver description languages, platforms, common services and software development tools. These will enable rapid development of Grid services and applications by providing a standard framework for describing and composing Web services and Grid services. They will also facilitate the commercialisation of the products from e-Science research.

An important feature of WS is the emergence of languages for describing aspects of the components they integrate that are independent from the implementation and platform technologies. They draw heavily on the power of XML Schema. For example, the Web Services Description Language (WSDL) is used to describe the function and interfaces (portTypes) of Web services and the Web Services Inspection Language (WSIL) is used to support simple registration and discovery systems. Simple Object Access Protocol (SOAP) is a common denominator interconnection language that transmits structured data across representational boundaries. There is currently considerable activity proposing revisions of these standards and additional languages for describing the integration and the coordination of WS, for describing quality-of-service properties and for extending Web service semantics to incorporate state, more sophisticated types for ports and transactions. It is uncertain what will emerge, though it is clear that the already strong support for distributed system integration will be strengthened. This will be useful for many of the integration tasks required to support e-Science.

Inevitably, the products lag behind the aspirations of the standards proposals and vary significantly. Nevertheless, they frequently include sophisticated platforms to support operations combined with powerful development tools. It is important that developers of e-Science applications take advantage of these. Consequently, the integration architectures used by e-Science should remain compatible with Web services and e-Science developers should consider carefully before they develop alternatives.

7.3.2 The Open Grid Services Architecture

As other chapters describe OGSA (see Chapter 8), it receives only minimal description here, mainly to introduce vocabulary for later sections. A system compliant with OGSA is built by composing GS. Each Grid service is also a Web service and is described by WSDL. Certain extensions to WSDL are proposed to allow Grid-inspired properties to be described, and these may be adopted for wider use in forthcoming standards. This extended version of WSDL is called *Grid Services Description Language* (GSDL).

To be a Grid service the component must implement certain portTypes, must comply with certain lifetime management requirements and must be uniquely identifiable by a Grid Service Handle (GSH) throughout its lifetime. The lifetime management includes a soft-state model to limit commitments, to avoid permanent resource loss when partial failures occur and to guarantee autonomy. In addition, evolution of interfaces and function are supported via the Grid Service Record (GSR). This is obtainable via a mapping from a GSH, and has a time to live so that contracts that use it must be renewed. These properties are important to support long-running scalable distributed systems.

A Grid service may present some of its properties via SDE. These SDE may be static or dynamic. Those that are static are invariant for the lifetime of the Grid service they describe, and so may also be available via an encoding in an extension of WSDL in the GSR. Those that are dynamic present aspects of a Grid service's state. The SDE may be used for introspection, for example, by tools that generate glue code, and for monitoring to support functions such as performance and progress analysis, fault diagnosis and accounting. The SDE are described by XML Schema and may be queried by a simple tag, by a value pair model and by more advanced query languages. The values may not be stored as XML but synthesised on demand.

An event notification, publish and subscribe, mechanism is supported. This is associated with the SDE, so that the query languages may be used to specify interest.

The functions supported through the mandatory portTypes include authentication and registration/discovery.

7.4 CASE FOR OGSA

The authors of OGSA [2] expect the first implementation, Globus Toolkit 3 (GT3), to faithfully reproduce the semantics and the APIs of the current GT2, in order to minimise the perturbation of current projects. However, the influence of the thinking and the industrial momentum behind WS, and the need to achieve regularities that can be exploited by tools, will surely provoke profound changes in Grid implementations of the future. Indeed, OGSA is perceived as a good opportunity to restructure and re-engineer the Globus foundation technology. This will almost certainly be beneficial, but it will also surely engender semantically significant changes.

Therefore, because of the investment in existing Grid technology (e.g. GT2) by many application projects, the case for a major change, as is envisaged with OGSA, has to be compelling. The arguments for adopting OGSA as the direction in which to focus the development of future Grid technology concern three factors: politics, commerce and technology.

The *political case* for OGSA is that it brings together the efforts of the e-Science pioneers and the major software companies. This is essential for achieving widely accepted standards and the investment to build and sustain high-quality, dependable Grid infrastructure. Only with the backing of major companies will we meet the challenges of

- installing widespread support in the network and the operating system infrastructures,
- developing acceptance of *general* mechanisms for interconnection across boundaries between different authorities and
- obtaining interworking agreements between nations permitting the exchange of significant data via the Grid.

The companies will expect from the e-Science community a contribution to the political effort particularly through compelling demonstrations.

The *commercial case* is the route to a sustainable Grid infrastructure and adequate Grid programming tools, both of which are missing for the Grid at present because the e-Science community's resources are puny compared to the demands of building *and sustaining* comprehensive infrastructure and tool sets. If convergence can be achieved between the technology used in commercial applications for distributed software integration and that used for scientific applications, then a common integration platform can be jointly constructed and jointly maintained. As commerce is ineluctably much larger than the science base alone, this amortises those costs over a much larger community. Commerce depends on rapid deployment and efficient use of many application developers who are rarely experts in distributed systems. Yet it also depends on a growing number of ever more sophisticated distributed systems. It therefore has strong incentives to build tool sets and encapsulated services that would also benefit scientists if we share infrastructure, as we do today for computers, operating systems, compilers and network Internet protocol (IP) stacks.

A further commercial advantage emerges from the proposed convergence. It will be easier to rapidly transfer e-Science techniques to commerce and industry. Using a common platform, companies will have less novel technology to learn about, and therefore less assimilation costs and risks when they take up the products of e-Science research.

The *technological case* for OGSA is largely concerned with software engineering issues. The present set of components provided by the Grid has little structure to guide the application developers. This lack of explicit structure may also increase the costs of maintaining and extending the existing Grid infrastructure.

The discipline of defining Grid services in terms of a language (GSDL) and of imposing a set of common requirements on each Grid service should significantly improve the ease and the accuracy with which components can be composed. Those same disciplines will help Grid service developers to think about relevant issues and to deliver dependable components. We expect significant families of GS that adopt additional constraints on their definition and address a particular domain. Such families will have improved compositional properties, and tools that exploit these will be a natural adjunct.

Dynamic binding and rebinding with soft state are necessary for large-scale, long-running systems that are also flexible and evolvable. The common infrastructure and disciplines will be an appropriate foundation from which to develop

- tools, subsystems and portals to facilitate e-Science application development, taking advantage of the richer information available from the metadata describing Grid services,
- advances in the precision and detail of the infrastructure and the disciplines to yield dependable, predictable and trustworthy services.

7.5 THE CHALLENGE OF OGSA

To deliver the potential of OGSA many challenges have to be met. Sustaining the effort to achieve widely adopted standards that deliver the convergence of the WS and the Grid and rallying the resources to build high-quality implementations are obvious international challenges. Here we focus on more technical issues.

1. The types commonly needed for e-Science applications and for database integration need to be defined as XML Schema namespaces. If this is not done, different e-Science application groups will develop their own standards and a babel of types will result.
2. The precision required in GSDL definitions will need to be specified so that it is sufficient to support the planned activities. A succession of improved standards should be investigated, so that progressively more of the assembly of Grid services can be automated or at least automatically verified.
3. Standard platforms on which Grid services and their clients are developed.
4. The infrastructure for problem diagnosis and response (e.g. detecting, reporting, localising and recovering from partial failures) has to be defined.
5. The infrastructure for accounting within an assembly of Grid services.
6. The infrastructure for management and evolution. This would deliver facilities for limiting and controlling the behaviour of Grid services and facilities for dynamically replacing Grid service instances in an extensively distributed and continuously operating system.
7. Coordination services that some Grid services can participate in.
8. Definitions and compliance testing mechanisms so that it is possible to establish and monitor quality and completeness standards for Grid services.
9. Programming models that establish and support good programming practice for this scale of integration.
10. Support for Grid service instance migration to permit operational, organisational and administrative changes.
11. Support for intermittently connected mobile Grid services to enable the use of mobile computing resources by e-Scientists.

These issues already exist as challenges for the 'classical' Grid architecture. In some cases, Global Grid Forum (GGF) working groups are already considering them. OGSA provides an improved framework in which they may be addressed. For example, interfaces can be defined using GSDL to more precisely delimit a working group's area of activity.

7.6 PLANNING THE UNITED KINGDOM'S OGSA CONTRIBUTIONS

The UK e-Science Core Programme should coordinate middleware development to align with, influence and develop the OGSA. This will inevitably be a dynamic process; that is, an initial plan will need to be monitored and modified in response to contributions by other countries and by companies. The UK Grid middleware community must work closely with pilot projects to explore the potential of OGSA, to conduct evaluations and to share implementation effort.

Our initial plan proposed work on a number of sub-themes.

Phase I: Current actions to position the UK e-Science Core Programme middleware effort.

1. Understanding, validating and refining OGSA concepts and technical design.
2. Establishing a common context, types and a baseline set of Grid services.
3. Defining and prototyping baseline database access technology [21].
4. Initiating a Grid service validation and testing process.
5. Establishing baseline logging to underpin accounting functions.

Phase II: Advanced development and research.

1. Refining GSDL, for example specifying semantics, and developing tools that use it.
2. Pioneering higher-level data integration services en route to the semantic Grid.
3. Pioneering database integration technology.
4. Developing advanced forms of Grid Economies.

The work in Phases I and II must take into account a variety of engineering and design issues that are necessary to achieve affordable, viable, maintainable and trustworthy services. These include performance engineering, dependable engineering, engineering for change, manageability and operations support, and privacy, ethical and legal issues.

The following sections of the chapter expand some of the topics in this plan.

7.7 ESTABLISHING COMMON INFRASTRUCTURE

There are three parts to this: the common Grid services, namely computational context, the standard set of e-Science and Grid service types and the minimal set of Grid service primitives.

Developers building new Grid services or applications require to know which operations are always supported by a hosting environment. As code portability can only pertain within a single hosting environment, these operations may be specific to that hosting environment. For example, the operations for developers working within a J2EE hosting environment need not be the same as those for developers using C. However, there will be a pervasive baseline functionality, which will have various syntactic forms. The physiology chapter (see Chapter 8) describes it thus:

*... implementation of Grid services can be facilitated by specifying baseline charac-
teristics that all hosting environments must possess, defining the 'internal' interface
from the service implementation to the global Grid environment. These characteristics would then be rendered into different implementation technologies (e.g. J2EE or
shared libraries).*

Whilst traditional Grid implementations have mapped such requirements directly to the
native operating system or to the library code, GS are likely to be significantly influenced
by the platforms used in Web service hosting.

7.7.1 Grid services hosting environment

We are familiar with the power of well-defined computational contexts, for example, that
defined by the Java Virtual Machine [22] and that defined for Enterprise Java Beans [23].
Designing such a context for GS requires a balance between the following:

- Parsimony of facilities to minimise transition and learning costs.
- Complete functionality to provide rich resources to developers.

A common issue is a standard representation of a computation's history, sometimes called
its *context*. This context contains information that must be passed from one service to
the next as invocations occur. Examples include the subject on behalf of whom the
computation is being conducted (needed for authorisation), the transaction identifier if
this computation is part of a transaction and so on. An example of such a context is
`MessageContext` in Apache Axis.

7.7.2 Standard types

The SOAP definition [24] defines a set of types including primitive types and the recursive
composition of these as structures and arrays, based on namespaces and notations of the
XML Schema definitions [25, 26]. The applications and libraries of e-Science use many
standard types, such as complex numbers, diagonalised matrices, triangular matrices and
so on. There will be a significant gain if widely used types are defined and named early,
so that the same e-Science-oriented namespaces can be used in many exchange protocols,
port definitions, components and services. The advantages include

1. simplification of interworking between components that adopt these standards,
2. better amortisation of the cost of type design,
3. early validation of the use of WSDL for these aspects of e-Science and
4. simplification of the task of providing efficient mappings for serialising and deserial-
   ising these structures by avoiding multiple versions.

Many communities, such as astronomers, protein crystallographers, bioinformaticians and
so on, are developing standards for their own domains of communication and for curated
data collections. The e-Science core programme can build on and facilitate this process

by developing component types that can be reused across domains. As higher-level types are standardised and reused, it becomes easier to move activities towards the information and knowledge layers to which the semantic Grid aspires.

7.8 BASELINE DATABASE ACCESS

This is primarily the responsibility of the UK Database Task Force [12] and now the GGF Database Access and Integration Services (DAIS) working group [21]. A Centre Project, OGSA-DAI based on a consortium of EPCC, IBM, NEeSC, NeSC, NWeSC and Oracle, is developing a set of components to serve this function and to contribute to the standards. This is an illustration of using OGSA and so it is presented as an example.

The suite of middleware envisaged is complementary to that produced by data management projects, such as the European Data Grid (see Chapter 15) and distributed file and replication management, such as GIGGLE [27]. Their primary concern is to manage reliable storage, distributed naming, replication and movement of large collections of data that are under their management. They operate largely without concern for the structure and the interpretation of data within their containers (mostly files and collections of files).

Database Access and Integration (DAI) components permit access to data, which is usually stored in standard Database Management Systems (DBMS), which is often managed autonomously and provide database operations, such as query and update, for the data held in these databases. The challenges include establishing connections, establishing authority and handling the variety of forms of data. At present DAI aspires to handle distributed database operations applied to collections of data in relational databases, XML collections and files whose structure is adequately described. A current description may be found in Reference [21] and in other papers prepared for GGF5 (http://www.cs.man.ac.uk/grid-db/).

7.8.1 An overview of the OGSA-DAI architecture

DAIcomponent categories: There are four categories of component. Each may have a variety of detailed forms. Their function and range of forms are introduced here.

Grid database services (GDS): These are either a structured data store or (more often) a proxy for a structured data store, such as an instance of a DBMS. In either case they provide access to the stored data through a standard portType. In later versions of DAI they will also represent federations. To accommodate the variety of data models, query languages, data description languages and proprietary languages, this API includes explicit identification of the language that is used to specify each (database) operation that is to be applied. This flexibility also allows for GDS that reveal proprietary operations of specific DBMS, allows batches of operations to be optimised and processed by interpreting a Grid Job Control Language and provides for extension for future data models. Developers may restrict themselves to standard widely supported languages, such as SQL'92 and Xpath, in order to achieve platform independence.

Grid database service factories (GDSF): These may be associated with an instance of a DBMS or they may be associated with a particular DBMS type or data model. In the former case, the GDS that are generated by this factory will act as proxies for the underlying database. In the latter case, the factory will produce data storage systems managed according to the associated DBMS type, for example Oracle 9i, or according to the associated model, for example XML. The result may then be a GDSF, which produces proxies to that storage system, or a GDS that allows direct access on that storage system. The API will again provide control and flexibility by explicitly defining the language used.

Grid data transport vehicles (GDTV): These provide an abstraction over bulk data transmission systems, such as GridFTP, MPICH-G, Unix Pipes and so on. They provide two APIs, one for a data producer and one for a data consumer.[1] These may then be kept invariant, while performance, synchronisation and reliability properties may be adjusted. For example, the data may be delivered immediately or stored for later collection, it may be transferred in bulk or as a stream, it may be sent to a third party within a specified time and so on. A GDTV may be used to supply data to an operation, for example to a distributed join or to a bulk load, or to transfer data from an operation, for example as a result set. Instances of GDTV may not be services, in order to avoid redundant data movement. These will be bindings to libraries supporting the APIs, hence 'vehicle' rather than 'service'. Others, for example those that store results for later collection, will be services.

Grid data service registries (GDSR): These allow GDS and GDSF to register and then allow client code to query GDSR to find data sources or GDS/F that match their requirements according to data content, operations, resources and so forth. They will be a special form of registry in the OGSA sense, providing support for data content and structure searches. These depend on a metadata infrastructure.

Grid data metadata: These metadata may be distinct from metadata used by applications and from metadata in database schemas, though, in some cases they may be derived from these. The metadata will include the following aspects of DAI components:

- The types of data model supported, for example, Relational, XML and so forth.
- The languages supported by this GDS, SQL'92, Xpath, Xquery and so forth.
- The operations supported, for example, query, bulk load, insert, update, delete, schema edit and so forth.
- Grid Data Transport Vehicles supported.
- The data content that is accessible.
- Resources and restrictions.
- Capacity and performance information.
- Access policies.
- Charging policies.

---

[1] There will normally be another API to permit progress monitoring.

7.8.1.1 Example showing execution model

A diagram of an application using OGSA-DAI is shown in Figure 7.1. It illustrates a scenario in which a sequence of steps takes place involving a client application, five OGSA-DAI components and four GDTV. Control messages are not shown. The notation uses yellow ellipses for OGSA-DAI components and a blue rectangle for a client. It uses various forms of dotted open arrow for invocation messages and solid thick arrows for the applications of GDTVs. The scenario presumes that the client wishes to achieve two things using data integrated from three sources managed by GDS:

- Obtain data combined from $database_1$ and $database_3$, perhaps to verify some aspect of the composite operation.
- Send data combined from $database_1$, $database_2$ and $database_3$ as a stream to a specified third party, for example a data-mining tool (not shown).

In the scenario, we envisage the client using OGSA-DAI as follows:

1. It knows the GSH of one of the GDSR and sends to that GDSR a description of the GDS it requires. The description specifies the three databases that must be accessed and the operations (query and bulk/stream transfer) that are required. The GDSR may use a peer-to-peer protocol to refer the request to other GDSR in a large system.
2. The GDSR replies with an indication that the required GDS do not exist. It provides a list of GDSFs (as their GSHs) that can generate the required GDS.
3. The client chooses one of these, perhaps after exercising some dialogues to determine more about the offered GDSF, and constructs a script requesting three GDS. These

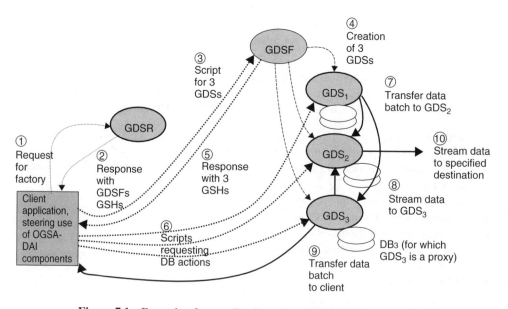

Figure 7.1   Example of an application using OGSA-DAI components.

will be described using the same notation[2] as was used in the request to the GDSR but probably with additional information.

4. The GDSF probably responds with an immediate confirmation that it has understood the script and will make the GDS (this message is not shown). The GDSF then schedules the construction and initialisation of the three GDS, presumably making each an OGSA-DAI-compliant proxy for the three databases of interest.

5. When they have all reported that they are initialised (not shown), the GDSF sends a composite message to the client providing their identities as GSH.

6. After a dialogue with each GDS to determine the finer details of its capabilities (not shown), the client sends a script to each GDS indicating the graph of tasks each is required to undertake. The script sent to GDS_1 indicated that it should transport a batch of data to a task in the script sent to GDS_2 and send a stream of data to a task identified in a script sent to GDS_3. The script sent to GDS_2 indicated that it should expect a batch of data from a task sent to GDS_1 and a stream of data from a task sent to GDS_3. It was required to run a task that established the flow of a stream of data to a specified third party. The script sent to GDS_3 contained a task that was required to send a batch of data to the client. For each of these data transfers there would be a description allowing an appropriate GDTV to be constructed and/or used.

7. GDS_1 uses a GDTV to send the specified *batch* of data to GDS_2 and another GDTV to send a *stream* of data to GDS_3.

8. GDS_3 combines that incoming data stream with its own data and uses a GDTV to send the result data as a stream to GDS_2.

9. GDS_3 uses the incoming data stream, its own data and a GDTV to send a batch of data to the client.

10. GDS_2 combines the incoming batch of data, its own data and the incoming stream of data to construct a stream of data for the third party that is delivered using a GDTV.

This scenario does not illustrate all possible relationships. For example, a GDSR may use a GDS to support its own operation, and a script may require data transport between tasks within the same GDS.

The first steps in data integration are distributed query systems in which the schemas are compatible. This has already been prototyped [28]. Subsequent stages require interposition of data transformations. Tools may be developed to help in the formulation of scripts and transformations, in order to render consistent data from heterogeneous data sources [29].

7.9 BASELINE LOGGING INFRASTRUCTURE

The community of e-Scientists uses a large variety of facilities, such as compute resources, storage resources, high-performance networks and curated data collections. Our goal is to establish a culture in UK science where the e-Science techniques, which depend on these resources, are widely used. To make this possible, accounting mechanisms must be

---

[2] It is important to avoid introducing extra notations, as that generates problems understanding and maintaining their equivalences.

effective. To support the resource sharing within virtual organisations and between real organisations it is essential that resource usage be recorded. The units that need to be recorded will differ between services but could include

- number of bytes crossing various boundaries, such as a gateway on a network, the main memory of a computer, the fibres from disk and so on;
- number of byte seconds of storage use for various stores;
- number of CPU-hours;
- number of records examined in a database;
- uses of licensed software.

There are clearly many potential units that could be the basis of charging, but for a logging infrastructure to be scalable and usable a relative small number must be agreed upon and understood. These units may be accumulated or the critical recorded unit may be based on peak rates delivered. If resource owners are to contribute their resources, these units also have to approximately reflect the origins of their costs within their organisation.

The OGSA needs to intercept service usage as part of its core architecture and record this information through a logging service. This service should provide reliable mechanisms to distribute this data to other organisations.

It is unlikely that a single charging model or a single basis for charging will emerge in a diverse community. There therefore needs to be a mechanism in which different charging policies and different clients can meet. We refer to it as a *market*.

The creation of a sustainable long-term economic model that will attract independent service providers requires

1. a logging infrastructure that reliably and economically records resource usage,
2. a commonly understood means of describing charges,
3. a mechanism to negotiate charges between the consumer and the provider,
4. a secure payment mechanism.

A project to develop such a Grid market infrastructure for OGSA has been proposed.

7.10 SUMMARY AND CONCLUSIONS

This chapter has recorded the UK e-Science commitment to the OGSA and explained the rationale for that commitment. It has illustrated the consequences of this commitment by presenting the steps that are necessary to augment the baseline OGSA middleware with other common facilities that are needed by e-Science projects.

The focus on access and integration of structured data, typically held in databases, was motivated by the prevalence of data integration within those projects. The OGSA-DAI project's plans for GS provided an illustration of the ways in which this requirement maps onto OGSA.

Many other middleware functions can be extracted and developed as GS. One example is accounting and the infrastructure for a Grid 'market'. This is identified as urgently required by many UK projects and by those who provide computation resources.

The OGSA infrastructure and the componentisation of e-Science infrastructure is expected to have substantial long-term benefits.

- It assists in the dynamic composition of components and makes it more likely that tools to support safe composition will be developed.
- It increases the chances of significant contribution to the required infrastructure from industry.
- It improves the potential for meeting challenges, such as agreements about interchange across political and organisational boundaries.
- By providing a description regime, it provides the basis for improved engineering and better partitioning of development tasks.

However, it is not a panacea. There are still many functionalities required by e-Scientists that are yet to be implemented; only two of them were illustrated above. It depends on WS, which are an excellent foundation for distributed heterogeneous system integration. But these are still the subject of vigorous development of standards and platforms. The outcome is very likely to be beneficial, but the journey to reach it may involve awkward revisions of technical decisions. As the e-Science community stands to benefit from effective WS and effective GS, it should invest effort in developing applications using these new technologies and use that experience to influence the design and ensure the compatibility of these foundation technologies.

ACKNOWLEDGEMENTS

This work was commissioned by the UK e-Science Core Programme Directorate and funded by the UK Department of Trade and Industry and by the Engineering and Physical Sciences Research Council. The chapter presents the joint work of the whole Architectural Task Force, namely,

Malcolm Atkinson (chair)	National e-Science Centre
Jon Crowcroft	Cambridge University
David De Roure	Southampton University
Vijay Dialani	Southampton University
Andrew Herbert	Microsoft Research Cambridge
Ian Leslie	Cambridge University
Ken Moody	Cambridge University
Steven Newhouse	Imperial College, London
Tony Storey	IBM Hursley Laboratory, UK

REFERENCES

1. Atkinson, M. P. *et al.* (2002) *UK Role in Open Grid Services Architecture*, Report for the UK e-Science Programme, April, 2002.

2. Tuecke, S., Czajkowski, K., Foster, I., Frey, J., Graham, S., Kesselman, C. and Nick, J. (2002) *Grid Service Specification*, Presented at GGF4, February, 2002. http://www.globus.org/research/papers.html.

3. Abiteboul, S., Buneman, P. and Suciu, D. *Data on the Web: From Relations to Semistructured data and XML*. San Francisco, CA: Morgan Kaufmann, 1999.

4. Atkinson, M. P., Dmitriev, M., Hamilton, C. and Printezis, T. (2000) Scalable and recoverable implementation of object evolution for the PJama platform. *Proc. of the Ninth International Workshop on Persistent Object Systems*, 2000, pp. 255–268.

5. Atkinson, M. P. and Jordan, M. J., *A Review of the Rationale and Architectures of PJama: A Durable, Flexible, Evolvable and Scalable Orthogonally Persistent Programming Platform*, Technical Report TR-2000-90, Sun Microsystems Laboratories, Palo Alto, CA, p. 103.

6. Atkinson, M. P. Persistence and Java – a balancing act. (2000) *Proceedings of the ECOOP Symposium on Objects and Databases*, 2000, pp. 1–32, LNCS Number 1944 (invited paper).

7. Buneman, P., Davidson, S., Hart, K., Overton, C. and Wong, L. (1995) A data transformation system for biological data sources. *Proceedings of the Twenty-first International Conference on Very Large Databases*, 1995.

8. Buneman, P., Khanna, S. and Tan, W.-C. (2000) *Data Provenance: Some Basic Issues,* in *Foundations of Software Technology and Theoretical Computer Science*, 2000.

9. Buneman, P., Khanna, S. and Tan, W.-C. (2001) Why and where – a characterization of data provenance. *Proceedings of the International Conference on Database Theory*, 2001.

10. Buneman, P., Khanna, S. and Tan, W.-C. (2002) *On propagation of deletions and annotations through views*. *Proceedings of the Conference on the Principles of Database Systems*, May, 2002; to appear.

11. Buneman, P., Khanna, S., Tajima, K. and Tan, W.-C. (2002) *Archiving Scientific Data*. *Proceedings of ACM SIGMOD*, 2002; to appear.

12. Paton, N. W., Atkinson, M. P., Dialani, V., Pearson, D., Storey, T. and Watson, P. (2002) *Database Access and Integration Services on the Grid*, UK DBTF working paper, January, 2002 (presented at GGF4), http://www.cs.man.ac.uk/grid-db/.

13. Pearson, D. (2002) *Data Requirements for the Grid: Scoping Study Report*, UK DBTF working paper, February, 2002 (presented at GGF4), http://www.cs.man.ac.uk/grid-db/.

14. Stevens, R., Goble, C., Paton, N., Bechhofer, S., Ng, G., Baker, P. and Brass, A. (1999) Complex Query Formulation Over Diverse Information Sources Using an Ontology. In *Proceedings of Workshop on Computation of Biochemical Pathways and Genetic Networks*, European Media Lab (EML), pp. 83–88.

15. Watson, P. (2002) *Databases and the Grid*, Version 3, Technical Report CS-TR-755, Newcastle University, Department of Computer Science, January, 2002, in this volume.

16. Gray, J. (2001) The World Wide Telescope: Mining the Sky SC2001, Denver Colorado, November 14, 2001, http://www.sc2001.org/plenary_gray.html

17. Szalay, A. S., Kuntz, P., Thacker, A., Gray, J., Slutz, D. and Brunner, R. J. *Designing and Mining Multi-Terabyte Astronomy Archives: The Sloan Digital Sky Survey*, Technical Report MR-TR-99-30, Microsoft Research, Revised 2000.

18. Kuntz, P. Z. (2002) *The Open Grid Services Architecture: A Summary and Evaluation*, Report for the UK e-Science Core Programme, April, 2002.

19. Gannon, D., Chiu, K., Chiu, Govindaraju, M. and Slominski, A. (2002). *An Analysis of the Open Grid Services Architecture*, Report for the UK e-Science Core Programme, April, 2002.

20. Graham, S., Simeonov, S., Boubez, T., Davis, D., Daniels, G., Nakamura, Y. and Nayama, R. *Building Web Services with Java: Making Sense of XML, SOAP, WSDL and UDDI*. Indianapolis, IN: Sams Publishing (2002).

21. Atkinson, M. P. *et al. Grid Database Access and Integration: Requirements and Functionalities*, (Presented at GGF5 July, 2002), http://www.cs.man.ac.uk/grid-db/.

22. Linholm, T. and Yellin, F. (1996) *The Java Virtual Machine*. Reading, MA: Addison-Wesley, 1996.

23. Mohan, *Application Servers and Associated Technologies*, Tutorials at EDBT, SIGMOD and VLDB, 2002, http://www.almaden.ibm.com/u/mohan/AppServersTutorial_SIGMOD2002.pdf and http://www.almaden.ibm.com/u/mohan/AppServersTutorial_VLDB2002_Slides.pdf.
24. Mitra, N. (2001) SOAP Version 1.2 Part 0: Primer, W3C Working Draft, December 17, 2001, www.w3.org/TR/2001/WD-soap12-part0-20011217/, supersedes Box+00.
25. W3C Working Draft, XML Schema Part 1: Structures, work in progress.
26. W3C Working Draft, XML Schema Part 2: Datatypes, work in progress.
27. Foster, I. *et al.* (2002) *GIGGLE: A Framework for Constructing Scalable Replica Location Services*, 2002.
28. Smith, J., Gounaris, A., Watson, P., Paton, N. W., Fernandes, A. A. A. and Sakellariou, R. *Distributed Query Processing on the Grid*, http://www.cs.man.ac.uk/grid-db/.
29. Melton, J., Michels, J.-E., Josifovski, V., Kulkarni, K., Schwarz, P. and Zeidenstein, K. (2002) *SQL and Management of External Data*, 2002.
30. Foster, I., Kesselman, C. and Tuecke, S. (2001) The Anatomy of the Grid: Enabling Virtual Organisations. *International Journal of Supercomputer Applications*, **15**(3), pp. 200–222.
31. Newhouse, S., Mayer, A., Furmento, N., McGough, S., Stanton, J. and Darlington, J. (2001) *Laying the Foundations for the Semantic Grid*. London, UK: London e-Science Centre, December, 2001, www-icpc.doc.ic.ac.uk/components/.

8

The physiology of the Grid

Ian Foster,[1,2] Carl Kesselman,[3] Jeffrey M. Nick,[4] and Steven Tuecke[1]

[1]*Argonne National Laboratory, Argonne, Illinois, United States,* [2]*University of Chicago, Chicago, Illinois, United States,* [3]*University of Southern California, Marina del Rey, California, United States,* [4]*IBM Corporation, Poughkeepsie, New York, United States*

8.1 INTRODUCTION

Until recently, application developers could often assume a target environment that was (to a useful extent) homogeneous, reliable, secure, and centrally managed. Increasingly, however, computing is concerned with collaboration, data sharing, and other new modes of interaction that involve distributed resources. The result is an increased focus on the interconnection of systems both within and across enterprises, whether in the form of intelligent networks, switching devices, caching services, appliance servers, storage systems, or storage area network management systems. In addition, companies are realizing that they can achieve significant cost savings by outsourcing nonessential elements of their IT environment to various forms of service providers.

These evolutionary pressures generate new requirements for distributed application development and deployment. Today, applications and middleware are typically developed

Grid Computing – Making the Global Infrastructure a Reality. Edited by F. Berman, A. Hey and G. Fox
© 2003 John Wiley & Sons, Ltd   ISBN: 0-470-85319-0

for a specific platform (e.g., Windows NT, a flavor of Unix, a mainframe, J2EE, Microsoft .NET) that provides a hosting environment for running applications. The capabilities provided by such platforms may range from integrated resource management functions to database integration, clustering services, security, workload management, and problem determination – with different implementations, semantic behaviors, and application programming interfaces (APIs) for these functions on different platforms. But in spite of this diversity, the continuing decentralization and distribution of software, hardware, and human resources make it essential that we achieve desired qualities of service (QoS) – whether measured in terms of common security semantics, distributed workflow and resource management performance, coordinated fail-over, problem determination services, or other metrics – on resources assembled dynamically from enterprise systems, SP systems, and customer systems. We require new abstractions and concepts that allow applications to access and share resources and services across distributed, wide-area networks.

Such problems have been for some time a central concern of the developers of distributed systems for large-scale scientific research. Work within this community has led to the development of *Grid technologies* [1, 2], which address precisely these problems and which are seeing widespread and successful adoption for scientific and technical computing.

In an earlier article, we defined Grid technologies and infrastructures as supporting the sharing and coordinated use of diverse resources in dynamic, distributed 'virtual organizations' (VOs) [2]. We defined essential properties of Grids and introduced key requirements for protocols and services, distinguishing among *connectivity* protocols concerned with communication and authentication, *resource* protocols concerned with negotiating access to individual resources, and *collective* protocols and services concerned with the coordinated use of multiple resources. We also described the Globus Toolkit™ ¹ [3], an open-source reference implementation of key Grid protocols that supports a wide variety of major e-Science projects.

Here we extend this argument in three respects to define more precisely how a Grid functions and how Grid technologies can be implemented and applied. First, while Reference [2] was structured in terms of the protocols required for interoperability among VO components, we focus here on the nature of the *services* that respond to protocol messages. We view a Grid as an extensible set of *Grid services* that may be aggregated in various ways to meet the needs of VOs, which themselves can be defined in part by the services that they operate and share. We then define the behaviors that such Grid services should possess in order to support distributed systems integration. By stressing functionality (i.e., 'physiology'), this view of Grids complements the previous protocol-oriented ('anatomical') description.

Second, we explain how Grid technologies can be aligned with Web services technologies [4, 5] to capitalize on desirable Web services properties, such as service description and discovery; automatic generation of client and server code from service descriptions; binding of service descriptions to interoperable network protocols; compatibility with emerging higher-level open standards, services and tools; and broad commercial support.

---

¹ Globus Project and Globus Toolkit are trademarks of the University of Chicago.

We call this alignment – and augmentation – of Grid and Web services technologies an *Open Grid Services Architecture* (OGSA), with the term *architecture* denoting here a well-defined set of basic interfaces from which can be constructed interesting systems and the term *open* being used to communicate extensibility, vendor neutrality, and commitment to a community standardization process. This architecture uses the Web Services Description Language (WSDL) to achieve self-describing, discoverable services and interoperable protocols, with extensions to support multiple coordinated interfaces and change management. OGSA leverages experience gained with the Globus Toolkit to define conventions and WSDL interfaces for a *Grid service*, a (potentially transient) stateful service instance supporting reliable and secure invocation (when required), lifetime management, notification, policy management, credential management, and virtualization. OGSA also defines interfaces for the discovery of Grid service instances and for the creation of transient Grid service instances. The result is a standards-based distributed service system (we avoid the term distributed object system owing to its overloaded meaning) that supports the creation of the sophisticated distributed services required in modern enterprise and interorganizational computing environments.

Third, we focus our discussion on commercial applications rather than the scientific and technical applications emphasized in References [1, 2]. We believe that the same principles and mechanisms apply in both environments. However, in commercial settings we need, in particular, seamless integration with existing resources and applications and with tools for workload, resource, security, network QoS, and availability management. OGSA's support for the discovery of service properties facilitates the mapping or *adaptation* of higher-level Grid service functions to such native platform facilities. OGSA's service orientation also allows us to *virtualize* resources at multiple levels, so that the same abstractions and mechanisms can be used both within distributed Grids supporting collaboration across organizational domains and within hosting environments spanning multiple tiers within a single IT domain. A common infrastructure means that differences (e.g., relating to visibility and accessibility) derive from policy controls associated with resource ownership, privacy, and security, rather than interaction mechanisms. Hence, as today's enterprise systems are transformed from separate computing resource islands to integrated, multitiered distributed systems, service components can be integrated dynamically and flexibly, both within and across various organizational boundaries.

The rest of this article is as follows. In Section 8.2, we examine the issues that motivate the use of Grid technologies in commercial settings. In Section 8.3, we review the Globus Toolkit and Web services, and in Section 8.4, we motivate and introduce our Open Grid Services Architecture. In Sections 8.5 to 8.8, we present an example and discuss protocol implementations and higher-level services. We discuss related work in Section 8.9 and summarize our discussion in Section 8.10.

We emphasize that the OGSA and associated Grid service specifications continue to evolve as a result of both standard work within the Global Grid Forum (GGF) and implementation work within the Globus Project and elsewhere. Thus the technical content in this article, and in an earlier abbreviated presentation [6], represents only a snapshot of a work in progress.

8.2 THE NEED FOR GRID TECHNOLOGIES

Grid technologies support the sharing and coordinated use of diverse resources in dynamic VOs – that is, the creation, from geographically and organizationally distributed components, of virtual computing systems that are sufficiently integrated to deliver desired QoS [2].

Grid concepts and technologies were first developed to enable resource sharing within far-flung scientific collaborations [1, 7–11]. Applications include collaborative visualization of large scientific datasets (pooling of expertise), distributed computing for computationally demanding data analyses (pooling of compute power and storage), and coupling of scientific instruments with remote computers and archives (increasing functionality as well as availability) [12]. We expect similar applications to become important in commercial settings, initially for scientific and technical computing applications (where we can already point to success stories) and then for commercial distributed computing applications, including enterprise application integration and business-to-business (B2B) partner collaboration over the Internet. Just as the World Wide Web began as a technology for scientific collaboration and was adopted for e-Business, we expect a similar trajectory for Grid technologies.

Nevertheless, we argue that Grid concepts are critically important for commercial computing, not primarily as a means of enhancing capability but rather as a solution to new challenges relating to the construction of reliable, scalable, and secure distributed systems. These challenges derive from the current rush, driven by technology trends and commercial pressures, to decompose and distribute through the network previously monolithic host-centric services, as we now discuss.

8.2.1 The evolution of enterprise computing

In the past, computing typically was performed within highly integrated host-centric enterprise computing centers. While sophisticated distributed systems (e.g., command and control systems, reservation systems, the Internet Domain Name System [13]) existed, these have remained specialized niche entities [14, 15].

The rise of the Internet and the emergence of e-Business have, however, led to a growing awareness that an enterprise's IT infrastructure also encompasses external networks, resources, and services. Initially, this new source of complexity was treated as a network-centric phenomenon, and attempts were made to construct 'intelligent networks' that intersect with traditional enterprise IT data centers only at 'edge servers': for example, an enterprise's Web point of presence or the virtual private network server that connects an enterprise network to SP resources. The assumption was that the impact of e-Business and the Internet on an enterprise's core IT infrastructure could thus be managed and circumscribed.

This attempt has, in general, failed because IT services decomposition is also occurring *inside* enterprise IT facilities. New applications are being developed to programming models (such as the Enterprise Java Beans component model [16]) that insulate the application from the underlying computing platform and support portable deployment across multiple platforms. This portability in turn allows platforms to be selected on the basis of price/performance and QoS requirements, rather than operating system supported. Thus, for example, Web serving and caching applications target commodity servers rather than traditional mainframe computing platforms. The resulting proliferation of Unix and NT servers necessitates distributed connections to legacy mainframe application and data assets. Increased load on those assets has caused companies to offload nonessential functions (such as query processing) from backend transaction-processing systems to midtier servers. Meanwhile, Web access to enterprise resources requires ever-faster request servicing, further driving the need to distribute and cache content closer to the edge of the network. The overall result is a decomposition of highly integrated internal IT infrastructure into a collection of heterogeneous and fragmented systems. Enterprises must then reintegrate (with QoS) these distributed servers and data resources, addressing issues of navigation, distributed security, and content distribution inside the enterprise, much as on external networks.

In parallel with these developments, enterprises are engaging ever more aggressively in e-Business and are realizing that a highly robust IT infrastructure is required to handle the associated unpredictability and rapid growth. Enterprises are also now expanding the scope and scale of their enterprise resource planning projects as they try to provide better integration with customer relationship management, integrated supply chain, and existing core systems. These developments are adding to the significant pressures on the enterprise IT infrastructure.

The aggregate effect is that *qualities of service traditionally associated with mainframe host-centric computing [17] are now essential to the effective conduct of e-Business across distributed compute resources, inside as well as outside the enterprise.* For example, enterprises must provide consistent response times to customers, despite workloads with significant deviations between average and peak utilization. Thus, they require flexible resource allocation in accordance with workload demands and priorities. Enterprises must also provide a secure and reliable environment for distributed transactions flowing across a collection of dissimilar servers, must deliver continuous availability as seen by end users, and must support disaster recovery for business workflow across a distributed network of application and data servers. Yet the current paradigm for delivering QoS to applications via the vertical integration of platform-specific components and services just does not work in today's distributed environment: the decomposition of monolithic IT infrastructures is not consistent with the delivery of QoS through vertical integration of services on a given platform. Nor are distributed resource management capabilities effective, being limited by their proprietary nature, inaccessibility to platform resources, and inconsistencies between similar resources across a distributed environment.

The result of these trends is that IT systems integrators take on the burden of reintegrating distributed compute resources with respect to overall QoS. However, without appropriate infrastructure tools, the management of distributed computing workflow becomes increasingly labor intensive, complex, and fragile as platform-specific operations staff watch for 'fires' in overall availability and performance and verbally collaborate on corrective actions across different platforms. This situation is not scalable, cost effective, or tenable in the face of changes to the computing environment and application portfolio.

8.2.2 Service providers and business-to-business computing

Another key trend is the emergence of service providers (SPs) of various types, such as Web-hosting SPs, content distribution SPs, applications SPs, and storage SPs. By exploiting economies of scale, SPs aim to take standard e-Business processes, such as creation of a Web-portal presence, and provide them to multiple customers with superior price/performance. Even traditional enterprises with their own IT infrastructures are offloading such processes because they are viewed as commodity functions.

Such emerging 'eUtilities' (a term used to refer to service providers offering continuous, on-demand access) are beginning to offer a model for carrier-grade IT resource delivery through metered usage and subscription services. Unlike the computing services companies of the past, which tended to provide off-line batch-oriented processes, resources provided by eUtilities are often tightly integrated with enterprise computing infrastructures and used for business processes that span both in-house and outsourced resources. Thus, a price of exploiting the economies of scale that are enabled by eUtility structures is a further decomposition and distribution of enterprise computing functions. Providers of eUtilities face their own technical challenges. To achieve economies of scale, eUtility providers require server infrastructures that can be easily customized on demand to meet specific customer needs. Thus, there is a demand for IT infrastructure that (1) supports dynamic resource allocation in accordance with service-level agreement policies, efficient sharing and reuse of IT infrastructure at high utilization levels, and distributed security from edge of network to application and data servers and (2) delivers consistent response times and high levels of availability, which in turn drives a need for end-to-end performance monitoring and real-time reconfiguration.

Still another key IT industry trend is cross-enterprise B2B collaboration such as multiorganization supply chain management, virtual Web malls, and electronic market auctions. B2B relationships are, in effect, virtual organizations, as defined above – albeit with particularly stringent requirements for security, audibility, availability, service-level agreements, and complex transaction processing flows. Thus, B2B computing represents another source of demand for distributed systems integration, characterized often by large differences among the information technologies deployed within different organizations.

8.3 BACKGROUND

We review two technologies on which we build to define the Open Grid Services Architecture: the Globus Toolkit, which has been widely adopted as a Grid technology solution for scientific and technical computing, and Web services, which have emerged as a popular standards-based framework for accessing network applications.

8.3.1 The Globus Toolkit

The Globus Toolkit [2, 3] is a community-based, open-architecture, open-source set of services and software libraries that support Grids and Grid applications. The toolkit addresses issues of security, information discovery, resource management, data management, communication, fault detection, and portability. Globus Toolkit mechanisms are in use at hundreds of sites and by dozens of major Grid projects worldwide.

The toolkit components that are most relevant to OGSA are the Grid Resource Allocation and Management (GRAM) protocol and its 'gatekeeper' service, which provides for secure, reliable service creation and management [18]; the Meta Directory Service (MDS-2) [19], which provides for information discovery through soft-state registration [20, 21], data modeling, and a local registry ('GRAM reporter' [18]); and the Grid security infrastructure (GSI), which supports single sign-on, delegation, and credential mapping. As illustrated in Figure 8.1, these components provide the essential elements of a service-oriented architecture, but with less generality than is achieved in OGSA.

The GRAM protocol provides for the reliable, secure remote creation and management of arbitrary computations: what we term in this article as *transient service instances*. GSI mechanisms are used for authentication, authorization, and credential delegation [22] to remote computations. A two-phase commit protocol is used for reliable invocation, based on techniques used in the Condor system [23]. Service creation is handled by a small, trusted 'gatekeeper' process (termed a *factory* in this article), while a GRAM reporter monitors and publishes information about the identity and state of local computations (*registry*).

MDS-2 [19] provides a uniform framework for discovering and accessing system configuration and status information such as compute server configuration, network status, or the locations of replicated datasets (what we term a *discovery* interface in this chapter). MDS-2 uses a soft-state protocol, the Grid Notification Protocol [24], for lifetime management of published information.

The public key-based GSI protocol [25] provides single sign-on authentication, communication protection, and some initial support for restricted delegation. In brief, *single sign-on* allows a user to authenticate once and thus create a proxy credential that a program can use to authenticate with any remote service on the user's behalf. *Delegation* allows for the creation and communication to a remote service of delegated proxy credentials that the remote service can use to act on the user's behalf, perhaps with various restrictions; this

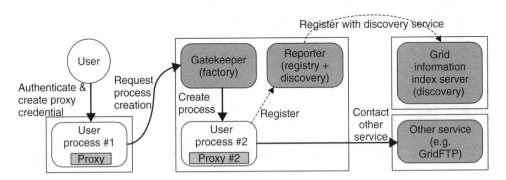

Figure 8.1 Selected Globus Toolkit mechanisms, showing initial creation of a proxy credential and subsequent authenticated requests to a remote gatekeeper service, resulting in the creation of user process #2, with associated (potentially restricted) proxy credential, followed by a request to another remote service. Also shown is soft-state service registration via MDS-2.

capability is important for nested operations. (Similar mechanisms can be implemented within the context of other security technologies, such as Kerberos [26], although with potentially different characteristics.)

GSI uses X.509 certificates, a widely employed standard for Public Key Infrastructure (PKI) certificates, as the basis for user authentication. GSI defines an X.509 proxy certificate [27] to leverage X.509 for support of single sign-on and delegation. (This proxy certificate is similar in concept to a Kerberos forwardable ticket but is based purely on public key cryptographic techniques.) GSI typically uses the Transport Layer Security (TLS) protocol (the follow-on to Secure Sockets Layer (SSL)) for authentication, although other public key-based authentication protocols could be used with X.509 proxy certificates. A remote delegation protocol of X.509 proxy certificates is layered on top of TLS. An Internet Engineering Task Force draft defines the X.509 Proxy Certificate extensions [27]. GGF drafts define the delegation protocol for remote creation of an X.509 proxy certificate [27] and Generic Security Service API (GSS-API) extensions that allow this API to be used effectively for Grid programming.

Rich support for restricted delegation has been demonstrated in prototypes and is a critical part of the proposed X.509 Proxy Certificate Profile [27]. Restricted delegation allows one entity to delegate just a subset of its total privileges to another entity. Such restriction is important to reduce the adverse effects of either intentional or accidental misuse of the delegated credential.

8.3.2 Web services

The term *Web services* describes an important emerging distributed computing paradigm that differs from other approaches such as DCE, CORBA, and Java RMI in its focus on simple, Internet-based standards (e.g., eXtensible Markup Language: XML [28, 29]) to address heterogeneous distributed computing. Web services define a technique for describing software components to be accessed, methods for accessing these components, and discovery methods that enable the identification of relevant SPs. Web services are programming language–, programming model–, and system software–neutral.

Web services standards are being defined within the W3C and other standards bodies and form the basis for major new industry initiatives such as Microsoft (.NET), IBM (Dynamic e-Business), and Sun (Sun ONE). We are particularly concerned with three of these standards: SOAP, WSDL, and WS-Inspection.

- The *Simple Object Access Protocol* (SOAP) [30] provides a means of messaging between a service provider and a service requestor. SOAP is a simple enveloping mechanism for XML payloads that defines a remote procedure call (RPC) convention and a messaging convention. SOAP is independent of the underlying transport protocol; SOAP payloads can be carried on HTTP, FTP, Java Messaging Service (JMS), and the like. We emphasize that Web services can describe multiple access mechanisms to the underlying software component. SOAP is just one means of formatting a Web service invocation.
- The *Web Services Description Language* (WSDL) [31] is an XML document for describing Web services as a set of *endpoints* operating on messages containing either document-oriented (messaging) or RPC payloads. Service interfaces are defined abstractly in terms

of message structures and sequences of simple message exchanges (or operations, in WSDL terminology) and then bound to a concrete network protocol and data-encoding format to define an endpoint. Related concrete endpoints are bundled to define abstract endpoints (services). WSDL is extensible to allow description of endpoints and the concrete representation of their messages for a variety of different message formats and network protocols. Several standardized binding conventions are defined describing how to use WSDL in conjunction with SOAP 1.1, HTTP GET/POST, and (MIME) Multimedia Internet Message Extensions.

• *WS-Inspection* [32] comprises a simple XML language and related conventions for locating service descriptions published by an SP. A WS-Inspection language (WSIL) document can contain a collection of service descriptions and links to other sources of service descriptions. A service description is usually a URL to a WSDL document; occasionally, a service description can be a reference to an entry within a Universal Description, Discovery, and Integration (UDDI) [33] registry. A link is usually a URL to another WS-Inspection document; occasionally, a link is a reference to a UDDI entry. With WS-Inspection, an SP creates a WSIL document and makes the document network accessible. Service requestors use standard Web-based access mechanisms (e.g., HTTP GET) to retrieve this document and discover what services the SP advertises. WSIL documents can also be organized in different forms of index.

Various other Web services standards have been or are being defined. For example, Web Services Flow Language (WSFL) [34] addresses Web services *orchestration*, that is, the building of sophisticated Web services by composing simpler Web services.

The Web services framework has two advantages for our purposes. First, our need to support the dynamic discovery and composition of services in heterogeneous environments necessitates mechanisms for registering and discovering interface definitions and endpoint implementation descriptions and for dynamically generating proxies based on (potentially multiple) bindings for specific interfaces. WSDL supports this requirement by providing a standard mechanism for defining interface definitions separately from their embodiment within a particular binding (transport protocol and data-encoding format). Second, the widespread adoption of Web services mechanisms means that a framework based on Web services can exploit numerous tools and extant services, such as WSDL processors that can generate language binding for a variety of languages (e.g., Web Services Invocation Framework: WSIF [35]), workflow systems that sit on top of WSDL, and hosting environments for Web services (e.g., Microsoft .NET and Apache Axis). We emphasize that the use of Web services does not imply the use of SOAP for all communications. If needed, alternative transports can be used, for example, to achieve higher performance or to run over specialized network protocols.

8.4 AN OPEN GRID SERVICES ARCHITECTURE

We have argued that within internal enterprise IT infrastructures, SP-enhanced IT infrastructures, and multiorganizational Grids, computing is increasingly concerned with the creation, management, and application of dynamic ensembles of resources and services (and people) – what we call *virtual organizations* [2]. Depending on the context, these ensembles can be small or large, short-lived or long-lived, single institutional or multi-institutional, and homogeneous or heterogeneous. Individual ensembles may be structured hierarchically from smaller systems and may overlap in membership.

We assert that regardless of these differences, developers of applications for VOs face common requirements as they seek to deliver QoS – whether measured in terms of common security semantics, distributed workflow and resource management, coordinated fail-over, problem determination services, or other metrics – across a collection of resources with heterogeneous and often dynamic characteristics.

We now turn to the nature of these requirements and the mechanisms required to address them in practical settings. Extending our analysis in Reference [2], we introduce an Open Grid Services Architecture that supports the creation, maintenance, and application of ensembles of services maintained by VOs.

We start our discussion with some general remarks concerning the utility of a service-oriented Grid architecture, the importance of being able to virtualize Grid services, and essential service characteristics. Then, we introduce the specific aspects that we standardize in our definition of what we call a *Grid service*. We present more technical details in Section 8.6 (and in Reference [36]).

8.4.1 Service orientation and virtualization

When describing VOs, we can focus on the physical resources being shared (as in Reference [2]) or on the services supported by these resources. (A *service* is a network-enabled entity that provides some capability. The term object could arguably also be used, but we avoid that term owing to its overloaded meaning.) In OGSA, we focus on *services*: computational resources, storage resources, networks, programs, databases, and the like are all represented as services.

Regardless of our perspective, a critical requirement in a distributed, multiorganizational Grid environment is for mechanisms that enable interoperability [2]. In a service-oriented view, we can partition the interoperability problem into two subproblems, namely, the definition of service interfaces and the identification of the protocol(s) that can be used to invoke a particular interface – and, ideally, agreement on a standard set of such protocols.

A service-oriented view allows us to address the need for standard interface definition mechanisms, local/remote transparency, adaptation to local OS services, and uniform service semantics. A service-oriented view also simplifies virtualization – that is, the encapsulation behind a common interface of diverse implementations. Virtualization allows for consistent resource access across multiple heterogeneous platforms with local or remote location transparency, and enables mapping of multiple logical resource instances onto the same physical resource and management of resources within a VO based on composition from lower-level resources. Virtualization allows the composition of services to form more sophisticated services – without regard for how the services being composed are implemented. Virtualization of Grid services also underpins the ability to map common service semantic behavior seamlessly onto native platform facilities.

Virtualization is easier if service functions can be expressed in a standard form, so that any implementation of a service is invoked in the same manner. WSDL, which we adopt for this purpose, supports a service interface *definition* that is distinct from the protocol bindings used for service *invocation*. WSDL allows for multiple bindings for a single interface, including distributed communication protocol(s) (e.g., HTTP) as well as locally optimized binding(s) (e.g., local Inter-Process Communication (IPC)) for interactions between request and service processes on the same host. Other binding properties may include reliability (and other forms of QoS) as well as authentication and delegation of credentials. The choice of binding should always be transparent to the requestor with respect to service invocation semantics – but not with respect to other things: for example, a requestor should be able to choose a particular binding for performance reasons.

The service interface definition and access binding are also distinct from the *implementation* of the functionality of the service. A service can support multiple implementations on different platforms, facilitating seamless overlay not only to native platform facilities but also, via the nesting of service implementations, to virtual ensembles of resources. Depending on the platform and context, we might use the following implementation approaches.

1. We can use a reference implementation constructed for full portability across multiple platforms to support the execution environment (container) for hosting a service.
2. On a platform possessing specialized native facilities for delivering service functionality, we might map from the service interface definition to the native platform facilities.
3. We can also apply these mechanisms recursively so that a higher-level service is constructed by the composition of multiple lower-level services, which themselves may either map to native facilities or decompose further. The service implementation then dispatches operations to lower-level services (see also Section 8.4.4).

As an example, consider a distributed trace facility that records trace records to a repository. On a platform that does not support a robust trace facility, a reference implementation can be created and hosted in a service execution environment for storing and retrieving trace records on demand. On a platform already possessing a robust trace facility, however, we can integrate the distributed trace service capability with the native platform trace mechanism, thus leveraging existing operational trace management tools, auxiliary off-load, dump/restore, and the like, while semantically preserving the logical trace stream through the distributed trace service. Finally, in the case of a higher-level service, trace records obtained from lower-level services would be combined and presented as the integrated trace facility for the service.

Central to this virtualization of resource behaviors is the ability to adapt to operating system functions on specific hosts. A significant challenge when developing these *mappings* is to enable exploitation of native capabilities – whether concerned with performance monitoring, workload management, problem determination, or enforcement of native platform security policy – so that the Grid environment does not become the least common denominator of its constituent pieces. Grid service discovery mechanisms are important in this regard, allowing higher-level services to discover what capabilities are supported by a particular implementation of an interface. For example, if a native platform supports reservation capabilities, an implementation of a resource management interface (e.g., GRAM [18, 37]) can exploit those capabilities.

Thus, our service architecture supports *local and remote transparency with respect to service location and invocation*. It also provides for *multiple protocol bindings* to facilitate localized optimization of services invocation when the service is hosted locally with the service requestor, as well as to enable protocol negotiation for network flows across organizational boundaries where we may wish to choose between several InterGrid protocols, each optimized for a different purpose. Finally, we note that an implementation of a particular Grid service interface may map to native, nondistributed platform functions and capabilities.

8.4.2 Service semantics: The Grid service

Our ability to virtualize and compose services depends on more than standard interface definitions. We also require standard semantics for service interactions so that, for example, different services follow the same conventions for error notification. To this end, OGSA defines what we call a *Grid service*: a Web service that provides a set of well-defined interfaces and that follows specific conventions. The interfaces address discovery, dynamic service creation, lifetime management, notification, and manageability; the conventions address naming and upgradability. We expect also to address authorization and concurrency control as OGSA evolves. Two other important issues, authentication and reliable invocation, are viewed as service protocol bindings and are thus external to the core Grid service definition but must be addressed within a complete OGSA implementation. This separation of concerns increases the generality of the architecture without compromising functionality.

The interfaces and conventions that define a Grid service are concerned, in particular, with behaviors related to the management of *transient service instances*. VO participants typically maintain not merely a static set of persistent services that handle complex activity requests from clients. They often need to instantiate new transient service instances dynamically, which then handle the management and interactions associated with the state of particular requested activities. When the activity's state is no longer needed, the service can be destroyed. For example, in a videoconferencing system, the establishment of a videoconferencing session might involve the creation of service instances at intermediate points to manage end-to-end data flows according to QoS constraints. Or, in a Web serving environment, service instances might be instantiated dynamically to provide for consistent user response time by managing application workload through dynamically added capacity. Other examples of transient service instances might be a query against a database, a data-mining operation, a network bandwidth allocation, a running data transfer, and an advance reservation for processing capability. (These examples emphasize that service instances can be extremely lightweight entities, created to manage even short-lived activities.) Transience has significant implications of how services are managed, named, discovered, and used.

8.4.2.1 Upgradeability conventions and transport protocols

Services within a complex distributed system must be independently *upgradable*. Hence, versioning and compatibility between services must be managed and expressed so that clients can discover not only specific service versions but also compatible services. Further, services (and the hosting environments in which they run) must be upgradable without disrupting the operation of their clients. For example, an upgrade to the hosting environment may change the set of network protocols that can be used to communicate with the service, and an upgrade to the service itself may correct errors or even enhance the interface. Hence, OGSA defines conventions that allow us to identify when a service changes and when those changes are backwardly compatible with respect to interface and semantics (but not necessarily network protocol). OGSA also defines mechanisms for refreshing a client's knowledge of a service, such as what operations it supports or what network protocols can be used to communicate with the service. A service's description indicates the protocol binding(s) that can be used to communicate with the service. Two properties will often be desirable in such bindings.

- *Reliable service invocation*: Services interact with one another by the exchange of messages. In distributed systems prone to component failure, however, one can never guarantee that a message has been delivered. The existence of internal state makes it important to be able to guarantee that a service has received a message either once or not at all. From this foundation one can build a broad range of higher-level per-operation semantics, such as transactions.
- *Authentication*: Authentication mechanisms allow the identity of individuals and services to be established for policy enforcement. Thus, one will often desire a transport protocol that provides for mutual authentication of client and service instance, as well as the delegation of proxy credentials. From this foundation one can build a broad range of higher-level authorization mechanisms.

8.4.2.2 Standard interfaces

The interfaces (in WSDL terms, portTypes) that define a Grid service are listed in Table 8.1, introduced here, and described in more detail in Section 8.6 (and in Reference [36]). Note that while OGSA defines a variety of behaviors and associated interfaces, all but one of these interfaces (*GridService*) are optional.

Discovery: Applications require mechanisms for discovering available services and for determining the characteristics of those services so that they can configure themselves and their requests to those services appropriately. We address this requirement by defining

- a standard representation for *service data*, that is, information about Grid service instances, which we structure as a set of named and typed XML elements called *service data elements*, encapsulated in a standard container format;
- a standard operation, FindServiceData (within the required *GridService* interface), for retrieving service data from individual Grid service instances ('pull' mode access; see the *NotificationSource* interface below for 'push' mode access); and
- standard interfaces for registering information about Grid service instances with registry services (*Registry*) and for mapping from 'handles' to 'references' (*HandleMap* – to be explained in Section 8.6, when we discuss naming).

Table 8.1 Proposed OGSA Grid service interfaces (see text for details). The names provided here are likely to change in the future. Interfaces for authorization, policy management, manageability, and other purposes remain to be defined

PortType	Operation	Description
GridService	FindServiceData	Query a variety of information about the Grid service instance, including basic introspection information (handle, reference, primary key, home handleMap: terms to be defined), richer per-interface information, and service-specific information (e.g., service instances known to a registry); extensible support for various query languages
	SetTerminationTime	Set (and get) termination time for Grid service instance
	Destroy	Terminate Grid service instance
NotificationSource	SubscribeTo-NotificationTopic	Subscribe to notifications of service-related events, based on message type and interest statement; allows for delivery via third-party messaging services.
NotificationSink	DeliverNotification	Carry out asynchronous delivery of notification messages
Registry	RegisterService	Conduct soft-state registration of Grid service handles
	UnregisterService	Deregister a Grid service handle
Factory	CreateService	Create a new Grid service instance
HandleMap	FindByHandle	Return Grid Service Reference currently associated with supplied Grid service handle

Dynamic service creation: The ability to dynamically create and manage new service instances is a basic tenet of the OGSA model and necessitates the existence of service creation services. The OGSA model defines a standard interface (*Factory*) and semantics that any service-creation service must provide.

Lifetime management: Any distributed system must be able to deal with inevitable failures. In a system that incorporates transient, stateful service instances, mechanisms must be provided for *reclaiming services and states associated with failed operations*. For example, termination of a videoconferencing session might also require the termination of services created at intermediate points to manage the flow. We address this requirement by defining two standard operations: *Destroy* and *SetTerminationTime* (within the required *GridService* interface), for explicit destruction and soft-state lifetime management of Grid service instances, respectively. (*Soft-state protocols* [20, 21] allow state established at a remote location to be discarded eventually, unless refreshed by a stream of subsequent 'keepalive' messages. Such protocols have the advantages of being both resilient to failure – a single lost message need not cause irretrievable harm – and simple: no reliable 'discard' protocol message is required.)

Notification: A collection of dynamic, distributed services must be able to notify each other asynchronously of interesting changes to their state. OGSA defines common abstractions and service interfaces for subscription to (*NotificationSource*) and delivery of (*NotificationSink*) such *notifications*, so that services constructed by the composition of simpler services can deal with notifications (e.g., for errors) in standard ways. The *Notification-Source* interface is integrated with service data, so that a notification request is expressed as a request for subsequent 'push' mode delivery of service data. (We might refer to the capabilities provided by these interfaces as an event service [38], but we avoid that term because of its overloaded meaning.)

Other interfaces: We expect to define additional standard interfaces in the near future, to address issues such as authorization, policy management, concurrency control, and the monitoring and management of potentially large sets of Grid service instances.

8.4.3 The role of hosting environments

OGSA defines the semantics of a Grid service instance: how it is created, how it is named, how its lifetime is determined, how to communicate with it, and so on. However, while OGSA is prescriptive on matters of basic behavior, it does not place requirements on what a service does or how it performs that service. In other words, OGSA does not address issues of implementation programming model, programming language, implementation tools, or execution environment.

In practice, Grid services are instantiated within a specific execution environment, or *hosting environment*. A particular hosting environment defines not only implementation programming model, programming language, development tools, and debugging tools but also how an implementation of a Grid service meets its obligations with respect to Grid service semantics.

Today's e-Science Grid applications typically rely on *native operating system processes* as their hosting environment, with, for example, creation of a new service instance involving the creation of a new process. In such environments, a service itself may be implemented in a variety of languages such as C, C++, Java, or Fortran. Grid semantics may be implemented directly as part of the service or provided via a linked library [39]. Typically semantics are not provided via external services beyond those provided by the operating system. Thus, for example, lifetime management functions must be addressed within the application itself, if required.

Web services, on the other hand, may be implemented on more sophisticated *container or component-based* hosting environments such as J2EE, WebSphere, .NET, and Sun One. Such environments define a framework (container) within which components adhering to environment-defined interface standards can be instantiated and composed to build complex applications. Compared with the low levels of functionality provided by native hosting environments, container/component hosting environments tend to offer superior programmability, manageability, flexibility, and safety. Consequently, component/container-based hosting environments are seeing widespread use for building e-Business services. In the OGSA context, the container (hosting environment) has primary responsibility for ensuring that the services it supports adhere to Grid service semantics, and thus OGSA may motivate modifications or additions to the container/component interface.

By defining service semantics, OGSA specifies interactions between services in a manner independent of any hosting environment. However, as the above discussion highlights, successful implementation of Grid services can be facilitated by specifying baseline characteristics that all hosting environments must possess, defining the 'internal' interface from the service implementation to the global Grid environment. These characteristics would then be rendered into different implementation technologies (e.g., J2EE or shared libraries).

A detailed discussion of hosting environment characteristics is beyond the scope of this article. However, we can expect a hosting environment to address mapping of Grid-wide names (i.e., Grid service handles) into implementation-specific entities (C pointers, Java object references, etc.); dispatch of Grid invocations and notification events into implementation-specific actions (events, procedure calls); protocol processing and the formatting of data for network transmission; lifetime management of Grid service instances; and interservice authentication.

8.4.4 Using OGSA mechanisms to build VO structures

Applications and users must be able to create transient services and to discover and determine the properties of available services. The OGSA *Factory, Registry, GridService,* and *HandleMap* interfaces support the creation of transient service instances and the discovery and characterization of the service instances associated with a VO. (In effect, a registry service – a service instance that supports the *Registry* interface for registration and the *GridService* interface's FindServiceData operation, with appropriate service data, for discovery – defines the service set associated with a VO.) These interfaces can be used to construct a variety of VO service structures, as illustrated in Figure 8.2 and described in the following text.

Simple hosting environment: A simple execution environment is a set of resources located within a single administrative domain and supporting native facilities for service management, for example, a J2EE application server, Microsoft .NET system, or Linux cluster. In OGSA, the user interface to such an environment will typically be structured as a registry, one or more factories, and a handleMap service. Each factory is recorded in the registry, to enable clients to discover available factories. When a factory receives a client request to create a Grid service instance, the factory invokes hosting-environment-specific capabilities to create the new instance, assigns it a handle, registers the instance with the registry, and makes the handle available to the handleMap service. The implementations of these various services map directly into local operations.

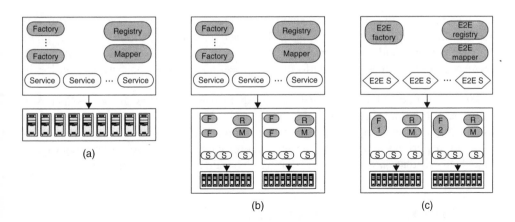

Figure 8.2 Three different VO structures, as described in the text: (a) simple hosting environment, (b) virtual hosting environment, and (c) collective services.

Virtual hosting environment: In more complex environments, the resources associated with a VO will span heterogeneous, geographically distributed 'hosting environments'. (e.g., in Figure 8.2, these resources span two simple hosting environments.) Nevertheless, this 'virtual hosting environment' (which corresponds, perhaps, to the set of resources associated with a B2B partnership) can be made accessible to a client via exactly the same interfaces as were used for the hosting environment just described. We create one or more 'higher-level' factories that delegate creation requests to lower-level factories. Similarly, we create a higher-level registry that knows about the higher-level factories and the service instances that they have created, as well any VO-specific policies that govern the use of VO services. Clients can use the VO registry to find factories and other service instances associated with the VO, and then use the handles returned by the registry to talk directly to those service instances. The higher-level factories and registry implement standard interfaces and so, from the perspective of the user, are indistinguishable from any other factory or registry.

Note that here, as in the previous example, the registry handle can be used as a globally unique name for the service set maintained by the VO. Resource management policies can be defined and enforced on the platforms hosting VO services, targeting the VO by this unique name.

Collective operations: We can also construct a 'virtual hosting environment' that provides VO participants with more sophisticated, virtual, 'collective,' or 'end-to-end' services. In this case, the registry keeps track of and advertises factories that create higher-level service instances. Such instances are implemented by asking lower-level factories to create multiple service instances and by composing the behaviors of those multiple lower-level service instances into that single, higher-level service instance.

These three examples, and the preceding discussion, illustrate how Grid service mechanisms can be used to integrate distributed resources both across virtual multiorganizational boundaries and within internal commercial IT infrastructures. In both cases, a collection of Grid services registered with appropriate discovery services can support functional capabilities delivering QoS interactions across distributed resource pools. Applications and middleware can exploit these services for distributed resource management across heterogeneous platforms with local and remote transparency and locally optimized flows.

Implementations of Grid services that map to native platform resources and APIs enable seamless integration of higher-level Grid services such as those just described with underlying platform components. Furthermore, service sets associated with multiple VOs can map to the same underlying physical resources, with those services represented as logically distinct at one level but sharing physical resource systems at lower levels.

8.5 APPLICATION EXAMPLE

We illustrate in Figure 8.3 the following stages in the life of a data-mining computation, which we use to illustrate the working of basic remote service invocation, lifetime management, and notification functions.

1. The environment initially comprises (from left to right) four simple hosting environments: one that runs the user application, one that encapsulates computing and storage resources (and that supports two factory services, one for creating storage reservations and the other for creating mining services), and two that encapsulate database services. The 'R's represent local registry services; an additional VO registry service presumably provides information about the location of all depicted services.
2. The user application invokes 'create Grid service' requests on the two factories in the second hosting environment, requesting the creation of a 'data-mining service' that will perform the data-mining operation on its behalf, and an allocation of temporary storage for use by that computation. Each request involves mutual authentication of the user and the relevant factory (using an authentication mechanism described in the factory's service description) followed by authorization of the request. Each request is successful and results in the creation of a Grid service instance with some initial lifetime. The new data-mining service instance is also provided with delegated proxy credentials that allow it to perform further remote operations on behalf of the user.
3. The newly created data-mining service uses its proxy credentials to start requesting data from the two database services, placing intermediate results in local storage. The data-mining service also uses notification mechanisms to provide the user application with periodic updates on its status. Meanwhile, the user application generates periodic 'keepalive' requests to the two Grid service instances that it has created.
4. The user application fails for some reason. The data-mining computation continues for now, but as no other party has an interest in its results, no further keepalive messages are generated.
5. Because of the application failure, keepalive messages cease, and so the two Grid service instances eventually time out and are terminated, freeing the storage and computing resources that they were consuming (not shown in figure).

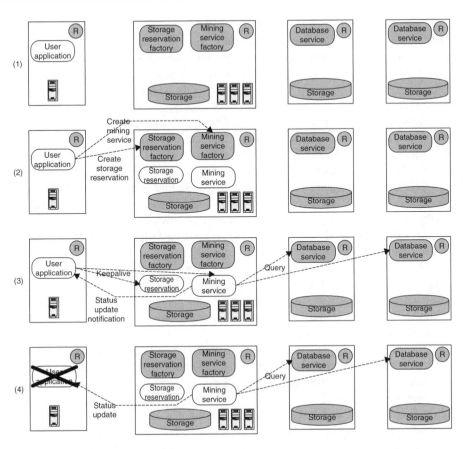

Figure 8.3 An example of Grid services at work. See text for details.

8.6 TECHNICAL DETAILS

We now present a more detailed description of the Grid service abstraction and associated interfaces and conventions.

8.6.1 The OGSA service model

A basic premise of OGSA is that everything is represented by a *service*: a network-enabled entity that provides some capability through the exchange of messages. Computational resources, storage resources, networks, programs, databases, and so forth are all services. This adoption of a uniform service-oriented model means that all components of the environment are virtual.

More specifically, OGSA represents everything as a *Grid service*: a Web service that conforms to a set of conventions and supports standard interfaces for such purposes as lifetime management. This core set of consistent interfaces, from which all Grid services are implemented, facilitates the construction of higher-order services that can be treated in a uniform way across layers of abstraction.

Grid services are characterized (*typed*) by the capabilities that they offer. A Grid service implements one or more *interfaces*, where each interface defines a set of operations that are invoked by exchanging a defined sequence of messages. Grid service interfaces correspond to portTypes in WSDL. The set of portTypes supported by a Grid service, along with some additional information relating to versioning, are specified in the Grid service's *serviceType*, a WSDL extensibility element defined by OGSA.

Grid services can maintain internal state for the lifetime of the service. The existence of state distinguishes one *instance* of a service from another that provides the same interface. We use the term *Grid service instance* to refer to a particular instantiation of a Grid service.

The protocol binding associated with a service interface can define a delivery semantics that addresses, for example, reliability. Services interact with one another by the exchange of messages. In distributed systems prone to component failure, however, one can never guarantee that a message that is sent has been delivered. The existence of internal state can make it important to be able to guarantee that a service has received a message once or not at all, even if failure recovery mechanisms such as retry are in use. In such situations, we may wish to use a protocol that guarantees exactly-once delivery or some similar semantics. Another frequently desirable protocol binding behavior is mutual authentication during communication.

OGSA services can be created and destroyed dynamically. Services may be destroyed explicitly, or may be destroyed or become inaccessible as a result of some system failure such as operating system crash or network partition. Interfaces are defined for managing service lifetime.

Because Grid services are dynamic and stateful, we need a way to distinguish one dynamically created service instance from another. Thus, every Grid service instance is assigned a globally unique name, the *Grid service handle* (*GSH*), that distinguishes a specific Grid service instance from all other Grid service instances that have existed, exist now, or will exist in the future. (If a Grid service fails and is restarted in such a way as to preserve its state, then it is essentially the same instance, and the same GSH can be used.)

Grid services may be upgraded during their lifetime, for example, to support new protocol versions or to add alternative protocols. Thus, the GSH carries no protocol- or instance-specific information such as network address and supported protocol bindings. Instead, this information is encapsulated, along with all other instance-specific information required to interact with a specific service instance, into a single abstraction called a *Grid service reference* (GSR). Unlike a GSH, which is invariant, the GSR(s) for a Grid service instance can change over that service's lifetime. A GSR has an explicit expiration time or may become invalid at any time during a service's lifetime, and OGSA defines mapping mechanisms, described below, for obtaining an updated GSR.

The result of using a GSR whose lifetime has expired is undefined. Note that holding a valid GSR does not guarantee access to a Grid service instance: local policy or access control constraints (e.g., maximum number of current requests) may prohibit servicing a request. In addition, the referenced Grid service instance may have failed, preventing the use of the GSR.

As everything in OGSA is a Grid service, there must be Grid services that manipulate the Grid service, handle, and reference abstractions that define the OGSA model. Defining a specific set of services would result in a specific rendering of the OGSA service model. We therefore take a more flexible approach and define a set of basic OGSA interfaces (i.e., WSDL portTypes) for manipulating service model abstractions. These interfaces can then be combined in different ways to produce a rich range of Grid services. Table 8.1 presents names and descriptions for the Grid service interfaces defined to date. Note that *only the* GridService *interface must be supported by all Grid services.*

8.6.2 Creating transient services: Factories

OGSA defines a class of Grid services that implement an interface that creates new Grid service instances. We call this the *Factory* interface, and a service that implements this interface a *factory*. The *Factory* interface's CreateService operation creates a requested Grid service and returns the GSH and initial GSR for the new service instance.

The *Factory* interface does not specify how the service instance is created. One common scenario is for the Factory interface to be implemented in some form of hosting environment (such as .NET or J2EE) that provides standard mechanisms for creating (and subsequently managing) new service instances. The hosting environment may define how services are implemented (e.g., language), but this is transparent to service requestors in OGSA, which see only the Factory interface. Alternatively, one can construct 'higher-level' factories that create services by delegating the request to other factory services (see Section 8.4.4). For example, in a Web serving environment, a new computer might be integrated into the active pool by asking an appropriate factory service to instantiate a 'Web serving' service on an idle computer.

8.6.3 Service lifetime management

The introduction of transient service instances raises the issue of determining the service's lifetime: that is, determining when a service can or should be terminated so that associated resources can be recovered. In normal operating conditions, a transient service instance is created to perform a specific task and terminates either on completion of this task or via an explicit request from the requestor or from another service designated by the requestor. In distributed systems, however, components may fail and messages may be lost. One result is that a service may never see an expected explicit termination request, thus causing it to consume resources indefinitely.

OGSA addresses this problem through a soft-state approach [21, 40] in which Grid service instances are created with a specified lifetime. The initial lifetime can be extended by a specified time period by explicit request of the client or another Grid service acting on the client's behalf (subject of course to the policy of the service). If that time period expires without having received a reaffirmation of interest from a client, either the hosting environment or the service instance itself is at liberty to terminate the service instance and release any associated resources.

Our approach to Grid service lifetime management has two desirable properties:

- A client knows, or can determine, when a Grid service instance will terminate. This knowledge allows the client to determine reliably when a service instance has terminated and hence its resources have been recovered, even in the face of system faults (e.g., failures of servers, networks, clients). The client knows exactly how long it has in order to request a final status from the service instance or to request an extension to the service's lifetime. Moreover, it also knows that if system faults occur, it need not continue attempting to contact a service after a known termination time and that any resources associated with that service would be released after that time – unless another client succeeded in extending the lifetime. In brief, lifetime management enables robust termination and failure detection, by clearly defining the lifetime semantics of a service instance.

- A hosting environment is guaranteed such that resource consumption is bounded, even in the face of system failures outside of its control. If the termination time of a service is reached, the hosting environment can reclaim all associated resources.

We implement soft-state lifetime management via the SetTerminationTime operation within the required *GridService* interface, which defines operations for negotiating an initial lifetime for a new service instance, for requesting a lifetime extension, and for harvesting a service instance when its lifetime has expired. We describe each of these mechanisms in turn.

Negotiating an initial lifetime: When requesting the creation of a new Grid service instance through a factory, a client indicates minimum and maximum acceptable initial lifetimes. The factory selects an initial lifetime and returns this to the client.

Requesting a lifetime extension: A client requests a lifetime extension via a SetTerminationTime message to the Grid service instance, which specifies a minimum and maximum acceptable new lifetime. The service instance selects a new lifetime and returns this to the client. Note that a keepalive message is effectively idempotent: the result of a sequence of requests is the same, even if intermediate requests are lost or reordered, as long as not so many requests are lost that the service instance's lifetime expires.

The periodicity of keepalive messages can be determined by the client based on the initial lifetime negotiated with the service instance (and perhaps renegotiated via subsequent keepalive messages) and knowledge about network reliability. The interval size allows trade-offs between currency of information and overhead.

We note that this approach to lifetime management provides a service with considerable autonomy. Lifetime extension requests from clients are not mandatory: the service can apply its own policies on granting such requests. A service can decide at any time to extend its lifetime, either in response to a lifetime extension request by a client or for any other reason. A service instance can also cancel itself at any time, for example, if resource constraints and priorities dictate that it relinquishes its resources. Subsequent client requests that refer to this service will fail.

The use of absolute time in the *SetTerminationTime* operation – and, for that matter, in Grid service information elements, and commonly in security credentials – implies the existence of a global clock that is sufficiently well synchronized. The Network Time Protocol (NTP) provides standardized mechanisms for clock synchronization and can typically synchronize clocks within at most tens of milliseconds, which is more than adequate for the purposes of lifetime management. Note that we are not implying by these statements a requirement for ordering of events, although we expect to introduce some such mechanisms in future revisions.

8.6.4 Managing handles and references

As discussed above, the result of a factory request is a GSH and a GSR. While the GSH is guaranteed to reference the created Grid service instance in perpetuity, the GSR is created with a finite lifetime and may change during the service's lifetime. While this strategy has the advantage of increased flexibility from the perspective of the Grid service provider, it introduces the problem of obtaining a valid GSR once the GSR returned by the service creation operation expires. At its core, this is a bootstrapping problem: how does one establish communication with a Grid service given only its GSH? We describe here how these issues are addressed in the Grid service specification as of June 2002, but note that this part of the specification is likely to evolve in the future, at a minimum to support multiple handle representations and handle mapping services.

The approach taken in OGSA is to define a handle-to-reference mapper interface (*HandleMap*). The operations provided by this interface take a GSH and return a valid GSR. Mapping operations can be access controlled, and thus a mapping request may be denied. An implementation of the *HandleMap* interface may wish to keep track of what Grid service instances are actually in existence and not return references to instances that it knows have terminated. However, possession of a valid GSR does not assure that a Grid service instance can be contacted: the service may have failed or been explicitly terminated between the time the GSR was given out and the time that it was used. (Obviously, if termination of a service is scheduled, it is desirable to represent this in the GSR lifetime, but it is not required.)

By introducing the *HandleMap* interface, we partition the general problem of obtaining a GSR for an arbitrary service into two more specific subproblems:

1. identifying a handleMap service that contains the mapping for the specified GSH, and
2. contacting that handleMap to obtain the desired GSR.

We address these two subproblems in turn. To ensure that we can always map a GSH to a GSR, we require that every Grid service instance be registered with at least one handleMap, which we call the *home* handleMap. By structuring the GSH to include the home handleMap's identity, we can easily and scalably determine which handleMap to contact to obtain a GSR for a given GSH. Hence, unique names can be determined locally, thus avoiding scalability problems associated with centralized name allocation services – although relying on the Domain Name System [13]. Note that GSH mappings can also live in other handleMaps. However, every GSH must have exactly one home handleMap.

How do we identify the home handleMap within a GSH? Any service that implements the *HandleMap* interface is a Grid service, and as such will have a GSH. If we use this name in constructing a GSH, however, then we are back in the same position of trying to obtain a GSR from the handleMap service's GSH. To resolve this bootstrapping problem, we need a way to obtain the GSR for the handleMap without requiring a handleMap! We accomplish this by requiring that all home handleMap services be identified by a URL and support a bootstrapping operation that is bound to a single, well-known protocol, namely, HTTP (or HTTPS). Hence, instead of using a GSR to describe what protocols should be used to contact the handleMap service, an HTTP GET operation is used on the URL that points to the home handleMap, and the GSR for the handleMap, in WSDL form, is returned.

Note that a relationship exists between services that implement the *HandleMap* and *Factory* interfaces. Specifically, the GSH returned by a factory request must contain the URL of the home handleMap, and the GSH/GSR mapping must be entered and updated into the handleMap service. The implementation of a factory must decide what service to use as the home handleMap. Indeed, a single service may implement both the *Factory* and *HandleMap* interfaces.

Current work within GGF is revising this Grid service component to allow for other forms of handles and mappers/resolvers and/or to simplify the current handle and resolver.

8.6.5 Service data and service discovery

Associated with each Grid service instance is a set of *service data*, a collection of XML elements encapsulated as service data elements. The packaging of each element includes a name that is unique to the Grid service instance, a type, and time-to-live information that a recipient can use for lifetime management.

The obligatory *GridService* interface defines a standard WSDL operation, FindService-Data, for querying and retrieving service data. This operation requires a simple 'by name' query language and is extensible to allow for the specification of the query language used, which may be, for example, Xquery [41].

The Grid service specification defines for each Grid service interface a set of zero or more service data elements that must be supported by any Grid service instance that supports that interface. Associated with the *GridService* interface, and thus obligatory for any Grid service instance, is a set of elements containing basic information about a Grid service instance, such as its GSH, GSR, primary key, and home handleMap.

One application of the *GridService* interface's FindServiceData operation is service discovery. Our discussion above assumed that one has a GSH that represents a desired service. But how does one obtain the GSH in the first place? This is the essence of *service discovery*, which we define here as the process of identifying a subset of GSHs from a specified set based on GSH attributes such as the interfaces provided, the number of requests that have been serviced, the load on the service, or policy statements such as the number of outstanding requests allowed.

A Grid service that supports service discovery is called a *registry*. A registry service is defined by two things: the *Registry* interface, which provides operations by which GSHs can be registered with the registry service, and an associated service data element used to contain information about registered GSHs. Thus, the *Registry* interface is used to register a GSH, and the *GridService* interface's FindServiceData operation is used to retrieve information about registered GSHs.

The *Registry* interface allows a GSH to register with a registry service to augment the set of GSHs that are considered for subsetting. As in MDS-2 [19], a service (or VO) can use this operation to notify interested parties within a VO of its existence and the service(s) that it provides. These interested parties typically include various forms of service discovery services, which collect and structure service information in order to respond efficiently to service discovery requests. As with other stateful interfaces in OGSA, GSH registration is a soft-state operation and must be periodically refreshed, thus allowing discovery services to deal naturally with dynamic service availability.

We note that specification of the attributes associated with a GSH is not tied to the registration of a GSH to a service implementing the *GridService* interface. This feature is important because attribute values may be dynamic and there may be a variety of ways in which attribute values may be obtained, including consulting another service implementing the *GridService* interface.

8.6.6 Notification

The OGSA notification framework allows clients to register interest in being notified of particular messages (the *NotificationSource* interface) and supports asynchronous, one-way delivery of such notifications (*NotificationSink*). If a particular service wishes to support subscription of notification messages, it must support the *NotificationSource* interface to manage the subscriptions. A service that wishes to receive notification messages must implement the *NotificationSink* interface, which is used to deliver notification messages. To start notification from a particular service, one invokes the subscribe operation on the notification source interface, giving it the service GSH of the notification sink. A stream of notification messages then flow from the source to the sink, while the sink sends periodic keepalive messages to notify the source that it is still interested in receiving notifications. If reliable delivery is desired, this behavior can be implemented by defining an appropriate protocol binding for this service.

An important aspect of this notification model is a close integration with service data: a subscription operation is just a request for subsequent 'push' delivery of service data that meet specified conditions. (Recall that the FindServiceData operation provides a 'pull' model.)

The framework allows both for direct service-to-service notification message delivery, and for integration with various third-party services, such as messaging services commonly used in the commercial world, or custom services that filter, transform, or specially deliver notification messages on behalf of the notification source. Notification semantics are a property of the protocol binding used to deliver the message. For example, a SOAP/HTTP protocol or direct User Datagram Protocol (UDP) binding would provide point-to-point, best-effort notification, while other bindings (e.g., some proprietary message service) would provide better than best-effort delivery. A multicast protocol binding would support multiple receivers.

8.6.7 Change management

In order to support *discovery* and *change management* of Grid services, Grid service interfaces must be globally and uniquely named. In WSDL, an interface is defined by a portType and is globally and uniquely named by the portType's qname (i.e., an XML namespace as defined by the targetNamespace attribute in the WSDL document's definitions element, and a local name defined by the portType element's name attribute). Any changes made to the definition of a Grid service, either by changing its interface or by making semantically significant implementation changes to the operations, must be reflected through new interface names (i.e., new portTypes and/or serviceTypes). This feature allows clients that require Grid Services with particular properties (either particular interfaces or implementation semantics) to discover compatible services.

8.6.8 Other interfaces

We expect in the future to define an optional *Manageability* interface that supports a set of manageability operations. Such operations allow potentially large sets of Grid service instances to be monitored and managed from management consoles, automation tools, and the like. An optional *Concurrency* interface will provide concurrency control operations.

8.7 NETWORK PROTOCOL BINDINGS

The Web services framework can be instantiated on a variety of different protocol bindings. SOAP + HTTP with TLS for security is one example, but others can and have been defined. Here we discuss some issues that arise in the OGSA context.

In selecting network protocol bindings within an OGSA context, we must address four primary requirements:

- *Reliable transport*: As discussed above, the Grid services abstraction can require support for reliable service invocation. One way to address this requirement is to incorporate appropriate support within the network protocol binding, as, for example, in HTTP-R.
- *Authentication and delegation*: As discussed above, the Grid services abstraction can require support for communication of proxy credentials to remote sites. One way to address this requirement is to incorporate appropriate support within the network protocol binding, as, for example, in TLS extended with proxy credential support.
- *Ubiquity*: The Grid goal of enabling the dynamic formation of VOs from distributed resources means that, in principle, it must be possible for any arbitrary pair of services to interact.
- *GSR format*: Recall that the Grid Service Reference can take a binding-specific format. One possible GSR format is a WSDL document; CORBA IOR is another.

The successful deployment of large-scale interoperable OGSA implementations would benefit from the definition of a small number of standard protocol bindings for Grid service discovery and invocation. Just as the ubiquitous deployment of the Internet Protocol allows essentially any two entities to communicate, so ubiquitous deployment of such 'Inter-Grid' protocols will allow any two services to communicate. Hence, clients can be particularly simple, since they need to know about only one set of protocols. (Notice that the definition of such standard protocols does not prevent a pair of services from using an alternative protocol, if both support it.) Whether such InterGrid protocols can be defined and gain widespread acceptance remains to be seen. In any case, their definition is beyond the scope of this article.

8.8 HIGHER-LEVEL SERVICES

The abstractions and services described in this article provide building blocks that can be used to implement a variety of higher-level Grid services. We intend to work closely with the community to define and implement a wide variety of such services that will, collectively, address the diverse requirements of e-Business and e-Science applications. These are likely to include the following:

- *Distributed data management services*, supporting access to and manipulation of distributed data, whether in databases or files [42]. Services of interest include database access, data translation, replica management, replica location, and transactions.
- *Workflow services*, supporting the coordinated execution of multiple application tasks on multiple distributed Grid resources.
- *Auditing services*, supporting the recording of usage data, secure storage of that data, analysis of that data for purposes of fraud and intrusion detection, and so forth.
- *Instrumentation and monitoring services*, supporting the discovery of 'sensors' in a distributed environment, the collection and analysis of information from these sensors, the generation of alerts when unusual conditions are detected, and so forth.
- *Problem determination services for distributed computing,* including dump, trace, and log mechanisms with event tagging and correlation capabilities.
- *Security protocol mapping services,* enabling distributed security protocols to be transparently mapped onto native platform security services for participation by platform resource managers not implemented to support the distributed security authentication and access control mechanism.

The flexibility of our framework means that such services can be implemented and composed in a variety of different ways. For example, a coordination service that supports the simultaneous allocation and use of multiple computational resources can be instantiated as a service instance, linked with an application as a library, or incorporated into yet higher-level services.

It appears straightforward to reengineer the resource management, data transfer, and information service protocols used within the current Globus Toolkit to build on these common mechanisms (see Figure 8.4). In effect, we can *refactor* the design of those protocols, extracting similar elements to exploit commonalities. In the process, we enhance

Figure 8.4 (a) Some current Globus Toolkit protocols and (b) a potential refactoring to exploit OGSA mechanisms.

the capabilities of the current protocols and arrive at a common service infrastructure. This process will produce Globus Toolkit 3.0.

8.9 RELATED WORK

We note briefly some relevant prior and other related work, focusing in particular on issues relating to the secure and reliable remote creation and management of transient, stateful services.

As discussed in Section 8.3.1, many OGSA mechanisms derive from the Globus Toolkit v2.0: in particular, the factory (GRAM gatekeeper [18]), registry (GRAM reporter [18] and MDS-2 [19]), use of soft-state registration (MDS-2 [19]), secure remote invocation with delegation (GSI [25]), and reliable remote invocation (GRAM [18]). The primary differences relate to how these different mechanisms are integrated, with OGSA refactoring key design elements so that, for example, common notification mechanisms are used for service registration and service state.

OGSA can be viewed as a distributed object system [43], in the sense that each Grid service instance has a unique identity with respect to the other instances in the system, and each instance can be characterized as state coupled with behavior published through type-specific operations. In this respect, OGSA exploits ideas developed previously in systems such as Eden [44], Argus [45], CORBA [46], SOS [47], Spring [48], Globe [49], Mentat [50], and Legion [51, 52]. In contrast to CORBA, OGSA, like Web services, addresses directly issues of secure interoperability and provides a richer interface definition language. In Grid computing, the Legion group has promoted the use of object models, and we can draw parallels between certain OGSA and Legion constructs, in particular the factory ('Class Object'), handleMap ('Binding Agent'), and timeouts on bindings. However, we also note that OGSA is nonprescriptive on several issues that are often viewed as central to distributed object systems, such as the use of object technologies in implementations, the exposure of inheritance mechanisms at the interface level, and hosting technology.

Soft-state mechanisms have been used for management of specific state in network entities within Internet protocols [21, 40, 53] and (under the name 'leases') in RMI and Jini [54]. In OGSA, all services and information are open to soft-state management. We prefer soft-state techniques to alternatives such as distributed reference counting [55] because of their relative simplicity.

Our reliable invocation mechanisms are inspired by those used in Condor [23, 56, 57], which in turn build on much prior work in distributed systems.

As noted in Section 8.4.3, core OGSA service behaviors will, in general, be supported via some form of hosting environment that simplifies the development of individual components by managing persistence, security, life cycle management, and so forth. The notion of a hosting environment appears in various operating systems and object systems.

The application of Web services mechanisms to Grid computing has also been investigated and advocated by others (e.g., [58, 59]), with a recent workshop providing overviews of a number of relevant efforts [60]. Gannon *et al.* [59] discuss the application of various contemporary technologies to e-Science applications and propose 'application factories' (with WSDL interfaces) as a means of creating application services dynamically. De Roure *et al.* [61] propose a 'Semantic Grid', by analogy to the Semantic Web [62], and propose a range of higher-level services. Work on service-oriented interfaces to numerical software in NetSolve [63, 64] and Ninf [65] is also relevant.

Sun Microsystems' JXTA system [66] addresses several important issues encountered in Grids, including discovery of and membership in virtual organizations – what JXTA calls 'peer groups'. We believe that these abstractions can be implemented within the OGSA framework.

There are connections to be made with component models for distributed and high-performance computing [67–69], some implementations of which build on Globus Toolkit mechanisms.

8.10 SUMMARY

We have defined an Open Grid Services Architecture that supports, via standard interfaces and conventions, the creation, termination, management, and invocation of *stateful, transient services as named, managed entities with dynamic, managed lifetime.*

Within OGSA, everything is represented as a *Grid service*, that is, a (potentially transient) service that conforms to a set of conventions (expressed using WSDL) for such purposes as lifetime management, discovery of characteristics, notification, and so on. Grid service implementations can target native platform facilities for integration with, and of, existing IT infrastructures. Standard interfaces for creating, registering, and discovering Grid services can be configured to create various forms of VO structure.

The merits of this service-oriented model are as follows: All components of the environment are virtualized. By providing a core set of consistent interfaces from which all Grid services are implemented, we facilitate the construction of hierarchal, higher-order services that can be treated in a uniform way across layers of abstraction. Virtualization also enables mapping of multiple logical resource instances onto the same physical resource, composition of services regardless of implementation, and management of resources within a VO based on composition from lower-level resources. It is virtualization of Grid services that underpins the ability for mapping common service semantic behavior seamlessly onto native platform facilities.

The development of OGSA represents a natural evolution of the Globus Toolkit 2.0, in which the key concepts of factory, registry, reliable and secure invocation, and so on exist, but in a less general and flexible form than here, and without the benefits of a uniform interface definition language. In effect, OGSA refactors key design elements so that, for example, common notification mechanisms are used for service registration and service state. OGSA also further abstracts these elements so that they can be applied at any level to virtualize VO resources. The Globus Toolkit provides the basis for an open-source OGSA implementation, Globus Toolkit 3.0, that supports existing Globus APIs as well as WSDL interfaces, as described at www.globus.org/ogsa.

The development of OGSA also represents a natural evolution of Web services. By integrating support for transient, stateful service instances with existing Web services technologies, OGSA extends significantly the power of the Web services framework, while requiring only minor extensions to existing technologies.

ACKNOWLEDGMENTS

We are pleased to acknowledge the many contributions to the Open Grid Services Architecture of Karl Czajkowski, Jeffrey Frey, and Steve Graham. We are also grateful to numerous colleagues for discussions on the topics covered here and/or for helpful comments on versions of this article, in particular, Malcolm Atkinson, Brian Carpenter, David De Roure, Andrew Grimshaw, Marty Humphrey, Keith Jackson, Bill Johnston, Kate Keahey, Gregor von Laszewski, Lee Liming, Miron Livny, Norman Paton, Jean-Pierre Prost, Thomas Sandholm, Peter Vanderbilt, and Von Welch.

This work was supported in part by the Mathematical, Information, and Computational Sciences Division subprogram of the Office of Advanced Scientific Computing Research, U.S. Department of Energy, under Contract W-31-109-Eng-38; by the National Science Foundation; by the NASA Information Power Grid program; and by IBM.

REFERENCES

1. Foster, I. and Kesselman, C. (eds) (1999) *The Grid: Blueprint for a New Computing Infrastructure*. San Francisco, CA: Morgan Kaufmann Publishers, 1999.
2. Foster, I., Kesselman, C. and Tuecke, S. (2001) The anatomy of the grid: Enabling scalable virtual organizations. *International Journal of High Performance Computing Applications*, **15**(3), 200–222, www.globus.org/research/papers/anatomy.pdf.
3. Foster, I. and Kesselman, C. (1999) Globus: A toolkit-based grid architecture, in Foster, I. and Kesselman, C. (eds) *The Grid: Blueprint for a New Computing Infrastructure*. San Francisco, CA: Morgan Kaufmann Publishers, 1999, pp. 259–278.
4. Graham, S., Simeonov, S., Boubez, T., Daniels, G., Davis, D., Nakamura, Y. and Neyama, R. (2001) *Building Web Services with Java: Making Sense of XML, SOAP, WSDL, and UDDI*. Indianapolis, IN: Sams Publishing, 2001.
5. Kreger, H. (2001) *Web Services Conceptual Architecture*, IBMTechnical Report WCSA 1.0.
6. Foster, I., Kesselman, C., Nick, J. M. and Tuecke, S. (2002) Grid services for distributed systems integration. *IEEE Computer*, **35**(6), 37–46.
7. Catlett, C. (1992) In search of gigabit applications. *IEEE Communications Magazine*, April, pp. 42–51.
8. Catlett, C. and Smarr, L. Metacomputing. (1992) *Communications of the ACM*, **35**(6), 44–52.

9. Foster, I. (2002) The grid: A new infrastructure for 21st century science. *Physics Today*, **55**(2), 42–47.
10. Johnston, W. E., Gannon, D. and Nitzberg, B. (1999) Grids as production computing environments: the engineering aspects of NASA's Information Power Grid, *Proc. 8th IEEE Symposium on High Performance Distributed Computing*. IEEE Press, 1999.
11. Stevens, R., Woodward, P., DeFanti, T. and Catlett, C. (1997) From the I-WAY to the National Technology Grid. *Communications of the ACM*, **40**(11), 50–61.
12. Johnston, W. E. (1999) Realtime widely distributed instrumentation systems, in Foster, I. and Kesselman, C. (eds) *The Grid: Blueprint for a New Computing Infrastructure*. San Francisco, CA: Morgan Kaufmann Publishers, 1999, pp. 75–103.
13. Mockapetris, P. V. and Dunlap, K. (1988) Development of the domain name system, *Proceedings of SIGCOMM '88, Computer Communication Review*, **18**(4), 123–133.
14. Bal, H. E., Steiner, J. G. and Tanenbaum, A. S. (1989) Programming languages for distributed computing systems. *ACM Computing Surveys*, **21**(3), 261–322.
15. Mullender, S. (ed.) (1989) *Distributed Systems*. New York: Addison-Wiley, A89.
16. Thomas, A. (1998) *Enterprise Java Beans Technology: Server Component Model for the Java Platform*, http://java.sun.com/products/ejb/white_paper.html.
17. Nick, J. M., Moore, B. B., Chung, J.-Y. and Bowen, N. S. (1997) S/390 cluster technology: Parallel sysplex. *IBM Systems Journal*, **36**(2), 172–201.
18. Czajkowski, K., Foster, I., Karonis, N., Kesselman, C., Martin, S., Smith, W. and Tuecke, S. (1998) A resource management architecture for metacomputing systems, *4th Workshop on Job Scheduling Strategies for Parallel Processing*. New York: Springer-Verlag, 1998, pp. 62–82.
19. Czajkowski, K., Fitzgerald, S., Foster, I. and Kesselman, C. (2001) Grid information services for distributed resource sharing, *10th IEEE International Symposium on High Performance Distributed Computing*. New York: IEEE Press, 2001, pp. 181–184.
20. Raman, S. and McCanne, S. (1999) A model, analysis, and protocol framework for soft state-based communication. *Computer Communication Review*, **29**(4), 15–25.
21. Zhang, L., Braden, B., Estrin, D., Herzog, S. and Jamin, S. (1993) RSVP: A new resource ReSerVation protocol. *IEEE Network*, **7**(5), 8–18.
22. Gasser, M. and McDermott, E. (1990) An architecture for practical delegation in a distributed system, *Proc. 1990 IEEE Symposium on Research in Security and Privacy*. IEEE Press, pp. 20–30.
23. Livny, M. (1999) High-throughput resource management, in Foster, I. and Kesselman, C. (eds) *The Grid: Blueprint for a New Computing Infrastructure*. San Francisco, CA: Morgan Kaufmann Publishers, 1999, pp. 311–337.
24. Gullapalli, S., Czajkowski, K., Kesselman, C. and Fitzgerald, S. (2001) The Grid Notification Framework, Global Grid Forum, Draft GWD-GIS-019.
25. Foster, I., Kesselman, C., Tsudik, G. and Tuecke, S. (1998) A security architecture for computational grids. *ACM Conference on Computers and Security*, 1998, pp. 83–91.
26. Steiner, J., Neuman, B. C. and Schiller, J. (1988) Kerberos: An authentication system for open network systems. *Proc. Usenix Conference*, 1988, pp. 191–202.
27. Tuecke, S., Engert, D., Foster, I., Thompson, M., Pearlman, L. and Kesselman, C. (2001) Internet X.509 Public Key Infrastructure Proxy Certificate Profile. IETF, Draft draft-ietf-pkix-proxy-01.txt.
28. Bray, T., Paoli, J. and Sperberg-McQueen, C. M. (1998) The Extensible Markup Language (XML) 1.0.
29. Fallside, D. C. (2001) XML Schema Part 0: Primer W3C, Recommendation, http://www.w3.org/TR/xmlschema-0/.
30. Simple Object Access Protocol (SOAP) 1.1, W3C Note 8, 2000.
31. Christensen, E., Curbera, F., Meredith, G. and Weerawarana, S. (2001) Web Services Description Language (WSDL) 1.1, W3C Note 15, www.w3.org/TR/wsdl.
32. Brittenham, P. (2001) An Overview of the Web Services Inspection Language, www.ibm.com/developerworks/webservices/library/ws-wsilover.
33. UDDI: Universal Description, Discovery and Integration, www.uddi.org.
34. Web Services Flow Language, www-4.ibm.com/software/solutions/webservices/pdf/WSFL.pdf.

35. Mukhi, N. (2001) Web Service Invocation Sans SOAP,
    http://www.ibm.com/developerworks/library/ws-wsif.html.
36. Tuecke, S., Czajkowski, K., Foster, I., Frey, J., Graham, S. and Kesselman, C. (2002) *Grid Services Specification*, www.globus.org/research/papers/gsspec.pdf.
37. Foster, I., Kesselman, C., Lee, C., Lindell, R., Nahrstedt, K. and Roy, A. (1999) A distributed resource management architecture that supports advance reservations and co-allocation. *Proc. International Workshop on Quality of Service*, 1999, pp. 27–36.
38. Barrett, D. J., Clarke, L. A., Tarr, P. L. and Wise, A. E. (1996) A framework for event-based software integration. *ACM Transactions on Software Engineering and Methodology*, **5**(4), 378–421.
39. Getov, V., Laszewski, G. V., Philippsen, M. and Foster, I. (2001) Multiparadigm communications in Java for Grid computing. *Communications of the ACM*, **44**(10), 118–125.
40. Clark, D. D. (1988) The design philosophy of the DARPA internet protocols, *SIGCOMM Symposium on Communications Architectures and Protocols*. Atlantic City, NJ: ACM Press, 1988, pp. 106–114.
41. Chamberlin, D. (2001) Xquery 1.0: An XML Query Language, W3C Working Draft 07.
42. Paton, N. W., Atkinson, M. P., Dialani, V., Pearson, D., Storey, T. and Watson, P. (2002) *Database Access and Integration Services on the Grid*. Manchester: Manchester University.
43. Chin, R. S. and Chanson, S. T. (1991) Distributed object-based programming systems. *ACM Computing Surveys*, **23**(1), 91–124.
44. Almes, G. T., Black, A. P., Lazowska, E. D. and Noe, J. D. (1985) The Eden system: A technical review. *IEEE Transactions on Software Engineering*, **SE-11**(1), 43–59.
45. Liskov, B. (1988) Distributed programming in Argus. *Communications of the ACM*, **31**(3), 300–312.
46. Common Object Request Broker: Architecture and Specification, Revision 2.2. Object Management Group Document 96.03.04, 1998.
47. Shapiro, M. (1989) SOS: An object oriented operating system – assessment and perspectives. *Computing Systems*, **2**(4), 287–337.
48. Mitchell, J. G. *et al.* (1994) *An overview of the spring system*. COMPCON, 1994, pp. 122–131
49. Steen, M. V., Homburg, P., Doorn, L. V., Tanenbaum, A. and Jonge, W. D. Towards object-based wide area distributed systems, in Carbrera, L.-F. and Theimer, M. (eds) *International Workshop on Object Orientation in Operating Systems*. 1995, pp. 224–227.
50. Grimshaw, A. S., Ferrari, A. J. and West, E. A. (1997) *Mentat: Parallel Programming Using C++*. Cambridge, MA: MIT Press, 1997, pp. 383–427.
51. Grimshaw, A. S., Ferrari, A., Knabe, F. C. and Humphrey, M. (1999) Wide-area computing: resource sharing on a large scale. *IEEE Computer*, **32**(5), 29–37.
52. Grimshaw, A. S. and Wulf, W. A. (1997) The Legion vision of a worldwide virtual computer. *Communications of the ACM*, **40**(1), 39–45.
53. Sharma, P., Estrin, D., Floyd, S. and Jacobson, V. (1997) Scalable timers for soft state protocols, *IEEE Infocom '97*. IEEE Press, 1997.
54. Oaks, S. and Wong, H. (2000) *Jini in a Nutshell*. O'Reilly; www.oreilly.com.
55. Bevan, D. I. (1987) Distributed garbage collection using reference counting, *PARLE Parallel Architectures and Languages Europe*. New York: Springer-Verlag, LNCS 259, pp. 176–187.
56. Frey, J., Tannenbaum, T., Foster, I., Livny, M. and Tuecke, S. (2001) Condor-G: A computation management agent for multi-institutional grids, *10th International Symposium on High Performance Distributed Computing*. IEEE Press, pp. 55–66.
57. Litzkow, M. and Livny, M. (1990) Experience with the Condor distributed batch system. *IEEE Workshop on Experimental Distributed Systems*, 1990.
58. Fox, G., Balsoy, O., Pallickara, S., Uyar, A., Gannon, D. and Slominski, A. Indian University in Bloomington, IN (2002) *Community Grids*. Community Grid Computing Laboratory, Indiana University, 2002.
59. Gannon, D. *et al.* (2001) Programming the grid: Distributed software components, P2P, and Grid Web services for scientific applications. *Grid 2001*, 2001.

60. Grid Web Services Workshop. (2001)
https://gridport.npaci.edu/workshop/webserv01/agenda.html.
61. De Roure, D., Jennings, N. and Shadbolt, N. (2002) *Research Agenda for the Semantic Grid: A Future e-Science Infrastructure*. Edinburgh, UK: UK National eScience Center, www.semanticgrid.org.
62. Berners-Lee, T., Hendler, J. and Lassila, O. (2001) The semantic Web. *Scientific American*, May 2001, **284**(5) pp. 34–43.
63. Casanova, H. and Dongarra, J. (1997) NetSolve: A network server for solving computational science problems. *International Journal of Supercomputer Applications and High Performance Computing*, **11**(3), 212–223.
64. Casanova, H., Dongarra, J., Johnson, C. and Miller, M. (1999) Application-specific tools, in Foster, I. and Kesselman, C. (eds) *The Grid: Blueprint for a New Computing Infrastructure*. San Francisco, CA: Morgan Kaufmann Publishers, pp. 159–180.
65. Nakada, H., Sato, M. and Sekiguchi, S. (1999) Design and implementations of Ninf: Towards a global computing infrastructure. *Future Generation Computing Systems*, **15**(5–6), 649–658.
66. JXTA, www.jxta.org.
67. Armstrong, R., Gannon, D., Geist, A., Keahey, K., Kohn, S., McInnes, L. and Parker, S. (1999) *Toward a common component architecture for high performance scientific computing*. Proc. 8th IEEE Symp. on High Performance Distributed Computing, 1999.
68. Bramley, R. *et al.* (1998) Component architectures for distributed scientific problem solving. *IEEE Computational Science and Engineering*, **5**(2), 50–63.
69. Villacis, J. *et al.* (1999) CAT: A high performance, distributed component architecture toolkit for the grid. *IEEE Intl. Symp. on High Performance Distributed Computing*, 1999.

9

Grid Web services and application factories

Dennis Gannon, Rachana Ananthakrishnan, Sriram Krishnan, Madhusudhan Govindaraju, Lavanya Ramakrishnan, and Aleksander Slominski

Indiana University, Bloomington, Indiana, United States

9.1 INTRODUCTION

A Grid can be defined as a layer of networked services that allow users single sign-on access to a distributed collection of compute, data, and application resources. The Grid services allow the entire collection to be seen as a seamless information processing system that the user can access from any location. Unfortunately, for application developers, this Grid vision has been a rather elusive goal. The problem is that while there are several good frameworks for Grid architectures (Globus [1] and Legion/Avaki [18]), the task of application development and deployment has not become easier. The heterogeneous nature of the underlying resources remains a significant barrier. Scientific applications often require extensive collections of libraries that are installed in different ways on different platforms. Moreover, Unix-based default user environments vary radically between different users and even between the user's interactive environment and the default environment provided in a batch queue. Consequently, it is almost impossible for one application developer to

Grid Computing – Making the Global Infrastructure a Reality.   Edited by F. Berman, A. Hey and G. Fox
© 2003 John Wiley & Sons, Ltd   ISBN: 0-470-85319-0

hand an execution script and an executable object code to another user and to expect the second user to be able to successfully run the program on the same machine, let alone a different machine on the Grid. The problem becomes even more complex when the application is a distributed computation that requires a user to successfully launch a heterogeneous collection of applications on remote resources. Failure is the norm and it can take days, if not weeks, to track down all the incorrectly set environment variables and path names.

A different approach, and the one advocated in this paper, is based on the Web services model [2–5], which is quickly gaining attention in the industry. The key idea is to isolate the responsibility of deployment and instantiation of a component in a distributed computation from the user of that component. In a Web service model, the users are only responsible for accessing running services. The Globus Toolkit provides a service for the remote execution of a job, but it does not attempt to provide a standard hosting environment that will guarantee that the job has been executed correctly. That task is left to the user. In a Web service model, the job execution and the lifetime becomes the responsibility of the service provider.

The recently proposed OGSA [6, 7] provides a new framework for thinking about and building Grid applications that are consistent with this service model view of applications. OGSA specifies three things that a Web service must have before it qualifies as a Grid services. First, it must be an instance of a service implementation of some service type as described above. Second, it must have a Grid Services Handle (GSH), which is a type of Grid Universal Resource Identifier (URI) for the service instance. The third property that elevates a Grid service above a garden-variety Web service is the fact that each Grid service instance must implement a port called *GridService*, which provides any client access to service metadata and service state information. In the following section of this paper we will describe the role that the GridService port can play in a distributed component system.

OGSA also provides several other important services and port types. Messaging is handled by the NotificationSource and the NotificationSink ports. The intent of this service is to provide a simple publish-subscribe system similar to JMS [8], but based on XML messages. A Registry service allows other services to publish service metadata and to register services. From the perspective of this paper, a very important addition is the OGSA concept of a Factory service, which is used to create instances of other services.

In this paper, we describe an implementation of an Application Factory Service that is designed to create instances of distributed applications that are composed of well-tested and deployed components each executing in a well-understood and predictable hosting environment. In this model both the executing component instances and the composite application are Web services. We also describe how some important features of OGSA can be used to simplify client access to the running application from a conventional Web portal. We also describe a simple security model for the system that is designed to provide both authentication and simple authorization. We conclude with a discussion of how the factory service can be used to isolate the user from the details of resource selection and management in Grid environments.

9.1.1 An overview of the application factory service

The concept of a *factory service* is not new. It is an extension of the Factory Design Pattern [9] to the domain of distributed system. A factory service is a secure and a stateless persistent service that knows how to create an instance of transient, possibly stateful, service. Clients contact the factory service and supply the needed parameters to instantiate the application instance. It is the job of the service to invoke exactly one instance of the application and return a Web Service Description Language (WSDL) document that clients can use to access the application. OGSA has a standard port type for factory services, which has the same goal as the one described here but the details differ in some respects.

To illustrate the basic concept we begin with an example (see Figure 9.1). Suppose a scientist at a location X has a simulation code that is capable of doing some interesting computation provided it is supplied with useful initial and bound conditions. A supplier at another location Y may have a special data archive that describes material properties that define possible boundary or initial conditions for this simulation. For example, these may be aerodynamic boundary conditions such as fluid temperature and viscosity used in a simulation of turbulence around a solid body or process parameters used in a simulation of a semiconductor manufacturing facility. Suppose the supplier at Y would like to provide users at other locations with access to the application that uses the data archive at Y to drive the simulation at X. Furthermore, suppose that the scientist at location X is willing to allow others to execute his application on his resources, provided he authorizes them to do so.

To understand the requirements for building such a grid simulation service, we can follow a simple use-case scenario.

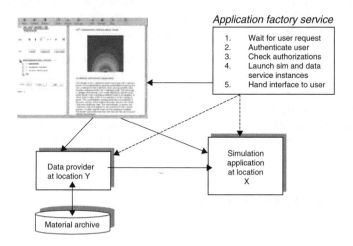

Figure 9.1 High-level view of user/application factory service. User contacts the persistent factory service from a Web interface. Factory service handles authentication and authorization and then creates an instance of the distributed application. A handle to the distributed application instance is returned to the user.

- The user would contact the factory service through a secure Web portal or a direct secure connection from a factory service client. In any case, the factory service must be able to authenticate the identity of the user.
- Once the identity of the user has been established, the factory service must verify that the user is authorized to run the simulation service. This authorization may be as simple as checking an internal access control list, or it may involve consulting an external authorization service.
- If the authorization check is successful, the factory service can allow the user to communicate any basic configuration requirements back to the factory service. These configuration requirements may include some basic information such as estimates of the size of the computation or the simulation performance requirements that may affect the way the factory service selects resources on which the simulation will run.
- The factory service then starts a process that creates running instances of a data provider component at Y and a simulation component at X that can communicate with each other. This task of activating the distributed application may require the factory service to consult resource selectors and workload managers to optimize the use of compute and data resources. For Grid systems, there is an important question here: under whose ownership are these two remote services run? In a classic grid model, we would require the end user to have an account on both the X and the Y resources. In this model, the factory service would now need to obtain a proxy certificate from the user to start the computations on the user's behalf. However, this delegation is unnecessary if the resource providers trust the factory service and allow the computations to be executed under the service owner's identity. The end users need not have an account on the remote resources and this is a much more practical service-oriented model.
- Access to this distributed application is then passed from the factory service back to the client. The easiest way to do this is to view the entire distributed application instance as a transient, stateful Web service that belongs to the client.
- The factory service is now ready to interact with another client.

In the sections that follow, we describe the basic technology used to build such a factory service. The core infrastructure used in this work is based on the eXtreme Component Architecture Toolkit (XCAT) [10, 11], which is a Grid-level implementation of the Common Component Architecture (CCA) [12] developed for the US Department of Energy. XCAT can be thought of as a tool to build distributed application-oriented Web services. We also describe how OGSA-related concepts can be used to build active control interfaces to these distributed applications.

9.2 XCAT AND WEB SERVICES

In this section, we describe the component model used by XCAT and discuss its relation to the standard Web service model and OGSA. XCAT components are software modules that provide part of a distributed application's functionality in a manner similar to that of a class library in a conventional application. A running instance of an XCAT component is a Web service that has two types of ports. One type of port, called a *provides-port*, is

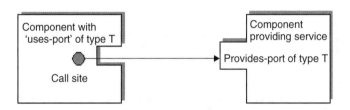

Figure 9.2 CCA composition model. A uses-port, which represents a proxy for an invocation of a remote service, may be bound at run time to any provides-port of the same type on another component.

essentially identical to a normal Web service port. A provides-port is a service provided by the component. The second type of port is called a *uses-port*. These are ports that are 'outgoing only' and they are used by one component to invoke the services of another or, as will be described later, to send a message to any waiting listeners. Within the CCA model, as illustrated in Figure 9.2, a uses-port on one component may be *connected* to a provides-port of another component if they have the same port interface type.

Furthermore, this connection is dynamic and it can be modified at run time. The provides-ports of an XCAT component can be described by the WebService Description Language (WSDL) and hence can be accessed by any Web service client that understands that port type. [A library to generate WSDL describing any remote reference is included as a part of XSOAP [13], which is an implementation of Java Remote Method Protocol (JRMP) in both C++ and Java with Simple Object Access Protocol (SOAP) as the communication protocol. Since, in XCAT a provides-port is a remote reference, the XSOAP library can be used to obtain WSDL for any provides-port. Further, a WSDL describing the entire component, which includes the WSDL for each provides-port, can be generated using this library.] The CCA/XCAT framework allows

- any component to create instances of other components on remote resources where it is authorized to do so (in XCAT this is accomplished using Grid services such as Globus),
- any component to connect together the uses-/provides-ports of other component instances (when it is authorized to do so), and
- a component to create new uses- and provides-ports as needed dynamically.

These dynamic connection capabilities make it possible to build applications in ways not possible with the standard Web services model. To illustrate this, we compare the construction of a distributed application using the CCA/XCAT framework with Web services using the Web Services Flow Language (WSFL) [5], which is one of the leading approaches to combining Web services into composite applications.

Typically a dynamically created and connected set of component instances represents a distributed application that has been invoked on behalf of some user or group of users. It is stateful and, typically, not persistent. For example, suppose an engineering design team wishes to build a distributed application that starts with a database query that provides initialization information to a data analysis application that frequently needs information found in a third-party information service. An application coordinator component (which

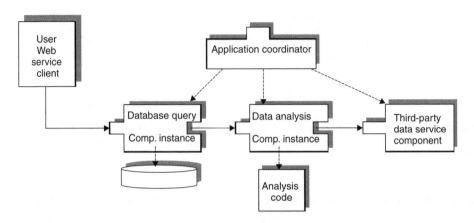

Figure 9.3   A data analysis application. An application coordinator instantiates three components: a database query component, a data analysis component that manages a legacy application, and a third-party data service (which may be a conventional Web service).

will be described later in greater detail) can be written that instantiates an instance of a database query component, a specialized legacy program driver component, and a component that consults a third-party data service, all connected as shown in Figure 9.3. Suppose the operation of this data analysis application is as follows. The database query component provides a Web service interface to users and when invoked by a user, it consults the database and contacts the analysis component. The analysis component, when receiving this information, interacts periodically with the data service and eventually returns a result to the database component, which returns it to the user.

This entire ensemble of connected component instances represents a distributed, transient service that may be accessed by one user or group of users and may exist for only the duration of a few transactions.

In the case above, the use of the application controller component to instantiate and connect together a chain of other components is analogous to a workflow engine executing a WSFL script on a set of conventional Web services. As shown in Figure 9.4, the primary advantage of the CCA component model is that the WSFL engine must intermediate at each step of application sequence and relay the messages from one service to the next.

If the data traffic between the services is heavy, it is probably not best to require it to go through a central flow engine. Furthermore, if logic that describes the interaction between the data analysis component and the third-party data service is complex and depends upon the application behavior, then putting it in the high-level workflow may not work. This is an important distinction between application-dependent flow between components and service mediation at the level of workflow.

In our current implementation, each application is described by three documents.

- The *Static Application Information* is an XML document that describes the list of components used in the computation, how they are to be connected, and the ports of the ensemble that are to be exported as application ports.

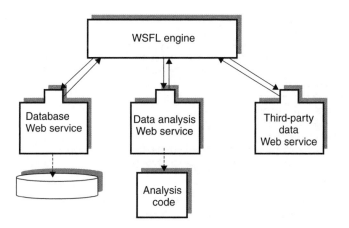

Figure 9.4 Standard Web service linking model using a Web service flow language document to drive a WSFL engine.

- The *Dynamic Application Information* is another XML document that describes the bindings of component instances to specific hosts and other initialization data.
- The *Component Static Information* is an XML document that contains basic information about the component and all the details of its execution environment for each host on which it has been deployed. This is the information that is necessary for the application coordinator component to create a running instance of the component. The usual way to obtain this document is through a call to a simple directory service, called the *Component Browser*, which allows a user to browse components by type name or to search for components by other attributes such as the port types they support.

To illustrate the way the Static and the Dynamic Application Information is used, consider the small example of the data analysis application above. The static application information, shown below, just lists the components and the connections between their ports. Each component is identified both by the type and by the component browser from which its static component information is found.

```
<application appName="Data Analysis Application">
   <applicationCoordinator>
      <component name="application coordinator">
         <compID>AppCoordinator</compID>
         <directoryService>uri-for-comp-browser</directoryService>
      </component>
   </applicationCoordinator>
   <applicationComponents>
      <component name="Database Query">
         <compID>DBQuery</compID>
         <directoryService>uri-for-comp-browser</directoryService>
      </component>
      <component name="Data Analysis">
         <compID>DataAnalysis</compID>
```

```
            <directoryService>uri-for-comp-browser</directoryService>
         </component>
         <component name="Joe's third party data source">
            <compID>GenericDataService</compID>
            <directoryService>uri-for-comp-browser</directoryService>
         </component>
    </applicationComponents>
    <applicationConnections>
         <connection>
             <portDescription>
                 <portName>DataOut</portName>
                 <compName>Database Query</compName>
             </portDescription>
              <portDescription>
                 <portName>DataIn</portName>
                 <compName>Data Analysis</compName>
             </portDescription>
         </connection>
          <connection>
             <portDescription>
                 <portName>Fetch Data</portName>
                 <compName>Data Analysis</compName>
             </portDescription>
              <portDescription>
                 <portName>Data Request</portName>
                 <compName>Joe's third party data source</compName>
             </portDescription>
         </connection>
    </applicationConnections>
</application>
```

The dynamic information document simply binds components to hosts based on availability of resources and authorization of user. For the example described above, it may look like the form shown below:

```
<applicationInstance name="Data Analysis Application">
   <appCoordinatorHost>application coordinator</appCoordinatorHost>
   <compInfo>
      <componentName>Data Query</componentName>
      <hostName>rainier.extreme.indiana.edu</hostName>
   </compInfo>
   <compInfo>
      <componentName>Data Analysis</componentName>
      <hostName>modi4.csrd.uiuc.edu</hostName>
   </compInfo>
   <compInfo>
      <componentName>Joe's third party data source</componentName>
      <hostName>joes_data.com</hostName>
   </compInfo>
</applicationInstance>
```

An extension to this dynamic information application instance document provides a way to supply any initial configuration the parameters that are essential for the operation of the component.

These documents are used by the application coordinator to instantiate the individual components. The way in which these documents are created and passed to the coordinator is described in detail in the next two sections.

9.2.1 The OGSA Grid services port and standard CCA ports

To understand how the CCA/XCAT component framework relates to the Open Grid Services Architecture, one must look at the required features of an OGSA service. In this paper we focus on one aspect of this question. The Open Grid Services Architecture requires that each service that is a fully qualified OGSA service must have a port that implements the GridService port. This port implements four operations. Three of these operations deal with service lifetime and one, *findServiceData*, is used to access service metadata and state information. The message associated with the findService-Data operation is a simple query to search for serviceData objects, which take the form shown below.

```
<gsdl:serviceData name="nmtoken"? globalName="qname"? type="qname"
      goodFrom="xsd:dateTime"? goodUntil="xsd:dateTime"?
      availableUntil="xsd:dateTime"?>
  <-- content element --> *
</gsdl:serviceData>
```

The *type* of a serviceData element is the XML schema name for the content of the element. Hence almost any type of data may be described here. OGSA defines about 10 different required serviceData types and we agree with most of them. There are two standard default serviceData search queries: finding a serviceData element by name and finding all serviceData elements that are of a particular type. This mechanism provides a very powerful and uniform way to allow for service reflection/introspection in a manner consistent with the Representational State Transfer model [15]. Though not implemented as a standard port in the current XCAT implementation, it will be added in the next release. The XCAT serviceData elements will contain the static component information record associated with its deployment. An important standard XCAT component serviceData element is a description of each port that the component supports. This includes the port name, the WSDL port type, whether it is a provides-port or a uses-port and if it is a uses-port whether it is currently connected to a provides-port. Often component instances publish event streams that are typed XML messages. An important serviceData element contains a list of all the event types and the handle for the persistent event channel that stores the published events.

Many application components also provide a custom control port that allows the user to directly interact with the running instance of the component. In this case a special ControlDocument serviceData element can be used to supply a user with a graphical user interface to the control port. This user interface can be either a downloadable appletlike program or a set of Web pages and execution scripts that can be dynamically loaded into a portal server such as the XCAT science portal [10]. As illustrated in Figure 9.5, this allows the user control of the remote component from a desktop Web browser.

Within the Web services community there is an effort called *Web Services for Remote Portals* (WSRP), which is attempting to address this problem [14]. The goal of this effort

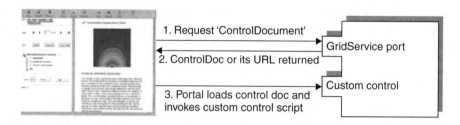

Figure 9.5   Portal interaction with GridService port.

is to provide a generic portlet, which runs in a portal server and acts as a proxy for a service-specific remote portlet. As this effort matures we will incorporate this standard into our model for component interaction.

Each XCAT component has another standard provides-port called the *Go* port. The life cycle of a CCA component is controlled by its Go. There are three standard operations in this port

- *sendParameter*, which is used to set initialization parameters of the component. The argument is an array of type Object, which is then cast to the appropriate specific types by the component instance. A standard serviceData element for each component is 'parameters', which provides a list of tuples of the form (name, type, default, current) for each component parameter.
- *start*, which causes the component to start running. In the CCA model, a component is first instantiated. At this point only the GridService, the start, and the sendParameter port are considered operational. Once parameters are set by the sendParameter method (or defaults are used) and the start method has been invoked, the other ports will start accepting calls. These rules are only used for stateful components or stateless components that require some initial, constant state.
- *kill*, which shuts down a component. If the start method registered the component with an information service, this method will also unregister the instance. In some cases, kill will also disconnect child components and kill them.

(It should be noted that in the CCA/XCAT framework we have service/component lifetime management in the Go port, while in OGSA it is part of the GridService port. Also, the OGSA lifetime management model is different.)

9.2.1.1 Application coordinator

Each XCAT component has one or more additional ports that are specific to its function. The Application Coordinator, discussed in the multicomponent example above, has the following port.

ACCreationProvidesPort provides the functionality to create applications by instantiating the individual components that make up the application, connecting the uses-

and provides-port of the different components and generating WSDL describing the application. This functionality is captured in the following two methods:

- *createApplication*, which accepts the static and the dynamic XML describing the application and the parameters. The XML strings are parsed to obtain installation information and host name for each component. The components are launched using the XCAT creation service and the appropriate execution environment values. Further, information about the connections between the different components is extracted from the static XML, and using the XCAT connection service the connections are established. The static XML also has information about the various ports of the components that need to be exposed to the clients. These ports, which belong to different components, are exported as ports of the Application Coordinator. In the next version, the static XML will also contain a list of bootstrap methods that need to be invoked on the individual components to start up the application. Once the application has been successfully instantiated, bootstrap methods, if any, are invoked.
- *generateWSDL*, which returns the WSDL describing the application as a whole. The ports of the Application Coordinator component and the exported ports are included as a part of this WSDL.

Another port, the *ACMonitorProvidesPort*, has methods such as 'killApplication' and 'pingApplication' for monitoring the application. The 'killApplication' method parses the static XML to retrieve information about the methods that need to be invoked on each of the components that the application is composed of, so as to kill the complete application. Similarly, the 'pingApplication' method invokes relevant methods on each of the components to get information about the status of the application.

9.3 THE APPLICATION FACTORY SERVICE

The application coordinator component described above is designed to be capable of launching and coupling together the components necessary to build a distributed application and shutting them down when the user is finished. It also provides the WSDL document that defines the ensemble service. It can also provide state information about the running application to the user. It is the job of the application factory service to launch an instance of the application coordinator.

A 'generic' application factory service (GAFS) is a stateless component that can be used to launch many types of applications. The GAFS accepts requests from a client that consists of the static and the dynamic application information. The GAFS authenticates the request and verifies that the user is authorized to make the request. Once authentication and authorization are complete, the GAFS launches an application coordinator. The GAFS passes the static and the dynamic application information to the application coordinator by invoking its createApplication method. Once the application component instances have been created, connected, and initialized, the application coordinator builds a WSDL document of the ensemble application and returns that to the GAFS, which returns it to the client. When the GAFS comes up, it generates a WSDL describing itself and binds the WSDL to a registry to facilitate discovery of this service. The service maintains no state information about any of the application instances it launches.

There are two ways to make the GAFS into an application-specific service. The first method is to build a Web portlet that contains the necessary user interface to allow the user to supply any needed initialization parameters. The portlet then constructs the needed dynamic and static application XML documents and sends them to the GAFS to use to instantiate the application coordinator. The other method is to attach an application-specific component to the GAFS, which contains the application-specific XML.

The complete picture of the application factory service and the application life cycle is shown in Figure 9.6. The basic order of events is listed below.

1. When the factory service comes up, it registers itself with a directory/discovery service.
2. A Web portlet can be used to discover the service.
3. The Web portlet contacts the factory service and initial configuration parameters are supplied.
4. The factory service contacts the authorization service.
5. If the user is authorized, it contacts the resource broker that is used to compose the dynamic application information.
6. The factory service instantiates the application coordinator that then instantiates the rest of the application components and connects them. It also builds the WSDL document for the distributed application that describes the exported ports of the ensemble.
7. The WSDL is returned to the factory, which returns it to the portal.

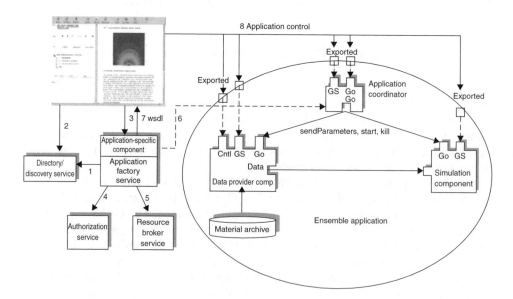
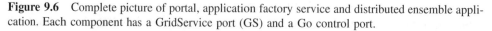

Figure 9.6 Complete picture of portal, application factory service and distributed ensemble application. Each component has a GridService port (GS) and a Go control port.

8. Using the ensemble application WSDL, the user can contact the application and interact with it.

9.4 CONCLUSIONS

The prototype XCAT application factory service components described here are available for download at http://www.extreme.indiana.edu/afws. The version to be released in November 2002 will contain the full OGSA-compliant components and contain several example service factories. The application factory service described here provides a Web service model for launching distributed Grid applications. There are several topics that have not been addressed in this document. Security in XCAT is based on Secure Sockets Layer (SSL) and Public Key Infrastructure (PKI) certificates. The authorization model is based on a very simple access control list. Other, more sophisticated authorization policies are under investigation.

The other area not discussed here is the component event system. This is based on a simple XML/SOAP-based messaging and event system and is available at http://www.extreme.indiana.edu/xgws. Any component may either publish or subscribe to event streams generated by other components. A persistent event channel serves as a publication/subscription target.

REFERENCES

1. Foster, I. and Kesselman, C. (eds) (1998) *The Grid: Blueprint for a New Computing Infrastructure*. San Francisco, CA: Morgan Kaufman Publishers, See also, Argonne National Lab, Math and Computer Science Division, http://www.mcs.anl.gov/globus.
2. Web Services Description Language (WSDL) 1.1, W3C, http://www.w3.org/TR/wsdl.
3. UDDI: Universal Description, Discover and Integration of Business for the Web, http://www.uddi.org.
4. Web Services Inspection Language (WSIL), http://xml.coverpages.org/IBM-WS-Inspection-Overview.pdf.
5. Web Services Flow Language (WSFL), http://www-4.ibm.com/software/solutions/webservices/pdf/WSFL.pdf.
6. Foster, I., Kesselman, C., Nick, J. and Tuecke, S. (2002) *The Physiology of the Grid: An Open Grid Services Architecture for Distributed Systems Integration*, Chapter 8 of this book.
7. Tuecke, S., Czajkowski, K., Foster, I., Frey, J., Graham, S. and Kesselman, C. (2002) *Grid Service Specification*, February, 2002, hhtp://www.gridforum.org/ogsi-wg/.
8. Sun Microsystems Inc. *Java Message Service Specification*, http://java.sun.com/products/jms/docs/html.
9. Gamma, E., Helm, R., Johnson, R. and Vlissides, J. (1995) *Design Patterns*. Reading, MA: Addison-Wesley.
10. Krishnan, S. *et al.* (2001) The XCAT science portal. *Proceedings SC 2001*, 2001.
11. Bramley, R. *et al.* (2000) A component based services architecture for building distributed applications. *Proceedings of HPDC*, 2000.
12. Armstrong, R. *et al.* (1999) Toward a common component architecture for high-performance scientific computing. Proceedings, *High Performance Distributed Computing Conference*, 1999.
13. Slominski, A., Govindaraju, M., Gannon, D. and Bramley, R. (2001) Design of an XML based interoperable RMI system: soapRMI C++/Java 1.1. *Proceedings of International Conference*

on Parallel and Distributed Processing Techniques and Applications, Las Vegas, June 25–28, 2001, pp. 1661–1667.

14. Web Services for Remote Portals, http://www.oasis-open.org/committees/wsrp/.
15. Fielding, R.T. (2000) *Styles and the Design of Network-based Software Architectures*, Architectural Ph.D. Dissertation, University of California, Irvine, 2000.
16. Grimshaw *et al. From Legion to Avaki: The Persistence of Vision*, Chapter 10 of this volume.

10

From Legion to Avaki: the persistence of vision*

Andrew S. Grimshaw,[1,2] **Anand Natrajan,**[2] **Marty A. Humphrey,**[1]
Michael J. Lewis,[3] **Anh Nguyen-Tuong,**[2] **John F. Karpovich,**[2]
Mark M. Morgan,[2] **and Adam J. Ferrari**[4]

[1]*University of Virginia, Charlottesville, Virginia, United States,* [2]*Avaki Corporation, Cambridge, Massachusetts, United States,* [3]*State University of New York at Binghamton, Binghamton, New York, United States,* [4]*Endeca Technologies Inc., Cambridge, Massachusetts, United States*

10.1 GRIDS ARE HERE

In 1994, we outlined our vision for wide-area distributed computing [1]:

For over thirty years science fiction writers have spun yarns featuring worldwide networks of interconnected computers that behave as a single entity. Until recently such science fiction fantasies have been just that. Technological changes are now occurring which may expand computational power in the same way that the invention of desktop calculators and personal computers did. In the near future computationally

* This work partially supported by DARPA (Navy) contract #N66001-96-C-8527, DOE grant DE-FG02-96ER25290, DOE contract Sandia LD-9391, Logicon (for the DoD HPCMOD/PET program) DAHC 94-96-C-0008, DOE D459000-16-3C, DARPA (GA) SC H607305A, NSF-NGS EIA-9974968, NSF-NPACI ASC-96-10920, and a grant from NASA-IPG.

demanding applications will no longer be executed primarily on supercomputers and single workstations using local data sources. Instead enterprise-wide systems, and someday nationwide systems, will be used that consist of workstations, vector super-computers, and parallel supercomputers connected by local and wide area networks. Users will be presented the illusion of a single, very powerful computer, rather than a collection of disparate machines. The system will schedule application components on processors, manage data transfer, and provide communication and synchroniza-tion in such a manner as to dramatically improve application performance. Further, boundaries between computers will be invisible, as will the location of data and the failure of processors.

The future is now; after almost a decade of research and development by the Grid community, we see Grids (then called *metasystems* [2]) being deployed around the world in both academic and commercial settings.

This chapter describes one of the major Grid projects of the last decade, Legion, from its roots as an academic Grid project [3–5] to its current status as the only commercial complete Grid offering, Avaki, marketed by a Cambridge, Massachusetts company called *AVAKI Corporation.* We begin with a discussion of the fundamental requirements for any Grid architecture. These fundamental requirements continue to guide the evolution of our Grid software. We then present some of the principles and philosophy underlying the design of Legion. Next, we present briefly what a Legion Grid looks like to adminis-trators and users. We introduce some of the architectural features of Legion and delve slightly deeper into the implementation in order to give an intuitive understanding of Grids and Legion. Detailed technical descriptions are available in References [6–12]. We then present a brief history of Legion and Avaki in order to place the preceding discussion in context. We conclude with a look at the future and how Legion and Avaki fit in with emerging standards such as Open Grid Services Infrastructure (OGSI) [13].

10.2 GRID ARCHITECTURE REQUIREMENTS

Of what use is a Grid? What is required of a Grid? Before we answer these questions, let us step back and define a Grid and its essential attributes.

Our definition, and indeed a popular definition is that a Grid *system* is a collection of distributed resources connected by a network. A Grid system, also called a *Grid*, gathers resources – desktop and handheld hosts, devices with embedded processing resources such as digital cameras and phones or tera-scale supercomputers – and makes them accessible to users and applications in order to reduce overhead and to accelerate projects. A Grid *application* can be defined as an application that operates in a Grid environment or is 'on' a Grid system. Grid *system software* (or middleware) is software that facilitates writing Grid applications and manages the underlying Grid infrastructure.

The resources in a Grid typically share at least some of the following characteristics:

- they are numerous;
- they are owned and managed by different, potentially mutually distrustful organizations and individuals;

- they are potentially faulty;
- they have different security requirements and policies;
- they are heterogeneous, that is, they have different CPU architectures, are running different operating systems, and have different amounts of memory and disk;
- they are connected by heterogeneous, multilevel networks;
- they have different resource management policies; and
- they are likely to be geographically separated (on a campus, in an enterprise, on a continent).

A Grid enables users to collaborate securely by sharing *processing, applications* and *data* across systems with the above characteristics in order to facilitate collaboration, faster application execution and easier access to data. More concretely this means being able to do the following:

Find and share data: When users need access to data on other systems or networks, they should simply be able to access it like data on their own system. System boundaries that are not useful should be invisible to users who have been granted legitimate access to the information.

Find and share applications: The leading edge of development, engineering and research efforts consists of custom applications – permanent or experimental, new or legacy, public-domain or proprietary. Each application has its own requirements. Why should application users have to jump through hoops to get applications together with the data sets needed for analysis?

Share computing resources: It sounds very simple – one group has computing cycles; some colleagues in another group need them. The first group should be able to grant access to its own computing power without compromising the rest of the network.

Grid computing is in many ways a novel way to construct applications. It has received a significant amount of recent press attention and been heralded as the next wave in computing. However, under the guises of 'peer-to-peer systems', 'metasystems' and 'distributed systems', Grid computing requirements and the tools to meet these requirements have been under development for decades. Grid computing requirements address the issues that frequently confront a developer trying to construct applications for a Grid. The novelty in Grids is that these requirements are addressed by the Grid infrastructure in order to reduce the burden on the application developer. The requirements are as follows:

- *Security*: Security covers a gamut of issues, including authentication, data integrity, authorization (access control) and auditing. If Grids are to be accepted by corporate and government information technology (IT) departments, a wide range of security concerns must be addressed. Security mechanisms must be integral to applications and capable of supporting diverse policies. Furthermore, we believe that security must be firmly built in from the beginning. Trying to patch security in as an afterthought (as some systems are attempting today) is a fundamentally flawed approach. We also believe that no single security policy is perfect for all users and organizations. Therefore, a Grid system must have mechanisms that allow users and resource owners to select policies that fit particular security and performance needs, as well as meet local administrative requirements.

- *Global namespace*: The lack of a global namespace for accessing data and resources is one of the most significant obstacles to wide-area distributed and parallel processing. The current multitude of disjoint namespaces greatly impedes developing applications that span sites. All Grid objects must be able to access (subject to security constraints) any other Grid object *transparently* without regard to location or replication.
- *Fault tolerance*: Failure in large-scale Grid systems is and will be a fact of life. Hosts, networks, disks and applications frequently fail, restart, disappear and behave otherwise unexpectedly. Forcing the programmer to predict and handle all these failures significantly increases the difficulty of writing reliable applications. Fault-tolerant computing is a known, very difficult problem. Nonetheless, it must be addressed, or businesses and researchers will not entrust their data to Grid computing.
- *Accommodating heterogeneity*: A Grid system must support interoperability between heterogeneous hardware and software platforms. Ideally, a running application should be able to migrate from platform to platform if necessary. At a bare minimum, components running on different platforms must be able to communicate transparently.
- *Binary management*: The underlying system should keep track of executables and libraries, knowing which ones are current, which ones are used with which persistent states, where they have been installed and where upgrades should be installed. These tasks reduce the burden on the programmer.
- *Multilanguage support*: In the 1970s, the joke was 'I don't know what language they'll be using in the year 2000, but it'll be called *Fortran*.' Fortran has lasted over 40 years, and C for almost 30. Diverse languages will always be used and legacy applications will need support.
- *Scalability*: There are over 400 million computers in the world today and over 100 million network-attached devices (including computers). Scalability is clearly a critical necessity. Any architecture relying on centralized resources is doomed to failure. A successful Grid architecture must strictly adhere to the distributed systems principle: the service demanded of any given component must be independent of the number of components in the system. In other words, the service load on any given component must not increase as the number of components increases.
- *Persistence*: I/O and the ability to read and write persistent data are critical in order to communicate between applications and to save data. However, the current files/file libraries paradigm should be supported, since it is familiar to programmers.
- *Extensibility*: Grid systems must be flexible enough to satisfy current user demands and unanticipated future needs. Therefore, we feel that mechanism and policy must be realized by replaceable and extensible components, including (and especially) core system components. This model facilitates development of improved implementations that provide value-added services or site-specific policies while enabling the system to adapt over time to a changing hardware and user environment.
- *Site autonomy*: Grid systems will be composed of resources owned by many organizations, each of which desire to retain control over their own resources. For each resource, the owner must be able to limit or deny use by particular users, specify when it can be used and so on. Sites must also be able to choose or rewrite an implementation of

each Legion component as best suited to its needs. A given site may trust the security mechanisms of one particular implementation over those of another so it should freely be able to use that implementation.

- *Complexity management*: Finally, but importantly, complexity management is one of the biggest challenges in large-scale Grid systems. In the absence of system support, the application programmer is faced with a confusing array of decisions. Complexity exists in multiple dimensions: heterogeneity in policies for resource usage and security, a range of different failure modes and different availability requirements, disjoint namespaces and identity spaces and the sheer number of components. For example, professionals who are not IT experts should not have to remember the details of five or six different file systems and directory hierarchies (not to mention multiple user names and passwords) in order to access the files they use on a regular basis. Thus, providing the programmer and system administrator with clean abstractions is critical to reducing the cognitive burden.

Solving these requirements is the task of a Grid infrastructure. An architecture for a Grid based on well-thought principles is required in order to address each of these requirements. In the next section, we discuss the principles underlying the design of one particular Grid system, namely, Legion.

10.3 LEGION PRINCIPLES AND PHILOSOPHY

Legion is a Grid architecture as well as an operational infrastructure under development since 1993 at the University of Virginia. The architecture addresses the requirements of the previous section and builds on lessons learned from earlier systems. We defer a discussion of the history of Legion and its transition to a commercial product named *Avaki* to Section 10.7. Here, we focus on the design principles and philosophy of Legion, which can be encapsulated in the following 'rules':

- *Provide a single-system view*: With today's operating systems, we can maintain the illusion that our local area network is a single computing resource. But once we move beyond the local network or cluster to a geographically dispersed group of sites, perhaps consisting of several different types of platforms, the illusion breaks down. Researchers, engineers and product development specialists (most of whom do not want to be experts in computer technology) must request access through the appropriate gatekeepers, manage multiple passwords, remember multiple protocols for interaction, keep track of where everything is located and be aware of specific platform-dependent limitations (e.g. this file is too big to copy or to transfer to one's system; that application runs only on a certain type of computer). Recreating the illusion of a single computing resource for heterogeneous distributed resources reduces the complexity of the overall system and provides a single namespace.
- *Provide transparency as a means of hiding detail*: Grid systems should support the traditional distributed system transparencies: access, location, heterogeneity, failure, migration, replication, scaling, concurrency and behavior [7]. For example, users and

programmers need not have to know where an object is located in order to use it (access, location and migration transparency), nor should they need to know that a component across the country failed – they want the system to recover automatically and complete the desired task (failure transparency). This is the traditional way to mask various aspects of the underlying system. Transparency addresses fault tolerance and complexity.

- *Provide flexible semantics*: Our overall objective was a Grid architecture that is suitable to as many users and purposes as possible. A rigid system design in which policies are limited, trade-off decisions are preselected, or all semantics are predetermined and hard-coded would not achieve this goal. Indeed, if we dictated a single system-wide solution to almost any of the technical objectives outlined above, we would preclude large classes of potential users and uses. Therefore, Legion allows users and programmers as much flexibility as possible in their applications' semantics, resisting the temptation to dictate solutions. Whenever possible, users can select both the *kind* and the *level* of functionality and choose their own trade-offs between function and cost. This philosophy is manifested in the system architecture. The Legion object model specifies the functionality but not the implementation of the system's core objects; the core system therefore consists of extensible, replaceable components. Legion provides default implementations of the core objects, although users are not obligated to use them. Instead, we encourage users to select or construct object implementations that answer their specific needs.

- *By default the user should not have to think*: In general, there are four classes of users of Grids: end users of applications, applications developers, system administrators and managers who are trying to accomplish some mission with the available resources. We believe that users want to focus on their jobs, that is, their applications, and not on the underlying Grid plumbing and infrastructure. Thus, for example, to run an application a user may type *legion_run my_application my_data* at the command shell. The Grid should then take care of all the messy details such as finding an appropriate host on which to execute the application, moving data and executables around and so on. Of course, the user may as an option be aware of and specify or override certain behaviors, for example, specify an architecture on which to run the job, or name a specific machine or set of machines or even replace the default scheduler.

- *Reduce activation energy*: One of the typical problems in technology adoption is getting users to use it. If it is difficult to shift to a new technology, then users will tend to not take the effort to try it unless their need is immediate and extremely compelling. This is not a problem unique to Grids – it is human nature. Therefore, one of our most important goals is to make using the technology easy. Using an analogy from chemistry, we keep the activation energy of adoption as low as possible. Thus, users can easily and readily realize the benefit of using Grids – and get the reaction going – creating a self-sustaining spread of Grid usage throughout the organization. This principle manifests itself in features such as 'no recompilation' for applications to be ported to a Grid, and support for mapping a Grid to a local operating system's file system. Another variant of this concept is the motto 'no play, no pay'. The basic idea is that if you do not need a feature, for example, encrypted data streams, fault resilient files or strong access control, you should not have to pay the overhead for using it.

- *Do not change host operating systems*: Organizations will not permit their machines to be used if their operating systems must be replaced. Our experience with Mentat [14] indicates, though, that building a Grid on top of host operating systems is a viable approach.
- *Do not change network interfaces*: Just as we must accommodate existing operating systems, we assume that we cannot change the network resources or the protocols in use.
- *Do not require Grids to run in privileged mode*: To protect their objects and resources, Grid users and sites will require Grid software to run with the lowest possible privileges.

Although we focus primarily on technical issues in this chapter, we recognize that there are also important political, sociological and economic challenges in developing and deploying Grids, such as developing a scheme to encourage the participation of resource-rich sites while discouraging free-riding by others. Indeed, politics can often overwhelm technical issues.

10.4 USING LEGION IN DAY-TO-DAY OPERATIONS

Legion is comprehensive Grid software that enables efficient, effective and secure sharing of data, applications and computing power. It addresses the technical and administrative challenges faced by organizations such as research, development and engineering groups with computing resources in disparate locations, on heterogeneous platforms and under multiple administrative jurisdictions. Since Legion enables these diverse, distributed resources to be treated as a single virtual operating environment with a single file structure, it drastically reduces the overhead of sharing data, executing applications and utilizing available computing power regardless of location or platform. The central feature in Legion is the single global namespace. Everything in Legion has a name: hosts, files, directories, groups for security, schedulers, applications and so on. The same name is used regardless of where the name is used and regardless of where the named object resides at any given point in time.

In this and the following sections, we use the term 'Legion' to mean both the academic project at the University of Virginia as well as the commercial product, Avaki, distributed by AVAKI Corp.

Legion helps organizations create a *compute Grid*, allowing processing power to be shared, as well as a *data Grid*, a virtual single set of files that can be accessed without regard to location or platform. Fundamentally, a compute Grid and a data Grid are the same product – the distinction is solely for the purpose of exposition. Legion's unique approach maintains the security of network resources while reducing disruption to current operations. By increasing sharing, reducing overhead and implementing Grids with low disruption, Legion delivers important efficiencies that translate to reduced cost.

We start with a somewhat typical scenario and how it might appear to the end user. Suppose we have a small Grid as shown below with four sites – two different departments in one company, a partner site and a vendor site. Two sites are using load management systems; the partner is using Platform Computing™ Load Sharing Facility (LSF) software and one department is using Sun™ Grid Engine (SGE). We will assume that there is a mix of hardware in the Grid, for example, Linux hosts, Solaris hosts, AIX hosts, Windows 2000 and Tru64 Unix. Finally, there is data of interest at three different sites. A user then sits down at a terminal, authenticates to Legion (logs in) and runs the command *legion_run my_application my_data*. Legion will then *by default*, determine the binaries available, find and select a host on which to execute *my_application*, manage the secure transport of credentials, interact with the local operating environment on the selected host (perhaps an SGE™ queue), create accounting records, check to see if the current version of the application has been installed (and if not install it), move all the data around as necessary and return the results to the user. The user does not need to know where the application resides, where the execution occurs, where the file *my_data* is physically located or any of the other myriad details of what it takes to execute the application. Of course, the user may choose to be aware of, and specify or override, certain behaviors, for example, specify an architecture on which to run the job, or name a specific machine or set of machines or even replace the default scheduler. In this example, the user exploits key features:

- *Global namespace*: Everything the user specifies is in terms of a global namespace that names everything: processors, applications, queues, data files and directories. The same name is used regardless of the location of the user of the name or the location of the named entity.
- *Wide-area access to data*: All the named entities, including files, are mapped into the local file system directory structure of the user's workstation, making access to the Grid transparent.
- *Access to distributed and heterogeneous computing resources*: Legion keeps track of binary availability and the current version.
- *Single sign-on*: The user need not keep track of multiple accounts at different sites. Indeed, Legion supports policies that do not require a local account at a site to access data or execute applications, as well as policies that require local accounts.
- *Policy-based administration of the resource base*: Administration is as important as application execution.
- *Accounting both for resource usage information and for auditing purposes*: Legion monitors and maintains a Relational Database Management System (RDBMS) with accounting information such as who used what application on what host, starting when and how much was used.
- Fine-grained security that protects both the user's resources and that of the others.
- Failure detection and recovery.

10.4.1 Creating and administering a Legion Grid

Legion enables organizations to collect resources – applications, computing power and data – to be used as a single virtual operating environment as shown in Figure 10.1. This set of shared resources is called a *Legion Grid*. A Legion Grid can represent resources from homogeneous platforms at a single site within a single department, as well as resources from multiple sites, heterogeneous platforms and separate administrative domains.

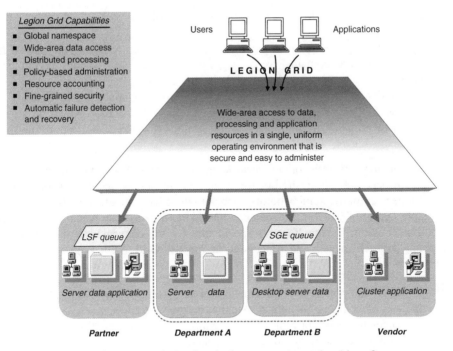

Figure 10.1   Example Legion deployment and associated benefits.

Legion ensures secure access to resources on the Grid. Files on participating computers become part of the Grid only when they are *shared* or explicitly made available to the Grid. Further, even when shared, Legion's fine-grained access control is used to prevent unauthorized access. Any subset of resources can be shared, for example, only the processing power or only certain files or directories. Resources that have not been shared are not visible to Grid users. By the same token, a user of an individual computer or network that participates in the Grid is not automatically a Grid user and does not automatically have access to Grid files. Only users who have explicitly been granted access can take advantage of the shared resources. Local administrators may retain control over who can use their computers, at what time of day and under which load conditions. Local resource owners control access to their resources.

Once a Grid is created, users can think of it as one computer with one directory structure and one batch processing protocol. They need not know where individual files are located physically, on what platform type or under which security domain. A Legion Grid can be administered in different ways, depending on the needs of the organization.

1. *As a single administrative domain*: When all resources on the Grid are owned or controlled by a single department or division, it is sometimes convenient to administer them centrally. The administrator controls which resources are made available to the Grid and grants access to those resources. In this case, there may still be separate

administrators at the different sites who are responsible for routine maintenance of the local systems.

2. *As a federation of multiple administrative domains*: When resources are part of multiple administrative domains, as is the case with multiple divisions or companies cooperating on a project, more control is left to administrators of the local networks. They each define which of their resources will be made available to the Grid and who has access. In this case, a team responsible for the collaboration would provide any necessary information to the system administrators, and would be responsible for the initial establishment of the Grid.

With Legion, there is little or no intrinsic need for central administration of a Grid. Resource owners are administrators for their own resources and can define who has access to them. Initially, administrators cooperate in order to create the Grid; after that, it is a simple matter of which management controls the organization wants to put in place. In addition, Legion provides features specifically for the convenience of administrators who want to track queues and processing across the Grid. With Legion, they can do the following:

• Monitor local and remote load information on all systems for CPU use, idle time, load average and other factors from any machine on the Grid.
• Add resources to queues or remove them without system interruption and dynamically configure resources based on policies and schedules.
• Log warnings and error messages and filter them by severity.
• Collect all resource usage information down to the user, file, application or project level, enabling Grid-wide accounting.
• Create scripts of Legion commands to automate common administrative tasks.

10.4.2 Legion Data Grid

Data access is critical for any application or organization. A Legion Data Grid [2] greatly simplifies the process of interacting with resources in multiple locations, on multiple platforms or under multiple administrative domains. Users access files by name – typically a pathname in the Legion virtual directory. *There is no need to know the physical location of the files.*

There are two basic concepts to understand in the Legion Data Grid – how the data is accessed and how the data is included into the Grid.

10.4.2.1 Data access

Data access is through one of three mechanisms: a Legion-aware NFS server called a *Data Access Point* (*DAP*), a set of command line utilities or Legion I/O libraries that mimic the C *stdio* libraries.

DAP access: The DAP provides a standards-based mechanism to access a Legion Data Grid. It is a commonly used mechanism to access data in a Data Grid. The DAP is a server that responds to NFS 2.0/3.0 protocols and interacts with the Legion system. When an NFS client on a host mounts a DAP, it effectively maps the Legion global namespace into the local host file system, providing completely transparent access to data throughout the Grid without even installing Legion software.

However, the DAP is not a typical NFS server. First, it has no actual disk or file system behind it – it interacts with a set of resources that may be distributed, be owned by multiple organizations, be behind firewalls and so on. Second, the DAP supports the Legion security mechanisms – access control is with signed credentials, and interactions with the data Grid can be encrypted. Third, the DAP caches data aggressively, using configurable local memory and disk caches to avoid wide-area network access. Further, the DAP can be modified to exploit semantic data that can be carried in the metadata of a file object, such as 'cacheable', 'cacheable until' or 'coherence window size'. In effect, DAP provides a highly secure, wide-area NFS.

To avoid the rather obvious hot spot of a single DAP at each site, Legion encourages deploying more than one DAP per site. There are two extremes: one DAP per site and one DAP per host. Besides the obvious trade-off between scalability and the shared cache effects of these two extremes, there is also an added security benefit of having one DAP per host. NFS traffic between the client and the DAP, typically unencrypted, can be restricted to one host. The DAP can be configured to only accept requests from a local host, eliminating the classic NFS security attacks through network spoofing.

Command line access: A Legion Data Grid can be accessed using a set of command line tools that mimic the Unix file system commands such as *ls*, *cat* and so on. The Legion analogues are *legion_ls*, *legion_cat* and so on. The Unix-like syntax is intended to mask the complexity of remote data access by presenting familiar semantics to users.

I/O libraries: Legion provides a set of I/O libraries that mimic the *stdio* libraries. Functions such as *open* and *fread* have analogues such as *BasicFiles_open* and *BasicFiles_fread*. The libraries are used by applications that need stricter coherence semantics than those offered by NFS access. The library functions operate directly on the relevant file or directory object rather than operating via the DAP caches.

10.4.2.2 Data inclusion

Data inclusion is through one of three mechanisms: a 'copy' mechanism whereby a copy of the file is made in the Grid, a 'container' mechanism whereby a copy of the file is made in a container on the Grid or a 'share' mechanism whereby the data continues to reside on the original machine, but can be accessed from the Grid. Needless to say, these three inclusion mechanisms are completely orthogonal to the three access mechanisms discussed earlier.

Copy inclusion: A common way of including data into a Legion Data Grid is by copying it into the Grid with the *legion_cp* command. This command creates a Grid object or service that enables access to the data stored in a copy of the original file. The copy of the data may reside anywhere in the Grid, and may also migrate throughout the Grid.

Container inclusion: Data may be copied into a Grid container service as well. With this mechanism the contents of the original file are copied into a container object or service that enables access. The container mechanism reduces the overhead associated with having one service per file. Once again, data may migrate throughout the Grid.

Share inclusion: The primary means of including data into a Legion Data Grid is with the *legion_export_dir* command. This command starts a daemon that maps a file or rooted directory in Unix or Windows NT into the data Grid. For example, *legion_export_dir C:\data/home/grimshaw/share-data* maps the directory *C:\data* on a Windows machine into the data Grid at */home/grimshaw/share-data*. Subsequently, files and subdirectories in *C:\data* can be accessed directly in a peer-to-peer fashion from anywhere else in the data Grid, subject to access control, without going through any sort of central repository. A Legion share is independent of the implementation of the underlying file system, whether a direct-attached disk on Unix or NT, an NFS-mounted file system or some other file system such as a hierarchical storage management system.

Combining shares with DAPs effectively federates multiple directory structures into an overall file system, as shown in Figure 10.2. Note that there may be as many DAPs as needed for scalability reasons.

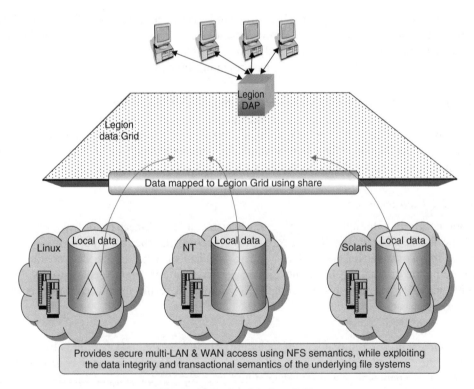

Figure 10.2   Legion data Grid.

10.4.3 Distributed processing

Research, engineering and product development depend on intensive data analysis and large simulations. In these environments, much of the work still requires computation-intensive data analysis – executing specific applications (that may be complex) with specific input data files (that may be very large or numerous) to create result files (that may also be large or numerous). For successful job execution, the data and the application must be available, sufficient processing power and disk storage must also be available and the application's requirements for a specific operating environment must be met. In a typical network environment, the user must know where the file is where the application is, and whether the resources are sufficient to complete the work. Sometimes, in order to achieve acceptable performance, the user or administrator must move data files or applications to the same physical location.

With Legion, users do not need to be concerned with these issues in most cases. Users have a single point of access to an entire Grid. Users log in, define application parameters and submit a program to run on available resources, which may be spread across distributed sites and multiple organizations. Input data is read securely from distributed sources without necessarily being copied to a local disk. Once an application is complete, computational resources are cleared of application remnants and the output is written to the physical storage resources available in the Grid. Legion's distributed processing support includes several features, listed below.

10.4.3.1 Automated resource matching and file staging

A Legion Grid user executes an application, referencing the file and application by name. In order to ensure secure access and implement necessary administrative controls, predefined policies govern where applications may be executed or which applications can be run on which data files. Avaki matches applications with queues and computing resources in different ways:

- *Through access controls*: For example, a user or application may or may not have access to a specific queue or a specific host computer.
- *Through matching of application requirements and host characteristics*: For example, an application may need to be run on a specific operating system, or require a particular library to be installed or require a particular amount of memory.
- *Through prioritization*: For example, on the basis of policies and load conditions.

Legion performs the routine tasks needed to execute the application. For example, Legion will move (or *stage*) data files, move application binaries and find processing power as needed, as long as the resources have been included into the Grid and the policies allow them to be used. If a data file or application must be migrated in order to execute the job, Legion does so automatically; the user does not need to move the files or know where the job was executed. Users need not worry about finding a machine for the application to run on, finding available disk space, copying the files to the machine and collecting results when the job is done.

10.4.3.2 Support for legacy applications – no modification necessary

Applications that use a Legion Grid can be written in any language, do not need to use a specific Application Programming Interface (API) and can be run on the Grid without source code modification or recompilation. Applications can run anywhere on the Grid without regard to location or platform as long as resources are available that match the application's needs. This is critical in the commercial world where the sources for many third-party applications are simply not available.

10.4.3.3 Batch processing – queues and scheduling

With Legion, users can execute applications interactively or submit them to a queue. With queues, a Grid can group resources in order to take advantage of shared processing power, sequence jobs based on business priority when there is not enough processing power to run them immediately and distribute jobs to available resources. Queues also permit allocation of resources to groups of users. Queues help insulate users from having to know where their jobs are physically run. Users can check the status of a given job from anywhere on the Grid without having to know where it was actually processed. Administrator tasks are also simplified because a Legion Grid can be managed as a single system. Administrators can do the following tasks:

- Monitor usage from anywhere on the network.
- Preempt jobs, reprioritize and requeue jobs, take resources off the Grid for maintenance or add resources to the Grid – all without interrupting work in progress.
- Establish policies based on time windows, load conditions or job limits. Policies based on time windows can make some hosts available for processing to certain groups only outside business hours or outside peak load times. Policies based on the number of jobs currently running or on the current processing load can affect load balancing by offloading jobs from overburdened machines.

10.4.3.4 Unifying processing across multiple heterogeneous networks

A Legion Grid can be created from individual computers and local area networks or from clusters enabled by software such as the LSF™ or SGE™. If one or more queuing systems, load management systems or scheduling systems are already in place, Legion can interoperate with them to create virtual queues, thereby allowing resources to be shared across the clusters. This approach permits unified access to disparate, heterogeneous clusters, yielding a Grid that enables an increased level of sharing and allows computation to scale as required to support growing demands.

10.4.4 Security

Security was designed in the Legion architecture and implementation from the very beginning [8]. Legion's robust security is the result of several separate capabilities, such as authentication, authorization and data integrity that work together to implement and enforce security policy. For Grid resources, the goal of Legion's security approach is to eliminate the need for any other software-based security controls, substantially reducing the overhead of sharing resources. With Legion security in place, users need to know only their sign-on protocol and the Grid pathname where their files are located. Accesses to all resources are mediated by access controls set on the resources. We will describe the security implementation in more detail in Section 10.5.3. Further details of the Legion security model are available in References [7, 9].

10.4.5 Automatic failure detection and recovery

Service downtimes and routine maintenance are common in a large system. During such times, resources may be unavailable. A Legion Grid is robust and fault tolerant. If a computer goes down, Legion can migrate applications to other computers based on predefined deployment policies as long as resources are available that match application requirements. In most cases, migration is automatic and unobtrusive. Since Legion is capable of taking advantage of all the resources on the Grid, it helps organizations optimize system utilization, reduce administrative overhead and fulfill service-level agreements.

- *Legion provides fast, transparent recovery from outages*: In the event of an outage, processing and data requests are rerouted to other locations, ensuring continuous operation. Hosts, jobs and queues automatically back up their current state, enabling them to restart with minimal loss of information if a power outage or other system interruption occurs.
- *Systems can be reconfigured dynamically*: If a computing resource must be taken offline for routine maintenance, processing continues using other resources. Resources can also be added to the Grid, or their access changed, without interrupting operations.
- *Legion migrates jobs and files as needed*: If a job's execution host is unavailable or cannot be restarted, the job is automatically migrated to another host and restarted. If a host must be taken off the Grid for some reason, administrators can migrate files to another host without affecting pathnames or users. Requests for that file are rerouted to the new location automatically.

10.5 THE LEGION GRID ARCHITECTURE: UNDER THE COVERS

Legion is an object-based system composed of independent objects, disjoint in logical address space that communicate with one another by method invocation.[1] Method calls are nonblocking and may be accepted in any order by the called object. Each method has a signature that describes its parameters and return values (if any). The complete set of signatures for an object describes that object's interface, which is determined by its class. Legion class interfaces are described in an Interface Description Language (IDL), three of which are currently supported in Legion: the CORBA IDL [15], MPL [14] and

---

[1] The fact that Legion is object-based does not preclude the use of non-object-oriented languages or non-object-oriented implementations of objects. In fact, Legion supports objects written in traditional procedural languages such as C and Fortran as well as object-oriented languages such as C++, Java and Mentat Programming Language (MPL, a C++ dialect with extensions to support parallel and distributed computing).

BFS [16] (an object-based Fortran interface language). Communication is also supported for parallel applications using a Legion implementation of the Message Passing Interface (MPI) libraries. The Legion MPI supports cross-platform, cross-site MPI applications.

All things of interest to the system – for example, files, applications, application instances, users and groups – are objects. All Legion objects have a name, state (which may or may not persist) and metadata (<name, valueset> tuples) associated with their state and an interface.[2] At the heart of Legion is a three-level global naming scheme with security built into the core. The top-level names are human names. The most commonly used are *path names*, for example, */home/grimshaw/my_sequences/seq_1*. Path names are all that most people ever see. The path names map to location-independent identifiers called *Legion Object Identifier* (LOIDs). LOIDs are extensible self-describing data structures that include as a part of the name, an RSA public key (the private key is part of the object's state). Consequently, objects may authenticate one another and communicate securely without the need for a trusted third party. Finally, LOIDs map to object addresses (OAs), which contain location-specific current addresses and communication protocols. An OA may change over time, allowing objects to migrate throughout the system, even while being used.

Each Legion object belongs to a class and each class is by itself a Legion object. All objects export a common set of object-mandatory member functions (such as deactivate(), ping() and getInterface()), which are necessary to implement core Legion services. Class objects export an additional set of class-mandatory member functions that enable them to manage their instances (such as createInstance() and deleteInstance()).

Much of the Legion object model's power comes from the role of Legion classes, since a sizeable amount of what is usually considered system-level responsibility is delegated to user-level class objects. For example, classes are responsible for creating and locating their instances and for selecting appropriate security and object placement policies. Legion core objects provide mechanisms that allow user-level classes to implement chosen policies and algorithms. We also encapsulate system-level policy in extensible, replaceable class objects, supported by a set of primitive operations exported by the Legion core objects. This effectively eliminates the danger of imposing inappropriate policy decisions and opens up a much wider range of possibilities for the application developer.

10.5.1 Naming with context paths, LOIDs and object addresses

As in many distributed systems over the years, access, location, migration, failure and replication transparencies are implemented in Legion's naming and binding scheme. Legion has a three-level naming scheme. The top level consists of human-readable namespaces. There are currently two namespaces, context space and an attribute space. A context is a directory-like object that provides mappings from a user-chosen string to a Legion object identifier, called a *LOID*. A LOID can refer to a context object, file object, host object or any other type of Legion object. For direct object-to-object communication, a LOID must be bound to its low-level OA, which is meaningful within the transport

---

[2] Note the similarity to the OGSA model, in which everything is a resource with a name, state and an interface.

protocol used for communication. The process by which LOIDs are mapped to OAs is called the *Legion binding process*.

Contexts: Contexts are organized into a classic directory structure called *context space*. Contexts support operations that lookup a single string, return all mappings, add a mapping and delete a mapping. A typical context space has a well-known root context that in turn 'points' to other contexts, forming a directed graph (Figure 10.3). The graph can have cycles, though it is not recommended.

We have built a set of tools for working in and manipulating context space, for example, *legion_ls*, *legion_rm*, *legion_mkdir*, *legion_cat* and many others. These are modeled on corresponding Unix tools. In addition, there are libraries that mimic the *stdio* libraries for doing all the same sort of operations on contexts and files (e.g. `creat()`, `unlink()`, etc.).

Legion object IDentifiers: Every Legion object is assigned a unique and immutable LOID upon creation. The LOID identifies an object to various services (e.g. method invocation). The basic LOID data structure consists of a sequence of variable length binary string fields. Currently, four of these fields are reserved by the system. The first three play an important role in the LOID-to-object address binding mechanism: the first field is the *domain identifier*, used in the dynamic connection of separate Legion systems; the second is the *class identifier*, a bit string uniquely identifying the object's class within its domain; the third is an *instance number* that distinguishes the object from other instances of its class. LOIDs with an instance number field of length zero are defined to refer to class objects. The fourth and final field is reserved for security purposes. Specifically, this field contains an RSA public key for authentication and encrypted communication with the named object. New LOID types can be constructed to contain additional security information, location hints and other information in the additional available fields.

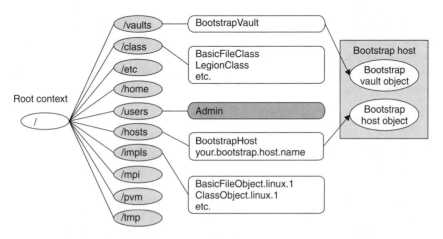

Figure 10.3 Legion Grid directory structure (context space).

Object addresses: Legion uses standard network protocols and communication facilities of host operating systems to support interobject communication. To perform such communication, Legion converts location-independent LOIDs into location-dependent communication system-level OAs through the Legion binding process. An OA consists of a list of *object address elements* and an *address semantic* field, which describes how to use the list. An OA element contains two parts, a 32-bit *address type* field indicating the type of address contained in the OA and the address itself, whose size and format depend on the type. The address semantic field is intended to express various forms of multicast and replicated communication. Our current implementation defines two OA types, the first consists of a single OA element containing a 32-bit IP address, 16-bit port number and 32-bit unique id (to distinguish between multiple sessions that reuse a single IP/port pair). This OA is used by our User Datagram Protocol (UDP)-based data delivery layer. The other is similar and uses TCP. Associations between LOIDs and OAs are called *bindings*, and are implemented as three-tuples. A binding consists of a LOID, an OA and a field that specifies the time at which the binding becomes invalid (including never). Bindings are first-class entities that can be passed around the system and cached within objects.

10.5.2 Metadata

Legion *attributes* provide a general mechanism to allow objects to describe themselves to the rest of the system in the form of metadata. An attribute is an *n*-tuple containing a *tag* and a list of *values*; the tag is a character string, and the values contain data that varies by tag. Attributes are stored as part of the state of the object they describe and can be dynamically retrieved or modified by invoking object-mandatory functions. In general, programmers can define an arbitrary set of attribute tags for their objects, although certain types of objects are expected to support certain standard sets of attributes. For example, host objects are expected to maintain attributes describing the architecture, configuration and state of the machine(s) they represent.

Attributes are collected together into metadata databases called *collections*. A collection is an object that 'collects' attributes from a set of objects and can be queried to find all the objects that have to meet some specification. Collections can be organized into directed acyclic graphs, for example, trees, in which collection data 'flows' up and is aggregated in large-scale collections. Collections can be used extensively for resource discovery in many forms, for example, for scheduling *(retrieve hosts where the one-minute load is less than some threshold)*.

10.5.3 Security

Security is an important requirement for Grid infrastructures as well as for clients using Grids. The definitions of security can vary depending on the resource accessed and the level of security desired. In general, Legion provides mechanisms for security while allowing Grid administrators to configure policies that use these mechanisms. Legion provides mechanisms for authentication, authorization and data integrity/confidentiality. The mechanisms are flexible enough to accommodate most policies desired by administrators. Legion explicitly accepts and supports that Grids may comprise multiple organizations that have different policies and that wish to maintain independent control of all their resources rather than trust a single global administrator.

In addition, Legion operates with existing network security infrastructures such as Network Address Translators (NATs) and firewalls. For firewall software, Legion requires port-forwarding to be set up. After this setup, all Legion traffic is sent through a single bi-directional UDP port opened in the firewall. The UDP traffic sent through the firewall is transmitted using one of the data protection modes described in Section 10.5.3.3. In addition, Legion provides a sandboxing mechanism whereby different Legion users are mapped to different local Unix/Windows users on a machine and thus isolated from each other, as well as from the Grid system on that machine. The sandboxing solution also permits the use of 'generic' IDs so that Legion users who do not have accounts on that machine can continue to access resources on that machine using generic accounts.

Recall that LOIDs contain an RSA public key as part of their structure. Thus, naming and identity have been combined in Legion. Since public keys are part of object names, any two objects in Legion can communicate securely and authenticate without the need for a trusted third party. Further, any object can generate and sign a credential. Since all Grid components, for example, files, users, directories, applications, machines, schedulers and so on are objects, security policies can be based not just on which person, but what component, is requesting an action.

10.5.3.1 Authentication

Users authenticate themselves to a Legion Grid with the login paradigm. Logging in requires specifying a user name and password. The user name corresponds to an authentication object that is the user's proxy to the Grid. The authentication object holds the user's private key, encrypted password and user profile information as part of its persistent state. The password supplied during login is compared to the password in the state of the authentication object in order to permit or deny subsequent access to the Grid. The state of the authentication object is stored on disk and protected by the local operating system. In other words, Legion depends on the local operating system to protect identities. Consequently, it requires careful configuration to place authentication objects on reliable machines. In a typical configuration, authentication objects for each administrative domain are restricted to a small number of trusted hosts.

A Legion system is neutral to the authentication mechanism. Past implementations have used the possession of a valid Kerberos credential as evidence of identity (as opposed to the password model above). In principle, any authentication mechanism, for example, NIS, Lightweight Directory Access Protocol (LDAP), Kerberos, RSA SecurID and so on, can be used with Legion. Upon authentication, the authentication object generates and signs an encrypted (non-X.509) credential that is passed back to the caller. The credential is stored either in a temporary directory similar to the approach of Kerberos or stored as part of the internal state of the DAP. In either case, the credential is protected by the security of the underlying operating system.

Although login is the most commonly used method for users to authenticate themselves to a Grid, it is not the only method. A Legion Data Access (DAP) enables access to a data Grid using the NFS protocol. The DAP maintains a mapping from the local Unix ID to the

LOID of an authentication object. The DAP communicates directly with authentication objects to obtain credentials for Unix users. Subsequently, the DAP uses the credentials to access data on the user's behalf. With this access method, users may be unaware that they are using Legion's security system to access remote data. Moreover, clients of the DAP typically do not require installation of Legion software.

Credentials are used for subsequent access to Grid resources. When a user initiates an action, the user's credentials are securely propagated along the chain of resources accessed in the course of the action using a mechanism called *implicit parameters*. Implicit parameters are the Grid equivalent of environment variables – they are tuples of the form <name, value> and are used to convey arbitrary information to objects.

10.5.3.2 Authorization

Authorization addresses the issue of who or what can do which operations on what objects. Legion is policy-neutral with respect to authorization, that is, a large range of policies is possible. Legion/Avaki ships with an access control list (ACL) policy, which is a familiar mechanism for implementing authorization. For each action (or function) on an object, for example, a 'read' or a 'write' on a file object, there exists an 'allow' list and a 'deny' list. Each list can contain the names of other objects, typically, but not restricted to authentication objects such as */users/jose*. A list may also contain a directory, which in turn contains a list of objects. If the directory contains authentication objects, then it functions as a group. Arbitrary groups can be formed by selecting desired authentication objects and listing them in directories. The allow list for an action on an object thus identifies a set of objects (often, authentication objects, which are proxies for humans) that are allowed to perform the action. Likewise, the deny list identifies a set of objects that are not allowed to perform the action. If an object is present in the allow list as well as the deny list, we choose the more restrictive option, that is, the object is denied permission. An ACL can be manipulated in several ways; one way provided in Legion is with a 'chmod' command that maps the intuitive '+/−' and 'rwx' options to the appropriate allow and deny lists for sets of actions.

ACLs can be manipulated to permit or disallow arbitrary sharing policies. They can be used to mimic the usual read-write-execute semantics of Unix, but can be extended to include a wider variety of objects than just files and directories. Complex authorizations are possible. For example, if three mutually distrustful companies wish to collaborate in a manner wherein one provides the application, another provides data and the third provides processing power, they can do so by setting the appropriate ACLs in a manner that permits only the actions desired and none other. In addition, the flexibility of creating groups at will is a benefit to collaboration. Users do not have to submit requests to create groups to system administrators – they can create their own groups by linking in existing users into a directory and they can set permissions for these groups on their own objects.

10.5.3.3 Data integrity and data confidentiality

Data integrity is the assurance that data has not been changed, either maliciously or inadvertently, either in storage or in transmission, without proper authorization. Data confidentiality (or privacy) refers to the inability of an unauthorized party to view/read/understand a particular piece of data, again either in transit or in storage. For on-the-wire protection, that is, protection during transmission, Legion follows the lead of almost all contemporary security systems in using cryptography. Legion transmits data in one of three modes: *private, protected* and *none*. Legion uses the OpenSSL libraries for encryption. In the private mode, all data is fully encrypted. In the protected mode, the sender computes a checksum (e.g. MD5) of the actual message. In the third mode, 'none', all data is transmitted in the clear except for credentials. This mode is useful for applications that do not want to pay the performance penalty of encryption and/or checksum calculations. The desired integrity mode is carried in an implicit parameter that is propagated down the call chain. If a user sets the mode to *private,* then all subsequent communication made by the user or on behalf of the user will be fully encrypted. Since different users may have different modes of encryption, different communications with the same object may have different encryption modes. Objects in the call chain, for example, the DAP, may override the mode if deemed appropriate.

For on-the-host data protection, that is, integrity/confidentiality during storage on a machine, Legion relies largely on the underlying operating system. Persistent data is stored on local disks with the appropriate permissions such that only appropriate users can access them.

10.6 CORE LEGION OBJECTS

In this section, we delve deeper into the Legion implementation of a Grid. In particular, we discuss some of the core objects that comprise a Grid. The intention of this section is to emphasize how a clean architectural design encourages the creation of varied services in a Grid. In the case of Legion, these services are implemented naturally as objects. We do not intend this section to be a catalogue of all Legion objects for two reasons: one, space considerations prevent an extensive catalogue and two, an abiding design principle in Legion is extensibility, which in turn means that a complete catalogue is intentionally impossible. Also, in this section, we stop at the object level. All objects in turn use a common communication substrate in order to perform their tasks. This communication is based on dataflow graphs, with each graph encapsulating a complex, reflection-based, remote method invocation. In turn, the method invocation involves a flow-controlled message-passing protocol. Each of these topics is interesting in itself; however, we refer the reader to literature on Legion as well as other systems regarding these topics [7, 17].

10.6.1 Class objects

Class objects manage particular instances. For example, a class object may manage all running instances of an application. Managing instances involves creating them on demand, destroying them when required, managing their state and keeping track of them.

Every Legion object is defined and managed by its class object. Class objects are *managers* and *policy makers* and have systemlike responsibility for creating new instances, activating and deactivating them, and providing bindings for clients. Legion encourages users to define and build their own class objects. These two features – class object management of the class' instances and applications programmers' ability to construct new classes – provide flexibility in determining how an application behaves and further support the Legion philosophy of enabling flexibility in the kind and level of functionality.

Class objects are ideal for exploiting the special characteristics of their instances. We expect that a vast majority of programmers will be served adequately by existing *metaclasses. Metaclass objects* are class objects whose instances are by themselves class objects. Just as a normal class object maintains implementation objects for its instances, so does a metaclass object. A metaclass object's implementation objects are built to export the class-mandatory interface and to exhibit a particular class functionality behind that interface. To use one, a programmer simply requests a 'create' on the appropriate metaclass object, and configures the resulting class object with implementation objects for the application in question. The new class object then supports the creation, migration, activation and location of these application objects in the manner defined by its metaclass object.

For example, consider an application that requires a user to have a valid software license in order to create a new object, for example, a video-on-demand application in which a new video server object is created for each request. To support this application, the developer could create a new metaclass object for its video server classes, the implementation of which would add a license check to the object creation method.

10.6.2 Hosts

Legion host objects abstract processing resources in Legion. They may represent a single processor, a multiprocessor, a Sparc, a Cray T90 or even an aggregate of multiple hosts. A host object is a machine's representative to Legion: it is responsible for executing objects on the machine, protecting the machine from Legion users, protecting user objects from each other, reaping objects and reporting object exceptions. A host object is also ultimately responsible for deciding which objects can run on the machine it represents. Thus, host objects are important points of security policy encapsulation.

Apart from implementing the host-mandatory interface, host object implementations can be built to adapt to different execution environments and to suit different site policies and underlying resource management interfaces. For example, the host-object implementation for an interactive workstation uses different process creation mechanisms when compared with implementations for parallel computers managed by batch queuing systems. In fact, the standard host object has been extended to submit jobs to a queuing system. Another extension to the host object has been to create container objects that permit execution of only one type of object on a machine. An implementation to address the performance problems might use threads instead of processes. This design improves the performance of object activation, and also reduces the cost of method invocation between objects on the same host by allowing shared address space communication.

While host objects present a uniform interface to different resource environments, they also (and more importantly) provide a means for resource providers to enforce security and resource management policies within a Legion system. For example, a host object implementation can be customized to allow only a restricted set of users, access to a resource.

Alternatively, host objects can restrict access on the basis of code characteristics (e.g. accepting only object implementations that contain a proof-carrying code demonstrating certain security properties, or rejecting implementations containing certain 'restricted' system calls).

We have implemented a spectrum of host object choices that trade off risk, system security, performance, and application security. An important aspect of Legion site autonomy is the freedom of each site to select the existing host object implementation that best suits its needs, to extend one of the existing implementations to suit local requirements, or to implement a new host object starting from the abstract interface. In selecting and configuring host objects, a site can control the use of its resources by Legion objects.

10.6.3 Vaults

Vault objects are responsible for managing other Legion objects' persistent representations (OPRs). In the same way that hosts manage active objects' direct access to processors, vaults manage inert objects on persistent storage. A vault has direct access to a storage device (or devices) on which the OPRs it manages are stored. A vault's managed storage may include a portion of a Unix file system, a set of databases or a hierarchical storage management system. The vault supports the creation of OPRs for new objects, controls access to existing OPRs and supports the migration of OPRs from one storage device to another. Manager objects manage the assignment of vaults to instances: when an object is created, its vault is chosen by the object's manager. The selected vault creates a new, empty OPR for the object and supplies the object with its state. When an object migrates, its manager selects a new target vault for its OPR.

10.6.4 Implementation objects

Implementation objects encapsulate Legion object executables. The executable itself is treated very much like a Unix file (i.e. as an array of bytes) so the implementation object interface naturally is similar to a Unix file interface. Implementation objects are also write-once, read-many objects – no updates are permitted after the executable is initially stored.

Implementation objects typically contain an executable code for a single platform, but may in general contain any information necessary to instantiate an object on a particular host. For example, implementations might contain Java byte code, Perl scripts or high-level source code that requires compilation by a host. Like all other Legion objects, implementation objects describe themselves by maintaining a set of attributes. In their attributes, implementation objects specify their execution requirements and characteristics, which may then be exploited during the scheduling process [18]. For example, an implementation object may record the type of executable it contains, its minimum target machine requirements, performance characteristics of the code, and so on.

Class objects maintain a complete list of (possibly very different) acceptable implementation objects appropriate for their instances. When the class (or its scheduling agent) selects a host and implementation for object activation, it selects them on the basis of the attributes of the host, the instance to be activated and the implementation object.

Implementation objects allow classes a large degree of flexibility in customizing the behavior of individual instances. For example, a class might maintain implementations with different time/space trade-offs and select between them depending on the currently available resources. To provide users with the ability to select their cost/performance trade-offs, a class might maintain both a slower, low-cost implementation and faster, higher-cost implementation. This approach is similar to abstract and concrete types in Emerald.

10.6.5 Implementation caches

Implementation caches avoid storage and communication costs by storing implementations for later reuse. If multiple host objects share access to some common storage device, they may share a single cache to further reduce copying and storage costs. Host objects invoke methods on the implementation caches in order to download implementations. The cache object either finds that it already has a cached copy of the implementation, or it downloads and caches a new copy. In either case, the cache object returns the executable's path to the host. In terms of performance, using a cached binary results in object activation being only slightly more expensive than running a program from a local file system.

Our implementation model makes the invalidation of cached binaries a trivial problem. Since class objects specify the LOID of the implementation to be used on each activation request, a class need only change its list of binaries to replace the old implementation LOID with the new one. The new version will be specified with future activation requests, and the old implementation will no longer be used and will time-out and be discarded from caches.

10.7 THE TRANSFORMATION FROM LEGION TO AVAKI

Legion began in late 1993 with the observation that dramatic changes in wide-area network bandwidth were on the horizon. In addition to the expected vast increases in bandwidth, other changes such as faster processors, more available memory, more disk space, and so on were expected to follow in the usual way as predicted by *Moore's Law*. Given the dramatic changes in bandwidth expected, the natural question was, how will this bandwidth be used? Since not just bandwidth will change, we generalized the question to, 'Given the expected changes in the physical infrastructure – what sorts of applications will people want, and given that, what is the system software infrastructure that will be needed to support those applications?' The Legion project was born with the determination to build, test, deploy and ultimately transfer to industry, a robust, scalable, Grid computing software infrastructure. We followed the classic design paradigm of first determining requirements, then completely designing the system architecture on paper after numerous design meetings, and finally, after a year of design work, coding. We made a decision to write from scratch rather than to extend and modify our existing system (Mentat [14]), which we had been using as a prototype. We felt that only by starting from scratch could we ensure adherence to our architectural principles. First funding was obtained in early 1996, and the first line of Legion code was written in June 1996.

The basic architecture was driven by the principles and requirements described above, and by the rich literature in distributed systems. The resulting architecture was reflective, object-based to facilitate encapsulation, extensible and was in essence an operating system for Grids. We felt strongly that having a common, accepted underlying architecture and a set of services that can be assumed to be critical for success in Grids. In this sense, the Legion architecture anticipated the drive to Web Services and OGSI. Indeed, the Legion architecture is very similar to OGSI [13].

By November 1997, we were ready for our first deployment. We deployed Legion at UVa, SDSC, NCSA and UC Berkeley for our first large-scale test and demonstration at Supercomputing 1997. In the early months, keeping the mean time between failures (MTBF) over 20 h under continuous use was a challenge. This is when we learned several valuable lessons. For example, we learned that the world is not 'fail-stop'. While we intellectually knew this, it was really brought home by the unusual failure modes of the various hosts in the system.

By November 1998, we had solved the failure problems and our MTBF was in excess of one month, and heading toward three months. We again demonstrated Legion – now on what we called *NPACI-Net*. NPACI-Net consisted of hosts at UVa, Caltech, UC Berkeley, IU, NCSA, the University of Michigan, Georgia Tech, Tokyo Institute of Technology and the Vrije Universiteit, Amsterdam. By that time, dozens of applications had been ported to Legion from areas as diverse as materials science, ocean modeling, sequence comparison, molecular modeling and astronomy. NPACI-Net continues today with additional sites such as the University of Minnesota, SUNY Binghamton and PSC. Supported platforms include Windows 2000, the Compaq iPaq, the T3E and T90, IBM SP-3, Solaris, Irix, HPUX, Linux, True 64 Unix and others.

From the beginning of the project a 'technology transfer' phase had been envisioned in which the technology would be moved from academia to industry. We felt strongly that Grid software would move into mainstream business computing only with commercially supported software, help lines, customer support, services and deployment teams. In 1999, Applied MetaComputing was founded to carry out the technology transition. In 2001, Applied MetaComputing raised $16 M in venture capital and changed its name to AVAKI. The company acquired legal rights to Legion from the University of Virginia and renamed *Legion* to *Avaki*. Avaki was released commercially in September 2001. Avaki is an extremely hardened, trimmed-down, focused-on-commercial-requirements version of Legion. While the name has changed, the core architecture and the principles on which it operates remain the same.

10.7.1 Avaki today

The first question that must be answered before looking at Avaki is 'Why commercialize Grids?' Many of the technological challenges faced by companies today can be viewed as variants of the requirements of Grid infrastructures. Today, science and commerce is increasingly a global enterprise, in which people collaborating on a single research project, engineering project or product development effort may be located over distances spanning a country or the world. The components of the project or product – data, applications, processing power and users – may be in distinct locations from one another. This scenario is particularly true in the life sciences, in which there are large amounts of data generated by many different organizations, both public and private, around the world. The attractive model of accessing all resources as if they were local runs into several immediate challenges. Administrative controls set up by organizations to prevent unauthorized access to resources hinder authorized access as well. Differences in platforms, operating systems, tools, mechanisms for running jobs, data organizations and so on impose a heavy cognitive burden on users. Changes in resource usage policies and security policies affect the day-to-day actions of users. Finally, large distances act as barriers to the quick communication necessary for collaboration. Consequently, users spend too much time on the procedures for accessing a resource and too little time on using the resource itself. These challenges lower productivity and hinder collaboration.

Grids offer a promise to solve the challenges facing collaboration by providing the mechanisms for easy and secure access to resources. Academic and government-sponsored Grid infrastructures, such as Legion, have been used to construct long-running Grids accessing distributed, heterogeneous and potentially faulty resources in a secure manner. There are clear benefits in making Grids available to an audience wider than academia or government. However, clients from industry make several demands that are not traditionally addressed by academic projects. For example, industry clients typically demand a product that is supported by a company whose existence can be assured for a reasonably long period. Moreover, the product must be supported by a professional services team that understands how to deploy and configure the product. Clients demand training engagements, extensive documentation, always-available support staff and a product roadmap that includes their suggestions regarding the product. Such clients do not necessarily view open source or free software as requisites; they are willing to pay for products and services that will improve the productivity of their users.

A successful technology is one that can transition smoothly from the comfort and confines of academia to the demanding commercial environment. Several academic projects are testament to the benefits of such transitions; often, the transition benefits not just the user community but the quality of the product as well. We believe Grid infrastructures are ready to make such a transition. Legion has been tested in nonindustry environments from 1997 onwards during which time we had the opportunity to test the basic model, scalability, security features, tools and development environment rigorously. Further improvement required input from a more demanding community with a vested interest in using the technology for their own benefit. The decision to commercialize Grids, in the form of the Avaki 2.x product and beyond was inevitable.

10.7.2 How are Grid requirements relevant to a commercial product?

Despite changes to the technology enforced by the push to commercialization, the basic technology in Avaki 2.x remains the same as Legion. All the principles and architectural features discussed earlier continue to form the basis of the commercial product. As a result, the commercial product continues to meet the requirements outlined in the introduction.

These requirements follow naturally from the challenges faced by commercial clients who attempt to access distributed, heterogeneous resources in a secure manner.

Consider an everyday scenario in a typical life sciences company as a use case. Such a company may have been formed by the merger of two separate companies, with offices located in different cities. The original companies would have purchased their IT infrastructure, that is, their machines, their operating systems, their file systems and their tools independent of each other. The sets of users at the two companies would overlap to a very small extent, often not at all. After the merger, the companies may be connected by virtual private network (VPN) software, but their traffic goes over the public network. Users would expect to use the resources of the merged companies shortly after the merger. Moreover, they may expect to use public databases as well as databases managed by other partner companies.

The above scenario is common in several companies not only in life sciences but also other verticals, such as finance, engineering design and natural sciences. Any solution that attempts to solve this scenario must satisfy several, if not all the Grid requirements. For example, the different IT infrastructures lead to the requirement of heterogeneity. The desire to access resources located remotely and with different access policies leads to complexity. Since resources are accessed over a public network, fault tolerance will be expected. Besides, as resources are increased, the mean time to failure of the infrastructure as a whole increases surprisingly rapidly even if the individual resources are highly reliable. Users in the merged company will expect the merged IT infrastructure to be secure from external threats, but in addition, they may desire security from each other, at least until enough trust has been built between users of the erstwhile companies. Accessing remote computing resources leads to binary management and to support for applications written in any language. The inclusion of private, public and partnered databases in accesses demands a scalable system that can grow as the numbers and sizes of the databases increase. Managers of these databases will expect to continue controlling the resources they make available to others, as will managers of other resources, such as system administrators. Users will expect to access resources in as simple a manner as possible – specifying just a name is ideal, but specifying any location, method, behavior, and so on is too cumbersome. Finally, users will expect to reuse these solutions. In other words, the infrastructure must continue to exist beyond one-time sessions. Any software solution that addresses all these requirements is a Grid infrastructure.

10.7.3 What is retained, removed, reinforced?

Avaki 2.x retains the vision, the philosophy, the principles and the underlying architecture of Legion. It eliminates some of the more esoteric features and functions present in Legion. It reinforces the robustness of the infrastructure by adding more stringent error-checking and recommending safer configurations. Moreover, it increases the usability of the product by providing extensive documentation, configuration guidelines and additional tools.

As a company, AVAKI necessarily balanced retaining only those features and functions of the product that addressed immediate revenue against retaining all possible features in Legion. For example, with Legion, we created a new kind of file object called a *2D file* partly to support applications that access large matrices and partly to demonstrate the ability to create custom services in Legion. Since a demand for 2D files was low in the commercial environment, Avaki 2.x does not support this object, although the code base includes it. Likewise, until clients request Avaki's heterogeneous MPI tools, support for this feature is limited.

The Legion vision of a single-system view of Grid components is a compelling one, and we expect it to pervade future releases of Avaki and products from AVAKI. In addition, we expect that several philosophical aspects such as the object model, naming and well-designed security will continue. Underlying principles such as 'no privileged user requirement' have been popular with users, and we expect to continue them. The underlying architecture may change, especially with the influence of web services or other emerging technologies. Likewise, the implementation of the product may change, but that change is an irrelevant detail to Grid customers.

The commercial product reinforces many of the strengths of Legion. For example, with the commercial product, installation and Grid creation is easier than before – these actions require a total of two commands. Several commands have been simplified while retaining their Unix-like syntax. Configuring the Grid in the myriad ways possible has been simplified. The underlying communication protocol as well as several tools and services have been made more robust. Error messaging has been improved, and services have been made more reliable. Support for several of the strengths of Legion – transparent data access, legacy application management, queuing, interfaces to third-party queuing and MPI systems and parameter-space studies – have been improved.

10.8 MEETING THE GRID REQUIREMENTS WITH LEGION

Legion continues to meet the technical Grid requirements outlined in Section 10.2. In addition, it meets commercial requirements for Grids as well. In this section, we discuss how Legion and Avaki meet the technical Grid requirements by revisiting each requirement identified in Section 10.2:

- *Security*: As described in Section 10.5.3 and in References [8, 9], security is a core aspect of both the Legion architecture and the Legion implementation. Legion addresses authentication, authorization, data integrity and firewalls in a flexible manner that separates policy and mechanism clearly.
- *Global namespace*: A global namespace is fundamental to this model and is realized by context space and LOIDs.
- *Fault tolerance*: Fault tolerance remains one of the most significant challenges in Grid systems. We began with a belief that there is no good one-size-fits-all solution to the fault-tolerance problem in Grids [10, 19]. Legion addresses fault tolerance for limited classes of applications. Specifically stateless application components, jobs in high-throughput computing, simple K-copy objects that are resilient to K-1 host failures and MPI applications that have been modified to use Legion save-and-restore state functions. There remains, however, a great deal of work to be done in the area of fault tolerance for Grids. Indeed, this is the thrust of the IBM efforts in autonomic computing.

- *Accommodating heterogeneity*: Operating system and architecture heterogeneity is actually one of the easier problems to solve, and a problem for which good solutions have been available for over two decades. In general, there are two basic challenges to overcome: data format conversions between heterogeneous architectures (e.g. those that arise from a cross-architecture remote procedure call (RPC) or reading binary data that was written on another architecture), and executable format differences including different instruction sets (you cannot use the same executable on all platforms[3]).

Data format conversions have traditionally been done in one of two ways. In the first technique, exemplified by XDR (External Data Representation), all data is converted from native format to a standard format when sent, and then converted back to the native format when received. The second technique has been called *receiver makes right*. The data is sent in its native format, along with a metadata tag that indicates which format the data is in. Upon receipt the receiver checks the tag. If the data is already of the native format, no conversion is necessary. If the data is of a different format, then the data is converted on the fly. Legion uses the second technique. Typed data is 'packed' into buffers. When data is 'unpacked', we check the metadata tag associated with the buffer and convert the data if needed. We have defined packers and unpackers (including all pairwise permutations) for Intel (32 bit), Sun, SGI (32 and 64 bit), HP, DEC Alpha (32 and 64 bit), IBM RS6000 and Cray T90 and C90.

- *Binary management*: The Legion run time ensures that the appropriate binaries, if they exist, are available on a host whenever needed. Each class object maintains a list of implementations available to the class and each implementation is by itself a Legion object with methods to read the implementation data. At run time the schedulers will ensure that only hosts for which binaries are available are selected. Once a host has been selected and an appropriate implementation chosen, the class object asks the host to instantiate an object instance. The host asks the implementation cache for a path to the executable, and the cache downloads the implementation if necessary. No user-level code is involved in this process.
- *Multilanguage support*: Avaki is completely agnostic to programming language. Applications have been written in C, C++, Fortran, SNOBOL, Perl, Java and shell-scripting languages.
- *Scalability*: A truly scalable system can keep growing without performance degradation. In order to have a scalable system, there must be no hot spots. System components should not become overloaded as the system scales to millions of hosts and billions of objects. This means that as the number of users and hosts increases the number of 'servers' that provide basic services must increase as well without requiring a superlinear increase in the amount of traffic.

Before we begin, let us examine where hot spots would be most likely to occur. The first and most obvious place is the root of the class structure and hence the binding structure.

---

[3] There is a long history of languages for which this is not true. They are all interpreted languages in one form or another, either directly as in Perl, Python and shell scripts or compiled into some form of byte code, which is then executed by a virtual machine or interpreter, for example, UCSD Pascal and Java.

A second point is near the top of the context space tree, at the root and perhaps the top level or two of directories. Once again, these are globally shared, and frequently accessed data structures.

We achieve scalability in Legion through three aspects of the architecture: hierarchical binding agents, distribution of naming and binding to multiple class objects and contexts, and replication and cloning of classes and objects.

A single binding agent would be no more scalable than a single class object. Therefore, we provide for a multitude of binding agents, typically one per host that are arranged in a tree and behave like a software-combining tree in high-performance shared memory architectures. The effect is to reduce the maximum lookup rate.

In the case of the LOID to OA mapping, Legion detects stale bindings (allowing safe aggressive caching, discussed above), by distributing the ultimate binding authority to class objects. There is a different class object for every type in the system, for example, for program1, for BasicFileObjects and so on.

In order to prevent frequently-used classes, for example BasicFileClass, from becoming hot spots, Legion supports *class cloning* and object replication. When a class is cloned, a new class object is instantiated. The new class has a different LOID, but all of the same implementations. When a new instance is created, one of the clones is selected (typically round-robin) to perform the instantiation and to manage the instance in the future.

- *Persistence*: Persistence in Legion is realized with Vaults (Section 10.6.3), and as far as most users are concerned, with the Legion Data Grid described above.
- *Extensibility*: No matter how carefully a system is designed and crafted, it will not meet the needs of all users. The reason is that different users, applications and organizations have different requirements and needs, for example, some applications require a coherent distributed file system, while others have weaker semantic requirements. Some organizations require the use of Kerberos credentials, while others do not. The bottom line is that a restrictive system will be unable to meet all current and future requirements.

Legion was designed to be customized and tailored to different application or organizational requirements. These customizations can take place either in the context of a closed system, for example, a total system change that affects all objects or in the context of a local change that affects only a subset of objects and does not affect the interoperability of the modified objects and other objects in the system.

Broadly, there are three places where customization can occur: modification of core daemon objects, definition of new metaclasses to change metabehaviours for a class of objects (Interobject Protocols), and changes in the Legion run-time libraries (LRTL) to change how Legion is implemented inside an object (Intraobject Protocols).

The core daemons, that is, hosts, vaults, classes, contexts and implementation objects and schedulers can be modified and extended by system administrators to override the default behaviors.

Legion class objects manage their instances and determine how their instances are instantiated. For example, the default, vanilla class object behavior instantiates its instances with one instance per address space, and the class object gives out the actual OA to binding agents. Similarly, the state of the instances is stored in vaults, and there is one copy of the state. In Legion, the metaobject protocols and interactions can be selectively modified on a class-by-class basis. This is done by overriding class-object behaviors with a new implementation. The result is a new metaclass object.

The final way to modify and extend the system is by modifying the LRTL [20], or by changing the configuration of the event handlers. Legion is a reflective system, so the internal representation of Legion in objects is exposed to programmers. Modifications to that representation will change behavior. The LRTL was designed to be changed, and to have the event 'stack' modified. To date we have used this primarily for modifications to the security layer and to build performance-optimized implementations for our MPI implementation.

- *Site autonomy*: Site autonomy is critical for Grid systems. We believe that very few individuals, and no enterprises, will agree to participate in a large-scale Grid system if it requires giving up control of their resources and control of local policy decisions such as who can do what on their resources. Site autonomy in Legion is guaranteed by two mechanisms: protecting the physical resources and setting the security policies of the objects used by local users.

  Host and vault objects represent a site's physical resources. They are protected by their respective ACLs. As discussed earlier, the access control policy can be overridden by defining a new host or vault class. The result is a change in local policy. In general, the hosts and vault policies can be set to whatever is needed by the enterprise.

- *Complexity management*: Complexity is addressed in Legion by providing a complete set of high-level services and capabilities that mask the underlying infrastructure from the user so that, by default the user does not have to think. At the level of programming the Grid, Legion takes a classic approach of using the object paradigm to encapsulate complexity in objects whose interfaces define behavior but not implementation.

10.9 EMERGING STANDARDS

Recently, at the Global Grid Forum an Open Grid Services Architecture (OGSA) [13] was proposed by IBM and the Globus PIs. The proposed architecture has many similarities to the Avaki architecture. One example of this congruence is that all objects (called *Grid Resources* in OGSA) have a name, an interface, a way to discover the interface, metadata and state, and are created by factories (analogous to Avaki class objects). The primary differences in the core architecture lie in the RPC model, the naming scheme and the security model.

The RPC model differences are of implementation – not of substance. This is a difference that Avaki intends to address by becoming Web Services compliant, that is, by supporting XML/SOAP (Simple Object Access Protocol) and WSDL (Web Services Description Language).

As for the naming models, both offer two low-level name schemes in which there is an immutable, location-independent name (Grid Services Handle (GSH) in OGSA, LOID in Avaki) and a lower-level 'address' (a WSDL Grid Service Reference (GSR) in OGSA and an OA in Avaki). The differences though are significant. First, OGSA names have no security information in them at all, requiring the use of an alternative mechanism to bind name and identity. We believe this is critical as Grids become widespread and consist of thousands of diverse organizations. Second, binding resolvers in OGSA currently are location- and protocol-specific, severely reducing the flexibility of the name-resolving process. To address these issues, Avaki has proposed the Secure Grid Naming Protocol (SGNP) [11] to the Global Grid Forum (GGF) as an open standard for naming in Grids. SGNP fits quite well with OGSA, and we are actively working with IBM and others within the GGF working group process to find the best solution for naming.

It is interesting to note the similarity of Legion to OGSI. Both architectures have at their core the notion of named entities that interact using method calls. Both use multilayer naming schemes and reflective interface discovery and defer to user-defined objects how those names will be bound, providing for a wide variety of implementations and semantics. The similarity is not a surprise as both draw on the same distributed systems literature.

As for security, at this juncture there is no security model in OGSA. This is a shortcoming that we certainly expect to remedy soon – at least as far as authentication and data integrity. The Globus group [21] has a long history in this area [22]. The Avaki security model was designed to be flexible, and has identity included in names. Further, the Avaki model has a notion of access control through replaceable modules called *MayI* that implement access control. The default policy is access control lists.

There are many nonarchitectural differences between Avaki and OGSA – Avaki is a complete, implemented system with a wealth of services. OGSA is a core architectural proposal – not a complete system. It is expected that higher-level interfaces and components will be added to the basic OGSA substrate over a period of time at GGF.

Finally, at Avaki we are fully committed to OGSI moving forward. We feel strongly that a foundation of good standards will accelerate the development of the Grid market and benefit everybody in the community, both users and producers of Grid software.

10.10 SUMMARY

The Legion project was begun in late 1993 to construct and deploy large-scale metasystems for scientific computing, though with a design goal to be a general-purpose metaoperating system. Since then, 'metacomputing' has become 'Grid computing' and the whole concept of Grid computing has begun to move into the mainstream. From the beginning we argued that Grid systems should be designed and engineered like any other large-scale software system – first by examining the requirements, then by designing an architecture to meet the requirements, and finally by building, deploying and testing the resulting software.

Thus Legion was born. Today, almost a decade later, not only have the basic design principles and architectural features stood the test of time but we have a production operational Grid system that has demonstrated utility not only in academia but in industry as well where robust software and hard return-on-investment are requirements.

In this chapter, we presented what we believe are some of the fundamental architectural requirements of Grids, as well as our design philosophy that we used in building Legion. We also presented a glimpse of how Legion is used in the field, as well as some basic aspects of the Legion architecture. We concluded with a discussion of the transformation of Legion into Avaki, and how we see Avaki fitting into the context of OGSI. The presentation was, of necessity, rather brief. We encourage the interested reader to follow references embedded in the text for much more complete descriptions.

REFERENCES

1. Grimshaw, A. S. (1994) Enterprise-wide computing. *Science*, **256**, 892–894.
2. Smarr, L. and Catlett, C. E. (1992) Metacomputing. *Communications of the ACM*, **35**(6), 44–52.
3. Grimshaw, A. S. and Wulf, W. A. (1997) The Legion vision of a worldwide virtual computer. *Communications of the ACM*, **40**(1), 39–45.
4. Grimshaw, A. S., Ferrari, A. J., Lindahl, G. and Holcomb, K. (1998) Metasystems. *Communications of the ACM*, **41**(11), 46–55.
5. Grimshaw, A. S., Ferrari, A. J., Knabe, F. C. and Humphrey, M. A. (1999) Wide-area computing: Resource sharing on a large scale. *IEEE Computer*, **32**(5), 29–37.
6. Grimshaw, A. S. *et al.* (1998) *Architectural Support for Extensibility and Autonomy in Wide-Area Distributed Object Systems*, Technical Report CS-98-12, Department of Computer Science, University of Virginia, June, 1998.
7. Grimshaw, A. S., Lewis, M. J., Ferrari, A. J. and Karpovich, J. F. (1998) *Architectural Support for Extensibility and Autonomy in Wide-Area Distributed Object Systems*, Technical Report CS-98-12, Department of Computer Science, University of Virginia, June, 1998.
8. Chapin, S. J., Wang, C., Wulf, W. A., Knabe, F. C. and Grimshaw, A. S. (1999) A new model of security for metasystems. *Journal of Future Generation Computing Systems*, **15**, 713–722.
9. Ferrari, A. J., Knabe, F. C., Humphrey, M. A., Chapin, S. J. and Grimshaw, A. S. (1999) A flexible security system for metacomputing environments. *7th International Conference on High-Performance Computing and Networking Europe (HPCN '99)*, Amsterdam, April, 1999, pp. 370–380.
10. Nguyen-Tuong, A. and Grimshaw, A. S. (1999) Using reflection for incorporating fault-tolerance techniques into distributed applications. *Parallel Processing Letters*, **9**(2), 291–301.
11. Apgar, J., Grimshaw, A. S., Harris, S., Humphrey, M. A. and Nguyen-Tuong, A. *Secure Grid Naming Protocol: Draft Specification for Review and Comment*, http://sourceforge.net/projects/sgnp.
12. Grimshaw, A. S., Ferrari, A. J., Knabe, F. C. and Humphrey, M. A. (1999) Wide-area computing: Resource sharing on a large scale. *IEEE Computer*, **32**(5), 29–37.
13. Foster, I., Kesselman, C., Nick, J. and Tuecke, S. *The Physiology of the Grid: An Open Grid Services Architecture for Distributed Systems Integration*, http://www.Gridforum.org/drafts/ogsi-wg/ogsa_draft2.9_2002-06-22.pdf.
14. Grimshaw, A. S., Weissman, J. B. and Strayer, W. T. (1996) Portable run-time support for dynamic object-oriented parallel processing. *ACM Transactions on Computer Systems*, **14**(2). 139–170.
15. Humphrey, M. A., Knabe, F. C., Ferrari, A. J. and Grimshaw, A. S. (2000) Accountability and control of process creation in metasystems. *Proceedings of the 2000 Network and Distributed Systems Security Conference (NDSS '00)*, San Diego, CA, February, 2000.
16. Ferrari, A. J. and Grimshaw, A. S. (1998) *Basic Fortran Support in Legion*, Technical Report CS-98-11, Department of Computer Science, University of Virginia, March, 1998.
17. Nguyen-Tuong, A., Chapin, S. J., Grimshaw, A. S. and Viles, C. *Using Reflection for Flexibility and Extensibility in a Metacomputing Environment*, Technical Report 98-33, University of Virginia, Department of Computer Science, November 19, 1998.
18. Chapin, S. J., Katramatos, D., Karpovich, J. F. and Grimshaw, A. S. (1999) Resource management in Legion. *Journal of Future Generation Computing Systems*, **15**, 583–594.
19. Nguyen-Tuong, A. *et al.* (1996) Exploiting data-flow for fault-tolerance in a wide-area parallel system. *Proceedings of the 15th International Symposium on Reliable and Distributed Systems (SRDS-15)*, pp. 2–11, 1996.

20. Viles, C. L. *et al.* (1997) Enabling flexibility in the Legion run-time library. *International Conference on Parallel and Distributed Processing Techniques (PDPTA '97)*, Las Vegas, NV, 1997.
21. Foster, I. and Kesselman, C. (1997) Globus: A metacomputing infrastructure toolkit. *International Journal of Supercomputing Applications*, **11**(2), 115–128, 1997.
22. Butler, R., Engert, D., Foster, I., Kesselman, C., Tuecke, S., Volmer, J. and Welch, V. (2000) A National-scale authentication infrastructure. *IEEE Computer*, **33**(12), 60–66.
23. Jin, L. J. and Grimshaw, A. S. (2000) From metacomputing to metabusiness processing. *Proceedings IEEE International Conference on Cluster Computing – Cluster 2000*, Saxony, Germany, December, 2000.
24. Natrajan, A., Humphrey, M. A. and Grimshaw, A. S. (2001) Capacity and capability computing using Legion. *Proceedings of the 2001 International Conference on Computational Science*, San Francisco, CA, May, 2001.
25. Natrajan, A., Fox, A. J., Humphrey, M. A., Grimshaw, A. S., Crowley, M. and Wilkins-Diehr, N. (2001) Protein folding on the grid: Experiences using CHARMM under Legion on NPACI resources. *10th International Symposium on High Performance Distributed Computing (HPDC)*, San Francisco, CA, August 7–9, 2001.
26. White, B. S., Walker, M. P., Humphrey, M. A. and Grimshaw, A. S. (2001) LegionFS: A secure and scalable file system supporting cross-domain high-performance applications. *Proceedings of Supercomputing 2001*, Denver, CO, 2001, www.sc2001.org/papers/pap.pap324.pdf.
27. Mullender, S. (ed.) (1989) *Distributed Systems*, ISBN number: 0-201-41660-3, ACM Press.

11

Condor and the Grid

Douglas Thain, Todd Tannenbaum, and Miron Livny

University of Wisconsin-Madison, Madison, Wisconsin, United States

11.1 INTRODUCTION

Since the early days of mankind the primary motivation for the establishment of communities has been the idea that by being part of an organized group the capabilities of an individual are improved. The great progress in the area of intercomputer communication led to the development of means by which stand-alone processing subsystems can be integrated into multicomputer communities.

– Miron Livny, *Study of Load Balancing Algorithms for Decentralized Distributed Processing Systems,* Ph.D. thesis, July 1983.

Ready access to large amounts of computing power has been a persistent goal of computer scientists for decades. Since the 1960s, visions of computing utilities as pervasive and as simple as the telephone have motivated system designers [1]. It was recognized in the 1970s that such power could be achieved inexpensively with collections of small devices rather than expensive single supercomputers. Interest in schemes for managing distributed processors [2, 3, 4] became so popular that there was even once a minor controversy over the meaning of the word 'distributed' [5].

Grid Computing – Making the Global Infrastructure a Reality.   Edited by F. Berman, A. Hey and G. Fox
© 2003 John Wiley & Sons, Ltd   ISBN: 0-470-85319-0

As this early work made it clear that distributed computing was *feasible*, theoretical researchers began to notice that distributed computing would be *difficult*. When messages may be lost, corrupted, or delayed, precise algorithms must be used in order to build an understandable (if not controllable) system [6, 7, 8, 9]. Such lessons were not lost on the system designers of the early 1980s. Production systems such as Locus [10] and Grapevine [11] recognized the fundamental tension between consistency and availability in the face of failures.

In this environment, the Condor project was born. At the University of Wisconsin, Miron Livny combined his 1983 doctoral thesis on cooperative processing [12] with the powerful Crystal Multicomputer [13] designed by DeWitt, Finkel, and Solomon and the novel Remote UNIX [14] software designed by Litzkow. The result was Condor, a new system for distributed computing. In contrast to the dominant centralized control model of the day, Condor was unique in its insistence that every participant in the system remain free to contribute as much or as little as it cared to.

Modern processing environments that consist of large collections of workstations interconnected by high capacity network raise the following challenging question: can we satisfy the needs of users who need extra capacity without lowering the quality of service experienced by the owners of under utilized workstations? . . . The Condor scheduling system is our answer to this question.

– Michael Litzkow, Miron Livny, and Matt Mutka, *Condor: A Hunter of Idle Workstations,* IEEE 8th Intl. Conf. on Dist. Comp. Sys., June 1988.

The Condor system soon became a staple of the production-computing environment at the University of Wisconsin, partially because of its concern for protecting individual interests [15]. A production setting can be both a curse and a blessing: The Condor project learned hard lessons as it gained real users. It was soon discovered that inconvenienced machine owners would quickly withdraw from the community, so it was decreed that owners must maintain control of their machines at any cost. A fixed schema for representing users and machines was in constant change and so led to the development of a schema-free resource allocation language called ClassAds [16, 17, 18]. It has been observed [19] that most complex systems struggle through an adolescence of five to seven years. Condor was no exception.

The most critical support task is responding to those owners of machines who feel that Condor is in some way interfering with their own use of their machine. Such complaints must be answered both promptly and diplomatically. Workstation owners are not used to the concept of somebody else using their machine while they are away and are in general suspicious of any new software installed on their system.

– Michael Litzkow and Miron Livny, *Experience With The Condor Distributed Batch System,* IEEE Workshop on Experimental Dist. Sys., October 1990.

The 1990s saw tremendous growth in the field of distributed computing. Scientific interests began to recognize that coupled commodity machines were significantly less expensive than supercomputers of equivalent power [20]. A wide variety of powerful batch execution systems such as LoadLeveler [21] (a descendant of Condor), LSF [22], Maui [23], NQE [24], and PBS [25] spread throughout academia and business. Several high-profile distributed computing efforts such as SETI@Home and Napster raised the public consciousness about the power of distributed computing, generating not a little moral and legal controversy along the way [26, 27]. A vision called grid computing began to build the case for resource sharing across organizational boundaries [28].

Throughout this period, the Condor project immersed itself in the problems of production users. As new programming environments such as PVM [29], MPI [30], and Java [31] became popular, the project added system support and contributed to standards development. As scientists grouped themselves into international computing efforts such as the Grid Physics Network [32] and the Particle Physics Data Grid (PPDG) [33], the Condor project took part from initial design to end-user support. As new protocols such as Grid Resource Access and Management (GRAM) [34], Grid Security Infrastructure (GSI) [35], and GridFTP [36] developed, the project applied them to production systems and suggested changes based on the experience. Through the years, the Condor project adapted computing structures to fit changing human communities.

Many previous publications about Condor have described in fine detail the features of the system. In this chapter, we will lay out a broad history of the Condor project and its design philosophy. We will describe how this philosophy has led to an organic growth of *computing communities* and discuss the *planning* and the *scheduling* techniques needed in such an uncontrolled system. Our insistence on dividing responsibility has led to a unique model of cooperative computing called *split execution*. We will conclude by describing how real users have put Condor to work.

11.2 THE PHILOSOPHY OF FLEXIBILITY

As distributed systems scale to ever-larger sizes, they become more and more difficult to control or even to describe. International distributed systems are heterogeneous in every way: they are composed of many types and brands of hardware, they run various operating systems and applications, they are connected by unreliable networks, they change configuration constantly as old components become obsolete and new components are powered on. Most importantly, they have many owners, each with private policies and requirements that control their participation in the community.

Flexibility is the key to surviving in such a hostile environment. Five admonitions outline our philosophy of flexibility.

Let communities grow naturally: Humanity has a natural desire to work together on common problems. Given tools of sufficient power, people will organize the computing structures that they need. However, human relationships are complex. People invest their time and resources into many communities with varying degrees. Trust is rarely complete or symmetric. Communities and contracts are never formalized with the same level of precision as computer code. Relationships and requirements change over time. Thus, we aim to build structures that permit but do not require cooperation. Relationships, obligations, and schemata will develop according to user necessity.

Plan without being picky: Progress requires optimism. In a community of sufficient size, there will always be idle resources available to do work. But, there will also always be resources that are slow, misconfigured, disconnected, or broken. An overdependence on the correct operation of any remote device is a recipe for disaster. As we design software, we must spend more time contemplating the consequences of failure than the potential benefits of success. When failures come our way, we must be prepared to retry or reassign work as the situation permits.

Leave the owner in control: To attract the maximum number of participants in a community, the barriers to participation must be low. Users will not donate their property to the common good unless they maintain some control over how it is used. Therefore, we must be careful to provide tools for the owner of a resource to set use policies and even instantly retract it for private use.

Lend and borrow: The Condor project has developed a large body of expertise in distributed resource management. Countless other practitioners in the field are experts in related fields such as networking, databases, programming languages, and security. The Condor project aims to give the research community the benefits of our expertise while accepting and integrating knowledge and software from other sources.

Understand previous research: We must always be vigilant to understand and apply previous research in computer science. Our field has developed over many decades and is known by many overlapping names such as operating systems, distributed computing, metacomputing, peer-to-peer computing, and grid computing. Each of these emphasizes a particular aspect of the discipline, but is united by fundamental concepts. If we fail to understand and apply previous research, we will at best rediscover well-charted shores. At worst, we will wreck ourselves on well-charted rocks.

11.3 THE CONDOR PROJECT TODAY

At present, the Condor project consists of over 30 faculties, full time staff, graduate and undergraduate students working at the University of Wisconsin-Madison. Together the group has over a century of experience in distributed computing concepts and practices, systems programming and design, and software engineering.

Condor is a multifaceted project engaged in five primary activities.

Research in distributed computing: Our research focus areas and the tools we have produced, several of which will be explored below and are as follows:

1. Harnessing the power of opportunistic and dedicated resources. *(Condor)*
2. Job management services for grid applications. *(Condor-G, DaPSched)*

3. Fabric management services for grid resources. *(Condor, Glide-In, NeST)*
4. Resource discovery, monitoring, and management. *(ClassAds, Hawkeye)*
5. Problem-solving environments. *(MW, DAGMan)*
6. Distributed I/O technology. *(Bypass, PFS, Kangaroo, NeST)*

Participation in the scientific community: Condor participates in national and international grid research, development, and deployment efforts. The actual development and deployment activities of the Condor project are a critical ingredient toward its success. Condor is actively involved in efforts such as the Grid Physics Network (GriPhyN) [32], the International Virtual Data Grid Laboratory (iVDGL) [37], the Particle Physics Data Grid (PPDG) [33], the NSF Middleware Initiative (NMI) [38], the TeraGrid [39], and the NASA Information Power Grid (IPG) [40]. Further, Condor is a founding member in the National Computational Science Alliance (NCSA) [41] and a close collaborator of the Globus project [42].

Engineering of complex software: Although a research project, Condor has a significant software production component. Our software is routinely used in mission-critical settings by industry, government, and academia. As a result, a portion of the project resembles a software company. Condor is built every day on multiple platforms, and an automated regression test suite containing over 200 tests stresses the current release candidate each night. The project's code base itself contains nearly a half-million lines, and significant pieces are closely tied to the underlying operating system. Two versions of the software, a stable version and a development version, are simultaneously developed in a multiplatform (Unix and Windows) environment. Within a given stable version, only bug fixes to the code base are permitted – new functionality must first mature and prove itself within the development series. Our release procedure makes use of multiple test beds. Early development releases run on test pools consisting of about a dozen machines; later in the development cycle, release candidates run on the production UW-Madison pool with over 1000 machines and dozens of real users. Final release candidates are installed at collaborator sites and carefully monitored. The goal is that each stable version release of Condor should be proven to operate in the field before being made available to the public.

Maintenance of production environments: The Condor project is also responsible for the Condor installation in the Computer Science Department at the University of Wisconsin-Madison, which consist of over 1000 CPUs. This installation is also a major compute resource for the Alliance Partners for Advanced Computational Servers (PACS) [43]. As such, it delivers compute cycles to scientists across the nation who have been granted computational resources by the National Science Foundation. In addition, the project provides consulting and support for other Condor installations at the University and around the world. Best effort support from the Condor software developers is available at no charge via ticket-tracked e-mail. Institutions using Condor can also opt for contracted support – for a fee, the Condor project will provide priority e-mail and telephone support with guaranteed turnaround times.

Education of students: Last but not the least, the Condor project trains students to become computer scientists. Part of this education is immersion in a production system. Students graduate with the rare experience of having nurtured software from the chalkboard all the way to the end user. In addition, students participate in the academic community by designing, performing, writing, and presenting original research. At the time of this writing, the project employs 20 graduate students including 7 Ph.D. candidates.

11.3.1 The Condor software: Condor and Condor-G

When most people hear the word 'Condor', they do not think of the research group and all of its surrounding activities. Instead, usually what comes to mind is strictly the *software* produced by the Condor project: the *Condor High Throughput Computing System*, often referred to simply as Condor.

11.3.1.1 Condor: a system for high-throughput computing

Condor is a specialized job and *resource management system* (RMS) [44] for compute-intensive jobs. Like other full-featured systems, Condor provides a job management mechanism, scheduling policy, priority scheme, resource monitoring, and resource management [45, 46]. Users submit their jobs to Condor, and Condor subsequently chooses when and where to run them based upon a policy, monitors their progress, and ultimately informs the user upon completion.

While providing functionality similar to that of a more traditional batch queueing system, Condor's novel architecture and unique mechanisms allow it to perform well in environments in which a traditional RMS is weak – areas such as sustained *high-throughput computing* and *opportunistic computing*. The goal of a high-throughput computing environment [47] is to provide large amounts of fault-tolerant computational power over prolonged periods of time by effectively utilizing all resources available to the network. The goal of opportunistic computing is the ability to utilize resources whenever they are available, *without requiring* 100% availability. The two goals are naturally coupled. High-throughput computing is most easily achieved through opportunistic means.

Some of the enabling mechanisms of Condor include the following:

- *ClassAds*: The ClassAd mechanism in Condor provides an extremely flexible and expressive framework for matching resource requests (e.g. jobs) with resource offers (e.g. machines). ClassAds allow Condor to adopt to nearly any desired resource utilization policy and to adopt a *planning* approach when incorporating Grid resources. We will discuss this approach further in a section below.
- *Job checkpoint and migration*: With certain types of jobs, Condor can transparently record a checkpoint and subsequently resume the application from the checkpoint file. A periodic checkpoint provides a form of fault tolerance and safeguards the accumulated computation time of a job. A checkpoint also permits a job to migrate from

one machine to another machine, enabling Condor to perform low-penalty *preemptive-resume scheduling* [48].

- *Remote system calls*: When running jobs on remote machines, Condor can often preserve the local execution environment via remote system calls. Remote system calls is one of Condor's *mobile sandbox* mechanisms for redirecting all of a jobs I/O-related system calls back to the machine that submitted the job. Therefore, users do not need to make data files available on remote workstations before Condor executes their programs there, even in the absence of a shared file system.

With these mechanisms, Condor can do more than effectively manage dedicated compute clusters [45, 46]. Condor can also scavenge and manage wasted CPU power from otherwise idle desktop workstations across an entire organization with minimal effort. For example, Condor can be configured to run jobs on desktop workstations only when the keyboard and CPU are idle. If a job is running on a workstation when the user returns and hits a key, Condor can migrate the job to a different workstation and resume the job right where it left off. Figure 11.1 shows the large amount of computing capacity available from idle workstations.

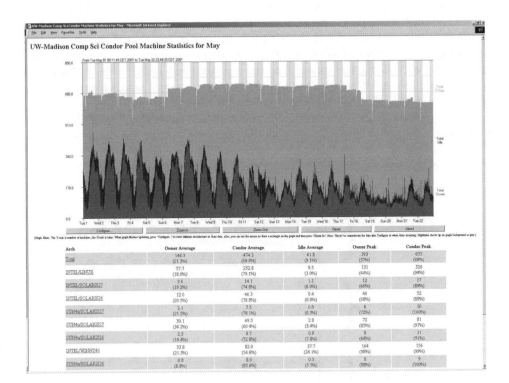

Arch	Owner Average	Condor Average	Idle Average	Owner Peak	Condor Peak
Total	146.3 (21.3%)	474.3 (69.6%)	61.8 (9.1%)	393 (57%)	635 (99%)
INTEL/LINUX	57.7 (18.0%)	252.8 (79.1%)	9.5 (3.0%)	131 (44%)	320 (94%)
INTEL/SOLARIS27	3.6 (19.2%)	14.1 (74.8%)	1.1 (6.0%)	12 (66%)	17 (89%)
INTEL/SOLARIS26	12.0 (20.5%)	46.3 (78.8%)	0.4 (0.8%)	40 (68%)	52 (88%)
SUN4x/SOLARIS27	2.1 (21.5%)	7.5 (78.1%)	0.0 (0.3%)	6 (72%)	10 (100%)
SUN4x/SOLARIS27	30.1 (36.2%)	49.5 (60.4%)	2.8 (3.4%)	72 (85%)	81 (97%)
SUN4x/SOLARIS26	2.3 (19.4%)	8.7 (72.8%)	0.9 (7.8%)	8 (66%)	11 (91%)
INTEL/WINNT40	33.8 (21.3%)	82.0 (54.6%)	37.7 (24.1%)	164 (98%)	156 (99%)
SUN4x/SOLARIS26	0.8 (8.8%)	8.0 (85.6%)	0.5 (5.5%)	8 (88%)	9 (100%)

Figure 11.1 The available capacity of the UW-Madison Condor pool in May 2001. Notice that a significant fraction of the machines were available for batch use, even during the middle of the work day. This figure was produced with CondorView, an interactive tool for visualizing Condor-managed resources.

Moreover, these same mechanisms enable preemptive-resume scheduling of dedicated compute cluster resources. This allows Condor to cleanly support priority-based scheduling on clusters. When any node in a dedicated cluster is not scheduled to run a job, Condor can utilize that node in an opportunistic manner – but when a schedule reservation requires that node again in the future, Condor can preempt any opportunistic computing job that may have been placed there in the meantime [30]. The end result is that Condor is used to seamlessly combine all of an organization's computational power into one resource.

The first version of Condor was installed as a production system in the UW-Madison Department of Computer Sciences in 1987 [14]. Today, in our department alone, Condor manages more than 1000 desktop workstation and compute cluster CPUs. It has become a critical tool for UW researchers. Hundreds of organizations in industry, government, and academia are successfully using Condor to establish compute environments ranging in size from a handful to thousands of workstations.

11.3.1.2 Condor-G: a computation management agent for Grid computing

Condor-G [49] represents the marriage of technologies from the Globus and the Condor projects. From Globus [50] comes the use of protocols for secure interdomain communications and standardized access to a variety of remote batch systems. From Condor comes the user concerns of job submission, job allocation, error recovery, and creation of a friendly execution environment. The result is very beneficial for the end user, who is now enabled to utilize large collections of resources that span across multiple domains as if they all belonged to the personal domain of the user.

Condor technology can exist at both the frontends and backends of a middleware environment, as depicted in Figure 11.2. Condor-G can be used as the reliable submission and job management service for one or more sites, the Condor High Throughput Computing system can be used as the fabric management service (a grid 'generator') for one or

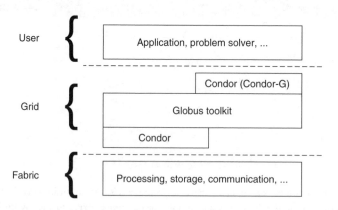

Figure 11.2 Condor technologies in Grid middleware. Grid middleware consisting of technologies from both Condor and Globus sit between the user's environment and the actual fabric (resources).

more sites, and finally Globus Toolkit services can be used as the bridge between them. In fact, Figure 11.2 can serve as a simplified diagram for many emerging grids, such as the USCMS Test bed Grid [51], established for the purpose of high-energy physics event reconstruction.

Another example is the European Union DataGrid [52] project's Grid Resource Broker, which utilizes Condor-G as its job submission service [53].

11.4 A HISTORY OF COMPUTING COMMUNITIES

Over the history of the Condor project, the fundamental structure of the system has remained constant while its power and functionality has steadily grown. The core components, known as the *kernel*, are shown in Figure 11.3. In this section, we will examine how a wide variety of *computing communities* may be constructed with small variations to the kernel.

Briefly, the kernel works as follows: The user submits jobs to an *agent*. The agent is responsible for remembering jobs in persistent storage while finding *resources* willing to run them. Agents and resources advertise themselves to a *matchmaker*, which is responsible for introducing potentially compatible agents and resources. Once introduced, an agent is responsible for contacting a resource and verifying that the match is still valid. To actually execute a job, each side must start a new process. At the agent, a *shadow* is responsible for providing all of the details necessary to execute a job. At the resource, a *sandbox* is responsible for creating a safe execution environment for the job and protecting the resource from any mischief.

Let us begin by examining how agents, resources, and matchmakers come together to form *Condor pools*. Later in this chapter, we will return to examine the other components of the kernel.

The initial conception of Condor is shown in Figure 11.4. Agents and resources independently report information about themselves to a well-known matchmaker, which then

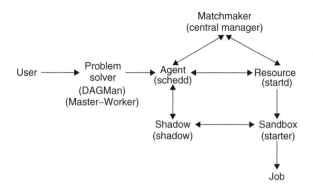

Figure 11.3 The Condor Kernel. This figure shows the major processes in a Condor system. The common generic name for each process is given in large print. In parentheses are the technical Condor-specific names used in some publications.

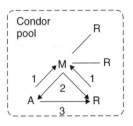

Figure 11.4 A Condor pool ca. 1988. An agent (A) is shown executing a job on a resource (R) with the help of a matchmaker (M). Step 1: The agent and the resource advertise themselves to the matchmaker. Step 2: The matchmaker informs the two parties that they are potentially compatible. Step 3: The agent contacts the resource and executes a job.

makes the same information available to the community. A single machine typically runs both an agent and a resource daemon and is capable of submitting and executing jobs. However, agents and resources are logically distinct. A single machine may run either or both, reflecting the needs of its owner. Furthermore, a machine may run more than one instance of an agent. Each user sharing a single machine could, for instance, run its own personal agent. This functionality is enabled by the agent implementation, which does not use any fixed IP port numbers or require any superuser privileges.

Each of the three parties – agents, resources, and matchmakers – are independent and individually responsible for enforcing their owner's policies. The agent enforces the submitting user's policies on what resources are trusted and suitable for running jobs. The resource enforces the machine owner's policies on what users are to be trusted and serviced. The matchmaker is responsible for enforcing community policies such as admission control. It may choose to admit or reject participants entirely on the basis of their names or addresses and may also set global limits such as the fraction of the pool allocable to any one agent. Each participant is autonomous, but the community as a single entity is defined by the common selection of a matchmaker.

As the Condor software developed, pools began to sprout up around the world. In the original design, it was very easy to accomplish resource sharing in the context of one community. A participant merely had to get in touch with a single matchmaker to consume or provide resources. However, a user could only participate in one community: that defined by a matchmaker. Users began to express their need to share across organizational boundaries.

This observation led to the development of *gateway flocking* in 1994 [54]. At that time, there were several hundred workstations at Wisconsin, while tens of workstations were scattered across several organizations in Europe. Combining all of the machines into one Condor pool was not a possibility because each organization wished to retain existing community policies enforced by established matchmakers. Even at the University of Wisconsin, researchers were unable to share resources between the separate engineering and computer science pools.

The concept of gateway flocking is shown in Figure 11.5. Here, the structure of two existing pools is preserved, while two gateway nodes pass information about participants

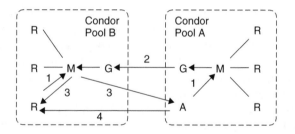

Figure 11.5 Gateway flocking ca. 1994. An agent (A) is shown executing a job on a resource (R) via a gateway (G). Step 1: The agent and resource advertise themselves locally. Step 2: The gateway forwards the agent's unsatisfied request to Condor Pool B. Step 3: The matchmaker informs the two parties that they are potentially compatible. Step 4: The agent contacts the resource and executes a job via the gateway.

between the two pools. If a gateway detects idle agents or resources in its home pool, it passes them to its peer, which advertises them in the remote pool, subject to the admission controls of the remote matchmaker. Gateway flocking is not necessarily bidirectional. A gateway may be configured with entirely different policies for advertising and accepting remote participants. Figure 11.6 shows the worldwide Condor flock in 1994.

The primary advantage of gateway flocking is that it is completely transparent to participants. If the owners of each pool agree on policies for sharing load, then cross-pool matches will be made without any modification by users. A very large system may be grown incrementally with administration only required between adjacent pools.

There are also significant limitations to gateway flocking. Because each pool is represented by a single gateway machine, the accounting of use by individual remote users

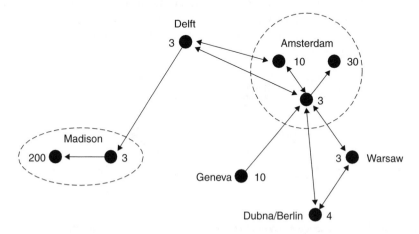

Figure 11.6 Worldwide Condor flock ca. 1994. This is a map of the worldwide Condor flock in 1994. Each dot indicates a complete Condor pool. Numbers indicate the size of each Condor pool. Lines indicate flocking via gateways. Arrows indicate the direction that jobs may flow.

is essentially impossible. Most importantly, gateway flocking only allows sharing at the organizational level – it does not permit an individual user to join multiple communities. This became a significant limitation as distributed computing became a larger and larger part of daily production work in scientific and commercial circles. Individual users might be members of multiple communities and yet not have the power or need to establish a formal relationship between both communities.

This problem was solved by *direct flocking*, shown in Figure 11.7. Here, an agent may simply report itself to multiple matchmakers. Jobs need not be assigned to any individual community, but may execute in either as resources become available. An agent may still use either community according to its policy while all participants maintain autonomy as before.

Both forms of flocking have their uses, and may even be applied at the same time. Gateway flocking requires agreement at the organizational level, but provides immediate and transparent benefit to all users. Direct flocking only requires agreement between one individual and another organization, but accordingly only benefits the user who takes the initiative.

This is a reasonable trade-off found in everyday life. Consider an agreement between two airlines to cross-book each other's flights. This may require years of negotiation, pages of contracts, and complex compensation schemes to satisfy executives at a high level. But, once put in place, customers have immediate access to twice as many flights with no inconvenience. Conversely, an individual may take the initiative to seek service from two competing airlines individually. This places an additional burden on the customer to seek and use multiple services, but requires no Herculean administrative agreement.

Although gateway flocking was of great use before the development of direct flocking, it did not survive the evolution of Condor. In addition to the necessary administrative complexity, it was also technically complex. The gateway participated in every interaction in the Condor kernel. It had to appear as both an agent and a resource, communicate with the matchmaker, and provide tunneling for the interaction between shadows and sandboxes. Any change to the protocol between any two components required a change

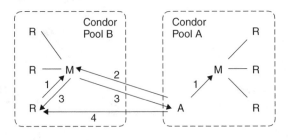

Figure 11.7 Direct flocking ca. 1998. An agent (A) is shown executing a job on a resource (R) via direct flocking. Step 1: The agent and the resource advertise themselves locally. Step 2: The agent is unsatisfied, so it also advertises itself to Condor Pool B. Step 3: The matchmaker (M) informs the two parties that they are potentially compatible. Step 4: The agent contacts the resource and executes a job.

to the gateway. Direct flocking, although less powerful, was much simpler to build and much easier for users to understand and deploy.

About 1998, a vision of a worldwide computational Grid began to grow [28]. A significant early piece in the Grid computing vision was a uniform interface for batch execution. The Globus Project [50] designed the GRAM protocol [34] to fill this need. GRAM provides an abstraction for remote process queuing and execution with several powerful features such as strong security and file transfer. The Globus Project provides a server that speaks GRAM and converts its commands into a form understood by a variety of batch systems.

To take advantage of GRAM, a user still needs a system that can remember what jobs have been submitted, where they are, and what they are doing. If jobs should fail, the system must analyze the failure and resubmit the job if necessary. To track large numbers of jobs, users need queueing, prioritization, logging, and accounting. To provide this service, the Condor project adapted a standard Condor agent to speak GRAM, yielding a system called Condor-G, shown in Figure 11.8. This required some small changes to GRAM such as adding durability and two-phase commit to prevent the loss or repetition of jobs [55].

The power of GRAM is to expand the reach of a user to any sort of batch system, whether it runs Condor or not. For example, the solution of the NUG30 [56] quadratic assignment problem relied on the ability of Condor-G to mediate access to over a thousand hosts spread across tens of batch systems on several continents. We will describe NUG30 in greater detail below.

The are also some disadvantages to GRAM. Primarily, it couples resource allocation and job execution. Unlike direct flocking in Figure 11.7, the agent must direct a particular job, with its executable image and all, to a particular queue without knowing the availability of resources behind that queue. This forces the agent to either oversubscribe itself by submitting jobs to multiple queues at once or undersubscribe itself by submitting jobs to potentially long queues. Another disadvantage is that Condor-G does not support all of the varied features of each batch system underlying GRAM. Of course, this is a necessity: if GRAM included all the bells and whistles of every underlying system, it

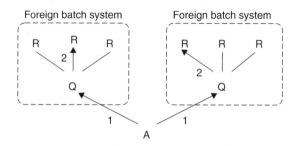

Figure 11.8   Condor-G ca. 2000. An agent (A) is shown executing two jobs through foreign batch queues (Q). Step 1: The agent transfers jobs directly to remote queues. Step 2: The jobs wait for idle resources (R), and then execute on them.

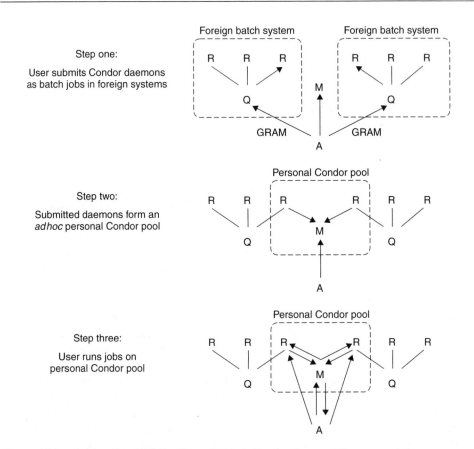

Figure 11.9   Condor-G and Gliding In ca. 2001. A Condor-G agent (A) executes jobs on resources (R) by gliding in through remote batch queues (Q). Step 1: A Condor-G agent submits the Condor daemons to two foreign batch queues via GRAM. Step 2: The daemons form a personal Condor pool with the user's personal matchmaker (M). Step 3: The agent executes jobs as in Figure 11.4.

would be so complex as to be unusable. However, a variety of useful features, such as the ability to checkpoint or extract the job's exit code are missing.

This problem is solved with a technique called *gliding in*, shown in Figure 11.9. To take advantage of both the powerful reach of Condor-G and the full Condor machinery, a personal Condor pool may be carved out of remote resources. This requires three steps. In the first step, a Condor-G agent is used to submit the standard Condor daemons as jobs to remote batch systems. From the remote system's perspective, the Condor daemons are ordinary jobs with no special privileges. In the second step, the daemons begin executing and contact a personal matchmaker started by the user. These remote resources along with the user's Condor-G agent and matchmaker form a personal Condor pool. In step three, the user may submit normal jobs to the Condor-G agent, which are then matched to and executed on remote resources with the full capabilities of Condor.

To this point, we have defined communities in terms of such concepts as responsibility, ownership, and control. However, communities may also be defined as a function of more tangible properties such as location, accessibility, and performance. Resources may group themselves together to express that they are 'nearby' in measurable properties such as network latency or system throughput. We call these groupings *I/O communities*.

I/O communities were expressed in early computational grids such as the Distributed Batch Controller (DBC) [57]. The DBC was designed in 1996 for processing data from the NASA Goddard Space Flight Center. Two communities were included in the original design: one at the University of Wisconsin and the other in the District of Columbia. A high-level scheduler at Goddard would divide a set of data files among available communities. Each community was then responsible for transferring the input data, performing computation, and transferring the output back. Although the high-level scheduler directed the general progress of the computation, each community retained local control by employing Condor to manage its resources.

Another example of an I/O community is the *execution domain*. This concept was developed to improve the efficiency of data transfers across a wide-area network. An execution domain is a collection of resources that identify themselves with a checkpoint server that is close enough to provide good I/O performance. An agent may then make informed placement and migration decisions by taking into account the rough physical information provided by an execution domain. For example, an agent might strictly require that a job remain in the execution domain that it was submitted from. Or, it might permit a job to migrate out of its domain after a suitable waiting period. Examples of such policies expressed in the ClassAd language may be found in Reference [58].

Figure 11.10 shows a deployed example of execution domains. The Istituto Nazionale de Fisica Nucleare (INFN) Condor pool consists of a large set of workstations spread across Italy. Although these resources are physically distributed, they are all part of a national organization, and thus share a common matchmaker in Bologna, which enforces institutional policies. To encourage local access to data, six execution domains are defined within the pool, indicated by dotted lines. Each domain is internally connected by a fast network and shares a checkpoint server. Machines not specifically assigned to an execution domain default to the checkpoint server in Bologna.

Recently, the Condor project developed a complete framework for building general-purpose I/O communities. This framework permits access not only to checkpoint images but also to executables and run-time data. This requires some additional machinery for all parties. The storage device must be an appliance with sophisticated naming and resource management [59]. The application must be outfitted with an interposition agent that can translate application I/O requests into the necessary remote operations [60]. Finally, an extension to the ClassAd language is necessary for expressing community relationships. This framework was used to improve the throughput of a high-energy physics simulation deployed on an international Condor flock [61].

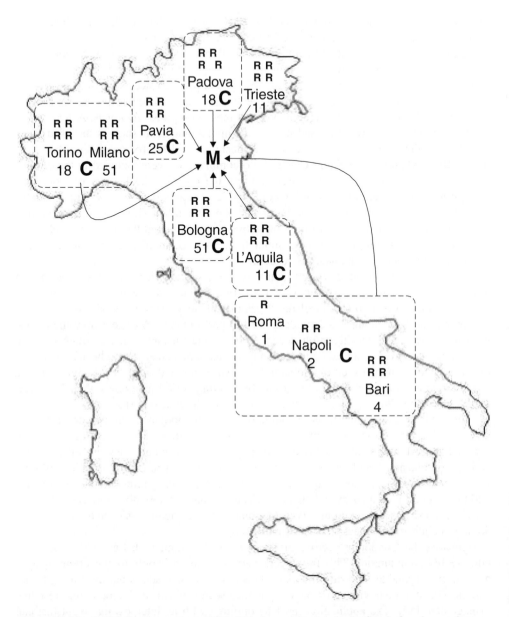

Figure 11.10 INFN Condor pool ca. 2002. This is a map of a single Condor pool spread across Italy. All resources (R) across the country share the same matchmaker (M) in Bologna. Dotted lines indicate execution domains in which resources share a checkpoint server (C). Numbers indicate resources at each site. Resources not assigned to a domain use the checkpoint server in Bologna.

11.5 PLANNING AND SCHEDULING

In preparing for battle I have always found that plans are useless, but planning is indispensable.

– Dwight D. Eisenhower (1890–1969)

The central purpose of distributed computing is to enable a community of users to perform work on a pool of shared resources. Because the number of jobs to be done nearly always outnumbers the available resources, somebody must decide how to allocate resources to jobs. Historically, this has been known as *scheduling*. A large amount of research in scheduling was motivated by the proliferation of massively parallel processor (MPP) machines in the early 1990s and the desire to use these very expensive resources as efficiently as possible. Many of the RMSs we have mentioned contain powerful scheduling components in their architecture.

Yet, Grid computing cannot be served by a centralized scheduling algorithm. By definition, a Grid has multiple owners. Two supercomputers purchased by separate organizations with distinct funds will never share a single scheduling algorithm. The owners of these resources will rightfully retain ultimate control over their own machines and may change scheduling policies according to local decisions. Therefore, we draw a distinction based on the ownership. Grid computing requires both *planning* and *scheduling*.

Planning is the acquisition of resources by users. Users are typically interested in increasing personal metrics such as response time, turnaround time, and throughput of their own jobs within reasonable costs. For example, an airline customer performs planning when she examines all available flights from Madison to Melbourne in an attempt to arrive before Friday for less than $1500. Planning is usually concerned with the matters of *what* and *where*.

Scheduling is the management of a resource by its owner. Resource owners are typically interested in increasing system metrics such as efficiency, utilization, and throughput without losing the customers they intend to serve. For example, an airline performs scheduling when its sets the routes and times that its planes travel. It has an interest in keeping planes full and prices high without losing customers to its competitors. Scheduling is usually concerned with the matters of *who* and *when*.

Of course, there is feedback between planning and scheduling. Customers change their plans when they discover a scheduled flight is frequently late. Airlines change their schedules according to the number of customers that actually purchase tickets and board the plane. But both parties retain their independence. A customer may purchase more tickets than she actually uses. An airline may change its schedules knowing full well it will lose some customers. Each side must weigh the social and financial consequences against the benefits.

The challenges faced by planning and scheduling in a Grid computing environment are very similar to the challenges faced by cycle-scavenging from desktop workstations.

The insistence that each desktop workstation is the sole property of one individual who is in complete control, characterized by the success of the personal computer, results in distributed ownership. Personal preferences and the fact that desktop workstations are often purchased, upgraded, and configured in a haphazard manner results in heterogeneous resources. Workstation owners powering their machines on and off whenever they desire creates a dynamic resource pool, and owners performing interactive work on their own machines creates external influences.

Condor uses *matchmaking* to bridge the gap between planning and scheduling. Matchmaking creates opportunities for planners and schedulers to work together while still respecting their essential independence. Although Condor has traditionally focused on producing robust planners rather than complex schedulers, the matchmaking framework allows both parties to implement sophisticated algorithms.

Matchmaking requires four steps, shown in Figure 11.11. In the first step, agents and resources advertise their characteristics and requirements in *classified advertisements* (ClassAds), named after brief advertisements for goods and services found in the morning newspaper. In the second step, a *matchmaker* scans the known ClassAds and creates pairs that satisfy each other's constraints and preferences. In the third step, the matchmaker informs both parties of the match. The responsibility of the matchmaker then ceases with respect to the match. In the final step, *claiming*, the matched agent and the resource establish contact, possibly negotiate further terms, and then cooperate to execute a job. The clean separation of the *claiming* step has noteworthy advantages, such as enabling the resource to independently authenticate and authorize the match and enabling the resource to verify that match constraints are still satisfied with respect to current conditions [62].

A ClassAd is a set of uniquely named expressions, using a semistructured data model, so no specific schema is required by the matchmaker. Each named expression is called an *attribute*. Each attribute has an *attribute name* and an *attribute value*. In our initial ClassAd implementation, the attribute value could be a simple integer, string, floating point value, or expression composed of arithmetic and logical operators. After gaining more experience, we created a second ClassAd implementation that introduced richer attribute value types and related operators for records, sets, and tertiary conditional operators similar to C.

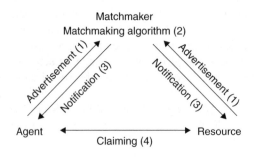

Figure 11.11 Matchmaking.

Because ClassAds are schema-free, participants in the system may attempt to refer to attributes that do not exist. For example, a job may prefer machines with the attribute (Owner == ''Fred''), yet some machines may fail to define the attribute Owner. To solve this, ClassAds use *three-valued logic* that allows expressions to be evaluated to either true, false, or undefined. This explicit support for missing information allows users to build robust requirements even without a fixed schema.

The Condor matchmaker assigns significance to two special attributes: Requirements and Rank. Requirements indicates a constraint and Rank measures the desirability of a match. The matchmaking algorithm requires that for two ClassAds to match, both of their corresponding Requirements must evaluate to true. The Rank attribute should evaluate to an arbitrary floating point number. Rank is used to choose among compatible matches: Among provider advertisements matching a given customer advertisement, the matchmaker chooses the one with the highest Rank value (noninteger values are treated as zero), breaking ties according to the provider's Rank value.

ClassAds for a job and a machine are shown in Figure 11.12. The Requirements state that the job must be matched with an Intel Linux machine that has enough free disk space (more than 6 MB). Out of any machines that meet these requirements, the job prefers a machine with lots of memory, followed by good floating point performance. Meanwhile, the machine advertisement Requirements states that this machine is not willing to match with any job unless its load average is low and the keyboard has been idle for more than 15 min. In other words, it is only willing to run jobs when it would otherwise sit idle. When it is willing to run a job, the Rank expression states it prefers to run jobs submitted by users from its own department.

11.5.1 Combinations of planning and scheduling

As we mentioned above, planning and scheduling are related yet independent. Both planning and scheduling can be combined within one system.

Condor-G, for instance, can perform *planning around a schedule*. Remote site schedulers control the resources, and once Condor-G submits a job into a remote queue, when

Job ClassAd	Machine ClassAd
[	[
MyType = "Job"	MyType = "Machine"
TargetType = "Machine"	TargetType = "Job"
Requirements =	Machine = "nostos.cs.wisc.edu"
((other.Arch=="INTEL"&&	Requirements =
other.OpSys=="LINUX")	(LoadAvg <= 0.300000) &&
&& other.Disk > my.DiskUsage)	(KeyboardIdle > (15 * 60))
Rank = (Memory * 10000) + KFlops	Rank = other.Department==self.Department
Cmd = "/home/tannenba/bin/sim-exe"	Arch = "INTEL"
Department = "CompSci"	OpSys = "LINUX"
Owner = "tannenba"	Disk = 3076076
DiskUsage = 6000	]
]	

Figure 11.12 Two sample ClassAds from Condor.

it will actually run is at the mercy of the remote scheduler (see Figure 11.8). But if the remote scheduler publishes information about its timetable or workload priorities via a ClassAd to the Condor-G matchmaker, Condor-G could begin making better choices by planning where it should submit jobs (if authorized at multiple sites), when it should submit them, and/or what types of jobs to submit. In fact, this approach is currently being investigated by the PPDG [33]. As more information is published, Condor-G can perform better planning. But even in a complete absence of information from the remote scheduler, Condor-G could still perform planning, although the plan may start to resemble 'shooting in the dark'. For example, one such plan could be to submit the job once to each site willing to take it, wait and see where it completes first, and then upon completion, delete the job from the remaining sites.

Another combination is *scheduling within a plan*. Consider as an analogy a large company that purchases, in advance, eight seats on a Greyhound bus each week for a year. The company does not control the bus schedule, so they must plan how to utilize the buses. However, after purchasing the tickets, the company is free to decide to send to the bus terminal whatever employees it wants in whatever order it desires. The Condor system performs scheduling within a plan in several situations. One such situation is when Condor schedules parallel jobs on compute clusters [30]. When the matchmaking framework offers a match to an agent and the subsequent claiming protocol is successful, the agent considers itself the owner of that resource until told otherwise. The agent then creates a schedule for running tasks upon the resources that it has claimed via planning.

11.5.2 Matchmaking in practice

Matchmaking emerged over several versions of the Condor software. The initial system used a fixed structure for representing both resources and jobs. As the needs of the users developed, these structures went through three major revisions, each introducing more complexity in an attempt to retain backwards compatibility with the old. This finally led to the realization that no fixed schema would serve for all time and resulted in the development of a C-like language known as *control expressions* [63] in 1992. By 1995, the expressions had been generalized into *classified advertisements* or ClassAds [64]. This first implementation is still used heavily in Condor at the time of this writing. However, it is slowly being replaced by a new implementation [16, 17, 18] that incorporated lessons from language theory and database systems.

A stand-alone open source software package for manipulating ClassAds is available in both Java and C++ [65]. This package enables the matchmaking framework to be used in other distributed computing projects [66, 53]. Several research extensions to matchmaking have been built. *Gang matching* [17, 18] permits the coallocation of more than once resource, such as a license and a machine. *Collections* provide persistent storage for large numbers of ClassAds with database features such as transactions and indexing. *Set matching* [67] permits the selection and claiming of large numbers of resource using a very compact expression representation. *Indirect references* [61] permit one ClassAd to refer to another and facilitate the construction of the I/O communities mentioned above.

In practice, we have found matchmaking with ClassAds to be very powerful. Most RMSs allow customers to set provide requirements and preferences on the resources they wish. But the matchmaking framework's ability to allow resources to impose constraints on the customers they wish to service is unique and necessary for preserving distributed ownership. The clean separation between matchmaking and claiming allows the match-maker to be blissfully ignorant about the actual mechanics of allocation, permitting it to be a general service that does not have to change when new types of resources or customers are added. Because stale information may lead to a bad match, a resource is free to refuse a claim even after it has been matched. Matchmaking is capable of repre-senting wildly divergent resources, ranging from electron microscopes to storage arrays because resources are free to describe themselves without a schema. Even with similar resources, organizations track different data, so no schema promulgated by the Condor software would be sufficient. Finally, the matchmaker is stateless and thus can scale to very large systems without complex failure recovery.

11.6 PROBLEM SOLVERS

We have delved down into the details of planning and execution that the user relies upon, but may never see. Let us now move up in the Condor kernel and discuss the environment in which a user actually works.

A *problem solver* is a higher-level structure built on top of the Condor agent. Two problem solvers are provided with Condor: *master–worker* (MW) and the *directed acyclic graph manager* (5). Each provides a unique programming model for managing large numbers of jobs. Other problem solvers are possible and may be built using the public interfaces of a Condor agent.

A problem solver relies on a Condor agent in two important ways. A problem solver uses the agent as a service for reliably executing jobs. It need not worry about the many ways that a job may fail in a distributed system, because the agent assumes all responsibility for hiding and retrying such errors. Thus, a problem solver need only concern itself with the application-specific details of ordering and task selection. The agent is also responsible for making the problem solver itself reliable. To accomplish this, the problem solver is presented as a normal Condor job that simply executes at the submission site. Once started, the problem solver may then turn around and submit subjobs back to the agent.

From the perspective of a user or a problem solver, a Condor agent is identical to a Condor-G agent. Thus, any of the structures we describe below may be applied to an ordinary Condor pool or to a wide-area Grid computing scenario.

11.6.1 Master–Worker

Master–Worker (MW) is a system for solving a problem of indeterminate size on a large and unreliable workforce. The MW model is well-suited for problems such as parameter

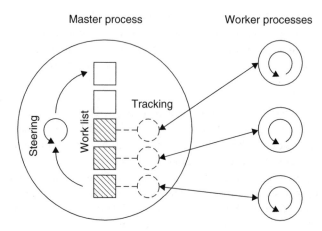

Figure 11.13   Structure of a Master–Worker program.

searches where large portions of the problem space may be examined independently, yet the progress of the program is guided by intermediate results.

The MW model is shown in Figure 11.13. One master process directs the computation with the assistance of as many remote workers as the computing environment can provide. The master itself contains three components: a *work list*, a *tracking* module, and a *steering* module. The work list is simply a record of all outstanding work the master wishes to be done. The tracking module accounts for remote worker processes and assigns them uncompleted work. The steering module directs the computation by examining results, modifying the work list, and communicating with Condor to obtain a sufficient number of worker processes.

Of course, workers are inherently unreliable: they disappear when machines crash and they reappear as new resources become available. If a worker should disappear while holding a work unit, the tracking module simply returns it to the work list. The tracking module may even take additional steps to replicate or reassign work for greater reliability or simply to speed the completion of the last remaining work units.

MW is packaged as source code for several C++ classes. The user must extend the classes to perform the necessary application-specific worker processing and master assignment, but all of the necessary communication details are transparent to the user.

MW is the result of several generations of software development. It began with Pruyne's doctoral thesis [64], which proposed that applications ought to have an explicit interface to the system responsible for finding resources and placing jobs. Such changes were contributed to PVM release 3.3 [68]. The first user of this interface was the Worker Distributor (WoDi or 'Woody'), which provided a simple interface to a work list processed by a large number of workers. The WoDi interface was a very high-level abstraction that presented no fundamental dependencies on PVM. It was quickly realized that the same functionality could be built entirely without PVM. Thus, MW was born [56]. MW provides an interface similar to WoDi, but has several interchangeable implementations.

Today, MW can operate by communicating through PVM, through a shared file system, over sockets, or using the standard universe (described below).

11.6.2 Directed Acyclic Graph Manager

The Directed Acyclic Graph Manager (DAGMan) is a service for executing multiple jobs with dependencies in a declarative form. DAGMan might be thought of as a distributed, fault-tolerant version of the traditional make. Like its ancestor, it accepts a declaration that lists the work to be done and the constraints on its order. Unlike make, it does not depend on the file system to record a DAG's progress. Indications of completion may be scattered across a distributed system, so DAGMan keeps private logs, allowing it to resume a DAG where it left off, even in the face of crashes and other failures.

Figure 11.14 demonstrates the language accepted by DAGMan. A JOB statement associates an abstract name (A) with a file (a.condor) that describes a complete Condor job. A PARENT-CHILD statement describes the relationship between two or more jobs. In this script, jobs B and C are may not run until A has completed, while jobs D and E may not run until C has completed. Jobs that are independent of each other may run in any order and possibly simultaneously.

In this script, job C is associated with a PRE and a POST program. These commands indicate programs to be run before and after a job executes. PRE and POST programs are not submitted as Condor jobs but are run by DAGMan on the submitting machine. PRE programs are generally used to prepare the execution environment by transferring or uncompressing files, while POST programs are generally used to tear down the environment or to evaluate the output of the job.

DAGMan presents an excellent opportunity to study the problem of multilevel error processing. In a complex system that ranges from the high-level view of DAGs all the way down to the minutiae of remote procedure calls, it is essential to tease out the source of an error to avoid unnecessarily burdening the user with error messages.

Jobs may fail because of the nature of the distributed system. Network outages and reclaimed resources may cause Condor to lose contact with a running job. Such failures are not indications that the job itself has failed, but rather that the system has failed.

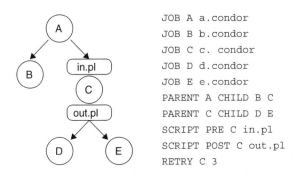

Figure 11.14   A Directed Acyclic Graph.

Such situations are detected and retried by the agent in its responsibility to execute jobs reliably. DAGMan is never aware of such failures.

Jobs may also fail of their own accord. A job may produce an ordinary error result if the user forgets to provide a necessary argument or input file. In this case, DAGMan is aware that the job has completed and sees a program result indicating an error. It responds by writing out a rescue DAG and exiting with an error code. The *rescue DAG* is a new DAG listing the elements of the original DAG left unexecuted. To remedy the situation, the user may examine the rescue DAG, fix any mistakes in submission, and resubmit it as a normal DAG.

Some environmental errors go undetected by the distributed system. For example, a corrupted executable or a dismounted file system *should* be detected by the distributed system and retried at the level of the agent. However, if the job was executed via Condor-G through a foreign batch system, such detail beyond 'job failed' may not be available, and the job will appear to have failed of its own accord. For these reasons, DAGMan allows the user to specify that a failed job be retried, using the RETRY command shown in Figure 11.14.

Some errors may be reported in unusual ways. Some applications, upon detecting a corrupt environment, do not set an appropriate exit code, but simply produce a message on the output stream and exit with an indication of success. To remedy this, the user may provide a POST script that examines the program's output for a valid format. If not found, the POST script may return failure, indicating that the job has failed and triggering a RETRY or the production of a rescue DAG.

11.7 SPLIT EXECUTION

So far, this chapter has explored many of the techniques of getting a job to an appropriate execution site. However, that only solves part of the problem. Once placed, a job may find itself in a hostile environment: it may be without the files it needs, it may be behind a firewall, or it may not even have the necessary user credentials to access its data. Worse yet, few resources sites are uniform in their hostility. One site may have a user's files yet not recognize the user, while another site may have just the opposite situation.

No single party can solve this problem. No process has all the information and tools necessary to reproduce the user's home environment. Only the execution machine knows what file systems, networks, and databases may be accessed and how they must be reached. Only the submission machine knows at run time what precise resources the job must actually be directed to. Nobody knows in advance what names the job may find its resources under, as this is a function of location, time, and user preference.

Cooperation is needed. We call this cooperation *split execution*. It is accomplished by two distinct components: the *shadow* and the *sandbox*. These were mentioned in Figure 11.3. Here we will examine them in detail.

The *shadow* represents the user to the system. It is responsible for deciding exactly what the job must do as it runs. The shadow provides absolutely everything needed to specify the job at run time: the executable, the arguments, the environment, the input files, and so on. None of this is made known outside of the agent until the actual moment of execution. This allows the agent to defer placement decisions until the last possible moment. If the agent submits requests for resources to several matchmakers, it may award the highest priority job to the first resource that becomes available without breaking any previous commitments.

The *sandbox* is responsible for giving the job a safe place to play. It must ask the shadow for the job's details and then create an appropriate environment. The sandbox really has two distinct components: the *sand* and the *box*. The sand must make the job feel at home by providing everything that it needs to run correctly. The box must protect the resource from any harm that a malicious job might cause. The box has already received much attention [69, 70, 71, 72], so we will focus here on describing the sand.[1]

Condor provides several *universes* that create a specific job environment. A universe is defined by a matched sandbox and shadow, so the development of a new universe necessarily requires the deployment of new software modules at both sides. The matchmaking framework described above can be used to select resources equipped with the appropriate universe. Here, we will describe the oldest and the newest universes in Condor: the standard universe and the Java universe.

11.7.1 The standard universe

The standard universe was the only universe supplied by the earliest versions of Condor and is a descendant of the Remote UNIX [14] facility.

The goal of the standard universe is to faithfully reproduce the user's home POSIX environment for a single process running at a remote site. The standard universe provides emulation for the vast majority of standard system calls including file I/O, signal routing, and resource management. Process creation and interprocess communication are not supported and users requiring such features are advised to consider the MPI and PVM universes or the MW problem solver, all described above.

The standard universe also provides *checkpointing*. This is the ability to take a snapshot of a running process and place it in stable storage. The snapshot may then be moved to another site and the entire process reconstructed and then resumed right from where it left off. This may be done to migrate a process from one machine to another or it may be used to recover failed processes and improve throughput in the face of failures.

Figure 11.15 shows all of the components necessary to create the standard universe. At the execution site, the sandbox is responsible for creating a safe and usable execution environment. It prepares the machine by creating a temporary directory for the job, and then fetches all of the job's details – the executable, environment, arguments, and so on – and places them in the execute directory. It then invokes the job and is responsible for monitoring its health, protecting it from interference, and destroying it if necessary.

At the submission site, the shadow is responsible for representing the user. It provides all of the job details for the sandbox and makes all of the necessary policy decisions about the job as it runs. In addition, it provides an I/O service accessible over a secure remote procedure call (RPC) channel. This provides remote access to the user's home storage device.

---

[1] The Paradyn Project has explored several variations of this problem, such as attacking the sandbox [73], defending the shadow [74], and hijacking the job [75].

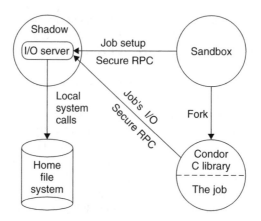

Figure 11.15   The standard universe.

To communicate with the shadow, the user's job must be relinked with a special library provided by Condor. This library has the same interface as the standard C library, so no changes to the user's code are necessary. The library converts all of the job's standard system calls into secure remote procedure calls back to the shadow. It is also capable of converting I/O operations into a variety of remote access protocols, including HTTP, GridFTP [36], NeST [59], and Kangaroo [76]. In addition, it may apply a number of other transformations, such as buffering, compression, and speculative I/O.

It is vital to note that the shadow remains in control of the entire operation. Although both the sandbox and the Condor library are equipped with powerful mechanisms, neither is authorized to make decisions without the shadow's consent. This maximizes the flexibility of the user to make run-time decisions about exactly what runs where and when.

An example of this principle is the *two-phase open*. Neither the sandbox nor the library is permitted to simply open a file by name. Instead, they must first issue a request to map a logical file name (the application's argument to open) into a physical file name. The physical file name is similar to a URL and describes the actual file name to be used, the method by which to access it, and any transformations to be applied.

Figure 11.16 demonstrates two-phase open. Here the application requests a file named alpha. The library asks the shadow how the file should be accessed. The shadow responds that the file is available using remote procedure calls, but is compressed and under a different name. The library then issues an open to access the file.

Another example is given in Figure 11.17. Here the application requests a file named beta. The library asks the shadow how the file should be accessed. The shadow responds that the file is available using the NeST protocol on a server named nest.wisc.edu. The library then contacts that server and indicates success to the user's job.

The mechanics of checkpointing and remote system calls in Condor are described in great detail by Litzkow *et al.* [77, 78]. We have also described Bypass, a stand-alone system for building similar split execution systems outside of Condor [60].

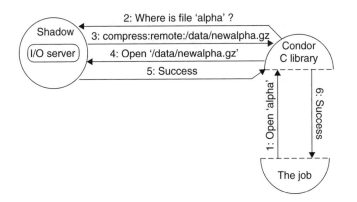

Figure 11.16   Two-phase open using the shadow.

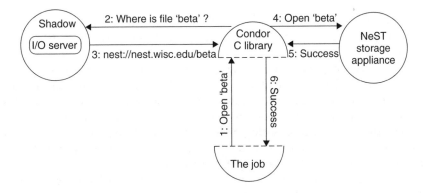

Figure 11.17   Two-phase open using a NeST.

11.7.2 The Java universe

A universe for Java programs was added to Condor in late 2001. This was due to a growing community of scientific users that wished to perform simulations and other work in Java. Although such programs might run slower than native code, such losses were offset by faster development times and access to larger numbers of machines. By targeting applications to the Java Virtual Machine (JVM), users could avoid dealing with the time-consuming details of specific computing systems.

Previously, users had run Java programs in Condor by submitting an entire JVM binary as a standard universe job. Although this worked, it was inefficient in two ways: the JVM binary could only run on one type of CPU, which defied the whole point of a universal instruction set, and the repeated transfer of the JVM and the standard libraries was a waste of resources on static data.

A new Java universe was developed which would raise the level of abstraction to create a complete Java environment rather than a POSIX environment. The components of the

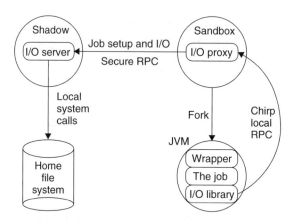

Figure 11.18   The Java universe.

new Java universe are shown in Figure 11.18. The responsibilities of each component are the same as other universes, but the functionality changes to accommodate the unique features of Java.

The sandbox is responsible for creating a safe and comfortable execution environment. It must ask the shadow for all of the job's details, just as in the standard universe. However, the location of the JVM is provided by the local administrator, as this may change from machine to machine. In addition, a Java program consists of a variety of run-time components, including class files, archive files, and standard libraries. The sandbox must place all of these components in a private execution directory along with the user's credentials and start the JVM according to the local details.

The I/O mechanism is somewhat more complicated in the Java universe. The job is linked against a Java I/O library that presents remote I/O in terms of standard interfaces such as `InputStream` and `OutputStream`. This library does not communicate directly with any storage device, but instead calls an I/O proxy managed by the sandbox. This unencrypted connection is secure by making use of the loopback network interface and presenting a shared secret. The sandbox then executes the job's I/O requests along the secure RPC channel to the shadow, using all of the same security mechanisms and techniques as in the standard universe.

Initially, we chose this I/O mechanism so as to avoid reimplementing all of the I/O and security features in Java and suffering the attendant maintenance work. However, there are several advantages of the I/O proxy over the more direct route used by the standard universe. The proxy allows the sandbox to pass through obstacles that the job does not know about. For example, if a firewall lies between the execution site and the job's storage, the sandbox may use its knowledge of the firewall to authenticate and pass through. Likewise, the user may provide credentials for the sandbox to use on behalf of the job without rewriting the job to make use of them.

The Java universe is sensitive to a wider variety of errors than most distributed computing environments. In addition to all of the usual failures that plague remote execution, the Java environment is notoriously sensitive to installation problems, and many jobs and sites are unable to find run-time components, whether they are shared libraries, Java classes, or the JVM itself. Unfortunately, many of these environmental errors are presented to the job itself as ordinary exceptions, rather than expressed to the sandbox as an environmental failure. To combat this problem, a small Java wrapper program is used to execute the user's job indirectly and analyze the meaning of any errors in the execution. A complete discussion of this problem and its solution may be found in Reference [31].

11.8 CASE STUDIES

Grid technology, and Condor in particular, is working today on real-world problems. The three brief case studies presented below provide a glimpse on how Condor and Condor-G are being used in production not only in academia but also in industry. Two commercial organizations, with the foresight to embrace the integration of computational Grids into their operations, are presented.

11.8.1 Micron Technology, Inc.

Micron Technology, Inc., has established itself as one of the leading worldwide providers of semiconductor solutions. Micron's quality semiconductor solutions serve customers in a variety of industries including computer and computer-peripheral manufacturing, consumer electronics, CAD/CAM, telecommunications, office automation, networking and data processing, and graphics display.

Micron's mission is to be the most efficient and innovative global provider of semiconductor solutions. This mission is exemplified by short cycle times, high yields, low production costs, and die sizes that are some of the smallest in the industry. To meet these goals, manufacturing and engineering processes are tightly controlled at all steps, requiring significant computational analysis.

Before Condor, Micron had to purchase dedicated compute resources to meet peak demand for engineering analysis tasks. Condor's ability to consolidate idle compute resources across the enterprise offered Micron the opportunity to meet its engineering needs without incurring the cost associated with traditional, dedicated compute resources. With over 18 000 employees worldwide, Micron was enticed by the thought of unlocking the computing potential of its desktop resources.

So far, Micron has set up two primary Condor pools that contain a mixture of desktop machines and dedicated compute servers. Condor manages the processing of tens of thousands of engineering analysis jobs per week. Micron engineers report that the analysis jobs run faster and require less maintenance. As an added bonus, dedicated resources that were formerly used for both compute-intensive analysis and less intensive reporting tasks can now be used solely for compute-intensive processes with greater efficiency.

Advocates of Condor at Micron especially like how easy it has been to deploy Condor across departments, owing to the clear model of resource ownership and sandboxed environment. Micron's software developers, however, would like to see better integration of Condor with a wider variety of middleware solutions, such as messaging or CORBA.

11.8.2 C.O.R.E. Digital Pictures

C.O.R.E. Digital Pictures is a highly successful Toronto-based computer animation studio, cofounded in 1994 by William Shatner (of film and television fame) and four talented animators.

Photo-realistic animation, especially for cutting-edge film special effects, is a compute-intensive process. Each frame can take up to an hour, and 1 s of animation can require 30 or more frames. When the studio was first starting out and had only a dozen employees, each animator would handle their own render jobs and resources by hand. But with lots of rapid growth and the arrival of multiple major motion picture contracts, it became evident that this approach would no longer be sufficient. In 1998, C.O.R.E. looked into several RMS packages and settled upon Condor.

Today, Condor manages a pool consisting of 70 Linux machines and 21 Silicon Graphics machines. The 70 Linux machines are all dual-CPU and mostly reside on the desktops of the animators. By taking advantage of Condor ClassAds and native support for multiprocessor machines, one CPU is dedicated to running Condor jobs, while the second CPU only runs jobs when the machine is not being used interactively by its owner.

Each animator has his own Condor queuing agent on his own desktop. On a busy day, C.O.R.E. animators submit over 15 000 jobs to Condor. C.O.R.E. has done a significant amount of vertical integration to fit Condor transparently into their daily operations. Each animator interfaces with Condor via a set of custom tools tailored to present Condor's operations in terms of a more familiar animation environment (see Figure 11.19). C.O.R.E. developers created a session metascheduler that interfaces with Condor in a manner similar to the DAGMan service previously described. When an animator hits the 'render' button, a new session is created and the custom metascheduler is submitted as a job into Condor. The metascheduler translates this session into a series of rendering jobs that it subsequently submits to Condor, asking Condor for notification on their progress. As Condor notification events arrive, this triggers the metascheduler to update a database and perhaps submit follow-up jobs following a DAG.

C.O.R.E. makes considerable use of the schema-free properties of ClassAds by inserting custom attributes into the job ClassAd. These attributes allow Condor to make planning decisions based upon real-time input from production managers, who can tag a project, or a shot, or an individual animator with a priority. When jobs are preempted because of changing priorities, Condor will preempt jobs in such a way that minimizes the loss of forward progress as defined by C.O.R.E.'s policy expressions.

To date, Condor has been used by C.O.R.E. for many major productions such as *X-Men*, *Blade II*, *Nutty Professor II*, and *The Time Machine*.

11.8.3 NUG30 Optimization Problem

In the summer of year 2000, four mathematicians from Argonne National Laboratory, University of Iowa, and Northwestern University used Condor-G and several other technologies discussed in this document to be the first to solve a problem known as NUG30 [79].

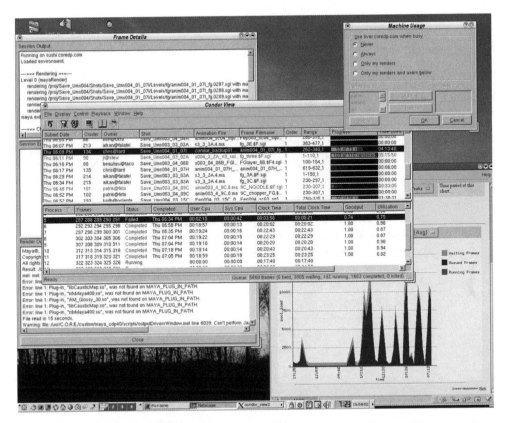

Figure 11.19 Vertical integration of Condor for computer animation. Custom GUI and database integration tools sitting on top of Condor help computer animators at C.O.R.E. Digital Pictures.

NUG30 is a quadratic assignment problem that was first proposed in 1968 as one of the most difficult combinatorial optimization challenges, but remained unsolved for 32 years because of its complexity.

In order to solve NUG30, the mathematicians started with a sequential solver based upon a *branch-and-bound* tree search technique. This technique divides the initial search space into smaller pieces and bounds what could be the best possible solution in each of these smaller regions. Although the sophistication level of the solver was enough to drastically reduce the amount of compute time it would take to determine a solution, the amount of time was still considerable: over seven years with the best desktop workstation available to the researchers at that time (a Hewlett Packard C3000).

To combat this computation hurdle, a parallel implementation of the solver was developed which fit the master–worker model. The actual computation itself was managed by Condor's *Master–Worker* (MW) problem-solving environment. MW submitted work to Condor-G, which provided compute resources from around the world by both *direct flocking* to other Condor pools and by *gliding in* to other compute resources accessible via the Globus GRAM protocol. *Remote System Calls*, part of Condor's *standard universe*, was used as the I/O service between the master and the workers. *Checkpointing* was performed every fifteen minutes for fault tolerance. All of these technologies were introduced earlier in this chapter.

The end result: a solution to NUG30 was discovered utilizing Condor-G in a computational run of less than one week. During this week, over 95 000 CPU hours were used to solve the over 540 billion linear assignment problems necessary to crack NUG30. Condor-G allowed the mathematicians to harness over 2500 CPUs at 10 different sites (8 Condor pools, 1 compute cluster managed by PBS, and 1 supercomputer managed by LSF) spanning 8 different institutions. Additional statistics about the NUG30 run are presented in Table 11.1.

Table 11.1 NUG30 computation statistics. Part A lists how many CPUs were utilized at different locations on the grid during the seven day NUG30 run. Part B lists other interesting statistics about the run

Part A Number	Architecture	Location
1024	SGI/Irix	NCSA
414	Intel/Linux	Argonne
246	Intel/Linux	U. of Wisconsin
190	Intel/Linux	Georgia Tech
146	Intel/Solaris	U. of Wisconsin
133	Sun/Solaris	U. of Wisconsin
96	SGI/Irix	Argonne
94	Intel/Solaris	Georgia Tech
54	Intel/Linux	Italy (INFN)
45	SGI/Irix	NCSA
25	Intel/Linux	U. of New Mexico
16	Intel/Linux	NCSA
12	Sun/Solaris	Northwestern U.
10	Sun/Solaris	Columbia U.
5	Intel/Linux	Columbia U.

Part B	
Total number of CPUs utilized	2510
Average number of simultaneous CPUs	652.7
Maximum number of simultaneous CPUs	1009
Running wall clock time (sec)	597 872
Total CPU time consumed (sec)	346 640 860
Number of times a machine joined the computation	19 063
Equivalent CPU time (sec) on an HP C3000 workstation	218 823 577

11.9 CONCLUSION

Through its lifetime, the Condor software has grown in power and flexibility. As other systems such as Kerberos, PVM, and Java have reached maturity and widespread deployment, Condor has adjusted to accommodate the needs of users and administrators without sacrificing its essential design. In fact, the Condor kernel shown in Figure 11.3 has not changed at all since 1988. Why is this?

We believe the key to lasting system design is to outline structures first in terms of *responsibility* rather than expected *functionality*. This may lead to interactions that, at first blush, seem complex. Consider, for example, the four steps to matchmaking shown in Figure 11.11 or the six steps to accessing a file shown in Figures 11.16 and 11.17. Yet, every step is necessary for discharging a component's responsibility. The matchmaker is responsible for enforcing community policies, so the agent cannot claim a resource without its blessing. The shadow is responsible for enforcing the user's policies, so the sandbox cannot open a file without its help. The apparent complexity preserves the independence of each component. We may update one with more complex policies and mechanisms without harming another.

The Condor project will also continue to grow. The project is home to a variety of systems research ventures in addition to the flagship Condor software. These include the Bypass [60] toolkit, the ClassAd [18] resource management language, the Hawkeye [80] cluster management system, the NeST storage appliance [59], and the Public Key Infrastructure Lab [81]. In these and other ventures, the project seeks to gain the hard but valuable experience of nurturing research concepts into production software. To this end, the project is a key player in collaborations such as the National Middleware Initiative (NMI) [38] that aim to harden and disseminate research systems as stable tools for end users. The project will continue to train students, solve hard problems, and accept and integrate good solutions from others. We look forward to the challenges ahead!

ACKNOWLEDGMENTS

We would like to acknowledge all of the people who have contributed to the development of the Condor system over the years. They are too many to list here, but include faculty and staff, graduates and undergraduates, visitors and residents. However, we must particularly recognize the first core architect of Condor, Mike Litzkow, whose guidance through example and advice has deeply influenced the Condor software and philosophy.

We are also grateful to Brooklin Gore and Doug Warner at Micron Technology, and to Mark Visser at C.O.R.E. Digital Pictures for their Condor enthusiasm and for sharing their experiences with us. We would like to thank Jamie Frey, Mike Litzkow, and Alain Roy provided sound advice as this chapter was written.

This research was made possible by the following grants: Department of Energy awards DE-FG02-01ER25443, DE-FC02-01ER25464, DE-FC02-01ER25450, and DE-FC02-01ER25458; European Commission award 18/GL/04/2002; IBM Corporation awards MHVU5622 and POS996BK874B; and National Science Foundation awards 795ET-21076A, 795PACS1077A, 795NAS1115A, 795PACS1123A, and 02-229 through the University of Illinois, NSF awards UF00111 and UF01075 through the University of Florida, and NSF award 8202-53659 through Johns Hopkins University. Douglas Thain is supported by a Lawrence Landweber NCR fellowship and the Wisconsin Alumni Research Foundation.

REFERENCES

1. Organick, E. I. (1972) *The MULTICS system: An examination of its structure*. Cambridge, MA, London, UK: The MIT Press.
2. Stone, H. S. (1977) Multiprocessor scheduling with the aid of network flow algorithms. *IEEE Transactions of Software Engineering*, **SE-3**(1), 95–93.
3. Chow, Y. C. and Kohler, W. H. (1977) Dynamic load balancing in homogeneous two-processor distributed systems. *Proceedings of the International Symposium on Computer Performance, Modeling, Measurement and Evaluation*, Yorktown Heights, New York, August, 1977, pp. 39–52.
4. Bryant, R. M. and Finkle, R. A. (1981) A stable distributed scheduling algorithm. Proceedings of the Second International Conference on Distributed Computing Systems, Paris, France, April, 1981, pp. 314–323.
5. Enslow, P. H. (1978) What is a distributed processing system? *Computer*, **11**(1), 13–21.
6. Lamport, L. (1978) Time, clocks, and the ordering of events in a distributed system. *Communications of the ACM*, **7**(21), 558–565.
7. Lamport, L., Shostak, R. and Pease, M. (1982) The byzantine generals problem. *ACM Transactions on Programming Languages and Systems*, **4**(3), 382–402.
8. Chandy, K. and Lamport, L. (1985) Distributed snapshots: determining global states of distributed systems. *ACM Transactions on Computer Systems*, **3**(1), 63–75.
9. Needham, R. M. (1979) Systems aspects of the Cambridge Ring. *Proceedings of the Seventh Symposium on Operating Systems Principles*, 0-89791-009-5, Pacific Grove, CA, USA, 1979, pp. 82–85.
10. Walker, B., Popek, G., English, R., Kline, C. and Thiel, G. (1983) The LOCUS distributed operating system. *Proceedings of the 9th Symposium on Operating Systems Principles (SOSP)*, November, 1983, pp. 49–70.
11. Birrell, A. D., Levin, R., Needham, R. M. and Schroeder, M. D. (1982) Grapevine: an exercise in distributed computing. *Communications of the ACM*, **25**(4), 260–274.
12. Livny, M. (1983) *The Study of Load Balancing Algorithms for Decentralized Distributed Processing Systems*. Weizmann Institute of Science.
13. DeWitt, D., Finkel, R. and Solomon, M. (1984) The CRYSTAL multicomputer: design and implementation experience. *IEEE Transactions on Software Engineering*, 553 UW-Madison Comp. Sci. Dept., September, 1984.
14. Litzkow, M. J. (1987) Remote UNIX – Turning Idle Workstations into Cycle Servers. Proceedings of USENIX, Summer, 1987, pp. 381–384.
15. Litzkow, M. and Livny, M. (1990) Experience with the condor distributed batch system. *Proceedings of the IEEE Workshop on Experimental Distributed Systems*, October, 1990.
16. Raman, R., Livny, M. and Solomon, M. (1998) Matchmaking: distributed resource management for high throughput computing. *Proceedings of the Seventh IEEE International Symposium on High Performance Distributed Computing (HPDC7)*, July, 1998.
17. Raman, R., Livny, M. and Solomon, M. (2000) Resource management through multilateral matchmaking. *Proceedings of the Ninth IEEE Symposium on High Performance Distributed Computing (HPDC9)*, Pittsburgh, PA, August, 2000, pp. 290–291.
18. Raman, R. (2000) *Matchmaking Frameworks for Distributed Resource Management*. University of Wisconsin WI.

19. Lauer, H. C. (1981) Observations on the Development of an Operating System, *Proceedings of the 8th Symposium on Operating Systems Principles (SOSP), ACM Operating Systems Review*, **15**, 30–36.

20. Sterling, T., Savarese, D., Becker, D. J., Dorband, J. E., Ranawake, U. A. and Packer, C. V. (1995) BEOWULF: a parallel workstation for scientific computation. *Proceedings of the 24th International Conference on Parallel Processing*, Oconomowoc, WI, pp. 11–14.

21. IBM Corporation. (1993) *IBM Load Leveler: User's Guide*. IBM Corporation.

22. Zhou, S. (1992) LSF: Load sharing in large-scale heterogenous distributed systems. *Proceedings of the Workshop on Cluster Computing*, 1992.

23. Jackson, D., Snell, Q. and Clement, M. (2001) Core algorithms of the Maui scheduler. *Proceedings of the 7th Workshop on Job Scheduling Strategies for Parallel Processing*, 2001.

24. Cray Inc. (1997) Introducing NQE, Technical Report 2153_2.97, Cray Inc., Seattle, WA, 1997.

25. Henderson, R. and Tweten, D. (1996) *Portable Batch System: External reference Specification*. NASA, Ames Research Center.

26. Anderson, D., Bowyer, S., Cobb, J., Gedye, D., Sullivan, W. T. and Werthimer, D. (1997) Astronomical and Biochemical Origins and the Search for Life in the Universe, *Proceedings of the 5th International Conference on Bioastronomy*. Bologna, Italy: Editrice Compositori.

27. Stern, R. (2000) Micro law: Napster: a walking copyright infringement? *IEEE Microsystems*, **20**(6), 4–5.

28. Foster, I. and Kesselman, C. (1998) *The Grid: Blueprint for a New Computing Infrastructure*. San Francisco, CA: Morgan Kaufmann Publishers.

29. Pruyne, J. and Livny, M. (1994) Providing resource management services to parallel applications. *Proceedings of the Second Workshop on Environments and Tools for Parallel Scientific Computing*, May, 1994.

30. Wright, D. (2001) Cheap cycles from the desktop to the dedicated cluster: combining opportunistic and dedicated scheduling with Condor. Conference on Linux Clusters: The HPC Revolution, Champaign-Urbana, IL, June, 2001.

31. Thain, D. and Livny, M. (2002) Error scope on a computational grid: theory and practice. *Proceedings of the 11th IEEE Symposium on High Performance Distributed Computing (HPDC)*, July, 2002.

32. The Grid Physics Network (GriPhyN), http://www.griphyn.org, August, 2002.

33. Particle Physics Data Grid (PPDG), http://www.ppdg.net, August, 2002.

34. Czajkowski, K., Foster, I., Karonis, N., Kesselman, C., Martin, S., Smith, W. and Tuecke, S. (1988) A resource management architecture for metacomputing systems. *Proceedings of the IPPS/SPDP Workshop on Job Scheduling Strategies for Parallel Processing*, 1988, pp. 62–82.

35. Foster, I., Kesselman, C., Tsudik, G. and Tuecke, S. (1998) A security architecture for computational grids. *Proceedings of the 5th ACM Conference on Computer and Communications Security Conference*, 1998, pp. 83–92.

36. Allcock, W., Chervenak, A., Foster, I., Kesselman, C. and Tuecke, S. (2000) Protocols and services for distributed data-intensive science. *Proceedings of Advanced Computing and Analysis Techniques in Physics Research (ACAT)*, 2000, pp. 161–163.

37. The International Virtual Data Grid Laboratory (iVDGL), http://www.ivdgl.org, August, 2002.

38. NSF Middleware Initiative (NMI), http://www.nsf-middleware.org, August, 2002.

39. TeraGrid Project, http://www.teragrid.org, August, 2002.

40. NASA Information Power Grid, http://www.ipg.nasa.gov/, August, 2002.

41. The National Computational Science Alliance, http://www.ncsa.uiuc.edu/About/Alliance/, August, 2002.

42. Foster, I. and Kesselman, C. (1998) The globus project: a status report. *Proceedings of the Seventh Heterogeneous Computing Workshop*, March 4–19, 1998, citeseer.nj.nec.com/foster98globus.html.

43. Alliance Partners for Advanced Computational Services (PACS), http://www.ncsa.uiuc.edu/About/Alliance/Teams, 2002.

44. Ferstl, F. (1999) Job and resource management systems, in Buyya, R. (ed.) *High Performance Cluster Computing: Architectures and Systems.* Vol. 1. Upper Saddle River NJ: Prentice Hall PTR.
45. Tannenbaum, T., Wright, D., Miller, K. and Livny, M. (2001) Condor – a distributed job Scheduler, in Sterling, T. (ed.) *Beowulf Cluster Computing with Linux.* Cambridge, MA; MIT Press.
46. Tannenbaum, T., Wright, D., Miller, K. and Livny, M. (2001) Condor – a distributed job scheduler, in Sterling, T. (ed.) *Beowulf Cluster Computing with Windows.* Cambridge, MA; MIT Press.
47. Basney, J. and Livny, M. (1999) Deploying a high throughput computing cluster, in eds-Buyya, R. (ed.) *High Performance Cluster Computing: Architectures and Systems.* Vol. 1. Upper Saddle River NJ: Prentice Hall PTR.
48. Krueger, P. E. (1988) Distributed Scheduling for a Changing Environment, Technical Report UW-CS-TR-780, University of Wisconsin – Madison, Computer Sciences Department, June, 1988.
49. Frey, J., Tannenbaum, T., Foster, I., Livny, M. and Tuecke, S. (2001) Condor-G: a computation management agent for multi-institutional grids. *Proceedings of the Tenth IEEE Symposium on High Performance Distributed Computing (HPDC),* San Francisco, CA, August, 7–9, 2001.
50. Foster, I. and Kesselman, C. (1997) Globus: a metacomputing infrastructure toolkit. *International Journal of Supercomputer Applications,* **11**(2), 115–128.
51. The Compact Muon Solenoid Collaboration, http://uscms.fnal.gov., August, 2002.
52. European Union DataGrid Project, http://www.eu-datagrid.org.
53. Anglano, C. *et al.* (2001) Integrating grid tools to build a computing resource broker: activities of datagrid WP1. *Proceedings of the Conference on Computing in High Energy Physics 2001 (CHEP01),* Beijing, September, 3–7.
54. Epema, D. H. J., Livny, M., van Dantzig, R., Evers, X. and Pruyne, J. (1996) A worldwide flock of condors: load sharing among workstation clusters. *Future Generation Computer Systems,* **12**, 53–65.
55. Frey, J., Tannenbaum, T., Foster, I., Livny, M. and Tuecke, S. (2002) Condor-G: a computation management agent for multi-institutional grids. *Cluster Computing,* **5**, 237–246.
56. Linderoth, J., Kulkarni, S., Goux, J.-P. and Yoder, M. (2000) An enabling framework for master-worker applications on the computational grid. *Proceedings of the Ninth IEEE Symposium on High Performance Distributed Computing (HPDC9),* Pittsburgh, PA, August, 2000, pp. 43–50.
57. Chen, C., Salem, K. and Livny, M. (1996) The DBC: processing scientific data over the internet. 16th International Conference on Distributed Computing Systems, May, 1996.
58. Basney, J., Livny, M. and Mazzanti, P. (2001) Utilizing widely distributed computational resources efficiently with execution domains. *Computer Physics Communications,* **140**, 252–256, 2001.
59. Bent, J. *et al.* (2002) Flexibility, manageability, and performance in a grid storage appliance. *Proceedings of the Eleventh IEEE Symposium on High Performance Distributed Computing,* Edinburgh, Scotland, July, 2002.
60. Thain, D. and Livny, M. (2001) Multiple bypass: interposition agents for distributed computing. *Journal of Cluster Computing,* **4**, 39–47.
61. Thain, D., Bent, J., Arpaci-Dusseau, A., Arpaci-Dusseau, R. and Livny, M. (2001) Gathering at the well: creating communities for grid I/O. *Proceedings of Supercomputing 2001,* Denver, CO, November, 2001.
62. Livny, M. and Raman, R. (1998) High-throughput resource management, in Foster, I. and Kesselman, C. *The Grid: Blueprint for a New Computing Infrastructure.* San Francisco, CA: Morgan Kaufmann Publishers.
63. Bricker, A., Litzkow, M. and Livny, M. (1992) Condor Technical Summary, Technical Report 1069, University of Wisconsin, Computer Sciences Department, January, 1992.
64. Pruyne, J. C. (ed.) (1996) *Resource Management Services for Parallel Applications.* University of Wisconsin WI.
65. Condor Manual, Condor Team, 2001, Available from http://www.cs.wisc.edu/condor/manual.

66. Vazhkudai, S., Tuecke, S. and Foster, I. (2001) Replica selection in the globus data grid, *IEEE International Symposium on Cluster Computing and the Grid (CCGrid)*. Brisbane, Australia: IEEE Computer Society Press, pp. 106–113.
67. Angulo, D., Foster, I., Liu, C. and Yang, L. (2002) Design and evaluation of a resource selection framework for grid applications. *Proceedings of the 11th IEEE Symposium on High Performance Distributed Computing (HPDC-11)*. Edinburgh, Scotland, July, 2002.
68. Pruyne, J. and Livny, M. (1996) Interfacing Condor and PVM to harness the cycles of workstation clusters. *Future Generation Computer Systems*, **12**, 67–86.
69. Lampson, B. (1971, 1974) Protection. *Proceedings of the 5th Princeton Symposium on Information Sciences and Systems, ACM Operating Systems Review*, **8**, 18–21.
70. Saltzer, J. H. and Schroeder, M. D. (1975) The protection of information in computer systems. *Proceedings of the IEEE*, **63**(9), 1278–1308.
71. Bershad, B. N. *et al.* (1995) Extensibility, Safety, and Performance in the SPIN Operating System. *Proceedings of the 15th ACM Symposium on Operating Systems Principles*, December, 1995, pp. 251–266.
72. Seltzer, M. I., Endo, Y., Small, C. and Smith, K. A. (1996) Dealing with disaster: surviving misbehaved kernel extensions. *Proceedings of the 2nd USENIX Symposium on Operating Systems Design and Implementation (OSDI)*, October, 1996, pp. 213–227.
73. Miller, B. P., Christodorescu, M., Iverson, R., Kosar, T., Mirgordskii, A. and Popovici, F. (2001) Playing inside the black box: using dynamic instrumentation to create security holes. *Parallel Processing Letters*, **11**(2/3), 267–280.
74. Giffin, J., Jha, S. and Miller, B. (2002) Detecting manipulated remote call streams. *Proceedings of the 11th USENIX Security Symposium*, San Francisco, CA, August, 2002.
75. Zandy, V. C., Miller, B. P. and Livny, M. (1999) Process hijacking. *Proceedings of the 8th IEEE International Symposium on High Performance Distributed Computing (HPDC8)*, 1999.
76. Thain, D., Basney, J., Son, S.-C. and Livny, M. (2001) The kangaroo approach to data movement on the grid. *Proceedings of the Tenth IEEE Symposium on High Performance Distributed Computing (HPDC10)*, San Francisco, CA, August, 7–9, 2001.
77. Solomon, M. and Litzkow, M. (1992) Supporting checkpointing and process migration outside the UNIX Kernel. *Proceedings of USENIX*, Winter, 1992, pp. 283–290.
78. Litzkow, M., Tannenbaum, T., Basney, J. and Livny, M. (1997) Checkpoint and Migration of UNIX Processes in the Condor Distributed Processing System, Technical Report UW-CS-TR-1346, University of Wisconsin – Madison, Computer Sciences Department, April, 1997.
79. Anstreicher, K., Brixius, N., Goux, J.-P. and Linderoth, J. (2002) Solving large quadratic assignment problems on computational grids, *Mathematical Programming*. Series B, **91**, 563–588.
80. HawkEye: A Monitoring and Management Tool for Distributed Systems, 2002, http://www.cs.wisc.edu/condor/hawkeye.
81. Public Key Infrastructure Lab (PKI-Lab), http://www.cs.wisc.edu/pkilab, August, 2002.

Architecture of a commercial enterprise desktop Grid: the Entropia system

Andrew A. Chien

Entropia, Inc., San Diego, California, United States
University of California, San Diego, California, United States

12.1 INTRODUCTION

For over four years, the largest computing systems in the world have been based on 'distributed computing', the assembly of large numbers of PCs over the Internet. These 'Grid' systems sustain multiple teraflops continuously by aggregating hundreds of thousands to millions of machines, and demonstrate the utility of such resources for solving a surprisingly wide range of large-scale computational problems in data mining, molecular interaction, financial modeling, and so on. These systems have come to be called 'distributed computing' systems and leverage the unused capacity of high performance desktop PCs (up to 2.2-GHz machines with multigigaOP capabilities [1]), high-speed local-area networks (100 Mbps to 1 Gbps switched), large main memories (256 MB to 1 GB configurations), and large disks (60 to 100 GB disks). Such 'distributed computing'

Grid Computing – Making the Global Infrastructure a Reality. Edited by F. Berman, A. Hey and G. Fox
© 2003 John Wiley & Sons, Ltd   ISBN: 0-470-85319-0

or desktop Grid systems leverage the installed hardware capability (and work well even with much lower performance PCs), and thus can achieve a cost per unit computing (or return-on-investment) superior to the cheapest hardware alternatives by as much as a factor of five or ten. As a result, distributed computing systems are now gaining increased attention and adoption within the enterprises to solve their largest computing problems and attack new problems of unprecedented scale. For the remainder of the chapter, we focus on enterprise desktop Grid computing. We use the terms *distributed computing, high throughput computing*, and *desktop Grids* synonymously to refer to systems that tap vast pools of desktop resources to solve large computing problems, both to meet deadlines or to simply tap large quantities of resources.

For a number of years, a significant element of the research and now commercial computing community has been working on technologies for Grids [2–6]. These systems typically involve servers and desktops, and their fundamental defining feature is to share resources in new ways. In our view, the Entropia system is a desktop Grid that can provide massive quantities of resources and will naturally be integrated with server resources into an enterprise Grid [7, 8].

While the tremendous computing resources available through distributed computing present new opportunities, harnessing them in the enterprise is quite challenging. Because distributed computing exploits existing resources, to acquire the most resources, capable systems must thrive in environments of extreme heterogeneity in machine hardware and software configuration, network structure, and individual/network management practice. The existing resources have naturally been installed and designed for purposes other than distributed computing, (e.g. desktop word processing, web information access, spreadsheets, etc.); the resources must be exploited without disturbing their primary use.

To achieve a high degree of utility, distributed computing must capture a large number of valuable applications – it must be easy to put an application on the platform – and secure the application and its data as it executes on the network. And of course, the systems must support large numbers of resources, thousands to millions of computers, to achieve their promise of tremendous power, and do so without requiring armies of IT administrators.

The Entropia system provides solutions to the above desktop distributed computing challenges. The key advantages of the Entropia system are the ease of application integration, and a new model for providing security and unobtrusiveness for the application and client machine. Applications are integrated using binary modification technology without requiring any changes to the source code. This binary integration automatically ensures that the application is unobtrusive, and provides security and protection for both the client machine and the application's data. This makes it easy to port applications to the Entropia system. Other systems require developers to change their source code to use custom Application Programming Interfaces (APIs) or simply provide weaker security and protection [9–11]. In many cases, application source code may not be available, and recompiling and debugging with custom APIs can be a significant effort.

The remainder of the chapter includes

- an overview of the history of distributed computing (desktop Grids);
- the key technical requirements for a desktop Grid platform: efficiency, robustness, security, scalability, manageability, unobtrusiveness, and openness/ease of application integration;
- the Entropia system architecture, including its key elements and how it addresses the key technical requirements;
- a brief discussion of how applications are developed for the system; and
- an example of how Entropia would be deployed in an enterprise IT environment.

12.2 BACKGROUND

The idea of distributed computing has been described and pursued as long as there have been computers connected by networks. Early justifications of the ARPANET [12] described the sharing of computational resources over the national network as a motivation for building the system. In the mid 1970s, the Ethernet was invented at Xerox PARC, providing high-bandwidth local-area networking. This invention combined with the Alto Workstation presented another opportunity for distributed computing, and the PARC Worm [13] was the result. In the 1980s and early 1990s, several academic projects developed distributed computing systems that supported one or several Unix systems [11, 14–17]. Of these, the Condor Project is best known and most widely used. These early distributed computing systems focused on developing efficient algorithms for scheduling [28], load balancing, and fairness. However, these systems provided no special support for security and unobtrusiveness, particularly in the case of misbehaving applications. Further, they do not manage dynamic desktop environments, limit what is allowed in application execution, and have significant per machine management effort.

In the mid-1980s, the parallel computing community began to leverage first Unix workstations [18], and in the late 1990s, low-cost PC hardware [19, 20]. Clusters of inexpensive PCs connected with high-speed interconnects were demonstrated to rival supercomputers. While these systems focused on a different class of applications, tightly coupled parallel, these systems provided clear evidence that PCs could deliver serious computing power.

The growth of the Worldwide Web (WWW) [21] and exploding popularity of the Internet created a new much larger scale opportunity for distributed computing. For the first time, millions of desktop PCs were connected to wide-area networks both in the enterprise and in the home. The number of machines potentially accessible to an Internet-based distributed computing system grew into the tens of millions of systems for the first time. The scale of the resources (millions), the types of systems (windows PCs, laptops), and the typical ownership (individuals, enterprises) and management (intermittent connection, operation) gave rise to a new explosion of interest in a new set of technical challenges for distributed computing.

In 1996, Scott Kurowski partnered with George Woltman to begin a search for large prime numbers, a task considered synonymous with the largest supercomputers. This effort, the 'Great Internet Mersenne Prime Search' or GIMPS [22, 23], has been running continuously for more than five years with more than 200 000 machines, and has discovered the 35th, 36th, 37th, 38th, and 39th Mersenne primes – the largest known prime numbers. The most recent was discovered in November 2001 and is more than 4 million digits.

The GIMPS project was the first project taken on by Entropia, Inc., a startup commercializing distributed computing. Another group, distributed.net [24], pursued a number of cryptography-related distributed computing projects in this period as well. In 1999, the best-known Internet distributed computing project SETI@home [25] began and rapidly grew to several million machines (typically about 0.5 million active). These early Internet distributed computing systems showed that aggregation of very large scale resources was possible and that the resulting system dwarfed the resources of any single supercomputer, at least for a certain class of applications. But these projects were single-application systems, difficult to program and deploy, and very sensitive to the communication-to-computation ratio. A simple programming error could cause network links to be saturated and servers to be overloaded.

The current generation of distributed computing systems, a number of which are commercial ventures, provide the capability to run multiple applications on a collection of desktop and server computing resources [9, 10, 26, 27]. These systems are evolving towards a general-use compute platform. As such, providing tools for application integration and robust execution are the focus of these systems.

Grid technologies developed in the research community [2, 3] have focused on issues of security, interoperation, scheduling, communication, and storage. In all cases, these efforts have been focused on Unix servers. For example, the vast majority if not all Globus and Legion activity has been done on Unix servers. Such systems differ significantly from Entropia, as they do not address issues that arise in a desktop environment, including dynamic naming, intermittent connection, untrusted users, and so on. Further, they do not address a range of challenges unique to the Windows environment, whose five major variants are the predominant desktop operating system.

12.3 REQUIREMENTS FOR DISTRIBUTED COMPUTING

Desktop Grid systems begin with a collection of computing resources, heterogeneous in hardware and software configuration, distributed throughout a corporate network and subject to varied management, and use regimens and aggregate them into an easily manageable and usable single resource. Furthermore, a desktop Grid system must do this in a fashion that ensures that there is little or no detectable impact on the use of the computing resources for other purposes. For end users of distributed computing, the aggregated resources must be presented as a simple to use, robust resource. On the basis of our experience with corporate end users, the following requirements are essential for a viable enterprise desktop Grid solution:

Efficiency: The system harvests virtually all the idle resources available. The Entropia system gathers over 95% of the desktop cycles unused by desktop user applications.

Robustness: Computational jobs must be completed with predictable performance, masking underlying resource failures.

Security: The system must protect the integrity of the distributed computation (tampering with or disclosure of the application data and program must be prevented). In addition, the desktop Grid system must protect the integrity of the desktops, preventing applications from accessing or modifying desktop data.

Scalability: Desktop Grids must scale to the 1000s, 10 000s, and even 100 000s of desktop PCs deployed in enterprise networks. Systems must scale both upward and downward – performing well with reasonable effort at a variety of system scales.

Manageability: With thousands to hundreds of thousands of computing resources, management and administration effort in a desktop Grid cannot scale up with the number of resources. Desktop Grid systems must achieve manageability that requires no incremental human effort as clients are added to the system. A crucial element is that the desktop Grid cannot increase the basic desktop management effort.

Unobtrusiveness: Desktop Grids share resources (computing, storage, and network resources) with other usage in the corporate IT environment. The desktop Grid's use of these resources should be unobtrusive, so as not to interfere with the primary use of desktops by their primary owners and networks by other activities.

Openness/Ease of Application Integration: Desktop Grid software is a platform that supports applications, which in turn provide value to the end users. Distributed computing systems must support applications developed with varied programming languages, models, and tools – all with minimal development effort.

Together, we believe these seven criteria represent the key requirements for distributed computing systems.

12.4 ENTROPIA SYSTEM ARCHITECTURE

The Entropia system addresses the seven key requirements by aggregating the raw desktop resources into a single logical resource. The aggregate resource is reliable, secure, and predictable, despite the fact that the underlying raw resources are unreliable (machines may be turned off or rebooted), insecure (untrusted users may have electronic and physical access to machines), and unpredictable (machines may be heavily used by the desktop user at any time). The logical resource provides high performance for applications through parallelism while always respecting the desktop user and his or her use of the desktop machine. Furthermore, the single logical resource can be managed from a single administrative console. Addition or removal of desktop machines is easily achieved, providing a simple mechanism to scale the system as the organization grows or as the need for computational cycles grows.

   To support a large number of applications, and to support them securely, we employ a proprietary binary sandboxing technique that enables any Win32 application to be deployed in the Entropia system without modification and without any special system support. Thus, end users can compile their own Win32 applications and deploy them in a matter of minutes. This is significantly different from the early large-scale distributed computing systems that required extensive rewriting, recompilation, and testing of the application to ensure safety and robustness.

12.5 LAYERED ARCHITECTURE

The Entropia system architecture consists of three layers: physical management, scheduling, and job management (see Figure 12.1). The base, the *physical node management* layer, provides basic communication and naming, security, resource management, and application control. The second layer is *resource scheduling*, providing resource matching, scheduling, and fault tolerance. Users can interact directly with the resource scheduling layer through the available APIs, or alternatively through the third layer, *job management*, which provides management facilities for handling large numbers of computations and files. Entropia provides a job management system, but existing job management systems can also be used.

Physical node management: The desktop environment presents numerous unique challenges to reliable computing. Individual client machines are under the control of the desktop user or IT manager. As such, they can be shutdown, rebooted, reconfigured, and be disconnected from the network. Laptops may be off-line or just off for long periods of time. The physical node management layer of the Entropia system manages these and other low-level reliability issues.

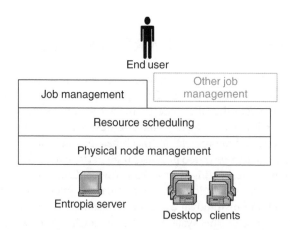

Figure 12.1 Architecture of the Entropia distributed computing system. The physical node management layer and resource scheduling layer span the servers and client machines. The job management layer runs only on the servers. Other (non-Entropia) job management systems can be used with the system.

The physical node management layer provides naming, communication, resource management, application control, and security. The resource management services capture a wealth of node information (e.g. physical memory, CPU speed, disk size and free space, software version, data cached, etc.), and collect it in the system manager.

This layer also provides basic facilities for process management including file staging, application initiation and termination, and error reporting. In addition, the physical node management layer ensures node recovery, terminating runaway, and poorly behaving applications.

The security services employ a range of encryption and binary sandboxing technologies to protect both distributed computing applications and the underlying physical node. Application communications and data are protected with high quality cryptographic techniques. A binary sandbox controls the operations and resources that are visible to distributed applications on the physical nodes, controlling access to protect the software and hardware of the underlying machine.

Finally, the binary sandbox also controls the usage of resources by the distributed computing application. This ensures that the application does not interfere with the primary users of the system – it is unobtrusive – without requiring a rewrite of the application for good behavior.

Resource scheduling: A desktop Grid system consists of resources with a wide variety of configurations and capabilities. The resource scheduling layer accepts units of computation from the user or job management system, matches them to appropriate client resources, and schedules them for execution. Despite the resource conditioning provided by the physical node management layer, the resources may still be unreliable (indeed the application software itself may be unreliable in its execution to completion). Therefore, the resource scheduling layer must adapt to changes in resource status and availability, and to high failure rates. To meet these challenging requirements the Entropia system can support multiple instances of heterogeneous schedulers.

This layer also provides simple abstractions for IT administrators, which automate the majority of administration tasks with reasonable defaults, but allow detailed control as desired.

Job management: Distributed computing applications often involve large overall computation (thousands to millions of CPU hours) submitted as a single large job. These jobs consist of thousands to millions of smaller computations and often arise from statistical studies (i.e. Monte Carlo or Genetic algorithm), parameter sweep, or database search (bioinformatics, combinatorial chemistry, etc.). Because so many computations are involved, tools to manage the progress and status of each piece, in addition to the performance of the aggregate job in order to provide short, predictable turnaround times are provided by the job management layer. The job manager provides simple abstractions for end users, delivering a high degree of usability in an environment in which it is easy to drown in the data, computation, and the vast numbers of activities.

Entropia's three-layer architecture provides a wealth of benefits in system capability, ease of use by end users and IT administrators, and for internal implementation. The modularity provided by the Entropia system architecture allows the physical node layer to contain many of the challenges of the resource-operating environment. The physical node layer manages many of the complexities of the communication, security, and management, allowing the layers above to operate with simpler abstractions. The resource scheduling layer deals with unique challenges of the breadth and diversity of resources, but need not deal with a wide range of lower level issues. Above the resource scheduling layer, the job management layer deals with mostly conventional job management issues. Finally, the higher-level abstractions presented by each layer support the easy enabling of applications. This process is highlighted in the next section.

12.6 PROGRAMMING DESKTOP GRID APPLICATIONS

The Entropia system is designed to support easy application enabling. Each layer of the system supports higher levels of abstraction, hiding more of the complexity of the under-lying resource and execution environment while providing the primitives to get the job done. Applications can be enabled without the knowledge of low-level system details, yet can be run with high degrees of security and unobtrusiveness. In fact, unmodified application binaries designed for server environments are routinely run in production on desktop Grids using the Entropia technology. Further, desktop Grid computing versions of applications can leverage existing job coordination and management designed for existing cluster systems because the Entropia platform provides high capability abstractions, similar to those used for clusters. We describe two example application-enabling processes:

Parameter sweep (single binary, many sets of parameters)

1. Process application binary to wrap in Entropia virtual machine, automatically providing security and unobtrusiveness properties
2. Modify your scripting (or frontend job management) to call Entropia job submission comment and catch completion notification
3. Execute large parameter sweep jobs on 1000 to 100 000 nodes
4. Execute millions of subjobs

Data parallel (single application, applied to parts of a database)

1. Process application binaries to wrap in Entropia virtual machine, automatically pro-viding security and unobtrusiveness properties
2. Design database-splitting routines and incorporate in Entropia Job Manager System
3. Design result combining techniques and incorporate in Entropia Job Manager System
4. Upload your data into the Entropia data management system
5. Execute your application exploiting Entropia's optimized data movement and caching system
6. Execute jobs with millions of subparts

12.7 ENTROPIA USAGE SCENARIOS

The Entropia system is designed to interoperate with many computing resources in an enterprise IT environment. Typically, users are focused on integrating desktop Grid capabilities with other large-scale computing and data resources, such as Linux clusters, database servers, or mainframe systems. We give two example integrations below:

Single submission: Users often make use of both Linux cluster and desktop Grid systems, but prefer not to manually select resources as delivered turnaround time depends critically on detailed dynamic information, such as changing resource configurations, planned maintenance, and even other competing users. In such situations, a single submission interface, in which an intelligent scheduler places computations where the best turnaround time can be achieved, gives end users the best performance.

Large data application: For many large data applications, canonical copies of data are maintained in enhanced relational database systems. These systems are accessed via the network, and are often unable to sustain the resulting data traffic when computational rates are increased by factors of 100 to 10 000. The Entropia system provides for data copies to be staged and managed in the desktop Grid system, allowing the performance demands of the desktop Grid to be separated from the core data infrastructure (see Figure 12.2). A key benefit is that the desktop Grid can then provide maximum computational speedup.

12.8 APPLICATIONS AND PERFORMANCE

Early adoption of distributed computing technology is focused on applications that are easily adapted, and whose high demands cannot be met by traditional approaches whether for cost or technology reasons. For these applications, sometimes called 'high throughput' applications, very large capacity provides a new kind of capability.

The applications exhibit large degrees of parallelism (thousands to even hundreds of millions) with little or no coupling, in stark contrast to traditional parallel applications that are more tightly coupled. These high throughput-computing applications are the only

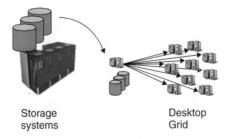

Storage
systems

Desktop
Grid

Figure 12.2   Data staging in the Entropia system.

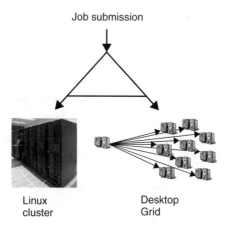

Figure 12.3   Single submission to multiple Grid systems.

ones capable of not being limited by Amdahl's law. As shown in Figure 12.3, these applications can exhibit excellent scaling, greatly exceeding the performance of many traditional high-performance computing platforms.

We believe the widespread availability of distributed computing will encourage reevaluation of many existing algorithms to find novel uncoupled approaches, ultimately increasing the number of applications suitable for distributed computing. For example, Monte Carlo or other stochastic methods that are very inefficient using conventional computing approaches may prove attractive when considering time to solution.

Four application types successfully using distributed computing include virtual screening, sequence analysis, molecular properties and structure, and financial risk analysis [29–51]. We discuss the basic algorithmic structure from a computational and concurrency perspective, the typical use and run sizes, and the computation/communication ratio. A common characteristic of all these applications is the independent evaluation requiring several minutes or more of CPU time, and using at most a few megabytes of data.

12.9 SUMMARY AND FUTURES

Distributed computing has the potential to revolutionize how much of large-scale computing is achieved. If easy-to-use distributed computing can be seamlessly available and accessed, applications will have access to dramatically more computational power to fuel increased functionality and capability. The key challenges to acceptance of distributed computing include robustness, security, scalability, manageability, unobtrusiveness, and openness/ease of application integration.

Entropia's system architecture consists of three layers: a physical node management layer, resource scheduling, and job scheduling. This architecture provides a modularity that allows each layer to focus on a smaller number of concerns, enhancing overall system capability and usability. This system architecture provides a solid foundation to meet the

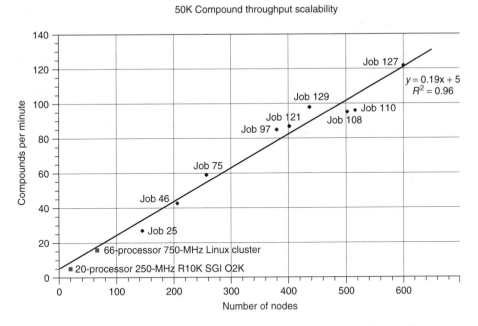

Figure 12.4   Scaling of Entropia system throughput on virtual screening application.

technical challenges as the use of distributed computing matures; it enables a broadening class of computations by supporting an increasing breadth of computational structure, resource usage, and ease of application integration.

We have described the architecture of the Entropia system and how to apply it – both to applications and in an enterprise IT environment. The system is applicable to a large number of applications, and we have discussed virtual screening, sequence analysis, molecular modeling, and risk analysis in this chapter. For all these application domains, excellent linear scaling has been demonstrated for large distributed computing systems (see Figure 12.4). We expect to extend these results to a number of other domains in the near future.

Despite the significant progress documented here, we believe we are only beginning to see the mass use of distributed computing. With robust commercial systems such as Entropia only recently available, widespread industry adoption of the technology is only beginning. At the time of writing this, we were confident that within a few years, distributed computing will be deployed and in use in production within a majority of large corporations and research sites.

ACKNOWLEDGEMENTS

We gratefully acknowledge the contributions of the talented team at Entropia to the design and implementation of this desktop Grid system. We specifically acknowledge the contributions of Kenjiro Taura, Scott Kurowski, Brad Calder, Shawn Marlin, Wayne Schroeder, Jon Anderson, Karan Bhatia, Steve Elbert, and Ed Anady to the definition and development of the system architecture. We also acknowledge Dean Goddette, Wilson Fong, and Mike May for their contributions to applications benchmarking and understanding performance data.

The author is supported in part by the Defense Advanced Research Projects Administration through United States Air Force Rome Laboratory Contracts AFRL F30602-99-1-0534 and the National Science Foundation through the National Computational Science Alliance and NSF EIA-99-75020 GrADS.

REFERENCES

1. Lyman, J. (2002) Intel Debuts 2.2 ghz Pentium 4 Chip, *The News Factor*,
   http://www.newsfactor.com/perl/story/17627.html.
2. Foster, I. and Kesselman, C. (1998) The Globus project: A status report. Presented at *IPPS/SPDP '98 Heterogeneous Computing Workshop*, 1998.
3. Grimshaw, A. and Wulf, W. (1997) The legion vision of a worldwide virtual computer. *Communications of the ACM*, **40**(1): 39–45.
4. Barkai, D. (2001) *Peer-to-Peer Computing: Technologies for Sharing and Collaborating on the Net*. Intel Press,
   http://www.intel.com/intelpress/index.htm.
5. Sun Microsystems, http://www.jxta.org.
6. Foster, I. (1998) *The Grid: Blueprint for a New Computing Infrastructure*. San Francisco, CA: Morgan Kaufmann.
7. Foster, I., Kesselman, C. and Tuecke, S. (2001) The anatomy of the grid: enabling scalable virtual organizations. *International Journal of Supercomputer Applications*, **15**.
8. Entropia Inc. (2002) *Entropia Announces Support for Open Grid Services Architecture*. Entropia Inc., Press Release Feb 21, 2002,
   http://www.entropia.com/.
9. United Devices, http://www.ud.com.
10. Platform Computing, http://www.platform.com.
11. Bricker, A., Litzkow, M. and Livny, M. (1992) *Condor Technical Summary*, Technical Report 1069, Department of Computer Science, University of Wisconsin, Madison, WI, January, 1992.
12. Heart, F., McKenzie, A., McQuillian, J. and Walden, D. (1978) *ARPANET Completion Report*, Technical Report 4799, BBN, January, 1978.
13. Schoch, J. F. and Hupp, J. A. (1982) The 'Worm' programs-early experience with a distributed computation. *Communications of the ACM*, **25**, 172–180.
14. Waldsburger, C. A., Hogg, T., Huberman, B. A., Kephart, J. O. and Stornetta, W. S. (1992) Spawn: A distributed computational economy. *IEEE Transactions on Software Engineering*, **18**, 103–117.
15. Litzkow, M. J. (1987) Remote Unix turning idle workstations into cycle servers. Presented at *Proceedings of the Summer 1987 USENIX Conference*, Phoenix, AZ, USA, 1987.
16. Litzkow, M. J., Livny, M. and Mutka, M. W. (1988) Condor-a hunter of idle workstations. Presented at *8th International Conference on Distributed Computing Systems*, San Jose, CA, USA, 1988.
17. Songnian, Z., Xiaohu, Z., Jingwen, W. and Delisle, P. (1993) Utopia: A load sharing facility for large, heterogeneous distributed computer systems. *Software – Practice and Experience*, **23**, 1305–1336.
18. Sunderam, V. S. (1990) PVM: A framework for parallel distributed computing. *Concurrency: Practice and Experience*, **2**, 315–339.

19. Chien, A. *et al.* (1999) Design and evaluation of HPVM-based Windows Supercomputer. *International Journal of High Performance Computing Applications*, **13**, 201–219.
20. Sterling, T. (2001) *Beowulf Cluster Computing with Linux*. Cambridge, MA: The MIT Press.
21. Gray, M. *Internet Growth Summary*. MIT,
    http://www.mit.edu/people.McCray/net/internet-growth-summary.html.
22. Entropia Inc. (2001) *Researchers Discover Largest Multi-Million-Digit Prime Using Entropia Distributed Computing Grid*. Entropia Inc., Press Release Dec. 2001.
23. Woltman, G. *The Great Internet Mersenne Prime Search*, http://www.mersenne.org/.
24. Distributed.net, *The Fastest Computer on Earth*, http://www.distributed.net/.
25. Sullivan, W. T., Werthimer, D., Bowyer, S. (1997) A new major Seti project based on project Serendip data and 100 000 personal computers, Astronomical and Biochemical Origins and the Search for Life in the Universe, in Cosmovici, C.B., Bowyer, S. and Werthimer, D. (eds) *Proc. of the Fifth Intl. Conf. on Bioastronomy, IAU Colloq. No. 161* Bologna, Italy: Editrice Compositori.
26. Entropia Inc., http://www.entropia.com.
27. DataSynapse Inc., http://www.datasynapse.com.
28. Veridian Systems, *Portable Batch System*, http://www.openpbs.org.
29. Kramer, B., Rarey, M. and Lengauer, T. (1999) Evaluation of the flex incremental construction algorithm for protein-ligand docking. *PROTEINS: Structure, Functions, and Genetics*, **37**, 228–241.
30. McGann, M. *Fred: Fast Rigid Exhaustive Docking*, OpenEye,
    http://www.eyesopen.com/fred.html.
31. Morris, G. M. *et al.* (1998) Automated docking using a Lamarckian genetic algorithm and empirical binding free energy function. *Journal of Computational Chemistry*, **19**, 1639–1662.
32. Ewing, T. J. A. and Kuntz, I. D. (1997) Critical evaluation of search algorithms for automated molecular docking and database screening. *Journal of Computational Chemistry*, **9**, 1175–1189.
33. Jones, G., Willett, P. and Glen, R. C. (1995) Molecular recognition of receptor sites using a genetic algorithm with a description of desolvation. *Journal of Molecular Biology*, **245**, 43–53.
34. Davies, K. *Think*, Department of Chemistry, Oxford University, Oxford,
    http://www.chem.ox.ac.uk/cancer/thinksoftware.html.
35. Eldridge, M. D., Murray, C. W., Auton, T. R., Paolini, G. V. and Mee, R. P. (1997) Empirical scoring functions: i. the development of a fast empirical scoring function to estimate the binding affinity of ligands in receptor complexes. *Journal of Computer-Aided Molecular Design*, **11**, 425–445.
36. Katchalski-Katzir, E. *et al.* (1992) Molecular surface recognition: determination of geometric fit between proteins and their ligands by correlation techniques. *Proceedings of the National Academy of Sciences of the United States of America*, **89**, 2195–2199.
37. Eyck, L. F. T., Mandell, J., Roberts, V. A. and Pique, M. E. (1995) Surveying molecular interactions with dot. Presented at *Supercomputing 1995*, San Diego, 1995.
38. Bohm, H. J. (1996) Towards the automatic design of synthetically accessible protein ligands: peptides, amides and peptidomimetics. *Journal of Computer-Aided Molecular Design*, **10**, 265–272.
39. Altschul, S. F., Gish, W., Miller, W., Myers, E. W., and Lipman, D. J. (1990) Basic local alignment search tool. *Journal of Molecular Biology*, **215**, 403–410.
40. Altschul, S. F. *et al.* (1997) Gapped blast and Psi-blast: A new generation of protein database search programs. *Nucleic Acids Research*, **25**, 3389–3402.
41. Gish, W. and States, D. J. (1993) Identification of protein coding regions by database similarity search. *Nature Genetics*, **3**, 266–272.
42. Madden, T. L., Tatusov, R. L. and Zhang, J. (1996) Applications of network blast server. *Methods of Enzymology*, **266**, 131–141.
43. Zhang, J. and Madden, T. L. (1997) Powerblast: A new network blast application for interactive or automated sequence analysis and annotation. *Genome Research*, **7**, 649–656.
44. Eddy, S. R. *Hmmer: Profile Hidden Markov Models for Biological Sequence Analysis*,
    http://hmmer.wustl.edu/.

45. Thompson, J. D., Higgins, D. G. and Gibson, T. J. (1994) Clustal W: improving the sensitivity of progressive multiple sequence alignment through sequence weighting, positions-specific gap penalties and weight matrix choice. *Nucleic Acids Research*, **22**, 4673–4680.
46. Pearson, W. R. and Lipman, D. J. (1988) Improved tools for biological sequence comparison. *Proceedings of the National Academy of Sciences of the United States of America*, **85**, 2444–2448.
47. Birney, E. *Wise2: Intelligent Algorithms for DNA Searches*, http://www.sanger.ac.uk/Software/Wise2/.
48. Smith, T. F. and Waterman, M. S. (1981) Comparison of biosequences. *Advance in Applied Mathematics*, **2**, 482–489.
49. Frisch, M. J. *et al.* (2001) *Gaussian 98*. Pittsburgh, PA: Gaussian Inc.
50. Schmidt, M. W. *et al.* (1993) General atomic and molecular electronic structure system. *Journal of Computational Chemistry*, **14**, 1347–1363.
51. Schrödinger. *Jaguar*, http://www.schrodinger.com/Products/jaguar.html.

13

Autonomic computing and Grid

Pratap Pattnaik, Kattamuri Ekanadham, and Joefon Jann

Thomas J. Watson Research Center, Yorktown Heights, New York, United States

13.1 INTRODUCTION

The goal of autonomic computing is the reduction of complexity in the management of large computing systems. The evolution of computing systems faces a continuous growth in the number of degrees of freedom the system must manage in order to be efficient. Two major factors contribute to the increase in the number of degrees of freedom: Historically, computing elements, such as CPU, memory, disks, network and so on, have nonuniform advancement. The disparity between the capabilities/speeds of various elements opens up a number of different strategies for a task depending upon the environment. In turn, this calls for a dynamic strategy to make judicious choices for achieving targeted efficiency. Secondly, the systems tend to have a global scope in terms of the demand for their services and the resources they employ for rendering the services. Changes in the demands/resources in one part of the system can have a significant effect on other parts of the system. Recent experiences with Web servers (related to popular events such as the Olympics) emphasize the variability and unpredictability of demands and the need to rapidly react to the changes. A system must perceive the changes in the environment and must be ready with a variety of choices, so that suitable strategies can be quickly selected for the new environment. The autonomic computing approach is to orchestrate

Grid Computing – Making the Global Infrastructure a Reality. Edited by F. Berman, A. Hey and G. Fox
© 2003 John Wiley & Sons, Ltd   ISBN: 0-470-85319-0

the management of the functionalities, efficiencies and the qualities of services of large computing systems through logically distributed, autonomous controlling elements, and to achieve a harmonious functioning of the global system within the confines of its stipulated behavior, while individual elements make locally autonomous decisions. In this approach, one moves from a *resource/entitlement model* to a *goal-oriented model*. In order to significantly reduce system management complexity, one must clearly delineate the boundaries of these controlling elements. The reduction in complexity is achieved mainly by making a significant amount of decisions locally in these elements. If the local decision process is associated with a smaller time constant, it is easy to revise it, before large damage is done globally.

Since Grid Computing, by its very nature, involves the controlled sharing of computing resources across distributed, autonomous systems, *we believe* that there are a number of synergistic elements between Grid computing and autonomic computing and that the advances in the architecture in either one of these areas will help the other. In Grid computing also, local servers are responsible for enforcing local security objectives and for managing various queuing and scheduling disciplines. Thus, the concept of cooperation in a federation of several autonomic components to accomplish a global objective is a common theme for both autonomic computing and Grid computing. As the architecture of Grid computing continues to improve and rapidly evolve, as expounded in a number of excellent papers in this issue, we have taken the approach of describing the autonomic server architecture in this paper. We make some observations on the ways we perceive it to be a useful part of the Grid architecture evolution.

The choice of the term *autonomic* in autonomic computing is influenced by an analogy with biological systems [1, 2]. In this analogy, a component of a system is like an organism that survives in an environment. A vital aspect of such an organism is a symbiotic relationship with others in the environment – that is, it renders certain services to others in the environment and it receives certain services rendered by others in the environment. A more interesting aspect for our analogy is its adaptivity – that is, it makes constant efforts to change its behavior in order to fit into its environment. In the short term, the organism perseveres to perform its functions despite adverse circumstances, by readjusting itself within the degrees of freedom it has. In the long term, evolution of a new species takes place, where environmental changes force permanent changes to the functionality and behavior. While there may be many ways to perform a function, an organism uses its local knowledge to adopt a method that economizes its resources. Rapid response to external stimuli in order to adapt to the changing environment is the key aspect we are attempting to mimic in autonomic systems.

The autonomic computing paradigm imparts this same viewpoint to the components of a computing system. The environment is the collection of components in a large system. The services performed by a component are reflected in the advertised methods of the component that can be invoked by others. Likewise, a component receives the services of others by invoking their methods. The semantics of these methods constitute the behavior that the component attempts to preserve in the short term. In the long term, as technology progresses new resources and new methods may be introduced. Like organisms, the components are not perfect. They do not always exhibit the advertised behavior exactly. There can be errors, impreciseness or even cold failures. An autonomic component watches for these variations in the behavior of other components that it interacts with and adjusts to the variations.

Reduction of complexity is not a new goal. During the evolution of computing systems, several concepts emerged that help manage the complexity. Two notable concepts are particularly relevant here: object-oriented programming and fault-tolerant computing. Object-oriented designs introduced the concept of abstraction, in which the interface specification of an object is separated from its implementation. Thus, implementation of an object can proceed independent of the implementation of dependent objects, since it uses only their interface specifications. The rest of the system is spared from knowing or dealing with the complexity of the internal details of the implementation of the object. Notions of hierarchical construction, inheritance and overloading render easy development of different functional behaviors, while at the same time enabling them to reuse the common parts. An autonomic system takes a similar approach, except that the alternative implementations are designed for improving the performance, rather than providing different behaviors. The environment is constantly monitored and suitable implementations are dynamically chosen for best performance.

Fault-tolerant systems are designed with additional support that can detect and correct any fault out of a predetermined set of faults. Usually, redundancy is employed to overcome faults. Autonomic systems generalize the notion of fault to encompass any behavior that deviates from the expected or the negotiated norm, including performance degradation or change-of-service costs based on resource changes. Autonomic systems do not expect that other components operate correctly according to stipulated behavior. The input–output responses of a component are constantly monitored and when a component's behavior deviates from the expectation, the autonomic system readjusts itself either by switching to an alternative component or by altering its own input–output response suitably.

Section 13.2 describes the basic structure of a typical autonomic component, delineating its behavior, observation of environment, choices of implementation and an adaptive strategy. While many system implementations may have these aspects buried in some detail, it is necessary to identify them and delineate them, so that the autonomic nature of the design can be improved in a systematic manner. Section 13.3 illustrates two speculative methodologies to collect environmental information. Some examples from server design are given to illustrate them. Section 13.4 elaborates on the role of these aspects in a Grid computing environment.

13.2 AUTONOMIC SERVER COMPONENTS

The basic structure of any Autonomic Server Component, C, is depicted in Figure 13.1, in which all agents that interact with C are lumped into one entity, called the *environment*. This includes clients that submit input requests to C, other components whose services can be invoked by C and resource managers that control the resources for C. An autonomic component has four basic specifications:

```
AutonomicComp ::= BehaviorSpec, StateSpec, MethodSpec, StrategySpec
BehaviorSpec  ::= InputSet Σ, OutputSet Φ, ValidityRelation β ⊆  Σ × Φ
```

```
StateSpec      ::= InternalState Ψ, EstimatedExternalState ξ̄
MethodSpec     ::= MethodSet Π, each π∈Π  :  Σ × Ψ × ξ̄ → Φ × Ψ × ξ̄
StrategySpec   ::= Efficiency η, Strategy α  :  Σ × Ψ × ξ̄ → Π
```

The functional behavior of C is captured by a relation, $\beta \subseteq \Sigma \times \Phi$, where Σ is the input alphabet, Φ is the output alphabet and β is a relation specifying valid input–output pair. Thus, if C receives an input $u \in \Sigma$, it delivers an output $v \in \Phi$, satisfying the relation $\beta(u, v)$. The output variability permitted by the relation β (as opposed to a function) is very common to most systems. As illustrated in Figure 13.1, a client is satisfied to get any one of the many possible outputs $(v, v', \ldots)$ for a given input u, as long as they satisfy some property specified by β. All implementations of the component preserve this functional behavior.

The state information maintained by a component comprises two parts: internal state Ψ and external state $\bar{\xi}$. Internal state, Ψ, contains the data structures used by an implementation and any other variables used to keep track of input–output history and resource utilization. The external state ξ is an abstraction of the environment of C and includes information on the input arrival process, the current level of resources available for C and the performance levels of other components of the system whose services are invoked by C. The component C has no control over the variability in the ingredients of ξ, as they are governed by agents outside C. The input arrival process is clearly outside C. We assume an external global resource manager that may supply or withdraw resources from C dynamically. Finally, the component C has no control over how other components are performing and must expect arbitrary variations (including failure) in their health. Thus the state information, ξ, is dynamically changing and is distributed throughout the system.

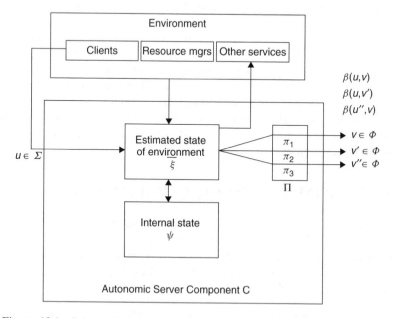

Figure 13.1 Schematic view of an autonomic component and its environment.

C cannot have complete and accurate knowledge of ξ at any time. Hence, the best C can do is to keep an estimate, $\bar{\xi}$, of ξ at any time and periodically update it as and when it receives correct information from the appropriate sources.

An implementation, π, is the usual input–output transformation based on state π : $\Sigma \times \Psi \times \bar{\xi} \to \Phi \times \Psi \times \bar{\xi}$, where an input–output pair $u \in \Sigma$ and $v \in \Phi$ produced will satisfy the relation $\beta(u, v)$. There must be many implementations, $\pi \in \Pi$, available for the autonomic component in order to adapt to the situation. A single implementation provides no degree of freedom. Each implementation may require different resources and data structures. For any given input, different implementations may produce different outputs (of different quality), although all of them must satisfy the relation β.

Finally, the intelligence of the autonomic component is in the algorithm α that chooses the best implementation for any given input and state. Clearly switching from one implementation to another might be expensive as it involves restructuring of resources and data. The component must establish a cost model that defines the efficiency, η, at which the component is operating at any time. The objective is to maximize η. In principle, the strategy, α, evaluates whether it is worthwhile to switch the current implementation for a given input and state, based on the costs involved and the benefit expected. Thus, the strategy is a function of the form $\alpha : \Sigma \times \Psi \times \bar{\xi} \to \Pi$. As long as the current implementation is in place, the component continues to make *local* decisions based on its estimate of the external state. When actual observation of the external state indicates significant deviations (from the estimate), an evaluation is made to choose the right implementation, to optimize η. This leads to the following two aspects that can be studied separately.

Firstly, given that the component has up-to-date and accurate knowledge of the state of the environment, it must have an algorithm to determine the best implementation to adapt. This is highly dependent upon the system characteristics, the costs associated and the estimated benefits from different implementations. An interesting design criterion is to choose the time constants for change of implementation, so that the system enters a stable state quickly. Criteria and models for such designs are under investigation and here we give a few examples.

Secondly, a component may keep an estimate of the external state (which is distributed and dynamically changing) and must devise a means to correct its estimate periodically, so that the deviation from the actual state is kept within bounds. We examine this question in the next section.

13.3 APPROXIMATION WITH IMPERFECT KNOWLEDGE

A general problem faced by all autonomic components is the maintenance of an estimate, $\bar{\xi}$, of a distributed and dynamically changing external state, ξ, as accurately as possible. We examine two possible ways of doing this: by self-observation and by collective observation.

13.3.1 Self-observation

Here a component operates completely autonomously and does not receive any explicit external state information from its environment. Instead, the component deduces information on its environment solely from its own interactions with the environment. This is indeed the way organisms operate in a biological environment. (No one explicitly tells an animal that there is a fire on the east side. It senses the temperatures as it tries to move around and organizes in its memory the gradients and if lucky, moves west and escapes the fire.) Following the analogy, an autonomic component keeps a log of the input–output history with its clients, to track both the quality that it is rendering to its clients as well as the pattern of input arrivals. Similarly, it keeps the history of its interactions with each external service that it uses and tracks its quality. On the basis of these observations, it formulates the estimate, $\bar{\xi}$, of the state of its environment, which is used in its local decisions to adapt suitable implementations. The estimate is constantly revised as new inputs arrive. This strategy results in a very independent component that can survive in any environment. However, the component cannot quickly react to the rapidly changing environment. It takes a few interactions before it can assess the change in its environment. Thus, it will have poor impulse response, but adapts very nicely to gradually changing circumstances. We illustrate this with the example of a memory allocator.

13.3.1.1 Example 1. Memory allocator

This simple example illustrates how an autonomic server steers input requests with frequently observed characteristics to implementations that specialize in efficient handling of those requests. The allocator does not require any resources or external services. Hence, the only external state it needs to speculate upon, $\bar{\xi}$, is the pattern of inputs – specifically how frequently a particular size is being requested in the recent past.

The behavior, (Σ, Φ, β), of a memory allocator can be summarized as follows: The input set Σ has two kinds of inputs: alloc(n) and free(a); the output set Φ has three possible responses: null, error and an address. Alloc(n) is a request for a block of n bytes. The corresponding output is an address of a block or an error indicating inability to allocate. The relation β validates any block, as long as it has the requested number of free bytes in it. Free(a) returns a previously allocated block. The system checks that the block is indeed previously allocated and returns null or error accordingly.

The quality of service, η, must balance several considerations: A client expects quick response time and also that its request is never denied. A second criterion is locality of allocated blocks. If the addresses are spread out widely in the address space, the client is likely to incur more translation overheads and prefers all the blocks to be within a compact region of addresses. Finally, the system would like to minimize fragmentation and avoid keeping a large set of noncontiguous blocks that prevent it from satisfying requests for large blocks.

We illustrate a Pi that has two implementations: The first is a linked-list allocator, which keeps the list of the addresses and sizes of the free blocks that it has. To serve a new allocation request, it searches the list to find a block that is larger than (or equal to) the requested size. It divides the block if necessary and deletes the allocated block from the list and returns its address as the output. When the block is returned, it searches the list again and tries to merge the block with any free adjacent portions in the free list. The second strategy is called *slab allocation*. It reserves a contiguous chunk of memory, called *slab*, for each size known to be frequently used. When a slab exists for the requested size, it peals off a block from that slab and returns it. When a block (allocated from a slab) is returned to it, it links it back to the slab. When no slab exists for a request, it fails to allocate.

The internal state, Ψ, contains the data structures that handle the linked-list and the list of available slabs. The estimated environmental state, $\overline{\xi}$, contains data structures to track the frequency at which blocks of each size are requested or released. The strategy, α, is to choose the slab allocator when a slab exists for the requested size. Otherwise the linked-list allocator is used. When the frequency for a size (for which no slab exists) exceeds a threshold, a new slab is created for it, so that subsequent requests for that size are served faster. When a slab is unused for a long time, it is returned to the linked-list. The cost of allocating from a slab is usually smaller than the cost of allocating from a linked-list, which in turn, is smaller than the cost of creating a new slab. The allocator sets the thresholds based on these relative costs. Thus, the allocator autonomically reorganizes its data structures based on the pattern of sizes in the inputs.

13.3.2 Collective observation

In general, a system consists of a collection of components that are interconnected by the services they offer to each other. As noted before, part of the environmental state, ξ, that is relevant to a component, C, is affected by the states of other components. For instance, if D is a component that provides services for C, then C can make more intelligent decisions if it has up-to-date knowledge of the state of D. If C is periodically updated about the state of D, the performance can be better than what can be accomplished by self-observation. To elaborate on this, consider a system of n interacting components, $C_i, i = 1, n$. Let $S_{ii}(t)$ denote the portion of the state of C_i at time t, that is relevant to other components in the system. For each $i \neq j$, C_i, keeps an estimate, $S_{ij}(t)$, of the corresponding state, $S_{jj}(t)$, of C_j. Thus, each component has an accurate value of its own state and an estimated value of the states of other components. Our objective is to come up with a communication strategy that minimizes the norm $\Sigma_{i,j} |S_{ij}(t) - S_{ii}(t)|$, for any time t. This problem is similar to the time synchronization problem and the best solution is for all components to broadcast their states to everyone after every time step. But since the broadcasts are expensive, it is desirable to come up with a solution that minimizes the communication unless the error exceeds certain chosen limits. For instance, let us assume that each component can estimate how its state is going to change in the near future. Let Δ_i^t be the estimated derivative of $S_{ii}(t)$, at time t – that is, the estimated value of $S_{ii}(t + dt)$ is given by $S_{ii}(t) + \Delta_i^t(dt)$. There can be two approaches to using this information.

13.3.2.1 Subscriber approach (push paradigm)

Suppose a component C_j is interested in the state of C_i. Then C_j will *subscribe* to C_i and obtains a tuple of the form, $\langle t, S_{ii}(t), \Delta_i^t \rangle$, which is stored as part of its estimate of the external state, $\overline{\xi}$. This means that at time t, the state of C_i was $S_{ii}(t)$ and it grows at the rate of Δ_i^t, so that C_j can estimate the state of C_i at future time, $t + \delta t$, as $S_{ii}(t) + \Delta_i^t * \delta t$. Component, C_i, constantly monitors its own state, $S_{ii}(t)$, and whenever the value $|S_{ii}(t) + \Delta_i^t - S_{ii}(t + \delta t)|$ exceeds a tolerance limit, it computes a new gradient, $\Delta_i^{t+\delta t}$ and sends to all its subscribers the new tuple $\langle t + \delta t, S_{ii}(t + \delta t), \Delta_i^{t+\delta t} \rangle$. The subscribers, replace the tuple in their $\overline{\xi}$ with the new information. Thus, the bandwidth of updates is proportional to the rate at which states change. Also, depending upon the tolerance level, the system can have a rapid impulse response.

13.3.2.2 Enquirer approach (pull paradigm)

This is a simple variation of the above approach, where an update is sent only upon explicit request from a subscriber. Each subscriber may set its own tolerance limit and monitor the variation. If the current tuple is $\langle t, S_{ii}(t), \Delta_i^t \rangle$, the subscriber requests for a new update when the increment $\Delta_i^t * \delta t$ exceeds its tolerance limit. This relieves the source component the burden of keeping track of subscribers and periodically updating them. Since all information flow is by demand from a requester, impulse response can be poor if the requester chooses poor tolerance limit.

13.3.2.3 Example 2. Routing by pressure propagation

This example abstracts a common situation that occurs in Web services. It illustrates how components communicate their state to each other, so that each component can make decisions to improve the overall quality of service. The behavior, β, can be summarized as follows: The system is a collection of components, each of which receives transactions from outside. Each component is capable of processing any transaction, regardless of where it enters the system. Each component maintains an input queue of transactions and processes them sequentially. When a new transaction arrives at a component, it is entered into the input queue of a selected component. This selection is the autonomic aspect here and the objective is to minimize the response time for each transaction.

Each component is initialized with some constant structural information about the system, $\langle \mu_i, \tau_{ij} \rangle$, where μ_i is the constant time taken by component C_i to process any transaction and τ_{ij} is the time taken for C_i to send a transaction to C_j. Thus, if a transaction that entered C_i was transferred and served at C_j, then its total response time is given by $\tau_{ij} + (1 + Q_j) * \mu_j$, where Q_j is the length of the input queue at C_j, when the transaction entered the queue there. In order to give best response to the transaction, C_i chooses to forward it to C_j, which minimizes $[\tau_{ij} + (1 + Q_j) * \mu_j]$, over all possible j. Since C_i has no precise knowledge of Q_j, it must resort to speculation, using the collective observation scheme.

As described in the collective observation scheme, each component, C_i, maintains the tuple $\langle t, \Delta_j^t, Q_j^t \rangle$, from which the queue size of C_j at time $t + \delta t$ can be estimated as $Q_j^t + \Delta_j^t * \delta t$. When a request arrives at C_l at time $t + \delta t$, it computes the target j, which minimizes $[\tau_{ij} + (1 + Q_j^t + \Delta_j^t * \delta t) * \mu_j]$, over all possible j. The request is sent to be queued at C_j. Each component, C_j, broadcasts a new tuple, $\langle t + \delta t, \Delta_j^{t+\delta t}, Q_j^{t+\delta t} \rangle$, to all other components whenever the quantity $|Q_j^t + \Delta_j^t * \delta t - Q_j^{t+\delta t}|$ exceeds a tolerance limit.

13.4 GRID COMPUTING

The primary objective of *Grid computing* [3] is to facilitate controlled sharing of resources and services that are made available in a heterogeneous and distributed system. Both heterogeneity and distributedness force the interactions between entities to be based on *protocols* that specify the exchanges of information in a manner that is independent of how a specific resource/service is implemented. Thus, a protocol is independent of details such as the libraries, language, operating system or hardware employed in the implementation. In particular, implementation of a protocol communication between two heterogeneous entities will involve some changes in the types and formats depending upon the two systems. Similarly, implementation of a protocol communication between two distributed entities will involve some marshaling and demarshaling of information and instantiation of local stubs to mimic the remote calls. The *fabric* layer of Grid architecture defines some commonly used protocols for accessing resources/services in such a system. Since the interacting entities span multiple administrative domains, one needs to put in place protocols for authentication and security. These are provided by the *connectivity* layer of the Grid architecture. A *Service* is an abstraction that guarantees a specified *behavior*, if interactions adhere to the protocols defined for the service. Effort is under way for standardization of the means in which a behavior can be specified, so that clients of the services can plan their interactions accordingly, and the implementers of the services enforce the behavior. The *resource* layer of Grid architecture defines certain basic protocols that are needed for acquiring and using the resources available. Since there can be a variety of ways in which resource sharing can be done, the next layer, called the *collective* layer, describes protocols for discovering available services, negotiating for desired services, and initiating, monitoring and accounting of services chosen by clients.

13.4.1 Synergy between the two approaches

The *service* abstraction of the Grid architecture maps to the notion of a *component* of autonomic computing described in Section 13.2. As we noted with components, the implementation of a high-level service for a virtual organization often involves several other resources/services, which are heterogeneous and distributed. The behavior of a service is the BehaviorSpec of a component in Section 13.2 and an implementation must ensure that they provide the advertised behavior, under all conditions. Since a service depends upon other services and on the resources that are allocated for its implementation, prudence dictates that its design be autonomic. Hence, it must monitor the behavior of its dependent services, its own level of resources that may be controlled by other agents and the quality of service it is providing to its clients. In turn, this implies that a service implementation must have a strategy such as α of Section 13.2, which must adapt to the changing environment and optimize the performance by choosing appropriate resources. Thus, all the considerations we discussed under autonomic computing apply to this situation. In particular, there must be general provisions for the maintenance of accurate estimates of global states as discussed in Section 13.3, using either the self-observation or collective observation method. A specialized protocol in the collective layer of the Grid architecture could possibly help this function.

Consider an example of a data-mining service offered on a Grid. There may be one or more implementations of the data-mining service and each of them requires database services on the appropriate data repositories. All the implementations of a service form a collective and they can coordinate to balance their loads, redirecting requests for services arriving at one component to components that have lesser loads. An autonomic data-mining service implementation may change its resources and its database services based on its performance and the perceived levels of service that it is receiving. Recursively the database services will have to be autonomic to optimize the utilization of their services. Thus, the entire paradigm boils down to designing each service from an autonomic perspective, incorporating logic to monitor performance, discover resources and apply them as dictated by its objective function.

13.5 CONCLUDING REMARKS

As systems get increasingly complex, natural forces will automatically eliminate interactions with components whose complexity has to be understood by an interactor. The only components that survive are those that hide the complexity, provide a simple and stable interface and possess the intelligence to perceive the environmental changes, and struggle to fit into the environment. While facets of this principle are present in various degrees in extant designs, explicit recognition of the need for being autonomic can make a big difference, and thrusts us toward designs that are robust, resilient and innovative. In the present era, where technological changes are so rapid, this principle assumes even greater importance, as adaptation to changes becomes paramount.

The first aspect of autonomic designs that we observe is the clear delineation of the interface of how a client perceives a server. Changes to the implementation of the service should not compromise this interface in any manner. The second aspect of an autonomic server is the need for monitoring the varying input characteristics of the clientele as well as the varying response characteristics of the servers on which this server is dependent. In the present day environment, demands shift rapidly and cannot be anticipated most of the time. Similarly, components degrade and fail, and one must move away from deterministic behavior to fuzzy behaviors, where perturbations do occur and must be observed and acted upon. Finally, an autonomic server must be prepared to quickly adapt to the observed changes in inputs as well as dependent services. The perturbations are not only due to failures of components but also due to performance degradations due to changing demands. Autonomic computing provides a unified approach to deal with both. A collective of services can collaborate to provide each other accurate information so that local decisions by each service contribute to global efficiency.

We observe commonalities between the objectives of Grid and autonomic approaches. We believe that they must blend together and Grid architecture must provide the necessary framework to facilitate the design of each service with an autonomic perspective. While we outlined the kinds of protocols and mechanisms that may be supported for this purpose, there is more work to be done in the area of formulating models that capture the stability characteristics of the algorithms that govern the autonomic behavior. Invariably they must map to economic models that involve costs of resources, rewards and penalties for keeping or breaking advertised behaviors and more importantly the temporal aspects of these variations, as time is of essence here. This is the subject of our future study.

REFERENCES

1. Horn, P. Autonomic Computing, http://www.research.ibm.com/autonomic.
2. Wladawsky-Berger, I. Project Eliza, http://www-1.ibm.com/servers/eserver/introducing/eliza.
3. Foster, I., Kesselman, C. and Tuecke, S. (2001) The anatomy of the grid. *International Journal of Supercomputing Applications*, **15**(3), 200–222.

14

Databases and the grid

Paul Watson

University of Newcastle, Newcastle upon Tyne, United Kingdom

14.1 INTRODUCTION

This chapter examines how databases can be integrated into the Grid [1]. Almost all early Grid applications are file-based, and so, to date, there has been relatively little effort applied to integrating databases into the Grid. However, if the Grid is to support a wider range of applications, both scientific and otherwise, then database integration into the Grid will become important. For example, many applications in the life and earth sciences and many business applications are heavily dependent on databases.

The core of this chapter considers how databases can be integrated into the Grid so that applications can access data from them. It is not possible to achieve this just by adopting or adapting the existing Grid components that handle files as databases offer a much richer set of operations (for example, queries and transactions), and there is greater heterogeneity between different database management systems (DBMSs) than there is between different file systems. Not only are there major differences between database paradigms (e.g. object and relational) but also within one paradigm, different database products (e.g. Oracle and DB2) vary in their functionality and interfaces. This diversity makes it more difficult to design a single solution for integrating databases into the Grid, but the alternative of requiring every database to be integrated into the Grid in a bespoke

Grid Computing – Making the Global Infrastructure a Reality.   Edited by F. Berman, A. Hey and G. Fox
© 2003 John Wiley & Sons, Ltd   ISBN: 0-470-85319-0

fashion would result in a much-wasted effort. Managing the tension between the desire to support the full functionality of different database paradigms, while also trying to produce common solutions to reduce effort, is key to designing ways of integrating databases into the Grid.

The diversity of DBMSs also has other important implications. One of the main hopes for the Grid is that it will encourage the publication of scientific data in a more open manner than is currently the case. If this occurs, then it is likely that some of the greatest advances will be made by combining data from separate, distributed sources to produce new results. The data that applications wish to combine would have been created by a set of different researchers who would often have made local, independent decisions about the best database paradigm and design for their data. This heterogeneity presents problems when data is to be combined. If each application has to include its own, bespoke solutions to federating information, then similar solutions will be reinvented in different applications, and will be a waste of effort. Therefore, it is important to provide generic middleware support for federating Grid-enabled databases.

Yet another level of heterogeneity needs to be considered. While this chapter focuses on the integration of structured data into the Grid (e.g. data held in relational and object databases), there will be the need to build applications that also access and federate other forms of data. For example, semi-structured data (e.g. XML) and relatively unstructured data (e.g. scientific papers) are valuable sources of information in many fields. Further, this type of data will often be held in files rather than in a database. Therefore, in some applications there will be a requirement to federate these types of data with structured data from databases.

There are therefore two main dimensions of complexity to the problem of integrating databases into the Grid: implementation differences between server products within a database paradigm and the variety of database paradigms. The requirement for database federation effectively creates a problem space whose complexity is abstractly the product of these two dimensions. This chapter includes a proposal for a framework for reducing the overall complexity.

Unsurprisingly, existing DBMSs do not currently support Grid integration. They are, however, the result of many hundreds of person-years of effort that allows them to provide a wide range of functionality, valuable programming interfaces and tools and important properties such as security, performance and dependability. As these attributes will be required by Grid applications, we strongly believe that building new Grid-enabled DBMSs from scratch is both unrealistic and a waste of effort. Instead we must consider how to integrate existing DBMSs into the Grid. As described later, this approach does have its limitations, as there are some desirable attributes of Grid-enabled databases that cannot be added in this way and need to be integrated in the underlying DBMS itself. However, these are not so important as to invalidate the basic approach of building on existing technology.

The danger with this approach is when a purely short-term view is taken. If we restrict ourselves to considering only how existing databases servers can be integrated with existing Grid middleware, then we may lose sight of long-term opportunities for more powerful connectivity. Therefore, we have tried to identify both the limitations of what can be achieved in the short term solely by integrating existing components and by identifying cases in which developments to the Grid middleware and database server components themselves will produce long-term benefits. An important aspect of this will occur naturally if the Grid becomes commercially important, as the database vendors will then wish to provide 'out-of-the-box' support for Grid integration, by supporting the emerging Grid standards. Similarly, it is vital that those designing standards for Grid middleware take into account the requirements for database integration. Together, these converging developments would reduce the amount of 'glue' code required to integrate databases into the Grid.

This chapter addresses three main questions: what are the requirements of Grid-enabled databases? How far do existing Grid middleware and database servers go towards meeting these requirements? How might the requirements be more fully met? In order to answer the second question, we surveyed current Grid middleware. The Grid is evolving rapidly, and so the survey should be seen as a snapshot of the state of the Grid as it was at the time of writing. In addressing the third question, we focus on describing a framework for integrating databases into the Grid, identifying the key functionalities and referencing relevant work. We do not make specific proposals at the interface level in this chapter – this work is being done in other projects described later.

The structure of the rest of the chapter is as follows. Section 14.2 defines terminology and then Section 14.3 briefly lists the possible range of uses of databases in the Grid. Section 14.4 considers the requirements of Grid-connected databases and Section 14.5 gives an overview of the support for database integration into the Grid offered by current Grid middleware. As this is very limited indeed, we go on to examine how the requirements of Section 14.4 might be met. This leads us to propose a framework for allowing databases to be fully integrated into the Grid, both individually (Section 14.6) and in federations (Section 14.7). We end by drawing conclusions in Section 14.8.

14.2 TERMINOLOGY

In this section, we briefly introduce the terminology that will be used through the chapter.

A *database* is a collection of related data. A *database management system* (DBMS) is responsible for the storage and management of one or more databases. Examples of DBMS are Oracle 9i, DB2, Objectivity and MySQL. A DBMS will support a particular database *paradigm*, for example, relational, object-relational or object. A DBS is created, using a DBMS, to manage a specific database. The DBS includes any associated application software.

Many Grid applications will need to utilise more than one DBS. An application can access a set of DBS individually, but the consequence is that any integration that is required (e.g. of query results or transactions) must be implemented in the application. To reduce the effort required to achieve this, *federated* databases use a layer of middleware running on top of autonomous databases to present applications with some degree of integration. This can include integration of schemas and query capability.

DBS and DBMS offer a set of *services* that are used to manage and to access the data. These include query and transaction services. A service provides a set of related *operations*.

14.3 THE RANGE OF USES OF DATABASES ON THE GRID

As well as the storage and retrieval of the data itself, databases are suited to a variety of roles within the Grid and its applications. Examples of the potential range of uses of databases in the Grid include the following:

Metadata: This is data about data, and is important as it adds context to the data, aiding its identification, location and interpretation. Key metadata includes the name and location of the data source, the structure of the data held within it, data item names and descriptions. There is, however, no hard division between data and metadata – one application's metadata may be another's data. For example, an application may combine data from a set of databases with metadata about their locations in order to identify centres of expertise in a particular category of data (e.g. a specific gene). Metadata will be of vital importance if applications are to be able to discover and automatically interpret data from large numbers of autonomously managed databases. When a database is 'published' on the Grid, some of the metadata will be installed into a catalogue (or catalogues) that can be searched by applications looking for relevant data. These searches will return a set of links to databases whose additional metadata (not all the metadata may be stored in catalogues) and data can then be accessed by the application. The adoption of standards for metadata will be a key to allowing data on the Gird to be discovered successfully. Standardisation efforts such as *Dublin Core* [2], along with more generic technologies and techniques such as *rdf* [3] and ontologies, will be as important for the Grid as they are expected to become to the *Semantic Web* [4]. Further information on the metadata requirements of early Grid applications is given in Reference [5].

Provenance: This is a type of metadata that provides information on the history of data. It includes information on the data's creation, source, owner, what processing has taken place (including software versions), what analyses it has been used in, what result sets have been produced from it and the level of confidence in the quality of information. An example would be a pharmaceutical company using provenance data to determine what analyses have been run on some experimental data, or to determine how a piece of derived data was generated.

Knowledge repositories: Information on all aspects of research can be maintained through knowledge repositories. This could, for example, extend provenance by linking research projects to data, research reports and publications.

Project repositories: Information about specific projects can be maintained through project repositories. A subset of this information would be accessible to all researchers through the knowledge repository. Ideally, knowledge and project repositories can be used to link data, information and knowledge, for example, raw data $\rightarrow$ result sets $\rightarrow$ observations $\rightarrow$ models and simulations $\rightarrow$ observations $\rightarrow$ inferences $\rightarrow$ papers.

In all these examples, some form of data is 'published' so that it can be accessed by Grid applications. There will also be Grid components that use databases internally, without directly exposing their contents to external Grid applications. An example would be a performance-monitoring package that uses a database internally to store information. In these cases, Grid integration of the database is not a requirement and so does not fall within the scope of this chapter.

14.4 THE DATABASE REQUIREMENTS OF GRID APPLICATIONS

A typical Grid application, of the sort with which this chapter is concerned, may consist of a computation that queries one or more databases and carries out further analysis on the retrieved data. Therefore, database access should be seen as being only one part of a wider, distributed application. Consequently, if databases are to be successfully integrated into Grid applications, there are two sets of requirements that must be met: firstly, those that are generic across all components of Grid applications and allow databases to be 'first-class components' within these applications, and secondly, those that are specific to databases and allow database functionality to be exploited by Grid applications. These two categories of requirements are considered in turn in this section.

If computational and database components are to be seamlessly combined to create distributed applications, then a set of agreed standards will have to be defined and will have to be met by all components. While it is too early in the lifetime of the Grid to state categorically what all the areas of standardisation will be, work on existing middleware systems (e.g. CORBA) and emerging work within the Global Grid Forum, suggest that security [6], accounting [7], performance monitoring [8] and scheduling [9] will be important. It is not clear that database integration imposes any additional requirements in the areas of accounting, performance monitoring and scheduling, though it does raise implementation issues that are discussed in Section 14.6. However, security is an important issue and is now considered.

An investigation into the security requirements of early data-oriented Grid applications [5] shows the need for great flexibility in access control. A data owner must be able to grant and revoke access permissions to other users, or delegate this authority to a trusted third party. It must be possible to specify all combinations of access restrictions (e.g. read, write, insert, delete) and to have fine-grained control over the granularity of the data against which they can be specified (e.g. columns, sets of rows). Users with access rights must themselves be able to delegate access rights to other users or to an application. Further, they must be able to restrict the rights they wish to delegate to a subset of the rights they themselves hold. For example, a user with read and write permission to a dataset may wish to write and distribute an application that has only read access to the data. Role-based access, in which access control is based on user role as well as on named individuals, will be important for Grid applications that support collaborative working. The user who performs a role may change over time, and a set of users may adopt the same role concurrently. Therefore, when a user or an application accesses a database they must be able to specify the role that they wish to adopt. All these requirements can be met 'internally' by existing database server products. However, they must also be supported by any Grid-wide security system if it is to be possible to write Grid applications all of whose components exist within a single unified security framework.

Some Grid applications will have extreme performance requirements. In an application that performs CPU-intensive analysis on a huge amount of data accessed by a complex query from a DBS, achieving high performance may require utilising high-performance servers to support the query execution (e.g. a parallel database server) and the computation (e.g. a powerful compute server such as a parallel machine or cluster of workstations). However, this may still not produce high performance, unless the communication between the query and analysis components is optimised. Different communication strategies will be appropriate in different circumstances. If all the query results are required before analysis can begin, then it may be best to transfer all the results efficiently in a single block from the database server to the compute server. Alternatively, if a significant computation needs to be performed on each element of the result set, then it is likely to be more efficient to stream the results from the DBS to the compute server as they are produced. When streaming, it is important to optimise communication by sending data in blocks, rather than as individual items, and to use flow control to ensure that the consumer is not swamped with data. The designers of parallel database servers have built up considerable experience in designing these communications mechanisms, and this knowledge can be exploited for the Grid [10–12].

If the Grid can meet these requirements by offering communications mechanisms ranging from fast large file transfer to streaming with flow control, then how should the most efficient mechanism be selected for a given application run? Internally, DBMSs make decisions on how best to execute a query through the use of cost models that are based on estimates of the costs of the operations used within queries, data sizes and access costs. If distributed applications that include database access are to be efficiently mapped onto Grid resources, then this type of cost information needs to be made available by the DBMS to application planning and scheduling tools, and not just used internally. Armed with this information a planning tool can not only estimate the most efficient communication mechanism to be used for data flows between components but also decide what network and computational resources should be acquired for the application. This will be particularly important where a user is paying for the resources that the application consumes: if high-performance platforms and networks are underutilised then money is wasted, while a low-cost, low-performance component that is a bottleneck may result in the user's performance requirements not being met.

If cost information was made available by Grid-enabled databases, then this would enable a potentially very powerful approach to writing and planning distributed Grid applications that access databases. Some query languages allow user-defined operation calls in queries, and this can allow many applications that combine database access and computation to be written as a single query (or if not then at least parts of them may be written in this way). The Object Database Management Group (ODMG) Object Query Language (OQL) is an example of one such query language [13]. A compiler and optimiser could then take the query and estimate how best to execute it over the Grid, making decisions about how to map and schedule the components of such queries onto the Grid, and the best ways to communicate data between them. To plan such queries efficiently requires estimates of the cost of operation calls. Mechanisms are therefore required for these to be provided by users, or for predictions to be based on measurements collected at run time from previous calls (so reinforcing the importance of performance-monitoring for Grid applications). The results of work on compiling and executing OQL queries on parallel object database servers can fruitfully be applied to the Grid [12, 14].

We now move beyond considering the requirements that are placed on all Grid middleware by the need to support databases, and consider the requirements that Grid applications will place on the DBMSs themselves. Firstly, there appears to be no reason Grid applications will not require at least the same functionality, tools and properties as other types of database applications. Consequently, the range of facilities already offered by existing DBMSs will be required. These support both the management of data and the management of the computational resources used to store and process that data. Specific facilities include

- query and update facilities
- programming interface
- indexing
- high availability
- recovery
- replication
- versioning
- evolution
- uniform access to data and schema
- concurrency control
- transactions
- bulk loading
- manageability
- archiving
- security
- integrity constraints
- change notification (e.g. triggers).

Many person-years of effort have been spent embedding this functionality into existing DBMS, and so, realistically, integrating databases into the Grid must involve building on existing DBMS, rather than on developing completely new, Grid-enabled DBMS from scratch. In the short term, this may place limitations on the degree of integration that is possible (an example is highlighted in Section 14.6), but in the longer term, there is the possibility that the commercial success of the Grid will remove these limitations by encouraging DBMS producers to provide built-in support for emerging Grid standards.

We now consider whether Grid-enabled databases will have requirements beyond those typically found in existing systems. The Grid is intended to support the wide-scale sharing of large quantities of information. The likely characteristics of such systems may be expected to generate the following set of requirements that Grid-enabled databases will have to meet:

Scalability: Grid applications can have extremely demanding performance and capacity requirements. There are already proposals to store petabytes of data, at rates of up to 1 terabyte per hour, in Grid-accessible databases [15]. Low response times for complex queries will also be required by applications that wish to retrieve subsets of data for further processing. Another strain on performance will be generated by databases that are accessed by large numbers of clients, and so will need to support high access throughput. Popular, Grid-enabled information repositories will fall into this category.

Handling unpredictable usage: The main aim of the Grid is to simplify and promote the sharing of resources, including data. Some of the science that will utilise data on the Grid will be explorative and curiosity-driven. Therefore, it will be difficult to predict in advance the types of accesses that will be made to Grid-accessible databases. This differs from most existing database applications in which types of access can be predicted. For example, many current e-Commerce applications 'hide' a database behind a Web interface that only supports limited types of access. Further, typical commercial 'line-of-business' applications generate a very large number of small queries from a large number of users, whereas science applications may generate a relatively small number of large queries, with much greater variation in time and resource usage. In the commercial world, data warehouses may run unpredictable workloads, but the computing resources they use are deliberately kept independent of the resources running the 'line-of-business' applications from which the data is derived. Providing open, ad hoc access to scientific databases, therefore, raises the additional problem of DBMS resource management. Current DBMSs offer little support for controlling the sharing of their finite resources (CPU, disk IOs and main memory cache usage). If they were exposed in an open Grid environment, little could be done to prevent deliberate or accidental denial of service attacks. For example, we want to be able to support a scientist who has an insight that running a particular complex query on a remote, Grid-enabled database could generate exciting new results. However, we do not want the execution of that query to prevent all other scientists from accessing the database for several hours.

Metadata-driven access: It is already generally recognised that metadata will be very important for Grid applications. Currently, the use of metadata in Grid applications tends to be relatively simple – it is mainly for mapping the logical names for datasets into the physical locations where they can be accessed. However, as the Grid expands into new application areas such as the life sciences, more sophisticated metadata systems and tools will be required. The result is likely to be a *Semantic Grid* [16] that is analogous to the *Semantic Web* [4]. The use of metadata to locate data has important implications for integrating databases into the Grid because it promotes a two-step access to data. In step one, a search of Metadata catalogues is used to locate the databases containing the data required by the application. That data is then accessed in the second step. A consequence of two-step access is that the application writer does not know the specific DBS that will be accessed in the second step. Therefore, the application must be general enough to connect and interface to any of the possible DBSs returned in step one. This is straightforward if all are built from the same DBMS, and so offer the same interfaces to the application, but more difficult if these interfaces are heterogeneous. Therefore, if it is to be successful, the two-step approach requires that all DBS should, as far as possible, provide a standard interface. It also requires that all data is held in a common format, or that the metadata that describes the data is sufficient to allow applications to understand the formats and interpret the data. The issues and problems of achieving this are discussed in Section 14.6.

Multiple database federation: One of the aims of the Grid is to promote the open publication of scientific data. A recent study of the requirements of some early Grid applications concluded that 'The prospect exists for literally billions of data resources and petabytes of data being accessible in a Grid environment' [5]. If this prospect is realised, then it is expected that many of the advances to flow from the Grid will come from applications that can combine information from multiple data sets. This will allow researchers to combine different types of information on a single entity to gain a more complete picture and to aggregate the same types of information about different entities. Achieving this will require support for integrating data from multiple DBS, for example, through distributed query and transaction facilities. This has been an active research area for several decades, and needs to be addressed on multiple levels. As was the case for metadata-driven access, the design of federation middleware will be made much more straightforward if DBS can be accessed through standard interfaces that hide as much of their heterogeneity as possible. However, even if APIs are standardised, this still leaves the higher-level problem of the semantic integration of multiple databases, which has been the subject of much attention over the past decades [17, 18]. In general, the problem complexity increases with the degree of heterogeneity of the set of databases being federated, though the provision of ontologies and metadata can assist. While there is much existing work on federation on which to build, for example, in the area of query processing [19, 20], the Grid should give a renewed impetus to research in this area because there will be clear benefits from utilising tools that can combine data over the Grid from multiple, distributed repositories. It is also important that the middleware that supports distributed services across federated databases meets the other Grid requirements. For example, distributed queries that run across the Grid may process huge amounts of data, and so the performance requirements on the middleware may, in some cases, exceed the requirements on the individual DBS.

In summary, there are a set of requirements that must be met in order to support the construction of Grid applications that access databases. Some are generic across all Grid application components, while others are database specific. It is reasonable to expect that Grid applications will require at least the functionality provided by current DBMSs. As these are complex pieces of software, with high development costs, building new, Grid-enabled DBMS from scratch is not an option. Instead, new facilities must be added by enhancing existing DBMSs, rather than by replacing them. The most commonly used DBMSs are commercial products that are not open-source, and so enhancement will have to be achieved by wrapping the DBMS externally. It should be possible to meet almost all the requirements given above in this way, and methods of achieving this are proposed in Sections 14.6 and 14.7. In the longer term, it is to be hoped that, if the Grid is a commercial success, then database vendors will wish to provide 'out-of-the-box' support for Grid integration, by supporting Grid requirements. Ideally, this would be encouraged by the definition of open standards. If this was to occur, then the level of custom wrapping required to integrate a database into the Grid would be considerably reduced.

The remainder of this chapter investigates how far current Grid middleware falls short of meeting the above requirements, and then proposes mechanisms for satisfying them more completely.

14.5 THE GRID AND DATABASES: THE CURRENT STATE

In this section, we consider how the current Grid middleware supports database integration. We consider Globus, the leading Grid middleware before looking at previous work on databases in Grids. As the Grid is evolving rapidly, this section should be seen as a snapshot taken at the time of writing.

The dominant middleware used for building computational grids is Globus, which provides a set of services covering grid information, resource management and data management [21]. Information Services allow owners to register their resources in a directory, and provide, in the Monitoring and Discovery Service (MDS), mechanisms through which they can be dynamically discovered by applications looking for suitable resources on which to execute. From MDS, applications can determine the configuration, operational status and loading of both computers and networks. Another service, the Globus Resource Allocation Manager (GRAM) accepts requests to run applications on resources, and manages the process of moving the application to the remote resource, scheduling it and providing the user with a job control interface.

An orthogonal component that runs through all Globus services is the Grid Security Infrastructure (GSI). This addresses the need for secure authentication and communications over open networks. An important feature is the provision of 'single sign-on' access to computational and data resources. A single X.509 certificate can be used to authenticate a user to a set of resources, thus avoiding the need to sign-on to each resource individually.

The latest version of Globus (2.0) offers a core set of services (called the Globus Data Grid) for file access and management. There is no direct support for database integration and the emphasis is instead on the support for very large files, such as those that might be used to hold huge datasets resulting from scientific experiments. GridFTP is a version of file transfer protocol (FTP) optimised for transferring files efficiently over high-bandwidth wide-area networks and it is integrated with the GSI. Globus addresses the need to have multiple, possibly partial, copies of large files spread over a set of physical locations by providing support for replica management. The Globus Replica Catalogue holds the location of a set of replicas for a logical file, so allowing applications to find the physical location of the portion of a logical file they wish to access. The Globus Replica Management service uses both the Replica Catalogue and GridFTP to create, maintain and publish the physical replicas of logical files.

There have been recent moves in the Grid community to adopt Web Services [22] as the basis for Grid middleware, through the definition of the Open Grid Services Architecture (OGSA) [23]. This will allow the Grid community to exploit the high levels of investment in Web Service tools and components being developed for commercial computing. The move also reflects the fact that there is a great deal of overlap between the Grid vision of supporting scientific computing by sharing resources, and the commercial vision of enabling Virtual Organisations – companies combining information, resources and processes to build new distributed applications.

Despite lacking direct support for database integration, Globus does have services that can assist in achieving this. The GSI could be used as the basis of a system that provides a *single sign-on* capability, removing the need to individually connect to each database with a separate username and password (which would not easily fit into the two-step access method described in Section 14.4). However, mechanisms for connecting a user or application to the database in a particular role and for delegating restricted access rights are required, as described in Section 14.4, but are not currently directly supported by GSI. A recent development – the Community Authorisation Service [24] – does offer restricted delegation, and so may offer a way forward. Other Globus components could also be harnessed in order to support other aspects of database integration into the Grid. For example, GridFTP could be used both for bulk database loading and, where efficient, for the bulk transfer of query results from a DBS to another component of an application. The MDS and GRAM services can be used to locate and run database federation middleware on appropriate computational resources, as will be discussed in Section 14.7. In the long term, the move towards an OGSA service-based architecture for Globus is in line with the proposed framework for integrating databases into the Grid that will be described in Sections 14.6 and 14.7.

Having examined Globus, the main generic Grid middleware project, we now describe two existing projects that include work on Grids and databases.

Spitfire [25], an European Data Grid project, has developed an infrastructure that allows a client to query a relational database over GSI-enabled Hypertext Transfer Protocol (HTTP) (S). An XML-based protocol is used to represent the query and its result. The system supports role-based security: clients can specify the role they wish to adopt for a query execution, and a mapping table in the server checks that they are authorised to take on this role.

The Storage Request Broker (SRB) is a middleware that provides uniform access to datasets on a wide range of different types of storage devices [26] that can include file systems and archival resources. The SRB's definition of dataset is 'stream-of-bytes', and so the primary focus is on files and collections of files rather than on the structured data held in databases that is the focus of this chapter. The SRB provides location transparency through the use of a metadata catalogue that allows access to datasets by logical names, or other metadata attributes. SRB installations can be federated, such that any SRB in the federation can accept client calls and forward them to the appropriate SRB. The SRB also provides replica management for datasets, providing fault tolerance by redirecting client requests to a replica if the primary storage system is unavailable. The SRB supports a variety of authentication protocols for clients accessing data, including the GSI. While the focus of the SRB is on file-based data, it does offer some limited capabilities for accessing data held in databases. Datasets can be held as Large OBjects (LOBs) in databases (in which case they are basically treated as files stored in the database), or a standard query language (SQL) query can be registered as an object with the SRB – the query is executed whenever the object is retrieved.

In summary, while there are aspects of existing Grid middleware that can contribute to integrating databases into the Grid, very few of the requirements of Section 14.4 have yet been met, and so we now move on to describe a service framework within which they could be achieved, and identify the key functionalities of the required services.

14.6 INTEGRATING DATABASES INTO THE GRID

In this section, we describe a framework for integrating databases into Grid applications and for identifying the main functionality that must be provided if the requirements of Section 14.4 are to be met. We do not make specific proposals at the interface level in this chapter; many different interfaces could be designed to meet the requirements within the proposed framework, though we hope that work within the Global Grid Forum will lead to the definition of interface standards.

The proposed framework is service-based. Each Grid-enabled DBS would offer a set of services covering the areas identified in the requirements given in Section 14.4. Where possible, individual operations offered by these services would be standardized to increase portability and reduce the effort required to build applications that interact with multiple DBSs. Standardisation would be done by adding wrapper code to map the service operation interface to the vendor-specific interface beneath. However, it is impossible to standardise all services: for example, different database paradigms support different types of query languages, and these cannot all be reconciled into a Standard Query Language (even within the set of relation databases there are variations). One advantage of a service-based framework is, however, that each DBS made available on the Grid can provide a metadata service that gives information on the range of services and the operations that it supports. For example, a DBS may describe itself as offering a query service that supports SQL-92. This service metadata would give application builders the information required to exploit whatever facilities were available, and is an important pre-requisite to the proposal for database federation given in Section 14.7.

Figure 14.1 shows the service-based framework, with a service wrapper placed between the Grid and the DBS (we deliberately refer to DBS here rather than DBMS, as the owner of the database can choose which services are to be made available on the Grid and who is allowed to access them). Initially, the service wrappers will have to be custom produced, but, in the future, if the commercial importance of the Grid increases, and standards are defined, then it is to be hoped that DBMS vendors will offer Grid-enabled service interfaces as an integral part of their products.

We now discuss each of the services shown in Figure 14.1:

Metadata: This service provides access to technical metadata about the DBS and the set of services that it offers to Grid applications. Examples include the logical and physical

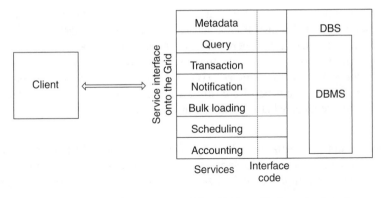

Figure 14.1 A database system with a Grid-enabled service interface.

name of the DBS and its contents, ownership, version numbers, the database schema and information on how the data can be accessed. The service description metadata would, for each service, describe exactly what functionality is offered. This would be used by Grid application builders and by tools that need to know how to interface to the DBS. It is particularly important for applications that are dynamically constructed – the two-step access to data described in Section 14.4 means that the databases that are to take part in an application are not known until some preliminary processing of metadata has taken place. Each run of such applications may result in the need to access a different set of databases, and so mechanisms are required to dynamically construct interfaces to those DBS – if they are not all able to offer completely standard interfaces, then the metadata can be accessed to determine their functionality and interfaces, so that they can be dynamically incorporated into the application.

Query: Query languages differ across different DBMSs, though the core of SQL is standard across most relational DBMSs. It is therefore important that the service metadata defines the type and level of query language that is supported. To provide input to scheduling decisions and to enable the efficient planning of distributed Grid applications, an operation that provides an estimate of the cost of executing a query is highly desirable. As described in the requirements section, the query service should also be able to exploit a variety of communications mechanisms in order to transfer results over the Grid, including streaming (with associated flow control) and transfer as a single block of data. Finally, it is important that the results of a query can be delivered to an arbitrary destination, rather than just to the sender of the query. This allows the creation of distributed systems with complex communications structures, rather than just simple client-server request-response systems.

Transaction: These operations would support transactions involving only a single DBS (e.g. operations to Begin, Commit and Rollback transactions), and also allow a DBS to participate in application-wide distributed transactions, where the DBS supports it. There are a variety of types of transactions that are supported by DBMSs (for example, some but not all support nested transactions), and so a degree of heterogeneity between DBSs is inevitable. In the longer term, there may also be a need for loosely coordinated, long-running transactions between multiple enterprises, and so support for alternative protocols [e.g. the Business Transaction Protocol (BTP)] [27] may become important. Given the variety of support that could be offered by a transaction service, the service description metadata must make clear what is available at this DBS.

Bulk loading: Support for the bulk loading of data over the Grid into the database will be important in some systems. For large amounts of data, the service should be able to exploit Grid communication protocols that are optimised for the transfer of large datasets (e.g. GridFTP).

Notification: This would allow clients to register some interest in a set of data and receive a message when a change occurs. Supporting this function requires both a mechanism that allows the client to specify exactly what it is interested in (e.g. additions, updates, deletions, perhaps further filtered by a query) and a method for notifying the client of a change. Implementing this service is made much simpler if the underlying DBMS provides native support, for example, through triggers. When a notification is generated, it would be fed into a generic Grid event service to determine what action is to be taken. For example, a user may be directly informed by e-mail, or an analysis computation may be automatically run on new data.

Scheduling: This would allow users to schedule the use of the DBS. It should support the emerging Grid-scheduling service [9], for example, allowing a DBS and a supercomputer to be co-scheduled, so that large datasets retrieved from the DBS can be processed by the supercomputer. Bandwidth on the network connecting them might also need to be pre-allocated. As providing exclusive access to a DBS is impractical, mechanisms are needed to dedicate sufficient resources (disks, CPUs, memory, network) to a particular task. This requires the DBS to provide resource pre-allocation and management, something that is not well supported by existing DBMSs and that cannot be implemented by wrapping the DBMS and controlling the resources at the operating system level. This is because DBMS, like most efficiently designed servers, run as a set of processes that are shared among all the users, and the management of sharing is not visible or controllable at the operating system process level.

Accounting: The DBS must be able to provide the necessary information for whatever accounting and payment scheme that emerges for the Grid. This service would monitor performance against agreed service levels and enable users to be charged for resource usage. The data collected would also provide valuable input for application capacity planning and for optimising the usage of Grid resources. As with scheduling, as a DBS is a shared server it is important that accounting is done in terms of the individual users (or groups) use of the DBS, and not just aggregated across all users.

We do not claim that this list of services is definitive. It is based on our experience of building systems that utilise databases, but the need for new services may emerge as more experience is gained with Grid applications. It is also the case that specific types of DBMSs may require other services – for example, a navigation service may be required for Object Oriented Database Management System (ODBMS).

There is no separate security service. This is because access control is needed on all the operations of all services, so DBS owners can choose what each user (or group of users) is allowed to do.

The above services are all at a relatively low level and are very generic: they take no account of the meaning of the stored data. Higher-level, semantics-based services will also be required for Grid applications, and these will sit above, and utilise, the lower-level services described in this chapter. For example, the need for a generic Provenance service might be identified, implemented once and used by a variety of applications. It may, for example, offer operations to locate data with a particular provenance, or identify the provenance of data returned by a query. Identifying these higher-level services and designing them to be as general as possible, will be important for avoiding duplicated effort in the construction of Grid applications.

14.7 FEDERATING DATABASE SYSTEMS ACROSS THE GRID

Section 14.4 stressed the importance of being able to combine data from multiple DBSs. The ability to generate new results by combining data from a set of distributed resources is one of the most exciting opportunities that the Grid will offer. In this section, we consider how the service-based framework introduced in Section 14.6 can help achieve this.

One option is for a Grid application to interface directly to the service interfaces of each of the set of DBSs whose data it wishes to access. This approach is illustrated in Figure 14.2. However, this forces application writers to solve federation problems within the application itself. This would lead to great application complexity and to the duplication of effort.

To overcome these problems we propose an alternative, in which Grid-enabled middleware is used to produce a single, federated 'virtual database system' to which the application interfaces. Given the service-based approach proposed in Section 14.6, federating a set of DBS reduces to federating each of the individual services (query, transaction etc.). This creates a Virtual DBS, which has exactly the same service interface as the DBS described in the previous section but does not actually store any data (advanced versions could however be designed to cache data in order to increase performance). Instead, calls made to the Virtual DBS services are handled by service federation middleware that interacts with the service interfaces of the individual DBS that are being federated in order to compute the result of the service call. This approach is shown in Figure 14.3. Because the Virtual DBS has an identical service interface to the 'real' DBS, it is possible for a Virtual DBS to federate the services of both the 'real' DBS and other Virtual DBSs.

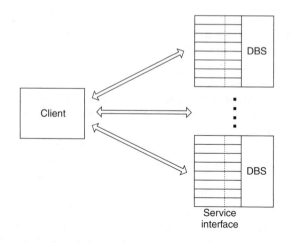

Figure 14.2 A Grid application interfacing directly to a set of DBS.

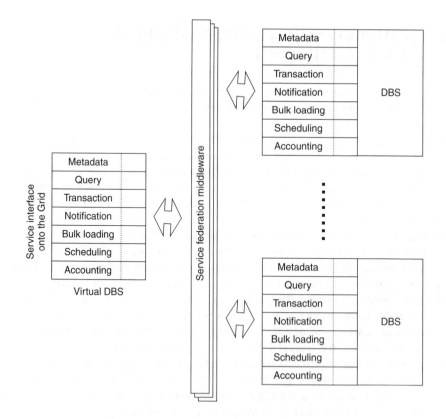

Figure 14.3 A virtual database system on the Grid.

Two different scenarios can be envisaged for the creation of a Virtual DBS:

1. A user decides to create a Virtual DBS that combines data and services from a specific set of DBSs that he/she wishes to work with. These may, for example, be well known as the standard authorities in his/her field.
2. A user wishes to find and work with data on a subject of his/her interest, but he/she does not know where it is located, for example, a biologist may want information on *Bacillus subtilis* 168. A metadata query would be used to locate appropriate datasets. These would then be federated to create a Virtual DBS that could then be queried. At the end of the work session, the Virtual DBS could be saved for future use. For example, the notification service might be configured to inform the user of interesting new data. Alternatively, the virtual DBS might not be required after the current session is ended, and so could be destroyed.

How can the Virtual DBS be created? The ideal situation would be for a tool to take a set of DBS and automatically create the Virtual DBS. At the other end of the scale, a set of bespoke programs could be written to implement each service of the Virtual DBS. Obviously, the former is preferable, especially if we wish to dynamically create Virtual DBS as in the second scenario above. Bearing this in mind, we now consider the issues in federating services.

The service-based approach proposed in Section 14.6 assists in the process of federating services, by encouraging standardisation. However, it will not be possible to fully standardise all services, and it is the resulting heterogeneity that causes problems. A tool could attempt to create a Virtual DBS automatically as follows. For each service, the tool would query the metadata service of each of the DBS being federated in order to determine their functionality and interface. Knowing the integration middleware that was available for the service and the requirements that this middleware had for the underlying services, the tool would determine the options for federation. If there were more than one option, then one would be selected (possibly taking into account application or user preferences). If no options were available, then the application or user would be informed that no integration of this service was possible. In this case, the user would either not be able to use the service, or would need to write new federation middleware to effect the integration, if that were possible.

Integrating each of the services proposed in Section 14.6 raises specific issues that are now described:

Query: Ideally, this would present to the user a single integrated schema for the virtual DBS and accept queries against it. A compiler and optimiser would determine how to split up the query across the set of DBS, and then combine the results of these sub-queries. The major relational DBMS products already offer 'Star' tools that implement distributed query middleware. Grid applications do, however, introduce new requirements, in particular, the need for conformance with Grid standards and the ability to query across dynamically changing sets of databases. The service-based approach to Grid-enabling databases simplifies the design of federation middleware as it promotes the standardisation of interfaces, but as was stated in the requirements section, it does not address the higher-level problem of the semantic integration of multiple databases, which has been the subject of much attention over the past decades [17]. It is to be hoped that the challenges of Grid applications give a further impetus to research in this area within the database community, as the results of this work are likely to be of great benefit to those building data-intensive Grid applications. Previous work on distributed query processing for parallel systems [10, 11] is very relevant to the Grid, which has a potential need for very high performance – distributed queries across large datasets may require huge joins that would benefit from parallelisation. The nature of the Grid does, however, offer some interesting new opportunities for distributed query processing [12]. Once a query has been compiled, Grid resources could be acquired on demand for running the distributed query execution middleware. The choice of resources could be made on the basis of the response time and price requirements of the user. For example, if a join operator was the bottleneck in a query and performance was important, then multiple compute nodes could be acquired and utilised to run that part of the query in parallel. If the user was charged for time on the compute nodes, then a trade-off between price and performance would need to be made. Further, because query optimisers can only estimate the cost of a query before it is run, queries sometimes take much longer than expected, perhaps because a filter or join in the middle of a query has produced more data than expected. An option here for Grid-based distributed query execution is to monitor the performance at run time and acquire more resources dynamically in order to meet the performance requirements of the user. There has been previous work in this area of dynamic query adaptation [28], and the Grid, which offers the ability to acquire and discard resources dynamically, could exploit this.

Transaction: The basic transaction service described in Section 14.6 already supports the creation of distributed transactions across multiple databases.

Bulk loading: This could be implemented by middleware that takes a load file, splits it into separate files for each DBS and uses the bulk load service of each individual DBMS to carry out the loading.

Notification: A client would register an interest in the virtual DBS. Middleware would manage the distribution of the notification operations: registration, filtering and notification, across the DBS. This should ideally be done using a generic Grid-enabled event service so that a database specific federation solution is not required.

Metadata: This would be a combination of the metadata services of the federated databases, and would describe the set of services offered by the Virtual DBS. At the semantic, data-description level (e.g. providing a unified view of the combined schema) the problems are as described above for the query service.

Scheduling: This would provide a common scheduling interface for the virtual DBS. When generic, distributed scheduling middleware is available for the Grid, the implementation of a federated service should be relatively straightforward (though, as described in Section 14.6, there is a major problem in controlling scheduling within the individual DBMS).

Accounting: This would provide a combined accounting service for the whole virtual DBS. As a Grid accounting service will have to support distributed components, the implementation of this service should be straightforward once that Grid accounting middleware is available.

As has been seen, the complexity of the service federation middleware will vary from service to service, and will, in general, increase as the degree of heterogeneity of the services being federated increases. However, we believe that the service-based approach to federating services provides a framework for the incremental development of a suite of federation middleware by more than one supplier. Initially, it would be sensible to focus on the most commonly required forms of service federation. One obvious candidate is query integration across relational DBMS. However, over time, applications would discover the need for other types of federation. When this occurs, then the aim is that the solution would be embodied in service federation middleware that fits into the proposed framework described above, rather than it being buried in the application-specific code. The former approach has the distinct advantage of allowing the federation software to be reused by other Grid applications. Each integration middleware component could be registered in a catalogue that would be consulted by tools attempting to integrate database services. The process of writing integration components would also be simplified by each taking a set of standard service interfaces as 'inputs' and presenting a single, standard federated service as 'output'. This also means that layers of federation can be created, with virtual databases taking other virtual databases as inputs.

While the focus in this document is on federating structured data held in databases, Section 14.4 also identified the requirement to federate this type of data with file-based, semi-structured and relatively unstructured data [18]. The framework for federation described above can also be used to support this through the use of special federation middleware. To enable any meaningful forms of federation, at least some basic services would have to be provided for the file-based data, for example, a simple query service. This could be achieved by either the owner, or the consumer of the data, providing a query wrapper that would be accessed by the middleware. With this in place, service federation middleware could be written that interfaces to and federates both file and database services for structured and less-structured data. The adoption of a service-based framework does not, however, provide any solutions for the semantic integration of data and metadata. That remains an open research area.

14.8 CONCLUSIONS

This chapter identified a set of requirements for Grid databases and showed that existing Grid middleware does not meet them. However, some of it, especially the Globus middleware, can be utilised as lower-level services on which database integration middleware can be built.

The chapter proposed a set of services that should be offered by a Grid-integrated DBS. This service-based approach will simplify the task of writing applications that access databases over the Grid. The services themselves vary considerably in the degree of complexity required to implement them. Some – *Transactions, Query and Bulk Loading* – already exist in most current DBMS, but work is needed both to integrate them into the emerging Grid standards and services and to introduce a level of interface standardisation where appropriate. Another, *Accounting*, should be relatively straightforward to implement once a Grid-wide accounting framework is in place. The effort required to develop a *Notification* service will depend on whether the underlying DBMS provides native support for it, and work is also required to integrate emerging Grid event and workflow services. Finally, *Scheduling* is the most problematic, as to do this properly requires a level of resource management that is not found in existing DBMSs, and this functionality cannot be added externally.

The chapter showed how the service-based approach to making databases available on the Grid could be used to structure and simplify the task of writing applications that need to combine information from more than one DBS. This is achieved by federating services to produce a single Virtual DBS with which the application interacts. The effort required to federate services will vary depending on the type of service and the degree of heterogeneity of the DBS being federated.

The service-based approach advocated in this chapter is independent of any particular implementation technology. However, the recent moves in Grid development towards a service-based middleware for the Grid through the OGSA [23] will clearly simplify the process of building distributed applications that access databases through service interfaces. While this chapter has focussed on describing a framework for integrating databases into the Grid and identifying required functionalities without making proposals at the detailed interface level, the UK core e-Science programme's OGSA – Database Access and Integration project (OGSA-DAI) is currently designing and building wrappers for relational databases and XML repositories so that they offer Grid-enabled services conforming to the OGSA framework. In conjunction with this, the Polar* project is researching into parallel query processing on the Grid [12].

To conclude, we believe that if the Grid is to become a generic platform, able to support a wide range of scientific and commercial applications, then the ability to publish and access databases on the Grid will be of great importance. Consequently, it is vitally important that, at this early stage in the Grid's development, database requirements are taken into account when Grid standards are defined and middleware is designed. In the short term, integrating databases into Grid applications will involve wrapping existing DBMSs in a Grid-enabled service interface. However, if the Grid becomes a commercial success then it is to be hoped that the DBMS vendors will Grid-enable their own products by adopting emerging Grid standards.

ACKNOWLEDGEMENTS

This chapter was commissioned by the UK e-Science Core Programme. Work to design and implement versions of the services described in the chapter is being carried out in the OGSA-DAI project (a collaboration between IBM Hursley, Oracle UK, EPCC and the Universities of Manchester and Newcastle) funded by that programme, and within the EPSRC funded Polar* project (at the Universities of Manchester and Newcastle). I would like to thank the other members of the UK Grid Database Taskforce (Malcolm Atkinson, Vijay Dialani, Norman Paton, Dave Pearson and Tony Storey) for many interesting discussions held over the last year. I am also grateful to the following people for their constructive comments on earlier versions of this chapter: Ian Foster, Alex Gray, Jim Gray, Tony Hey, Inderpal Narang, Pete Lee, Clive Page, Guy Rixon, David Skillicorn, Anne Trefethen and the anonymous referees.

REFERENCES

1. Foster, I. and Kesselman, C. (eds) (1999) *The Grid: Blueprint for a New Computing Infrastructure*. San Francisco, CA: Morgan Kaufmann Publishers.
2. Hillmann D. (2001) Using Dublin core, *Dublin Core Metadata Initiative*, dublincore.org/documents/usageguide.
3. Lassila, O. and Swick, R. (1999) *Resource Description Framework (RDF) Model and Syntax Specification*, W3C, www.w3c.org.REC-rdf-syntax-19990222.
4. Berners-Lee, T., Hendler, J. and Lassila, O. (2001) The semantic web. *Scientific American*, May, 34–43.
5. Pearson, D. (2002) *Data Requirements for the Grid: Scoping Study Report*. UK Grid Database Taskforce.
6. Global Grid Forum, Global Grid Forum Security Working Group, www.gridforum.org, 2002.
7. Global Grid Forum, Accounting Models Research Group, www.gridforum.org, 2002.
8. Tierney, B., Aydt, R., Gunter, D., Smith, W., Taylor, V., Wolski, R. and Swany, M. (2002) *A Grid Monitoring Architecture*, Global Grid Forum, GWD-Perf-16-2.
9. Global Grid Forum, Scheduling and Resource Management Area, www.gridforum.org, 2002.
10. Graefe, G. (1990) Encapsulation of parallelism in the volcano query processing system, *SIGMOD Conference*. Atlantic City, NJ: ACM Press.
11. Smith, J., Sampaio, S. D. F. M., Watson, P. and Paton, N. W. (2000) Polar: An architecture for a parallel ODMG compliant object database, *ACM CIKM International Conference on Information and Knowledge Management*. McLean, VA: ACM Press.
12. Smith, J., Gounaris, A., Watson, P., Paton, N. W., Fernandes, A. A. A. and Sakellariou, R. (2002) Distributed query processing on the grid. *Proceedings of the 3rd International Workshop on Grid Computing (GRID 2002)*, in LNCS 2536, Springer-Verlag, pp. 279–290.
13. Cattell, R. (1997) *Object Database Standard: ODMG 2.0*. San Francisco, CA: Morgan Kaufmann Publishers.
14. Smith, J., Watson, P., Sampaio, S. D. F. M. and Paton, N. W. (2002) Speeding up navigational requests in a parallel object database system. *EuroPar 2002*, LNCS 2400, Paderborn, Germany, pp. 332–341, 2002.
15. Shiers, J. (1998) Building a multi-petabyte database: The RD45 project at CERN, in Loomis, M. E. S. and Chaudhri, A. B. (eds) *Object Databases in Practice*. New York: Prentice Hall, pp. 164–176.
16. De Roure, D., Jennings, N. and Shadbolt, N. (2001) *Research Agenda for the Semantic Grid: A Future e-Science Infrastructure*. UKeS-2002-02, Edinburgh, UK: UK National e-Science Centre.
17. Sheth, A. P. and Larson, J. A. (1990) Federated database systems for managing distributed, heterogeneous and autonomous databases. *ACM Computing Surveys*, **22**(3), 183–236.
18. Lahiri, T., Abiteboul, S. and Widom, J. (1999) Ozone: Integrating structured and semistructured data, *Seventh Intl. Workshop on Database Programming Languages*. Scotland: Kinloch Rannoch 1999.

19. Yu, C. T. and Chang, C. C. (1984) Distributed query processing. *ACM Computing Surveys*, **16**(4), 399–433.
20. Suciu, D. (2002) Distributed query evaluation on semistructured data. *ACM Transactions on Database Systems*, **27**(1), 1–62.
21. Foster, I., Kesselman, C. and Tuecke, S. (2001) The anatomy of the grid: enabling scalable virtual organizations. *International Journal of Supercomputer Applications*, **15**(3), 200–222.
22. Graham, S., Simeonov, S., Boubez, T., Daniels, G., Davis, D., Nakamura, Y. and Neyama, R. (2001) *Building Web Services with Java: Making Sense of XML, SOAP, WSDL and UDDI*. Indianapolis, IN: Sams Publishing.
23. Foster, I., Kesselman, C., Nick, J. and Tuecke, S. (2002) *The Physiology of the Grid: An Open Grid Services Architecture for Distributed Systems Integration*, The Globus Project, www.globus.org.
24. Pearlman, L., Welch, V., Foster, I. and Kesselman, C. (2001) A community authorization service for group communication. Submitted to the *3rd Intl. Workshop on Policies for Distributed Systems and Networks*, 2001.
25. Hoschek, W. and McCance, G. (2001) *Grid Enabled Relational Database Middleware*. Global Grid Forum, Frascati, Italy, www.gridforum.org/1_GIS/RDIS.htm.
26. Rajasekar, A., Wan, M. and Moore, R. (2002) MySRB & SRB – Components of a data grid. *11th International Symposium on High Performance Distributed Computing (HPDC-11)*, Edinburgh, 2002.
27. OASIS Committee. (2002) Business Transaction Protocol Version 1.0.
28. Gounaris, A., Paton, N. W., Fernandes, A. A. A. and Sakellariou, R. (2002) Adaptive query processing: a survey, *British National Conference on Databases (BNCOD)*. 2002. Sheffield, UK: Springer.

The Open Grid Services Architecture, and data Grids

Peter Z. Kunszt and Leanne P. Guy

CERN, Geneva, Switzerland

15.1 INTRODUCTION

Data Grids address computational and data intensive applications that combine very large datasets and a wide geographical distribution of users and resources [1, 2]. In addition to computing resource scheduling, Data Grids address the problems of storage and data management, network-intensive data transfers and data access optimization, while maintaining high reliability and availability of the data (see References [2, 3] and references therein).

The Open Grid Services Architecture (OGSA) [1, 4] builds upon the anatomy of the Grid [5], where the authors present an open Grid Architecture, and define the technologies and infrastructure of the Grid as 'supporting the sharing and coordinated use of diverse resources in dynamic distributed Virtual Organizations (VOs)'. OGSA extends and complements the definitions given in Reference [5] by defining the architecture in terms of Grid services and by aligning it with emerging Web service technologies.

Web services are an emerging paradigm in distributed computing that focus on simple standards-based computing models. OGSA picks the Web Service Description Language (WSDL) [6], the Simple Object Access Protocol (SOAP) [7], and the Web Service

Grid Computing – Making the Global Infrastructure a Reality. Edited by F. Berman, A. Hey and G. Fox
© 2003 John Wiley & Sons, Ltd   ISBN: 0-470-85319-0

Introspection Language (WSIL) [8] from the set of technologies offered by Web services and capitalizes especially on WSDL. This is a very natural approach since in the distributed world of Data Grid computing the same problems arise as on the Internet concerning the description and discovery of services, and especially the heterogeneity of data [9].

In this article the application of the OGSA is discussed with respect to Data Grids. We revisit the problem that is being addressed, the vision and functionality of Data Grids and of OGSA. We then investigate OGSA's benefits and possible weaknesses in view of the Data Grid problem. In conclusion, we address what we feel are some of the shortcomings and open issues that expose potential areas of future development and research.

The European Data Grid project (EU Data Grid project) aims to build a computational and data-intensive grid to facilitate the analysis of millions of Gigabytes of data generated by the next generation of scientific exploration. The main goal is to develop an infrastructure that will provide all scientists participating in a research project with the best possible access to data and resources, irrespective of geographical location. All the Data Grid discussion presented in this chapter draws from existing concepts in the literature, but is strongly influenced by the design and experiences with data management in the context of the EU Data Grid project [10]. In particular, data management in the EU Data Grid project has focused primarily on file-based data management [11, 12]. However, the usage of the Web services paradigm to build a Grid infrastructure was always considered in the architecture and has lead to the definition of the Web Service discovery architecture WSDA [13].

15.1.1 The vision

In this section we summarize the Data Grid problem and the vision of the Grid as presented in the literature [1, 5, 14].

15.1.2 Desirable features of Grids

In the scientific domain there is a need to share resources to solve common problems and to facilitate collaboration among many individuals and institutes. Likewise, in commercial enterprise computing there is an increasing need for resource sharing, not just to solve common problems but to enable enterprise-wide integration and business-to-business (B2B) partner collaboration

To be successful on a large scale, the sharing of resources should be

FLEXIBLE, SECURE, COORDINATED, ROBUST, SCALABLE,

UBIQUITOUSLY ACCESSIBLE, MEASURABLE (QOS METRICS),

TRANSPARENT TO THE USERS

The distributed resources themselves should have the following qualities:

INTEROPERABLE, MANAGEABLE, AVAILABLE AND EXTENSIBLE

Grid computing attempts to address this very long list of desirable properties. The concept of the Virtual Organization (VO) gives us the first necessary semantics by which to address these issues systematically and to motivate these properties in detail.

15.1.3 Virtual organizations

A Virtual Organization (VO) is defined as a group of individuals or institutes who are geographically distributed but who appear to function as one single unified organization. The members of a VO usually have a common focus, goal or vision, be it a scientific quest or a business venture. They collectively dedicate resources, for which a well-defined set of rules for sharing and quality of service (QoS) exists, to this end. Work is coordinated through electronic communication. The larger the VO, the more decentralized and geographically distributed it tends to be in nature.

The concept of the VO is not a new one; VOs have existed for many years in the fields of business and engineering, in which consortia are assembled to bid for contracts, or small independent companies temporarily form a VO to share skills, reduce costs, and expand their market access. In the commercial context, VOs may be set up for anything from a single enterprise entity to complex B2B relationships.

In scientific communities such as High Energy Physics (HEP), astrophysics or medical research, scientists form collaborations, either individually or between their respective institutes, to work on a common set of problems or to share a common pool of data. The term *Virtual Organization* has only recently been coined to describe such collaborations in the scientific domain.

VOs may be defined in a very flexible way so as to address highly specialized needs and are instantiated through the set of services that they provide to their users/applications. Ultimately, the organizations, groups and individuals that form a VO also need to pay for the resources that they access in some form or another – either by explicitly paying money for services provided by Grid service providers or by trading their own resources. Grid resource and service providers may choose to which VOs they can offer their resources and services. The models for paying and trading of resources within a VO will probably be vastly different in the scientific and business domains.

In the vision of the Grid, end users in a VO can use the shared resources according to the rules defined by the VO and the resource providers. Different VOs may or may not share the same resources. VOs can be dynamic, that is, they may change in time both in scope and in extension by adjusting the services that are accessible by the VO and the constituents of the VO. The Grid needs to support these changes such that they are transparent to the users in the VO.

15.1.4 Motivation for the desirable features in Data Grids

In the specific case of Data Grids, the members of a VO also share their collective data, which is potentially distributed over a wide area (SCALABLE). The data needs to be accessible from anywhere at anytime (FLEXIBLE, UBIQUITOUS ACCESS); however, the user is not necessarily interested in the exact location of the data (TRANSPARENT). The access to the data has to be secured, enabling different levels of security according to the needs of the VO in question, while at the same time maintaining the manageability of the data and its accessibility (SECURE, COORDINATED, MANAGEABLE). The reliability and robustness requirements in the existing communities interested in Data Grids are high, since the computations are driven by the data – if the data are not accessible, no computation is possible (ROBUST, AVAILABLE).

In many VOs, there is a need to access existing data – usually through interfaces to legacy systems and data stores. The users do not want to be concerned with issues of data interoperability and data conversion (INTEROPERABLE, EXTENSIBLE). In addition, some Data Grids also introduce the concept of Virtual Data, [15] in which data needs to be generated by a series of computations before it is accessible to the user.

15.2 THE OGSA APPROACH

The OGSA views the Grid as an extensible set of Grid services. A Grid service is defined as 'a Web service that provides a set of well-defined interfaces and that follows specific conventions' [14]. Virtual Organizations are defined by their associated services, since the set of services 'instantiate' a VO. In OGSA each VO builds its infrastructure from existing Grid service components and has the freedom to add custom components that have to implement the necessary interfaces to qualify as a Grid service. We summarize the requirements that motivate the OGSA as stated in References [1, 4, 14] and the properties it provides to the Grid (see previous section for the discussion of properties) in Table 15.1.

Table 15.1 Requirements mentioned in References [1, 4, 14] for OGSA. We try to map each requirement to the properties that grid systems are supposed to have on the basis of the requirements given in Section 15.2. The primary property is highlighted. An analysis of these requirements is given in the next section

	Requirement	Property
1	Support the creation, maintenance and application of ensembles of services maintained by VOs.	COORDINATED, TRANSPARENT, SCALABLE, EXTENSIBLE, *MANAGEABLE*
2	Support local and remote transparency with respect to invocation and location.	*TRANSPARENT*, FLEXIBLE
3	Support multiple protocol bindings (enabling e.g. localized optimization and protocol negotiation).	*FLEXIBLE*, TRANSPARENT, INTEROPERABLE, EXTENSIBLE, UBIQ.ACCESSIBLE
4	Support virtualization in order to provide a common interface encapsulating the actual implementation.	*INTEROPERABLE*, TRANSPARENT, MANAGEABLE
5	Require mechanisms enabling interoperability.	*INTEROPERABLE*, UBIQ.ACCESSIBLE
6	Require standard semantics (like same conventions for error notification).	*INTEROPERABLE*
7	Support transience of services.	SCALABLE, MANAGEABLE, AVAILABLE, *FLEXIBLE*
8	Support upgradability without disruption of the services.	MANAGEABLE, FLEXIBLE, *AVAILABLE*

Table 15.2 Current elements of OGSA and elements that have been mentioned by OGSA but are said to be dealt with elsewhere. We also refer to the requirement or to the Grid capability that the elements address. Please be aware that this is a snapshot and that the current OGSA service specification might not be reflected correctly

Responsibility	Interfaces	Requirements addressed
Information & discovery	GridService.FindServiceData Registry.(Un)RegisterService HandleMap.FindByHandle *Service Data Elements Grid Service Handle and Grid Service Reference*	1,2,4,5,6
Dynamic service creation	Factory.CreateService	1,7
Lifetime management	GridService.SetTerminationTime, GridService.Destroy	1,7
Notification	NotificationSource. (Un)SubscribeToNotificationTopic NotificationSink.DeliverNotification	5
Change management	*CompatibilityAssertion data element*	8
Authentication	Impose on the hosting environment	SECURE
Reliable invocation	Impose on hosting environment	ROBUST
Authorization and policy management	Defer to later	SECURE
Manageability	Defer to later	MANAGEABLE

As mentioned before, OGSA focuses on the nature of services that makes up the Grid. In OGSA, existing Grid technologies are aligned with Web service technologies in order to profit from the existing capabilities of Web services, including

- service description and discovery,
- automatic generation of client and server code from service descriptions,
- binding of service descriptions to interoperable network protocols,
- compatibility with higher-level open standards, services and tools, and
- broad industry support.

OGSA thus relies upon emerging Web service technologies to address the requirements listed in Table 15.1. OGSA defines a preliminary set of Grid Service interfaces and capabilities that can be built upon. All these interfaces are defined using WSDL to ensure a distributed service system based on standards. They are designed such that they address the responsibilities dealt with in Table 15.2.

The GridService interface is imposed on all services. The other interfaces are optional, although there need to be many service instances present in order to be able to run a VO. On the basis of the services defined, higher-level services are envisaged such as data management services, workflow management, auditing, instrumentation and monitoring, problem determination and security protocol mapping services. This is a nonexhaustive list and is expected to grow in the future.

OGSA introduces the concept of Grid Service Handles (GSHs) that are unique to a service and by which the service may be uniquely identified and looked up in the Handle Map. GSHs are immutable entities. The Grid Service Reference (GSR) is then the description of the service instance in terms of (although not necessarily restricted to) WSDL and gives the user the possibility to refer to a mutable running instance of this service. OGSA envisages the possibility of a hierarchical structure of service hosting environments, as they demonstrate in their examples.

Another essential building block of OGSA is the standardized representation for service data, structured as a set of named and typed XML fragments called *service data elements* (SDEs). The SDE may contain any kind of information on the service instance allowing for basic introspection information on the service. The SDE is retrievable through the GridService port type's FindServiceData operation implemented by all Grid services. Through the SDE the Grid services are stateful services, unlike Web services that do not have a state. This is an essential difference between Grid services and Web services: Grid services are stateful and provide a standardized mechanism to retrieve their state.

15.3 DATA GRID SERVICES

There has been a lot of effort put into the definition of Data Grid services in the existing Grid projects. Figure 15.1 shows the Global Grid Forum GGF Data Area concept space,

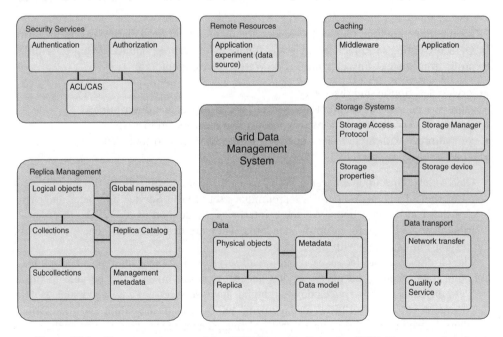

Figure 15.1 The concept space of the GGF data area (from the GGF data area website).

namely, their domains of interest. In this article we address the same concept space. Examples of Data Grid architectures are given in References [2, 15–21].

15.3.1 The data

In most of the existing architectures, the data management services are restricted to the handling of files. However, for many VOs, files represent only an intermediate level of data granularity. In principle, Data Grids need to be able to handle data elements – from single bits to complex collections and even virtual data, which must be generated upon request. All kinds of data need to be identifiable through some mechanism – a logical name or ID – that in turn can be used to locate and access the data. This concept is identical to the concept of a GSH in OGSA [14], so in order to be consistent with the OGSA terminology we will call this logical identifier the Grid Data Handle (GDH) (see below).

The following is a nondefinitive list of the kinds of data that are dealt with in Data Grids.

- *Files*: For many of the VOs the only access method to data is file-based I/O, data is kept only in files and there is no need for other kind of data granularity. This simplifies data management in some respect because the semantics of files are well understood. For example, depending on the QoS requirements on file access, the files can be secured through one of the many known mechanisms (Unix permissions, Access Control Lists (ACLs), etc.). Files can be maintained through directory services like the Globus Replica Catalog [3], which maps the Logical File Name (LFN) to their physical instances, and other metadata services that are discussed below.
- *File collections*: Some VOs want to have the possibility of assigning logical names to file collections that are recognized by the Grid. Semantically there are two different kinds of file collections: *confined* collections of files in which all files making up the collection are always kept together and are effectively treated as one – just like a *tar* or *zip* archive – and *free* collections that are composed of files as well as other collections not necessarily available on the same resource – they are effectively a bag of logical file and collection identifiers. Confined collections assure the user that all the data are always accessible at a single data source while free collections provide a higher flexibility to freely add and remove items from the collection, but don't guarantee that all members are accessible or even valid at all times. Keeping free collections consistent with the actual files requires additional services.
- *Relational databases*: The semantics of data stored in a Relational Database Management System (RDBMS) are also extremely well understood. The data identified by a GDH may correspond to a database, a table, a view or even to a single row in a table of the RDBMS. Distributed database access in the Grid is an active research topic and is being treated by a separate article in this volume (see Chapter 14).
- *XML databases and semistructured data*: Data with loosely defined or irregular structure is best represented using the semistructured data model. It is essentially a dynamically typed data model that allows a 'schema-less' description format in which the data is less constrained than in relational databases [22]. The XML format is the standard in which the semistructured data model is represented. A GDH may correspond to any XML data object identifiable in an XML Database.

- *Data objects*: This is the most generic form of a single data instance. The structure of an object is completely arbitrary, so the Grid needs to provide services to describe and access specific objects since the semantics may vary from object type to object type. VOs may choose different technologies to access their objects – ranging from proprietary techniques to open standards like CORBA/IIOP and SOAP/XML.
- *Virtual data*: The concept of virtualized data is very attractive to VOs that have large sets of secondary data that are derived from primary data using a well-defined set of procedures and parameters. If the secondary data are more expensive to store than to regenerate, they can be virtualized, that is, only created upon request. Additional services are required to manage virtual data.
- *Data sets*: Data sets differ from free file collections only in that they can contain any kind of data from the list above in addition to files. Such data sets are useful for archiving, logging and debugging purposes. Again, the necessary services that track the content of data sets and keep them consistent need to be provided.

15.3.1.1 The Grid data handle GDH

The common requirement for all these kinds of data is that the logical identifier of the data, the GDH, be globally unique. The GDH can be used to locate the data and retrieve information about it. It can also be used as a key by which more attributes can be added to the data through dedicated application-specific catalogs and metadata services. There are many possibilities to define a GDH, the simplest being the well-known Universal Unique IDentifier (UUID) scheme [23]. In order to be humanly readable, a uniform resource identifier (URI)-like string might be composed, specifying the kind of data instead of the protocol, a VO namespace instead of the host and then a VO-internal unique identifier, which may be a combination of creation time and creation site and a number (in the spirit of UUID), or another scheme that may even be set up by the VO. The assignment and the checking of the GDH can be performed by the Data Registry (DR) (see below).

Similar to the GSH, a semantic requirement on the GDH is that it must not change over time once it has been assigned. The reason for this requirement is that we want to enable automated Grid services to be able to perform all possible operations on the data – such as location, tracking, transmission, and so on. This can easily be achieved if we have a handle by which the data is identified. One of the fundamental differences between Grids and the Web is that for Web services, ultimately, there always is a human being at the end of the service chain who can take corrective action if some of the services are erroneous. On the Grid, this is not the case anymore. If the data identifier is neither global nor unique, tracking by automated services will be very difficult to implement – they would need to be as 'smart' as humans in identifying the data based on its content.

15.3.1.2 The Grid data reference

For reasons of performance enhancement and scalability, a single logical data unit (any kind) identified by a GDH may have several physical instances – replicas – located throughout the Grid, all of which also need to be identified. Just like the GSH, the GDH carries no protocol or instance-specific information such as supported protocols to access the data. All this information may be encapsulated into a single abstraction – in analogy to the Grid Service Reference – into a Grid Data Reference (GDR). Each physical data instance has a corresponding GDR that describes how the data can be accessed. This description may be in WSDL, since a data item also can be viewed as a Grid resource. The elements of the GDR include physical location, available access protocols, data lifetime, and possibly other metadata such as size, creation time, last update time, created by, and so on. The exact list and syntax would need to be agreed upon, but this is an issue for the standardization efforts in bodies like GGF. The list should be customizable by VOs because items like update time may be too volatile to be included in the GDR metadata.

In the OGSA context, the metadata may be put as Service Data Elements in the Data Registry.

15.3.1.3 The data registry

The mapping from GDH to GDR is kept in the Data Registry. This is the most basic service for Data Grids, which is essential to localize, discover, and describe any kind of data. The SDEs of this service will hold most of the additional information on the GDR to GDH mapping, but there may be other more dedicated services for very specific kinds of GDH or GDR metadata.

To give an example, in Data Grids that only deal with files – as is currently the case within the EU DataGrid project – the GDH corresponds to the LFN and the GDR is a combination of the physical file name and the transport protocol needed to access the file, sometimes also called *Transport File Name* (TFN) [16–18]. The Data Registry corresponds to the Replica Catalog [24] in this case.

15.3.2 The functionality and the services

In this section we discuss the functionalities of Data Grids that are necessary and/or desirable in order to achieve the Data Grid vision.

15.3.2.1 VO management

Virtual Organizations can be set up, maintained and dismantled by Grid services. In particular, there needs to be functionality provided for bootstrapping, user management, resource management, policy management and budget management. However, these functionalities are generic to all kinds of Grids, and their detailed discussion is beyond the scope of this article. Nevertheless, it is important to note that since VOs form the organic organizational unit in Grids, VO management functionalities need to be supported by all services. Some of these issues are addressed by the services for control and monitoring (see below).

15.3.2.2 Data transfer

The most basic service in Data Grids is the data transfer service. To ensure reliable data transfer and replication, data validation services need to be set up. Data validation after transfer is a QoS service that might not be required by all VOs. For file-based Data Grids, there is already an ongoing effort to provide a reliable file transfer service for the Grid [3].

15.3.2.3 Data storage

Data Grid storage services will be provided by many resources. There will be very primitive storage systems with limited functionality as well as high-level systems for data management such as RDBMSs or Storage Resource Managers [18]. All may provide high-quality Grid services, but with vast range of available functionality. The requirements on the data store depend very much on the kind of data (see Section 15.3.1). Two common QoS additions are

Resource allocation: The ability to reserve directly (through advance reservation) or indirectly (through quotas) sufficient space for data to be transferred to the store later.

Lifetime management: The ability to specify the lifetime of the data directly (timestamps and expiration times), or indirectly (through data status flags like permanent, durable, volatile).

15.3.2.4 Data management

In all existing Data Grids, data management is one of the cornerstones of the architecture. The data management services need to be very flexible in order to accommodate the peculiarities and diverse requirements on QoS of the VOs and their peculiarities with respect to different kinds of data (see previous section) and data access.

In the vision of a Data Grid, the data management services maintain, discover, store, validate, transfer and instantiate the data for the user's applications transparently. Because of the multitude of different kinds of data, it is essential that the data can be described and validated in the Data Grid framework.

In the previous section, we introduced the concept of GDHs and GDRs as well as a service for their mapping, the Data Registry. These are the core concepts of data management. In addition, we have the following data management functionalities in Data Grids:

Registration of data: The first obvious functionality is to register new or existing data on the Grid. The data will receive a GDH; first instances will have to be described through their GDR, and the corresponding mappings need to be registered in the DR. This functionality may be part of the DR or set up as a separate data registration service.

Materialization of virtual data: This functionality of the data management services will materialize virtual data according to a set of materialization instructions. These instructions include references to the executables and any input datasets required, as well as to any additional job parameters. All information pertaining to the materialization of virtual data is stored in a catalog. After materialization, physical copies of the data exist and the corresponding catalogs need to be updated and a GDR assigned.

GDH assignment and validation: The uniqueness of the GDH can only be assured by the data management services themselves. The VOs may have their own GDH generation and validation schemes but in order to be certain, those schemes should be pluggable and complementary to the generic GDH validation scheme of the Data Grid. In order to scale well, the GDH validation and generation functionality should not be assured by a central service. Optimally, it should even work on disconnected Grid nodes, that is, a local service should be able to decide whether a new GDH is indeed unique or not.

Data location based on GDH or metadata: The DR is the service that provides the mapping functionality between GDH and GDR. It should be a distributed service in the spirit of Reference [24] that enables the location of local data even if the other Grid sites are unreachable. In the case of the location of data based on metadata, there will be higher-level services, probably databases and not just registries that can execute complex queries to find data (see Chapter 14). It is also necessary to be able to do the reverse mapping from GDR to GDH. There is no reason that a single GDR might not be referenced by more than one GDH.

Pre- and post-processing before and after data transfer: In addition to data validation, there might be necessary pre- and post-processing steps to be executed before and after data transfer, respectively. For example, the data might need to be extracted, transformed into a format that can be transmitted, stripped of confidential data, and so on, before transfer. After transfer, there may be additional steps necessary to store the data in the correct format in its designated store.

Replica management: The replication process including processing, transfer, registration and validation should appear as a single atomic operation to the user. In order to assure correct execution of all steps, a replica manager (RM) service is necessary that orchestrates and logs each step and can take corrective action in the case of failures. It also needs to query the destination data store to determine whether the application or user initiating the replication has the required access rights to the resource, and whether enough storage space is available. All replication requests should be addressed to the RM service.

Replica selection: The job scheduler will need to know about all replicas of a given GDH in order to make a qualified choice as to where to schedule jobs, or whether replication or materialization of data needs to be initiated. The replica selection functionality of the data management services can select the optimal replica for this purpose.

Subscription to data: Automatic replication may be initiated by a simple subscription mechanism. This is very useful in VOs in which there is a single data source but many analysts all around the world. Data appearing at the data source adhering to certain criteria can be automatically replicated to remote sites.

Replica load balancing: On the basis of access patterns of frequently used data, this service can initiate replication to improve load balancing of the Data Grid. This service should be configurable so that many different strategies can be used to do automated replication [25].

Data management consistency: The data in all the metadata catalogs and the actual data store might get out of synchronization due to failures, or because the data was accessed outside the Grid context, and so on. There could be a service enhancing this QoS aspect of data management that periodically checks the consistency of data in store with the data management service metadata.

Data update: If data can be updated, there are many different strategies regarding how to propagate the updates to the replicas. The strategy to be chosen is again dependent on the QoS requirements of the VO in question: how much latency is allowed, requirements on consistent state between replicas, and so on. Of great interest are also the access patterns: are there many concurrent updates or only sporadic updates? The data management services need to be generic enough to accommodate many update strategies.

15.3.2.5 Metadata

Metadata can be classified into two groups according to their usage and content. Metadata associated with the operation of the system or any of its underlying components is designated 'technical metadata', whereas those metadata accumulated by the users of the system that pertain to the usage or classification of the corresponding data, is designated 'application metadata'. Technical metadata is treated as metadata by the system and is essential to its successful operation. Application metadata are metadata managed by the Grid on behalf of the users and are treated simply as data by the Grid system. The boundary between these two groups is not always discernible or unambiguous.

The GDR should contain most of the necessary technical metadata on the given physical instance of the data. This will be stored in the Data Registry. There is, however, a need for metadata services that store both technical and application metadata on all the other aspects of the data management services.

Security metadata: These include local authorization policies to be used by all data management services, and specific authorization data for specialized high-level services. The details of the security infrastructure are not worked out yet, but these and other security metadata will certainly have to be stored in the Grid. See the next section for more on security.

GDH metadata and data relations: As mentioned before, the GDH may serve as the primary key to other catalogs to add more metadata on the data elements.

Virtual data generation catalogs: The metadata stored in these catalogs should describe the programs to be invoked and their necessary parameters in order to materialize a virtual data set.

Replication progress data: As described above, the RM service acts on behalf of users or other services as an orchestrator of replication. It needs to control each step in the replication process and needs to make sure that failures are dealt with properly; hence the need for progress data. This may be stored in a dedicated metadata store or just log files, depending on the QoS requirements.

Subscription configurations: This is metadata to be used by the subscription service, defining the source and destination of the automated replication process. Also, data on triggers like initiation time or quota usage level are stored here, and so on.

Data access patterns: The automated load-balancing mechanisms need to store monitoring data and other computed statistics in order to optimize data access.

Provenance data: A record of how data was created or derived may be generated automatically by the Data Grid technology, as well as by users or by the application code. Such data can be classified as both technical and application metadata.

Durability and lifetime data: Data indicating how long data should last, and whether its intermediate states should be recoverable.

Type and format data: Tools used for the diagnostics and validation of data would benefit from metadata that describes the type and format of data.

15.3.2.6 Security

Data security is another one of the cornerstones of Data Grids. Certain industrial and research domains, such as biomedical research and the pharmaceutical industry, will impose very strict QoS requirements on security. Domains that deal with sensitive or proprietary data, as well as data that may lead to patents will probably develop their own security infrastructure to replace that developed by the Grid community.

In Grids the most basic functionality for security is authentication of users and groups using single-sign-on techniques and the possibility of delegating certain privileges to Grid services to act on the user's behalf. In Data Grids there is a very strong emphasis on authorization, especially fine-grained authorization on data access and data update.

The resource providers also need to have an instrument by which they can enforce local policies and are able to block malicious users locally without the need to contact a remote service. On the other hand, users need to be assured that their confidential data cannot be compromised by local site administrators. This calls for two-way authentication: the service needs to be authenticated with the user as well. As we will see later, security is a component that has received insufficient recognition in the OGSA framework to date.

There are services that address security already such as the Community Authorization Service (CAS) by Globus [26], but they are not yet exploited in the Grid services context.

15.3.2.7 Control and monitoring

Most of the data management services need to be monitored and controlled by other services or users. The status of the services needs to be published, statistics need to be kept for optimization and accounting purposes, services need to be deployed or terminated upon request, and so on.

The services that we can identify here are service registries and factories, as proposed by the OGSA. For gathering statistics, there needs to be additional services for monitoring; one possibility is to set up services in the scheme of the Grid monitoring architecture that follows a consumer-producer model [27].

These services are essential for service and VO maintenance, as well as logging, accounting, billing and auditing. But the security services will also need a way to shut another service down if they detect that the service has been compromised. The OGSA mechanisms for lifetime management adequately address this issue.

15.3.2.8 Reliability and fault tolerance

In addition to control and monitoring services, there need to be additional services that enhance the provided QoS with respect to reliability and fault tolerance. Upon occurrences of unpredictable failures, such as the sudden unavailability of resources, Grid nodes, networks, and so on, dedicated services may choose to act in a predefined manner, failover to other sites or services for example.

In terms of fault tolerance, all services need to be implemented using robust protocols and mechanisms. OGSA does not enforce, but also does not hinder the usage of such protocols.

15.3.3 Data Grid and OGSA

In this section we investigate the steps that need to be taken to provide OGSA versions of the services described above. OGSA introduces several service concepts that need to be adopted in order to qualify as Grid services. Necessary components are factories, registries, and handle maps. Additional mechanisms that can be used to build our Data Grid are notification and lifetime management.

15.3.3.1 Factories

Factories have a very positive impact on the robustness and availability of services: if a service fails, it can be restarted by higher-level controlling services by calling on the service factory. Factories themselves may act as smart controlling services. This is also a very useful pattern for manageability; only certain factories need to be contacted to bootstrap or to disband a VO.

Transactions, which are very important in Data Grids, may be implemented more easily in higher-level services by building upon the functionality of the factories. As an example, consider the RM service. The operation of replication involves data access checking, space reservation at the remote site, data transfer and lookup and update operations in many

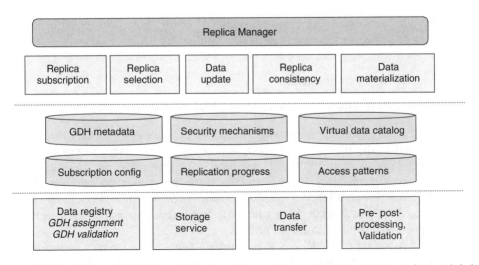

Figure 15.2 Elements of the data storage, transfer, management, and metadata services and their logical layering. The top layer contains the high-level services with the replica manager being the 'chief orchestrator'. The middle layer shows the logical metadata services. The lowest layer contains the basic services.

catalogs. To a high-level client these steps should be atomic: the RM needs to roll back upon failure or to retry on a different path if possible. The factory model facilitates the implementation of such transactional services considerably.

All the services in Figure 15.2 could be instantiated through a set of factories; we could even have dedicated factories for each component. One attractive possibility is to build the replica manager with the factory interface to many of its fellow high-level services so that it can instantiate any of the services, providing the functionality upon request.

15.3.3.2 Registries

How should the registry interface be deployed in order to make service discovery possible? One trivial possibility is to implement the registry interface in all services so that they always have access to a registry to discover other services. The problem with such a model is scalability: if all services need to be notified of a service joining or leaving the Grid, the number of messages will increase exponentially with the number of services. So it is much more sensible to keep just a few registries for a VO. They should be instantiated at Grid nodes that adhere to a higher QoS in terms of stability and accessibility of the node (for example it should be run at a large computing facility with good local support and not on somebody's desktop machine in a university).

It also does not make sense to add the registry functionality to any other service because of the vital role of the registry. It should not be destroyed 'by accident' because the Grid service that hosts it was destroyed for some other reason. Nor should there be services running that have additional functionality just because their registry functionality is still needed by other services – this opens up potential security holes. So registries are best deployed as dedicated services on high-quality Grid nodes.

15.3.3.3 Service lifetime management

Let us consider three simple models for service lifetime management.

Model 1: A very natural possibility is to keep a set of factories with a very long lifetime (possibly equal to the lifetime of the VO) that create the necessary services to assure the QoS requirements of the VO and keep them alive themselves by using the OGSA lifetime-extension mechanism. If a service fails, it should be restarted automatically.

Model 2: Another possibility is to set up factories that create new services upon application demand, in which case the applications would be responsible for keeping the services alive until they are no longer needed.

Model 3: A variant of Model 2 is to not create a new service for each application but to redirect incoming applications to existing services if they are not already overloaded and only to create new services for load-balancing purposes.

The preferred model depends on the applications and the usage patterns of the VOs. If there are a lot of parallel applications that can easily share the same resource, Model 1 or 3 is probably preferable – Model 2 would generate too many services. If there are only a small number of very intensive applications, Model 2 might be preferable. In the case of Model 2 and 3, it is important that applications do not set very long lifetimes on the services, since in the case of the failure of the applications the services would remain too long.

Of course, the different models may be mixed within a VO. The metadata services (middle layer in Figure 15.2) are less dynamic since they have persistent data associated with them, which is usually associated with a database or other data store. The VO would not want to let them expire if no application has used them for a while, hence Model 1 is the most appropriate for these services.

The higher-level data management services as well as the lowest-level services may be created upon demand; they are much more lightweight, so Models 2 or 3 might be appropriate, depending on the VO.

15.3.3.4 Data lifetime management

Managing the data lifetime is orthogonal to managing the lifetime of the Data Grid services. Persistency requirements on certain kinds of data are so strong that their lifetime will most certainly exceed the lifetime of Grid service instances (identified by GSRs) – they depend on their hosting environment that evolves over time. Nevertheless, the GSH, the service that is associated with accessing persistent data need not change even if the implementation of the service has changed radically.

It is not up to the Grid to solve the issues of migrating data that has to be kept 'forever' to new storage technologies, but it should provide an access point that does not change over time – the GDH. The GDR may also change radically, the schema of the data it

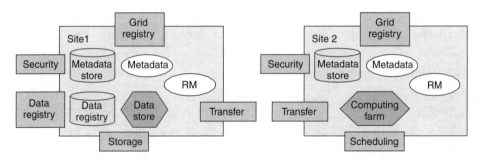

Figure 15.3 A simple VO providing services at two sites, one site capable of only storage and another capable of only computing. Ovals represent factories, rectangles represent exposed services. The replica manager (RM) factory may spawn other high-level data management services, see Figure 15.2.

points to may have changed, but semantically it should still be the same data as it was 'forever before'. Information on migration of the data might be kept in the metadata catalogs as well if it is desirable.

15.3.3.5 A VO example

What is the minimal set of services that a Grid node should have? It depends of course on what kind of resources the node site offers. If the site has only storage capabilities, it will need most of the data management services discussed above. If it only has computing capabilities but no storage capabilities, then it needs only a subset of those, but it will need job management services instead. A site offering both storage and computing capabilities will require both data and job management services.

The example shown in Figure 15.3 shows a simple VO layout with two sites and a possible set of services.

Of course, there may be many other services such as monitoring and management services that are not shown in the picture.

15.4 ISSUES

It is understood that the OGSA work has just begun and there is a long way to go before we have a complete architectural specification in which all the desired properties of Grids are addressed. This can only happen by having reference implementations and deployments of OGSA-compliant Grid middleware that will eventually expose the strengths and weaknesses of the architecture. OGSA will have to be refined and adjusted iteratively, but this is a natural and healthy process, and a very important fact is that the first steps have been taken.

In this section, shortcomings and open issues are discussed that indicate potential areas of future development. The current state of OGSA is analyzed by addressing each of the

Grid properties we listed in Section 15.2. We identify the areas that we feel are not adequately addressed by OGSA at this stage by referring to Tables 15.1 and 15.2.

15.4.1 Availability and robustness

By introducing the Factory pattern, the availability and robustness of a service can be greatly improved in the context of data management services, as discussed in Section 15.3. We have discussed how higher-level services might deal with failing or unavailable instances and start up new ones automatically.

SDEs that deal with failing instances and policies on how to restart them need to be introduced. Possible items include policies that would answer the following questions: What happens if a service has been unavailable for a given time? How is service overload dealt with? What if the network of a VO becomes partitioned? What happens if the Factory or the Registry is suddenly unavailable? This also touches somewhat on the desired property of scalability.

15.4.2 Scalability

By designing OGSA to explicitly support transient services and dynamic service creation and by using soft state updates, scalability seems to be assured. In Table 15.1, scalability is dealt with in requirements 1 and 7, which in turn are addressed by several interfaces (see Table 15.2). We will have to see in future implementations and deployments of OGSA where the bottlenecks are and whether any architectural issues arise due to an eventual limitation in scalability. As we have mentioned before, the robustness of the system needs to be ensured also in a scalable manner.

15.4.3 Monitorability

In the introduction of Reference [14], it is mentioned that each VO needs certain levels of QoS to be achieved and that they may be measured in many different ways. The ability to set up VOs that fulfill many different QoS requirements is highlighted as one of the most desirable properties of Grids. OGSA does not elaborate further on QoS metrics: The Grid property MEASURABLE is not addressed by any OGSA requirement in Table 15.1. We agree that this area is fundamental to the success of Grids in general and needs to be dealt with in the very near future.

There needs to be not only agreed metrics of QoS, but definitions from each Grid service on how it will enhance or decrease certain QoS metrics. We have outlined in the previous section how QoS might be measured for data management services; this needs to be done for all Grid services in detail. There might be the need to define a QoS namespace to be able to query this property of services more explicitly in the WSDL description of the GSR. Each service also needs to declare its own internal QoS metrics and to give a value in a specific instance in the case in which different instances of the same service can be set up such that the given metrics can change.

Measurability is also very important when a VO defines its own set of necessary services, or when it analyzes its own state, and optimizes its performance due to changes in its nature or requirements. In order to define, bootstrap and optimize a VO, it is essential to have QoS metrics by which the VO can measure itself and by which it can be measured by others, especially for billing purposes.

15.4.4 Integration

There is a need to integrate services not just within but also across VOs. OGSA solves this problem by defining the GridService interface and requiring all services to implement it. A lot of effort still needs to be put into the exact mechanisms and definition of common semantics in order that the integration of services (across VOs) may be achieved. In Data Grids, the integration of data access is also essential; hence our introduction of the concept of Grid Data References and the Data Registry.

15.4.5 Security

In the OGSA framework, the hosting environment has the burden of authentication – which is reasonable – but there is no discussion on how local and VO-wide security policies are enforced. Is there the need for a Grid service that deals with these issues, or should each of the services have an interface addressing this, making it part of the GridService base interface? By deferring this problem to a later time, the design decision is made that security needs to be dealt with at the protocol layer or by the higher-level services.

By relying on Web services, the strong industrial drive to come to a solution in this area will help speed up the process to design a suitable security infrastructure. Recent press releases by Microsoft and IBM have indicated the industry's commitment in this area. The release of WS-Security [28], however, only deals with SOAP-based protocols; there is no effort to set up a security infrastructure spanning all possible protocols. There needs to be a lot of effort put into this domain also from the Grid community to check how existing Grid security infrastructures might interoperate with Web service security mechanisms.

For Data Grids, authorization and accounting will be particularly complex for certain VOs. Experience shows that global authorization schemes do not work because local resources refuse to trust the global authorization authority to perform the authorization in accordance with the local policies – which may change over time. Also for logging, auditing, and accounting purposes the local resource managers will always want to know exactly who has done what to their resource. An issue is how to delegate rights to automated Grid services that need to use the resources on behalf of the user even if the user did not initiate their usage explicitly.

Security will have to be dealt with very soon within OGSA since it will depend on the success of the underlying security framework. Open questions include: How are VO-wide policies applied? How are local security policies enforced? What is the role of hosting environments? How is an audit performed? Can a user belong to more than one VO and use both resources even if the security mechanisms differ?

15.4.6 Interoperability and compatibility

Interoperability is explicitly mentioned as a requirement and is one of the driving concepts behind OGSA. Web services, including Grid services, are designed such that the modules are highly interoperable. There is no uniform protocol required that each service has to speak. WSDL descriptions are there to ensure interoperability.

The notification framework for passing messages between Grid services is addressing this property explicitly. It resembles the concepts of the Grid monitoring architecture [27], but without the possibility of registering notification sources with the target (sink) of the notification event. Interoperability is very closely related to discovery because services that need to interoperate have to discover among other things, which common protocols they can use and whether there are issues of compatibility.

15.4.7 Service discovery

As mentioned before, users of many services and services that want to interoperate need to get hold of the service descriptions to discover which services meet their needs or which services are still missing to achieve a given QoS. But how does the user know how to get hold of these descriptions? The OGSA's answer is the Registry and the HandleMap. The Registry needs to be searched to find the GSHs of the services that fulfill the user requirements – formulated in a query if necessary. The HandleMap then can be contacted to retrieve the detailed description of the services in question.

By holding a GSH one can get at the corresponding (WSDL) description and the HandleMap is bound to the HTTP(S) protocol to assure availability of the description without another necessary discovery step.

This whole mechanism, however, leads us back to the service discovery problem: How do we get the GSH of the relevant registries in the first place? There has been significant effort in the P2P community to provide robust and scalable protocols addressing this issue, like Chord [29].

This touches again on the issue of QoS metrics and SDEs. How is it possible to get the handles of the registries that we can query to get a set of services that we might be interested in, that is, how do we find the registry or registries relevant to a given query? How is a query formulated to do so? OGSA considers using XQuery [30, 31] to query the Service Data of the Registry, but then we need SDEs defining QoS. We tried to give a few examples for data management.

15.4.8 Manageability

Manageability is not dealt with explicitly in OGSA. The idea of unified monitoring and controlling mechanisms is there, but not further exploited. We have discussed the manageability issues before with respect to VO management.

15.5 SUMMARY

OGSA defines and standardizes a set of (mostly) orthogonal multipurpose communication primitives that can be combined and customized by specific clients and services to yield powerful behavior. OGSA defines standard interfaces (portTypes in WSDL terminology) for basic Grid services.

OGSA is an implementation independent specification. Several implementations of this specification may be offered. For example, a future Globus distribution will offer one or more reference implementations.

As with all complex systems, the devil is in the details. OGSA touches on the points of scalability, manageability and interoperability but there are many questions that remain unanswered or are deferred to a later time. How are QoS metrics defined and described? How does a VO get bootstrapped? How are VO-wide and local security policies enforced? How to deal with requirements of high availability and consistency? How do local resources – hosting environments – control and monitor their services and guard themselves against malicious abuse? It is understandable that OGSA is still in a very fluid state in which lots of ideas are tested and evaluated before making it into the specification.

While being incomplete for the time being, OGSA provides a solid basis for future Grid infrastructures. The basic elements of OGSA have been analyzed and summarized in this article and we have tried to point out its current strengths and limitations.

We gave an overview on Data Grids and the kinds of data they manage and have tried to motivate many basic services from first principles: data transfer, storage, management and metadata services. We have introduced the notion of the GDH, which is the unique logical identifier of the data and the GDR pointing to a physical instance and describing the data if necessary – in analogy to the GSH and GSR in OGSA. One of the real issues is how the data is actually located; we have introduced the notion of the Data Registry, which holds the GDH to GDR mappings. There are many issues with respect to data localization that we did not touch upon, but which have been dealt with elsewhere in the literature (see References [13, 31, 32–34] and references therein).

We have tried to apply OGSA to Data Grids and have seen that all the services can be deployed in the OGSA framework, which proves to be extensible and flexible enough to accommodate many different QoS requirements. All this is a very preliminary, high-level view of Data Grids and OGSA, which has to be refined and tested by actual implementations.

ACKNOWLEDGEMENTS

This work was inspired by many people, first and foremost by our friends and colleagues at CERN: Wolfgang Hoschek, Erwin Laure, Ben Segal, Heinz Stockinger, and Kurt Stockinger. We would also like to acknowledge Ann Chervenak, Ewa Deelman, Ian Foster, Carl Kesselman, Miron Livny, Arie Shoshani and Mike Wilde for many interesting discussions and insights.

Peter is most grateful to his family – Janka, Talia and Andor, for their incredible support.

REFERENCES

1. Foster, I. and Kesselman, C. (eds) (1999) *The Grid: Blueprint for a New Computing Infrastructure*. San Francisco, CA: Morgan Kaufmann Publishers.
2. Chervenak, A., Foster, I., Kesselman, C., Salisbury, C. and Tuecke, S. (2001) The data grid: towards an architecture for the distributed management and analysis of large scientific datasets. *Journal of Network and Computer Applications*, **23**, 187–200.

3. Allcock, W. *et al.* (2002) Data management and transfer in high performance computational grid environments. *Parallel Computing*, **28**(5), 749–771.
4. Tuecke, S., Czajkowski, K., Foster, I., Frey, J., Graham, S. and Kesselman, C. (2002) *Grid Service Specifications*, February 15, 2002, http://www.globus.org/ogsa.
5. Foster, I., Kesselman, C. and Tuecke, S. (2001) The anatomy of the grid. *International Journal of Supercomputer Applications*, **15**(3).
6. Christensen, E., Curbera, F., Meredith, G. and Weerawarana, S. (2001) Web Services Description Language 1.1, W3C Note 15, http://www.w3.org/TR/wsdl.
7. World Wide Web Consortium, Simple Object Access Protocol (SOAP) 1.1, W3C Note 8, 2000.
8. Brittenham, P. (2001) An Overview of the Web Services Inspection Language, http://www.ibm.com/developerworks/webservices/library/ws-wsilover.
9. Gannon, D. *et al. The Open Grid Services Architecture and Web Services*; Chapter 9 of this book.
10. EU DataGrid Data Management Workpackage WP2, http://cern.ch/grid-data-management/.
11. Guy, L., Laure, E., Kunszt, P., Stockinger, H. and Stockinger, K. *Data Replication in Data Grids*, Informational document submitted to GGF5, Replication-WG, http://cern.ch/grid-data-management/docs/ReplicaManager/ReptorPaper.pdf.
12. Stockinger, H., Samar, A., Allcock, B., Foster, I., Holtman, K. and Tierney, B. (2001) File and object replication in data grids, *Proceedings of the Tenth International Symposium on High Performance Distributed Computing (HPDC-10)*. San Francisco: IEEE Press.
13. Hoschek, W. (2002) The web service discovery architecture. *Proc. of the Intl. IEEE Supercomputing Conference (SC 2002)*, Baltimore, USA, November, 2002, see also his article in this volume; to appear.
14. Foster, I., Kesselman, C., Nick, J. M. and Tuecke, S. The physiology of the grid, *An Open Grid Services Architecture for Distributed Systems Integration*. http://www.globus.org/ogsa/, Chapter 8 of this book.
15. The Griphyn Project Whitepaper, http://www.griphyn.org/documents/white_paper/white_paper4.html.
16. Jones, B. (ed.), *The EU DataGrid Architecture*, EDG Deliverable 12.4, http://edms.cern.ch/document/333671.
17. Hoschek, W., Jaen-Martinez, J., Kunszt, P., Segal, B., Stockinger, H., Stockinger, K. and Tierney, B. *Data Management (WP2) Architecture Report*, EDG Deliverable 2.2, http://edms.cern.ch/document/332390.
18. Shoshani, A. *et al.* (2002) *SRM Joint Functional Design – Summary of Recommendations*, GGF4 Informational Document Toronto 2002, http://www.zib.de/ggf/data/Docs/GGF4-PA-Arie_Shoshani-SRM_Joint_func_design.doc.
19. Foster, I. and Kesselman, C. *A Data Grid Reference Architecture*, GRIPHYN-2001-12, http://www.griphyn.org/documents/document_server/technical_reports.html.
20. Baru, C., Moore, R., Rajasekar, A. and Wan, M. (1998) The SDSC storage resource broker. *Proc. CASCON '98 Conference*, 1998.
21. Luekig, L. *et al.* (2001) The D0 experiment data grid – SAM, *Proceedings of the Second International Workshop on Grid Computing GRID2001*. Denver: Springer-Verlag, pp. 177–184.
22. Abiteboul, S., Buneman, P. and Suciu, D. (1999) *Data on the Web: From Relations to Semistructured Data and XML*. San Francisco, CA: Morgan Kaufmann.
23. Leach, P. and Salz, R. (1997) *UUIDs and GUIDs*, Internet-Draft, ftp://ftp.isi.edu/internet-drafts/draft-leach-uuids-guids-00.txt, February 24, 1997.
24. Chervenak, A. *et al.* (2001) Giggle: A framework for constructing scalable replica location services. Presented at *Global Grid Forum*, Toronto, Canada, 2001.
25. Carman, M., Zini, F., Serafini, L. and Stockinger, K. (2002) Towards an economy-based optimisation of file access and replication on a data grid, *International Workshop on Agent based Cluster and Grid Computing at International Symposium on Cluster Computing and the Grid (CCGrid 2002)*. Berlin, Germany: IEEE Computer Society Press.

26. Pearlman, L., Welch, V., Foster, I., Kesselman, C. and Tuecke, S. (2001) A community authorization service for group collaboration. Submitted to *IEEE 3rd International Workshop on Policies for Distributed Systems and Networks*, 2001.

27. Tierney, B., Aydt, R., Gunter, D., Smith, W., Taylor, V., Wolski, R. and Swany, M. (2002) *A Grid Monitoring Architecture*, Grid Forum Working Draft, GWD-Perf-16-2, January, 2002.

28. Aitkinson, B. *et al.* (2002) Web Services Security (WS-Security), Version 1.0, April 5, 2002, http://www.verisign.com/wss/wss.pdf.

29. Stoica, I. *et al.* (2001) Chord: A scalable peer-to-peer lookup service for Internet applications, *Proc. ACM SIGCOMM*. New York: ACM Press, pp. 149–160.

30. World Wide Web Consortium, XQuery 1.0: An XML Query Language, W3C Working Draft, December, 2001.

31. Hoschek, W. (2002) *A Unified Peer-to-Peer Database Framework for XQueries over Dynamic Distributed Content and its Application For Scalable Service Discovery*, Ph.D. thesis, Technical University of Vienna, Austria, http://edms.cern.ch/file/341826/1/phd2002-hoschek.pdf.

32. Hoschek, W. (2002) Dynamic timeouts and neighbor selection queries in peer-to-peer networks. *Intl. Conf. on Networks, Parallel and Distributed Processing and Applications (NPDPA 2002)*, Tsukuba, Japan, October, 2002; to appear.

33. Aberer, K., Hauswirth, M., Punceva, M. and Schmidt, R. (2002) Improving data access in P2P systems. *IEEE Internet Computing*, **6**(1), 58–67.

34. Shoshani, A., Sim, A. and Gu, J. (2002) Storage resource managers: middleware components for grid storage. *Proceedings of the 19th IEEE Symposium on Mass Storage Systems*, 2002.

Virtualization services for Data Grids

Reagan W. Moore and Chaitan Baru

University of California, San Diego, California, United States

16.1 INTRODUCTION

The management of data within Grids is a challenging problem. It requires providing easy access to distributed, heterogeneous data that may reside in different 'administrative domains' and may be represented by heterogeneous data formats, and/or have different semantic meaning. Since applications may access data from a variety of storage repositories, for example, file systems, database systems, Web sites, document management systems, scientific databases, and so on, there is a need to define a higher-level abstraction for data organization. This is generally referred to as a *data collection*. A data collection contains named entities that may in actuality be stored in any of the repositories mentioned above. Data collections can be organized hierarchically, so that a collection may contain subcollections, and so on. An important issue in building collections is *naming*. Typically, the naming conventions are dependent upon the type of digital entities. While there are technologies to build uniform namespaces, they are typically different for each type of digital entity. For example, a Geographic Information System may use one approach to organize and name spatial data, database systems may use another approach for structured

Grid Computing – Making the Global Infrastructure a Reality.   Edited by F. Berman, A. Hey and G. Fox
© 2003 John Wiley & Sons, Ltd   ISBN: 0-470-85319-0

data, and file systems may use yet another for files. Data Grids must provide the capability to assemble (and, further, *integrate*) such disparate data into coherent collections.

This problem is further exacerbated owing to the fact that data collections can persist longer than their supporting software systems. Thus, the organization of data collections may need to be preserved across multiple generations of supporting infrastructure. There is a need for technologies that allow a data collection to be preserved while software evolves [1]. Data Grids must provide services, or mechanisms, to address both the data naming and the data persistence issues.

Data Grids provide a set of *virtualization services* to enable management and integration of data that are distributed across multiple sites and storage systems. These include services for organization, storage, discovery, and knowledge-based retrieval of digital entities – such as output from word processing systems, sensor data, and application output – and associated information. Some of the key services are *naming, location transparency, federation, and information integration*. Data Grid applications may extract information from data residing at multiple sites, and even different sets of information from the same data. Knowledge representation and management services are needed to represent the different semantic relationships among information repositories.

This chapter provides a survey of data management and integration concepts used in Data Grids.

Further details can be found in Reference [2, 3]. In the following sections, we define 'digital entities' in the Grid as combinations of data, information, and knowledge, and define the requirements for persistence (Section 16.2). We discuss issues in managing data collections at the levels of data, information, and knowledge. The state of the art in Data Grid technology is discussed, including the design of a persistent archive infrastructure, based upon the convergence of approaches across several different extant Data Grids (Section 16.3). Approaches to information integration are also described on the basis of data warehousing, database integration, and semantic-based data mediation (Section 16.4). We conclude in Section 16.5 with a statement of future research challenges.

16.2 DIGITAL ENTITIES

Digital entities are bit streams that can only be interpreted and displayed through a supporting infrastructure. Examples of digital entities include sensor data, output from simulations, and even output from word processing programs. Sensor data usually are generated by instruments that convert a numerical measurement into a series of bits (that can be interpreted through a data model). The output from simulation codes must also be interpreted, typically by applying a data format; the output from word processing programs requires interpretation by the program that generated the corresponding file. Digital entities inherently are composed of data, information (metadata tags), and knowledge in the form of logical relationships between metadata tags or structural relationships defined by the data model. More generally, it is possible for these relationships to be statistical, based, for example, on the results of knowledge discovery and data mining techniques.

Digital entities reside within a supporting infrastructure that includes software systems, hardware systems, and encoding standards for semantic tags, structural data models, and presentation formats. *The challenge in managing digital entities is not just the management of the data bits but also the management of the infrastructure* required to interpret, manipulate, and display these entities or images of reality [4].

In managing digital entities, we can provide Grid support for the procedures that generate these digital entities. Treating the processes used to generate digital entities as first-class objects gives rise to the notion of 'virtual' digital entities, or *virtual data*. This is similar to the notion of a *view* in database systems. Rather than actually retrieving a digital entity, the Grid can simply invoke the process for creating that entity, when there is a request to access such an entity. By careful management, it is possible to derive multiple digital entities from a single virtual entity. Managing this process is the focus of *virtual Data Grid* projects [5].

The virtual data issue arises in the context of long-term data persistence as well, when a digital entity may need to be accessed possibly years after its creation. A typical example is the preservation of the engineering design drawings for each aeroplane that is in commercial use. The lifetime of an aeroplane is measured in decades, and the design drawings typically have to be preserved through multiple generations of software technology. Either the application that was used to create the design drawings is preserved, in a process called *emulation*, or the information and the knowledge content is preserved in a process called *migration*. We can maintain digital entities by either preserving the processes used to create their information and knowledge content or by explicitly characterizing their information and knowledge content, and then preserving the characterization.

There is strong commonality between the software systems that implement *virtual Data Grids* and *emulation* environments. These approaches to data management focus on the ability to maintain the process needed to manipulate a digital entity, either by characterization of the process or by wrapping of the original application. There is also a strong commonality between the software systems that implement *Data Grids* and *migration* environments. In this discussion, we will show that similar infrastructure can be used to implement both emulation and migration environments. In emulation, a level of abstraction makes it possible to characterize the presentation application independent of the underlying operating system. Similarly in migration, for a given data model it is possible to characterize the information and knowledge content of a digital entity independent of the supporting software infrastructure. We will also look at the various extraction-transformation-load (ETL) issues involved in information integration across such multiple archives, across subdisciplines and disciplines. In the next section, we discuss the notion of long-term persistence of digital entities.

16.2.1 Long-term persistence

Digital entity management would be an easier task if the underlying software and hardware infrastructure remained invariant over time. With technological innovation, new infrastructure may provide lower cost, improved functionality, or higher performance. New systems appear roughly every 18 months. In 10 years, the underlying infrastructure may have evolved through six generations. Given that the infrastructure continues to evolve, one approach to digital entity management is to try to keep the interfaces between the infrastructure components invariant.

Standards communities attempt to encapsulate infrastructure through definition of an interface, data model, or protocol specification. Everyone who adheres to the specification uses the defined standard. When new software is written that supports the standard, all applications that also follow the standard can manipulate the digital entity. *Emulation* specifies a mapping from the original interface (e.g. operating system calls) to the new interface. Thus, emulation is a mapping between interface standards. *Migration* specifies a mapping from the original encoding format of a data model to a new encoding format. Thus, migration is a mapping between encoding format standards. Preservation can thus be viewed as the establishment of a mechanism to maintain mappings from the current interface or data model standard to the oldest interface or data model standard that is of interest.

We can characterize virtual Data Grids and digital libraries by the mappings they provide between either the digital entities and the generating application or the digital entity file name and the collection attributes. Each type of mapping is equivalent to a level of abstraction. A data management system can be specified by the levels of abstraction that are needed to support interoperability across all the software and hardware infrastructure components required to manipulate digital entities. Specification of the infrastructure components requires a concise definition of what is meant by data, information, and knowledge.

16.3 DATA, INFORMATION, AND KNOWLEDGE

It is possible to use computer science–based specifications to describe what data, information, and knowledge represent [6]. In the simplest possible terms, they may be described as follows:

- *Data* corresponds to the bits (zeroes and ones) that comprise a digital entity.
- *Information* corresponds to any semantic tag associated with the bits. The tags assign semantic meaning to the bits and provide context. The semantically tagged data can be extracted as attributes that are managed in a database as metadata.
- *Knowledge* corresponds to any relationship that is defined between information attributes or that is inherent within the data model. The types of relationships are closely tied to the data model used to define a digital entity. At a minimum, semantic/logical relationships can be defined between attribute tags and spatial/structural, temporal/procedural, and systemic/epistemological relationships can be used to characterize data models.

Data and information that are gathered and represented, as described above, at the subdisciplinary and disciplinary levels must then be integrated in order to create cross-disciplinary information and knowledge. Thus, information integration at all levels becomes a key issue in distributed data management.

16.3.1 A unifying abstraction

Typically, data are managed as files in a storage repository, information is managed as metadata in a database, and knowledge is managed as relationships in a knowledge repository. This is done at a subdisciplinary or disciplinary level (in science) or at an agency or department level (in government or private sector organizations). For each type of repository, mechanisms are provided for organizing and manipulating the associated digital entity. Files are manipulated in file systems, information in digital libraries, and knowledge in inference engines. A Data Grid defines the interoperability mechanisms for interacting with multiple versions of each type of repository. These levels of interoperability can be captured in a single diagram that addresses the ingestion, management, and access of data, information, and knowledge [7]. Figure 16.1 shows an extension of the two-dimensional diagram with a third dimension to indicate the requirement for information integration across various boundaries, for example, disciplines and/or organizations.

In two dimensions, the diagram is a 3 × 3 data management matrix that characterizes the data handling systems in the lowest row, the information handling systems in the middle row, and the knowledge handling systems in the top row. The ingestion mechanisms used to import digital entities into management systems are characterized in the left column. Management systems for repositories are characterized in the middle column and access systems are characterized in the right column. Each ingestion system, management system, and access system is represented as a rectangle. The interoperability mechanisms for mapping between data, information, and knowledge systems are represented by the Grid that interconnects the rectangles.

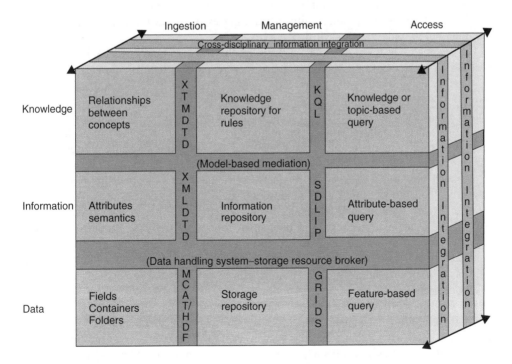

Figure 16.1 Characterization of digital entity management systems.

The rectangles in the left column of the data management matrix represent mechanisms for organizing digital entities. Data may be organized as fields in a record, aggregated into containers such as tape archive (tar) files, and then sorted into folders in a logical name-space. Information may be characterized as attributes, with associated semantic meanings that are organized in a schema. Knowledge is represented as relationships between con-cepts that also have associated semantic meanings that may be organized in a relationship schema or ontology [8]. The third dimension repeats the 3 × 3 matrix, but for different disciplinary databases or for information systems from different organizations. A major challenge is to define techniques and technologies for integration across layers in the third dimension, as discussed in Section 16.4.

The rectangles in the middle column represent instances of repositories. Data are stored as files in storage repositories such as file systems and archives and in databases as binary large objects or 'blobs'. Information is stored as metadata attributes in databases. Knowl-edge is stored as relationships in a knowledge repository. There are many possible choices for each type of repository. For instance, file systems tightly couple the management of attributes about files (location on disk, length, last update time, owner) with the man-agement of the files, and store the attributes as i-nodes intermixed with the data files. On the other hand, Data Grids need to use a logical namespace to organize and man-age attributes about digital entities stored at multiple sites [9]. The attributes could be stored in a database, with the digital entities stored in a separate storage system, such as an archive [10], with referential integrity between the two. The ability to separate the management of the attributes (information) from the management of the data makes it possible for Data Grids to build uniform namespaces that span multiple storage systems and, thereby, provide location virtualization services.

The rectangles in the right column represent the standard query mechanisms used for discovery and access to each type of repository. Typically, files in storage systems are accessed by using explicit file names. The person accessing a file system is expected to know the names of all of the relevant files in the file system. Queries against the files require some form of feature-based analysis performed by executing an application to determine if a particular characteristic is present in a file. Information in a database is queried by specifying operations on attribute–value pairs and is typically written in SQL. The person accessing the database is expected to know the names and meaning of all of the attributes and the expected ranges of attribute values. Knowledge-based access may rely upon the concepts used to describe a discipline or a business. An example is the emerging topic map ISO standard [11], which makes it possible to define terms that will be used in a business, and then map from these terms to the attribute names implemented within a local database. In this case, the knowledge relationships that are exploited are logical relationships between semantic terms. An example is the use of 'is a' and 'has a' logical relationships to define whether two terms are semantically equivalent or subordinate [12–15].

The Grid in the data management matrix that interconnects the rectangles represents the interoperability mechanisms that make up the levels of abstraction. The lower row of the Grid represents a data handling system used to manage access to multiple storage systems [16]. An example of such a system is the San Diego Supercomputer Center (SDSC) Storage Resource Broker (SRB) [17, 18]. The system provides a uniform storage system abstraction for accessing data stored in files' systems, archives, and binary large objects (blobs) in databases. The system uses a logical namespace to manage attributes about the data. This is similar to the 'simple federation' mentioned in Reference [2].

The upper row of the Grid in the data management matrix represents the mediation systems used to map from the concepts described in a knowledge space to the attributes used in a collection. An example is the SDSC model-based mediation system used to interconnect multiple data collections [19]. The terms used by a discipline can be organized in a concept space that defines their semantic or logical relationships. The concept space is typically drawn as a directed graph with the links representing the logical relationships and the nodes representing the terms. Links are established between the concepts and the attributes used in the data collections. By mapping from attribute names used in each collection to common terms defined in the concept space, it is possible to define the equivalent semantics between the attributes used to organize disparate collections. Queries against the concept space are then automatically mapped into SQL queries against the data collections, enabling the discovery of digital entities without having to know the names of the data collection attributes.

The left column of the Grid in the data management matrix represents the encoding standards used for each type of digital entity. For data, the Hierarchical Data Format version 5 (HDF) [20] may be used to annotate the data models that organize bits into files. The Metadata CATalog system (MCAT) [21] provides an encoding standard for aggregating files into containers before storage in an archive. For information annotation, the XML (extensible Markup Language) syntax [22] provides a standard markup language. The information annotated by XML can be organized into an XML schema. For relationships, there are multiple choices for a markup language. The Resource Description Framework (RDF) [23] may be used to specify a relationship between two terms. The ISO 13250 Topic Map standard [11] provides a way to specify typed associations (relationships) between topics (concepts) and occurrences (links) to attribute names in collections. Again the relationships can be organized in an XML Topic Map Document Type Definition.

The right column of the Grid in the data management matrix represents the standard access mechanisms that can be used to interact with a repository. For data access, a standard set of operations used by Data Grids is the Unix file system access operations (open, close, read, write, seek, stat, sync). Grid data access mechanisms support these operations on storage systems. For information access, the Simple Digital Library Interoperability Protocol (SDLIP) [24] provides a standard way to retrieve results from search engines. For knowledge access, there are multiple existing mechanisms, including the Knowledge Query Manipulation Language (KQML) [25], for interacting with a knowledge repository.

The interoperability mechanisms between information and data and between knowledge and information represent the levels of abstraction needed to implement a digital entity management system. Data, information, and knowledge management systems have been integrated into distributed data collections, persistent archives, digital libraries, and Data Grids. Each of these systems spans a slightly different portion of the 3 × 3 data management matrix.

Digital libraries are focused on the central row of the data management matrix and the manipulation and presentation of information related to a collection [26]. Digital libraries are now extending their capabilities to support the implementation of logical namespaces, making it possible to create a personal digital library that points to material that is stored somewhere else on the Web, as well as to material that is stored on your local disk. Web interfaces represent one form of knowledge access systems (the upper right hand corner of the Grid), in that they organize the information extraction methods, and organize the presentation of the results for a particular discipline. Examples are portals, such as the Biology Workbench [27], that are developed to tie together interactions with multiple Web sites and applications to provide a uniform access point.

Data Grids are focused on the lower row of the Grid and seek to tie together multiple storage systems and create a logical namespace [28]. The rows of the data management matrix describe different naming conventions. At the lowest level, a data model specifies a 'namespace' for describing the structure within a digital entity. An example is the METS standard [29] for describing composite objects, such as multimedia files. At the next higher level, storage systems use file names to identify all the digital entities on their physical media. At the next higher level, databases use attributes to characterize digital entities, with the attribute names forming a common namespace. At the next higher level, Data Grids use a logical namespace to create global, persistent identifiers that span multiple storage systems and databases. Finally, at the knowledge level, concept spaces are used to span multiple Data Grids [30].

16.3.2 Virtualization and levels of data abstraction

The management of persistence of data has been discussed through two types of mechanisms, emulation of the viewing application and migration of the digital entity to new data formats. Both approaches can be viewed as parts of a continuum, based upon the choice for data abstraction. From the perspective of emulation, software infrastructure is characterized not by the ability to annotate data, information, and knowledge but by the systems used to interconnect an application to storage and display systems. Figure 16.2 shows the levels of interoperability for emulation. Note that emulation must address not only possible changes in operating systems but also changes in storage systems and display systems.

Since any component of the hardware and software infrastructure may change over time, emulation needs to be able to deal with changes not only in the operating system calls but also in the storage and the display systems. An application can be wrapped to map the original operating system calls used by the application to a new set of operating system calls. Conversely, an operating system can be wrapped, by adding support for the old system calls, either as issued by the application or as used by old storage and display systems. Finally, the storage systems can be abstracted through use of a data handling system, which maps from the protocols used by the storage systems to the protocols required by the application. A similar approach can be used to build a display system abstraction that maps from the protocols required by the display to the protocols used by the application. Thus, the choice of abstraction level can be varied from the application, to the operating system, to the storage and display systems. Migration puts the abstraction

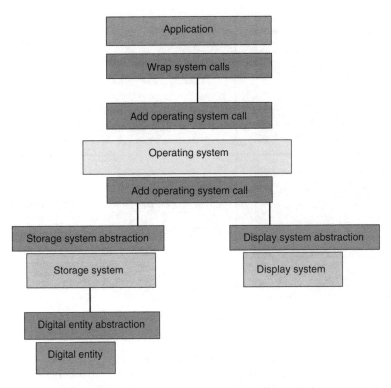

Figure 16.2 Levels of interoperability for emulation.

level at the data model. In this context, the transformative migration of a digital entity to a new encoding format is equivalent to wrapping the digital entity to create a new digital object.

Once an abstraction level is chosen, software infrastructure is written to manage the associated mappings between protocols. The abstraction level typically maps from an original protocol to a current protocol. When new protocols are developed, the abstraction level must be modified to correctly interoperate with the new protocol. Both the migration approach and the emulation approach require that the chosen level of abstraction be migrated forward in time, as the underlying infrastructure evolves. The difference in approach between emulation and migration is mainly concerned with the choice of the desired level of abstraction. We can make the choice more concrete by explicitly defining abstraction levels for data and knowledge. Similar abstraction levels can be defined for information catalogs in information repositories.

A Data Grid specifies virtualization services and a set of abstractions for interoperating with multiple types of storage systems. It is possible to define an abstraction for storage that encompasses file systems, databases, archives, Web sites, and essentially all types of storage systems, as shown in Figure 16.3.

Abstraction levels differentiate between the digital entity and the infrastructure that is used to store or manipulate the digital entity. In Figure 16.3, this differentiation is

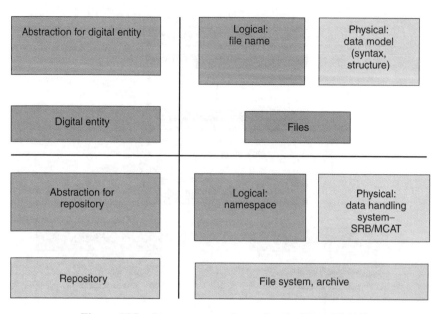

Figure 16.3   Storage system abstraction for Data Grids.

explicitly defined. The abstraction for the storage repository is presented in the bottom row of the 2 × 2 matrix. The abstraction for the digital entity that is put into the storage repository is presented in the upper row of the 2 × 2 matrix. The storage repository has a physical instantiation represented as the box labeled 'File System, Archive'. The abstraction for the physical repository has both a logical namespace used to reference items deposited into the storage as well as a set of physical operations that can be performed upon the items. Similarly, the digital entity has a physical instantiation, a logical namespace, and an associated set of physical operations. The name used to label the physical entity does not have to be the same as the name used to manage the physical entity within the storage system. The set of operations that can be performed upon the digital entity are determined by its data model and are typically supported by aggregating multiple operations within the physical storage system. This characterization of abstraction levels can be made concrete by separately considering the mechanisms used to support Data Grids that span multiple storage systems and the knowledge management systems that span multiple knowledge repositories.

The storage system abstraction for a Data Grid uses a logical namespace to reference digital entities that may be located on storage systems at different sites. The logical namespace provides a global identifier and maintains the mapping to the physical file names. Each of the data, the information, and the knowledge abstractions in a Data Grid introduces a new namespace for characterizing digital entities. For example, consider the following levels of naming:

- *Document level*: Definition of the structure of multimedia or compound documents through use of an [31] Archival Information Packet (Open Archival Information System

(OAIS)), or digital library Metadata Encoding and Transformation Standard (METS) attribute set, or Hierarchical Data Format version 5 (HDF5) file.

- *Data Grid level*: Mapping from local file names to global persistent *identifiers*.
- *Collection level*: Mapping from global persistent identifiers to collection attributes that describe provenance- and discipline-specific metadata.
- *Ontology level*: Mapping from collection attributes to discipline concepts.
- *Multidiscipline level*: Mapping between discipline concepts.

Each time a new namespace is introduced, access is provided to a wider environment, either to data spread across multiple storage systems, or to data stored in multiple collections, or to data organized by multiple disciplines. Each naming level specifies a set of relationships that can be captured in an ontology. The challenge is to provide the capability to map across multiple ontology levels in such a hierarchy. The use of multiple levels of naming conventions is quite powerful. Note that the mapping from logical namespace to physical namespace may be one to many in the case in which replicas of the file exist on multiple storage systems.

The Data Grid defines and implements a standard set of operations for the storage systems that are managed by the Grid. In doing so, it needs to develop 'wrappers' for these systems to ensure that all systems can respond to Grid requests. As an example, in the SRB, the set of operations are an extension of those provided by the Unix file system. A similar set of operations are being defined for the XIO Grid I/O interface for the Globus GridFTP transport program [32]. New capabilities, such as making a replica of a digital entity on a second storage system, creating a version of a digital entity, and aggregating digital entities into physical containers, have been added for performance as well as functionality to the SRB, for example, to facilitate long-term preservation.

It is essential to identify *core* versus *extended* virtualization services and ensure that the most general possible set of capabilities are provided. As discussed here, the use of a logical namespace to define unique identifiers is a key component. This makes it possible to separate manipulation of the namespace from interactions with the storage systems. The logical namespace can be extended to support registration of arbitrary digital entities. For example, in the Storage Resource Broker, digital entities that can be registered include files, directories, URLs, SQL command strings, and database tables [10].

A second observation of this level of abstraction is that the management of storage systems requires the use of information [16]. A logical file name needs to be assigned, in addition to the physical file name. The logical file name makes it possible to create a global, persistent identifier. Addition of capabilities such as replication, version numbers, access control, audit trails, and containers can be thought of as administration metadata that are managed as information attributes in the logical namespace. It is also possible to add structural metadata, defining how a digital entity is encoded (Multipurpose Internet Mail Extensions (MIME type)), preservation metadata, describing the provenance of the digital entity, and discipline metadata, describing the context-specific attributes that justify formation of a unique collection. These attributes are most effectively managed as information in a database.

A registration and discovery service is needed to enable registration of new catalogs and searching of catalogs to find information and data. The schema of the corresponding information repository does not have to be identical to the schema of the catalog. The information repository needs to be extensible, in order to allow the registration of new schemas and support addition of new attributes and mapping of attributes to new table structures. It is now possible to arrange digital entities in collections and to assemble a hierarchy of collections and subcollections, in which each subcollection is given a different set of attributes. It is also possible to define attributes that are unique to a single digital entity. The discovery service must be capable of accessing multiple database systems and querying multiple schemas.

The integration of multiple information repositories can be accomplished by defining semantic relationships between the attributes used in each collection. These relationships are managed in a knowledge repository that specifies rules for integration. In order to use multiple instantiations of knowledge repositories, a knowledge repository abstraction level is needed, as shown in Figure 16.4.

We face the same challenges when dealing with multiple knowledge repositories as we had with multiple storage systems. A major research challenge is the creation of a consensus on the management of relationships. As before, a logical description of a set of relationships can be defined as a schema. However, the syntax to use for annotating relationships currently has many contending standards. The Resource Description Framework (RDF) may be an adequate choice. The relationships are organized in a concept space, which describes the relationships between the concepts used as terms to describe a collection. An example is the development of a semantic domain map to describe how attribute names are logically related ('is a' and 'has a' relationships) to other attribute names.

Knowledge management is very closely tied to the data model used to describe the digital entity. Thus, if the digital entity includes English language terms, a concept space can be developed that describes the semantic relationships between the words present within

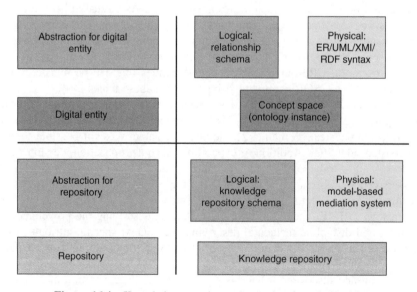

Figure 16.4 Knowledge repository abstraction for relationships.

the collection. If the digital entity includes spatial information, an atlas can be developed that describes the spatial relationships between the digital entities in the collection. If the digital entity includes temporal information, a process map can be developed that describes the procedural steps through which the digital entity was created. The management of knowledge relationships can be generalized by typing both the concepts (nodes of the graph) and the relationships (links between the nodes).

A knowledge management service needs to support

- annotation syntax for relationships,
- management system for organizing relationships,
- mapping from relationships to inference rules that can be applied to digital entities to discover if the relationship is present,
- logic (inference) engine for applying the inference rules.

The challenge is that standard services have not yet been defined for this area. In addition, the knowledge management community needs to agree upon a standard syntax for annotating relationships to enable development of widely used knowledge support utilities.

16.3.3 Data Grid infrastructure

The Data Grid community is developing a consensus on the fundamental capabilities that should be provided by Data Grids [33]. In addition, the Persistent Archive Research Group of the Global Grid Forum [34] is developing a consensus on the additional capabilities that are needed in Data Grids to support the implementation of a persistent archive. Distributed data management has been largely solved by the Data Grid community. Data Grids provide the data handling systems needed to manage a logical namespace and implement a storage system abstraction.

For example, the SDSC Storage Resource Broker is a data handling environment that is integrated with an information repository management system, the Extensible MCAT. Together, the systems implement a logical namespace, provide attribute management within the logical namespace, and manage digital entities across file systems (Windows, Unix, Linux, Mac OS X), archives (HPSS, DMF, ADSM, UniTree), databases (Oracle, DB2, Sybase), and FTP sites. The system is shown in Figure 16.5.

A storage abstraction is used to implement a common set of operations across archives, Hierarchical Resource Managers, file systems, and databases. The supported operations against files and database blobs include the standard Unix file system operations (create, open, close, unlink, read, write, seek, sync, stat, fstat, mkdir, rmdir, chmod, opendir, closedir, and readdir). The set of operations has been extended to support capabilities required by digital libraries and persistent archives. The extensions include aggregation of files into containers, replication of digital entities, and support for version numbers, audit trails, and access controls on each digital entity.

The SRB organizes the attributes for each digital entity within a logical namespace that is implemented in an object-relational database. The logical namespace supports organization of digital entities as collections/subcollections and supports soft links between subcollections. The types of digital entities that can be registered include files, URLs, SQL command strings, and databases.

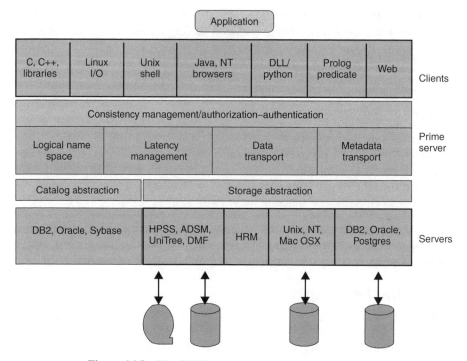

Figure 16.5   The SDSC storage resource broker Data Grid.

The SRB/MCAT system has been used to build distributed data collections, digital libraries, and persistent archives. In each case, the data management system has relied upon the ability of the underlying Data Grid technology to provide the levels of abstraction needed to handle multiple storage systems and heterogeneous information repositories. For example, a persistent archive can use replication to migrate data from an old storage system to a new storage system, and thus can use a Data Grid as the migration environment.

The European Data Grid project (EDG) [35] has similar goals for the management of distributed data. They implement a logical namespace that can be used to map global names to local file names, support replication of data files across multiple sites, provide query mechanisms to discover data by attribute values, and provide multiple latency management mechanisms ranging from caching to data streaming. The European Data Grid is primarily used as a federation mechanism to integrate multiple storage systems into a common data handling environment.

16.3.4 Data Grid projects

The Data Grid community has been developing software infrastructure to support distributed data collections for many scientific disciplines, including high-energy physics, chemistry, biology, Earth systems science, and astronomy. These systems are in production use, managing data collections that contain millions of digital entities and aggregate terabytes in size. It is noteworthy that across the many implementations, a common approach is emerging. There are a number of important projects in the areas of Data Grids and Grid toolkits. They include the SRB Data Grid described here, the European DataGrid replication environment (based upon GDMP, a project in common between the European DataGrid and the Particle Physics Data Grid, and augmented with an additional product of the European DataGrid for storing and retrieving metadata in relational databases called Spitfire and other components) (EDG), the Scientific Data Management [36] Data Grid from Pacific Northwest Laboratory (PNL) (SDM), the Globus toolkit [37], the SAM Sequential Access using Metadata Data Grid from Fermi National Accelerator Laboratory [38, 39], the Magda data management system from Brookhaven National Laboratory [40], and the JASMine Data Grid from Jefferson National Laboratory [41]. These systems have evolved as the result of input from their user communities for the management of data across heterogeneous, distributed storage resources.

EGP, SAM, Magda, and JASMine Data Grids support high-energy physics data. In addition, the EDG provides access to distributed data for Earth observation and computational biology communities. The SAM Data Grid is a file-based data management and access layer between storage and data processing layers that is used to support the D-Zero high energy physics experiment at Fermi National Accelerator Laboratory. Magda is a distributed data management system used to support the ATLAS high-energy physics experiment. Magda serves as a file catalog and an automated file replication tool between CERN (European Organization for Nuclear Research) and Brookhaven National Laboratory mass stores and the United States ATLAS Grid test bed sites. JASMine is a Mass Storage System Manager that is designed to meet the needs of current and future high data rate experiments at Jefferson Laboratory.

The SDM system provides a digital library interface to archived data for PNL and manages data from multiple scientific disciplines. The Globus toolkit provides services that can be composed to create a Data Grid. The SRB data handling system is used in projects for multiple US federal agencies, including the NASA Information Power Grid (digital library front end to archival storage), the DOE Accelerated Strategic Computing Initiative (collection-based data management), the National Library of Medicine Visible Embryo project (distributed data collection), the National Archives Records Administration (persistent archive), the National Science Foundation (NSF) National Partnership for Advanced Computational Infrastructure [42] (distributed data collections for astronomy, Earth systems science, and neuroscience), the Joint Center for Structural Genomics (Data Grid), and the National Institute of Health Biomedical Informatics Research Network (BIRN) (Data Grid).

The seven data-management systems listed above included not only Data Grids but also distributed data collections, digital libraries, and persistent archives. However, at the core of each system was a Data Grid that supported access to distributed data. By examining the capabilities provided by each system for data management, we can define the core capabilities that all the systems have in common. The systems that have the most diverse set of user requirements tend to have the largest number of features. Across the seven data management systems, a total of 152 capabilities were identified. Ninety percent of the capabilities are implemented in the SRB Data Grid. Over three-quarters of the capabilities

(120) were implemented in at least two of the seven data management systems. Two-thirds of the capabilities were either implemented or are planned for development in both the SRB and the EDG Data Grids. About one-third of the capabilities (50) have been implemented in at least five of the Data Grids. These 50 capabilities comprise the current core features of Data Grids [34].

The common Data Grid capabilities that are emerging across all of the Data Grids include implementation of a logical namespace that supports the construction of a uniform naming convention across multiple storage systems. The logical namespace is managed independently of the physical file names used at a particular site, and a mapping is maintained between the logical file name and the physical file name. Each Data Grid has added attributes to the namespace to support location transparency (access without knowing the physical location of the file), file manipulation, and file organization. Most of the Grids provide support for organizing the data files in a hierarchical directory structure within the logical namespace and support for ownership of the files by a community or collection ID.

The logical namespace attributes typically include the replica storage location, the local file name, and the user-defined attributes. Mechanisms are provided to automate the generation of attributes such as file size and creation time. The attributes are created synchronously when the file is registered into the logical namespace, but many of the Grids also support asynchronous registration of attributes.

Most of the Grids support synchronous replica creation and provide data access through parallel I/O. The Grids check transmission status and support data transport restart at the application level. Writes to the system are done synchronously, with standard error messages returned to the user. The error messages provided by the Grids to report problems are quite varied. Although all of the Grids provided some form of an error message, the number of error messages varied from less than 10 to over 1000 for the SRB. The Grids statically tuned the network parameters (window size and buffer size) for transmission over wide-area networks. Most of the Grids provide interfaces to the GridFTP transport protocol.

The most common access APIs to the Data Grids are a C++ I/O library, a command line interface, and a Java interface. The Grids are implemented as distributed client server architectures. Most of the Grids support federation of the servers, enabling third-party transfer. All of the Grids provide access to storage systems located at remote sites including at least one archival storage system. The Grids also currently use a single catalog server to manage the logical namespace attributes. All of the Data Grids provide some form of latency management, including caching of files on disk, streaming of data, and replication of files.

16.4 INFORMATION INTEGRATION

As described earlier, Data Grid technologies are becoming available for creating and managing data collections, information repositories, and knowledge bases. However, a new requirement emerges for a virtualization service that integrates information from different collections, repositories, and knowledge bases. Such a virtualization service provides a (virtual) single site view over a set of distributed, heterogeneous information sources. The motivation arises from the fact that human endeavors – whether in science, commerce, or security – are inherently becoming multidisciplinary in nature. 'Next-generation' applications in all of these areas require access to information from a variety of distributed information sources. The virtualization service may support *simple federation*, that is, federation of a set of distributed files and associated metadata or it may provide *complex semantic integration* for data from multiple domains.

The need for such integration was realized several years ago in some domains, for example, commercial and government applications, owing to the existence of multiple legacy, 'stovepiped' systems in these domains. Lacking a Data Grid infrastructure at the time, an approach that was followed was to create a separate resource, namely, the *data warehouse*, where data from multiple independent sources was 'physically' brought together into a single system [43]. Creating such warehouses is generally possible – though still difficult – when it is done within the bounds of a single organization or enterprise, for example, within a Fortune company, within a laboratory, or even within science disciplines. However, owing to organizational issues, creating data warehouses becomes difficult when the original sources are owned by different organizations, jurisdictions, or disciplines.

Data warehouses typically suffer from a 'latency' problem, that is, the data in the warehouse may be out of date with respect to the data at the source, depending on the nature (synchronous vs asynchronous) and frequency of updates. An alternate approach is to integrate information across a collection of sources by 'federating' such distributed information sources. While the data remains at its source, queries or requests from the client are sent to the appropriate sources and the integrated result is returned to the client. *Database integration* provides one approach to federation based on using database mediation software. A formal database model is employed for integration, including the use of a well-defined data model for representing data at the sources and a corresponding query language for integrating and querying this data [44].

Another approach to information integration is the *application integration* approach. While the objective is similar to database integration, that is, to integrate data and information from disparate, distributed sources, the techniques are based on a programming language approach rather than a database approach. While this affords more flexibility in integration and allows for greater degree of *ad hoc* integration of data, it also typically requires the use of an object model – to provide the necessary modeling capability – and a programming language (e.g. the Java object model and Java), rather than a data model and query algebra, as in the case of database integration.

Database integration approaches have traditionally focused on schema integration issues at the 'structural' level of the schema. For example, two different databases may contain the same concept of, say, an *Employee*, and may even employ the same name to describe this concept, except the concept may have been implemented as an entity in one database but as an attribute (of an entity) in the other. However, there are well-known approaches for handling some aspects of database integration at the semantic level as well. For example, in the above example, one database may have implemented the concept of *Employee,* while another may only contain the concept of *Graduate Student*. In order to integrate information from these two sources, one would then model the fact that *Graduate*

Student is a *specialization* of *Employee*. The concepts of *generalization* and *specialization* in data modeling were introduced more than two decades ago [45]. For example, in the above case, 'graduate student research assistant' *is_an* employee (specialization). In another instance, say, automobiles, ships, and aeroplanes can be generalized to the concept of 'vehicles'.

Nonetheless, the extant schema integration efforts focus on relatively simple contexts, for example, all employees in an organization (where an employee_ID or SSN can be used to cross-relate information from multiple databases), or the 'universe' of all types of transportation vehicles (cross-related on the basis of Vehicle_ID). More complex integration scenarios arise commonly in scientific applications. For example, geoscientists wish to ask a query about a geologic phenomenon (say, *plutons*) that requires integration of information from multiple disciplines, for example, petrology, geophysics, and stratigraphy [46]. Similarly, neuroscientists may wish to integrate structural brain information with functional brain information [47]. While the information integration request is across seemingly unrelated sources, integration is still possible on the basis of using a well-defined higher-level abstraction or concept, for example, 'the solid Earth' or 'the brain', which enables the integration of the lower-level sources. While the data from each of these sources is quite different in its details, conceptually, it is simple to understand that both sets of data are 'about' the solid Earth or about the brain.

In these scenarios, the source data usually originates from very different communities, for example, different labs, different subdisciplines and disciplines, or different jurisdictions and organizational entities. While the information is about the same higher-level abstraction, the originating communities that have created the sources have all typically evolved independently of one another and have not traditionally had much communication with each other. As a result, each may have developed terminologies that are quite distinct and different from the others even though they all refer to the same high-level concept or object. Therefore, the key requirement here is *semantic data integration*. A typical approach to semantic data integration is to define formal terminology or ontology structures and then define mappings from one structure to another using knowledge representation techniques and logic rules.

Logic-based, semantic data integration may be viewed as a 'horizontal integration' approach, since the integration is across databases that are all 'about' the same high-level concept, but employ disparate terminology, format, and type of content, for example, geologic maps, geochemistry data, gravity maps, and geochronology tables for plutons. An additional level of complexity is introduced in scenarios that require 'vertical integration' of data. There are increasingly more cases in which there is need to integrate data across scale. In bioinformatics applications, this includes integrating data from the molecular level to sequence, genome, protein, cell, tissue, organ, and so on. In these scenarios, the integration occurs across different concepts (molecule, genome, protein, cell). In geoinformatics, this includes integration of data from local and regional scale to continental and planetary scale. Another example is integration of social science data from local, city, state, to federal and international levels. Such scenarios require domain models that describe causal relationships, statistical linkages, and the various interactions that occur across these levels, to enable integration of information. Indeed, it would be impossible to integrate the information in the absence of such models. We refer to this approach as *model-based integration*, which requires statistical and probabilistic techniques and the ability to handle probabilistic relationships among data from different sources rather than just logical relationships.

16.4.1 Data warehousing

As mentioned earlier, data warehousing has been in existence for almost 10 years. Since all the data are brought into one physical location, it becomes possible to exploit this feature by developing efficient storage and indexing schemes for the centralized data. Early data warehouse software (e.g. Redbrick) demonstrated superiority over traditional relational database management systems by providing novel indexing schemes that supported efficient *on-line analytical processing (OLAP)* of large databases. This included support for basic forms of exploratory data analysis using techniques such as *drill-down* and *roll-up* of data along multiple data *dimensions*.

To create a warehouse, data is physically exported from source data systems. Thus, a copy of the necessary data is sent for incorporation into the warehouse. Typically, a defined subset, or 'view', of the source data is exported. Data warehousing is fundamentally an implementation approach. At the conceptual level, it is still necessary to address the issue of schema heterogeneity – which is also encountered in database integration (see below). A number of metadata management tools have been created to help with the process of data warehousing. For example, such tools would allow a warehouse designer to specify that, say, the 'Address' field is a single, text field (i.e. unstructured) in one data source, while it has been partially expanded (i.e. semistructured) in another source, and fully expanded (structured) in yet another source. In another instance, the designer may have to deal with data that is aggregated differently in different data sources, for example, sales data that is aggregated by shift, by day, or by week. The data warehouse technology platforms were initially specially designed software systems (e.g. the Redbrick Systems, which was subsequently acquired by Informix Corporation) that supported a relational data model interface and SQL enhanced with special operations. Subsequently, relational database systems themselves began to support the warehouse capability, including extensions to the SQL language to support OLAP operations.

In the scientific applications domain, there are several important examples of centralized data warehouses. These include the Protein Data Bank (PDB) [48], the Alliance for Cell Signaling (AFCS) [49], the Interuniversity Consortium for Political and Social Research (ICPSR) – which warehouses demographic survey data [50], and the Incorporated Research Institutions for Seismology (IRIS) [51]. An emerging approach is to package the data warehouse in a 'storage appliance' (say, IBM's Shark or Fast-T platform containing a set of FCS-attached Redundant Array of Inexpensive Disks (RAID) disks along with a set of processors and fast communications link) [52] and a software 'stack', for example, containing OS, database system, system management software, and Data Grid software (e.g. the SRB), and offer this package as a building block for a Data Grid, that is, a *GridBrick*.

16.4.2 Database and application integration

One of the drawbacks of warehouses is that the data can be out of date with respect to the actual data at the sources. Moreover, the warehouse model requires a higher level of commitment from each participating source, since each source is required to take on the additional task of periodically exporting its data to the warehouse.

Database integration and mediation techniques focus on integration of data from multiple sources without assuming that the data is exported from the sources into a single location. Instead, the database integration or mediation middleware performs the task of integrating the data across distributed sources. The database integration technique is predicated on the assumption that all data sources subscribe to a common data model (e.g. relational, object-oriented) or can at least export a *view* of their local data in a common model. The integration middleware is then used to define *integrated views*, which are published to the end user. In reality, since different sources may utilize different data models internally, it may become necessary to implement 'wrappers' at each source to convert source data into the common, mediator data model.

Thus, the key difference here is that sources export *views* rather than *data*. However, many of the conceptual issues in integrating data may be quite similar to the data warehouse case. Different types of schema heterogeneity need to be addressed, which can be done by transformations defined at both the wrapper level and the view integration level. The IBM DiscoveryLinks product supports database integration based on the relational data model [53]. Where necessary, wrappers are defined to convert local data into a relational (table-oriented) view. The *Xmediator* system from Enosys employs an XML Schema-based model for integration [54]. Wrappers map local data into XML schemas and the mediator employs an XML query language (XQuery) to define integrated views and query the database.

The database integration approach extends database technology to support information integration by exploiting the well-known notions of data models and query languages and is able to reuse a significant base of relational technology. The task of a 'database integrator' is, therefore, not much more complex than that of a 'traditional' database designer.

An example of database integration is in astronomy. A variety of sky surveys are being conducted to record and identify objects in the sky, for example, Sloan Digital Sky Survey, Digital Palomar Observatory Sky Survey, 2-micron All Sky Survey (SDSS, DPOSS, 2MASS) [55–57]. In each case, a sky catalog is constructed of all objects that have been identified in the particular survey. Objects are indexed in space either using a standard coordinate system or by mapping from the local coordinate system into a standard system. Users may be interested in issuing a query that selects a subset of objects, given coordinate ranges, and cross-correlates these objects across catalogs. Thus, a selection operation followed by a database join operation is needed to process this query, and an integrated view can be created that hides this detail from the end user.

Application integration techniques are also designed to provide an integrated view of disparate data sources and applications, to the end user. They differ in approach from database integration techniques in that they employ a programming language and associated 'data model' for integration. An object-oriented language, such as Java, is commonly used for this purpose. In this case, sources are 'wrapped' to return well-defined objects in the language model. Once the source data is represented as (Java) objects, arbitrary manipulation of these objects is possible using the programming language. The *Extensible Information Systems (XIS) platform* from *Polexis* provides this approach for information integration [58]. When wrapping source data, object behaviors as well as associated metadata can be defined for each type of object. These are referred to as *InfoBeans* in XIS. The software infrastructure can then manipulate and integrate the objects using the metadata and the object methods/behaviors. This approach may provide more flexibility than the database integration model since it does not depend on preexisting database views and employs a programming language. Thus, this approach may be more suited to *ad hoc* or *on-the-fly* integration of data. Owing to the programming language-based approach (versus database-based approach), it is generally possible to integrate a greater variety of data sources. There is no prerequisite to implement wrappers to convert to a common data model and support query processing. However, there is the requirement to write (Java) programs to integrate data, rather than just defining database views and using query languages. Thus, the integration task may take more effort and may not be as easily extensible.

The Information Integration Testbed [59] project at SDSC employs XML-based database integration techniques, using the Enosys XMediator along with information integration technology being developed at SDSC, which is focused on extensions to database mediation technology necessary for dealing with scientific databases, for example, differences in scale and resolution (e.g. for geospatial information), accuracy, statistical bases, ontologies, and semantics. The I2T test bed wraps sources in the form of Web services (see Figure 16.6). Thus, rather than exporting database views and query capabilities, sources publish one or more Web services. The Web Service Description Language (WSDL) standard is used to describe services and the Simple Object Access Protocol (SOAP) is used for service invocation. The Web services approach has some advantages since it provides a uniform interface to both data and computational services.

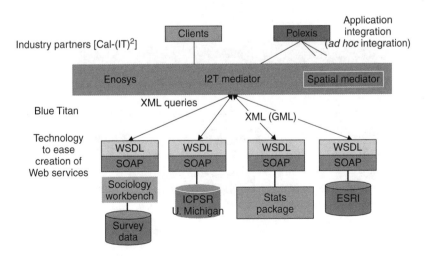

Figure 16.6   The Information integration test bed [59].

It can also be used to better control the types of queries/requests accepted by a source, thereby providing better control of the resources consumed at the source. The I2T system also provides application integration capabilities using the Polexis XIS system, to enable *ad hoc* integration of data. In many situations, end users may not know the particular integrated view of information they need until they have arrived at a particular computation. Since it is not possible to define all possible integrated database views in advance, there is a need for on-the-fly creation of integrated views.

The I2T test bed is being developed in cooperation with industry partners Enosys, Polexis, and Blue Titan. Blue Titan provides software infrastructure to help in wrapping sources using Web services. They also provide a common Web services development and implementation platform, that is, the equivalent of a Software Development Kit (SDK) for developing 'Web programs' [60].

Database mediation and information integration are very active areas of research with many research papers on the topic. In addition to the I2T system described here, another research project that is developing software in this area is Active Mediators for Information Integration (AMOS)s, which employs a novel approach of a 'web of mediators' [61]. As mentioned above, there are a limited number of industry products in this area (e.g. IBM DiscoveryLinks and Enosys XMediator), though this will probably be an area of active development in the future.

The I2T system is being used in the ClimDB project to integrate climatological data from several distributed field stations sites in the United States [62]. As Figure 16.7 indicates, each site may have a different method for storing information, and may even be collecting different information. Each source is wrapped using a uniform Web services interface. Individual sites typically also use different data formats. However, the site wrapper ensures that all data is returned in a standardized XML format, namely, the Ecological Metadata Language (EML) [63].

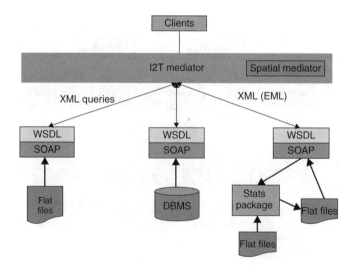

Figure 16.7   Application of I2T technology in the ClimDB project.

16.4.3 Semantic data integration

Semantic data integration is necessary in cases in which information is integrated across sources that have differing terminologies or ontologies. For example, in the Geosciences there is a need to integrate geophysics data with, say, stratigraphy, geochronology, and gravitation data. In the BIRN project, there is a requirement to integrate information from multiple human brain and mouse brain laboratories, and also to integrate human brain with mouse brain databases, from structural MRI images. There is also the requirement to integrate structural MRI images with functional MRI data. While all of the information relates to the brain, different human brain labs may be focused on different aspects of the human brain and, thus, employ disjoint terminologies. For example, one lab may represent an aspect of the brain in more detail than another. In the case of interspecies integration, the terminologies used to describe mouse brains and human brains may be quite different, even though the same brain structure is being described, since the brain researchers belong to different subdisciplines and each may have evolved their own terminologies.

Information integration in this case is based on developing 'conceptual models' for each source and linking these models to a global knowledge representation structure, which represents the encyclopedic knowledge in the domain (or an appropriate subset thereof). Formal methods of knowledge representation would enable the system to deduce linkages between seemingly disparate terminologies and ontologies using logic-based and rule-based systems. Figure 16.8 illustrates the semantic data integration infrastructure used

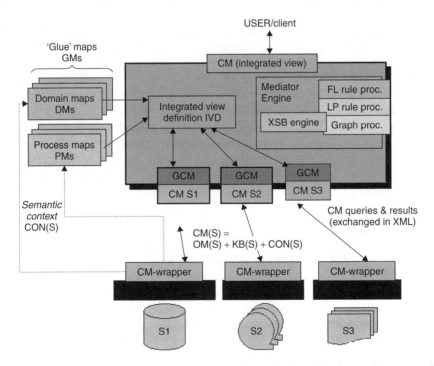

Figure 16.8 Architecture of a logic (or knowledge)-based information integration system [19].

to integrate neuroscience databases [19, 64, 65]. While Ludaescher [19] refers to this as 'model-based mediation', the term 'model-based' is used to describe logic-based models. In our nomenclature, we refer to this simply as 'semantic data integration', understanding that such integration requires logic-based (or knowledge-based) techniques and reserve the term 'model-based' for techniques that employ *domain models* for data integration.

16.4.4 Model-based integration

There is an increasing need in some scientific disciplines to integrate information 'across scale'. In bioinformatics applications, scientists are interested in integrating information from the molecular level to the sequence, genome, protein, cell, tissue, and even organ level. For example, there is interest in annotating (integrating) individual genes with related protein information. This type of integration goes beyond the capabilities of logic-based knowledge representation systems. Indeed, integration of information at this level may typically require the execution of computationally intensive models and analysis packages to find, say, sequence similarities, approximate matches, and causal and functional relationships among data at different levels of abstraction (e.g. genes and proteins). Model-based integration is an emerging requirement in many science application areas and requires the use of computationally intensive (scientific) models, statistical analysis, and data mining. Many examples can be found in the bioinformatics area, where pipelined processing of experimental data is the standard approach. An example of a project that integrates multilevel data (genomic and proteomic) using such pipelines is the *Encyclopedia of Life* project described in Chapter 40 in this book.

Supporting analysis pipelines requires a virtualization service that can support robust workflows in the Data Grid. Efforts are under way to standardize Web workflow specifications [66]. Given such a specification, the workflow virtualization service should be able to schedule the workflow processes on the Grid, taking care of prefetching data where necessary, reserving the necessary resources, and so on. Complex analysis pipelines will contain decision points, iterators, and loops, which are similar to control structures in a programming language. Thus, the virtualization service is essentially capable of scheduling and managing the execution of (distributed) Grid programs. Increasingly, the processes in the workflow/pipeline are Web services, so managing the execution of a pipeline is essentially the same as the problem of managing, or 'orchestrating', the execution of a set of Web services.

16.5 SUMMARY

Virtualization services are designed to hide the complexities of data management and data integration in Data Grids. They provide better usability and also support extensibility, since they hide the details of data formats, information structures, and knowledge representation from the Data Grid application. Thus, through the use of levels of abstraction for describing interactions with storage repositories, information repositories, and knowledge repositories, it is possible to implement data handling systems that span multiple sites, multiple administration domains, and even multiple disciplines. We note that the implementation of a level of abstraction can be done at any level of the data management software infrastructure. Thus, approaches can be created for persistence that range from process management (virtual Data Grids) to data migration. Similarly, approaches for the management of information can be created that range from centralized data warehouses (effectively data migration) to mediation systems that wrap existing information sources (effectively process management). The 'deep data modeling' required by scientific applications, that is, the modeling of implicit and hidden domain semantics, requires advanced data models and knowledge representation techniques in order to develop semantic structures that span subdisciplines and disciplines.

ACKNOWLEDGMENTS

The ideas presented here were developed by members of the Data and Knowledge Systems group at the San Diego Supercomputer Center. Michael Wan developed the data management systems, Arcot Rajasekar developed the information management systems, Bertram Ludaescher and Amarnath Gupta developed the logic-based integration and knowledge management systems, and Ilya Zaslavsky developed the spatial information integration and knowledge management systems. Richard Marciano created digital library and persistent archive prototypes. The data management system characterizations were only possible with the support of Igor Terekhov (Fermi National Accelerator Laboratory), Torre Wenaus (Brookhaven National Laboratory), Scott Studham (Pacific Northwest Laboratory), Chip Watson (Jefferson Laboratory), Heinz Stockinger and Peter Kunszt (CERN), Ann Chervenak (Information Sciences Institute, University of Southern California), and Arcot Rajasekar (San Diego Supercomputer Center). This research was supported by NSF NPACI ACI-9619020 (NARA supplement), NSF NSDL/UCAR Subaward S02-36645, NSF I2T EIA9983510, DOE SciDAC/SDM DE-FC02-01ER25486, and NIH BIRN-CC3~P41~RR08605-08S1. The views and conclusions contained in this document are those of the authors and should not be interpreted as representing the official policies, either expressed or implied, of the National Science Foundation, the National Archives and Records Administration, or the US government.

REFERENCES

1. Rajasekar, A., Marciano, R. and Moore, R. (1999) Collection based persistent archives. *Proceedings of the 16th IEEE Symposium on Mass Storage Systems*, March, 1999, pp. 176–184.
2. Moore, R., Baru, C., Rajasekar, A., Marciano, R. and Wan M. (1999) Data intensive computing, in Foster, I. and Kesselman, C. (eds) *The Grid: Blueprint for a New Computing Infrastructure*. San Francisco, CA: Morgan Kaufmann Publishers.
3. Raman, V., Narang, I., Crone, C., Haas, L., Malaika, S., Mukai, T., Wolfson, D. and Baru, C. (2002) *Data Access and Management Services on the Grid*, Technical Report, submission to Global Grid Forum 5, Edinburgh, Scotland, July 21–26, 2002, http://www.cs.man.ac.uk/grid-db/papers/dams.pdf.
4. Moore, R. *et al.* (2000) *Collection-based persistent digital archives – Parts 1& 2*, D-Lib Magazine, April/March, 2000, http://www.dlib.org/.
5. Grid Physics Network Project, http://www.griphyn.org/index.php.

6. Boisvert, R. and Tang, P. (2001) The architecture of scientific software, *Data Management Systems for Scientific Applications*. Norwell, MA: Kluwer Academic Publishers, pp. 273–284.
7. Moore, R. (2000) Knowledge-based persistent archives. *Proceedings of La Conservazione Dei Documenti Informatici Aspetti Organizzativi E Tecnici*, Rome, Italy, October, 2000.
8. Chen, C. (2001) Global digital library development, *Knowledge-based Data Management for Digital Libraries*. Beijing, China: Tsinghua University Press, pp. 197–204.
9. Stockinger, H., Rana, O., Moore, R. and Merzky, A. (2001) Data management for grid environments. *High Performance Computing and Networking (HPCN 2001)*, Amsterdam, Holland, June, 2001, pp. 151–160.
10. Rajasekar, A. and Moore, R. (2001) Data and metadata collections for scientific applications. *High Performance Computing and Networking (HPCN 2001)*, Amsterdam, NL, June, 2001, pp. 72–80.
11. ISO/IEC FCD 13250, 1999, http://www.ornl.gov/sgml/sc34/document/0058.htm.
12. Kifer, M., Lausen, G. and Wu, J. (1995) Logical foundations of object-oriented and frame-based languages. *Journal of the ACM*, **42**(4): 741, 843.
13. Baru, C., Chu, V., Gupta, A., Ludaescher, B., Marciano, R., Papakonstantinou, Y. and Velikhov, P. (1999) XML-Based information mediation for digital libraries. *ACM Conf. On Digital Libraries (exhibition program)*, 1999.
14. Paton, N. W., Stevens, R., Baker, P. G., Goble, C. A., Bechhofer, S. and Brass, A. (1999) Query processing in the TAMBIS bioinformatics source integration system, Ozsoyoglo, Z. M., *et al.* (eds), *Proc. 11th Intl. Conf. on Scientific and Statistical Database Management (SSDBM)*. Piscataway, NJ: IEEE Press, pp. 138–147.
15. Ludäscher, B., Marciano, R. and Moore, R. (2001) *Towards Self-Validating Knowledge-Based Archives, 11th Workshop on Research Issues in Data Engineering (RIDE)*, Heidelberg, Germany: IEEE Computer Society, pp. 9–16, RIDE-DM.
16. Moore, R., Baru, C., Bourne, P., Ellisman, M., Karin, S., Rajasekar, A. and Young S. (1997) Information based computing. *Proceedings of the Workshop on Research Directions for the Next Generation Internet*, May, 1997.
17. The Storage Resource Broker Web Page, http://www.npaci.edu/DICE/SRB/.
18. Baru, C., Moore, R., Rajasekar, A. and Wan, M. (1998) The SDSC storage resource broker. *Proc. CASCON '98 Conference*, November 30–December 3, Toronto, Canada, 1998.
19. Ludäscher, B., Gupta, A. and Martone, M. E. (2000) Model-based information integration in a neuroscience mediator system demonstration track, *26th Intl. Conference on Very Large Databases (VLDB)*, Cairo, Egypt, September, 2000.
20. Hierarchical Data Format, http://hdf.ncsa.uiuc.edu/.
21. The Metadata Catalog, http://www.npaci.edu/DICE/SRB/mcat.html.
22. Extensible Markup Language, http://www.w3.org/XML/.
23. Resource Description Framework (RDF). W3C Recommendation, www.w3.org/TR/REC-rdf-syntax, February, 1999.
24. Simple Digital Library Interoperability Protocol, http://www-diglib.stanford.edu/~testbed/doc2/SDLIP/.
25. Knowledge Query Manipulation Language, http://www.cs.umbc.edu/kqml/papers/.
26. Brunner, R., Djorgovski, S. and Szalay, A. (2000) *Virtual Observatories of the Future*. Astronomical Society of the Pacific Conference Series, Vol. 225. pp. 257–264.
27. The Biology Workbench, http://www.ncsa.uiuc.edu/News/Access/Releases/96Releases/960516.BioWork.html.
28. Baru, C. *et al.* (1998) A data handling architecture for a prototype federal application. *Sixth Goddard Conference on Mass Storage Systems and Technologies*, March, 1998.
29. Metadata Encoding and Transmission Standard, http://www.loc.gov/standards/mets/.
30. Moore, R. (2001) Knowledge-based grids. *Proceedings of the 18th IEEE Symposium on Mass Storage Systems and Ninth Goddard Conference on Mass Storage Systems and Technologies*, San Diego, April, 2001.
31. *Reference Model for an Open Archival Information System (OAIS)*, submitted as ISO Draft, http://www.ccsds.org/documents/pdf/CCSDS-650.0-R-1.pdf, 1999.

32. Allcock, W. E. (2002) Argonne National Laboratory.
33. Grid Forum Remote Data Access Working Group, http://www.sdsc.edu/GridForum/RemoteData/.
34. Moore, R. (2002) *Persistent Archive Concept Paper*, Global Grid Forum 5, Edinburgh, Scotland, July 21–26, 2002.
35. European Data Grid, http://eu-datagrid.web.cern.ch/eu-datagrid/.
36. Scientific Data Management in the Environmental Molecular Sciences Laboratory, http://www.computer.org/conferences/mss95/berard/berard.htm.
37. The Globus Toolkit, http://www.globus.org/toolkit/.
38. Terekhov, I. and White, V. (2000). Distributed data access in the sequential access model in the D0 run II data handling at Fermilab. *Proceedings of the 9th IEEE International Symposium on High Performance Distributed Computing*, Pittsburgh, PA, August, 2000.
39. Sequential Data Access Using Metadata, http://d0db.fnal.gov/sam/.
40. Manager for Distributed Grid-Based Data, http://atlassw1.phy.bnl.gov/magda/info.
41. Jefferson Laboratory Asynchronous Storage Manager, http://cc.jlab.org/scicomp/JASMine/.
42. NPACI Data Intensive Computing Environment Thrust Area, http://www.npaci.edu/DICE/.
43. Inmon, W. H. (2002) *Building the Data Warehouse*, 3rd ed., John Wiley & Sons.
44. Wiederhold, G. (1992) Mediators in the architecture of future information systems. *IEEE Computer*, **25**(8), 38–49 (Original Report.); reprinted in Michael Huhns and Munindar Singh: Readings in Agents; Morgan Kaufmann, San Francisco, CA, October, 1997, pp. 185–196.
45. Smith, J. M. and Smith, D. C. P. (1977) Database abstractions: aggregation and generalization, *ACM Transactions on Database Systems (TODS)*, **2**(2), 105–133.
46. Baru, C. and Sinha, K. (2002) *GEON: A Research Project to Create Cyberinfrastructure for the Geosciences*, SDSC; under preparation.
47. The Biomedical Informatics Research Network, http://www.nbirn.net.
48. The Protein Data Bank, http://www.rcsb.org/pdb.
49. Alliance for Cell Signaling, http://www.afcs.org.
50. Inter-University Consortium for Political and Social Research, http://www.icspr.umich.edu.
51. Incorporated Research Institutions for Seismology, http://www.iris.edu.
52. IBM TotalStorage Enterprise Storage Server, http://www.storage.ibm.com/hardsoft/products/ess/ess.htm.
53. IBM DiscoveryLink, http://www.ibm.com/solutions/lifesciences/discoverylink.hmtl.
54. Enosys XMediator, http://www.enosyssoftware.com.
55. Two Micron All Sky Survey, http://www.ipac.caltech.edu/2mass/.
56. Sloan Digital Sky Survey SkyServer, http://skyserver.sdss.org.
57. Digital Palomar Sky Survey, http://www.sdss.jhu.edu/~rrg/science/dposs/.
58. Polexis XIS, http://www.polexis.com.
59. The Information Integration Testbed Project, http://www.sdsc.edu/DAKS/I2T.
60. Blue Titan Software, www.bluetitan.com.
61. Active Mediators for Information Integration, http://www.dis.uu.se/~udbl/amos/amoswhite.html.
62. The Climate Database Project, http://www.fsl.orst.edu/climdb/.
63. Ecological Metadata Language, http://knb.ecoinformatics.org/software/eml/.
64. Gupta, A., Ludäscher, B., Martone, M. E. (2000) Knowledge-based integration of neuroscience data sources, *12th Intl. Conference on Scientific and Statistical Database Management (SSDBM)*. Berlin, Germany: IEEE Computer Society Press.
65. Ludäscher, B., Gupta, A. and Martone, M. E. (2001) Model-based mediation with domain maps, *17th Intl. Conference on Data Engineering (ICDE)*. Heidelberg, Germany: IEEE Computer Society Press.
66. Leymann, F. (2001) The Web Services Flow Language, http://www-3.ibm.com/software/solutions/webservices/pdf/WSFL.pdf, May, 2001.

The Semantic Grid: a future e-Science infrastructure

David De Roure, Nicholas R. Jennings, and Nigel R. Shadbolt

University of Southampton, Southampton, United Kingdom

17.1 INTRODUCTION

Scientific research and development has always involved large numbers of people, with different types and levels of expertise, working in a variety of roles, both separately and together, making use of and extending the body of knowledge. In recent years, however, there have been a number of important changes in the nature and the process of research. In particular, there is an increased emphasis on collaboration between large teams, an increased use of advanced information processing techniques, and an increased need to share results and observations between participants who are not physically co-located. When taken together, these trends mean that researchers are increasingly relying on computer and communication technologies as an intrinsic part of their everyday research activity. At present, the key communication technologies are predominantly e-mail and the Web. Together these have shown a glimpse of what is possible; however, to more fully support the e-Scientist, the next generation of technology will need to be much richer, more flexible and much easier to use. Against this background, this chapter focuses on the requirements, the design and implementation issues, and the

Grid Computing – Making the Global Infrastructure a Reality. Edited by F. Berman, A. Hey and G. Fox
© 2003 John Wiley & Sons, Ltd ISBN: 0-470-85319-0

research challenges associated with developing a computing infrastructure to support future e-Science.

The computing infrastructure for e-Science is commonly referred to as *the Grid* [1] and this is, therefore, the term we will use here. This terminology is chosen to connote the idea of a 'power grid': that is, that e-Scientists can plug into the e-Science computing infrastructure like plugging into a power grid. An important point to note, however, is that the term 'Grid' is sometimes used synonymously with a networked, high-performance computing infrastructure. While this aspect is certainly an important enabling technology for future e-Science, it is only a part of a much larger picture that also includes information handling and support for knowledge processing within the e-Scientific process. It is this broader view of the e-Science infrastructure that we adopt in this document and we refer to this as the *Semantic Grid* [2]. Our view is that as the Grid is to the Web, so the Semantic Grid is to the Semantic Web [3, 4]. Thus, the Semantic Grid is characterised as an open system in which users, software components and computational resources (all owned by different stakeholders) come and go on a continual basis. There should be a high degree of automation that supports flexible collaborations and computation on a global scale. Moreover, this environment should be personalised to the individual participants and should offer seamless interactions with both software components and other relevant users.[1]

The Grid metaphor intuitively gives rise to the view of the e-Science infrastructure as a set of services that are provided by particular individuals or institutions for consumption by others. Given this, and coupled with the fact that many research and standards activities are embracing a similar view [5], we adopt a *service-oriented view* of the Grid throughout this document (see Section 17.3 for a more detailed justification of this choice). This view is based upon the notion of various *entities* (represented as software agents) providing *services* to one another under various forms of *contract* (or service level agreement) in various forms of *marketplace*.

Given the above view of the scope of e-Science, it has become popular to characterise the computing infrastructure as consisting of three conceptual layers:[2]

- *Data/computation*: This layer deals with the way that computational resources are allocated, scheduled and executed and the way in which data is shipped between the various processing resources. It is characterised as being able to deal with large volumes of data, providing fast networks and presenting diverse resources as a single metacomputer. The data/computation layer builds on the physical 'Grid fabric', that is, the underlying network and computer infrastructure, which may also interconnect scientific equipment. Here data is understood as uninterpreted bits and bytes.
- *Information*: This layer deals with the way that information is represented, stored, accessed, shared and maintained. Here information is understood as data equipped with

---

[1] Our view of the Semantic Grid has many elements in common with the notion of a 'collaboratory' [58]: a centre without walls, in which researchers can perform their research without regard to geographical location – interacting with colleagues, accessing instrumentation, sharing data and computational resource, and accessing information in digital libraries. We extend this view to accommodate 'information appliances' in the laboratory setting, which might, for example, include electronic logbooks and other portable devices.

[2] The three-layer Grid vision is attributed to Keith G. Jeffery of CLRC, who introduced it in a paper for the UK Research Councils Strategic Review in 1999.

meaning. For example, the characterisation of an integer as representing the temperature of a reaction process, the recognition that a string is the name of an individual.

• *Knowledge*: This layer is concerned with the way that knowledge is acquired, used, retrieved, published, and maintained to assist e-Scientists to achieve their particular goals and objectives. Here knowledge is understood as information applied to achieve a goal, solve a problem or enact a decision. In the Business Intelligence literature, knowledge is often defined as actionable information. For example, the recognition by a plant operator that in the current context a reaction temperature demands shutdown of the process.

There are a number of observations and remarks that need to be made about this layered structure. Firstly, all Grids that have or will be built have some element of all three layers in them. The degree to which the various layers are important and utilised in a given application will be domain dependent – thus, in some cases, the processing of huge volumes of data will be the dominant concern, while in others the knowledge services that are available will be the overriding issue. Secondly, this layering is a conceptual view of the system that is useful in the analysis and design phases of development. However, the strict layering may not be carried forward to the implementation for reasons of efficiency. Thirdly, the service-oriented view applies at all the layers. Thus, there are services, producers, consumers, and contracts at the computational layer, at the information layer, and at the knowledge layer (Figure 17.1).

Although this view is widely accepted, to date most research and development work in this area has concentrated on the data/computation layer and on the information layer. While there are still many open problems concerned with managing massively distributed computations in an efficient manner and in accessing and sharing information from heterogeneous sources (see Chapter 3 for more details), we believe the full potential of Grid computing can only be realised by fully exploiting the functionality and capabilities provided by knowledge layer services. This is because it is at this layer that the reasoning necessary for seamlessly automating a significant range of the actions and interactions takes place. Thus, this is the area we focus on most in this chapter.

The remainder of this chapter is structured in the following manner. Section 17.2 provides a motivating scenario of our vision for the Semantic Grid. Section 17.3 provides a justification of the service-oriented view for the Semantic Grid. Section 17.4 concentrates

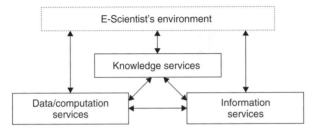

Figure 17.1 Three-layered architecture viewed as services.

on knowledge services. Section 17.5 concludes by presenting the main research challenges that need to be addressed to make the Semantic Grid a reality.

17.2 A SEMANTIC GRID SCENARIO

To help clarify our vision of the Semantic Grid, we present a motivating scenario that captures what we believe are the key characteristics and requirements of future e-Science environments. We believe this is more instructive than trying to produce an all-embracing definition.

This scenario is derived from talking with e-Scientists across several domains including the physical sciences. It is not intended to be domain-specific (since this would be too narrow) and at the same time it cannot be completely generic (since this would not be detailed enough to serve as a basis for grounding our discussion). Thus, it falls somewhere in between. Nor is the scenario science fiction – these practices exist today, but on a restricted scale and with a limited degree of automation. The scenario itself (Figure 17.2) fits with the description of Grid applications as 'coordinated resource sharing and problem solving among dynamic collections of individuals' [6].

The sample arrives for analysis with an ID number. The technician logs it into the database and the information about the sample appears (it had been entered remotely when the sample was taken). The appropriate settings are confirmed and the sample is placed with the others going to the analyser (a piece of laboratory equipment). The analyser runs automatically and the output of the analysis is stored together with a record of the parameters and laboratory conditions at the time of analysis.

The analysis is automatically brought to the attention of the company scientist who routinely inspects analysis results such as these. The scientist reviews the results from

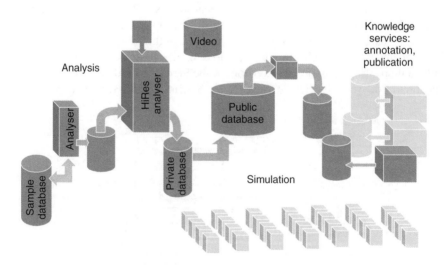

Figure 17.2 Workflow in the scenario.

their remote office and decides the sample needs further investigation. They request a booking to use the High Resolution Analyser and the system presents configurations for previous runs on similar samples; given this previous experience the scientist selects appropriate parameters. Prior to the booking, the sample is taken to the analyser and the equipment recognizes the sample identification. The sample is placed in the equipment which configures appropriately, the door is locked and the experiment is monitored by the technician by live video then left to run overnight; the video is also recorded, along with live data from the equipment. The scientist is sent a URL to the results.

Later the scientist looks at the results and, intrigued, decides to replay the analyser run, navigating the video and associated information. They then press the query button and the system summarises previous related analyses reported internally and externally, and recommends other scientists who have published work in this area. The scientist finds that their results appear to be unique.

The scientist requests an agenda item at the next research videoconference and publishes the experimental information for access by their colleagues (only) in preparation for the meeting. The meeting decides to make the analysis available for the wider community to look at, so the scientist then logs the analysis and associated metadata into an international database and provides some covering information. Its provenance is recorded. The availability of the new information prompts other automatic processing and a number of databases are updated; some processing of this new information occurs.

Various scientists who had expressed interest in samples or analyses fitting this description are notified automatically. One of them decides to run a simulation to see if they can model the sample, using remote resources and visualizing the result locally. The simulation involves the use of a problem-solving environment (PSE) within which to assemble a range of components to explore the issues and questions that arise for the scientist. The parameters and results of the simulations are made available via the public database. Another scientist adds annotation to the published information.

This scenario draws out a number of underlying assumptions and raises a number of requirements that we believe are broadly applicable to a range of e-Science applications:

- *Storage*: It is important that the system is able to store and process potentially huge volumes of content in a timely and efficient fashion.
- *Ownership*: Different stakeholders need to be able to retain ownership of their own content and processing capabilities, but there is also a need to allow others access under the appropriate terms and conditions.
- *Provenance*: Sufficient information is stored so that it is possible to repeat the experiment, reuse the results, or provide evidence that this information was produced at this time (the latter may involve a third party).
- *Transparency*: Users need to be able to discover, transparently access and process relevant content wherever it may be located in the Grid.
- *Communities*: Users should be able to form, maintain and disband communities of practice with restricted membership criteria and rules of operation.

- *Fusion*: Content needs to be able to be combined from multiple sources in unpredictable ways according to the users' needs; descriptions of the sources and content will be used to combine content meaningfully.
- *Conferencing*: Sometimes it is useful to see the other members of the conference, and sometimes it is useful to see the artefacts and visualisations under discussion.
- *Annotation*: From logging the sample through to publishing the analysis, it is necessary to have annotations that enrich the description of any digital content. This metacontent may apply to data, information or knowledge and depends on agreed interpretations.
- *Workflow*: To support the process enactment and automation, the system needs descriptions of processes. The scenario illustrates workflow both inside and outside the company.
- *Notification*: The arrival of new information prompts notifications to users and initiates automatic processing.
- *Decision support*: The technicians and scientists are provided with relevant information and suggestions for the task at hand.
- *Resource reservation*: There is a need to ease the process of resource reservation. This applies to experimental equipment, collaboration (the conference), and resource scheduling for the simulation.
- *Security*: There are authentication, encryption and privacy requirements, with multiple organisations involved, and a requirement for these to be handled with minimal manual intervention.
- *Reliability*: The systems appear to be reliable but in practice there may be failures and exception handling at various levels, including the workflow.
- *Video*: Both live and stored video have a role, especially where the video is enriched by associated temporal metacontent (in this case to aid navigation).
- *Smart laboratory*: For example, the equipment detects the sample (e.g. by barcode or RFID tag), the scientist may use portable devices for note taking, and visualisations may be available in the lab.
- *Knowledge*: Knowledge services are an integral part of the e-Science process. Examples include: finding papers, finding people, finding previous experimental design (these queries may involve inference), annotating the uploaded analysis, and configuring the lab to the person.
- *Growth*: The system should support evolutionary growth as new content and processing techniques become available.
- *Scale*: The scale of the scientific collaboration increases through the scenario, as does the scale of computation, bandwidth, storage, and complexity of relationships between information.

17.3 A SERVICE-ORIENTED VIEW

This section expands upon the view of the Semantic Grid as a service-oriented architecture in which *entities* provide *services* to one another under various forms of *contract*.[3] Thus,

---

[3] This view pre-dates the work of Foster *et al.* on the Open Services Grid Architecture [59]. While Foster's proposal has many similarities with our view, he does not deal with issues associated with developing services through autonomous agents, with

as shown in Figure 17.1, the e-Scientist's environment is composed of data/computation services, information services, and knowledge services. However, before we deal with the specifics of each of these different types of service, it is important to highlight those aspects that are common since this provides the conceptual basis and rationale for what follows. To this end, Section 17.3.1 provides the justification for a *service-oriented* view of the different layers of the Semantic Grid. Section 17.3.2 then addresses the technical ramifications of this choice and outlines the key technical challenges that need to be overcome to make service-oriented Grids a reality. The section concludes (Section 17.3.3) with the e-Science scenario of Section 17.2 expressed in a service-oriented architecture.

17.3.1 Justification of a service-oriented view

Given the set of desiderata and requirements from Section 17.2, a key question in designing and building Grid applications is what is the most appropriate conceptual model for the system? The purpose of such a model is to identify the key constituent components (abstractions) and specify how they are related to one another. Such a model is necessary to identify generic Grid technologies and to ensure that there can be reuse between different Grid applications. Without a conceptual underpinning, Grid endeavours will simply be a series of handcrafted and *ad hoc* implementations that represent point solutions.

To this end, an increasingly common way of viewing many large systems (from governments, to businesses, to computer systems) is in terms of the *services* that they provide. Here a service can simply be viewed as an abstract characterization and encapsulation of some content or processing capabilities. For example, potential services in our exemplar scenario could be the equipment automatically recognising the sample and configuring itself appropriately, the logging of information about a sample in the international database, the setting up of a video to monitor the experiment, the locating of appropriate computational resources to support a run of the High Resolution Analyser, the finding of all scientists who have published work on experiments similar to those uncovered by our e-Scientist, and the analyser raising an alert whenever a particular pattern of results occurs (see Section 17.3.3 for more details). Thus, services can be related to the domain of the Grid, the infrastructure of the computing facility, or the users of the Grid – that is, at the data/computation layer, at the information layer, or at the knowledge layer (as per Figure 17.1). In all these cases, however, it is assumed that there may be multiple versions of broadly the same service present in the system.

Services do not exist in a vacuum; rather they exist in a particular institutional context. Thus, all services have an owner (or set of owners). The owner is the body (individual or institution) that is responsible for offering the service for consumption by others. The owner sets the terms and conditions under which the service can be accessed. Thus, for example, the owner may decide to make the service universally available and free to all on a first-come, first-served basis. Alternatively, the owner may decide to limit access to particular classes of users, to charge a fee for access and to have priority-based access. All options between these two extremes are also possible. It is assumed that in a given

---

the issue of dynamically forming service level agreements, nor with the design of marketplaces in which the agents trade their services.

system there will be multiple service owners (each representing a different stakeholder) and that a given service owner may offer multiple services. These services may correspond to genuinely different functionality or they may vary in the way that broadly the same functionality is delivered (e.g. there may be a quick and approximate version of the service and one that is more time consuming and accurate).

In offering a service for consumption by others, the owner is hoping that it will indeed attract consumers for the service. These consumers are the entities that decide to try and invoke the service. The purpose for which this invocation is required is not of concern here: it may be for their own private use, it may be to resell to others, or it may be to combine with other services.

The relationship between service owner and service consumer is codified through a service contract. This contract specifies the terms and conditions under which the owner agrees to provide the service to the consumer. The precise structure of the contract will depend upon the nature of the service and the relationship between the owner and the provider. However, examples of relevant attributes include the price for invoking the service, the information the consumer has to provide to the provider, the expected output from the service, an indication about when this output can be expected, and the penalty for failing to deliver according to the contract. Service contracts can be established by either an off-line or an on-line process depending on the prevailing context.

The service owners and service producers interact with one another in a particular environmental context. This environment may be common to all entities in the Grid (meaning that all entities offer their services in an entirely open marketplace). In other cases, however, the environment may be closed and the entrance may be controlled (meaning that the entities form a private club).[4] In what follows, a particular environment will be called a *marketplace* and the entity that establishes and runs the marketplace will be termed the *market owner*. The rationale for allowing individual marketplaces to be defined is that they offer the opportunity to embed interactions in an environment that has its own set of rules (both for membership and ongoing operation) and they allow the entities to make stronger assumptions about the parties with which they interact (e.g. the entities may be more trustworthy or cooperative since they are part of the same club). Such marketplaces may be appropriate, for example, if the nature of the domain means that the services are particularly sensitive or valuable. In such cases, the closed nature of the marketplace will enable the entities to interact more freely because of the rules of membership.

To summarise, the key components of a service-oriented architecture are as follows (Figure 17.3): service owners (rounded rectangles) that offer services (filled circles) to service consumers (filled triangles) under particular contracts (solid links between producers and consumers). Each owner-consumer interaction takes place in a given marketplace (denoted by ovals) whose rules are set by the market owner (filled cross). The market owner may be one of the entities in the marketplace (either a producer or a consumer) or it may be a neutral third party.

---

[4] This is analogous to the notion of having a virtual private network overlaid on top of the Internet. The Internet corresponds to the open marketplace in which anybody can participate and the virtual private network corresponds to a closed club that can interact under its own rules.

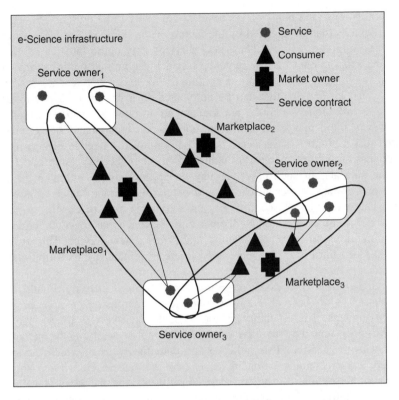

Figure 17.3 Service-oriented architecture: key components.

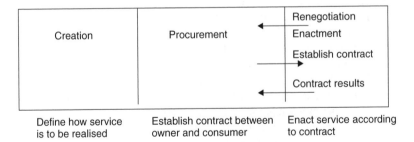

Figure 17.4 Service life cycle.

Given the central role played by the notion of a service, it is natural to explain the operation of the system in terms of a *service life cycle* (Figure 17.4). The first step is for service owners to define a service they wish to make available to others. The reasons for wanting to make a service available may be many and varied – ranging from altruism, through necessity, to commercial benefit. It is envisaged that in a given Grid application all three motivations (and many others besides) are likely to be present, although perhaps to varying degrees that are dictated by the nature of the domain. *Service creation* should be seen as an ongoing activity. Thus, new services may come into the environment at any time and existing ones may be removed (service decommissioning) at any time. This means that the system is in a state of continual flux and never reaches a steady state. Creation is also an activity that can be automated to a greater or lesser extent. Thus, in some cases, all services may be put together in an entirely manual fashion. In other cases, however, there may be a significant automated component. For example, it may be decided that a number of services should be combined, either to offer a new service (if the services are complementary in nature) or to alter the ownership structure (if the services are similar). In such cases, it may be appropriate to automate the processes of finding appropriate service providers and of getting them to agree to new terms of operation. This dynamic service composition activity is akin to creating a new *virtual organisation*: a number of initially distinct entities can come together, under a set of operating conditions, to form a new entity that offers a new service. This grouping will then stay in place until it is no longer appropriate to remain in this form, whereupon it will disband.

The *service creation* process covers three broad types of activity. Firstly, specifying how the service is to be realized by the service owner using an appropriate service description language. These details are not available externally to the service consumer (i.e. they are encapsulated by the service owner). Secondly, specifying the metainformation associated with the service. This indicates the potential ways in which the service can be procured. This metainformation indicates who can access the service and what are the likely contract options for procuring it. Thirdly, making the service available in the appropriate marketplace. This requires appropriate service advertising and registration facilities to be available in the marketplace.

The *service procurement* phase is situated in a particular marketplace and involves a service owner and a service consumer establishing a contract for the enactment of the service according to a particular set of terms and conditions. There are a number of points to note about this process. Firstly, it may fail. That is, for whatever reason, a service owner may be unable or unwilling to provide the service to the consumer. Secondly, in most cases, the service owner and the service consumer will represent different and autonomous stakeholders. Thus, the process by which contracts are established will be some form of *negotiation* – since the entities involved need to come to a mutually acceptable agreement on the matter. If the negotiation is successful (i.e. both parties come to an agreement), then the outcome of the procurement is a contract between the service owner and the service consumer. Thirdly, this negotiation may be carried out off-line by the respective service owners or it may be carried out at run time. In the latter case, the negotiation may be automated to a greater or lesser extent – varying from the system merely by automatically flagging the fact that a new service contract needs to be established to automating the entire negotiation process.[5]

---

[5] Automated negotiation technology is now widely used in many e-Commerce applications [60]. It encompasses various forms of auctions (a one-to-many form of negotiation) as well as bi-lateral negotiations. Depending on the negotiation protocol that is in place, the negotiation can be concluded in a single round or it may last for many rounds. Thus negotiation need not be a lengthy process; despite the connotation from human interactions that it may be!

The final stage of the service life cycle is *service enactment*. Thus, after having established a service contract, the service owner has to undertake the necessary actions in order to fulfil its obligations as specified in the contract. After these actions have been performed, the owner needs to fulfil its reporting obligations to the consumer with respect to the service. This may range from a simple inform indicating that the service has been completed, to reporting back complex content that represents the results of performing the service. The above assumes that the service owner is always able to honour the contracts that it establishes. However, in some cases the owner may not be able to stick to the terms specified in the contract. In such cases, it may have to renegotiate the terms and conditions of the contract, paying any penalties that are due. This enforcement activity is undertaken by the market owner and will be covered by the terms and conditions that the service providers and consumers sign up when they enter the marketplace.

Having described the key components of the service-oriented approach, we return to the key system-oriented desiderata noted in Section 17.2. From the above discussion, it can be seen that a service-oriented architecture is well suited to Grid applications:

- Able to store and process huge volumes of content in a timely fashion.
  - *The service-oriented model offers a uniform means of describing and encapsulating activities at all layers in the Grid. This model then needs to be underpinned by the appropriate processing and communication infrastructure to ensure it can deliver the desired performance.*
- Allow different stakeholders to retain ownership of their own content and processing capabilities, but to allow others access under the appropriate terms and conditions.
  - *Each service owner retains control over the services that they make available to others. They determine how the service is realized and set the policy for accessing the service.*
- Allow users to discover, transparently access and process relevant content wherever it may be located in the Grid.
  - *The overall system is simply viewed as a number of service marketplaces. Any physical distribution and access problems are masked via the service interface and the service contract. The marketplace itself has advertisement and brokering mechanisms to ensure appropriate service owners and consumers are put together.*
- Allow users to form, maintain, and disband communities of practice with restricted membership criteria and rules of operation.
  - *Each community can establish its own marketplace. The marketplace owner defines the conditions that have to be fulfilled before entities can enter, defines the rules of interaction for the marketplace once operational, and enforces the rules through appropriate monitoring.*
- Allow content to be combined from multiple sources in unpredictable ways according to the users' needs.
  - *It is impossible to a priori predict how the users of a system will want to combine the various services contained within it. Thus services must be such that they can be composed in flexible ways. This is achieved by negotiation of appropriate contracts. This composition can be done on a one-off basis or may represent a more permanent*

binding into a new service that is offered on an ongoing basis (as in the establishment of a new virtual organisation).

- Support evolutionary growth as new content and processing techniques become available.
  - *Services represent the unit of extension of the system. Thus, as new content or processing techniques become available they are simply represented as new services and placed in a marketplace(s). Also, new marketplaces can be added as new communities of practice emerge.*

17.3.2 Key technical challenges

The previous section outlined the service-oriented view of the Semantic Grid. Building upon this, this section identifies the key technical challenges that need to be overcome to make such architectures a reality. To this end, Table 17.1 represents the key functionality of the various components of the service-oriented architecture, each of which is then described in more detail in the remainder of this section.

17.3.2.1 Service owners and consumers as autonomous agents

A natural way to conceptualise the service owners and the service consumers are as autonomous agents. Although there is still some debate about exactly what constitutes agenthood, an increasing number of researchers find the following characterisation useful [7]:

An agent is an encapsulated computer system that is situated in some environment and that is capable of flexible, autonomous action in that environment in order to meet its design objectives.

There are a number of points about this definition that require further explanation. Agents are [8]: (1) clearly identifiable problem-solving entities with well-defined boundaries and interfaces, (2) situated (embedded) in a particular environment – they receive inputs related to the state of their environment through sensors and they act on the environment through effectors, (3) designed to fulfill a specific purpose – they have particular objectives (goals) to achieve, (4) autonomous – they have control both over their internal state and over their own behaviour,[6] and (5) capable of exhibiting flexible problem-solving

Table 17.1 Key functions of the service-oriented architecture components

Service owner	Service consumer	Marketplace
Service creation	Service discovery	Owner and consumer registration
Service advertisement		Service registration
Service contract creation	Service contract creation	Policy specification
Service delivery	Service result reception	Policy monitoring and enforcement

---

[6] Having control over their own behaviour is one of the characteristics that distinguishes agents from objects. Although objects encapsulate state and behaviour (more accurately behaviour realization), they fail to encapsulate behaviour activation or action choice. Thus, any object can invoke any publicly accessible method on any other object at any time. Once the method is invoked, the corresponding actions are performed. In this sense, objects are totally obedient to one another and do not have autonomy over their choice of action.

behaviour in pursuit of their design objectives – they need to be both reactive (able to respond in a timely fashion to changes that occur in their environment) and proactive (able to act in anticipation of future goals).

Thus, each service owner will have one or more agents acting on its behalf. These agents will manage access to the services for which they are responsible and will ensure that the agreed contracts are fulfilled. This latter activity involves the scheduling of local activities according to the available resources and ensuring that the appropriate results from the service are delivered according to the contract in place. Agents will also act on behalf of the service consumers. Depending on the desired degree of automation, this may involve locating appropriate services, agreeing to contracts for their provision, and receiving and presenting any received results.

17.3.2.2 Interacting agents

Grid applications involve multiple stakeholders interacting with one another in order to procure and deliver services. Underpinning the agents' interactions is the notion that they need to be able to interoperate in a meaningful way. Such semantic interoperation is difficult to obtain in Grids (and all other open systems) because the different agents will typically have their own individual information models. Moreover, the agents may have a different communication language for conveying their own individual terms. Thus, meaningful interaction requires mechanisms by which this basic interoperation can be effected (see Section 17.4.2 for more details).

Once semantic interoperation has been achieved, the agents can engage in various forms of interaction. These interactions can vary from simple information interchanges, to requests for particular actions to be performed and on to cooperation, coordination and negotiation in order to arrange interdependent activities. In all these cases, however, there are two points that qualitatively differentiate agent interactions from those that occur in other computational models. Firstly, agent-oriented interactions are conceptualised as taking place at the *knowledge level* [9]. That is, they are conceived in terms of which goals should be followed, at what time, and by whom. Secondly, as agents are flexible problem solvers, operating in an environment over which they have only partial control and observability, interactions need to be handled in a similarly flexible manner. Thus, agents need the computational apparatus to make *run-time* decisions about the nature and scope of their interactions and to initiate (and respond to) interactions that were not foreseen at design time (cf. the hard-wired engineering of such interactions in extant approaches).

The subsequent discussion details what would be involved if all these interactions were to be automated and performed at run time. This is clearly the most technically challenging scenario and there are a number of points that need to be made. Firstly, while such automation is technically feasible, in a limited form, using today's technology, this is an area that requires more research to reach the desired degree of sophistication and maturity. Secondly, in some cases, the service owners and consumers may not wish to automate all these activities since they may wish to retain a degree of human control over these decisions. Thirdly, some contracts and relationships may be set up at design time rather than being established at run time. This can occur when there are well-known links and dependencies between particular services, owners and consumers.

The nature of the interactions between the agents can be broadly divided into two main camps. Firstly, those that are associated with making service contracts. This will typically be achieved through some form of automated negotiation since the agents are autonomous [10]. When designing these negotiations, three main issues need to be considered:

- *The Negotiation Protocol*: The set of rules that govern the interaction. This covers the permissible types of participants (e.g. the negotiators and any relevant third parties), the negotiation states (e.g. accepting bids, negotiation closed), the events that cause negotiation states to change (e.g. no more bidders, bid accepted), and the valid actions of the participants in particular states (e.g. which messages can be sent by whom, to whom, at what stage).
- *The negotiation object*: The range of issues over which agreement must be reached. At one extreme, the object may contain a single issue (such as price), while on the other hand it may cover hundreds of issues (related to price, quality, timings, penalties, terms and conditions, etc.). Orthogonal to the agreement structure, and determined by the negotiation protocol, is the issue of the types of operation that can be performed on agreements. In the simplest case, the structure and the contents of the agreement are fixed and participants can either accept or reject it (i.e. a take it or leave it offer). At the next level, participants have the flexibility to change the values of the issues in the negotiation object (i.e. they can make counter-proposals to ensure the agreement better fits their negotiation objectives). Finally, participants might be allowed to dynamically alter (by adding or removing issues) the structure of the negotiation object (e.g. a car salesman may offer one year's free insurance in order to clinch the deal).
- *The agent's decision-making models*: The decision-making apparatus that the participants employ so as to act in line with the negotiation protocol in order to achieve their objectives. The sophistication of the model, as well as the range of decisions that have to be made are influenced by the protocol in place, by the nature of the negotiation object, and by the range of operations that can be performed on it. It can vary from the very simple, to the very complex.

In designing any automated negotiation system, the first thing that needs to be established is the protocol to be used (this is called the *mechanism design problem*). In this context, the protocol will be determined by the market owner. Here the main consideration is the nature of the negotiation. If it is a one-to-many negotiation (i.e. one buyer and many sellers or one seller and many buyers), then the protocol will typically be a form of auction. Although there are thousands of different permutations of auction, four main ones are typically used. These are English, Dutch, Vickrey, and First-Price Sealed Bid. In an English auction, the auctioneer begins with the lowest acceptable price and bidders are free to raise their bids successively until there are no more offers to raise the bid. The winning bidder is the one with the highest bid. The Dutch auction is the converse of the English one; the auctioneer calls for an initial high price, which is then lowered progressively until there is an offer from a bidder to claim the item. In the first-priced sealed bid, each bidder submits his/her offer for the item independently without any knowledge of the other bids. The highest bidder gets the item and they pay a price equal to their bid amount. Finally, a Vickrey auction is similar to a first-price sealed bid auction, but the item is awarded to the highest bidder at a price equal to the second highest bid. More complex forms of auctions exist to deal with the cases in which there are multiple buyers and sellers that wish to trade (these are called double auctions) and with cases in which agents wish to purchase multiple interrelated goods at the same time (these are called combinatorial auctions). If it is a one-to-one negotiation (one buyer and one seller), then a form of heuristic model is needed (e.g. [11, 12]). These models vary depending upon the nature of the negotiation protocol and, in general, are less well developed than those for auctions.

Having determined the protocol, the next step is to determine the nature of the contract that needs to be established. This will typically vary from application to application and again it is something that is set by the market owner. Given these two, the final step is to determine the agent's reasoning model. This can vary from the very simple (bidding truthfully) to the very complex (involving reasoning about the likely number and nature of the other bidders).

The second main type of interaction is when a number of agents decide to come together to form a new virtual organisation. This involves determining the participants of the coalition and determining their various roles and responsibilities in this new organisational structure. Again, this is typically an activity that will involve negotiation between the participants since they need to come to a mutually acceptable agreement about the division of labour and responsibilities. Here there are a number of techniques and algorithms that can be employed to address the coalition formation process [13, 14] although this area requires more research to deal with the envisaged scale of Grid applications.

17.3.2.3 Marketplace structures

It should be possible to establish marketplaces by any agent(s) in the system (including a service owner, a service consumer or a neutral third party). The entity that establishes the marketplace is here termed the market owner. The owner is responsible for setting up, advertising, controlling and disbanding the marketplace. In order to establish a marketplace, the owner needs a representation scheme for describing the various entities that are allowed to participate in the marketplace (terms of entry), a means of describing how the various allowable entities are allowed to interact with one another in the context of the marketplace, and what monitoring mechanisms (if any) are to be put in place to ensure that the marketplace's rules are adhered to.

17.3.3 A service-oriented view of the scenario

The first marketplace is the one connected with the scientist's own lab. This marketplace has agents to represent the humans involved in the experiment; thus, there is a *scientist agent* (SA) and a *technician agent* (TA). These are responsible for interacting with the scientist and the technician, respectively, and then for enacting their instructions in the Grid. These agents can be viewed as the computational proxies of the humans they represent – endowed with their personalised information about their owner's preferences and objectives. These personal agents need to interact with other (artificial) agents in the marketplace in order to achieve their objectives. These other agents include an *analyser agent* (AA) (that is responsible for managing access to the analyser itself), the *analyser database agent* (ADA) (that is responsible for managing access to the database containing information about the analyser), and the *high resolution analyser agent* (HRAA) (that is responsible for managing access to the high resolution analyser). There is also an *interest notification agent* (INA) (that is responsible for recording which scientists in the lab are interested in which types of results and for notifying them when appropriate results are generated) and an *experimental results agent* (ERA) (that can discover similar analyses of data or when similar experimental configurations have been used in the past). The services provided by these agents are summarised in Table 17.2.

The operation of this marketplace is as follows. The technician uses the `logSample` service to record data about the sample when it arrives and the `setAnalysisConfiguration` service to set the appropriate parameters for the forthcoming experiment. The technician then instructs the TA to book a slot on the analyser using the `bookSlot` service. At the appropriate time, the ADA informs the AA of the settings that it should adopt (via the `configureParameters` service) and that it should now run the experiment

Table 17.2 Services in the scientist's lab marketplace

Agent	Services offered	Services consumed by
Scientist agent (SA)	resultAlert	Scientist
	reportAlert	Scientist
Technician agent (TA)	MonitorAnalysis	Technician
Analyser agent (AA)	configureParameters	ADA
	runSample	ADA
Analyser database agent (ADA)	logSample	Technician
	setAnalysisConfiguration	Technician
	bookSlot	TA
	recordAnalysis	AA
High resolution analyser agent (HRAA)	bookSlot	SA
	configureParameters	Scientist
	runAnalysis	Scientist
	videoAnalysis	Scientist,
	monitorAnalysis	Technician
	reportResults	Technician
	replayExperiment	SA
	suggestRelatedConfigurations	Scientist
		Scientist
Interest notification agent (INA)	registerInterest	Scientists,
	notifyInterestedParties	Technicians
	findInterestedParties	ADA
		Scientist
Experimental results agent (ERA)	FindSimilarExperiments	HRAA

(via the `runSample` service). As part of the contract for the `runSample` service, the AA informs the ADA of the results of the experiment and these are logged along with the appropriate experimental settings (using the `recordAnalysis` service). Upon receipt of these results, the ADA informs the INA about them. The INA then disseminates the results (via the `notifyInterestedParties` service) to scientists who have registered an interest in results of that kind (achieved via the `registerInterest` service).

When interesting results are received, the SA alerts the scientist (via the `resultAlert` service). The scientist then examines the results and decides that they are of interest and that further analysis is needed. The scientist then instructs the SA to make a booking on the High Resolution Analyser (via the `bookSlot` service). When the booking is made, the HRAA volunteers information to the scientist about the configurations of similar experiments that have previously been run (via the `suggestRelatedConfigurations` service). Using this information, the scientist sets the appropriate configurations (via the `configureParameters` service). At the appropriate time, the experiment is started (via the `runAnalysis` service). As part of the contract for this service, the experiment is videoed (via the `videoAnalysis` service), monitoring information is sent to the technician (via the `monitorAnalysis` service) and a report is prepared and sent to the SA (via the `reportResults` service). In preparing this report, the HRAA interacts with the ERA to discover if related experiments and results have already been undertaken (achieved via the `findSimilarExperiments` service).

The scientist is alerted to the report by the SA (via the `reportAlert` service). The scientist decides the results may be interesting and decides to replay some of the key segments of the video (via the `replayExperiment` service). The scientist decides the results are indeed interesting and so asks for relevant publications and details of scientists who have published any material on this topic. This latter activity is likely to be provided through an external marketplace that provides this service for the wider community (see Table 17.3). In such a marketplace, there may be multiple *Paper Repository Agents*

Table 17.3 Services in the general scientific community marketplace

	Services offered	Services consumed by
International sample database agent (ISDA)	logSample	Scientist
	registerInterests	Scientist
	disseminateInformation	SAs
Paper repository agent (PRA)	FindRelatedPapers	SAs
	FindRelatedAuthors	SAs
Scientist agent (SA)	ReceiveRelevantData	Scientist
	ArrangeSimulation	Scientist
Simulation provider agent (SPA)	offerSimulationResource	SA
	utiliseSimulationResource	SA
Problem-solving environment agent (PSEA)	WhatSimulationTools	Scientist
	simulationSettingInfo	Scientist
	analyseResults	Scientist

Table 17.4 Services in the scientist's organisation marketplace

Agent	Services offered	Service consumed by
Research meeting convener agent (RMCA)	`arrangeMeeting`	SAs
	`setAgenda`	Scientist
	`disseminateInformation`	SAs
Scientist agent (SA)	`arrangeMeeting`	RMCA
	`receiveInformation`	RMCA

(*PRAs*) that offer the same broad service (`findRelatedPapers` and `findRelated Authors`) but to varying degrees of quality, coverage, and timeliness.

Armed with all this information, the scientist decides that the results should be discussed within the wider organisation context. This involves interacting in the Scientist's Organisation Marketplace. The agents involved in this marketplace are the *research meeting convener agent* (RMCA) (responsible for organising research meetings) and the various *scientist agents* that represent the relevant scientists. The services provided by these agents are given in Table 17.4. The RMCA is responsible for determining when research meetings should take place; this is achieved via the `arrangeMeeting` service through interaction with the SAs of the scientists involved. The scientist requests a slot to discuss the latest experimental findings (via the `setAgenda` service) and provides the appropriate data for discussion to the RMCA that disseminates it to the SA's of the relevant participants (via the `disseminateInformation` service). As a consequence of the meeting, it is decided that the results are appropriate for dissemination into the scientific community at large.

The general scientific community is represented by a series of distinct marketplaces that are each responsible for different aspects of the scientific process. As decided upon at the organisation's meeting, the sample data is logged in the appropriate international database (using the `logSample` service). This database has an attached notification service at which individual scientists can register their interests in particular types of data (via the `registerInterests` service). Scientists will then be informed, via their SA, when new relevant data is posted (via the `disseminateInformation` service).

One of the scientists who receives notification of the new results believes that they should be investigated further by undertaking a new round of simulations. The scientist instructs the SA to arrange for particular simulations to be arranged. The SA enters a marketplace where providers of processing capabilities offer their resources (via the `offerSimulationResource` service). The SA will arrange for the appropriate amount of resource to be made available at the desired time such that the simulations can be run. Once these contracts have been established, the SA will invoke the simulation (via the `utiliseSimulationResource` service). During the course of these simulations, the scientist will make use of the Problem-Solving Environment Agent (PSEA) to assist in the tasks of determining what simulation tools to exploit (via the `whatSimulation-Tools` service), setting the simulation parameters appropriately for these tools (via the `simulationSettingInfo` service), and analysing the results (via the `analyseRe-sults` service).

This then characterises our scenario as an active marketplace of agents offering and consuming services. As already indicated, we do not expect that this complete set of interactions will be dealt with seamlessly by computational agents in the near future. However, it provides a level of abstraction and defines capabilities that we claim is important to aspire to if the full potential of the Semantic Grid is to be realised.

17.4 THE KNOWLEDGE LAYER

The aim of the knowledge layer is to act as an infrastructure to support the management and application of scientific knowledge to achieve particular types of goal and objective. In order to achieve this, it builds upon the services offered by the data/computation and information layers (see Chapter 3 for more details of the services and technologies at these layers).

The first thing to reiterate with respect to this layer is the problem of the sheer scale of content we are dealing with. We recognise that the amount of data that the data Grid is managing is likely to be huge. By the time that data is equipped with meaning and turned into information, we can expect order of magnitude reductions in the amount. However, what remains will certainly be enough to present us with the problem of *infosmog* – the condition of having too much information to be able to take effective action or apply it in an appropriate fashion to a specific problem. Once information is delivered that is destined for a particular purpose, we are in the realm of the knowledge Grid. Thus, at this level we are fundamentally concerned with abstracted and annotated content and with the management of scientific knowledge.

We can see this process of scientific knowledge management in terms of a life cycle of knowledge-oriented activity that ranges over knowledge acquisition and modelling, knowledge retrieval and reuse, knowledge publishing and knowledge maintenance (Section 17.4.1). Next we discuss the fundamental role that ontologies will play in providing the underpinning semantics for the knowledge layer (Section 17.4.2). Section 17.4.3 then considers the knowledge services aspects of our scenario. Finally, we review the research issues associated with our requirements for a knowledge Grid (Section 17.4.4).

17.4.1 The knowledge life cycle

The knowledge life cycle can be regarded as a set of challenges as well as a sequence of stages. Each stage has variously been seen as a bottleneck. The effort of acquiring knowledge was one bottleneck recognised early [15]. But so too are modelling, retrieval, reuse, publication and maintenance. In this section, we examine the nature of the challenges at each stage in the knowledge life cycle and review the various methods and techniques at our disposal.

Although we often suffer from a deluge of data and too much information, all too often what we have is still insufficient or too poorly specified to address our problems, goals, and objectives. In short, we have insufficient knowledge. *Knowledge acquisition* sets the challenge of getting hold of the information that is around, and turning it into knowledge by making it *usable*. This might involve, for instance, making tacit knowledge explicit, identifying gaps in the knowledge already held, acquiring and integrating knowledge from multiple sources (e.g. different experts, or distributed sources on the Web), or acquiring knowledge from unstructured media (e.g. natural language or diagrams).

A range of techniques and methods has been developed over the years to facilitate knowledge acquisition. Much of this work has been carried out in the context of attempts to build knowledge-based or expert systems. Techniques include varieties of interview, different forms of observation of expert problem-solving, methods of building conceptual maps with experts, various forms of document and text analysis, and a range of machine learning methods [16]. Each of these techniques has been found to be suited to the elicitation of different forms of knowledge and to have different consequences in terms of the effort required to capture and model the knowledge [17, 18]. Specific software tools have also been developed to support these various techniques [19] and increasingly these are now Web enabled [20].

However, the process of explicit knowledge acquisition from human experts remains a costly and resource intensive exercise. Hence, the increasing interest in methods that can (semi-) automatically elicit and acquire knowledge that is often implicit or else distributed on the Web [21]. A variety of information extraction tools and methods are being applied to the huge body of textual documents that are now available [22]. Examples include programs to extract information about protein function from various scientific papers, abstracts and databases that are increasingly available on-line. Another style of automated acquisition consists of systems that observe user behaviour and infer knowledge from that behaviour. Examples include recommender systems that might look at the papers downloaded by a researcher and then detect themes by analysing the papers using methods such as term frequency analysis [23]. The recommender system then searches other literature sources and suggests papers that might be relevant or else of interest to the user.

Methods that can engage in the sort of background knowledge acquisition described above are still in their infancy but with the proven success of pattern directed methods in areas such as data mining, they are likely to assume a greater prominence in our attempts to overcome the knowledge acquisition bottleneck.

Knowledge modelling bridges the gap between the acquisition of knowledge and its use. Knowledge models must be able *both* to act as straightforward placeholders for the acquired knowledge, *and* to represent the knowledge so that it can be used for problem solving. Knowledge representation technologies have a long history in Artificial Intelligence. There are numerous languages and approaches that cater for different knowledge types; structural forms of knowledge, procedurally oriented representations, rule-based characterisations and methods to model uncertainty, and probabilistic representations [24].

Most large applications require a range of knowledge representation formats. CommonKADS [25], one of the most comprehensive methodologies for the development of knowledge intensive systems, uses a range of modelling methods and notations – including logic and structured objects. It also factors out knowledge into various types and identifies recurrent patterns of inference and knowledge type that denote characteristic problem solvers. These patterns are similar to design patterns in software engineering [26] and attempt to propose a set of components out of which problem-solving architectures can be composed. One of the major constituents of the models built in CommonKADS is domain ontologies, which we discuss in the next section.

Recently, with the explosion of content on the Web there has arisen the recognition of the importance of metadata. Any kind of content can be 'enriched' by the addition of annotations about what the content is about [27]. Such semantic metadata is an important additional element in our modelling activity. It may indicate the origin of content, its provenance, value or longevity. It may associate other resources with the content such as the rationale as to why the content is in the form it is in and so on.

Certainly, given the sheer amount of content available in a Grid context, it is crucial to have some technical support for metadata 'enrichment'. To this end a number of systems are now under development that aim to take given metadata structures and help annotate, tag, or associate content with that metadata [28, 29].

In any modelling exercise, it is important to recognise that the modelling reflects a set of interests and perspectives. These may be made more or less explicit but they are always present. It is also important to recognise that models may be more or less formal and aspire to various degrees of precision and accuracy. The model is, of course, not the object or process, rather it is an artefact built with a particular set of goals and intentions in mind.

Once knowledge has been acquired and modelled, it needs to be stored or hosted somewhere so that it can be retrieved efficiently. In this context, there are two related problems to do with *knowledge retrieval*. First, there is the issue of finding knowledge again once it has been stored. And second, there is the problem of retrieving the subset of content that is relevant to a particular problem. This will set particular problems for a knowledge retrieval system in which content alters rapidly and regularly.

Technologies for information retrieval exist in many forms [30]. They include methods that attempt to encode structural representations about the content to be retrieved such as explicit attributes and values. Varieties of matching algorithm can be applied to retrieve cases that are similar to an example or else a partial set of attributes presented to the system. Such explicit Case-Based Reasoning [31] and Query engines have been widely adopted. They suffer from the problem of content encoding – the ease with which new content and examples can be represented in the required structural format. There are also perennial issues about the best measures of similarity to use in these systems.

Other retrieval methods are based on statistical encoding of the objects to be retrieved. These might be as vectors representing the frequency of the terms in a document or other piece of content. Retrieval is a matter of matching a query of an example piece of content against these stored representations and generating the closest matches [32].

Search engines such as Google that are manifestly capable of scaling and also demonstrate good retrieval performance rely on concepts such as relevance ranking. Here, given any set of terms to search, Google looks at the interconnected nature of content and the frequency of its being accessed to help determine in part the rank of how good a match to the material sought it is likely to be.

In the general field of content retrieval there is no one dominant paradigm – it can occur at the fine-grained level at which point it is a form of information extraction, or else at the level of complete documents or even work flows or data logs that might encode entire experimental configurations and subsequent runs.

One of the most serious impediments to the cost-effective use of knowledge is that too often knowledge components have to be constructed afresh. There is little *knowledge reuse*. This arises partly because knowledge tends to require different representations depending on the problem solving that it is intended to do. We need to understand how to find patterns in knowledge, to allow for its storage so that it can be reused when circumstances permit. This would save a good deal of effort in reacquiring and restructuring the knowledge that had already been used in a different context.

We have already alluded to the form of reuse embodied in methodologies such as CommonKADS. Here, a problem-solving template for monitoring might be used in one domain and its general structure reused elsewhere. The actual ontology of components or processes might be another candidate for reuse. Complete problem-solving runs or other results might offer the chance to reuse previously solved problems in areas that are similar. Workflows themselves might be reused. Technical support in the area of reuse tends to be focused on the type of product being reused. At one end of the spectrum we have reuse of ontologies in tools such as Protégé [33], at the other end there are tools to facilitate the reuse of complete problem-solving architectures [34–36]. Obstacles to reuse include the very real possibility that it is sometimes easier to reconstruct the knowledge fragment than to hunt for it. Even when it is found it is often necessary to modify it to suit the current context. Some knowledge is so difficult to model in a reusable fashion that an explicit decision is made to reacquire when needed.

Having acquired knowledge, modelled and stored it, the issue then arises as to how to get that knowledge to the people who subsequently need it. The challenge of *knowledge publishing* or disseminating can be described as getting the right knowledge, in the right form, to the right person or system, at the right time. Different users and systems will require knowledge to be presented and visualised in different ways. The quality of such presentation is not merely a matter of preference. It may radically affect the utility of the knowledge. Getting presentation right involves understanding the different perspectives of people with different agendas and systems with different requirements. An understanding of knowledge content will help ensure that important related pieces of knowledge get published at the appropriate time.

Technologies to help publish content in fast and flexible ways are now starting to appear. One such is the distributed link service (DLS). This is a method for associating hyperlinks with content in such a way that the link is held separate from the content and not represented in the content itself. This means that different link structures or *link bases* can be associated with the same content. This allows very different hypertext structures to be associated with the same content and supports very different styles of publishing and subsequently navigating content [37]. More recently, DLS systems have been built that generate links that can be switched in and out depending on the ontology or conceptualization in play at the time [38]. Ontologies can also act as filters on portals. By looking at the metadata associated with content, the portal can elect to show various content in different ways to different users [http://www.ontoportal.org.uk/].

Given accumulated fragments of knowledge, methods now exist to thread this information together and to generate connected text to explain or to present the fragments [39, 40].

Some publication models seek to generate extended narratives from harvested Web content [41]. Publishing services extend to concepts such as the Open Archives Initiative [42] and e-Prints [43]. In these models, individuals deposit their papers with associated metadata. The e-Prints system, for example, can then offer the basis for additional services running on a significant publication base. For example, it currently runs on the Los Alamos Physics Archive consisting of some 100 000 documents and offers citation and automatic cross-indexing services [http://opcit.eprints.org/].

Problems with publication include the fact that it has to be timely and it should not overwhelm the recipient with detail nor content that is not of interest. Related to these last two issues we find technologies under development to carry out summarisation [44] of texts and subject content identification [45, 46].

Finally, having acquired and modelled the knowledge, and having managed to retrieve and disseminate it appropriately, the last challenge is to keep the knowledge content current – *knowledge maintenance*. This may involve the regular updating of content as knowledge changes. Some content has considerable longevity, while other knowledge dates quickly. If knowledge is to remain useful over a period of time, it is essential to know which parts of the knowledge base must be updated or else discarded and when. Other problems involved in maintenance include verifying and validating the content, and certifying its safety.

Historically, the difficulty and expense of maintaining large software systems has been underestimated. Where that information and knowledge content is to be maintained in a distributed fashion the problem would appear to be even more acute. Whether it is a repository full of documents or databases full of experimental data, the problem of curation needs to be addressed early in the system design process. Moreover, it needs to be tackled early in the knowledge life cycle. When content is acquired and modelled, metadata regarding its provenance, quality and value ought to be captured too. Otherwise one has little evidence about what is important to maintain and what are the likely consequences if it is changed or removed.

Technologies have been developed to look at the effects of refining and maintaining knowledge bases [47]. These technologies attempt to implement a range of checking algorithms to see if altering the knowledge base leads to cyclic reasoning behaviour or else disables or enables new classes of inference or behaviours. A different type of maintenance relates to the domain descriptions or conceptualizations themselves. Again, it is important that at the point at which the ontology is designed, careful thought is given to those parts of the conceptualization that are likely to remain stable as opposed to areas where it is recognised that change and modification is likely to happen. Once built, an ontology is typically populated with instances to produce the knowledge bases over which processing occurs. Populating ontologies with instances is a constant process of maintenance and whenever it is carried out there can be much post processing to eliminate, for example, duplicate instances from the knowledge base [48].

As with so many aspects of the knowledge life cycle, effective maintenance will also depend on socio-technical issues having to do with whether there are clear owners and stakeholders whose primary function is content and knowledge management.

We have already indicated that if the knowledge intensive activities described above are to be delivered effectively in the Semantic Grid context then a crucial step is to establish a basic level of semantic interoperation (Section 17.3.2.2). This requires the development of a shared vocabulary, description or conceptualization for the particular domain of interest. It is to this ontological engineering that we now turn.

17.4.2 Ontologies and the knowledge layer

The concept of an *ontology* is necessary to capture the expressive power that is needed for modelling and reasoning with knowledge. Generally speaking, an ontology determines the extension of terms and the relationships between them. However, in the context of knowledge and Web engineering, an ontology is simply a published, more or less agreed, conceptualization of an area of content. The ontology may describe objects, processes, resources, capabilities or whatever.

Recently, a number of languages have appeared that attempt to take concepts from the knowledge representation languages of AI and extend the expressive capability of those of the Web (e.g. RDF and RDF Schema). Examples include SHOE [49], DAML [50], and OIL [51]. Most recently there has been an attempt to integrate the best features of these languages in a hybrid called DAML+OIL. As well as incorporating constructs to help model ontologies, DAML+OIL is being equipped with a logical language to express rule-based generalizations.

However, the development of the Semantic Grid is not simply about producing machine-readable languages to facilitate the interchange and integration of heterogeneous information. It is also about the elaboration, enrichment and annotation of that content. To this end, the list below is indicative of how rich annotation can become. Moreover, it is important to recognize that enrichment or metatagging can be applied at any conceptual level in the three-tier Grid of Figure 17.1. This yields the idea of metadata, metainformation, and metaknowledge.

- *Domain ontologies*: Conceptualizations of the important objects, properties, and relations between those objects. Examples would include an agreed set of annotations for medical images, an agreed set of annotations for climate information, and a controlled set of vocabulary for describing significant features of engineering design.
- *Task ontologies*: Conceptualizations of tasks and processes, their interrelationships and properties. Examples would include an agreed set of descriptors for the stages of a synthetic chemistry process, an agreed protocol for describing the dependencies between optimisation methods, and a set of descriptions for characterizing the enrichment or annotation process when describing a complex medical image.
- *Quality ontologies*: Conceptualizations of the attributes that knowledge assets possess and their interrelationships. Examples would include annotations that would relate to the expected error rates in a piece of medical imaging, the extent to which the quality of a result from a field geologist depended on their experience and qualifications, and whether results from particular scientific instruments were likely to be superseded by more accurate devices.
- *Value ontologies*: Conceptualizations of those attributes that are relevant to establishing the value of content. Examples would include the cost of obtaining particular physics data, the scarcity of a piece of data from the fossil record, and how widely known a particular metabolic pathway was.

- *Personalization ontologies*: Conceptualizations of features that are important to establishing a user model or perspective. Examples would include a description of the prior familiarity that a scientist had with particular information resources, the amount of detail that the user was interested in, and the extent to which the user's current e-Science activities might suggest other content of interest.
- *Argumentation ontologies*: A wide range of annotations can relate to the *reasons why* content was acquired, why it was modelled in the way it was, and who supports or dissents from it. This is particularly powerful when extended to the concept of associating discussion threads with content. Examples are the integration of authoring and reviewing processes in on-line documents. Such environments allow structured discussions of the evolution and development of an idea, paper, or concept. The structured discussion is another annotation that can be held in perpetuity. This means that the reason for a position in a paper or a design choice is linked to the object of discussion itself.

The benefits of an ontology include improving communication between systems whether machines, users, or organizations. They aim to establish an agreed and perhaps normative model. They endeavour to be consistent and unambiguous, and to integrate a range of perspectives. Another benefit that arises from adopting an ontology is interoperability and this is why they figure largely in the vision for the Semantic Web [4]. An ontology can act as an interlingua, it can promote reuse of content, ensure a clear specification of what content or a service is about, and increase the chance that content and services can be successfully integrated.

A number of ontologies are emerging as a consequence of commercial imperatives where vertical marketplaces need to share common descriptions. Examples include the Common Business Library (CBL), Commerce XML (cXML), ecl@ss, the Open Applications Group Integration Specification (OAGIS), Open Catalog Format (OCF), the Open Financial Exchange (OFX), Real Estate Transaction Markup Language (RETML), RosettaNet, UN/SPSC (see www.diffuse.org), and UCEC.

We can see examples of ontologies built and deployed in a range of traditional knowledge-intensive applications ranging from chemical processing [52] through to engineering plant construction [53]. Moreover, there are a number of large-scale ontology initiatives under way in specific scientific communities. One such is in the area of genetics where a great deal of effort has been invested in producing common terminology and definitions to allow scientists to manage their knowledge [http://www.geneontology.org/]. This effort provides a glimpse of how ontologies will play a critical role in sustaining the e-Scientist.

This work can also be exploited to facilitate the sharing, reuse, composition, mapping, and succinct characterizations of (Web) services. In this vein, [54] McIlraith *et al.* exploit a Web service mark-up that provides an agent-independent *declarative API* that is aimed at capturing the data and metadata associated with a service together with specifications of its properties and capabilities, the interface for its execution, and the prerequisites and consequences of its use. A key ingredient of this work is that the mark-up of Web content exploits ontologies. They have used DAML for *semantic mark-up* of Web Services. This provides a means for agents to populate their local knowledge bases so that they can reason about Web services to perform automatic Web service discovery, execution, composition, and interoperation.

It can be seen that ontologies clearly provide a basis for the communication, integration, and sharing of content. But they can also offer other benefits. An ontology can be used for improving search accuracy by removing ambiguities and spotting related terms, or by associating the information retrieved from a page with other information. They can act as the backbone for accessing information from a community Web portal [55]. Moreover, Internet reasoning systems are beginning to emerge that exploit ontologies to extract and generate annotations from the existing Web [56].

Given the developments outlined in this section, a general process that might drive the emergence of the knowledge Grid would comprise the following:

- The development, construction, and maintenance of application (specific and more general areas of science and engineering) and community (sets of collaborating scientists) based ontologies.
- The large-scale annotation and enrichment of scientific data, information and knowledge in terms of these ontologies.
- The exploitation of this enriched content by knowledge technologies.

There is a great deal of activity in the whole area of ontological engineering at the moment. In particular, the World Wide Web Consortium (W3C) has a working group developing a language to describe ontologies on the Web; this Web Ontology language, which is known as OWL, is based on DAML+OIL. The development and deployment of ontologies is a major topic in the Web services world and is set to assume an important role in Grid computing.

17.4.3 Knowledge layer aspects of the scenario

Let us now consider our scenario in terms of the opportunities it offers for knowledge services (see Table 17.5). We will describe the knowledge layer aspects in terms of the agent-based service-oriented analysis developed in Section 17.3.3. Important components of this conceptualization were the software proxies for human agents such as the *scientist agent* and the *technician agent*. These software agents will interact with their human counterparts to elicit preferences, priorities, and objectives. The software proxies will then realise these elicited items on the Grid. This calls for knowledge acquisition services. As indicated in Section 17.4.1, a range of methods could be used. Structured interview methods invoke templates of expected and anticipated information. Scaling and sorting methods enable humans to rank their preferences according to relevant attributes that can either be explicitly elicited or pre-enumerated. The laddering method enables users to construct or select from ontologies. Knowledge capture methods need not be explicit – a range of pattern detection and induction methods exist that can construct, for example, preferences from past usage.

One of the most pervasive knowledge services in our scenario is the partial or fully automated annotation of scientific data. Before it can be used as knowledge, we need to equip the data with meaning. Thus, agents require capabilities that can take data streaming

Table 17.5 Example knowledge services in the scenario

Agent requirements	Knowledge technology services
Scientist agent (SA)	Knowledge acquisition of scientist profile
	Ontology service
Technician agent (TA)	Knowledge acquisition of technician profile
	Ontology service
	Knowledge-based scheduling service to book analyser
Analyser agent (AA)	Annotation and enrichment of instrument streams
	Ontology service
Analyser database agent (ADA)	Annotation and enrichment of databases
	Ontology service
High resolution analyser agent (HRAA)	Annotation and enrichment of media
	Ontology service
	Language generation services
	Internet reasoning services
Interest notification agent (INA)	Knowledge publication services
	Language generation services
	Knowledge personalization services
	Ontology service
Experimental results agent (ERA)	Language generation services
	Result clustering and taxonomy formation
	Knowledge and data mining service
	Ontology service
Research meeting convener agent (RMCA)	Constraint based scheduling service
	Knowledge personalization service
	Ontology service
International sample database agent (ISDA)	Result Clustering and taxonomy formation
	Knowledge and data mining services
	Ontology service
Paper repository agent (PRA)	Annotation and enrichment of papers
	Ontology service
	Dynamic link service
	Discussion and argumentation service
Problem-solving environment agent (PSEA)	Knowledge-based configuration of PSE components
	Knowledge-based parameter setting and input selection
	Ontology service

from instruments and annotate it with meaning and context. Example annotations include the experimental context of the data (where, when, what, why, which, how). Annotation may include links to other previously gathered information or its contribution and relevance to upcoming and planned work. Such knowledge services will certainly be one of the main functions required by the Analyser Agent and analyser database agent. In the case of the High Resolution Analyser Agent, we have the additional requirement to enrich a range of media types with annotations. In the original scenario, this included video of the actual experimental runs.

These acquisition and annotation services along with many others will be underpinned by ontology services that maintain agreed vocabularies and conceptualizations of the scientific domain. These are the names and relations that hold between the objects and processes of interest to us. Ontology services will also manage the mapping between ontologies that will be required by agents with differing interests and perspectives.

Personalization services will also be invoked by a number of agents in the scenario. These might interact with the annotation and ontology services already described so as to customize the generic annotations with personal mark-ups – the fact that certain types of data are of special interest to a particular individual. Personal annotations might reflect genuine differences of terminology or perspective – particular signal types often have local vocabulary to describe them. Ensuring that certain types of content are noted as being of particular interest to particular individuals brings us on to services that notify and push content in the direction of interested parties. The INA and the RMCA could both be involved in the publication of content either customized to individual or to group interests. Portal technology can support the construction of dynamic content to assist the presentation of experimental results.

Agents such as the High Resolution Analyser (HRAA) and Experimental Results Analyser (ERA) have interests in classifying or grouping certain information and annotation types together. Examples might include all signals collected in a particular context, or sets of signals collected and sampled across contexts. This, in turn, provides a basis for knowledge discovery and the mining of patterns in the content. Should such patterns arise, they might be further classified against existing pattern types held in international databases – in our scenario this is managed in marketplaces by agents such as the International Sample Database Agent (ISDA).

At this point, agents are invoked whose job it is to locate other systems or agents that might have an interest in the results. Negotiating the conditions under which the results can be released, determining the quality of results, might all be undertaken by agents that are engaged to provide result brokering and result update services.

Raw results are unlikely to be especially interesting so that the generation of natural language summaries of results will be important for many of the agents in our scenario. Results that are published this way will also want to be linked and threaded to existing papers in the field and made available in ways that discussion groups can usefully comment on. Link services are one type of knowledge technology that will be ubiquitous here – this is the dynamic linking of content in documents in such a way that multiple markups and hyperlink annotations can be simultaneously maintained. Issue tracking and design rationale methods allow multiple discussion threads to be constructed and followed through documents. In our scenario, the PRA will not only retrieve relevant papers but also mark them up and thread them in ways that reflect the personal interests and conceptualizations (ontologies) of individuals or research groups.

The use of PSEAs in our simulation of experimentally derived results presents us with classic opportunities for knowledge intensive configuration and processing. Once again, these results may be released to communities of varying size with their own interests and viewpoints.

Ultimately it will be up to application designers to determine if the knowledge services described in this scenario are invoked separately or else as part of the inherent competencies of the agents described earlier. Whatever the design decisions, it is clear that knowledge services will play a fundamental role in realizing the potential of the Semantic Grid for the e-Scientist.

17.4.4 Research issues

The following is a list of the key research issues that remain for exploiting knowledge services in the Semantic Grid. In many cases there are already small-scale exemplars for most of these services; consequently, many of the issues relate to the problems of scale and distribution

- Languages and infrastructures are needed to describe, advertise and locate knowledge services. We need the means to invoke and communicate the results of such services. This is the sort of work that is currently under way in the Semantic Web effort of DAML-S [57]. However, it is far from clear how this work will interface with that of the agent-based computing, Web services and Grid communities.
- Methods are required to build large-scale ontologies and tools deployed to provide a range of ontology services.
- Annotation services are required that will run over large corpora of local and distributed data. In some cases, for example, the annotation and cleaning of physics data, this process will be iterative and will need to be near real time as well as supporting fully automatic and mixed initiative modes. These annotation tools are required to work with mixed media.
- Knowledge capture tools are needed that can be added as plug-ins to a wide variety of applications and that draw down on ontology services. This will include a clearer understanding of profiling individual and group e-Science perspectives and interests.
- Dynamic linking, visualization, navigation, and browsing of content from many perspectives over large content sets.
- Retrieval methods based on explicit annotations.
- Construction of repositories of solution cases with sufficient annotation to promote reuse as opposed to discovering the solution again because the cost of finding the reusable solution is too high.
- Deployment of routine natural language processing as Internet services. Capabilities urgently required include tagging and mark-up of documents, discovering different linguistic forms of ontological elements, and providing language generation and summarisation methods for routine scientific reporting.
- Deployment of Internet-based reasoning services – whether as particular domain PSEs or more generic problem solvers such as scheduling and planning systems.
- Provision of knowledge discovery services with standard input/output APIs to ontologically mapped data.
- Understanding how to embed knowledge services in ubiquitous and pervasive devices.

17.5 CONCLUSIONS

This paper has outlined our vision of the Semantic Grid as a future e-Science infrastructure in which there is a high degree of easy-to-use and seamless automation and in which there are flexible collaborations and computations on a global scale. We have argued that this infrastructure should be conceptualised and implemented as a service-oriented architecture in which agents interact with one another in various types of information marketplace. Moreover, we have highlighted the importance of knowledge services in this vision and have outlined the key research challenges that need to be addressed at this level.

In order to make the Semantic Grid a reality, a number of research challenges need to be addressed. These include (in no particular order):

- *Smart laboratories*: We believe that for e-Science to be successful and for the Grid to be effectively exploited much more attention needs to be focused on how laboratories need to be instrumented and augmented. For example, infrastructure that allows a range of equipment to advertise its presence, be linked together, annotate and mark-up content it is receiving or producing.
- *Service-oriented architectures*: Research the provision and implementation of Grid facilities in terms of service-oriented architectures. Also, research into service description languages as a way of describing and integrating the Grid's problem-solving elements.
- *Agent-based approaches*: Research the use of agent-based architectures and interaction languages to enable e-Science marketplaces to be developed, enacted and maintained.
- *Trust and provenance*: Further research is needed to understand the processes, methods and techniques for establishing computational trust and determining the provenance and quality of content in Grid systems. This extends to the issue of digital rights management in making content available.
- *Metadata and annotation*: Whilst the basic metadata infrastructure already exists in the shape of RDF, metadata issues have not been fully addressed in current Grid deployments. It is relatively straightforward to deploy some of the technology in this area, and this should be promoted. RDF, for example, is already encoding metadata and annotations as shared vocabularies or ontologies. However, there is still a need for extensive work in the area of tools and methods to support the design and deployment of e-Science ontologies. Annotation tools and methods need to be developed so that emerging metadata and ontologies can be applied to the large amount of content that will be present in Grid applications.
- *Knowledge technologies*: In addition to the requirement for the research in metadata and annotation, there is a need for a range of other knowledge technologies to be developed and customised for use in e-Science contexts. These include knowledge capture tools and methods, dynamic content linking, annotation based search, annotated reuse repositories, natural language processing methods (for content tagging, mark-up, generation and summarisation), data mining, machine learning and Internet reasoning services. These technologies will need shared ontologies and service description languages if they are to be integrated into the e-Science workflow. These technologies will also need to be incorporated into the pervasive devices and smart laboratory contexts that will emerge in e-Science.
- *Integrated media*: Research into incorporating a wide range of media into the e-Science infrastructure. This will include video, audio, and a wide range of imaging methods. Research is also needed into the association of metadata and annotation with these various media forms.

- *Content presentation*: Research is required into methods and techniques that allow content to be visualized in ways consistent with the e-Science collaborative effort. This will also involve customizing content in ways that reflect localised context and should allow for personalization and adaptation.
- *e-Science workflow and collaboration*: Much more needs to be done to understand the workflow of current and future e-Science collaborations. Users should be able to form, maintain and disband communities of practice with restricted membership criteria and rules of operation. Currently, most studies focus on the e-Science infrastructure behind the socket on the wall. However, this infrastructure will not be used unless it fits in with the working environment of the e-Scientists. This process has not been studied explicitly and there is a pressing need to gather and understand these requirements. There is a need to collect real requirements from users, to collect use cases and to engage in some evaluative and comparative work. There is also a need to more fully understand the process of collaboration in e-Science.
- *Pervasive e-Science*: Currently, most references and discussions about Grids imply that their primary task is to enable global access to huge amounts of computational power. Generically, however, we believe Grids should be thought of as the means of providing seamless and transparent access from and to a diverse set of networked resources. These resources can range from PDAs to supercomputers and from sensor's and smart laboratories to satellite feeds.
- *e-Anything*: Many of the issues, technologies, and solutions developed in the context of e-Science can be exploited in other domains in which groups of diverse stakeholders need to come together electronically and interact in flexible ways. Thus, it is important that relationships are established and exploitation routes are explored with domains such as e-Business, e-Commerce, e-Education, and e-Entertainment.

REFERENCES

1. Foster, I. and Kesselman, C. (eds) (1998) *The Grid: Blueprint for a New Computing Infrastructure*. San Francisco, CA: Morgan Kaufmann Publishers.
2. De Roure, D., Jennings, N. R. and Shadbolt, N. R. (2001) *Research Agenda for the Semantic Grid: A Future e-Science Infrastructure*, UKeS-2002-02, Technical Report of the National e-Science Centre, 2001.
3. Berners-Lee, T. and Fischetti, M. (1999) *Weaving the Web: The Original Design and Ultimate Destiny of the World Wide Web by its Inventor*. San Francisco: Harper Collins
4. Berners-Lee, T., Hendler, J. and Lassila, O. (2001) The semantic web. *Scientific American*, **284**(5), 34–43.
5. *Proceedings of W3C Web Services Workshop*, April 11, 12, 2001, http://www.w3.org/2001/03/wsws-program
6. Foster, I., Kesselman, C. and Tuecke, S. (2001) The anatomy of the grid: enabling scalable virtual organizations. *International Journal of Supercomputer Applications and High Performance Computing*, 2001.
7. Wooldridge, M. 1997 Agent-based software engineering. *IEE Proceedings on Software Engineering*, **144**(1), 26–37.
8. Jennings, N. R. (2000) On agent-based software engineering. *Artificial Intelligence*, **117**, 277–296.
9. Newell, A. (1982) The knowledge level. *Artificial Intelligence*, **18**, 87–127.

10. Jennings, N. R., Faratin, P., Lomuscio, A. R., Parsons, S., Sierra, C. and Wooldridge, M. (2001) Automated negotiation: prospects, methods and challenges. *International Journal of Group Decision and Negotiation*, **10**(2), 199–215.

11. Faratin, P., Sierra, C. and Jennings, N. R. (1999) Negotiation decision functions for autonomous agents. *International Journal of Robotics and Autonomous Systems*, **24**(3–4), 159–182.

12. Kraus, S. (2001) *Strategic Negotiation in Multi-agent Environments*. Cambridge, MA: MIT Press.

13. Sandholm, T. (2000) Agents in electronic commerce: component technologies for automated negotiation and coalition formation. *Autonomous Agents and Multi-Agent Systems*, **3**(1), 73–96.

14. Shehory, O. and Kraus, S. (1998) Methods for task allocation via agent coalition formation. *Artificial Intelligence*, **101**(1–2), 165–200.

15. Hayes-Roth, F., Waterman, D. A. and Lenat, D. B. (1983) *Building Expert Systems*. Reading, Mass.: Addison-Wesley.

16. Shadbolt, N. R. and Burton, M. (1995) Knowledge elicitation: a systematic approach, in evaluation of human work, in Wilson, J. R. and Corlett, E. N. (eds) *A Practical Ergonomics Methodology*. ISBN-07484-0084-2 London, UK: Taylor & Francis, pp. 406–440.

17. Hoffman, R., Shadbolt, N. R., Burton, A. M. and Klein, G. (1995) Eliciting knowledge from experts: a methodological analysis. *Organizational Behavior and Decision Processes*, **62**(2), 129–158.

18. Shadbolt, N. R., O'Hara, K. and Crow, L. (1999) The experimental evaluation of knowledge acquisition techniques and methods: history, problems and new directions. *International Journal of Human Computer Studies*, **51**(4), 729–755.

19. Milton, N., Shadbolt, N., Cottam, H. and Hammersley, M. (1999). Towards a knowledge technology for knowledge management. *International Journal of Human Computer Studies*, **51**(3), 615–64.

20. Shaw, M. L. G. and Gaines, B. R. (1998) WebGrid-II: developing hierarchical knowledge structures from flat grids. In *Proceedings of the 11th Knowledge Acquisition Workshop (KAW '98)*, Banff, Canada, April, 1998, pp. 18–23, Available at http://repgrid.com/reports/KBS/WG/.

21. Crow, L. and Shadbolt, N. R. (2001) Extracting focused knowledge from the semantic web. *International Journal of Human Computer Studies*, **54**(1), 155–184.

22. Ciravegna, F.(2001) Adaptive information extraction from text by rule induction and generalisation. *Proceedings of 17th International Joint Conference on Artificial Intelligence (IJCAI2001)*, Seattle, August, 2001.

23. Middleton, S. E., De Roure, D. and Shadbolt, N. R. (2001) Capturing knowledge of user preferences: ontologies in recommender systems, *Proceedings of the First International Conference on Knowledge Capture, K-CAP2001*. New York: ACM Press.

24. Brachman, R. J. and Levesque, H. J. (1983) *Readings in Knowledge Representation*. San Mateo, CA: Morgan Kaufmann Publishers.

25. Schreiber, G., Akkermans, H., Anjewierden, A., de Hoog, R., Shadbolt, N. R., Van de Velde, W. and Wielinga, B. (2000) *Knowledge Engineering and Management*. Cambridge, MA: MIT Press.

26. Gamma, E., Helm, R., Johnson, R. and Vlissides, J. (1995) *Design Patterns: Elements of Reusable Object-Oriented Software*. Reading, MA: Addison-Wesley.

27. Motta, E., Buckingham Shum, S. and Domingue, J. (2001) Ontology-driven document enrichment: principles. *Tools and Applications. International Journal of Human Computer Studies*, **52**(5), 1071–1109.

28. Motta, E., Vargas-Vera, M., Domingue, J., Lanzoni, M., Stutt, A., and Ciravegna, F. (2002) MnM: ontology driven semi-automatic and automatic support for semantic Markup. *13th International Conference on Knowledge Engineering and Knowledge Management (EKAW2002)*, Sigüenza, Spain, 2002.

29. Handschuh, S., Staab, S. and Ciravegna, F. (2002) S-CREAM – Semi-automatic CREAtion of Metadata. *13th International Conference on Knowledge Engineering and Knowledge Management (EKAW2002)*, Sigüenza, Spain, 2002.

30. Sparck-Jones, K. and Willett, P. (1997) *Readings In Information Retrieval*. Mountain View, CA: Morgan Kaufmann Publishers.
31. Lenz, M., Bartsch-Spörl, B., Burkhard, H. and Wess, S. (eds) (1998) *Case-Based Reasoning Technology – From Foundations to Applications*. Lecture Notes in Artificial Intelligence 1400. Heidelberg, Springer-Verlag.
32. Croft, W. B. (2000) Information retrieval based on statistical language models. *ISMIS 2000*, 2000, pp. 1–11.
33. Schreiber, G., Crubezy, M. and Musen, M. A. (2000) A case study in using protege-2000 as a tool for CommonKADS. *12th International Conference on Knowledge Engineering and Knowledge Management (EKAW2000)*. France, Juan-les-Pins, pp. 33–48, Springer LNAI.
34. Motta, E., Fensel, D., Gaspari, M. and Benjamins, R. (1999) specifications of knowledge components for reuse. *Eleventh International Conference on Software Engineering and Knowledge Engineering (SEKE '99)*, June, 1999.
35. Fensel, D., Benjamins, V. R., Motta, E. and Wielinga, B. (1999). UPML: A framework for knowledge system reuse. *Proceedings of the International Joint Conference on AI (IJCAI-99)*, Stockholm, Sweden, July 31–August 5, 1999.
36. Crubézy, M., Lu, W., Motta, E. and Musen, M. A. (2002) Configuring online problem-solving resources with the internet reasoning service, *Conference on Intelligent Information Processing (IIP 2002) of the International Federation for Information Processing World Computer Congress (WCC 2002)*. Montreal, Canada: Kluwer.
37. Carr, L., De Roure, D., Davis, H. and Hall, W. (1998) Implementing an open link service for the world wide web. *World Wide Web Journal*, **1**(2), 61–71, Baltzer.
38. Carr, L., Hall, W., Bechhofer, S. and Goble, C. (2001) Conceptual linking: oncology-based open hypermedia. *Proceedings of the Tenth International World Wide Web Conference*, Hong Kong, May 1–5, pp. 334–342.
39. Bontcheva, K. (2001) Tailoring the content of dynamically generated explanations, in Bauer, M., Gmytrasiewicz, P. J. and Vassileva, J. (eds) *User Modelling 2001:8th International Conference, UM2001*. Lecture Notes in Artificial Intelligence 2109, Heidelberg, Springer-Verlag.
40. Bontcheva, K. and Wilks, Y. (2001) Dealing with dependencies between content planning and surface realisation in a pipeline generation architecture. *Proceedings of International Joint Conference in Artificial Intelligence (IJCAI '01)*, Seattle, August 7–10, 2001.
41. Sanghee, K., Alani, H., Hall, W., Lewis, P., Millard, D., Shadbolt, N., Weal, M. (2002) Artequakt: generating tailored biographies with automatically annotated fragments from the Web. *Proceedings Semantic Authoring, Annotation and Knowledge Markup Workshop in the 15th European Conference on Artificial Intelligence*, Lyon, France, 2002.
42. Harnad, S. (2001) The self-archiving initiative. *Nature*, **410**, 1024–1025.
43. Hitchcock, S., Carr, L., Jiao, Z., Bergmark, D., Hall, W., Lagoze, C. and Harnad, S. (2000) Developing services for open eprint archives: globalisation, integration and the impact of links. *Proceedings of the 5th ACM Conference on Digital Libraries*, San Antonio, TX, June, 2000, pp. 143–151.
44. Knight, K. and Marcu, D. (2000) Statistics-Based Summarization – Step One: Sentence Compression. *Proceedings of National Conference on Artificial Intelligence (AAAI)*, 2000.
45. Landauer, T. K. and Dumais, S. T. (1997) A solution to Plato's problem: the latent semantic analysis theory of the acquisition, induction, and representation of knowledge. *Psychological Review*, **104**, 211–240.
46. Landauer, T. K., Foltz, P. W. and Laham, D. (1998) Introduction to latent semantic analysis. *Discourse Processes*, **25**, 259–284.
47. Carbonara, L. and Sleeman, D. (1999) Effective and efficient knowledge base refinement. *Machine Learning*, **37**, 143–181.
48. Alani, H. *et al.* (2002) Managing reference: ensuring referential integrity of ontologies for the semantic web. *14th International Conference on Knowledge Engineering and Knowledge Management*, Spain, October, 2002.
49. Luke, S. and Heflin,J. (2000) SHOE 1.1. Proposed Specification, www.cs.umd.edu/projects/plus/SHOE/spec1.01.html, 2000 (current 20 Mar. 2001).

50. Hendler, J. and McGuinness, D. (2000) The DARPA agent markup language. *IEEE Intelligent Systems*, **15**(6), 72, 73.
51. van Harmelen, F. and Horrocks, I. (2000) FAQs on oil: the ontology inference layer. *IEEE Intelligent Systems*, **15**(6), 69–72.
52. Lopez, M. F. *et al.* (1999) Building a chemical ontology using methontology and the ontology design environment. *IEEE Intelligent Systems*, **14**(1), 37–46.
53. Mizoguchi, R., Kozaki, K., Sano, T. and Kitamura, Y. (2000) Construction and deployment of a plant ontology. *12th International Conference on Knowledge Engineering and Knowledge Management*, French Riviera, Juan-les-Pins, October, 2000.
54. McIlraith, S. A., Son, T. C. and Zeng, H. (2001) Semantic web services. *IEEE Intelligent Systems*, **16**(2), 46–53.
55. Staab, S. *et al.* (2000) Semantic community Web portal. *Proc. of WWW-9*, Amsterdam, 2000.
56. Decker, S., Erdmann, M., Fensel, D. and Studer, R. (1999) Ontobroker: ontology-based access to distributed and semi-structured information, in Meersman, R. (ed.) *Semantic Issues in Multimedia Systems: Proceedings of DS-8*. Boston: Kluwer Academic Publishers, pp. 351–369.
57. DAML Services Coalition (alphabetically Ankolenkar, A. *et al.*) (2002) DAML-S: Web service description for the semantic web. *The First International Semantic Web Conference (ISWC)*, June, 2002, pp. 348–363.
58. Cerf, V. G. *et al.* (1993) *National Collaboratories: Applying Information Technologies for Scientific Research*. Washington, D.C.: National Academy Press.
59. Foster, I., Kesselman, C., Nick, J. and Tuecke, S. (2002) *The Physiology of the Grid: Open Grid Services Architecture for Distributed Systems Integration*, presented at GGF4, February, 2002 http://www.globus.og/research/papers/ogsa.pdf.
60. Guttman, R. H., Moukas, A. G. and Maes, P. (1998) Agent-mediated electronic commerce: a survey. *The Knowledge Engineering Review*, **13**(2), 147–159.

18

Peer-to-peer Grids

Geoffrey Fox,[1] Dennis Gannon,[1] Sung-Hoon Ko,[1] Sangmi-Lee,[1,3] Shrideep Pallickara,[1] Marlon Pierce,[1] Xiaohong Qiu,[1,2] Xi Rao,[1] Ahmet Uyar,[1,2] Minjun Wang,[1,2] and Wenjun Wu[1]

[1]*Indiana University, Bloomington, Indiana, United States*
[2]*Syracuse University, Syracuse, New York, United States*
[3]*Florida State University, Tallahassee, Florida, United States*

18.1 PEER-TO-PEER GRIDS

There are no crisp definitions of Grids [1, 2] and Peer-to-Peer (P2P) Networks [3] that allow us to unambiguously discuss their differences and similarities and what it means to integrate them. However, these two concepts conjure up stereotype images that can be compared. Taking 'extreme' cases, Grids are exemplified by the infrastructure used to allow seamless access to supercomputers and their datasets. P2P technology is exemplified by Napster and Gnutella, which can enable *ad hoc* communities of low-end clients to advertise and access the files on the communal computers. Each of these examples offers services but they differ in their functionality and style of implementation. The P2P example could involve services to set up and join peer groups, to browse and access files on a peer, or possibly to advertise one's interest in a particular file. The 'classic' grid could support job submittal and status services and access to sophisticated data management systems.

Grid Computing – Making the Global Infrastructure a Reality. Edited by F. Berman, A. Hey and G. Fox
© 2003 John Wiley & Sons, Ltd   ISBN: 0-470-85319-0

Grids typically have structured robust security services, while P2P networks can exhibit more intuitive trust mechanisms reminiscent of the 'real world'. Again, Grids typically offer robust services that scale well in preexisting hierarchically arranged organizations; P2P networks are often used when a best-effort service is needed in a dynamic poorly structured community. If one needs a particular 'hot digital recording', it is not necessary to locate all sources of this; a P2P network needs to search enough plausible resources that success is statistically guaranteed. On the other hand, a 3D simulation of the universe might need to be carefully scheduled and submitted in a guaranteed fashion to one of the handful of available supercomputers that can support it.

In this chapter, we explore the concept of a P2P Grid with a set of services that include the services of Grids and P2P networks and support naturally environments that have features of both limiting cases. We can discuss two examples in which such a model is naturally applied. In High Energy Physics data analysis (e-Science [4]) problem discussed in Chapter 39, the initial steps are dominated by the systematic analysis of the accelerator data to produce summary events roughly at the level of sets of particles. This Gridlike step is followed by 'physics analysis', which can involve many different studies and much debate among involved physicists as to the appropriate methods to study the data. Here we see some Grid and some P2P features. As a second example, consider the way one uses the Internet to access information – either news items or multimedia entertainment. Perhaps the large sites such as Yahoo, CNN and future digital movie distribution centers have Gridlike organization. There are well-defined central repositories and high-performance delivery mechanisms involving caching to support access. Security is likely to be strict for premium channels. This structured information is augmented by the P2P mechanisms popularized by Napster with communities sharing MP3 and other treasures in a less organized and controlled fashion. These simple examples suggest that whether for science or for commodity communities, information systems should support both Grid and Peer-to-Peer capabilities [5, 6].

In Section 18.2, we describe the overall architecture of a P2P Grid emphasizing the role of Web services and in Section 18.3, we describe the event service appropriate for linking Web services and other resources together. In the following two sections, we describe how collaboration and universal access can be incorporated in this architecture. The latter includes the role of portals in integrating the user interfaces of multiple services. Chapter 22 includes a detailed description of a particular event infrastructure.

18.2 KEY TECHNOLOGY CONCEPTS FOR P2P GRIDS

The other chapters in this book describe the essential architectural features of Web services and we first contrast their application in Grid and in P2P systems. Figure 18.1 shows a traditional Grid with a Web [Open Grid Services Architecture (OGSA)] middleware mediating between clients and backend resources. Figure 18.2 shows the same capabilities but arranged democratically as in a P2P environment. There are some 'real things' (users, computers, instruments), which we term external resources – these are the outer band around the 'middleware egg'. As shown in Figure 18.3, these are linked by

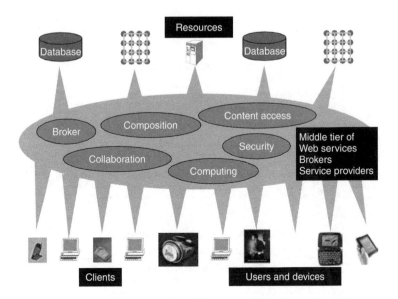

Figure 18.1 A Grid with clients accessing backend resources through middleware services.

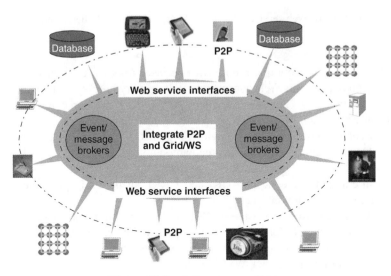

Figure 18.2 A Peer-to-peer Grid.

a collection of Web services [7]. All entities (external resources) are linked by messages whose communication forms a distributed system integrating the component parts.

Distributed object technology is implemented with objects defined in an XML-based IDL (Interface Definition Language) called WSDL (Web Services Definition Language). This allows 'traditional approaches' such as CORBA or Java to be used 'under-the-hood'

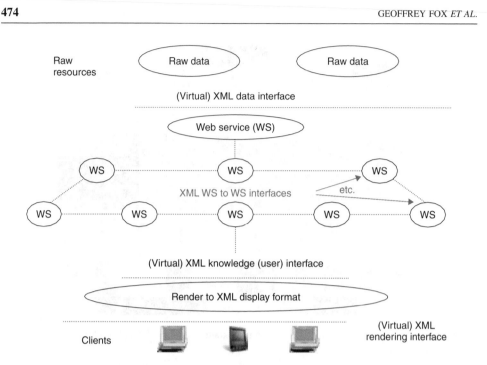

Figure 18.3 Role of Web services (WS) and XML in linkage of clients and raw resources.

with an XML wrapper providing a uniform interface. Another key concept – that of the resource – comes from the Web consortium W3C. Everything – whether an external or an internal entity – is a resource labeled by a Universal Resource Identifier (URI), a typical form being *escience://myplace/mything/mypropertygroup/leaf*. This includes not only macroscopic constructs like computer programs or sensors but also their detailed properties. One can consider the URI as the barcode of the Internet – it labels everything. There are also, of course, Universal Resource Locations (URLs) that tell you where things are. One can equate these concepts (URI and URL) but this is in principle inadvisable, although of course a common practice.

Finally, the environments of Figures 18.1 to 18.3 are built with a service model. A service is an entity that accepts one or more inputs and gives one or more results. These inputs and results are the messages that characterize the system. In WSDL, the inputs and the outputs are termed ports and WSDL defines an overall structure for the messages. The resultant environment is built in terms of the composition of services.

In summary, everything is a resource. The basic macroscopic entities exposed directly to users and to other services are built as distributed objects that are constructed as services so that capabilities and properties are accessed by a message-based protocol. Services contain multiple properties, which are themselves individual resources. A service corresponds roughly to a computer program or a process; the ports (interface of a communication channel with a Web service) correspond to subroutine calls with input parameters and returned data. The critical difference from the past is that one assumes that each service runs on a different computer scattered around the globe. Typically services can be dynamically migrated between computers. Distributed object technology allows us to properly encapsulate the services and provide a management structure. The use of XML and standard interfaces such as WSDL give a universality that allows the interoperability of services from different sources. This picture is consistent with that described throughout this book with perhaps this chapter emphasizing more on the basic concept of resources communicating with messages.

There are several important technology research and development areas on which the above infrastructure builds:

1. Basic system capabilities packaged as Web services. These include security, access to computers (job submittal, status etc.) and access to various forms of databases (information services) including relational systems, Lightweight Directory Access Protocol (LDAP) and XML databases/files. Network wide search techniques about Web services or the content of Web services could be included here. In Section 18.1, we described how P2P and Grid systems exhibited these services but with different trade-offs in performance, robustness and tolerance of local dynamic characteristics.
2. The messaging subsystem between Web services and external resources addressing functionality, performance and fault tolerance. Both P2P and Grids need messaging, although if you compare JXTA [8] as a typical P2P environment with a Web service–based Grid you will see important differences described in Section 18.3. Items 3 to 7 listed below are critical e-Science [4] capabilities that can be used more or less independently.
3. Toolkits to enable applications to be packaged as Web services and construction of 'libraries' or more precisely components. Near-term targets include areas like image processing used in virtual observatory projects or gene searching used in bioinformatics.
4. Application metadata needed to describe all stages of the scientific endeavor.
5. Higher-level and value-added system services such as network monitoring, collaboration and visualization. Collaboration is described in Section 18.4 and can use a common mechanism for both P2P and Grids.
6. What has been called the Semantic Grid [9] or approaches to the representation of and discovery of knowledge from Grid resources. This is discussed in detail in Chapter 17.
7. Portal technology defining user-facing ports on Web services that accept user control and deliver user interfaces.

Figure 18.3 is drawn as a classic three-tier architecture: client (at the bottom), backend resource (at the top) and multiple layers of middleware (constructed as Web services). This is the natural virtual machine seen by a given user accessing a resource. However, the implementation could be very different. Access to services can be mediated by 'servers in the core' or alternatively by direct P2P interactions between machines 'on the edge'. The distributed object abstractions with separate service and message layers allow either P2P or server-based implementations. The relative performance of each approach (which could reflect computer/network horsepower as well as existence of firewalls) would be used in deciding on the implementation to use. P2P approaches best support local dynamic interactions; the server approach scales best globally but cannot easily manage the rich

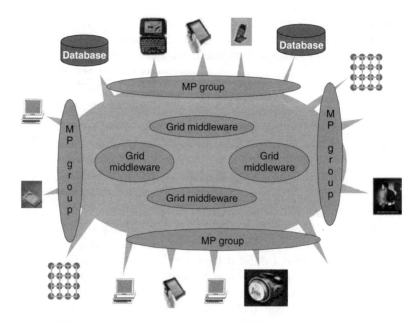

Figure 18.4   Middleware Peer (MP) groups of services at the 'edge' of the Grid.

structure of transient services, which would characterize complex tasks. We refer to our architecture as a P2P grid with peer groups managed locally arranged into a global system supported by core servers. Figure 18.4 redraws Figure 18.2 with Grids controlling central services, while 'services at the edge' are grouped into less organized 'middleware peer groups'. Often one associates P2P technologies with clients but in a unified model, they provide services, which are (by definition) part of the middleware. As an example, one can use the JXTA search technology [8] to federate middle-tier database systems; this dynamic federation can use either P2P or more robust Grid security mechanisms. One ends up with a model shown in Figure 18.5 for managing and organizing services. There is a mix of structured (Gridlike) and unstructured dynamic (P2P-like) services.

We can ask if this new approach to distributed system infrastructure affects key hardware, software infrastructure and their performance requirements. First we present some general remarks. Servers tend to be highly reliable these days. Typically they run in controlled environments but also their software can be proactively configured to ensure reliable operation. One can expect servers to run for months on end and often one can ensure that they are modern hardware configured for the job at hand. Clients on the other hand can be quite erratic with unexpected crashes and network disconnections as well as sporadic connection typical of portable devices. Transient material can be stored by clients but permanent information repositories must be on servers – here we talk about 'logical' servers as we may implement a session entirely within a local peer group of 'clients'. Robustness of servers needs to be addressed in a dynamic fashion and on a scale greater than in the previous systems. However, traditional techniques of replication and careful transaction processing probably can be extended to handle servers and the

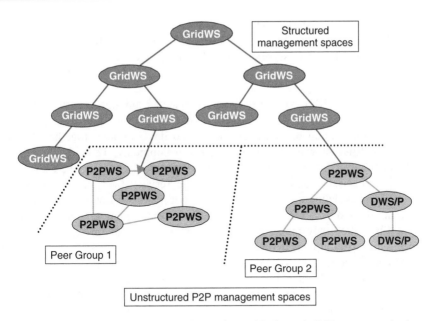

Figure 18.5 A hierarchy of Grid (Web) services with dynamic P2P groups at the leaves.

Web services that they host. Clients realistically must be assumed to be both unreliable and sort of outside our control. Some clients will be 'antiques' and underpowered and are likely to have many software, hardware and network instabilities. In the simplest model, clients 'just' act as a vehicle to render information for the user with all the action on 'reliable' servers. Here applications like Microsoft Word 'should be' packaged as Web services with message-based input and output. Of course, if you have a wonderful robust PC you can run both server(s) and thin client on this system.

18.3 PEER-TO-PEER GRID EVENT SERVICE

Here we consider the communication subsystem, which provides the messaging between the resources and the Web services. Its characteristics are of a Jekyll and Hyde nature. Examining the growing power of optical networks, we see the increasing universal bandwidth that in fact motivates the thin client and the server-based application model. However, the real world also shows slow networks (such as dial-ups), links leading to a high fraction of dropped packets and firewalls stopping our elegant application channels dead in their tracks. We also see some chaos today in the telecom industry that is stunting somewhat the rapid deployment of modern 'wired' (optical) and wireless networks. We suggest that the key to future e-Science infrastructure will be messaging subsystems that manage the communication between external resources, Web services and clients to achieve the highest possible system performance and reliability. We suggest that this problem is sufficiently hard and that we only need to solve this problem 'once', that is, that all communication – whether TCP/IP, User Datagram Protocol (UDP), RTP, RMI, XML or so forth – be handled by a single messaging or event subsystem. Note that this implies that we would tend to separate control and high-volume data transfer, reserving specialized protocols for the latter and more flexible robust approaches for setting up the control channels.

As shown in Figure 18.6, we see the event service as linking all parts of the system together and this can be simplified further as in Figure 18.7 – the event service is to provide the communication infrastructure needed to link resources together. Messaging is addressed in different ways by three recent developments. There is Simple Object Access Protocol (SOAP) messaging [10] discussed in many chapters, the JXTA peer-to-peer protocols [8] and the commercial Java Message Service (JMS) message service [11]. All these approaches define messaging principles but not always at the same level of the Open Systems Interconnect (OSI) stack; further, they have features that sometimes can be compared but often they make implicit architecture and implementation assumptions that hamper interoperability and functionality. SOAP 'just' defines the structure of the message content in terms of an XML syntax and can be clearly used in both Grid and P2P networks. JXTA and other P2P systems mix transport and application layers as the message routing, advertising and discovery are intertwined. A simple example of this is publish–subscribe systems like JMS in which general messages are not sent directly but queued on a broker that uses somewhat *ad hoc* mechanisms to match publishers and subscribers. We will see an important example of this in Section 18.4 when we discuss collaboration; here messages are not unicast between two designated clients but rather shared between multiple clients. In general, a given client does not know the locations of

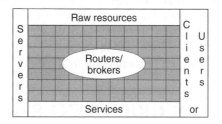

Figure 18.6   One view of system components with event service represented by central mesh.

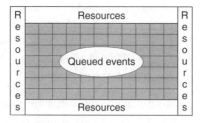

Figure 18.7   Simplest view of system components showing routers of event service supporting queues.

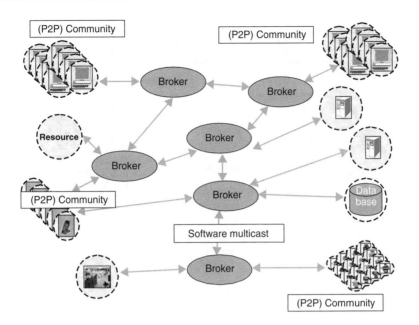

Figure 18.8 Distributed brokers implementing event service.

those other collaborators but rather establishes a criterion for collaborative session. Thus, as in Figure 18.8, it is natural to employ routers or brokers whose function is to distribute messages between the raw resources, clients and servers of the system. In JXTA, these routers are termed *rendezvous peers*.

We consider that the servers provide services (perhaps defined in the WSDL [7] and related XML standards [10]) and do *not* distinguish at this level between what is provided (a service) and what is providing it (a server). Note that we do not distinguish between events and messages; an event is defined by some XML Schema including a time stamp but the latter can of course be absent to allow a simple message to be thought of as an event. Note that an event is itself a resource and might be archived in a database raw resource. Routers and brokers actually provide a service – the management of queued events and so these can themselves be considered as the servers corresponding to the event or message service. This will be discussed a little later as shown in Figure 18.9. Here we note that we design our event systems to support some variant of the publish–subscribe mechanism. Messages are queued from 'publishers' and then clients subscribe to them. XML tag values are used to define the 'topics' or 'properties' that label the queues.

Note that in Figure 18.3, we call the XML Interfaces 'virtual'. This signifies that the interface is logically defined by an XML Schema but could in fact be implemented differently. As a trivial example, one might use a different syntax with say *<sender>meoryou</sender>* replaced by *sender:meoryou*, which is an easier-to-parse-but-less-powerful notation. Such simpler syntax seems a good idea for 'flat' schemas that can be mapped into it. Less trivially, we could define a linear algebra Web service in WSDL

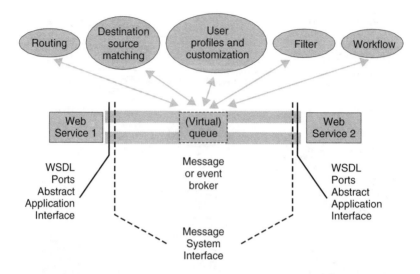

Figure 18.9 Communication model showing subservices of event service.

but compile it into method calls to a Scalapack routine for high-performance implementation. This compilation step would replace the XML SOAP-based messaging [10]
with serialized method arguments of the default remote invocation of this service by
the natural in-memory stack-based use of pointers to binary representations of the arguments. Note that we like publish–subscribe messaging mechanisms but this is sometimes
unnecessary and indeed creates unacceptable overhead. We term the message queues in
Figures 18.7 and 18.9 *virtual* to indicate that the implicit publish–subscribe mechanism
can be bypassed if this agreed in the initial negotiation of communication channel. The use
of virtual queues and virtual XML specifications could suggest the interest in new run-time
compilation techniques, which could replace these universal but at times unnecessarily
slow technologies by optimized implementations.

   We gather together all services that operate on messages in ways that are largely independent of the process (Web service) that produced the message. These are services that
depend on 'message header' (such as destination), message format (such as multimedia
codec) or message process (as described later for the publish–subscribe or workflow
mechanism). Security could also be included here. One could build such capabilities into
each Web service but this is like 'inlining' (more efficient but a job for the run-time
compiler we mentioned above). Figure 18.9 shows the event or message architecture,
which supports communication channels between Web services that can either be direct
or pass through some mechanism allowing various services on the events. These could
be low-level such as routing between a known source and destination or the higher-level
publish–subscribe mechanism that identifies the destinations for a given published event.
Some routing mechanisms in P2P systems in fact use dynamic strategies that merge these
high- and low-level approaches to communication. Note that the messages must support
multiple interfaces: as a 'physical' message it should support SOAP and above this the

event service should support added capabilities such as filtering, publish–subscribe, collaboration, workflow that corresponds to changing message content or delivery. Above this there are application and service standards. All of these are defined in XML, which can be virtualized. As an example, consider an audio–video-conferencing Web service [12, 13]. It could use a simple publish/subscribe mechanism to advertise the availability of some video feed. A client interested in receiving the video would negotiate (using the Session Initiation Protocol (SIP) perhaps) the transmission details. The video could either be sent directly from publisher to subscriber or alternatively from publisher to Web service and then from Web service to subscriber; as a third option, we could send it from the Web service to the client but passing it through a filter that converted one codec into another if required. In the last case, the location of the filter would be negotiated on the basis of computer/network performance issues – it might also involve proprietary software available only at special locations. The choice and details of these three different video transport and filtering strategies would be chosen at the initial negotiation and one would at this stage 'compile' a generic interface to its chosen form. One could of course allow dynamic 'runtime compilation' when the event processing strategy needs to change during a particular stream. This scenario is not meant to be innovative but rather to illustrate the purpose of our architecture building blocks in a homely example. Web services are particularly attractive owing to their support of interoperability, which allows the choices described.

We have designed and implemented a system, NaradaBrokering, supporting the model described here with a dynamic collection of brokers supporting a generalized publish–subscribe mechanism. As described elsewhere [5, 6, 14–16], this can operate either in a client–server mode like JMS or in a completely distributed JXTA-like peer-to-peer mode. By combining these two disparate models, NaradaBrokering can allow optimized performance-functionality trade-offs for different scenarios. Note that typical overheads for broker processing are around 1 ms. This is acceptable for real-time collaboration [6, 13, 17] and even for audio–video-conferencing in which each frame takes around 30 ms. We have demonstrated that such a general messaging system can be applied to real-time synchronous collaboration using the commercial Anabas infrastructure [17, 18].

18.4 COLLABORATION IN P2P GRIDS

Both Grids and P2P networks are associated with collaborative environments. P2P networks started with *ad hoc* communities such as those sharing MP3 files; Grids support virtual enterprises or organizations – these are unstructured or structured societies, respectively. At a high level, collaboration involves sharing and in our context this is sharing of Web services, objects or resources. These are in principle essentially the same thing, although today sharing 'legacy applications' like Microsoft Word is not so usefully considered as sharing Web services. Nevertheless, we can expect that Web service interfaces to 'everything' will be available and will take this point of view later where Word, a Web Page, a computer visualization or the audio–video (at say 30 frames per second) from some videoconferencing system will all be viewed as objects or resources with a known Web service interface. Of course, if you want to implement collaborative systems today, then one must use Microsoft COM as the object model for Word but architecturally at least, this is similar to a Web service interface.

There are many styles and approaches to collaboration. In asynchronous collaboration, different members of a community access the same resource; the Web has revolutionized asynchronous collaboration in its simplest form: one member posting or updating a Web page and others accessing it. Asynchronous collaboration has no special time constraint and typically each community member can access the resource in their own fashion; objects are often shared in a coarse grain fashion with a shared URL pointing to a large amount of information. Asynchronous collaboration is quite fault-tolerant as each user can manage their access to the resource and accommodate difficulties such as poor network connectivity; further, well-established caching techniques can usually be used to improve access performance as the resource is not expected to change rapidly. Synchronous collaboration at a high level is no different from the asynchronous case except that the sharing of information is done in real time. The 'real-time' constraint implies delays of around 10 to 1000 ms per participant or rather 'jitter in transit delays' of a 'few' milliseconds. Note that these timings can be compared to the second or so it takes a browser to load a new page, the several seconds it takes a lecturer to gather thoughts at the start of a new topic (new PowerPoint slide) and the 30-ms frame size natural in audio/video transmission. These numbers are much longer than the parallel computing Message Passing Interface (MPI) message latency measured in microsecond(s) and even the 0.5- to 3-ms typical latency of a middle-tier broker. Nevertheless, synchronous collaboration is much harder than the asynchronous case for several reasons. The current Internet has no reliable Quality of Service (QoS) and so it is hard to accommodate problems coming from unreliable networks and clients. If the workstation crashes during an asynchronous access, one just needs to reboot and restart one's viewing at the point of interruption; unfortunately in the synchronous case, after recovering from an error, one cannot resume from where one lost contact because the rest of the collaborators have moved on. Further, synchronizing objects among the community must often be done at a fine grain size. For asynchronous education, the teacher can share a complete lecture, whereas in a synchronous session we might wish to share a given page in a lecture with a particular scrolling distance and particular highlighting. In summary, synchronous and asynchronous collaboration both involve object sharing but the former is fault-sensitive, has modest real-time constraints and requires fine grain object state synchronization.

The sharing mechanism can be roughly the same for both synchronous and asynchronous case. One needs to establish communities (peer groups) by either direct (members join a session) or indirect (members express interest in topics and are given an opportunity to satisfy this interest) mechanism. The indirect mechanism is the most powerful and is deployed in P2P systems such as JXTA by using XML-expressed advertisements to link together those interested in a particular topic. Audio–video-conferencing systems typically have a direct method with perhaps e-mail used to alert potential attendees of an upcoming session. Commercial Web-conferencing systems such as WebEx and Placeware use this approach. In asynchronous collaboration, one typically 'just' has notification mechanisms for object availability and update. Systems such as CVS (version control) and WebDAV (distributed authoring) are particularly sophisticated in this regard. However, such sharing environments do not usually support the 'microscopic' events as, say, one

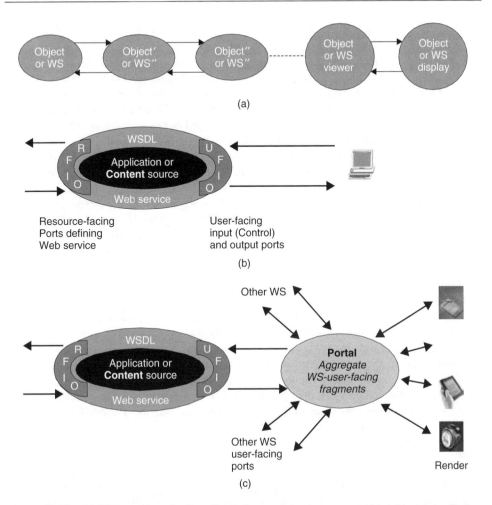

Figure 18.10 (a) Web service pipeline (flow) from originating resource(s) (objects) to display, (b) Web services can have resource-facing and user-facing ports and (c) portal as part of display 'Web service' aggregating user-facing fragments from multiple Web services.

user edits a file; rather 'major events' (check-in and check-out) are communicated between participants. Nevertheless, it is worth stressing that asynchronous collaboration can be supported through the generalized publish–subscribe mechanism with events (messages, advertisements) being linked together to define the collaboration. In order to describe a general approach to collaboration, we need to assume that every Web service has one or more ports in each of the three classes shown in Figure 18.10. The first class is (resource-facing) input ports that supply the information needed to define the state of the Web service; these may be augmented by user-facing input port(s) that allow control information to be passed by the user. The final class is user-facing output ports that supply information needed to construct the user interface. Asynchronous collaboration can share the data (e.g. URL for a Web page or body of an e-mail message) needed to define a Web service (display Web page or browse e-mail in examples).

Let us now consider synchronous collaboration. If one examines an object, then there is typically some pipeline as seen in Figure 18.10 from the original object to the eventual displayed user interface; as described above, let us assume that each stage of the pipeline is a Web service with data flowing from one to another. Other chapters have discussed this composition or flow problem for Web services and our discussion is not sensitive to details of linkages between Web services. Rather than a simple pipeline, one can have a complex dynamic graph linking services together. Figure 18.10 shows the role of portals in this approach and we will return to this in Section 18.5. One can get different types of sharing depending on which 'view' of the basic object one shares, that is, where one intercepts the pipeline and shares the flow of information after this interception. We can identify three particularly important cases illustrated in Figures 18.11 to 18.13; these are shared display, shared input port and shared user-facing output port. The shared input port case is usually called 'shared event', but in fact in a modern architecture with all resources communicating by messages, all collaboration modes can be implemented with a similar event mechanism. The commercial Anabas environment uses the JMS to handle all collaboration modes and we have successfully used NaradaBrokering to replace JMS. In each collaboration mode we assume there is a single 'master' client that 'controls' the Web service; one can in fact have much more complex scenarios with simultaneous and interchangeable control. However, in all cases, there is instantaneously one 'master' and one must transmit the state as seen by this system to all other participants.

In shared display model of Figure 18.11, one shares the bitmap (vector) display and the state is maintained between the clients by transmitting (with suitable compression) the changes in the display. As with video compression like Moving Picture Experts Group

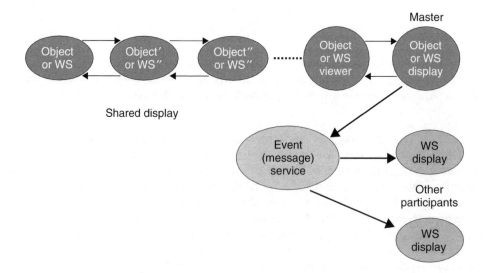

Figure 18.11   Shared display collaboration.

Shared input port (replicated WS) collaboration

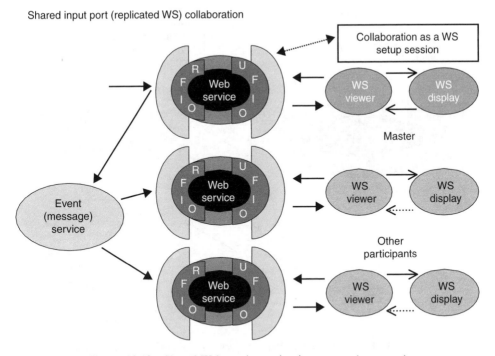

Figure 18.12   Shared Web services using input ports (messages).

(MPEG), one uses multiple event types with some defining full display and others just giving updates. Obviously, the complete display requires substantial network bandwidth but it is useful every now and then, so that one can support clients joining throughout a session, has more fault tolerance and can define full display update points (major events) where asynchronous clients can join a recording. Supporting heterogeneous clients requires that sophisticated shared display environments automatically change size and color resolution to suit each community member. Shared display has one key advantage – it can immediately be applied to all shared objects; it has two obvious disadvantages – it is rather difficult to customize and requires substantial network bandwidth.

In the shared input port (or input message) model of Figure 18.12, one replicates the Web service to be shared with one copy for each client. Then sharing is achieved by intercepting the pipeline before the master Web service and directing copies of the messages on each input port of the 'master' Web service to the replicated copies. Only the user-facing ports in this model are typically partially shared with data from the master transmitted to each replicated Web service, but in a way that can be overridden on each client. We can illustrate this with a more familiar PowerPoint example. Here all the clients have a copy of the PowerPoint application and the presentation to be shared. On the master client, one uses some sort of COM wrapper to detect PowerPoint change events such as slide and animation changes. These 'change' events are sent to all participating clients. This model is not usually termed *shared input ports* but that is just because PowerPoint

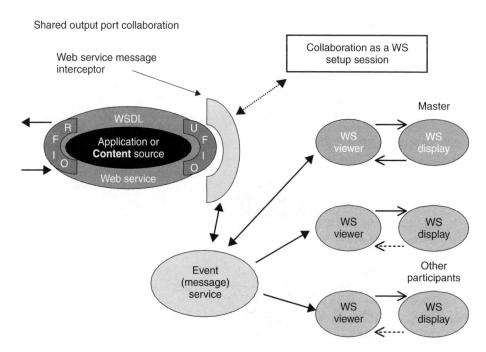

Figure 18.13 Collaborative Web services using shared user-facing ports.

as currently shipped is not set up as a Web service with a messaging interface. One can build a similar shared Web browser and for some browsers (such as that for SVG from Apache) one can in fact directly implement the Web service model. There is a variant here as one can either trap the internal events (such as slide changes in PowerPoint or text area changes in a browser) or the external mouse and keyboard events that generated them. We once developed a sophisticated shared browser using the JavaScript event model to trap user input to a browser. These events were transmitted directly to participating clients to implement such a shared input port model with the user interface playing the role of input ports. We can hope that developments [19] such as Web services for Remote Portals (WSRP) and Web services for Interactive Applications (WSIA) will define the user-facing message ports and the interactions of Web services with input devices so that a coherent systematic approach can be given for replicated Web services with shared input ports.

The shared output port model of Figure 18.13 only involves a single Web service with user-facing ports giving a messaging interface to the client. As in the next section and Figure 18.10, a portal could manage these user-facing ports. Since (by definition) the user-facing ports of a Web service define the user interface, this mechanism simply gives a collaborative version of any Web service. One simple example can be built around any content or multimedia server with multicast output stream(s). This method naturally gives, like shared display, an identical view for each user but with the advantage of typically less network bandwidth since the bitmap display usually is more voluminous than the data transmitted to the client to define the display. In the next section, we discuss user interfaces and suggest that the user-facing ports should not directly define the interface but a menu allowing the client to choose the interface style. In such a case, one can obtain from the shared user-facing model, collaborative views that are customized for each user. Of course, the replicated Web service model of Figure 18.12 offers even greater customizability as each client has the freedom to accept or reject data defining the shared Web service.

Here we have discussed how to make general applications collaborative. Figures 18.12 and 18.13 point out that collaboration is itself a Web service [12, 13] with ports allowing sessions to be defined and to interact with the event service. This collaboration Web service can support asynchronous and all modes of synchronous collaboration.

We proposed that Web services interacting with messages unified P2P and Grid architectures. Here we suggest that sharing either the input or the user-facing ports of a Web service allows one to build flexible environments supporting either the synchronous or the asynchronous collaboration needed to support the communities built around this infrastructure.

18.5 USER INTERFACES AND UNIVERSAL ACCESS

There are several areas where the discussion of Section 18.4 is incomplete and here we clarify some user interface issues that we discuss in the context of universal accessibility, that is, ensuring that any Web service can be accessed by any user irrespective of their physical capability or their network/client connection. Universal access requires that the user interface be defined intelligently by an interaction between the user 'profile' (specifying user and client capabilities and preferences) and the semantics of the Web service [20]. Only the service itself can in general specify what is essential about its user-facing view. This equates to some ways in which the output should not be modified as it is adapted for situations with varying device capabilities, environmental interference and user needs and preferences. In the negotiation between service and client, both sides have a veto and the ability to give hints. The final rendering is in general an iterative optimization process between client and service. This principle implies that the modular pipeline of Figure 18.10 is deficient in the sense that there must be a clear flow not only from the 'basic Web service' to the user but also back again. This can be quite complicated and it is not clear how this is achieved in general as the pipeline from Web service to the user can include transformations, which are not reversible. We can consider the output of each stage of the Web service as a 'document' – each with its own document object model – preferably different instances of the W3C DOM are possible. The final user interface could be a pure audio rendering for a blind user or a bitmap transmitted to a primitive client (perhaps a cell phone) not able to perform full browser functions. Our user-facing port must transmit information in a way such that user interactions can be properly passed back from the user with the correct semantic meaning. Current browsers and transformation tools (such as XSLT) do not appear to address this. At this stage we must assume that this problem is solved perhaps with a back channel communicating directly between the user interface and the Web service. In general, a direct back channel does not appear to be the usual solution but rather as a mix of transport backwards through reversible transformations in the pipeline and direct communication around stages (such as the portal controller) where necessary. In any case some virtual back channel must exist that translates user interaction into an appropriate response by the Web service on the user-facing ports. Actually there are three key user-facing sets of ports:

1. The main user-facing specification output ports that in general do *not* deliver the information defining the display but rather a menu that defines many possible views. A *selector* in Figure 18.14 combines a user profile from the client (specified on a special profile port) with this menu to produce the 'specification of actual user output' that is used by a *portal*, which aggregates many user interface components (from different Web services) into a single view. The result of the transformer may just be a handle that points to a user-facing customized output port.
2. Customized user-facing output port that delivers the selected view from step 1 from the Web service to the client. This in general need not pass through the portal as this only needs the specification of the interface and not the data defining the interface. For multimedia content, step 1 could involve a choice of codec and step 2 could involve the transmission of the chosen codec. The conversion between codecs could in fact involve a general *filter* capability of the event service as another Web filter service as illustrated in Figure 18.7. It seems appropriate to consider interactions with user profiles and filters as outside the original Web service as they can be defined as interacting with the message using a general logic valid for many originating Web services.
3. User-facing input/output port, which is the control channel shown in Figure 18.14.

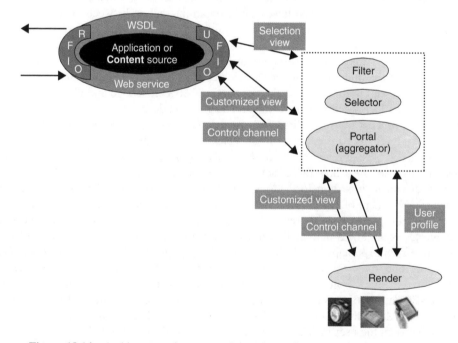

Figure 18.14   Architecture of event service and portal to support universal access.

Note that in Figure 18.14 we have lumped a portal (such as Jetspeed [21, 22] from Apache) as part of the 'event service' as it provides a general service (aggregating user interface components) for all applications (Web services). This packaging may not be very convenient but architecturally portals share features with workflow, filters and collaboration. These are services that operate on message streams produced by Web services. Considering universal access in this fashion could make it easier to provide better customizable interfaces and help those for whom the current display is unsuitable. We expect that more research is needed in areas like the DOM to allow universal interfaces to be generated for general Web services. We would like to thank Al Gilman and Gregg Vanderheiden from the Wisconsin Trace Center for discussions in this area.

ACKNOWLEDGEMENTS

This publication is made possible through partial support provided by DoD High Performance Computing Modernization Program (HPCMP), Programming Environment & Training (PET) activities through Mississippi State University under the terms of Agreement No. # GS04T01BFC0060. The University of Illinois also provided support through the PACI Partners program funded by NSF. The opinions expressed herein are those of the authors and do not necessarily reflect the views of the sponsors.

REFERENCES

1. The Grid Forum, http://www.gridforum.org.
2. Globus Grid Project, http://www.globus.org.
3. Oram, A. (ed.) (2001) *Peer-to-Peer: Harnessing the Power of Disruptive Technologies.* Sebastapol, California: O'Reilly.
4. United Kingdom e-Science Activity, http://www.escience-grid.org.uk/.
5. Bulut, H. *et al.* (2002) An architecture for e-Science and its implications. Presented at *2002 International Symposium on Performance Evaluation of Computer and Telecommunication Systems (SPECTS 2002)*, July 17, 2002, http://grids.ucs.indiana.edu/ptliupages/publications/spectsescience.pdf.
6. Fox, G., Balsoy, O., Pallickara, S., Uyar, A., Gannon, D. and Slominski, A. (2002) Community grids. Invited talk at *The 2002 International Conference on Computational Science*, Amsterdam, The Netherlands, April 21–24, 2002, http://grids.ucs.indiana.edu/ptliupages/publications/iccs.pdf.
7. Web Services Description Language (WSDL) Version 1.1, http://www.w3.org/TR/wsdl.
8. Sun Microsystems JXTA Peer to Peer Technology, http://www.jxta.org.
9. W3C Semantic Web, http://www.w3.org/2001/sw/.
10. XML Based Messaging and Protocol Specifications SOAP, http://www.w3.org/2000/xp/.
11. Sun Microsystems. Java Message Service, http://java.sun.com/products/jms.
12. Fox, G., Wu, W., Uyar, A. and Bulut, H. (2002) A web services framework for collaboration and audio/videoconferencing. *Proceedings of 2002 International Conference on Internet Computing IC '02*, Las Vegas, USA, June 24–27, 2002, http://grids.ucs.indiana.edu/ptliupages/publications/avwebserviceapril02.pdf.
13. Bulut, H., Fox, G., Pallickara, S., Uyar, A. and Wu, W. (2002) Integration of NaradaBrokering and audio/video conferencing as a Web service. *Proceedings of the IASTED International*

Conference on Communications, Internet, and Information Technology, St.Thomas, US Virgin Islands, November 18–20, 2002.

14. Fox, G. and Pallickara, S. (2002) The NaradaBrokering event brokering system: overview and extensions. *Proceedings of the 2002 International Conference on Parallel and Distributed Processing Techniques and Applications (PDPTA '02)*, 2002, http://grids.ucs.indiana.edu/ptliupages/projects/NaradaBrokering/papers/NaradaBrokeringBrokeringSystem.pdf.

15. Fox, G. and Pallickara, S. (2002) JMS compliance in the NaradaBrokering event brokering system. To appear in the *Proceedings of the 2002 International Conference on Internet Computing (IC-02)*, 2002, http://grids.ucs.indiana.edu/ptliupages/projects/NaradaBrokering/papers/JMSSupportInNaradaBrokering.pdf.

16. Fox, G. and Pallickara, S. (2002) Support for peer-to-peer interactions in web brokering systems. *ACM Ubiquity*, **3**(15).

17. Fox, G. *et al.* (2002) Grid services for earthquake science. *Concurrency and Computation: Practice and Experience*, in ACES Special Issue, **14**, 371–393, http://aspen.ucs.indiana.edu/gemmauisummer2001/resources/gemandit7.doc.

18. Anabas Collaboration Environment, http://www.anabas.com.

19. OASIS Web Services for Remote Portals (WSRP) and Web Services for Interactive Applications (WSIA), http://www.oasis-open.org/committees/.

20. Fox, G., Ko, S.-H., Kim, K., Oh, S. and Lee, S. (2002) Integration of hand-held devices into collaborative environments. *Proceedings of the 2002 International Conference on Internet Computing (IC-02)*, Las Vegas, USA, June 24–27, 2002, http://grids.ucs.indiana.edu/ptliupages/publications/pdagarnetv1.pdf.

21. Jetspeed Portal from Apache, http://jakarta.apache.org/jetspeed/site/index.html.

22. Balsoy, O. *et al.* (2002) The online knowledge center: building a component based portal. *Proceedings of the International Conference on Information and Knowledge Engineering*, 2002, http://grids.ucs.indiana.edu:9000/slide/ptliu/research/gateway/Papers/OKCPaper.pdf.

19

Peer-to-Peer Grid Databases for Web Service Discovery

Wolfgang Hoschek

CERN IT Division, European Organization for Nuclear Research, Switzerland

19.1 INTRODUCTION

The fundamental value proposition of computer systems has long been their potential to automate well-defined repetitive tasks. With the advent of distributed computing, the Internet and World Wide Web (WWW) technologies in particular, the focus has been broadened. Increasingly, computer systems are seen as enabling tools for effective long distance communication and collaboration. Colleagues (and programs) with shared interests can work better together, with less respect paid to the physical location of themselves and the required devices and machinery. The traditional departmental team is complemented by cross-organizational virtual teams, operating in an open, transparent manner. Such teams have been termed *virtual organizations* [1]. This opportunity to further extend knowledge appears natural to science communities since they have a deep tradition in drawing their strength from stimulating partnerships across administrative boundaries. In particular, Grid Computing, Peer-to-Peer (P2P) Computing, Distributed Databases, and Web Services introduce core concepts and technologies for *Making the Global Infrastructure a Reality*. Let us look at these in more detail.

Grid Computing – Making the Global Infrastructure a Reality. Edited by F. Berman, A. Hey and G. Fox
© 2003 John Wiley & Sons, Ltd ISBN: 0-470-85319-0

Grids: Grid technology attempts to support flexible, secure, coordinated information sharing among dynamic collections of individuals, institutions, and resources. This includes data sharing as well as access to computers, software, and devices required by computation and data-rich collaborative problem solving [1]. These and other advances of distributed computing are necessary to increasingly make it possible to join loosely coupled people and resources from multiple organizations. Grids are collaborative distributed Internet systems characterized by large-scale heterogeneity, lack of central control, multiple autonomous administrative domains, unreliable components, and frequent dynamic change.

For example, the scale of the next generation Large Hadron Collider project at CERN, the European Organization for Nuclear Research, motivated the construction of the European DataGrid (EDG) [2], which is a global software infrastructure that ties together a massive set of people and computing resources spread over hundreds of laboratories and university departments. This includes thousands of network services, tens of thousands of CPUs, WAN Gigabit networking as well as Petabytes of disk and tape storage [3]. Many entities can now collaborate among each other to enable the analysis of High Energy Physics (HEP) experimental data: the HEP user community and its multitude of institutions, storage providers, as well as network, application and compute cycle providers. Users utilize the services of a set of remote application providers to submit jobs, which in turn are executed by the services of compute cycle providers, using storage and network provider services for I/O. The services necessary to execute a given task often do not reside in the same administrative domain. Collaborations may have a rather static configuration, or they may be more dynamic and fluid, with users and service providers joining and leaving frequently, and configurations as well as usage policies often changing.

Services: Component oriented software development has advanced to a state in which a large fraction of the functionality required for typical applications is available through third-party libraries, frameworks, and tools. These components are often reliable, well documented and maintained, and designed with the intention to be reused and customized. For many software developers, the key skill is no longer hard-core programming, but rather the ability to find, assess, and integrate building blocks from a large variety of third parties.

The software industry has steadily moved towards more software execution flexibility. For example, dynamic linking allows for easier customization and upgrade of applications than static linking. Modern programming languages such as Java use an even more flexible link model that delays linking until the last possible moment (the time of method invocation). Still, most software expects to link and run against third-party functionality installed on the local computer executing the program. For example, a word processor is locally installed together with all its internal building blocks such as spell checker, translator, thesaurus, and modules for import and export of various data formats. The network is not an integral part of the software execution model, whereas the local disk and operating system certainly are.

The maturing of Internet technologies has brought increased ease-of-use and abstraction through higher-level protocol stacks, improved APIs, more modular and reusable server frameworks, and correspondingly powerful tools. The way is now paved for the next step toward increased software execution flexibility. In this scenario, some components are network-attached and made available in the form of network *services* for use by the general public, collaborators, or commercial customers. Internet Service Providers (ISPs) offer to run and maintain reliable services on behalf of clients through hosting environments. Rather than invoking functions of a local library, the application now invokes functions on remote components, in the ideal case, to the same effect. Examples of a service are as follows:

- A replica catalog implementing an interface that, given an identifier (logical file name), returns the global storage locations of replicas of the specified file.
- A replica manager supporting file replica creation, deletion, and management as well as remote shutdown and change notification via publish/subscribe interfaces.
- A storage service offering GridFTP transfer, an explicit TCP buffer-size tuning interface as well as administration interfaces for management of files on local storage systems. An auxiliary interface supports queries over access logs and statistics kept in a registry that is deployed on a centralized high-availability server, and shared by multiple such storage services of a computing cluster.
- A gene sequencing, language translation or an instant news and messaging service.

Remote invocation is always necessary for some demanding applications that cannot (exclusively) be run locally on the computer of a user because they depend on a set of resources scattered over multiple remote domains. Examples include computationally demanding gene sequencing, business forecasting, climate change simulation, and astronomical sky surveying as well as data-intensive HEP analysis sweeping over terabytes of data. Such applications can reasonably only be run on a remote supercomputer or several large computing clusters with massive CPU, network, disk and tape capacities, as well as an appropriate software environment matching minimum standards.

The most straightforward but also most inflexible configuration approach is to hard wire the location, interface, behavior, and other properties of remote services into the local application. Loosely coupled decentralized systems call for solutions that are more flexible and can seamlessly adapt to changing conditions. For example, if a user turns out to be less than happy with the perceived quality of a word processor's remote spell checker, he/she may want to plug in another spell checker. Such dynamic plug-ability may become feasible if service implementations adhere to some common interfaces and network protocols, and if it is possible to match services against an interface and network protocol specification. An interesting question then is: *What infrastructure is necessary to enable a program to have the capability to search the Internet for alternative but similar services and dynamically substitute these?*

Web Services: As communication protocols and message formats are standardized on the Internet, it becomes increasingly possible and important to be able to describe communication mechanisms in some structured way. A service description language addresses this need by defining a grammar for describing Web services as collections of service interfaces capable of executing operations over network protocols to end points. Service descriptions provide documentation for distributed systems and serve as a recipe for automating the details involved in application communication [4]. In contrast to popular belief, a Web Service is neither required to carry XML messages, nor to be bound to Simple Object Access Protocol (SOAP) [5] or the HTTP protocol, nor to run within a .NET hosting environment, although all of these technologies may be helpful for implementation. For clarity, service descriptions in this chapter are formulated in the Simple Web Service Description Language (SWSDL), as introduced in our prior studies [6]. SWSDL describes the interfaces of a distributed service object system. It is a compact pedagogical vehicle trading flexibility for clarity, not an attempt to replace the Web Service Description Language (WSDL) [4] standard. As an example, assume we have a simple scheduling service that offers an operation submitJob that takes a job description as argument. The function should be invoked via the HTTP protocol. A valid SWSDL service description reads as follows:

```
<service>
  <interface type = "http://gridforum.org/Scheduler-1.0">
    <operation>
      <name>void submitJob(String jobdescription)</name>
      <allow> http://cms.cern.ch/everybody </allow>
      <bind:http verb= "GET" URL="https://sched.cern.ch/submitjob"/>
    </operation>
  </interface>
</service>
```

It is important to note that the concept of a service is a logical rather than a physical concept. For efficiency, a *container* of a virtual hosting environment such as the Apache Tomcat servlet container may be used to run more than one service or interface in the same process or thread. The service interfaces of a service may, but need not, be deployed on the same host. They may be spread over multiple hosts across the LAN or WAN and even span administrative domains. This notion allows speaking in an abstract manner about a coherent interface bundle without regard to physical implementation or deployment decisions. We speak of a *distributed (local) service*, if we know and want to stress that service interfaces are indeed deployed across hosts (or on the same host). Typically, a service is persistent (long-lived), but it may also be transient (short-lived, temporarily instantiated for the request of a given user).

The next step toward increased execution flexibility is the (still immature and hence often hyped) *Web Services* vision [6, 7] of distributed computing in which programs are no longer configured with static information. Rather, the promise is that programs are made more flexible, adaptive, and powerful by querying Internet databases (registries) at run time in order to discover information and network-attached third-party building blocks. Services can advertise themselves and related metadata via such databases, enabling the assembly of distributed higher-level components. While advances have recently been made in the field of Web service specification [4], invocation [5], and registration [8], the problem of how to use a rich and expressive general-purpose query language to discover services that offer functionality matching a detailed specification has so far received little attention. A natural question arises: *How precisely can a local application discover relevant remote services?*

For example, a data-intensive HEP analysis application looks for remote services that exhibit a suitable combination of characteristics, including appropriate interfaces, operations, and network protocols as well as network load, available disk quota, access rights, and perhaps quality of service and monetary cost. It is thus of critical importance to develop capabilities for rich service discovery as well as a query language that can support advanced resource brokering. What is more, it is often necessary to use several services in combination to implement the operations of a request. For example, a request may involve the combined use of a file transfer service (to stage input and output data from remote sites), a replica catalog service (to locate an input file replica with good data locality), a request execution service (to run the analysis program), and finally again a file transfer service (to stage output data back to the user desktop). In such cases, it is often helpful to consider correlations. For example, a scheduler for data-intensive requests may look for input file replica locations with a fast network path to the execution service where the request would consume the input data. If a request involves reading large amounts of input data, it may be a poor choice to use a host for execution that has poor data locality with respect to an input data source, even if it is very lightly loaded. *How can one find a set of correlated services fitting a complex pattern of requirements and preferences?*

If one instance of a service can be made available, a natural next step is to have more than one identical distributed instance, for example, to improve availability and performance. Changing conditions in distributed systems include latency, bandwidth, availability, location, access rights, monetary cost, and personal preferences. For example, adaptive users or programs may want to choose a particular instance of a content download service depending on estimated download bandwidth. If bandwidth is degraded in the middle of a download, a user may want to switch transparently to another download service and continue where he/she left off. *On what basis could one discriminate between several instances of the same service?*

Databases: In a large heterogeneous distributed system spanning multiple administrative domains, it is desirable to maintain and query dynamic and timely information about the active participants such as services, resources, and user communities. Examples are a (worldwide) service discovery infrastructure for a DataGrid, the Domain Name System (DNS), the e-mail infrastructure, the World Wide Web, a monitoring infrastructure, or an instant news service. The shared information may also include quality-of-service description, files, current network load, host information, stock quotes, and so on. However, the set of information tuples in the universe is partitioned over one or more database nodes from a wide range of system topologies, for reasons including autonomy, scalability, availability, performance, and security. As in a data integration system [9, 10, 11], the goal is to exploit several independent information sources as if they were a single source. This enables queries for information, resource and service discovery, and collective collaborative functionality that operate on the system as a whole, rather than on a given part of it. For example, it allows a search for descriptions of services of a file-sharing system, to determine its total download capacity, the names of all participating organizations, and so on.

However, in such large distributed systems it is hard to keep track of metadata describing participants such as services, resources, user communities, and data sources. Predictable, timely, consistent, and reliable global state maintenance is infeasible. The information to be aggregated and integrated may be outdated, inconsistent, or not available at all. Failure, misbehavior, security restrictions, and continuous change are the norm rather than the exception. The problem of how to support expressive general-purpose discovery queries over a view that integrates autonomous dynamic database nodes from a wide range of distributed system topologies has so far not been addressed. Consider an instant news service that aggregates news from a large variety of autonomous remote data sources residing within multiple administrative domains. New data sources are being integrated frequently and obsolete ones are dropped. One cannot force control over multiple administrative domains. Reconfiguration or physical moving of a data source is the norm rather than the exception. The question then is *How can one keep track of and query the metadata describing the participants of large cross-organizational distributed systems undergoing frequent change?*

Peer-to-peer networks: It is not obvious how to enable powerful discovery query support and collective collaborative functionality that operate on the distributed system as a whole, rather than on a given part of it. Further, it is not obvious how to allow for search results that are fresh, allowing time-sensitive dynamic content. Distributed (relational) database systems [12] assume tight and consistent central control and hence are infeasible in Grid environments, which are characterized by heterogeneity, scale, lack of central control, multiple autonomous administrative domains, unreliable components, and frequent dynamic change. It appears that a P2P database network may be well suited to support dynamic distributed database search, for example, for service discovery.

In systems such as Gnutella [13], Freenet [14], Tapestry [15], Chord [16], and Globe [17], the overall P2P idea is as follows: rather than have a centralized database, a distributed framework is used where there exist one or more autonomous database nodes, each maintaining its own, potentially heterogeneous, data. Queries are no longer posed to a central database; instead, they are recursively propagated over the network to some or all database nodes, and results are collected and sent back to the client. A node holds a set of tuples in its database. Nodes are interconnected with links in any arbitrary way. A link enables a node to query another node. A *link topology* describes the link structure among nodes. The centralized model has a single node only. For example, in a service discovery system, a link topology can tie together a distributed set of administrative domains, each hosting a registry node holding descriptions of services local to the domain. Several link topology models covering the spectrum from centralized models to fine-grained fully distributed models can be envisaged, among them single node, star, ring, tree, graph, and hybrid models [18]. Figure 19.1 depicts some example topologies.

In any kind of P2P network, nodes may publish themselves to other nodes, thereby forming a topology. In a P2P network for service discovery, a *node* is a service that exposes *at least* interfaces for publication and P2P queries. Here, nodes, services, and other content providers may publish (their) service descriptions and/or other metadata to one or more nodes. Publication enables distributed node topology construction (e.g. ring, tree, or graph) and at the same time constructs the federated database searchable by queries. In other examples, nodes may support replica location [19], replica management, and optimization [20, 21], interoperable access to Grid-enabled relational databases [22], gene sequencing or multilingual translation, actively using the network to discover services such as replica catalogs, remote gene mappers, or language dictionaries.

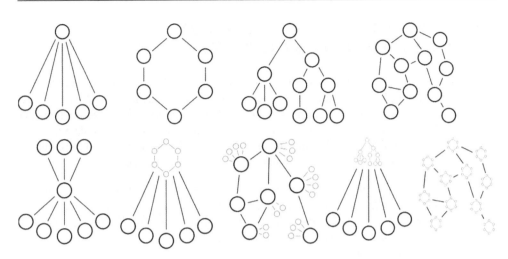

Figure 19.1   Example link topologies [18].

Organization of this chapter: This chapter distills and generalizes the essential properties of the discovery problem and then develops solutions that apply to a wide range of large distributed Internet systems. It shows how to support expressive general-purpose queries over a view that integrates autonomous dynamic database nodes from a wide range of distributed system topologies. We describe the first steps toward the convergence of Grid computing, P2P computing, distributed databases, and Web services. The remainder of this chapter is organized as follows:

Section 2 addresses the problems of maintaining dynamic and timely information populated from a large variety of unreliable, frequently changing, autonomous, and heterogeneous remote data sources. We design a database for XQueries over dynamic distributed content – the so-called *hyper registry*.

Section 3 defines the *Web Service Discovery Architecture (WSDA)*, which views the Internet as a large set of services with an extensible set of well-defined interfaces. It specifies a small set of orthogonal multipurpose communication primitives (building blocks) for discovery. These primitives cover service identification, service description retrieval, data publication as well as minimal and powerful query support. WSDA promotes interoperability, embraces industry standards, and is open, modular, unified, and simple yet powerful.

Sections 4 and 5 describe the *Unified Peer-to-Peer Database Framework (UPDF)* and corresponding *Peer Database Protocol (PDP)* for general-purpose query support in large heterogeneous distributed systems spanning many administrative domains. They are unified in the sense that they allow expression of specific discovery applications for a wide range of data types, node topologies, query languages, query response modes, neighbor selection policies, pipelining characteristics, time-out, and other scope options.

Section 6 discusses related work. Finally, Section 7 summarizes and concludes this chapter. We also outline interesting directions for future research.

19.2 A DATABASE FOR DISCOVERY OF DISTRIBUTED CONTENT

In a large distributed system, a variety of information describes the state of autonomous entities from multiple administrative domains. Participants frequently join, leave, and act on a best-effort basis. Predictable, timely, consistent, and reliable global state maintenance is infeasible. The information to be aggregated and integrated may be outdated, inconsistent, or not available at all. Failure, misbehavior, security restrictions, and continuous change are the norm rather than the exception. The key problem then is

How should a database node maintain information populated from a large variety of unreliable, frequently changing, autonomous, and heterogeneous remote data sources? In particular, how should it do so without sacrificing reliability, predictability, and simplicity? How can powerful queries be expressed over time-sensitive dynamic information?

A type of database is developed that addresses the problem. A database for XQueries over dynamic distributed content is designed and specified – the so-called *hyper registry*. The registry has a number of key properties. An XML data model allows for structured and semistructured data, which is important for integration of heterogeneous content. The XQuery language [23] allows for powerful searching, which is critical for nontrivial applications. Database state maintenance is based on soft state, which enables reliable, predictable, and simple content integration from a large number of autonomous distributed content providers. Content link, content cache, and a hybrid pull/push communication model allow for a wide range of dynamic content freshness policies, which may be driven by all three system components: content provider, registry, and client.

A hyper registry has a database that holds a set of tuples. A *tuple* may contain a piece of arbitrary *content*. Examples of content include a service description expressed in WSDL [4], a quality-of-service description, a file, file replica location, current network load, host information, stock quotes, and so on. A tuple is annotated with a *content link* pointing to the authoritative data source of the embedded content.

19.2.1 Content link and content provider

Content link: A *content link* may be any arbitrary URI. However, most commonly, it is an HTTP(S) URL, in which case it points to the content of a content provider, and an HTTP(S) GET request to the link must return the current (up-to-date) content. In other words, a simple hyperlink is employed. In the context of service discovery, we use the term *service link* to denote a content link that points to a service description. Content links can freely be chosen as long as they conform to the URI and HTTP URL specification [24]. Examples of content links are

```
urn:/iana/dns/ch/cern/cn/techdoc/94/1642-3
urn:uuid:f81d4fae-7dec-11d0-a765-00a0c91e6bf6
http://sched.cern.ch:8080/getServiceDescription.wsdl
https://cms.cern.ch/getServiceDesc?id=4712&cache=disable
```

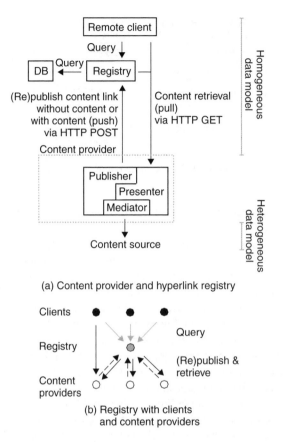

(a) Content provider and hyperlink registry

(b) Registry with clients
and content providers

Figure 19.2 (a) Content provider and hyper registry and (b) registry with clients and content providers.

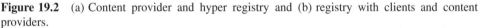

```
http://phone.cern.ch/lookup?query="select phone from book where phone=4711"
http://repcat.cern.ch/getPFNs?lfn="myLogicalFileName"
```

Content provider: A *content provider* offers information conforming to a homogeneous global data model. In order to do so, it typically uses some kind of internal mediator to transform information from a local or proprietary data model to the global data model. A content provider can be seen as a gateway to heterogeneous content sources. A content provider is an umbrella term for two components, namely, a presenter and a publisher. The *presenter* is a service and answers HTTP(S) GET content retrieval requests from a registry or client (subject to local security policy). The *publisher* is a piece of code that publishes content link, and perhaps also content, to a registry. The publisher need not be a service, although it uses HTTP(S) POST for transport of communications. The structure of a content provider and its interaction with a registry and a client are depicted in Figure 19.2(a). Note that a client can bypass a registry and directly pull

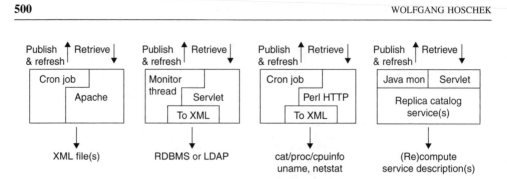

Figure 19.3 Example content providers.

current content from a provider. Figure 19.2(b) illustrates a registry with several content providers and clients.

Just as in the dynamic WWW that allows for a broad variety of implementations for the given protocol, it is left unspecified how a presenter computes content on retrieval. Content can be static or dynamic (generated on the fly). For example, a presenter may serve the content directly from a file or database or from a potentially outdated cache. For increased accuracy, it may also dynamically recompute the content on each request. Consider the example providers in Figure 19.3. A simple but nonetheless very useful content provider uses a commodity HTTP server such as Apache to present XML content from the file system. A simple `cron` job monitors the health of the Apache server and publishes the current state to a registry. Another example of a content provider is a Java servlet that makes available data kept in a relational or LDAP database system. A content provider can execute legacy command line tools to publish system-state information such as network statistics, operating system, and type of CPU. Another example of a content provider is a network service such as a replica catalog that (in addition to servicing replica look up requests) publishes its service description and/or link so that clients may discover and subsequently invoke it.

19.2.2 Publication

In a given context, a content provider can publish content of a given type to one or more registries. More precisely, a content provider can publish a dynamic pointer called a *content link*, which in turn enables the registry (and third parties) to retrieve the current (up-to-date) content. For efficiency, the `publish` operation takes as input a set of zero or more tuples. In what we propose to call the *Dynamic Data Model (DDM)*, each XML tuple has a content link, a type, a context, four soft-state time stamps, and (optionally) metadata and content. A tuple is an annotated multipurpose soft-state data container that may contain a piece of arbitrary *content* and allows for refresh of that content at any time, as depicted in Figures 19.4 and 19.5.

- *Link*: The content link is an URI in general, as introduced above. If it is an HTTP(S) URL, then the current (up-to-date) content of a content provider can be retrieved (pulled) at any time.

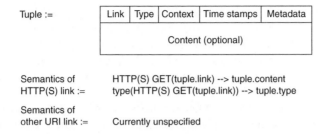

Figure 19.4 Tuple is an annotated multipurpose soft-state data container, and allows for dynamic refresh.

```
<tupleset>
    <tuple link="http://registry.cern.ch/getDescription" type="service" ctx="parent"
            TS1="10" TC="15" TS2="20" TS3="30">
        <content>
            <service>
                <interface type="http://cern.ch/Presenter-1.0">
                    <operation>
                        <name>XML getServiceDescription()</name>
                        <bind:http verb="GET" URL="https://registry.cern.ch/getDesc"/>
                    </operation>
                </interface>

                <interface type = "http://cern.ch/XQuery-1.0">
                    <operation>
                        <name> XML query(XQueryquery)</name>
                        <bind:beep URL="beep://registry.cern.ch:9000"/>
                    </operation>
                </interface>
            </service>
        </content>

        <metadata> <owner name="http://cms.cern.ch"/> </metadata>
    </tuple>

    <tuple link="http://repcat.cern.ch/getDesc?id=4711" type="service" ctx="child"
            TS1="30" TC="0" TS2="40" TS3="50">
    </tuple>

    <tuple link="urn:uuid:f81d4fae-11d0-a765-00a0c91e6bf6"
            type="replica" TC="65" TS1="60" TS2="70" TS3="80">
        <content>
            <replicaSet LFN="urn:/iana/dns/ch/cern/cms/higgs-file" size="10000000" type="MySQL/ISAM">
                <PFN URL="ftp://storage.cern.ch/file123" readCount="17"/>
                <PFN URL="ftp://se01.infn.it/file456"    readCount="1"/>
            </replicaSet>
        </content>
    </tuple>

    <tuple link="http://monitor.cern.ch/getHosts" type="hosts" TC="65" TS1="60" TS2="70" TS3="80">
        <content>
            <hosts>
                <host name="fred01.cern.ch" os="redhat 7.2" arch="i386" mem="512M" MHz="1000"/>
                <host name="fred02.cern.ch" os="solaris 2.7" arch="sparc" mem="8192M" MHz="400"/>
            </hosts>
        </content>
    </tuple>
</tupleset>
```

Figure 19.5 Example tuple set from dynamic data model.

- *Type*: The type describes *what* kind of content is being published (e.g. `service`, `application/octet-stream`, `image/jpeg`, `networkLoad`, `hostinfo`).
- *Context*: The context describes *why* the content is being published or *how* it should be used (e.g. `child`, `parent`, `x-ireferral`, `gnutella`, `monitoring`). Context and type allow a query to differentiate on crucial attributes even if content caching is not supported or authorized.
- *Time stamps TS1, TS2, TS3, TC*: On the basis of embedded soft-state time stamps defining lifetime properties, a tuple may eventually be discarded unless refreshed by a stream of timely confirmation notifications. The time stamps allow for a wide range of powerful caching policies, some of which are described below in Section 2.5.
- *Metadata*: The optional metadata element further describes the content and/or its retrieval beyond what can be expressed with the previous attributes. For example, the metadata may be a secure digital XML signature [25] of the content. It may describe the authoritative content provider or owner of the content. Another metadata example is a Web Service Inspection Language (WSIL) document [26] or fragment thereof, specifying additional content retrieval mechanisms beyond HTTP content link retrieval. The metadata argument is an extensibility element enabling customization and flexible evolution.
- *Content*: Given the link the current (up-to-date) content of a content provider can be retrieved (pulled) at any time. Optionally, a content provider can also include a copy of the current content as part of publication (push). Content and metadata can be structured or semistructured data in the form of any arbitrary well-formed XML document or fragment.[1] An individual element may, but need not, have a schema (XML Schema [28]), in which case it must be valid according to the schema. All elements may, but need not, share a common schema. This flexibility is important for integration of heterogeneous content.

The publish operation of a registry has the signature `void publish(XML tuple-set)`. Within a tuple set, a tuple is uniquely identified by its *tuple key*, which is the pair (`content link, context`). If a key does not already exist on publication, a tuple is inserted into the registry database. An existing tuple can be updated by publishing other values under the same tuple key. An existing tuple (key) is 'owned' by the content provider that created it with the first publication. It is recommended that a content provider with another identity may not be permitted to publish or update the tuple.

19.2.3 Query

Having discussed the data model and how to publish tuples, we now consider a query model. It offers two interfaces, namely, `MinQuery` and `XQuery`.

---

[1] For clarity of exposition, the content is an XML element. In the general case (allowing nontext-based content types such as `image/jpeg`), the content is a MIME [27] object. The XML-based publication input tuple set and query result tuple set is augmented with an additional MIME multipart object, which is a list containing all content. The content element of a tuple is interpreted as an index into the MIME multipart object.

MinQuery: The `MinQuery` interface provides the simplest possible query support (*'select all'*-style). It returns tuples including or excluding cached content. The `getTuples()` query operation takes no arguments and returns the full set of all tuples 'as is'. That is, query output format and publication input format are the same (see Figure 19.5). If supported, output includes cached content. The `getLinks()` query operation is similar in that it also takes no arguments and returns the full set of all tuples. However, it always substitutes an empty string for cached content. In other words, the content is omitted from tuples, potentially saving substantial bandwidth. The second tuple in Figure 19.5 has such a form.

XQuery: The `XQuery` interface provides powerful XQuery [23] support, which is important for realistic service and resource discovery use cases. XQuery is the standard XML query language developed under the auspices of the W3C. It allows for powerful searching, which is critical for nontrivial applications. Everything that can be expressed with SQL can also be expressed with XQuery. However, XQuery is a more expressive language than SQL, for example, because it supports path expressions for hierarchical navigation. Example XQueries for service discovery are depicted in Figure 19.6. A detailed discussion of a wide range of simple, medium, and complex discovery queries and their representation in the XQuery [23] language is given in Reference [6]. XQuery can dynamically integrate external data sources via the `document(URL)` function, which can be used to process the XML results of remote operations invoked over HTTP. For example, given

- *Find all (available) services.*

  ```
  RETURN /tupleset/tuple[@type="service"]
  ```

- *Find all services that implement a replica catalog service interface that CMS members are allowed to use, and that have an HTTP binding for the replica catalog operation "XML getPFNs(String LFN).*

  ```
  LET $repcat := "http://cern.ch/ReplicaCatalog-1.0"
  FOR $tuple in /tupleset/tuple[@type="service"]
  LET $s := $tuple/content/service
  WHERE SOME $op in $s/interface[@type = $repcat]/operation SATISFIES
      $op/name="XML getPFNs(StringLFN)" AND $op/bindhttp/@verb="GET" AND contains($op/allow, "cms.cern.ch")
  RETURN $tuple
  ```

- *Find all replica catalogs and return their physical file names (PFNs) for a given logical file name (LFN); suppress PFNs not starting with "ftp://".*

  ```
  LET $repcat := "http://cern.ch/ReplicaCatalog-1.0"
  LET $s := /tupleset/tuple[@type="service"]/content/service[interface@type = $repcat]
  RETURN
      FOR $pfn IN invoke($s, $repcat, "XML getPFNs(StringLFN)", "http://myhost.cern.ch/myFile")/tupleset/PFN
      WHERE starts-with($pfn, "ftp://")
      RETURN $pfn
  ```

- *Return the number of replica catalog services.*

  ```
  RETURN count(/tupleset/tuple/content/service[interface/@type="http://cern.ch/ReplicaCatalog-1.0"])
  ```

- *Find all (execution service, storage service) pairs where both services of a pair live within the same domain.(Job wants to read and write locally).*

  ```
  LET $executorType:="http://cern.ch/executor-1.0"
  LET $storageType:="http://cern.ch/storage-1.0"
  FOR $executorIN /tupleset/tuple[content/service/interface/@type=$executorType],
      $storage IN /tupleset/tuple[content/service/interface/@type=$storageType
                              AND domainName(@link) = domainName($executor/@link)]
  RETURN <pair> {$executor} {$storage} </pair>
  ```

Figure 19.6   Example XQueries for service discovery.

a service description with a `getPhysicalFileNames(LogicalFileName)` operation, a query can match on values dynamically produced by that operation. The same rules that apply to minimalist queries also apply to XQuery support. An implementation can use a modular and simple XQuery processor such as `Quip` for the operation `XML query(XQuery query)`. Because not only content but also content link, context, type, time stamps, metadata, and so on are part of a tuple, a query can also select on this information.

19.2.4 Caching

Content *caching* is important for client efficiency. The registry may not only keep content links but also a copy of the current content pointed to by the link. With caching, clients no longer need to establish a network connection for each content link in a query result set in order to obtain content. This avoids prohibitive latency, in particular, in the presence of large result sets. A registry may (but need not) support caching. A registry that does not support caching ignores any content handed from a content provider. It keeps content links only. Instead of cached content, it returns empty strings (see the second tuple in Figure 19.5 for an example). Cache coherency issues arise. The query operations of a caching registry may return tuples with stale content, that is, content that is out of date with respect to its master copy at the content provider.

A caching registry may implement a *strong* or *weak cache coherency policy*. A strong cache coherency policy is *server invalidation* [29]. Here a content provider notifies the registry with a publication tuple whenever it has locally modified the content. We use this approach in an adapted version in which a caching registry can operate according to the client push pattern (*push registry*) or server pull pattern (*pull registry*) or a hybrid thereof. The respective interactions are as follows:

- *Pull registry*: A content provider publishes a content link. The registry then pulls the current content via content link retrieval into the cache. Whenever the content provider modifies the content, it notifies the registry with a publication tuple carrying the time the content was last modified. The registry may then decide to pull the current content again, in order to update the cache. It is up to the registry to decide if and when to pull content. A registry may pull content at any time. For example, it may dynamically pull fresh content for tuples affected by a query. This is important for frequently changing dynamic data like network load.
- *Push registry*: A publication tuple pushed from a content provider to the registry contains not only a content link but also its current content. Whenever a content provider modifies content, it pushes a tuple with the new content to the registry, which may update the cache accordingly.
- *Hybrid registry*: A hybrid registry implements both pull and push interactions. If a content provider merely notifies that its content has changed, the registry may choose to pull the current content into the cache. If a content provider pushes content, the cache may be updated with the pushed content. This is the type of registry subsequently assumed whenever a caching registry is discussed.

A noncaching registry ignores content elements, if present. A publication is said to be *without content* if the content is not provided at all in the tuple. Otherwise, it is said to be *with content*. Publication without content implies that no statement at all about cached content is being made (neutral). It does *not* imply that content should not be cached or invalidated.

19.2.5 Soft state

For reliable, predictable, and simple distributed state maintenance, a registry tuple is maintained as *soft state*. A tuple may eventually be discarded unless refreshed by a stream of timely confirmation notifications from a content provider. To this end, a tuple carries time stamps. A tuple is expired and removed unless explicitly renewed via timely periodic publication, henceforth termed *refresh*. In other words, a refresh allows a content provider to cause a content link and/or cached content to remain present for a further time.

The strong cache coherency policy *server invalidation* is extended. For flexibility and expressiveness, the ideas of the Grid Notification Framework [30] are adapted. The publication operation takes four absolute time stamps TS1, TS2, TS3, TC per tuple. The semantics are as follows: The content provider asserts that its content was last modified at time TS1 and that its current content is expected to be valid from time TS1 until at least time TS2. It is expected that the content link is alive between time TS1 and at least time TS3. Time stamps must obey the constraint $TS1 \leq TS2 \leq TS3$. TS2 triggers expiration of cached content, whereas TS3 triggers expiration of content links. Usually, TS1 equals the time of last modification or first publication, TS2 equals TS1 plus some minutes or hours, and TS3 equals TS2 plus some hours or days. For example, TS1, TS2, and TS3 can reflect publication time, 10 min, and 2 h, respectively.

A tuple also carries a time stamp TC that indicates the time when the tuple's embedded content (not the provider's master copy of the content) was last modified, typically by an intermediary in the path between client and content provider (e.g. the registry). If a content provider publishes with content, then we usually have TS1=TC. TC must be zero-valued if the tuple contains no content. Hence, a registry not supporting caching always has TC set to zero. For example, a highly dynamic network load provider may publish its link without content and TS1=TS2 to suggest that it is inappropriate to cache its content. Constants are published with content and TS2=TS3=infinity, TS1=TC=currentTime. Time stamp semantics can be summarized as follows:

```
TS1 = Time content provider last modified content
TC  = Time embedded tuple content was last modified (e.g. by intermediary)
TS2 = Expected time while current content at provider is at least valid
TS3 = Expected time while content link at provider is at least valid (alive)
```

Insert, update, and delete of tuples occur at the time stamp–driven state transitions summarized in Figure 19.7. Within a tuple set, a tuple is uniquely identified by its *tuple key*, which is the pair (content link, context). A tuple can be in one of three states: *unknown, not cached,* or *cached*. A tuple is unknown if it is not contained in the registry (i.e. its key does not exist). Otherwise, it is known. When a tuple is assigned *not cached* state, its last internal modification time TC is (re)set to zero and the cache is

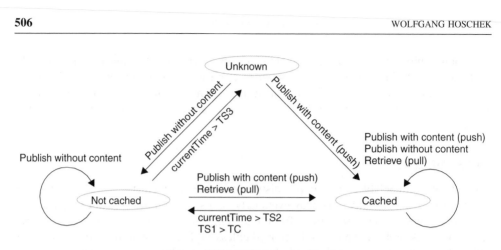

Figure 19.7   Soft state transitions.

deleted, if present. For a *not cached* tuple, we have `TC < TS1`. When a tuple is assigned *cached* state, the content is updated and `TC` is set to the current time. For a *cached* tuple, we have `TC ≥ TS1`.

A tuple moves from *unknown* to *cached* or *not cached* state if the provider publishes with or without content, respectively. A tuple becomes *unknown* if its content link expires (`currentTime > TS3`); the tuple is then deleted. A provider can force tuple deletion by publishing with `currentTime > TS3`. A tuple is upgraded from *not cached* to *cached* state if a provider push publishes with content or if the registry pulls the current content itself via retrieval. On content pull, a registry may leave `TS2` unchanged, but it may also follow a policy that extends the lifetime of the tuple (or any other policy it sees fit). A tuple is degraded from *cached* to *not cached* state if the content expires. Such expiry occurs when no refresh is received in time (`currentTime > TS2`) or if a refresh indicates that the provider has modified the content (`TC < TS1`).

19.2.6   Flexible freshness

Content link, content cache, a hybrid pull/push communication model, and the expressive power of XQuery allow for a wide range of dynamic content freshness policies, which may be driven by all three system components: content provider, registry, and client. All three components may indicate how to manage content according to their respective notions of freshness. For example, a content provider can model the freshness of its content via pushing appropriate time stamps and content. A registry can model the freshness of its content via controlled acceptance of provider publications and by actively pulling fresh content from the provider. If a result (e.g. network statistics) is up to date according to the registry, but out of date according to the client, the client can pull fresh content from providers as it sees fit. However, this is inefficient for large result sets. Nevertheless, it is important for clients that query results are returned according to their notion of freshness, in particular, in the presence of frequently changing dynamic content.

Recall that it is up to the registry to decide to what extent its cache is stale, and if and when to pull fresh content. For example, a registry may implement a policy that dynamically pulls fresh content for a tuple whenever a query touches (affects) the tuple. For example, if a query interprets the content link as an identifier within a hierarchical namespace (e.g. as in LDAP) and selects only tuples within a subtree of the namespace, only these tuples should be considered for refresh.

Refresh-on-client-demand: So far, a registry must guess what a client's notion of freshness might be, while at the same time maintaining its decisive authority. A client still has no way to indicate (as opposed to force) its view of the matter to a registry. We propose to address this problem with a simple and elegant *refresh-on-client-demand* strategy under control of the registry's authority. The strategy exploits the rich expressiveness and dynamic data integration capabilities of the XQuery language. The client query may itself inspect the time stamp values of the set of tuples. It may then decide itself to what extent a tuple is considered interesting yet stale. If the query decides that a given tuple is stale (e.g. if type="networkLoad" AND TC < currentTime() - 10), it calls the XQuery document (URL contentLink) function with the corresponding content link in order to pull and get handed fresh content, which it then processes in any desired way.

This mechanism makes it unnecessary for a registry to guess what a client's notion of freshness might be. It also implies that a registry does not require complex logic for query parsing, analysis, splitting, merging, and so on. Moreover, the fresh results pulled by a query can be reused for subsequent queries. Since the query is executed within the registry, the registry may implement the document function such that it not only pulls and returns the current content but as a side effect also updates the tuple cache in its database. A registry retains its authority in the sense that it may apply an authorization policy, or perhaps ignore the query's refresh calls altogether and return the old content instead. The refresh-on-client-demand strategy is simple, elegant, and controlled. It improves efficiency by avoiding overly eager refreshes typically incurred by a guessing registry policy.

19.3 WEB SERVICE DISCOVERY ARCHITECTURE

Having defined all registry aspects in detail, we now proceed to the definition of a Web service layer that promotes interoperability for Internet software. Such a layer views the Internet as a large set of services with an extensible set of well-defined interfaces. A Web service consists of a set of interfaces with associated operations. Each operation may be bound to one or more network protocols and end points. The definition of interfaces, operations, and bindings to network protocols and end points is given as a service description. A discovery architecture defines appropriate services, interfaces, operations, and protocol bindings for discovery. The key problem is

> Can we define a discovery architecture that promotes interoperability, embraces industry standards, and is open, modular, flexible, unified, nondisruptive, and simple yet powerful?

We propose and specify such an architecture, the so-called *Web Service Discovery Architecture (WSDA)*. WSDA subsumes an array of disparate concepts, interfaces, and protocols under a single semitransparent umbrella. It specifies a small set of orthogonal multipurpose communication primitives (building blocks) for discovery. These primitives cover service identification, service description retrieval, data publication as well as minimal and powerful query support. The individual primitives can be combined and plugged together by specific clients and services to yield a wide range of behaviors and emerging synergies.

19.3.1 Interfaces

The four WSDA interfaces and their respective operations are summarized in Table 19.1. Figure 19.8 depicts the interactions of a client with implementations of these interfaces. Let us discuss the interfaces in more detail.

Table 19.1 WSDA interfaces and their respective operations

Interface	Operations	Responsibility
Presenter	XML getServiceDescription()	Allows clients to retrieve the current description of a service and hence to bootstrap all capabilities of a service.
Consumer	(TS4,TS5) publish(XML tupleset)	A content provider can publish a dynamic pointer called a *content link*, which in turn enables the consumer (e.g. registry) to retrieve the current content. Optionally, a content provider can also include a copy of the current content as part of publication. Each input tuple has a content link, a type, a context, four time stamps, and (optionally) metadata and content.
MinQuery	XML getTuples() XML getLinks()	Provides the simplest possible query support (*'select all'*-style). The getTuples operation returns the full set of all available tuples 'as is'. The minimal getLinks operation is identical but substitutes an empty string for cached content.
XQuery	XML query(XQuery query)	Provides powerful XQuery support. Executes an XQuery over the available tuple set. Because not only content, but also content link, context, type, time stamps, metadata and so on are part of a tuple, a query can also select on this information.

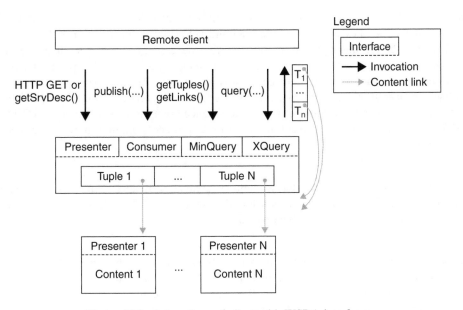

Figure 19.8   Interactions of client with WSDA interfaces.

Presenter: The `Presenter` interface allows clients to retrieve the current (up-to-date) service description. Clearly, clients from anywhere must be able to retrieve the current description of a service (subject to local security policy). Hence, a service needs to present (make available) to clients the means to retrieve the service description. To enable clients to query in a global context, some identifier for the service is needed. Further, a description retrieval mechanism is required to be associated with each such identifier. Together these are the bootstrap key (or handle) to all capabilities of a service.

In principle, identifier and retrieval mechanisms could follow any reasonable convention, suggesting the use of any arbitrary URI. In practice, however, a fundamental mechanism such as service discovery can only hope to enjoy broad acceptance, adoption, and subsequent ubiquity if integration of legacy services is made easy. The introduction of service discovery as a new and additional auxiliary service capability should require as little change as possible to the large base of valuable existing legacy services, preferably no change at all. It should be possible to implement discovery-related functionality without changing the core service. Further, to help easy implementation the retrieval mechanism should have a very narrow interface and be as simple as possible.

Thus, for generality, we define that an identifier may be any URI. However, in support of the above requirements, the identifier is most commonly chosen to be a URL [24], and the retrieval mechanism is chosen to be HTTP(S). If so, we define that an HTTP(S) GET request to the identifier must return the current service description (subject to local security policy). In other words, a simple hyperlink is employed. In the remainder of this chapter, we will use the term *service link* for such an identifier enabling service description retrieval. Like in the WWW, service links (and content links, see below) can freely be chosen as long as they conform to the URI and HTTP URL specification [24].

Because service descriptions should describe the essentials of the service, it is recommended[2] that the service link concept be an integral part of the description itself. As a result, service descriptions may be retrievable via the `Presenter` interface, which defines an operation `getServiceDescription()` for this purpose. The operation is identical to service description retrieval and is hence bound to (invoked via) an HTTP(S) GET request to a given service link. Additional protocol bindings may be defined as necessary.

Consumer: The `Consumer` interface allows content providers to publish a tuple set to a consumer. The publish operation has the signature `(TS4, TS5) publish(XML tuple set)`. For details, see Section 2.2.

MinQuery: The `MinQuery` interface provides the simplest possible query support (*'select all'*-style). The `getTuples()` and `getLinks()` operations return tuples including and excluding cached content, respectively. For details, see Section 2.3.

Advanced query support can be expressed on top of the minimal query capabilities. Such higher-level capabilities conceptually do not belong to a consumer and minimal query interface, which are only concerned with the fundamental capability of making a content link (e.g. service link) *reachable*[3] for clients. As an analogy, consider the related but distinct concepts of Web hyperlinking and Web searching: Web hyperlinking is a fundamental capability without which nothing else on the Web works. Many different kinds of Web search engines using a variety of search interfaces and strategies can and are layered on top of Web linking. The kind of XQuery support we propose below is certainly not the only possible and useful one. It seems unreasonable to assume that a single global standard query mechanism can satisfy all present and future needs of a wide range of communities. Many such mechanisms should be able to coexist. Consequently, the consumer and query interfaces are deliberately separated and kept as minimal as possible, and an additional interface type (`XQuery`) for answering XQueries is introduced.

XQuery: The greater the number and heterogeneity of content and applications, the more important expressive general-purpose query capabilities become. Realistic ubiquitous service and resource discovery *stands and falls* with the ability to express queries in a rich general-purpose query language [6]. A query language suitable for service and resource discovery should meet the requirements stated in Table 19.2 (in decreasing order of significance). As can be seen from the table, LDAP, SQL, and XPath do not meet a number of essential requirements, whereas the XQuery language meets all requirements

---

[2] In general, it is not mandatory for a service to implement any 'standard' interface.

[3] *Reachability* is interpreted in the spirit of garbage collection systems: A content link is reachable for a given client if there exists a direct or indirect retrieval path from the client to the content link.

Table 19.2 Capabilities of XQuery, XPath, SQL, and LDAP query languages

Capability	XQuery	XPath	SQL	LDAP
Simple, medium, and complex queries over a set of tuples	yes	no	yes	no
Query over structured and semi-structured data	yes	yes	no	yes
Query over heterogeneous data	yes	yes	no	yes
Query over XML data model	yes	yes	no	no
Navigation through hierarchical data structures (path expressions)	yes	yes	no	exact match only
Joins (combine multiple data sources into a single result)	yes	no	yes	no
Dynamic data integration from multiple heterog. sources such as databases, documents, and remote services	yes	yes	no	no
Data restructuring patterns (e.g. SELECT-FROM-WHERE in SQL)	yes	no	yes	no
Iteration over sets (e.g. FOR clause)	yes	no	yes	no
General-purpose predicate expressions (WHERE clause)	yes	no	yes	no
Nesting several kinds of expressions with full generality	yes	no	no	no
Binding of variables and creating new structures from bound variables (LET clause)	yes	no	yes	no
Constructive queries	yes	no	no	no
Conditional expressions (IF ... THEN ... ELSE)	yes	no	yes	no
Arithmetic, comparison, logical, and set expressions	yes, all	yes	yes, all	log. & string
Operations on data types from a type system	yes	no	yes	no
Quantified expressions (e.g. SOME, EVERY clause)	yes	no	yes	no
Standard functions for sorting, string, math, aggregation	yes	no	yes	no
User defined functions	yes	no	yes	no
Regular expression matching	yes	yes	no	no
Concise and easy to understand queries	yes	yes	yes	yes

and desiderata posed. The operation XML query(XQuery query) of the XQuery interface is detailed in Section 2.3.

19.3.2 Network protocol bindings and services

The operations of the WSDA interfaces are bound to (carried over) a default transport protocol. The XQuery interface is bound to the *Peer Database Protocol (PDP)* (see Section 5). PDP supports database queries for a wide range of database architectures and response models such that the stringent demands of ubiquitous Internet discovery infrastructures in terms of scalability, efficiency, interoperability, extensibility, and reliability can be met. In particular, it allows for high concurrency, low latency, pipelining as well as early and/or partial result set retrieval, both in pull and push mode. For all other operations and arguments, we assume for simplicity HTTP(S) GET and POST as transport, and XML-based parameters. Additional protocol bindings may be defined as necessary.

We define two kinds of example registry services: The so-called *hypermin registry* must (at least) support the three interfaces Presenter, Consumer, and MinQuery (excluding XQuery support). A *hyper registry* must (at least) support these interfaces plus the XQuery interface. Put another way, any service that happens to support, among others, the respective interfaces qualifies as a hypermin registry or hyper registry. As usual, the interfaces may have end points that are hosted by a single container, or they may be spread across multiple hosts or administrative domains.

It is by no means a requirement that only dedicated hyper registry services and hypermin registry services may implement WSDA interfaces. Any arbitrary service may decide to offer and implement none, some or all of these four interfaces. For example, a job scheduler may decide to implement, among others, the MinQuery interface to indicate a simple means to discover metadata tuples related to the current status of job queues and the supported Quality of Service. The scheduler may not want to implement the Consumer interface because its metadata tuples are strictly read-only. Further, it may not want to implement the XQuery interface, because it is considered overkill for its purposes. Even though such a scheduler service does not qualify as a hypermin or hyper registry, it clearly offers useful added value. Other examples of services implementing a subset of WSDA interfaces are consumers such as an instant news service or a cluster monitor. These services may decide to implement the Consumer interface to invite external sources for data feeding, but they may not find it useful to offer and implement any query interface.

In a more ambitious scenario, the example job scheduler may decide to publish its local tuple set also to an (already existing) remote hyper registry service (i.e. with XQuery support). To indicate to clients how to get hold of the XQuery capability, the scheduler may simply copy the XQuery interface description of the remote hyper registry service and advertise it as its own interface by including it in its own service description. This kind of *virtualization* is not a 'trick', but a feature with significant practical value, because it allows for minimal implementation and maintenance effort on the part of the scheduler.

Alternatively, the scheduler may include in its local tuple set (obtainable via the `getLinks()` operation) a tuple that refers to the service description of the remote hyper registry service. An interface referral value for the context attribute of the tuple is used, as follows:

```
<tuple link="https://registry.cern.ch/getServiceDescription"
       type="service" ctx="x-ireferral://cern.ch/XQuery-1.0"
       TS1="30" TC= "0" TS2="40" TS3="50">
</tuple>
```

19.3.3 Properties

WSDA has a number of key properties:

- *Standards integration*: WSDA embraces and integrates solid and broadly accepted industry standards such as XML, XML Schema [28], the SOAP [5], the WSDL [4], and XQuery [23]. It allows for integration of emerging standards such as the WSIL [26].
- *Interoperability*: WSDA promotes an interoperable Web service layer on top of Internet software, because it defines appropriate services, interfaces, operations, and protocol bindings. WSDA does not introduce new Internet standards. Rather, it judiciously combines existing interoperability-proven open Internet standards such as HTTP(S), URI [24], MIME [27], XML, XML Schema [28], and BEEP [33].
- *Modularity*: WSDA is modular because it defines a small set of orthogonal multi-purpose communication primitives (building blocks) for discovery. These primitives cover service identification, service description retrieval, publication, as well as minimal and powerful query support. The responsibility, definition, and evolution of any given primitive are distinct and independent of that of all other primitives.
- *Ease-of-use and ease-of-implementation*: Each communication primitive is deliberately designed to avoid any unnecessary complexity. The design principle is to *'make simple and common things easy, and powerful things possible'*. In other words, solutions are rejected that provide for powerful capabilities yet imply that even simple problems are complicated to solve. For example, service description retrieval is by default based on a simple HTTP(S) GET. Yet, we do not exclude, and indeed allow for, alternative identification and retrieval mechanisms such as the ones offered by Universal Description, Discovery and Integration (UDDI) [8], RDBMS or custom Java RMI registries (e.g. via tuple metadata specified in WSIL [26]). Further, tuple content is by default given in XML, but advanced usage of arbitrary MIME [27] content (e.g. binary images, files, MS-Word documents) is also possible. As another example, the minimal query interface requires virtually no implementation effort on the part of a client and server. Yet, where necessary, also powerful XQuery support may, but need not, be implemented and used.
- *Openness and flexibility*: WSDA is open and flexible because each primitive can be used, implemented, customized, and extended in many ways. For example, the interfaces of a service may have end points spread across multiple hosts or administrative domains. However, there is nothing that prevents all interfaces to be colocated on the same host or implemented by a single program. Indeed, this is often

a natural deployment scenario. Further, even though default network protocol bindings are given, additional bindings may be defined as necessary. For example, an implementation of the Consumer interface may bind to (carry traffic over) HTTP(S), SOAP/BEEP [34], FTP or Java RMI. The tuple set returned by a query may be maintained according to a wide variety of cache coherency policies resulting in static to highly dynamic behavior. A consumer may take any arbitrary custom action upon publication of a tuple. For example, it may interpret a tuple from a specific schema as a command or an active message, triggering tuple transformation, and/or forwarding to other consumers such as loggers. For flexibility, a service maintaining a WSDA tuple set may be deployed in any arbitrary way. For example, the database can be kept in a XML file, in the same format as returned by the getTuples query operation. However, tuples can also be dynamically recomputed or kept in a relational database.

- *Expressive power*: WSDA is powerful because its individual primitives can be combined and plugged together by specific clients and services to yield a wide range of behaviors. Each single primitive is of limited value all by itself. The true value of simple orthogonal multipurpose communication primitives lies in their potential to generate powerful emerging synergies. For example, combination of WSDA primitives enables building services for replica location, name resolution, distributed auctions, instant news and messaging, software and cluster configuration management, certificate and security policy repositories, as well as Grid monitoring tools. As another example, the consumer and query interfaces can be combined to implement a P2P database network for service discovery (see Section 19.4). Here, a node of the network is a service that exposes *at least* interfaces for publication and P2P queries.

- *Uniformity*: WSDA is unified because it subsumes an array of disparate concepts, interfaces, and protocols under a single *semi-transparent* umbrella. It allows for multiple competing distributed systems concepts and implementations to coexist and to be integrated. Clients can dynamically adapt their behavior based on rich service introspection capabilities. Clearly, there exists no solution that is optimal in the presence of the heterogeneity found in real-world large cross-organizational distributed systems such as DataGrids, electronic marketplaces and instant Internet news and messaging services. Introspection and adaptation capabilities increasingly make it unnecessary to mandate a single global solution to a given problem, thereby enabling integration of collaborative systems.

- *Non-disruptiveness*: WSDA is nondisruptive because it offers interfaces but does not mandate that every service in the universe must comply to a set of 'standard' interfaces.

19.4 PEER-TO-PEER GRID DATABASES

In a large cross-organizational system, the set of information tuples is partitioned over many distributed nodes, for reasons including autonomy, scalability, availability, performance, and security. It is not obvious how to enable powerful discovery query support and collective collaborative functionality that operate on the distributed system as a whole, rather than on a given part of it. Further, it is not obvious how to allow for search results that are fresh, allowing time-sensitive dynamic content. It appears that a P2P database network may be well suited to support dynamic distributed database search, for example, for service discovery. The key problems then are

- What are the detailed architecture and design options for P2P database searching in the context of service discovery? What response models can be used to return matching query results? How should a P2P query processor be organized? What query types can be answered (efficiently) by a P2P network? What query types have the potential to immediately start piping in (early) results? How can a maximum of results be delivered reliably within the time frame desired by a user, even if a query type does not support pipelining? How can loops be detected reliably using timeouts? How can a query scope be used to exploit topology characteristics in answering a query?
- Can we devise a unified P2P database framework for general-purpose query support in large heterogeneous distributed systems spanning many administrative domains? More precisely, can we devise a framework that is unified in the sense that it allows expression of specific discovery applications for a wide range of data types, node topologies, query languages, query response modes, neighbor selection policies, pipelining characteristics, time-out, and other scope options?

In this section, we take the first steps toward unifying the fields of Database Management Systems (DBMSs) and P2P computing, which so far have received considerable, but separate, attention. We extend database concepts and practice to cover P2P search. Similarly, we extend P2P concepts and practice to support powerful general-purpose query languages such as XQuery [23] and SQL [35]. As a result, we propose the *Unified Peer-to-Peer Database Framework (UPDF)* and corresponding *Peer Database Protocol (PDP)* for general-purpose query support in large heterogeneous distributed systems spanning many administrative domains. They are unified in the sense that they allow expression of specific discovery applications for a wide range of data types, node topologies, query languages, query response modes, neighbor selection policies, pipelining characteristics, timeout, and other scope options.

19.4.1 Routed versus direct response, metadata responses

When any *originator* wishes to search a P2P network with some query, it sends the query to an *agent node*. The node applies the query to its local database and returns matching results; it also forwards the query to select *neighbor nodes*. These neighbors return their local query results, they also forward the query to select neighbors, and so on. We propose to distinguish four techniques of return matching query results to an originator: *Routed Response, Direct Response, Routed Metadata Response*, and *Direct Metadata Response*, as depicted in Figure 19.9. Let us examine the main implications with a Gnutella use case. A typical Gnutella query such as *Like a virgin* is matched by some hundreds of files, most of them referring to replicas of the very same music file. Not all matching files are identical because there exist multiple related songs (e.g. remixes, live recordings) and multiple versions of a song (e.g. with different sampling rates). A music file has a size

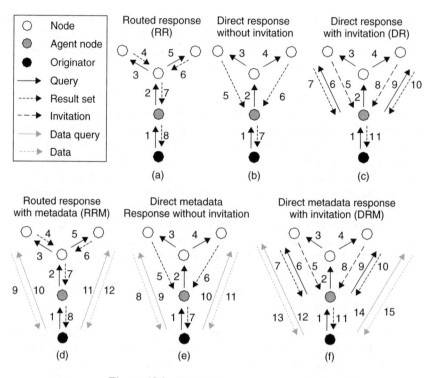

Figure 19.9   Peer-to-peer response modes.

of at least several megabytes. Many thousands of concurrent users submit queries to the Gnutella network.

- *Routed response*: (Figure 19.9a). Results are propagated back into the originator along the paths on which the query flowed outwards. Each (passive) node returns to its (active) client not only its own local results but also all remote results it receives from neighbors. Routing messages through a logical overlay network of P2P nodes is much less efficient than routing through a physical network of IP routers [36]. Routing back even a single Gnutella file (let alone all results) for each query through multiple nodes would consume large amounts of overall system bandwidth, most likely grinding Gnutella to a screeching halt. As the P2P network grows, it is fragmented because nodes with low bandwidth connections cannot keep up with traffic [37]. Consequently, routed responses are not well suited for file-sharing systems such as Gnutella. In general, *overall economics* dictate that routed responses are not well suited for systems that return many and/or large results.
- *Direct response*: With and without invitation. To better understand the underlying idea, we first introduce the simpler variant, which is Direct Response Without Invitation (Figure 19.9b). Results are not returned by routing back through intermediary nodes. Each (active) node that has local results sends them directly to the (passive) agent, which combines and hands them back to the originator. Response traffic does

not travel through the P2P system. It is offloaded via individual point-to-point data transfers on the edges of the network. Let us examine the main implications with a use case. As already mentioned, a typical Gnutella query such as *Like a virgin* is matched by some hundreds of files, most of them referring to replicas of the very same music file. For Gnutella users, it would be sufficient to receive just a small subset of matching files. Sending back *all* such files would unnecessarily consume large amounts of direct bandwidth, most likely restricting Gnutella to users with excessive cheap bandwidth at their disposal. Note, however, that the overall Gnutella system would be only marginally affected by a single user downloading, say, a million music files, because the largest fraction of traffic does not travel through the P2P system itself. In general, *individual economics* dictate that direct responses without invitation are not well suited for systems that return many equal and/or large results, while a small subset would be sufficient. A variant based on invitation (Figure 19.9c) softens the problem by inverting control flow. Nodes with matching files do not blindly push files to the agent. Instead, they invite the agent to initiate downloads. The agent can then act as it sees fit. For example, it can filter and select a subset of data sources and files and reject the rest of the invitations. Owing to its inferiority, the variant without invitation is not considered any further. In the remainder of this chapter, we use the term *Direct Response* as a synonym for Direct Response With Invitation.

- *Routed metadata response and direct metadata response*: Here, interaction consists of two phases. In the first phase, routed responses (Figure 19.9d) or direct responses (Figure 19.9e,f) are used. However, nodes do not return data results in response to queries, but only small metadata results. The metadata contains just enough information to enable the originator to retrieve the data results and possibly to apply filters before retrieval. In the second phase, the originator selects, on the basis of the metadata, which data results are relevant. The (active) originator directly connects to the relevant (passive) data sources and asks for data results. Again, the largest fraction of response traffic does not travel through the P2P system. It is offloaded via individual point-to-point data transfers on the edges of the network.

The routed metadata response approach is used by file-sharing systems such as Gnutella. A Gnutella query does not return files; it just returns an annotated set of HTTP URLs. The originator connects to a subset of these URLs to download files as it sees fit. Another example is a service discovery system in which the first phase returns a set of service links instead of full service descriptions. In the second phase, the originator connects to a subset of these service links to download service descriptions as it sees fit. Another example is a *referral* system in which the first phase uses routed metadata response to return the service links of the set of nodes having local matching results (*Go ask these nodes for the answer*). In the second phase, the originator or agent connects directly to a subset of these nodes to query and retrieve result sets as it sees fit. This variant avoids the 'invitation storm' possible under Direct Response. Referrals are also known as *redirections*. A metadata response mode with a radius scope of zero can be used to implement the referral behavior of the DNS.

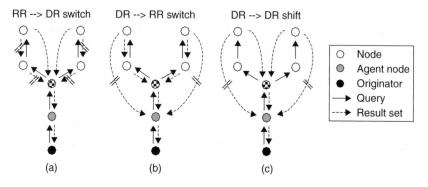

Figure 19.10   Response mode switches and shifts (RR...routed response, DR...direct response).

For a detailed comparison of the properties of the various response models, see our prior studies [38]. Although from the functional perspective all response modes are equivalent, no mode is optimal under all circumstances. The question arises as to what extent a given P2P network must mandate the use of any particular response mode throughout the system. Observe that nodes are autonomous and defined by their interface only. Consequently, we propose that response modes can be mixed by *switches* and *shifts*, in arbitrary permutations, as depicted in Figure 19.10. The response flows that would have been taken are shown crossed out. It is useful to allow specifying as part of the query a hint that indicates the preferred response mode (`routed` or `direct`).

19.4.2   Query processing

In a distributed database system, there exists a single local database and zero or more neighbors. A classic centralized database system is a special case in which there exists a single local database and zero neighbors. From the perspective of query processing, a P2P database system has the same properties as a distributed database system, in a recursive structure. Hence, we propose to organize the P2P query engine like a general distributed query engine [39, 12]. A given query involves a number of operators (e.g. SELECT, UNION, CONCAT, SORT, JOIN, SEND, RECEIVE, SUM, MAX, IDENTITY) that may or may not be exposed at the query language level. For example, the SELECT operator takes a set and returns a new set with tuples satisfying a given predicate. The UNION operator computes the union of two or more sets. The CONCAT operator concatenates the elements of two or more sets into a multiset of arbitrary order (without eliminating duplicates). The IDENTITY operator returns its input set unchanged. The semantics of an operator can be satisfied by several operator implementations, using a variety of algorithms, each with distinct resource consumption, latency, and performance characteristics. The query optimizer chooses an efficient query execution plan, which is a tree plugged together from operators. In an execution plan, a parent operator consumes results from child operators.

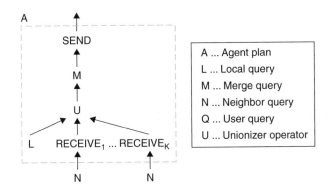

Figure 19.11 Template execution plan.

Template query execution plan: *Any* query Q within our query model can be answered by an agent with the *template execution plan* A depicted in Figure 19.11. The plan applies a local query L against the tuple set of the local database. Each neighbor (if any) is asked to return a result set for (the same) neighbor query N. Local and neighbor result sets are unionized into a single result set by a unionizer operator U that must take the form of either UNION or CONCAT. A merge query M is applied that takes as input the result set and returns a new result set. The final result set is sent to the client, that is, another node or an originator.

Centralized execution plan: To see that indeed any query against any kind of database system can be answered within this framework, we derive a simple *centralized execution plan* that always satisfies the semantics of any query Q. The plan substitutes specific sub-plans into the template plan A, leading to distinct plans for the agent node (Figure 19.12a) and neighbors nodes (Figure 19.12b). In the case of XQuery and SQL, parameters are substituted as follows:

XQuery	SQL
A: M=Q, U=UNION, L="RETURN /", N' =N	A: M=Q, U=UNION, L="SELECT *", N' =N
N: M=IDENTITY, U=UNION, L= "RETURN /", N' =N	N: M=IDENTITY, U=UNION, L= "SELECT *", N' =N

In other words, the agent's plan A fetches all raw tuples from the local and remote databases, unionizes the result sets, and then applies the query Q. Neighbors are handed a rewritten neighbor query N that recursively fetches all raw tuples, and returns their union. The neighbor query N is recursively partitionable (see below).

The same centralized plan works for routed and direct response, both with and without metadata. Under direct response, a node does forward the query N, but does not attempt

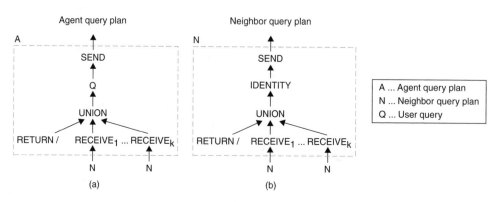

Figure 19.12   Centralized execution plan.

to receive remote result sets (conceptually empty result sets are delivered). The node does not send a result set to its predecessor, but directly back to the agent.

The centralized execution plan can be inefficient because potentially large amounts of base data have to be shipped to the agent before locally applying the user's query. However, sometimes this is the only plan that satisfies the semantics of a query. This is always the case for a complex query. A more efficient execution plan can sometimes be derived (as proposed in the following section). This is always the case for a simple and medium query.

Recursively partitionable query: A P2P network can be efficient in answering queries that are recursively partitionable. A query Q is *recursively partitionable* if, for the template plan A, there exists a merge query M and a unionizer operator U to satisfy the semantics of the query Q assuming that L and N are chosen as L = Q and N = A. In other words, a query is recursively partitionable if the very same execution plan *can* be recursively applied at every node in the P2P topology. The corresponding execution plan is depicted in Figure 19.13.

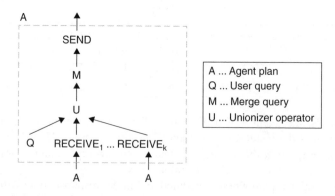

Figure 19.13   Execution plan for recursively partitionable query.

The input and output of a merge query have the same form as the output of the local query L. Query processing can be parallelized and spread over all participating nodes. Potentially very large amounts of information can be searched while investing little resources such as processing time per individual node. The recursive parallel spread of load implied by a recursively partitionable query is the basis of the massive P2P scalability potential. However, query performance is not necessarily good, for example, because of high network I/O costs.

Now we are in a position to clarify the definition of simple, medium, and complex queries.

- *Simple query*: A query is *simple* if it is recursively partitionable using M = IDEN-TITY, U = UNION. An example is *Find all (available) services*.
- *Medium query*: A query is a *medium* query if it is not simple, but it is recursively partitionable. An example is *Return the number of replica catalog services*.
- *Complex query*: A query is *complex* if it is not recursively partitionable. An example is *Find all (execution service, storage service) pairs in which both services of a pair live within the same domain*.

For simplicity, in the remainder of this chapter we assume that the user explicitly provides M and U along with a query Q. If M and U are not provided as part of a query to any given node, the node acts defensively by assuming that the query is not recursively partitionable. Choosing M and U is straightforward for a human being. Consider, for example, the following medium XQueries.

- *Return the number of replica catalog services*: The merge query computes the sum of a set of numbers. The unionizer is CONCAT.

```
Q = RETURN
        <tuple>
          count(/tupleset/tuple/content/service[interface/@type="repcat"])
        </tuple>
M = RETURN
        <tuple>
          sum(/tupleset/tuple)
        </tuple>
U = CONCAT
```

- *Find the service with the largest uptime*:

```
Q=M= RETURN (/tupleset/tuple[@type="service "] SORTBY (./@uptime)) [last()]
U = UNION
```

Note that the query engine always encapsulates the query output with a tuple set root element. A query need not generate this root element as it is implicitly added by the environment.

Pipelining: The success of many applications depends on how fast they can start producing initial/relevant portions of the result set rather than how fast the entire result set is produced [40]. Often an originator would be happy to already work with one or a few *early results*, as long as they arrive quickly and reliably. Results that arrive later can be handled later or are ignored anyway. This is particularly often the case in distributed systems in which many nodes are involved in query processing, each of which may be unresponsive for many reasons. The situation is even more pronounced in systems with loosely coupled autonomous nodes.

Operators of any kind have a uniform iterator interface, namely, the three methods `open()`, `next()`, and `close()`. For efficiency, the method `next()` can be asked to deliver several results at once in a so-called *batch*. Semantics are as follows: *Give me a batch of at least N and at most M results* (less than N results are delivered when the entire query result set is exhausted). For example, the SEND and RECEIVE network communication operators typically work in batches.

The monotonic semantics of certain operators such as SELECT, UNION, CONCAT, SEND, and RECEIVE allow that operator implementations consume just one or a few child results on `next()`. In contrast, the nonmonotonic semantics of operators such as SORT, GROUP, MAX, some JOIN methods, and so on require that operator implementations consume *all* child results already on `open()` in order to be able to deliver a result on the first call to `next()`. Since the output of these operators on a subset of the input is not, in general, a subset of the output on the whole input, these operators need to see all of their input before they produce the correct output. This does not break the iterator concept but has important latency and performance implications. Whether the root operator of an agent exhibits a short or long latency to deliver to the originator the first result from the result set depends on the query operators in use, which in turn depend on the given query. In other words, for some query types the originator has the potential to immediately start piping in results (at moderate performance rate), while for other query types it must wait for a long time until the first result becomes available (the full result set arrives almost at once, however).

A query (an operator implementation) is said to be *pipelined* if it can already produce at least one result tuple before all input tuples have been seen. Otherwise, a query (an operator) is said to be *nonpipelined*. Simple queries do support pipelining (e.g. Gnutella queries). Medium queries may or may not support pipelining, whereas complex queries typically do not support pipelining. Figure 19.14 depicts pipelined and nonpipelined example queries.

19.4.3 Static loop time-out and dynamic abort time-out

Clearly, there comes a time when a user is no longer interested in query results, no matter whether any more results might be available. The query roaming the network and its response traffic should fade away after some time. In addition, P2P systems are well advised to attempt to limit resource consumption by defending against *runaway* queries roaming forever or producing gigantic result sets, either unintended or malicious. To address these problems, an absolute *abort time-out* is attached to a query, as it travels across hops. An abort time-out can be seen as a deadline. Together with the query, a node tells a neighbor *I will ignore (the rest of) your result set if I have not received it before 12:00:00 today*. The problem, then, is to ensure that a maximum of results can be delivered reliably within the time frame desired by a user. The value of a *static time-out*

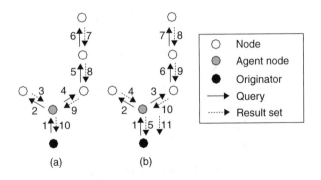

Figure 19.14   (a) Nonpipelined and (b) Pipelined query.

remains unchanged across hops, except for defensive modification in flight triggered by runaway query detection (e.g. infinite time-out). In contrast, it is intended that the value of a *dynamic time-out* be decreased at each hop. Nodes further away from the originator may time out earlier than nodes closer to the originator.

Dynamic abort time-out: A static abort time-out is entirely unsuitable for nonpipelined result set delivery, because it leads to a serious reliability problem, which we propose to call *simultaneous abort time-out*. If just one of the many nodes in the query path fails to be responsive for whatever reasons, all other nodes in the path are waiting, and eventually time-out and attempt to return at least a partial result set. However, it is impossible that any of these partial results ever reach the originator, because all nodes time-out *simultaneously* (and it takes some time for results to flow back).

To address the simultaneous abort timeout problem, we propose dynamic abort time-outs. Under *dynamic abort time-out*, nodes further away from the originator time out earlier than nodes closer to the originator. This provides some safety time window for the partial results of any node to flow back across multiple hops to the originator. Intermediate nodes can and should adaptively decrease the time-out value as necessary, in order to leave a large enough time window for receiving and returning partial results subsequent to timeout.

Observe that the closer a node is to the originator, the more important it is (if it cannot meet its deadline, results from a large branch are discarded). Further, the closer a node is to the originator, the larger is its response and bandwidth consumption. Thus, as a good policy to choose the safety time window, we propose *exponential decay with halving*. The window size is halved at each hop, leaving large safety windows for important nodes and tiny window sizes for nodes that contribute only marginal result sets. Also, taking into account network latency and the time it takes for a query to be locally processed, the timeout is updated at each hop N according to the following recurrence formula:

$$\text{timeout}_N = \text{currenttime}_N + \frac{\text{timeout}_{N-1} - \text{currenttime}_N}{2}$$

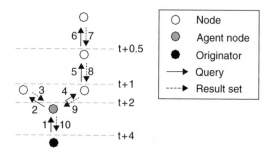

Figure 19.15 Dynamic abort time-out.

Consider, for example, Figure 19.15. At time t the originator submits a query with a dynamic abort time-out of t+4 s. In other words, it warns the agent to ignore results after time t+4. The agent in turn intends to safely meet the deadline and so figures that it needs to retain a safety window of 2 s, already starting to return its (partial) results at time t+2. The agent warns its own neighbors to ignore results after time t+2. The neighbors also intend to safely meet the deadline. From the 2 s available, they choose to allocate 1 s, and leave the rest to the branch remaining above. Eventually, the safety window becomes so small that a node can no longer meet a deadline on time-out. The results from the unlucky node are ignored, and its partial results are discarded. However, other nodes below and in other branches are unaffected. Their results survive and have enough time to hop all the way back to the originator before time t+4.

Static loop time-out: The same query may arrive at a node multiple times, along distinct routes, perhaps in a complex pattern. For reliable loop detection, a query has an identifier and a certain lifetime. To each query, an originator attaches a *loop time-out* and a different *transaction identifier*, which is a universally unique identifier (UUID). A node maintains a state table of transaction identifiers and returns an error when a query is received that has already been seen and has not yet timed-out. On loop time-out, a node may 'forget' about a query by deleting it from the state table. To be able to reliably detect a loop, a node must not forget a transaction identifier before its loop time-out has been reached. Interestingly, a static loop time-out is required in order to fully preserve query semantics. Otherwise, a problem arises that we propose to call *nonsimultaneous loop time-out*. The nonsimultaneous loop time-out problem is caused by the fact that some nodes still forward the query to other nodes when the destinations have already forgotten it. In other words, the problem is that loop time-out does not occur simultaneously everywhere. Consequently, a loop time-out must be static (does not change across hops) to guarantee that loops can reliably be detected. Along with a query, an originator not only provides a dynamic abort time-out, but also a static loop time-out. Initially at the originator, both values must be identical (e.g. t+4). After the first hop, both values become unrelated.

To summarize, we have abort timeout ≤ loop timeout. To ensure reliable loop detection, a loop time-out must be static, whereas an abort time-out may be static or dynamic. Under nonpipelined result set delivery, dynamic abort time-out using *exponential decay with halving* ensure that a maximum of results can be delivered reliably within the time frame desired by a user. We speculate that dynamic timeouts could also incorporate sophisticated cost functions involving latency and bandwidth estimation and/or economic models.

19.4.4 Query scope

As in a data integration system, the goal is to exploit several independent information sources as if they were a single source. This is important for distributed systems in which node topology or deployment model change frequently. For example, cross-organizational Grids and P2P networks exhibit such a character. However, in practice, it is often sufficient (and much more efficient) for a query to consider only a subset of all tuples (service descriptions) from a subset of nodes. For example, a typical query may only want to search tuples (services) within the scope of the domain cern.ch and ignore the rest of the world. To this end, we cleanly separate the concepts of (logical) *query* and (physical) *query scope*. A query is formulated against a global database view and is insensitive to link topology and deployment model. In other words, to a query the set of tuples appears as a single homogenous database, even though the set may be (recursively) partitioned across many nodes and databases. This means that in a relational or XML environment, at the global level, the set of all tuples *appears* as a single, very large, table or XML document, respectively. The query scope, on the other hand, is used to navigate and prune the link topology and filter on attributes of the deployment model. Conceptually, the scope is the input fed to the query. The query scope is a set and may contain anything from all tuples in the universe to none. Both query and scope can prune the search space, but they do so in a very different manner. A query scope is specified either *directly* or *indirectly*. One can distinguish scopes based on neighbor selection, timeout, and radius.

Neighbor selection: For simplicity, all our discussions so far have implicitly assumed a *broadcast* model (on top of TCP) in which a node forwards a query to all neighbor nodes. However, in general, one can select a subset of neighbors and forward concurrently or sequentially. Fewer query forwards lead to less overall resource consumption. The issue is critical because of the snowballing (epidemic, flooding) effect implied by broadcasting. Overall bandwidth consumption grows exponentially with the query radius, producing enormous stress on the network and drastically limiting its scalability [41, 36].

Clearly selecting a neighbor subset can lead to incomplete coverage, missing important results. The best policy to adopt depends on the context of the query and the topology. For example, the scope can select only neighbors with a service description of interface type 'Gnutella'. In an attempt to explicitly exploit topology characteristics, a virtual organization of a Grid may deliberately organize global, intermediate and local job schedulers into a treelike topology. Correct operation of scheduling may require reliable discovery of all or at least most relevant schedulers in the tree. In such a scenario, random selection of half of the neighbors at each node is certainly undesirable. A policy that selects all child nodes and ignores all parent nodes may be more adequate. Further, a node may maintain statistics about its neighbors. One may only select neighbors that meet minimum requirements in terms of latency, bandwidth, or historic query outcomes (`maxLatency`, `minBandwidth`, `minHistoricResult`). Other node properties such as host name, domain name, owner, and so on can be exploited in query scope guidance, for example, to implement security policies. Consider an example in which the scheduling system may only trust nodes from a select number of security domains. Here a query should never be forwarded to nodes not matching the trust pattern.

Further, in some systems, finding a single result is sufficient. In general, a user or any given node can guard against unnecessarily large result sets, message sizes, and resource consumption by specifying the maximum number of result tuples (`maxResults`) and bytes (`maxResultsBytes`) to be returned. Using sequential propagation, depending on the number of results already obtained from the local database and a subset of the selected neighbors, the query may no longer need to be forwarded to the rest of the selected neighbors.

Neighbor selection query: For flexibility and expressiveness, we propose to allow the user to specify the selection policy. In addition to the normal query, the user defines a *neighbor selection query* (XQuery) that takes the tuple set of the current node as input and returns a subset that indicates the nodes selected for forwarding. For example, a neighbor query implementing broadcasting selects all services with registry and P2P query capabilities, as follows:

```
RETURN /tupleset/tuple[@type= "service"
  AND content/service/interface[@type= "Consumer-1.0"]
  AND content/service/interface[@type= "XQuery-1.0"]]
```

A wide range of policies can be implemented in this manner. The neighbor selection policy can draw from the rich set of information contained in the tuples published to the node. Further, recall that the set of tuples in a database may not only contain service descriptions of neighbor nodes (e.g. in WSDL [4] or SWSDL [6]), but also other kind of (soft state) content published from any kind of content provider. For example, this may include the type of queries neighbor nodes can answer, descriptions of the kind of tuples they hold (e.g. their types), or a compact summary or index of their content. Content available to the neighbor selection query may also include host and network information as well as statistics that a node periodically publishes to its immediate neighbors. A neighbor selection query enables group communication to all nodes with certain characteristics (e.g. the same group ID). For example, broadcast and random selection can be expressed with a neighbor query. One can select nodes that support given interfaces (e.g. Gnutella [13], Freenet [14] or job scheduling). In a tree topology, a policy can use the tuple `context` attribute to select all `child` nodes and to ignore all `parent` nodes. One can implement domain filters and security filters (e.g. `allow`/`deny` regular expressions as used in the Apache HTTP server if the tuple set includes metadata such as host name and node owner. Power law policies [42] can be expressed if metadata includes the number of neighbors to the nth radius. To summarize, a neighbor selection query can be used to implement *smart dynamic routing*.

Radius: The *radius* of a query is a measure of path length. More precisely, it is the maximum number of hops a query is allowed to travel on any given path. The radius is decreased by one at each hop. The roaming query and response traffic must fade away upon reaching a radius of less than zero. A scope based on radius serves similar purposes as a time-out. Nevertheless, timeout and radius are complementary scope features. The radius can be used to indirectly limit result set size. In addition, it helps to limit latency and bandwidth consumption and to guard against runaway queries with infinite lifetime. In Gnutella and Freenet, the radius is the primary means to specify a query scope. The radius is termed *time-to-live* (TTL) in these systems. Neither of these systems support timeouts.

For maximum result, set size limiting, a timeout, and/or radius can be used in conjunction with neighbor selection, routed response, and perhaps sequential forward to implement the *expanding ring* [43] strategy. The term stems from IP multicasting. Here an agent first forwards the query to a small radius/timeout. Unless enough results are found, the agent forwards the query again with increasingly large radius/timeout values to reach further into the network, at the expense of increasingly large overall resource consumption. On each expansion, radius/timeout is multiplied by some factor.

19.5 PEER DATABASE PROTOCOL

In this section, we summarize how the operations of the UPDF and XQuery interface from Section 19.3.1 are carried over (bound to) our *Peer Database Protocol* (PDP) [6, 44]. PDP supports centralized and P2P database queries for a wide range of database architectures and response models such that the stringent demands of ubiquitous Internet infrastructures in terms of scalability, efficiency, interoperability, extensibility, and reliability can be met. Any client (e.g. an originator or a node) can use PDP to query the P2P network and to retrieve the corresponding result set in an iterator style. While the use of PDP for communication between nodes is mandatory to achieve interoperability, any arbitrary additional protocol and interface may be used for communication between an originator and a node (e.g. a simple stateless SOAP/HTTP request-response or shared memory protocol). For flexibility and simplicity, and to allow for gatewaying, mediation, and protocol translation, the relationship between an originator and a node may take any arbitrary form, and is therefore left unspecified.

The high-level messaging model employs four request messages (QUERY, RECEIVE, INVITE, and CLOSE) and a response message (SEND). A *transaction* is a sequence of one or more message exchanges between two peers (nodes) for a given query. An example transaction is a QUERY-RECEIVE-SEND-RECEIVE-SEND-CLOSE sequence. A peer can concurrently handle multiple independent transactions. A transaction is identified by a transaction identifier. Every message of a given transaction carries the same transaction identifier.

A QUERY message is asynchronously forwarded along hops through the topology. A RECEIVE message is used by a client to request query results from another node. It requests the node to respond with a SEND message, containing a batch of at least N and at most M results from the (remainder of the) result set. A client may issue a CLOSE request to inform a node that the remaining results (if any) are no longer needed and can safely be discarded. Like a QUERY, a CLOSE is asynchronously forwarded. If the local result set is not empty under direct response, the node directly contacts the agent with an INVITE message to solicit a RECEIVE message. A RECEIVE request can ask to deliver SEND messages in either synchronous (pull) or asynchronous (push) mode. In the synchronous mode, a single RECEIVE request must precede every single SEND response. An example sequence is RECEIVE-SEND-RECEIVE-SEND. In the asynchronous mode, a single RECEIVE request asks for a sequence of successive SEND responses. A client need not explicitly request more results, as they are automatically pushed in a sequence of zero or more SENDs. An example sequence is RECEIVE-SEND-SEND-SEND. Appropriately sized batched delivery greatly reduces the number of hops incurred by a single RECEIVE. To reduce latency, a node may prefetch query results.

Discrete messages belong to well-defined message exchange patterns. For example, the pattern of synchronous exchanges (one-to-one, pull) as well as the pattern of asynchronous exchanges (one-to-many, push) is supported. For example, the response to a MSG RECEIVE may be an *error* (ERR), a *reply* (RPY SEND) or a sequence of zero or more *answers* (ANS SEND), followed by a *null terminator* message (NULL). The RPY OK and ERR message type are introduced because any realistic messaging model must deal with acknowledgments and errors. The following message exchanges are permitted:

```
MSG_QUERY    --> RPY_OK | ERR
MSG_RECEIVE  --> RPY_SEND | (ANS_SEND [0:N], NULL) | ERR
MSG_INVITE   --> RPY_OK | ERR
MSG_CLOSE    --> RPY_OK | ERR
```

For simplicity and flexibility, PDP uses straightforward XML representations for messages, as depicted in Figure 19.16. Without loss of generality, example query expressions (e.g. user query, merge query, and neighbor selection query) are given in the XQuery language [23]. Other query languages such as XPath, SQL, LDAP [45] or subscription interest statements could also be used. Indeed, the messages and network interactions required to support efficient P2P publish-subscribe and event trigger systems do not differ at all from the ones presented earlier.

PDP has a number of key properties. It is applicable to any node topology (e.g. centralized, distributed or P2P and to multiple P2P response modes (routed response and direct response, both with and without metadata modes). To support loosely coupled autonomous Internet infrastructures, the model is connection-oriented (ordered, reliable, congestion sensitive) and message-oriented (loosely coupled, operating on structured data). For efficiency, it is stateful at the protocol level, with a transaction consisting of one or more discrete message exchanges related to the same query. It allows for low latency, pipelining, early and/or partial result set retrieval due to synchronous pull, and result set delivery in one or more variable-sized batches. It is efficient, because of asynchronous push with delivery of multiple results per batch. It provides for resource consumption and flow control on a per query basis, due to the use of a distinct channel per transaction. It is scalable, because of application multiplexing, which allows for very high query concurrency and very low latency, even in the presence of secure TCP connections. To encourage interoperability and extensibility it is fully based on the BEEP [33] Internet

```
<MSG_QUERY transactionID = "12345">
  <query>
        <userquery> RETURN /tupleset/tuple </userquery>
        <mergequery unionizer="UNION"> RETURN /tupleset/tuple </mergequery>
  </query>
  <scope loopTimeout = "2000000000000" abortTimeout = "1000000000000"
        logicalRadius = "7" physicalRadius = "4"
        maxResults = "100" maxResultsBytes = "100000">
        <neighborSelectionQuery>          <!-- implements broadcasting -->
           RETURN /tupleset/tuple[@type="service"
              AND content/service/interface[@type="Consumer-1.0"]
              AND content/service/interface[@type="XQuery-1.0"]]
        </neighborSelectionQuery>
  </scope>
  <options>
        <responseMode> routed </responseMode>
        <originator> fred@example.com </originator>
  </options>
</MSG_QUERY>

<MSG_RECEIVE transactionID =  "12345">
  <modeminResults =  "1" maxResults = "10"> synchronous </mode>
</MSG_RECEIVE>

<RPY_SEND transactionID =  "12345">
  <data nonBlockingResultsAvailable =  "-1" estimatedResultsAvailable =  "-1">
    <tupleset TS4="100">
       <tuple link="http://sched.infn.it:8080/pub/getServiceDescription"
              type="service" ctx="child" TS1="20" TC="25" TS2="30" TS3="40">
          <content>
             <service> service description B goes here </service>
          </content>
       </tuple>
       ... more tuples can go here...
    </tupleset>
  </data>
</RPY_SEND>

<ANS_SEND transactionID = "12345">
    structure is identical to RPY_SEND (seeabove)...
</ANS_SEND>

<MSG_INVITE transactionID =  "12345">
  <avail nonBlockingResultsAvailable="50" estimatedResultsAvailable="100"/>
</MSG_INVITE>

<MSG_CLOSE transactionID =  "12345" code="555"> maximum idle time exceeded </MSG_CLOSE>

<RPY_OK transactionID =  "12345"/>
<ERR transactionID =  "12345" code ="550"> transaction identifier unknown </ERR>
```

Figure 19.16 Example messages of peer database protocol.

Engineering Task Force (IETF) standard, for example, in terms of asynchrony, encoding, framing, authentication, privacy, and reporting. Finally, we note that SOAP can be carried over BEEP in a straightforward manner [34] and that BEEP, in turn, can be carried over any reliable transport layer (TCP is merely the default).

19.6 RELATED WORK

RDBMS: (Distributed) Relational database systems [12] provide SQL as a powerful query language. They assume tight and consistent central control and hence are infeasible in Grid environments, which are characterized by heterogeneity, scale, lack of central control, multiple autonomous administrative domains, unreliable components, and frequent dynamic change. They do not support an XML data model and the XQuery language. Further, they do not provide dynamic content generation, softstate-based publication and content caching. RDBMS are not designed for use in P2P systems. For example, they do not support asynchronous push, invitations, scoping, neighbor selection, and dynamic timeouts. Our work does not compete with an RDBMS, though. A node may as well internally use an RDBMS for data management. A node can accept queries over an XML view and internally translate the query into SQL [46, 47]. An early attempt towards a WAN distributed DBMS was Mariposa [48], designed for scalability to many cooperating sites, data mobility, no global synchronization, and local autonomy. It used an economic model and bidding for adaptive query processing, data placement, and replication.

ANSA and CORBA: The ANSA project was an early collaborative industry effort to advance distributed computing. It defined trading services [49] for advertisement and discovery of relevant services on the basis of service type and simple constraints on attribute/value pairs. The CORBA Trading service [50] is an evolution of these efforts.

UDDI: Universal Description, Discovery and Integration (UDDI) [8] is an emerging industry standard that defines a business-oriented access mechanism to a centralized registry holding XML-based WSDL service description. UDDI is a definition of a specific service class, not a discovery architecture. It does not offer a dynamic data model. It is not based on soft state, which limits its ability to dynamically manage and remove service descriptions from a large number of autonomous third parties in a reliable, predictable, and simple way. Query support is rudimentary. Only key lookups with primitive qualifiers are supported, which is insufficient for realistic service discovery use cases.

Jini, SLP, SDS, INS: The centralized Jini Lookup Service [51] is located by Java clients via a UDP multicast. The network protocol is not language-independent because it relies on the Java-specific object serialization mechanism. Publication is based on soft state. Clients and services must renew their leases periodically. Content freshness is not addressed. The query 'language' allows for simple string matching on attributes and is even less powerful than LDAP.

The Service Location Protocol (SLP) [52] uses multicast, soft state, and simple filter expressions to advertise and query the location, type, and attributes of services. The query 'language' is simpler than Jini's. An extension is the Mesh Enhanced Service Location Protocol (mSLP) [53], increasing scalability through multiple cooperating directory agents. Both assume a single administrative domain and hence do not scale to the Internet and Grids.

The Service Discovery Service (SDS) [54] is also based on multicast and soft state. It supports a simple XML-based exact match query type. SDS is interesting in that it mandates secure channels with authentication and traffic encryption, and privacy and authenticity of service descriptions. SDS servers can be organized in a distributed hierarchy. For efficiency, each SDS node in a hierarchy can hold an index of the content of its subtree. The index is a compact aggregation and is custom tailored to the narrow type of query SDS can answer. Another effort is the Intentional Naming System [55]. Like SDS, it integrates name resolution and routing.

JXTA: The goal of the JXTA P2P network [56, 57, 58] is to have peers that can cooperate to form self-organized and self-configured peer groups independent of their position in the network and without the need of a centralized management infrastructure. JXTA defines six stateless best-effort protocols for *ad hoc*, pervasive, and multihop P2P computing. These are designed to run over unidirectional, unreliable transports. Because of this ambitious goal, a range of well-known higher-level abstractions (e.g. bidirectional secure messaging) are (re)invented from first principles.

The Endpoint Routing Protocol allows discovery of a route (sequence of hops) from one peer to another peer, given the destination peer ID. The Rendezvous Protocol offers publish-subscribe functionality within a peer group. The Peer Resolver Protocol and Peer Discovery Protocol allow for publication of advertisements and *simple* queries that are unreliable, stateless, nonpipelined, and nontransactional. We believe that this limits scalability, efficiency, and applicability for service discovery and other nontrivial use cases. Lacking expressive means for query scoping, neighbor selection, and timeouts, it is not clear how chained rendezvous peers can form a search network. We believe that JXTA Peer Groups, JXTA search, and publish/subscribe can be expressed within our UPDF framework, but not vice versa.

GMA: The Grid Monitoring Architecture (GMA) [59, 60] is intended to enable efficient monitoring of distributed components, for example, to allow for fault detection and performance prediction. GMA handles performance data transmitted as time-stamped *events*. It consists of three types of components, namely, Directory Service, Producer, and Consumer. Producers and Consumers publish their existence in a centralized directory service. Consumers can use the directory service to discover producers of interest, and vice versa. GMA briefly sketches three interactions for transferring data between producers and consumers, namely, publish/subscribe, query/response, and notification. Both consumers and producers can initiate interactions.

GMA neither defines a query language, nor a data model, nor a network protocol. It does not consider the use of multihop P2P networks, and hence does not address loop detection, scoping, timeouts, and neighbor selection. Synchronous multimessage exchanges and routed responses are not considered. Event data is always asynchronously pushed from a producer directly to a consumer. GMA like server-initiated interactions could be offered via the INVITE message of the PDP, but currently we do not see enough compelling reasons for doing so. For a comparison of various response modes, see our prior studies [38]. We believe that GMA can be expressed within our framework, but not vice versa.

OGSA: The independently emerging *Open Grid Services Architecture (OGSA)* [31, 32] exhibits striking similarities with WSDA, in spirit and partly also in design. However, although it is based on soft state, OGSA does not offer a DDM allowing for dynamic refresh of content. Hence it requires trust delegation on publication, which is problematic for security-sensitive data such as detailed service descriptions. Further, the data model and publication is not based on sets, resulting in scalability problems in the presence of large numbers of tuples. The absence of set semantics also seriously limits the potential for query optimization. The use of arbitrary MIME content is not foreseen, reducing applicability. The concepts of notification and registration are not unified, as in the WSDA Consumer interface. OGSA does define an interesting interface for publish/subscribe functionality, but no corresponding network protocol (e.g. such as the PDP). OGSA mandates that every service in the universe must comply with the GridService interface, unnecessarily violating the principle of nondisruptiveness. It is not clear from the material whether OGSA intends in the future to support either or both XQuery, XPath, or none. Finally, OGSA does not consider that the set of information tuples in the universe is partitioned over multiple autonomous nodes. Hence, it does not consider P2P networks, and their (nontrivial) implications, for example, with respect to Query-processing, timeouts, pipelining, and network protocols. For a detailed comparison of WSDA and OGSA, see Reference [61].

DNS: Distributed databases with a hierarchical namespace such as the Domain Name System [62] can efficiently answer *simple* queries of the form *'Find an object by its full name'*. Queries are not forwarded (routed) through the (hierarchical) link topology. Instead, a node returns a *referral* message that redirects an originator to the next closer node. The originator explicitly queries the next node, is referred to yet another closer node, and so on. To support neighbor selection in a hierarchical namespace within our UPDF framework, a node can publish to its neighbors not only its service link but also the namespace it manages. The DNS referral behavior can be implemented within UPDF by using a radius scope of zero. The same holds for the LDAP referral behavior (see below).

LDAP and MDS: The Lightweight Directory Access Protocol (LDAP) [45] defines a network protocol in which clients send requests to and receive responses from LDAP servers. LDAP is an extensible network protocol, not a discovery architecture. It does not offer a DDM, is not based on soft state, and does not follow an XML data model.

The expressive power of the LDAP query language is insufficient for realistic service discovery use cases [6]. Like DNS, it supports referrals in a hierarchical namespace but not query forwarding.

The Metacomputing Directory Service (MDS) [63] inherits all properties of LDAP. As a result its query language is insufficient for service discovery, and it does not follow an XML data model. MDS does not offer a dynamic data model, limiting cache freshness steering. However, it is based on soft state. MDS is not a Web service, because it is not specified by a service description language. It does not offer interfaces and operations that may be bound to multiple network protocols. However, it appears that MDS is being recast to fit into the OGSA architecture. Indeed, the OGSA registry and notification interfaces could be seen as new and abstracted clothings for MDS. Beyond LDAP, MDS offers a simple form of query forwarding that allows for multilevel hierarchies but not for arbitrary topologies. It does not support radius and dynamic abort timeout, pipelined query execution across nodes as well as direct response and metadata responses.

Query processing: *Simple* queries for lookup by key are assumed in most P2P systems such as DNS [62], Gnutella [13], Freenet [14], Tapestry [15], Chord [16], and Globe [17], leading to highly specialized *content-addressable* networks centered around the theme of distributed hash table lookup. *Simple* queries for exact match (i.e. given a flat set of attribute values find all tuples that carry exactly the same attribute values) are assumed in systems such as SDS [54] and Jini [51]. Our approach is distinguished in that it not only supports all of the above query types, but it also supports queries from the rich and expressive general-purpose query languages XQuery [23] and SQL.

Pipelining: For a survey of adaptive query processing, including pipelining, see the special issue of Reference [64]. Reference [65] develops a general framework for producing partial results for queries involving any nonmonotonic operator. The approach inserts update and delete directives into the output stream. The Tukwila [66] and Niagara projects [67] introduce data integration systems with adaptive query processing and XML query operator implementations that efficiently support pipelining. Pipelining of hash joins is discussed in References [68, 69, 70]. Pipelining is often also termed *streaming* or *nonblocking* execution.

Neighbor selection: *Iterative deepening* [71] is a similar technique to *expanding ring* in which an optimization is suggested that avoids reevaluating the query at nodes that have already done so in previous iterations. Neighbor selection policies that are based on randomness and/or historical information about the result set size of prior queries are simulated and analyzed in Reference [72]. An efficient neighbor selection policy is applicable to simple queries posed to networks in which the number of links of nodes exhibits a power law distribution (e.g. Freenet and Gnutella) [42]. Here, most (but not all) matching results can be reached with few hops by selecting just a very small subset of neighbors (the neighbors that themselves have the most neighbors to the nth radius). Note, however, that the policy is based on the assumption that not all results must be found and that all query results are equally relevant. These related works discuss in isolation neighbor selection techniques for a particular query type, without the context of a framework for comprehensive query support.

19.7 CONCLUSIONS

This chapter distills and generalizes the essential properties of the discovery problem and then develops solutions that apply to a wide range of large distributed Internet systems. It shows how to support expressive general-purpose queries over a view that integrates autonomous dynamic database nodes from a wide range of distributed system topologies. We describe the first steps towards the convergence of Grid Computing, P2P Computing, Distributed Databases, and Web services, each of which introduces core concepts and technologies necessary for *making the global infrastructure a reality.*

Grids are collaborative distributed Internet systems characterized by large-scale heterogeneity, lack of central control, multiple autonomous administrative domains, unreliable components, and frequent dynamic change. We address the problems of maintaining dynamic and timely information populated from a large variety of unreliable, frequently changing, autonomous, and heterogeneous remote data sources by designing a database for XQueries over dynamic distributed content – the so-called *hyper registry*. The registry has a number of key properties. An XML data model allows for structured and semistructured data, which is important for integration of heterogeneous content. The XQuery language allows for powerful searching, which is critical for nontrivial applications. Database state maintenance is based on soft state, which enables reliable, predictable, and simple content integration from a large number of autonomous distributed content providers. Content link, content cache and a hybrid pull/push communication model allow for a wide range of dynamic content freshness policies, which may be driven by all three system components: content provider, registry, and client.

We propose and specify an open discovery architecture, the *Web Service Discovery Architecture (WSDA)*. WSDA views the Internet as a large set of services with an extensible set of well-defined interfaces. It has a number of key properties. It promotes an interoperable Web service layer on top of Internet software, because it defines appropriate services, interfaces, operations, and protocol bindings. It embraces and integrates solid industry standards such as XML, XML Schema, SOAP, WSDL, and XQuery. It allows for integration of emerging standards like the WSIL. It is modular because it defines a small set of orthogonal multipurpose communication primitives (building blocks) for discovery. These primitives cover service identification, service description retrieval, data publication, as well as minimal and powerful query support. Each communication primitive is deliberately designed to avoid any unnecessary complexity. WSDA is open and flexible because each primitive can be used, implemented, customized, and extended in many ways. It is powerful because the individual primitives can be combined and plugged together by specific clients and services to yield a wide range of behaviors and emerging synergies. It is unified because it subsumes an array of disparate concepts, interfaces and protocols under a single semitransparent umbrella.

We take the first steps towards unifying the fields of DBMSs and P2P computing, which so far have received considerable, but separate, attention. We extend database concepts and practice to cover P2P search. Similarly, we extend P2P concepts and practice to support powerful general-purpose query languages such as XQuery and SQL. As a result, we propose the *Unified Peer-to-Peer Database Framework (UPDF)* and corresponding *Peer Database Protocol (PDP)* for general-purpose query support in large heterogeneous distributed systems spanning many administrative domains. They are unified in the sense that they allow expression of specific discovery applications for a wide range of data types, node topologies (e.g. ring, tree, graph), query languages (e.g. XQuery, SQL), query response modes (e.g. Routed, Direct, and Referral Response), neighbor selection policies (in the form of an XQuery), pipelining characteristics, timeout, and other scope options.

The uniformity and wide applicability of our approach is distinguished from related work, which (1) addresses some but not all problems and (2) does not propose a unified framework.

The results presented in this chapter open four interesting research directions.

First, it would be interesting to extend further the unification and extension of concepts from DBMSs and P2P computing. For example, one could consider the application of database techniques such as buffer cache maintenance, view materialization, placement, and selection as well as query optimization for use in P2P computing. These techniques would need to be extended in the light of the complexities stemming from autonomous administrative domains, inconsistent, and incomplete (soft) state, dynamic, and flexible cache freshness policies and, of course, tuple updates. An important problem left open in our work is whether a query processor can automatically determine if a correct merge query and unionizer exist, and if so, how to choose them (we require a user to explicitly provide these as parameters). Here approaches from query rewriting for heterogeneous and homogenous relational database systems [39, 73] should prove useful. Further, database resource management and authorization mechanisms might be worthwhile to consider for specific flow control policies per query or per user.

Second, it would be interesting to study and specify in more detail specific cache freshness interaction policies between content provider, hyper registry and client (query). Our specification allows expressing a wide range of policies, some of which we outline, but do not evaluate in detail the merits and drawbacks of any given policy.

Third, it would be valuable to rigourously assess, review, and compare the WSDA and the OGSA in terms of concepts, design, and specifications. A strong goal is to achieve convergence by extracting best-of-breed solutions from both proposals. Future collaborative work could further improve current solutions, for example, in terms of simplicity, orthogonality, and expressiveness. For practical purposes, our pedagogical service description language (SWSDL) could be mapped to WSDL, taking into account the OGSA proposal. This would allow to use SWSDL as a tool for greatly improved clarity in high-level architecture and design discussions, while at the same time allowing for painstakingly detailed WSDL specifications addressing ambiguity and interoperability concerns.

We are working on a multipurpose interface for persistent XQueries (i.e. server-side trigger queries), which will roughly correspond to the OGSA publish-subscribe interface, albeit in a more general and powerful manner. The PDP already supports, in a unified manner, all messages and network interactions required for efficient implementations of P2P publish-subscribe and event-trigger interfaces (e.g. synchronous pull and asynchronous push, as well as invitations and batching).

Fourth, Tim Berners-Lee designed the World Wide Web as a consistent interface to a flexible and changing heterogeneous information space for use by CERN's staff, the HEP community, and, of course, the world at large. The WWW architecture [74] rests on four simple and orthogonal pillars: URIs as identifiers, HTTP for retrieval of content pointed to by identifiers, MIME for flexible content encoding, and HTML as the primus-inter-pares (MIME) content type. On the basis of our Dynamic Data Model, we hope to proceed further towards a self-describing meta content type that retains and wraps all four WWW pillars 'as is', yet allows for flexible extensions in terms of identification, retrieval and caching of content. Judicious combination of the four Web pillars, DDM, the WSDA, the Hyper Registry, the UPDF and its corresponding PDP are used to define how to bootstrap, query, and publish to a dynamic and heterogeneous information space maintained by self-describing network interfaces.

ACKNOWLEDGMENTS

This chapter is dedicated to Ben Segal, who supported this work with great integrity, enthusiasm, and, above all, in a distinct spirit of humanity and friendship. Gerti Kappel, Erich Schikuta and Bernd Panzer-Steindel patiently advised, suggesting what always turned out to be wise alleys. Testing ideas against the solid background of all members of the EDG WP2 (Grid Data Management) team proved an invaluable recipe in separating wheat from chaff. This work was carried out in the context of a Ph.D. thesis [6] for the European Data Grid project (EDG) at CERN, the European Organization for Nuclear Research, and supported by the Austrian Ministerium für Wissenschaft, Bildung und Kultur.

REFERENCES

1. Foster, I., Kesselman, C. and Tuecke, S. (2001) The Anatomy of the Grid: enabling scalable virtual organizations. *Int'l. Journal of Supercomputer Applications*, **15**(3).
2. Hoschek, W., Jaen-Martinez, J., Samar, A., Stockinger, H. and Stockinger, K. (2000) Data management in an international data grid project, In *1st IEEE/ACM Int'l. Workshop on Grid Computing (Grid'2000)*. Bangalore, India, December.
3. Bethe, S., Hoffman, H. *et al* (2001) Report of the LHC Computing Review, Technical Report, CERN/LHCC/2001-004, CERN, Switzerland, April 2001, http://cern.ch/lhc-computing-review-public/Public/Report_final.PDF.
4. Christensen, E., Curbera, F., Meredith, G. and Weerawarana. S. (2001) Web Services Description Language (WSDL) 1.1. *W3C Note 15*, 2001. http://www.w3.org/TR/wsdl.
5. Box, D. *et al*, (2000) World Wide Web Consortium, Simple Object Access Protocol (SOAP) 1.1. *W3C Note 8*, 2000.
6. Hoschek, W. (2002) *A Unified Peer-to-Peer Database Framework for XQueries over Dynamic Distributed Content and its Application for Scalable Service Discovery*. Ph.D. Thesis Austria: Technical University of Vienna, March, 2002.

7. Cauldwell P. *et al.* (2001) *Professional XML Web Services.* ISBN 1861005091, Chicago, IL: Wrox Press.
8. UDDI Consortium. UDDI: Universal Description, Discovery and Integration. http://www.uddi.org.
9. Ullman, J. D. (1997) Information integration using logical views. In *Int'l. Conf. on Database Theory (ICDT)*, Delphi, Greece, 1997.
10. Florescu, D., Manolescu, I., Kossmann, D. and Xhumari, F. (2000) Agora: Living with XML and Relational. In *Int'l. Conf. on Very Large Data Bases (VLDB)*, Cairo, Egypt, February 2000.
11. Tomasic, A., Raschid, L. and Valduriez, P. (1998) Scaling access to heterogeneous data sources with DISCO. *IEEE Transactions on Knowledge and Data Engineering*, **10**(5), 808–823.
12. Tamer Özsu, M. and Valduriez, P. (1999) *Principles of Distributed Database Systems.* New York: Prentice Hall.
13. Gnutella Community. Gnutella Protocol Specification v0.4. dss.clip2.com/GnutellaProtocol04.pdf.
14. Clarke, I., Sandberg, O., Wiley, B. and Hong, T. (2000) Freenet: A distributed anonymous information storage and retrieval system, In *Workshop on Design Issues in Anonymity and Unobservability.* 2000.
15. Zhao, B., Kubiatowicz, J. and Joseph, A. (2001) Tapestry: An infrastructure for fault-resilient wide-area location and routing. Technical report.U.C. Berkeley UCB//CSD-01-1141, 2001.
16. Stoica, I., Morris, R., Karger, D., Kaashoek, M. and Balakrishnan, H. (2001) Chord: A scalable peer-to-peer lookup service for internet applications. In *ACM SIGCOMM*, 2001.
17. van Steen, M. Homburg, P. and Tanenbaum, A. A wide-area distributed system. *IEEE Concurrency*, 1999.
18. Minar, N. (2001) Peer-to-Peer is Not Always Decentralized. In *The O'Reilly Peer-to-Peer and Web Services Conference*, Washington, D.C., November, 2001.
19. Chervenak, A. *et al.* (2002) Giggle: A Framework for Constructing Scalable Replica Location Services, In *Proc. of the Int'l. IEEE/ACM Supercomputing Conference (SC 2002).* Baltimore, USA: IEEE Computer Society Press, November.
20. Guy, L., Kunszt, P., Laure, E., Stockinger, H. and Stockinger, K. 2002 Replica Management in Data Grids. Technical report, Global Grid Forum Informational Document, GGF5, Edinburgh, Scotland, July 2002.
21. Stockinger, H., Samar, A., Mufzaffar, S. and Donno, F. (2002) Grid Data Mirroring Package (GDMP). *Journal of Scientific Computing* **10**(2), 121–134.
22. Bell, W., Bosio, D., Hoschek, W., Kunszt, P., McCance, G. and Silander, M. (2002) Project Spitfire - Towards Grid Web Service Databases. Technical report, Global Grid Forum Informational Document, GGF5, Edinburgh, Scotland, July 2002.
23. World Wide Web Consortium. XQuery 1.0: An XML Query Language. *W3C Working Draft*, December, 2001.
24. Berners-Lee, T., Fielding, R. and Masinter, L.. Uniform Resource Identifiers (URI): Generic Syntax. *IETF RFC 2396.*
25. World Wide Web Consortium. XML-Signature Syntax and Processing. *W3C Recommendation*, February 2002.
26. Brittenham, P. An Overview of the Web Services Inspection Language, 2001. www.ibm.com/developerworks/webservices/library/ws-wsilover.
27. Freed, N. and Borenstein, N. (1996) Multipurpose Internet Mail Extensions (MIME) Part One: Format of Internet Message Bodies. *IETF RFC 2045*, November, (1996).
28. World Wide Web Consortium. XML Schema Part 0: Primer. *W3C Recommendation*, May 2001.
29. Wang, J. (1999) A survey of web caching schemes for the Internet. *ACM Computer Communication Reviews*, **29**(5), October.
30. Gullapalli, S., Czajkowski, K., Kesselman, C. and Fitzgerald, S. (2001) The grid notification framework. Technical report, Grid Forum Working Draft GWD-GIS-019, June, 2001. http://www.gridforum.org.
31. Rose, M. (2001) The Blocks Extensible Exchange Protocol Core. *IETF RFC 3080*, March, 2001.

32. O'Tuathail, E. and Rose, M. (2002) Using the Simple Object Access Protocol (SOAP) in Blocks Extensible Exchange Protocol (BEEP). *IETF RFC 3288*, June, 2002.

33. International Organization for Standardization (ISO). Information Technology-Database Language SQL. *Standard No. ISO/IEC 9075:1999*, 1999.

34. Ripeanu M. (2001) Peer-to-Peer Architecture Case Study: Gnutella Network. In *Int'l. Conf. on Peer-to-Peer Computing (P2P2001)*, Linkoping, Sweden, August 2001.

35. Clip2Report. Gnutella: To the Bandwidth Barrier and Beyond. http://www.clip2.com/gnutella.html.

36. Foster, I., Kesselman, C., Nick, J. and Tuecke, Steve The Physiology of the Grid: An Open Grid Services Architecture for Distributed Systems Integration, January, 2002. http://www.globus.org.

37. Tuecke, S., Czajkowski, K., Foster, I., Frey, J., Graham, S. and Kesselman, C. (2002) Grid Service Specification, February, 2002. http://www.globus.org.

38. Hoschek, W. (2002) A Comparison of Peer-to-Peer Query Response Modes. In *Proc. of the Int'l. Conf. on Parallel and Distributed Computing and Systems (PDCS 2002)*, Cambridge, USA, November 2002.

39. Kossmann, D. (2000) The state of the art in distributed query processing. *ACM Computing Surveys* **32**(4), 422–469.

40. Urhan, T. and Franklin, M. (2001) Dynamic Pipeline Scheduling for Improving Interactive Query Performance. *The Very Large Database (VLDB) Journal* 2001: 501–510.

41. Ritter, J. Why Gnutella Can't Scale. No, Really. http://www.tch.org/gnutella.html.

42. Puniyani, A., Huberman, B., Adamic, L., Lukose, R. (2001) Search in power-law networks. *Phys. Rev, E*, **64**, 46135.

43. Deering, S. E. *Multicast Routing in a Datagram Internetwork.* Ph.D. thesis, Stanford University, Stanford 1991.

44. Hoschek, W. (2002) A Unified Peer-to-Peer Database Protocol. Technical report, DataGrid-02-TED-0407, April, 2002.

45. Yeong, W., Howes, T. and Kille, S. 1995 Lightweight Directory Access Protocol. *IETF RFC 1777*, March, 1995.

46. Fernandez, M., Atsuyuki, M., Suciu, D. and Wang-Chiew, T. (2001) Publishing relational data in xml: the silkroute approach. *IEEE Data Engineering Bulletin*, **24**(2), 12–19.

47. Florescu, D., Manolescu, I. and Kossmann, D. (2001) Answering XML queries over heterogeneous data sources, International Conference on Very Large Data Bases (VLDB), Roma, Italy, September, 2001.

48. Stonebraker, M., *et al.* (1996) Mariposa: a wide-area distributed database system. *The Very Large Database (VLDB) Journal*, **5**(1), 48–63.

49. Beitz, A., Bearman, M. and Vogel. A. (1995) Service Location in an Open Distributed Environment, Proceedings of the International Workshop on Services in Distributed and Networked Environments, Whistler, Canada, June 1995.

50. Object Management Group, Trading Object Service, (1996) OMG RPF5 Submission, May, 1996.

51. Waldo, J. (1999) The Jini architecture for network-centric computing. *Communications of the ACM*, **42**(7), 76–82.

52. Guttman, E. (1999) Service location protocol: automatic discovery of ip network services. *IEEE Internet Computing Journal*, **3**(4).

53. Zhao, W., Schulzrinne, H. and Guttman, E. (2000) mSLP - mesh enhanced service location protocol, Proceedings of the IEEE International Conference on Computer Communications and Networks (ICCCN'00), Las Vegas, USA, October 2000.

54. Czerwinski, S. E., Zhao, B. Y., Hodes, T., Joseph, A. D. and Katz, R. (1999) An architecture for a secure service discovery service, Fifth Annual International Conference on Mobile Computing and Networks (MobiCOM '99), Seattle, WA, August 1999.

55. Adjie-Winoto, W., Schwartz, E., Balakrishnan, H. and Lilley, J. (1999) The design and implementation of an intentional naming system, Proceedings of the Symposium on Operating Systems Principles, Kiawah Island, USA, December 1999.

56. Traversat, B., Abdelaziz, M., Duigou, M., Hugly, J.-C., Pouyoul, E. and Yeager, B. (2002) Project JXTA Virtual Network, White Paper, http://www.jxta.org.
57. Waterhouse, S. (2001) JXTA Search: Distributed Search for Distributed Networks, White Paper, http://www.jxta.org.
58. Project JXTA. (2002) JXTA v1.0 Protocols Specification, http://spec.jxta.org.
59. Tierney, B., Aydt, R., Gunter, D., Smith, W., Taylor, V., Wolski, R. and Swany, M. (2002) A grid monitoring architecture. technical report, Global Grid Forum Informational Document, January, http://www.gridforum.org.
60. Lee, J., Gunter, D., Stoufer, M. and Tierney, B. (2002) Monitoring Data Archives for Grid Environments, Proceedings of the International IEEE/ACM Supercomputing Conference (SC 2002), Baltimore, USA: IEEE Computer Society Press, November 2002.
61. Hoschek, W. (2002) The web service discovery architecture, Proceedings of the International IEEE/ACM Supercomputing Conference (SC 2002), Baltimore, USA: IEEE Computer Society Press, November 2002.
62. Mockapetris, P. (1987) Domain names – implementation and specification. *IETF RFC 1035*, November 1987.
63. Czajkowski, K., Fitzgerald, S., Foster, I. and Kesselman, C. (2001) Grid information services for distributed resource sharing, *Tenth IEEE International Symposium on High-Performance Distributed Computing (HPDC-10)*, San Francisco, California, August, 2001.
64. IEEE Computer Society. (2000) *Data Engineering Bulletin*, **23**(2).
65. Shanmugasundaram, J., Tufte, K., DeWitt, D. J., Naughton, J. F. and Maier, D. (2000) Architecting a network query engine for producing partial results. *WebDB 2000*.
66. Ives, Z. G., Halevy, A. Y. and Weld. D. S. (2001) Integrating network-bound xml data. *IEEE Data Engineering Bulletin*, **24**(2), 20–26.
67. Naughton, J. F. *et al.* (2001) The Niagara internet query system. *IEEE Data Engineering Bulletin*, **24**(2), 27–33.
68. Wilschut, A. N. and Apers, P. M. G. (1991) Dataflow query execution in a parallel main-memory environment, First International Conference on Parallel and Distributed Information Systems, December 1991.
69. Ives, Z. G., Florescu, D., Friedman, M. T., Levy, A. Y. and Weld. D. S. (1999) An adaptive query execution system for data integration, ACM SIGMOD Conf. On Management of Data, 1999.
70. Urhan, T. and Franklin, M. J. (2000) Xjoin, A reactively-scheduled pipelined join operator. *IEEE Data Engineering Bulletin*, **23**(2), 27–34.
71. Yang, B. and Garcia-Molina, H. (2002) Efficient Search in Peer-to-Peer Networks, 22nd International Conference on Distributed Computing Systems, Vienna, Austria, July 2002.
72. Iamnitchi, A. and Foster, I. (2001) On Fully Decentralized Resource Discovery in Grid Environments, International IEEE Workshop on Grid Computing, Denver, Colorado, November, 2001.
73. Papakonstantinou, Y. and Vassalos, V. (1999) Query rewriting for semistructured data, ACM SIGMOD Conference On Management of Data, 1999.
74. Ph.D. Thesis Roy Thomas Fielding, (2000) Architectural Styles and the Design of Network-based Software Architectures, Ph.D. Thesis, University of California, Irvine, 2000.

PART C

Grid computing environments

Overview of Grid computing environments

Geoffrey Fox,[1] Dennis Gannon,[1] and Mary Thomas[2]

[1]*Indiana University, Bloomington, Indiana, United States*
[2]*The University of Texas at Austin, Austin, Texas, United States*

20.1 INTRODUCTION

This short chapter summarizes the current status of Grid Computational and Programming environments. It puts the corresponding section of this book in context and integrates a survey of a set of 28 chapters gathered together by the Grid Computing Environment (GCE) group of the Global Grid Forum, which is being published in 2002 as a special issue of *Concurrency and Computation: Practice and Experience*. Several of the chapters here are extensions or reprints of those papers.

We can define a GCE as a set of tools and technologies that allows users 'easy' access to Grid resources and applications. Often it takes the form of a Web portal that provides the user interface to a multitier Grid application development stack, but it may also be as simple as a GridShell that allows a user access to and control over Grid resources in the same way a conventional shell allows the user access to the file system and process space of a regular operating system.

Grid Computing – Making the Global Infrastructure a Reality.   Edited by F. Berman, A. Hey and G. Fox
© 2003 John Wiley & Sons, Ltd   ISBN: 0-470-85319-0

20.2 OVERALL CLASSIFICATION

Grid Computing Environments can be classified in several different ways. One straight-forward classification is in terms of the technologies used. The different projects differ in terms of languages used, nature of treatment of objects (if any), use of particular technology such as Java servlets, the Globus Toolkit, or GridFTP, and other implementation issues. Some of these issues are important for performance or architecture, but often can look to the user as not so important. For instance, there is a trend to use more heavily Java, XML, and Web Services, but this will only be interesting if the resultant systems have important properties such as better customizability, sustainability, and ease of use, without sacrificing too much in areas such as performance. The ease of development using modern technologies often yields greater functionality in the GCE for a given amount of implementation effort. Technology differences in the projects are important, but more interesting at this stage are the differences in capabilities and the model of computing explicit or implicit in the GCE.

All GCE systems assume that there are some backend remote resources (the Grid), and endeavor to provide convenient access to their capabilities. This implies that one needs some sort of model for 'computing'. At the simplest this is running a job, which already has nontrivial consequences as data usually needs to be properly set up, and access is required for the running job status and final output. More complex examples require coordinated gathering of data, many simulations (either linked at a given time or following each other), visualization, analysis of results and so on. Some of these actions require substantial collaboration between researchers, and sharing of results and ideas. This leads to the concept of GCE collaboratories supporting sharing among scientific teams working on the same problem area.

We can build a picture of different GCE approaches by viewing the problem as some sort of generalization of the task of computing on a single computer. So we can highlight the following classes of features:

1. Handling the basic components of a distributed computing system – files, computing and data resources, programs, and accounts. The GCE will typically interface with an environment like Globus or a batch scheduler like PBS (Portable Batch System) to actually handle the backend resources. However, the GCE will present the user interfaces to handle these resources. This interface can be simple or complex and often constructed hierarchically to reflect tools built in such a fashion. We can follow the lead of UNIX [and Legion (Chapter 10) in its distributed extension], and define a basic GCEShell providing access to the core distributed computing functions. For example, JXTA [1] also builds Grid-like capabilities with a UNIX shell model. GCEShell would support running and compiling jobs, moving among file systems, and so on. GCEShell can have a command line or more visually appealing graphical user interface.

2. The three-tier model, which is typically used for most systems, implies that any given capability (say run a matrix inversion program) can appear at multiple levels. Maybe there is a backend parallel computer running an MPI (Message Passing Interface) job; this is frontended perhaps as a service by some middle-tier component running on a totally different computer, which could even be in a different security domain. One can

'interact' with this service at either level; a high performance I/O transfer at the parallel computing level, and/or by a slower middle-tier protocol such as Simple Object Access Protocol (SOAP) at the service level. These two (or more) calls (component interactions) can represent different functions or the middle-tier call can be coupled with a high performance mirror; typically the middle-tier provides control and the backend provides 'raw data transfer'. The resultant rather complicated model is shown in Figure 20.1. We have each component (service) represented in both middle and High Performance Computing (HPC) (raw or native) tiers. Intratier and intertier linkage is shown. Chapter 21 has an excellent review of the different programming models for the Grid.

3. One broadly important general-purpose feature is security (authentication, authorization, and privacy), which is addressed in some way or the other by essentially all environments.

4. Data management is another broadly important topic, which gets even more important on a distributed system than it is on single machines. It includes file manipulation, databases, and access to raw signals from instruments such as satellites and accelerators.

5. One augments the basic GCEShell with a library of other general-purpose tools, and this can be supported by the GCE. Such tools include (Grid)FTP, (Grid)MPI, parameter sweep and more general workflow, and the composition of GCEShell primitives.

6. Other higher-level tools are also important, and many tend to be rather application dependent; visualization and intelligent-decision support as to what type of algorithm to use can be put here.

7. Looking at commercial portals, one finds that they usually support sophisticated user interfaces with multiple subwindows aggregated in the user interface. The Apache Jetspeed project is a well-known toolkit supporting this [2]. This user interface aggregation

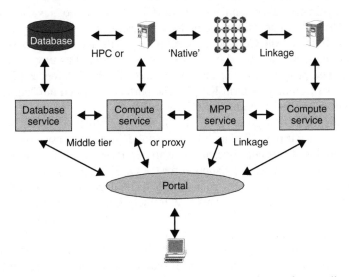

Figure 20.1 Middle-tier and raw (HPC) linked components of an application.

is often supported by a GCE. This aggregation is not stressed in any paper in this special issue although it is provided implicitly.

Apart from particular features, a GCE usually implies a particular computing model for the Grid, and this model is reflected in the GCE architecture and the view of the Grid presented to the user. For example, object models for applications are very popular, and this object view is reflected in the view of the Grid presented to the user by the GCE. Note the programming model for a GCE is usually the programming of rather large objects – one can describe programs and hardware resources as objects without this object model necessarily changing the software model used in applications.

With this preamble, we can now classify the papers in this special issue. There are, as always, no absolute classifications for a complex topic such as distributed Grid systems. Hence it is often the case that these projects can be looked at from many overlapping points of view.

20.3 SUMMARY OF GCE PROJECTS AND FEATURES

20.3.1 Technology for building GCE systems

In the previous section of this book we have described the basic architecture and technologies needed to build a Grid, and we have described the basic component for the different types of GCEs mentioned in the previous section. Chapter 21 provides an excellent overview of many of the backend application programming issues.

The Globus Toolkit [3] is the most widely used Grid middleware system, but it does not provide much direct support for building GCEs. References [4–7] and Chapter 26 describe, respectively, Java, CORBA, Python and Perl Commodity Grid interfaces to the Globus Toolkit. These provide the basic building blocks of full GCEs. Chapter 27, describes the Grid Portal Development Toolkit (GPDK), a suite of JavaBeans suitable for Java-based GCE environments; the technology is designed to support Java Server Pages (JSP) displays. Together, the Commodity Grid (COG) Kits and GPDK constitute the most widely used frameworks for building GCEs that use the Globus environment for basic Grid services. The problem-solving environments (PSEs)in References [8–10] are built on top of the Java Commodity Grid Kit [4]. The portals described in Chapter 28 are built directly on top of the Perl Commodity Grid Kit [7].

Another critical technology for building GCEs is a notification/event service. Reference [11] notes that current Grid architectures build more and more on message-based middleware, and this is particularly clear for Web Services; this chapter designs and prototypes a possible event or messaging support for the Grid. Chapter 22 describes the Narada Brokering system, which leverages peer-to-peer technology to provide a framework for routing messages in the wide area. This is extremely important in cases where the GCE must cross the trust boundaries between the users environment and the target Grid.

Reference [12] provides C support for interfacing to the Globus Toolkit, and portals exposing the toolkit's capabilities can be built on the infrastructure of this paper. Reference [13] proposes an interesting XML-based technology for supporting the run-time coupling of multidisciplinary applications with matching geometries. Reference [14] describes a rather different technology; that is, a Grid simulator aimed at testing new scheduling algorithms.

20.3.2 Largely problem-solving environments

We have crudely divided those GCEs offering user interfaces into two classes. One class focuses on a particular application (set), which is sometimes called application portals or Problem-Solving Environments (PSEs). The second class offers generic application capabilities and has been termed as *user portals*; in our notation introduced above, we can call them *GCEShell portals*. Actually one tends to have a hierarchy with PSEs building on GCEShell portals; the latter building on middleware such as GPDK; GPDK builds on the Java CoG Kit [4], which by itself builds on the Globus Toolkit that finally builds on the native capabilities of the Grid component resources. This hierarchy is for one set of technologies and architecture, but other approaches are similarly built in a layered fashion.

Several chapters in this issue include discussions of Grid PSEs. Reference [15] has an interesting discussion of the architectural changes to a 'legacy' PSE consequent to switching to a Grid Portal approach. Reference [16] illustrates the richness of PSE with a survey of several operational systems; these share a common heritage with the PSEs of Reference [17], although the latter paper is mainly focused on a recommended tool described later.

Five further papers describe PSEs that differ in terms of GCE infrastructure used and applications addressed. Reference [8] describes two PSEs built on top of a GCEShell portal with an object computing model. A similar portal is the XCAT Science portal [18], which is based on the concept of application Notebooks that contain web pages, Python scripts and control code specific to an application. In this case, the Python script code plays the role of the GCEShell. The astrophysics collaboratory [10] includes the Globus Toolkit link via Java [4] and the GPDK [19]; it also interfaces to the powerful Cactus distributed environment [20]. Reference [21] and Chapter 34 presents a portal for computational physics using Web services – especially for data manipulation services. The Polder system [22] and SCIRun [23] offer rich visualization capabilities within several applications including biomedicine. SCIRun has been linked to several Grid technologies including NetSolve [24], and it supports a component model (the common component architecture (CCA) [25], which is described in Chapter 9) with powerful workflow capabilities.

The Discover system described in Chapter 31 describes a PSE framework that is built to enable computational steering of remote Grid applications. This is also an important objective of the work on Cactus described in Chapter 23.

20.3.3 Largely basic GCEShell portals

Here we describe the set of portals designed to support generic computing capabilities on the Grid. Reference [26] is interesting as it is a Grid portal designed to support the stringent requirements of DoE's ASCI (Accelerated Strategic Computing Initiative) program. This reflects not only security and performance issues but the particular and well-established computing model for the computational physicists using the ASCI machines.

Reference [27] describes a portal interface of the very sophisticated Legion Grid, which has through the Legion Shell a powerful generic interface to the shared object (file) system supported by Legion [Chapter 10]. This paper also describes how specific PSEs can be built on top of the basic GCEShell portal.

Unicore [28] was one of the pioneering full-featured GCEShell portals developed originally to support access to a specific set of European supercomputers, but recently has been interfaced to the Globus Toolkit, and as described in Chapter 29, to the Open Grid Services Architecture described in Chapter 8. Unicore has developed an interesting abstract job object (AJO) with full workflow support.

Chapter 30 and References [8, 29] describe well-developed GCEShell portals technology on which several application specific PSEs have been built. Chapter 28 describes the NPACI Grid Portal toolkit, GridPort, which is middleware using the Perl Community Grid Kit [7] to access the Globus Toolkit. This chapter and Reference [30] also describes HotPage, a GCEShell portal built on top of GridPort.

20.3.4 Security

One of the primary tasks of any Grid portal is to manage secure access to Grid resources.

Consequently, security is discussed in most papers on this topic. The GCEs based on Globus and Legion use the Public Key Infrastructure. Kerberos is required by some installations (DoD and DoE, for instance, in the United States), and Grid Computing Environments developed for them [8, 26, 29] are based on this security model.

20.3.5 Workflow

Workflow corresponds to composing a complete job from multiple distributed components. This is broadly important and is also a major topic within the commercial Web service community. It is also inherently a part of a GCEShell or PSE, since these systems are compositions of specific sequences of tasks. Several projects have addressed this but currently there is no consensus on how workflow should be expressed, although several groups have developed visual user interfaces to define the linkage between components. Workflow is discussed in References [9, 13, 23, 26, 28]. The latter integrates Grid workflow with the dataflow paradigm, which is well established in the visualization community. Reference [13] has stressed the need for powerful run time to support the coupling of applications, and this is implicit in other papers including Reference [9]. Business Process Execution Language for Web Services (BPEL4WS) is a major commercial initiative in this area led by BEA Systems IBM and Microsoft (http://www-106.ibm.com/developerworks/webservices/library/ws-bpel/). We expect this to be a major topic of study in the future.

20.3.6 Data management

Data intensive applications are expected to be critical on the Grid but support of this is not covered in the papers of this special issue. Interfaces with file systems, databases and data transfer through mechanisms like GridFTP are covered in several papers. This is primarily because of the fact that data management software is still relatively new on the Grid.

Chapter 34 describes a SOAP-based web service and a portal interface for managing data used within a large scientific data Grid project. This basic model may become the standard for GCE data management tools.

20.3.7 GCEShell tools

In our GCE computing model, one expects a library of tools to be built up that add value to the basic GCEShell capabilities. The previous two sections describe two tools – workflow and data management of special interest, and here we present a broad range of other tools that appeared in several chapters in this special issue.

NetBuild [31] supports distributed libraries with automatic configuration of software on the wide variety of target machines on the Grids of growing heterogeneity.

NetSolve [24] in Chapter 24 pioneered the use of agents to aid the mapping of appropriate Grid resources to client needs. Reference [17] describes a recommendation system that uses detailed performance information to help users on a PSE choose the best algorithms to address their problem.

Many projects have noted the importance of 'parameter sweep' problems where a given application is run many times with different input parameters. Such problems are very suitable for Grid environments, and Reference [32] describes a particular parameter sweep system Nimrod-G. This paper focuses on a different tool – that is, a novel-scheduling tool based on an economic model of Grid suppliers and consumers. Chapter 33 describes another well-regarded parameter sweep system APST that builds on the AppLeS application level scheduling system.

HotPage, described in Chapter 28 and Reference [30] is well known for pioneering the provision of job-status information to portals; such a tool is clearly broadly important.

Finally, we should stress visualization as a critical tool for many users and here References [13, 23] describe this area. There are many other important tools like data-mining that fall into this category.

20.3.8 GCE computing model

In the preamble we suggested that it was interesting to consider the computing model underlying GCEs. This refers to the way we think about the world of files, computers, databases and programs exposed through a portal. NetSolve described in Chapter 24 [24], together with the Ninf effort in Chapter 25 and Reference [33], has developed the important Network Service model for distributed computing. Rather than each user downloading a library to solve some part of their problem, this task is dispatched to a Network resource providing this computational service. Both Ninf and NetSolve support the new GridRPC remote procedure call standard, which encapsulates a key core part of their Grid computing model described in Chapter 21. GridRPC supports scientific data structures as well as Grid-specific security and resource management.

Reference [34] describes Grid implementations of MPI (message passing standard for parallel computing), which addresses the incompatibilities between MPI implementations and binary representations on different parallel computers. Note that in the notation of Figure 20.1, MPI is at the 'HPC backend linkage' layer and not at the middleware layer. Reference [10] supports the Cactus environment [20], Chapter 23] that has well-developed support for Grid computing at the HPC layer, that is, it supports backend programming interfaces and not the middle-tier GCEShell environment. The astrophysics problem-solving environment of Reference [10] augments Cactus with a full middle-tier environment.

Legion, described in Chapter 10 and Reference [27], built a very complete Grid object model. Reference [9] describes a CORBA distributed object model for the Grid, and Chapter 31 and Reference [35] describe the surprisingly hard issues involved in providing interoperability between multiple CORBA GCEs. We can hope that Web services can be easily made interoperable, as the technology used (XML, SOAP) is more open than CORBA, which has evolved with several often incompatible implementations as listed in Reference [5].

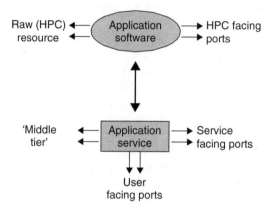

Figure 20.2 A proxy service programming model showing four types of interactions to and from users (portal interface), between proxy and raw resource, other middle-tier components and between other raw (HPC) resources.

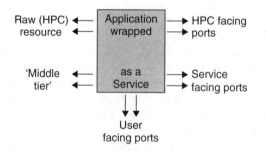

Figure 20.3 A wrapped application programming model showing three types of interactions to and from users (portal interface), to and from other middle-tier components, and between other raw (HPC) resources.

Chapters 9, 29, 30, References [8, 28, 29] and the XCAT Science Portal [18], also present an object model for GCE computing, but with one critical feature – that is, the middle-tier objects are always proxies that hold the metadata, which describe 'real resources' that operate in conventional environments. This proxy strategy appears useful for many Grid resources although the true Network service model of NetSolve is also essential. Let us give a simple example from UNIX and suppose one wanted to send data between two programs (in different machines). One could choose the mechanism within the program and use a simple socket or FTP or Java Remote Method Invocation (RMI) interaction mechanism. Alternatively, the programs could be written generically with output and input or 'standard I/O'. The programs could then have the output of one 'piped' to the input of the other from a UNIX shell command. Such a hybrid-programming model with actions partly specified internally and partly specified at the service level is important to the success of the Grid and should be built into programming models for it.

Any GCE computing model should support both the metadata and the wrapped styles of Grid objects. Actually going back to point 2 in Section 20.2, the proxy and NetSolve models are not really different as indicated in Figures 20.2. and 20.3. Both models effectively wrap application (software) resources as objects. In the proxy model, one exposes the interaction between middle-tier and backend. In the wrapped service model of NetSolve and Ninf, one presents a single entity to the user. In both cases, one can have separate middle-tier and HPC ('real', 'raw' or 'native') communication. To complicate the classification, there can of course be a difference between programming model abstraction (proxy or not) and implementation. In the XCAT model, a software component system [Chapter 9] is used, which implements the wrapped service or proxy model. The component system is based on Web Service standards, so it is possible that the wrapped service components may be arbitrary Grid services.

An additional aspect of the computing model that must be addressed by GCE systems is the way in which resources are managed. In Chapter 32, the authors present the case for an economic model of resource allocation and provisioning. While this is a concept that is not used in any of the current systems described here, there is a good chance it will be used as we scale Grid system to very large sizes.

REFERENCES

1. JXTA Peer-to-Peer Environment, http://www.jxta.org.
2. Apache Jetspeed Portal, http://jakarta.apache.org/jetspeed/site/index.html.
3. The Globus Grid Project, http://www.globus.org.
4. von Laszewski, G., Gawor, J., Lane, P., Rehn, N. and Russell, M. (2002) Features of the Java commodity grid kit. *Concurrency and Computation: Practice and Experience*, **14**, Grid Computing Environments Special Issue 13–14.
5. von Laszewski, G., Parashar, M., Verma, S., Gawor, J., Keahey, K. and Rehn, N. (2002) A CORBA commodity grid kit. *Concurrency and Computation: Practice and Experience*, **14**, Grid Computing Environments Special Issue 13–14.
6. Jackson, K. (2002) pyGlobus: A python interface to the globus toolkit. *Concurrency and Computation: Practice and Experience*, **14**, Grid Computing Environments Special Issue 13–14.
7. Mock, S., Dahan, M., Thomas, M. and von Lazewski, G. (2000) The perl commodity grid toolkit. *Concurrency and Computation: Practice and Experience*, **14**, Grid Computing Environments Special Issue 13–14.

8. Haupt, T., Bangalore, P. and Henley, G. (2002) Mississippi computational web portal. *Concurrency and Computation: Practice and Experience*, **14**, Grid Computing Environments Special Issue 13–14.

9. Schreiber, A. (2002) The integrated simulation environment TENT. *Concurrency and Computation: Practice and Experience*, **14**, Grid Computing Environments Special Issue 13–14.

10. von Laszewski, G. *et al.* (2000) Community software development with the astrophysics simulation collaboratory. *Concurrency and Computation: Practice and Experience*, **14**, Grid Computing Environments Special Issue 13–14.

11. Fox, G. and Pallickara, S. (2000) An event service to support grid computational environments. *Concurrency and Computation: Practice and Experience*, **14**, Grid Computing Environments Special Issue 13–14.

12. Aloisio, G. and Cafaro, M. (2002) Web-based access to the grid using the grid resource broker portal. *Concurrency and Computation: Practice and Experience*, **14**, Grid Computing Environments Special Issue 13–14.

13. Clarke, J. A. and Namburu, R. R. (2002) A distributed computing environment for interdisciplinary applications. *Concurrency and Computation: Practice and Experience*, **14**, Grid Computing Environments Special Issue 13–14.

14. Murshed, M., Buyya, R. and Abramson, D. (2000) GridSim: A toolkit for the modeling and simulation of distributed resource management and scheduling for Grid Computing. *Concurrency and Computation: Practice and Experience*, **14**, Grid Computing Environments Special Issue 13–14.

15. Schuchardt, K., Didier, B. and Black, G. (2002) Ecce – A problem solving environment's evolution toward grid services and a web architecture. *Concurrency and Computation: Practice and Experience*, **14**, Grid Computing Environments Special Issue 13–14.

16. Ramakrishnan, N., Watson, L. T., Kafura, D. G., Ribbens, C. J. and Shaffer, C. A. (2002) Programming environments for multidisciplinary grid communities. *Concurrency and Computation: Practice and Experience*, **14**, Grid Computing Environments Special Issue 13–14.

17. Houstis, E., Catlin, A. C., Dhanjani, N., Rice, J. R., Ramakrishnan, N. and Verykios, V. (2002) MyPYTHIA: A recommendation portal for scientific software and services. *Concurrency and Computation: Practice and Experience*, **14**, Grid Computing Environments Special Issue 13–14.

18. Krishnan, S. *et al.* (2001) The XCAT science portal. *Proceedings SC2001*, Denver, November, 2001.

19. Novotny, J. (2002) The grid portal development kit. *Concurrency and Computation: Practice and Experience*, **14**, Grid Computing Environments Special Issue 13–14, This article is reprinted in this Volume.

20. Cactus Grid Computational Toolkit, http://www.cactuscode.org.

21. Watson III, W. A., Bird, I., Chen, J., Hess, B., Kowalski, A. and Chen, Y. (2002) A web services data analysis grid. *Concurrency and Computation: Practice and Experience*, **14**, Grid Computing Environments Special Issue 13–14.

22. Iskra, K. A. *et al.* (2000) The polder computing environment: a system for interactive distributed simulation. *Concurrency and Computation: Practice and Experience*, **14**, Grid Computing Environments Special Issue 13–14.

23. Johnson, C. Parker, S. and Weinstein, D. (2000) Component-based problem solving environments for large-scale scientific computing. *Concurrency and Computation: Practice and Experience*, **14**, Grid Computing Environments Special Issue 13–14.

24. Arnold, D., Casanova, H. and Dongarra, J. (2002) Innovations of the NetSolve grid computing system. *Concurrency and Computation: Practice and Experience*, **14**, Grid Computing Environments Special Issue 13–14.

25. Common Component Architecture, http://www.cca-forum.org/.

26. Rheinheimer, R., Humphries, S. L., Bivens, H. P. and Beiriger, J. I. (2002) The ASCI computational grid: initial deployment. *Concurrency and Computation: Practice and Experience*, **14**, Grid Computing Environments Special Issue 13–14.

27. Natrajan, A., Nguyen-Tuong, A. Humphrey, M. A. and Grimshaw, A. S. (2002) The legion grid portal. *Concurrency and Computation: Practice and Experience*, **14**, Grid Computing Environments Special Issue 13–14.
28. Erwin, D. W. (2000) UNICORE – A grid computing environment. *Concurrency and Computation: Practice and Experience*, **14**, Grid Computing Environments Special Issue 13–14.
29. Pierce, M. E., Youn, C. and Fox, G. C. (2002) The gateway computational web portal. *Concurrency and Computation: Practice and Experience*, **14**, Grid Computing Environments Special Issue 13–14.
30. Dahan, M., Mueller, K., Mock, S., Mills, C. and Thomas, M. (2000) Application portals: practice and experience. *Concurrency and Computation: Practice and Experience*, **14**, Grid Computing Environments Special Issue 13–14.
31. Moore, K. and Dongarra, J. (2002) NetBuild: transparent cross-platform access to computational software libraries. *Concurrency and Computation: Practice and Experience*, **14**, Grid Computing Environments Special Issue 13–14.
32. Buyya, R., Abramson, D., Giddy, J. and Stockinger, H. (2000) Economics paradigm for resource management and scheduling in grid computing. *Concurrency and Computation: Practice and Experience*, **14**, Grid Computing Environments Special Issue 13–14.
33. Ninf network Server Project, http://ninf.apgrid.org/.
34. Mueller, M., Gabriel, E. and Resch, M. (2002) A software development environment for grid computing. *Concurrency and Computation: Practice and Experience*, **14**, Grid Computing Environments Special Issue 13–14.
35. Mann, V. and Parashar, M. (2002) Engineering an interoperable computational collaboratory on the grid. *Concurrency and Computation: Practice and Experience*, **14**, Grid Computing Environments Special Issue 13–14.

Grid programming models: current tools, issues and directions

Craig Lee[1] and Domenico Talia[2]

[1]*The Aerospace Corporation, California, United States,*
[2]*Universitá della Calabria, Rende, Italy*

21.1 INTRODUCTION

The main goal of Grid programming is the study of programming models, tools, and methods that support the effective development of portable and high-performance algorithms and applications on Grid environments. Grid programming will require capabilities and properties beyond that of simple sequential programming or even parallel and distributed programming. Besides orchestrating simple operations over private data structures, or orchestrating multiple operations over shared or distributed data structures, a Grid programmer will have to manage a computation in an environment that is typically open-ended, heterogeneous, and dynamic in composition with a deepening memory and bandwidth/latency hierarchy. Besides simply operating over data structures, a Grid programmer would also have to design the interaction between remote services, data sources, and hardware resources. While it may be possible to build Grid applications with current programming tools, there is a growing consensus that current tools and languages are insufficient to support the effective development of efficient Grid codes.

Grid Computing – Making the Global Infrastructure a Reality. Edited by F. Berman, A. Hey and G. Fox
© 2003 John Wiley & Sons, Ltd ISBN: 0-470-85319-0

Grid applications will tend to be heterogeneous and dynamic, that is, they will run on different types of resources whose configuration may change during run time. These dynamic configurations could be motivated by changes in the environment, for example, performance changes or hardware failures, or by the need to flexibly compose *virtual organizations* [1] from any available Grid resources. Regardless of their cause, can a programming model or tool give those heterogeneous resources a common 'look-and-feel' to the programmer, hiding their differences while allowing the programmer some control over each resource type if necessary? If the proper abstraction is used, can such transparency be provided by the run-time system? Can *discovery* of those resources be assisted or hidden by the run-time system?

Grids will also be used for large-scale, high-performance computing. Obtaining high performance requires a balance of computation and communication among all resources involved. Currently, this is done by managing computation, communication, and data locality using message passing or remote method invocation (RMI) since they require the programmer to be aware of the marshalling of arguments and their transfer from source to destination. To achieve petaflop rates on tightly or loosely coupled Grid clusters of gigaflop processors, however, applications will have to allow extremely large granularity or produce upwards of approximately 10^8-way parallelism such that high latencies can be tolerated. In some cases, this type of parallelism, and the performance delivered by it in a heterogeneous environment, will be manageable by hand-coded applications.

In light of these issues, we must clearly identify where current programming models are lacking, what new capabilities are required, and whether they are best implemented at the language level, at the tool level, or in the run-time system. The term *programming model* is used here since we are not just considering programming languages. A programming model can be present in many different forms, for example, a language, a library API, or a tool with extensible functionality. Hence, programming models are present in frameworks, portals, and problem-solving environments, even though this is typically not their main focus. The most successful programming models will enable both high performance and the flexible composition and management of resources. Programming models also influence the entire software life cycle: design, implementation, debugging, operation, maintenance, and so on. Hence, successful programming models should also facilitate the effective use of all types of development tools, for example, compilers, debuggers, performance monitors, and so on.

First, we begin with a discussion of the major issues facing Grid programming. We then take a short survey of common programming models that are being used or proposed in the Grid environment. We next discuss programming techniques and approaches that can be brought to bear on the major issues, perhaps using the existing tools.

21.2 GRID PROGRAMMING ISSUES

There are several general properties that are desirable for all programming models. Properties for parallel programming models have also been discussed in Reference [2]. Grid programming models inherit all these properties. The Grid environment, however, will shift the emphasis on these properties dramatically to a degree not seen before and present several major challenges.

21.2.1 Portability, interoperability, and adaptivity

Current high-level languages allowed codes to be processor-independent. Grid programming models should enable codes to have similar portability. This could mean *architecture independence* in the sense of an interpreted virtual machine, but it can also mean the ability to use different prestaged codes or services at different locations that provide equivalent functionality. Such portability is a necessary prerequisite for coping with dynamic, heterogeneous configurations.

The notion of using different but equivalent codes and services implies *interoperability* of programming model implementations. The notion of an *open and extensible Grid architecture* implies a distributed environment that may support protocols, services, application programming interface, and software development kits in which this is possible [1]. Finally, portability and interoperability promote *adaptivity*. A Grid program should be able to adapt itself to different configurations based on available resources. This could occur at start time, or at run time due to changing application requirements or fault recovery. Such adaptivity could involve simple restart somewhere else or actual process and data migration.

21.2.2 Discovery

Resource discovery is an integral part of Grid computing. Grid codes will clearly need to discover suitable hosts on which to run. However, since Grids will host many *persistent services*, they must be able to discover these services and the interfaces they support. The use of these services must be programmable and composable in a uniform way. Therefore, programming environments and tools must be aware of available discovery services and offer a user explicit or implicit mechanisms to exploit those services while developing and deploying Grid applications.

21.2.3 Performance

Clearly, for many Grid applications, performance will be an issue. Grids present heterogeneous bandwidth and latency hierarchies that can make it difficult to achieve high performance and good utilization of coscheduled resources. The communication-to-computation ratio that can be supported in the typical Grid environment will make this especially difficult for tightly coupled applications.

For many applications, however, *reliable* performance will be an equally important issue. A dynamic, heterogeneous environment could produce widely varying performance results that may be unacceptable in certain situations. Hence, in a shared environment, *quality of service* will become increasingly necessary to achieve reliable performance for a given programming construct on a given resource configuration. While some users may require an actual deterministic performance model, it may be more reasonable to provide reliable performance within some statistical bound.

21.2.4 Fault tolerance

The dynamic nature of Grids means that some level of fault tolerance is necessary. This is especially true for highly distributed codes such as Monte Carlo or parameter sweep applications that could initiate thousands of similar, independent jobs on thousands of hosts. Clearly, as the number of resources involved increases, so does the probability that some resource will fail during the computation. Grid applications must be able to check run-time faults of communication and/or computing resources and provide, at the program level, actions to recover or react to faults. At the same time, tools could assure a minimum level of reliable computation in the presence of faults implementing run-time mechanisms that add some form of reliability of operations.

21.2.5 Security

Grid codes will commonly run across multiple administrative domains using shared resources such as networks. While providing strong authentication between two sites is crucial, in time, it will not be uncommon that an application will involve multiple sites all under program control. There could, in fact, be call trees of arbitrary depth in which the selection of resources is dynamically decided. Hence, a security mechanism that provides authentication (and privacy) must be integral to Grid programming models.

21.2.6 Program metamodels

Beyond the notion of just interface discovery, complete Grid programming will require models about the programs themselves. Traditional programming with high-level languages relies on a compiler to make a translation between two programming models, that is, between a high-level language, such as Fortran or C, and the hardware instruction set presented by a machine capable of applying a sequence of functions to data recorded in memory. Part of this translation process can be the construction of a number of models concerning the semantics of the code and the application of a number of enhancements, such as optimizations, garbage-collection, and range checking. Different but analogous *metamodels* will be constructed for Grid codes. The application of enhancements, however, will be complicated by the distributed, heterogeneous Grid nature.

21.3 A BRIEF SURVEY OF GRID PROGRAMMING TOOLS

How these issues are addressed will be tempered by both current programming practices and the Grid environment. The last 20 years of research and development in the areas of parallel and distributed programming and distributed system design has produced a body of knowledge that was driven by both the most feasible and effective hardware architectures and by the desire to be able to build systems that are more 'well-behaved' with properties such as improved maintainability and reusability. We now provide a brief survey of many specific tools, languages, and environments for Grids. Many, if not most, of these systems have their roots in 'ordinary' parallel or distributed computing and are being applied in Grid environments because they are established programming methodologies. We discuss both programming models and tools that are actually available today, and those that are being proposed or represent an important set of capabilities that will eventually be needed. Broader surveys are available in References [2] and [3].

21.3.1 Shared-state models

Shared-state programming models are typically associated with tightly coupled, synchronous languages and execution models that are intended for shared-memory machines or distributed memory machines with a dedicated interconnection network that provides very high bandwidth and low latency. While the relatively low bandwidths and deep, heterogeneous latencies across Grid environments will make such tools ineffective, there are nonetheless programming models that are essentially based on shared state where the producers and consumers of data are decoupled.

21.3.1.1 JavaSpaces

JavaSpaces [4] is a Java-based implementation of the Linda tuplespace concept, in which tuples are represented as serialized objects. The use of Java allows heterogeneous clients and servers to interoperate, regardless of their processor architectures and operating systems. The model used by JavaSpaces views an application as a collection of processes communicating between them by putting and getting objects into one or more *spaces*. A *space* is a shared and persistent object repository that is accessible via the network. The processes use the repository as an exchange mechanism to get coordinated, instead of communicating directly with each other. The main operations that a process can do with a *space* are to *put, take*, and *read* (copy) objects. On a *take* or *read* operation, the object received is determined by an *associative matching* operation on the type and arity of the objects put into the space. A programmer that wants to build a space-based application should design *distributed data structures* as a set of objects that are stored in one or more *spaces*. The new approach that the JavaSpaces programming model gives to the programmer makes building distributed applications much easier, even when dealing with such dynamic, environments. Currently, efforts to implement JavaSpaces on Grids using Java toolkits based on Globus are ongoing [5, 6].

21.3.1.2 Publish/subscribe

Besides being the basic operation underlying JavaSpaces, *associative matching* is a fundamental concept that enables a number of important capabilities that cannot be accomplished in any other way. These capabilities include *content-based routing, event services*, and *publish/subscribe* communication systems [7]. As mentioned earlier, this allows the producers and consumers of data to coordinate in a way in which they can be decoupled and may not even know each other's identity.

Associative matching is, however, notoriously expensive to implement, especially in wide-area environments. On the other hand, given the importance of *publish/subscribe*

to basic Grid services, such as event services that play an important role in supporting fault-tolerant computing, such a capability will have to be available in some form.
Significant work is being done in this area to produce implementations with acceptable
performance, perhaps by constraining individual instantiations to a single application's
problem space. At least three different implementation approaches are possible [8]:

- *Network of servers*: This is the traditional approach for many existing, distributed services. The Common Object Request Broker Architecture (CORBA) Event Service [9]
  is a prime example, providing decoupled communication between producers and consumers using a hierarchy of clients and servers. The fundamental design space for
  server-based event systems can be partitioned into (1) the local matching problem and
  (2) broker network design [10].
- *Middleware*: An advanced communication service could also be encapsulated in a
  layer of middleware. A prime example here is *A Forwarding Layer for Application-level
  Peer-to-Peer Services (FLAPPS* [11]). *FLAPPS* is a routing and forwarding middleware
  layer in user-space interposed between the application and the operating system. It is
  composed of three interdependent elements: (1) peer network topology construction
  protocols, (2) application-layer routing protocols, and (3) explicit request forwarding.
  FLAPPS is based on the store-and-forward networking model, in which messages and
  requests are relayed hop-by-hop from a source peer through one or more transit peers
  en route to a remote peer. Routing behaviors can be defined over an application-defined
  namespace that is hierarchically decomposable such that collections of resources and
  objects can be expressed compactly in routing updates.
- *Network overlays*: The topology construction issue can be separated from the server/
  middleware design by the use of *network overlays*. Network overlays have generally
  been used for *containment, provisioning*, and *abstraction* [12]. In this case, we are
  interested in abstraction, since network overlays can make isolated resources appear
  to be virtually contiguous with a specific topology. These resources could be service
  hosts, or even active network routers, and the communication service involved could
  require and exploit the virtual topology of the overlay. An example of this is a communication service that uses a tree-structured topology to accomplish time management in
  distributed, discrete-event simulations [13].

21.3.2 Message-passing models

In message-passing models, processes run in disjoint address spaces, and information is
exchanged using message passing of one form or another. While the explicit parallelization with message passing can be cumbersome, it gives the user full control and is thus
applicable to problems where more convenient semiautomatic programming models may
fail. It also forces the programmer to consider exactly where a potential expensive communication must take place. These two points are important for single parallel machines,
and even more so for Grid environments.

21.3.2.1 MPI and variants

The Message Passing Interface (MPI) [14, 15] is a widely adopted standard that defines
a two-sided message passing library, that is, with matched sends and receives, that is

well-suited for Grids. Many implementations and variants of MPI have been produced. The most prominent for Grid computing is MPICH-G2.

MPICH-G2 [16] is a Grid-enabled implementation of the MPI that uses the Globus services (e.g. job start-up, security) and allows programmers to couple multiple machines, potentially of different architectures, to run MPI applications. MPICH-G2 automatically converts data in messages sent between machines of different architectures and supports multiprotocol communication by automatically selecting TCP for intermachine messaging and vendor-supplied MPI for intramachine messaging. MPICH-G2 alleviates the user from the cumbersome (and often undesirable) task of learning and explicitly following site-specific details by enabling the user to launch a multimachine application with the use of a single command, mpirun. MPICH-G2 requires, however, that Globus services be available on all participating computers to contact each remote machine, authenticate the user on each, and initiate execution (e.g. fork, place into queues, etc.).

The popularity of MPI has spawned a number of variants that address Grid-related issues such as dynamic process management and more efficient collective operations. The MagPIe library [17], for example, implements MPI's collective operations such as broadcast, barrier, and reduce operations with optimizations for wide-area systems as Grids. Existing parallel MPI applications can be run on Grid platforms using MagPIe by relinking with the MagPIe library. MagPIe has a simple API through which the under-lying Grid computing platform provides the information about the number of clusters in use, and which process is located in which cluster. PACX-MPI [18] has improvements for collective operations and support for intermachine communication using TCP and SSL. Stampi [19] has support for MPI-IO and MPI-2 dynamic process management. MPI_Connect [20] enables different MPI applications, under potentially different vendor MPI implementations, to communicate.

21.3.2.2 One-sided message-passing

While having matched send/receive pairs is a natural concept, *one-sided communication* is also possible and included in MPI-2 [15]. In this case, a *send* operation does not necessarily have an explicit *receive* operation. Not having to match sends and receives means that irregular and asynchronous communication patterns can be easily accommodated. To implement one-sided communication, however, means that there is usually an *implicit* outstanding receive operation that *listens* for any incoming messages, since there are no remote memory operations between multiple computers. However, the one-sided communication semantics as defined by MPI-2 can be implemented on top of two-sided communications [21].

A number of one-sided communication tools exist. One that supports multiprotocol communication suitable for Grid environments is Nexus [22]. In Nexus terminology, a *remote service request (RSR)* is passed between *contexts*. Nexus has been used to build run-time support for languages to support parallel and distributed programming, such as Compositional C++ [23], and also MPI.

21.3.3 RPC and RMI models

Message-passing models, whether they are point-to-point, broadcast, or associatively addressed, all have the essential attribute of explicitly marshaled arguments being sent to a matched receive that unmarshalls the arguments and decides the processing, typically based on message type. The semantics associated with each message type is usually defined statically by the application designers. One-sided message-passing models alter this paradigm by not requiring a matching receive and allowing the sender to specify the type of remote processing. Remote Procedure Call (RPC) and Remote Method Invocation (RMI) models provide the same capabilities as this, but structure the interaction between sender and receiver more as a language construct, rather than a library function call that simply transfers an uninterpreted buffer of data between points A and B. RPC and RMI models provide a simple and well-understood mechanism for managing remote computations. Besides being a mechanism for managing the flow of control and data, RPC and RMI also enable some checking of argument type and arity. RPC and RMI can also be used to build higher-level models for Grid programming, such as components, frameworks, and network-enabled services.

21.3.3.1 Grid-enabled RPC

GridRPC [24] is an RPC model and API for Grids. Besides providing standard RPC semantics with asynchronous, coarse-grain, task-parallel execution, it provides a convenient, high-level abstraction whereby the many details of interacting with a Grid environment can be hidden. Three very important Grid capabilities that GridRPC could transparently manage for the user are as follows:

- *Dynamic resource discovery and scheduling*: RPC services could be located anywhere on a Grid. Discovery, selection, and scheduling of remote execution should be done on the basis of user constraints.
- *Security*: Grid security via GSI and X.509 certificates is essential for operating in an open environment.
- *Fault tolerance*: Fault tolerance via automatic checkpoint, rollback, or retry becomes increasingly essential as the number of resources involved increases.

The management of interfaces is an important issue for all RPC models. Typically this is done in an *Interface Definition Language* (IDL). GridRPC was also designed with a number of other properties in this regard to both improve usability and ease implementation and deployment:

- *Support for a 'scientific IDL'*: This includes large matrix arguments, shared-memory matrix arguments, file arguments, and call-by-reference. Array strides and sections can be specified such that communication demand is reduced.
- *Server-side-only IDL management*: Only GridRPC servers manage RPC stubs and monitor task progress. Hence, the client-side interaction is very simple and requires very little client-side state.

Two fundamental objects in the GridRPC model are *function handles* and the *session IDs*. GridRPC function names are mapped to a server capable of computing the function. This mapping is subsequently denoted by a function handle. The GridRPC model does not specify the mechanics of resource discovery, thus allowing different implementations to use different methods and protocols. All RPC calls using a function handle will be executed on the server specified by the handle. A particular (nonblocking) RPC call is denoted by a session ID. Session IDs can be used to check the status of a call, wait for completion, cancel a call, or check the returned error code.

It is not surprising that GridRPC is a straightforward extension of network-enabled service concept. In fact, prototype implementations exist on top of both Ninf [25] and NetSolve [26]. The fact that server-side-only IDL management is used means that deployment and maintenance is easier than other distributed computing approaches, such as CORBA, in which clients have to be changed when servers change. We note that other RPC mechanisms for Grids are possible. These include SOAP [27] and XML-RPC [28] which use XML over HTTP. While XML provides tremendous flexibility, it currently has limited support for scientific data, and a significant encoding cost [29]. Of course, these issues could be rectified with support for, say, double-precision matrices, and binary data fields. We also note that GridRPC could, in fact, be hosted on top of Open Grid Services Architecture (OGSA) [30].

21.3.3.2 Java RMI

Remote invocation or execution is a well-known concept that has been underpinning the development of both – originally RPC and then Java's RMI. Java Remote Method Invocation (RMI) enables a programmer to create distributed Java-based applications in which the methods of remote Java objects can be invoked from other Java virtual machines, possibly on different hosts. RMI inherits basic RPC design in general; it has distinguishing features that reach beyond the basic RPC. With RMI, a program running on one Java virtual machine (JVM) can invoke methods of other objects residing in different JVMs. The main advantages of RMI are that it is truly object-oriented, supports all the data types of a Java program, and is garbage collected. These features allow for a clear separation between caller and callee. Development and maintenance of distributed systems becomes easier. Java's RMI provides a high-level programming interface that is well suited for Grid computing [31] that can be effectively used when efficient implementations of it will be provided.

21.3.4 Hybrid models

The inherent nature of Grid computing is to make all manner of hosts available to Grid applications. Hence, some applications will want to run both within and across address spaces, that is to say, they will want to run perhaps multithreaded within a shared-address space and also by passing data and control between machines. Such a situation occurs in *clumps* (clusters of symmetric multiprocessors) and also in Grids. A number of programming models have been developed to address this issue.

21.3.4.1 OpenMP and MPI

OpenMP [32] is a library that supports parallel programming in shared-memory parallel machines. It has been developed by a consortium of vendors with the goal of producing a standard programming interface for parallel shared-memory machines that can be used within mainstream languages, such as Fortran, C, and C++. OpenMP allows for the parallel execution of code (*parallel DO loop*), the definition of shared data (*SHARED*), and the synchronization of processes.

The combination of both OpenMP and MPI within one application to address the clump and Grid environment has been considered by many groups [33]. A prime consideration in these application designs is 'who's on top'. OpenMP is essentially a multithreaded programming model. Hence, OpenMP on top of MPI requires MPI to be thread-safe, or requires the application to explicitly manage access to the MPI library. (The MPI standard claims to be 'thread-compatible' but the thread-safety of a particular implementation is another question.) MPI on top of OpenMP can require additional synchronization and limit the amount of parallelism that OpenMP can realize. Which approach actually works out best is typically application-dependent.

21.3.4.2 OmniRPC

OmniRPC [34] was specifically designed as a thread-safe RPC facility for clusters and Grids. OmniRPC uses OpenMP to manage thread-parallel execution while using Globus to manage Grid interactions, rather than using message passing between machines; however, it provides RPC. OmniRPC is, in fact, a layer on top of Ninf. Hence, it uses the Ninf machinery to discover remote procedure names, associate them with remote executables, and retrieve all stub interface information at run time. To manage multiple RPCs in a multithreaded client, OmniRPC maintains a queue of outstanding calls that is managed by a scheduler thread. A calling thread is put on the queue and blocks until the scheduler thread initiates the appropriate remote call and receives the results.

21.3.4.3 MPJ

All these programming concepts can be put into one package, as is the case with message-passing Java, or MPJ [35]. The argument for MPJ is that many applications naturally require the symmetric message-passing model, rather than the asymmetric RPC/RMI model. Hence, MPJ makes multithreading, RMI and message passing available to the application builder. MPJ message-passing closely follows the MPI-1 specification.

This approach, however, does present implementation challenges. Implementation of MPJ on top of a native MPI library provides good performance, but breaks the Java security model and does not allow applets. A native implementation of MPJ in Java, however, usually provides slower performance. Additional compilation support may improve overall performance and make this single-language approach more feasible.

21.3.5 Peer-to-peer models

Peer-to-peer (P2P) computing [36] is the sharing of computer resources and services by direct exchange between systems. Peer-to-peer computing takes advantage of existing desktop computing power and networking connectivity, allowing economical clients to leverage their collective power to benefit the entire enterprise. In a P2P architecture, computers that have traditionally been used solely as clients communicate directly among themselves and can act as both clients and servers, assuming whatever role is most efficient for the network. This reduces the load on servers and allows them to perform specialized services (such as mail-list generation, billing, etc.) more effectively. As computers become ubiquitous, ideas for implementation and use of P2P computing are developing rapidly and gaining importance. Both peer-to-peer and Grid technologies focus on the flexible sharing and innovative use of heterogeneous computing and network resources.

21.3.5.1 JXTA

A family of protocols specifically designed for P2P computing is JXTA [37]. The term JXTA is derived from 'juxtapose' and is simply meant to denote that P2P computing is juxtaposed to client/server and Web-based computing. As such, JXTA is a set of open, generalized P2P protocols, defined in XML messages, that allows any connected device on the network ranging from cell phones and wireless PDAs to PCs and servers to communicate and collaborate in a P2P manner. Using the JXTA protocols, peers can cooperate to form self-organized and self-configured peer groups independent of their positions in the network (edges, firewalls), and without the need for a centralized management infrastructure. Peers may use the JXTA protocols to advertise their resources and to discover network resources (services, pipes, etc.) available from other peers. Peers form and join peer groups to create special relationships. Peers cooperate to route messages allowing for full peer connectivity. The JXTA protocols allow peers to communicate without the need to understand or manage the potentially complex and dynamic network topologies that are becoming common. These features make JXTA a model for implementing P2P Grid services and applications [6].

21.3.6 Frameworks, component models, and portals

Besides these library and language-tool approaches, entire programming environments to facilitate the development and deployment of distributed applications are available. We can broadly classify these approaches as frameworks, component models, and portals. We review a few important examples.

21.3.6.1 Cactus

The Cactus Code and Computational Toolkit [38] provides a modular framework for computational physics. As a framework, Cactus provides application programmers with a high-level API for a set of services tailored for computational science. Besides support for services such as parallel I/O and parallel check pointing and restart, there are services for computational steering (dynamically changing parameters during a run) and remote visualization. To build a Cactus application, a user builds modules, called *thorns*, that are plugged into the framework *flesh*. Full details are available elsewhere in this book.

21.3.6.2 CORBA

The Common Object Request Broker Architecture (CORBA) [9] is a standard tool in which a metalanguage interface is used to manage interoperability among objects. Object member access is defined using the IDL. An Object Request Broker (ORB) is used to provide resource discovery among client objects. While CORBA can be considered middleware, its primary goal has been to manage interfaces between objects. As such, the primary focus has been on client-server interactions within a relatively static resource environment. With the emphasis on flexibly managing interfaces, implements tend to require layers of software on every function call resulting in performance degradation.

To enhance performance for those applications that require it, there is work being done on High-Performance CORBA [39]. This endeavors to improve the performance of CORBA not only by improving ORB performance but also by enabling 'aggregate' processing in clusters or parallel machines. Some of this work involves supporting parallel objects that understand how to communicate in a distributed environment [40].

21.3.6.3 CoG kit

There are also efforts to make CORBA services directly available to Grid computations. This is being done in the CoG Kit project [41] to enable 'Commodity Grids' through an interface layer that maps Globus services to a CORBA API. Full details are available elsewhere in this book.

21.3.6.4 Legion

Legion [42] provides objects with a globally unique (and opaque) identifier. Using such an identifier, an object, and its members, can be referenced from anywhere. Being able to generate and dereference globally unique identifiers requires a significant distributed infrastructure. We note that all Legion development is now being done as part of the AVAKI Corporation [43].

21.3.6.5 Component architectures

Components extend the object-oriented paradigm by enabling objects to manage the interfaces they present and discover those presented by others [44]. This also allows implementation to be completely separated from definition and version. Components are required to have a set of *well-known ports* that includes an *inspection* port. This allows one component to query another and discover what interfaces are supported and their exact specifications. This capability means that a component must be able to provide metadata about its interfaces and also perhaps about its functional and performance properties. This capability also supports software reuse and composibility.

A number of component and component-like systems have been defined. These include COM/DCOM [45], the CORBA 3 Component Model [9], Enterprise Java Beans and Jini [46, 47], and the Common Component Architecture [48]. Of these, the Common Component Architecture includes specific features for high-performance computing, such as *collective ports* and *direct connections*.

21.3.6.6 Portals

Portals can be viewed as providing a Web-based interface to a distributed system. Commonly, portals entail a *three-tier architecture* that consists of (1) a first tier of clients, (2) a middle tier of brokers or servers, and (3) a third tier of object repositories, compute servers, databases, or any other resource or service needed by the portal. Using this general architecture, portals can be built that support a wide variety of application domains, for example, science portals, compute portals, shopping portals, education portals, and so on. To do this effectively, however, requires a set of portal building tools that can be customized for each application area.

A number of examples are possible in this area. One is the Grid Portal Toolkit, also known as *GridPort* [49]. The GridPort Toolkit is partitioned into two parts: (1) the client interface tools and (2) the Web portal services module. The client interface tools enable customized portal interface development and do not require users to have any specialized knowledge of the underlying portal technology. The Web portal services module runs on commercial Web servers and provides authenticated use of Grid resources.

Another very important example is the XCAT Science Portal [50]. In this effort, portals are designed using a *notebook* of typical Web pages, text, graphics, forms, and executable scripts. Notebooks have an interactive script/forms editor based on JPython that allows access to other tool kits such as CoG Kit and the XCAT implementation of the Common Component Architecture. This coupling of portals and components will facilitate ease of use by the user and the dynamic composition of Grid codes and services in a way that will provide the best of both worlds.

More information on portals is available elsewhere in this book.

21.3.7 Web service models

21.3.7.1 OGSA

Grid technologies are evolving toward an Open Grid Services Architecture (OGSA) ([51] and elsewhere in this book) in which a Grid provides an extensible set of services that virtual organizations can aggregate in various ways. OGSA defines a uniform exposed service semantics (the so-called Grid service) based on concepts and technologies from both the Grid and Web services communities. OGSA defines standard mechanisms for creating, naming, and discovering transient Grid service instances, provides location transparency and multiple protocol bindings for service instances, and supports integration with underlying native platform facilities.

The OGSA effort aims to define a common resource model that is an abstract representation of both real resources, such as nodes, processes, disks, file systems, and logical resources. It provides some common operations and supports multiple underlying resource models representing resources as service instances. OGSA abstractions and services provide building blocks that developers can use to implement a variety of higher-level Grid services, but OGSA services are in principle programming language- and programming model-neutral. OGSA aims to define the semantics of a Grid service instance: how it is created, how it is named, how its lifetime is determined, how to communicate with it, and so on.

OGSA does not, however, address issues of implementation programming model, programming language, implementation tools, or execution environment. OGSA definition and implementation will produce significant effects on Grid programming models because these can be used to support and implement OGSA services, and higher-level models could incorporate OGSA service models offering high-level programming mechanisms to use those services in Grid applications. The Globus project is committed to developing an open source OGSA implementation by evolving the current Globus Toolkit towards an OGSA-compliant Globus Toolkit 3.0. This new release will stimulate the research community in developing and implementing OGSA-oriented programming models and tools.

21.3.8 Coordination models

The purpose of a coordination model is to provide a means of integrating a number of possibly heterogeneous components together, by interfacing with each component in such a way that the collective set forms a single application that can execute on parallel and distributed systems [52]. Coordination models can be used to distinguish the computational concerns of a distributed or parallel application from the cooperation ones, allowing separate development but also the eventual fusion of these two development phases.

The concept of coordination is closely related to that of heterogeneity. Since the coordination interface is separate from the computational one, the actual programming languages used to write the computational code play no important role in setting up the coordination mechanisms. Furthermore, since the coordination component offers a homogeneous way for interprocess communication and abstracts from the architecture details, coordination encourages the use of heterogeneous ensembles of machines.

A coordination language offers a composing mechanism and imposes some constraints on the forms of parallelism and on the interfacing mechanisms used to compose an application. Coordination languages for Grid computing generally are orthogonal to sequential or parallel code used to implement the single modules that must be executed, but provide a model for composing programs and should implement inter-module optimizations that take into account machine and interconnection features for providing efficient execution on Grids. Some recent research activities in this area use XML-based [53, 54] or skeleton-based models for Grid programming. Another potential application domain for Grid coordination tools is *workflow* [55], a model of enterprise work management in which work units are passed between *processing points* based on procedural rules.

21.4 ADVANCED PROGRAMMING SUPPORT

While these programming tools and models are extremely useful (and some are actually finding wide use), they may be underachieving in the areas of both performance and flexibility. While it may be possible to hand code applications using these models and low-level, common Grid services that exhibit good performance and flexibility, we also have the goal of making these properties as easy to realize as possible. We discuss several possibilities for advanced programming support.

21.4.1 Traditional techniques

While tightly coupled applications that have been typically supported by a shared-memory abstraction will not be effective in Grids, there are a number of traditional performance-enhancing techniques that can be brought to bear in Grid codes. Work reported in

Reference [56] describes many of these techniques, all applied to a single, tightly coupled MPI solver code run between two institutions separated by roughly 2000 km.

- *Overlapping computation with communication*: This requires a Grid-aware communication schedule such that it is known when boundary data can be exchanged while computation is done on the interior.
- *Shadow arrays*: The use of overlapping 'ghostzones' allows more latency to be tolerated at the expense of some redundant computation.
- *Aggregated communication*: Communication efficiency can be improved by combining many smaller messages into fewer larger messages.
- *Compression*: With the smooth data in this physics simulation, very good compression ratios were achieved such that latency, and not bandwidth, became more of a problem.
- *Protocol tuning*: By tuning communication protocol parameters, such as the TCP window size, applications can realize better communication performance.

With all these well-known techniques, respectable scaling is claimed (88% and 63%) for the problem size and resources used. The outstanding issue here is how well these techniques can be incorporated into programming models and tools such that they can be transparently applied to Grid applications.

21.4.2 Data-driven techniques

Besides improving communication performance, the techniques in the previous section are also oriented towards providing a more loosely coupled execution of the MPI code. How can a more asynchronous, loosely coupled execution model be realized to support programming models? Clearly *data-driven* programming techniques can facilitate this. While such models can suffer from excessive operand matching and scheduling overheads, restricted, coarse-grain forms can realize significant net benefits. *Workflow* is an instance of this model. Another instance is *stream programming*.

As exemplified by the DataCutter framework [57, 58], stream programming can be used to manage the access to large data stores and associate processing with communication in a distributed manner. Data sets that are too large to easily copy or move can be accessed through upstream filters that can do spatial filtering, decimation, corner-turns, or caching copies 'closer to home'. The coallocation of filters and streams in a Grid environment is an important issue. Stream programming can also have a wide variety of semantics and representations, as catalogued in Reference [59]. This is also closely related to the notion of advanced communication services discussed below.

21.4.3 Speculative or optimistic techniques

Another method for producing a more asynchronous, loosely coupled execution is that of *speculative* or *optimistic computing*. By direct analogy with optimistic discrete-event simulation, speculative or optimistic computing is the relaxation of synchronization and communication requirements by allowing speculative execution among multiple hosts with the probability that some work optimistically computed would have to be discarded when it is determined to be incompatible or redundant. The goal is to control the level of optimism such that the benefits of loosely coupled execution are maximized while the overhead of wasted computation is minimized, thus hitting a 'sweetspot'.

An example of the use of optimistic computation in an application domain is *optimistic mesh generation*. The Parallel Bowyer–Watson method [60] for mesh generation allows an implementation in which boundary cavity generation can be computed optimistically with a control parameter for the level of optimism. This should enable the generation code to stay in an 'operating region' where reasonable performance and utilization is realized.

21.4.4 Distributed techniques

Yet another method is the distribution of processing over the data. In a Grid's deep, heterogeneous latency hierarchy, synchronous data-parallel language approaches will clearly be inappropriate. Assuming that synchronization and intercommunication requirements are not excessively dense, distributed techniques can achieve very high aggregate bandwidths between local data and processing.

This basic technique has been applied in contexts other than Grids. The *macroserver* model [61] uses a coarse-grain, message-driven approach and was developed to support the *processor-in-memory* (PIM) technology in the Hybrid Technology MultiThreaded Architecture (HTMT) architecture design. Besides executing code 'at the sense amps', *parcels* containing code, data, and environment *percolate* through the machine to the PIM where they execute. The goal, of course, is to hide all latency.

An analogous approach can be taken in Grid environments, but on a completely different scale. The Grid Datafarm architecture [62] is designed to exploit access locality by scheduling programs across a large-scale distributed disk farm that has processing close to the storage. To promote tight coupling between storage and processing, the *owner computes* rule (as devised for data-parallel languages) was adopted here. The Grid Datafarm also provides a parallel I/O API.

21.4.5 Grid-aware I/O

While I/O systems may concentrate on the movement of data, they can certainly have a large effect on how programs are written. Grid Datafarm files are distributed across disks but they can be opened, read, and written as a single, logical file. For communication within a program, the KeLP system [63, 64] uses a notion of *structural abstraction* and an associated *region calculus* to manage message passing, thread scheduling, and synchronization. While developed to support Parallel Object-Oriented Methods and Applications (POOMA) [65], Shared-Memory Asynchronous Run-Time System (SMARTS) [66] may have applicability in the Grid environment. SMARTS uses macrodataflow scheduling to manage coarse-grain data-parallel operations and to hide latency. It would be interesting to determine if this approach is scalable to the Grid environment.

21.4.6 Advanced communication services

Feasible programming models may depend on the infrastructure support that is available. Advanced communication services are part of this infrastructure. What is meant here by 'advanced communication services' is essentially any type of semantics associated with communication beyond the simple, reliable unicast transfer of data from point A to point B, or even the multicast of data from one to many. Hence, what constitutes an advanced communication service can be broadly defined and can be motivated by different factors.

In Grid computations, understanding and utilizing the network topology will be increasingly important since overall Grid communication performance will be increasingly dominated by propagation delays, that is to say, in the next five to ten years and beyond, network 'pipes' will be getting fatter (as bandwidths increase) but not commensurately shorter (because of latency limitations) [67]. To maintain performance, programming tools such as MPI will have to become *topology-aware*. An example of this is MagPIe [68]. MagPie transparently accommodates wide-area clusters by minimizing the data traffic for collective operations over the slow links. Rather than being governed by $O(nlogn)$ messages across the diameter of the network, as is typical, topology-aware collective operations could be governed by just the average diameter of the network.

Another motivating factor for advanced communication services is the need for fundamentally different communication properties. Such is the case for *content-based* or *policy-based* routing. A traditional multicast group, for example, builds relevant routing information driven by the physical network topology. Content-based routing would enable an application to control the communication scheduling, routing, and filtering on the basis of the application's dynamic communication requirements within a given multicast group, rather than always having to use point-to-point communication. Of course, this requires topology-awareness at some level.

Hence, advanced communication services can be classified into several broad categories. Some of these services are simply more efficiently implemented when topology-aware, while others are not possible in any other way [69].

- *Augmented communication semantics*: Rather than changing fundamental routing behaviors, and so on, most of the communication could simply be augmented with additional functionality. Common examples of this include caching (Web caching), filtering, compression, encryption, quality of service, data-transcoding, or other user-defined functions.
- *Collective operations*: Applications may require synchronous operations, such as barriers, scans and reductions. These operations are typically implemented with a communication topology based on point-to-point operations. For performance in a wide-area network, it is crucial to match these operations to the topology defined by the physical or virtual network.
- *Content-based and policy-based routing*: Content-based routing is a fundamentally different paradigm that enables a host of important capabilities. By allowing applications to determine routing based on application-defined fields in the data payload, this enables capabilities such as publish/subscribe for interest management, event services, and even tuple spaces. Policy-based routing is also possible. Examples of this include routing to meet QoS requirements and message consistency models in which a policy must be enforced on the message-arrival order across some set of end hosts.
- *Communication scope*: Some communication services could be expensive to implement, especially on a large scale. Hence, if applications could define their own *scope*

for the service, then they could keep the problem size to a minimum, thereby helping the service to remain feasible. Communication scope could be associated with a *named topology* such that multiple scopes can be managed simultaneously for the same or separate applications.

Many of these services are suitable for the implementation approaches discussed in Section 21.3.1.2.

21.4.7 Security

Grid applications may want authentication, authorization, integrity checking, and privacy. In the context of a programming model, this carries additional ramifications. Basic, point-to-point security can be accomplished by integrating a security mechanism with a programming construct. An example of this is the integration of SOAP with GSI [70]. In the large context, however, such RMI or RPC calls could exist in a *call tree*. Supporting security along a call tree requires the notion of *delegation of trust*. We note that cancellation of a secure call could require the revocation of delegated trust [71].

Signing and checking certificates on an RPC also represents an overhead that must be balanced against the amount of work represented by the RPC. Security overheads could be managed by establishing secure, trusted domains. RPCs within a domain could dispense with certificates; RPCs that cross domains would have to use them. Trusted domains could be used to limit per-RPC security overheads in favor of the one-time cost of establishing the domain.

21.4.8 Fault tolerance

Reliability and fault tolerance in Grid programming models/tools are largely unexplored, beyond simple check pointing and restart. Certain application domains are more amendable to fault tolerance than others, for example, parameter sweep or Monte Carlo simulations that are composed of many independent cases in which a case can simply be redone if it fails for any reason. The issue here, however, is how to make Grid programming models and tools inherently more reliable and fault tolerant. Clearly, a distinction exists between reliability and fault tolerance in the application versus that in the programming model/tool versus that in the Grid infrastructure itself. An argument can be made that reliability and fault tolerance have to be available at all lower levels to be possible at the higher levels.

A further distinction can be made between fault detection, fault notification, and fault recovery. In a distributed Grid environment, simply being able to detect when a fault has occurred is crucial. Propagating notification of that fault to relevant sites is also critical. Finally these relevant sites must be able to take action to recover from or limit the effects of the fault.

These capabilities require that *event models* be integral to Grid programming models and tools [71]. Event models are required for many aspects of Grid computing, such as a performance-monitoring infrastructure. Hence, it is necessary that a widely deployed Grid event mechanism become available. The use of such a mechanism will be a key element for reliable and fault-tolerant programming models.

As a case in point, consider a Grid RPC mechanism. An RPC typically has a *call* and a *return* in a call tree, but it can also have a *cancellation* or *rejection*. For a chain of synchronous RPCs, cancellation or rejection must flow along one linear path. For multiple, asynchronous RPCs, however, cancellations and rejections may have to flow along multiple branches. Rejections may also precipitate cancellations on other branches.

Hence, a Grid RPC mechanism clearly needs an event service to manage cancellation and rejection. This is critical to designing and implementing an RPC mechanism that is fault tolerant, that is, a mechanism by which any abnormal operation is detected within a bounded length of time and reliable signaling occurs whereby the RPC service cleans up any obsolete state. Reliability of any cancel and reject event is critical to achieving any fault tolerance.

While the simplest (and probably the most common) case of cancellation will involve one RPC handle that is carried on a single event delivered point-to-point, it may be useful to cancel RPCs en masse. In this case, RPCs could be identified as members of a process group. Such a process group may include (1) the single active branch of a call tree, (2) a parallel call tree with a single root, or (3) one or more branches of a call tree in which the parent or root node is not a member.

Cancellation of an entire process group could be accomplished by point-to-point events. However, one-to-many or some-to-all event notification would enable the entire group to be cancelled more quickly by 'short-circuiting' the call tree topology. Such group event notification could be accomplished by membership in the group (as in membership in a multicast group) or by a publish/subscribe interface whereby remote RPC servers subscribe to cancellation events for the process groups of the RPCs that they are hosting.

21.4.9 Program metamodels and Grid-aware run-time systems

Another serious issue in Grid programming models is the concept of *program metamodels* and their use by Grid-aware run-time systems. Regardless of how Grids are ultimately deployed, they will consist of components and services that are either persistent or can be instantiated. Some of these components and services will become widely used and commonly available. Hence, many applications will be built, in part or in whole, through the composition of components and services.

How can such a composition be accomplished automatically such that characteristics like performance are understood and maintained, in addition to maintaining properties like security and fault tolerance? This can only be done by producing *metamodels* that define a component's characteristics and properties. Metamodels could be produced by hand, but they could also be produced automatically.

Hence, compilers and composition tools could be responsible for producing metamodels and using them to identify and enforce valid compositions. In this context, 'valid' can mean more than just whether the interface arguments are compatible. Valid could mean preserving performance characteristics, security properties, or fault tolerance. On the basis of a high-level program description (as in a portal scripting language, for instance), a 'compiler' could map higher-level semantics to lower-level components and services. This raises the possibility of the definition of a 'Grid compiler target', not in the traditional sense of a machine instruction set but rather as a set of commonly available services.

Preliminary work has been done in this area by the Grid Application Development Software (GrADS) project [72]. The GrADS approach is based on (1) the GrADS Program Preparation System and (2) the GrADS Program Execution System. The Program Preparation System takes user input, along with reusing components and libraries, to produce a *configurable object program*. These objects are annotated with their resource requirements and predicted performance. The execution environment uses this information to select appropriate resources for execution, and also to monitor the application's compliance with a performance 'contract'.

This type of approach is a cornerstone for the Dynamic Data-Driven Application Systems (DDDAS) concept [73] being developed by the National Science Foundation. DDDAS promotes the notion of a *run-time compiling system* that accomplishes *dynamic application composition*. The ultimate goal of DDDAS, however, is to enable dynamic applications that can discover and ingest new data on the fly and automatically form new collaborations with both computational systems and physical systems through a network of sensors and actuators.

21.5 CONCLUSION

We have considered programming models for Grid computing environments. As with many fundamental areas, what will comprise a successful Grid-programming model consists of many aspects. To reiterate, these include portability, interoperability, adaptivity, and the ability to support discovery, security, and fault tolerance while maintaining performance. We identified a number of topics in which further work is needed to realize important capabilities, such as data-driven and optimistic programming techniques, advanced communication and I/O services, and finally program metamodels.

Regardless of these insights, however, the programming models and tools that get developed will largely depend on which models and tools are considered to be the *dominant paradigm*. Enhancements to these models and tools will have a lower *barrier to acceptance* and will be perceived as potentially having a *broader impact* on a larger community of users. There is also a distinction to be made between the capabilities supported in the common infrastructure and the programming models and tools built on top of them.

With regard to common infrastructure, the tremendous commercial motivation for the development of Web services means that it would be a mistake not to leverage these capabilities for scientific and engineering computation. Hence, for these practical reasons, we will most probably see continued development of the Open Grid Services Architecture.

With regards to programming models and tools, and for the same practical reasons, we will also most probably see continued development in MPI. MPI is a standard with an established user base. Many of the potential enhancements discussed earlier could be provided in MPI with minimal changes to the API.

Other models and tools, however, will see increasing development. Frameworks that provide a rich set of services on top of common Grid services, such as Cactus and XCAT, will incorporate many of the capabilities that we have described. For other applications, however, a fundamental programming construct that is Grid-aware, such as GridRPC, will be completely sufficient for their design and implementation.

Finally, we discuss the issue of *programming style*. This evolution in available computing platforms, from single machines to parallel machines to Grids, will precipitate a corresponding evolution in how programming is done to solve computational problems. Programmers, by nature, will adapt their codes and programming style to accommodate the available infrastructure. They will strive to make their codes more loosely coupled. 'Problem architectures' will be conceived in a way to make them better suited to the Grid environment.

This raises a concomitant issue. Besides the tremendous flexibility that Grids will offer for virtual organizations, what will be their limits for computational science? Will computational science be limited to the size of 'single-chassis' machines, such as the ASCI machines [74] and the HTMT [75]? Or can the problem architectures for science and engineering, and their associated computational models, be made sufficiently *Grid-friendly* such that increasingly large problems can be solved? Much work remains to be done.

REFERENCES

1. Foster, I., Kesselman, C. and Tuecke, S. (2001) The anatomy of the grid: enabling scalable virtual organizations. *International Journal of Supercomputer Applications*, **15**, 200–222, available at www.globus.org/research/papers/anatomy.pdf.
2. Skillicorn, D. and Talia, D. (1998) Models and languages for parallel computation. *ACM Computing Surveys*, **30**(2), 123–169.
3. Lee, C., Matsuoka, S., Talia, D., Sussman, A., Mueller, M., Allen, G. and Saltz, J. (2001) A Grid Programming Primer, submitted to the Global Grid Forum, August, 2001, http://www.gridforum.org/7_APM/APS.htm http://www.eece.unm.edu/~apm/docs/APM_Primer_0801.pdf.
4. Freeman, E., Hupfer, S. and Arnold, K. (1999) *JavaSpaces: Principles, Patterns, and Practice*. Reading, MA: Addison-Wesley.
5. Saelee, D. and Rana, O. F. (2001) Implementing services in a computational Grid with Jini and Globus. *First EuroGlobus Workshop*, 2001, www.euroglobus.unile.it.
6. Rana, O. F., Getov, V. S., Sharakan, E., Newhouse, S. and Allan, R. (2002) Building Grid Services with Jini and JXTA, February, 2002, GGF2 Working Document.
7. Lee, C. and Michel, B. S. (2002) The use of content-based routing to support events, coordination and topology-aware communication in wide-area Grid environments, in Marinescu, D. and Lee, C. (eds) *Process Coordination and Ubiquitous Computing*. Boca Raton, FL: CRC Press.
8. Lee, C., Coe, E., Michel, B. S., Solis, I., Clark, J. M., and Davis, B. (2003) Using topology-aware communication services in grid environments. *Workshop on Grids and Advanced Networks*, co-located with the *IEEE International Symposium on Cluster Computing and the Grid (CCGrid 2003)*, May 2003, to appear.
9. CORBA 3 Release Information, The Object Management Group, 2000, www.omg.org/technology/corba/corba3releaseinfo.htm.
10. Jacobsen, H.-A. and Llirbat, F. Publish/subscribe systems. *17th International Conference on Data Engineering*, 2001; Tutorial.
11. Michel, B. S. and Reiher, P. (2001) Peer-to-Peer Internetworking. *OPENSIG*, September, 2001.
12. Touch, J. Dynamic internet overlay deployment and management using the x-bone. *Computer Networks*, July 2001, 117–135.
13. Lee, C. *et al.* (2002) Scalable time management algorithms using active networks for distributed simulation. *DARPA Active Network Conference and Exposition*, May 29–30, 2002.
14. Message Passing Interface Forum, MPI: A Message Passing Interface Standard, June, 1995, www.mpi-forum.org/.

15. Message Passing Interface Forum, MPI-2: Extensions to the Message Passing Interface, July, 1997, www.mpi-forum.org/.
16. Foster, I. and Karonis, N. T. (1998) A grid-enabled MPI: message passing in heterogeneous distributed computing systems. *Supercomputing*. IEEE, November, 1998, www.supercomp.org/sc98.
17. Kielmann, T., Hofman, R. F. H., Bal, H. E., Plaat, A. and Bhoedjang, R. A. F. (1999) MagPIe: MPI's collective communication operations for clustered wide area systems. *Proc. Seventh ACM SIGPLAN Symposium on Principles and Practice of Parallel Programming (PPoPP'99)*, Atlanta, GA, May 4–6, 1999 pp. 131–140.
18. Gabriel, E., Resch, M., Beisel, T. and Keller, R. (1998) Distributed computing in a Heterogeneous Computing Environment, in Alexandrov, V. and Dongarra, J. (eds) *Recent Advances in Parallel Virtual Machine and Message Passing Interface. 5th European PVM/MPI Users' Group Meeting*, Springer, 180–188, Liverpool, UK.
19. Imamura, T., Tsujita, Y., Koide, H. and Takemiya, H. (2000) An architecture of Stampi: MPI library on a cluster of parallel computers, in Dongarra, J., Kacsuk, P. and Podhorszki, N. (eds) *Recent Advances in Parallel Virtual Machine and Message Passing Interface*. Lecture Notes In Computer Science, Vol. 1908. Springer, pp. 200–207, *7th European PVM/MPI Users' Group Meeting*, Lake Balaton, Hungary.
20. Fagg, G. E., London, K. S. and Dongarra, J. J. (1998) MPI_Connect: managing heterogeneous MPI applications interoperation and process control, in Alexandrov, V. and Dongarra, J. (eds) *Recent Advances in Parallel Virtual Machine and Message Passing Interface*. Lecture Notes in Computer Science, Vol. 1497. Springer, pp. 93–96, *5th European PVM/MPI Users' Group Meeting*, Liverpool, UK.
21. Booth, S. and Mourao, E. (2000) Single sided MPI implementations for SUN MPI. *Proceedings of Supercomputing 2000*, Dallas, November, 2000.
22. Foster, I., Geisler, J., Kesselman, C. and Tuecke, S. (1997) Managing multiple communication methods in high-performance networked computing systems. *Journal of Parallel and Distributed Computing*, **40**, 35–48.
23. Chandy, K. and Kesselman, C. (1993) Compositional C++: compositional parallel programming. Languages and Compilers for Parallel Computing, *6th International Workshop Proceedings*, 1993, pp. 124–144.
24. Nakada, H., Matsuoka, S., Seymour, K., Dongarra, J., Lee, C. and Casanova, H. (2002) Overview of GridRPC: A Remote Procedure Call API for Grid Computing. *3rd International Workshop on Grid Computing*, LNCS, Vol. 2536, November, 2002, pp. 274–278.
25. Nakada, H., Sato, M. and Sekiguchi, S. (1999) Design and implementations of Ninf: towards a global computing infrastructure. *Future Generation Computing Systems*, Metacomputing Issue, **15**(5–6), 649–658.
26. Arnold, D. *et al.* (2001) Users' Guide to NetSolve V1.4, Technical Report CS-01-467, Computer Science Department, University of Tennessee, Knoxville, TN, July, 2001.
27. SOAP, Simple Object Access Protocol (SOAP) 1.1, http://www.w3.org/TR/SOAP/, W3C Note, May, 2000.
28. XML-RPC, XML-RPC, http://www.xml-rpc.com/.
29. Govindaraju, M., Slominski, A., Choppella, V., Bramley, R. and Gannon, D. (2000) Requirements for and evaluation of RMI protocols for scientific computing. *Proceedings of SC 2000*, Dallas, TX, 2000.
30. Shirasuna, S., Nakada, H., Matsuoka, S. and Sekiguchi, S. (2002) Evaluating Web services based implementations of GridRPC. *HPDC-11*, 2002, pp. 237–245.
31. Getov, V., von Laszewski, G., Philippsen, M. and Foster, I. (2001) Multi-paradigm communications in java for grid computing. *CACM* **44**(10), 118–125 (2001).
32. OpenMP Consortium, OpenMP C and C++ Application Program Interface, Version 1.0, 1997.
33. Smith, L. and Bull, M. (2000) Development of Mixed Mode MPI/OpenMP Applications. *WOMPAT*, 2000.

34. Sato, M., Hirono, M., Tanaka, Y. and Sekiguchi, S. (2001) OmniRPC: a grid RPC facility for cluster and global computing in OpenMP, *WOMPAT*. LNCS, Vol. 2104. West Lafayette, IN: Springer-Verlag, pp. 130–136.

35. Carpenter, B. *et al.* (2000) MPJ: MPI-like message-passing for java. *Concurrency: Practice and Experience*, **12**(11), 1019–1038.

36. Gong, L. (ed.) (2002) *IEEE Internet computing*, **6**(1), Peer-to-Peer Networks in Action, http://dsonline.computer.org/0201/ic/w102gei.htm.

37. Gong, L. (2001) JXTA: A Network Programming Environment. *IEEE Internet Computing*, **5**(3), 88–95.

38. Cactus Webmeister, The Cactus Code Website, www.CactusCode.org, 2000.

39. Jacobsen, H. A. *et al.* (2001) High Performance CORBA Working Group, www.omg.org/realtime/working_groups/ high_performance_corba.html.

40. Denis, A., Pérez, C., Priol, T. and Ribes, A. (2002) Programming the grid with distributed objects, in Marinescu, D. and Lee, C. (eds) *Process Coordination and Ubiquitous Computing*. Boca Raton, FL: CRC Press.

41. Verma, S., Gawor, J., von Laszewski, G. and Parashar, M. (2001) A CORBA commodity grid kit. *Grid 2001*, November, 2001, pp. 2–13.

42. Lewis, M. and Grimshaw, A. (1995) The Core Legion Object Model, Technical Report TR CS-95-35, University of Virginia, 1995.

43. AVAKI, The AVAKI Corporation, www.avaki.com, 2001.

44. Szyperski, C. (1999) *Component Software: Beyond Object-oriented Programming*. Reading, MA: Addison-Wesley.

45. Sessions, R. (1997) *COM and DCOM: Microsoft's Vision for Distributed Objects*. New York: John Wiley & Sons.

46. Englander, R. (1997) *Developing Java Beans*. O'Reilly.

47. The Jini Community (2000) The Community Resource for Jini Technology, www.jini.org.

48. Gannon, D. *et al.* (2000) CCAT: The Common Component Architecture Toolkit, www.extreme.indiana.edu/ccat.

49. Thomas, M. *et al.* (2001) The Grid Portal Toolkit, gridport.npaci.edu.

50. Krishnan, S. *et al.* (2001) The XCAT science portal. *Supercomputing*, November, 2001.

51. Foster, I., Kesselman, C., Nick, J. M. and Tuecke, S. (2002) Grid services for distributed system integration. *IEEE Computer*, **June** 37–46.

52. Marinescu, D. and Lee, C. (eds) (2002) *Process Coordination and Ubiquitous Computing*. Boca Raton, FL: CRC Press.

53. Tolksdorf, R. (2002) Models of coordination and web-based systems, in Marinescu, D. and Lee, C. (eds) *Process Coordination and Ubiquitous Computing*. Boca Raton, FL: CRC Press.

54. Furmento, N., Mayer, A., McGough, S., Newhouse, S., Field, T. and Darlington, J. (2001) An integrated grid environment for component applications. *Second International Workshop on Grid Computing (Grid 2001)*, LNCS, Vol. 2242, pp. 26–37, 2001.

55. Fischer, L. (2002) *The Workflow Handbook 2002*. Lighthouse Point, FL: Future Strategies, Inc.

56. Allen, G. *et al.* (2001) Supporting efficient execution in heterogeneous distributed computing environments with cactus and globus. *Supercomputing*, November, 2001.

57. Beynon, M. D., Ferreira, R., Kurc, T., Sussman, A. and Saltz, J. (2000) DataCutter: middleware for filtering very large scientific datasets on archival storage systems. *MASS2000*, March, 2000, pp. 119–133, National Aeronautics and Space Administration, NASA/CP 2000-209888.

58. Beynon, M., Kurc, T., Sussman, A. and Saltz, J. (2001) Optimizing execution of component-based applications using group instances. *IEEE International Symposium on Cluster Computing and the Grid (CCGrid 2001)*, Brisbane, Australia, May, 2001, pp. 56–63.

59. Lee, C. (2001) Stream Programming: In Toto and Core Behavior, www.eece.unm.edu/~apm/WhitePapers/stream.pdf, September, 2001.

60. Chrisochoides, N. and Nave, D. (2001) Parallel delaunay mesh generation kernel. *International Journal for Numerical Methods in Engineering*, **1**(1).

61. Zima, H. P. and Sterling, T. L. (2000) Macroservers: an object-based programming and execution model for processor-in-memory arrays. *Proc. International Symposium on High Performance Computing (ISHPC2K)*, October, 2000.
62. Tatebe, O., Morita, Y., Matsuoka, S., Soda, N., and Sekiguchi, S. (2002) Grid datafarm architecture for petascale data intensive computing. *Proceedings of the 2nd IEEE/ACM International Symposium on Cluster Computing and the Grid (CCGrid 2002)*, 2002, pp. 102–110.
63. Fink, S. and Baden, S. (1997) Runtime support for multi-tier programming of block-structured applications on SMP clusters. *International Scientific Computing in Object-Oriented Parallel Environments Conference (ISCOPE '97)*, December, 1997, available at www-cse.ucsd.edu/groups/hpcl/scg/kelp/pubs.html.
64. Baden, S. and Fink, S. (1999) The data mover: a machine-independent abstraction for managing customized data motion. *12th Workshop on Languages and Compilers for Parallel Computing*, August, 1999.
65. LANL, POOMA: Parallel Object-Oriented Methods and Applications, www.acl.lanl.gov/PoomaFramework, 2000.
66. Vajracharya, S. *et al.* (1999) SMARTS: exploiting temporal locality and parallelism through vertical execution. *International Conference on Supercomputing*, 1999.
67. Lee, C. and Stepanek, J. (2001) On future global Grid communication performance. *10th IEEE Heterogeneous Computing Workshop*, May, 2001.
68. Kielmann, T. *et al.* (1999) MagPIe: MPI's collective communication operations for clustered wide area systems. *Symposium on Principles and Practice of Parallel Programming*, Atlanta, GA, May, 1999, pp. 131–140.
69. Lee, C. (2000) On active grid middleware. *Second Workshop on Active Middleware Services*, August 1, 2000.
70. Jackson, K. (2002) pyGlobus: a python interface to the Globus Toolkit. *Concurrency and Computation: Practice and Experience*, Vol. 14, Grid Computing Environments, Special Issue 13–14, 2002, available from http://www.cogkits.org/papers/c545python-cog-cpe.pdf; to appear.
71. Lee, C. (2001) Grid RPC, Events and Messaging, www.eece.unm.edu/~apm/WhitePapers/APM_Grid_RPC _0901.pdf, September, 2001.
72. Kennedy, K. *et al.* (2002) Toward a framework for preparing and executing adaptive grid programs. *Proceedings of NSF Next Generation Systems Program Workshop (International Parallel and Distributed Processing Symposium)*, April, 2002.
73. Darema, F. (2002) Dynamic data-driven application systems, in Marinescu, D. and Lee, C. (eds) *Process coordination and ubiquitous computing*. Boca Raton, FL: CRC Press.
74. Subcommittee on Computing, Information, and Communications R&D, DOE, DOE's ASCI Program, Technical Report DOE, 2000, www.hpcc.gov/pubs/blue00/asci.html.
75. Sterling, T. and Bergman, L. (1999) Design analysis of a hybrid technology multithreaded architecture for petaflops scale computation. *International Conference on Supercomputing*, June, 1999, pp. 286–293.

NaradaBrokering: an event-based infrastructure for building scalable durable peer-to-peer Grids

Geoffrey Fox and Shrideep Pallickara

Indiana University, Bloomington, Indiana, United States

22.1 INTRODUCTION

The peer-to-peer (P2P) style interaction [1] model facilitates sophisticated resource sharing environments between 'consenting' peers over the 'edges' of the Internet; the 'disruptive' [2] impact of which has resulted in a slew of powerful applications built around this model. Resources shared could be anything – from CPU cycles, exemplified by SETI@home (extraterrestrial life) [3] and Folding@home (protein folding) [4] to files (Napster and Gnutella [5]). Resources in the form of direct human presence include collaborative systems (Groove [6]) and Instant Messengers (Jabber [7]). Peer 'interactions' involve advertising resources, search and subsequent discovery of resources, request for access to these resources, responses to these requests and exchange of messages between peers. An overview of P2P systems and their deployments in distributed computing and

Grid Computing – Making the Global Infrastructure a Reality. Edited by F. Berman, A. Hey and G. Fox
© 2003 John Wiley & Sons, Ltd ISBN: 0-470-85319-0

collaboration can be found in Reference [8]. Systems tuned towards large-scale P2P systems include *Pastry* [9] from Microsoft, which provides an efficient location and routing substrate for wide-area P2P applications. Pastry provides a self-stabilizing infrastructure that adapts to the arrival, departure and failure of nodes. FLAPPS [10], a Forwarding Layer for Application-level Peer-to-Peer Services, is based on the general 'peer internetworking' model in which routing protocols propagate availability of shared resources exposed by remote peers. File replications and hoarding services are examples in which FLAPPS could be used to relay a source peer's request to the closest replica of the shared resource. The JXTA [11] (from *juxta*position) project at Sun Microsystems is another research effort that seeks to provide such large-scale P2P infrastructures. Discussions pertaining to the adoption of event services as a key building block supporting P2P systems can be found in References [8, 12].

We propose an architecture for building a scalable, durable P2P Grid comprising resources such as relatively static clients, high-end resources and a dynamic collection of multiple P2P subsystems. Such an infrastructure should draw upon the evolving ideas of computational Grids, distributed objects, Web services, peer-to-peer networks and message-oriented middleware while seamlessly integrating users to themselves and to resources, which are also linked to each other. We can abstract such environments as a distributed system of 'clients', which consist either of 'users' or 'resources' or proxies thereto. These clients must be linked together in a flexible, fault-tolerant, efficient, high-performance fashion. We investigate the architecture, comprising a distributed brokering system that will support such a hybrid environment. In this chapter, we study the event brokering system – NaradaBrokering – that is appropriate to link the clients (both users and resources of course) together. For our purposes (registering, transporting and discovering information), events are just messages – typically with time stamps. The event brokering system NaradaBrokering must scale over a wide variety of devices – from handheld computers at one end to high-performance computers and sensors at the other extreme. We have analyzed the requirements of several Grid services that could be built with this model, including computing and education and incorporated constraints of collaboration with a shared event model. We suggest that generalizing the well-known publish–subscribe model is an attractive approach and this is the model that is used in NaradaBrokering. Services can be hosted on such a P2P Grid with peer groups managed locally and arranged into a global system supported by core servers. Access to services can then be mediated either by the 'broker middleware' or alternatively by direct P2P interactions between machines 'on the edge'. The relative performance of each approach (which could reflect computer/network cycles as well as the existence of firewalls) would be used in deciding on the implementation to use. P2P approaches best support local dynamic interactions; the distributed broker approach scales best globally but cannot easily manage the rich structure of transient services, which would characterize complex tasks. We use our research system NaradaBrokering as the distributed brokering core to support such a hybrid environment. NaradaBrokering is designed to encompass both P2P and the traditional centralized middle-tier style of interactions. This is needed for robustness (since P2P interactions are unreliable and there are no guarantees associated with them) and dynamic resources (middle-tier style interactions are not natural for very dynamic clients and resources). This chapter describes the support for these interactions in NaradaBrokering.

There are several attractive features in the P2P model, which motivate the development of such hybrid systems. Deployment of P2P systems is entirely user-driven obviating the need for any dedicated management of these systems. Peers expose the resources that they are willing to share and can also specify the security strategy to do so. Driven entirely on demand a resource may be replicated several times; a process that is decentralized and one over which the original peer that advertised the resource has sometimes little control. Peers can form groups with the fluid group memberships. In addition, P2P systems tend to be very dynamic with peers maintaining an intermittent digital presence. P2P systems incorporate schemes for searching and subsequent discovery of resources. Communication between a requesting peer and responding peers is facilitated by peers en route to these destinations. These intermediate peers are thus made aware of capabilities that exist at other peers constituting dynamic real-time knowledge propagation. Furthermore, since peer interactions, in most P2P systems, are XML-based, peers can be written in any language and can be compiled for any platform. There are also some issues that need to be addressed while incorporating support for P2P interactions. P2P interactions are self-attenuating with interactions dying out after a certain number of hops. These attenuations in tandem with traces of the peers, which the interactions have passed through, eliminate the continuous echoing problem that results from loops in peer connectivity. However, attenuation of interactions sometimes prevents peers from discovering certain services that are being offered. This results in P2P interactions being very localized. These attenuations thus mean that the P2P world is inevitably fragmented into many small subnets that are not connected. Peers in P2P systems interact directly with each other and sometimes use other peers as intermediaries in interactions. Specialized peers are sometimes deployed to enhance routing characteristics. Nevertheless, sophisticated routing schemes are seldom in place and interactions are primarily through simple forwarding of requests with the propagation range being determined by the attenuation indicated in the message.

NaradaBrokering must support many different frameworks including P2P and centralized models. Though native NaradaBrokering supports this flexibility we must also expect that realistic scenarios will require the integration of multiple brokering schemes. NaradaBrokering supports this hybrid case through gateways to the other event worlds. In this chapter we look at the NaradaBrokering system and its standards-based extensions to support the middle-tier style and P2P style interactions. This chapter is organized as follows; in Section 22.2 we provide an overview of the NaradaBrokering system. In Section 22.3 we outline NaradaBrokering's support for the Java Message Service (JMS) specification. This section also outlines NaradaBrokering's strategy for replacing single-server JMS systems with a distributed broker network. In Section 22.4 we discuss NaradaBrokering's support for P2P interactions, and in Section 22.5 we discuss NaradaBrokering's integration with JXTA.

22.2 NARADABROKERING

NaradaBrokering [13–18] is an event brokering system designed to run a large network of cooperating broker nodes while incorporating capabilities of content-based routing and publish/subscribe messaging. NaradaBrokering incorporates protocols for organizing broker nodes into a cluster-based topology. The topology is then used for incorporating efficient calculation of destinations, efficient routing even in the presence of failures, provisioning of resources to clients, supporting application defined communications scope and incorporating fault-tolerance strategies. Strategies for adaptive communication scheduling based on QoS requirements, content type, networking constraints (such as presence of firewalls, MBONE [19] support or the lack thereof) and client-processing capabilities (from desktop clients to Personal Digital Assistant (PDA) devices) are currently being incorporated into the system core. Communication within NaradaBrokering is asynchronous, and the system can be used to support different interactions by encapsulating them in specialized events. Events are central in NaradaBrokering and encapsulate information at various levels as depicted in Figure 22.1. Clients can create and publish events, specify interests in certain types of events and receive events that conform to specified templates. Client interests are managed and used by the system to compute destinations associated with published events. Clients, once they specify their interests, can disconnect and the system guarantees the delivery of matched events during subsequent reconnects. Clients reconnecting after prolonged disconnects, connect to the local broker instead of the remote broker that it was last attached to. This eliminates bandwidth degradations caused by heavy concentration of clients from disparate geographic locations accessing a certain known remote broker over and over again. The delivery guarantees associated with individual events and clients are met even in the presence of failures. The approach adopted by the Object Management Group (OMG) is one of establishing event channels and registering suppliers and consumers to those channels. The channel approach in the CORBA Event Service [20] could however entail clients (consumers) to be aware of a large number of event channels.

22.2.1 Broker organization and small worlds behavior

Uncontrolled broker and connection additions result in a broker network that is susceptible to network partitions and that is devoid of any logical structure making the creation of

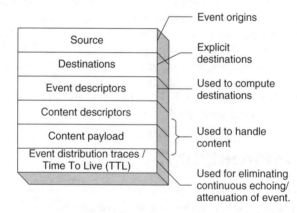

Figure 22.1   Event in NaradaBrokering.

efficient broker network maps (BNM) an arduous if not impossible task. The lack of this knowledge hampers development of efficient routing strategies, which exploits the broker topology. Such systems then resort to 'flooding' the entire broker network, forcing clients to discard events they are not interested in. To circumvent this, NaradaBrokering incorporates a broker organization protocol, which manages the addition of new brokers and also oversees the initiation of connections between these brokers. The node organization protocol incorporates Internet protocol (IP) discriminators, geographical location, cluster size and concurrent connection thresholds at individual brokers in its decision-making process.

In NaradaBrokering, we impose a hierarchical structure on the broker network, in which a broker is part of a cluster that is part of a super-cluster, which in turn is part of a super-super-cluster and so on. Clusters comprise strongly connected brokers with multiple links to brokers in other clusters, ensuring alternate communication routes during failures. This organization scheme results in 'small world networks' [21, 22] in which the average communication 'pathlengths' between brokers increase logarithmically with geometric increases in network size, as opposed to exponential increases in uncontrolled settings. This distributed cluster architecture allows NaradaBrokering to support large heterogeneous client configurations that scale to arbitrary size. Creation of BNMs and the detection of network partitions are easily achieved in this topology. We augment the BNM hosted at individual brokers to reflect the cost associated with traversal over connections, for example, intracluster communications are faster than intercluster communications. The BNM can now not only be used to compute valid paths but also to compute shortest paths. Changes to the network fabric are propagated only to those brokers that have their broker network view altered. Not all changes alter the BNM at a broker and those that do result in updates to the routing caches, containing shortest paths, maintained at individual brokers.

22.2.2 Dissemination of events

Every event has an implicit or explicit destination list, comprising clients, associated with it. The brokering system as a whole is responsible for computing broker destinations (targets) and ensuring efficient delivery to these targeted brokers en route to the intended client(s). Events as they pass through the broker network are to be updated to snapshot its dissemination within the network. The event dissemination traces eliminate continuous echoing and in tandem with the BNM – used for computing shortest paths – at each broker, is used to deploy a near optimal routing solution. The routing is near optimal since for every event the associated targeted set of brokers are usually the only ones involved in disseminations. Furthermore, every broker, either targeted or en route to one, computes the shortest path to reach target destinations while employing only those links and brokers that have not failed or have not been failure-suspected. In the coming years, increases in communication bandwidths will not be matched by commensurately reduced communication latencies [23]. Topology-aware routing and communication algorithms are needed for efficient solutions. Furthermore, certain communication services [24] are feasible only when built on top of a topology-aware solution. NaradaBrokering's routing solution thus provides a good base for developing efficient solutions.

22.2.3 Failures and recovery

In NaradaBrokering, stable storages existing in parts of the system are responsible for introducing state into the events. The arrival of events at clients advances the state associated with the corresponding clients. Brokers do not keep track of this state and are responsible for ensuring the most efficient routing. Since the brokers are stateless, they can fail and remain failed forever. The guaranteed delivery scheme within NaradaBrokering does not require every broker to have access to a stable store or database management system (DBMS). The replication scheme is flexible and easily extensible. Stable storages can be added/removed and the replication scheme can be updated. Stable stores can fail but they do need to recover within a finite amount of time. During these failures, the clients that are affected are those that were being serviced by the failed storage.

22.2.4 Support for dynamic topologies

Support for local broker accesses, client roams and stateless brokers provide an environment extremely conducive to dynamic topologies. Brokers and connections could be instantiated dynamically to ensure efficient bandwidth utilizations. These brokers and connections are added to the network fabric in accordance with rules that are dictated by the agents responsible for broker organization. Brokers and connections between brokers can be dynamically instantiated on the basis of the concentration of clients at a geographic

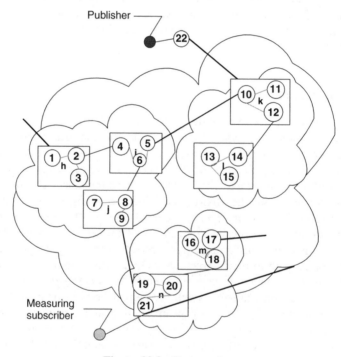

Figure 22.2 Test topology.

location and also on the basis of the content that these clients are interested in. Similarly, average pathlengths for communication could be reduced by instantiating connections to optimize clustering coefficients within the broker network. Brokers can be continuously added or can fail and the broker network can undulate with these additions and failures of brokers. Clients could then be induced to roam to such dynamically created brokers for optimizing bandwidth utilization. A strategy for incorporation of dynamic self-organizing overlays similar to MBONE [19] and X-Bone [25] is an area for future research.

22.2.5 Results from the prototype

Figure 22.3 illustrates some results [14, 17] from our initial research in which we studied the message delivery time as a function of load. The results are from a system comprising 22 broker processes and 102 clients in the topology outlined in Figure 22.2. Each broker node process is hosted on one physical Sun SPARC Ultra-5 machine (128 MB RAM, 333 MHz), with no SPARC Ultra-5 machine hosting two or more broker node processes. The publisher and the *measuring* subscriber reside on the same SPARC Ultra-5 machine. In addition, there are 100 subscribing client processes with 5 client processes attached to every other broker node (broker nodes 22 and 21 do not have any other clients besides the publisher and the measuring subscriber, respectively) within the system. The 100 client node processes all reside on a SPARC Ultra-60 (512 MB RAM, 360 MHz)

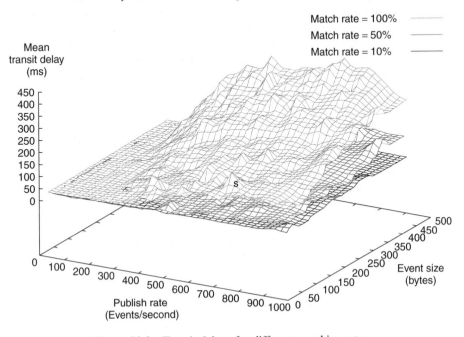

Figure 22.3   Transit delays for different matching rates.

machine. The run-time environment for all the broker node and client processes is Solaris JVM (JDK 1.2.1, native threads, JIT). The three matching values correspond to the percentages of messages that are delivered to any given subscriber. The 100% case corresponds to systems that would flood the broker network. The system performance improves significantly with increasing selectivity from subscribers. We found that the distributed network scaled well with adequate latency (2 ms per broker hop) unless the system became saturated at very high publish rates. We do understand how a production version of the NaradaBrokering system could give significantly higher performance – about a factor of 3 lower in latency than the prototype. By improving the thread scheduling algorithms and incorporating flow control (needed at high publish rates), significant gains in performance can be achieved. Currently, we do not intend to incorporate any non-Java modules.

22.3 JMS COMPLIANCE IN NARADABROKERING

Industrial strength solutions in the publish/subscribe domain include products like *TIB/Rendezvous* [26] from TIBCO and *SmartSockets* [27] from Talarian. Other related efforts in the research community include *Gryphon* [28], *Elvin* [29] and *Sienna* [30]. The push by Java to include publish–subscribe features into its messaging middleware include efforts such as Jini and JMS. One of the goals of JMS is to offer a unified Application Programming Interface (API) across publish–subscribe implementations. The JMS specification [31] results in JMS clients being vendor agnostic and interoperating with any service provider; a process that requires clients to incorporate a few vendor specific initialization sequences. JMS does not provide for interoperability between JMS providers, though interactions between clients of different providers can be achieved through a client that is connected to the different JMS providers. Various JMS implementations include solutions such as SonicMQ [32] from Progress, JMQ from iPlanet and FioranoMQ from Fiorano. Clients need to be able to invoke operations as specified in the specification; expect and partake from the logic and the guarantees that go along with these invocations. These guarantees range from receiving only those events that *match* the specified subscription to receiving events that were published to a given topic irrespective of the failures that took place or the duration of client disconnect. Clients are built around these calls and the guarantees (implicit and explicit) that are associated with them. Failure to conform to the specification would result in clients expecting certain sequences/types of events and not receiving those sequences, which in turn lead to deviations that could result in run-time exceptions.

22.3.1 Rationale for JMS compliance in NaradaBrokering

There are two objectives that we meet while providing JMS compliance within NaradaBrokering:

Providing support for JMS clients within the system: This objective provides for JMS-based systems to be replaced transparently by NaradaBrokering and also for NaradaBrokering clients (including those from other frameworks supported by NaradaBrokering such as P2P via JXTA) to interact with JMS clients. This also provides NaradaBrokering access to a plethora of applications developed around JMS.

To bring NaradaBrokering functionality to JMS clients/systems developed around it: This approach (discussed in Section 22.3.3) will transparently replace single-server or limited-server JMS systems with a very large scale distributed solution, with failure resiliency, dynamic real-time load balancing and scaling benefits.

22.3.2 Supporting JMS interactions

NaradaBrokering provides clients with connections that are then used for communications, interactions and any associated guarantees that would be associated with these interactions. Clients specify their interest, accept events, retrieve lost events and publish events over this connection. JMS includes a similar notion of connections. To provide JMS compliance we write a bridge that performs all the operations that are required by NaradaBrokering connections in addition to supporting operations that would be performed by JMS clients. Some of the JMS interactions and invocations are either supported locally or are mapped to corresponding NaradaBrokering interactions initiated by the connections. Each connection leads to a separate instance of the bridge. In the distributed JMS strategy it is conceivable that a client, with multiple connections and associated sessions, would not have all of its connections initiated to the same broker. The bridge instance per connection helps every connection to be treated independent of the others.

In addition to connections, JMS also provides the notion of sessions that are registered to specific connections. There can be multiple sessions on a given connection, but any given session can be registered to only one connection. Publishers and subscribers are registered to individual sessions. Support for sessions is provided locally by the bridge instance associated with the connection. For each connection, the bridge maintains the list of registered sessions, and the sessions in turn maintain a list of subscribers. Upon the receipt of an event over the connection, the corresponding bridge instance is responsible for forwarding the event to the appropriate sessions, which then proceed to deliver the event to the listeners associated with subscribers having subscriptions matching the event. In NaradaBrokering, each connection has a unique ID and guarantees are associated with individual connections. This ID is contained within the bridge instance and is used to deal with recovery and retrieval of events after prolonged disconnects or after induced roam due to failures.

We also need to provide support for the creation of different message types and assorted operations on these messages as dictated by the JMS specification, along with serialization and deserialization routines to facilitate transmission and reconstruction. In NaradaBrokering, events are routed as streams of bytes, and as long as we provide mar-shalling–unmarshalling operations associated with these types there are no issues regarding support for these message types. We also make use of the JMS selector mechanism implemented in OpenJMS [33]. The JMS subscription request is mapped to the corresponding NaradaBrokering profile propagation request and propagated through the system. The bridge maps persistent/transient subscriptions to the corresponding NaradaBrokering subscription types. JMS messages that are published are routed through the NaradaBrokering broker as a NaradaBrokering event. The anatomy of a Narada/JMS event, encapsulating the JMS messages, is shown in Figure 22.4. Events are routed on the basis of

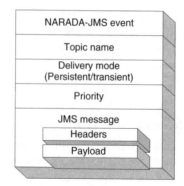

Figure 22.4   Narada-JMS event.

the mapped JMS. The topic name is contained in the event. Storage to databases is done on the basis of the delivery mode indicator in the event. Existing JMS applications in which we successfully replaced the JMS provider with NaradaBrokering include the multimedia-intensive distance education audio/video/text/application conferencing system [34] by Anabas Inc. and the Online Knowledge Center (OKC) [35] developed at IU Grid Labs. Both these applications were based on SonicMQ.

22.3.3 The distributed JMS solution

By having individual brokers interact with JMS clients, we have made it possible to replace the JMS provider's broker instance with a NaradaBrokering broker instance. The features in NaradaBrokering are best exploited in distributed settings. However, the distributed network should be transparent to the JMS clients and these clients should not be expected to keep track of broker states, failures and associated broker network partitions and so on. Existing systems built around JMS should be easily replaced with the distributed model with minimal changes to the client. In general, setups on the client side are to be performed in a transparent manner. The solution to the transparent distributed JMS solution would allow for any JMS-based system to benefit from the distributed solution. Applications would be based on source codes conforming to the JMS specification, while the scaling benefits, routing efficiencies and failure resiliency accompanying the distributed solution are all automatically inherited by the integrated solution.

To circumvent the problem of discovering valid brokers, we introduce the notion of *broker locators*. The broker locators' primary function is the discovery of brokers that a client can connect to. Clients thus do not need to keep track of the brokers and their states within the broker network. The broker locator has certain properties and constraints based on which it arrives at the decision regarding the broker that a client would connect to as follows:

Load balancing: Connection requests are always forked off to the best available broker based on broker metrics (Section 22.3.3.1). This enables us to achieve distributed dynamic real-time load balancing.

Incorporation of new brokers: A newly added broker is among the best available brokers to handle new connection requests. Clients thus incorporate these brokers faster into the routing fabric.

Availability: The broker locator itself should not constitute a single point of failure nor should it be a bottleneck for clients trying to utilize network services. The NaradaBrokering topology allows brokers to be part of domains. There could be more than one broker locator for a given administrative domain.

Failures: The broker locator does not maintain active connections to any element within the NaradaBrokering system and its loss does not affect processing pertaining to any other node.

22.3.3.1 Metrics for decision making

To determine the best available broker to handle the connection request, the metrics that play a role in the broker locator's decision include the IP-address of the requesting client, the number of connections still available at the brokers that are best suited to handle the connection, the number of connections that currently exist, the computing capabilities and finally, the availability of the broker (a simple ping test). Once a valid broker has been identified, the broker locator also verifies if the broker process is currently up and running. If the broker process is not active, the computed broker is removed from the list of available brokers and the broker locator computes the best broker from the current list of available brokers. If the computed broker is active, the broker locator proceeds to route broker information to the client. The broker information propagated to the client includes the hostname/IP-address of the machine hosting the broker, the port number on which it listens for connections/communications and the transport protocol that is used for communication. The client then uses this information to establish a communication channel with the broker. Once it is connected to a NaradaBrokering broker, the JMS client can proceed with interactions identical to those in the single broker case.

22.3.4 JMS performance data

To gather performance data, we run an instance of the SonicMQ (Version 3.0) broker and NaradaBrokering broker on the same dual CPU (Pentium-3, 1 GHz, 256 MB) machine. We then set up 100 subscribers over 10 different JMS TopicConnections on another dual CPU (Pentium-3, 866 MHz, 256 MB) machine. A *measuring* subscriber and a publisher are then set up on a third dual CPU (Pentium-3, 866 MHz, 256 MB RAM) machine. Setting up the measuring subscriber and publisher on the same machine enables us to obviate the need for clock synchronizations and differing clock drifts while computing delays. The three machines involved in the benchmarking process have Linux (Version 2.2.16) as their operating system. The run-time environment for the broker, publisher and subscriber processes is Java 2 JRE (Java-1.3.1, Blackdown-FCS, mixed mode). Subscribers subscribe to a certain topic and the publisher publishes to the same topic. Once the publisher starts issuing messages, the factor that we are most interested in is the *transit delay* in the receipt of these messages at the subscribers. We measure this delay at the measuring subscriber while varying the publish rates and payload sizes of the messages being published. For a sample

Transit delays for message samples in Narada and SonicMQ

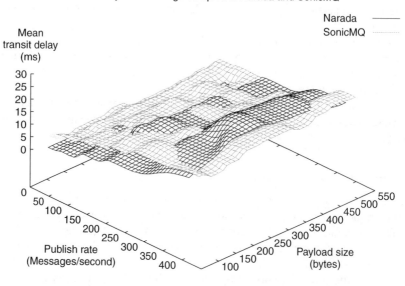

Figure 22.5 Transit delays – higher publish rates smaller payloads.

Standard deviation in the message samples–Narada and SonicMQ

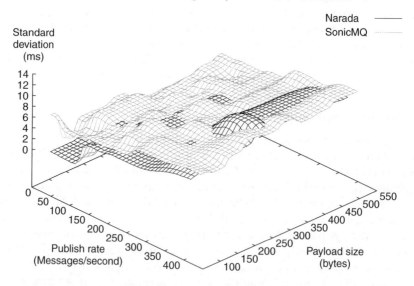

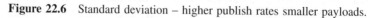

Figure 22.6 Standard deviation – higher publish rates smaller payloads.

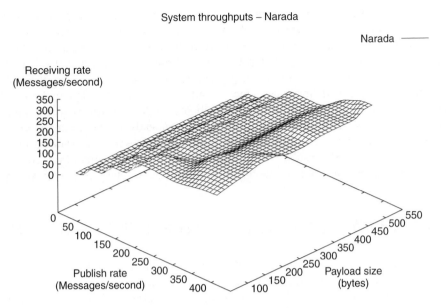

Figure 22.7   System throughputs (Narada) – higher publish rates smaller payloads.

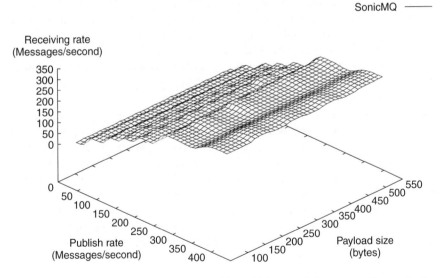

Figure 22.8   System throughputs (SonicMQ) – higher publish rates smaller payloads.

of messages received at the measuring subscriber, we calculate the *mean transit delay* and the *standard deviation* within this sample. We also calculate the *system throughput* measured in terms of the number of messages received per second at the measuring subscriber. Figure 22.5 depicts the mean transit delay for JMS clients under NaradaBrokering and SonicMQ for varying publish rates and payload sizes. Figure 22.6 depicts the standard deviation associated with message samples under conditions depicted in Figure 22.5. Figures 22.7 and 22.8 depict the system throughputs for NaradaBrokering and SonicMQ clients, respectively. As can be seen from the results, NaradaBrokering compares very well with SonicMQ. Comprehensive results comparing NaradaBrokering and SonicMQ can be found in Reference [18]. Results for NaradaBrokering's User Datagram Protocol (UDP)-based JMS solution's use in multimedia applications can be found in Reference [36].

22.4 NARADABROKERING AND P2P INTERACTIONS

Issues in P2P systems pertaining to the discovery of services and intelligent routing can be addressed very well in the NaradaBrokering brokering system. The broker network would be used primarily as a delivery engine, and a pretty efficient one at that, while locating peers and propagating interactions to relevant peers. The most important aspect in P2P systems is the satisfaction of peer requests and discovery of peers and associated resources that could handle these requests. The broker network forwards these requests only to those peers that it believes can handle the requests. Peer interactions in most P2P systems are achieved through XML-based data interchange. XML's data description and encapsulation properties allow for ease of accessing specific elements of data. Individual brokers routing interactions could access relevant elements, cache this information and use it subsequently to achieve the best possible routing characteristics. The brokering system, since it is aware of advertisements, can also act as a hub for search and discovery operations. These advertisements, when organized into 'queryspaces', allow the integrated system to respond to search operations more efficiently.

Resources in NaradaBrokering are generally within the purview of the broker network. P2P systems replicate resources in an *ad hoc* fashion, the availability of which is dependent on the peer's active digital presence. Some resources, however, are best managed by the brokering system rather than being left to the discretion of peers who may or may not be present at any given time. An understanding of the network topology and an ability to pinpoint the existence of peers interested in that resource are paramount for managing the efficient replications of a resource. The distributed broker network, possessing this knowledge, best handles this management of resources while ensuring that these replicated resources are 'closer' and 'available' at locations with a high interest in that resource. Furthermore, the broker network is also better suited, than a collection of peers, to eliminate race conditions and deadlocks that could exist because of a resource being accessed simultaneously by multiple peers. The broker network can also be responsive to changes in peer concentrations, volumes of peer requests and resource availability. Brokers and associated interconnections can be dynamically instantiated or purged to compensate for affected routing characteristics due to changes in peer interactions.

As mentioned earlier, P2P systems fragment into multiple disconnected subsystems. NaradaBrokering could also be used to connect islands of peers together. Peers that are not directly connected through the peer network could be indirectly connected through the broker network. Peer interactions and resources in the P2P model are traditionally unreliable, with interactions being lost or discarded because of peer failures or absences, overloading of peers and queuing thresholds being reached. Guaranteed delivery properties existing in NaradaBrokering can augment peer behavior to provide a notion of reliable peers, interactions and resources. Such an integrated brokering solution would also allow for hybrid interaction schemes to exist alongside each other. Applications could be built around hybrid-clients that would exhibit part peer behavior and part traditional client behavior (e.g. JMS). P2P communications could be then used for traffic where loss of information can be sustained. Similarly, hybrid-clients needing to communicate with each other in a 'reliable' fashion could utilize the brokering system's capabilities to achieve that. Hybrid-clients satisfy each other's requests whenever they can, obviating the need for funneling interactions through the broker network. The broker merely serves as an efficient conduit for supporting interaction between different applications (clients, peers or hybrid).

22.4.1 JXTA

JXTA is a set of open, generalized protocols to support peer-to-peer interactions and core P2P capabilities such as indexing, file sharing, searching, peer grouping and security. The JXTA peers and rendezvous peers (specialized routers) rely on a simple forwarding of interactions for disseminations and on time-to-live (TTL) indicators and peer traces to attenuate interaction propagations. However, JXTA interactions are unreliable and tend to be very localized. Figure 22.9 depicts the protocols that comprise the XML-encoded JXTA protocol suite [37]. JXTA is independent of transport protocols and can be implemented on top of TCP/IP, hypertext transfer protocol (HTTP), TLS, Bluetooth, HomePNA and many other protocols. JXTA provides features such as dynamic discovery and a rich search mechanism while allowing peers to communicate across NAT, DHCP and firewall

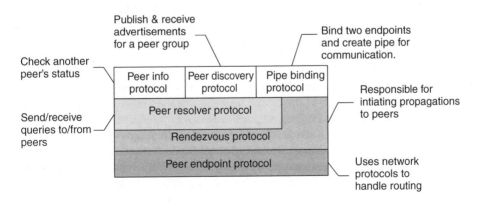

Figure 22.9   The JXTA protocol suite.

boundaries. In JXTA, a peer is any node that supports JXTA protocols and could be any digital device. Peers that seek to collaborate could come together to form a peer group. Peers within a peer group can identify each other, agree on group memberships and exchange information with each other. Peers publish the existence of a resource through an advertisement, which is simply an XML document describing the resource. Peers locate other peers, peer groups and properties pertaining to them. Once a peer joins a JXTA group, JXTA's discovery capabilities support queries for services, resources and other peers. The queries could be centralized or decentralized involving the entire peer group.

JXTA is also programming language–independent; the C language binding for JXTA core was released earlier this year. Implementation of the core JXTA protocols in Perl 5, Object C, Ruby, Smalltalk and Python are currently under way. It is expected that existing P2P systems would either support JXTA or have bridges initiated to it from JXTA thus enabling us to leverage other existing P2P systems along with applications built around these systems. To extend JXTA support to lightweight devices, projects such as *pocketjxta* [38] have been started to implement the JXTA platform and build applications for PDAs. NaradaBrokering's support for JXTA, in addition to the support for JMS, would result in interactions that are robust and dynamic while being supported by a scalable and highly available system. One good example of a dynamic 'peer' group is a set of Grid/Web services [38–41] generated dynamically when a complex task runs – here existing registration/discovery mechanisms are unsuitable. A JXTA-like discovery strategy within such a dynamic group combined with NaradaBrokering's JMS mode between groups seems attractive. These 'peers' can of course be in 'middle tier' – so such a model can be applied in the Internet universe where we have 'clusters' (In our analogy, JXTA runs galaxies while JMS runs the universe). We intend to investigate this model of dynamic resource management in later chapters.

22.5 JXTA INTEGRATION IN NARADABROKERING

In our strategy for providing support for P2P interactions within NaradaBrokering, we need to ensure minimal or zero changes to the NaradaBrokering system core and the associated protocol suites. We also make no changes to the JXTA core and the associated protocols. We do make additions to the rendezvous layer for integration purposes. Peers do not communicate directly with the NaradaBrokering system. Furthermore, this integration should entail neither any changes to the peers nor a straitjacketing of the interactions that these peers could have had prior to the integration. The integration is based on the proxy model, which essentially acts as the bridge between the NaradaBrokering system and the JXTA. The Narada-JXTA proxy, operating inside the JXTA rendezvous layer, serves a dual role as both a rendezvous peer and as a NaradaBrokering client providing a bridge between NaradaBrokering and JXTA. NaradaBrokering could be viewed as a service by JXTA. The discovery of this service is automatic and instantaneous as a result of the Narada-JXTA proxy's integration inside the rendezvous layer. Any peer can utilize NaradaBrokering as a service so long as it is connected to a Narada-JXTA proxy. Nevertheless, peers do not know that the NaradaBrokering broker network is routing some of their interactions.

Furthermore, these Narada-JXTA proxies, since they are configured as clients within the NaradaBrokering system, inherit all the guarantees that are provided to clients within NaradaBrokering.

22.5.1 The interaction model

Different JXTA interactions are queued at the queues associated with the relevant layers comprising the JXTA protocol suite [37]. Each layer performs some operations including the addition of additional information. The rendezvous layer processes information arriving at its input queues from the peer-resolving layer and the pipe-binding layer. Since the payload structure associated with different interactions is different, we can easily identify the interaction types associated with the payloads. Interactions pertaining to discovery/search or communications within a peer group would be serviced both by JXTA rendezvous peers and also by Narada-JXTA proxies. The interactions that peers have with the Narada-JXTA proxies are what are routed through the NaradaBrokering system. JXTA peers can continue to interact with each other and of course, some of these peers can be connected to pure JXTA rendezvous peers. Peers have multiple routes to reach each other and some of these could include the NaradaBrokering system, while others need not. Such peers can interact directly with each other during the request/response interactions. Figure 22.10 outlines the NaradaBrokering JXTA interaction model.

22.5.2 Interaction disseminations

Peers can create a peer group, request to be part of a peer group and perform search/request/discovery all with respect to a specific targeted peer group. Peers always issue requests/responses to a specific peer group and sometimes to a specific peer. Peers and peer

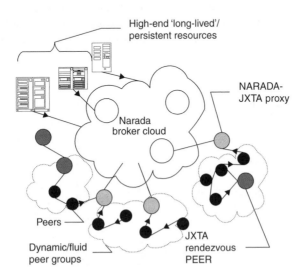

Figure 22.10   The Narada-JXTA interaction model.

groups are identified by Universal Unique IDentifier (UUID) [43] [Internet Engineering Task Force (IETF) specification guarantees uniqueness until 3040 A.D.] based identifiers. Every peer generates its own peer id while the peer that created the peer group generates the associated peer group id. Each rendezvous peer keeps track of multiple peer groups through the peer group advertisements that it receives. Any given peer group advertisement could of course be received at multiple rendezvous peers. These rendezvous peers are then responsible for forwarding interactions, if it had received an advertisement for the peer group contained in these interactions.

Narada-JXTA proxies are initialized both as rendezvous peers and also as NaradaBrokering clients. During its initialization as a NaradaBrokering client, every proxy is assigned a unique connection ID by the NaradaBrokering system, after which the proxy subscribes to a topic identifying itself as a Narada-JXTA proxy. This enables NaradaBrokering to be aware of all the Narada-JXTA proxies that are present in the system. The Narada-JXTA proxy in its role as a rendezvous peer to peers receives

- peer group advertisements,
- requests from peers to be part of a certain peer group and responses to these requests and
- messages sent to a certain peer group or a targeted peer.

To ensure the efficient dissemination of interactions, it is important to ensure that JXTA interactions that are routed by NaradaBrokering are delivered only to those Narada-JXTA proxies that should receive them. This entails that the Narada-JXTA proxy performs a sequence of operations, based on the interactions that it receives, to ensure selective delivery. The set of operations that the Narada-JXTA proxy performs comprise gleaning relevant information from JXTA's XML encapsulated interactions, constructing an event (depicted in Figure 22.11) based on the information gleaned and finally, in its role as a NaradaBrokering client subscribing (if it chooses to do so) to a topic to facilitate selective delivery. By subscribing to relevant topics and creating events targeted to specific topics, each proxy ensures that the broker network is not flooded with interactions routed by them. The events constructed by the Narada-JXTA proxies include the entire interaction as the

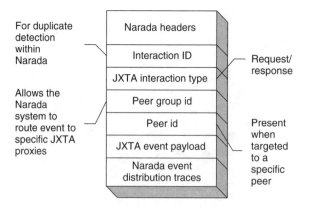

Figure 22.11   The Narada-JXTA event.

event's payload. Upon receipt at a proxy, this payload is deserialized and the interaction is propagated as outlined in the proxy's dual role as a rendezvous peer. Events constructed from interactions need to have a unique identifier associated with them. Advertisements, since, they encapsulate information pertaining to a uniquely identifiable resource can use the UUID associated with the advertised resource as the interaction identifier of the constructed event. The interaction type along with the interaction identifier allows us to uniquely identify each event. In the case of JXTA messages, the unique interaction identifier is constructed on the basis of the peer id of the peer issuing the message and the timestamp in milliseconds (based on the system clock at the peer node) associated with the message. We now proceed to outline the sequence of operations associated with different JXTA interactions.

As opposed to the simple forwarding of interactions, the intelligent routing in NaradaBrokering in tandem with the duplicate detection scheme [44] in our solution ensures faster disseminations and improved communication latencies for peers. Furthermore, targeted peer interactions traversing along shortest paths within the broker network obviate the need for a peer to maintain dedicated connections to a lot of peers. Dynamic discovery of peers is significantly faster using NaradaBrokering's JXTA proxy strategy. The scheme also allows us to connect islands of peers and rendezvous peers together, allowing for a greater and richer set of interactions for these clients. NaradaBrokering, since it is JMS-compliant, also opens up possibilities for JMS clients interacting with JXTA peers. Details pertaining to the JXTA integration can be found in Reference [44].

22.5.3 JXTA applications and NaradaBrokering

JXTA applications that were tested within the integrated Narada-JXTA environment included the *JXTA shell*, which provides a command line interface to JXTA functionality and *myjxta* (also known as InstantP2P), which also includes an instant messaging environment. Work is currently under way to integrate *myjxta* into the Anabas Conferencing system in which NaradaBrokering would provide backend support for both JMS and JXTA interactions. In Reference [45], we described the Garnet Message Service (based on the Anabas-shared display infrastructure) as a demonstration that JMS and millisecond latency for small messages can support major collaborative functions – Shared Display, Whiteboard, Polls and Text chat among others. The goal of our effort is to explore the use of Garnet with hybrid messaging and bring Garnet/NaradaBrokering functionality to JXTA.

22.5.4 NaradaBrokering-JXTA systems

The NaradaBrokering JXTA integration service scales naturally since NaradaBrokering provides dynamic 'long-distance' support while JXTA provides for dynamic localized supports. In the combined global scale infrastructure, NaradaBrokering works best for 'long-lived' and persistent resources that would be efficiently replicated within NaradaBrokering. This integrated model provides efficient search/discovery not only for static activity but also for dynamic activity while allowing JXTA interactions at the edges. The resultant system also scales with multiple JXTA Peer Groups linked by NaradaBrokering, which can dynamically instantiate new brokers to balance the load. As opposed to the simple forwarding of interactions, the intelligent routing in NaradaBrokering in tandem with the duplicate detection scheme in our solution ensures faster disseminations and improved communication latencies for peers. Furthermore, targeted peer interactions traversing along shortest paths within the broker network obviates the need for a peer to maintain dedicated connections to a lot of peers. Proxies, owing to their initialization as clients within NaradaBrokering, inherit all the guarantees accorded to clients within the NaradaBrokering such as guaranteed delivery in the presence of failures and fast disseminations. Discovery of rendezvous peer in JXTA is a slow process. A rendezvous peer generally downloads a list of other rendezvous peers from a server, not all of which would be up and running at that time. We allow for dynamic discovery of Narada-JXTA proxies, which need not be directly aware of each other, but do end up receiving interactions sent to a peer group if they had both received peer group advertisements earlier. The scheme also allows us to connect islands of peers and rendezvous peers together, allowing for a greater and richer set of interactions between these peers.

A typical application of the hybrid NaradaBrokering/JXTA/JMS technology is distance education. This often consists of multiple linked classrooms in which the participants in each classroom are individually linked to the collaborative environment. Here a peer-to-peer model (such as JXTA) can be used in a classroom to give fast dynamic response to shared input control, while the JMS style global NaradaBrokering capability is used to multicast between classrooms. More generally, this combination of globally structured and locally dynamic messaging scales to support general applications. We can package the JXTA/JMS/NaradaBrokering hybrid as an event or messaging Web service whose different modes could correspond to different ports in the Web Services Description Language (WSDL) [39]. These ports trade off scaling, performance and reliability in a fashion that can be chosen by the user. Alternatively Web services communicate via channels and we could use the technology of this chapter to flexibly link different services together with a dynamic choice of the service provider. Other applications we are pursuing include workflows in which administrative agents control the traffic between different Web services.

22.6 CONCLUSION

In this chapter, we presented our strategy for a scalable, durable P2P Grid. In NaradaBrokering, we have based support for P2P interactions through JXTA. We also enumerated the benefits that can be accrued, by both NaradaBrokering and P2P systems such as JXTA, through such integrations. The successful integration of NaradaBrokering with JXTA can be combined with our JMS integration demonstrate such that one can both support very different brokering models within the same system and deploy hybrid systems with NaradaBrokering linking different environments. We believe that such an environment is appropriate for building scalable, durable P2P Grids supporting both dynamic local and long-range static resources. We believe that these integrations make us well positioned to Web service 'enable' NaradaBrokering.

REFERENCES

1. openp2p P2P Web Site from O'Reilly, http://www.openp2p.com.
2. Oram, A. (eds) (2001) *Peer-to-Peer: Harnessing the Power of Disruptive Technologies.* Sebastapol, California: O'Reilly.
3. SETI@home Project, http://setiathome.ssl.berkeley.edu.
4. Folding@home Project, http://www.stanford.edu/group/pandegroup/Cosm.
5. Gnutella, http://gnutella.wego.com.
6. Groove Network, http://www.groove.net.
7. Jabber, http://www.jabber.org.
8. Fox, G. (2001) Peer-to-Peer Networks. *Computing in Science & Engineering*, **3**(3), pp. 75–77.
9. Antony, R. and Druschel, P. (2001) Pastry: scalable, decentralized object location and routing for large-scale peer-to-peer systems. *Proceedings of Middleware 2001*, 2001.
10. Michel, S. and Reiher, P. (2001) Peer-to-Peer Internetworking. *OPENSIG 2001 Workshop*, Imperial College, London, September, 2001.
11. Sun Microsystems, *The JXTA Project and Peer-to-Peer Technology*, http://www.jxta.org.
12. Fox, G., Balsoy, O., Pallickara, S., Uyar, A., Gannon, S. and Slominski, A. (2002) Community grids. *Proceedings of the International Conference on Computational Science (ICCS 2002)*, Netherlands, April, 2002.
13. The NaradaBrokering System, http://www.naradabrokering.org.
14. Fox, G. and Pallickara, S. An event service to support grid computational environments. *Concurrency and Computation: Practice and Experience*, Special Issue on Grid Computing Environments; to be published.
15. Fox, G. and Pallickara, S. (2001) An approach to high performance distributed web brokering. *ACM Ubiquity*, **2**(38).
16. Pallickara, S. (2001) *A Grid Event Service*, Ph.D. thesis, Syracuse University, Syracuse, New York.
17. Fox, G. C. and Pallickara, S. (2002) The Narada event brokering system: overview and extensions. *Proceedings of the 2002 International Conference on Parallel and Distributed Processing Techniques and Applications*, 2002.
18. Fox, G. C. and Pallickara, S. (2002) JMS compliance in the Narada event brokering system. Proceedings of the 2002 International Conference on Internet Computing (IC-02), 2002.
19. Eriksson, H. (1994) MBone: the multicast backbone. *Communications of the ACM*, **37**, 54–60.
20. The Object Management Group, (2002) The CORBA Event Service, Version 1.1., http://www.omg.org/technology/ documents/formal/event_service.htm, 2002.
21. Watts, D. J. and Strogatz, S. H. (1998) Collective dynamics of small-world networks. *Nature*, **393**, 440.
22. Albert, R., Jeong, H. and Barabasi, A. (1999) Diameter of the world wide web. *Nature*, **401**, 130.
23. Lee, C. and Stepanek, J. (2001) On future global grid communication performance. *10th IEEE Heterogeneous Computing Workshop*, May, 2001.
24. Lee, C. (2000) On active grid middleware. *Second Workshop on Active Middleware Services*, August 1, 2000.
25. Touch, J. (2001) Dynamic Internet overlay deployment and management using the X-bone. *Computer Networks*, **36**(2–3), 117–135.
26. TIBCO Corporation, (1999) TIB/Rendezvous White Paper, http://www.rv.tibco.com/whitepaper.html. June.
27. Talarian Corporation, *SmartSockets: Everything You Need to Know About Middleware: Mission Critical Interprocess Communication*, Technical Report, http://www.talarian.com/products/smartsockets.
28. Aguilera, M., Strom, R., Sturman, D., Astley, M. and Chandra, T. (1999) Matching events in a content-based subscription system. *Proceedings of the 18th ACM Symposium on Principles of Distributed Computing*, May, 1999.

29. Segall, B. and Arnold, D. (1997) Elvin has left the building: a publish/subscribe notification service with quenching. *Proceedings AUUG '97*, September, 1997.
30. Carzaniga, A., Rosenblum, D. and Wolf, A. (2000) Achieving scalability and expressiveness in an internet-scale event notification service. *Proceedings of 19th ACM Symposium on Principles of Distributed Computing*, July, 2000.
31. Happner, M., Burridge, R. and Sharma, R. Sun Microsystems. (2000) Java Message Service Specification, http://java.sun.com/products/jms.
32. SonicMQ JMS Server, http://www.sonicsoftware.com/.
33. The OpenJMS Project, http://openjms.exolab.org/.
34. The Anabas Conferencing System, http://www.anabas.com.
35. The Online Knowledge Center (OKC) Web Portal, http://judi.ucs.indiana.edu/okcportal/index.jsp.
36. Bulut, H., Fox, G., Pallickara, S., Uyar, A. and Wu, W. (2002) Integration of NaradaBrokering and Audio/Video Conferencing as a Web Service. *Proceedings of the IASTED International Conference on Communications, Internet, and Information Technology*, St. Thomas, US Virgin Islands, November 18–20, 2002.
37. The JXTA Protocol Specifications, http://spec.jxta.org/v1.0/docbook/JXTAProtocols.html.
38. PocketJXTA Project: Porting the JXTA-c Platform to the PocketPC and Similar Devices, http://pocketjxta.jxta.org/.
39. Web Services Description Language (WSDL) 1.1, http://www.w3.org/TR/wsdl.
40. Kuno, H., Lemon, M., Karp, A. and Beringer, D. WSCL – Web Services Conversational Language, Conversations + Interfaces = Business Logic, http://www.hpl.hp.com/techreports/2001/HPL-20 01127.html.
41. Semantic Web from W3C: Self organizing Intelligence from enhanced web resources, http://www.w3c.org/ 2001/sw/.
42. Berners-Lee, T., Hendler, J. and Lassila, O. (2001) The semantic web. *Scientific American*: **284**(5), 34–43.
43. Leach, P. J. and Salz, R. Network Working Group. (1998) *UUIDs and GUIDs*, February, 1998.
44. Fox, G., Pallickara, S. and Rao, Xi. A scaleable event infrastructure for peer to peer Grids. *Proceedings of ACM Java Grande ISCOPE Conference 2002*, Seattle, Washington, November, 2002.
45. Fox, G. *et al*. Grid services for earthquake science. *Concurrency & Computation: Practice and Experience*, ACES Computational Environments for Earthquake Science Special Issue. **14**(6–7, 371–393.

Classifying and enabling Grid applications

Gabrielle Allen,[1] Tom Goodale,[1] Michael Russell,[1] Edward Seidel,[1] and John Shalf[2]

[1]*Max-Planck-Institut für Gravitationsphysik, Golm, Germany,* [2]*Lawrence Berkeley National Laboratory, Berkeley, California, United States*

23.1 A NEW CHALLENGE FOR APPLICATION DEVELOPERS

Scientific and engineering applications have driven the development of high-performance computing (HPC) for several decades. Many new techniques have been developed over the years to study increasingly complex phenomena using larger and more demanding jobs with greater throughput, fidelity, and sophistication than ever before. Such techniques are implemented as hardware, as software, and through algorithms, including now familiar concepts such as vectorization, pipelining, parallel processing, locality exploitation with memory hierarchies, cache use, and coherence.

As each innovation was introduced, at either the hardware, operating system or algorithm level, new capabilities became available – but often at the price of rewriting applications. This often slowed the acceptance or widespread use of such techniques. Further,

Grid Computing – Making the Global Infrastructure a Reality. Edited by F. Berman, A. Hey and G. Fox
© 2003 John Wiley & Sons, Ltd ISBN: 0-470-85319-0

when some novel or especially disruptive technology was introduced (e.g. MPPs programmed using message passing) or when an important vendor disappeared (e.g. Thinking Machines), entire codes had to be rewritten, often inducing huge overheads and painful disruptions to users.

As application developers and users who have witnessed and experienced both the promise and the pain of so many innovations in computer architecture, we now face another revolution, *the Grid*, offering the possibility of aggregating the capabilities of the multitude of computing resources available to us around the world. However, like all revolutions that have preceded it, along with the fantastic promise of this new technology, we are also seeing our troubles multiply. While the Grid provides platform-neutral protocols for fundamental services such as job launching and security, it lacks sufficient abstraction at the application level to accommodate the continuing evolution of individual machines. The application developer, already burdened with keeping abreast of evolution in computer architectures, operating systems, parallel paradigms, and compilers, must simultaneously consider how to assemble these rapidly evolving, heterogeneous pieces, into a useful collective computing resource atop a dynamic and rapidly evolving Grid infrastructure.

However, despite such warnings of the challenges involved in migrating to this potentially hostile new frontier, we are very optimistic. We strongly believe that *the Grid can be tamed and will enable new avenues of exploration for science and engineering, which would remain out of reach without this new technology.* With the ability to build and deploy applications that can take advantage of the distributed resources of Grids, we will see truly novel and very dynamic applications. Applications will use these new abilities to acquire and release resources on demand and according to need, notify and interact with users, acquire and interact with data, or find and interact with other Grid applications. Such a world has the potential to fundamentally change the way scientists and engineers think about their work. While the Grid offers the ability to attack much larger scale problems with phenomenal throughput, new algorithms will need to be developed to handle the kinds of parallelism, memory hierarchies, processor and data distributions found on the Grid. Although there are many new challenges in such an environment, many familiar concepts in parallel and vector processing remain present in a Grid environment, albeit under a new guise. Many decades-old strategies that played a role in the advancement of HPC, will find a new life and importance when applied to the Grid.

23.2 APPLICATIONS MUST BE THE LIFEBLOOD!

Grids are being engineered and developed to be *used*; thus attention to application needs is crucial if Grids are to evolve and be widely embraced by users and developers. What must happen before this new Grid world is used effectively by the application community? First, the underlying Grid infrastructure must mature and must be widely and *stably* deployed and supported. Second, different virtual organizations must possess the appropriate mechanisms for both co-operating and interoperating with one another. We are still, however, missing a crucial link: *applications need to be able to take advantage of this infrastructure*. Such applications will not appear out of thin air; they must be developed, and developed on top of an increasingly complex fabric of heterogeneous resources, which in the Grid world may take on different incarnations day-to-day and hour-to-hour. Programming applications to exploit such an environment without burdening users with the true Grid complexity is a challenge indeed!

Of many problems, three major challenges emerge: (1) Enabling application developers to incorporate the abilities to harness the Grid, so that new application classes, like those described in this chapter, can be realized; (2) Abstracting the various Grid capabilities sufficiently so that they may be accessed easily from within an application, without requiring detailed knowledge about the underlying fabric that will be found at run time; and (3) Posing these abstractions to match application-level needs and expectations. While current Grid abstractions cover extremely low-level capabilities such as job launching, information services, security, and file transfer (the *'Grid assembly language'*), applications require higher-level abstractions such as checkpointing, job migration, distributed data indices, distributed event models for interactive applications, and collaborative interfaces (both on-line and off-line).

In the experience of communities developing applications to harness the power of computing, *frameworks* are an effective tool to deal with the complexity and heterogeneity of today's computing environment, and an important insurance policy against disruptive changes in future technologies. A properly designed framework allows the application developer to make use of APIs that encode simplified abstractions for commonly-used operations such as creation of data-parallel arrays and operations, ghostzone synchronization, I/O, reductions, and interpolations. The framework communicates directly with the appropriate machine-specific libraries underneath and this abstraction allows the developer to have easy access to complex libraries that can differ dramatically from machine to machine, and also provides for the relatively seamless introduction of new technologies. Although the framework itself will need to be extended to exploit the new technology, a well-designed framework will maintain a constant, unchanged interface to the application developer. If this is done, the application will still be able to run, and even to take advantage of new capabilities with little, if any, change. As we describe below, one such framework, called *Cactus* [1], has been particularly successful in providing such capabilities to an active astrophysics and relativity community – enabling very sophisticated calculations to be performed on a variety of changing computer architectures over the last few years.

The same concepts that make Cactus and other frameworks so powerful on a great variety of machines and software infrastructures will also make them an important and powerful methodology for harnessing the capabilities of the Grid. A Grid application framework can enable scientists and engineers to write their applications in a way that frees them from many details of the underlying infrastructure, while still allowing them the power to write fundamentally new types of applications, and to exploit still newer technologies developed in the future without disruptive application rewrites. In particular, we discuss later an important example of an abstracted Grid development toolkit with precisely these goals. The *Grid Application Toolkit*, or GAT, is being developed to enable generic applications to run in any environment, without change to the application code itself, to discover Grid and other services at runtier, and to enable scientists and engineers themselves to develop their applications to fulfil this vision of the Grid of the future.

23.3 CASE STUDY: REAL-WORLD EXAMPLES WITH THE CACTUS COMPUTATIONAL TOOLKIT

Several application domains are now exploring Grid possibilities (see, e.g. the GriPhyN, DataGrid, and the National Virtual Observatory projects). Computational framework and infrastructure projects such as Cactus, Triana, GrADs, NetSolve, Ninf, MetaChaos, and others are developing the tools and programming environments to entice a wide range of applications onto the Grid. It is crucial to learn from these early Grid experiences with real applications. Here we discuss some concrete examples provided by one specific programming framework, Cactus, which are later generalized to more generic Grid operations.

Cactus is a generic programming framework, particularly suited (by design) for developing and deploying large scale applications in diverse, dispersed collaborative environments. From the outset, Cactus has been developed with Grid computing very much in mind; both the framework and the applications that run in it have been used and extended by a number of Grid projects. Several basic tools for remote monitoring, visualization, and interaction with simulations are commonly used in production simulations [2]. Successful prototype implementations of Grid scenarios, including job migration from one Grid site to another (perhaps triggered by 'contract violation', meaning a process run more slowly than contracted at one site, so another more suitable site was discovered and used); task spawning, where parts of a simulation are 'outsourced' to a remote resource; distributed computing with dynamic load balancing, in which multiple machines are used for a large distributed simulation, while various parameters are adjusted during execution to improve efficiency, depending on intrinsic and measured network and machine characteristics [3, 4, 5], have all shown the potential benefits and use of these new technologies. These specific examples are developed later into more general concepts.

These experiments with Cactus and Grid computing are not being investigated out of purely academic interest. Cactus users, in particular, those from one of its primary user domains in the field of numerical relativity and astrophysics, urgently require for their science more and larger computing resources, as well as easier and more efficient use of these resources. To provide a concrete example of this need, numerical relativists currently want to perform large-scale simulations of the spiraling coalescence of two black holes, a problem of particular importance for interpreting the gravitational wave signatures that will soon be seen by new laser interferometric detectors around the world. Although they have access to the largest computing resources in the academic community, no single machine can supply the resolution needed for the sort of high-accuracy simulations necessary to gain insight into the physical systems. Further, with limited computing cycles from several different sites, the physicists have to work daily in totally different environments, working around the different queue limitations, and juggling their joint resources for best effect.

Just considering the execution of a single one of their large-scale simulations shows that a functioning Grid environment implementing robust versions of our prototypes would provide large benefits: appropriate initial parameters for the black hole simulations are usually determined from a large number of smaller scale test runs, which could be automatically staged to appropriate resources (*task farming for parameter surveys*). An intelligent module could then interpret the collected results to determine the optimal parameters for the real simulation. This high-resolution simulation could be automatically staged across suitable multiple machines (*resource brokering* and *distributed computing*). As it runs, independent, yet computationally expensive, tasks could be separated and moved to cheaper machines (*task spawning*). During the big simulation, additional lower resolution parameter surveys could be farmed to determine the necessary changes to parameters governing the simulation, *parameter steering*, providing the mechanism for communicating these changes back to the main simulation. Since these long-running simulations usually require run times longer than queue times, the entire simulation could be automatically moved to new resources when needed, or when more appropriate machines are located (*job migration*). Throughout, the physicists would monitor, interact with, and visualize the simulation.

Implementing such composite scenarios involves many different underlying Grid operations, each of which must function robustly, interoperating to good effect with many other components. The potential complexity of such systems motivates us to step back and consider more general ways to describe and deliver these requirements and capabilities.

23.4 STEPPING BACK: A SIMPLE MOTIVATION-BASED TAXONOMY FOR GRID APPLICATIONS

The Grid is becoming progressively better defined [6]. Bodies such as the Global Grid Forum are working to refine the terminology and standards required to understand and communicate the infrastructure and services being developed. For Grid developers to be able to ensure that their new technologies satisfy general application needs, we need to apply the same diligence in classifying applications; what type of applications will be using the Grid and how will they be implemented; what kinds of Grid operations will they require, and how will they be accessed; what limitations will be placed by security and privacy concerns or by today's working environments; how do application developers and users want to use the Grid, and what new possibilities do they see.

In the following sections we make a first pass at categorizing the kinds of applications we envisage wanting to run on Grids and the kinds of operations we will need to be available to enable our scenarios.

23.4.1 Generic types of Grid applications

There are many ways to classify Grid applications. Our taxonomy divides them here into categories based on their primary driving reasons for using the Grid. Current Grid applications can be quite easily placed in one or another of these categories, but in the future it is clear that they will become intricately intertwined, and such a classification scheme will need to be extended and refined.

1. *Community-centric*: These are applications that attempt to bring people or communities together for collaborations of various types. Examples range from the Access Grid, allowing interactive video presentation and conferencing from many sites simultaneously, to distributed musical concerts, and to supporting collaborations of dozens of scientists, engineers, and mathematicians around the world, needed to perform complex simulations of cosmic events such as supernova explosions, on demand, as data from the events are pouring into detectors of various kinds. Workflow management is also a strong component of this paradigm in which the flow straddles many fields of expertise, organizational boundaries, and widely separated resources.

2. *Data-centric*: Data is the primary driving force behind the Grid at present, and will become even more so in the future. Not limited to particle and astrophysics experiments, which themselves will be generating multiterabyte data sets each day, sensors for everything from precise maps of the earth's crust to highly localized weather data will be feeding the Grid with large quantities of data from sources around the world. Storing, transferring, managing, and mining these data for content quickly becomes impossible without rapidly improving Grid technology.

3. *Computation-centric*: These are the traditional HPC applications, common in astrophysics (e.g. simulations of a supernova explosion or black-hole collision), automotive/aerospace industry (e.g. simulations of a car crash or a rocket engine), climate modeling (e.g. simulations of a tornado or prediction of the earth's climate for the next century), economics (e.g. modeling the world economy), and so on. Typically, the models are simplified to the extent that they *can be* computed on presently available machines; usually many important effects are left out because the computational power is not adequate to include them. Capturing the true complexity of nature (or mankind!) is simply beyond reach at present. Just as such applications turned to parallel computing to overcome the limitations of a single processor, many of them will turn to Grid computing to overcome the limitations of parallel computers!

4. *Interaction-centric*: Finally, there are applications that require, or are enhanced by, real-time user interaction. This interaction can be of many forms, ranging from decision-making to visualization. The requirements for responsiveness are often in direct contradiction to the high average throughput and load-balancing needs of typical batch-oriented HPC systems. Furthermore, the existing Grid lacks standards for event management that could possibly support the sort of real-time interaction required for effective real-time data analysis.

23.4.2 Grid operations for applications

What kinds of Grid operations and processes will developers want to use in their current and future applications? Two extremes of Grid use that have been commonly discussed in recent years are: (1) task farming of many (hundreds, thousands, millions) independent jobs, across workstations or PCs around the world (e.g. Monte Carlo techniques or

Table 23.1 Tabulation of the Grid operations described in this section

Basic	Information & interaction	Compound
Resource selection	Application monitoring	Migration
Job initialization	Notification	Spawning
Data transfer	Interaction	Task farming

SETI@Home), where only small amounts of data need to be retrieved from each job and (2) coupled simulations, in which multiple supercomputers are harnessed to carry out tightly coupled, *distributed* simulations that cannot be easily done on a single resource (e.g. the collision of two neutron stars or the simulation of the earth's ocean and atmosphere). Both types of scenarios have proven effective for a limited class of problems, and we consider them below. But we expect that future Grid applications will go far beyond such 'simple' present-day examples, which are merely building blocks for much more sophisticated scenarios, as we sketch them in Table 23.1.

We begin with some simple scenarios involving operations that can immediately aid users in present-day tasks without necessarily modifying applications. These could be used individually to good effect, making use of new technologies as they emerge, archiving output data for the simulations of a whole collaboration consistently in one location or the remote monitoring of a simulation to isolate performance bottlenecks. We then illustrate, how with appropriate Grid services in place, these can in principle be extended and combined to create extremely complex, and even autonomous hierarchies of Grid processes for future applications.

Here we only have room to describe a small subset of envisaged operations, and even these operations must now be more rigorously defined in terms of inputs, outputs, and functionality. Such an endeavor, as is now ongoing in, for example, the GridLab project, will require input and standardization across the whole community, but will result in the formal definition of the set of operations which any Grid-enabling technology should provide.

23.4.2.1 Basic operations

First we consider *basic* Grid operations, such as resource selection, job initiation, and data transfer. Properly implemented, transparent, and automated services to handle these operations could provide huge benefits to the simplest of Grid applications. Many research groups, organizations, and companies have access to large numbers of computational resources scattered across the world, usually with different characteristics, loads, and user accounts. Selection of the most appropriate resource for a given job taking into account factors such as cost, CPU speed, and availability is a difficult issue. Tools to automatically find remote machines, and start jobs, would provide a consistent and easy to use interface to facilitate a currently tedious and prone-to-error task. Data transfer tools could automatically stage the required executables and input files to the correct location on a machine, and archive the resulting output files.

Even this type of Grid computing, although imminent, is yet to be very widely used, because the necessary infrastructure to support it at the various sites, is not fully developed or deployed. Furthermore, user interfaces, such as portals to facilitate user interaction with the various resources, are still in their infancy. For examples of the more advanced portal development efforts, providing Web-based interfaces to computing resources, see References [7, 8].

Such basic operations are, however, just the start. Advanced versions of these operations could include many more features: resource selection could include both application-specific (performance on a given machine, data transfer overheads) and community-specific (balancing of cost and resource usage for a whole group of users) factors; job initiation could be extended to include staging simulations across multiple machines for increased resources or quicker turnaround time; data transfer could include replication and searching.

23.4.2.2 Information and interaction operations

The collection, organization, and distribution of information is fundamental to fully functioning Grids. Information about resources, software, people, jobs, data, and running applications, are just a few potential sources. Flexible operations for *information retrieval* will be needed by all applications.

A further set of operations providing *application monitoring* can be divided into two cases: (1) information about running applications can be archived, eventually providing a database of characteristics such as performance on different machines, typical run times, and resource usage (CPU, memory, file space) and (2) monitoring information can be made available interactively, and accessible, for example, by an entire collaboration, while an application is running.

Related to application monitoring is *notification* and *interaction*. Applications could have the ability to notify users (or whole collaborations) of important events, both resource related *'You will fill the current hard disk in half an hour!'* and concerning the application itself *'Found new phenomenon – contact the news agencies!'*. Notification could use a variety of media, from now traditional email to the latest mobile devices and protocols (WAP, SMS, MMS, imode), and include where relevant attachments such as images and recordings. Interaction with running applications provides users with more control over the use of resources, and provides new insights into their results. Interacting with applications includes remote steering and visualization.

23.4.2.3 Compound operations

The most exciting Grid operations would occur when all these pieces are put together, and applications are able to make *dynamic* use of their resources, either self-deterministically or controlled by remote services. Here we have room to discuss only a couple of important possibilities, but clearly there are many more potential uses of Grids for applications.

Migration: Let us first consider a combination of three basic Grid operations: resource selection, information retrieval/publication, and data transfer. These can be combined into a new operation that we call *migration*, in which a running process is transferred from one resource to another for some reason. Migration can have many uses. For example, when a job is initiated, a resource selector may find that the most appropriate resource is busy, and will not be available for several hours, but that a second choice, although not optimal, is available immediately. The job could then be started on the second choice and then migrated when the better resource becomes available. The migration could be triggered when the primary resource becomes available by writing a platform-independent checkpoint file, transferring it to the remote resource, staging the executable to, or building it on, the same resource and continuing the job there.

But there are more interesting and more dynamic possibilities: a simulation running somewhere may find that its resource needs a change dramatically. Perhaps it needs an order of magnitude more memory, or less, or it requires interaction with a large dataset located on a distant machine or set of machines. Such needs may not be known at the compile time, but are only discovered sometime during the execution of the job. Given appropriate access to resources, information services to determine availability, network capabilities, and so on, there is no reason Grid processes cannot migrate themselves from one collection of resources to another, triggered by a set of users who have been notified of the need, by an external master process, or by the process itself. The migration may involve processes that are distributed across multiple resources. Once a migration has occurred, the old process must be shut down (if it was not already), and the new process must register its new location(s) and status in an information server that users or other processes use to track it.

Spawning: Related to migration is another operation that we call *spawning*. Rather than migrating an entire process from one set of resources to another, particular subprocesses may be 'outsourced' as needed. For example, a parallel simulation of a black hole collision may involve analysis of the data for the gravitational waves emitted, or locating the horizon surfaces of the holes. These operations can be very time consuming, do not necessarily feed back into the main evolution loop, or may not even be done with parallel algorithms. In such cases, in the Grid world it would be sensible to seek out other resources, during the execution, to which these tasks could be spawned. Typically, a basic spawning operation would involve a request for and acquisition of resources, transfer of data from the main process to the remote resource(s), initiation of the remote process (which may be very different from the main process), and return of a result, possibly to the main process. If a spawning request cannot be met, the task can simply be carried out inline with the main process (as it usually is now), but at the cost of holding it up while the unspawned task is completed.

Spawned tasks may themselves spawn other tasks, or request migrations, as their needs change, or as their environment changes (e.g. network, file system, or computational capacities and loads change). Such operations could be combined to create complex Grid pipeline, workflow, and vector type operations. Each process in these potentially very complex hierarchies should be able to publish its state to an information server, so that its status can be monitored. Furthermore, each process will have the need for data management, archiving, and so on.

Task farming: Another related operation, *task farming*, builds upon other operations such as spawning and resource finding. In task farming, a master process spawns off multiple slave processes, which do some jobs on its behalf. When these processes are completed, the master process creates more processes until the master has finished its task. A typical use of this is for a parameter search. In this scenario the master process is tasked with exploring one or more parameter ranges, and spawns a slave process for each distinct parameter combination. Obviously, the real operation is more complex than this, as the number of parameter combinations may be extremely large, but the master process would only be able to spawn a small number of slave processes at any one time, and so would start a new process only when one finishes. This scenario could also be hierarchical in nature whereby the master delegates sets of parameter ranges to secondary masters that then perform the task farming, and so on.

This scenario requires resource management to find the hosts to spawn jobs to, and a spawning facility as described above.

These application-level capabilities that are being built on top of basic Grid infrastructure go way beyond existing Grid protocol and service standards, but are needed in a similar form by a wide variety of emerging applications. And yet these services are immensely complicated to implement and need to be re-implemented by each different application team in much the same way as basic Grid services that were re-implemented prior to the emergence of toolkits like Globus and standards bodies like the Global Grid Forum. Furthermore, such services remain extremely fragile when implemented in this manner, as they cannot withstand even minor revisions or nonuniformity in the Grid infrastructure.

Even assuming that basic Grid services needed to support these kinds of operations are fully deployed, the question is how to prepare and develop applications for the Grid so that they may take advantage of these technologies? The basic services are likely to differ in their implementation and version from site to site, or multiple services may exist that provide the same basic capability, but with different performance characteristics, and further, they may come and go and resources may fail or come on-line at any time. Further, the applications must be able to run in many different environments, from laptops to supercomputers, and yet the application programmer cannot develop and maintain different versions of routines for each of these environments. Creating an environment that enables the application developer and user to take advantage of such new Grid capabilities, through abstraction and discovery of various services, which may or may not be available, without having to change the application code for the different possibilities, is the subject of the following sections.

23.5 THE WAY FORWARD: GRID PROGRAMMING ENVIRONMENTS AND THEIR EFFECTIVE USE

To enable applications to take advantage of the potential of Grid computing, it will be imperative that the new capabilities it offers be both easy to use and robust. Furthermore, by its construction the Grid is extremely heterogeneous, with different networks, machines, operating systems, file systems, and just as importantly, different versions of the underlying Grid infrastructure. Applications will need to work seamlessly in a variety of different environments. Even worse, although many services may be deployed, there is no guarantee that they will be operational at run time! Application developers and users must be able to operate in an environment in which they cannot be sure of what services will exist in advance, and applications must fail gracefully by falling back on alternative service implementations or strategies when the intended service is unavailable. For this reason, it is not only convenient, but rather it is imperative, that applications can be written and developed free from the details of where specific Grid services are located and the individual mechanisms and protocols used for accessing them.

We have, in the first section of this chapter, already discussed how the techniques that have been successfully implemented in HPC programming frameworks should be now used to build frameworks for programming on the Grid. Frameworks and environments for Grid applications are currently being discussed and developed in several projects and more broadly in the different groups of the Global Grid Forum (in particular, the Advanced Programming Models and Applications Research Groups). See Reference [9] for a detailed review of current Grid-programming tools and the issues associated with them.

Previous work has shown repeatedly that it is essential to develop a flexible API for Grid operations, which insulates application developers and users from the details of the underlying Grid infrastructure and its deployment. Such an API will allow developers to write and utilize software today, which will then seamlessly make use of more advanced infrastructure as it matures. The GridLab project aims to develop such an API and its accompanying infrastructure.

The authors of this paper are among the instigators and researchers of the European GridLab project – *A Grid Application Toolkit and Testbed*. The GridLab project brings together Grid infrastructure and service developers with those experienced in developing successful frameworks and toolkits for HPC.

The basic architecture of the GridLab project splits the system into two basic sets of components: user space components, which the developer is aware of and are deployed locally, and components that provide some functionality, which may be deployed remotely or locally, We refer to these functionality components as 'capability providers' (we avoid the much-overloaded word 'service' as these capability providers may be local libraries).

The user-space components consist of the application itself, linked to a library that provides the Grid Application Toolkit API (GAT-API). When a user requests a Grid operation through the GAT-API, the library checks a database or uses some discovery mechanism to find an appropriate capability provider, and dispatches the operation to it. Thus, to develop an application the programmer merely needs to make GAT-API calls for required Grid operations. These then make use of whatever Grid infrastructure is actually deployed to perform these operations. Note that since the Grid infrastructure is highly dynamic the set of capability providers may vary even within one run of an application.

This architecture is very flexible, and allows the capability providers to be a simple library call, a CORBA application, or an OGSA service, or anything else, which may be developed in the future. Thus, we may leverage off the huge investment that has been made in the business community in developing Web services and related technologies, while not being tied to any specific technology.

There are still many issues to be clarified, such as security, discovery mechanisms, and so on; however, these have no impact on the code that an application developer needs to write to use the GAT-API, and thus the developer is insulated from changes in how Grid services are deployed, or which technologies are used to contact them.

The effectiveness of a particular application on a Grid is determined both by the nature of the application itself and the functionality of the code that implements it. For example, Monte Carlo type schemes are embarrassing parallel, highly conducive to task farming scenarios, whereas tightly coupled finite difference applications require new algorithms and techniques to run efficiently across loosely coupled machines.

Improvements in technology may enable compilers or automated tools to analyze applications and determine the best way to utilize the Grid, in the same way that compiler technology has allowed automatic parallelization of certain programs. However, just as with automatic parallelization technology, automatic Grid-enabling technologies will take years to develop, and will almost certainly require the user to embed some form of directives in the source code to give hints to the process, in the same way that parallelizing compilers require today. It is possible that an equivalent of the OpenMP standard could be developed to allow compatibility between different compiler or tool vendors; such a standard would be essential for truly heterogeneous environments.

Even if such enabling technologies become available, there are things that application developers will always need to be aware of and functionality that their codes must provide. For example, to make full use of an operation such as migration, an application must be able to checkpoint its current state in a manner independent of machine architecture and the number of processors it is running on. One advantage of programming frameworks such as Cactus is that a lot of this required functionality is automatically and transparently available.

One of the dangers of middleware development is that it cannot be used effectively by applications until the development is nearly completed. Otherwise the applications must suffer extreme disruptions while the API is reorganized during the course of development or while the fundamental bugs are uncovered. However, we want to use the Grid now, before such technologies are fully developed, both to exploit the parts that are complete and also to begin building our applications to incorporate the appropriate scenarios and operations. Currently, application developers must either Grid-enable their applications by hand, or use an enabling framework, such as Cactus, to do so. Grid application framework solutions like the GAT offer a way to break this cycle of dependence.

23.6 IN CONCLUSION

This chapter barely scratches the surface in addressing how applications can interact with, and benefit from, the new emerging Grid world. We have shown how real-world examples of computational needs can lead to new and very different ways of carrying out computational tasks in the new environment provided by the Grid. Much more than merely convenient access to remote resources, we expect that very dynamic Grid-enabled processes, such as on-demand task farming, job migration, and spawning, will be automated and intermixed, leading to complex Grid processes matching the particular needs of individual applications. These applications will be able to interact with each other, with data, and with users and developers across the Grid.

However, in order to enable these future Grid applications, in the face of myriad, increasingly complex technologies and software infrastructure, higher-level services must be abstracted and provided to application developers in such a way that they can be understood and used without a knowledge of the details of Grid technologies.

We have focused on computation-centric requirements for applications in the HPC world, and although such applications are likely to be among the first real benefactors of these new technologies, a mature Grid environment will impact *all classes of applications* each with their own special needs. Much further work is now needed, in fully classifying and abstracting applications and their required operations, and in developing a usable API to encode these abstractions. Such a program is under way, in particular, within the GridLab project, whose mission is to deliver a GAT that can be used by generic applications to access the fully available functionality of the dynamic Grid.

ACKNOWLEDGMENTS

We acknowledge the input, both in ideas and implementations, from Werner Benger, Thomas Dramlitsch, Gerd Lanfermann, and Thomas Radke. We warmly thank Denis Pollney and others in the AEI numerical relativity group for their suggestions, as well as their careful testing, invaluable feedback, and endless patience with our new technologies. We thank the Max Planck Gesellschaft and Microsoft for their financial support of the Cactus project. This material is based in part upon the work supported by EU GridLab project (IST-2001-32133) and the German DFN-Verein TiKSL project.

REFERENCES

1. Cactus Code homepage, http://www.cactuscode.org, July 10, 2002.
2. Allen, G. *et al.* (2001) Cactus grid computing: review of current development, in Sakellariou, R., Keane, J., Gurd, J. and Freeman, L. (eds) *Euro-Par 2001: Parallel Processing, Proceedings of 7th International Euro-Par Conference Manchester*, UK: Springer.
3. Allen, G. *et al.* (2001) Supporting efficient execution in heterogeneous distributed computing environments with Cactus and Globus. Proceedings of SC 2001, Denver, CO, November 10–16, 2001.
4. Liu, C., Yang, L., Foster I. and Angulo, D. (2002) Design and evaluation of a resource selection framework for grid applications HPDC-11, Edinburgh, Scotland July, 2002.
5. Allen, G., Angulo, D., Foster, I., Lanfermann, G., Liu, C., Radke, T. and Seidel, E. (2001) The cactus worm: experiments with dynamic resource discovery and allocation in a grid environment. *International Journal of High Performance Computing Applications*, **15**(4), 345–358.
6. Foster, I., Kesselman, C. and Tuecke, S. 2001 The anatomy of the grid: enabling scalable virtual organizations. *International Journal of Supercomputing Applications*, **15**(3).
7. Thomas, M., Mock S., Boisseau, J., Dahan, M., Mueller, K. and Sutton, D. (2001) The GridPort toolkit architecture for building grid portals, Proceedings of the 10th IEEE Intl. Symp. on High Performance Distributed Computing, August, 2001.

8. Russell, M. *et al.* (2002) The astrophysics simulation collaborary: a science portal enabling community software development. *Cluster Computing*, **5**(3), 297–304.

9. Lee, C., Matsuoka, S., Talia, D., Sussman, A., Mueller, M., Allen, G. and Saltz, J. (2002) A Grid Computing Primer (Global Grid Forum document),
http://www.eece.unm.edu/~apm/docs/APM_Primer_0801.pdf, July 10, 2002.

NetSolve: past, present, and future – a look at a Grid enabled server[1]

Sudesh Agrawal, Jack Dongarra, Keith Seymour, and Sathish Vadhiyar

University of Tennessee, Tennessee, United States

24.1 INTRODUCTION

The emergence of Grid computing as the prototype of a next-generation cyber infrastructure for science has excited high expectations for its potential as an accelerator of discovery, but it has also raised questions about whether and how the broad population of research professionals, who must be the foundation of such productivity, can be motivated to adopt this new and more complex way of working. The rise of the new era of scientific modeling and simulation has, after all, been precipitous, and many science and engineering professionals have only recently become comfortable with the relatively simple world of uniprocessor workstations and desktop scientific computing tools. In this world, software packages such as Matlab and Mathematica, and languages such as C and Fortran represent general-purpose scientific computing environments that enable users – totaling more than a million worldwide – to solve a wide variety of problems

[1] Work supported in part by the NSF/NGS GRANT #NSF EIA-9 975 015, and NSF GRANT ACI-9 876 895

Grid Computing – Making the Global Infrastructure a Reality. Edited by F. Berman, A. Hey and G. Fox
© 2003 John Wiley & Sons, Ltd   ISBN: 0-470-85319-0

through flexible user interfaces that can model in a natural way the mathematical aspects of many different problem domains. Moreover, the ongoing, exponential increase in the computing resources supplied by the typical workstation makes these scientific computing environments more and more powerful, and thereby tends to reduce the need for the kind of resource sharing that represents a major strength of Grid computing. Certainly, there are various forces now urging collaboration across disciplines and distances, and the burgeoning Grid community, which aims to facilitate such collaboration, has made significant progress in mitigating the well-known complexities of building, operating, and using distributed computing environments. But it is unrealistic to expect the transition of research professionals to the Grid to be anything but halting and slow if it means abandoning the scientific computing environments that they rightfully view as a major source of their productivity.

The NetSolve project [1] addresses this difficult problem directly: the purpose of Net-Solve is to create the middleware necessary to provide a seamless bridge between the simple, standard programming interfaces and desktop systems that dominate the work of computational scientists, and the rich supply of services supported by the emerging Grid architecture, so that the users of the former can easily access and reap the benefits (shared processing, storage, software, data resources, etc.) of using the latter. This vision of the broad community of scientists, engineers, research professionals, and students, working with the powerful and flexible tool set provided by their familiar desktop computing environment, and yet being able to easily draw on the vast, shared resources of the Grid for unique or exceptional resource needs, or to collaborate intensively with colleagues in other organizations and locations, is the vision that NetSolve is designed to realize.

24.2 HOW NetSolve WORKS TODAY

Currently, we have released Version 1.4.1 of NetSolve. The NetSolve homepage, located at http://icl.cs.utk.edu/NetSolve/, contains detailed information and the source code.

In any network-based system, we can distinguish three main paradigms: *proxy computing*, *code shipping* [2], and *remote computing*. These paradigms differ in the way they handle the user's data and the program that operates on this data.

In *proxy computing*, the data and the program reside on the user's machine and are both sent to a server that runs the code on the data and returns the result. In *code shipping*, the program resides on the server and is downloaded to the user's machine, where it operates on the data and generates the result on that machine. This is the paradigm used by Java applets within Web browsers, for example. In the third paradigm, *remote computing*, the program resides on the server. The user's data is sent to the server, where the programs or numerical libraries operate on it; the result is then sent back to the user's machine. NetSolve uses the third paradigm.

There are three main components in the NetSolve system: the agent, the server, and the client. These components are described below.

Agent: The agent represents the gateway to the NetSolve system. It maintains a database of NetSolve servers along with their capabilities and dynamic usage statistics for use in scheduling decisions. The NetSolve agent attempts to find the server that will service the request, balance the load amongst its servers, and keep track of failed servers. Requests are directed away from failed servers. The agent also adds fault-tolerant heuristics that attempt to use every likely server until it finds one that successfully services the request.

Server: The NetSolve server is the computational backbone of the system. It is a daemon process that awaits client requests. The server can run on single workstations, clusters of workstations, symmetric multiprocessors (SMPs), or massively parallel processors (MPPs). One key component of the server is the ability to wrap software library routines into NetSolve software services by using an Interface Definition Language (IDL) facility called the *NetSolve Problem Description File* (PDF).

Client: A NetSolve client user accesses the system through the use of simple and intuitive application programming interfaces (APIs). The NetSolve client uses these APIs to make a request to the NetSolve system, with the specific details required with the request. This call automatically contacts the NetSolve system through the agent, which in turn returns to the server, which can service the request. The client then contacts the server to start running the job with the input data. Figure 24.1 shows this organization.

As mentioned earlier, NetSolve employs fault tolerance to fulfill client requests. The NetSolve system ensures that a user request will be completed unless every single resource capable of solving the problem has failed. When a client sends a request to a NetSolve agent, it receives a sorted list of computational servers to contact. When one of these servers has been successfully contacted, the numerical computation starts. If the contacted server fails during the computation, another server is contacted, and the computation is restarted. Table 24.1 shows how calls are made to NetSolve through Matlab, C, and Fortran client interfaces.

There are a number of topics that we will not be able to cover fully in this chapter. The list includes the following:

- Security
  Uses Kerberos V5 for authentication
- Separate server characteristics
  Prototype implementation of hardware and software servers
- Hierarchy of agents
  Providing a more scalable configuration
- Monitor NetSolve network
  Track and monitor usage
- Network status
  Use of network weather service
- Internet Backplane protocol
  Middleware for managing and using remote storage
- Fault tolerance
- The dynamic nature of servers

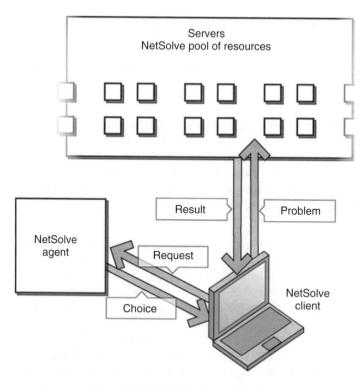

Figure 24.1   NetSolve's organization.

Table 24.1 Solving a linear system A x = b with NetSolve

'A' is an $m \times m$ matrix and 'b' is an $m \times n$ matrix
In C and Fortran, the right-hand side is overwritten by the solution

From MATLAB:
x = netsolve('linsol', A,b)

From C:
netsl("linsol",nsinfo,m,n,A,lda,b,ldb)

From FORTRAN:
CALL NETSL("linsol", NSINFO,M,N,A,LDA,B,LDB)

- Automated adaptive algorithm selection
  Dynamically determine the best algorithm based on the system status and the nature of user problem.

Additional details can be found on the NetSolve [3] Web site.

24.3 NetSolve IN SCIENTIFIC APPLICATIONS

NetSolve has been used in a number of applications for resource management purposes, to enable parallelism in the applications and to help users avoid installation of cumbersome software. Following Sections detail some of the applications that reap the benefits of using NetSolve.

24.3.1 IPARS (integrated parallel accurate reservoir simulators)

IPARS [4], developed under the directorship of Mary Wheeler at the Center for Subsurface Modeling, at the University of Texas' Institute for Computational and Applied Mathematics, TICAM, is a framework for developing parallel models of subsurface flow and transport through porous media. IPARS can simulate single-phase (water only), two-phase (water and oil), or three-phase (water, oil, and gas) flow through a multiblock 3D porous medium. IPARS can be applied to model water table decline due to overproduction near urban areas, or enhanced oil and gas recovery in industrial applications.

A NetSolve interface to the IPARS system that allows users to access the full functionality of IPARS was constructed. Accessing the system via the MATLAB, C, Mathematica, or FORTRAN interfaces automatically executes simulations on a cluster of dual-node workstations that allow for much quicker execution than what would be possible on a single local machine. The NetSolve system also does the postprocessing of the output to use the third-party software, TECPLOT, to render the 3D output images. Among other things, NetSolve provides a gateway to the IPARS system without downloading and installing the IPARS code. This means it can even be used on platforms that it has not yet been ported to. The interface was further enhanced by embedding it in HTML form [5] within a Web browser so that with just access to a Web browser one can enter input parameters and submit a request for execution of the IPARS simulator on a NetSolve system. The output images are then brought back and displayed to the Web browser. This interaction showed how the NetSolve system can be used to create a robust Grid computing environment in which powerful modeling software, like IPARS, becomes both easier to use and to administer.

24.3.2 MCell

MCell [6] is a general Monte Carlo simulator of cellular microphysiology. MCell uses Monte Carlo diffusion and chemical reaction algorithms in 3D to simulate the complex biochemical interactions of molecules inside and outside of living cells. NetSolve is used as a resource management framework to manage the execution of a large number of MCell simulations on a large number of resources in the NetSolve Grid. One of the central pieces of the framework is a scheduler that takes advantage of MCell input data requirements to minimize turnaround time. This scheduler is part of the larger AppLeS at the University of California, San Diego. The use of robust, coherent, and fault-tolerant NetSolve pool of resources allowed the MCell researchers to implement parallelism in the simulations with simple sequential calls to NetSolve.

24.3.3 SARA3D

SARA3D [7] is a code developed by BBN Acoustic Technologies that is used to solve structural acoustic problems for finite-element structures emerging in a user-defined fluid. The SARA3D application accepts as input a file describing the structure, the materials involved, the fluid properties, and the actual analyses to be performed. As output, the application generates a variable number of files that contain different types of data calculated during the analyses. These output files can be further processed to compute quantities such as radiated power, near and far-filled pressures, and so on. The SARA3D application also consists of a large number of phases. The output files produced by one phase of the application will serve as inputs to the next phase of the application. These intermediate files are of no particular interest to the end user. Also, it is desirable that the final output files produced by the application be shared by a large number of users.

For these purposes, SARA3D was integrated into the NetSolve framework utilizing the data staging capability of NetSolve. Different NetSolve servers implementing the different phases of the application were started. The user caches his input data near the servers using the data staging facility in NetSolve. The user also indicates to the servers to output the intermediate results and the final outputs to data caches. By this framework, only relevant and sharable data are transmitted to the end users.

24.3.4 Evolutionary farming

Evolutionary Farming (EvCl) is a joint effort between the Department of Mathematics and the Department of Ecology and Evolutionary Biology at the University of Tennessee. The goals of the EvCl research are to develop a computer-based model simulating evolution and diversification of metapopulations in a special setting and to explore relationships between various parameters affecting speciation dynamics. EvCl consists of two main phases, *evolve*, which simulates the evolution and *cluster*, which is used for species determination. Since the problem involves separate runs of the same parameters due to the stochastic nature of simulation, the NetSolve task-farming interface was used to farm these separate simulations onto different machines. The use of NetSolve helped in significantly reducing the cumbersome management of disparate runs for the problem.

24.3.5 LSI-based conference organizer

Latent Semantic Indexing (LSI) [8] is an information retrieval method that organizes information into a semantic structure that takes advantage of some of the implicit higher-order associations of words with text objects. The LSI method is based on singular value decomposition (SVD) of matrices consisting of documents and query terms. Currently, LSI is being used at the University of Tennessee to construct a Web-based conference organizer [9] that organizes conference sessions based on submitted abstracts or full-text documents. The computationally intensive SVD decompositions will be implemented by the NetSolve servers allowing the developers of the Conference Organizer to concentrate their research efforts in other parts of the organizer, namely, the Web interfaces. Integrating the Conference Organizer into the NetSolve system also allows the users of the organizer to avoid installing the SVD software in their system.

24.4 FUTURE WORK

Over time, many enhancements have been made to NetSolve to extend its functionality or to address various limitations. Some examples of these enhancements include task farming, request sequencing, and security. However, some desirable enhancements cannot be easily implemented within the current NetSolve framework. Thus, future work on NetSolve will involve redesigning the framework from the ground up to address some of these new requirements.

On the basis of our experience in developing NetSolve, we have identified several requirements that are not adequately addressed in the current NetSolve system. These new requirements – coupled with the requirements for the original NetSolve system – will form the basis for the next generation of NetSolve.

The overall goal is to address three general problems: ease of use, interoperability, and scalability. Improving ease of use primarily refers to improving the process of integrating user codes into a NetSolve server. Interoperability encompasses several facets, including better handling of different network topologies, better support for parallel libraries and parallel architectures, and better interaction with other Grid computing systems such as Globus [10] and Ninf [11]. Scalability in the context used here means that the system performance does not degrade as a result of adding components to the NetSolve system.

The sections below describe some of the specific solutions to the general problems discussed earlier.

24.4.1 Network address translators

As the rapid growth of the Internet began depleting the supply of IP addresses, it became evident that some immediate action would be required to avoid complete IP address depletion. The IP Network Address Translator [12] is a short-term solution to this problem. Network Address Translation (NAT) allows reuse of the same IP addresses on different subnets, thus reducing the overall need for unique IP addresses.

As beneficial as NATs may be in alleviating the demand for IP addresses, they pose many significant problems to developers of distributed applications such as NetSolve [13]. Some of the problems pertaining to NetSolve include the following:

- *IP addresses are not unique*: In the presence of a NAT, a given IP address may not be globally unique. Typically the addresses used behind the NAT are from one of the several blocks of IP addresses reserved for use in private networks, though this is not strictly required. Consequently, any system that assumes that an IP address can serve as the unique identifier for a component will encounter problems when used in conjunction with a NAT.
- *IP address-to-host bindings may not be stable*: This has similar consequences to the first issue in that NetSolve can no longer assume that a given IP address corresponds uniquely to a certain component. This is because, among other reasons, the NAT may change the mappings.
- *Hosts behind the NAT may not be contactable from outside*: This currently prevents all NetSolve components from existing behind a NAT because they must all be capable of accepting incoming connections.

- *NATs may increase network failures*: This implies that NetSolve needs more sophisticated fault-tolerance mechanisms to cope with the increased frequency of failures in a NAT environment.

To address these issues we are currently investigating the development of a new communications framework for NetSolve. To avoid problems related to potential duplication of IP addresses, the NetSolve components will be identified by a globally unique identifier, for example, a 128-bit random number. The mapping between the component identifier and a real host will not be maintained by the NetSolve components themselves, rather there will be a discovery protocol to locate the actual machine running the NetSolve component with the given identifier. In a sense, the component identifier is a network address that is layered on top of the real network address such that a component identifier is sufficient to uniquely identify and locate any NetSolve component, even if the real network addresses are not unique. This is somewhat similar to a machine having an IP address layered on top of its MAC address in that the protocol to obtain the MAC address corresponding to a given IP address is abstracted in a lower layer.

An important aspect to making this new communications model work is the *relay*, which is a component that will allow servers to exist behind a NAT. Since a server cannot accept unsolicited connections from outside the private network, it must first register with a relay. The relay acts on behalf of the component behind the NAT by establishing connections with other components or by accepting incoming connections. The component behind the NAT keeps the connection with the relay open as long as possible since it can only be contacted by other components while it has a control connection established with the relay. To maintain good performance, the relay will only examine the header of the messages that it forwards, and it will use a simple table-based lookup to determine where to forward each message. Furthermore, to prevent the relay from being abused, authentication will be required.

Since NATs may introduce more frequent network failures, we must implement a protocol to allow NetSolve components to reconnect to the system and resume the data transfer if possible. We are still investigating the specifics of this protocol, but at the least it should allow the servers to store the results of a computation to be retrieved at some time later when the network problem has been resolved. Additionally, this would allow a client to submit a problem, break the connection, and reconnect later at a more convenient time to retrieve the results.

24.4.2 Resource selection criteria

In the current NetSolve system, the only parameter that affects the selection of resources is the problem name. Given the problem name, the NetSolve agent selects the 'best' server to solve that problem. However, the notion of which server is best is entirely determined by the agent. In the next generation of NetSolve, we plan to extend this behavior in two ways. First, we should allow the user to provide constraints on the selection process. These selection constraints imply that the user has some knowledge of the characteristics that will lead to a better solution to the problem (most probably in terms of speed). Second, we should allow the service providers (that is, those organizations that provide NetSolve servers) to specify constraints on the clients that can access that service. For example, an organization may want to restrict access to a certain group of collaborators. We are currently examining the use of XML as a resource description language.

24.5 CONCLUSIONS

We continue to evaluate the NetSolve model to determine how we can architect the system to meet the needs of our users. Our vision is that NetSolve will be used mostly by computational scientists who are not particularly interested in the mathematical algorithms used in the computational solvers or the details of network-based computing. NetSolve is especially helpful when the software and hardware resources are not available at hand. With Grid computing, there exists many interesting areas of research to explore and much room for improvement. We envision future work in features like dynamically extensible servers whose configuration can be modified on the fly. The new strategy will be to implement a just-in-time binding of the hardware and software service components, potentially allowing servers to dynamically download software components from service repositories. Parallel libraries could be better supported by data distribution/collection schemes that will marshal input data directly from the client to all computational nodes involved and collect results in a similar fashion. Efforts also need to be made so that clients can solve jobs with large data sets on parallel machines; the current implementation requires this data to be passed in the call since the calling sequence expects a reference to the data and not a reference via a file pointer, and this may not be possible.

As researchers continue to investigate feasible ways to harness computational resources, the NetSolve system will continue to emerge as a leading programming paradigm for Grid technology. Its lightweight and ease of use make it an ideal candidate for middleware, and as we discover the needs of computational scientists, the NetSolve system will be extended to become applicable to an even wider range of applications.

REFERENCES

1. Casanova, H. and Dongarra, J. (1997) NetSolve: A network-enabled server for solving computational science problems. *The International Journal of Supercomputer Applications and High Performance Computing*, **11**(3), 212–223.
2. Agrawal, S. and Dongarra, J. J. (2002) Hardware Software Server in NetSolve, UT-CS-02-480, University of Tennessee, Computer Science Department, Knoxville, 2002.
3. NetSolve Web site http://icl.cs.utk.edu/netsolve/.
4. IPARS Web site http://www.ticam.utexas.edu/CSM/ACTI/ipars.html.
5. TICAM Web site http://www.ticam.utexas.edu/~ut/webipars/AIR.
6. MCell Web site http://www.mcell.cnl.salk.edu.
7. Allik, H., Dees, R., Moore, S. and Pan, D. (1995) SARA-3D User's Manual, BBN Acoustic Technologies.
8. Berry, M. W., Dumais, S. T. and O'Brien, G. W. (1995) Using linear algebra for intelligent information retrieval. *SIAM Review*, **37**(4), 573–595.
9. COP Web site http://shad.cs.utk.edu/cop.

10. Foster, I. and Kesselman, C. (1997) Globus: A metacomputing infrastructure toolkit. *The International Journal of Supercomputer Applications and High Performance Computing*, **11**(2), 115–128, citeseer.nj.nec.com/foster96globu.html.
11. Nakada, H., Sato, M. and Sekiguchi, S. (1999) Design and implementations of Ninf: towards a global computing infrastructure. *Future Generation Computing Systems*, Metacomputing Issue, **15**(5–6), 649–658.
12. Egevang, K. and Francis, P. (1994) The IP Network Address Translator (NAT), RFC Tech Report 1631, May 1994.
13. Moore, K. (2002) Recommendations for the Design and Implementation of NAT-Tolerant Applications, private communication, February 2002.

Ninf-G: a GridRPC system on the Globus toolkit

Hidemoto Nakada,[1,2] Yoshio Tanaka,[1] Satoshi Matsuoka,[2] and Satoshi Sekiguchi[1]

[1]*National Institute of Advanced Industrial Science and Technology (AIST), Grid Technology Research Center, Tsukuba, Ibaraki, Japan,* [2]*Tokyo Institute of Technology, Global Scientific Information and Computing Center, Tokyo, Japan*

25.1 INTRODUCTION

Recent developments in high-speed networking enables collective use of globally distributed computing resources as a huge single problem-solving environment, also known as the Grid.

The Grid not only presents a new, more difficult degree of inherent challenges in distributed computing such as heterogeneity, security, and instability, but will also require the constituent software substrates to be seamlessly interoperable across the network. As such, software layers are constructed so that higher-level middleware sits on some common, lower-level software layer, just as it is with single-box computers in which most applications running on top of it share a common operating system and standard libraries.

Grid Computing – Making the Global Infrastructure a Reality.   Edited by F. Berman, A. Hey and G. Fox
© 2003 John Wiley & Sons, Ltd   ISBN: 0-470-85319-0

Currently, the Globus Toolkit [1] serves as the '*de facto* standard' lower-level substrate for the Grid. Globus provides important Grid features such as authentication, authorization, secure communication, directory services, and so on that the software above can utilize so as not to replicate the lower-level programming efforts, as well as provide interoperability between the middleware mentioned above. However, the Globus Toolkit alone is insufficient for programming on the Grid. The abstraction level of the Globus, being a lower-level substrate, is terse and primitive in a sense; this is fine for the intended use of the Toolkit, but nevertheless higher-level programming layers would be absolutely necessary for most users. This is analogous to the situation in programming, say, parallel programming on a Linux cluster; one would certainly program with higher-level parallel programming systems such as MPI and OpenMP, rather than by using the raw TCP/IP socket interface of the operating system.

Over the years there have been several active research programs in programming middlewares for the Grid environment, including MPICH-G [2] and Nimrod/G [3], Ninf [4], and NetSolve [5]. MPICH-G and MPICH-G2 are MPI systems achieving security and cross-organization MPI interoperability using Globus Toolkit, although as a programming model for the Grid the utility is somewhat less significant compared to tightly coupled environments. Nimrod-G is a high-throughput computing system implemented on Globus toolkit. It automatically invokes existing applications on many computing resources based on deadline scheduling. While the system is quite useful for some applications, the applications area is largely limited to parameter-sweep applications and not general task-parallel programming.

Ninf and NetSolve are implementations of the GridRPC [6] programming model, providing simple yet powerful server-client-based framework for programming on the Grid. GridRPC facilitates an easy-to-use set of APIs, allowing easy construction of globally distributed parallel applications without complex learning curves. Both systems have seen successful usages in various Grid application projects.

On the other hand, Ninf and NetSolve themselves are mutually interoperable, but they do not interoperate well with other Grid tools on Globus such as GridFTP. The reason is that, because the first versions of both Ninf and NetSolve were essentially developed at the same time as Globus, both systems were built rather independently without fully utilizing the Globus features or taking into account their mutual interoperability.

To resolve the situation, we redesigned the whole GridRPC system in collaboration with the NetSolve team at the University of Tennessee, Knoxville. The redesign has been extensive, including the software architecture to fully utilize the Globus features, user-level API specifications that retain the simplicity but generalize call contexts for more flexible Grid-level programming, various changes in the protocols, and so on. The redesign, especially the API specification, has been carefully done so that each group can independently produce respective implementations of GridRPC.

Our implementation is Ninf-G, which is in effect a full reimplementation of Ninf on top of Globus. Compared to older implementations, Ninf-G fully utilizes and effectively sits on top of Globus. The result is that applications constructed with Ninf-G can take advantage of any modules implemented for the Globus Toolkit, such as special job managers or GridFTP servers.

For the remainder of this chapter, we discuss the implementation of the Ninf-G system, and demonstrate the usage with a simple parallel programming example with the Ninf-G, along with its performance. In Section 25.2 we briefly introduce the Globus Toolkit; in Section 25.3 we discuss the overall design of GridRPC, and how each feature can be mapped on to lower-level Globus Toolkit features. Section 25.4 outlines the implementation of the Ninf-G, while Sections 25.5 and 25.6 illustrate a typical usage scenario along with its performance evaluation. Section 25.7 will conclude and hint at future directions.

25.2 GLOBUS TOOLKIT

The Globus Toolkit is a collection of modules that provides standardized lower-level features for implementing a distributed system on the Grid. Table 25.1 covers the services provided by Globus, in which each service can be used independently when needed. We give brief descriptions of the most relevant modules for GridRPC.

GSI: Grid Security Infrastructure (GSI) serves as the common authentication facility underlying all the features of Globus. It enables single sign-on using certificate delegation based on PKI.

GRAM: Globus Resource Allocation Manager (GRAM) is a 'secure inetd' that authenticates clients using GSI-based certificates, maps to the local user account, and invokes executable files.

MDS: Monitoring and Discovering Service (MDS) is a directory service that provides resource information within the Grid. MDS consists of two layers of Lightweight Directory Access Protocol (LDAP) servers, Grid Index Information Service (GIIS), which manages project-wide information, and Grid Resource Information Service (GRIS), which is responsible for site local information.

Globus-I/O: Globus-I/O enables secure communication between Grid peers using GSI, providing standard read/write APIs, plus nonblocking I/O that integrates with the Globus Threads library.

GASS: Global Access to Secondary Storage (GASS) provides easy-to-use file transfer facility across Grid peers. GASS can be used in conjunction with GRAM to stage client side files.

Table 25.1 Globus services

Service	Module name	Description
Security	GSI	Single sign-on, authentication
Information infrastructure	MDS/GRIS/GIIS	Information service
Resource management	GRAM/DUROC	Computing resource management
Data management	GASS/Grid FTP	Remote data access
Communication	Nexus/Globus-I/O	Communication service
Portability		Portability library (libc, pthread)

25.3 DESIGN OF NINF-G

25.3.1 GridRPC system

GridRPC is a programming model based on client-server remote procedure call (RPC), with features added to allow easy programming and maintenance of code for scientific applications on the Grid. Application programmers write task-parallel client programs using simple and intuitive GridRPC APIs that hide most of the complexities involving Grid programming. As a result, programmers lacking experience on parallel programming, let alone the Grid, could still construct Grid applications effortlessly.

At the server side, computation routines are encapsulated into an executable component called the *Remote Executable*. Client program basically invokes Remote Executables to request computation to be done on the server; asynchronous parallel executions of multiple Remote Executables on different servers result in simple fork-join parallel execution, while nested GridRPC invocations with complex synchronizations resulting in generalized task-parallel computations are also possible if the programmers have control over the Remote Executables on all the constituent nodes.

GridRPC system generally consists of the following four components:

Client component: Client components are programs that issue requests for GridRPC invocations. Each component consists of the user's main program and the GridRPC library.

Server component: Server component invokes Remote Executables as described below,

Remote executable: Remote Executables perform the actual computation at the Server. Each component consists of user-supplied server-side compute routine, system generated stub main program, and system supplied communication library.

Information service: Information service provides various information for the client component to invoke and to communicate with the Remote Executable component.

25.3.2 Ninf-G implementation on the Globus toolkit

As mentioned earlier, in contrast to previous incarnations of Ninf in which the Globus interface was added as an afterthought, various features of Ninf-G integrates Globus components directly to implement the GridRPC features. More specifically, Ninf-G employs the following components from the Globus Toolkit as shown in Figure 25.1.

GRAM: Serves the role of the server in the old Ninf system.

MDS: Publishes interface information and pathname of GridRPC components.

Globus-I/O: Client and remote executable communicate with each other using Globus-I/O.

GASS: Redirects stdout and stderr of the GridRPC component to the client console.

25.3.3 API of Ninf-G

Ninf-G has two categories of APIs. One is the Ninf-compatible 'legacy' API, and the other is the new (low-level) GridRPC API that has been subjected to collaborative standardization activities at the Global Grid Forum with the NetSolve team from UTK. Ninf-G

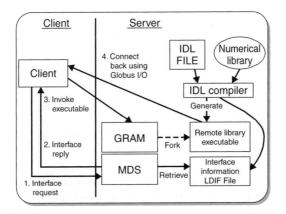

Figure 25.1 Overview of Ninf-G.

Table 25.2 GridRPC API principal functions

Function	Description
```int grpc_function_handle_init (grpc_function_handle_t * handle, char * host_name, int port, char * func_name);```	Initializes function handle using the provided information.
```int grpc_call (grpc_function_handle_t *, ...);```	Performs a blocking RPC with the function handle and arguments. This call blocks until RPC completes.
```int grpc_call_async (grpc_function_handle_t *, ...);```	Performs a nonblocking(asynchronous) RPC with the function handle and arguments. Returns sessionID as a future reference to the session.
```int grpc_wait(int sessionID);```	Wait for completion of the session specified by the sessionID.
```int grpc_wait_any (int * idPtr);```	Wait for completion of one of the RPCs invoked by grpc_call_async beforehand.

serves as one of the reference implementations of the GridRPC API. Principal functions of GridRPC API are shown in Table 25.2.

25.3.4 Server side IDL

In order to 'gridify' a library, the Ninf library provider describes the interface of the library function using the Ninf IDL to publish his library function, which is only manifested and handled at the server side. The Ninf IDL supports datatypes mainly tailored for

```
Module sample;
Define mmul(IN   int    N,
            IN   double A[N*N],
            IN   double B[N*N],
            OUT  double C[N*N])
Required "mmul_lib.o"
Calls "C" mmul(N, A, B, C);
```

Figure 25.2   An example of Ninf IDL file.

serving numerical applications: for example, the basic datatypes are largely scalars and their multidimensional arrays. On the other hand, there are special provisions such as support for expressions involving input arguments to compute array sizes, designation of temporary array arguments that need to be allocated on the server side but not transferred, and so on.

This allows direct 'gridifying' of existing libraries that assumes array arguments to be passed by call-by-reference (thus requiring shared-memory support across nodes via the software), and supplementing the information lacking in the C and Fortran-type systems regarding array sizes, array stride usage, array sections, and so on.

As an example, interface description for the matrix multiply is shown in Figure 25.2, in which the access specifiers IN and OUT specify whether the argument is read or written within the Gridified library. Other IN arguments can specify array sizes, strides, and so on with size expressions. In this example, the value of N is referenced to calculate the size of the array arguments A, B, C. In addition to the interface definition of the library function, the IDL description contains the information needed to compile and link the necessary libraries. Ninf-G tools allow the IDL files to be compiled into stub main routines and makefiles, which automates compilation, linkage and registration of Gridified executables.

25.4 NINF-G IMPLEMENTATION

We now describe the implementation of the Ninf-G in more detail.

25.4.1 'Gridifying' a library or an application using GridRPC

Using Ninf-G, a user needs to merely take a few simple steps to make his application 'Gridified' in the following manner on the server side (we note again that no IDL handling is necessary on the client side, as opposed to traditional RPC systems such as CORBA):

1. Describe the interface of a library function or an application with Ninf IDL.
2. Process the Ninf IDL description file, generate stub main routine for the remote executable and a makefile as described above.
3. Link the stub main routine with the remote library, obtain the 'Gridified' executable.

4. Register the 'Gridified' executable into the MDS. (Steps 3 and 4 are automated by the makefile).

To register information into the MDS, the program that acts as the information provider outputs data complying with the LDAP Data Interchange Format (LDIF); moreover, the program itself is registered within the MDS setup file. Ninf-G places all such relevant information to be registered as a file under a specific directory (${GLOBUS_LOCATION}/var/gridrpc/) readily formatted as a valid LDIF file, and also provides and registers a filter program that performs appropriate filtering as described below. For this purpose, Ninf-G adds the lines as shown in Figure 25.3 into the 'grid-info-resource-ldif.conf' file.

25.4.1.1 Generating the LDIF file

The Ninf-G IDL compiler also generates interface information source files, which are utilized upon 'make' to automatically generate interface information files in XML as well as in LDIF formats. Both embody interface information (in XML), pathname of the remote executable, and the signature of the remote library.

Figure 25.4 shows a sample LDIF file generated from an IDL file in Figure 25.2. (Note that the XML-represented interface information is base64 encoded because of its length.) Here, '__ROOT_DN__' in the first line will be replaced with the proper root-distinguished name by the information provider filter program described above.

25.4.1.2 Registration into the MDS

In order to register the LDIF files generated in the current directory into the MDS, we merely copy them into the ${GLOBUS_LOCATION}/var/gridrpc/ directory. Note that this step is also automated by the makefile.

```
dn: Mds-Software-deployment=GridRPC-Ninf-G, Mds-Host-hn=brain-n.a02.aist.go.jp, \
Mds-Vo-name=local,o=grid
objectclass: GlobusTop
objectclass: GlobusActiveObject
objectclass: GlobusActiveSearch
type: exec
path: /usr/local/globus/var/gridrpc
base: catldif
args: -devclassobj -devobjs \
      -dn Mds-Host-hn=brain-n.a02.aist.go.jp, Mds-Vo-name=local,o=grid \
      -validto-secs 900 -keepto-secs  900
cachetime: 60
timelimit: 20
sizelimit: 10
```

Figure 25.3 Addition to the grid-info-resource-ldif.conf.

```
dn: GridRPC-Funcname=sample/mmul, Mds-Software-deployment=GridRPC-Ninf-G, __ROOT_DN__
objectClass: GlobusSoftware
objectClass: MdsSoftware
objectClass: GridRPCEntry
Mds-Software-deployment: GridRPC-Ninf-G
GridRPC-Funcname: sample/mmul
GridRPC-Module: sample
GridRPC-Entry: mmul
GridRPC-Path: /home/ninf/tests/sample/_stub_mmul
GridRPC-Stub:: PGZlbmN0aW9uICB2ZXJzaW9uPSIyMjEuMDAwMDAwMDAwIiA+PGZlbmN0aW9udX2
 PSJwZXZJmIiBlbnRyeT0icGluZ3Bvbmci IC8+IDxhcmcgZGF0YV90eXBlPSJpbnQiIGlvZGVf
 ··· (The rest is omitted)
```

Figure 25.4 LDIF file for matrix multiply.

25.4.2 Performing GridRPC

Now we are ready to make the actual Ninf-G GridRPC call that can be broken down into the steps as shown in Figure 25.1.

1. The client requests interface information and executable pathname from the MDS.
2. MDS sends back the requested information.
3. Client requests the GRAM gatekeeper to invoke the 'Gridified' remote executable, passing on the necessary information described below.
4. The remote executable connects back to the client using Globus-I/O for subsequent parameter transfer, and so on.

25.4.2.1 Retrieval of interface information and executable pathname

The client retrieves the interface information and executable pathname registered within the MDS using the library signature as a key. The retrieved information is cached in the client program to reduce the MDS retrieval overhead.

25.4.2.2 Invoking the remote executable

The client invokes the remote executable (done by Ninf-G via the Globus GRAM), specifying the remote executable path obtained from the MDS and a port address that accepts the callback from the remote executable. Here, the accepting port authenticates its peer using Globus-I/O, preventing malicious third-party attacks; this is because the party that owns the proper Globus proxy certificates derived from the client user certificate can connect to the port.

25.4.2.3 Remote executable callbacks to the client

The remote executable obtains the client address and the port from argument list and connects back to the client using Globus-I/O. Subsequent remote executable communication with the client will use this port.

25.5 USAGE SCENARIO

Here, we show a sample deployment of a distributed parallel application using Ninf-G. As a sample application, consider computing an approximation of the value of *pi* using a simple Monte Carlo method.[1] The Monte Carlo method generates a huge amount of random numbers, converts them into meaningful input parameters, performs some calculations using the parameters, and statistically processes the results of the calculation to arrive at a meaningful result. Because each computation will largely be independent as a result of independent processing on each randomly generated number, Monte-Carlo is known to be quite suitable for distributed computing.

For this example, we assume an environment shown in Figure 25.5. There are four Ninf-G server hosts named Host0, Host1, Host2, and Host3 that run Globus GRAM. A GIIS server runs on HostIS. All the GRIS servers on others hosts register themselves with the GIIS. The Client Program looks up the GIIS server as an MDS server, and performs GridRPC onto the GRAM server on each host.

25.5.1 Setup remote executable

To setup the servers, the Monte Carlo calculation is defined as a standard C function, and an IDL description of the function is also defined. Figures 25.6 and 25.7 show the function and its IDL description, respectively. The function receives a seed for random number generation and a number of Monte Carlo trials, and returns the count of how many points have fallen within the circle.

25.5.2 Client program

The client program performs GridRPCs onto the servers in parallel using the asynchronous invocation API. The core fragment of the client program is shown in Figure 25.8, demonstrating the relative ease of parallel application construction in Ninf-G.

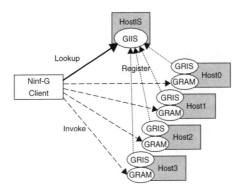

Figure 25.5   Usage scenario environment.

---

[1] This is one of the simplest Monte Carlo applications. Define a circle and a square that encompasses the circle. A randomly generated point in the square would fall inside the circle with probability '$\frac{\pi}{4}$'. In reality, problems are much more complex

```
long pi_trial(int seed, long time){
  long 1, counter = 0;
  srandom(seed);
  for (1 = 0; 1 < times; 1++){
    double x = (double)random() / RAND_MAX;
    double y = (double)random() / RAND_MAX;
    if (x * x + y * y < 1.0)
      counter++;
  }
  return counter;
}
```

Figure 25.6 Monte Carlo PI trials.

```
Module pi;

Define pi_trial (IN int seed, IN long times,
                 OUT long * count)
"monte carlo pi computation"
Required "pi_trial.o"
{
   long counter;
   counter = pi_trial(seed, times);
   *count = counter;
}
```

Figure 25.7 IDL for the PI trial function.

```
/* Initialize handles. */
for (i = 0; i < NUM_HOSTS; i++) {
  if (grpc_function_handle_init(&handles[i],
               hosts[i], port, "pi/pi_trial")
                              == GRPC_ERROR){
      grpc_perror("handle_init");
      exit(2);
  }
}
for (i = 0; i < NUM_HOSTS; i++){
  /* (Parallel non-blocking remote
     function invocation. */
  if ((ids[i] =
    grpc_call_async(&handles[i], i,
                    times, &count[i]))
                            == GRPC_ERROR){
    grpc_perror("pi_trial");
    exit(2);
  }
}
/* Synchronize on the result return */
if (grpc_wait_all() == GRPC_ERROR){
  grpc_perro("wait_all");
  exit(2);
}
```

Figure 25.8 Client PI program.

25.6 PRELIMINARY EVALUATION

To examine the basic performance of Ninf-G, we have preliminarily measured (1) the MDS lookup cost, (2) the GRAM invocation cost, and (3) the communication throughput, in both the WAN and the LAN environment.

25.6.1 Experimental setup

Our experiment was performed using two remote sites AIST and TITECH. AIST is located in Tsukuba, a city Northeast of Tokyo, while TITECH is located in the Southern residential area of Tokyo. The distance between AIST and TITECH is approximately 50 miles (80 km). Although most of the network paths between two sites go through fast National backbones, the state of the network upgrading was in flux and there still remained some 10 Mbps bottlenecks at the time of the measurement.

As the server machine, we deployed a machine at AIST, namely, SunFire 280 with 2 GB of memory and two UltraSPARC-III+ (900 MHz) CPUs. For clients, we deployed a Sun Workstation with UltraSPARC-III(Client A) and a Linux box at TITECH (Client B). Figure 25.9 shows the setup and FTP throughput and ping latency between the clients and the server. First, we used a simple ping-pong RPC program, which just sends bulk data and receives it back as it is. The communication time was measured at the server side.

25.6.2 Results and discussion

Table 25.3 shows the MDS query cost and the GRAM invocation cost. MDS lookup cost is categorized into expired and cached. MDS gathers information invoking 'information-providers', and caches the information for a specified period. When MDS finds that the requested data had already expired, it reinvokes the relevant information provider

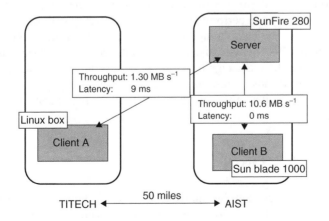

Figure 25.9 Experiment setup.

Table 25.3 MDS lookup cost and GRAM invocation cost

| | MDS lookup [s] | | GRAM |
	Expired	Cached	invocation [s]
Client A(LAN)	2.053	0.020	1.247
Client B(WAN)	1.851	0.046	1.192

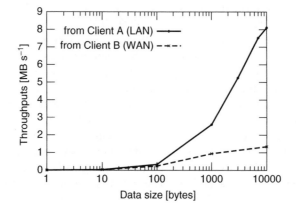

Figure 25.10 Ninf-G throughput.

to reobtain the information. As such the variance of MDS lookup cost is significant, depending on whether the data is expired or not – approximately 2 s if expired and 0.2 to 0.5 s if cached.

Here, we observe that both MDS and GRAM costs are independent of the network latency between the client and the server. The costs are not trivial, but we believe it is acceptable on the Grid since MDS query result is cached and reused, and remote executable once invoked can perform computation repeatedly without reinvocation.

Figure 25.10 shows the data throughput difference between the clients and the server in both the LAN and WAN environments. For small-sized data, throughputs are more or less equivalent, while there is a significant difference for large-sized data. This is largely due to the startup overheads as described above. Overall, throughput is similar to that of FTP for large-sized data, demonstrating that Globus-I/O is reasonably efficient for remote data transfer.

25.7 CONCLUSION

We described the concepts and the implementation of a Globus-based GridRPC system Ninf-G. Thanks to the Globus Toolkit, Ninf-G now provides a secure and interoperable GridRPC framework, compared to its predecessor. Modular design of the Globus Toolkit allowed us to conduct step-by-step deployment of the components, facilitating productive development, verifying the claim made by the Globus development team.

Preliminary evaluation shows that the initial overhead (involving mostly MDS lookup and GRAM invocation cost) is not trivial but acceptable, especially when the data size is large. We should, however, devise ways to reduce this cost by extensive caching of GridRPC states in the future versions of Ninf-G.

Ninf-G has just been released as of July 1, 2002, with both Globus 1.1.3 and 2.0 support. The Java client is under development and will be released this summer.

REFERENCES

1. Foster, I. and Kesselman, C. (1997) Globus: a metacomputing infrastructure toolkit. *International Journal of Supercomputer Applications*, 115–128.
2. Foster, I. and Karonis, N. (1998) A grid-enabled mpi: message passing in heterogeneous distributed computing systems. *Proc. 1998 SC Conference*, November, 1998.
3. Abramson, D., Foster, I., Giddy, J., Lewis, A., Sosic, R., Sutherst, R. and White, N. (1997) The Nimrod computational workbench: a case study in desktop metacomputing. *Proceedings of ACSC '97*, 1997.
4. Sato, M., Nakada, H., Sekiguchi, S., Matsuoka, S., Nagashima, U. and Takagi, H. (1997) Ninf: a network based information library for a global world-wide computing infrastructure. *Proc. of HPCN '97 (LNCS-1225)*, 1997, pp. 491–502.
5. Casanova, H. and Dongarra, J. (1996) NetSolve: a network server for solving computational science problems. *Proceedings of Super Computing '96*, 1996.
6. Seymour, K., Nakada, H., Matsuoka, S., Dongarra, J., Lee, C. and Casanova, H. (2002) GridRPC a remote procedure call API for Grid computing. *Grid2002*, 2002; submitted.

26

Commodity Grid kits – middleware for building Grid computing environments

Gregor von Laszewski,[1] Jarek Gawor,[1] Sriram Krishnan,[1,3] and Keith Jackson[2]

[1]*Argonne National Laboratory, Argonne, Illinois, United States,*
[2]*Lawrence Berkeley National Laboratory, Berkeley, California, United States,*
[3]*Indiana University, Bloomington, Indiana, United States*

26.1 INTRODUCTION

Over the past few years, various international groups have initiated research in the area of parallel and distributed computing in order to provide scientists with new programming methodologies that are required by state-of-the-art scientific application domains. These methodologies target collaborative, multidisciplinary, interactive, and large-scale applications that access a variety of high-end resources shared with others. This research has resulted in the creation of computational Grids.

The term *Grid* has been popularized during the past decade and denotes an integrated distributed computing infrastructure for advanced science and engineering applications.

Grid Computing – Making the Global Infrastructure a Reality. Edited by F. Berman, A. Hey and G. Fox
© 2003 John Wiley & Sons, Ltd   ISBN: 0-470-85319-0

The concept of the Grid is based on coordinated resource sharing and problem solving in dynamic multi-institutional virtual organizations [1]. In addition to providing access to a diverse set of remote resources located at different organizations, Grid computing is required to accommodate numerous computing paradigms, ranging from client-server to peer-to-peer computing. High-end applications using such computational Grids include data-, compute-, and network-intensive applications. Application examples range from nanomaterials [2], structural biology [3], and chemical engineering [4], to high-energy physics and astrophysics [5]. Many of these applications require the coordinated use of real-time large-scale instrument and experiment handling, distributed data sharing among hundreds or even thousands of scientists [6], petabyte distributed storage-facilities, and teraflops of compute power. Common to all these applications is a complex infrastructure that is difficult to manage [7]. Researchers therefore have been developing basic and advanced services, and portals for these services, to facilitate the realization of such complex environments and to hide the complexity of the underlying infrastructure. The Globus Project [8] provides a set of basic Grid services, including authentication and remote access to resources, and information services to discover and query such remote resources. However, these services may not be available to the end user at a level of abstraction provided by the commodity technologies that they use for their software development.

To overcome these difficulties, the Commodity Grid project is creating as a community effort what we call Commodity Grid Toolkits (CoG Kits) that define mappings and interfaces between Grid services and particular commodity frameworks. Technologies and frameworks of interest currently include Java [9, 10], Python [11], CORBA [12], Perl [13], and Web Services.

In the following sections, we elaborate on our motivation for the design of CoG Kits. First, we define what we understand by terms such as Grid Computing Environments (GCEs) and Portals. We then illustrate the creation of a GCE with the help of commodity technologies provided through the Java framework. Next, we outline differences from other CoG Kits and provide an overview of ongoing research in the Java CoG Kit Project, which is part of the Globus Project.

26.2 GRID COMPUTING ENVIRONMENTS AND PORTALS

GCEs [14] are aimed at providing scientists and other Grid users with an environment that accesses the Grid by using a coherent and interoperable set of frameworks that include Portals, Problem-Solving Environments, and Grid and Commodity Services. This goal is achieved by developing Grid and commodity standards, protocols, APIs, SDKs, and methodologies, while reusing existing ones.

We define the term *Grid Computing Environment* as follows:

An integrated set of tools that extend the user's computing environment in order to provide access to Grid Services.

GCEs include portals, shells, and collaborative and immersive environments running on the user's desktop on common operating systems such as Windows and Linux or on

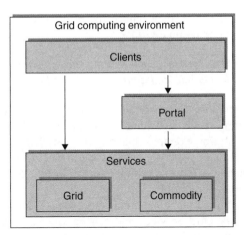

Figure 26.1   A Grid computing environment hides many of the complex interactions between the accessible services.

specialized devices ranging from Personal Digital Assistants (PDAs) to virtual reality environments such as stereographic devices or even CAVEs.

The architecture of a GCE can be represented as a multitier model. The components of this architecture are shown in Figure 26.1. Clients access the services through a portal or communicate with them directly. The user is oblivious of the fact that a service may engage other services on his or her behalf.

The term *Portal* is not defined uniformly within the computer science community. Sometimes it represents integrated desktops, electronic marketplaces, or information hubs [15, 16, 17] We use the term here in the more general sense of a community access point to information and services. Hence, we define the term as follows:

A community service with a single point of entry to an integrated system providing access to information, data, applications, and services.

In general, a portal is most useful when designed with a particular community in mind. Today, most *Web Portals* build on the current generation of Web-based commodity technologies, based on the HTTP protocol for accessing the information through a browser.

A Web Portal is a portal providing users ubiquitous access, with the help of Web-based commodity technologies, to information, data, applications, and services.

A *Grid portal* is a specialized portal useful for users of computational Grids. A Grid portal provides information about the status of the Grid resources and services. Commonly this information includes the status of batch queuing systems, load, and network performance between the resources. Furthermore, the Grid portal may provide a targeted access point to useful high-end services, such as a compute and data-intensive parameter study for climate change. Grid portals provide communities another advantage: they hide much of the complex logic to drive Grid-related services with simple interaction through the portal interface. Furthermore, they reduce the effort needed to deploy software for accessing resources on computational Grids.

A Grid Portal is a specialized portal providing an entry point to the Grid to access applications, services, information, and data available within a Grid.

In contrast to Web portals, Grid portals may not be restricted to simple browser technologies but may use specialized plug-ins or executables to handle the data visualization requirements of, for example, macromolecular displays or three-dimensional high-resolution weather data displays. These custom-designed visual components are frequently installed outside a browser, similar to the installation of MP3 players, PDF browsers, and videoconferencing tools.

Figure 26.2 presents a more elaborate architecture [18, 7] for representing a GCE that integrates many necessary Grid Services and can be viewed as a basis for many Grid portal activities. We emphasize that special attention must be placed on deployment and administrative services, which are almost always ignored in common portal activities [19]. As shown in the Figure 26.2, users are interested in services that deal with advanced job management to interface with existing batch queuing systems, to execute jobs in a fault-tolerant and reliable way, and to initiate workflows. Another useful service is reliable data management that transfers files between machines even if a user may not be logged in. Problem session management allows the users to initiate services, checkpoint them,

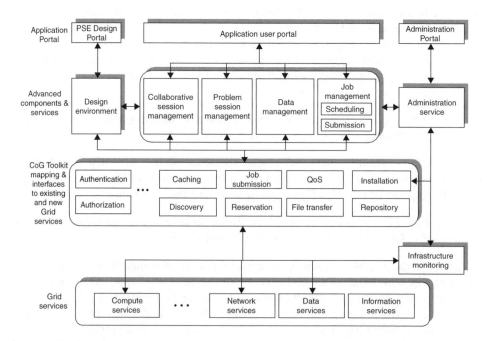

Figure 26.2 An example of a Grid computing environment that integrates basic and advanced Grid and commodity services.

and check on their status at a later time. All of these services are examples of the many possible services in a GCE and are based on the most elementary Grid services. The availability of commodity solutions for installation and rapid prototyping is of utmost importance for acceptance within the demanding user communities.

A Grid portal may deal with different user communities, such as developers, application scientists, administrators, and users. In each case, the portal must support a personal view that remembers the preferred interaction with the portal at the time of entry. To meet the needs of this diverse community, sophisticated Grid portals (currently under development) are providing commodity collaborative tools such as newsreaders, e-mail, chat, videoconferencing, and event scheduling. Additionally, some Grid portal developers are exploiting commodity technologies such as JavaBeans and Java Server Pages (JSP), which are already popular in Web portal environments.

Researchers interested in GCEs and Portals can participate in the GCE working group [14], which is part of the Global Grid Forum [20]. The origins of this working group can be traced back to the Desktop Access to Remote Resources organization that was later renamed to ComputingPortals.org and are spin-offs from the Java Grande Forum efforts [21].

26.3 COMMODITY TECHNOLOGIES

GCEs are usually developed by reusing a number of commodity technologies that are an integral part of the target environment. For example, a GCE implementing a Web Portal may require the use of protocols such as HTTPS and TCP/IP. It may make use of APIs such as CGI, SDKs such as JDK1.4, and commercial products such as Integrated Development Environments (IDEs) to simplify the development of such an environment. The Grid community has so far focused mostly on the development of protocols and development kits with the goal of defining a standard. This effort has made progress with the introduction of the Global Grid Forum and pioneering projects such as the Globus Project. So far the activities have mostly concentrated on the definition of middleware that is intended to be reused in the design of Grid applications. We believe that it is important to learn from these early experiences and to derive a middleware toolkit for the development of GCEs. This is where CoG Kits come into the picture.

CoG Kits play the important role of enabling access to the Grid functionality from within the commodity technology chosen to build a GCE. Because of the use of different commodity technologies as part of different application requirements, a variety of CoG Kits must be supported. In Table 26.1, we list a subset of commodity technologies that we have found useful to develop GCEs.

The availability of such CoG Kits is extremely helpful for the Grid application developers as they do not have to worry about the tedious details of interfacing the complex Grid services into the desired commodity technology. As good examples, we present the Java and the Python CoG Kits for the Globus Toolkit, known as Java CoG and pyGlobus, respectively. Both have been used in several GCE developments. However, it is important to recognize the different approaches the Java and the Python CoG Kit pursue.

While the Python CoG Kit interfaces with the Globus Toolkit on an API-based level, the Java CoG Kit interfaces with Globus services on a protocol level. The Python CoG

Table 26.1 A subset of commodity technologies used to develop Grid computing environments

	Languages	APIs	SDKs	Protocols	Hosting Environments	Methodologies
Web portals	Java, Perl, Python	CGI	JDK1.4	HTTPS, TCP/IP, SOAP	JVM, Linux, Windows	OO and procedural
Desktops	C, C + +, VisualBasic, C#	KParts, GTK	KDE, GNOME.NET	CORBA DCOM	Linux, Windows	OO and procedural
Immersive environments	C + +	CaveLib	Viz5D	TCP/IP	Linux	OO

Kit assumes the availability of precompiled Globus Toolkit libraries on the current hosting system, while the Java CoG Kit is implemented in pure Java and does not rely on the C-based Globus Toolkit. Both approaches provide a legitimate approach to achieve Globus Toolkit compliance. Each approach has advantages and disadvantages that are independent from the language chosen. Since the Python interface is generated by using the Simplified Wrapper and Interface Generator (SWIG) [22], it is far easier and faster to provide adaptations to a possibly changing toolkit such as the Globus Toolkit. Nevertheless, the price is that the Globus Toolkit libraries must be tightly integrated in the hosting environment in which the Python interpreter is executed. The first version of the Java CoG Kit was based on Java Native Interface (JNI) wrappers for the Globus Toolkit APIs. This approach, however, severely restricted the usage of the Java CoG Kit for developing pure Java clients and portals that are to be executed as part of browser applets. Hence, we implemented the protocols and some major functionality in pure Java in order to provide compliance with the Globus Toolkit. The availability of the functionality of the Globus Toolkit in another language has proved valuable in providing portability and assurance of code quality through protocol compliance.

Both the Python and Java CoG Kits provide additional value to Grids over and above a simple implementation of the Globus Toolkit APIs. The use of the commodity technologies such as object orientation, stream management, sophisticated exception, and event handling enhances the ability to provide the next generation of Grid services. Moreover, in many cases we find it inappropriate to develop such advanced services from scratch if other commodity technologies can be effectively used. A good example is the abstraction found in Java that hides access to databases or directories in general class libraries such as Java Database Connector (JDBC) and Java Naming and Directory Interface (JNDI); the absence of such abstractions in other languages might make it more complicated to implement the requisite functionality in such languages.

The availability of a variety of CoG Kits targeting different commodity technologies provides a great deal of flexibility in developing complicated services. We now focus on the Java CoG Kit as an example CoG Kit, and illustrate how it can be used to effectively build components that can be reused in the implementation of a GCE.

26.4 OVERVIEW OF THE JAVA COG KIT

Several factors make Java a good choice for GCEs. Java is a modern, object-oriented programming language that makes software engineering of large-scale distributed systems much easier. Thus, it is well suited as a basis for an interoperability framework and for exposing the Grid functionality at a higher level of abstraction than what is possible with the C Globus Toolkit. Numerous factors such as platform independence, a rich set of class libraries, and related frameworks make Grid programming easier. Such libraries and frameworks include JAAS [23], JINI [24], JXTA [25], JNDI [26], JSP [27], EJBs [28], and CORBA/IIOP [29]. We have depicted in Figure 26.3 a small subset of the Java technology that can be used to support various levels of the Grid architecture [1]. The Java CoG Kit builds a bridge between existing Grid technologies and the Java framework while enabling each to use the other's services to develop Grid services based on Java technology and to expose higher-level frameworks to the Grid community while providing interoperability [9]. The Java CoG Kit provides convenient access to the functionality of the Grid through client side and a limited set of server-side classes and components.

Furthermore, Java is well suited as a development framework for Web applications. Accessing technologies such as XML [30], XML schema [31], SOAP [32], and WSDL [33] will become increasingly important for the Grid community. We are currently investigating these and other technologies for Grid computing as part of the Commodity Grid projects to prototype a new generation of Grid services.

Because of these advantages, Java has received considerable attention by the Grid community in the area of application integration and portal development. For example,

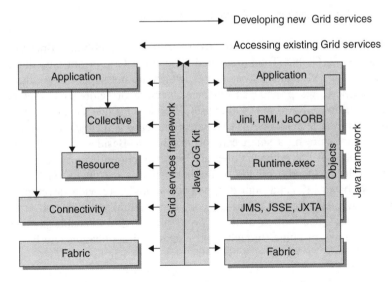

Figure 26.3 The Java CoG Kit allows users to access Grid services from the Java framework and enables application and Grid developers to use a higher level of abstraction for developing new Grid services and GCEs.

the EU DataGrid effort recently defined Java, in addition to C, as one of their target implementation languages. Additional motivation for choosing Java for Grid computing can be found in Reference [34].

The Java CoG Kit is general enough to be used in the design of a variety of advanced Grid applications with different user requirements. The Java CoG Kit integrates Java and Grid components and services within one toolkit, as a bag of services and components. In general, each developer chooses the components, services, and classes that ultimately support his or her development requirements. The goal of the Java CoG Kit is to enable Grid developers to use much of the Globus Toolkit functionality and to have access to the numerous additional libraries and frameworks developed by the Java community, allowing network, Internet, enterprise, and peer-to-peer computing. Since the Java CoG Kit strives to be only protocol compliant, it does not provide a simple one-to-one mapping between the C Globus Toolkit and Java CoG Kit API. Instead, it uses the more advanced features of Java, such as the sophisticated Java events and exception handling, rather than using the archaic C-based functions. It provides client-side access to the following Grid services:

- An information service compatible with the Globus Toolkit Metacomputing Directory Service (MDS) [35] implemented using JNDI.
- A security infrastructure compatible with the Globus Toolkit Grid Security Infrastructure (GSI) implemented with the IAIK security library [36].
- A data transfer compatible with a subset of the Globus Toolkit GridFTP [37] and/or GSIFTP [38].
- Resource management and job submission to the Globus Resource Access Manager (GRAM) [39].
- A certificate store based on the MyProxy server [40].

Additionally, the Java CoG Kit contains a set of command-line scripts that provide convenient access to Globus Toolkit-enabled production Grids from the client. This set includes support for MS Windows batch files, which are not supported by the C Globus Toolkit. Furthermore, we provide an enhanced version of 'globusrun' that allows the submission of multiple GRAM jobs. Other useful services include the ability to access Java smart card or iButton technology [41] to perform secure authentication with a possible multiple credential store on a smart card or an iButton. Besides these elementary Grid services and tools, several other features and services currently not provided by the C Globus Toolkit are included explicitly or implicitly within the Java CoG Kit.

The Java Webstart [42] and signed applet technologies provide developers with an advanced service to simplify code start-up, code distribution, and code update. Java Webstart allows the easy distribution of the code as part of downloadable jar files that are installed locally on a machine through a browser or an application interface. We have demonstrated the use of Webstart within the Java CoG Kit by installing sophisticated Graphical User Interface (GUI) applications on client machines. Component frameworks, such as JavaBeans, and the availability of commercial integrated development environments (IDEs) enable the Grid developer to use IDEs as part of rapid Grid prototyping while enabling code reuse in the attempt to reduce development costs.

Thus, our goal of developing collaborative scientific problem-solving environments and portals, based on the combined strength of the Java and the Grid technologies, is well substantiated by the Java CoG Kit. In the past, we had proposed portal architectures similar to the one depicted in Figure 26.2, in which the Java CoG Kit is used as an elementary middleware to integrate Grid services within portals and applications. We expect that advanced services will be integrated in future releases within the Java CoG Kit or as extension packages. Additionally, it is possible to implement several core Grid services, currently provided as part of the C Globus Toolkit, in pure Java while exposing the service through the Web Services Framework proposed recently by W3C. This possibility has been demonstrated for file transfer and for job execution. The availability of these services and protocol handlers in pure Java will make future portal development and the integration with the existing production Grid far easier. We have provided example programs using advanced GUI components in Java as part of the Java CoG Kit. These examples include a setup component for the Java CoG Kit, a form-based job submission component, a drag-and-drop-based submission component similar to a Windows desktop, an information service browser, and search queries. We hope that the community will contribute more components so that the usefulness of the Java CoG Kit will increase.

26.5 CURRENT WORK

Our current work is focused on the creation of an extended execution service and the integration of Web services in our CoG Kit efforts. Although these are currently prototyped in Java, it is easily possible to provide implementations in other languages like C and C++.

26.5.1 InfoGram

An important result from this prototyping has been the development of the 'InfoGram' service, which integrates a job submission service and an information service into a single service while reducing the development complexity. This InfoGram service has been described in more detail in Reference [43] outlining extensions to the Globus Resources Specification Language (RSL) [44] and the integration of checkpointing. Currently, we are also exploring the use of the InfoGram Service as part of 'Sporadic Grids', which are computational Grids dealing with sporadically available resources such as a computer at a beamline or a computer donated for a short period of time to a compute cluster. The InfoGram service can enable a SETI@home type of service, which can be used to integrate machines running on a cluster of MS Windows machines. Besides executing processes outside of the JVM, we have enhanced the security model for Grid computing while reusing Java's security model to, for example, restrict access to machine resources and prevent Trojan programs.

26.5.2 Web services

The Web services approach is quickly gaining popularity in the industry. Web services are designed to provide application integration via the use of standard mechanisms to

```
<implMap>
     <mapping>
          <source portName="CMCSPortType" operation="qEngine" />
          <target command="/bin/QEngine" />
     </mapping>
     <mapping>
          <source portName="CMCSPortType" operation="polyFit" />
          <target command="/bin/PolyFit" />
     </mapping>
</implMap>
```

Figure 26.4 XML mapping file for the command to Web services converter.

describe, publish, discover, invoke, and compose themselves. Moreover, Web services are platform- and implementation-independent. In other words, Web services written in a certain language can be accessed and invoked by clients written in other languages, executing under different environments. This capability is highly appealing to the scientific community, as it enables a high level of collaboration between various pieces of software written by different organizations in different languages. Despite all the advantages of the Web service technology, currently there are only limited Web service development environments, especially in languages other than Java. In such a scenario, it would be very convenient if there existed a tool that would be able to wrap an existing scientific application and expose it as a Web service. We are exploring the viability of this idea, using a prototypical implementation of a command to Web service converter. This converter is built by using Apache Axis [45] as the development environment. The converter takes as input the service description in the form of a WSDL document as well as an XML-encoded mapping between the operations exported in the WSDL and the target executables that they map to. The converter generates client- and server-side code for the target Web service using the standard Axis WSDL2Java converter, as well as the code for the actual implementation of the Web service using the XML-based mapping that has been provided.

An example of the mapping, which has been used as part of the CMCS project [4], is shown in the Figure 26.4. The qEngine operation maps to the executable '/bin/Qengine', while the polyFit operation maps to the executable '/bin/PolyFit'. The scientific codes can then be converted into Web services by automatic generation of wrapper code using the information defined in the XML format. These Web services can then be deployed, so that remote clients can have access to these codes over the network. We are currently analyzing patterns that would be appropriate for code generation. Such patterns have to be suitably captured in the XML mapfile and understood by the code generator so as to generate appropriate glue code.

26.6 ADVANCED COG KIT COMPONENTS

Now that we have illustrated the usefulness of CoG Kits, using the example of the Java CoG Kit, we demonstrate how we use it to provide clients with access to advanced services. As we have seen in Figure 26.2, we desire to implement services related to job, data, and workflow management. We have developed prototypes of advanced services and client interfaces that address these issues. Together these components can be used as part of a GCE. Other suggestions for components and services are listed in References [18] and [4].

26.6.1 Sample components

The first component models a desktop in which the user can create job specifications and machine representations through simple icons. Dragging a job onto a machine will automatically start the execution of this job on the remote machine. The user is able to monitor all jobs submitted to a particular machine by double-clicking on the machine icon. The associated output of the remote job can be downloaded by clicking on the appropriate file descriptor in the monitoring component. The specification of the icons and the associations to jobs and machines are represented in XML format. Figure 26.5 shows a screenshot of this component.

The second component is an interface to file transfers based on various protocols such as ftp [46], gsiftp [38], gridftp [37], and Reliable File Transfer (RFT) [47]. It is a drag-and-drop component allowing the user to conveniently use third-party file transfers between different Globus ftp servers by using either the gridftp or the gsiftp protocols. While using the RFT protocol, the user can also monitor the progress of RFTs that are executing in parallel. Figure 26.6 shows the snapshot for this component.

The third component is a workflow component that is currently used to define the workflow of an application in a graphical fashion, with the possibility to define dependencies between tasks as a hypergraph while using a graph data structure in recursive fashion. This feature allows the user to conveniently define large graphs hierarchically, thus increasing the readability. Such a tool could also be modified to create graph representations used by other projects such as Condor-G [48] and OGSA [49] while specifying dependencies between Grid Services. Therefore, the usefulness of such a component goes beyond the simple use as part of a dependency graph creation for simple job executions. Figure 26.7 shows how a workflow can be defined using this component.

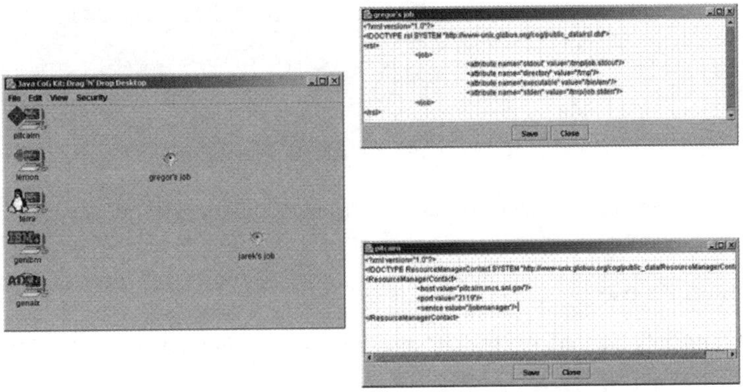

Figure 26.5   A prototypical GUI component performing job management for the GCE using the Java CoG Kit.

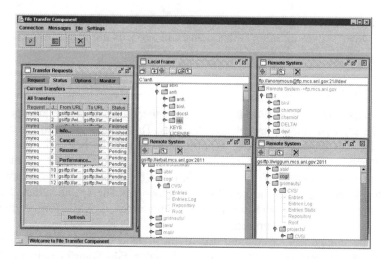

Figure 26.6 A prototypical GUI performing data management for the GCE using the Java CoG Kit.

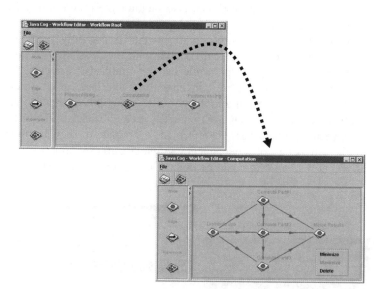

Figure 26.7 A prototypical component using the Java CoG Kit to perform workflow for the GCE.

26.6.2 Community use

The user community served by the Java CoG Kit is quite diverse. The Java CoG Kit allows

- *middleware developers* to create new middleware components that depend on the Java CoG Kit;

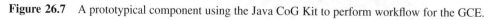

- *portal developers* to create portals that expose transparently the Grid functionality as part of a portal service; and
- *application developers* for the use of Grid services within the application portal.

A subset of projects currently using the Java CoG Kit for accessing Grid functionality includes the following:

- CoGBox [50] provides a simple GUI for much of the client-side functionality such as file transfer and job submission.
- CCAT [51] and XCAT [52] provide an implementation of a standard suggested by the Common Component Architecture Forum, defining a minimal set of standard features that a high-performance component framework has to provide, or can expect, in order to be able to use components developed within different frameworks.
- Grid Portal Development Kit (GPDK) [53] provides access to Grid services by using JSP and JavaBeans using Tomcat, a Web application server.
- JiPANG (Jini-based Portal AugmeNting Grids) [54] is a computing portal system that provides uniform access layer to a large variety of Grid services including other Problem-Solving Environments, libraries, and applications.
- The NASA IPG LaunchPad [55] uses the GPDK based on the Java CoG Kit. The tool consists of easy-to-use windows for users to input job information, such as the amount of memory and number of processors needed.
- The NCSA Science Portal [56] provides a personal Web server that the user runs on a workstation. This server has been extended in several ways to allow the user to access Grid resources from a Web browser or from desktop applications.
- The Astrophysics Simulation Code Portal (ASC Portal) [57] is building a computational collaboratory to bring the numerical treatment of the Einstein Theory of General Relativity to astrophysical problems.
- TENT [58] is a distributed simulation and integration system used, for example, for airplane design in commercial settings.
- ProActive [59] is a Java library for parallel, distributed, and concurrent computing and programming. The library is based on a reduced set of rather simple primitives and supports an active object model. It is based on the standard Java RMI library. The CoG Kit provides access to the Grid.
- DISCOVER [60] is developing a generic framework for interactive steering of scientific applications and collaborative visualization of data sets generated by such simulations. Access to the Grid will be enabled through the CORBA and Java CoG Kits.
- The Java CORBA CoG Kit [61] provides a simple Grid domain that can be accessed from CORBA clients. Future implementations in C++ are possible.
- The UNICORE [62] project as part of the Grid Interoperability Project (GRIP) [63] uses the Java CoG Kit to interface with Globus.

Additionally, work is currently performed as part of the Globus Project to provide a reference implementation of the Open Grid Services Architecture (OGSA) proposed through the Global Grid Forum. The current technology preview uses the Java CoG Kit's GSI security implementation and a modified version of the Java CoG Kit's GRAM

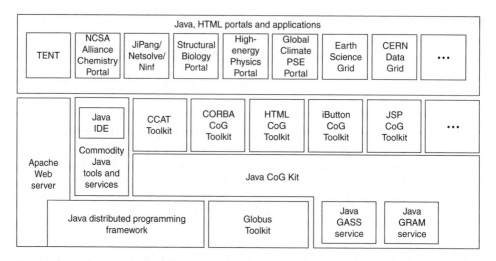

Figure 26.8 The Java CoG Kit builds a solid foundation for developing Grid applications based on the ability to combine Grid and Web technologies.

Table 26.2 Examples of community CoG Kits

Language	Name	Globus Compatibility	Web link
Perl	Perl CoG	API-based	gridport.npaci.edu/cog/
Python	pyGlobus	API-based	www-itg.lbl.gov/gtg/projects/pyGlobus/
Java	Java CoG Kit	Protocol-based	www.globus.org/cog
JSP	GPDK	Through Java CoG Kit	doesciencegrid.org/projects/GPDK/
CORBA	CORBA CoG	Through Java CoG Kit	www.caip.rutgers.edu/TASSL/Projects/CorbaCoG/

gatekeeper. The role of the Java CoG Kit for some of these projects is depicted in Figure 26.8.

A regularly updated list of such projects can be found at http://www.cogkits.org. We encourage the users to notify us of additional projects using CoG Kits, so that we can receive feedback about the requirements of the community. We also like to document the use and existence of other CoG Kits. In Table 26.2, we list a number of successfully used CoG Kits.

26.7 CONCLUSION

Commodity distributed computing technologies enable the rapid construction of sophisticated client-server applications. Grid technologies provide advanced network services for large-scale, wide area, multi-institutional environments and for applications that require the coordinated use of multiple resources. In the Commodity Grid project, we bridge these two worlds so as to enable advanced applications that can benefit from both Grid services and sophisticated commodity technologies and development environments. Various Commodity Grid projects are creating such a bridge for different commodity technologies. As part of the Java and Python Commodity Grid project, we provide an elementary set of classes that allow the Java and Python programmers to access basic Grid services, as well as enhanced services suitable for the definition of desktop problem-solving environments. Additionally, we provided the Globus Toolkit with an independent set of client tools that was able to increase the code quality of the C Globus Toolkit and the productivity of the end user.

Our future work will involve the integration of more advanced services into the Java CoG Kit and the creation of other CoG Kits and the integration of Web services technologies. We hope to gain a better understanding of where changes to commodity or Grid technologies can facilitate interoperability and how commodity technologies can be exploited in Grid environments. We believe that it is important to develop middleware for creating GCEs. We emphasize that a CoG Kit provides more than just an API to existing Grid services. Indeed, it brings the modalities and the unique strength of the appropriate commodity technology to the Grid as the Grid brings its unique strengths to the commodity users. This relationship is summarized in Figure 26.9, which shows the modified Grid architecture [64], which is introduced in Reference [7] with the explicit vertical support for a variety of commodity and high-end technologies into the Grid architecture.

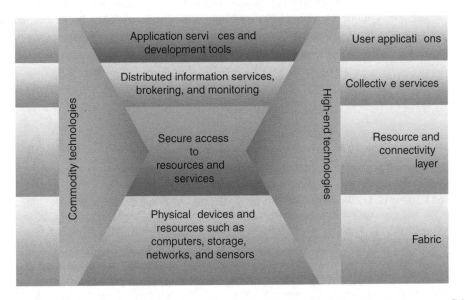

Figure 26.9 Commodity and high-end technologies bring enhanced value to the core Grid architecture.

26.8 AVAILABILITY

The Java CoG Kit closely monitors the development within the Globus Project to ensure that interoperability is maintained. The CoG Kit development team continues to keep track of projects that use the Java CoG Kit and documents the requirements of the community, in order to feed this information back to the Globus development team and to develop new features within the Java CoG Kit. For up-to-date release notes, readers should refer to the Web page at http://www.globus.org/cog, where the Java CoG kit is available for download. New releases are announced to the mailing list at cog-news@globus.org. Information about other CoG Kits such as Python, Perl, and CORBA can also be obtained from this Web page. We welcome contributions and feedback from the community.

ACKNOWLEDGEMENTS

This work is supported by the Mathematical, Information, and Computational Science Division subprogram of the Office of Advanced Scientific Computing Research, US Department of Energy, under Contract W-31-109-Eng-38. DARPA, DOE, and NSF supported Globus Toolkit research and development. We thank Ian Foster, Keith Jackson, Geoffrey C. Fox, Dennis Gannon, Shawn Hampton, Manish Parashar, Snigdha Verma, Mary Thomas, and Jay Alameda for the valuable discussions during the course of the ongoing CoG Kit development. We thank Nell Rehn, Peter Lane, Pawel Plaszczak, Mike Russell, Jason Novotny, Ravi Madduri, Benjamin Temko, Shamsuddin Ladha, Beulah Alunkal, and all others who have contributed to the Java CoG Kit. We thank Gail Pieper for her valuable comments. This work would not have been possible without the help of the Globus Project team. The Java CoG Kit is developed at Argonne National Laboratory as part of the Globus Project.

REFERENCES

1. Foster, I., Kesselman, C. and Tuecke, S. (2001) The anatomy of the grid: enabling scalable virtual organizations. *International Journal of Supercomputer Applications*, **15**(3), 200–222, http://www.globus.org/research/papers/anatomy.pdf.
2. Foster, I., Insleay, J., von Laszewski, G., Kesselman, C. and Thiebaux, M. (1999) Data visualization: data exploration on the grid. *IEEE Computer*, **14**, 36–41.
3. von Laszewski, G., Westbrook, M., Foster, I., Westbrook, E. and Barnes, C. (2000) Using computational grid capabilities to enhance the ability of an x-ray source for structural biology. *Cluster Computing*, **3**(3), 187–199, ftp://info.mcs.anl.gov/pub/tech_reports/P785.ps.Z.
4. von Laszewski, G. *et al.* (2002) A Grid Service Based Active Thermochemical Table Framework, In Preprint ANL/MCS-P972-0702.
5. Russell, M. *et al.* (2002) The astrophysics simulation collaboratory: a science portal enabling community software development. *Journal on Cluster Computing*, **5**(3), 297–304.
6. Allcock, W., Chervenak, A., Foster, I., Kesselman, C., Salisbury, C. and Tuecke, S. (2001) The data grid: towards an architecture for the distributed management and analysis of large scientific datasets. *Journal of Network and Computer Applications*, **23**, 187–200.

7. von Laszewski, G., Pieper, G. and Wagstrom, P. Gestalt of the grid, *Performance Evaluation and Characterization of Parallel and Distributed Computing Tools*. Wiley Book Series on Parallel and Distributed Computing, http://www.mcs.anl.gov/ gregor/bib to be published.
8. The Globus Project WWW Page, 2001, http://www.globus.org/.
9. von Laszewski, G., Foster, I., Gawor, J. and Lane, P. (2001) A java commodity grid kit. *Concurrency and Computation: Practice and Experience*, **13**(8,9), 643–662, http://www.globus.org/cog/documentation/papers/cog-cpe-final.pdf.
10. Novotny, J. (2001) The Grid Portal Development Kit, http://dast.nlanr.net/Features/GridPortal/.
11. Jackson, K. (2002) pyGlobus – A CoG Kit for Python, http://www-itg.lbl.gov/gtg/projects/pyGlobus/.
12. Verma, S., Parashar, M., Gawor, J. and von Laszewski, G. (2001) Design and implementation of a CORBA commodity grid kit, in Lee, C. A. (ed.) *Second International Workshop on Grid Computing – GRID 2001*. Number 2241 in Lecture Notes in Computer Science, Denver: Springer, pp. 2–12, In conjunction with SC'01, http://www.caip.rutgers.edu/TASSL/CorbaCoG/CORBACog.htm.
13. Thomas, M., Mock, S. and von Laszewski, G. A perl commodity grid kit. *Concurrency and Computation: Practice and Experience*, accepted, http://gridport.npaci.edu/cog/ and http://www.cogkits.org/papers/CPE_Perl_CoG_submitted.pdf.
14. Global Grid Forum Grid Computing Environments Working Group, www.computingportals.org.
15. Fox, G. C. (2000) Portals for Web Based Education and Computational Science.
16. Smarr, L. (2001) Infrastructures for Science Portals, http://www.computer.org/internet/v4n1/smarr.htm.
17. Fox G. C. and Furmanski, W. (1999) High performance commodity computing, in Foster I. and Kesselman, C. (eds) *The Grid: Blueprint for a New Computing Infrastructure*. San Francisco, CA: Morgan Kaufmann Publishers.
18. von Laszewski, G., Foster, I., Gawor, J., Lane, P., Rehn, N. and Russell, M. (2001) Designing grid-based problem solving environments and portals. *34th Hawaiian International Conference on System Science*, Hawaii, Maui, January, 2001, pp. 3–6, http://www.mcs.anl.gov/laszewsk/papers/cog-pse-final.pdf, http://computer.org/Proceedings/hicss/0981/volume.
19. von Laszewski, G., Blau, E., Bletzinger, M., Gawor, J., Lane, P., Martin, S. and Russell, M. (2002) Software, component, and service deployment in computational grids, in Bishop J., (ed.) *IFIP/ACM Working Conference on Component Deployment*, volume 2370 of Lecture Notes in Computer Science, Berlin, Germany: Springer, pp. 244–256, http://www.globus.org/cog.
20. Global Grid Forum, www.gridforum.org.
21. Java Grande Forum, www.javagrande.org.
22. Beazley, D. M. (1997) SWIG, a simple wrapper interface generator, http://www.swig.org/doc.html.
23. Sun Microsystems, Java Authentication and Authorization Service (JAAS), (2001), http://java.sun.com/products/jaas/.
24. Edwards, W. (2000) *Core JINI*. 2nd edn. NJ: Prentice Hall.
25. Gong, L. (2001) Get connected with Jxta, Sun MicroSystems, Java One, June 2001.
26. Lee, R. (2000) *Jndi api Tutorial and Reference: Building Directory-Enabled Java Applications*. Reading, MA: Addison-Wesley.
27. Hall, M. (2000) *Core Servlets and JavaServer Pages (JSP)*. 1st edn. NJ: Prentice Hall/Sun Microsystems Press.
28. Monson-Haefel, R. (2001) *Enterprise JavaBeans*. 3rd edn. Cambridge MA: O'Reilly.
29. Siegel, J. (2000) *Corba 3: Fundamentals and Programming*. 2nd edn. Indianapolis, IN: John Wiley & Sons.
30. Holzner, S. (2000) *Inside XML*. 1st edn. Indianapolis, IN: New Riders Publishing.
31. XML Schema, Primer 0–3, 2001, http://www.w3.org/XML/Schema.
32. Box, D., *et al.* (2000) Simple Object Access Protocol (SOAP) 1.1, http://www.w3.org/TR/SOAP.
33. Christensen, E., Curbera, F., Meredith, G. and Weerawarana. S. (2001) Web Services Description Language (WSDL) 1.1, http://www.w3.org/TR/wsdl.

34. Getov, V., von Laszewski, G., Philippsen, M. and Foster, I. (2001) Multi-paradigm communications in java for grid computing. *Communications of ACM*, **44**(10), 119–125, http://www.globus.org/cog/documentation/papers/.

35. von Laszewski, G., Fitzgerald, S., Foster, I., Kesselman, C., Smith, W. and Tuecke, S. (1997) A directory service for configuring high-performance distributed computations. *Proceedings of the 6th IEEE Symposium on High-Performance Distributed Computing*, August 5–8, 1997, pp. 365–375.

36. Grid Security Infrastructure, http://www.globus.org/security.

37. Grid FTP, 2002, http://www.globus.org/datagrid/gridftp.html.

38. GSI Enabled FTP, 2002, http://www.globus.org/datagrid/deliverables/gsiftp-tools.html.

39. Czajkowski, K., Foster, I. and Kesselman, C. (1999) Resource co-allocation in Computational Grids.

40. Novotny, J., Tuecke, S. and Welch, V. (2001) An online credential repository for the grid: Myproxy, *Proceedings of the Tenth International Symposium on High Performance Distributed Computing (HPDC-10)*. San Francisco: IEEE Press.

41. iButton Web Page, 2001, http://www.ibutton.com/.

42. Java Web Start, Version 1.0.1 Edition, 2001, http://www.sun.com/products/javawebstart/.

43. von Laszewski, G., Gawor, Jarek, Peña, C. J. and Foster, I. (2002) InfoGram: a peer-to-peer information and job submission service. *Proceedings of the 11th Symposium on High Performance Distributed Computing*, Scotland, Edinburgh, July, 2002, pp. 24–26, http://www.globus.org/cog.

44. Resource Specification Language, 2002, http://www.globus.org/gram/gram_rsl_parameters.html.

45. Apache Axis, 2002, http://xml.apache.org/axis/.

46. File Transfer Protocol, 2002, http://www.w3.org/Protocols/rfc959/Overview.html.

47. Reliable File Transfer Service, 2002, http://www-unix.mcs.anl.gov/madduri/RFT.html.

48. Frey, J., Tannenbaum, T., Foster, I., Livny, M. and Tuecke, S. (2001) Condor-G: a computation management agent for multi-institutional grids, *Proceedings of the Tenth IEEE Symposium on High Performance Distributed Computing (HPDC10)*, San Francisco: IEEE Press.

49. The Physiology of the Grid: An Open Grid Services Architecture for Distributed Systems Integration, http://www.globus.org/research/papers/ogsa.pdf.

50. Temko, B. The CoGBox Home Page, http://www.extreme.indiana.edu/btemko/cogbox/.

51. Indiana CCAT Home Page, 2001, http://www.extreme.indiana.edu/ccat/.

52. Govindaraju, M., Krishnan, S., Chiu, K., Slominski, A., Gannon, D. and Bramley, R. (2002) XCAT 2.0 : a component based programming model for grid web services. Submitted to Grid 2002, 3rd International Workshop on Grid Computing, 2002.

53. The Grid Portal Development Kit, 2000. http://dast.nlanr.net/Projects/GridPortal/.

54. Suzumura, T., Matsuoka, S. and Nakada, H. (2001) A Jini-based Computing Portal System, http://matsu-www.is.titech.ac.jp/suzumura/jipang/.

55. Launching into Grid Space with the NASA IPG LaunchPad, 2001, http://www.nas.nasa.gov/Main/Features/2001/Winter/launchpad.html.

56. Krishnan, S. *et al.* (2001) The XCAT science portal. *Proceedings of SC2001*, 2001.

57. The Astrophysics Simulation Collaboratory: A Laboratory for Large Scale Simulation of Relativistic Astrophysics, 2001, http://www.ascportal.org/.

58. German Air and Space Agency (DLR), TENT Home Page, 2001, http://www.sistec.dlr.de/tent/.

59. Caromel, D. ProActive Java Library for Parallel, Distributed and Concurrent Programming, 2001, http://www-sop.inria.fr/oasis/ProActive/.

60. Parashar M. *et al.* DISCOVER, 2001, http://www.discoverportal.org/.

61. CORBA CoG Kits, 2002, http://www.globus.org/cog/corba/index.html/.

62. UNICORE, 2001, http://www.unicore.de/.

63. Grid Interoperability Project, http://www.grid-interoperability.org/.

64. Foster, I. (2002) The grid: a new infrastructure for 21st century science. *Physics Today*, **55**(22), 42, http://www.aip.org/pt/vol-55/iss-2/p42.html.

Reprint from *Concurrency and Computation: Practice and Experience* © 2002 John Wiley & Sons, Ltd.
Minor changes to the original have been made to conform with house style.

27

The Grid portal development kit

Jason Novotny

Lawrence Berkeley National Laboratory, Berkeley, California, United States

27.1 INTRODUCTION

Computational Grids [1] have emerged as a distributed computing infrastructure for providing pervasive, ubiquitous access to a diverse set of resources ranging from high-performance computers (HPC), tertiary storage systems, large-scale visualization systems, expensive and unique instruments including telescopes and accelerators. One of the primary motivations for building Grids is to enable large-scale scientific research projects to better utilize distributed, heterogeneous resources to solve a particular problem or set of problems. However, Grid infrastructure only provides a common set of services and capabilities that are deployed across resources and it is the responsibility of the application scientist to devise methods and approaches for accessing Grid services.

Unfortunately, it still remains a daunting task for an application scientist to easily 'plug into' the computational Grid. While command line tools exist for performing atomic Grid operations, a truly usable interface requires the development of a customized problem solving environment (PSE). Traditionally, specialized PSE's were developed in the form of higher-level client side tools that encapsulate a variety of distributed Grid operations such as transferring data, executing simulations and post-processing or visualization of data across heterogeneous resources. A primary barrier in the widespread acceptance of monolithic client side tools is the deployment and configuration of specialized software.

Grid Computing – Making the Global Infrastructure a Reality. Edited by F. Berman, A. Hey and G. Fox
© 2003 John Wiley & Sons, Ltd   ISBN: 0-470-85319-0

Scientists and researchers are often required to download and install specialized software libraries and packages. Although client tools are capable of providing the most direct and specialized access to Grid enabled resources, we consider the web browser itself to be a widely available and generic problem solving environment when used in conjunction with a Grid portal. A Grid portal is defined to be a web based application server enhanced with the necessary software to communicate to Grid services and resources. A Grid portal provides application scientists a customized view of software and hardware resources from a web browser.

Furthermore, Grid Portals can be subdivided into application-specific and user-specific portal categories. An application specific portal provides a specialized subset of Grid operations within a specific application domain. Examples of application specific portals include the Astrophysics Simulation Collaboratory [2] and the Diesel Combustion Collaboratory [3]. User portals generally provide site specific services for a particular community or research center. The HotPage user portal [4], the Gateway project [5], and UNICORE [6] are all examples of user portals that allow researchers to seamlessly exploit Grid services via a browser-based view of a well defined set of Grid resources.

The Grid Portal Development Kit [7] seeks to provide generic user and application portal capabilities and was designed with the following criteria:

- The core of GPDK should reside in a set of generic, reusable, common components to access those Grid services that are supported by the Globus toolkit [8] including the Grid Security Infrastructure (GSI) [9]. As Globus [10] becomes a *de facto* standard for Grid middleware and gains support within the Global Grid Forum [11], the GPDK shall maintain Globus compatibility through the use of the Java Commodity Grid (CoG) kit [12]. An enumeration and description of the Grid services is provided in the next section.
- Provide a customizable user profile that contains user specific information such as past jobs submitted, resource and application information, and any other information that is of interest to a particular user. GPDK User profiles are intended to be extensible allowing for the easy creation of application portal specific profiles as well as serializable such that users' profiles are persistent even if the application server is shutdown or crashes.
- Provide a complete development environment for building customized application specific portals that can take advantage of the core set of GPDK Grid service components. The true usefulness of the Grid Portal Development Kit is in the rapid development and deployment of specialized application or user portals intended to provide a base set of Grid operations for a particular scientific community. The GPDK shall provide both an extensible library and a template portal that can be easily extended to provide specialized capabilities.
- The GPDK should leverage commodity and open source software technologies to the highest degree possible. Technologies such as Java beans and servlets and widespread protocols such as HTTP and LDAP provide interoperability with many existing internet applications and services. Software libraries used by the GPDK should be freely available and ideally provide open source implementations for both extensibility and for the widespread acceptance and adoption within the research community.

The following sections explain the design and architecture of the Grid Development Kit with an emphasis on implementation and the technologies used. The advanced portal development capabilities of the Grid Portal Development Kit and future directions will also be discussed.

27.2 OVERVIEW OF THE GRID PORTAL DEVELOPMENT KIT

The Grid Portal Development Kit is based on the standard 3-tier architecture adopted by most web application servers as shown in Figure 27.1. Tiers represent physical and administrative boundaries between the end user and the web application server. The client tier is represented as tier 1 and consists of the end-user's workstation running a web browser. The only requirements placed upon the client tier is a secure (SSL-capable) web browser that supports DHTML/Javascript for improved interactivity, and cookies to allow session data to be transferred between the client and the web application server.

The second tier is the web application server and is responsible for handling HTTP requests from the client browser. The application server is necessarily multi-threaded and must be able to support multiple and simultaneous connections from one or more client browsers. The Grid Portal Development Kit augments the application server with Grid enabling software and provides multi-user access to Grid resources. All other resources accessed by the portal including any databases used for storing user profiles, online credential repositories or additional resources forms the third tier, known as the back-end. Back-end resources are generally under separate administrative control from the web application server and subject to different policies and use conditions. The GPDK has been specially tailored to provide access to Grid resources as the back-end resources. It is generally assumed that Grid resources understand a subset of defined Grid and Internet protocols.

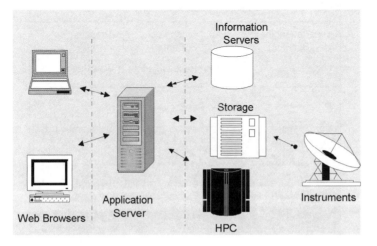

Figure 27.1   Standard 3-tier web architecture.

27.3 GRID PORTAL ARCHITECTURE

The Grid Portal Development Kit provides Grid enabling middleware for the middle-tier and aids in providing a Grid enabled application server. The GPDK is part of a complex vertical software stack as shown in Figure 27.2. At the top of the stack is a secure high-performance web server capable of handling multiple and simultaneous HTTPS requests. Beneath the web server is an application server that provides generic object invocation capabilities and offers support for session management. The deployed GPDK template portal creates a web application that is managed by the application server and provides the necessary components for accessing Grid services.

The Grid Portal Development Kit uses the Model-View-Controller (MVC) design pattern [13] to separate control and presentation from the application logic required for invoking Grid services. The GPDK is composed of three core components that map to the MVC paradigm. The Portal Engine (PE), provides the control and central organization of the GPDK portal in the form of a Java servlet that forwards control to the Action Page Objects (APO) and the View Pages (VP). The Action Page Objects form the 'model' and provide encapsulated objects for performing various portal operations. The View Pages are executed after the Action Page Objects and provide a user and application specific display (HTML) that is transmitted to the client's browser.

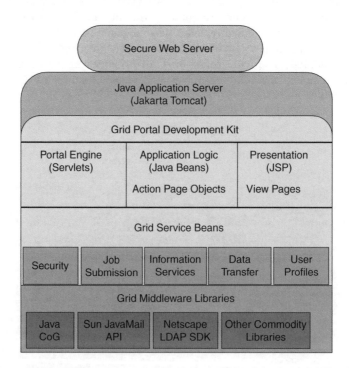

Figure 27.2 GPDK architecture and vertical stack of services and libraries.

The Grid service beans form the foundation of the GPDK and are used directly by the Portal Engine, Action Page Objects and View Pages. The Grid service beans are reusable Java components that use lower-level Grid enabling middleware libraries to access Grid services. Each Grid service bean encapsulates some aspect of Grid technology including security, data transfer, access to information services, and resource management. Commodity technologies are used at the lowest level to access Grid resources. The Java CoG Toolkit, as well as other commodity software APIs from Sun and Netscape, provide the necessary implementations of Grid services to communicate a subset of Grid protocols used by the GPDK service beans.

The modular and flexible design of the GPDK core services led to the adoption of a servlet container for handling more complex requests versus the traditional approach of invoking individual CGI scripts for performing portal operations. In brief, a servlet is a Java class that implements methods for handling HTTP protocol requests in the form of GET and POST. Based on the request, the GPDK servlet can be used as a controller to forward requests to either another servlet or a Java Server Page (JSP). Java Server Pages provides a scripting language using Java within an HTML page that allows for the instantiation of Java objects, also known as beans. The result is the dynamic display of data created by a Java Server Page that is compiled into HTML.

Figure 27.3 shows the sequence of events associated with performing a particular portal action. Upon start-up, the GPDK Servlet (GS) performs several key initialization steps including the instantiation of a Portal Engine (PE) used to initialize and destroy resources that are used during the operation of the portal. The PE performs general portal functions including logging, job monitoring, and the initialization of the portal informational database used for maintaining hardware and software information. The Portal Engine is also responsible for the authorizing users and managing users' credentials used to securely access Grid services. When a client sends an HTTP/HTTPS request to the application server, the GS is responsible for invoking an appropriate Action Page (AP) based on the 'action value' received as part of the HTTP header information. The Page Lookup Table is a plaintext configuration file that contains mappings of 'action values' to the appropriate Action Page Objects and View Pages. An AP is responsible for performing the logic of a particular portal operation and uses the GPDK service beans to execute the required operations. Finally, the GS forwards control to a View Page, a Java Server Page, after the AP is executed. The view page formats the results of an AP into a layout that is compiled dynamically into HTML and displayed in a client's browser.

27.4 GPDK IMPLEMENTATION

While a web server is unnecessary for the development of project specific portals using the GPDK, a secure web server is needed for the production deployment of a GPDK based portal. A production web server should offer maximum flexibility including the configuration of the number of supported clients, network optimization parameters, as well as support for 56 or 128-bit key based SSL authentication and support for a Java application server. The GPDK has been successfully deployed using the Apache [14] web

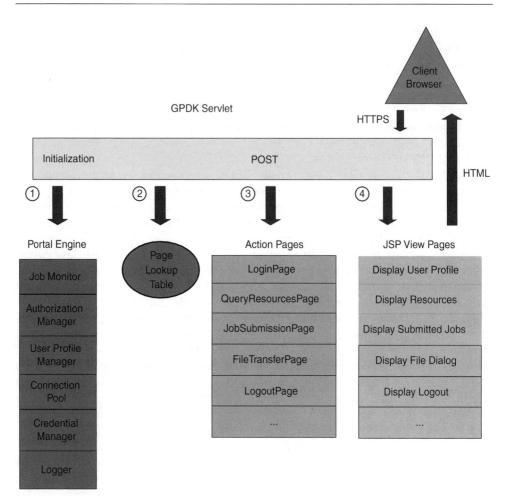

Figure 27.3 The GPDK event cycle.

server, a free, open source web server that provides a high-level of scalability and SSL support using the modSSL [15] package.

As mentioned previously, the GPDK relies on commodity Java technologies including Java beans, servlets and Java Server Pages for its general framework. The Grid Portal Development Kit was developed under the open source Tomcat application server available from the Jakarta Apache project [16]. The Tomcat [17] Application server was chosen as it is freely and widely available and implements the latest JSP and Servlet specifications from Sun and is included as part of the Java Enterprise Edition (J2EE) production application server.

The Java Commodity Grid (CoG) toolkit provides most of the functionality required to implement the various Grid services that have been encapsulated by the core GPDK service beans. The Java CoG Kit was developed to provide compliance with Globus Grid services and compatibility with the C reference implementation of Globus. The Java CoG toolkit provides the following Grid protocols and services that are used by the GPDK:

- The Grid security infrastructure (GSI) provides a secure communication protocol that uses X.509 certificates and SSL to perform mutual authentication. The Java CoG Toolkit implements GSI using the IAIK [18] security libraries. Although the IAIK libraries are proprietary and not open source, they remain free for research and academic use. Implementation of GSI using other Java security libraries such as Sun's Java Secure Sockets Extensions [19] (JSSE) is being investigated.
- The Globus resource and management (GRAM) [20] protocol is used to submit jobs to a Globus gatekeeper, a standard authentication and job spawning service provided by Globus enabled resources.
- The Grid FTP [21] protocol provides an optimized data transfer library and a specialized set of FTP commands for performing data channel authentication, third-party file transfers, and partial file transfers.
- The Myproxy [22] service provides an online credential repository for securely storing users' delegated credentials. The Java CoG provides a client API for communicating to a Myproxy certificate repository for retrieval of a user's security credential.

One of the powerful features of Java beans in the context of web application servers and the GPDK service beans is *bean scope*. Bean scope refers to the level of persistence offered by a bean within the servlet container. For instance, beans may have session, application, or request scope. Session scope implies that the bean persists for the duration of a user's session, typically determined by the servlet container. For instance, user profiles are represented as session beans and persist until a user decides to log out of the portal or their session times out as determined by the servlet container. Session scoped beans rely on the use of cookies used by most web browsers to retain state information on a particular client connection overcoming the inherent lack of state in the HTTP protocol. A bean with application scope persists for the complete duration of the servlet container and provides a persistent object used to store application specific static data that can be referenced by any Java Server Page on behalf of any user. The addition of collaborative capabilities such as a whiteboard or chat room, for instance, requires that messages be maintained with application scope, so logged in clients can see others' messages. A Bean with request scope persists only for the duration of a client HTTP request and is destroyed after the JSP page is processed into an HTML response for the client.

The GPDK has been developed and tested under Windows, Linux and Solaris platforms using the various JDK's provided by Sun in conjunction with the Apache web server available on both Windows and Unix platforms.

27.5 GPDK SERVICES

The usefulness of the GPDK as a portal development framework rests on the currently supported set of common Grid operations that are required for a typical scientific collaboration. A scientific collaboration may involve one or more of the following capabilities:

- Submission, cancellation and monitoring of specialized programs (serial and/or parallel) to a variety of compute resources including those requiring batch (non-interactive) submission.
- The ability to store and retrieve data accumulated either from experiment or simulation to a variety of storage resources.
- Use resource discovery mechanisms to enable the discovery of hardware and software resources that are available to a particular scientific collaboration.
- The ability to perform the above operations securely by allowing scientists to authenticate to remote resources as required by the remote site administrators.
- Application specific profile information including user preferences, submitted jobs, files transferred and other information that scientists may wish to archive along with results obtained from computational simulations or laboratory experiments.

Within the GPDK framework, the above requirements have been encapsulated into one or more GPDK service beans. As discussed in the following sections, the GPDK service beans are organized according to Grid services in the areas of security, job submission, file transfer and information services. The deployed GPDK demo portal highlights the capabilities of the supported Grid services through template web and JSP pages.

27.5.1 Security

The security working group of Grid Forum has been actively promoting the Grid Security Infrastructure (GSI) [23] as the current best practice for securely accessing Grid services. GSI is based upon public key infrastructure (PKI) and requires users' to possess a private key and an X.509 certificate used to authenticate to Grid resources and services. A key feature of GSI is the ability to perform delegation, the creation of a temporary private key and certificate pair known as a proxy that is used to authenticate to Grid resources on a users behalf. The GSI has been implemented over the Secure Sockets Layer (SSL) and is incorporated in the Globus and Java CoG toolkit. One of the key difficulties in developing a portal to access Grid services is providing a mechanism for users to delegate their credentials to the portal since current web browsers and servers do not support the concept of delegation. Past solutions have involved the storage of users' long-lived keys and certificates on the portal. A user would then provide their long-term pass phrase to the portal, which creates a valid proxy that can be used on the user's behalf. The danger in this approach, however, is the risk of the web server being broken into and having possibly many users' long term private keys compromised. For this reason, the Myproxy service [22] was developed to provide an online certificate repository where users can delegate temporary credentials that can be retrieved securely by the user from the portal. Briefly, a user delegates a credential to the Myproxy server with a chosen lifetime and user name and pass phrase. The user would enter the same user name and pass phrase from their browser over an HTTPS connection and the portal would retrieve a newly delegated credential valid for a chosen amount of time. Currently, GPDK doesn't enforce any maximum lifetime for the credential delegated to the portal, but when a user logs off, the proxy is destroyed reducing any potential security risk of their delegated credential being compromised on the portal. The portal retrieves credentials from the Myproxy Server

Figure 27.4 GPDK demo pages clockwise from top left (a) login page, (b) user profile page, (c) resources page, and (d) file transfer page.

using the GPDK security component, the MyproxyBean. The MyproxyBean component is actually a wrapper around the CoG toolkit client API to the Myproxy server. For users that have their delegated credential local to their workstation, the GPDK template portal allows them to upload the proxy to the portal directly using standard file upload mechanisms over HTTPS. In addition, the JMyproxy package [24] provides a Java GUI that can create a proxy locally and delegate a credential to a Myproxy server. The initial login page that displays the Myproxy interface to the demo portal is shown in Figure 27.4(a). In the current implementation, all users that can either supply a delegated credential or retrieve one from the Myproxy server are authorized portal users. However, if the portal administrator wished to further restrict access, an encrypted password file on the portal or a secure back-end database could also be used to determine authorization information.

27.5.2 Job submission

The GPDK Job Submission beans provide two different secure mechanisms for executing programs on remote resources. A GSI enhanced version of the Secure Shell (SSH) [9] software enables interactive jobs to be submitted to Grid resources supporting the GSI enabled SSH daemon. For all other job submissions, including batch job submissions, the Globus GRAM protocol is used and jobs are submitted to Globus gatekeepers deployed on Grid resources. Briefly, the GRAM protocol enables resource submission to a variety of resource scheduling systems using the Resource Specification Language (RSL) [20], allowing various execution parameters to be specified, for example, number of processors, arguments, wall clock or CPU time.

The primary GPDK components used to submit jobs are the JobBean, the JobSubmissionBean and the JobInfoBean. The JobBean provides a description of the job to be submitted, and encapsulates RSL by including methods for setting and returning values for the executable, additional arguments passed to the executable, number of processors for parallel jobs, batch queue if submitting a batch mode and more. The JobSubmissionBean is actually an abstract class that is sub-classed by the GramSubmissionBean in the case of submitting a job to a Globus gatekeeper or a GSISSHSubmissionBean using the GSI enhanced SSH client [9]. The GramSubmissionBean capabilities are provided once again by the Java CoG library.

Once a job has been successfully submitted, a JobInfoBean is created containing a time stamp of when the job was submitted and other useful information about the job, including a GRAM URL that can be used to query on the status of the job.

Job monitoring of submitted jobs is provided through the JobMonitorBean, a component initialized at start-up by the Portal Engine. The JobMonitorBean periodically queries the GRAM URL's on behalf of a user to keep track of job status based on the GRAM job status codes, for example, active, running, or failed. Because The JobMonitorBean has application scope, it can save job status to a user's profile even if the user has logged out.

27.5.3 File transfer

The GPDK file transfer beans encapsulate the GridFTP [21] API implemented as part of the CoG toolkit and provide file transfer capabilities, including third-party file transfer between GSI enabled FTP servers, as well as file browsing capabilities. The FileTransferBean provides a generic file transfer API that is extended by the GSIFTPTransferBean and the GSISCPTransferBean, an encapsulation of file transfer via the GSI enhanced scp command tool. The GSIFTPServiceBean provides a session scoped bean that manages multiple FTP connections to GSI enabled FTP servers. The GSIFTPServiceBean allows users to browse multiple GSI FTP servers simultaneously and a separate thread monitors server time-outs. The GSIFTPViewBean is an example view bean used by a JSP to display the results of browsing a remote GSI FTP server. Figure 27.4 shows the demo file browsing and transferring page.

27.5.4 Information services

The Grid Forum Information Services working group has proposed the Grid Information Services (GIS) architecture for deploying information services on the Grid and supported the Lightweight Directory Access Protocol (LDAP) as the communication protocol used to query information services. Information services on the Grid are useful for obtaining both static and dynamic information on software and hardware resources. The Globus toolkit provides an implementation of a Grid Information Service, known as the Metacomputing Directory Service using OpenLDAP, an open source LDAP server. Although, the Java CoG toolkit provides support for LDAP using the Java Naming and Directory Interface

(JNDI), GPDK uses the open source Netscape/Mozilla Directory SDK [25] as it proved easier to use in practice and also provides support for developing a connection pool for maintaining multiple connections to several Grid Information service providers, thus eliminating the need for clients to re-connect during each query. However, this model will need to be re-evaluated with the widespread deployment of the MDS-2 architecture that includes GSI enhancements making it necessary for clients to re-authenticate to the MDS for each query. GPDK provides an MDSQueryBean and MDSResultsBean for querying and formatting results obtained from the MDS. Currently GDK supports querying the MDS for hardware information such as CPU type, number of processors and other details as well as CPU load and queue information that can be used by the user to make more effective job scheduling decisions.

27.5.5 GPDK user profiles

User profiles are an integral part of the portal architecture as they enable the customization of a particular set of resources (hardware and software) by portal users. In addition, user profiles create a 'value-added' feature of using the portal to perform common Grid operations since a user history is maintained in a user profile allowing users to keep track of past jobs submitted and results obtained. A user profile is a persistent session bean that is either created the first time a user logs into the portal or is loaded (de-serialized) from a file or database if one exists already. User profiles are serialized when a user logs out of the portal or the profile has been modified. Application portals built on top of the GPDK can subclass the GPDK user profile bean to provide additional application specific information for a particular class of users.

27.6 GPDK AS A PORTAL DEVELOPMENT ENVIRONMENT

Computational science collaborations can benefit significantly by providing web access to software, data and expertise. The Grid Portal Development Kit is designed to be a development environment that enables rapid development of application specific portals by leveraging off core GPDK service beans and the MVC architectural model incorporated by the GPDK. The core set of services remain generic enough to be useful to any community interested in staging data, executing codes, and providing a customizable, commonly accessible, collaborative environment.

The GPDK is packaged with a demo portal to showcase the functionality of the service beans and provide several common portal operations. During installation, the Grid Portal Development Kit deploys the library of core service beans as well as a central servlet and a collection of fully functional demo template web pages to the Java application server. The template web pages include HTML and Java Server Pages intended to demonstrate the GPDK service beans. The template source code contains the central servlet class used by the application specific portal as well as sub-classes for project specific user profiles and a project specific Portal Engine (PE) used to initialize and shutdown any resources required by the new project portal. The goal of the template portal is not to provide a fully

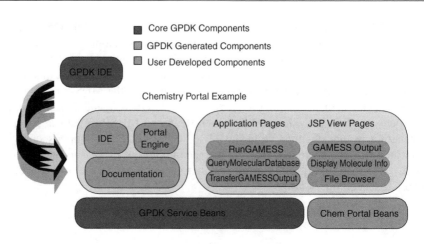

Figure 27.5   Creating a new portal.

polished interface ready for production deployment, but rather a demonstration of GPDK capabilities that can be easily extended and specialized for providing an application group with a common subset of Grid capabilities. Figure 27.5 shows the generalized GPDK architecture organized into core GPDK components, GPDK generated components, and application specific user developed components.

The GPDK template source code also contains all the Action Page Objects necessary for performing the various portal operations that demonstrate core GPDK services and beans. The APs are composed of a LoginPage, UpdateProfilePage and LogoutPage that demonstrates the retrieval of credentials from the Myproxy server and the loading, editing and saving of user profiles. The JobSubmissionPage and FileTransferPage demonstrate the GPDK service beans in the areas of job submission and file transfer. The template Java Server Pages are displayed upon the successful completion of the Page object operations. If an error occurs during the execution of an AP, an exception is thrown with a relevant error message that is displayed to the client by a specialized error JSP.

The Grid Portal Development Kit relies on ANT [26] for cross-platform project compilation and deployment. ANT provides an XML based Java build tool that is analogous to the Makefile process and provides support for common operations such as invoking the Java compiler, the Java archiver tool used for creating libraries, the Javadoc tool used for creating API documentation from commented source code, as well as file copying, and directory creation/deletion operations.

The following steps are followed to create a new application specific portal called **ChemPortal** using the GPDK:

- Compile core GPDK service beans and portal objects: When GPDK is compiled for the first time using the ANT build script, for example, build all, all of the necessary classes are compiled and placed into the GPDK Java Archive Repository (JAR) file.
- Invoke the ANT build tool to create a new portal with the desired name, for example, build new ChemPortal: By specifying a new project target, for example, ChemPortal,

when invoking the GPDK build script, a new portal sub project, ChemPortal, is created containing the pre-processed template portal pages. The template portal pages are composed of functional Java source code, template action page classes and view pages in JSP in addition to Javadoc generated API documentation for the generated source code. File pre-processing capabilities that allow the GPDK to generate a complete template portal project is made possible by the substitution capabilities provided by the ANT build tool. Ant pre-processes the template file names as well as the contents of the file to provide project specific Java code. For example, the template GPDK Servlet (GS) is pre-processed to create a ChemPortalServlet. The Portal Engine (PE) template is pre-processed to create a ChemPortal object that uses the core GPDK Portal object for initialization and shutdown routines. Similarly, the GPDK User Profile object is sub-classed to create a ChemPortal User Profile object that can be edited to contain data pertinent to the users of the ChemPortal.

The specified project namespace also provides a packaging namespace for portal source code. Core GPDK objects and beans are packaged under the org.gpdk namespace as per the Sun source code packaging convention and the pre-processed template project code is packaged under a new project namespace, for example, org.chemportal. This logical separation of core GPDK service beans and project specific classes allows for the creation and deployment of multiple portal projects to the same application server that can make use of the standard set of GPDK service beans.

- Modify and create new Action Pages (AP) and JSP view pages for project specific portal operations: As new portal operations are needed, ChemPortal developers would create a new Action Page object that implements the Page interface. The Page interface defines only one method, execute, that must be filled in by the developer. A correspond view page written as a JSP would need to be developed to return a display of the operation results to the client. As an example, imagine a new capability that allows ChemPortal users to run a GAMESS [27] simulation with a particular set of parameters. (GAMESS is a scientific code used for performing *ab-initio* chemistry simulations). The ChemPortal developer would create a RunGAMESSPage class that would use the GPDK Job Submission beans to initiate a GAMESS simulation with the appropriate run parameters. Furthermore, the GPDK file transfer beans could be used to move any generated output files to a storage server for permanent archiving. Finally, the developer would create a GAMESSOutput JSP that would be used to provide information on a completed run to the end user.

- Deploy new portal project to application server: Simply invoking the ANT build script, for example, build.pl all will package all of the web application files, HTML, java classes and JSP in the form of a web application repository (WAR) file and deployed to the Tomcat application server, where it can be accessed by a web browser.

The steps presented above allow an application developer to create a fully functional portal complete with template code in a matter of minutes. The build environment provides easy to use compilation and deployment capabilities, eliminating the need for additional Makefiles or project specific build scripts. In fact, most of the work done in developing a new portal is in the action pages and JSPs that are project specific.

27.6.1 Application portals based on GPDK

The Grid Portal Development Kit has proven successful in the creation of application specific portals under development by other research groups. The following list briefly describes various ongoing portal projects that have been developed using the Grid Portal Development Kit framework: The NASA Launchpad [28] user portal seeks to provide web based access to users of the NASA Information Power Grid (IPG) [29]. Launchpad is based entirely on GPDK and takes advantage of the GPDK service beans to allow IPG users access to high-performance computer resources and IPG Grid Information Services. The NASA Nebula portal provides a Web-based interface to the Nebula simulation code. Scientists use the Nebula code to study the atomic composition of interstellar clouds. The Nebula portal allows researchers to create an appropriate list of input parameters to the Nebula simulation and submit their job to NASA IPG resources. The Astrophysics Simulation Collaboratory Portal (ASC) [30] is an application portal developed to provide astrophysics researchers web based access to the Cactus computational toolkit. Cactus is a framework for solving various wave equations with an emphasis on astrophysical simulations of colliding neutron stars and black holes. The ASC portal allows users to remotely checkout and compile Cactus with specialized options as well as submit various Cactus simulations to a distributed set of HPC resources. The NCSA portal development effort is focused on the development of several application portals for computational chemistry and particle physics simulations. GridGaussian, a computational chemistry portal allows users to provide input files to the popular Gaussian chemistry package and easily submit simulations via the web. Similarly, the MIMD Lattice Computation (MILC) portal is aimed at providing particle physicists access to a popular community code used to understand elementary particle interactions. Both portals rely on GPDK for job submission and file staging capabilities.

27.7 RELATED WORK

While the GPDK demo portal is very similar to many other portal projects, the ability to develop new portals using the GPDK framework offers a new capability to the scientific community and is one of the first of its kind. As a Software Development Kit (SDK), the GPDK is most similar to the GridPort [31] Toolkit developed by the San Diego Supercomputer Center (SDSC) to facilitate the development of application specific portals. The GridPort toolkit is implemented in Perl and makes use of the existing HotPage [4] technology for providing access to Grid services. GridPort supports many of the same Grid services as GPDK including the Myproxy service, GRAM, and GIS.

Application portal development using GridPort is enabled in two ways: The first approach requires that Globus software tools be installed because the GridPort scripts wrap the C Globus command line tools in the form of Perl CGI scripts. The second method of developing a portal using GridPort does not require Globus, but relies on the CGI scripts that have been configured to use a primary GridPort portal as a proxy for access to GridPort Services such as authentication, job submission, file transfer, etc. In the second approach, an application scientist can quickly deploy a web server configured with a set of GridPort CGI scripts to perform very generic portal operations. However, the ease of deployment comes at a cost of portal customizability. Because the HTTP response from the proxy GridPort server contains the HTML to be displayed, the graphical interface displayed to portal users is the same as the base GridPort server. In addition, both logic and design are encapsulated in the GridPort Perl CGI scripts making code maintainability and customization more difficult over the approach taken by GPDK which separates the graphical interface provided by template JSP pages and the logic contained within the GPDK beans.

27.8 CONCLUSION AND FUTURE WORK

The architecture of the GPDK is highly modular and provides easy re-use of Grid components and a useful framework for the development and deployment of application specific portals. Based on the Model View Controller design paradigm, the GPDK has proven to be an extensible and flexible software package in use by several other portal building efforts. The GPDK makes use of commodity technologies including the open-source servlet container Tomcat and the Apache web server. The Grid Portal Development Kit makes use of the Java Commodity Grid (CoG) toolkit for its pure Java implementation of client side Globus Grid services as well as other widely available, commodity Java libraries. Future work on the Grid Portal Development Kit involves development in the following areas:

- Improve and enhance the functionality of GPDK service beans to support emerging Grid services and develop new portal capabilities. The GPDK information service beans will be enhanced to support the improved and secure MDS-2 infrastructure as it becomes widely deployed. Other future developments include the integration of a secure database to store user profiles rather than maintaining them on the portal. Additional application information may also be stored in the database or in the information servers, for access by the GPDK information service beans. Task composition from a portal has become increasingly important as portal operations become more complex. Portal users would like to specify Grid or portal operations as tasks and be able to combine tasks together to create a work flow system for an entire calculation involving staging data, running a simulation and migrating output data to a storage system. Emerging Grid monitoring and event technologies [32] will provide useful information on network and computer utilization and performance allowing portal users and scheduling software alike to make better informed resource scheduling decisions. As these technologies become widely deployed, components for accessing these services will be developed in the GPDK.
- Investigate and prototype new distributed computing technologies including web services. Web services have become increasingly important in the enterprise community and many new standards and implementations are emerging from Sun, Microsoft and IBM. The Web Services Definition Language (WSDL) [33] permits services to be defined by a standard interface and registered and discovered using the Universal Description, Discovery and Integration (UDDI) [34] specification. The Simple Object

Access Protocol (SOAP) [35] provides a standard for communicating structured information using XML. GPDK client web services beans will be developed that are capable of exchanging SOAP messages with Grid enabled web services as they become widespread.

ACKNOWLEDGEMENTS

We are grateful to many colleagues for discussions on portal development and working with early incarnations of GPDK to improve on its usefulness and robustness. In particular, we wish to thank Jarek Gawor, Gregor Laszewski, Nell Rehn, George Myers, Mark Wallace, Farah Hasnat, Yinsyi Hung, John Lehman, Jeffrey Becker, Michael Russell, Shawn Hampton, Scott Koranda and John White.

REFERENCES

1. Foster, I. and Kesselman, C. (eds) (1998) *The Grid: Blueprint for a New Computing Infrastructure*. San Francisco, CA: Morgan Kaufmann.
2. The Astrophysics Simulation Collaboratory, http://www.ascportal.org, November 22, 2001.
3. Pancerella, C. M., Rahn, L. A. and Yang, C. L. (1999) The diesel combustion collaboratory: combustion researchers collaborating over the internet. *Proc. of IEEE Conference on High Performance Computing and Networking*, November, 1999.
4. Boisseau, J., Mock, S. and Thomas, M. (2000) Development of Web toolkits for computational science portals: the NPACI HotPage. *Proc. of the 9th IEEE Intl. Symp. on High Perf. Dist. Comp*, 2000.
5. Akarsu, E. Fox, G., Haupt, T. and Youn, C. (2000) The gateway system: uniform web based access to remote resources. *IEEE Concurrency: Practice and Experience*, July.
6. Romberg, M. (1999) The UNICORE architecture. *Proc. of the 8th IEEE Intl. Symp. on High Perf. Dist. Comp.*, 1999.
7. The GPDK Project Page, http://www.itg-lbl.gov/Grid/projects/GPDK/index.html, November 22, 2001.
8. Globus Web Site, http://www.globus.org, November 22, 2001.
9. GSI Software Information, http://www.globus.org/security, November 22, 2001.
10. Foster, I. and Kesselman, C. (1997) Globus: a metacomputing infrastructure toolkit. *International Journal of Supercomputing Applications*.
11. Grid Forum Web Site, http://www.gridforum.org, November 22, 2001.
12. Laszewski, G., Foster, I. and Gawor, J. (2000) CoG kits: a bridge between commodity distributed computing and high-performance grids. *Proceedings of the ACM Java Grande Conference*, 2000.
13. Gamma, E., Helm, R., Johnson, R. and Vlissides, J. (1995) *Design Patterns: Elements of Reusable Object Oriented Software*. Reading, MA: Addison-Wesley.
14. Apache Webserver Project, http://www.apache.org, November 22, 2001.
15. ModSSL, http://www.modssl.org, November 22, 2001.
16. Jakarta Apache Project, http://jakarta.apache.org, November 22, 2001.
17. Tomcat Open-Source Servlet Container, http://jakarta.apache.org/tomcat, November 22, 2001.
18. IAIK Security Libraries, http://jcewww.iaik.at/, November 22, 2001.
19. Sun Java Secure Sockets Extension, http://java.sun.com/products/jsse, November 22, 2001.
20. Czajkowski, K., Foster, I., Karonis, N., Kesselman, C., Martin, S., Smith, W. and Tuecke, S. (1998) A resource management architecture for metacomputing systems. *Proc. IPPS/SPDP '98 Workshop on Job Scheduling Strategies for Parallel Processing*, 1998.

21. Allcock, W., Bester, J., Bresnahan, J., Chervenak, A., Liming, L. and Tuecke, S. (2001) GridFTP: Protocol Extensions to FTP for the Grid. *Grid Forum Working Draft*, March, http://www.gridforum.org.
22. Novotny, J., Tuecke, S. and Welch, V. (2001) An online credential repository for the grid: MyProxy. *Proc. 10th IEEE Symp. On High Performance Distributed Computing*, 2001.
23. Foster, I., Karonis, N., Kesselman, C., Koenig, G. and Tuecke, S. (1997) A secure communications infrastructure for high-performance distributed computing. *Proc. 6th IEEE Symp. on High Performance Distributed Computing*, 1997.
24. The JMyproxy Client, ftp://ftp.george.lbl.gov/pub/globus/jmyproxy.tar.gz, November 22, 2001.
25. Netscape Directory and LDAP Developer Central, http://developer.netscape.com/tech/directory/index.html, November 22, 2001.
26. The Jakarta ANT Project, http://jakarta.apache.org/ant/index.html, November 22, 2001.
27. The General Atomic and Molecular Electronic Structure System, http://www.msg.ameslab.gov/GAMESS/GAMESS.html, November 22, 2001.
28. The NASA Launchpad User Portal, http://www.ipg.nasa.gov/launchpad, November 22, 2001.
29. Johnston, W. E., Gannon, D. and Nitzberg, B. (1999) Grids as production computing environments: the engineering aspects of NASA's information power grid. *Proc. 8th IEEE Symp. on High Performance Distributed Computing*, 1999.
30. Allen, G. *et al.* (2001) The astrophysics simulation collaboratory portal: a science portal enabling community software development. *Proc. of the 10th IEEE Intl. Symp. on High Perf. Dist. Comp.*, 2001.
31. Boisseau, J., Dahan, M., Mock, S., Mueller, K., Sutton, D. and Thomas, M. (2001) The gridport toolkit: a system for building grid portals. *Proc. of the 10th IEEE Intl. Symp. on High Perf. Dist. Comp.*, 2001.
32. Grid Monitoring Architecture, http://www-didc.lbl.gov/GGF-PERF/GMA-WG/, November 22, 2001.
33. Web Services Definition Language, http://www.w3.org/TR/wsdl, November 22, 2001.
34. Universal Description, Discovery and Integration, http://www.uddi.org, November 22, 2001.
35. Simple Object Access Protocol, http://www.w3.org/TR/SOAP/, November 22, 2001.
36. Czajkowski, K., Fitzgerald, S., Foster, I. and Kesselman, C. (2001) Grid information services for distributed resource sharing. *Proc. 10th IEEE Symp. on High Performance Distributed Computing*, 2001.
37. Allen, G., Russell, M., Novotny, J., Daues, G. and Shalf, J. (2001) The astrophysics simulation collaboratory: a science portal enabling community software development. *Proc. 10th IEEE Symp. on High Performance Distributed Computing*, 2001.
38. Java Servlet 2.3 and Java Server Pages 1.2 Specifications, http://java.sun.com/products/servlets, November 22, 2001.

Building Grid computing portals: the NPACI Grid portal toolkit

Mary P. Thomas and John R. Boisseau

The University of Texas at Austin, Austin, Texas, United States

28.1 INTRODUCTION

In this chapter, we discuss the development, architecture, and functionality of the National Partnership for Advanced Computational Infrastructure NPACI Grid Portals project. The emphasis of this paper is on the NPACI Grid Portal Toolkit (GridPort); we also discuss several Grid portals built using GridPort including the NPACI HotPage. We discuss the lessons learned in developing this toolkit and the portals built from it, and finally we present our current and planned development activities for enhancing Grid-Port and thereby the capabilities, flexibility, and ease-of-development of portals built using GridPort.

28.1.1 What are Grid computing portals?

Web-based Grid computing portals, or *Grid portals* [1], have been established as effective tools for providing users of computational Grids with simple, intuitive interfaces for accessing Grid information and for using Grid resources [2]. Grid portals are now being

Grid Computing – Making the Global Infrastructure a Reality. Edited by F. Berman, A. Hey and G. Fox
© 2003 John Wiley & Sons, Ltd ISBN: 0-470-85319-0

developed, deployed, and used on large Grids including the National Science Foundation (NSF) Partnership for Advanced Computational Infrastructure (PACI) TeraGrid, the
NASA Information Power Grid, and the National Institute of Health (NIH) Biomedical
Informatics Research Network. Grid *middleware* such as the Globus Toolkit provides
powerful capabilities for integrating a wide variety of computing and storage resources,
instruments, and sensors, but Grid middleware packages generally have complex user
interfaces (UIs) and Application Programming Interfaces (APIs). Grid portals make these
distributed, heterogeneous compute and data Grid environments more accessible to users
and scientists by utilizing common Web and UI conventions. Grid portals, and other
Web-based portals, provide developers and users with the capabilities to customize the
content and presentation (e.g. page layout, level of detail) for the set of tools and services
presented. Grid portals can enable automated execution of specific applications, provide
explicit links to discipline-specific data collections, integrate (and hide) data workflow
between applications, and automate the creation of collections of application output files.
Portals can also provide a window to the underlying execution environment, reporting
the availability of resources, the status of executing jobs, and the current load on the
Grid resources.

The software used to build Grid portals must interact with the middleware running on
Grid resources, and in some cases it must provide missing functionality when the Grid
middleware is not available for a specific resource or it is lacking capabilities needed by
the Grid portal. The portal software must also be compatible with common Web servers
and browsers/clients. Several generalized Grid portal toolkits have emerged that help
simplify the portal developer's task of utilizing the complex Grid technologies used for
Grid services and making them available via a familiar Web interface. With the advent
of Web services, interoperable protocols and standards are now being developed for Grid
information and other Grid services. Web services will further simplify the use of Grids
and Grid technologies and will encourage the use and deployment of more general *Grid
applications* on the Grid. As Grid portal toolkits and the underlying Grid technologies
mature and as Web services standards become more common, Grid portals will become
easier to develop, deploy, maintain, and use.

Software for creating Grid portals must integrate a wide variety of other software and
hardware systems. Thus, portals represent an integration environment. This is part of the
unique role that portals play at the Grid middleware layer: portals drive the integration of
'lower' middleware packages and enforce the integration of other toolkits. Furthermore,
projects such as the GridPort [3], the Grid Portal Development Toolkit [4], and the Common Component Architecture project [5] have demonstrated that integrated toolkits can
be developed that meet the generalized needs of Grid applications as well as Web-based
Grid portals.

28.1.2 History and motivation

The NPACI, led by the San Diego Supercomputer Center (SDSC), was initiated in 1997
by the NSF PACI program [6]. NPACI is charged with developing, deploying, and supporting an advanced computational *infrastructure* – hardware, software, and support – to
enable the next generation of computational science. NPACI resources include diverse

high performance computing (HPC) architectures and storage systems at SDSC and at university partners. US academic researchers may apply for accounts on multiple resources at multiple sites, so NPACI must enable users to utilize this distributed collection of resources effectively.

The NPACI HotPage was developed to help facilitate and support usage of the NPACI resources. The HotPage initially served as an *informational* portal for HPC users, especially those with accounts on multiple NPACI systems [7, 8]. The World Wide Web had recently become established as a popular method for making information available over the Internet, so the HotPage was developed as a Web portal to provide information about the HPC and the archival storage resources operated by the NPACI partner sites (SDSC, the University of Texas at Austin, the University of Michigan, Caltech, and UC-Berkeley). As an informational service, the HotPage provided users with centralized access to technical documentation for each system. However, the HotPage also presented *dynamic* informational data for each system, including current operational status, loads, and status of queued jobs. The integration of this information into a single portal presented NPACI users with data to make decisions on where to submit their jobs. However, the goals of the HotPage included not only the provision of information but also the capability to use all NPACI resources interactively via a single, integrated Web portal. Grid computing technologies, which were supported as part of the NPACI program, were utilized to provide these functionalities. In 1999, a second version of the HotPage was developed that used the Globus Toolkit [9]. Globus capabilities such as the Grid Security Infrastructure (GSI) and the Globus Resource Allocation Manager (GRAM) enabled the HotPage to provide users with real time, secure access to NPACI resources. HotPage capabilities were added to allow users to manage their files and data and to submit and delete jobs. This version of the HotPage has been further enhanced and is in production for NPACI and for the entire PACI program. Versions of the HotPage are in use at many other universities and government laboratories around the world.

The mission of the HotPage project has always been to provide a Web portal that would present an integrated appearance and set of services to NPACI users: a *user portal*. This has been accomplished using many custom scripts and more recently by using Grid technologies such as Globus. The HotPage is still relatively 'low level', however, in that it enables NPACI users to manipulate files in each of their system accounts directly and to launch jobs on specific systems. It was apparent during the development of the HotPage that there was growing interest in developing higher-level *application portals* that launched specific applications on predetermined resources. These application portals trade low-level control of jobs, files, and data for an even simpler UI, making it possible for non-HPC users to take advantage of HPC systems as 'virtual laboratories'. Much of the functionality required to build these higher-level application portals had already been developed for the HotPage. Therefore, the subset of software developed for Hot-Page account management and resource usage functions was abstracted and generalized into GridPort. GridPort was then enhanced to support multiple application portals on a single Grid with a single-login environment. The usefulness of this system has been successfully demonstrated with the implementation and development of several *production* application portals.

The driving philosophy behind the design of the HotPage and GridPort is the conviction that many potential Grid users and developers will benefit from portals and portal technologies that provide universal, easy access to resource information and usage while requiring minimal work by Grid developers. Users of GridPort-based portals are not required to perform any software downloads or configuration changes; they can use the Grid resources and services via common Web browsers. Developers of Grid portals can avoid many of the complexities of the APIs of Grid middleware by using GridPort and similar toolkits. Design decisions were thus guided by the desire to provide a *generalized* infrastructure that is accessible to and useable by the computational science community. If every Grid portal developer or user were required to install Web technologies, portal software, and Grid middleware in order to build and use portals, there would be a tremendous duplication of effort and unnecessary complexity in the resulting network of connected systems.

GridPort attempts to address these issues by meeting several key design goals:

- *Universal access*: enables Web-based portals that can run anywhere and any time, that do not require software downloads, plug-ins, or helper applications, and that work with 'old' Web browsers that do not support recent technologies (e.g. client-side XML).
- *Dependable information services*: provide portals, and therefore users, with centralized access to comprehensive, accurate information about Grid resources.
- *Common Grid technologies and standards*: minimize impact on already burdened resource administrators by not requiring a proprietary GridPort daemon on HPC resources.
- *Scalable and flexible infrastructure*: facilitates adding and removing application portals, Grid software systems, compute and archival resources, services, jobs, users, and so on.
- *Security*: uses GSI, support HTTPS/SSL (Secure Sockets Layer) encryption at all layers, provide access control, and clean all secure data off the system as soon as possible.
- *Single login*: requires only a single login for easy access to and navigation between Grid resources.
- *Technology transfer*: develops a toolkit that portal developers can easily download, install, and use to build portals.
- *Global Grid Forum standards*: adhere to accepted standards, conventions, and best practices.
- *Support for distributed client applications and portal services*: enables scientists to build their own application portals and use existing portals for common infrastructure services.

Adhering to these design goals and using the lessons learned from building several production Grid portals resulted in a Grid portal toolkit that is generalized and scalable. The GridPort project has met all of the goals listed above with the exception of the last one. A Web services–based architecture, in which clients host Grid application portals on local systems and access distributed Grid Web services, will address the last design goal. As these Grid portal toolkits continue to evolve, they will enable developers, and even users, to construct more general Grid applications that use the Grid resources and services.

28.1.3 Grid portal users and developers

The Grid is just beginning the transition to deployment in production environments, so there are relatively few Grid users at this time. As Grids move into production, users will require much assistance in trying to develop applications that utilize them.

In the NPACI HPC/science environment, there are three general classes of potential Grid users. First, there are end users who only run prepackaged applications, most commonly launched in Unix shell windows on the HPC resources. Adapting these applications to the Grid by adding a simple Web interface to supply application-specific parameters and execution configuration data is straightforward. Therefore, this group will be easiest for transition to using Grids instead of individual resources, but most of the work falls on the Grid portal developers. These users generally know little about the HPC systems they currently use – just enough to load input data sets, start the codes, and collect output data files. This group includes users of community models and applications (e.g. GAMESS, NASTRAN, etc.). Many of the users in this group may never know (or care) anything about HPC or how parallel computers are running their code to accomplish their science. For them, the application code is a virtual laboratory. Additionally, there exists a large constituency that is absent from the HPC world because they find even this modest level of HPC knowledge to be intimidating and/or too time-consuming. However, with an intuitive, capable application portal, these researchers would not be exposed to the HPC systems or Grid services in order to run their application. Effective Grid portals can provide both novice and potential HPC users with simple, effective mechanisms to utilize HPC systems transparently to achieve their research goals.

The second group consists of researchers who are more experienced HPC users and who often have accounts on multiple HPC systems. Most current NPACI users fall into this category. For them, building HPC applications is challenging though tractable, but as scientists they prefer conducting simulations with production applications to developing new code. While this group will accept the challenges inherent in building parallel computing applications in order to solve their scientific problems, they are similar to the first group: their first interest is in solving scientific problems. For this group, a user portal like the HotPage is ideal: it provides information on all the individual systems on which they have accounts. It allows users to learn how to use each system, observe which systems are available, and make an assessment of which system is the best for running their next simulations. While they already have experience of using Unix commands and the commands native to each HPC system on which they have ported their codes, the HotPage allows them to submit jobs, archive and retrieve data, and manage files on *any* of these from a *single* location. For these users, a user portal like the HotPage cannot replace their use of the command line environment of each individual system during periods of code development and tuning, but it can augment their usage of the systems for production runs in support of their research.

The third group of HPC users in our science environment includes the researchers who are computational experts and invest heavily in evaluating and utilizing the latest computing technologies. This group is often at least as focused on computing technologies as on the applications science. This group has programmers who are 'early adopters', and so in some cases have already begun investigating Grid technologies. Users in this group may benefit from a user portal, but they are more likely to build a Grid application using their base HPC application and integrating Grid components and services directly. In addition, there are also Grid developers who develop portals, libraries, and applications. For these Grid users and for Grid developers, a Grid application toolkit is ideal: something like GridPort, but enhanced to provide greater flexibility for applications in general, not just portals.

28.2 THE GRID PORTAL TOOLKIT (GRIDPORT)

GridPort has been the portal software toolkit used for the PACI and NPACI HotPage user portals and for various application portals since 1999. It was developed to support NPACI scientific computing objectives by providing centralized services such as secure access to distributed Grid resources, account management, large file transfers, and job management via Web-based portals that operate on the NSF computational Grid.

Implementation of these services requires the integration and deployment of a large and diverse number of Web and Grid software programs and services, each with a different client and server software and APIs. Furthermore, the Web and the Grid are continually evolving, making this task not only challenging but also requiring constant adaptation to new standards and tools. GridPort evolved out of the need to simplify the integration of these services and technologies for portal developers. As additional application portals for the NPACI user community were constructed, the need for an architecture that would provide a single, uniform API to these technologies and an environment that would support multiple application portals emerged. GridPort was designed to provide a common shared instance of a toolkit and its associated services and data (such as user account information, session data, and other information), and to act as a mediator between client requests and Grid services.

28.2.1 GridPort architecture

The GridPort design is based on a multilayered architecture. On top there exists a client layer (e.g. Web browsers) and beneath it is a portal layer (the actual portals that format the content for the client). On the bottom is a backend services layer that connects to distributed resources via Grid technologies such as Globus. GridPort is a portal services layer that mediates between the backend and the portal layers (see Figure 28.1). GridPort modularizes each of the steps required to translate the portal requests into Grid service function calls, for example, a GRAM submission. In the context of the architecture of Foster *et al.* [10] that describes the 'Grid anatomy', GridPort represents an API that supports applications (portals) and interfaces to the Collective and Resources layers. In a sense, a producer/consumer model exists between each layer. Each layer represents a logical part of the system in which data and service requests flow back and forth and addresses some specific aspect or function of the GridPort portal system. For example, GridPort consumes Grid services from a variety of providers (e.g. via the Globus/GRAM Gatekeeper); as a service provider, GridPort has multiple clients such as the HotPage

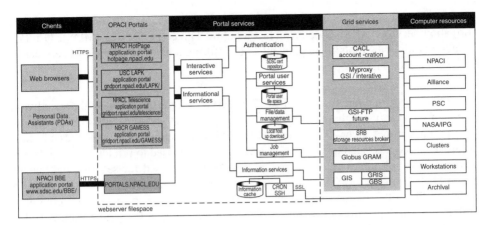

Figure 28.1   The diagram shows the layers used in the GridPort multilayered architecture. Each layer represents a logical part of the portal in which data and service requests flow back and forth. On top is the client layer and beneath it is a portal layer. On the bottom is the backend services layer that connects to distributed resources via Grid technologies. GridPort is the *portal services* layer that mediates between the backend and portal layers. GridPort modularizes each of the steps required to translate the portal requests into Grid service function calls, for example, a GRAM submission.

and other application portals that use the same instance of GridPort to submit a job. We describe each of the layers and their functions below.

Client layer: The client layer represents the consumers of Grid computing portals, typically Web browsers, Personal Digital Assistants (PDAs), or even applications capable of pulling data from a Web server. Typically, clients interact with the portal via HTML form elements and use HTTPS to submit requests. Owing to limitations in client-level security solutions, application portals running at different institutions other than the GridPort instance are not currently supported. This limitation can now be addressed, however, owing to the advent of Web services that are capable of proxy delegation and forwarding. The issue is discussed in further detail in Section 28.5.

Portals layer: The portals layer consists of the portal-specific code itself. Application portals run on standard Web servers and process the client requests and the responses to those requests. One instance of GridPort can support multiple concurrent application portals, but they must exist on the same Web server system in which they share the same instance of the GridPort libraries. This allows the application portals to share portal-related user and account data and thereby makes possible a single-login environment. These portals can also share libraries, file space, and other services. Application portals based on GridPort are discussed in more detail in Section 28.3.

Portal services layer: GridPort and other portal toolkits or libraries reside at the portal services layer. GridPort performs common services for application portals including management of session state, portal accounts, and file collections and monitoring of Grid information services (GIS) Globus Metacomputing Directory Service (MDS). GridPort provides portals with tools to implement both informational and interactive services as described in Section 28.1.

Grid services (technologies) layer: The Grid services layer consists of those software components and services that are needed to handle requests being submitted by software to the portal services layer. Wherever possible, GridPort employs simple, reusable middleware technologies, such as Globus/GRAM Gatekeeper, used to run interactive jobs and tasks on remote resources [9]; Globus GSI and MyProxy, used for security and authentication [11]; Globus GridFTP and the SDSC Storage Resource Broker (SRB), used for distributed file collection and management [12]; and GIS based primarily on proprietary GridPort information provider scripts and the Globus MDS 2.1 – Grid Resource Information System (GRIS). Resources running any of the above can be added to the set of Grid resources supported by GridPort by incorporating the data about the system into GridPort's configuration files.

Resources layer: GridPort-hosted portals can be connected to any system defined in the local configuration files, but interactive capabilities are only provided for GSI-enabled systems. For example, on the PACI HotPage [13], the following computational systems are supported: multiple IBM SPs and SGI Origins, a Compaq ES-45 cluster, an IBM Regatta cluster, a Sun HPC10000, a Cray T3E, a Cray SV1, an HP V2500, an Intel Linux Cluster, and a Condor Flock. Additionally, via the GSI infrastructure it is possible to access file archival systems running software compatible with the Public Key Infrastructure/Grid Security Infrastructure (PKI/GSI) certificate system, such as GridFTP or SRB. GridPort supports resources located across organizational and geographical locations, such as NASA/IPG, NPACI, and the Alliance.

GridPort is available for download from the project Website [8]. The latest version of GridPort has been rewritten as a set of Perl packages to improve modularity and is based on a model similar to the Globus Commodity Grid project that also provides CoG Kits in Java, Python, and CORBA [14, 15].

28.2.2 GridPort capabilities

GridPort provides the following capabilities for portals.

Portal accounts: All portal users must have a portal account and a valid PKI/GSI certificate. (Note that these accounts are not the same as the individual accounts a user needs on the resources.) The portal manages the user's account and keeps track of sessions, user preferences, and portal file space.

Authentication: Users may authenticate GridPort portals using either of two mechanisms: by authenticating against certificate data stored in the GridPort repository or by using a MyProxy server. GridPort portals can accept certificates from several sites; for example,

NPACI GridPort portals accept certificates from the Alliance, NASA/IPG, Cactus, and Globus as well as NPACI. Once a user is logged in to a GridPort portal and has been authenticated, the user has access to any other GridPort-hosted portal that is part of the single-login environment. Thus, a portal user can use any Grid resource with the same permissions and privileges as if he had logged into each directly, but now must only authenticate through the portal.

Jobs and command execution: Remote tasks are executed via the Globus/GRAM Gatekeeper, including compiling and running programs, performing process and job submission and deletion, and viewing job status and history. Additionally, users may execute simple Unix-type commands such as mkdir, ls, rmdir, cd, and pwd.

Data, file, and collection management: GridPort provides file and directory access to compute and archival resources and to portal file space. Using GSI-FTP (file transfer protocol) and SRB commands, GridPort enables file transfer between the local workstation and the remote Grid resources as well as file transfers between remote resources. GridPort also supports file and directory access to a user's file space on the Grid resource.

Information services: GridPort provides a set of information services that includes the status of all systems, node-level information, batch queue load, and batch queue job summary data. Data is acquired via the MDS 2.0 GIIS/GRIS or via custom information scripts for those systems without MDS 2.0 installed. Portals can use the GridPort GIS to gather information about resources on the Grid for display to the users (as in the HotPage portal) or use the data to influence decisions about jobs that the portal will execute.

28.2.3 GridPort security

Secure access at all layers is essential and is one of the more complex issues to address. All connections require secure paths (SSL, HTTPS) between all layers and systems involved, including connections between the client's local host and the Web server and between the Web server and the Grid resources that GridPort is accessing. Security between the client and the Web server is handled via SSL using an RC4-40 128-bit key and all connections to backend resources are SSL-encrypted and use GSI-enabled software wherever required. GridPort does not use an authorization service such as Akenti [16], but there are plans to integrate such services in the future versions.

The portal account creation process requires the user to supply the portal with a digital GSI certificate from a known Certificate Authority (CA). Once the user has presented this credential, the user will be allowed to use the portal with the digital identity contained within the certificate.

Logging into the portal is based on the Globus model and can be accomplished in one of two ways. The Web server may obtain a user ID and a passphrase and attempt to create a proxy certificate using *globus-proxy-init* based on certificate data stored in a local repository. Alternatively, the proxy may be generated from a proxy retrieved on behalf of the user from a MyProxy server.

If the proxy-init is successful, session state is maintained between the client and the Web server with a session cookie that is associated with a session file. Access restrictions for cookie retrieval are set to be readable by any portal hosted on the same domain. As a result, the user is now successfully authenticated for any portal that is hosted using the same domain, allowing GridPort to support a single login environment. Session files and sensitive data including user proxies are stored in a restricted access repository on the Web server. The repository directory structure is located outside Web server file space and has user and group permissions set such that no user except the Web server daemon may access these files and directories. Furthermore, none of the files in these directories are allowed to be executable, and the Web server daemon may not access files outside these directories. The session file also contains a time stamp that GridPort uses to expire user login sessions that have been inactive for a set period of time.

Grid task execution is accomplished using the GSI model: when a portal uses GridPort to make a request on behalf of the user, GridPort presents the user's credentials to the Grid resource, which decides, on the basis of the local security model, whether the request will be honored or denied. Thus, the portal acts as a proxy for executing requests on behalf of the user (on resources that the user is authorized to access) on the basis of the credentials presented by the user who created the portal account. Therefore, portal users have the same level of access to a particular resource through the portal as they would if they logged into the resource directly.

28.3 GRIDPORT PORTALS

As the number of portals using GridPort increased, the complexity of supporting them also increased. The redesign of GridPort as a shared set of components supporting centralized services (e.g. common portal accounts and common file space) benefited both developers and users: it became much simpler for developers to support multiple application portals because code, files, and user accounts were shared across all portals, while users only had to sign in once to a single account in order to gain access to all portals.

Figure 28.2 depicts the GridPort approach used for implementing portals. The diagram shows the relationships between multiple application portals residing on the same machine and accessing the same instance of GridPort. In this example, all the application portals are hosted on the *.npaci.edu domain. The Web servers for these URLs and the application portal software reside on the same physical machine and have access to the shared GridPort file space. Each portal also has its own file space containing specialized scripts and data. The portal developer incorporates the GridPort libraries *directly* into the portal code, making subroutine calls to GridPort software in order to access the functionality that GridPort provides. The application portal software is responsible for handling the HTTP/CGI (Common Gateway Interface) request, parsing the data, formatting the request, and invoking GridPort when appropriate. Furthermore, since GridPort is an open source, the application portal developers at a given domain may modify GridPort to suit their needs. GridPort intentionally decouples handling HTTP/CGI data so that clients can make GridPort requests using other languages (e.g. Java, PHP, etc.).

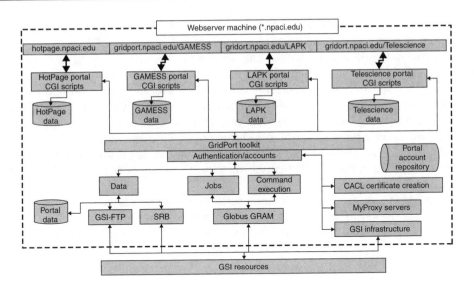

Figure 28.2 GridPort multiportal architecture diagram showing the method in which multiple portals can be installed on the same Web server machine. In this design, each portal has its own file space and shares the same instance of the GridPort modules as well as common portal account and authorization information. All access to the Grid is done through functions provided by GridPort and each request is authenticated against the account data.

Each portal request is first checked by authentication mechanisms built into GridPort; if the user is logged in, then the request will be passed on to the correct Grid service. Results are passed back to the requesting application portal, which has the responsibility of formatting and presenting results to the user.

There are several types of portals based on GridPort [17], examples of which include user portals, such as the NPACI and PACI HotPages, in which the user interacts directly with the HPC resources; community model application portals, such as the Laboratory for Applied Pharmacokinetics (LAPK) portal [18], which hide the fact that a Grid or an HPC resource is being used; remote instrument application portals, such as the UCSD Telescience portal, in which users control remote equipment and migrate data across institutional boundaries [19]; and systems of portals such as those being constructed by the NSF funded National Biomedical Computation Resource (NBCR) program [the Protein Data Bank (PDB) Combinatorial Extension (CE) portal, the GAMESS and Amber portals, and others] that share data to achieve a common research objective [20, 21]. In addition, there are a number of remote sites that have ported GridPort and installed local versions of the HotPage [22–25].

Each of these portal types demonstrates different motivations that computational scientists have for using computational Grid portals. Descriptions of examples of these portal types are given below. These examples illustrate some of the features and capabilities of GridPort. They also reveal some of the issues encountered in constructing Grid portals owing to limitations of GridPort and the Grid services employed by GridPort or to the fact that some necessary services have not yet been created. These limitations are summarized in Section 28.4 and plans to address them are discussed in Section 28.5.

Grid computing portals are popular and easy to use because Web interfaces are now common and well understood. Because they are Web-based, GridPort portals are accessible wherever there is a Web browser, regardless of the user's location. GridPort only requires that portals employ simple, commodity Web technologies, although more complex mechanisms for requesting and formatting data may be used if desired. For example, portals can be built using a minimal set of Web technologies including HTML 3.0, JavaScript 1.1, and the HTTPS protocol (Netscape Communicator or Microsoft Internet Explorer versions 4.0 or later), but GridPort portals may use technologies such as DHTML, Java, Python, XML, and so on.

On the server side, common Web server software should be adequate to host GridPort-based portals. To date, GridPort has been demonstrated to work on Netscape Enterprise Server running on Solaris and Apache running on Linux, but it should run on any Web server system that supports CGI and HTTPS [26].

28.3.1 HotPage user portal

The HotPage user portal is designed to be a single point of access to all Grid resources that are represented by the portal for both informational and interactive services and has a unique identity within the Grid community. The layout of the HotPage allows a user to view information about these resources from either the Grid or the individual resource perspectives in order to quickly determine system-wide information such as operational status, computational load, and available computational resources (see Figure 28.3). The interactive portion of the HotPage includes a file navigation interface and Unix shell-like tools for editing remote files, submitting jobs, archiving and retrieving files, and selecting multiple files for various common Unix operations. With the introduction of personalization in a new beta version of the HotPage, users can now choose which systems they wish to have presented for both informational and interactive services.

The informational services provide a user-oriented interface to NPACI resources and services. These services provide on-line documentation, static informational pages, and dynamic information, including

- summary data of all resources on a single page,
- basic user information such as documentation, training, and news,
- real-time information for each machine, including operational status and utilization of all resources,
- summaries of machine status, load, and batch queues,
- displays of currently executing and queued jobs,
- graphical map of running applications mapped to nodes,
- batch script templates for each resource.

Published as part of the NPACI Portal GIS, much of this data is generated via specialized custom scripts that are installed into the Globus Grid Resource Information Services (GRIS running on MDS 2.0 on each resource) and are also published via the HotPage.

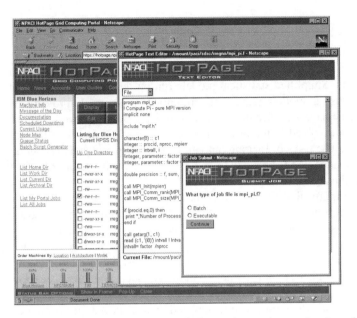

Figure 28.3 Screenshot of the NPACI HotPage, during an interactive session (user is logged in). In this example, the user is preparing to submit an MPI job to the Blue Horizon queue. Results of the job request will be displayed in a pop-up window.

The GIS service model allows us to share data with or to access data from other portal or Grid systems such as the Alliance/NCSA or the Cactus Grids [27, 28].

The HotPage interactive interface enables users to

- view directory listings and contents on all computational and storage systems on which they have accounts, including hyperlinked navigation through directories and display of text files;
- build batch job scripts to be submitted to the queues on the computational resources;
- execute interactive processes on the computational resources;
- perform file management operations on remote files (e.g. tar/untar, gzip/gunzip, etc.) and move files to and from the local host, HPC resources, archival storage resources, and SRB collections;
- manage NPACI accounts.

Each of the capabilities listed above is provided using Perl scripts that employ GridPort modules and can be extended by the portal application developer. Thus, the HotPage software consists of static HTML pages, server-side includes to simplify dynamic construction of other HTML pages, and Perl scripts to actually build the dynamic pages.

28.3.2 Laboratory for Applied Pharmacokinetics modeling portal

GridPort is being used to create a portal for medical doctors who run a drug dosage modeling package developed as a result of a collaboration between the Laboratory for

Applied Pharmacokinetics (LAPK), led by Roger Jelliffe, a physician and professor of medicine at the University of Southern California (USC), and Dr Robert Leary, a computational mathematician at the SDSC [18, 29]. For 30 years, Jelliffe and the LAPK have been developing software that enables physicians to tailor drug dosage regimens to individual patients. Simplified statistical models run on desktop computers, but newer, more accurate, nonlinear models require supercomputing power. Improvements in the algorithm by Leary have significantly reduced run times and improved portability. However, the physicians involved in the research have historically found working on HPC systems to be difficult. The tasks of logging onto an HPC system, uploading input data, submitting a job, and downloading results had to be kept extremely simple in order for the physicians to complete their jobs.

There was motivation to create a portal environment that physicians could use to run jobs on HPC resources without being exposed to the complexities of those computational resources. A key advance realized by the construction of the portal is that researchers can now run multiple tasks at the same time, whereas before they were limited to single job runs. The portal (see Figure 28.4) manages file names, moves data around for the clients, and allows researchers to vary simple parameters such as number of CPUs to utilize or job run time. Another significant capability allowed by the portal is the maintenance of an on-line 'notebook' where users can review and reuse data and files from previous sessions.

GridPort was chosen to build the LAPK portal based on developers' ability to use GridPort to rapidly deploy a prototype portal demonstrating the viability of the portal concept. Construction of the prototype portal took less than two weeks. Further development has been focused on adding more sophisticated interfaces and features, including

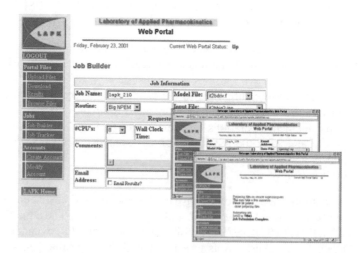

Figure 28.4 Screenshot of the LAPK Portal, in which users (medical researchers) utilize the portal without being exposed to the fact that HPC Grid and archival resources are being used. In this example, the three windows show the process by which the user created a new job (batch queue) in the back window, the job was submitted, and the results of the job submission information (front window).

job tracking and history, user access control, and user administrative pages. The portal also allows researchers to view 2D plots of their results and allows them to manage tests and results interactively. Most of the work on the portal is done by student interns at SDSC, and thus it is a relatively inexpensive project. Like the HotPage, the portal uses only frames and simple client-side JavaScript. The size of the portal software is relatively large, but not complex: there are approximately 50 cgi files, with an average size of about 200 lines each. Not all of these files are used to access the Grid. The LAPK portal system demonstrates one of the reasons GridPort has been successful: it is easy to program (the portal was developed by interns), easy to extend (the portal provides graphical images), and is extensible (LAPK has an additional authentication layer).

28.3.3 Telescience portal

The Telescience portal is a Web interface to tools and resources that are used in conducting biological studies involving electron tomography at the National Center for Microscopy and Imaging Research (NCMIR) at UCSD [19]. The basic architecture of the Telescience portal is shown in Figure 28.5: portal users have access to remote instrumentation (a microscope) via an Applet, an SRB collection for storage of raw results and analyzed data, portal tools for creating and managing projects, and remote visualization applications. The portal guides the user through the tomography process, from an applet-based data acquisition program to analysis.

A central component of the portal is a tool known as FastTomo, which enables microscopists to obtain near real-time feedback (in the range of 5 to 15 min) from the tomographic reconstruction process, on-line and without calibration, to aid in the selection of specific areas for tomographic reconstruction. By using SRB, the portal provides an infrastructure that enables secure and transparent migration of data between data sources,

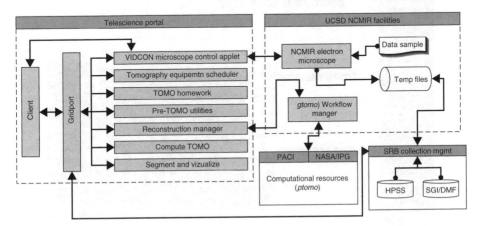

Figure 28.5 Basic architecture diagram of the Telescience portal, in which portal users will have access to remote instrumentation (microscope), automatic SRB collection access for storage of raw results and analyzed data, portal tools for creating and managing projects, and remote visualization applications.

processing sites, and long-term storage. Telescience users can perform computations and use SRB resources at NCMIR and SDSC and on NASA/IPG. The tomography process generates large quantities of data (several hundred megabytes per run, on average) and the portal moves data between these locations over high-speed network connections between compute and archival resources such as the Sun Ultra E10k and the High Performance Storage System (HPSS).

This portal demonstrates the capability within GridPort to integrate and interoperate with an existing portal system, to utilize Grid resources across organizations (both NPACI and IPG resources are used here), and to interact dynamically with sensors and systems.

28.3.4 NBCR computational portal environment

SDSC's NBCR project, funded by NIH-NCRR, is building software and portals to conduct, catalyze, and enable biomedical research by harnessing advanced computational technology [30]. The ultimate purpose of the resource is to facilitate biomedical research by making advanced computational data and visualization capabilities as easy to access and use as the World Wide Web, freeing researchers to focus on biology. The NBCR transparent supercomputing initiative was started five years ago and augments the work of researchers by providing Web-based access to computational tools without requiring training in supercomputer systems.

GAMESS/QMView (see Figure 28.6) is a computational chemistry package that provides all the tools necessary to build, launch, compute, and understand computational

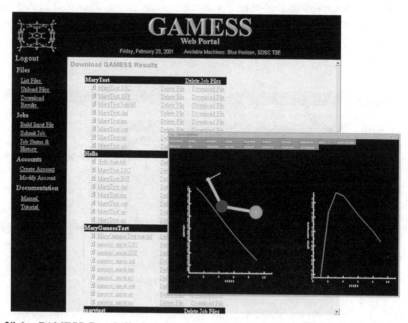

Figure 28.6 GAMESS Portal display showing a file listing of completed results that are stored and retrieved dynamically by GridPort from the user's SRB collection. Of interest, this view shows the 3D QMView visualization software that is available as a plug-in to the portal.

chemistry data [20]. While the original version of the GAMESS/QMView infrastructure provided a complete system for carrying out computational quantum chemistry on a variety of platforms, the system was not accessible through a secure and robust Web interface. The GAMESS portal was created to simplify the process of using the GAMESS program on sophisticated supercomputer architectures by enabling file transfer, input file creation, intelligent job building and submission, job history and tracking, and (in the near future) the ability to use the visualization component of the package (QMView) through the Web browser [31].

Scientists can use GAMESS from any Web browser. We have begun work on a version that is accessible from portable wireless devices including PDAs and mobile phones. We are also working to integrate the SRB to provide advanced file handling and storage management.

Another NBCR-funded project is the PDB CE Grid Computing Web Portal [32], which enables users to perform structural comparisons of proteins using the CE algorithm on HPC resources at NPACI. Like GAMESS, the CE portal can be accessed from any Web browser. The CE Portal gives users on-line capabilities to upload input files to query the CE database, submit searches to the CE database, display a status of the database searches, and view results of the searches. The portal enables users to upload their input files that contain representations of their proteins. The user can select a specific protein chain and query a CE database against it. The query looks for similar protein structures and aligns the proteins so that the user can view similarities and differences. It is distinct from the other portals because it allows two forms of login: anonymous and registered. Anonymous login gives users general functionality such as starting a CE search and viewing results of the search. By registering for a portal account, the portal user has additional benefits such as a personal and password-protected account to store queries and results, the ability to track the progress of searches after submission, and reduced restrictions on the usage of the HPC resources. The portal also supports molecular visualization programs such as the Compare 3D (which renders a simplified 3D view of a structural alignment with a Java applet), QMView, and Rasmol. Future plans include the addition of enhanced file handling capabilities using the SRB and automatic computational resource selection.

28.4 SUMMARY OF LESSONS LEARNED

Grid computing portals such as the HotPage and toolkits such as GridPort have become popular in the HPC community. The HotPage continues to maintain a position of high visibility because it provides the appropriate kind of information that a traditional HPC user requires. GridPort has proven to be useful and popular with application portal developers because it is based on simple technologies that are easy to deploy and use for development. In addition to NPACI, GridPort and the HotPage have been ported to several systems in the United States, the United Kingdom, and Asia and are used for multiple production applications. Science application portals based on GridPort have been implemented for the NASA Information Power Grid, the Department of Defense Naval Oceanographic Office Major Shared Resource Center (MSRC), the National Center for Microscopy and Imaging

Research (NCMIR), the University of Southern California (USC), the UK Daresbury Labs, and other locations [18–23, 25, 32].

User portals must enable the users to do more or better research in computational science. Traditional HPC users value *information* from user portals. For example, the NPACI HotPage is one of the most-visited Web pages in the NPACI domain. Even dedicated command line users find the current HotPage useful because it is quicker for determining which systems are up, how busy they are, where a job is likely to run fastest, why a job is not running somewhere, and so on. Grid information services are, relatively speaking, more mature than other Grid services, and this has enabled the development of useful informational user portals. However, experience has demonstrated that traditional HPC users have not fully adopted Grid portals for *interactive* services for three main reasons: user portals do not possess the full capabilities of the command line or mature user-oriented interfaces for the commands they do present, portal-interactive capabilities are not responsive as command-line actions, and Grid portals interactive responses are slower than traditional Web portals (e.g. Yahoo!) with which the users are familiar. These interactive issues are due to performance and capability limitations in the evolving set of Grid technologies, which are maturing but are certainly not complete (and are discussed in greater detail below). As Grid infrastructure matures – which is occurring rapidly because of massive government and industrial investment – Grid portals will meet the interactive performance and capability requirements of even the most demanding interactive users.

While some Grid application portals (such as those discussed in Section 28.3) have proven successful, the Grid is not yet being heavily used for application portals and general Grid applications. The reasons are both obvious and understandable: the Grid is still new and evolving, and much of the basic framework and infrastructure is still being developed or even conceived. With the basic infrastructure still being hardened (and in some cases developed!), the construction of Grid application portals and Grid applications that realize the benefits of the underlying network of Grid resources, and the toolkits being developed to simplify their construction, is difficult and in some cases impossible. In order to motivate scientists to further utilize the Grid, applications and application portals require the following: Grid resources must be transparent (users should not care where an application runs, but just that it got done), Grid tools must be easy to set up and use, it must be easy to get and use accounts and to configure and manage allocations, it must be easy to access and modify a user's Grid environment, and it must be easy to initialize Grid applications. A toolkit that makes these capabilities more easily accessible to developers of application portals and general applications will greatly accelerate effective utilization of Grid technologies and resources.

Furthermore, all these Grid capabilities must have the right focus: they must ease the task of code development/debugging/testing, they must couple data/knowledge searches with current research, they must support connection to physical experiments and management of data from sensors, and they must support distributed visualization tools. Portals must provide true connection to on-line knowledge centers, data repositories, and user data and make all data easy to incorporate into analytical processes, as well as enable enhanced usage of the available computational resources.

Many of these requirements cannot be satisfied until some basic Grid infrastructure issues have been resolved. The usefulness of Grid portals and applications, and therefore their rate of development and acceptance, will be limited until the following Grid tools, components, and capabilities are available:

- *Grid accounts and allocations*: Accounts and access should be managed at the organizational level instead of the machine level by allowing users to run on all systems under an umbrella account/allocation.
- *Single sign-on*: Users must have single sign-on for access to all Grid resources, from portals and for applications. GridPort provides a single sign-on method (see Section 28.2) and has a single sign-on mechanism that utilizes cookies, but this is limited to GridPort-hosted portals and does not extend to portals hosted on other domains or to general Grid applications.
- *Single logical file space*: Management and movement of data across Grid resources is challenging, but has been addressed in technologies such as the SRB. Registration and management of the executable binaries for each Grid compute architecture is even more difficult and must be automated. Avaki has accomplished much in this area, but it must become available for all Grids, not just Grids based on Avaki software.
- *Grid schedulers*: Effective utilization of Grid resources requires efficient Grid scheduling beyond advanced reservations. Portals and applications must be able to use the next available resource(s) and to utilize multiple resources for parallel applications and for workflow usage models. Grid schedulers must account not only for computational resource loads but also for network performance and data transfer times.
- *Advanced Grid Tools*: To fully utilize the Grid, additional tools are needed for development and optimized performance of Grid applications. This includes Grid-aware compilers for handling 'Grid binaries', Grid I/O libraries, and parallel libraries that interoperate with existing scheduling systems. Information services that contain archived historical data for prediction and performance analysis would also be useful for Grid scheduling.

In addition to the missing components and capabilities, there are known latencies associated with existing Grid services (e.g. authentication) that can drive users away. The latencies are due to factors such as network availability, encryption/decryption and key generation, and poor performance of Grid software itself. This creates a significant perception problem for portal users because they expect that events on the portal will happen instantaneously, which is not always the case. Additionally, portals themselves may induce latencies if not properly designed.

Grid computing portals have the potential to provide application users with the optimal environment to utilize Grid resources, and Grid portal toolkits have the potential to accelerate this utilization by driving the development of useful portals and applications. However, missing capabilities in both and performance latencies on both the portal toolkits and the underlying Grid technologies also have the largest capacity to drive users away if there are performance problems.

28.5 CURRENT AND FUTURE WORK

On the basis of the experiences discussed above and on the emergence of new portal and Web technologies, work has begun on a new version of GridPort that will have a 'components and tools' approach to facilitate dynamic composition of portal tasks. The scope of GridPort will be expanded to provide a toolkit not only for the development of portals but also for developing more general Grid applications and distributed Grid Web services. ('Grid Web services' are defined to be Grid services that are based on either standard Web services [33] or on the Open Grid Services Architecture (OGSA) [10, 34].) Plans also include improving the users' interactions with the HotPage through the use of portal technologies that support personalization and customization.

28.5.1 GridPort GCE architecture

The GridPort project has been expanded to provide a developer's framework that will support a variety of Grid computing environments (GCEs) including portals (user and application), Grid applications, legacy applications, problem solving environments, and shells [35]. The initial redesign phase will focus on Grid portals and the Grid Web services needed to support them. Subsequent efforts will further generalize GridPort for the development of Grid applications and other GCEs, after the Grid infrastructure needed to support Grid applications has matured.

As with the current version of GridPort, the layers in the new architecture are based on function and on what they produce or consume (see Figure 28.7). Each layer is described briefly below.

Client layer: As before, this will include consumers of applications such as Web browsers and PDAs. In addition, this layer has been expanded to include command line or shell interfaces and problem solving environments (PSEs).

Applications layer (enhancement of the current portals layer): This layer has been expanded to include applications as well as Grid computing portals. A significant change to the GridPort architecture is the fact that portals, and the systems on which they are hosted (the Web server machines), may now be physically separated from the machines on which GridPort application services are hosted through the use of technologies such as portlets and Grid Web services. Applications can now be located anywhere, providing the user community with the capability to develop and host science portals or applications on locally owned and controlled resources.

Application services layer (enhancement of the portal services layer – GridPort): GridPort components will be expanded to provide mediation services between Grid technologies and applications as well as portals. Components will be added to interoperate with common accounts, allocations, and authorization mechanisms; provide customization and personalization; track jobs and task history; collect and display network performance data (via Network Weather Service) and estimate data transfer times [36]; and interoperate with Grid schedulers. GridPort services can be local or remote. Grid Web services will be used to provide many of these capabilities. Portlets are included at this layer because they can deliver content to portals.

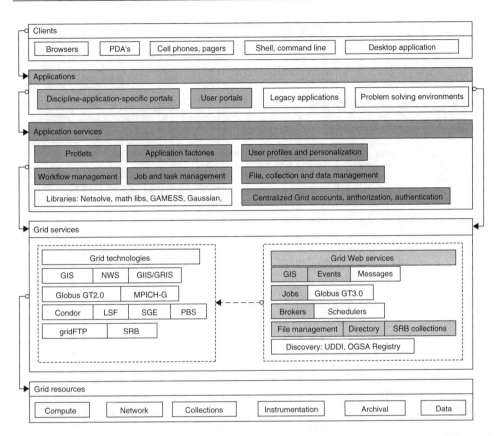

Figure 28.7 This diagram depicts the GridPort GCE architecture, which is based on a multilayered approach, as in Figure 28.1. The architecture has been extended to include applications (which include portals) and Grid Web services. Shaded boxes represent layers and types of services that will be implemented as part of the GridPort toolkit.

Grid services (technologies) layer: This layer represents the resource-level Grid. However, Grid Web services interfaces will be developed for many existing Grid technologies in the near future. This migration will take some time, and as a result GridPort will include both 'standard' and Web services interfaces to these technologies.

Resources layer: This is essentially the same resources layer that appears in the resources layer described in Reference [10] and contains compute, archival, networking, and data resources.

28.5.2 GridPort Grid Web services

A major focus for all future work on GridPort will be on the development and use of Grid Web services to facilitate dynamic operation of the Grid portals and applications. The

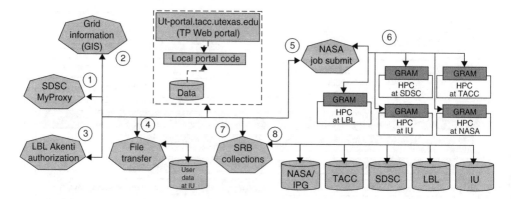

Figure 28.8 Diagram depicts a workflow scenario in which multiple sites support a Grid of resources, with sites hosting various centralized services. This example depicts job submission: (1) user points browser to the Texas Portal (TP), authenticates using MyProxy Web service; (2) TP checks status of HPC resources via GIS; (3) TP checks Akenti authorization service; (4) TP pulls input files from user's data file system; (5) TP submits request to JobSubmit Web Service (JSWS); (6) JSWS chooses HPC system, submits request to GRAM; (7) upon job completion, TP sends request to SRB Web service (SRB-WS) to migrate files into user's virtual collection; and (8) SRB-WS moves physical file to physical location.

impact of the Web services architecture on the development of the Grid is demonstrated in Figure 28.8. The diagram conceptualizes a simple workflow example of how an application portal might utilize distributed Grid Web services to submit a job to an HPC resource. The environment is similar to the Interoperable Web services test bed [37] being developed by the GCE Research Group of the Global Grid Forum (GGF), where a BatchScript Generator Web service between SDSC and IU has been successfully demonstrated [38]. The conceptual scenario is that TACC, SDSC, LBL, Indiana University (IU), and NASA have cooperated to develop a Grid connecting their various resources, with different sites hosting the various centralized services (authorization, authentication, a particular portal, etc.) and *all* sites contributing HPC systems and storage resources managed by SRB. The steps are as follows:

1. User points Web browser to the Texas Portal (TP) and authenticates using MyProxy service hosted at SDSC as a Web Service;
2. TP checks status of HPC resources and sees that two HPC systems are available, so request gets processed;
3. TP checks AKENTI authorization service hosted at LBL and sees that user is authorized to use portal to run applications on both available HPC systems;
4. TP pulls necessary input files from user's data from file system (non-SRB) at IU;
5. TP submits request to JobSubmit Web Service (JSWS) hosted at NASA;
6. JSWS chooses HPC system at SDSC and submits request to GRAM Gatekeeper;
7. Upon job completion, TP sends request to SRB Web service (SRB-WS) to migrate files into user's virtual collection; and
8. SRB-WS moves physical file to physical location.

Note that in this design Web services are also able to function as clients of other Web services. Not shown in this diagram are event services, which let the portal, and potential user, know the status of operations and pending actions. Event services would be useful at several steps in this workflow, for example, after steps 4, 5, 7, and 8.

There are many clear advantages to this system. All tasks are performed as messages based on a simple protocol, and Grid software systems can be encapsulated within this protocol. A Web service can be both a provider and a consumer of other Web services. Using discovery services such as UDDI (Universal Description Discovery and Integration) or the OGSA registry, Grid and Web services can dynamically choose services or Grid resources. The portal developer only needs to program the Web services protocol and support the Web services software on the local Web system. The implementation details such as programming language on which Grid software is used becomes irrelevant as long as the protocols can be agreed upon. Finally, the application programmer is not required to install, update, and maintain a large number of Grid service software components (Globus, NWS, LSF, Sun Grid Engine, Condor, MPICH-G); those details are left to the service providers.

The initial set of Web services planned for GridPort will include *information* Web services (load, node, status, queue usage, other MDS 2.1 information, network performance data, etc.) that are both secure and public. The set of *interactive* Web services will be based on those supported by the current HotPage (job submit, batch job submit, job cancelation, file transfer, data collection management, etc.). The Web service protocols currently being used include Simple Object Access Protocol (SOAP) [39], Web Services Description Language (WSDL) [40], UDDI [41], and others. SOAP and WSDL are standards supported by the World Wide Web Consortium (W3C) [42]. SOAP is used for remote method invocation between services, WSDL to describe the method interfaces, and UDDI as the service repository and discovery system. GridPort Grid Web services system will interface with the OGSA services.

Although these services have the potential to enhance and simplify Grid computing capabilities, the Grid community will need to address basic research questions about the scalability of Web services, how well they interact on the Grid, and how to determine the logical order in which these services and tasks should be done.

28.5.3 Grid portals and applications

The introduction of Web services provides the Grid portal community with the opportunity to redefine what is meant by a Grid computing portal or application. The introduction of new portal technologies such as portlets further enables portal service providers to develop and host a set of basic building blocks and tools for use by portals. Although there exist many commercial portal technologies, GridPort plans include investigation of Java-based technologies such as JetSpeed [43] and Uportal [44] because they are open-source and customizable, employ portlets, and would enable integration and adaptation to the unique requirements of Grid computing. Note that Perl does not support proxy delegation and forwarding (at the time of writing this article), further influencing the decision to migrate to Java as the primary programming language in GridPort.

With these capabilities, portals can provide a set of modular tools that the user can use to customize a view and to 'compose' a task, that is, to dynamically set up an application workflow. All sessions can be stored, so a user can reuse previous views and workflows. In this sense, the user can build a customized view or application 'on the fly', store the steps, come back and run it again later. This information can be incorporated into an on-line notebook or registered to be used by other applications. These new features and capabilities will be implemented in the next version of GridPort, enabling users to build their own customized views of the HotPage and their own application portals and applications. The system will support customization by virtual organization (e.g. NPACI, IPG, University of Texas), by machines, or by tasks that the user wishes to perform. The set of choices is left to the needs and imagination of the developers and users.

ACKNOWLEDGMENTS

The authors would like to thank their funding agencies for support of work, namely, the NSF PACI Program and the NPACI (NSF-ACI-975249), the NASA Information Power Grid (IPG) program (NPACI–NSF-NASA IPG Project), the Naval Oceanographic Office Major Shared Resource Center (NAVO MSRC) Programming Environment and Training (PET) Program, the Pharmacokinetic Modeling Project (NCRR Grant No RR11526), NBCR Project (NIH/NCRR P41 RR08605-07), The SDSC, and the Texas Advanced Computing Center (TACC) at the University of Texas at Austin.

We would also like to thank our colleagues at the SDSC – Steve Mock, Cathie Mills, Kurt Mueller, Keith Thompson, Bill Link, and Josh Polterock – and our former students at SDSC: Ray Regno, Akhil Seth, Kathy Seyama, and Derek Trikarso. We would also like to thank our new colleagues at the Texas Advanced Computing Center (TACC) – Maytal Dahan, Rich Toscano, and Shyamal Mitra, and our current students: Eric Roberts and Jeson Martajaya. We also wish to acknowledge the contributions by our many collaborators at other institutions, especially Gregor von Laszewski, Jason Novotny, Geoffrey Fox, Dennis Gannon, Marlon Pierce, Bill Johnston, and Keith Jackson.

REFERENCES

1. Fox, G., Furmanski, W. (1998) High performance commodity computing, Chapter 10, in Foster, I. and Kesselman, C. (eds) *The Grid: Blueprint for a New Computing Infrastructure*. San Francisco, CA: Morgan Kaufmann Publishers.
2. Foster, I. and Kesselman, C. (eds) (1998) *The Grid: Blueprint for a New Computing Infrastructure*. San Francisco, CA: Morgan Kaufmann Publishers.
3. Thomas, M. P., Mock, S., Boisseau, J. (2000) Development of web toolkits for computational science portals: the NPACI hotpage. *Proceedings of the Ninth IEEE International Symposium on High Performance Distributed Computing*, August, 2000, Project Website last accessed on 7/1/02 at http://hotpage.npaci.edu.
4. Novotny, J. (2002) Grid portal development toolkit (GPDK). *Concurrency and Computation: Practice and Experiencer*, Special Edition on Grid Computing Environments, Spring 2002 (to be published).

5. Bramley, R. *et al.* (2000) A component based services system for building distributed applications. *Proceedings of the Ninth IEEE International Symposium on High Performance Distributed Computing*, August, 2000.
6. National Partnership for Advanced Computational Infrastructure (NPACI), Project Website last accessed on 7/1/02 at http://www.npaci.edu.
7. NPACI HotPage Grid Computing Portal, Project Website last accessed on 7/1/02 at http://hotpage.paci.org.
8. Thomas, M. P., Mock, S., Dahan, M., Mueller, K. and Sutton, D. (2001) The GridPort toolkit: a system for building grid portals. *Proceedings of the Tenth IEEE International Symposium on High Performance Distributed Computing*, August, 2001, Project Website last accessed on 7/1/02 at http://gridport.npaci.edu.
9. Foster, I. and Kesselman, C. (1998) Globus: a metacomputing infrastructure toolkit. *International Journal of Supercomputer Applications*, **11**(2), 115–129, Globus Project Website last accessed on 7/1/02 at http://www.globus.org.
10. Foster, I., Kesselman, C. and Tuecke, S. (2001) The anatomy of the grid: enabling scalable virtual organizations. *International Journal of High Performance Computing Applications*, **15**(3), 200–222.
11. Novotny, J., Tuecke, S., Welch, V. (2001) An Online Credential Repository for the Grid: MyProxy. *Proceedings of the Tenth International Symposium on High Performance Distributed Computing (HPDC-10)*, IEEE Press, August 2001.
12. Baru, C., Moore, R., Rajasekar, A. and Wan, M. (1998) (1998) The SDSC storage resource broker. *Proc. CASCON '98 Conference*, Toronto, Canada, November 30-December 3, 1998, SRB Project Website last accessed on 7/2/02 at http://www.npaci.edu/SRB.
13. PACI HotPage Grid Computing Portal, Project Website last accessed on 7/1/02 at http://hotpage.paci.org.
14. Mock, S., Dahan, M., von Laszewski, G. and Thomas, M. (2002) The perl commodity grid toolkit. *Concurrency and Computation: Practice and Experience*, Special Edition on Grid Computing Environments, Winter, 2002. Project Website last accessed on 7/1/02 at http://gridport.npaci.edu/cog; to be published.
15. Globus Commodity Grid (CoG) Kits, Project Website last accessed on 7/1/02 at http://www.globus.org/cog.
16. Johnston, W., Mudumbai, S. and Thompson, M. (1998) Authorization and attribute certificates for widely distributed access control. *Proceedings of IEEE 7th International Workshops on Enabling Technologies: Infrastructures for Collaborative Enterprises – WETICE '98*, 1998, Website last accessed on 7/1/02 at http://www-itg.lbl.gov/Akenti.
17. Thomas, M. P., Dahan, M., Mueller, K., Mock, S., Mills, C. and Regno, R. (2002) Application portals: practice and experience. *Concurrency and Computation: Practice and Experience*, Special Edition on Grid Computing Environments, Spring 2002 (to be published).
18. LAPK Portal: The Laboratory for Pharmacokinetic Modeling, Website last accessed on 7/1/02 at: http://gridport.npaci.edu/LAPK.
19. Telescience for Advanced Tomography Applications Portal, Project Website last accessed on 7/1/02 at http://gridport.npaci.edu/Telescience/.
20. GAMESS Portal: The Laboratory for Pharmacokinetic Modeling, Website last accessed on 7/1/02 at https://gridport.npaci.edu/GAMESS.
21. AMBER (Assisted Model Building with Energy Refinement) Portal, Website last accessed on 7/1/02 at https://gridport.npaci.edu/Amber
22. Allan, R. J., Daresbury Laboratory, CLRC e-Science Centre, HPCGrid Services Portal, Last accessed on 1/1/02 at http://esc.dl.ac.uk/HPCPortal/.
23. Asian Pacific Grid (ApGrid) Testbed, Project Website last accessed on 7/1/02 at http://www.apgrid.org.
24. NASA Information Power Grid, Project Website last accessed on 7/1/02 at http://www.ipg.nasa.gov/.
25. NCSA Alliance User Portal, Project Website last accessed on 7/1/02 at http://aup.ncsa.uiuc.edu.
26. Apache Software Foundation, Website last accessed on 7/1/02 at http://www.apache.org.

27. Allen, G. *et al.* (1999) The cactus code: a problem solving environment for the grid. *Proceedings of the Eight IEEE International Symposium on High Performance Distributed Computing*, August, 1999, Cactus problem solving environment project. Website last accessed on 7/1/02 at http://www.cactuscode.org.

28. National Computational Science Alliance (Alliance), Project Website last accessed on 7/1/02 at http://www.ncsa.uiuc.edu/About/Alliance/.

29. Botnen, A., Wang, X., Jelliffe, R., Thomas, M. and Hoem, N. (2001) *Population Pharmacokinetic/Dynamic (PK/PD) Modeling via the World Wide Web*. National Library of Medicine Online Publications. Website last accessed http://www.nlm.nih.gov/

30. National Biomedical Computation Resource (NBCR), Project Website last accessed on 7/1/02 at http://nbcr.sdsc.edu/.

31. QMView: Quantum Mechanical Viewing Tool, Project Website last accessed on 7/1/02 at http://www.sdsc.edu/QMView/.

32. Combinatorial Extension (CE), Grid Computing Web Portal, Project Website last accessed on 7/1/02 at https://gridport.npaci.edu/CE/.

33. W3C: Web Services Activity, Project Website last accessed on 7/1/02 at http://www.w3.org/2002/ws/.

34. Foster, I., Kesselman, C., Nick, J. and Tuecke, S. *The Physiology of the Grid: An Open Grid Services Architecture for Distributed Systems Integration*. Paper available from Globus Project Website.

35. Global Grid Forum (GGF), Grid Computing Environments Research Group (GCE), Website last accessed on 7/1/02 at http://www.computingportals.org/GCE.

36. Network Weather Service (NWS), Project Website last accessed on 7/1/02 at: http://nws.cs.ucsb.edu/.

37. GCE Interoperable Web Services Testbed Project, Website last accessed on 7/1/02 at http://www.computingportals.org/gce-testbed.

38. Mock, S., Mueller, K., Pierce, M., Youn, C., Fox, G. and Thomas, M. (2002) A batch script generator web service for computational portals. *Proceedings of the International Multiconference in Computer Science*, July, 2002.

39. SOAP: Simple Object Access Protocol, Project Website last accessed on 7/1/02 at http://www.w3.org/TR/SOAP.

40. W3C: Web Services Description Language (WSDL) 1.1, Project Website last accessed on 7/1/02 at http://www.w3c.org/TR/wsdl.

41. Universal Description, Discovery, and Integration, Project Website last accessed on 7/1/02 at http://www.uddi.org.

42. Jetspeed Enterprise Information Portal, Project Website last accessed on 7/1/02 at http://jakarta.apache.org/jetspeed.

43. JavaSIG Uportal, Project Website last accessed on 7/1/02 at http://www.mis2.udel.edu/ja-sig/portal.html.

44. W3C: World Wide Web Consortium. Project website last accessed on 12/1/02 at: http://www.w3c.org.

Unicore and the Open Grid Services Architecture

David Snelling

Fujitsu Laboratories of Europe, Hayes, Middlesex, United Kingdom

29.1 INTRODUCTION

This chapter describes a GridService demonstrator built around the Unicore Grid environment, its architectural design and implementation [1]. It then examines some lessons learned from the process of developing an implementation of a family of GridServices that conforms to the Open Grid Services Architecture (OGSA) [2] and the Grid Service Specification [3].

The goals of this project were two fold. Primarily, it is only through implementation that complexities such as those that arise in OGSA can be fully understood and analyzed. Secondly, Unicore (www.unicore.org) is a complete production Grid environment that seemed to conform extremely well to the OGSA. The author wanted to test this hypothesis and see how difficult it would be to construct an OGSA-hosting environment on top of the Unicore infrastructure.

By way of a brief summary, both goals were met easily and with satisfaction. The OGSA model is remarkably clean and the abstractions well matched to what is needed at the implementation level. The lessons learned reveal where cracks exist and, in many

Grid Computing – Making the Global Infrastructure a Reality.   Edited by F. Berman, A. Hey and G. Fox
© 2003 John Wiley & Sons, Ltd   ISBN: 0-470-85319-0

cases, support the changes currently under discussion in the Open Grid Services Infrastructure (OGSI) working group of the Global Grid Forum (GGF). The reader should note that this chapter and the implementation on which it is based are targeted at the first draft (2/15/2002) of the Grid service specification [3]. By the time of going to the press, many of the issues discussed here and elsewhere will have become part of the next version.

With respect to the second goal, the implementation of a complete OGSA GridService environment that could be used to deploy task specific GridServices was significantly easier than expected initially. The combination of the Unicore Grid infrastructure with a flexible Web Services tooling environment (GLUE from www.themindelectric.com) proved to be effective and easy to use.

29.1.1 Caveats

The implementation discussed here conforms to the OGSA model, but not to the specifications as such. The specification is still under development and requires extensions to the Web Services Definition Language (WSDL) [4]. It will be some time before tooling platforms are available that support these extensions. Therefore, our approach has been to develop the architecture described by OGSA, but to work within the constraints of existing Web Services. The OGSA abstractions (serviceTypes in particular) are exposed in Java implementation, and then merged into a 'flat' WSDL representation, as the various portTypes are aggregated into a single Web service. This restriction has very little impact on our ability to meet the goals of this project. In fact, the extent to which this project was possible within the Web Services framework raises questions about the need for extensions, see Section 29.3.1 below.

29.2 IMPLEMENTATION

This section outlines the overall architecture and implementation of this GridService demonstrator. On top of a 'hosting environment', the demonstrator supports a generalized, distributed, application-steering GridService. This section also includes a summary of this application-steering GridService.

29.2.1 Infrastructure and architecture

In 'The Physiology of the Grid' [2], Foster *et al.*, describe a *hosting environment* as follows:

> *In practice, Grid services are instantiated within a specific execution environment or hosting environment. A particular hosting environment defines not only implementation programming model, programming language, development tools, and debugging tools, but also how an implementation of a Grid service meets its obligation with respect to Grid service semantics.*

The *platform* component of this demonstrator fits easily into this definition; however, the overall environment is more complex than this definition indicates, see Figure 29.1.

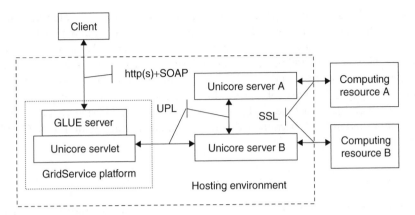

Figure 29.1 Architecture of Unicore GridService hosting environment.

Implicit in the above definition, a hosting environment is based on a single programming model/language; it implies that the GridService client interacts with a single 'program'. Although true of this demonstrator's platform component (i.e. the client interacts with a Java program constructed as described below), this simplification belies the complexity of the Unicore infrastructure that interacts with the platform and the distributed computing resources. The Unicore *hosting environment* supports applications written in a variety of programming languages and run on many different computer architectures. Nonetheless, the notion of Unicore deployed in this way is well described as a hosting environment.

This hosting environment is based on two primary building blocks, the Web Services platform GLUE from The Mind Electric, (www.themindelectric.com) and the Unicore server implementation developed by Fujitsu Laboratories of Europe for Unicore-related projects (www.unicore.org). Both environments are freely available for downloading along with the source code of this demonstrator.[1] All components are written in Java. Only a small part of the Unicore environment requires Perl.

29.2.1.1 Client

The user accesses several GridServices from a single client application, linked with the client side classes from GLUE. The client included in the demonstration download is a crude, command line client supporting all the GridServices provided, for example, Factory, Registry, HandleMap, a NotificationService, and various JobGridServices. More realistic and user friendly clients could be written using any Web Service environment, for example, .NET.

The client connects to the hosting environment's platform component, via http or https, with Simple Object Access Protocol (SOAP) bindings, as specified in the WSDL

---

[1] Note the demonstrator download does not include the Unicore or GLUE infrastructure components. To experiment with this demonstrator, the GLUE Web Services environment is required and must be obtained from The Mind Electric. A 'DummyGrid-Service' is included in the demonstrator that does not require a complete Unicore Grid. This demonstrator will interact with the demonstration Unicore Grid at (www.fz-juelich.de/unicore-test), which supports some of the demonstration applications.

sent from the platform. For https, a pair of certificates and the associated key stores are required for the client and server. The use of an https binding automatically activates a password-based authorization step. In the http mode, no user authentication is needed.

With the simple http binding, the 'console' supported by GLUE can be started (see GLUE documentation for details). The console introspects the WSDL and generates dynamic Web pages for most of the GridService operations. The exceptions are those that return structured data types, which the simple console interface cannot introspect; however, no fundamental barriers exist to automatic tooling within this GridService environment.

29.2.1.2 Platform

On the server side, the GLUE Web Services server is integrated with a servlet package that supports the Unicore protocols. From the servlet, Unicore Abstract Job Objects (AJOs) can be consigned to Unicore sites, monitored, controlled, and the results retrieved. Within a Unicore Grid, this servlet can be used to construct lightweight clients. In this implementation, the *GridService* package implements the basic OGSA portTypes and a collection of JobGridServices. The *servlet* package manages the interaction between the *GridService* package and the Unicore Grid. This servlet also manages Unicore security, see section 29.2.1.4.

29.2.1.3 Unicore server

From the servlet, Unicore sites are contacted via the Unicore Protocols Layer (UPL). The Unicore Server receives AJOs from Unicore clients, the GridService platform, and other Unicore Servers. As the various workflow tasks of the AJO are processed, they are converted from their abstract representation in the AJO to platform-specific forms by Unicore mechanisms. All aspects of the Unicore Grid, including the AJO, are of course hidden from the OGSA user. Further details of the Unicore side are not central to this chapter. See Unicore download documentation and Reference [5] for more information.

29.2.1.4 Security framework

At the time of writing, the Grid Service Specification [3] only indicated that security was a binding time property. As a result, there are two separate levels to the security framework, one at the https binding between the client and the GridService platform and another between the platform and the rest of the Unicore hosting environment.

Https is used to connect the client to the platform component of the GridService. The use of https triggers the need for simple authentication using user names and passwords.[2] Once the https connection is established, a Unicore certificate, loaded by the servlet when it starts up, is used to authenticate and authorize work by all users of the GridService. In

---

[2] The user name/password step is redundant, given that the authentication can be taken from the https connection. However, we chose to leave it in place, since the normal practice for an https connection assumes the use of a generic client certificate (e.g. as supplied with most browsers) and a separate login to sites requiring authentication.

this case, the GridService is acting like an Application Service Provider (ASP). The user authenticates the ASP, using user name and password over https, and the ASP authenticates the supercomputer centres, using Unicore authentication over UPL. In this demonstrator, the computer centres do not know the identity of the end user, only that of the ASP.

Other security and trust models are clearly possible as detailed below:

1. Globus style proxy credentials could be generated by the Web Services client and sent to the platform. This proxy certificate could then be used by Unicore to identify the user to the computer centres.
2. The servlet can be run on the client side. In this case, Unicore authentication and authorization are used throughout the process. The Web Services authentication can be omitted since this connection is local. This approach has little to offer byway of practical GridServices beyond providing a GridService client to existing Unicore users.
3. The Unicore sites could accept authentication according to the ASP model as described above, but the Unicore sites can be configured to authorize on the basis of the login used at the Web Services point of entry.
4. As OGSA security plans evolve, other approaches will be considered.

29.2.2 Supported Grid Services

The basic set of portTypes, as outlined in Reference [3], has been implemented. By aggregating and extending these, a collection of practical GridServices has been implemented.

29.2.2.1 Basic GridService portTypes

All the basic portTypes are coded as Java interfaces. The WSDL representation of each is generated automatically by the GLUE environment. However, GLUE has not been extended to support the serviceType extension; hence, the aggregation of portTypes into services takes place in the Java implementation and the complete Web service description for each GridService is created automatically from the Java implementation.

Given that the Grid Service Specification indicates that all GridServices should support the GridService portType, all other portTypes are constructed using interface extensions from the GridService interface. The basic GridService interface hierarchy is therefore as follows:

```
interface com.fujitsu.arcon.gridservice.IGridService
interface com.fujitsu.arcon.gridservice.IGridServiceFactory
interface com.fujitsu.arcon.gridservice.IHandleMap
interface com.fujitsu.arcon.gridservice.IJobGridService
interface com.fujitsu.arcon.gridservice.ISteeringGridService
interface com.fujitsu.arcon.gridservice.INotificationService
interface com.fujitsu.arcon.gridservice.INotificationSink
interface com.fujitsu.arcon.gridservice.INotificationSource
interface com.fujitsu.arcon.gridservice.IRegistry
```

29.2.2.2 Grid Service implementations

The implementation hierarchy parallels the interface hierarchy; there is an implementation of the GridService portType that all other portTypes extend. Although a GridService that implemented only the GridService portType could be created and used, it performs no useful function and therefore, no factory has been written for it. The following GridServices can be created (or are created automatically) and accessed from the demonstrator's client application.

RegistryAndMapper: The HandleMap and Registry portTypes are implemented as a single GridService. This GridService is created automatically when the service starts up as a function of the hosting environment. GridServices created by the hosting environment are registered automatically with both the HandleMap and Registry when they are created. Also, there is a Registry portType for registering new, external GridServices with the Registry.

GridServiceFactory: The factory is started automatically, but the client must enquire of the Registry to find the Grid Service Handle (GSH) of the factory and then contact the HandleMap to obtain a Grid Service Reference (GSR) with which to bind to the factory. This process could be streamlined; however, the goal was to expose the full functioning structure of the OGSA. The factory can construct a variety of JobGridServices and a NotificationService, as outlined below.

NotificationService: Given a Grid Service Handle, this service will wait for any notified state change in the target GridService. The NotificationService implements the NotificationSink portType and requires that the target GridService implements the NotificationSource portType.

DummyJobGridService: This service mimics the functions of the JobGridService, but does not require access to a Unicore enabled site, nor does it support the NotificationSource portType.

JobGridService: This GridService provides support for running a 'black box' application at a Unicore site. The serviceData describing these applications is instance specific, but made available as part of the factory's serviceData. Each Unicore site may support many different applications, and the factory may query the Unicore sites for this serviceData. Note that, in this implementation, the factory's version of this serviceData element is hard coded, so that the DummyJobGridService can function without reference to a Unicore site. The dynamic information from the Unicore site is available in the GridServiceDescription of the JobGridService after it has been created.

Each 'black box' application accepts an input file and writes an output file. The names of these files are part of the GridServiceDescription, and the users may specify file names of their choosing. The GridService manages the renaming automatically. The application may also access other resources at the site, but these are configured by the site and invisible to the user (effectively inside the 'black box'). The input file may come from any Unicore site or the client and the output file may be sent to any Unicore site or returned to the client.

SteeringGridService: This GridService directly extends the JobGridService with operations that allow the user to fetch a sample file back to the client and write a control file to be read by the running application. The assumption is that the application occasionally inspects the control file and modifies its behaviour accordingly. The sample file may be written explicitly as a result of fields in the control file or according to some other policy, for example, every ten iterations. The client, included with the download of this demonstrator, assumes that the sample file is in GIF format and starts a viewer to display the contents. The same flexibility with respect to file names exists for the sample and control files as it does for the input and output files.

29.3 LESSONS

This section briefly outlines the author's thoughts on some issues that arose out of the process of implementing this OGSA interface to Unicore. Although arrived at independently, many of these ideas are not unique. Following the release of the first draft of the Grid Service Specification, a significant and active community of informal reviewers was formed. Many of the ideas and perspectives listed here can also be seen in the proceedings of this community, see Reference [6].

29.3.1 Are GridService extensions to WSDL really needed?

The author certainly feels that some of the proposed extensions are desirable, but from both the server and client-development points of view, many of the proposed extensions seem unnecessary. This project was carried out with only standard WSDL facilities on both the client and server sides and has resulted in a working demonstrator that could be deployed in a variety of contexts. However, several aspects of the OGSA model are missing.

29.3.1.1 ServiceType

The serviceType extension provides a mechanism for aggregating and naming a collection of portTypes. Multiple serviceTypes can be combined to create new serviceTypes, and multiple implementations of a given serviceType can be constructed. In this implementation, Java interfaces encapsulate the abstractions of the OGSA portTypes and therefore capture the serviceType structure, albeit in Java rather than (extended) WSDL. For each serviceType, the WSDL emitted by the platform is a single portType containing all the operations of the constituent portTypes (represented by Java interfaces) used to form the serviceType. The serviceType structure, retained at the Java level, is effectively flattened into a single portType at the WSDL level. For example, the RegistryAndHandleMap serviceType would contain three portTypes, that is, GridService, HandleMap, and Registry. The single WSDL portType constructed by the platform is semantically equivalent to what would be expressed in the extended serviceType representation, but the underlying structure is lost.

The author's belief is that a serviceType extension that supports serviceType inheritance should be a minimum extension to WSDL. Such an inheritance mechanism should support aggregation, versioning, extension (in the sense of adding additional operations through new portTypes), and implementation naming, see Section 29.3.1.2.

29.3.1.2 ServiceImplementation

The related extension, serviceImplementation, provides a mechanism for naming a particular implementation of a serviceType as part of the creation of a service. However, serviceImplementation extends the WSDL element service, which includes endpoint and binding information. The author favours separating the naming aspect from binding and endpoint aspects. Allowing serviceTypes to be extended, as recommended earlier, provides an effective naming mechanism for particular implementations, while the standard WSDL service element provides the binding and endpoint details. The author believes that serviceImplementation can be dropped from the specification.

29.3.1.3 ServiceData

Because serviceData can be obtained dynamically, through the findServiceData operation in the required GridService portType, there is no requirement to expose service-Data as an extension to the WSDL representation of a serviceType. The typical pattern adopted in this implementation is to include some (usually static) serviceData elements for one GridService as run-time serviceData elements in the Grid Service used to find (or create) the given GridService. In this way, serviceData that is needed to create a Grid Service, for example, is available from the factory as part of its run-time serviceData.

Although not required, there are good reasons for making some serviceData available through the WSDL. In particular, it allows serviceData to remain associated with the serviceType to which it refers. Clearly not all serviceData elements can be defined when the WSDL is generated, however, wherever possible there are clear advantages of having it available. For example, if the list of applications supported by a given JobGridService were available as part of the WSDL, the client could decide if this service supported the desired application before creating the service. It is possible to obtain this information from the serviceData of the factory, as in our demonstrator, but this separates this important information from the GridService.

Furthermore, serviceData is an ideal framework to capture increasing detail about a GridService as it evolves from interface, to implementation, to created instance, to running instance, to completed instance, and finally to destroyed GridService. In some cases the content changes as the state of the service changes; in other cases new serviceData elements are added. At the transition from implementation to instance, serviceData allows site-specific details to be provided by the Unicore hosting environment. In particular, the list of supported applications at a given site is included in the factory's serviceData. Also, details about the file names required by the application are provided to the client as the GridService is created.

However, there are several places in the architecture where care must be taken to keep serviceData consistent with other structures in the model, in particular, the GSR. There are three different sources for a GSR. The client may have a copy of the GSR that was obtained from the factory when the GridService was created, one may be obtained from a HandleMap, or the client may query the serviceData of the GridService itself. In most cases the consistency of the GSR is not important, but with serviceData being generally available from several sources, consistency can become an issue.

In general, serviceData is an essential part of the OGSA framework. The specification should make it clear what serviceData elements are required and when in the GridService life cycle they are relevant. Allowing serviceData as part of the WSDL of a serviceType is certainly desirable, but all serviceData should remain available dynamically from some source, be it the GridService itself or a registry or a factory.

29.3.1.4 Compatibility assertions

There are many issues surrounding compatibility assertions. Since this demonstrator is the first GridService of any type (that the author had access to), it could not have been compatible with another implementation. During the analysis and design phases for this effort, many questions arose. These were also echoed in the OGSI mailing list and are detailed as follows:

> *Who asserts the assertions?*
> *How do we believe the assertion?*
> *Does an assertion mean more semantically that 'This one is like that one.'?*
> *Is compatibility asserted at the operation, port, service, or implementation level?*
> *Are compatibility assertions object-oriented programming by the back door?*
> *How are compatibility assertions composed?*
> *What is the meaning of composition of compatibility assertions?*

Recent discussions in the OGSI Working Group indicate that compatibility assertions will be removed from the specification and discussion postponed to some later time. The author concurs with this strategy.

In the meantime, it is possible to experiment with compatibility issues in the context of GridServices. As part of GridService development, inclusion of a compatibility portType would provide information on the compatibility of one version to all earlier versions. There is no complex trust framework required, as the service provider is making assertions about his own implementation only. Similarly, as part of a registry or other discovery services, a compatibility portType could check that two distinct implementations were compatible.

Such a service could be supported and run by a VO, or a corporate consortium, covering their own members and implementations.

29.3.1.5 *PrimaryKey*

In this implementation no use could be found for the primaryKey portType. Initially, during development, it seemed to make sense as an internal handle, but it never made sense as an externally visible concept. As development proceeded, its internal role also vanished. The primaryKey portType should be removed from the specification.

29.3.2 Can a single GridService support multiple portTypes?

One aspect of GridServices in general, and of this implementation in particular, is that multiple portTypes can be combined into a single serviceType. In this demonstrator, the Registry and the HandleMap are combined into a single serviceType. However, a single service possessing two distinct functions creates some complexities in the development of a clean client demonstrator, as well as raising questions of identity and equality within the GridService context.

The bootstrap process for the client is based on knowing the address of an 'interesting' Registry. The Registry advertises a running factory for creating application steering Grid-Services. Following the procedure set out in the Grid Service Specification, the Registry's or GridService's GSH yields the handle for the home HandleMap. These two GSHs (the Registry and the HandleMap) and their mapped GSRs are different, but point to the same instance and therefore have the same serviceData. One of the serviceData elements is the GSH of the GridService. The GSH must appear exactly once in the serviceData and, in this case, can have two possible values. In the demonstrator, the serviceData contains the GSH of the Registry. In general, the issue of identity needs further discussion and some resolution needs to be arrived at in the context of GridServices.

29.3.3 Is the 'Push Only' notification framework sufficient?

The OGSA notification framework is based around a subscribe/deliver model. However, the underlying assumptions in Unicore are based on periodic polling for changes in state. This was largely motivated by the need to deal with clients sitting behind firewalls, the norm with the majority of Unicore-related partners.[3]

For this demonstrator, a NotificationService provides an operation that waits for a state change in a selected GridService and then returns that state to the client, that is, a 'pull'-based model. To implement 'push'-based notification back to the client, at least partial server-side functionality would need to be provided to the client, creating the same firewall issues that motivated the 'pull'-based model in Unicore.

This NotificationService (a NotificationSink) subscribes to state change notification with the nominated JobGridService (a NotificationSource), using the OGSA 'push'-based

---

[3] Incidentally, this is the only significant point where the architecture of OGSA and Unicore differ; otherwise a very direct mapping of concepts is possible.

model. Web Service-based bindings are used, even though both services are running in the same Java Virtual Machine (JVM) to demonstrate the principle. Note that within the JobGridService implementation, a 'pull' model exists to obtain up-to-date information from the remote Unicore sites. It seems likely that lightweight clients, running behind firewalls, will be needed in many instances. Therefore, some thought should be given to establishing a framework within the Grid service specification that supports a 'pull'-based model with the same scope as the 'push'-based model currently described.

29.3.4 Is security provision at binding level only adequate?

In the case of this demonstrator, it is clear that providing security binding at the protocol layer only is not adequate. The platform component must use its own credentials (rather than the user's) to authenticate to the Unicore sites. This is consistent with a Web Services model. A Web service that needs to use subsequent services authenticates under its own credentials as a service provider to the subsidiary service. Eventually, GridServices will need more flexibility than this. Some of the approaches outlined in Section 29.2.1.4 above would allow extended flexibility. There are many issues pertaining to security and trust frameworks that cannot be addressed in this chapter. When we have a chance to increase the flexibility of this demonstrator with respect to these issues, a better indication of the full scope of these issues should be more apparent.

29.4 CONCLUSIONS AND FUTURE DIRECTIONS

This project has produced what can best be described as a 'conceptual prototype', and therefore a fresh start is as probable as continued development. Among the issues that will be addressed in the next steps are the following:

Single sign-on: The current implementation does not support single sign-on. Both Unicore and Globus provide this functionality and thus it will be expected of GridServices. There are several approaches for achieving this, two of which are apparent in the differences between Unicore and Globus.

Workflow: Unicore includes a complete framework for workflow management in a batch job context. The extent to which GridServices can be controlled with similar mechanisms is a current area of investigation. A hosting environment that supports a lightweight workflow model would meet the needs of many users and help answer questions pertaining to wider issues in workflow management such as that might arise from the work on Web Services Flow Language (WSFL) [7].

Interoperability: The primary goal in standards development is interoperability. It is planned that within the context of the GRIP [8] it will be possible to explore issues of interoperability between this demonstrator and the early releases of the Globus Tool Kit version 3, which will include OGSA interfaces.

In closing, both GLUE and Unicore represent significant infrastructures in their own rights. The merging of the two was remarkably straightforward, and therefore, the development effort has been minimal. The total effort for this project so far, including the learning curve for Web Services, was not more than 2 person months (3 months elapsed). It is this accessibility of Web Services that we hope to see extended to GridServices. The key requirement for this will be standardization, and for this reason the author stresses the importance of limiting the number and complexity of extensions to WSDL.

REFERENCES

1. Unicore information is available on the following web sites: www.unicore.org – Unicore protocols and source code; www.unicore.de – The home page for the Unicore Plus project; www.fz-juelich.de/unicore-test – A complete Unicore environment in which users can download a client, obtain a certificate, and trial the Unicore environment.
2. Foster, I., Kesselman, C., Nick, J. and Tuecke, S. (2002) *The Physiology of the Grid*, Draft 2/3/2002, www.globus.org/research/papers/ogsa.pdf.
3. Tuecke, S., Czaijkowski, K., Foster, I., Kesselman, C., Frey, J. and Graham, A. (2002) *Grid Service Specification*, Draft of 2/15/ 2002, www.globus.org/research/papers/gsspec.pdf.
4. The Web Services Description Language, www.w3.org/TR/wsdl.
5. Erwin, D. and Snelling, D. (2001) *Unicore: A Grid Computing Environment*. LNCS 2150, Euro-Par 2001, Springer.
6. Global Grid Forum, OGSI Working Group, www.gridforum.org/ogsi-wg/.
7. Leymann, F., The Web Services Flow Language, www-4.ibm.com/software/solutions/webservices/pdf/WSFL.pdf.
8. Grid Interoperability Project, IST-2001-32257, www.grid-interoperability.org.

Distributed object-based Grid computing environments

Tomasz Haupt[1] and Marlon E. Pierce[2]

[1]*Mississippi State University, Starkville, Mississippi, United States,* [2]*Indiana University, Bloomington, Indiana, United States*

30.1 INTRODUCTION

Computational Grid technologies hold the promise of providing global scale distributed computing for scientific applications. The goal of projects such as Globus [1], Legion [2], Condor [3], and others is to provide some portion of the infrastructure needed to support ubiquitous, geographically distributed computing [4, 5]. These metacomputing tools provide such services as high-throughput computing, single login to resources distributed across multiple organizations, and common Application Programming Interfaces (APIs) and protocols for information, job submission, and security services across multiple organizations. This collection of services forms the backbone of what is popularly known as the computational Grid, or just the Grid.

The service-oriented architecture of the Grid, with its complex client tools and programming interfaces, is difficult to use for the application developers and end users. The perception of complexity of the Grid environment comes from the fact that often Grid

Grid Computing – Making the Global Infrastructure a Reality. Edited by F. Berman, A. Hey and G. Fox
© 2003 John Wiley & Sons, Ltd   ISBN: 0-470-85319-0

services address issues at levels that are too low for the application developers (in terms of API and protocol stacks). Consequently, there are not many Grid-enabled applications, and in general, the Grid adoption rate among the end users is low.

By way of contrast, industry has undertaken enormous efforts to develop easy user interfaces that hide the complexity of underlying systems. Through Web portals the user has access to a wide variety of services such as weather forecasts, stock market quotes and on-line trading, calendars, e-mail, auctions, air travel reservations and ticket purchasing, and many others yet to be imagined. It is the simplicity of the user interface, which hides all implementation details from the user, that has contributed to the unprecedented success of the idea of a Web browser.

Grid computing environments (GCEs) such as computational Web portals are an extension of this idea. GCEs are used for aggregating, managing, and delivering grid services to end users, hiding these complexities behind user-friendly interfaces. Computational Web portal takes advantage of the technologies and standards developed for Internet computing such as HTTP, HTML, XML, CGI, Java, CORBA [6, 7], and Enterprise JavaBeans (EJB) [8], using them to provide browser-based access to High Performance Computing (HPC) systems (both on the Grid and off). A potential advantage of these environments also is that they may be merged with more mainstream Internet technologies, such as information delivery and archiving and collaboration.

Besides simply providing a good user interface, computing portals designed around distributed object technologies provide the concept of persistent state to the Grid. The Grid infrastructure is implemented as a bag of services. Each service performs a particular transaction following a client-server model. Each transaction is either stateless or supports only a conversional state. This model closely resemble HTTP-based Web transaction model: the user makes a request by pointing the Web browser to a particular URL, and a Web server responds with the corresponding, possibly dynamically generated, HTML page. However, the very early Web developers found this model too restrictive. Nowadays, most Web servers utilize object- or component-oriented technologies, such as EJB or CORBA, for session management, multistep transaction processing, persistence, user profiles, providing enterprise-wide access to resources including databases and for incorporating third-party services. There is a remarkable similarity between the current capabilities of the Web servers (the Web technologies), augmented with Application Servers (the Object and Component Technologies), and the required functionality of a Grid Computing Environment.

This paper provides an overview of Gateway and Mississippi Computational Web Portal (MCWP). These projects are being developed separately at Indiana University and Mississippi State University, respectively, but they share a common design heritage. The key features of both MCWP and Gateway are the use of XML for describing portal metadata and the use of distributed object technologies in the control tier.

30.2 DEPLOYMENT AND USE OF COMPUTING PORTALS

In order to make concrete the discussion presented in the introduction, we describe below our deployed portals. These provide short case studies on the types of portal users and the services that they require.

30.2.1 DMEFS: an application of the Mississippi Computational Web Portal

The Distributed Marine Environment Forecast System (DMEFS) [9] is a project of the Mississippi State team that is funded by the Office of Naval Research. DMEFS's goal is to provide open framework to simulate the littoral environments across many temporal and spatial scales in order to accelerate the evolution of timely and accurate forecasting. DMEFS is expected to provide a means for substantially reducing the time to develop, prototype, test, validate, and transition simulation models to operation, as well as support a genuine, synergistic collaboration among the scientists, the software engineers, and the operational users. In other words, the resulting system must provide an environment for model development, including model coupling, model validation and data analysis, routine runs of a suite of forecasts, and decision support.

Such a system has several classes of users. The model developers are expected to be computer savvy domain specialists. On the other hand, operational users who routinely run the simulations to produce daily forecasts have only a limited knowledge on how the simulations actually work, while the decision support is typically interested only in accessing the end results. The first type of users typically benefits from services such as archiving and data pedigree as well as support for testing and validation. The second type of users benefits from an environment that simplifies the complicated task of setting up and running the simulations, while the third type needs ways of obtaining and organizing results.

DMEFS is in its initial deployment phase at the Naval Oceanographic Office Major Shared Resource Center (MSRC). In the next phase, DMEFS will develop and integrate metadata-driven access to heterogenous, distributed data sources (databases, data servers, scientific instruments). It will also provide support for data quality assessment, data assimilation, and model validation.

30.2.2 Gateway support for commodity codes

The Gateway computational Web portal is deployed at the Army Research Laboratory MSRC, with additional deployment approved for the Aeronautical Systems Center MSRC. Gateway's initial focus has been on simplifying access to commercial codes for novice HPC users. These users are assumed to understand the preprocessing and postprocessing tools of their codes on their desktop PC or workstation but not to be familiar with common HPC tasks such as queue script writing and job submission and management. Problems using HPC systems are often aggravated by the use of different queuing systems between and even within the same center, poor access for remote users caused by slow network speeds at peak hours, changing locations for executables, and licensing issues for commercial codes. Gateway attempts to hide or manage as much of these details as possible, while providing a browser front end that encapsulates sets of commands into relatively few portal actions. Currently, Gateway supports job creation, submission, monitoring, and archiving for ANSYS, ZNS, and Fluent, with additional support planned for CTH. Gateway interfaces to these codes are currently being tested by early users.

Because Gateway must deal with applications with restricted source codes, we wrap these codes in generic Java proxy objects that are described in XML. The interfaces for the invocation of these services likewise are expressed in XML, and we are in the process of converting our legacy service description to the Web service standard Web Services Description Language (WSDL) [10].

Gateway also provides secure file transfer, job monitoring and job management through a Web browser interface. These are currently integrated with the application interfaces but have proven popular on their own and so will be provided as stand-alone services in the future.

Future plans for Gateway include integration with the Interdisciplinary Computing Environment (ICE) [11], which provides visualization tools and support for light code coupling through a common data format. Gateway will support secure remote job creation and management for ICE-enabled codes, as well as secure, remote, sharable visualization services.

30.3 COMPUTING PORTAL SERVICES

One may build computational environments such as the one above out of a common set of core services. We list the following as the base set of abstract service definitions, which may be (but are not necessarily) implemented more or less directly with typical Grid technologies in the portal middle tier.

1. *Security*: Allow access only to authenticated users, give them access only to authorized areas, and keep all communications private.
2. *Information resources*: Inform the user about available codes and machines.
3. *Queue script generation*: On the basis of the user's choice of code and host, create a script to run the job for the appropriate queuing system.
4. *Job submission*: Through a proxy process, submit the job with the selected resources for the user.
5. *Job monitoring*: Inform the user of the status of his submitted jobs, and more generally provide events that allow loosely coupled applications to be staged.
6. *File transfer and management*: Allow the user to transfer files between his desktop computer and a remote system and to transfer files between remote systems.

Going beyond the initial core services above, both MCWP and Gateway have identified and have or are in the process of implementing the following GCE-specific services.

1. *Metadata-driven resource allocation and monitoring*: While indispensable for acquiring adequate resources for an application, allocation of remote resources adds to the complexity of all user tasks. To simplify this chore, one requires a persistent and platform-independent way to express computational tasks. This can be achieved by the introduction of application metadata. This user service combines standard authentication, information, resource allocation, and file transfer Grid services with GCE services: metadata discovery, retrieval and processing, metadata-driven Resource Specification Language (RSL) (or batch script) generation, resource brokerage, access to remote file systems and data servers, logging, and persistence.
2. *Task composition or workflow specification and management*: This user service automates mundane user tasks with data preprocessing and postprocessing, file transfers, format conversions, scheduling, and so on. It replaces the nonportable 'spaghetti' shell

scripts currently widely used. It requires task composition tools capable of describing the workflow in a platform-independent way, since some parts of the workflow may be preformed on remote systems. The workflow is built hierarchically from reusable modules (applications), and it supports different mechanisms for triggering execution of modules: from static sequences with branches to data flow to event-driven systems. The workflow manager combines information, resource brokers, events, resource allocation and monitoring, file transfer, and logging services.

3. *Metadata-driven, real-time data access service*: Certain simulation types perform assimilation of observational data or analyze experimental data in a real time. These data are available from many different sources in a variety of formats. Built on top of the metadata, file transfer and persistence services, this user service closely interacts with the resource allocation and monitoring or workflow management services.

4. *User space, persistency, and pedigree service*: This user service provides support for reuse and sharing of applications and their configuration, as well as for preserving the pedigree of all jobs submitted by the user. The pedigree information allows the user to reproduce any previous result on the one hand and to localize the product of any completed job on the other. It collects data generated by other services, in particular, by the resource allocation and workflow manager.

30.4 GRID PORTAL ARCHITECTURE

A computational Web portal is implemented as a multitier system composed of clients running on the users' desktops or laptops, portal servers providing user level services (i.e. portal middleware), and backend servers providing access to the computing resources.

30.4.1 The user interface

The user interacts with the portal through either a Web browser, a client application, or both. The central idea of both the Gateway and the MCWP user interfaces is to allow users to organize their work into problem contexts, which are then subdivided into *session contexts* in Gateway terminology, or projects and tasks using MCWP terms. Problems (or projects) are identified by a descriptive name handle provided by the user, with sessions automatically created and time-stamped to give them unique names. Within a particular session (or task), the user chooses applications to run and selects computing resources to use. This interface organization is mapped to components in the portal middleware (user space, persistency, and pedigree services) described below. In both cases, the Web browser–based user interface is developed using JavaServer Pages (JSP), which allow us to dynamically generate Web content and interface easily with our Java-based middleware.

The Gateway user interface provides three tracks: code selection, problem archive, and administration. The code selection track allows the user to start a new problem, make an initial request for resources, and submit the job request to the selected host's queuing system. The problem archive allows the user to revisit and edit old problem sessions so that he/she can submit his/her job to a different machine, use a different input file, and so forth. Changes to a particular session are stored in a newly generated session name. The administration track allows privileged users to add applications and host computers to the portal, modify the properties of these entities, and verify their installation. This information is stored in an XML data record, described below.

The MCWP user interface provides five distinct views of the system, depending on the user role: developer, analyst, operator, customer, and administrator. The developer view combines the selection and archive tracks. The analyst view provides tools for data selection and visualizations. The operator view allows for creating advance scheduling of tasks for routine runs (similar to creating a cron table). The customer view allows access to routinely generated and postprocessed results (plots, maps, and so forth). Finally, the administrator view allows configuration and controlling of all operations performed by the portal.

30.4.2 Component-based middleware

The portal middleware naturally splits into two layers: the actual implementation of the user services and the presentation layer responsible for providing mechanisms for the user interactions with the services. The presentation layer accepts the user requests and returns the service responses. Depending on the implementation strategy for the client, the services' responses are directly displayed in the Web browser or consumed by the client-side application.

A key feature of both Gateway and MCWP is that they provide a container-based middle tier that holds and manages the (distributed) proxy wrappers for basic services like those listed above. This allows us to build user interfaces to services without worrying about the implementation of those services. Thus, for example, we may implement the portal using standard service implementations from the Globus toolkit, we may implement some core services ourselves for stand-alone resources, or we may implement the portal as a mixture of these different service implementation styles.

The Gateway middle tier consists of two basic sections: a Web server running a servlet engine and a distributed CORBA-based middle tier (WebFlow). This is illustrated in Figure 30.1. The Web server typically runs a single Java Virtual Machine (JVM) on a single server host that contains local JavaBean components. These components may implement specific local services or they may act as proxies for WebFlow-distributed components running in different JVMs on a nest of host computers. WebFlow servers consist of a top-level master server and any number of child servers. The master server acts as a gatekeeper and manages the life cycle of the children. These child servers can in turn provide access to remote backend services such as HPCs running Portal Batch System (PBS) or Load Sharing Facility (LSF) queuing systems, a Condor flock, a Globus grid, and data storage devices. By running different WebFlow child servers on different hosts, we may easily span organizational barriers in a lightweight fashion. For more information on the WebFlow middleware, see References [12, 13, 14]. For a general overview of the role of commodity technologies in computational Grids, see Reference [15].

The MCWP application server is implemented using EJB. The user space is a hierarchy of entities: users, projects, tasks, and applications. The abstract application metadata tree is implemented as entity beans as well with the host-independent information as one database table and host-dependent information as another one. Finally, there are two entities related

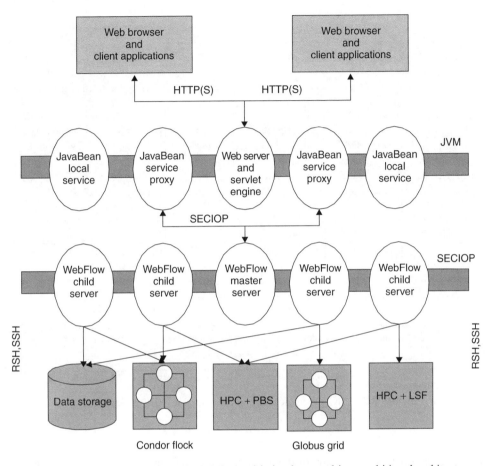

Figure 30.1 The Gateway computational portal is implemented in a multitiered architecture.

to job status: a job entity (with the unique jobId as the key in the job table) and a host that describes the target machines properties (metadata). It is important to note that all metadata beans (i.e. application, hosts, and data sets) are implemented using a hybrid technology: EJB and XML, that is, a database is used to store many short XML files.

The MCWP services are implemented as EJB session beans, and their relationship is depicted in Figure 30.2. The bottom-layer services are clients to the low-level Grid services, the upper-layer services are user level services, and the middle-layer services provides mapping between the two former ones. The task composition service provides a high-level interface for the metadata-driven resource allocation and monitoring. The knowledge about the configuration of each component of the computational task is encompassed in the application metadata and presented to the user in the form of a GUI. The user does not need to know anything about the low-level Globus interfaces, syntax of RSL or batch schedulers on the remote systems. In addition, the user is given either the default values of parameters for all constituent applications that comprise the task or the

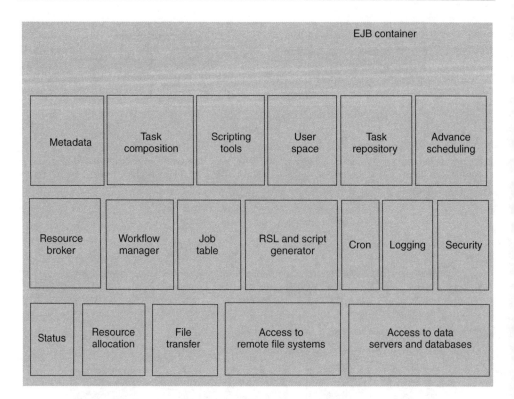

Figure 30.2 MCWP services are implemented as EJB session beans.

values of parameters used in any of the previous runs. The application metadata are accessible through the metadata service. A configured task, that is, application parameters for all components and relationship between the components (e.g. workflow specification) is transparently saved in the user space (or application context) for later reuse. Optionally, the user may choose to publish the configured task to be used by others through the task repository service.

The scripting tool is similar to the task composition service. If several steps are to be executed on the same machine running in a batch mode, it is much more efficient to generate a shell script that orchestrate these steps in a single run, rather than to submit several batch jobs under control of the workflow manager. The advance scheduling service allows an operator to schedule a selected application to run routinely at specified times, say everyday at 2 p.m. The services in the middle and bottom layers have self-describing names. The job table is an EJB entity that keeps track of all jobs submitted through MCWP and is used for reconnection, monitoring, and preserving the task pedigree. The cron service reproduces the functionality of the familiar Unix service to run commands at predefined times, and it is closely related to the advance scheduling user service. The security service is responsible for delegation of the user credentials. For Globus-based implementation, it is a client of the myProxy server [16], which stores the user's temporary certificates. For Kerberos-based systems, it serializes the user tickets.

For both MCWP and Gateway, it is natural to implement clients as stand-alone Java applications built as a collection of (Enterprise) JavaBean clients. However, this approach has several drawbacks if applied to the 'World Wide Grids'. CORBA and EJB technologies are well suited for distributed, enterprise-wide applications but not for the Internet. Going beyond the enterprise boundaries, there is always a problem with client software distribution and in particular the upgrades of the service interfaces. Secondly, the protocols employed for the client-server communication are not associated with standard ports and are often filtered by firewalls, making it impossible for the external users to access the services. Finally, in the case of EJB, currently available containers do not implement robust security mechanisms for extra-enterprise method invocation.

An alternative solution is to restrict the scope of client application to the application server and to provide access to it through the World Wide Web, as shown in Figure 30.1 for Gateway and Figure 30.3 for MCWP. Here, the clients are implemented as the server-side Java Beans and these beans are accessed by JSP to dynamically generate user interface as HTML forms. This approach solves the problem of the client software distribution as well as the problem of secure access to the Grid resources. With the Web browser–server communications secured using the HTTPS protocol, and using myProxy server to store the

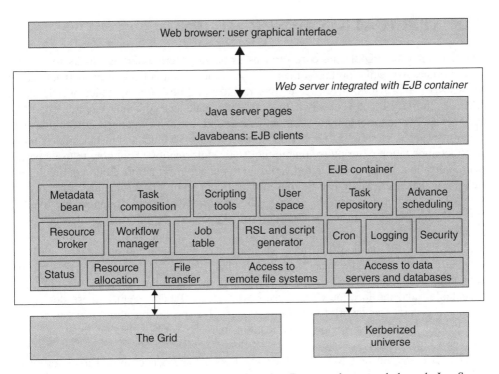

Figure 30.3 EJB clients may be implemented as JavaBeans and accessed through JavaServer Pages.

user's Globus (proxy) certificate, the MCWP services are capable of securely allocating services, transfer files, and access data using the Globus grid services. Finally, the server-side Java Beans acting as EJB clients can be easily converted into Web services (Simple Object Access Protocol/Web Services Description Language (SOAP/WSDL)) [17]. Therefore, the MCWP can be implemented as a stand-alone application, deployed using Java WebStart technology, acting as a WSDL client as opposed to EJB client.

30.4.3 Resource tier

Computing resources constitute the final tier of the portal. These again are accessed through standard protocols, such as the Globus protocols for Grid-enabled computing resources, and also including protocols such as Java Database Connectivity (JDBC) for database connections.

There is always the problem that the computing resource may not be using a grid service, so the transport mechanism for delivering commands from the middle tier to the backend must be pluggable. We implement this in the job submission proxy service in the middle tier, which constructs and invokes commands on the backend either through secure remote shell invocations or else through something such as a globusrun command. The actual command to use in a particular portal installation is configured.

30.5 APPLICATION DESCRIPTORS

One may view the middle tier core services as being generic building blocks for assembling portals. A specific portal on the other hand includes a collection of metadata about the services it provides. We refer to this metadata as *descriptors*, which we define in XML. Both MCWP and Gateway define these metadata as a container hierarchy of XML schema: applications contain host computing resources, which contain specific entities like queuing systems. Descriptors are divided into two types: abstract and instance descriptors. Abstract application descriptors contain the 'static' information about how to use a particular application. Instance descriptors are used to collect information about a particular run by a particular user, which can be reused later by an archiving service and for pedigree.

XML descriptors are used to describe data records that should remain long-lived or static. As an example, an application descriptor contains the information needed to run a particular code: the number of input and output files that must be specified on the command line, the method that the application uses for input and output, the machines that the code is installed on, and so on. Machine, or host, descriptors describe specific computing resources, including the queuing systems used, the locations of application executables, and the location of the host's workspace. Taken together, these descriptors provide a general framework for building requests for specific resources that can be used to generate batch queue scripts. Applications may be further chained together into a workflow.

The GCE Application Metadata Working Group has been proposed as a forum for different groups to exchange ideas and examples of using application metadata, which may potentially lead to a standardization of some of the central concepts.

30.6 SAMPLE SERVICE IMPLEMENTATIONS

In previous sections we outlined several cores services. In this section, we look in some detail at two service implementations: batch script generation and context management. The first serves as an exemplar for building proxy services to specific implementations and the second is an example of a portal-specific service.

30.6.1 Batch script generation

Portals are designed in part to aid users who are not familiar with HPC systems and so need tools that will assist them in creating job scripts to work with a particular queuing system. From our viewpoint as developers, it is also important to design an infrastructure that will allow us to support many different queuing systems at different sites. It was our experience that most queuing systems were quite similar (to first approximation) and could be broken down into two parts: a queuing system-specific set of header lines, followed by a block of script instructions that were queue-independent. We decided that a script generating service would best be implemented with a simple 'Factory' design pattern [18], with queue script generators for specific queues extending a common parent. This is illustrated in Figure 30.4. This allows us to choose the appropriate generator at run time. New generators can be added by extending a common parent.

Queue scripts are generated on the server, based on the user's choice of machine, application, memory requirements, and so on. This script is then moved (if necessary) to the selected remote host using a WebFlow module. The information needed to generate the queue script is stored as context data, described below. This allows the user to return to old sessions and revise portions of the resource requests. If this modified request is to be run on a different queuing system, then a new script can be generated using the slightly modified context (session) data.

Batch script generation is an excellent candidate for a simple Web service, and we have recently implemented this in collaboration with a team from the San Diego Supercomputer Center [19].

30.6.2 Context management

As described in previous sections, the user's interactions with either the MCWP or Gateway portals are mapped to nested containers (objects living in context). The session data associated with these containers is stored persistently and can be recovered later by the user for editing and resubmission. The base of each tree is the user context, with subdirectories for problems and subsubdirectories for sessions.

We refer to this data as context data. Context data minimally describes the parent and children of a particular node, but can be extended to include any arbitrary name-value pairs. This is useful for storing, for instance, HTTP GET or POST requests. In practice, we store all information gained from the user's interaction with the browser forms. For instance, the user's request to run a particular code on a certain host with a specified amount of memory, nodes, and wall time is represented as a series of linked hashtables

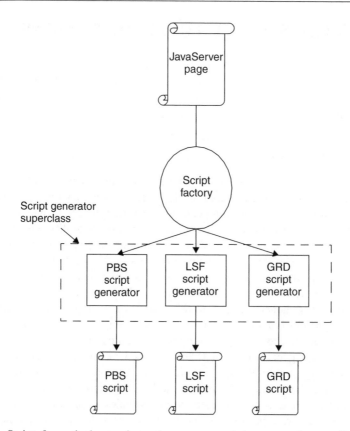

Figure 30.4   Scripts for particular queuing systems are generated on request from a calling JavaServer Page. A factory JavaBean creates an instance of the appropriate script generator. All script generators extend a common parent.

containing name-value pairs that is stored in a directory that is associated with the problem and session contexts in which the request was created.

30.7 KERBEROS SECURITY REQUIREMENTS IN MULTITIERED ARCHITECTURES

Security is of utmost importance when grid resources are made available through the Internet. Security solutions of commercial Web sites are typically inadequate for computational grids: customers and retailers are protected by third parties such as credit card companies and banks, the company sponsoring the site profits from the sale of goods rather than services, and the site has no need to allow users direct access to its computational resources. None of these conditions apply to centers running computational portals.

Portals must be compliant with the existing security requirements of the centers where they are run. Both the Gateway and the MCWP are funded in part to be deployed at

Department of Defense (DoD) computing centers that require Kerberos [20] for authentication, data transmission integrity, and privacy [21]. In the following section we describe some general security issues for portals as well as discuss the details of providing Kerberos support.

As can be seen in both Figures 30.1 and 30.3, our three-tiered Web portals require at least three layers of security: client-server and server-backend connections must be authenticated, authorized, and made private. This also applies to our distributed middle tier. The critical feature is that the control layer must be given delegation powers in order to make invocations of the backend services on behalf of the user. This is simple enough to implement with Kerberos tickets, but the key to doing this securely is to provide strong authentication between the client and the server. The server layer in the middle tier will be performing tasks on the backend as the user, on the basis of requests it receives from a browser purporting to be associated with a particular user.

One of the difficulties of using Kerberos as an authentication service for Web applications is that it is not compatible with existing browsers. Early implementations of Gateway attempted to solve this problem by using a Java applet to establish the secure connection and manage the client requests. This introduced new problems, however, as this required installing native libraries on the client. We were thus required to override the native browser's Java Runtime Environment with a plug-in, but we discovered in turn that this broke the applet-browser communication capabilities.

We have explored two solutions to this problem. The first solution is to tunnel client-to-middle tier HTTP requests through secure, Kerberos-authenticated proxies. Our implementation of this idea (which we call Charon) consists of a client-side proxy application and a server-side module that intercepts HTTP requests and responses between the user's browser and the Apache server. A user wishing to connect to the Web site must first get a Kerberos ticket-granting ticket (TGT) and then launch the client Charon application. The user then points his browser to a specified port on his local host. All traffic through this port is forwarded to the Charon server (a module running in the middle tier), which forwards the request to the local Web server. The response is collected by the Charon server and sent back to the Charon client, which sends it to the local port. All Web traffic is encrypted and message digested to preserve message integrity. Charon thus represents a fully kerberized method for authenticating HTTP clients and servers.

A drawback to the previous scheme is that it requires some installation on the user's desktop, and the dynamically linked native libraries require privileged access before they can be installed on some desktop machines. As an alternative solution, we provide a browser interface that will create a Kerberos TGT on the server for the user. The user is authenticated in the usual way, with password and passcode. After the ticket is created on the server, we use a message-digesting algorithm to create a unique cookie for the browser. This cookie and the client browser's IP address are used for future identification. Before the browser can view secured pages, these two identification pieces are verified against the session values stored on the server. All wire communications go over 128-bit encrypted Secure Socket Layer (SSL) [22] connections, and SSL session IDs provide a third method of identification. Client certificates can be used as a fourth means of identifying the client browser. The user can delete his server-side session tickets through the browser, and they will also be automatically deleted when the JSP session context expires. This is currently our preferred security mechanism as it requires no installation and software upgrades on the client. It can also make use of existing proxy servers to cross firewall boundaries, requiring no additional development on our part.

The distributed object servers represent the second layer that must be secured. One challenge here is that the servers typically do not run as privileged processes, whereas Kerberos makes the assumption that kerberized services are run as system level processes. Practically, this requires the service to have access to a system's keytab file, which has restricted access. In implementing kerberized servers, we have obtained special keytab files from the Kerberos administrator that are owned by the application account and that are tied to a specific host machine. The use of keytab files can be avoided by user-to-user authentication. This would allow a client and server to both authenticate with TGTs, instead of a TGT and keytab. User-to-user authentication is not supported by the standard Generic Security Service API (GSSAPI), although extensions have been implemented.

Finally, secure connections to the remote backend machines (such as HPCs or mass storage) must also be made. To accomplish this, we create the user's personal server in the middle tier with a forwardable Kerberos ticket. The server can then contact the remote backend by simply invoking an external mechanism such as a kerberized remote shell invocation. This allows commands to be run on the remote host through the preexisting, approved remote invocation method.

30.8 SUMMARY AND FUTURE WORK

The MCWP and Gateway portals provide examples of a coarse-grained approach to accessing remote, distributed high-performance computing resources. We have the capability to provide bridges to different grid infrastructure services, and where required implement these services ourselves.

The so-called Web services model and particularly the proposed Open Grid Services Architecture (OGSA) [23] represent an important future development for computing portals and their services. This is particularly true for both Gateway and MCWP, since Web services in some respects duplicate some of the basic ideas of both CORBA and EJB. In some respects, we are ahead of the game because our use of CORBA and EJB has always dictated the separation of portal service interface from implementation and access protocol. It becomes now a matter of bridging to our legacy portal service implementations through new WSDL interfaces and SOAP invocations. These protocol bridges are relatively easy to implement for specific cases: we have done so with legacy WebFlow services as an experiment, for example.

The fundamental impact of the OGSA services on the portal services described in this paper remains unclear, while the OGSA specification is still being developed. However, it is likely that the core service list will not change and the portal middle tier proxies to these services will just be clients to OGSA services, with the benefit that the service definition is in WSDL. Advanced services such as metadata-driven resource allocation, task composition, and user session persistency, and data pedigree access will remain important services that must be implemented at the application level, although they may use underlying core OGSA services.

Web services are an important development for computational portals because they promote the development of interoperable and reusable services with well-defined interfaces. The implication of this is that clients to Web services will be easy to build (by independent groups and through dynamic clients). We thus are moving toward a system in which each service (simple or complex) has its on user interface. Portals can now be thought of as an aggregation of the interfaces to these services. We believe that the efforts of the commercial world, including standard portlet APIs [24] and portlet-based Web service [25] should be followed, exploited, and extended by the computational Web portal community.

REFERENCES

1. The Globus Project, http://www.globus.org/, July 20, 2001.
2. Legion: A Worldwide Virtual Computer, http://www.cs.virginia.edu/ legion, July 20, 2001.
3. Condor: High Throughput Computing, http://www.cs.wisc.edu/condor, July 20, 2001.
4. Foster, I. and Kesselman, C. (eds) (1999) *The Grid: Blueprint for a New Computing Infrastructure*. San Francisco, CA: Morgan Kaufmann Publishers.
5. Global Grid Forum, http://www.gridforum.org, July 20, 2001.
6. Orfali, R. and Harkey, D. (1998) *Client/Server Programming with Java and CORBA*. New York: John Wiley & Sons.
7. CORBA/IIOP Specification, http://www.omg.org/technology/documents/formal/corba_iiop.htm, July 20, 2001.
8. Enterprise JavaBeans Technology, http://java.sun.com/products/ejb.
9. Haupt, T., Bangalore, P. and Henley, G. (2001) A computational web portal for distributed marine environment forecast system. *Proceedings of the High-Performance Computing and Networking*, HPCN-Europe 2001, 2001, pp. 104–114.
10. Web Services Description Language (WSDL) 1.1, http://www.w3c.org/TR/wsdl, July 8, 2002.
11. ICE Home Page, http://www.arl.hpc.mil/ice/, July 8, 2002.
12. Haupt, T., Akarsu, E., Fox, G. and Youn, C. (2000) The gateway system: uniform web based access to remote resources. *Concurrency and Computation: Practice and Experience*, **12**(8), 629–642.
13. Bhatia, D., Burzevski, V., Camuseva, M., Fox, G., Furmanski, W. and Premchandran, G. (1997) WebFlow – A visual programming paradigm for Web/Java based coarse grain distributed computing. *Concurrency and Computation: Practice and Experience*, **9**(6), 555–577.
14. Akarsu, E. (1999) Integrated Three-Tier Architecture for High-Performance Commodity Metacomputing, Ph.D. Dissertation, Syracuse University, Syracuse, 1999.
15. Fox, G. and Furmanski, W. (1999) High performance commodity computing, in Foster, I. and Kesselman, C. (eds) *The Grid: Blueprint for a New Computing Infrastructure*. San Francisco, CA: Morgan Kaufmann.
16. Novotny, J., Tuecke, S. and Welch, V. (2001) An online credential repository for the grid: myproxy, *Proceedings of the Tenth International Symposium on High Performance Distributed Computing*. IEEE Press.
17. SOAP Version 1.2, http://www.w3c.org/TR/soap12, July 20, 2001.
18. Gamma E., Helm, R., Johnson, R. and Vlissides, J. (1995) *Design Patterns: Elements of Reusable Object-Oriented Software*. Reading, MA: Addison-Wesley.
19. Mock S., Pierce, M., Youn, C., Fox G. and Thomas, M. (2002) A batch script generator Web service for computational portals. *Proceedings of the International Conference on Communications in Computing*, 2002.
20. Neuman C. and Tso, T. (1994) Kerberos: an authentication service for computer networks. *IEEE Communications*, **32**(9), 33–38.

21. Department of Defense High Performance Computing Modernization Program Security Issues, http://www.hpcmo.hpc.mil/Htdocs/Security.
22. SSL 3.0 Specification, http://home.netscape.com/eng/ssl3.
23. Foster, I., Kesselman, C., Nick, J. and Tuecke, S. The Physiology of the Grid, Draft available from http://www.globus.org/ogsa, July 8, 2002.
24. Java Specification Request (JSR) 162: Portlet API, http://www.jcp.org/jsr/detail/162.jsp, July 8, 2002.
25. OASIS Web Services for Remote Portals (WSRP), http://www.oasis-open.org/committees/wsrp, July 8, 2002.

DISCOVER: a computational collaboratory for interactive Grid applications[‡]

Vijay Mann and Manish Parashar[*,†]

Rutgers, The State University of New Jersey, Piscataway, New Jersey, United States

31.1 INTRODUCTION

A collaboratory is defined as a place where scientists and researchers work together to solve complex interdisciplinary problems, despite geographic and organizational boundaries [1]. The growth of the Internet and the advent of the computational 'Grid' [2, 3] have made it possible to develop and deploy advanced computational collaboratories [4, 5] that provide uniform (collaborative) access to computational resources, services, applications and/or data. These systems expand the resources available to researchers, enable

[‡] The DISCOVER collaboratory can be accessed at http://www.discoverportal.org/
[*] National Science Foundation (CAREERS, NGS, ITR) ACI9984357, EIA0103674, EIA0120934
[†] Department of Energy/California Institute of Technology (ASCI) PC 295251

Grid Computing – Making the Global Infrastructure a Reality. Edited by F. Berman, A. Hey and G. Fox
© 2003 John Wiley & Sons, Ltd ISBN: 0-470-85319-0

multidisciplinary collaborations and problem solving, accelerate the dissemination of knowledge, and increase the efficiency of research.

This chapter presents the design, implementation and deployment of the DISCOVER computational collaboratory that enables interactive applications on the Grid. High-performance simulations are playing an increasingly critical role in all areas of science and engineering. As the complexity and computational cost of these simulations grows, it has become important for scientists and engineers to be able to monitor the progress of these simulations and to control or steer them at run time. The utility and cost-effectiveness of these simulations can be greatly increased by transforming traditional batch simulations into more interactive ones. Closing the loop between the user and the simulations enables experts to drive the discovery process by observing intermediate results, by changing parameters to lead the simulation to more interesting domains, play what-if games, detect and correct unstable situations, and terminate uninteresting runs early. Furthermore, the increased complexity and multidisciplinary nature of these simulations necessitates a collaborative effort among multiple, usually geographically distributed scientists/engineers. As a result, collaboration-enabling tools are critical for transforming simulations into true research modalities.

DISCOVER [6, 7] is a virtual, interactive computational collaboratory that enables geographically distributed scientists and engineers to collaboratively monitor and control high-performance parallel/distributed applications on the Grid. Its primary goal is to bring Grid applications to the scientists/'engineers' desktop, enabling them to collaboratively access, interrogate, interact with and steer these applications using Web-based portals. DISCOVER is composed of three key components (see Figure 31.1):

1. *DISCOVER middleware substrate*, which enables global collaborative access to multiple, geographically distributed instances of the DISCOVER computational collaboratory and provides interoperability between DISCOVER and external Grid services.

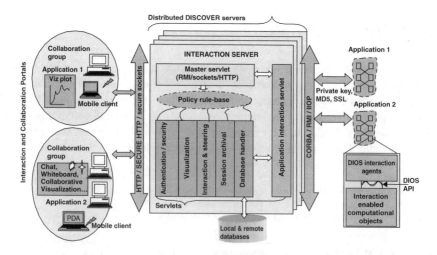

Figure 31.1 Architectural schematic of the DISCOVER computational collaboratory.

The middleware substrate enables DISCOVER interaction and collaboration servers to dynamically discover and connect to one another to form a peer network. This allows clients connected to their local servers to have global access to all applications and services across all servers based on their credentials, capabilities and privileges. The DISCOVER middleware substrate and interaction and collaboration servers build on existing Web servers and leverage commodity technologies and protocols to enable rapid deployment, ubiquitous and pervasive access, and easy integration with third party services.

2. *Distributed Interactive Object Substrate (DIOS)*, which enables the run-time monitoring, interaction and computational steering of parallel and distributed applications on the Grid. DIOS enables application objects to be enhanced with sensors and actuators so that they can be interrogated and controlled. Application objects may be distributed (spanning many processors) and dynamic (be created, deleted, changed or migrated at run time). A control network connects and manages the distributed sensors and actuators, and enables their external discovery, interrogation, monitoring and manipulation.

3. *DISCOVER interaction and collaboration portal*, which provides remote, collaborative access to applications, application objects and Grid services. The portal provides a replicated shared workspace architecture and integrates collaboration tools such as chat and whiteboard. It also integrates 'Collaboration Streams,' that maintain a navigable record of all client–client and client-application interactions and collaborations.

Using the DISCOVER computational collaboratory clients can connect to a local server through the portal and can use it to discover and access active applications and services on the Grid as long as they have appropriate privileges and capabilities. Furthermore, they can form or join collaboration groups and can securely, consistently and collaboratively interact with and steer applications based on their privileges and capabilities. DISCOVER is currently operational and is being used to provide interaction capabilities to a number of scientific and engineering applications, including oil reservoir simulations, computational fluid dynamics, seismic modeling, and numerical relativity. Furthermore, the DISCOVER middleware substrate provides interoperability between DISCOVER interaction and collaboration services and Globus [8] Grid services. The current DISCOVER server network includes deployments at CSM, University of Texas at Austin, and is being expanded to include CACR, California Institute of Technology.

The rest of the chapter is organized as follows. Section 31.2 presents the DISCOVER middleware substrate. Section 31.3 describes the DIOS interactive object framework. Section 31.3.4 presents the experimental evaluation. Section 31.4 describes the DISCOVER collaborative portal. Section 31.5 presents a summary of the chapter and the current status of DISCOVER.

31.2 THE DISCOVER MIDDLEWARE SUBSTRATE FOR GRID-BASED COLLABORATORIES

The proliferation of the computational Grid and recent advances in Grid technologies have enabled the development and deployment of a number of advanced problem-solving environments and computational collaboratories. These systems provide specialized services to their user communities and/or address specific issues in wide-area resource sharing and Grid computing [9]. However, solving real problems on the Grid requires combining these services in a seamless manner. For example, execution of an application on the Grid requires security services to authenticate users and the application, information services for resource discovery, resource management services for resource allocation, data transfer services for staging, and scheduling services for application execution. Once the application is executing on the Grid, interaction, steering and collaboration services allow geographically distributed users to collectively monitor and control the application, allowing the application to be a true research or instructional modality. Once the application terminates data storage and cleanup, services come into play. Clearly, a seamless integration and interoperability of these services is critical to enable global, collaborative, multi-disciplinary and multi-institutional, problem solving.

Integrating these collaboratories and Grid services presents significant challenges. The collaboratories have evolved in parallel with the Grid computing effort and have been developed to meet unique requirements and support specific user communities. As a result, these systems have customized architectures and implementations and build on specialized enabling technologies. Furthermore, there are organizational constraints that may prevent such interaction as it involves modifying existing software. A key challenge then is the design and development of a robust and scalable middleware that addresses interoperability and provides essential enabling services such as security and access control, discovery, and interaction and collaboration management. Such a middleware should provide loose coupling among systems to accommodate organizational constraints and an option to join or leave this interaction at any time. It should define a minimal set of interfaces and protocols to enable collaboratories to share resources, services, data and applications on the Grid while being able to maintain their architectures and implementations of choice.

The DISCOVER middleware substrate [10, 11] defines interfaces and mechanisms for a peer-to-peer integration and interoperability of services provided by domain-specific collaboratories on the Grid. It currently enables interoperability between geographically distributed instances of the DISCOVER collaboratory. Furthermore, it also integrates DISCOVER collaboratory services with the Grid services provided by the Globus Toolkit [8] using the CORBA Commodity Grid (CORBA CoG) Kit [12, 13]. Clients can now use the services provided by the CORBA CoG Kit to discover available resources on the Grid, to allocate required resources and to run applications on these resources, and use DISCOVER to connect to and collaboratively monitor, interact with, and steer the applications. The middleware substrate enables DISCOVER interaction and steering servers as well as Globus servers to dynamically discover and connect to one another to form a peer network. This allows clients connected to their local servers to have global access to all applications and services across all the servers in the network based on their credentials, capabilities and privileges.

31.2.1 DISCOVER middleware substrate design

The DISCOVER middleware substrate has a hybrid architecture, that is, it provides a client-server architecture from the users' point of view, while the middle tier has a peer-to-peer architecture. This approach provides several advantages. The middle-tier peer-to-peer network distributes services across peer servers and reduces the requirements of a server. As clients connect to the middle tier using the client-server approach, the number of peers in the system is significantly smaller than a pure peer-to-peer system. The smaller number of peers allows the hybrid architecture to be more secure and better managed as compared to a true peer-to-peer system and restricts the security and manageability concerns to the middle tier. Furthermore, this approach makes no assumptions about the capabilities of the clients or the bandwidth available to them and allows for very thin clients. Finally, servers in this model can be lightweight, portable and easily deployable and manageable, instead of being heavyweight (as in pure client-server systems). A server may be deployed anywhere there is a growing community of users, much like a HTTP Proxy server.

A schematic overview of the overall architecture is presented in Figure 31.2(a). It consists of (collaborative) client portals at the frontend, computational resources, services or applications at the backend, and the network of peer servers in the middle. To enable ubiquitous access, clients are kept as simple as possible. The responsibilities of the middleware include providing a 'repository of services' view to the client, providing controlled access to these backend services, interacting with peer servers and collectively managing and coordinating collaboration. A client connects to its 'closest' server and should have access to all (local and remote) backend services and applications defined by its privileges and capabilities.

Backend services can divided into two main classes – (1) resource access and management toolkits (e.g. Globus, CORBA CoG) providing access to Grid services and (2) collaboratory-specific services (e.g. high-performance applications, data archives and network-monitoring tools). Services may be specific to a server or may form a pool of services that can be accessed by any server. A service will be server specific if direct access to the service is restricted to the local server, possibly due to security, scalability or compatibility constraints. In either case, the servers and the backend services are accessed using standard distributed object technologies such as CORBA/IIOP [14, 15] and RMI [16]. XML-based protocols such as SOAP [17] have been designed considering the services model and are ideal candidates.

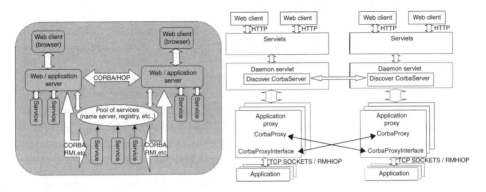

Figure 31.2   DISCOVER middleware substrate: (a) architecture and (b) implementation.

The middleware architecture defines three levels of interfaces for each server in the substrate. The level-one interfaces enable a server to authenticate with peer servers and query them for active services and users. The level-two interfaces are used for authenticating with and accessing a specific service at a server. The level-three interfaces (Grid Infrastructure Interfaces) are used for communicating with underlying core Grid services (e.g. security, resource access). The implementation and operation of the current DISCOVER middleware substrate is briefly described below. Details can be found in References [10, 18].

31.2.2 DISCOVER middleware substrate implementation

31.2.2.1 DISCOVER interaction and collaboration server

The DISCOVER interaction/collaboration servers build on commodity Web servers, and extend their functionality (using Java Servlets [19]) to provide specialized services for real-time application interaction and steering and for collaboration between client groups. Clients are Java applets and communicate with the server over HTTP using a series of HTTP *GET* and *POST* requests. Application-to-server communication either uses standard distributed object protocols such as CORBA [14] and Java RMI [16] or a more optimized, custom protocol over TCP sockets. An *ApplicationProxy* object is created for each active application/service at the server and is given a unique identifier. This object encapsulates the entire context for the application. Three communication channels are established between a server and an application: (1) a *MainChannel* for application registration and periodic updates, (2) a *CommandChannel* for forwarding client interaction requests to the application, and (3) a *ResponseChannel* for communicating application responses to interaction requests. At the other end, clients differentiate between the various messages (i.e. Response, Error or Update) using Java's reflection mechanism. Core service handlers provided by each server include the Master Handler, Collaboration Handler, Command Handler, Security/Authentication Handler and the Daemon Servlet that listens for application connections. Details about the design and implementation of the DISCOVER Interaction and Collaboration servers can be found in Reference [7].

31.2.2.2 DISCOVER middleware substrate

The current implementation of the DISCOVER middleware consists of multiple independent collaboratory domains, each consisting of one or more DISCOVER servers, applications/services connected to the server(s) and/or core Grid services. The middleware substrate builds on CORBA/IIOP, which provides peer-to-peer connectivity between servers within and across domains, while allowing them to maintain their individual architectures and implementations. The implementation is illustrated in Figure 31.2(b). It uses the level-one and level-two interfaces to construct a network of DISCOVER servers. A third level of interfaces is used to integrate Globus Grid Services [8] via the CORBA CoG [12, 13]. The different interfaces are described below.

DiscoverCorbaServer interface: The *DiscoverCorbaServer* is the level-one interface and represents a server in the system. This interface is implemented by each server and defines the methods for interacting with a server. This includes methods for authenticating with the server, querying the server for active applications/services and obtaining the list of users logged on to the server. A *DiscoverCorbaServer* object is maintained by each server's Daemon Servlet and publishes its availability using the CORBA trader service. It also maintains a table of references to *CorbaProxy* objects for remote applications/services.

CorbaProxy interface: The *CorbaProxy* interface is the level-two interface and represents an active application (or service) at a server. This interface defines methods for accessing and interacting with the application/service. The *CorbaProxy* object also binds itself to the CORBA naming service using the application's unique identifier as the name. This allows the application/service to be discovered and remotely accessed from any server. The *DiscoverCorbaServer* objects at servers that have clients interacting with a remote application maintain a reference to the application's *CorbaProxy* object.

Grid Infrastructure Interfaces: The level-three interfaces represent core Globus Grid Services. These include: (1) the *DiscoverGSI* interface that enables the creation and delegation of secure proxy objects using the Globus GSI Grid security service, (2) the *DiscoverMDS* that provides access to the Globus MDS Grid information service using Java Naming and Directory Interface (JNDI) [20] and enables users to securely connect to and access MDS servers, (3) the *DiscoverGRAM* interface that provides access to the Globus GRAM Grid resource management service and allows users to submit jobs on remote hosts and to monitor and manage these jobs using the CORBA Event Service [21], and (4) the *DiscoverGASS* interface that provides access to the Globus Access to Secondary Storage (GASS) Grid data access service and enables Grid applications to access and store remote data.

31.2.3 DISCOVER middleware operation

This section briefly describes key operations of the DISCOVER middleware. Details can be found in References [10, 18].

31.2.3.1 Security/authentication

The DISCOVER security model is based on the Globus GSI protocol and builds on the CORBA Security Service. The GSI delegation model is used to create and delegate an intermediary object (the CORBA GSI Server Object) between the client and the service. The process consists of three steps: (1) client and server objects mutually authenticate using the CORBA Security Service, (2) the client delegates the *DiscoverGSI* server object to create a proxy object that has the authority to communicate with other GSI-enabled Grid Services, and (3) the client can use this secure proxy object to invoke secure connections to the services.

Each DISCOVER server supports a two-level access control for the collaboratory services: the first level manages access to the server while the second level manages access to a particular application. Applications are required to be registered with a server and to provide a list of users and their access privileges (e.g. read-only, read-write). This information is used to create access control lists (ACL) for each user-application pair.

31.2.3.2 Discovery of servers, applications and resources

Peer DISCOVER servers locate each other using the CORBA trader services [22]. The CORBA trader service maintains server references as *service offer* pairs. All DISCOVER servers are identified by the service-id *DISCOVER*. The service offer contains the CORBA object reference and a list of properties defined as name-value pairs. Thus, the object can be identified on the basis of the service it provides or its properties. Applications are located using their globally unique identifiers, which are dynamically assigned by the DISCOVER server and are a combination of the server's IP address and a local count at the server. Resources are discovered using the Globus MDS Grid information service, which is accessed via the *MDSHandler* Servlet and the *DiscoverMDS* interface.

31.2.3.3 Accessing Globus Grid services: job submission and remote data access

DISCOVER middleware allows users to launch applications on remote resources using the Globus GRAM service. The clients invoke the *GRAMHandler* Servlet in order to submit a job. The *GRAMHandler* Servlet, using the delegated CORBA GSI Server Object, accesses the *DiscoverGRAM* server object to submit jobs to the Globus gatekeeper. The user can monitor jobs using the CORBA Event Service. Similarly, clients can store and access remote data using the Globus GASS service. The *GASSHandler* Servlet, using the delegated CORBA GSI Server Object, accesses the *DiscoverGASS* server object and the corresponding GASS service using the protocol specified by the client.

31.2.3.4 Distributed collaboration

The DISCOVER collaboratory enables multiple clients to collaboratively interact with and steer (local and remote) applications. The *collaboration handler* servlet at each server handles the collaboration on the server side, while a dedicated polling thread is used on the client side. All clients connected to an application instance form a collaboration group by default. However, as clients can connect to an application through remote servers, collaboration groups can span multiple servers. In this case, the *CorbaProxy* objects at the servers poll each other for updates and responses.

The peer-to-peer architecture offers two significant advantages for collaboration. First, it reduces the network traffic generated. This is because instead of sending individual collaboration messages to all the clients connected through a remote server, only one message is sent to that remote server, which then updates its locally connected clients. Since clients always interact through the server closest to them and the broadcast messages for collaboration are generated at this server, these messages do not have to travel large distances across the network. This reduces overall network traffic as well as client latencies, especially when the servers are geographically far away. It also leads to better scalability in terms of the number of clients that can participate in a collaboration session without overloading a server, as the session load now spans multiple servers.

31.2.3.5 Distributed locking and logging for interactive steering and collaboration

Session management and concurrency control is based on capabilities granted by the server. A simple locking mechanism is used to ensure that the application remains in a consistent state during collaborative interactions. This ensures that only one client 'drives' (issues commands) the application at any time. In the distributed server case, locking information is only maintained at the application's host server, that is, the server to which the application connects directly.

The session archival handler maintains two types of logs. The first log maintains all interactions between a client and an application. For remote applications, the client logs are maintained at the server where the clients are connected. The second log maintains all requests, responses and status messages for each application throughout its execution. This log is maintained at the application's host server (the server to which the application is directly connected).

31.2.4 DISCOVER middleware substrate experimental evaluation

This section gives a brief summary of the experimental evaluation of the DISCOVER middleware substrate. A more detailed description is presented in References [10, 18].

31.2.4.1 Evaluation of DISCOVER collaboratory services

This evaluation compared latencies for indirect (remote) accesses and direct accesses to DISCOVER services over a local area network (LAN) and a wide area network (WAN). The first set of measurements was for a 10-Mbps LAN and used DISCOVER servers at Rutgers University in New Jersey. The second set of measurements was for a WAN and used DISCOVER servers at Rutgers University and at University of Texas at Austin. The clients were running on the LAN at Rutgers University for both sets of measurements and requested data of different sizes from the application. Response times were measured for both, a direct access to the server where the application was connected and an indirect (remote) access through the middleware substrate. The time taken by the application to compute the response was not included in the measured time. Indirect (remote) access time included the direct access time plus the time taken by the server to forward the request to the remote server and to receive the result back from the remote server over IIOP. An average response time over 10 measurements was calculated for each response size.

The resulting response latencies for direct and indirect accesses measured on the LAN indicated that it is more efficient to directly access an application when it is on the same LAN. In contrast to the results for the LAN experiment, indirect access times measured on the WAN were of comparable order to direct access times. In fact, for small data sizes (1 KB, 10 KB and 20 KB) indirect access times were either equal to or smaller than direct access times. While these results might appear to be contradictory to expectations, the underlying communication for the two accesses provides an explanation. In the direct access measurement, the client was running at Rutgers and accessing the server at Austin over HTTP. Thus, in the direct access case, a large network path across the Internet was covered over HTTP, which meant that a new TCP connection was set up over the WAN for every request. In the indirect access case, however, the client at Rutgers accessed the local server at Rutgers over HTTP, which in turn accessed the server at Austin over IIOP. Thus, the path covered over HTTP was short and within the same LAN, while the larger network path (across the Internet) was covered over IIOP, which uses the same TCP connection for multiple requests. Since the time taken to set up a new TCP connection for every request over a WAN is considerably larger than that over a LAN, the direct access times are significantly larger. As data sizes increase, the overhead of connection set up time becomes a relatively smaller portion of the overall communication time involved. As a result, the overall access latency is dominated by the communication time, which is larger for remote accesses involving accesses to two servers. In both the cases, the access latency was less than a second.

31.2.4.2 Evaluation of DISCOVER Grid services

This experiment evaluated access to Grid services using the DISCOVER middleware substrate. The setup consisted of two DISCOVER server running on grid1.rutgers.edu and tassl-pc-2.rutgers.edu, connected via a 10-Mbps LAN. The Globus Toolkit was installed on grid1.rutgers.edu. The test scenario consisted of: (1) the client logging on to the Portal, (2) the client using the *DiscoverMDS* service to locate an appropriate resource, (3) the client using the *DiscoverGRAM* service to launch an application on the remote resource, (4) the client using the *DiscoverGASS* to transfer the output and error files produced by the application, (5) the client interacting and steering the application using the collaboratory services, and (6) the client terminating the application using the *DiscoverGRAM* service. The number of clients was varied up to a maximum of 25. The *DiscoverMDS* access time averaged around 250 ms. The total time for finding a resource also depends on the search criterion. We restricted our search criteria to memory size and available memory. The *DiscoverGASS* service was used to transfer files of various sizes. *DiscoverGASS* service performed well for small file sizes (below 10 MB) and deteriorated for larger files. The total time taken for the entire test scenario was measured for two cases: (1) the services were accessed locally at grid1.rutgers.edu and (2) the server at tassl-pc-2.rutgers accessed the Grid services provided by grid1.rutgers.edu. This time was further divided into five distinct time intervals: (1) time taken for resolving services, (2) time taken for delegation, (3) time taken for event channel creation to receive job updates, (4) time taken for unbinding the job, and (5) time taken for transferring the error file. The time taken was approximately 14.5 s in the first case and approximately 18 s in the second case. The additional time in the second case was spent in resolving the services not present locally.

31.3 DIOS: DISTRIBUTED INTERACTIVE OBJECT SUBSTRATE

DIOS is a distributed object infrastructure that enables the development and deployment of interactive application. It addresses three key challenges: (1) definition and deployment of interaction objects that extend distributed and dynamic computational objects with sensors and actuators for interaction and steering, (2) definition of a scalable control network that interconnects interaction objects and enables object discovery, interrogation and control, and (3) definition of an interaction gateway that enables remote clients to access, monitor and interact with applications. The design, implementation and evaluation of DIOS are presented below. DIOS is composed of two key components: (1) interaction objects that encapsulate sensors and actuators and (2) a hierarchical control network composed of *Discover Agents, Base Stations* and an *Interaction Gateway.*

31.3.1 Sensors, actuators and interaction objects

Interaction objects extend application computational objects with interaction and steering capabilities through embedded sensors and actuators. Computational objects are the objects (data structures, algorithms) used by the application for its computations.[1] In order to enable application interaction and steering, these objects must export interaction interfaces that enable their state to be externally monitored and changed. Sensors and actuators provide such an interface. Sensors provide an interface for viewing the current state of the object, while actuators provide an interface to process commands to modify the state. Note that the sensors and actuators need to be co-located in memory with the computational objects and have access to their internal state. Transforming computational objects into interaction objects can be a significant challenge. This is especially true when the computational objects are distributed across multiple processors and can be dynamically created, deleted, migrated and redistributed. Multiple sensors and actuators now have to coordinate in order to collectively process interaction requests.

DIOS provides application-level programming abstractions and efficient run-time support to support the definition and deployment of sensors and actuators for distributed and dynamic computational objects. Using DIOS, existing applications can be converted by deriving the computational objects from a DIOS virtual interaction base class. The derived objects can then selectively overload the base class methods to define their interaction interfaces as a set of views that they can provide and a set of commands that they can accept and process. Views represent sensors and define the type of information that the object can provide. For example, a mesh object might export views for its structure and distribution. Commands represent actuators and define the type of controls that can be applied to the object. Commands for the mesh object may include refine, coarsen and redistribute. The view and command interfaces may be guarded by access polices that define who can access the interfaces, how they can access them and when they can access them. This process requires minimal modification to original computational objects. Discover agents, which are a part of the DIOS control network, combine the individual interfaces and export them to the interaction server using an Interaction Interface Definition Language (IDL). The Interaction IDL contains metadata for interface discovery and access and is compatible with standard distributed object interfaces such as CORBA [14] and RMI [16]. In the case of applications written in non-object-oriented languages such as Fortran, application data structures are first converted into computation objects using a C++ wrapper object. Note that this has to be done only for those data structures that require interaction capabilities. These objects are then transformed to interaction objects as described above. DIOS interaction objects can be created or deleted

---

[1] Note that computational objects do not refer only to objects in an object-oriented implementation of an application but also to application data structures and operations on these data structures implemented in languages such as C and Fortran.

during application execution and can migrate between computational nodes. Furthermore, a distributed interaction object can modify its distribution at any time.

31.3.2 Local, global and distributed objects

Interaction objects can be classified on the basis of the address space(s) they span during the course of the computation as *local, global* and *distributed objects.* Local interaction objects are created in a processor's local memory. These objects may migrate to another processor during the lifetime of the application, but always exist in a single processor's address space at any time. Multiple instances of a local object could exist on different processors at the same time. Global interaction objects are similar to local objects, except that there can be exactly one instance of the object that is replicated on all processors at any time. A distributed interaction object spans multiple processors' address spaces. An example is a distributed array partitioned across available computational nodes. These objects contain an additional distribution attribute that maintains its current distribution type (e.g. blocked, staggered, inverse space filling curve-based, or custom) and layout. This attribute can change during the lifetime of the object. Like local and global interaction objects, distributed objects can be dynamically created, deleted or redistributed.

In order to enable interaction with distributed objects, each distribution type is associated with *gather* and *scatter* operations. Gather aggregates information from the distributed components of the object, while scatter performs the reverse operation. For example, in the case of a distributed array object, the gather operation would collate views generated from sub blocks of the array while the scatter operator would scatter a query to the relevant sub blocks. An application can select from a library of gather/scatter methods for popular distribution types provided by DIOS or can register gather/scatter methods for customized distribution types.

31.3.3 DIOS control network and interaction agents

The control network has a hierarchical 'cellular' structure with three components – Discover Agents, Base Stations and Interaction Gateway, as shown in Figure 31.3. Computational nodes are partitioned into interaction cells, with each cell consisting of a set of Discover Agents and a Base Station. The number of nodes per interaction cell is programmable. Discover Agents are present on each computational node and manage run-time references to the interaction objects on the node. The Base Station maintains information about interaction objects for the entire interaction cell. The highest level of the hierarchy is the Interaction Gateway that provides a proxy to the entire application. The control network is automatically configured at run time using the underlying messaging environment (e.g. Message Passing Interface (MPI) [23]) and the available number of processors.

31.3.3.1 Discover agents, base stations and interaction gateway

Each computation node in the control network houses a Discover Agent (DA). Each Discover Agent maintains a local interaction object registry containing references to all

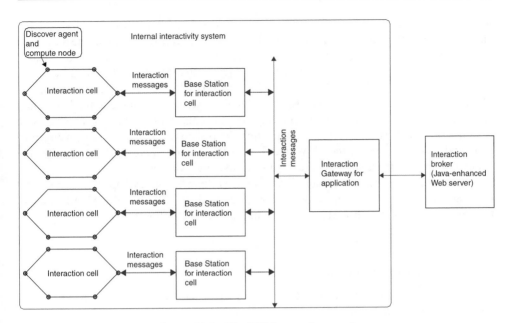

Figure 31.3 The DIOS control network.

interaction objects currently active and registered by that node and exports the interaction interfaces for these objects (using the Interaction IDL). Base Stations (BS) form the next level of control network hierarchy. They maintain interaction registries containing the Interaction IDL for all the interaction objects in the interaction cell and export this information to the Interaction Gateway. The Interaction Gateway (IG) represents an interaction proxy for the entire application. It manages a registry of interaction interfaces for all the interaction objects in the application and is responsible for interfacing with external interaction servers or brokers. The Interaction Gateway delegates incoming interaction requests to the appropriate Base Stations and Discover Agents and combines and collates responses. Object migrations and redistributions are handled by the respective Discover Agents (and Base Stations if the migration/redistribution is across interaction cells) by updating corresponding registries. The Discover Agent, Base Station and Interaction Gateway are all initialized on the appropriate processors during application start up. They execute in the same address space as the application and communicate using the application messaging environment, for example, MPI. A recent extension to DIOS allows clients to define and deploy rules to automatically monitor and control applications and/or application objects. The conditions and actions of the rules are composed using the exported view/command interfaces. A distributed rule engine is built in the control network that authenticates and validates incoming rules, decomposes the rules and distributes components to appropriate application objects and manages the execution of the rules.

In our implementation, interactions between an interaction server and the interaction gateway are achieved using two approaches. In the first approach, the Interaction Gateway serializes the interaction interfaces and associated metadata information for all registered interaction objects to the server. A set of Java classes at the server parse the serialized Interaction IDL stream to generate corresponding interaction object proxies. In the second approach, the Interaction Gateway initializes a Java Virtual Machine (JVM) and uses the Java Native Interface (JNI) [24] to create Java mirrors of registered interaction objects. These mirrors are registered with an RMI [16] registry service executing at the Interaction Gateway. This enables the Server to gain access to and control the interaction objects using the Java RMI API. We are currently evaluating the performance overheads of using Java RMI and JNI. The use of the JVM and JNI in the second approach assumes that the computing environment supports the Java run-time environment.

A more detailed description of the DIOS framework, including examples for converting existing applications into interactive ones, registering them with the DISCOVER interaction server and using web portals for monitoring and controlling them, can be found in References [25, 26].

31.3.4 Experimental evaluation

DIOS has been implemented as a C++ library and has been ported to a number of operating systems including Linux, Windows NT, Solaris, IRIX and AIX. This section summarizes an experimental evaluation of the DIOS library using the IPARS reservoir simulator framework on the Sun Starfire E10000 cluster. The E10000 configuration used consists of 64, 400 MHz SPARC processors, a 12.8 GB s^{-1} interconnect. IPARS is a Fortran-based framework for developing parallel/distributed oil reservoir simulators. Using DIOS/DISCOVER, engineers can interactively feed in parameters such as water/gas injection rates and well bottom hole pressure, and observe the water/oil ratio or the oil production rate. The transformation of IPARS using DIOS consisted of creating C++ wrappers around the IPARS well data structures and defining the appropriate interaction interfaces in terms of views and commands. The DIOS evaluation consists of five experiments:

Interaction object registration: Object registration (generating the Interaction IDL at the Discover Agents and exporting it to Base Station/Gateway) took 500 μs per object at each Discover Agent, 10 ms per Discover Agent in the interaction cell at the Base Station and 10 ms per Base Station in the control network at the Gateway. Note that this is a one-time cost.

Overhead of minimal steering: This experiment measured the run-time overheads introduced because of DIOS monitoring during application execution. In this experiment, the application automatically updated the DISCOVER server and connected clients with the current state of its interactive objects. Explicit command/view requests were disabled during the experiment. The application contained five interaction objects, two local objects and three global objects. The measurements showed that the overheads due to the DIOS run time are very small and typically within the error of measurement. In some cases, because of system load dynamics, the performance with DIOS was slightly better. Our observations have shown that for most applications the DIOS overheads are less that 0.2% of the application computation time.

Table 31.1 View and command processing times

View type	Data size (bytes)	Time taken	Command	Time taken
Text	65	0.4 ms	Stop, pause or resume	250 μs
Text	120	0.7 ms	Refine GridHierarchy	32 ms
Text	760	0.7 ms	Checkpoint	1.2 s
XSlice generation	1024	1.7 ms	Rollback	43 ms

View/command processing time: The query processing time depends on (1) the nature of interaction/steering requested, (2) the processing required at the application to satisfy the request and generate a response, and (3) type and size of the response. In this experiment, we measured time required for generating and exporting different views and commands. A sampling of the measured times for different scenarios is presented in Table 31.1.

DIOS control network overheads: This experiment consisted of measuring the overheads due to communication between the Discover Agents, Base Stations and the Interaction Gateway while processing interaction requests for local, global and distributed objects. As expected, the measurements indicated that the interaction request processing time is minimum when the interaction objects are co-located with the Gateway, and is the maximum for distributed objects. This is due to the additional communication between the different Discover Agents and the Gateway, and the `gather` operation performed at the Gateway to collate the responses. Note that for the IPARS application, the average interaction time was within 0.1 to 0.3% of the average time spent in computation during each iteration.

End-to-end steering latency: This measured the time to complete a round-trip steering operation starting with a request from a remote client and ending with the response delivered to that client. Remote clients executed within Web browsers on laptops/workstations on different subnets. These measurements of course depend on the state of the client, the server and the network interconnecting them. The DISCOVER system exhibits end-to-end latencies comparable to related steering systems, as reported in Reference [25].

31.4 THE COLLABORATIVE INTERACTION AND STEERING PORTAL

The DISCOVER collaborative computational portal provides scientists and engineers with an anytime/anywhere capability of collaboratively (and securely) launching, accessing, monitoring and controlling Grid applications. It integrates access to the collaboratory services and the Grid services provided by the DISCOVER middleware. A screen shot of the current DISCOVER portal is presented in Figure 31.4.

The DISCOVER portal consists of a virtual desktop with local and shared areas. The shared areas implement a replicated shared workspace and enable collaboration among dynamically formed user groups. Locking mechanisms are used to maintain consistency. The base portal, presented to user after authentication and access verification, is a control

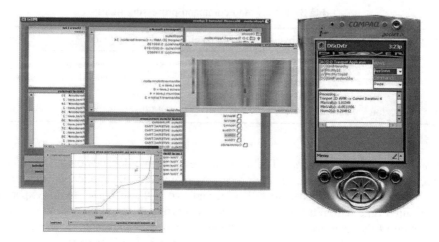

Figure 31.4 The DISCOVER collaborative interaction/steering portal.

panel. The control panel provides the user with a list of services and applications and is customized to match each user's access privileges. Once clients download the control panel they can launch any desired service such as resource discovery, application execution, application interrogation, interaction, collaboration, or application/session archival access. For application access, the desktop consists of (1) a list of interaction objects and their exported interaction interfaces (views and/or commands), (2) an information pane that displays global updates (current timestep of a simulation) from the application, and (3) a status bar that displays the current mode of the application (computing, interacting) and the status of issued command/view requests. The list of interaction objects is once again customized to match the client's access privileges. Chat and whiteboard tools can be launched from the desktop to support collaboration. View requests generate separate (possibly shared) panes using the corresponding view plug-in. All users choosing to steer a particular application form a collaboration group by default with a corresponding shared area on the virtual desktop. New groups can be formed or modified at any time. A separate application registration page is provided to allow superseders to register application, add application users and modify user capabilities.

31.5 SUMMARY AND CURRENT STATUS

In this chapter we presented an overview of the DISCOVER computational collaboratory for enabling interactive applications on the Grid. Its primary goal is to bring large distributed Grid applications to the scientists'/engineers' desktop and enable collaborative application monitoring, interaction and control. DISCOVER is composed of three key components: (1) a middleware substrate that integrates DISCOVER servers and enables interoperability with external Grid services, (2) an application control network consisting of sensors, actuators and interaction agents that enable monitoring, interaction and steering of distributed applications, and (3) detachable portals for collaborative access to Grid applications and services. The design, implementation, operation and evaluation of these components were presented. DISCOVER is currently operational and is being used to provide these capabilities to a number of application-specific PSEs including the IPARS oil reservoir simulator system at the Center for Subsurface Modeling, University of Texas at Austin, the virtual test facility at the ASCI/ASAP Center, California Institute of Technology, and the Astrophysical Simulation Collaboratory. Furthermore, the DISCOVER middleware integrates access to Globus Grid services. Additional information and an online demonstration are available at http://www.discoverportal.org.

ACKNOWLEDGEMENTS

We would like to thank the members of the DISCOVER team, V. Bhat, M. Dhillon, S. Kaur, H. Liu, V. Matossian, R. Muralidhar, A. Swaminathan and S. Verma for their contributions to this project. We would also like to thank J. Gawor and G. von Laszewski for their help with the CORBA CoG Kit.

REFERENCES

1. Kouzes, R. T., Myers, J. D. and Wulf, W. A. (1996) Collaboratories: doing science on the Internet. *IEEE Computer*, **29[A1]**, (8).
2. Foster, I. and Kesselman, C. (1998) *The Grid: Blueprint for a New Computing Infrastructure.* San Francisco, CA: Morgan Kaufmann.
3. Foster, I. (2000) Internet Computing and the Emerging Grid, Nature, Web Matters, http://www.nature.com /nature/webmatters/grid/grid.html, December, 2000.
4. The Global Grid Forum, http://www.gridforum.org.
5. Grid Computing Environments Working Group, Global Grid Forum, http://www.computingportals.org.
6. The DISCOVER Computational Collaboratory, http://www.discoverportal.org.
7. Kaur, S., Mann, V., Matossian, V., Muralidhar, R. and Parashar, M. (2001) Engineering a distributed computational collaboratory. 34th Hawaii Conference on System Sciences, January, 2001.
8. Foster, I. and Kesselman, C. (1997) Globus: a metacomputing infrastructure toolkit. *International Journal of Supercomputing Applications*, **11**(2), 115–128.
9. Foster, I., Kesselman, C. and Tuecke, S. (2001) The anatomy of the grid: enabling scalable virtual organizations. *International Journal of High Performance Computing Applications*, **15**(3), 200–222.
10. Mann, V. and Parashar, M. (2002) Engineering an interoperable computational collaboratory on the grid, Special Issue on Grid Computing Environments, *Concurrency and Computation: Practice and Experience.* John Wiley & Sons, to appear.
11. Mann, V. and Parashar, M. (2001) Middleware support for global access to integrated computational collaboratories, *Proceedings of 10th IEEE International Symposium on High Performance Distributed Computing.* San Francisco, CA, USA: IEEE Computer Society Press, pp. 35–46.
12. Parashar, M., von Laszewski, G., Verma, S., Gawor, J., Keahey, K. and Rehn, N. (2002) A CORBA commodity grid kit, Special Issue on Grid Computing Environments, *Concurrency and Computation: Practice and Experience.* John Wiley & Sons, to appear.

13. Verma, S., Parashar, M., Gawor, J. and von Laszewski, G. (2001) Design and implementation of a CORBA commodity grid kit. Second International Workshop on Grid Computing – GRID 2001, Denver, CO, USA, Springer LNCS 2242, November, 2001, pp. 2–12.
14. CORBA: Common Object Request Broker Architecture, http://www.corba.org.
15. CORBA/IIOP Specification, http://www.omg.org/technology/documents/formal/corbaiiop.html.
16. Java Remote Method Invocation, http://java.sun.com/products/jdk/rmi.
17. Simple Object Access Protocol (SOAP), http://www.w3.org/TR/SOAP.
18. Bhat, V., Mann, V. and Parashar, M. (2002) Integrating Grid Services using the DISCOVER Middleware Substrate, Technical Report 268, The Applied Software Systems Laboratory, CAIP, Rutgers University, September, 2002, http://www.caip.rutgers.edu/TASSL.
19. Hunter, J. (1998) *Java Servlet Programming*. 1st edn. Sebastopol, CA: O'Reilly & Associates.
20. Java Naming and Directory Interface, http://java.sun.com/products/jndi/libraries.
21. CORBA's Event Service Version 1.1, http://www.omg.org/technology/documents/formal/event_service.htm.
22. CORBA Trader Service Specification, ftp://ftp.omg.org/pub/docs/formal/97-07-26.pdf.
23. MPI Forum, MPI: Message Passing Interface, www.mcs.anl.gov/mpi.
24. Java Native Interface Specification, http://web2.java.sun.com/products/jdk/1.1/docs/guide/jni.
25. Muralidhar, R. and Parashar, M. (2001) A distributed object infrastructure for interaction and steering, in Sakellariou, R., Keane, J., Gurd, J. and Freeman, L. (eds) *Proceedings of the 7th International Euro-Par Conference (Euro-Par 2001)*. Lecture Notes in Computer Science, Vol. **2150**. Manchester, UK: Springer-Verlag, pp. 67–74.
26. Muralidhar, R. (2000) A Distributed Object Framework for the Interactive Steering of High-Performance Applications, MS thesis, Department of Electrical and Computer Engineering, The State University of New Jersey, Rutgers, October, 2000.

32

Grid resource allocation and control using computational economies

Rich Wolski,[1] **John Brevik,**[2] **James S. Plank,**[3] **and Todd Bryan**[1]

[1]*University of California, Santa Barbara, California, United States,* [2]*Wheaton College, Norton, Massachusetts, United States,* [3]*University of Tennessee, Knoxville, Tennessee, United States*

32.1 INTRODUCTION

Most, if not all, Computational Grid resource allocation and scheduling research espouses one of two paradigms: centralized omnipotent resource control [1–4] or localized application control [5–8]. The first is not a scalable solution either in terms of execution efficiency (the resource broker or scheduler becomes a bottleneck) or fault resilience (the allocation mechanism is a single point of failure). On the other hand, the second approach can lead to unstable resource assignments as 'Grid-aware' applications adapt to compete for resources. The complexity of the allocation problem and the dynamically changing performance characteristics of Grid resources (due to contention and resource failure) are such that automatic scheduling programs are needed to allocate and re-allocate resources. With resource allocation decisions under local control, the potential for instability exists

Grid Computing – Making the Global Infrastructure a Reality. Edited by F. Berman, A. Hey and G. Fox
© 2003 John Wiley & Sons, Ltd   ISBN: 0-470-85319-0

as competing application schedulers constantly adjust to load fluctuations they themselves induce.

As Grid systems are deployed, a variety of engineering approaches to Grid resource allocation have been and will continue to be used [1, 9–12]. However, almost all of them either rely on a centralized information base, or offer little assurance of allocation stability. There are two formal approaches to the Grid resource allocation problem, however, that specifically address questions of efficiency, stability, and scalability. They are *control theory* and *economics*. Control theory is a viable choice since it includes the notion of feedback explicitly, and system response can be represented by stochastic variables. Good stochastic models of Grid performance response, however, remain elusive, making it difficult to define formally tractable control theoretic mechanisms.

Economic systems, on the other hand, are attractive for several reasons. First, there is a considerable body of theory that attempts to explain the 'emergent' behavior of the overall system based on the presumed behavior of the constituents. Secondly, the concept of 'efficiency' is well defined, although it is different from the notion of efficiency typically understood in a computer performance evaluation setting. Finally, economic systems and the assumptions upon which they are based seem familiar, making common intuition a more valuable research asset. While it may be difficult to visualize the effect of a change in the covariance matrix (needed by many control theoretic systems), most people understand fairly quickly that an increase in price for a resource will squelch demand for it.

For these reasons, economic systems have garnered quite a bit of attention as an approach to Grid resource allocation under the heading *computational economy* [3, 13–15]. In this chapter, we will focus on computational economies for the Grid. It is important to note, however, that the term 'computational economy' is also used in e-Commerce settings to refer to the use of computers to solve difficult financial problems. For example, researchers are studying ways to resolve combinatorial auctions [16–19] optimally so that they may be used to distribute 'real-world' goods in different economic settings. The distinction largely hinges on whether the economy in question is the 'real' economy or an artificial one that has been set up to conform to a certain set of constraints. The approaches we will discuss are based on the latter model. That is, we will assume that the Grid system in question can be made to obey a certain set of economic principles *before* execution begins. Moreover, Grid computational economies can be restarted, whereas the 'real' economy cannot. This distinction turns out to be an important one (as we discuss in Section 32.5), but it necessarily implies that the economy in question is artificial.

Computational economies, defined in this way, can be broadly categorized into two types. *Commodities markets* treat equivalent resources as interchangeable. A purchaser of a resource (such as a CPU of a particular type) buys one of those that are available from a pool of equivalent choices without the ability to specify which resource exactly will be purchased. Alternatively, in *auction markets* purchasers (consumers) bid on and ultimately purchase specific resources provided by *producers*. Each formulation is characterized by theoretical as well as practical advantages and disadvantages. While it is an open question as to which is most suitable for future Grid settings, in this chapter we describe early work (presented more completely in Reference [20]) that addresses these characteristic differences.

The process of conducting computational economic research is, itself, an interesting research topic. With a few exceptions [21, 22], most investigations of various market formulations and mechanisms are simulated. Simulated consumer agents buy resources from simulated producers under various conditions so that the overall economy can be observed. The benefits of this approach are obvious (repeatability, controlled experimentation, etc.) but the model for the interaction of the agents with the system they are using is rarely included. That is, agents interact 'out-of-band' when they negotiate resource usage in most simulations. In Grid settings, the load introduced by agents in the market will itself affect resource availability. The extent of this effect is hard to model for the same reasons that control theory is difficult to apply in Grid settings: few stochastic models that match observed performance response are available.

Empirical studies offer an alternative to simulation, although they are labor intensive to conduct. More importantly, they are difficult to design. In particular, Grid resources and test beds are expensive to maintain, often supporting active 'production' users concurrently with Grid research. Additionally, a valid study must motivate users to attach real value to resources and applications. In most experimental environments [23], the subjects realize that they are under study and that they are operating in an artificial economy. While it possible to motivate them to compete economically, this motivation is often unconvincing. In this chapter, we also discuss our experiences pursuing empirical economic results in Grid settings. We have discovered several key issues that must be addressed to make such results convincing.

The remainder of this chapter is organized as follows. In the next section, we present a basic set of assumptions that must be made in order to support a Grid resource economy, and discuss the advantages and disadvantages of different market formulations. Section 32.4 covers the simulation of Grid economies and presents a methodology we have used to generate early results as a case study. In Section 32.5, we outline some of the issues associated with effective empirical studies, using our own work-in-progress as an example. Finally, in Section 32.6 we conclude and summarize as a way of pointing to future research.

32.2 COMPUTATIONAL ECONOMIES AND THE GRID

While a variety of schemes have been proposed as economic approaches to Grid resource allocation, there are a fundamental set of assumptions upon which any 'true' economic system must rest. It is often difficult to understand whether a particular resource allocation system

- can be predicted to behave as an economic system, and
- can be analyzed as an economic system.

The former characteristic is beneficial because it permits reasoning about the overall 'state' of Grid allocations and the likely changes to that state that may occur. The latter characteristic is desirable because rigorous analysis often leads to explanatory insight.

For a resource allocation mechanism to be economic in nature, it fundamentally must rest on the following assumptions. If it does not, or it is not clear whether these assumptions hold true, the system is not, at its core, a computational economy.

The most fundamental assumption concerns the relationship between supply, demand, and value.

Assumption #1: The relative worth of a resource must be determined by its supply and the demand for it.

In any economic system, the relationship of supply to demand determines value. It is interesting to note that we will not insist that these relationships be *monotonic* in the sense that raising the price will always cause supply to increase and demand to decrease, although this characteristic seems to be an intuitive and natural restriction. Philosophically, we do not include monotonicity as a condition because it is not necessary for the economic results to which we will appeal. The only condition of this type that is mathematically necessary is that if the price of a commodity is allowed to increase without bound, the supply will eventually overtake the demand. That is, from a theoretical point of view, supply must exceed demand for a sufficiently large price. More pragmatically, there are instances in which consumers may increase their demand in the face of rising prices (e.g. to purchase on the principle that 'you get what you pay for'). That must be modeled correctly.

A second, related assumption concerns the relationship of currency, price, and value.

Assumption #2: The price of a given resource is its worth relative to the value of a unit resource called currency.

There is nothing inherently special about money in an economic system with the possible exception of its portability. In fact, currency may be 'tied' to a specific commodity; for our purposes, however, we will assume that currency is a separate commodity that has utility value for all agents. If currency cannot be used to purchase resources universally (e.g. there is a different currency for each resource type and no way to translate between them), the system (however effective) is not an economic one. Taken in conjunction with assumption *#1*, the result is that price is a function of the relationship between supply and demand, and its units are the units of currency.

'Notice that systems that use terms from economics to describe the way allocation decisions are made are not necessarily economic systems. For example, consider a system based on lottery scheduling [24]. Lottery scheduling is a methodology in which resource providers hand out some number of 'tickets', each representing the right to use a fixed part of the resource for a given time duration. For each time period (e.g. time slice), a ticket is drawn at random, and its holder is permitted to use the resource. The proportion of time a particular consumer will be allocated is, therefore, proportional to the number of tickets that consumer possesses relative to other consumers.

It is tempting to think of tickets in a lottery-scheduling scheme as currency, and the proportion of tickets each consumer holds as a measure of relative wealth. Unless the proportions are determined through some aggregate function of supply and demand, this simple system is not an economic one under our stricter set of assumptions. The use of currency, by itself, does not necessarily imply economic behavior. Even if each user chooses only to 'bid' a fraction of its job's tickets, that fraction controls directly the proportion of the occupancy the job can expect. That is, lottery scheduling as described here is a powerful and effective mechanism for establishing execution priorities, but a system that uses it cannot be expected necessarily to act as an economic one.

This example also illustrates another potential point of confusion with respect to resource brokering schemes. In particular, the use of bidding as a negotiative mechanism does not, by itself, ensure that economic principles govern the behavior of an allocation scheme. Some resource brokering schemes [3, 25] are based on the idea that applications announce their computational requirements to a broker, and resource providers 'bid' (based on their processor type, load, etc.) for the outstanding jobs. The job is then 'awarded' to the resource that is best able to execute it according to some affinity function. Notice that the decision to acquire a resource is not dependent on its worth as measured by its price, nor does the decision to supply a resource hinge on the possible affinities it may have for some jobs.

However, a slight modification to the lottery-scheduling example results in a system that obeys (and can be analyzed using) economic principles. If each job is allocated some number of tickets (even based on who its user is) and the user is allowed to *retain* unused tickets from one job for use with another, then the overall system can be treated as an economic one. Supply (CPU time slices, in this case) is constant, but the fraction of CPU time that a job may purchase is now a function of the demand for that fixed supply.

Assumption #3: Relative worth is accurately measured only when market equilibrium is reached.

Another way of stating this assumption is that a given price is an accurate measure of value only when the available supply of a resource at that price is equal to the demand for it at that price. The relationship of equilibrium to value is important when evaluating the efficiency of a particular economic system. If the price of a resource is very high, causing little demand for it (i.e. the market is in equilibrium at a high price point), the allocation mechanism is working efficiently from an economic point of view even if many of the resources may be idle. In this case, price accurately captures the value of the resource as measured by the willingness of a producer to sell it *and* the willingness of a consumer to buy it. Now consider what happens if the same amount of resource is available, but the price that is specified (because of the workings of some price-setting mechanism – see the next subsection) is lower than the equilibrium price. In this case, demand would exceed supply, leaving some consumers unsatisfied.

32.2.1 Price-setting mechanisms: commodities markets and auctions

Broadly speaking, there are two categories of mechanism for setting prices: commodities markets and auctions. In both formulations, consumers and producers appeal to a trusted third party to mediate the necessary transactions. In a commodities market setting, the third party (often termed *the market*) sets a price for a resource and then queries both producers and consumers for a willingness to sell and buy respectively at that price. Those wishing to participate agree to transact business at the given price point, and an exchange of currency for resource takes place. The market observes the unsatisfied supply or demand and uses that information (as well as other inputs) to set a new price. Price setting and transactions may occur in distinct stages, or may be concurrent and asynchronous. The key feature, however, that distinguishes a commodity market from an individual auction is that the consumer does not purchase a 'specific' commodity, but rather takes one of many equivalents. For example, in the soybean market, a buyer does not purchase a specific lot of soybeans from a specific grower, but rather some quantity of soybeans of a given quality from the market.

Alternatively, prices may be set through an auction. In this case, the third party (termed *the auctioneer*) collects resources and bids, but determines the sale of an individual resource (or resource bundle) based on the bids. Only one bidder ('the winner') is awarded a resource per auction round, and the process is repeated for each available resource.

Taken another way, commodity markets and auctions represent two ends of a spectrum of market formulations. On the commodity market end, an attempt is made to satisfy *all* bidders and sellers at a given price. At the other end – the auction end – *one* bidder and seller is satisfied at a given price. Obviously, it is possible to consider market organizations that are between the extremes. While there may be limited theoretical value in doing so, from an implementation perspective it may be that satisfying some limited number of transactions between price-setting 'rounds' is a useful organization. While we have studied the feasibility of commodity markets and auctions in Grid settings (see Section 32.3) in the extreme, much work addressing the practical implications of different market organizations remains.

32.2.2 Pricing functions

The way in which a price is determined for a particular transaction is, in some sense, independent of how the market is organized. In auction settings, a variety of price determining rules can be used such as an *English auction, Dutch auction, First-price-sealed-bid, Vickery*, and so on. Each of these mechanisms is designed to maximize a different objective function with respect to the market. For example, the most familiar auction style is the English auction in which an auctioneer solicits ascending public bids for an item (i.e. each bid must be larger than the current largest) until only one bidder remains. This style of auction is designed to maximize the revenue available to the seller by drawing out the highest possible price for the item on sale. Notice, however, that revenue maximization in this way does not guarantee that the price paid accurately represents the value of the good. In particular, English auctions are susceptible to the 'winner's curse' in which the buyer typically overbids for the good. That is, statistically, the winning bid for a good is higher than the good is actually worth since the distribution of bids is likely to straddle the 'true' value.

In a Grid setting, the problem with this overpricing is that it may translate to inefficiency. Grid application users at a scientific computer center probably do not want their resource allocation mechanisms wasting resources to over bid each other. If currency can be exchanged for resources, than excess currency expended is equivalent to lost resource time. As a result, frugal users will either collude to keep prices low (which may extinguish supply) or will simply refuse to participate. In either case, users are 'encouraged' by the market to understate the value of the resource because of the possibility of overpaying. Worse, the market may fluctuate wildly as users search for the true value of a resource with their respective bids.

From an economic perspective, the term 'incentive compatibility' refers to the incentive a market participant has to state his or her valuations truthfully. English auctions are not incentive-compatible with respect to buyers. Second-price Vickery auctions, however, are buyer incentive-compatible. In a Second-price Vickery auction (or Vickery auction, for short) each buyer submits a single-sealed bid that is hidden from all other buyers. The winning bidder pays the price specified by the second-highest bid. For example, if a CPU were being auctioned to a set of Grid schedulers, and the highest bid were $G100 (100 'Grid bucks'), but the second-highest bid were $G90, the bidder who bid $G100 would win, but only pay $G90. It can be proved that this scheme induces buyers to bid truthfully since the possibility of winner's curse has been eliminated.

Unfortunately, it can also be proved that no auction price-setting scheme is incentive-compatible for both buyers and sellers. As such, proofs about the global stability of an auction-based Grid resource allocation scheme (or indeed any auction-based scheme, Grid or otherwise) remain elusive. Auctions 'work' in the sense that they are used in many different economic settings. They are attractive because they are easy to understand and efficient to implement (given a single, centralized auctioneer configuration). They are not, however, easy to analyze. Indeed, the problem of determining an optimal winner in combinatorial auction settings (where buyers and sellers trade currency for combinations of items) is, itself, NP-complete [19, 26]. By itself, this result does not rule out auction systems as a market formulation for Grid settings, but it is almost certainly the case that Grid resource allocators (human or automatic) will need to purchase different resources. Doing so from a number of single-unit auctions can lead to inefficiency, as we describe later. If combinatorial auctions are used, then the auction protocols that can be brought to bear offer only heuristic guarantees of optimality at best.

For commodity market formulations, the theoretical results are more attractive, but the implementation complexity is higher and the intuitive appeal less. Note that for our purposes, we will assume that *all agents, producers and consumers in the economy are price takers – that is, none of the agents represents a large enough market share to affect prices unilaterally, and therefore their decisions are made in response to a price that is given to them.* This assumption, also called *perfect competition* in the literature, is far from harmless, but it typically holds, at least approximately, in large-scale market economies.

The basic result of interest is a rather old one from general equilibrium theory. It says that, for a market in which multiple goods are exchanged, if the aggregate functions governing supply and demand are homogeneous, continuous, and obey Walras' Law, then there exists a way to assign a price to each good so that the *entire* market is brought into equilibrium. A more complete technical description of these assumptions can be found in Reference [20] – we will attempt only an outline here. Homogeneity is the property that the difference between demand and supply at any given price point (termed *excess demand*) is not a function of the units. (In other words, the excess demands will be the same whether the prices are reported in Grid bucks, Grid sawbucks, Grid talents, or Grid dimes.) Continuity is the mathematical definition. Finally, Walras' Law can be roughly paraphrased as 'the sum of the excess demands is zero.' That is, for any given set of prices, the total value supplied by all producers in the economy is equal to the total value demanded. (This may seem a bit strange if one does not bear in mind that we are treating our currency as just another commodity. Thus, a possible action of a 'consumer' is to provide a supply of currency whose value is equal to the sum of the values of the commodities that this consumer demands.)

Under these assumptions, several algorithmic methods have been proposed that provably will produce a sequence of prices converging to an equilibrium [27–30]. From the perspective of Grid resource allocation (but also, perhaps, in real-world markets) the biggest theoretical drawback is the need for the excess demand functions to be continuous. Clearly, supply, demand, and price are discrete quantities regardless of their units. Further, there is an implementation issue concerning the realization of commodities markets for the Computational Grid in that the price-setting algorithms assume access to the global excess demand functions and possibly their partial derivatives. The Newton-type algorithm proposed by Smale in Reference [30], which can compute an equilibrium price with very low computational complexity, for example, needs the partial derivatives for all excess demand functions as a function of price. In a Grid economy, it is unlikely that individual users will even be able to state their own excess demand functions reliably. If they could, it is not clear that these functions would be time invariant, or that they could be aggregated effectively into global functions for the market as a whole. For these reasons, little work has focused on the use of general equilibrium theory in computational economic settings.

On the other hand, one might reasonably expect that in a large Grid economy, the aggregate excess demand functions would become 'continuous enough' in the sense that the granularity caused by indivisibility of commodities is small compared to the magnitudes of these functions. Also, there are well-understood 'quasi-Newton' or secant methods, which have good numerical stability properties and operate in the absence of explicit knowledge about partial derivatives.

It is another question whether a given (approximate) equilibrium that we compute satisfies any desirable optimality conditions that can be formulated (or even observed). This is a difficult question, and even in an economy that perfectly satisfies our above criteria, the best that we can say is that the equilibrium is *Pareto optimal* in the sense that no change in price can occur without making at least one of the agents worse off (in terms of that agent's utility). Note then that the optimality properties of the market equilibrium tend to be 'spread out' over the whole economy rather than seeking, say, to maximize cash flow, which is in the interest of the producers. One might hope that a market equilibrium is one that maximizes total utility, but we cannot say this. In any event, the concept of 'total utility' is difficult to define in the absence of a well-defined unit of utility ('util'). Moreover, different equilibria will arise from different starting points (initial allocations) of the agents, and some of these may be judged under some criteria to be socially 'better' than others. This said, however, it is natural to look at market equilibrium, which after all arises from the desires of the agents, as a useful benchmark for the performance of an economy.

32.3 AN EXAMPLE STUDY: G-COMMERCE

As mentioned previously, using economic principles as the basis for Grid resource allocation has several advantages. In this section, we describe a previous study (detailed more completely in Reference [20]) that illustrates the opportunity to leverage rigorous economic theory profitably and the 'natural' fit resource economies seem to have intuitively for Grid systems.

The goal of this work is to investigate the feasibility and potential efficiency of different market formulations for Grid resource economies. *G-commerce* refers to the economic mechanisms *and* policies that will need to be put in place to create an effective Grid allocation system. Much of the work in Grid computing that comes before it or is concomitant with it [1, 3, 15, 25] has been focused on the mechanisms necessary to support computational Grid economies. While these mechanisms (many of which take the form of elaborate brokering systems) will assuredly play an important role, G-commerce attempts to study both the necessary mechanisms and the effect of different economic policies in Grid settings.

As an example, consider the potential trade-offs between an auction-based economy and one based on a commodities market approach. Auction systems have been widely studied in other computational settings [13, 21, 31], and as discussed previously, and much of what is known about them rigorously is restricted to a microeconomic level. Alternatively, a commodities market approach has strong macroeconomic properties, but important microeconomic characteristics such as incentive compatibility are often taken for granted.[1] The following study attempts to address two questions about Grid performance under different market formulations (i.e. auctions and commodity markets).

The first question is: *Given automatic program schedulers that can react at machine speeds such as those described in Reference [5, 6, 33], what is the effect on resource allocation stability of auctions versus commodities markets?* If a particular economic formulation results in allocators constantly changing their allocation decisions, the overall system will waste much of its available resource on the work necessary to constantly adjust resource allocations.

The second question is: *What is the effect of choosing a particular market formulation on resource utilization?* While it may be necessary for the performance evaluation community to develop new metrics for measuring Grid efficiency, and these metrics may ultimately rely on computational economic principles, it is certainly enlightening to assess the potential impact of a computational economy in terms of well-understood metrics like utilization.

---

[1] Generally, such microeconomic considerations are well behaved for an idealized market economy; in fact, it is proven in Reference [32] that a market mechanism, that is, a way of redistributing goods, is incentive-compatible for all agents if and only if it is perfectly competitive.

Our initial study uses simulation. While we are currently developing a methodology for validating these simulation results empirically, we motivate the credibility of the study in terms of the issues discussed in Section 32.2 of this chapter.

32.3.1 Producers and consumers

To compare the efficacy of commodities markets and auctions as Grid resource allocation schemes, we define a set of simulated Grid producers and consumers representing resource providers and applications respectively. We then use the same set of producers and consumers to compare commodity and auction-based market settings.

We simulate two different kinds of producers in this study: producers of CPUs and producers of disk storage. That is, from the perspective of a resource market, there are two kinds of resources within our simulated Grids: CPUs and disks (and therefore, counting currency, our market has three commodities). While the results should generalize to include a variety of other commodities, networks present a special problem. Our consumer model is that an application may request a specified amount of CPU and disk (the units of which we discuss below) and that these requests may be serviced by any provider, regardless of location or network connectivity. Since network links cannot be combined with other resources arbitrarily, they cannot be modeled as separate commodities. We believe that network cost can be represented in terms of 'shipping' costs in more complicated markets, but for the purposes of this study, we consider network connectivity to be uniform.

32.3.1.1 CPU producer model

In this study, a CPU represents a computational engine with a fixed dedicated speed. A CPU producer agrees to sell to the Grid some number of fixed 'shares' of the CPU it controls. The real-world scenario for this model is for CPU owners to agree to host a fixed number of processes from the Grid in exchange for Grid currency. Each process gets a fixed, predetermined fraction of the dedicated CPU speed, but the owner determines how many fractions or 'slots' he or she is willing to sell. For example, in our study, the fraction is 10% so each CPU producer agrees to sell a fixed number (less than 10) of 10%-sized slots to the Grid. When a job occupies a CPU, it is guaranteed to get 10% of the available cycles for each slot it consumes. Each CPU, however, differs in the total number of slots it is willing to sell.

To determine supply at a given price point, each CPU calculates

$$mean\_price = revenue/now/slots \tag{32.1}$$

where *revenue* is the total amount of Grid currency (hereafter referred to as \$G which is pronounced 'Grid bucks'), *now* is an incrementing clock, and *slots* is the total number of process slots the CPU owner is willing to support. The *mean_price* value is the average \$G per time unit per slot the CPU has made from selling to the Grid. In our study, CPU producers will only sell if the current price of a CPU slot exceeds the *mean_price* value, and when they sell, they sell all unoccupied slots. That is, the CPU will sell all of its available slots when it will turn a profit (per slot) with respect to the average profit over time.

32.3.1.2 Disk producer model

The model we use for a disk producer is similar to that for the CPU producer, except that disks sell some number of fixed-sized 'files' that applications may use for storage. The *mean_price* calculation for disk files is

$$mean\_price = revenue/now/capacity \qquad (32.2)$$

where *capacity* is the total number of files a disk producer is willing to sell to the Grid. If the current price for a file is greater than the *mean_price*, a disk producer will sell all of its available files.

Note that the resolution of CPU slots and file sizes is variable. It is possible to make a CPU slot equivalent to the duration of a single clock cycle, and limit the disk file to a single byte. Since our markets transact business at the commodity level, however, we hypothesize that any real implementation for the Grid will need to work with larger-scale aggregations of resources for reasons of efficiency. For the simulations described in Section 32.4, we choose values for these aggregations that we believe reflect a market formulation that is currently implementable.

32.3.1.3 Consumers and jobs

Consumers express their needs to the market in the form of jobs. Each job specifies both size and occupancy duration for each resource to be consumed. Each consumer also sports a budget of $G (pronounced 'Grid bucks') that it can use to pay for the resources needed by its jobs. Consumers are given an initial budget and a periodic allowance, but they are not allowed to hold $G over from one period until the next. This method of budget refresh is inspired by the allocation policies currently in use at the NSF Partnerships for Advanced Computational Infrastructure (PACIs) [34, 35]. At these centers, allocations are perishable.

When a consumer wishes to purchase resources for a job, it declares the size of the request for each commodity, but not the duration. Our model is that job durations are relatively long, and that producers allow consumers occupancy without knowing for how long the occupancy will last. At the time a producer agrees to sell to a consumer, a price is fixed that will be charged to the consumer for each simulated time unit until the job completes.

For example, consider a consumer wishing to buy a CPU slot for 100 minutes and a disk file for 300 minutes to service a particular job. If the consumer wishes to buy each for a particular price, it declares to the market a demand of 1 CPU slot and 1 disk slot, but does not reveal the 100 and 300 minute durations. A CPU producer wishing to sell at the CPU price agrees to accept the job until the job completes (as does the disk producer for the disk job). Once the sales are transacted, the consumer's budget is decremented by the agreed-upon price every simulated minute, and each producer's revenue account is incremented by the same amount. If the job completes, the CPU producer will have accrued 100 times the CPU price, the disk producer will have accrued 300 times the disk price, and the consumer's budget will have been decremented by the sum of 100 times the CPU price and 300 times the disk price.

In defining this method of conducting resource transactions, we make several assumptions. First, we assume that in an actual Grid setting, resource producers or suppliers will commit some fraction of their resources to the Grid, and that fraction is slowly changing. Once committed, the fraction 'belongs' to the Grid so producers are not concerned with occupancy. This assumption corresponds to the behavior of some batch systems in which, once a job is allowed to occupy its processors, it is allowed to run either until completion, or until its user's allocation is exhausted. Producers are concerned, in our models, with profit and they only sell if it is profitable on the average. By including time in the supply functions, producers consider past occupancy (in terms of profit) when deciding to sell. We are also assuming that neither consumers nor producers are malicious and that both honor their commitments (i.e. perfect competition). In practice, this requirement assuredly will be difficult to enforce. However, if consumers and producers must agree to use secure authentication methods and system-provided libraries to gain access to Grid resources, then it may be possible to approximate.

This last assumption, however, illustrates the weakness of a commodities market approach in terms of incentive compatibility. An incentive-compatible auction would naturally induce consumers to act 'fairly' because it would be in their best interest to do so. In a commodities market setting, the system would need to rely on external policies and mechanisms to enforce good consumer behavior (e.g. noncollusion). A key realization for the Grid, however, is that such a requirement already exists for security. For example, it is quite difficult to ensure that collaborating users do not share user-ids and passwords. The primary method of enforcement is policy based rather than mechanism based. That is, users have their access privileges revoked if they share logins as a matter of policy rather than by some automatic mechanism.

32.3.1.4 Consumer demand

The consumer demand function is more complex than the CPU and disk supply functions. Consumers must purchase enough CPU and disk resources for each job they wish to run. If they cannot satisfy the request for only one type, they do not express demand for the other. That is, the demand functions for CPU and disks are strongly correlated (although the supply functions are not). This relationship between supply and demand functions constitutes the most difficult of market conditions. Most theoretical market systems make weaker assumptions about the difference in correlation. By addressing the more difficult case, we believe our work more closely resembles what can be realized in practice.

To determine their demand at a given price, each consumer first calculates the average rate at which it would have spent $G for the jobs it has run so far if it had been charged the current price. It then computes how many $G it can spend per simulated time unit until the next budget refresh. That is, it computes

$$avg\_rate = \frac{\sum_i total\_work_i * price_i i}{now} \qquad (32.3)$$

$$capable\_rate = \frac{remaining\_budget}{(refresh - now)} \qquad (32.4)$$

where *total_work_i* is the total amount of work performed so far using commodity *i*, *price_i* is the current price for commodity *i*, *remaining_budget* is the amount left to spend before the budget refresh, *refresh* is the budget refresh time, and *now* is the current time. When *capable_rate* is greater than or equal to *avg_rate*, a consumer will express demand.

Unlike our supply functions, the consumer demand function does not consider past price performance directly when determining demand. Instead, consumers using this function act opportunistically on the basis of the money they have left to spend and when they will receive more. They use past behavior only as an indication of how much work they expect to introduce and buy when they believe they can afford to sustain this rate.

Consumers, in our simulations, generate work as a function of time. We arbitrarily fix some simulated period to be a 'simulated day.' At the beginning of each day, every consumer generates a random number of jobs. By doing so, we hope to model the diurnal user behavior that is typical in large-scale computational settings. In addition, each consumer can generate a single new job every time step with a pre-determined probability. Consumers maintain a queue of jobs waiting for service before they are accepted by producers. When calculating demand, they compute *avg_rate* and *capable_rate* and demand as many jobs from this queue as they can afford.

To summarize, for our G-commerce simulations:

- All entities, except the market-maker, act individually in their respective self-interests.
- Producers consider long-term profit and past performance when deciding to sell.
- Consumers are given periodic budget replenishments and spend opportunistically.
- Consumers introduce workloads in bulk at the beginning of each simulated day, and randomly throughout the day.

We believe that this combination of characteristics captures a reasonable set of producer and consumer traits as evidenced by the current allocation and usage patterns at large-scale compute centers. It does not, however, attempt to capture what the behavior Grid users in an e-Commerce or peer-to-peer setting. This narrowing of focus is an important one. It may be that different producer and consumer settings will require different market formulations – clearly an avenue for future study.

32.3.2 Commodities markets

Our model is an example of an *exchange economy*, that is, a system involving *agents* (producers and consumers), and several commodities. Each agent is assumed to control a sufficiently small segment of the market. In other words, the individual behavior of any one agent will not affect the system as a whole appreciably. In particular, prices will be regarded as beyond the control of the agents. Given a system of prices, then, each agent decides upon a course of action, which may consist of the sale of some commodities and the purchase of others with the proceeds. Thus, we define supply and demand functions for each commodity, which are functions of the aggregate behavior of all the agents. These are determined by the set of market prices for the various commodities.

The method for price determination is based on Smale's technique [30] for finding general market equilibria. A motivation for its choice and a complete explanation of its application are available from Reference [20]. Here, we limit or description to a discussion of the implementation issues as they pertain to the economic principles embodied by equilibrium theory.

In particular, it is not possible to use Smale's method directly for a number of reasons. First, any actual economy is inherently discrete, so the partial derivatives that the method requires do not exist, strictly speaking. Second, given the behavior of the producers and consumers described above, there are threshold prices for each agent that bring about sudden radical changes in behavior, so that a reasonable model for excess demand functions would involve sizeable jump discontinuities. Finally, the assumptions in Smale's model are that supply and demand are functions of price only and independent of time, whereas in practice there are a number of ways for supply and demand to change over time for a given price vector.

Obtaining the partial derivatives necessary to carry out Smale's process in an actual economy is impossible; however, within the framework of our simulated economy, we are able to get good approximations for the partials at a given price vector by polling the producers and consumers. Starting with a price vector, we find their preferences at price vectors obtained by fixing all but one price and varying the remaining price slightly, thus achieving a 'secant' approximation for each commodity separately; we then substitute these approximations for the values of the partial derivatives in the matrix $D_z(\mathbf{p})$, discretize with respect to time, solve for a price vector, and iterate. We will refer, conveniently but somewhat inaccurately, to this price adjustment scheme as *Smale's method*.

The First Bank of G: The drawback to the above scheme is that it relies on polling the entire market for aggregate supply and demand repeatedly to obtain the partial derivatives of the excess demand functions. If we were to try and implement Smale's method directly, each individual producer and consumer would have to be able to respond to the question 'how much of commodity x would you buy (sell) at price vector $\mathbf{p}$?' In practice, producers and consumers may not be able to make such a determination accurately for all possible values of $\mathbf{p}$. Furthermore, even if explicit supply and demand functions are made into an obligation that all agents must meet in order to participate in an actual Grid economy, the methodology clearly will not scale. For these reasons, in practice, we do not wish to assume that such polling information will be available.

A theoretically attractive way to circumvent this difficulty is to approximate each excess demand function z_i by a polynomial in $p_1, p_2, \ldots, p_n$ which fits recent price and excess demand vectors and to use the partial derivatives of these polynomials in place of the actual partials required by the method. In simulations, this method does not, in general, produce prices that approach equilibrium. The *First Bank of G* is a price adjustment scheme that both is practicable and gives good results; this scheme involves using *tâtonnement* [36] until prices get 'close' to equilibrium, in the sense that excess demands have sufficiently small absolute value, and then using the polynomial method for 'fine tuning.' Thus, the First Bank of G approximates Smale's method but is implementable in real-world Grid settings since it hypothesizes excess demand functions and need not poll the market for them. Our experience is that fairly high-degree polynomials are required to capture excess demand behavior with the sharp discontinuities described above. For all simulations described in Section 32.4, we use a degree 17 polynomial.

While this analysis may seem quite detailed, it illustrates the linkage between theoretical economic results and their potential application to Grid economies. At the same time, the work also provides a practical implementation of a price-setting mechanism. This combination of theory and practice is critical to the realization of effective Grid economies.

32.3.3 Auctions

Auctions have been extensively studied as resource allocation strategies for distributed computing systems (e.g. References [13, 31, 37, 38]). See Reference [26] for a particularly illucidating discussion of the relevant issues.

When consumers simply desire one commodity, for example, CPUs in Popcorn [37], auctions provide a convenient, straightforward mechanism for clearing the marketplace. However, the assumptions of a Grid Computing infrastructure pose a few difficulties to this model. First, when an application (the consumer in a Grid Computing scenario) desires multiple commodities, it must place simultaneous bids in multiple auctions, and may only be successful in a few of these. To do so, it must expend currency on the resources that it has obtained while it waits to obtain the others. This expenditure is wasteful, and the uncertain nature of auctions may lead to inefficiency for both producers and consumers.

Second, while a commodities market presents an application with a resource's worth in terms of its price, thus allowing the application to make meaningful scheduling decisions, an auction is more unreliable in terms of both pricing and the ability to obtain a resource, and may therefore result in poor scheduling decisions and more inefficiency for consumers.

To gain a better understanding of how auctions fare in comparison to commodities markets, we implement the following simulation of an auction-based resource allocation mechanism for Computational Grids. At each time step, CPU and disk producers submit their unused CPU and file slots to a CPU and a disk auctioneer. These are accompanied by a minimum selling price, which is the average profit per slot, as detailed in Section 3.1.1 above. Consumers use the demand function as described in Section 3.1.3 to define their bid prices, and as long as they have money to bid on a job, and a job for which to bid, they bid on each commodity needed by their oldest uncommenced job.

Once the auctioneers have received all bids for a time step, they cycle through all the commodities in a random order, performing one auction per commodity. In each auction, the highest-bidding consumer gets the commodity if the bid price is greater than the commodity's minimum price. If there is a second-highest bidder whose price is greater than the commodity's minimum price, then the price for the transaction is the second-highest bidder's price. If there is no such second-highest bidder, then the price of the commodity is the average of the commodity's minimum selling price and the consumer's bid price. When a consumer and commodity have been matched, the commodity is removed from the auctioneer's list of commodities, as is the consumer's bid. At that point, the consumer can submit another bid to that or any other auction, if desired. This situation occurs when a consumer has obtained all commodities for its oldest uncommenced job, and has another job to run. Auctions are transacted in this manner for every commodity, and the entire auction process is repeated at every time step.

Note that this structuring of the auctions means that each consumer may have at most one job for which it is currently bidding. When it obtains all the resources for that job, it immediately starts bidding on its next job. When a time step expires and all auctions for that time step have been completed, there may be several consumers whose jobs have some resources allocated and some unallocated, as a result of failed bidding. These consumers have to pay for their allocated resources while they wait to start bidding in the next time step.

While the auctions determine transaction prices based on individual bids, the supply and demand functions used by the producers and consumers to set ask and bid price are the same functions we use in the commodities market formulations. *Thus, we can compare the market behavior and individual producer and consumer behavior in both auction and commodity market settings.* It is this comparison that forms the basis of the economic study. Since the individual producers and consumers use the same rules to decide when to supply and demand in each market setting, we can observe the effects of market formulation on the overall economy.

32.4 SIMULATIONS AND RESULTS

We compare commodities markets and auctions using the producers and consumers described in Section 3.1. Recall that our markets do not include resale components. Consumers do not make money. Instead, $G are given to them periodically in much the same way that PACIs dole out machine-time allocations. Similarly, producers do not spend money. Once gathered, it is hoarded or, for the purposes of the economy, 'consumed.' We present the under-demand case that corresponds to a Grid economy in which the supply easily exceeds what is necessary to complete all the consumers' jobs. In the *over-demand* case (described in Reference [20]), consumers wish to buy more resources than are available. New jobs are generated fast enough to keep all producers almost completely busy, thereby creating a work backlog.

Using our simulated markets, we wish to investigate three questions with respect to commodities markets and auctions.

1. *Do the theoretical results from Smale's work [39] apply to plausible Grid simulations?*
2. *Can we approximate Smale's method with one that is practically implementable?*
3. *Are auctions or commodities markets a better choice for Grid computational economies?*

Question (1) is important because if Smale's results apply, they dictate that an equilibrium price point must exist (in a commodity market formulation), and they provide a methodology for finding those prices that make up the price point. Assuming the answer to question (1) is affirmative, we also wish to explore methodologies that achieve or approximate Smale's results, but which are implementable in real Grid settings. Lastly, recent work in Grid economies [1, 3, 25] and much previous work in computational economic settings [40–43] has centered on auctions as the appropriate market formulation. We wish to investigate question (3) to determine whether commodities markets are a viable alternative and how they compare to auctions as a market-making strategy.

32.4.1 Market conditions, under-demand case

We present only the under-demand case from Reference [20] (where maximum consumer demand is quickly extinguished by the maximum available supply) as an illustration of the results our simulation system has generated. A discussion of the over-demand case, particularly with respect to the presence of multiple equilibria, is important as well, but beyond the scope of this work. Again, the details are available from Reference [20].

Figure 32.1 shows the CPU and disk prices for an unrealistic-but-correct implementation of Smale's method in our simulated Grid economy over 10 000 time units. The implementation is unrealistic (even though it exists) because it relies on the ability to stop the entire economy and poll all producers and consumers completely for their individual supply and demand functions. It is, therefore, able to verify that the discretization that is necessary to represent Grid economies does not 'break' the effectiveness of Smale's method. It does not, however, imply that Smale's method is implementable in realistic way for Grid settings.

A diurnal cycle of consumer job submission that we deliberately introduce is evident from the price fluctuations. Every 1440 'minutes' each consumer generates between 1 and 100 new jobs causing demand and prices to spike. However, Smale's method is able to find an equilibrium price for both commodities quickly, as is evidenced in Figure 32.2.

Notice that the excess demand spikes in conjunction with the diurnal load, but is quickly brought near zero by the pricing shown in Figure 32.1 in which it hovers until the next cycle. The 'speed' with which Smale's method finds equilibrium is predicted

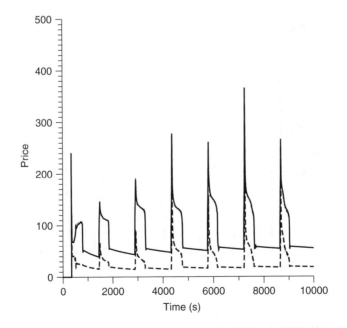

Figure 32.1   Smale's prices for the under-demand case. Solid line is CPU price, and dotted line is disk price in $G.

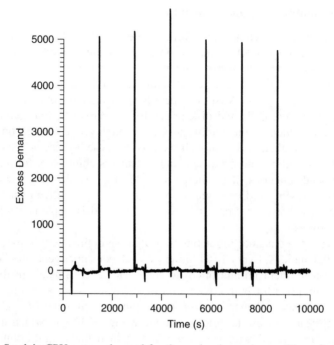

Figure 32.2   Smale's CPU excess demand for the under-demand case. The units are CPU slots.

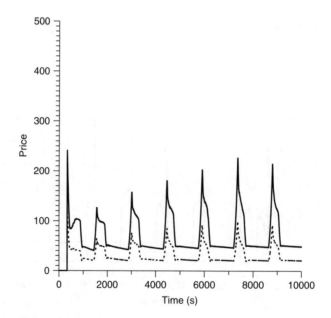

Figure 32.3   First Bank of G prices for the under-demand case. Solid line is CPU price, and dotted line is disk price in $G.

by his work, but the simulation verifies that discretization does not prevent the pricing mechanism from responding faster than the agents in the market.

In Figure 32.3, we show the pricing determined by our engineering approximation to Smale's method – the First Bank of G. The First Bank of G pricing closely approximates the theoretically achievable results generated by Smale's method in our simulated environment. The Bank, though, does not require polling to determine the partial derivatives for the aggregate supply and demand functions. Instead, it uses an iterative polynomial approximation that it derives from simple observations of purchasing and consumption. Thus, it is possible to implement the First Bank of G for use in a real Grid setting without polling Grid producers or consumers for their supply and demand functions explicitly.

The pricing determined by auctions is quite different, however, as depicted in Figures 32.4 and 32.5 (we show CPU and disk price separately as they are almost identical and obscure the graph when overlayed). In the figure, we show the average price paid by all consumers for CPU during each auction round. We use the average price for all auctions as being representative of the 'global' market price. Even though this price is smoothed as an average (some consumers pay more and some pay less during each time step), it shows considerably more variance than prices set by the commodities market. The spikes in workload are not reflected in the price, and the variance seems to increase (i.e. the price becomes less stable) over time.

The use of an average price in these results further illustrates the difficulty in analyzing auction systems at the macroeconomic level. Each consumer, potentially, pays a unique price, making overall supply and demand relationships difficult to gauge. This drawback by itself, however, does not imply that auctions are ineffective – only that they are harder to treat analytically on a global level. Given that the average (i.e. smoothed) price is significantly more unstable in an auction setting, however, indicates that equilibrium theory (despite its own inherent drawbacks) is a potentially useful tool with which to build a Grid economy.

32.4.2 Efficiency

While commodities markets using Smale's method of price determination appear to offer better theoretical and simulated economic properties (equilibrium and price stability) than auctions do, we also wish to consider the effect of the two pricing schemes on producer and consumer efficiency. To do so, we report the average percentage of time each resource is occupied as a utilization metric for suppliers, and the average number of jobs/minute each consumer was able to complete as a consumer metric. Table 32.1 summarizes these values the under-demand case.

In terms of efficiency, Smale's method is best and the First Bank of G achieves almost the same results. Both are significantly better than the auction in all metrics except disk utilization in the over-demanded case [20]. Since CPUs are the scarce resource, disk price may fluctuate through a small range without consequence when lack of CPU supply throttles the system. The auction seems to achieve slightly better disk utilization under these conditions. In general, however, Smale's method and the First Bank of G approximation both outperform the auction in the simulated Grid setting.

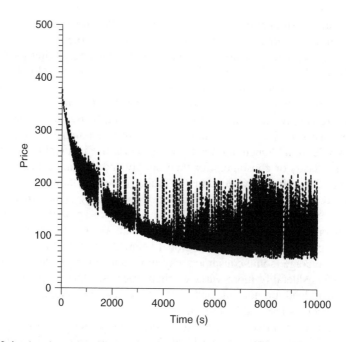

Figure 32.4   Auction prices for the under-demand case, average CPU price only, in $G.

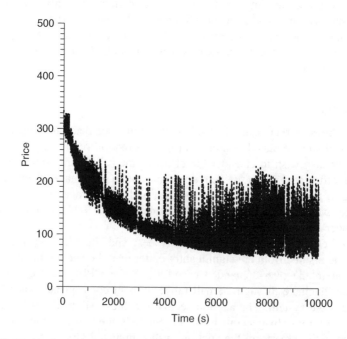

Figure 32.5   Auction prices for the under-demand case, average disk price only, in $G.

Table 32.1 Consumer and producer efficiencies

Efficiency metric	Under-demand
Smale consumer jobs/min	0.14 j/m
B of G consumer jobs/min	0.13 j/m
Auction consumer jobs/min	0.07 j/m
Smale CPU utilization %	60.7%
B of G CPU utilization %	60.4%
Auction CPU utilization %	35.2%
Smale disk utilization %	54.7%
B of G disk utilization %	54.3%
Auction disk utilization %	37.6%

32.5 DISCUSSION

In computational economic terms, this study has various strengths and weaknesses. One of its most obvious weaknesses is that it is unvalidated. At the time of this writing, there are no available infrastructures for implementing general Grid economies of which we are aware. Systems such as Nimrod-G [25] and GrADSoft [44] implement various components, but as of yet no mechanism exists for empirical studies of incentivized users using arbitrary programs. There are two immediate problems with building such a system that are particular to Grid economics.

The first problem is that it is difficult to incentivize credibly 'real' Grid consumers and producers in an experimental setting. Even if users are induced to participate, the decisions they make may be based on the inducement and not on their perceived value in the system. For example, in the Laboratory for **M**iddleware and **A**pplications **Y**ielding **H**eterogeneous **E**nvironments for **M**etacomputing (MAYHEM) [45] at the University of California, Santa Barbara, we have developed an infrastructure for implementing Grid resource economies that includes the First Bank of G as a price-setting mechanism. The system is compatible with a variety of Grid execution infrastructures including Globus [46], NetSolve [11], Condor [47], and Legion [2]. Designing effective empirical studies using the infrastructure has proved challenging. In particular, graduate students (when used as test subjects) are prone to skewing the results (either explicitly or inadvertently) according to the perceived wishes of their advisor. Our current work focuses on developing experiments that incentivize participants on the basis of the value of resources to the computations at hand and not the computational economics results that may be generated.

A second related problem concerns starting and restarting the economy. A key requirement for Grid economies in general, and Grid economics research in particular, is that Grid systems will need to be partially or entirely 'reset.' Software upgrades (particularly with respect to middleware services), security concerns, hardware maintenance, and so on can cause substantial reconfigurations. The economics research community has little guidance to offer with respect to starting an economy from scratch (or reset, in the case of the Grid). In an empirical setting, start-up can be critical since the amount of time that can be expected from test subjects may be limited. If the behavior under test will not emerge for months of years because the economy has been started in the wrong 'place', it may not be extractable. Again, our early attempts to design credible empirical experiments have been plagued by this problem.

Another important consideration in any economic study concerns the availability of what are usually hidden functions in a 'real-world' economy. In particular, some economic research hypothesizes the presence of utility functions that (in their simplest form) measure an agent's 'happiness' in terms of other economic quantities. These abstractions (sometimes termed *utility functions*) are useful for lending theoretical rigor, but are often difficult or impossible to determine in any practical way.

For G-commerce, these functions take the form of the consumer and producer decision-making algorithms that control supply and demand. Recall, that our simulated consumers used a stochastic model that was biased toward opportunistic purchasing while producers were stochastic bargain hunters. We designed them on the basis of our real-world experience with Grid users (and potential Grid users) in the PACI environment. Still, despite their non-linear and discontinuous definition, they remain a viable motivation for skepticism. Notice, though, that even though the individual producer and consumer functions are ill-behaved with respect to the assumptions underlying general equilibrium theory, they compose in a way that allows the basic equilibrium results to emerge. This observation about the composition of agent behavior is a strong point of the work.

32.6 CONCLUSIONS

In this chapter, we outline many of the features that characterize artificial resource economies as they pertain to the Computational Grid. Effective resource discovery and allocation mechanisms for Grid systems remain a critical area of research. By structuring such systems according to economic principles, it is possible to apply a considerable body of work from economics to analyze and predict Grid behavior. Doing so requires a keen understanding of the underlying assumptions governing allocation decisions.

In particular, it is important to understand, fundamentally, the relationship between price, supply, and demand that are induced as users acquire and release Grid resources. The choice of pricing protocol can also affect both the implementation complexity and the type of analysis that is possible. Auction systems, for example, are popular because they are easy to implement, as well as being well understood in terms of their effect on the individual market participants. They are difficult to analyze globally, however, and the optimal use of various formulations (e.g. combinatorial auctions) result in the need to solve NP-complete problems. Alternatively, commodity markets offer attractive global properties such as stability at the expense of useful features at the microeconomic level (e.g. incentive compatibility).

As an example illustrating many of these issues, we present G-commerce – a set of implementable policies and mechanisms economies for controlling resource allocation Computational Grid settings. Using simulated consumers and producers we investigate commodities markets and auctions as ways of price setting. Under similar conditions, we examine the overall price stability, market equilibrium, producer efficiency, and consumer efficiency. The results of this study, which appeal in large part of the work of Smale [39], show that commodity markets (despite their potential drawbacks) outperform auctions in our simulated Grid settings.

Designing and executing an empirical verification of this work remains a work in progress. Early experiences with accurately incentivized consumers illustrate several important research questions. In particular, unlike real-world economies, Grid resource economies may need to be restarted. Little formal theory exists for the analysis of the start-up and shut-down properties of economic systems. As a result, Grid economies may require new theoretical results in economics to be truly effective.

Despite this risk, the intuitive appeal of economically based resource allocation mechanisms combined with the theoretical results that are available make computational economies a viable and active research area. Clearly, there is much more work to do.

ACKNOWLEDGMENTS

This work was supported in part by NSF grants EIA-9975020, EIA-9975015, and ACI-9876895.

REFERENCES

1. Foster, I., Roy, A. and Winkler, L. (2000) A quality of service architecture that combines resource reservation and application adaptation. *Proceedings of TERENA Networking Conference, 2000.*
2. Grimshaw, A. S., Wulf, W. A., French, J. C., Weaver, A. C. and Reynolds, P. F. (1994) Legion: The Next Logical Step Toward a Nationwide Virtual Computer, Technical Report CS-94-21, University of Virginia, 1994, Charlotteselle, VA.
3. Buyya, R. EcoGRID Home Page, http://www.csse.monash.edu.au/~rajkumar/ecogrid/index.html.
4. Buyya, R. EconomyGRID Home Page, http://www.computingportals.org/projects/economyManager.xml.html.
5. Casanova, H., Obertelli, G., Berman, F. and Wolski, R. (2000) The AppLeS parameter sweep template: user-level middleware for the +grid. *Proceedings of SuperComputing 2000 (SC '00),* November, 2000.
6. Berman, F., Wolski, R., Figueira, S., Schopf, J. and Shao G. (1996). Application Level Scheduling on Distributed Heterogeneous Networks. *Proceedings of Supercomputing 1996,* 1996.
7. Arndt, O., Freisleben, B., Kielmann T. and Thilo, F. (1998) scheduling parallel applications in networks of mixed uniprocessor/multiprocessor workstations. *Proceedings of ISCA 11th Conference on Parallel and Distributed Computing,* September, 1998.
8. Gehrinf, J. and Reinfeld, A. (1996) MARS – a framework for minimizing the job execution time in a metacomputing environment. *Proceedings of Future general Computer Systems,* 1996.
9. Raman, R., Livny, M. and Solomon, M. (1998) Matchmaking: distributed resource management for high throughput computing. HPDC7, 1998.
10. globus-duroc, The Dynamically-Updated Request Online Coallocator (DUROC), 2002, Available from http://www.globus.org/duroc.
11. Casanova, H. and Dongarra, J. (1997) NetSolve: a network server for solving computational science problems. *The International Journal of Supercomputer Applications and High Performance Computing,* **11**(3), 212–223, Also in Proceedings of Supercomputing '96, Pittsburgh.

12. Nakada, H., Takagi, H., Matsuoka, S., Nagashima, U., Sato, M. and Sekiguchi, S. (1998) Utilizing the metaserver architecture in the ninf global computing system. *High-Performance Computing and Networking '98*, LNCS Vol. **1401**, 1998 pp. 607–616.
13. Chun, B. N. and Culler, D. E. (1999) Market-based Proportional Resource Sharing for Clusters, Millenium Project Research Report, http://www.cs.berkeley.edu/~plink/papers/market.pdf, September, 1999, Commerce Papers.
14. Wolski, R., Plank, J. S., Brevik, J. and Bryan, T. (2001) G-Commerce: market formulations controlling resource allocation on the computational grid. *International Parallel and Distributed Processing Symposium (IPDPS)*, April, 2001.
15. Buyya, R. (2002) Economic-based Distributed Resource Management and Scheduling for Grid Computing, Ph.D. thesis, Monash University, 2002, http://www.buyya.com/thesis/thesis.pdf.
16. Sandholm, T. and Suri, S. (2002) Optimal clearing of supply/demand curves. *Proceedings of 13th Annual International Symposium on Algorithms and Computation (ISAAC)*, 2002, an Explores complexity of clearing markets for a single commodity and supply and demand curves.
17. Sandholm, T., Suri, S., Gilpin, A. and Levine, D. (2001) CABOB: a fast optimal algorithm for combinatorial auctions. *Proc. of 17th International Joint Conference on Artificial Intelligence (IJCAI)*, 2001, an A fast and scalable algorithm to solve winner determination in comb auctions.
18. Parkes, D. and Ungar, L. (2000) Iterative combinatorial auctions: theory and practice. *Proceedings of 17th National Conference on Artificial Intelligence*, 2000.
19. Fujishima, Y., Leyton-Brown, K. and Shoham, Y. (1999) Taming the computational complexity of combinatorial auctions. *Proceedings of Sixteenth International Joint Conference on Artificial Intelligence*, 1999.
20. Wolski, R., Plank, J., Bryan, T. and Brevik, J. Analyzing market-based resource allocation strategies for the computational grid. *International Journal of High Performance Computing Applications*, **15**(3), 2001.
21. Regev, O. and Nisan, N. (1998) The popcorn market – online markets for computational resources. *First International Conference On Information and Computation Economies*, Charleston, SC, 1998, http:/www.cs.huji.ac.li/~popcorn/documentation/index.html.
22. Wellman, M. (1996) Market-oriented Programming: Some Early Lessons, Chapter 4, in Clearwater, S. (ed.) *Market-based Control: A Paradigm for Distributed Resource Allocation*, River Edge, NJ: World Scientific.
23. Varian, H. (2002) Broadband: Should We Regulate High-speed Internet Access? Alleman, J. and Crandall, R. (eds) *The Demand for Bandwidth: Evidence from the INDEX Experiment*, Washington, DC: Brookings Institute.
24. Waldspurger, C. A. and Weihl, W. E. Lottery scheduling: flexible proportional-share resource management. *First Symposium on Operating Systems Design and Implementation (OSDI '94)*, November, 1994, pp. 1–11, http://www.waldspurger.org/carl/papers.html.
25. Abramson, D., Giddy, J., Foster, I. and Kotler, L. (2000) High performance parametric modeling with nimrod/g: killer application for the global grid? *Proceedings of the International Parallel and Distributed Processing Symposium*, May, 2000.
26. Wellman, M., Walsh, W., Wurman, P. and MacKie-Mason, J. (1998) in Wellman, M. P., Walsh, W. E., Wurman, P. R. and MacKie-Mason, J. K. (eds) Auction Protocols for Decentralized Scheduling, Technical Report, University of Michigan, July, 1998, citeseer.nj.nec.com/article/wellman98auction.html.
27. Scarf, H. (1973) *The Computation of Economic Equilibria*. Yale University Press.
28. Merrill, O. H. (1972) Applications and Extensions of an Algorithm that Computes Fixed Points of Certain Upper Semi-Continuous Point to Set Mappings, Ph.D. Dissertation, Department of Industrial Engineering, University of Michigan, 1972.
29. Curtis Eaves, B. (1972) Homotopies for computation of fixed points. *Mathematical Programming*, **13**, 1–22.
30. Smale, S. (1975) Price adjustment and global Newton methods. *Frontiers of Quantitative Economics*, **IIIA**, 191–205.

31. Waldspurger, C. A., Hogg, T., Huberman, B., Kephart, J. O. and Stornetta, S. (1992) Spawn: A Distributed Computational Economy. *IEEE Transactions on Software Engineering*, **18**(2), 103–117, http://www.waldspurger.org/carl/papers.html.
32. Makowski, L., Ostroy, J. and Segal, U. (1995) Perfect Competition as the Blueprint for Efficiency and Incentive Compatibility, UCLA Economics Working Papers #745, October, 1995, http://ideas.uqam.ca/ideas/data/Papers/clauclawp745.htm.
33. Petitet, A., Blackford, S., Dongarra, J., Ellis, B., Fagg, G., Roche, K. and Vadhiyar, S. (2001) Numerical libraries and the grid. *Proceedings of SC '01*, November, 2001.
34. The National Partnership for Advanced Computational Infrastructure, http://www.npaci.edu.
35. The National Computational Science Alliance, http://www.ncsa.edu.
36. Walras, L. (1874) *Eléments d'Economie Politique Pure*. Corbaz.
37. Nisan, N., London, S., Regev, O. and Camiel, N. (1998) Globally distributed computation over the Internet – the POPCORN project. *International Conference on Distributed Computing Systems*, 1998, http:/www.cs.huji.ac.li/~popcorn/documentation/index.html.
38. Bredin, J., Kotz, D. and Rus, D. (1998) Market-based resource control for mobile agents, *Second International Conference on Autonomous Agents*. ACM Press, pp. 197–204, http://agent.cs.dartmouth.edu/papers/.
39. Smale, S. (1976) Dynamics in general equilibrium theory. *American Economic Review*, **66**(2), 284–294.
40. Chun, B. N. and Culler, D. E. (1999) Market-Based Proportional Resource Sharing for Clusters, Millenium Project Research Report, http://www.cs.berkeley.edu/~bnc/papers/market.pdf, September, 1999.
41. Regev, O. and Nisan, N. (1998) The popcorn market – online market for computational resources. *First International Conference On Information and Computation Economies*, Charleston, SC, 1998; to appear.
42. Bredin, J., Kotz, D. and Rus, D. (1997) Market-Based Resource Control for Mobile Agents, Technical Report PCS-TR97-326, Dartmouth College, Computer Science Department, Hanover, NH, November, 1997, ftp://ftp.cs.dartmouth.edu/TR/TR97-326.ps.Z.
43. Waldspurger, C. A., Hogg, T., Huberman, B. A., Kephart, J. O. and Stornetta, W. S. (1992) Spawn: a distributed computational economy. *IEEE Transactions on Software Engineering*, **18**(2), 103–117.
44. Berman, F. *et al.* (2000) The GrADS Project: Software Support for High-Level Grid Application Development, Technical Report Rice COMPTR00-355, Houston, TX: Rice University, February, 2000.
45. The MAYHEM Laboratory, UCSB, http://pompone.cs.ucsb.edu.
46. Foster, I. and Kesselman, C. (1997) Globus: a metacomputing infrastructure toolkit, *The International Journal of Supercomputer Applications and High Performance Computing IJSA*, **11**(2), 115–128, Summer 1997.
47. Tannenbaum, T. and Litzkow, M. (1995) Checkpoint and Migration of Unix Processes in the Condor Distributed Processing System, *Dr. Dobbs Journal*, **February**, 40–48.

Parameter sweeps on the Grid with APST

Henri Casanova[1] and Fran Berman[1]

[1]*University of California, San Diego, California, United States*

33.1 INTRODUCTION

Computational Grids [1, 2] are large collections of resources such as computers, networks, on-line instruments, or storage archives, and they are becoming popular platforms for running large-scale, resource-intensive applications. Many challenges exist in providing the necessary mechanisms for accessing, discovering, monitoring, and aggregating Grid resources. Consequently, a tremendous effort has been made to develop middleware technology to establish a Grid software infrastructure (GSI) [2–4]. Although middleware provides the fundamental building blocks, the APIs and access methods are often too complex for end users. Instead, there is a need for abstractions and tools that make it easy for users to deploy their applications. Several projects have addressed this need at various stages of application development and execution. For instance, the GrADS project [5] seeks to provide a comprehensive application-development system for Grid computing that integrates Grid-enabled libraries, application compilation, scheduling, staging of binaries and data, application launching, and monitoring of application execution progress.

Grid Computing – Making the Global Infrastructure a Reality. Edited by F. Berman, A. Hey and G. Fox
© 2003 John Wiley & Sons, Ltd ISBN: 0-470-85319-0

Another approach is to provide simple programming abstractions and corresponding runtime support for facilitating the development of Grid applications. For instance, a number of projects enable Remote Procedure Call (RPC) programming on the Grid [6, 7] and are currently collaborating to define GridRPC [8]. Alternatively, other projects have developed environments that deploy applications on Grid resources without involving the user in any Grid-related development effort. The AppLeS Parameter Sweep Template (APST) project presented here belongs in the last category and targets the class of parameter sweep applications.

Parameter sweep applications (PSAs) are structured as sets of computational tasks that are mostly *independent*: there are few task-synchronization requirements, or data dependencies, among tasks. In spite of its simplicity, this application model arises in many fields of science and engineering, including Computational Fluid Dynamics [9], Bio-informatics [10–12], Particle Physics [13, 14], Discrete-event Simulation [15, 16], Computer Graphics [17], and in many areas of Biology [18–20].

PSAs are commonly executed on a network of workstations. Indeed, it is straightforward for users to launch several independent jobs on those platforms, for instance, via *ad hoc* scripts. However, many users would like to scale up their PSAs and benefit from the vast numbers of resources available in Grid platforms. Fortunately, PSAs are not tightly coupled, as tasks do not have stringent synchronization requirements. Therefore, they can tolerate high network latencies such as the ones expected on wide-area networks. In addition, they are amenable to straightforward fault-tolerance mechanisms as tasks can be restarted from scratch after a failure. The ability to apply widely distributed resources to PSAs has been recognized in the Internet computing community (e.g. SETI@home [21]). There are two main challenges for enabling PSAs at such a wide scale: making application execution easy for the users, and achieving high performance. APST addresses those two challenges by providing transparent deployment and automatic scheduling of both data and computation.

We have published several research articles describing our work and results on APST. In Reference [22], we evaluated in simulation a number of scheduling algorithms and heuristics for scheduling PSAs onto a Grid platform consisting of a set of clusters. In Reference [23], we described the first APST prototype and presented experimental results obtained on a Grid platform spanning clusters in Japan, California, and Tennessee. In Reference [24], we described the use of APST for a computational neuroscience application. Our goal here is to briefly introduce APST, discuss its usability, and show that it is a powerful tool for running PSAs on small networks of workstations, large clusters of PCs, and Grid platforms. We focus APST v2.0, which includes many improvements compared to previous versions [23].

33.2 THE APST PROJECT

33.2.1 Motivation

The genesis of APST lies in our work on the AppLeS (Application-Level Scheduling) project [25]. The AppLeS work has been motivated by two primary goals: (1) to investigate and validate adaptive scheduling for Grid computing and (2) to apply our results to real applications in production environments and improve the performance experienced by end users. We have achieved these goals by combining static and dynamic resource information, performance predictions, application- and user-specific information, and by developing scheduling techniques that use that information to improve application performance. Using several applications, we demonstrated that adaptive scheduling is key for achieving high performance in Grid environments [25–32]. Each application was fitted with a customized scheduling agent that strives to improve application performance given the resources at hand, the structural characteristics of the application, and the user's performance goals.

During the course of the AppLeS project, we have often been approached by application developers asking for AppLeS code so that they could enhance their own applications. However, AppLeS agents are integrated pieces of software in which the application code and the agent are combined, and therefore are not easily separated for reuse. The next logical step, then, was to develop software environments that are usable for classes of applications. In that context, the APST project was established so that users can easily and effectively deploy PSAs.

One challenge is to *transparently deploy* applications on behalf of users. APST should handle most logistics of the deployment, which include discovering resources, performing application data transfers, keeping track of the application data, launching and monitoring computations on Grid resources, and detecting and recovering from failures. Many of those tasks can be implemented with middleware services invoked on behalf of the user. APST provides application deployment that is as transparent as possible, while letting the user control key aspects of deployment. Another challenge is that of *performance*. This can be achieved by developing scheduling algorithms that make decisions concerning where to transfer/download application data and where to start application tasks. Since PSAs are generally long running, these algorithms must refine decisions during application execution to tolerate changes in resource conditions. APST implements several such scheduling algorithms.

33.2.2 Principles and architecture

When designing and implementing APST, we focused on the following basic principles.

Ubiquitous deployment: We wish APST users to be able to deploy their applications on as many resources as possible. APST must therefore support a variety of middleware services (e.g. Grid services) for discovering, using, and monitoring storage, compute, and network resources. The design of APST must ensure that it is possible (and easy) to add support for such services as they become available. To that end, APST contains modules that abstract resource discovery and monitoring, job launching and monitoring, data movement and storage. Each module can be instantiated with several implementations that can be used simultaneously.

Opportunistic execution: Another principle behind APST is that no specific service is required. For instance, if services for resource monitoring are deployed and available to the user, then they can be used by a scheduler within APST for making more informed scheduling decisions. However, if no such service is available, APST will still function, but will probably achieve lower performance. Similarly, a user should be able to benefit from APST out-of-the-box by deploying his/her application on local resources with default services (e.g. ssh to start remote jobs). If needed, the user can incrementally acquire new resources that may require other services (e.g. Globus GRAM). This principle has proved very successful in getting users to adopt APST from the start and progressively scale up to Grid platforms.

Lightweight software: A big impediment to the acceptance of software by scientific user communities is the complexity of the compilation, installation, and deployment of that software. To that end, we use standard packaging technology for the APST software. Furthermore, the APST software only needs to be installed on a single host, typically the user's local machine. This is possible because APST reuses middleware services that are already deployed and can be used to access resources. This contributes to making the software lightweight and is critical for gaining acceptance from users.

Automation of user processes: We do not wish to change the way in which users run their applications, but rather automate the process by which they do it. In addition, APST generally does not require any change to the application code, provided that all I/O is done via files and command-line arguments (which is typical for most PSAs). This is another critical factor in getting users to adopt APST.

Simple user interface: When designing APST, we examined several alternatives for a user interface. We chose a simple, XML-based interface that can be used from the command-line or from scripts. Because it uses a well-defined XML-based protocol, our current interfaces can be easily integrated with more sophisticated interfaces such as the ones provided by Grid portals [33], ILAB [34], or Nimrod/G [35].

Resilience: Grid resources are shared and federated, and are therefore prone to failures and downtimes. APST must implement simple fault-detection restart mechanisms. Such mechanisms are already available in some middleware services and can be leveraged by APST. Since PSAs are typically long running, APST also implements a checkpointing mechanism to easily recover from crashes of APST itself with minimal loss for the application.

33.2.3 Software architecture

We designed the APST software to run as two distinct processes: a daemon and a client. The *daemon* is in charge of deploying and monitoring applications. The *client* is essentially a console that can be used periodically, either interactively or from scripts. The user can invoke the client to interact with the daemon to submit requests for computation and check on application progress.

We show the overall architecture of the APST software in Figure 33.1. The computing platform consists of storage, compute, and network resources depicted at the bottom of the

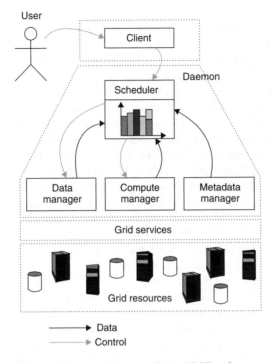

Figure 33.1 Architecture of the APST software.

figure. Those resources are accessible via deployed middleware services (e.g. Grid services as shown on the figure). The central component of the daemon is a *scheduler* that makes all decisions regarding the allocation of resources to application tasks and data. To implement its decisions, the scheduler uses a *data manager* and a *compute manager*. Both components use middleware services to launch and monitor data transfers and computations. In order to make decisions about resource allocation, the scheduler needs information about resource performance. As shown in the figure, the scheduler gathers information from 3 sources. The data manager and the compute manager both keep records of past resource performance and provide the scheduler with that historical information. The third source, the *metadata manager*, uses information services to actively obtain published information about available resources (e.g. CPU speed information from MDS [36]). A predictor, not shown on the figure, compiles information from those three sources and computes forecasts, using techniques from NWS [37]. Those forecasts are then used by APST's scheduling algorithms. The cycle of control and data between the scheduler and the three managers is key for adaptive scheduling of PSAs onto Grid platforms.

33.2.4 Scheduling

APST started as a research project in the area of Grid application scheduling. Therefore, most of our initial efforts were focused on the Scheduler component. On the basis of

AppLeS results, the scheduler uses static and dynamic information about resources, as well as application-level information (number of tasks, size of data files, etc.) in order to make scheduling decisions. Previous AppLeS work only addressed adaptive scheduling at the onset of the application, as opposed to during the execution. Departing from the AppLeS work, APST targets applications that are long running and therefore it must refine scheduling decisions throughout application execution.

In Reference [22], we presented an adaptive scheduling algorithm that refines the application schedule periodically. In particular, we focused on application scenarios in which potentially large input data files can be shared by several application tasks, which occurs for many PSAs. It is then critical to maximize the reuse of those files. This can be achieved via file replication and scheduling of computational tasks 'close' to relevant files. This scheduling problem is NP-complete and we employed list scheduling heuristics with dynamic priorities [38]. We hypothesized that this would be a good approach to our scheduling problem by identifying commonalities between the concept of task-host *affinities*, defined in Reference [39], and the notion of *closeness* of data to computation. We developed a new heuristic, XSufferage, and validated our scheduling approach in simulation. We showed that our adaptive scheduling algorithm tolerates the kind of performance prediction errors that are expected in Grid environments. We implemented our algorithm in the APST scheduler for four different heuristics. We also implemented a simple greedy algorithm that uses task duplication. We then compared these different scheduling approaches on a real Grid test bed in Reference [23].

Deciding which scheduling algorithm is appropriate for which situation is a difficult question. In Reference [22], we have seen that our XSufferage heuristic is effective when large input files are shared by several application tasks and when performance prediction errors are within reasonable bounds. However, in an environment in which resource availability varies significantly, thereby making performance unpredictable, a greedy algorithm may be more appropriate. Also, if the amount of application data is small, the algorithms presented in Reference [22] may not be effective and a greedy approach may be preferable. Currently, our results do not allow us to precisely decide which scheduling algorithm to employ on the fly. However, the APST design is amenable to experimentation for tackling that open research question. The scheduler is completely isolated from other components and can therefore be replaced easily. If services to monitor resources are available, then the scheduler can use resource information for making decisions. If no such service is available, then the scheduler uses estimates based solely on historical application behavior on the resources. The current APST implementation allows the user to choose which of the available scheduling algorithms to use; the scheduling process is completely transparent from then on.

33.2.5 Implementation

The current APST implementation can make use of a number of middleware services and standard mechanisms to deploy applications. We provide a brief description of those capabilities.

Launching application tasks: APST can launch application tasks on the local host using `fork`. Remote hosts can be accessed via `ssh`, Globus `GRAM` [40], and `NetSolve` [6].

The ssh mechanism allows for ssh-tunneling in order to go through firewalls and to private networks. APST inherits the security and authentication mechanisms available from those services (e.g. GSI [41]), if any. APST can launch applications directly on interactive resources and can start jobs via schedulers such as PBS [42], LoadLeveler [43], and Condor [44]. We are conducting research on the use of batch resources for the efficient deployment of PSAs. Batch schedulers are complex systems that are not adapted for applications that consist of large numbers of small, possibly sequential jobs. We are investigating several techniques that will adapt to the behavior of batch schedulers.

Moving and storing application data: APST can read, copy, transfer, and store application data among storage resources with the following mechanisms. It can use cp to copy data between the user's local host to storage resources that are on the same Network File System; data can also be used in place. APST can also use scp, FTP, GASS [45], GridFTP [46], and SRB [47]. Version 1.0 of APST also supported IBP [48] and we are currently planning an integration of IBP's newest set of tools into APST v2.0. APST inherits any security mechanisms provided by those services.

Discovering and monitoring resources: APST can obtain static and dynamic information from information services such as MDS [36] and NWS [37]. We also support the Ganglia [49] system that is increasingly popular on clusters. In addition, we have implemented a few straightforward mechanisms for obtaining information on resource. For instance, we use standard UNIX commands such as uptime for resources that are not registered with information services. APST also learns about available resources by keeping track of their past performance when computing application tasks or transferring application data.

We summarize the services that can be used by APST to deploy users' applications in Table 33.1. The software can be easily configured and installed to use one or more of those services. The default installation enables the use of fork, ssh, cp, and scp so that users can run applications immediately on resources in their laboratories. The software consists of about 10 000 lines of C code, uses the AppleSeeds library [50], and has been ported to most flavors of UNIX. An early prototype of the software was demonstrated at the SC'99 conference for a computational neuroscience application [20] and Version 1.1 was demonstrated at SC'01 for a volume rendering application [17]. APST is freely available at [51].

Table 33.1 Services usable by the current APST implementation

Functionality	Mechanisms
Computation	fork, GRAM, NetSolve, ssh, Condor, PBS, LoadLeveler
Data	cp, scp, FTP, GASS, GridFTP, SRB
Information	MDS, NWS, Ganglia

33.3 APST: USAGE AND APPLICATIONS

The APST software consists of a daemon, `apstd`, and a client, `apst`, which communicates with the daemon over a socket. The user can send a variety of commands to the client, either from the command-line or in interactive mode. Some of these commands require input (e.g. which tasks to run). That input is structured as XML files within `<apst>` and `</apst>` tags. We provide a description and examples of how APST is used in the next two sections.

33.3.1 APST and Grid resources

APST's logical view of available resources is depicted in Figure 33.2. The platform consists of *sites*, in which each site contains at least one storage resource and a number of compute resources that can read and write data on the storage resources. Sites are interconnected over the network and data can be moved from one site's storage to another site's storage. The client and daemon run on the user's local system.

The user must describe the available storage and compute resources in XML, denoting for each one which access mechanisms should be used. To form sites, the XML also describes which storage resource is associated with each compute resource. In addition, the XML can specify information sources (e.g. a MDS server). We show below a small example.

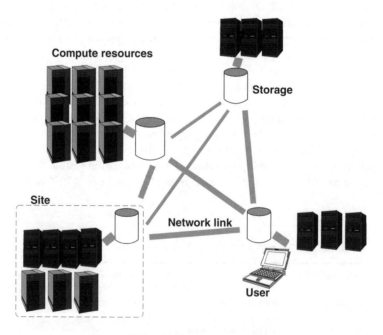

Figure 33.2   APST computing platform.

```
<apst>
  <storage>
    <disk id="DISK1" datadir="/home/data">
      <gridftp server="storage.site1.edu" />
    </disk>
    <disk id="DISK2" datadir="/usr/home/data">
      <scp server="storage.site2.edu" />
    </disk>
  </storage>

  <compute>
    <host id="HOST1" disk="DISK1">
      <globus server="host.site1.edu" />
    </host>
    <host id="HOST2" disk="DISK2">
      <ssh server="host.site2.edu" processors="40" />
      <condor/>
    </host>
  </compute>

  <gridinfo>
    <nws server="nws.site1.edu" />
    <mds server="mds.site2.edu" />
  </gridinfo>
</apst>
```

This XML section describes a compute platform that consists of two sites. The first site contains a GridFTP server and one host accessible via Globus. The second site consists of a Condor pool with 40 hosts, with Condor submissions done via ssh. Application data can be stored and retrieved from the second site via scp. In addition, resource information may be available from an NWS server and an MDS server. There are many options (and appropriate default values) available to the user for XML resource descriptions; all details are available in the APST documentation. The user can add resources at any time by sending XML descriptions to the APST daemon during application execution. We believe that APST resource descriptions strike a good balance between hiding needless details from the user while providing good control over resource usage.

33.3.2 Running applications with APST

The user must describe application tasks that are to be executed. Each task is specified by several parameters, including the name of an executable, command-line arguments, input and output files, and so forth. As PSAs consist of large numbers of tasks, users typically write their own scripts to generate XML task descriptions. Here is an example:

```
<tasks>
  <task executable="app" arguments="f1 g1" input="f1" output="g1" cost="1" />
  <task executable="app" arguments="f2 g2" input="f2" output="g2" cost="2" />
</tasks>
```

where each task runs the app executable with two different input files to generate two output files. In this case, the user also gives APST a hint that the second task requires twice as much computation as the first task, which can be used for scheduling decisions. APST checks dependencies among tasks in case a task's output is needed as input by another task. Note that the APST scheduler assumes that those dependencies are infrequent, meaning that we do not use sophisticated Directed Acyclic Graph (DAG) scheduling algorithms [52], but simply maintain a list of 'ready' tasks that are all independent.

The user can also specify application data that has been pre-staged on storage resources as follows:

```
<files>
  <file id="f1" size="500M">
    <copy disk="DISK1" />
  <file>
</files>
```

meaning that file f1 need not be transferred to DISK1 as a copy is already available at that site's storage resource.

The APST client provides ways for the user to submit XML (for resources, application tasks, and application files) to the APST daemon *at any time*. In addition, the client provides a number of ways for the user to check on progress and completion of tasks, check the utilization of resources, cancel tasks, enable/disable resources manually, download/upload files from remote storage manually, and be notified of task completions. The APST daemon periodically checkpoints its state (as an XML file) during execution. If the daemon crashes, it can be restarted from that state with minimal loss for the user.

33.3.3 Discussion

APST started as a research prototype for exploring adaptive scheduling of PSAs on the Grid platform. Since then, it has evolved into a usable software tool that is gaining popularity in several user communities. The first application to use APST in production was MCell [20], a computational neuroscience application developed at the Salk institute and the Pittsburgh Supercomputer Center. Since then, APST has been used for computer graphics applications [17], discrete-event simulations [16], and bio-informatics applications [12, 10, 11]. There is a growing interest in the bio-informatics community as biological sequence matching applications all fit under the PSA model. While interacting with users, we have learned the following lessons.

Many disciplinary scientists are still running their applications on single workstations. It was surprising to realize that, even for parallel applications as simple as PSAs, there are still many hurdles for users to overcome. APST provides a good solution because it does not require modification of the application, because it requires only a minimal understanding of XML, and because it can be used immediately with ubiquitous mechanisms (e.g. ssh and scp). In addition, users can easily and progressively transition to larger scale platforms on which more sophisticated Grid services are required. Moreover, the fact that the APST software needs to be installed only on the user's host makes it easier to adopt.

Our experience has been that most users find the current APST interface appropriate for their needs. In fact, they usually build simple, application-specific interfaces on top of the APST interface. Some scientists are considering building application-specific GUIs on top of APST. This is a fairly straightforward process as it is possible to communicate directly to the APST daemon over a socket, with a well-defined protocol that uses XML to structure messages. Also, generic interfaces such as the ones provided by Grid portals [33], ILAB [34], or Nimrod/G [35], could be easily adapted to be used with APST.

Another realization is that at this stage of Grid computing users are much more concerned with usability than with performance. Even though APST originated as a scheduling research project that focused primarily on application performance, recent developments have been focused on usability. The gap between disciplinary scientists and the Grid is still large, and having a usable tool that allows users to easily run applications is invaluable.

An issue is that of multiple users using APST simultaneously over a same set of resources. Competitive/cooperative resource sharing is a difficult problem from a scheduling standpoint, and it has been investigated for a number of objectives (e.g. fairness, resource utilization, etc.) by *batch schedulers* [42–44]. The goal of APST is not to replace batch schedulers. Instead, APST can use batch schedulers over shared resources in order to submit jobs on behalf of a user. Nevertheless, interesting issues need to be addressed for APST to interact with batch schedulers effectively.

33.4 RELATED WORK

A number of projects in the Grid community are related to APST. Like APST, ILAB [34] and Nimrod/G [35] target the deployment of PSAs on distributed resources. Both projects place a strong emphasis on user interface whereas APST provides only a basic interface. In terms of application deployment, APST provides more mechanisms for accessing resource than either ILAB or Nimrod/G. APST could be interfaced/integrated with parts of both Nimrod/G and ILAB in order to generate a sweep software environment with full-fledged user interfaces and richer functionalities. Related projects also include Condor [44] and various projects being developed in industry [53, 54, 55] that aim at deploying large number of jobs on widely distributed resources. We have already built an interface from APST to Condor [56], and similar work could be done for other projects. Our goal is to target as many resources as possible, while still providing a unified application execution environment to the user.

A great deal of research has investigated the question of scheduling independent tasks onto networks of heterogeneous processors. We have reviewed relevant work, including work we have leveraged for developing the APST scheduler, in Reference [22]. APST makes an important and practical contribution by providing adaptive scheduling of both application computation and data during application execution. An interesting aspect of Nimrod/G [35] is that its scheduling approach is based on a Grid economy model with deadlines. The Nimrod/G scheduler achieves trade-offs between performance and resource cost, whereas APST only considers performance but takes into account application data movements. We are currently working together with the Nimrod/G team to investigate how our scheduling approaches could be combined.

33.5 CONCLUSION AND FUTURE DEVELOPMENTS

The APST project started as a research project to investigate adaptive scheduling of PSAs on the Grid and evolved into a usable application execution environment. We have reported on our results on scheduling in previous papers [22–24]. In this chapter, we briefly introduced APST, focused on usability issues, and explained how APST enables disciplinary scientists to easily, and progressively, deploy their applications on the Grid.

We are currently pursuing a number of new development and research directions. Batch schedulers were not designed to support PSAs, and we are investigating ways for APST to use batch resources effectively. An ongoing process is to continue integrating APST with Grid technology. For instance, DataGrid technology [57] could be leveraged by APST in order to manage application data replicas. We are pursuing the integration of APST with the Grid Security Infrastructure [41] to allow multiple users to cooperatively share an APST daemon securely. We are also investigating scenarios in which the application's workload is divisible, meaning that the APST scheduler can decide the size of application tasks. This is relevant for applications in which the size of a base computational unit is many orders of magnitude lower than the entire application's workload, which is a valid assumption for many PSAs. Scheduling divisible workloads is a challenging question [58] which we will investigate by using APST as a research platform.

Further information on the APST project, documentation, and software is available at http://grail.sdsc.edu/projects/apst.

REFERENCES

1. Foster, I. and Kesselman, C. (1999) *The Grid: Blueprint for a New Computing Infrastructure.* San Francisco, CA, USA: Morgan Kaufmann Publishers.
2. Foster, I., Kesselman, C. and Tuecke, S. (2001) The anatomy of the grid: enabling scalable virtual organizations. *International Journal of High Performance Computing Applications*, **15**(3), 200–222.
3. Foster, I., Kesselman, C., Nick, J. and Tuecke, S. (2002) *The Physiology of the Grid: An Open Grid Services Architecture for Distributed Systems Integration*, Global Grid Forum, Open Grid Service Infrastructure WG available at http://www.globus.org.
4. Global Grid Forum Webpage, http://www.gridforum.org.
5. Berman, F. *et al.* (2001) The GrADS project: software support for high-level grid application development. *International Journal of High Performance Computing Applications*, **15**(4), 327–344.
6. Casanova, H. and Dongarra, J. (1997) NetSolve: a network server for solving computational science problems. *The International Journal of Supercomputer Applications and High Performance Computing*, **11**(3), 212–223.
7. Nakada, H., Sato, M. and Sekiguchi, S. (1999) Design and implementations of Ninf: towards a global computing infrastructure. *Future Generation Computing Systems*, Metacomputing Issue, **15**(5–6), 649–658.
8. Nakada, H., Matsuoka, S., Seymour, K., Dongarra, J., Lee, C. and Casanova, H. (2002) GridRPC: A Remote Procedure Call API for Grid Computing, Technical Report ICL-UT-02-06, Department of Computer Science, University of Tennessee, Knoxville, 2002.
9. Rogers, S. (1995) A comparison of implicit schemes for the incompressible Navier-Stokes equations with artificial compressibility. *AIAA Journal*, **33**(10), 2066–2072.

10. Altschul, S., Dish, W., Miller, W., Myers, E. and Lipman, D. (1990) Basic local alignment search tool. *Journal of Molecular Biology*, **215**, 403–410.
11. Altschul, S., Madden, T., Schäffer, A., Zhang, J., Zhang, Z., Miller, W. and Lipman, D. (1997) Gapped BLAST and PSI-BLAST: a new generation of protein database search programs. *Nucleic Acids Research*, **25**, 3389–3402.
12. Durbin, R., Eddy, S., Krogh, A. and Mitchison, G. (1998) *Biological Sequence Analysis: Probabilistic Models of Proteins and Nucleic Acids*. Cambridge, UK: Cambridge University Press.
13. Basney, J., Livny, M. and Mazzanti, P. (2000) Harnessing the capacity of computational grids for high energy physics. Conference on Computing in High Energy and Nuclear Physics, (2000).
14. Majumdar, A. (2000) Parallel performance study of Monte-Carlo photon transport code on shared-, distributed-, and distributed-shared-memory architectures. *Proceedings of the 14th Parallel and Distributed Processing Symposium, IPDPS '00*, May, 2000, pp. 93–99.
15. Casanova, H. (2001) Simgrid: a toolkit for the simulation of application scheduling. *Proceedings of the IEEE International Symposium on Cluster Computing and the Grid (CCGrid '01)*, May, 2001, pp. 430–437.
16. Takefusa, A., Matsuoka, S., Nakada, H., Aida, K. and Nagashima, U. (1999) Overview of a performance evaluation system for global computing scheduling algorithms. *Proceedings of the 8th IEEE International Symposium on High Performance Distributed Computing (HPDC)*, August, 1999, pp. 97–104.
17. NPACI Scalable Visualisation Tools Webpage, http://vistools.npaci.edu.
18. Berman, H. *et al.* (2000) The protein data bank. *Nucleic Acids Research*, **28**(1), 235–242.
19. Natrajan, A., Crowley, M., Wilkins-Diehr, N., Humphrey, M., Fox, A. and Grimshaw, A. (2001) Studying protein folding on the grid: experiences using CHARM on NPACI resources under Legion. *Proceedings of the Tenth IEEE International Symposium on High Performance Distributed Computing (HPDC-10)*, 2001.
20. Stiles, J. R., Bartol, T. M., Salpeter, E. E. and Salpeter, M. M. (1998) Monte Carlo simulation of neuromuscular transmitter release using MCell, a general simulator of cellular physiological processes, in Bower, J. M. (ed.) *Computational Neuroscience*. New York: Plenum Press, pp. 279–284.
21. SETI@home, http://setiathome.ssl.berkeley.edu, (2001).
22. Casanova, H., Legrand, A., Zagorodnov, D. and Berman, F. (2000) Heuristics for scheduling parameter sweep applications in grid environments. *Proceedings of the 9th Heterogeneous Computing Workshop (HCW '00)*, May, 2000, pp. 349–363.
23. Casanova, H., Obertelli, G., Berman, F. and Wolski, R. (2000) The AppLeS parameter sweep template: user-level middleware for the grid. *Proceedings of Supercomputing 2000 (SC '00)*, November, 2000.
24. Casanova, H., Bartol, T., Stiles, J. and Berman, F. (2001) Distributing MCell simulations on the grid. *International Journal of High Performance Computing Applications*, **14**(3), 243–257.
25. Berman, F., Wolski, R., Figueira, S., Schopf, J. and Shao, G. (1996) Application level scheduling on distributed heterogeneous networks. *Proceedings of Supercomputing '96*, 1996.
26. Spring, N. and Wolski, R. (1998) Application level scheduling of gene sequence comparison on metacomputers. *Proceedings of the 12th ACM International Conference on Supercomputing*, Melbourne, Australia, July, 1998.
27. Su, A., Berman, F., Wolski, R. and Strout, M. (1999) Using AppLeS to schedule simple SARA on the computational grid. *The International Journal of High Performance Computing Applications*, **13**(3), 253–262.
28. Smallen, S. *et al.* (2000) Combining workstations and supercomputers to support grid applications: the parallel tomography experience. *Proceedings of the 9th Heterogeneous Computing Workshop*, May, (2000), pp. 241–252.
29. Cirne, W. and Berman, F. (2000) Adaptive selection of partition size for supercomputer requests. *Proceedings of the 6th Workshop on Job Scheduling Strategies for Parallel Processing*, May, 2000.

30. Dail, H., Obertelli, G., Berman, F., Wolski, R. and Grimshaw, A. (2000) Application-aware scheduling of a magnetohydrodynamics application in the Legion metasystem. *Proceedings of the 9th Heterogeneous Computing Workshop (HCW '00)*, May, 2000.
31. Schopf, J. and Berman, F. (1999) Stochastic scheduling. *Proceedings of Supercomputing '99*, 1999.
32. Schopf, J. and Berman, F. (2001) Using stochastic information to predict application behavior on contended resources. *International Journal of Foundations of Computer Science*, **12**(3), 341–363.
33. Thomas, M., Mock, S., Boisseau, J., Dahan, M., Mueller, K. and Sutton, D. (2001) The Grid-Port toolkit architecture for building grid portals. *Proceedings of the 10th IEEE International Symposium on High Performance Distributed Computing (HPDC-10)*, August, 2001.
34. Yarrow, M., McCann, K., Biswas, R. and Van der Wijngaart, R. (2000) An advanced user interface approach for complex parameter study process specification on the information power grid. *GRID 2000*, Bangalore, India, December, 2000.
35. Abramson, D., Giddy, J. and Kotler, L. (2000) High performance parametric modeling with Nimrod/G: killer application for the global grid? *Proceedings of the International Parallel and Distributed Processing Symposium (IPDPS)*, Cancun, Mexico, May, (2000), pp. 520–528.
36. Czajkowski, K., Fitzgerald, S., Foster, I. and Kesselman, C. (2001) Grid information services for distributed resource sharing. *Proceedings of the 10th IEEE Symposium on High-Performance Distributed Computing*, 2001; to appear.
37. Wolski, R., Spring, N. and Hayes, J. (1999) The network weather service: a distributed resource performance forecasting service for metacomputing. *Future Generation Computer Systems*, **15**(5–6), 757–768.
38. Li, K. (1999) Analysis of the list scheduling algorithm for precedence constrained parallel tasks. *Journal of Combinatorial Optimization*, **3**, 73–88.
39. Maheswaran, M., Ali, S., Siegel, H. J., Hensgen, D. and Freund, R. (1999) Dynamic matching and scheduling of a class of independent tasks onto heterogeneous computing systems. *8th Heterogeneous Computing Workshop (HCW '99)*, April, 1999, pp. 30–44.
40. Czajkowski, K., Foster, I., Karonis, N., Kesselman, C., Martin, S., Smith, W. and Tuecke, S. (1998) A resource management architecture for metacomputing systems. *Proceedings of IPPS/SPDP '98 Workshop on Job Scheduling Strategies for Parallel Processing*, 1998.
41. Foster, I., Kesselman, C., Tsudik, G. and Tuecke, S. (1998) A security architecture for computational grids. *Proceedings of the 5th ACM Conference on Computer and Communications Security*, 1998, pp. 83–92.
42. The Portable Batch System Webpage, http://www.openpbs.com.
43. IBM Corporation, IBM LoadLeveler User's Guide, (1993).
44. Litzkow, M., Livny, M. and Mutka, M. (1988) Condor – a hunter of idle workstations. *Proceedings of the 8th International Conference of Distributed Computing Systems*, June, 1988, pp. 104–111.
45. Foster, I., Kesselman, C., Tedesco, J. and Tuecke, S. (1999) GASS: a data movement and access service for wide area computing systems, *Proceedings of the Sixth workshop on I/O in Parallel and Distributed Systems*, May, 1999.
46. Allcock, W., Bester, J., Bresnahan, J., Chervenak, A., Liming, L. and Tuecke, S. (2001) GridFTP: Protocol Extension to FTP for the Grid, Grid Forum Internet-Draft, March, 2001
47. The Storage Resource Broker, http://www.npaci.edu/dice/srb, (2002).
48. Plank, J., Beck, M., Elwasif, W., Moore, T., Swany, M. and Wolski, R. (1999) The internet backplane protocol: storage in the network. *Proceedings of NetSore '99: Network Storage Symposium*, Internet2, 1999.
49. Ganglia Cluster Toolkit, http://ganglia.sourceforge.net, (2002).
50. AppleSeeds Webpage, http://grail.sdsc.edu/projects/appleseeds.
51. APST Webpage, http://grail.sdsc.edu/projects/apst.
52. Kwok, Y. and Ahmad, I. (1999) Benchmarking and comparison of the task graph scheduling algorithms. *Journal of Parallel and Distributed Computing*, **59**(3), 381–422.
53. Sun Microsystems Grid Engine, http://www.sun.com/gridware/.

54. Entropia Inc., Entropia, http://www.entropia.com.
55. United Device Inc., http://www.ud.com.
56. Williams, D. (2002) APST-C: Expanding APST to target Condor, Master's thesis, University of California San Diego, June, 2002.
57. Chervenak, A., Foster, I., Kesselman, C., Salisbury, C. and Tuecke, S. (2000) The data grid: towards an architecture for the distributed management and analysis of large scientific datasets. *Journal of Network and Computer Applications*; **23**, 187–200.
58. Bharadwaj, V., Ghose, D., Mani, V., T. G. and Robertazzi (1996) *Scheduling Divisible Loads in Parallel and Distributed Systems*. Los Alamitos, CA: IEEE computer society press.

34

Storage manager and file transfer Web services[1]

William A. Watson III, Ying Chen, Jie Chen, and Walt Akers

Thomas Jefferson National Accelerator Facility, Newport News, Virginia, United States

34.1 INTRODUCTION

Built upon a foundation of Simple Object Access Protocol (SOAP), Web Services Description Language (WSDL) and Universal Description Discovery and Integration (UDDI) technologies, Web services have become a widely accepted industry standard in the last few years [1, 2]. Because of their platform independence, universal compatibility, and network accessibility, Web services will be at the heart of the next generation of distributed systems. As more vendors offer SOAP tools and services, the advantages of using SOAP and Web services as an integration point will become even more pronounced.

The Grid computing community has also recognized the importance of using Web services. The Globus project has proposed adding a Web service layer to its existing infrastructure, and has proposed an Open Grid Services Architecture (OGSA) [3]. The Unicore project has already developed an OGSA-compliant Web services wrapper for

---

[1] Work supported by the Department of Energy, contract DE-AC05-84ER40150.

Grid Computing – Making the Global Infrastructure a Reality.   Edited by F. Berman, A. Hey and G. Fox
© 2003 John Wiley & Sons, Ltd   ISBN: 0-470-85319-0

their Grid Toolkit [4]. Other forums are likewise engaged in developing Web services for Grids and portals [5].

In the Fall of 2000, ahead of the major projects' move to Web services, Jefferson Lab started to investigate the feasibility of providing Grid capabilities using XML-based Web services. These XML services were later migrated to SOAP-based Web services. The goal of this ongoing work is to present all available resources in the Grid to the users as a single virtual system, a computing metafacility. To achieve this, many loosely coupled Web systems, such as a distributed data Grid, a distributed batch system, and so on are required. In this chapter we will emphasize file management services, focusing upon two of the services and summarizing several others.

The requirements of the Jefferson Lab distributed data analysis are as described in [6]. In that chapter, the emphasis was on the laboratory's existing computing infrastructure and Web services layer architecture. This chapter will concentrate more on the implementation details and the lessons learned from the first prototypes of the data management Web services.

34.1.1 Motivation

The Web has been overwhelmingly successful at providing seamless access to a wide range of information, and at providing a basic level of interaction with distributed systems (e.g. electronic commerce). Web services seeks to exploit this successful distributed architecture by adding a messaging layer (SOAP) and a syntax for constructing messages (XML, WSDL) on top of the Web protocols. This allows one to exploit all of the existing features such as secure http to build a loosely coupled, distributed system.

As an example of this leveraging, it is very simple for a user with valid X.509 credentials to use resources across the world without having to log into individual systems, because https defines a mechanism for using Secure Sockets Layer (SSL) to connect to a server using these credentials. Using SOAP over https, one can submit a batch request to a metafacility which automatically and optimally assigns the job to a particular compute site. This activity would include staging input files, running the job, and then automatically archiving the outputs from the job. Just as the Web is easily extended by adding a new Web server, it will be easy to add a new compute or storage node to this Web-based metafacility.

34.1.2 Architecture

The Web services Grid we envisage contains many components arranged in three layers (Figure 34.1). The lowest level contains site-specific backend services to manage disks and tertiary storage (a silo) and to schedule batch jobs onto local compute nodes. The middle layer provides a standard interface to these resources. It may be a thin layer (protocol translation and standardization) or may contain considerable business or management logic (as in our implementation). On top of these layers are the user interfaces and other client applications. The top layer needs no knowledge of the often complicated and site-specific lowest level (abstraction), so our design goal is to hide this lower layer and only expose an easy-to-use Web services middle layer to all top-level applications.

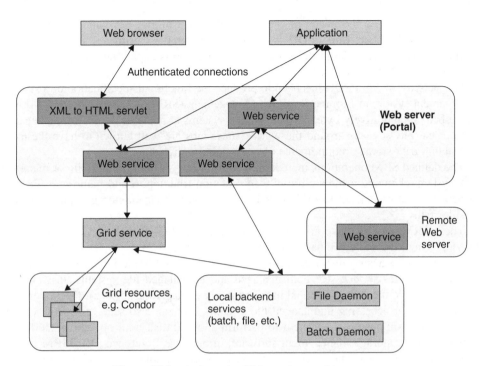

Figure 34.1   A three tier Web services architecture.

The only exception to this principle is bulk data transfer, where the application can directly contact a file transfer daemon. This multiprotocol approach follows the Web in general where many file transfer and streaming protocols exist alongside http.

The foundation of Web services is SOAP – XML messaging over standard Web protocols such as http. This lightweight communication mechanism ensures that any programming language, middleware, or platform can be easily integrated. Using SOAP, exchanging structured information in a decentralized, distributed environment is straightforward. SOAP mandates an XML vocabulary that is used for representing method parameters, return values, and exceptions. Although SOAP is not yet an Internet standard, the W3C Working Draft does provide a way to explore the protocol and decide whether it is appropriate to solve the task of deploying a metafacility.

34.2 DATA GRID WEB SERVICES

34.2.1 Data Grids

Jefferson Lab is part of a collaboration funded by the Department of Energy's Scientific Discovery through Advanced Computing (SciDAC) program. The Particle Physics Data Grid (PPDG) collaboratory [7] seeks to exploit Grid technologies within high energy and nuclear physics experiments worldwide. PPDG is working closely with counterparts in Europe to address the needs of the Large Hadron Collider now under construction at CERN. In the short run, PPDG is using current large experiments as test beds for production data Grids. Within this context, Jefferson Lab is deploying a test bed for data Grids based on Web services.

One activity within PPDG has been to produce the functional specification of a storage management system or Storage Resource Management (SRM). This specification [8] was developed collaboratively by experts in storage systems at many of the major high energy and nuclear physics labs around the world. These labs are well known to have the most demanding data storage requirements of any scientific discipline.

The defined SRM operations include moving files into or out of a tape silo or managed disk system, changing a file's eligibility for deletion (pin/unpin) and marking a file permanent (cannot be deleted from disk until it exists on permanent storage), and reporting the file status.

These SRM services have now been implemented as SOAP Web services, using Java Servlet technology. Here, we name our resource management Web service JSRM, for Java Storage Resource Management. JSRM implements a superset of the PPDG SRM functionality and provides services for both managed and unmanaged file systems (in which by 'unmanaged' we mean conventional user and group managed disk space). JSRM contains functionality for accessing unmanaged file systems, and can also serve as a front end or gateway to a (simpler) SRM system to perform silo and disk pool operations (Jefferson Lab's JASMine storage management software, in our case Reference [9]). JSRM is thus a Web-entry point to all the resources of a single site, including a user's home directory. A collection of such sites (when linked with additional services) forms a data Grid.

The majority of the service request/response data structures of JSRM are defined according to the SRM functional specification document. Additional operations not defined in the current SRM specification may be added in a future version.

The remaining discussion of functionality uses the following terms from the SRM spec:

GFN: Global file name; the name by which a file is known across the Grid. This file is typically the primary index in a catalog, and contains no location information
SFN: Site file name; the name by which a file is known by a data Grid node or site
SURL: Site URL (Uniform Resource Locator), a concatenation of the SOAP endpoint used to communicate with a site and the SFN
TURL: Transfer URL, a standard Web URL that includes a protocol specification for transferring a file, a file server hostname, and a local file path; for example: ftp://hpcfs1.jlab.org/some-path/some-filename

Generally, the SURL is persistent, but if the local site migrates a file from one server to another, the TURL can change.

34.2.2 File/dataset catalogs

In the PPDG model, a GFN can either be known in advance, or can be discovered by consulting an *application metadata catalog*, which allows mapping from domain specific metadata (such as beam energy or target specification for Jefferson Lab) into a set of file names (GFNs). The set of GFN's forms a namespace similar to the namespace of a Unix file system (recursive directories, etc.).

For optimal performance in a wide-area distributed system, a particular dataset or file may exist at multiple locations on the data Grid, and so a *replica catalog* is used to map between the global name and location information, in other words to convert each GFN into one or more SURLs. Jefferson Lab has implemented a ReplicaCatalog Web service to hold the GFN namespace, and to map between GFN and SURLs. This service is implemented as a combination of a Java servlet and a Standard Query Language (SQL) database (the persistence layer for the namespace), and is accessible both as a SOAP Web service, and as browsable Web pages (via style sheets).

For our replica catalog, we have also implemented the notion of softlinks, which allows a user to create a directory holding (links to) a set of files located in various other directories within the replica catalog. Links can also point to directories, or even to other links.

Because the replica catalog holds a file-system-like namespace, many of the operations supported by this service are designed to be identical to the operations supported by the storage manager Web service (described below). This set of common operations includes *list, mkdir, rmdir, delete, status*. Through this design, it is possible to treat the replica catalog and the storage manager Web service polymorphically, as is done by the Grid File Manager application (described below).

We have not yet implemented an application metadata catalog, but it, too, is likely to be implemented with an SQL database as the persistence layer. There may be advantages to colocating the application metadata catalog and the replica catalog so that the replica catalog becomes just a view of a larger set of data management information, and this will be explored in future work.

34.2.3 Storage resource management Web service – JSRM

The first Web service component we implemented is an interface to the software responsible for the site's storage resource management. At Jefferson Lab, the various experiments are able to produce up to one terabyte of data daily, and this data is held in a 12 000 slot StorageTek silo. In addition, a group of disk pools exists to store calibration and analysis data, and data files currently in use. These disk pools are managed by a daemon (a disk cache manager) with site-specific file deletion policies. We use the term 'managed file system' to refer to these disk pools. Each lab typically develops its own software to manage all its storage resources including the mass storage system and cache disk pools. Although these software packages may be written in different languages and with different designs, they are performing the same task – Storage Resource Management (SRM).

The JSRM component provides the following major services:

1. Directory listings, including file metadata (size, owner, cached, pinned, permanent, etc.).
2. Mapping from SFN to TURL to read/write files from/to this site, including protocol negotiation and space allocation. The implementation allows for pluggable protocols, and the current implementation supports http, ftp, and jparss (see below).

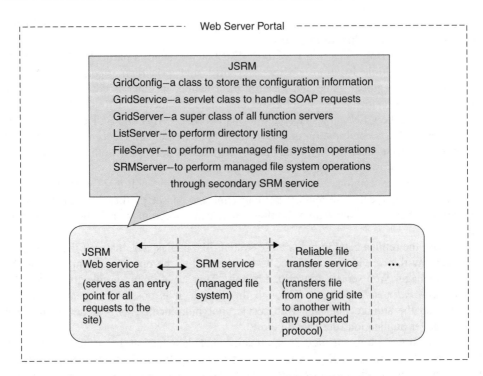

Figure 34.2   The architecture of the JSRM Web service.

3. Translation between SFN, or SURL, and a local file path, allowing local applications to find a Grid file on a locally mounted disk without needing to know any local mount point naming conventions.
4. File status change operations, such as stage a file, pin a file, migrate or copy a file to the silo.
5. File system management (copy, delete, make directory, etc., operations), including operations on managed and unmanaged areas.

SOAP is used as the service protocol: requests and responses are in XML, which is humanly readable, and can be described by WSDL. The WSDL for JSRM is available online [10].

Because of the platform independence and widely available tools and packages [11, 12], Java has been selected as the implementation language. The architecture of JSRM is illustrated in Figure 34.2. The functionality is distributed among several Java classes: one Java servlet is responsible for getting the SOAP requests from the user, validating the SOAP message, and passing it to an appropriated server object. Site-specific information (such as the set of available file systems) is configured via an XML configuration file, which is loaded into a Java object and passed to the individual function server by the Java servlet. With this design, an installation can be configured differently according to local policy (such as allowing access to the user's home directory) and file storage structure.

SRM operations for unmanaged disk areas are handled directly by JSRM. This allows JSRM to be deployed at a site that has only user-managed disk space (no disk management system, no tertiary storage system).

JSRM can also be layered above an existing storage manager. Certain operations on the managed disk areas and silo are then performed by deferring the requests to a separate object or SRM, which in itself can be a Web service or a Java class. As long as every site uses the same SRM or Java interface, JSRM can be deployed as a gateway, adding functionality without modifying the secondary SRM's code.

The Java API for XML Messaging (JAXM) is used to build, send, receive, and decompose the SOAP messages in this work. The current version of JAXM implements SOAP 1.1 with Attachments messaging. It takes care of all low-level XML communications – JSRM servers focus only on providing the service the user has requested.

34.2.4 Reliable file transfer

The next essential component for a data Grid is a service to move datasets from one Grid node (site) to another, otherwise known as third party file transfers. This service must be reliable, gracefully handling intermittent network or node unavailability problems.

A reliable file transfer service could be included in the site JSRM service (additional methods of the same Web service) However, in this project, we decided to implement it as an independent Java server class with a Web service interface. In this way, Reliable File Transfer functionality can be easily invoked by JSRM or serve as a user callable Web service. This demonstrates the power and simplicity of Web services: any Web service written in any language can interact with any other Web service.

To ensure the reliability of the transfer request, a MySQL database is used to store every valid request. The request status, any error that occurred during the transfer, the time stamp of each step, and other relevant data are saved in a database table. Once an operation completes, the summary history of the transfer is also kept in the database for statistical use in the future. The architecture of the Reliable File Transfer Web service is illustrated in Figure 34.3.

When the JRFT server receives a valid SOAP request, it first adds the request to the queue, and then returns a response to the user. This response only contains one element – transfer request-id – and with this the user can later check the transfer status. The decision to implement a pull architecture for such long-lived transactions is intentional – it allows the initiating client to exit, and then comes back later, perhaps long after the transfer has completed, to check for errors. This loosely coupled style simplifies system construction and improves flexibility, with only a minor increase in overhead to poll (which in any case is vanishingly small compared to the cost of moving large datasets).

Once a request is queued, there is a background thread running constantly checking the queue to find files to transfer. For each request, it must first negotiate the transfer protocol with the source site's JSRM and destination site's JSRM. If there is an agreement between both sites, the static method getTransferClient (String protocol_name) in the TransferClientFactory class will be called to obtain the corresponding transfer client to perform the real file transfer.

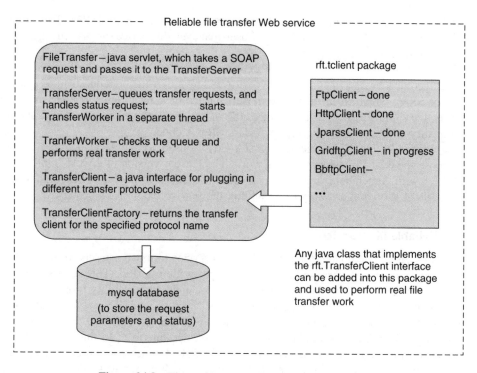

Figure 34.3 The architecture of the JRFT Web service.

In order to discover the set of transfer clients (protocols) automatically, all available transfer clients must implement TransferClient interface and follow a certain naming convention, for example, they must be named as *ProtocolnameClient.java* (only the first letter of the protocol name is capital and the remaining letters of the name are in lower case). So far there are three transfer protocol clients implemented in the rft.tclient package. Ftp-Client and HttpClient are used to transfer files between sites for which authentication is not needed. JparssClient uses the Java Parallel Secure Stream (JPARSS) package developed by Jefferson Lab [13] to perform authenticated parallel file transfers. Additional transfer clients using different protocols can be plugged in dynamically without any modification to the existing code (additional protocols will be added as time permits).

Depending upon the underlying file transfers capabilities, the Reliable File Transfer RFT service might be at a different site from both the source and destination sites. For simple protocols like ftp and http, only pull is implemented, and so the destination must be the same as the RFT site. (Request forwarding from one RFT instance to another is foreseen to deal with these protocol restrictions). Each protocol's capabilities are described in XML, so the RFT service discovers these restrictions dynamically, and uses them in the protocol negotiation.

When a file transfer process has failed, SenderException or ReceiverException will be thrown by the TransferClient. SenderException is thrown only when a fatal error occurs, for example, when the source SURL is invalid, or authentication has failed. When the transfer fails because of a possibly transient error, for example, database error, network problem, and so on, ReceiverException will be thrown and the request will be added back to the queue and will be tried again later.

After submitting a file transfer request, a user can check the transfer status. If the transfer has already started, additional information such as the number of bytes transferred and average bandwidth obtained will be included in the SOAP response.

34.2.5 Security

The security mode used for these Web services is certificate-based browser-model security, which means all privileged services are provided only through https/ssl connections. A permanent X.509 user certificate or a temporary proxy certificate [14] is required to use these services. The server (using a map file) maps the subject of a user's certificate to the local user account to perform privileged operations using operating system access control.

In some cases, we have found it useful to allow some limited superuser operations on the Grid, analogous to those performed by privileged daemons on a local system. As one example, the replica catalog allows entries to be made by certain privileged users (daemons using server certificates known to the replica catalog) to have the owner field set to a name other than the name of the calling user. In this way we have built automatic file registration 'spiders', which do not need an individual user's proxy certificate to function. In each such case, the extended privileges just allow the special caller to do operations as if he were a different user.

34.3 GRID FILE MANAGER

34.3.1 Grid file interface – a Java API

To make these file management Web services easier to use, it has been found to be helpful to develop a client library to hide the somewhat complicated SOAP request and response operations from a typical user. A Grid File Interface has been designed to wrap the communication between the service providers into a simple Java Application Programming Interface (API) (Figure 34.4). A SOAP implementation of this Grid File Interface, based upon JSRM, JRFT, and the ReplicaCatalog has been done.

The GridNode object represents a Grid node or physical site. It has a name, a URL, and other properties. A static method, getGridNode in GridNodeFactory will return a corresponding GridNode object with a specified URL. By using the list method in a GridNode class you can obtain a pointer to a given directory, file, or link. GridDirectory class, like File in the java.io package, provides file system manipulation methods, such as delete, link, mkdir, addFile, list, and so on. The client classes implemented with these interfaces then deal with the SOAP service request and response. In the addFile implementation, the Reliable File Transfer Web service is used to perform the data movement when the source and destination are on different sites, and the corresponding JSRM service is used when a file copy within a site is requested. Using the classes defined in this interface,

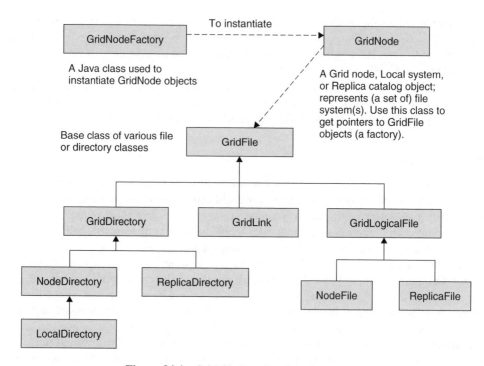

Figure 34.4 Grid file interface inheritance tree.

managing the resource on the Grid is as easy as managing a local file system (as for local Java clients with the java.io.File class).

While the current implementation of this client library is done in Java, it would be straightforward to also produce a corresponding C++ client library, by building upon existing C++ SOAP implementations.

34.3.2 Grid File Manager

A Grid File Manager user interface is developed on top of the Grid file API. This is a Java application client and is used to manage the resources located on different systems within the Grid, in the same or different organizations. An XML configuration file (accessed over the network via a standard URL) configures the systems that the Grid File Manager tool manages. In our version 1.0 release, the user is able to choose either his configuration file or the default one. Included in this configuration file are pointers to the replica catalog and a set of initial Grid nodes, so that this application can be deployed without a Web services directory service. At some point the URL to a configuration file could instead be a URL to a UDDI repository (directory).

Any server site that provides JSRM and JRFT services can be manipulated via the Grid File Manager. In addition, the tool can also manage the user's desktop computer (local system) without the need for locally installed Web services.

Grid File Manager provides a graphical interface for the user to request services. All underlying communications with service providers are hidden. With this tool, the user can browse and manage the supported file system within the Grid, and his entire file system on the local computer, including deleting files and directories, making new directories, obtaining file or directory metadata, and so on. To transfer a file or an entire directory tree from one site to another is as easy as a single click.

There are two identical file browser panes in the Grid File Manager interface, as shown in Figure 34.5. They serve as source and destination when you request a file copy/transfer.

In addition to the Grid node and local system, Grid File Manager can also browse the Jefferson Lab Replica Catalog system [15], but manipulations of the replica catalog other than 'list' are not implemented in current release. It is foreseen that 'drag-N-drop' operations from the ReplicaCatalog to a Grid node or local system will imply a hidden look-up operation on the file followed by a transfer from a site holding the file, and operations 'copying' a file to the ReplicaCatalog will imply registering the file in the catalog. These operations will be added in a future release.

We use Java Web Start [16], an application deployment technology from SUN, to launch and manage the resources of Grid File Manager. By clicking on a Web page link, Java Web Start automatically downloads all necessary application files to the user's computer if the application is not present. These files are cached on the user's computer so it is always ready to be relaunched anytime, either from an icon on the desktop or from

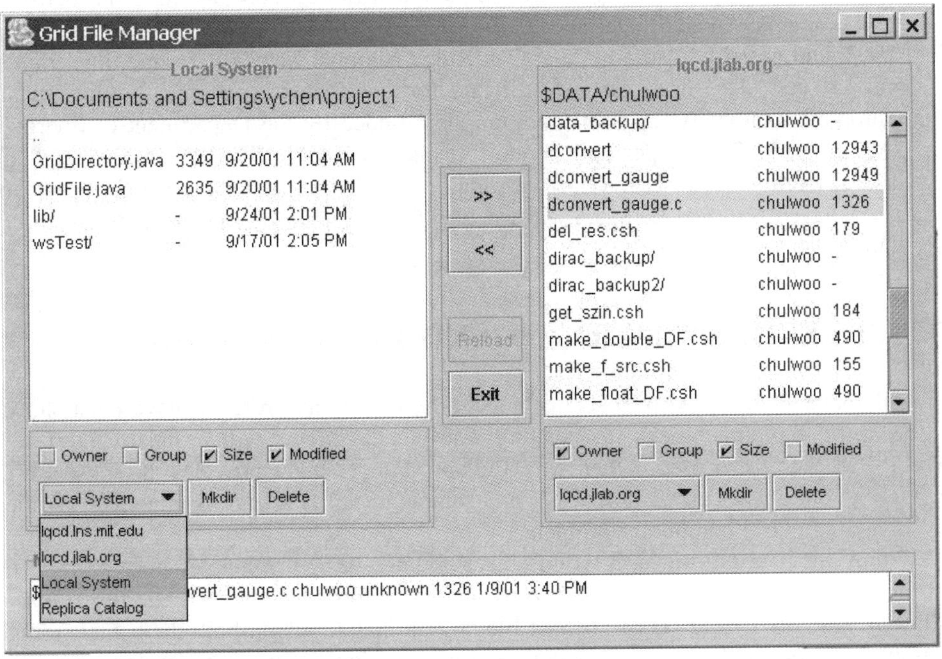

Figure 34.5 Grid file manage interface.

the browser link. Each time the application is launched, Java Web Start will check the server for a new version of the application, and automatically download it if a new version is available. If not yet installed by the user, the user will be prompted to download the Java Web Start tool and browser plug-in. This technology ensures that users are always using the latest, correct version of software, and eases the deployment workload.

The security model used for this application is X.509 credential based, with short lived credentials (proxies) been given to a server to allow it to act on a client's behalf. For authentication purposes, a Jefferson Lab or DOE Science Grid certificate is needed to use the Grid File Manager against the Jefferson Lab nodes. When launching an application, the user will be prompted for the pass phrase of his local certificate repository if no valid proxy exists, and a 24-h life proxy certificate will be generated. This proxy will be used to connect to the server for privileged service operations. Within the 24-h period, the user can use the application without entering a pass phrase again.

34.4 THE LATTICE PORTAL

The components described above are currently being deployed onto a tested consisting of Jefferson Lab and the MIT Lab for Nuclear Science, for use by the Lattice Hadron Physics Collaboration. This small not-quite-metafacility includes a cluster of 48 processors at MIT (alphas), a small cluster of 40 alphas at Jefferson Lab, and a cluster of 128 Pentium-4 nodes at Jefferson Lab (to be upgraded in the Fall of 2002 to at least 256). The data Grid tools are being used for moving lattice simulation files between the sites, and between sites and desktops. An additional node at the University of Maryland will be added during Fall 2002, and possibly also a node at Oak Ridge National Lab in the same time frame.

This test bed also serves to test software for use within a national Lattice (quantum chromo-dynamics (QCD) collaboration also funded by the Department of Energy's SciDAC program. This larger collaboration plans to deploy a distributed 20+ teraflops (sustained) facility within four years with major installations at Jefferson Lab, Fermilab, and Brookhaven National Lab, and will use data Grid technology to manage the resulting simulation datasets.

34.5 LESSONS LEARNED AND FUTURE PLANS

A well-constructed set of Web services, using SOAP as the protocol, have been shown to be useful building blocks for creating distributed systems with standard interfaces. The JSRM Web service combined with Reliable File Transfer provides the most useful services to our cluster users. It provides an example of utilizing cutting-edge technologies to solve a key Grid computing problem.

Within this data Grid project, good object-oriented design principles have allowed us to incorporate multiple file transfer protocols (via interface implementation), and have allowed us to do a large set of abstract file system operations on both the actual storage resource and the replica catalog polymorphically, easing the work of implementing the Grid File Manager.

As we have begun to deploy these tools into a production environment, the value of privileged processes on the Grid has become clear, and such processes may provide an appropriate mechanism for dealing with the problem of expired user proxies associated with long running batch jobs or file transfer jobs.

In the next step of this prototyping work, we will investigate a different security mode, IETF XML signature (Digital Signature). Such signatures could allow for finer grained authorization of delegated work, and could help limit the scope of privileged Grid processes so as to prevent a compromise of the Grid at one point from propagating to the entire Grid.

So far, the JSRM implementation is being used primarily with unmanaged file systems, with limited interaction with the JASMine cache manager and silo through a Java interface. JASMine is currently being modified to have an SRM interface (conforming to the PPDG spec), and when that is finished JSRM will be converted to forward managed file system operations to a secondary SRM Web service, with JASMine as the first implementation to be tested.

For the Grid File Manager, we will add the remaining Replica Catalog functions: virtual read (look up and fetch), write (publish, perhaps including a push to a Grid node if the data source is the local desktop), and delete (unpublish). We are also producing a number of command line tools for fetching and publishing data sets.

REFERENCES

1. Glass G. (2002). The Web Services (R)evolution: Part 1, http://www-106.ibm.com/developerworks/library/ws-peer1.html, July 5, 2002.
2. Waters J. (2002) Web services: The Next Big Thing? http://www.adtmag.com/article.asp?id =6124, July 5, 2002.
3. http://www.globus.org/research/papers/ogsa.pdf, July 5, 2002.
4. The Unicore Project, http://www.unicore.org/, July 5, 2002. The Web services implementation can be found at http://www.unicore.org/downloads.htm looking in the index for OGSA.
5. http://gridport.npaci.edu/pubs/workshops/gce/WebservMay02/, July 5, 2002.
6. Watson, W., Bird, I., Chen, J., Hess, B., Kowalski, A. and Chen, Y. (2002) A Web Service Data Analysis Grid, *Concurrency and Computation: Practice and Experience*, 14:1–9 (DOI: 10.1002/cpe.686).
7. Particle Physics Data Grid (PPDG), http://www.ppdg.net/ July 5, 2002.
8. Shoshani, A., Sim, A. and Gu, J. (2001) Storage resource managers: middleware components for grid storage. Proceedings of the 18th IEEE Symposium on Mass Storage Systems, 2001.
9. Bird, I., Hess, B. and Kowalski, A. (2001) Building the Mass Storage System at Jefferson Lab, Proceedings of the 18th IEEE Symposium on Mass Storage Systems, 2001.
10. http://lqcd.jlab.org/wsdl/jsrm.wsdl, July 5, 2002.
11. Apache Java Project, http://java.apache.org/, July 5, 2002.
12. Sun xml Web site, http://java.sun.com/xml/, July 5, 2002.
13. Chen, J. and Watson, W. (2002) JPARSS: A Java parallel network package for grid computing. Proceedings of the 2nd International Workshop on Multimedia and Intelligent Networks, Raleigh, NC, March, 2002, p. 944.
14. Foster, I., Kesselman, C., Tsudik G. and Tueche, S. (1998) A security architecture for computational grids. Proceedings 5th ACM Conference on Computer and Communications Security Conference, 1998, pp. 83–92.
15. Jefferson Lab Replica Catalog, Unpublished Work, http://www.jlab.org/datagrid/, July 5, 2002; for additional information.
16. Java Web Start Web site, http://java.sun.com/products/javaWebstart/, July 5, 2002.

PART D

Grid applications

35

Application overview for the book: Grid computing – making the global infrastructure a reality

Fran Berman,[1,2] Geoffrey Fox,[3] and Tony Hey[4,5]

[1]*San Diego Supercomputer Center, and Department of Computer Science and Engineering, University of California, San Diego, California, United States,* [2]*Indiana University, Bloomington, Indiana, United States,* [3]*EPSRC, Swindon, United Kingdom,* [4]*University of Southampton, Southampton, United Kingdom*

35.1 INTRODUCTION

This book, *Grid Computing: Making the Global Infrastructure a Reality*, is divided into four parts. This short chapter introduces the last part, Part D, on applications for the Grid. All the chapters in the book contain material relevant for Grid applications, but in this part the focus is the applications themselves. Some of the previous chapters also cover applications as part of an overview or to illustrate a technological issue.

Rather than merely reviewing a list of applications in this introduction, we abstract some general principles about the features of different types of applications well-suited for the Grid. Note that in addition to Chapters 37 to 43 devoted to applications in Part D,

Grid Computing – Making the Global Infrastructure a Reality. Edited by F. Berman, A. Hey and G. Fox
© 2003 John Wiley & Sons, Ltd ISBN: 0-470-85319-0

applications are found in Chapters 1, 2, 11, 12, 16, 23, 24, 28, 30 and 33 from Parts A, B and C and we abstract this material as well here.

35.2 GRID APPLICATIONS

Exactly what types of applications are suitable for Grid computing is still an active research area but some initial discussion is provided in Chapters 1 (Section 1.5), 2 (introduction), and 23 (Section 23.4). One can identify three broad problem architectures for which Grids have been successful:

1. *Megacomputing problems*: These correspond to problems that can be divided up into large numbers of independent parts and are often termed *pleasingly or embarrassingly parallel* in the parallel computing domain [1]. This area is addressed in detail with several examples in Chapters 1, 11, 12, 24 and 33. Further the data analysis part of particle physics in Chapter 39 and many biological computations (Chapters 40 and 41) fall into this class. Chapter 12 discusses in detail bioinformatics, molecular modeling and finance applications of this architecture. This class also covers 'parameter sweep' applications, an important and popular application paradigm for the Grid. Such applications can use tools such as Condor and AppLeS Parameter Sweep Template (APST), discussed in Chapters 11 and 33.
2. *Mega and seamless access problems*: These correspond to use of Grids to integrate the access and use of multiple data and compute resources. This underlies all the 'data deluge' applications such as those of Chapters 36 and 38 to 43. The scientific collections described in Chapters 16 and 36 are of this class.
3. *Loosely coupled nets*: These correspond to functionally decomposed problems (such as get data, compute, visualize or simulate ocean and simulate atmosphere) where synchronized (possibly pipelined) operation on the Grid is possible. See discussion of category 6 below.

In Chapter 23, a rather different view is taken of Grid problem classification. This chapter uses the motivation or style of computing involved in the Grid to identify four categories of Grid applications:

4. *Community centric*: The Grid is used to knit organizations together for collaboration and would often correspond to problem architecture 2 above. Education in Chapter 43 is clearly of this class. The virtual observatory in Chapter 38 has features of this as it integrates observations from different instruments.
5. *Data-centric*: This corresponds closely to problem architecture 2 above and reflects the 'data deluge'.
6. *Compute-centric*: This case is limited in applicability owing to the high latency of Grid connections, but certain loosely coupled applications and seamless access to multiple back-end hosts (architectures 3 and 2 above) make the Grid attractive for this category. There have been several problem-solving environments (see Chapters 24, 28 and 30 for example) built using Grid portals that support the many different loosely coupled stages of scientific computation with linked data and compute modules.
7. *Interaction-centric*: This corresponds to problems requiring real-time response and is not an area where there is much experience so far except perhaps in the real-time military simulations illustrated by Synthetic Forces Express [2] discussed in Chapter 1.

Further applications involving data-compute-visualization pipelines (Section 1.1 of Chapter 1) are of this type as are the data navigation examples in Chapter 37. Control of scientific instruments (Chapters 1, 28 and 37) also has this flavor.

We stress that the above categories overlap and our examples given above are not meant as precise classifications. For example, seamless access and community integration (Classes 2 and 4 above) are often supported in conjunction with other cases.

Chapter 1 describes several applications including a general discussion of the e-Science applications. These include synchrotron data analysis, astronomical virtual observatory, megacomputing, aircraft design and real-time engine data analysis, satellite operation, particle physics, combinatorial chemistry and bioinformatics. Chapter 37 gives a historical perspective with the 1992 vision of this area and gives visualization pipeline, instrument control and data navigation examples.

As in categories 2 and 5 above, data-intensive applications are expected to be of growing importance in the next decade as new instruments come online in a variety of fields. These include basic research in high-energy physics and astronomy, which are perhaps leading the use of the Grid for coping with the so-called data deluge described in Chapter 36. This chapter also describes data-centric applications in bioinformatics, environmental science, medicine and health, social sciences and digital libraries. The virtual observatory of Chapter 38 describes a new type of astronomy using the Grid to analyze the data from multiple instruments observing at different wavelengths. High-energy physics described in Chapter 39 is preparing for the wealth of data (100 petabytes by 2007) expected from the new Large Hadron Collider at CERN with a careful designed distributed computer and data architecture.

Biology and chemistry, as we have discussed, may actually stress the Grid even more with the growing number of commodity instruments, with ultimately, for instance, personal gene measurements enabling new approaches to healthcare. Aspects of this grand biology/chemistry vision are described in Chapters 40, 41 and 42. Chapter 40 describes a variety of biology applications involving both distributed gene sequencing and parameter sweep style simulations. Chapter 41 describes a different important 'e-health-care' problem – using the Grid to manage a federated database of distributed mammograms. The use of metadata and the implications for Grid-enabled medicine are stressed. Chapter 42 describes the importance of Grid for combinatorial chemistry – with new instruments producing and analyzing compounds in parallel. Here the Grid will manage both individual laboratories and enable their world- or company-wide integration.

Early Grid applications are naturally focused primarily on academic research but the Grid will soon be valuable in supporting enterprise IT systems and make an impact in the broader community. Chapter 43 describes the Grid supporting the collaboration between teachers, students and the public – a community Grid for education and outreach. It introduces this vision with a general discussion of the impact of web services on enterprise computing.

All of these applications indicate both the current and future promise of the Grid. As Grid software becomes more robust and sophisticated, applications will be able to better utilize the Grid for adaptive applications, real-time data analysis and interaction, more tightly coupled applications and 'poly-applications' that can adapt algorithm structure of individual application components to available Grid resources. Ultimately, applications are key to the perceived success or failure of Grid technologies and are critical to drive technology forward. In Part D of this book, we describe current applications visions enabled by the Grid.

REFERENCES

1. Fox, G. (2002) Chapter 4, in Dongarra, J., Foster, I., Fox, G., Gropp, W., Kennedy, K., Torczon, L. and White, A. (eds) *The Sourcebook of Parallel Computing*. San Francisco: Morgan Kaufmann Publishers, ISBN 1-55860-871-0.
2. Synthetic Forces Express, http://www.cacr.caltech.edu/SFExpress/.

36

The data deluge: an e-Science perspective

Tony Hey[1,2] and Anne Trefethen[1]

[1]*EPSRC, Swindon, United Kingdom,* [2]*University of Southampton, Southampton,*
United Kingdom

36.1 INTRODUCTION

There are many issues that should be considered in examining the implications of the imminent flood of data that will be generated both by the present and by the next generation of global 'e-Science' experiments. The term *e-Science* is used to represent the increasingly global collaborations – of people and of shared resources – that will be needed to solve the new problems of science and engineering [1]. These e-Science problems range from the simulation of whole engineering or biological systems, to research in bioinformatics, proteomics and pharmacogenetics. In all these instances we will need to be able to pool resources and to access expertise distributed across the globe. The information technology (IT) infrastructure that will make such collaboration possible in a secure and transparent manner is referred to as the *Grid* [2]. Thus, in this chapter the term *Grid* is used as a shorthand for the middleware infrastructure that is currently being developed to support global e-Science collaborations. When mature, this Grid middleware

Grid Computing – Making the Global Infrastructure a Reality. Edited by F. Berman, A. Hey and G. Fox
© 2003 John Wiley & Sons, Ltd   ISBN: 0-470-85319-0

will enable the sharing of computing resources, data resources and experimental facilities in a much more routine and secure fashion than is possible at present. Needless to say, present Grid middleware falls far short of these ambitious goals. Both e-Science and the Grid have fascinating sociological as well as technical aspects. We shall consider only technological issues in this chapter.

The two key technological drivers of the IT revolution are Moore's Law – the exponential increase in computing power and solid-state memory – and the dramatic increase in communication bandwidth made possible by optical fibre networks using optical amplifiers and wave division multiplexing. In a very real sense, the actual cost of any given amount of computation and/or sending a given amount of data is falling to zero. Needless to say, whilst this statement is true for any fixed amount of computation and for the transmission of any fixed amount of data, scientists are now attempting calculations requiring orders of magnitude more computing and communication than was possible only a few years ago. Moreover, in many currently planned and future experiments they are also planning to generate several orders of magnitude more data than has been collected in the whole of human history.

The highest performance supercomputing systems of today consist of several thousands of processors interconnected by a special-purpose, high-speed, low-latency network. On appropriate problems it is now possible to achieve sustained performance of several teraflop per second – a million million floating-point operations per second. In addition, there are experimental systems under construction aiming to reach petaflop per second speeds within the next few years [3, 4]. However, these very high-end systems are, and will remain, scarce resources located in relatively few sites. The vast majority of computational problems do not require such expensive, massively parallel processing but can be satisfied by the widespread deployment of cheap clusters of computers at university, department and research group level.

The situation for data is somewhat similar. There are a relatively small number of centres around the world that act as major repositories of a variety of scientific data. Bioinformatics, with its development of gene and protein archives, is an obvious example. The Sanger Centre at Hinxton near Cambridge [5] currently hosts 20 terabytes of key genomic data and has a cumulative installed processing power (in clusters – not a single supercomputer) of around $1/2$ teraflop s^{-1}. Sanger estimates that genome sequence data is increasing at a rate of four times each year and that the associated computer power required to analyse this data will 'only' increase at a rate of two times per year – still significantly faster than Moore's Law. A different data/computing paradigm is apparent for the particle physics and astronomy communities. In the next decade we will see new experimental facilities coming on-line, which will generate data sets ranging in size from hundreds of terabytes to tens of petabytes per year. Such enormous volumes of data exceed the largest commercial databases currently available by one or two orders of magnitude [6]. Particle physicists are energetically assisting in building Grid middleware that will not only allow them to distribute this data amongst the 100 or so sites and the 1000 or so physicists collaborating in each experiment but will also allow them to perform sophisticated distributed analysis, computation and visualization on all or subsets of the data [7–11]. Particle physicists envisage a data/computing model with a hierarchy of data centres with associated computing resources distributed around the global collaboration.

The plan of this chapter is as follows: The next section surveys the sources and magnitudes of the data deluge that will be imminently upon us. This survey is not intended to be exhaustive but rather to give numbers that will illustrate the likely volumes of scientific data that will be generated by scientists of all descriptions in the coming decade. Section 36.3 discusses issues connected with the annotation of this data with metadata as well as the process of moving from data to information and knowledge. The need for metadata that adequately annotates distributed collections of scientific data has been emphasized by the Data Intensive Computing Environment (DICE) Group at the San Diego Supercomputer Center [12]. Their Storage Resource Broker (SRB) data management middleware addresses many of the issues raised here. The next section on Data Grids and Digital Libraries argues the case for scientific data digital libraries alongside conventional literature digital libraries and archives. We also include a brief description of some currently funded UK e-Science experiments that are addressing some of the related technology issues. In the next section we survey self-archiving initiatives for scholarly publications and look at a likely future role for university libraries in providing permanent repositories of the research output of their university. Finally, in Section 36.6 we discuss the need for 'curation' of this wealth of expensively obtained scientific data. Such digital preservation requires the preservation not only of the data but also of the programs that are required to manipulate and visualize it. Our concluding remarks stress the urgent need for Grid middleware to be focused more on data than on computation.

36.2 THE IMMINENT SCIENTIFIC DATA DELUGE

36.2.1 Introduction

There are many examples that illustrate the spectacular growth forecast for scientific data generation. As an exemplar in the field of engineering, consider the problem of health monitoring of industrial equipment. The UK e-Science programme has funded the DAME project [13] – a consortium analysing sensor data generated by Rolls Royce aero-engines. It is estimated that there are many thousands of Rolls Royce engines currently in service. Each trans-Atlantic flight made by each engine, for example, generates about a gigabyte of data per engine – from pressure, temperature and vibration sensors. The goal of the project is to transmit a small subset of this primary data for analysis and comparison with engine data stored in three data centres around the world. By identifying the early onset of problems, Rolls Royce hopes to be able to lengthen the period between scheduled maintenance periods thus increasing profitability. The engine sensors will generate many petabytes of data per year and decisions need to be taken in real time as to how much data to analyse, how much to transmit for further analysis and how much to archive. Similar (or larger) data volumes will be generated by other high-throughput sensor experiments in fields such as environmental and Earth observation, and of course human health-care monitoring.

A second example from the field of bioinformatics will serve to underline the point [14]. It is estimated that human genome DNA contains around 3.2 Gbases that translates to only about a gigabyte of information. However, when we add to this gene sequence data, data on the 100 000 or so translated proteins and the 32 000 000 amino acids, the relevant data volume expands to the order of 200 GB. If, in addition, we include X-ray structure measurements of these proteins, the data volume required expands dramatically to several petabytes, assuming only one structure per protein. This volume expands yet again when we include data about the possible drug targets for each protein – to possibly as many as 1000 data sets per protein. There is still another dimension of data required when genetic variations of the human genome are explored. To illustrate this bioinformatic data problem in another way, let us look at just one of the technologies involved in generating such data. Consider the production of X-ray data by the present generation of electron synchrotron accelerators. At 3 s per image and 1200 images per hour, each experimental station generates about 1 terabyte of X-ray data per day. At the next-generation 'DIA-MOND' synchrotron currently under construction [15], the planned 'day 1' beamlines will generate many petabytes of data per year, most of which will need to shipped, analysed and curated.

From these examples it is evident that e-Science data generated from sensors, satellites, high-performance computer simulations, high-throughput devices, scientific images and so on will soon dwarf all of the scientific data collected in the whole history of scientific exploration. Until very recently, commercial databases have been the largest data collections stored electronically for archiving and analysis. Such commercial data are usually stored in Relational Database Management Systems (RDBMS) such as Oracle, DB2 or SQLServer. As of today, the largest commercial databases range from 10 s of terabytes up to 100 terabytes. In the coming years, we expect that this situation will change dramatically in that the volume of data in scientific data archives will vastly exceed that of commercial systems. Inevitably this watershed will bring with it both challenges and opportunities. It is for this reason that we believe that the data access, integration and federation capabilities of the next generation of Grid middleware will play a key role for both e-Science and e-Business.

36.2.2 Normalization

To provide some sort of normalization for the large numbers of bytes of data we will be discussing, the following rough correspondences [16] provide a useful guide:

A large novel	1 Mbyte
The Bible	5 Mbytes
A Mozart symphony (compressed)	10 Mbytes
OED on CD	500 Mbytes
Digital movie (compressed)	10 Gbytes
Annual production of refereed journal literature	1 Tbyte
(∼20 k journals; ∼2 M articles)	
Library of Congress	20 Tbytes
The Internet Archive (10 B pages)	100 Tbytes
(From 1996 to 2002) [17]	
Annual production of information (print, film,	1500 Pbytes
optical & magnetic media) [18]	

Note that it is estimated that printed information constitutes only 0.003% of the total stored information content [18].

36.2.3 Astronomy

The largest astronomy database at present is around 10 terabytes. However, new telescopes soon to come on-line will radically change this picture. We list three types of new 'e-Astronomy' experiments now under way:

1. *Virtual observatories*: e-Science experiments to create 'virtual observatories' containing astronomical data at many different wavelengths are now being funded in the United States (NVO [19]), in Europe (AVO [20]) and in the United Kingdom (Astro-Grid [21]). It is estimated that the NVO project alone will store 500 terabytes per year from 2004.
2. *Laser Interferometer Gravitational Observatory (LIGO)*: LIGO is a gravitational wave observatory and it is estimated that it will generate 250 terabytes per year beginning in 2002 [22].
3. *VISTA*: The VISTA visible and infrared survey telescope will be operational from 2004. This will generate 250 GB of raw data per night and around 10 terabytes of stored data per year [23]. By 2014, there will be several petabytes of data in the VISTA archive.

36.2.4 Bioinformatics

There are many rapidly growing databases in the field of bioinformatics [5, 24]:

1. *Protein Data Bank (PDB)*: This is a database of 3D protein structures. At present there are around 20 000 entries and around 2000 new structures are being added every 12 months. The total database is quite small, of the order of gigabytes.
2. *SWISS-PROT*: This is a protein sequence database currently containing around 100 000 different sequences with knowledge abstracted from around 100 000 different scientific articles. The present size is of the order of tens of gigabytes with an 18% increase over the last 8 months.
3. *TrEMBL*: This is a computer-annotated supplement to SWISS-PROT. It was created to overcome the time lag between submission and appearance in the manually curated SWISS-PROT database. The entries in TrEMBL will eventually move to SWISS-PROT. The current release has over 600 000 entries and is updated weekly. The size is of the order of hundreds of gigabytes.
4. *MEDLINE*: This is a database of medical and life sciences literature (Author, Title, Abstract, Keywords, Classification). It is produced by the National Library of Medicine in the United States and has 11.3 M entries. The size is of the order of hundreds of gigabytes.
5. *EMBLnucleotide sequence database*: The European Bioinformatics Institute (EBI) in the United Kingdom is one of the three primary sites for the deposition of nucleotide sequence data. It contains around 14 M entries of 15 B bases. A new entry is received

every 10 s and data at the 3 centres – in the United States, United Kingdom and Japan – is synchronized every 24 h. The European Molecular Biology Laboratory (EMBL) database has tripled in size in the last 11 months. About 50% of the data is for human DNA, 15% for mouse and the rest for a mixture of organisms. The total size of the database is of the order of terabytes.

6. *GeneExpression database*: This is extremely data-intensive as it involves image data produced from DNA chips and microarrays. In the next few years we are likely to see hundreds of experiments in thousands of laboratories worldwide. Data storage requirements are predicted to be in the range of petabytes per year.

These figures give an indication of the volume and the variety of data that is currently being created in the area of bioinformatics. The data in these cases, unlike in some other scientific disciplines, is a complex mix of numeric, textual and image data. Hence mechanisms for curation and access are necessarily complicated. In addition, new technologies are emerging that will dramatically accelerate this growth of data. Using such new technologies, it is estimated that the human genome could be sequenced in days rather than the years it actually took using older technologies [25].

36.2.5 Environmental science

The volume of data generated in environmental science is projected to increase dramatically over the next few years [26]. An example from the weather prediction community illustrates this point.

The European Centre for Medium Range Weather Forecasting (ECMWF) in Reading, United Kingdom, currently has 560 active users and handles 40 000 retrieval requests daily involving over 2 000 000 meteorological fields. About 4 000 000 new fields are added daily, amounting to about 0.5 terabytes of new data. Their cumulative data store now contains 3×10^9 meteorological fields and occupies about 330 terabytes. Until 1998, the increase in the volume of meteorological data was about 57% per year; since 1998, the increase has been 82% per year. This increase in data volumes parallels the increase in computing capability of ECMWF supercomputers.

This pattern is mirrored in the United States and elsewhere. Taking only one agency, NASA, we see predicted rises of data volumes of more than tenfold in the five-year period from 2000 to 2005. The Eros Data Center (EDC) predicts that their data holdings will rise from 74 terabytes in 2000 to over 3 petabytes by 2005. Similarly, the Goddard Space Flight Center (GSFC) predicts that its holdings will increase by around a factor of 10, from 154 terabytes in 2000 to about 1.5 petabytes by 2005. Interestingly, this increase in data volumes at EDC and GSFC is matched by a doubling of their corresponding budgets during this period and steady-state staffing levels of around 100 at each site It is estimated that NASA will be producing 15 petabytes of data by 2007. The NASA EOSDIS data holdings already total 1.4 petabytes.

In Europe, European Space Agency (ESA) satellites are currently generating around 100 GB of data per day. With the launch of Envisat and the forthcoming launches of the Meteosat Second Generation satellite and the new MetOp satellites, the daily data volume generated by ESA is likely to increase at an even faster rate than that of the NASA agencies.

36.2.6 Particle physics

The BaBar experiment has created what is currently the world's largest database: this is 350 terabytes of scientific data stored in an Objectivity database [27]. In the next few years these numbers will be greatly exceeded when the Large Hadron Collider (LHC) at CERN in Geneva begins to generate collision data in late 2006 or early 2007 [28]. The ATLAS and CMS experiments at the LHC each involve some 2000 physicists from around 200 institutions in Europe, North America and Asia. These experiments will need to store, access and process around 10 petabytes per year, which will require the use of some 200 teraflop s^{-1} of processing power. By 2015, particle physicists will be using exabytes of storage and petaflops per second of (non-Supercomputer) computation. At least initially, it is likely that most of this data will be stored in a distributed file system with the associated metadata stored in some sort of database.

36.2.7 Medicine and health

With the introduction of electronic patient records and improvements in medical imaging techniques, the quantity of medical and health information that will be stored in digital form will increase dramatically. The development of sensor and monitoring techniques will also add significantly to the volume of digital patient information. Some examples will illustrate the scale of the problem.

The company InSiteOne [29] is a US company engaged in the storage of medical images. It states that the annual total of radiological images for the US exceeds 420 million and is increasing by 12% per year. Each image will typically constitute many megabytes of digital data and is required to be archived for a minimum of five years.

In the United Kingdom, the e-Science programme is currently considering funding a project to create a digital mammographic archive [30]. Each mammogram has 100 Mbytes of data and must be stored along with appropriate metadata (see Section 36.3 for a discussion on metadata). There are currently about 3 M mammograms generated per year in the United Kingdom. In the United States, the comparable figure is 26 M mammograms per year, corresponding to many petabytes of data.

A critical issue for such medical images – and indeed digital health data as a whole – is that of data accuracy and integrity. This means that in many cases compression techniques that could significantly reduce the volume of the stored digital images may not be used. Another key issue for such medical data is security – since privacy and confidentiality of patient data is clearly pivotal to public confidence in such technologies.

36.2.8 Social sciences

In the United Kingdom, the total storage requirement for the social sciences has grown from around 400 GB in 1995 to more than a terabyte in 2001. Growth is predicted in the next decade but the total volume is not likely to exceed 10 terabytes by 2010 [31]. The ESRC Data Archive in Essex, the MIMAS service in Manchester [32] and the EDINA service in Edinburgh [33] have experience in archive management for social science. The MIMAS and EDINA services provide access to UK Census statistics, continuous government surveys, macroeconomic time series data banks, digital map datasets, bibliographical databases and electronic journals. In addition, the Humanities Research Board and JISC organizations in the UK jointly fund the Arts and Humanities Data Service [34]. Some large historical databases are now being created. A similar picture emerges in other countries.

36.3 SCIENTIFIC METADATA, INFORMATION AND KNOWLEDGE

Metadata is data about data. We are all familiar with metadata in the form of catalogues, indices and directories. Librarians work with books that have a metadata 'schema' containing information such as Title, Author, Publisher and Date of Publication at the minimum. On the World Wide Web, most Web pages are coded in HTML. This 'HyperText Markup Language' (HTML) contains instructions as to the appearance of the page – size of headings and so on – as well as hyperlinks to other Web pages. Recently, the XML markup language has been agreed by the W3C standards body. XML allows Web pages and other documents to be tagged with computer-readable metadata. The XML tags give some information about the structure and the type of data contained in the document rather than just instructions as to presentation. For example, XML tags could be used to give an electronic version of the book schema given above.

More generally, information consists of semantic tags applied to data. Metadata consists of semantically tagged data that are used to describe data. Metadata can be organized in a schema and implemented as attributes in a database. Information within a digital data set can be annotated using a markup language. The semantically tagged data can then be extracted and a collection of metadata attributes assembled, organized by a schema and stored in a database. This could be a relational database or a native XML database such as Xindice [35]. Such native XML databases offer a potentially attractive alternative for storing XML-encoded scientific metadata.

The quality of the metadata describing the data is important. We can construct search engines to extract meaningful information from the metadata that is annotated in documents stored in electronic form. Clearly, the quality of the search engine so constructed will only be as good as the metadata that it references. There is now a movement to standardize other 'higher-level' markup languages, such as DAML + OIL [36] that would allow computers to extract more than the semantic tags and to be able to reason about the 'meaning or semantic relationships' contained in a document. This is the ambitious goal of Tim Berners-Lee's 'semantic Web' [37].

Although we have given a simple example of metadata in relation to textual information, metadata will also be vital for storing and preserving scientific data. Such scientific data metadata will not only contain information about the annotation of data by semantic tags but will also provide information about its provenance and its associated user access controls. These issues have been extensively explored by Reagan Moore, Arcot Rajasekar and Mike Wan in the DICE group at the San Diego Supercomputer Center [38]. Their SRB middleware [39] organizes distributed digital objects as logical 'collections' distinct from the particular form of physical storage or the particular storage representation. A vital component of the SRB system is the metadata catalog (MCAT) that manages the attributes of the digital objects in a collection. Moore and his colleagues distinguish four types of metadata for collection attributes:

- Metadata for storage and access operations
- Provenance metadata based on the Dublin Core [40]
- Resource metadata specifying user access arrangements
- Discipline metadata defined by the particular user community.

In order for an e-Science project such as the Virtual Observatory to be successful, there is a need for the astronomy community to work together to define agreed XML schemas and other standards. At a recent meeting, members of the NVO, AVO and AstroGrid projects agreed to work together to create common naming conventions for the physical quantities stored in astronomy catalogues. The semantic tags will be used to define equivalent catalogue entries across the multiple collections within the astronomy community. The existence of such standards for metadata will be vital for the interoperability and federation of astronomical data held in different formats in file systems, databases or other archival systems. In order to construct 'intelligent' search engines, each separate community and discipline needs to come together to define generally accepted metadata standards for their community Data Grids. Since some disciplines already support a variety of existing different metadata standards, we need to develop tools that can search and reason across these different standards. For reasons such as these, just as the Web is attempting to move beyond information to knowledge, scientific communities will need to define relevant 'ontologies' – roughly speaking, relationships between the terms used in shared and well-defined vocabularies for their fields – that can allow the construction of genuine 'semantic Grids' [41, 42].

With the imminent data deluge, the issue of how we handle this vast outpouring of scientific data becomes of paramount importance. Up to now, we have generally been able to manually manage the process of examining the experimental data to identify potentially interesting features and discover significant relationships between them. In the future, when we consider the massive amounts of data being created by simulations, experiments and sensors, it is clear that in many fields we will no longer have this luxury. We therefore need to automate the discovery process – from data to information to knowledge – as far as possible. At the lowest level, this requires automation of data management with the storage and the organization of digital entities. At the next level we need to move towards automatic information management. This will require automatic annotation of scientific data with metadata that describes both interesting features of the data and of the storage and organization of the resulting information. Finally, we need to attempt to progress beyond structure information towards automated knowledge management of our scientific data. This will include the expression of relationships between information tags as well as information about the storage and the organization of such relationships.

In a small first step towards these ambitious goals, the UK GEODISE project [43] is attempting to construct a knowledge repository for engineering design problems. Besides traditional engineering design tools such as Computer Aided Design (CAD) systems, Computational Fluid Dynamics (CFD) and Finite Element Model (FEM) simulations on high-performance clusters, multi-dimensional optimization methods and interactive visualization techniques, the project is working with engineers at Rolls Royce and BAESystems to capture knowledge learnt in previous product design cycles. The combination of traditional engineering design methodologies together with advanced knowledge technologies makes for an exciting e-Science research project that has the potential to deliver significant industrial benefits. Several other UK e-Science projects – the myGrid project [44] and the Comb-*e*-Chem project [45] – are also concerned with automating some of the steps along the road from data to information to knowledge.

36.4 DATA GRIDS AND DIGITAL LIBRARIES

The DICE group propose the following hierarchical classification of scientific data management systems [46]:

1. *Distributed data collection*: In this case the data is physically distributed but described by a single namespace.
2. *Data Grid*: This is the integration of multiple data collections each with a separate namespace.
3. *Federated digital library*: This is a distributed data collection or Data Grid with services for the manipulation, presentation and discovery of digital objects.
4. *Persistent archives*: These are digital libraries that curate the data and manage the problem of the evolution of storage technologies.

In this chapter we shall not need to be as precise in our terminology but this classification does illustrate some of the issues we wish to highlight. Certainly, in the future, we envisage that scientific data, whether generated by direct experimental observation or by *in silico* simulations on supercomputers or clusters, will be stored in a variety of 'Data Grids'. Such Data Grids will involve data repositories together with the necessary computational resources required for analysis, distributed around the global e-Science community. The scientific data – held in file stores, databases or archival systems – together with a metadata catalogue, probably held in an industry standard relational database, will become a new type of distributed and federated digital library. Up to now the digital library community has been primarily concerned with the storage of text, audio and video data. The scientific digital libraries that are being created by global, collaborative e-Science experiments will need the same sort of facilities as conventional digital libraries – a set of services for manipulation, management, discovery and presentation. In addition, these scientific digital libraries will require new types of tools for data transformation, visualization and data mining. We return to the problem of the long-term curation of such data and its ancillary data manipulation programs below.

The UK e-Science programme is funding a number of exciting e-Science pilot projects that will generate data for these new types of digital libraries. We have already described both the 'AstroGrid' Virtual Observatory project [21] and the GridPP project [10] that will be a part of a worldwide particle physics Grid that will manage the flood of data to be generated by the CERN LHC accelerator under construction in Geneva. In other areas of science and engineering, besides the DAME [13] and e-Diamond [30] projects described above, there are three projects of particular interest for bioinformatics and drug discovery. These are the myGrid [44], the Comb-*e*-Chem [45] and the DiscoveryNet [47] projects. These projects emphasize data federation, integration and workflow and are concerned with the construction of middleware services that will automatically annotate the experimental data as it is produced. The new generation of hardware technology will generate data faster than humans can process it and it will be vital to develop software tools and middleware to support annotation and storage. A further project, RealityGrid [48], is concerned with supercomputer simulations of matter and emphasizes remote visualization and computational steering. Even in such a traditional High Performance Computing (HPC) project, however, the issue of annotating and storing the vast quantities of simulation data will be an important aspect of the project.

36.5 OPEN ARCHIVES AND SCHOLARLY PUBLISHING

In the United Kingdom, the Higher Education Funding Council, the organization that provides core funding for UK universities, is looking at the implications of the flood of e-Science data for libraries on a 10-year timescale. In such a 10-year time-frame, e-Science data will routinely be automatically annotated and stored in a digital library offering the 'usual' digital library services for management, searching and so on, plus some more specialized 'scientific data'–oriented services such as visualization, transformation, other types of search engines and so on. In addition, scientific research in many fields will require the linking of data, images and text so that there will be a convergence of scientific data archives and text archives. Scientific papers will also routinely have active links to such things as the original data, other papers and electronic theses. At the moment such links tend to be transitory and prone to breaking – perhaps the research group Web address '~tony' stops working when Tony leaves and so on. The Open Archive Initiative [49], which provides software and tools for self-archiving of their research papers by scientists, addresses this issue to some extent, but this is clearly a large issue with profound implications for the whole future of university libraries. On the matter of standards and interworking of scientific digital archives and conventional repositories of electronic textual resources, the recent move of Grid middleware towards Web services [50, 51] is likely to greatly facilitate the interoperability of these architectures.

Scholarly publishing will presumably eventually make a transition from the present situation – in which the publishers own the copyright and are therefore able to restrict the group of people who can read the paper – to a model in which publishers are funded not for the paper copy but for providing a refereeing service and a curated electronic journal archive with a permanent URL. The difference between this model (proposed by Stevan Harnad [52]) and Paul Ginsparg's 'Eprint' archive for physics papers [53] is that Ginsparg's model is central and discipline-based, whereas Harnad's is distributed and institution-based. Both models depend on publishers to implement the peer review for the papers. Peer review is essential in order to identify signal from noise in such public archives. In Harnad's model, researchers' institutions pay 'publishers' to organize the peer reviewing of their research output and to certify the outcome with their journal name and its established quality standard. The institutions' research output, both pre-peer review 'preprints' and post-peer review 'postprints', are archived in distributed, interoperable institutional Eprint archives. The Open Archives Initiative is providing a metadata harvesting protocol that could enable this interoperability. Using open source archiving software partly sponsored by the Budapest Open Access Initiative of the Soros Foundation, a growing number of universities in the United States and elsewhere are setting up Eprint Archives to provide permanent open access to their research. In addition to archiving their own research output, users also want to be able to search these archives for related works of others. Using the metadata associated with the archived paper, the OAI Metadata Harvesting Protocol [54] provides one solution to the problem of constructing suitable search engines. Any search engine produced in this manner will only be as good as the metadata associated with the papers [55], so strengthening and extending the metadata tagging and standards is a task of very high priority.

It seems just a question of time before scholarly publishing makes the 'Harnad Switch' – the outcome that Harnad has for a decade been describing as both optimal and inevitable. Authors actually want to maximize the impact and uptake of their research findings by making them accessible to as many would-be users as possible, rather than having them restricted, as they were in the paper era, to the minority of wealthy research libraries that can afford the access tolls. The Web has changed publishing forever and such a transition is inevitable. A similar transformation is likely to affect university libraries. The logical role for a university library in 10 years will surely be to become the responsible organization that hosts and curates (digitally) all the research papers produced by the university. It will be the university library that is responsible for maintaining the digital archive so that the '~tony' link continues to work for posterity. The Caltech Library System Digital Collections project [56] and the MIT DSpace project with HP [57] are two interesting exemplars of such an approach. There is also the interesting issue of how much responsibility individual universities would undertake for hosting and curating the scientific data produced by their researchers. Presumably, some universities would act as repositories for the scientific data for a number of university e-Science 'collaboratories', as well as acting as mirror sites for other organizations in the collaboration. Of course, particular communities will support specialized data archives – such as those of the EBI [24] and some national research organizations – and no doubt there will be commercial archives as well. An important issue not considered here is the question of ownership of data. Since much of the research in universities is funded by public bodies, there is clearly room for debate as to the ownership – and the curation costs!

36.6 DIGITAL PRESERVATION AND DATA CURATION

Generating the data is one thing, preserving it in a form so that it can be used by scientists other than the creators is entirely another issue. This is the process of 'curation'. For example, the SWISS-PROT database is generally regarded as the 'gold standard' for protein structure information [58]. Curation is done by a team of 25 full-time curators split between the Swiss Bioinformatics Institute and the EBI. This shows how expensive the curation process is and why it will be necessary to address this support issue – involving extreme levels of automated, semi-automated and manual annotation and data cleansing. In addition, preservation of the data will be a crucial aspect of the work of a data repository. A recent EU/US study [59] recommended the establishment of a 'Data Rescue Centre' that would be concerned with research into the longevity of electronic data archives. The report envisaged that such a centre would examine the issues concerned with the refreshment, replication, repackaging and transformation of data and become a centre of much-needed expertise in these technologies.

There are many technical challenges to be solved to ensure that the information generated today can survive long-term changes in storage media, devices and digital formats. An introduction to the issues surrounding this problem has been given by Rothenberg [60]. To illustrate these issues we shall briefly summarize a novel approach to long-term preservation recently suggested by Lorie [61]. Lorie distinguishes between the archiving of data files and the archiving of programs. The archiving of programs is necessary in order that their original behaviour with the original data set can be reproduced in the future. For example, it is likely that a significant percentage of the scientific digital data to be preserved will be generated directly via some program P. A simple example is a spreadsheet program. In order to make sense of the data in the future, we need to save the original program P that was used to create and manipulate the data along with the data itself. Of course, in one sense the program P is just a bit stream like the data it produces – but the important difference is that the machine and the operating system required to run P may no longer exist. Lorie discusses the pros and cons of two proposed solutions to this problem: 'conversion' – copying files and programs to each new system as new systems are introduced – and 'emulation' – saving the data and the program as a bit stream along with a detailed description of the original machine architecture and a textual description of what the original program P should do to the data. Lorie then proposes a third approach based on specifying the program P in terms of instructions for a 'Universal Virtual Computer' (UVC). When archiving data, the UVC would be used to archive the methods that are required to interpret the stored data stream. For archiving a program, the UVC would be used to specify the functioning of the original computer. It is not clear which of these three approaches will turn out to be most feasible or reliable. Needless to say, a solution to these problems is much more than just a technical challenge: all parts of the community from digital librarians and scientists to computer scientists and IT companies need to be involved.

36.7 CONCLUDING REMARKS

From the above discussion, it can be seen that the coming digital data deluge will have profound effects on much of the current scientific infrastructure. Data from a wide variety of new sources will need to be annotated with metadata, archived and curated so that both the data and the programs used to transform can be reproduced in the future. e-Scientists will want to search distributed sources of diverse types of data and co-schedule computation time on the nearest appropriate resource to analyse or visualize their results.

This vision of Grid middleware will require the present functionality of both SRB [39] and Globus [62] middleware systems and much more. The present move towards Grid Services and Open Grid Services Architecture represents a unique opportunity to exploit synergies with commercial IT suppliers and make such a Grid vision a reality.

ACKNOWLEDGEMENTS

The vision of the Grid described in this chapter – with its emphasis on the access and the integration of distributed data resources combined with that of remote access to distributed compute resources – owes much to discussions with many people. We would particularly like to acknowledge the contributions to our understanding of these issues from Jim Gray, Jeff Nick, Bill Johnstone, Reagan Moore, Paul Messina and the Globus team in the United States and Malcolm Atkinson, Stevan Harnad, Jessie Hey, Liz Lyon, Norman Paton and Paul Watson in the United Kingdom. We are also grateful to Malcolm Read of JISC for his ever innovative support, to Sir Brian Follet for his early insight into the implications of e-Science for libraries and for universities and to John Taylor for both his vision for e-Science and for obtaining funding for the UK e-Science programme.

The authors are also grateful to David Boyd, Reagan Moore and Stevan Harnad for some helpful detailed comments on an earlier version of this chapter.

REFERENCES

1. Taylor, J. M., http://www.e-science.clrc.ac.uk.
2. Foster, I. and Kesselman, C. (eds) (1999) *The Grid: Blueprint for a New Computing Infrastructure*. San Francisco, CA: Morgan Kaufmann Publishers.
3. Allen, F. *et al.* (2001) BlueGene: A vision for protein science using a petaflop computer. *IBM Systems Journal*, **40**(2), 310–327.
4. Sterling, T. (2002) The Gilgamesh MIND processor-in-memory architecture for petaflops-scale computing. *ISHPC Conference*, Kansai, Japan, May, 2002.
5. Sanger Institute, Hinxton, UK, http://www.sanger.ac.uk.
6. Gray, J. and Hey T. (2001) In search of petabyte databases. Talk at *2001 HPTS Workshop Asilomar*, 2001, www.research.microsoft/~gray.
7. EU DataGrid Project, http://eu-datagrid.web.cern.ch.
8. NSF GriPhyN Project, http://www.griphyn.org.
9. DOE PPDataGrid Project, http://www.ppdg.net.
10. UK GridPP Project, http://www.gridpp.ac.uk.
11. NSF iVDGL Project, http://www.ivdgl.org.
12. Rajasekar, A., Wan, M. and Moore, R. (2002) MySRB & SRB – components of a data grid. *11th International Symposium on High Performance Distributed Computing*, Edinburgh, Scotland, 2002.
13. DAME Project, www.cs.york.ac.uk/DAME.
14. Stuart, D. (2002) Presentation at *NeSC Workshop*, Edinburgh, June, 2002.
15. DIAMOND Project, http://www.diamond.ac.uk.
16. Lesk, M. (1997) *Practical Digital Libraries*. San Francisco, CA: Morgan Kaufmann Publishers.
17. Internet Archive, http:// www.archive.org.
18. Lyman, P. and Varian, H. R. (2000) *How Much Information?* UC Berkeley School of Information Management & Systems Report, http://www.sims.berkeley.edu/how-much-info.

19. NVO, http://www.nvo.org.
20. AVO, htpp://www.eso.org/avo.
21. AstroGrid, http://www.astrogrid.ac.uk.
22. LIGO, http://www.ligo.caltech.edu.
23. VISTA, http://www.vista.ac.uk.
24. European Bioinformatics Institute, http://www.ebi.ac.uk.
25. Hassard, J. (2002); private communication.
26. Gurney, R. (2002); private communication.
27. BaBar Experiment, www.slac.stanford.edu/BFROOT/.
28. LHC Computing Project, http://lhcgrid.web.cern.ch/LHCgrid.
29. InSiteOne Digital Image Storing and Archive Service, http://www.Insiteone.com.
30. Proposed e-Diamond Project, http://e-science.ox.ac.uk/. See also Chapter 41.
31. Neathey, J. (2002); private communication.
32. MIMAS Service, http://www.mimas.ac.uk.
33. EDINA Service, http://edina.ac.uk.
34. Arts and Humanities Data Service, http://www.ahds.ac.uk.
35. Xindice Native XML Database, http://xml.apache.org/xindice.
36. DAML+OIL, http://www.daml.org/2001/03/daml+oil-index.html.
37. Berners-Lee, T., Fischetti, M. (1999) *Weaving the Web.* New York: Harper Collins.
38. Rajasekar, A. K. and Moore, R. W. (2001) Data and metadata collections for scientific applications. *European High Performance Computing Conference*, Amsterdam, Holland, 2001.
39. The Storage Resource Broker, http://www.npaci.edu/DICE/SRB.
40. Dublin Core Metadata Initiative, http://Dublin core.org.
41. DeRoure, D., Jennings, N. and Shadbolt, N. Towards a semantic grid. *Concurrency & Computation*; (to be published) and in this collection.
42. Moore, R. W. (2001) knowledge-based grids. *Proceeding of the 18th IEEE Symposium on Mass Storage Systems and Ninth Goddard Conference on Mass Storage Systems and Technologies*, San Diego, April, 2001.
43. The GEODISE Project, http://www.geodise.org/.
44. The myGrid Project, http://mygrid.man.ac.uk.
45. The Comb-e-Chem Project, http://www.combechem.org.
46. Moore, R. W. (2001) Digital Libraries, Data Grids and Persistent Archives. Presentation at *NARA*, December, 2001.
47. The DiscoveryNet Project, http://www.discovery-on-the.net.
48. The RealityGrid Project, http://www.realitygrid.org.
49. Lagoze, C. and Van De Sompel, H. (2001) The open archives initiative: building a low-barrier interoperability framework, *JCDL '01*. Roanoke, Virginia: ACM Press, pp. 54–62.
50. Foster, I., Kesselman, C. and Nick, J. Physiology of the grid. *Concurrency and Computation*; (to be published) and in this collection.
51. Hey, T. and Lyon, L. (2002) Shaping the future? grids, web services and digital libraries. *International JISC/CNI Conference*, Edinburgh, Scotland, June, 2002.
52. Harnad, S. and Hey, J. M. N. (1995) Esoteric knowledge: the scholar and scholarly publishing on the Net, in Dempsey, L., Law, D. and Mowlat, I. (eds) *Proceedings of an International Conference on Networking and the Future of Libraries: Managing the Intellectual Record*. Bath, 19–21 April, 1995 London: Library Association Publications (November 1995), pp. 110–16.
53. Ginsparg e-Print Archive, http://arxiv.org.
54. Lynch, C. (2001) *Metadata Harvesting and the Open Archives Initiative*, ARL Bimonthly Report 217:1–9, 2001.
55. Lui, Z., Maly, K., Zubair, M. and Nelson, M. (2001) Arc – An OAI service provider for cross-archive searching, *JCDL'01*, Roanoke, VA: ACM Press, pp. 65–66.
56. The Caltech Library Systems Digital Collections Project, http://library.caltech.edu/digital/.
57. The MIT Dspace Project, http://www.mit.edu/dspace/.
58. SWISS-PROT, http://www.ebi.ac.uk/swissprot.
59. EU/US Workshop on Large Scientific Databases, http://www.cacr.caltech.edu/euus.

60. Rothenberg, J. (1995) Ensuring the longevity of digital documents. *Scientific American*, **272**(1), 42–7.
61. Lorie, R. A. (2001) Long Term Preservation of Digital Information. *JCDL '01*, Roanoke, VA, June, 2001.
62. The Globus Project, http://www.globus.org.

37

Metacomputing

Larry Smarr[1] and Charles E. Catlett[2]

[1]*Cal-(IT)², University of California, San Diego, California, United States, *[2]*Argonne
National Laboratory, Argonne, Illinois, United States*

From the standpoint of the average user, today's computer networks are extremely primitive compared to other networks. While the national power, transportation, and telecommunications networks have evolved to their present state of sophistication and ease of
use, computer networks are at an early stage in their evolutionary process. Eventually,
users will be unaware that they are using any computer but the one on their desk, because
it will have the capability to reach out across the national network and obtain whatever
computational resources that are necessary.

The computing resources transparently available to the user via this networked environment have been called a *metacomputer*. The metacomputer is a network of heterogeneous,
computational resources linked by software in such a way that they can be used as easily
as a personal computer. In fact, the PC can be thought of as a minimetacomputer, with a
general-purpose microprocessor, perhaps floating point-intensive coprocessor, a computer
to manage the I/O – or memory – hierarchy, and a specialized audio or graphics chip.
Like the metacomputer, the minimetacomputer is a heterogeneous environment of computing engines connected by communications links. Driving the software development

Grid Computing – Making the Global Infrastructure a Reality.   Edited by F. Berman, A. Hey and G. Fox
© 2003 John Wiley & Sons, Ltd   ISBN: 0-470-85319-0

and system integration of the National Center for Supercomputing Applications (NCSA) metacomputer are a set of 'probe' metaapplications.

The first stage in constructing a metacomputer is to create and harness the software to make the user's job of utilizing different computational elements easier. For any one project, a typical user might use a desktop workstation, a remote supercomputer, a mainframe supporting the mass storage archive, and a specialized graphics computer. Some users have worked in this environment for the past decade, using *ad hoc* custom solutions, providing specific capabilities at best, and in most cases moving data and porting applications by hand from machine to machine. The goal of building a metacomputer is elimination of the drudgery involved in carrying out a project on such a diverse collection of computer systems. This first stage is largely a software and hardware integration effort. It involves interconnecting all of the resources with high-performance networks, implementing a distributed file system, coordinating user access across the various computational elements, and making the environment seamless using existing technology. This stage is well under way at a number of federal agency supercomputer centers.

The next stage in metacomputer development moves beyond the software integration of a heterogeneous network of computers. The second phase involves spreading a single application across several computers, allowing a center's heterogeneous collection of computers to work in concert on a single problem. This enables users to attempt types of computing that are virtually impossible without the metacomputer. Software that allows this to be done in a general way (as opposed to one-time, *ad hoc* solutions) is just now emerging and is in the process of being evaluated and improved as users begin to work with it.

The evolution of metacomputing capabilities is constrained not only by software but also by the network infrastructure. At any one point in time, the capabilities available on the local area metacomputer are roughly 12 months ahead of those available on a wide-area basis. In general, this is a result of the difference between the network capacity of a local area network (LAN) and that of a wide-area network (WAN). While the individual capabilities change over time, this flow of capabilities from LAN to WAN remains constant.

The third stage in metacomputer evolution will be a transparent national network that will dramatically increase the computational and information resources available to an application. This stage involves more than having the local metacomputer use remote resources (i.e. changing the distances between the components). Stage three involves putting into place both adequate WAN infrastructure and developing standards at the administrative, file system, security, accounting, and other levels to allow multiple LAN metacomputers to cooperate. While this third epoch represents the five-year horizon, an early step toward this goal is the collaboration between the four National Science Foundation (NSF) supercomputer centers to create a 'national virtual machine room'. Ultimately, this will grow to a truly national effort by encompassing any of the attached National Research and Education Network (NREN) systems. System software must evolve to transparently handle the identification of these resources and the distribution of work.

In this article, we will look at the three stages of metacomputing, beginning with the local area metacomputer at the NCSA as an example of the first stage. The capabilities to be demonstrated in the SIGGRAPH'92 Showcase'92 environment represent the beginnings of the second stage in metacomputing. This involves advanced user interfaces that allow for participatory computing as well as examples of capabilities that would not be possible without the underlying stage one metacomputer. The third phase, a national metacomputer, is on the horizon as these new capabilities are expanded from the local metacomputer out onto gigabit per second network test beds.

37.1 LAN METACOMPUTER AT NCSA

Following the PC analogy, the hardware of the LAN metacomputer at NCSA consists of subcomponents to handle processing, data storage and management, and user interface with high-performance networks to allow communication between subcomponents [see Figure 37.1(a)]. Unlike the PC, the subsystems now are not chips or dedicated controllers but entire computer systems whose software has been optimized for its task and communication with the other components. The processing unit of the metacomputers is a collection of systems representing today's three major architecture types: massively parallel (Thinking Machines CM-2 and CM-5), vector multiprocessor (CRAY-2, CRAY Y-MP, and Convex systems), and superscalar (IBM RS/6000 systems and SGI VGX multiprocessors). Generally, these are differentiated as shared memory (Crays, Convex, and SGI) and distributed memory (CM-2, CM-5, and RS/6000 s) systems.

Essential to the Phase I LAN metacomputer is the development of new software allowing the program applications planner to divide applications into a number of components that can be executed separately, often in parallel, on a collection of computers. This requires both a set of primitive utilities to allow low-level communications between parts of the code or *processes* and the construction of a programming environment that takes available metacomputer resources into account during the design, coding, and execution phases of an application's development. One of the problems faced by the low-level communications software is that of converting data from one system's representation to that of a second system. NCSA has approached this problem through the creation of the Data Transfer Mechanism (DTM), which provides message-based interprocess communication and automatic data conversion to applications programmers and to designers of higher-level software development tools.[1]

At the level above interprocess communication, there is a need for standard packages that help the applications designer parallelize code, decompose code into functional units, and spread that distributed application onto the metacomputer. NCSA's approach to designing a distributed applications environment has been to acquire and evaluate several leading packages for this purpose, including Parallel Virtual Machine (PVM)[2] and *Express*,[3] both of which allow the programmer to identify subprocesses or subsections

---

[1] DTM was developed by Jeff Terstriep at NCSA as part of the BLANCA test bed efforts. NCSA's research on the BLANCA test bed is supported by funding from DARPA and NSF through the Corporation for National Research Initiatives.

[2] PVM was developed by a team at Oak Ridge National Laboratory, University of Tennessee, and Emory University. Also see A. Beguelin, J. Dongarra, G. Geist, R. Manchek, and V. Sunderam. Solving Computational Grand Challenges Using a Network of Supercomputers. In *Proceedings of the Fifth SIAM Conference on Parallel processing*, Danny Sorenson, Ed., SIAM, Philadelphia, 1991.

[3] Express was developed at Caltech and was subsequently distributed by ParaSoft. It is a suite of tools similar to PVM.

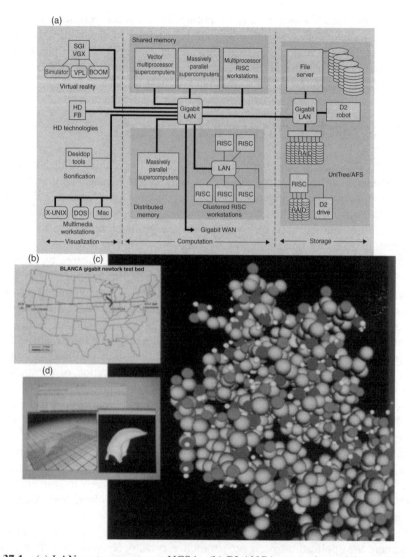

Figure 37.1 (a) LAN metacomputer at NCSA; (b) BLANCA research participants include the University of California – Berkeley, Lawrence Livermore National Laboratories, University of Wisconsin-Madison (CS, Physics, Space Science, and Engineering Center), and University of Illinois at Urbana-Champaign (CS, NCSA). Additional XUNET participants include Lawrence Livermore National Laboratories and Sandia. BLANCA uses facilities provided by the AT&T Bell Laboratories XUNET Communications Research Program in cooperation with Ameritech, Bell Atlantic, and Pacific Bell. Research on the BLANCA test bed is supported by the Corporation for National Research Initiatives with funding from industry, NSF, and DARPA. Diagram: Charles Catlett; (c) Three-dimensional image of a molecule modeled with molecular dynamics software. Credit: Klaus Schulten, NCSA visualization group; (d) Comparing video images (background) with live three-dimensional output from thunderstorm model using the NCSA digital library. Credit: Bob Wilhelmson, Jeff Terstriep.

of a dataset within the application and manage their distribution across a number of processors, either on the same physical system or across a number of networked computational nodes. Other software systems that NCSA is investigating include Distributed Network Queueing System (DNQS)[4] and Network Linda.[5] The goal of these efforts is to prototype distributed applications environments, which users can either use on their own LAN systems or use to attach NCSA computational resources when appropriate. Demonstrations in SIGGRAPH'92 Showcase will include systems developed in these environments.

A balanced system is essential to the success of the metacomputer. The network must provide connectivity at application-required bandwidths between computational nodes, information and data storage locations, and user interface resources, in a manner independent of geographical location.

The national metacomputer, being developed on gigabit network test beds such as the BLANCA test bed illustrated in Figure 37.1(b), will change the nature of the scientific process itself by providing the capability to collaborate with geographically dispersed researchers on Grand Challenge problems. Through heterogeneous networking technology, interactive communication in real time – from one-on-one dialogue to multiuser conferences – will be possible from the desktop. When the Internet begins to support capacities at 150 Mbit/s and above, commensurate with local area and campus area 100 Mbit/s FDDI networks, then remote services and distributed services will operate at roughly the same level as today's local services. This will result in the ability to extend local area metacomputers to the national scale.

37.2 METACOMPUTING AT SIGGRAPH'92 SHOWCASE'92

The following descriptions represent a cross section of a variety of capabilities to be demonstrated by application developers from many different institutions. These six applications also cut across three fundamental areas of computational science. Theoretical simulation can be thought of as using the metacomputers to solve scientific equations numerically. Instrument/sensor control can be thought of as using the metacomputer to translate raw data from scientific instruments and sensors into visual images, allowing the user to interact with the instrument or sensor in real time as well. Finally, Data Navigation can be thought of as using the metacomputer to explore large databases, translating numerical data into human sensory input.

37.2.1 Theoretical simulation

Theoretical simulation is the use of high-performance computing to perform numerical experiments, using scientific equations to create an artificial numerical world in the metacomputer memory where experiments take place without the constraints of space or time.

---

[4] 'DNQS, A Distributed Network Queueing System' and 'DQS, A Distributed Queueing System' are both 1991 papers by Thomas Green and Jeff Snyder from SCRI/FSU. DNQS was developed at Florida State University.

[5] Network Linda was developed at Yale University.

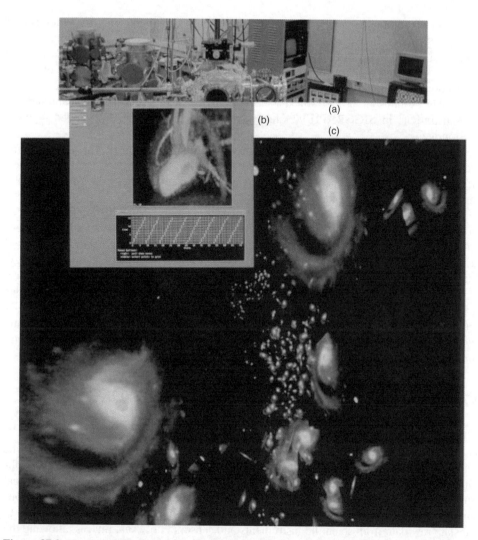

Figure 37.2 (a) Scanning Tunneling Microscopy Laboratory at the Beckman Institute for Advanced Science and Technology. Courtesy: Joe Lyding; (b) Volume rendering sequence using 'tiller' to view dynamic spatial reconstructor data of a heart of a dog. Credit: Pat Moran, NCSA; (c) Three-dimensional rendering of Harvard CFA galaxy redshift data. Credit: Margaret Geller, Harvard University, and NCSA visualization group.

One of these applications takes advantage of emerging virtual reality (VR) technologies to explore molecular structure, while the second theoretical simulation application we describe allows the user to explore the formation and dynamics of severe weather systems. An important capability these applications require of the metacomputer is to easily interconnect several computers to work on a single problem at the same time.

37.2.2 Molecular virtual reality

This project will demonstrate the interaction between a VR system and a molecular dynamics program running on a Connection Machine. Molecular dynamics models, developed by Klaus Schulten and his colleagues at the University of Illinois at Urbana-Champaign's Beckman Institute Center for Concurrent Biological Computing, are capable of simulating the ultrafast motion of macromolecular assemblies such as proteins [Figure 37.1(c)].[6] The new generation of parallel machines allows one to rapidly simulate the response of biological macromolecules to small structural perturbations, administered through the VR system, even for molecules of several thousand atoms.

Schulten's group, in collaboration with NCSA staff, developed a graphics program that collects the output of a separate program running on a Connection Machine and renders it on a SGI workstation. The imagery can be displayed on the Fake Space Labs boom display system, VPL's EyePhone head-mounted display, or the SGI workstation screen. The program provides the ability to interact with the molecule using a VPL DataGlove. The DataGlove communicates alterations of the molecular structure to the Connection Machine, restarting the dynamics program with altered molecular configurations.

This meta-application will provide the opportunity to use VR technology to monitor and control a simulation run on a Connection Machine stationed on the show floor. In the past, remote process control has involved starting, stopping, and changing the parameters of a numerical simulation. The VR user interface, on the other hand, allows the user to interact with and control the objects within the model – the molecules themselves – rather than just the computer running the model.

37.2.3 User-executed simulation/analysis of severe thunderstorm phenomena

In an effort to improve weather prediction, atmospheric science researchers are striving to better understand severe weather features. Coupled with special observing programs are intense numerical modeling studies that are being used to explore the relationship between these features and larger-scale weather conditions.[7] A supercomputer at NCSA will be used to run the model, and several workstations at both NCSA and Showcase will perform distributed visualization processing and user control. See Figure 37.1(d).

In Showcase'92, the visitor will be able to explore downburst evolution near the ground through coupled model initiation, simulation, analysis, and display modules. In this integrated, real-time environment, the analysis modules and visual display will be tied to new flow data as it becomes available from the model. This is a precursor to the kind of metacomputer forecasting environment that will couple observations, model simulations, and visualization together. The metacomputer is integral to the future forecasting environment for handling the large volumes of data from a variety of observational platforms and models being used to 'beat the real weather'. In the future, it is possible that real-time Doppler data will be used to initialize storm models to help predict the formation of tornadoes 20 to 30 min ahead of their actual occurrence.

---

[6] This research is by Mike Krogh, Rick Kufrin, William Humphrey and Klaus Schulten Department of Physics, National Center for Supercomputing Applications at Beckman Institute.

[7] This research is by Robert Wilhelmson, Crystal Shaw, Matthew Arrott, Gautum Mehrotra, and Jeff Thingvold, NCSA.

37.2.4 Instrument/sensor control

Whereas the numerical simulation data came from a computational model, the data in the following applications comes from scientific instruments. Now that most laboratory and medical instruments are being built with computers as control devices, remote observation and instrument control is possible using networks.

37.2.5 Interactive imaging of atomic surfaces

The scanning tunneling microscope (STM) has revolutionized surface science by enabling the direct visualization of surface topography and electronic structure with atomic spatial resolution. This project will demonstrate interactive visualization and distributed control of remote imaging instrumentation [Figure 37.2(a)].[8] Steering imaging experiments in real time is crucial as it enables the scientist to optimally utilize the instrument for data collection by adjusting observation parameters during the experiment. An STM at the Beckman Institute at the University of Illinois at Urbana-Champaign (UIUC) will be controlled remotely from a workstation at Showcase'92. The STM data will be sent as it is acquired to a Convex C3800 at NCSA for image processing and visualization. This process will occur during data acquisition. STM instrument and visualization parameters will be under user control from a workstation at Showcase'92. The user will be able to remotely steer the STM in Urbana from Chicago and visualize surfaces at the atomic level in real time.

The project will use AVS (Advanced Visualization System) for distributed components of the application between the Convex C3800 at NCSA and a Showcase'92 workstation. Viewit, a multidimensional visualization interface, will be used as the user interface for instrument control and imaging.

37.2.6 Data navigation

Data navigation may be regarded not only as a field of computational science but also as the method by which all computational science will soon be carried out. Both theoretical simulation and instrument/sensor control produce large sets of data that rapidly accumulate over time. Over the next several years, we will see an unprecedented growth in the amount of data that is stored as a result of theoretical simulation, instruments and sensors, and also text and image data produced by network-based publication and collaboration systems. While the three previous applications involve user interfaces to specific types of data, the three following applications address the problem faced by scientists who are searching through many types of data. Capabilities are shown for solving the problem of locating data as well as examining the data.

37.3 INTERACTIVE FOUR-DIMENSIONAL IMAGING

There are many different methods for visualizing biomedical image data sets. For instance, the Mayo Clinic Dynamic Spatial Reconstructor (DSR) is a CT scanner that can collect

---

[8] This research is by Clint Potter, Rachael Brady, Pat Moran, NCSA/Beckman Institute.

entire three-dimensional scans of a subject as quickly as 30 times per second. Viewing a study by examining individual two-dimensional plane images one at a time would take an enormous amount of time, and such an approach would not readily support identification of out-of-plane and/or temporal relationships.

The biomedical scientist requires computational tools for better navigation of such an 'ocean' of data. Two tools that are used extensively in the NCSA biomedical imaging activities are 'viewit' and 'tiller' [Figure 37.2(c)].[9] 'Viewit' is a multidimensional 'calculator' used for multidimensional image reconstruction and enhancement, and display preparation. It can be used to read instrument data, reconstruct, and perform volumetric projections saved in files as image frames. Each frame provides a view of the subject from a unique viewpoint at an instant in time. 'Tiller' collects frames generated by 'viewit', representing each frame as a cell on a two-dimensional Grid. One axis of the Grid represents a spatial trajectory and the other axis represents time. The user charts a course on this time–space map and then sets sail. A course specifies a frame sequence constructed on the fly and displayed interactively. This tool is particularly useful for exploring sets of precomputed volumetric images, allowing the user to move freely through the images by animating them.

At Showcase'92, interactive visualization of four-dimensional data will use an interface akin to that of 'Tiller'; however, the volumetric images will be generated on demand in real time, using the Connection Machine at NCSA. From a workstation at Showcase'92, the user will explore a large, four-dimensional data set stored at NCSA. A dog heart DSR data set from Eric Hoffman, University of Pennsylvania, will be used for the Showcase'92 demo.

37.4 SCIENTIFIC MULTIMEDIA DIGITAL LIBRARY

The Scientific Digital Library[10] will be available for browsing and data analysis at Showcase'92. The library contains numerical simulation data, images, and other types of data as well as software. To initiate a session, the participant will use a Sun or SGI workstation running the Digital Library user interface, to connect to a remote database located at NCSA. The user may then perform queries and receive responses from the database. The responses represent matches to specific queries about available data sets. After selecting a match, the user may elect to examine the data with a variety of scientific data analysis tools. The data is automatically retrieved from a remote system and presented to the researcher within the chosen tool.

One capability of the Digital Library was developed for radio astronomers. Data and processed images from radio telescopes are stored within the library and search mechanisms have been developed with search fields such as frequency and astronomical object names. This allows the radio astronomer to perform more specialized and comprehensive searches in the library based on the content of the data rather than simply by author or general subject.

---

[9] This research is by Clint Potter, Rachael Brady, Pat Moran, NCSA/Beckman Institute.

[10] The digital library architecture and development work at NCSA is led by Charlie Catlett and Jeff Terstriep.

The data may take the form of text, source code, data sets, images (static and animated), audio, and even supercomputer simulations and visualizations. The digital library thus aims to handle the entire range of multimedia options. In addition, its distributed capabilities allow researchers to share their findings with one another, with the results displayed on multiple workstations that could be located across the building or across the nation.

37.5 NAVIGATING SIMULATED AND OBSERVED COSMOLOGICAL STRUCTURES

The Cosmic Explorer[11] is motivated by Carl Sagan's imaginary spaceship in the Public Broadcasting System (PBS) series 'Cosmos', in which he explores the far corners of the universe. In this implementation, the user will explore the formation of the universe, the generation of astrophysical jets, and the colliding galaxies by means of numerical simulations and VR technology. The numerical simulations produce very large data sets representing the cosmic structures and events. It is important for the scientist not only to be able to produce images from this data but also to be able to animate events and view them from multiple perspectives.

Numerical simulations will be performed on supercomputers at NCSA and their resulting data sets will be stored at NCSA. Using the 45 Mbit/s NSFNET connection between Showcase'92 and NCSA, data from these simulations will be visualized remotely using the VR 'CAVE'.[12] The 'CAVE' will allow the viewer to 'walk around' in the data, changing the view perspective as well as the proximity of the viewer to the objects in the data.

Two types of simulation data sets will be used. The first is produced by a galaxy cluster formation model and consists of galaxy position data representing the model's predicted large-scale structure of the universe. The second is produced by a cosmological event simulator that produces data representing structures caused by the interaction of gases and objects in the universe.

Using the cosmic explorer and the 'CAVE', a user will be able to compare the simulated structure of the universe with the observed structure, using the Harvard CFA galaxy redshift database assembled by Margaret Geller and John Huchra. This will allow comparisons between the real and theoretical universes. The VR audience will be able to navigate the 'Great Wall' – a supercluster of galaxies over 500 million light years in length – and zoom in on individual galaxies. Similarly, the simulated event structures, such as gas jets and remains of colliding stars, will be compared with similar structures observed by radio telescopes. The radio telescope data, as mentioned earlier, has been accumulated within the scientific multimedia digital library. This combined simulation/observation environment will also allow the participant to display time sequences of the simulation data, watching the structures evolve and converge with the observed data.

---

[11] The Cosmic Explorer VR application software is based on software components already developed for VR and interactive graphic applications, including the Virtual Wind Tunnel developed by Steve Bryson of NASA Ames. Also integrated will be Mike Norman of NCSA for interactive visualization of numerical cosmology data bases, and the NCSA VR interface library developed by Mike McNeill.

[12] The CAVE, or "Cave Automated Virtual Environment," is a fully immersive virtual environment development by professor Tom DeFanti and his colleagues at the University of Illinois-Chicago Electronic Visualization Laboratory."

ACKNOWLEDGMENTS

This article is in part an expanded version of the NCSA newsletter Access special issue on the metacomputer, November/December 1991. Much information, text, and assistance was provided by Melanie Loots, Sara Latta, David Lawrence, Mike Krogh, Patti Carlson, Bob Wilhelmson, Clint Potter, Michael Norman, Jeff Terstriep, Klaus Schulten, Pat Moran, and Rachael Brady.

FURTHER READING

Becker, J. and Dagum, L. (1992) Distributed 3-D particle simulation using Cray Y-MP, CM-2. *NASA NAS, NASNEWS Numerical Aerodynamic Simulation Program Newsletter*, **6**, 10.

Catlett, C. E. (1992) In search of gigabit applications. *IEEE Communications*, **30**(4), 42–51.

Committee on Physical, Mathematical, and Engineering Sciences, Federal Coordinating Council for Science, Engineering, and Technology, Office of Science and Technology Policy. *Grand challenges: High Performance Computing and Communications*, Supplement to the President's Fiscal Year 1992 Budget.

Committee on Physical, Mathematical, and Engineering Sciences, Federal Coordinating Council for Science, Engineering, and Technology, Office of Science and Technology Policy. *Grand Challenges: High Performance Computing and Communications*, Supplement to the President's Fiscal Year 1993 Budget.

Corcoran, E. (1991) Calculating reality. *Scientific American*, **264**.

Corporation for National Research Initiatives, *1991 Annual Testbed Reports*, Reports prepared by project participants in each of five gigabit network testbeds.

Hibbard, W., Santek, D. and Tripoli, G. (1991) Interactive atmospheric data access via high speed networks. *Computer Networks and ISDN Systems*, **22**, 103–109.

Lederberg, J. and Uncapher, K. (1989) *Towards a National Collaboratory*. Report of an individual workshop at the Rockefeller University, March, 1989.

Lynch, D. (ed.) (1992) *Internet System Handbook*. New York: Manning Publications, Addison-Wesley.

Stix, G. (1990) Gigabit connection. *Scientific American*. **263**.

Supercomputers: Directions in Technology and Applications. ISBN 0309-04088-4, Washington, DC: National Academy Press, 1989.

Cohen, J. (1992) NCSA's metacomputer: a special report. Access: 5, *NCSA High-Performance Computing Newsletter*, **5**.

CR Categories and Subject Descriptors: C.2.4 [Computer-Communication Networks]: Distributed Systems – Distributed applications, Distributed databases; C.3 [Special-Purpose and Application-Based Systems]; 1.3.2 [Computer Graphics]: Graphics Systems – Distributed network graphics, remote systems; 1.6.3 [Simulation and modeling]: Applications; J.2 [Computer Applications]: Physical Sciences and Engineering; J.3 [Computer Applications]: Life and Medical Sciences.

38

Grids and the virtual observatory

Roy Williams

California Institute of Technology, California, United States

38.1 THE VIRTUAL OBSERVATORY

Astronomers have always been early adopters of technology, and information technology has been no exception. There is a vast amount of astronomical data available on the Internet, ranging from spectacular processed images of planets to huge amounts of raw, processed and private data. Much of the data is well documented with citations, instrumental settings, and the type of processing that has been applied. In general, astronomical data has few copyright, or privacy or other intellectual property restrictions in comparison with other fields of science, although fresh data is generally sequestered for a year or so while the observers have a chance to reap knowledge from it.

As anyone with a digital camera can attest, there is a vast requirement for storage. Breakthroughs in telescope, detector, and computer technology allow astronomical surveys to produce terabytes of images and catalogs (Figure 38.1). These datasets will cover the sky in different wavebands, from γ- and X rays, optical, infrared, through to radio. With the advent of inexpensive storage technologies and the availability of high-speed networks, the concept of multiterabyte on-line databases interoperating seamlessly is no longer outlandish [1, 2]. More and more catalogs will be interlinked, query engines will become more and more sophisticated, and the research results from on-line data will be

Grid Computing – Making the Global Infrastructure a Reality.   Edited by F. Berman, A. Hey and G. Fox
© 2003 John Wiley & Sons, Ltd   ISBN: 0-470-85319-0

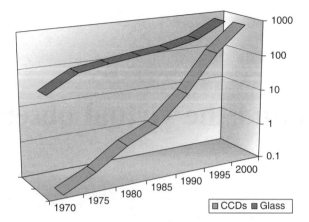

Figure 38.1   The total area of astronomical telescopes in m², and CCDs measured in gigapixels, over the last 25 years. The number of pixels and the data double every year.

just as rich as that from 'real' observatories. In addition to the quantity of data increasing exponentially, its heterogeneity – the number of data publishers – is also rapidly increasing. It is becoming easier and easier to put data on the web, and every scientist builds the service, the table attributes, and the keywords in a slightly different way. Standardizing this diversity without destroying it is as challenging as it is critical. It is also critical that the community recognizes the value of these standards, and agrees to spend time on implementing them.

Recognizing these trends and opportunities, the National Academy of Sciences Astronomy and Astrophysics Survey Committee, in its decadal survey [3] recommends, as a first priority, the establishment of a **National Virtual Observatory** (**NVO**), leading to US funding through the NSF. Similar programs have begun in Europe and Britain, as well as other national efforts, now unified by the International Virtual Observatory Alliance (IVOA). The Virtual Observatory (VO) will be a 'Rosetta Stone' linking the archival data sets of space- and ground-based observatories, the catalogs of multiwavelength surveys, and the computational resources necessary to support comparison and cross-correlation among these resources. While this project is mostly about the US effort, the emerging International VO will benefit the entire astronomical community, from students and amateurs to professionals.

We hope and expect that the fusion of multiple data sources will also herald a sociological fusion. Astronomers have traditionally specialized by wavelength, based on the instrument with which they observe, rather than by the physical processes actually occurring in the Universe: having data in other wavelengths available by the same tools, through the same kinds of services will soften these artificial barriers.

38.1.1 Data federation

Science, like any deductive endeavor, often progresses through *federation* of information: bringing information from different sources into the same frame of reference. The police detective investigating a crime might see a set of suspects with the motive to commit the crime, another group with the opportunity, and another group with the means. By federating this information, the detective realizes there is only one suspect in all three groups – this federation of information has produced knowledge. In astronomy, there is great interest in objects between large planets and small stars – the so-called brown dwarf stars. These very cool stars can be found because they are visible in the infrared range of wavelengths, but not at optical wavelengths. A search can be done by federating an infrared and an optical catalog, asking for sources in the former, but not in the latter.

The objective of the Virtual Observatory is to enable the federation of much of the digital astronomical data. A major component of the program is about efficient processing of large amounts of data, and we shall discuss projects that need Grid computing, first those projects that use images and then projects that use databases.

Another big part of the Virtual Observatory concerns standardization and translation of data resources that have been built by many different people in many different ways. Part of the work is to build enough metadata structure so that data and computing resources can be automatically connected in a scientifically valid fashion. The major challenge with this approach, as with any standards' effort, is to encourage adoption of the standard in the community. We can then hope that those in control of data resources can find it within them to expose it to close scrutiny, including all its errors and inconsistencies.

38.2 WHAT IS A GRID?

People often talk about the Grid, as if there is only one, but in fact *Grid* is a concept. In this paper, we shall think of a Grid in terms of the following criteria:

- *Powerful resources*: There are many Websites where clients can ask for computing to be done or for customized data to be fetched, but a true Grid offers sufficiently powerful resources that their owner does not want arbitrary access from the public Internet. Supercomputer centers will become delocalized, just as digital libraries are already.
- *Federated computing*: The Grid concept carries the idea of geographical distribution of computing and data resources. Perhaps a more important kind of distribution is human: that the resources in the Grid are managed and owned by different organizations, and have agreed to federate themselves for mutual benefit. Indeed, the challenge resembles the famous example of the federation of states – which is the United States.
- *Security structure*: The essential ingredient that glues a Grid together is security. A federation of powerful resources requires a superstructure of control and trust to limit uncontrolled, public use, but to put no barriers in the way of the valid users.

In the Virtual Observatory context, the most important Grid resources are data collections rather than processing engines. The Grid allows federation of collections without worry about differences in storage systems, security environments, or access mechanisms. There may be directory services to find datasets more effectively than the Internet search engines that work best on free text. There may be replication services that find the nearest copy of a given dataset. Processing and computing resources can be used through allocation services based on the batch queue model, on scheduling multiple resources for a given time, or on finding otherwise idle resources.

38.2.1 Virtual Observatory middleware

The architecture is based on the idea of *services*: Internet-accessible information resources with well-defined *requests* and consequent *responses*. There are already a large number of astronomical information services, but in general each is hand-made, with arbitrary request and response formats, and little formal directory structure. Most current services are designed with the idea that a human, not a computer, is the client, so that output comes back as HTML or an idiosyncratic text format. Furthermore, services are not designed with scaling in mind to gigabyte or terabyte result sets, with a consequent lack of authentication mechanisms that are necessary when resources become significant.

To solve the scalability problem, we are borrowing heavily from progress by information technologists in the Grid world, using GSI authentication [4], Storage Resource Broker [5], and GridFTP [6] for moving large datasets. In Sections 38.3 and 38.4, we discuss some of the applications in astronomy of this kind of powerful distributed computing framework, first for image computing, then for database computing. In Section 38.5, we discuss approaches to the semantic challenge in linking heterogeneous resources.

38.3 IMAGE COMPUTING

Imaging is a deep part of astronomy, from pencil sketches, through photographic plates, to the 16 gigapixel camera recently installed on the Hubble telescope. In this section, we consider three applications of Grid technology for federating and understanding image data.

38.3.1 Virtual Sky: multiwavelength imaging

The Virtual Sky project [7] provides seamless, federated images of the night sky; not just an album of popular places, but also the entire sky at multiple resolutions and multiple wavelengths (Figure 38.2). Virtual Sky has ingested the complete DPOSS survey (Digital Palomar Observatory Sky Survey [8]) with an easy-to-use, intuitive interface that anyone can use. Users can zoom out so the entire sky is on the screen, or zoom in, to a maximum resolution of 1.4 arcseconds per pixel, a magnification of 2000. Another theme is the Hubble Deep Field [9], a further magnification factor of 32. There is also a gallery of interesting places, and a blog (bulletin board) where users can record comments. Virtual Sky is a collaboration between the Caltech Center for Advanced Computing Research, Johns Hopkins University, the Sloan Sky Survey [10], and Microsoft Research. The image storage and display is based on the popular Terraserver [11].

Virtual Sky federates many different image sources into a unified interface. Like most federation of heterogeneous data sources, there is a loss of information – in this case because of resampling the original images – but we hope that the federation itself will provide a new insight to make up for the loss.

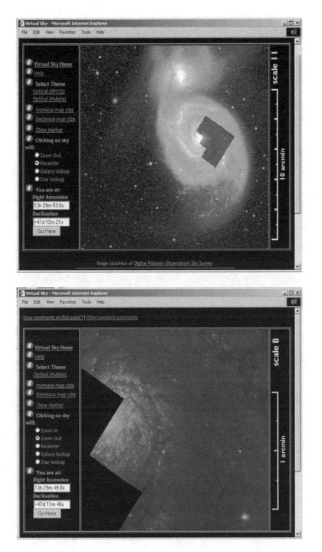

Figure 38.2 Two views from the Virtual Sky image federation portal. On the left is the view of the galaxy M51 seen with the DPOSS optical survey from Palomar. Overset is an image from the Hubble space telescope. At the right is the galactic center of M51 at eight times the spatial resolution. The panel on the left allows zooming and panning, as well as changing theme.

The architecture is based on a hierarchy of precomputed image tiles, so that response is fast. Multiple 'themes' are possible, each one being a different representation of the night sky. Some of the themes are as follows:

- Digital Palomar Observatory Sky Survey;
- Sloan Digital Sky Survey;

- A multi-scale star map from John Walker, based on the Yoursky server;
- The Hubble Deep Field.
- The 'Uranometria', a set of etchings from 1603 that was the first true star atlas;
- The ROSAT All Sky Survey in soft and hard X rays;
- The NRAO VLA Sky Survey at radio wavelengths (1.4 GHz);
- The 100 micron Dust Map from Finkbeiner *et al.*
- The NOAO Deep Wide Field survey.

All the themes are resampled to the same standard projection, so that the same part of the sky can be seen in its different representations, yet perfectly aligned. The Virtual Sky is connected to other astronomical data services, such as NASA's extragalactic catalog (NED [12]) and the Simbad star catalog at CDS Strasbourg [13]. These can be invoked simply by clicking on a star or galaxy, and a new browser window shows the deep detail and citations available from those sources.

Besides the education and outreach possibilities of this 'hyper-atlas' of the sky, another purpose is as an index to image surveys, so that a user can directly obtain the pixels of the original survey from a Virtual Sky page. A cutout service can be installed over the original data, so that Virtual Sky is used as a visual index to the survey, from which fully calibrated and verified *Flexible Image Transport Specification* (FITS) files can be obtained.

38.3.1.1 Virtual Sky implementation

When a telescope makes an image, or when a map of the sky is drawn, the celestial sphere is projected to the flat picture plane, and there are many possible mappings to achieve this. Images from different surveys may also be rotated or stretched with respect to each other. The Virtual Sky federates images by computationally stretching each one to a standard projection. Because all the images are on the same pixel Grid, they can be used for searches in multiwavelength space (see next section for scientific motivation). For the purposes of a responsive Website, however, the images are reduced in dynamic range and JPEG compressed before being loaded into a database. The sky is represented as 20 pages (like a star atlas), which has the advantage of providing large, flat pages that can easily be zoomed and panned. The disadvantage, of course, is distortion far from the center.

Thus, the chief computational demand of Virtual Sky is resampling the raw images. For each pixel of the image, several projections from pixel to sky and the same number of inverse projections are required. There is a large amount of I/O, with random access either on the input or output side. Once the resampled images are made at the highest resolution, a hierarchy is built, halving the resolution at each stage.

There is a large amount of data associated with a sky survey: the DPOSS survey is 3 Terabytes, the Two-Micron All Sky Survey (2MASS [14]) raw imagery is 10 Terabytes. The images were taken at different times, and may overlap. The resampled images built for Virtual Sky form a continuous mosaic with little overlap; they may be a fraction of these sizes, with the compressed tiles even smaller. The bulk of the backend processing has been done on an HP Superdome machine, and the code is now being ported to

Teragrid [15] Linux clusters. Microsoft SQL Server runs the Website on a dual-Pentium Dell Poweredge, at 750 MHz, with 250 GB of disks.

38.3.1.2 Parallel computing

Image stretching (resampling) (Figure 38.3) that is the computational backbone of Virtual Sky implies a mapping between the position of a point in the input image and the position of that point in the output. The resampling can be done in two ways:

- *Order by input*: Each pixel of the input is projected to the output plane, and its flux distributed there. This method has the advantage that each input pixel can be spread over the output such that total flux is preserved; therefore the brightness of a star can be accurately measured from the resampled dataset.
- *Order by output*: For each pixel of the output image, its position on the input plane is determined by inverting the mapping, and the color computed by sampling the input image. This method has the advantage of minimizing loss of spatial resolution. Virtual Sky uses this method.

If we order the computation by the input pixels, there will be random write access into the output dataset, and if we order by the output pixels, there will be random read access into the input images. This direction of projection also determines how the problem parallelizes. If we split the input files among the processors, then each processor opens one file at a time for reading, but must open and close output files arbitrarily, possibly leading to contention. If we split the data on the output, then processors are arbitrarily opening files from the input plane depending on where the output pixel is.

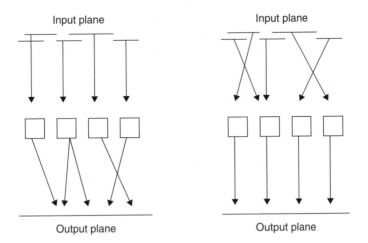

Figure 38.3 Parallelizing the process of image resampling. (a) The input plane is split among the processors, and data drops arbitrarily on the output plane. (b) The output plane is split among processors, and the arbitrary access is on the input plane.

38.3.2 MONTAGE: on-demand mosaics

Virtual Sky has been designed primarily as a delivery system for precomputed images in a fixed projection, with a resampling method that emphasizes spatial accuracy over flux conservation. The background model is a quadratic polynomial, with a contrast mapping that brings out fine detail, even though that mapping may be nonlinear.

The NASA-funded MONTAGE project [16] builds on this progress with a comprehensive mosaicking system that allows broad choice in the resampling and photometric algorithms, and is intended to be operated on a Grid architecture such as Teragrid. MONTAGE will operate as an on-demand system for small requests, up to a massive, wide-area data-computing system for large jobs. The services will offer simultaneous, parallel processing of multiple images to enable fast, deep, robust source detection in multiwavelength image space. These services have been identified as cornerstones of the NVO. We intend to work with both massive and diverse image archives: the 10 terabyte 2MASS (infrared [14]), the 3 terabyte DPOSS (optical [8]), and the much larger SDSS [10] optical survey as it becomes available. There are many other surveys of interest. MONTAGE is a joint project of the NASA Infrared Processing and Analysis Center (IPAC), the NASA Jet Propulsion Laboratory (JPL), and Caltech's Center for Advanced Computing Research (CACR).

38.3.3 Science with federated images

Modern sky surveys, such as 2MASS and Sloan provide small images ($\sim$1000 pixels on a side), so that it is difficult to study large objects and diffuse areas, for example, the Galactic Center. Another reason for mosaicking is to bring several image products from different instruments to the same projection, and thereby federate the data. This makes possible such studies as:

- *Stacking*: Extending source detection methods to detect objects an order of magnitude fainter than currently possible. A group of faint pixels may register in a single wavelength at the two-sigma level (meaning there may be something there, but it may also be noise). However, if the same pixels are at two-sigma in other surveys, then the overall significance may be boosted to five sigma – indicating an almost certain existence of signal rather than just noise. We can go fainter in image space because we have more photons from the combined images and because the multiple detections can be used to enhance the reliability of sources at a given threshold.
- *Spectrophotometry*: Characterizing the spectral energy distribution of the source through 'bandmerge' detections from the different wavelengths.
- *Extended sources*: Robust detection and flux measurement of complex, extended sources over a range of size scales. Larger objects in the sky (e.g. M31, M51) may have both extended structure (requiring image mosaicking) and a much smaller active center, or diffuse structure entirely. Finding the relationship between these attributes remains a scientific challenge. It will be possible to combine multiple-instrument imagery to build a multiscale, multiwavelength picture of such extended objects. It is also interesting to make statistical studies of less spectacular, but extended, complex sources that vary in shape with wavelength.

- *Image differencing*: Differences between images taken with different filters can be used to detect certain types of sources. For example, planetary nebulae (PNe) emit strongly in the narrow Hα band. By subtracting out a much wider band that includes this wavelength, the broad emitters are less visible and the PNe is highlighted.
- *Time federation*: A trend in astronomy is the *synoptic* survey, in which the sky is imaged repeatedly to look for time-varying objects. MONTAGE will be well placed for mining the massive data from such surveys. For more details, see the next section on the Quest project.
- *Essentially multiwavelength objects*: Multiwavelength images can be used to specifically look for objects that are not obvious in one wavelength alone. Quasars were discovered in this way by federating optical and radio data. There can be sophisticated, self-training, pattern recognition sweeps through the entire image data set. An example is a distant quasar so well aligned with a foreground galaxy to be perfectly gravitationally lensed, but where the galaxy and the lens are only detectable in images at different wavelengths.

38.3.4 MONTAGE architecture

The architecture will be based on the Grid paradigm, where data is fetched from the most convenient place, and computing is done at any available platform, with single sign-on authentication to make the process practical (Figure 38.4). We will also rely on the concept of 'virtual data', the idea that data requests can be satisfied transparently whether the data is available on some storage system or whether is needs to be computed in some way. With these architectural drivers, we will be able to provide customized, high-quality data, with great efficiency, to a wide spectrum of usage patterns.

At one end of the usage spectrum is the scientist developing a detailed, quantitative data pipeline to squeeze all possible statistical significance from the federation of multiple image archives, while maintaining parentage, rights, calibration, and error information. Everything is custom: the background estimation, with its own fitting function and masking, as well as cross-image correlation; projection from sky to pixel Grid, the details of the resampling and flux preservation; and so on. In this case, the scientist would have enough authorization that powerful computational resources can be brought to bear, each processor finding the nearest replica of its input data requirements and the output being hierarchically collected to a final composite. Such a product will require deep resources from the Teragrid [15], and the result will be published in a peer-reviewed journal as a scientifically authenticated, multiwavelength representation of the sky.

Other users will have less stringent requirements for the way in which image mosaics are generated. They will build on a derived data product such as described above, perhaps using the same background model, but with the resampling different, or perhaps just using the derived product directly. When providing users with the desired data, we want to be able to take advantage of the existing data products and produce only the necessary missing pieces. It is also possible, that it may take longer to access the existing data rather than performing the processing. These situations need to be analyzed in our system and appropriate decisions need to be made.

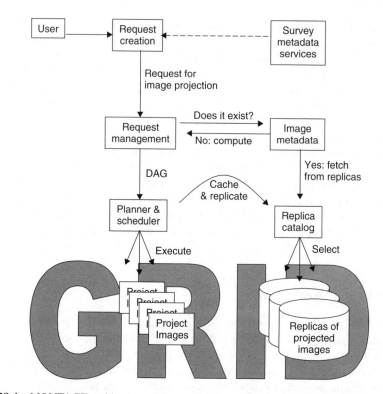

Figure 38.4 MONTAGE architecture. After a user request has been created and sent to the Request Manager, part of the request may be satisfied from existing (cached) data. The Image Metadata (IM) system looks for a suitable file, and if found, gets it from the distributed Replica Catalog (RC). If not found, a suitable computational graph Directed Acyclic Graph (DAG) is assembled and sent to be executed on Grid resources. Resulting products may be registered with the IM and stored in RC. The user is notified that the requested data is available until a specified expiry time.

38.3.4.1 Replica management

Management of replicas in a data pipeline means that intermediate products are cached for reuse: for example, in a pipeline of filters ABC, if the nature of the C filter is changed, then we need not recompute AB, but can use a cached result. Replica management can be smarter than a simple file cache: if we already have a mosaic of a certain part of the sky, then we can generate all subsets easily by selection. Simple transformations (like selection) can extend the power and reach of the replica software. If the desired result comes from a series of transformations, it may be possible to change the order of the transformations, and thereby make better use of existing replicas.

38.3.4.2 Virtual data

Further gains in efficiency are possible by leveraging the concept of 'virtual data' from the GriPhyN project [17]. The user specifies the desired data using domain specific attributes and not by specifying how to derive the data, and the system can determine how to efficiently build the desired result. Replica management is one strategy, choosing the appropriate computational platforms and input data locations is another.

The interface between an application programmer and the Virtual Data Management Systems (VDMS) now being deployed, is the definition of what is to be computed, which is called a Virtual Data Request (VDR). The VDR specifies what data is to be fetched – but not the physical location of the data, since there may be several copies and the VDMS should have the liberty to choose which copy to use. Similarly, the VDR specifies the set of objects to compute, but not the locations at which the computations will take place.

38.3.5 Quest: multitemporal imaging

A powerful digital camera is installed on the 48" Ochsin telescope of the Palomar observatory in California, and time is shared between the Quest [18] and the Near Earth Asteroid Tracking (NEAT [19]) projects. Each of these is reaching into a new domain of astronomy: systematic searching of the sky for transients and variability. Images of the same part of the sky are taken repeatedly, and computationally compared to pick out transient changes. A serendipitous example is shown in the image below, where a transient of unknown origin was caught in the process of making a sky survey.

The NEAT project is searching for asteroids that may impact the Earth, while Quest is directed to detect more distant events, for example:

- *Transient gravitational lensing*: As an object in our Galaxy passes between a distant star and us, it will focus and amplify the light of the background star over a period of days. Several events of this type have been observed by the MACHO project in the 1990s. These objects are thought to be brown dwarfs (stars not massive enough to ignite), or white dwarfs (stars already ignited and burned out). This will lead to a better understanding of the nature of the nonluminous mass of the Galaxy.
- *Quasar gravitational lensing*: At much larger scales than our Galaxy, the Quest team hopes to detect strong lensing of very remote objects such as quasars.
- *Supernovae*: The Quest system will be able to detect large numbers of very distant supernovae, leading to prompt follow-up observations, and a better understanding of supernova classification, as well as their role as standard candles for understanding the early Universe.
- *Gamma-ray burst (GRB) afterglows*: GRBs represent the most energetic processes in the Universe, and their nature is not clear. The GRB lasts for seconds, and may be detected by an orbiting observatory, but the optical afterglow may last much longer. Quest will search for these fading sources, and try to correlate them with known GRBs.

38.3.5.1 Data challenges from Quest

The Quest camera produces about 2 MB per second, which is stored on Digital Linear Tapes (DLT) at the observatory. A night of observing can produce about 50 GB of data, corresponding to 500 square degrees of sky in four filters (Figure 38.5). The tapes are

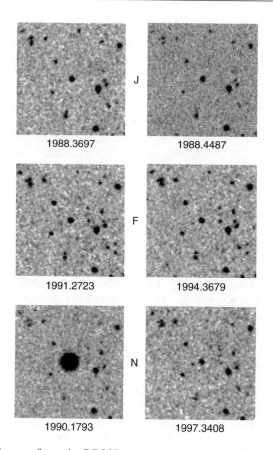

Figure 38.5 These images from the DPOSS survey were taken at different times, as indicated, and in three different color filters (J, F, N). The image at the bottom left, taken in 1990, is much brighter than the 1988 or 1991 images. The reason for this brightening is not known.

streamed by a microwave link from Palomar Mountain to San Diego Supercomputer Center (SDSC). Depending on whether the data is from NEAT or Quest, it may be sent to Berkeley (LBNL) and JPL, or to Caltech and Yale (Figure 38.6).

The objective of the Caltech team is prompt comparison processing with the previous coverage of that sky, and also with an average view of that sky. The comparison may be done directly with images, stretching the images to fit precisely over each other, correcting the sky background, then comparing new and old pixel values directly. Another way to do the comparison is through a database. Stars and galaxies are extracted from the images with source-finding software, and the resulting star catalog stored in the database. Then entries from different times can be compared with a fuzzy join, followed by picking out sources in which the brightness has changed. The QUEST team will be making use of Grid computing and database technology to extract the astronomical knowledge from the data stream.

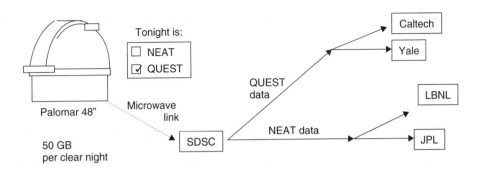

Figure 38.6 The telescope is switched between the Quest and NEAT projects. If it is Quest, the 50 GB result is transmitted by microwave to San Diego Supercomputer Center (SDSC), then by land line to Caltech and to Yale.

38.3.6 A galaxy morphology study

The classic Hubble classification relied on a few bright, nearby galaxies. It looks like a fork, with spiral galaxies, possibly barred, on the branches, and ellipticals on the 'handle' of the fork, plus a class of so-called 'irregulars'. But as our vision moves further out, and further back in time, there are more and more irregulars and the Hubble classification is no longer valid. A significant problem is that light is red-shifted with distance, so that optical output (in the rest frame of the galaxy) is shifted to infrared light as received on Earth. Since we want to compare images in the rest frame of the galaxies, it is necessary to federate infrared and optical imagery and then draw conclusions from differences between the younger (further, infrared) population and the older (nearer, optical) population. It is always difficult to compare results from different instruments, made by different groups of people.

Several investigators have attempted automatic classification [20] by computing numbers from galaxy images that measure morphology (shape), then using data-mining methods such as neural nets and unsupervised clustering to build a classification. These morphology coefficients include, for example, the concentration of light at the center, an asymmetry index, or the radial Fourier modes of the light distribution.

The objective of this study, one of the key science projects of the NVO, is to correlate galaxy morphology and X-ray surface brightness in clusters of galaxies. In particular, we will look for disturbed morphologies indicative of mergers or tidal perturbations.

From a catalog of clusters of galaxies, we select clusters, then use a catalog of galaxies to identify galaxies that are probably members of that cluster, and find both optical and X-ray images of each galaxy. A morphology program analyzes the pixels of each galaxy image, extracting morphology parameters. We run the morphology program on the optical images, and create a new galaxy catalog that combines cluster membership, X-ray flux, and morphological parameters.

The major initial problem is the federation of these databases and image libraries, building the data pipeline to read in these catalogs, correlate positions, contrast adjustment of images, a model of error propagation, and so on. Eventually, the pipeline will utilize powerful computing facilities for the image analysis. When it is time to prove a scientific hypothesis, there may be a comparison of the observed features of the galaxies to 'null hypothesis' models, meaning a lot of computing of the statistical features of synthetic galaxies.

38.4 DATABASE COMPUTING

Scientific data analysis is often called data mining because of its similarity to mining for precious metals or diamonds. The initial stages work coarsely with large quantities, and in the final stages we carefully separate small quantities of dross from small quantities of product. Often the initial stages of scientific data mining consist of *pattern-matching* of some kind: searching for signals in the data. In astronomy, the raw data is often images, and the patterns are the stars and galaxies. There may be dozens of attributes for each such source: magnitudes in several wavebands, some shape and morphology parameters such as those discussed above, error estimates, classification into different types of source, and so on. Such *catalogs of sources* are generally stored in a Relational Database Management System (RDBMS), where they are available for studies in statistical astronomy.

However, the source catalog may involve a large amount of data by itself. When complete, the Sloan Digital Sky Survey will have about 15 terabytes of image data, and this is reduced to about two terabytes in the RDBMS. With datasets of this size, the Grid community must pay serious attention to bulk database records as they do for large file-based data. In this section, we consider some projects that involve large quantities of catalog (RDBMS) data.

Often we hear the advice that the computing should be close to the data – which maximizes computing efficiency. However, in these days of cheap computers and fat data pipes, the efficiency may come by keeping code close to its developer: to minimize the amount of time spent reinstalling code from one machine to another, minimizing the maintenance of a rapidly developing code base on multiple platforms. Just as in the business world, we are moving from the old model of software installation on the desktop to a net-based model of services on demand.

Scientifically, this architecture could deliver a significant part of the promise of the NVO: data from the big sky surveys (SDSS, 2MASS, DPOSS, etc.) can be cross-matched and cached, run through statistical engines to build clusters and find outliers, and piped to visualization with flexible, innovative software on high-performance or desktop platforms. Outliers can be piped back to the original databases for further information, perhaps image cutouts.

Much of what is needed is already in place – the collaboration of astronomers and statisticians, the commitment to standards, namely, the NVO, and the recognition that big data is a gold mine and not a pile of dusty tapes. The format for transmitting these database records could be the VOTable standard, developed internationally under the VO umbrella, enabling separation of metadata from data, and high-performance binary transfer. One component that is missing is the 'driver' programs that affect the transfer, mediating between the applications and the Grid. Such a program implements the GSI (Globus) security model [4], and it can also package many database records into big files for transport over the Grid. Even though large files are being moved, perhaps asynchronously, each end of the pipe sees a continuous stream of records.

38.4.1 The VOTable XML proposal

The VOTable format [21] is a proposed XML standard for representing a table. In this context, a table is an unordered set of rows, each of a uniform format, as specified in the table metadata. Each row is a sequence of table cells, and each of these is either primitive data types, or an array of such primitives. The format is derived from the FITS Binary Table format [22]. The VOTable standard represents an initial success of the International VO community, with the initial effort spearheaded by F. Ochsenbein of Strasbourg and this author, joined by many colleagues from many countries, and sharpened and extended by an intensive on-line discussion group.

Astronomers have always been at the forefront of developments in information technology, and funding agencies across the world have recognized this by supporting the Virtual Observatory movement, in the hopes that other sciences and business can follow their lead in making on-line data both interoperable and scalable. VOTable is designed as a flexible storage and exchange format for tabular data, with particular emphasis on astronomical tables.

Interoperability is encouraged through the use of standards (XML); there are many compelling arguments for this, which will not be repeated here. However, another standard propagated by VOTable is about semantic labeling of data objects: physical quantities are tagged not only with units, but also through a common descriptive ontology that expresses the nature of the quantity (e.g. Gunn J magnitude, declination). For more on semantic interoperability, see Section 38.5.

VOTable has built-in features for big data and Grid computing. It allows metadata and data to be stored separately, with the remote data linked according to the Xlink model. Processes can then use metadata to 'get ready' for their full-sized input data, or to organize third-party or parallel transfers of the data. Remote data allow the metadata to be sent in e-mail and referenced in documents without pulling the whole dataset with it: just as we are used to the idea of sending a pointer to a document (URL) in place of the document, so we can now send metadata-rich pointers to data tables in place of the data itself. The remote data is referenced with the URL syntax protocol://location, meaning that arbitrarily complex protocols are allowed.

When we are working with very large tables in a distributed computing environment ('the Grid'), the data streams between processors, with flows being filtered, joined, and cached in different geographic locations. It would be very difficult if the number of rows of the table were required in the header – we would need to stream in the whole table into a cache, compute the number of rows, and then stream it again for the computation. In the Grid-data environment, the component in short supply is not the computers, but rather these very large caches! Furthermore, these remote data streams may be created dynamically by another process or cached in temporary storage: for this reason VOTable can express that remote data may not be available after a certain time (expires). Data on the net may require authentication for access, so VOTable allows expression of password or other identity information (the 'rights' attribute).

38.4.1.1 Data storage: flexible and efficient

The data part in a VOTable may be represented using one of three different formats:

- TABLEDATA is a pure XML format so that small tables can be easily handled in their entirety by XML tools.
- The FITS binary table format is well-known to astronomers, and VOTable can be used either to encapsulate such a file, or to re-encode the metadata; unfortunately it is difficult to stream FITS, since the dataset size is required in the header (NAXIS2 keyword), and FITS requires a specification up front of the maximum size of its variable-length arrays.
- The BINARY format is supported for efficiency and ease of programming: no FITS library is required, and the streaming paradigm is supported.

We expect that VOTable can be used in different ways, as a data storage and transport format, and also as a way to store metadata alone (table structure only). In the latter case, we can imagine a VOTable structure being sent to a server, which can then open a high-bandwidth connection to receive the actual data, using the previously digested structure as a way to interpret the stream of bytes from the data socket.

VOTable can be used for small numbers of small records (pure XML tables), or for large numbers of simple records (streaming data), or it can be used for small numbers of larger objects. In the latter case, there will be software to spread large data blocks among multiple processors on the Grid. Currently, the most complex structure that can be in a VOTable Cell is a multidimensional array.

38.4.1.2 Future

In future versions of the VOTable format, we expect to benefit from both experience and tool-building. Such tools include presentation and transformations of the metadata (and data too, when it is in XML), using XML transformation language and software: XSL and XSLT. We would like to migrate to the more powerful document validation provided by XSchema rather than the current Document Type Definition (DTD).

We also expect XSchema to allow better modularization of the document schema, so that, for example, users might put whatever serialized objects they wish into the table cells. In this way, we expect to use VOTable for handling the flow of large data objects through Grid resources, objects such as FITS files or XDF [23] documents. Also, it would mean, for example, that an XML definition of nonstandard astronomical coordinate systems could be seamlessly integrated.

We expect to add features for efficiency in the future also: to specify that the data stream has a particular sort order, to specify that a column in one table is a key into another table; to specify that one table is an index into another. The binary format will be extended to facilitate large-scale streaming.

38.4.2 Database mining and visualization

Databases accept queries, and produce a set of records as output. In the Grid era, we naturally think of combining data and applications in a distributed graph of modules and pipes. Moving along the pipes is a stream of database records. Objects that will be passed between modules are relational tables, meaning a combination of real-valued records and metadata. Specifically, we intend to exchange a Table object, which is defined as a combination of a metadata record and a collection of data records. The metadata defines the nature of the columns of the table, where each column is defined by a short name, with optional elements such as units, column type, minimum/maximum values, and description. The paradigm is quite similar to the discussion of VOTable.

Modules acting on the data stream can do such tasks as given below:

- Database modules advertise their tables and data dictionaries, and allow a user to put in a query. The single output pipe contains the results of this query.
- Storage modules for reading or writing tables into a file system.
- Joining modules have two input streams, and do database join. For astronomers, a fuzzy join is also necessary, so that a star at a particular position in one catalog is correctly matched with a star from another catalog if the positions are within astrometric error.
- Statistical and data-mining procedures such as unsupervised clustering, creation of a Bayes net, removal of columns that contain little information, principle component analysis, density estimation, outlier detection, and so on.
- Visualization modules that allow a diversity of geometric renderings of this high-dimensional space, include 3D scatter plots, parallel coordinates, and so on (Figure 38.7).

When a table is modified or transformed in structure, it is expected that its metadata is also transformed appropriately. For example, if the module takes in a table with columns Mg, Mr, and Mi, (magnitudes of a star in three wavebands) and the module output is the differences of these, it could output the metadata stream to read Mg-Mr, Mr-Mi, and Mi-Mg. Thus, there is processing not only of floating-point data, but also the attributes, elements, and properties of metadata processing.

38.5 A SEMANTIC WEB OF ASTRONOMICAL DATA

The NVO is about interoperability and federation of astronomical data resources. There are two major thrusts: Grid and Semantics. The Grid thrust is the main focus of this chapter, and it is concerned with moving large amounts of data between machines, about high-performance computing, about parallelism in processing and data movement, and about security and authentication. At the Grid level, a data object has the semantics of a file, or of a storage silo, or of a database.

Semantic interoperability, however, requires that a data object not be viewed as just bytes or records, but as scientific observations and all that is entailed: with physical units, data parentage, error estimates, and bibliography. The aim is to document data objects and the applications that manipulate them sufficiently well that a computer can connect them. The key concept for achieving this is an *ontology*, which is a shared vocabulary.

Astronomy already has several ontologies. The Library of Congress classification, for example, lists QB6 as 'Star Catalogs', and QB349-QB480 as 'Theoretical astronomy and

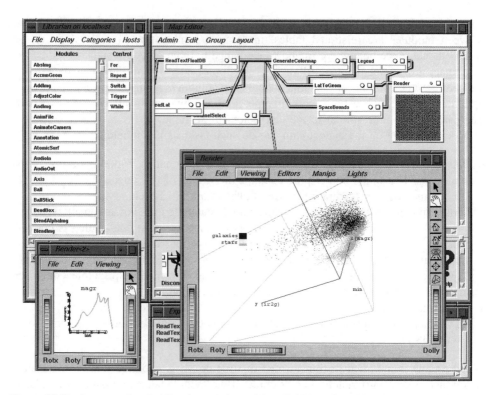

Figure 38.7 An example visualization session with a Grid-based multiparameter analysis suite. The image shows a 3D scatter plot of magnitude vs two measures of fuzziness: two populations are visible: stars and galaxies. Such a toolbox can be used to build an effective discriminator, or to find clusters or outliers.

astrophysics', or the American Physical Society PACS system lists 98.20.-d as 'Stellar clusters and associations'. These are ways of classifying books, articles, and other library materials, and were derived from analyzing a large number of such instances.

The ontology for the Virtual Observatory, by contrast, is about attributes of *data*, not books. Suppose a table has been published that contains stars, with position in the sky and brightness. Suppose further that there is a program that can draw charts of star positions as dots on the paper, and the brightness shown by the size of the dot (Figure 38.8). Explained in English, it is clear that the program can consume the table. How can we represent this knowledge formally so that the computer can also reach this conclusion?

The solution is some kind of standard ontology, and the difficult part is to decide how the ontology is to be used, created, published, and extended.

38.5.1 The Strasbourg ontology

In the NVO, we have decided to begin our ontology work with one that was generated at the astronomical data center at Strasbourg by F. Ochsenbein and colleagues, called Unified

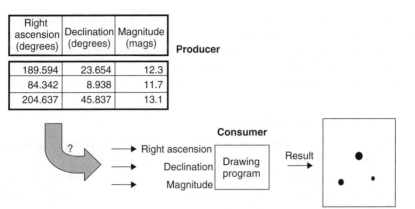

Figure 38.8 A human being can visualize that the data from the producer is usable by the consumer. How can the computer understand this?.

Content Descriptor (UCD) [24]. It was created by analyzing the attributes of each of the 5000 astronomical tables that are available at the Strasbourg center. Some examples from the UCD ontology are

- PHOT_INT-MAG_B   Integrated total blue magnitude
- ORBIT_ECCENTRICITY   Orbital eccentricity
- POS_EQ_RA_MAIN   Right ascension (main)
- INST_QE   Detector's quantum efficiency

In a UCD, the underscores represent hierarchy, so that POS_EQ (position – Equatorial coordinates) is a subclass of POS (position).

We can think of these in the same way that we think of physical units. In elementary science, students are told that they should not say 'the length is 42', but rather they should specify units: 'the length is 42 inches'. In a similar fashion, physical quantities should be augmented with the metadata expressing what sort of measurement the number represents.

The VOTable format, discussed earlier, supports labeling of parameters and table columns with UCD information. We have defined a new class of tables that have a certain type of positional information as those whose columns exhibit specific UCDs, and several data providers have provided services that return this type of table [25]. Referring back to Figure 38.8, a solution is to add UCD information to the table: the right ascension column has a UCD that is POS_EQ_RA_MAIN and the declination column has a UCD that is POS_EQ_DEC_MAIN, and similarly for the magnitude column. The drawing application specifies its requirement that suitable input tables must have these UCDs attached to columns. In this way, the data and the application can be made to 'plug and play'.

Thus, we have a first step towards the idea of publishing data in a NVO-compliant fashion. VOTable can publish tables with positional information, making sure that the coordinates are identified with labels selected from the standard ontology. We are in the process of building higher-level services with these positional tables, for cross-matching of tables (fuzzy join), for source density mapping, and others.

We can image other types of tables that are distinguished by the occurrence of certain UCDs, and applications that are able to detect these in the table metadata, and thereby process the dataset. There could be names for these different types of tables, and directories of services that list the UCD content of the results. In this way, we could use the VOTable format and the Strasbourg Ontology to create an interoperable matrix of services.

38.5.2 I want my own ontology

The Strasbourg Ontology is an excellent way to describe the meaning of data that comes from the Strasbourg holdings. But it would be naive to expect a fixed collection of descriptors to be capable of representing an expanding science. Suppose a new concept or a new type of instrument appears with its own vocabulary; how can the resulting data be described within the existing infrastructure?

The answer is to recognize that an ontology is another namespace, and the descriptors in the ontology are names in the namespace. A namespace is a collection of unique descriptors together with a syntax for expressing them. Consider the following sentence:

'We took the table and chair dimensions, and wrote them in a table'.

As a human, you probably realize the different meanings of the word 'table', but it might be a lot clearer to a machine if ambiguous terms were prefixed by their namespace:

'We took the (furniture) table and chair dimensions, and wrote them in a (word-processing) table'.

This kind of mechanism enables us to mix ontologies. In the networked world, a suitable way to ensure that namespaces are unique is to use Universal Resource Identifiers (URIs). The scheme relies on the uniqueness of Internet domain names. Thus, we could restate one of the names from the Strasbourg Ontology in a more general way:

- Namespace called CDS is shorthand for the URI which is http://astro.u-strasbg.fr/
- The UCD is then combined with the namespace as CDS:PHOT_INT-MAG_B

Thus anyone can define their own ontology, and produce data objects that are described in terms of such an ontology (Figure 38.9).

We have introduced flexibility and extensibility into the semantic framework, but unfortunately we have removed most of the interoperability! If everyone creates his or her own ontology, then nobody can understand anyone else's language!

38.5.3 Ontology mapping

Mapping one ontology to another can restore the interoperability. Referring back to the figure with the data producer and data consumer, let us suppose that the data requirements for the consuming application have been written with a different ontology, and let us suppose there is one called DRAW with descriptors X and Y for position and SIZE to indicate symbol size. In order to make these resources interoperate, we need a mapping between the two ontologies. Such a mapping may simply say that a name in one ontology

Right ascension CDS:POS_EQ_RA_MAIN	Declination CDS:POS_EQ_DEC_MAIN	Magnitude CDS:PHOT_MAG	Producer
189.594	23.654	12.3	
84.342	8.938	11.7	
204.637	45.837	13.1	

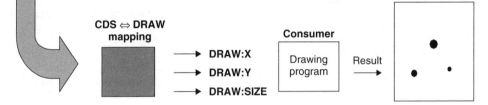

Figure 38.9 An ontology mapping.

is identical to a name in another, or it may say that one is an approximation of another, or that one name is an instance of the other. There may be a mathematical relationship – a sky projection in this case, or the conversion of the astronomical magnitude scale to a symbol size on the paper.

Representation of relationships between semantic entities is a subject of research in the computer science and digital library communities, through projects and protocols such as Resource Description Framework (RDF) [26], or Topic Maps [27].

38.6 CONCLUSION

The approaching tsunami of data and servers in the astronomical sciences will change the way astronomers discover knowledge about the Universe. It is not just the huge quantity of data that will forge the revolution, but also the large number of data collections that were built by different people, but now must operate seamlessly. Work on the new methods is well under way in the US NVO, as well as European, British, Canadian, Indian, and Australian VO projects. Close collaboration with the emerging standards of the Grid community will serve astronomers well in the emerging data-rich environment.

The author is grateful to the US National Science Foundation for funding under the NVO (Cooperative Agreement No. AST-0122449).

REFERENCES

1. Framework for the National Virtual Observatory, National Science Foundation, http://us-vo.org/.
2. Proceedings of the European-United States joint workshop on Large Scientific Databases, http://www.cacr.caltech.edu/euus.

3. Astronomy and Astrophysics in the New Millennium (Decadal Survey), National Academy of Science, Astronomy and Astrophysics Survey Committee, http://www.nap.edu/books/0309070317/html/.
4. Grid Security Infrastructure (GSI), http://www.globus.org/security/
5. Storage Resource Broker, San Diego Supercomputer Center, http://www.sdsc.edu/DICE/SRB/.
6. The GridFTP Protocol and Software, http://www.globus.org/datagrid/gridftp.html
7. Virtual Sky: Multi-Resolution, Multi-Wavelength Astronomical Images, http://VirtualSky.org.
8. Digital Palomar Observatory Sky Survey, http://www.astro.caltech.edu/~george/dposs/.
9. The Hubble Deep Field, http://www.stsci.edu/ftp/science/hdf/hdf.html.
10. The Sloan Digital Sky Survey, http://www.sdss.org.
11. Microsoft Terraserver, the original at http://terraserver.microsoft.com, and the commercial version at http://www.terraserver.com.
12. NASA Extragalactic Database (NED), http://ned.ipac.caltech.edu.
13. SIMBAD Astronomical Database, http://simbad.u-strasbg.fr.
14. Two-Micron All-Sky Survey, http://www.ipac.caltech.edu/2mass.
15. Teragrid, A Supercomputing Grid Comprising Argonne National Laboratory, California Institute of Technology, National Center for Supercomputing Applications, and San Diego Supercomputing Center, http://www.teragrid.org.
16. Montage, An Astronomical Image Mosaic Service for the National Virtual Observatory, http://montage.ipac.caltech.edu/.
17. GriPhyN, Grid Physics Network, http://www.griphyn.org.
18. QUEST: Synoptic Astronomical Imaging, http://hepwww.physics.yale.edu/www_info/astro/quest.html.
19. NEAT: Near Earth Asteroid Tracking, http://neat.jpl.nasa.gov/
20. Abraham, R. G. and van den Bergh, S. (2001) The morphological evolution of galaxies. *Science*, **293**, 1273–1278, http://xxx.lanl.gov/abs/astro-ph/0109358.
21. VOTable: A Proposed XML Format for Astronomical Tables, http://us-vo.org/VOTable and http://cdsweb.u-strasbg.fr/doc/VOTable/.
22. FITS: Flexible Image Transport Specification, Specifically the Binary Tables Extension, http://fits.gsfc.nasa.gov/.
23. XDF: Extensible Data format, http://xml.gsfc.nasa.gov/XDF/XDF_home.html.
24. Ochsenbein, F. *et al.* Unified Content Descriptors, CDS Strasbourg, http://vizier.u-strasbg.fr/doc/UCD.htx.
25. Simple Cone Search: A First Guide for Data Curators NVO Publishing, http://www.us-vo.org/metadata/conesearch/.
26. Resource Description Framework, see for example Bray, T., What is RDF? http://www.xml.com/lpt/a/2001/01/24/rdf.html.
27. Topic Maps, see for example Pepper, S., The TAO of Topic Maps, http://www.ontopia.net/topicmaps/materials/tao.html.

Data-intensive Grids for high-energy physics

Julian J. Bunn and Harvey B. Newman

California Institute of Technology, Pasadena, California, United States

39.1 INTRODUCTION: SCIENTIFIC EXPLORATION AT THE HIGH-ENERGY FRONTIER

The major high-energy physics (HEP) experiments of the next twenty years will break new ground in our understanding of the fundamental interactions, structures and symmetries that govern the nature of matter and space-time. Among the principal goals are to find the mechanism responsible for mass in the universe, and the 'Higgs' particles associated with mass generation, as well as the fundamental mechanism that led to the predominance of matter over antimatter in the observable cosmos.

The largest collaborations today, such as CMS [1] and ATLAS [2] who are building experiments for CERN's Large Hadron Collider (LHC) program [3], each encompass 2000 physicists from 150 institutions in more than 30 countries. Each of these collaborations include 300 to 400 physicists in the US, from more than 30 universities, as well as the major US HEP laboratories. The current generation of operational experiments at SLAC (BaBar [4]), Fermilab (D0 [5] and CDF [6]), as well as the experiments at the Relativistic

Grid Computing – Making the Global Infrastructure a Reality. Edited by F. Berman, A. Hey and G. Fox
© 2003 John Wiley & Sons, Ltd ISBN: 0-470-85319-0

Heavy Ion Collider (RHIC) program at Brookhaven National Laboratory (BNL) [7], face similar challenges. BaBar in particular has already accumulated datasets approaching a petabyte (1PB $= 10^{15}$ Bytes).

Collaborations on this global scale would not have been attempted if the physicists could not plan on excellent networks: to interconnect the physics groups throughout the life cycle of the experiment, and to make possible the construction of Data Grids capable of providing access, processing and analysis of massive datasets. These datasets will increase in size from petabytes to exabytes (1EB $= 10^{18}$ Bytes) within the next decade.

An impression of the complexity of the LHC data can be gained from Figure 39.1, which shows simulated particle trajectories in the inner 'tracking' detectors of CMS. The particles are produced in proton–proton collisions that result from the crossing of two proton bunches. A rare proton–proton interaction (approximately 1 in 10^{13}) resulting in the production of a Higgs particle that decays into the distinctive signature of four muons, is buried in 30 other 'background' interactions produced in the same crossing, as shown in the upper half of the figure. The CMS software has to filter out the background interactions by isolating the point of origin of the high momentum tracks in the interaction containing the Higgs. This filtering produces the clean configuration shown in the bottom half of the figure. At this point, the (invariant) mass of the Higgs can be measured from the shapes of the four muon tracks (labelled), which are its decay products.

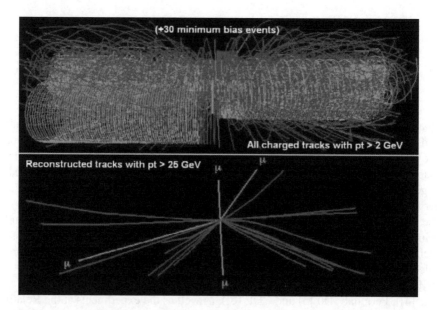

Figure 39.1 A simulated decay of the Higgs Boson into four muons. (a) The high momentum charged particles in the Higgs event and (b) how the event would actually appear in the detector, submerged beneath many other 'background' interactions.

39.2 HEP CHALLENGES: AT THE FRONTIERS OF INFORMATION TECHNOLOGY

Realizing the scientific wealth of these experiments presents new problems in data access, processing and distribution, and collaboration across national and international networks, on a scale unprecedented in the history of science. The information technology challenges include the following:

- Providing rapid access to data subsets drawn from massive data stores, rising from petabytes in 2002 to ~100 petabytes by 2007, and exabytes (10^{18} bytes) by approximately 2012 to 2015.
- Providing secure, efficient, and transparent managed access to heterogeneous worldwide-distributed computing and data-handling resources, across an ensemble of networks of varying capability and reliability.
- Tracking the state and usage patterns of computing and data resources in order to make possible rapid turnaround as well as efficient utilization of global resources.
- Matching resource usage to policies set by the management of the experiments' collaborations over the long term; ensuring that the application of the decisions made to support resource usage among multiple collaborations that share common (network and other) resources are internally consistent.
- Providing the collaborative infrastructure that will make it possible for physicists in all world regions to contribute effectively to the analysis and the physics results, particularly while they are at their home institutions.
- Building regional, national, continental, and transoceanic networks, with bandwidths rising from the gigabit per second to the terabit per second range over the next decade.[1]

All these challenges need to be met so as to provide the first integrated, managed, distributed system infrastructure that can serve 'virtual organizations' on the global scale.

39.3 MEETING THE CHALLENGES: DATA GRIDS AS MANAGED DISTRIBUTED SYSTEMS FOR GLOBAL VIRTUAL ORGANIZATIONS

The LHC experiments have thus adopted the 'Data Grid Hierarchy' model (developed by the MONARC[2] project) shown schematically in Figure 39.2. This five-tiered model shows data at the experiment being stored at the rate of 100 to 1500 MB s^{-1} throughout the year, resulting in many petabytes per year of stored and processed binary data, which are accessed and processed repeatedly by the worldwide collaborations searching for new physics processes. Following initial processing and storage at the 'Tier0' facility at the CERN laboratory site, the processed data is distributed over high-speed networks

---

[1] Continuing the trend of the last decade, where the affordable bandwidth increased by a factor of order 1000.

[2] http://www.cern.ch/MONARC. This project is described further in Section 39.4.3.

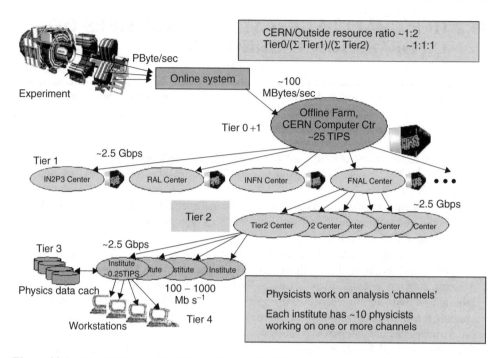

Figure 39.2 The LHC data Grid hierarchy model. This was first proposed by the MONARC collaboration in 1999.

to ~10 to 20 national 'Tier1' centers in the United States, leading European countries, Japan, and elsewhere.[3] The data is there further processed and analyzed and then stored at approximately 60 'Tier2' regional centers, each serving a small to medium-sized country, or one region of a larger country (as in the US, UK and Italy). Data subsets are accessed and further analyzed by physics groups using one of hundreds of 'Tier3' workgroup servers and/or thousands of 'Tier4' desktops.[4]

The successful use of this global ensemble of systems to meet the experiments' scientific goals depends on the development of Data Grids capable of managing and marshaling the 'Tier-N' resources, and supporting collaborative software development by groups of varying sizes spread around the globe. The modes of usage and prioritization of tasks need to ensure that the physicists' requests for data and processed results are handled within a reasonable turnaround time, while at the same time the collaborations' resources are used efficiently.

The GriPhyN [8], Particle Physics Data Grid (PPDG) [9], iVDGL [10], EU Data-Grid [11], DataTAG [12], the LHC Computing Grid (LCG) [13] and national Grid projects in Europe and Asia are working together, in multiyear R&D programs, to develop the

---

[3] At the time of this writing, a major Tier1 center in Rio de Janeiro Brazil is being planned.
[4] Tier4 also includes laptops, and the large number of handheld devices with broadband connections that are expected to come into use before LHC start-up in 2007.

necessary Grid systems. The DataTAG project is also working to address some of the network R&D issues and to establish a transatlantic test bed to help ensure that the US and European Grid systems interoperate smoothly.

The data rates and network bandwidths shown in Figure 39.2 are per LHC Experiment for the first year of LHC operation. The numbers shown correspond to a conservative 'baseline', formulated using a 1999–2000 evolutionary view of the advance of network technologies over the next five years [14]. The reason for this is that the underlying 'Computing Model' used for the LHC program assumes a very well-ordered, group-oriented and carefully scheduled approach to data transfers supporting the production processing and analysis of data samples. More general models supporting more extensive access to data samples on demand [15] would clearly lead to substantially larger bandwidth requirements.

39.4 EMERGENCE OF HEP GRIDS: REGIONAL CENTERS AND GLOBAL DATABASES

It was widely recognized from the outset of planning for the LHC experiments that the computing systems required to collect, analyze and store the physics data would need to be distributed and global in scope. In the mid-1990s, when planning for the LHC computing systems began, calculations of the expected data rates, the accumulated yearly volumes and the required processing power led many to believe that HEP would need a system whose features would not have looked out of place in a science fiction novel. However, careful extrapolations of technology trend lines, and detailed studies of the computing industry and its expected development [16] encouraged the experiments that a suitable system could be designed and built in time for the first operation of the LHC collider in 2005.[5] In particular, the studies showed that utilizing computing resources external to CERN, at the collaborating institutes (as had been done on a limited scale for the Large Electron Positron Collider (LEP) experiments) would continue to be an essential strategy, and that a global computing system architecture would need to be developed. (It is worthwhile noting that, at that time, the Grid was at an embryonic stage of development, and certainly not a concept the Experiments were aware of.) Accordingly, work began in each of the LHC experiments on formulating plans and models for how the computing could be done. The CMS experiment's 'Computing Technical Proposal', written in 1996, is a good example of the thinking that prevailed at that time. Because the computing challenges were considered so severe, several projects were instigated by the Experiments to explore various aspects of the field. These projects included RD45 [17], GIOD, MONARC and ALDAP, as discussed in the following sections.

39.4.1 The CMS computing model circa 1996

The CMS computing model as documented in the CMS 'Computing Technical Proposal' was designed to present the user with a simple logical view of all objects needed to

---

[5] At that time, the LHC planning specified a start of machine operations in 2005. The machine is now expected to come on-line in 2007.

perform physics analysis or detector studies. The word 'objects' was used loosely to refer to data items in files (in a traditional context) and to transient or persistent objects (in an object-oriented (OO) programming context). The proposal explicitly noted that, often, choices of particular technologies had been avoided since they depended too much on guesswork as to what would make sense or be available in 2005. On the other hand, the model explicitly assumed the use of OO analysis, design and programming. With these restrictions, the model's fundamental requirements were simply summed up as,

1. objects that cannot be recomputed must be stored somewhere;
2. physicists at any CMS institute should be able to query any objects (recomputable or not recomputable) and retrieve those results of the query which can be interpreted by a human;
3. the resources devoted to achieving 1 and 2 should be used as efficiently as possible.

Probably the most interesting aspect of the model was its treatment of how to make the CMS physics data (objects) persistent. The Proposal states '*at least one currently available ODBMS appears quite capable of handling the data volumes of typical current experiments and requires no technological breakthroughs to scale to the data volumes expected during CMS operation. Read performance and efficiency of the use of storage are very similar to Fortran/Zebra systems in use today. Large databases can be created as a federation of moderate (few GB) sized databases, many of which may be on tape to be recalled automatically in the event of an 'access fault'. The current product supports a geographically distributed federation of databases and heterogeneous computing platforms. Automatic replication of key parts of the database at several sites is already available and features for computing (computable) objects on demand are recognized as strategic developments.*'

It is thus evident that the concept of a globally distributed computing and data-serving system for CMS was already firmly on the table in 1996. The proponents of the model had already begun to address the questions of computing 'on demand', replication of data in the global system, and the implications of distributing computation on behalf of end-user physicists.

Some years later, CMS undertook a major requirements and consensus-building effort to modernize this vision of a distributed computing model to a Grid-based computing infrastructure. Accordingly, the current vision sees CMS computing as an activity that is performed on the 'CMS Data Grid System' whose properties have been described in considerable detail [18]. The CMS Data Grid System specifies a division of labor between the Grid projects (described in this chapter) and the CMS core computing project. Indeed, the CMS Data Grid System is recognized as being one of the most detailed and complete visions of the use of Grid technology among the LHC experiments.

39.4.2 GIOD

In late 1996, Caltech's HEP department, its Center for Advanced Computing Research (CACR), CERN's Information Technology Division, and Hewlett Packard Corporation initiated a joint project called 'Globally Interconnected Object Databases'. The GIOD

Project [19] was designed to address the key issues of wide-area network-distributed data access and analysis for the LHC experiments. It was spurred by the advent of network-distributed Object database management systems (ODBMSs), whose architecture held the promise of being scalable up to the multipetabyte range required by the LHC experiments. GIOD was set up to leverage the availability of a large (200 000 MIP) HP Exemplar supercomputer, and other computing and data-handling systems at CACR as of mid-1997. It addressed the fundamental need in the HEP community at that time to prototype object-oriented software, databases and mass storage systems, which were at the heart of the LHC and other (e.g. BaBar) major experiments' data analysis plans. The project plan specified the use of high-speed networks, including ESnet, and the transatlantic link managed by the Caltech HEP group, as well as next-generation networks (CalREN2 in California and Internet2 nationwide) which subsequently came into operation with speeds approaching those to be used by HEP in the LHC era.

The GIOD plan (formulated by Bunn and Newman in late 1996) was to develop an understanding of the characteristics, limitations, and strategies for efficient data access using the new technologies. A central element was the development of a prototype 'Regional Center'. This reflected the fact that both the CMS and ATLAS Computing Technical Proposals foresaw the use of a handful of such centers, in addition to the main center at CERN, with distributed database 'federations' linked across national and international networks. Particular attention was to be paid to how the system software would manage the caching, clustering and movement of collections of physics objects between storage media and across networks. In order to ensure that the project would immediately benefit the physics goals of CMS and US CMS while carrying out its technical R&D, it also called for the use of the CACR computing and data storage systems to produce terabyte samples of fully simulated LHC signal and background events that were to be stored in the Object database.

The GIOD project produced prototype database, reconstruction, analysis and visualization systems. This allowed the testing, validation and development of strategies and mechanisms that showed how the implementation of massive distributed systems for data access and analysis in support of the LHC physics program would be possible. Deployment and tests of the terabyte-scale GIOD database were made at a few US universities and laboratories participating in the LHC program. In addition to providing a source of simulated events for evaluation of the design and discovery potential of the CMS experiment, the database system was used to explore and develop effective strategies for distributed data access and analysis at the LHC. These tests used local, regional, national and international backbones, and made initial explorations of how the distributed system worked, and which strategies were most effective. The GIOD Project terminated in 2000, its findings documented [19], and was followed by several related projects described below.

39.4.3 MONARC

The MONARC[6] project was set up in 1998 to model and study the worldwide-distributed Computing Models for the LHC experiments. This project studied and attempted to optimize the site architectures and distribution of jobs across a number of regional computing

---

[6] Models of Networked Analysis at Regional Centers. http://www.cern.ch/MONARC

centers of different sizes and capacities, in particular, larger Tier-1 centers, providing a full range of services, and smaller Tier-2 centers. The architecture developed by MONARC is described in the final report [20] of the project.

MONARC provided key information on the design and operation of the Computing Models for the experiments, who had envisaged systems involving many hundreds of physicists engaged in analysis at laboratories and universities around the world. The models encompassed a complex set of wide-area, regional and local-area networks, a heterogeneous set of compute- and data-servers, and an undetermined set of priorities for group-oriented and individuals' demands for remote data and compute resources. Distributed systems of the size and complexity envisaged did not yet exist, although systems of a similar size were predicted by MONARC to come into operation and be increasingly prevalent by around 2005.

The project met its major milestones, and fulfilled its basic goals, including

- identifying first-round baseline computing models that could provide viable (and cost-effective) solutions to meet the basic simulation, reconstruction, and analysis needs of the LHC experiments,
- providing a powerful (CPU and time efficient) simulation toolset [21] that enabled further studies and optimization of the models,
- providing guidelines for the configuration and services of Regional Centers,
- providing an effective forum in which representatives of actual and candidate Regional Centers may meet and develop common strategies for LHC computing.

In particular, the MONARC work led to the concept of a Regional Center hierarchy, as shown in Figure 39.2, as the best candidate for a cost-effective and efficient means of facilitating access to the data and processing resources. The hierarchical layout was also believed to be well adapted to meet local needs for support in developing and running the software, and carrying out the data analysis with an emphasis on the responsibilities and physics interests of the groups in each world region. In the later phases of the MONARC project, it was realized that computational Grids, extended to the data-intensive tasks and worldwide scale appropriate to the LHC, could be used and extended (as discussed in Section 39.9) to develop the workflow and resource management tools needed to effectively manage a worldwide-distributed 'Data Grid' system for HEP.

39.4.4 ALDAP

The NSF funded three-year ALDAP project (which terminated in 2002) concentrated on the data organization and architecture issues for efficient data processing and access for major experiments in HEP and astrophysics. ALDAP was a collaboration between Caltech, and the Sloan Digital Sky Survey (SDSS[7]) teams at Johns Hopkins University and Fermilab. The goal was to find fast space- and time-efficient structures for storing

---

[7] The Sloan Digital Sky Survey (SDSS) will digitally map about half of the northern sky in five filter bands from UV to the near IR. SDSS is one of the first large physics experiments to design an archival system to simplify the process of 'data mining' and shield researchers from the need to interact directly with any underlying complex architecture.

large scientific data sets. The structures needed to efficiently use memory, disk, tape, local, and wide area networks, being economical on storage capacity and network bandwidth.

The SDSS is digitally mapping about half the northern sky in five filter bands from UV to the near IR. SDSS is one of the first large physics experiments to design an archival system to simplify the process of 'data mining' and shield researchers from the need to interact directly with any underlying complex architecture.

The need to access these data in a variety of ways requires it to be organized in a hierarchy and analyzed in multiple dimensions, tuned to the details of a given discipline. But the general principles are applicable to all fields. To optimize for speed and flexibility there needs to be a compromise between fully ordered (sequential) organization, and totally 'anarchic', random arrangements. To quickly access information from each of many 'pages' of data, the pages must be arranged in a multidimensional mode in a neighborly fashion, with the information on each page stored judiciously in local clusters. These clusters themselves form a hierarchy of further clusters. These were the ideas that underpinned the ALDAP research work.

Most of the ALDAP project goals were achieved. Besides them, the collaboration yielded several other indirect benefits. It led to further large collaborations, most notably when the ALDAP groups teamed up in three major successful ITR projects: GriPhyN, iVDGL and NVO. In addition, one of the ALDAP tasks undertaken won a prize in the Microsoft-sponsored student Web Services contest. The 'SkyServer'[22], built in collaboration with Microsoft as an experiment in presenting complex data to the wide public, continues to be highly successful, with over 4 million Web hits in its first 10 months.

39.5 HEP GRID PROJECTS

In this section we introduce the major HEP Grid projects. Each of them has a different emphasis: PPDG is investigating short-term infrastructure solutions to meet the mission-critical needs for both running particle physics experiments and those in active development (such as CMS and ATLAS). GriPhyN is concerned with longer-term R&D on Grid-based solutions for, collectively, Astronomy, Particle Physics and Gravity Wave Detectors. The international Virtual Data Grid Laboratory (iVDGL) will provide global test beds and computing resources for those experiments. The EU DataGrid has similar goals to GriPhyN and iVDGL, and is funded by the European Union. LCG is a CERN-based collaboration focusing on Grid infrastructure and applications for the LHC experiments. Finally, CrossGrid is another EU-funded initiative that extends Grid work to eleven countries not included in the EU DataGrid. There are several other smaller Grid projects for HEP, which we do not cover here because of space limitations.

39.5.1 PPDG

The Particle Physics Data Grid (www.ppdg.net) collaboration was formed in 1999 to address the need for Data Grid services to enable the worldwide-distributed computing model of current and future high-energy and nuclear physics (HENP) experiments. Initially funded from the Department of Energy's NGI program and later from the MICS[8] and HENP[9] programs, it has provided an opportunity for early development of the Data Grid architecture as well as for the evaluation of some prototype Grid middleware.

PPDG's second round of funding is termed the Particle Physics Data Grid Collaboratory Pilot. This phase is concerned with developing, acquiring and delivering vitally needed Grid-enabled tools to satisfy the data-intensive requirements of particle and nuclear physics. Novel mechanisms and policies are being vertically integrated with Grid middleware and experiment-specific applications and computing resources to form effective end-to-end capabilities. As indicated in Figure 39.3, PPDG is a collaboration of computer

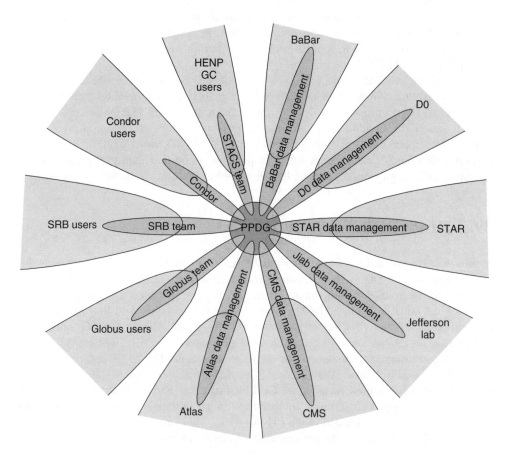

Figure 39.3 Showing the collaboration links between PPDG and the experiments and user communities.

---

[8] Mathematics, information, and computer sciences, through the SCIDAC (Scientific discovery through advanced computing) initiative.

[9] High energy and nuclear physics.

scientists with a strong record in distributed computing and Grid technology, and physicists with leading roles in the software and network infrastructures for major high-energy and nuclear experiments. A three-year program has been outlined for the project that takes full advantage of the strong driving force provided by currently operating physics experiments, ongoing Computer Science (CS) projects and recent advances in Grid technology. The PPDG goals and plans are ultimately guided by the immediate, medium-term and longer-term needs and perspectives of the physics experiments, and by the research and development agenda of the CS projects involved in PPDG and other Grid-oriented efforts.

39.5.2 GriPhyN

The GriPhyN (Grid Physics Network – http://www.griphyn.org) project is a collaboration of CS and other IT researchers and physicists from the ATLAS, CMS, Laser Interferometer Gravitational-wave Observatory (LIGO) and SDSSexperiments. The project is focused on the creation of Petascale Virtual Data Grids that meet the data-intensive computational needs of a diverse community of thousands of scientists spread across the globe. The concept of Virtual Data encompasses the definition and delivery to a large community of a (potentially unlimited) virtual space of data products derived from experimental data as shown in Figure 39.4. In this virtual data space, requests can be satisfied via direct access and/or computation, with local and global resource management, policy, and security constraints determining the strategy used. Overcoming this challenge and realizing the Virtual Data concept requires advances in three major areas:

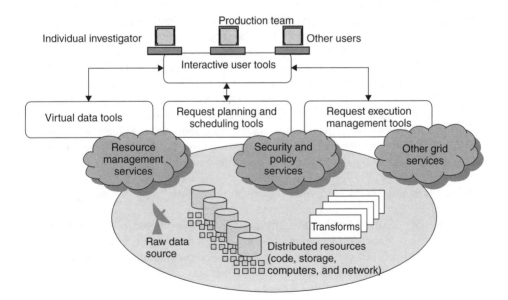

Figure 39.4 A production Grid, as envisaged by GriPhyN, showing the strong integration of data generation, storage, computing, and network facilities, together with tools for scheduling, management and security.

- *Virtual data technologies*: Advances are required in information models and in new methods of cataloging, characterizing, validating, and archiving software components to implement virtual data manipulations.
- *Policy-driven request planning and scheduling of networked data and computational resources*: Mechanisms are required for representing and enforcing both local and global policy constraints and new policy-aware resource discovery techniques.
- *Management of transactions and task execution across national-scale and worldwide virtual organizations*: New mechanisms are needed to meet user requirements for performance, reliability, and cost. Agent computing will be important to permit the Grid to balance user requirements and Grid throughput, with fault tolerance.

The GriPhyN project is primarily focused on achieving the fundamental IT advances required to create Petascale Virtual Data Grids, but is also working on creating software systems for community use, and applying the technology to enable distributed, collaborative analysis of data. A multifaceted, domain-independent Virtual Data Toolkit is being created and used to prototype the virtual Data Grids, and to support the CMS, ATLAS, LIGO and SDSS analysis tasks.

39.5.3 iVDGL

The international Virtual Data Grid Laboratory (iVDGL) (http://www.ivdgl.org) has been funded to provide a global computing resource for several leading international experiments in physics and astronomy. These experiments include the LIGO, the ATLAS and CMS experiments, the SDSS, and the National Virtual Observatory (NVO). For these projects, the powerful global computing resources available through the iVDGL should enable new classes of data-intensive algorithms that will lead to new scientific results. Other application groups affiliated with the NSF supercomputer centers and EU projects are also taking advantage of the iVDGL resources. Sites in Europe and the United States are, or soon will be, linked together by a multi-gigabit per second transatlantic link funded by a companion project in Europe. Management of iVDGL is integrated with that of the GriPhyN Project. Indeed, the GriPhyN and PPDG projects are providing the basic R&D and software toolkits needed for iVDGL. The European Union DataGrid (see the next section) is also a major participant and is contributing some basic technologies and tools. The iVDGL is based on the open Grid infrastructure provided by the Globus Toolkit and builds on other technologies such as the Condor resource management tools.

As part of the iVDGL project, a Grid Operations Center (GOC) has been created. Global services and centralized monitoring, management, and support functions are being coordinated by the GOC, which is located at Indiana University, with technical effort provided by GOC staff, iVDGL site staff, and the CS support teams. The GOC operates iVDGL just as a Network Operations Center (NOC) manages a network, providing a single, dedicated point of contact for iVDGL status, configuration, and management, and addressing overall robustness issues.

39.5.4 DataGrid

The European DataGrid (eu-datagrid.web.cern.ch) is a project funded by the European Union with the aim of setting up a computational and data-intensive Grid of resources for the analysis of data coming from scientific exploration. Next generation science will require coordinated resource sharing, collaborative processing and analysis of huge amounts of data produced and stored by many scientific laboratories belonging to several institutions.

The main goal of the DataGrid initiative is to develop and test the technological infrastructure that will enable the implementation of scientific 'collaboratories' where researchers and scientists will perform their activities regardless of geographical location. It will also allow interaction with colleagues from sites all over the world, as well as the sharing of data and instruments on a scale previously unattempted. The project is devising and developing scalable software solutions and test beds in order to handle many petabytes of distributed data, tens of thousand of computing resources (processors, disks, etc.), and thousands of simultaneous users from multiple research institutions.

The DataGrid initiative is led by CERN, together with five other main partners and fifteen associated partners. The project brings together the following European leading research agencies: the European Space Agency (ESA), France's Centre National de la Recherche Scientifique (CNRS), Italy's Istituto Nazionale di Fisica Nucleare (INFN), the Dutch National Institute for Nuclear Physics and High-Energy Physics (NIKHEF) and the UK's Particle Physics and Astronomy Research Council (PPARC). The fifteen associated partners come from the Czech Republic, Finland, France, Germany, Hungary, Italy, the Netherlands, Spain, Sweden and the United Kingdom.

DataGrid is an ambitious project. Its development benefits from many different kinds of technology and expertise. The project spans three years, from 2001 to 2003, with over 200 scientists and researchers involved.

The DataGrid project is divided into twelve Work Packages distributed over four Working Groups: Test bed and Infrastructure, Applications, Computational & DataGrid Middleware, Management and Dissemination. Figure 39.5 illustrates the structure of the project and the interactions between the work packages.

39.5.5 LCG

The job of CERN's LHC Computing Grid Project (LCG – http://lhcgrid.web.cern.ch) is to prepare the computing infrastructure for the simulation, processing and analysis of LHC data for all four of the LHC collaborations. This includes both the common infrastructure of libraries, tools and frameworks required to support the physics application software, and the development and deployment of the computing services needed to store and process the data, providing batch and interactive facilities for the worldwide community of physicists involved in the LHC (see Figure 39.6).

The first phase of the project, from 2002 through 2005, is concerned with the development of the application support environment and of common application elements, the development and prototyping of the computing services and the operation of a series of computing data challenges of increasing size and complexity to demonstrate the effectiveness of the software and computing models selected by the experiments. During this period, there will be two series of important but different types of data challenge under way: computing data challenges that test out the application, system software, hardware

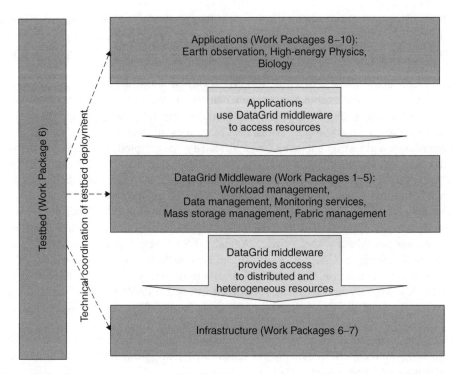

Figure 39.5   Showing the structure of the EU DataGrid project and its component work packages.

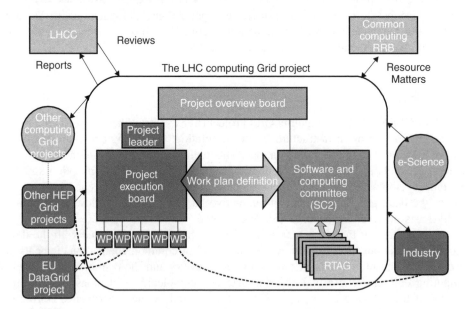

Figure 39.6   The organizational structure of the LHC computing Grid, showing links to external projects and industry.

and computing model, and physics data challenges aimed at generating data and analyzing it to study the behavior of the different elements of the detector and triggers. During this R&D phase, the priority of the project is to support the computing data challenges, and to identify and resolve problems that may be encountered when the first LHC data arrives. The physics data challenges require a stable computing environment, and this requirement may conflict with the needs of the computing tests, but it is an important goal of the project to arrive rapidly at the point where stability of the Grid prototype service is sufficiently good to absorb the resources that are available in Regional Centers and CERN for physics data challenges.

This first phase will conclude with the production of a Computing System Technical Design Report, providing a blueprint for the computing services that will be required when the LHC accelerator begins production. This will include capacity and performance requirements, technical guidelines, costing models, and a construction schedule taking account of the anticipated luminosity and efficiency profile of the accelerator.

A second phase of the project is envisaged, from 2006 through 2008, to oversee the construction and operation of the initial LHC computing system.

39.5.6 CrossGrid

CrossGrid (http://www.crossgrid.org) is a European project developing, implementing and exploiting new Grid components for interactive compute- and data-intensive applications such as simulation and visualization for surgical procedures, flooding crisis team

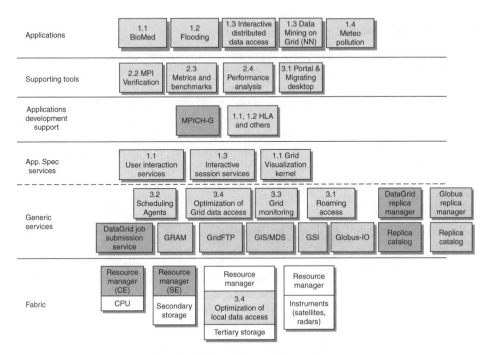

Figure 39.7 The CrossGrid architecture.

decision-support systems, distributed data analysis in high-energy physics, and air pollution combined with weather forecasting. The elaborated methodology, generic application architecture, programming environment, and new Grid services are being validated and tested on the CrossGrid test bed, with an emphasis on a user-friendly environment. Cross-Grid collaborates closely with the Global Grid Forum (GGF) and the DataGrid project in order to profit from their results and experience, and to ensure full interoperability (see Figure 39.7). The primary objective of CrossGrid is to further extend the Grid environment to a new category of applications of great practical importance. Eleven European countries are involved.

The essential novelty of the CrossGrid project consists in extending the Grid to a completely new and socially important category of applications. The characteristic feature of these applications is the presence of a person in a processing loop, with a requirement for real-time response from the computer system. The chosen interactive applications are both compute- and data-intensive.

39.6 EXAMPLE ARCHITECTURES AND APPLICATIONS

In this section we take a look at how HEP experiments are currently making use of the Grid, by introducing a few topical examples of Grid-based architectures and applications.

39.6.1 TeraGrid prototype

The beneficiary of NSF's Distributed Terascale Facility (DTF) solicitation was the TeraGrid project (www.teragrid.org), a collaboration between Caltech, SDSC, NCSA and Argonne. The project is supplementing the already powerful PACI[10] resources and services (e.g. the NPACI[11] leading-edge site, SDSC, has the world's largest High Performance Storage System (HPSS) managed archive, with over 200 TB of storage) with much greater emphasis on providing access to large scientific data collections. Massive compute power (13.6 TFLOPS), very large data caches (450 terabytes), expanded archival storage activities (HPSS), distributed visualization, ultra high-speed network connections among selected sites are features of the prototype.

TeraGrid is strongly supported by the physics community participating in the LHC, through the PPDG, GriPhyN and iVDGL projects, due to its massive computing capacity, leading-edge network facilities, and planned partnerships with distributed systems in Europe.

As part of the planning work for the TeraGrid proposal, a successful 'preview' of its potential use was made, in which a highly compute and data-intensive Grid task for the CMS experiment was distributed between facilities at Caltech, Wisconsin and NCSA (see Figure 39.8). The TeraGrid test runs were initiated at Caltech, by a simple script

---

[10] The US partnership for advanced computing infrastructure. See http://www.paci.org
[11] One of the two PACI programs, led by the San Diego Supercomputer Center.

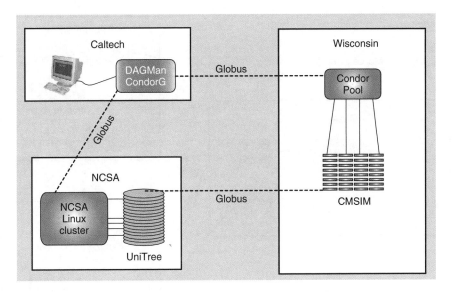

Figure 39.8 Showing the Grid-based production of Monte Carlo data for the CMS experiment. The setup, distributed between Caltech, Wisconsin, and NCSA, was an early demonstrator of the success of Grid infrastructure for HEP computing.

invocation. The necessary input files were automatically generated and, using Condor-G[12], a significant number of Monte Carlo simulation jobs were started on the Wisconsin Condor flock. Each of the jobs produced a data file that was then automatically transferred to a UniTree mass storage facility at NCSA. After all the jobs had finished at Wisconsin, a job at NCSA was automatically started to begin a further phase of processing. This being completed, the output was automatically transferred to UniTree and the run was completed.

39.6.2 MOP for Grid-enabled simulation production

The MOP[13] (short for 'CMS Monte Carlo Production') system was designed to provide the CMS experiment with a means for distributing large numbers of simulation tasks between many of the collaborating institutes. The MOP system as shown in Figure 39.9 comprises task description, task distribution and file collection software layers. The Grid Data Management Pilot (GDMP) system (a Grid-based file copy and replica management scheme using the Globus Toolkit) is an integral component of MOP, as is the Globus Replica Catalogue. Globus software is also used for task distribution. The task scheduler is the 'gridified' version of the Condor scheduler, Condor-G. In addition, MOP includes a set of powerful task control scripts developed at Fermilab.

---

[12] See http://www.cs.wisc.edu/condor/condorg/
[13] See http://grid.fnal.gov/test_beds/MOP_install.htm

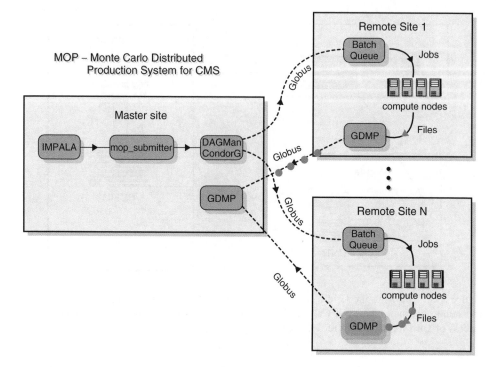

Figure 39.9 The MOP system, as demonstrated at SuperComputing 2001. In this schematic are shown the software components, and the locations at which they execute. Of particular note, is the use of the GDMP Grid tool.

The MOP development goal was to demonstrate that coordination of geographically distributed system resources for production was possible using Grid software. Along the way, the development and refinement of MOP aided the experiment in evaluating the suitability, advantages and shortcomings of various Grid tools. MOP developments to support future productions of simulated events at US institutions in CMS are currently underway.

39.6.3 GRAPPA

GRAPPA is an acronym for Grid Access Portal for Physics Applications. The preliminary goal of this project in the ATLAS experiment was to provide a simple point of access to Grid resources on the US ATLAS Testbed. GRAPPA is based on the use of a Grid-enabled portal for physics client applications. An initial portal prototype developed at the Extreme Computing Laboratory at Indiana University was the XCAT Science Portal (shown in Figure 39.10), which provided a script-based approach for building Grid Portals. This allowed users to build personal Grid Portals and was demonstrated with several applications. The ATLAS analysis and control framework, Athena, was used as the target application.

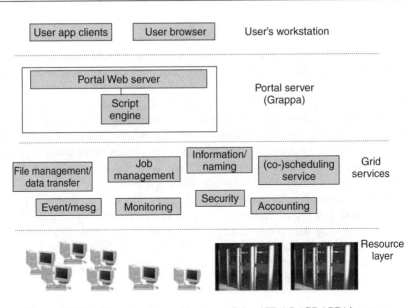

Figure 39.10 Showing the architecture of the ATLAS 'GRAPPA' system.

The GRAPPA[14] user authenticates to the portal using a GSI credential; a proxy credential is then stored so that the portal can perform actions on behalf of the user (such as authenticating jobs to a remote compute resource). The user can access any number of active notebooks within their notebook database. An active notebook encapsulates a session and consists of HTML pages describing the application, forms specifying the job's configuration, and Java Python scripts for controlling and managing the execution of the application. These scripts interface with Globus services in the GriPhyN Virtual Data Toolkit and have interfaces following the Common Component Architecture (CCA) Forum's specifications. This allows them to interact with and be used in high-performance computation and communications frameworks such as Athena.

Using the XCAT Science Portal tools, GRAPPA is able to use Globus credentials to perform remote task execution, store user's parameters for reuse or later modification, and run the ATLAS Monte Carlo simulation and reconstruction programs. Input file staging and collection of output files from remote sites is handled by GRAPPA. Produced files are registered in a replica catalog provided by the PPDGproduct MAGDA,[15] developed at BNL. Job monitoring features include summary reports obtained from requests to the Globus Resource Allocation Manager (GRAM[16]). Metadata from job sessions are captured to describe dataset attributes using the MAGDA catalog.

---

[14] See http://iuatlas.physics.indiana.edu/grappa/
[15] See http://atlassw1.phy.bnl.gov/magda/info.
[16] The Globus resource allocation manager. See http://www.globus.org/gram/

39.6.4 SAM

The D0 experiment's data and job management system software, sequential data access via metadata (SAM),[17] is an operational prototype of many of the concepts being developed for Grid computing (see Figure 39.11).

The D0 data-handling system, SAM, was built for the '*virtual organization*', D0, consisting of 500 physicists from 72 institutions in 18 countries. Its purpose is to provide a worldwide system of shareable computing and storage resources that can be brought to bear on the common problem of extracting physics results from about a petabyte of measured and simulated data. The goal of the system is to provide a large degree of transparency to the user who makes requests for datasets (collections) of relevant data and submits jobs that execute Monte Carlo simulation, reconstruction or analysis programs on available computing resources. Transparency in storage and delivery of data is currently in a more advanced state than transparency in the submission of jobs. Programs executed, in the context of SAM, transform data by consuming data file(s) and producing resultant data file(s) of different content, that is, in a different 'data tier'. Data files are read-only and are never modified, or versioned.

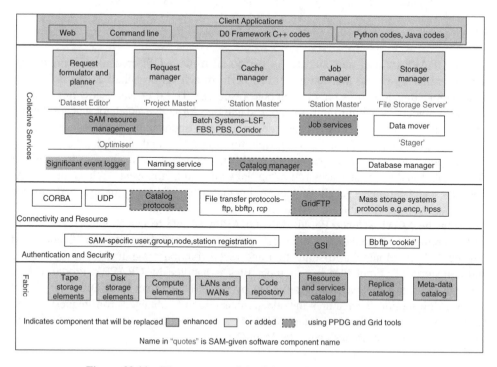

Figure 39.11   The structure of the D0 experiment's SAM system.

---

[17] See http://d0db.fnal.gov/sam/

The data-handling and job control services, typical of a data Grid, are provided by a collection of servers using CORBA communication. The software components are D0-specific prototypical implementations of some of those identified in Data Grid Architecture documents. Some of these components will be replaced by 'standard' Data Grid components emanating from the various Grid research projects, including PPDG. Others will be modified to conform to Grid protocols and APIs. Additional functional components and services will be integrated into the SAM system. (This work forms the D0/SAM component of the PPDG project.)

39.7 INTER-GRID COORDINATION

The widespread adoption by the HEP community of Grid technology is a measure of its applicability and suitability for the computing models adopted and/or planned by HEP experiments. With this adoption there arose a pressing need for some sort of coordination between all the parties concerned with developing Grid infrastructure and applications. Without coordination, there was a real danger that a Grid deployed in one country, or by one experiment, might not interoperate with its counterpart elsewhere. Hints of this danger were initially most visible in the area of conflicting authentication and security certificate granting methods and the emergence of several incompatible certificate granting authorities. To address and resolve such issues, to avoid future problems, and to proceed toward a mutual knowledge of the various Grid efforts underway in the HEP community, several inter-Grid coordination bodies have been created. These organizations are now fostering multidisciplinary and global collaboration on Grid research and development. A few of the coordinating organizations are described below.

39.7.1 HICB

The DataGrid, GriPhyN, iVDGL and PPDG, as well as the national European Grid projects in UK, Italy, Netherlands and France agreed to coordinate their efforts to design, develop and deploy a consistent open source standards-based global Grid infrastructure. The coordination body is HICB[18].

The consortia developing Grid systems for current and next generation High-Energy and Nuclear Physics (HENP) experiments, as well as applications in the earth sciences and biology, recognized that close collaboration and joint development is necessary in order to meet their mutual scientific and technical goals. A framework of joint technical development and coordinated management is therefore required to ensure that the systems developed will interoperate seamlessly to meet the needs of the experiments, and that no significant divergences preventing this interoperation will arise in their architecture or implementation.

To that effect, it was agreed that their common efforts would be organized in three major areas:

---

[18] See http://www.hicb.org

- An HENP Inter-Grid Coordination Board (HICB) for high-level coordination,
- A Joint Technical Board (JTB), and
- Common Projects and Task Forces to address needs in specific technical areas.

The HICB is thus concerned with ensuring compatibility and interoperability of Grid tools, interfaces and APIs, and organizing task forces, reviews and reporting on specific issues such as networking, architecture, security, and common projects.

39.7.2 GLUE

The Grid Laboratory Uniform Environment (GLUE[19]) collaboration is sponsored by the HICB, and focuses on interoperability between the US Physics Grid Projects (iVDGL, GriPhyN and PPDG) and the European physics Grid development projects (EDG, DataTAG etc.). The GLUE management and effort is provided by the iVDGL and DataTAG projects. The GLUE effort reports to and obtains guidance and oversight from the HICB and JTBs described in Section 39.7.1. The GLUE collaboration includes a range of subprojects to address various aspects of interoperability:

- Tasks to define, construct, test, and deliver interoperable middleware to and with the Grid projects.
- Tasks to help experiments with their intercontinental Grid deployment and operational issues; establishment of policies and procedures related to interoperability and so on.

Since the initial proposal for the GLUE project, the LCG Project Execution Board and SC2[20] have endorsed the effort as bringing benefit to the project goals of deploying and supporting global production Grids for the LHC experiments.

The GLUE project's work includes the following:

1. Definition, assembly and testing of core common software components of Grid middleware drawn from EU DataGrid, GriPhyN, PPDG, and others, designed to be part of the base middleware of the Grids that will be run by each project. GLUE will not necessarily assemble a complete system of middleware, but will choose components to work on that raise particular issues of interoperability. (Other projects may address some of these issues in parallel before the GLUE effort does work on them).
2. Ensuring that the EU DataGrid and GriPhyN/PPDG Grid infrastructure will be able to be configured as a single interoperable Grid for demonstrations and ultimately application use.
3. Experiments will be invited to join the collaboration to build and test their applications with the GLUE suite. GLUE will work with Grid projects to encourage experiments to build their Grids using the common Grid software components.

39.7.3 DataTAG

The main objective of the DataTAG (www.datatag.org) project is to create a large-scale intercontinental Grid test bed involving the EU DataGrid project, several national projects

---

[19] See http://www.hicb.org/glue/glue.htm.

[20] The software and computing steering committee of the LCG project. See http://sc2.web.cern.ch/sc2/

in Europe, and related Grid projects in the United States,. This will allow the exploration of advanced networking technologies and interoperability issues between different Grid domains.

DataTAG aims to enhance the EU programme of development of Grid-enabled technologies through research and development in the sectors relevant to interoperation of Grid domains on a global scale. In fact, a main goal is the implementation of an experimental network infrastructure for a truly high-speed interconnection between individual Grid domains in Europe and in the US, to be shared with a number of EU projects. However, the availability of a high-speed infrastructure is not sufficient, so DataTAG is proposing to explore some forefront research topics such as the design and implementation of advanced network services for guaranteed traffic delivery, transport protocol optimization, efficiency and reliability of network resource utilization, user-perceived application performance, middleware interoperability in multidomain scenarios, and so on.

The DataTAG project is thus creating a large-scale intercontinental Grid test bed that will link the Grid domains. This test bed is allowing the project to address and solve the problems encountered in the high-performance networking sector, and the interoperation of middleware services in the context of large-scale data-intensive applications.

39.7.4 Global Grid Forum

The Global Grid Forum (GGF – http://www.gridforum.org) is a group of individuals engaged in research, development, deployment, and support activities related to Grids in general (see Figure 39.12). The GGF is divided into working groups tasked with investigating a range of research topics related to distributed systems, best practices for the design and interoperation of distributed systems, and recommendations regarding the implementation of Grid software. Some GGF working groups have evolved to function as sets of related subgroups, each addressing a particular topic within the scope of the working group. Other GGF working groups have operated with a wider scope, surveying a broad range of related topics and focusing on long-term research issues. This situation

Current Group	Type
Accounting (ACCT)	Research
Advanced Collaborative Environments (ACE)	Research
Advanced Programming Models (APM)	Research
Applications & Testbeds (APPS)	Research
Grid Computing Environments (GCE)	Research
Grid Protocol Architecture (GPA)	Research
Grid User Services (GUS)	Research
Dictionary (DICT)	Working
Grid Performance (PERF)	Area
JINI	Working
Remote Data Access (DATA)	Area
Grid Information Services (GIS)	Area
Grid Security (SEC)	Area
Scheduling and Resource Management (SRM)	Area

Figure 39.12   Global Grid forum working groups, as defined in 2001.

has resulted in a different set of objectives, appropriate expectations, and operating styles across the various GGF working groups.

39.8 CURRENT ISSUES FOR HEP GRIDS

This section summarizes a number of critical issues and approaches that apply to the most data-intensive and/or extensive Grids, such as those being constructed and used by the major HEP experiments. While some of these factors appear to be special to HEP now (in 2002), it is considered likely that the development of petabyte-scale-managed Grids with high performance for data access, processing and delivery will have broad application within and beyond the bounds of scientific research in the next decade.

It should be noted that several of the Grid projects mentioned above, notably PPDG, iVDGL and DataTAG, are designed to address the issue of deploying and testing vertically integrated systems serving the major experiments. These projects are thus suitable testing grounds for developing the complex, managed Grid systems described in this section.

39.8.1 HEP Grids versus classical Grids

The nature of HEP Grids, involving processing and/or handling of complex terabyte-to-petabyte subsamples drawn from multipetabyte data stores, and many thousands of requests per day posed by individuals, small and large workgroups located around the world, raises a number of operational issues that do not appear in most of the Grid systems currently in operation or conceived.

While the ensemble of computational, data handling and network resources foreseen is large by present-day standards, it is going to be limited compared to the potential demands of the physics user community. Many large tasks will be difficult to service, as they will require the coscheduling of storage, computing and networking resources over hours and possibly days. This raises the prospect of task-redirection, checkpointing/resumption, and perhaps task reexecution on a substantial scale. The trade-off between high levels of utilization and turnaround time for individual tasks thus will have to be actively pursued and optimized with new algorithms adapted to increasingly complex situations, including an expanding set of failure modes if demands continue to outstrip the resources.[21]

Each physics collaboration, large as it is, has a well-defined management structure with lines of authority and responsibility.[22] Scheduling of resources and the relative priority among competing tasks becomes a matter of *policy* rather than *moment-to-moment technical capability* alone. The performance (efficiency of resource use; turnaround time) in completing the assigned range of tasks, and especially the weekly, monthly and annual partitioning of resource usage among tasks at different levels of priority must be tracked, and matched to the policy by *steering* the system as a whole. There will also be site-dependent policies on the use of resources at each facility, negotiated in advance between

---

[21] An initial approach to this optimization procedure is introduced in Section 39.8.5.

[22] While quite different in detail, this has many structural similarities to multinational corporations. Hence the conjecture that the solution to HEP's largest Grid-related problems will have broad applicability to industry and eventually commerce.

each site-facility and the Collaboration and Laboratory managements. These local and regional policies need to be taken into account in any of the instantaneous decisions taken as to where a task will run, and in setting its instantaneous priority.

So the net result is that the system's assignment of priorities and decisions will be both inherently *time-dependent* and *location-dependent.*

The relatively limited resources (compared to the potential demand) also leads to the potential for long queues, and to the need for *strategic* as well as *tactical* planning of resource allocations and task execution. The overall state of the complex system of site facilities and networks needs to be monitored in real time, tracked, and sometimes steered (to some degree). As some tasks or classes of tasks will take a long time to complete, long decision processes (hours to days) must be carried out. A strategic view of workload also has to be maintained, in which even the longest and lowest (initial) priority tasks are completed in a finite time.

For complex, constrained distributed systems of this kind, simulation and prototyping has a key role in the design, trial and development of effective management strategies, and for constructing and verifying the effectiveness and robustness of the Grid services themselves (see Section 39.8.5).

In contrast, most current Grid implementations and concepts have implicit assumptions of *resource-richness.* Transactions (request/handling/delivery of results) are assumed to be relatively short, the probability of success relatively high, and the failure modes and the remedial actions required relatively simple. This results in the 'classical' Grid (even if it involves some data) being a relatively simple system with little internal state, and simple scaling properties. The services to be built to successfully operate such a system are themselves relatively simple, since difficult (strategic) decisions in the scheduling and use of resources, and in the recovery from failure or the redirection of work away from a 'hot spot', rarely arise.

39.8.2 Grid system architecture: above (or within) the collective layer

The highest layer below the Applications layer specified in the current standard Grid architecture[23] is the Collective layer, that 'contains protocols and services (and APIs and SDKs) that are not associated with any one specific resource but rather are global in nature and capture interactions across collections of resources. ... Collective components ... can implement a wide variety of sharing behaviors without placing new requirements on the resources being shared'. Examples include workload management systems and collaboration frameworks, workload management systems, and so on.

Although the Collective layer includes some of the ideas required for effective operation with the experiments' application software, it is currently only defined at a conceptual level. Moreover, as discussed in Section 39.10.2, physicists deal with *object collections*, rather than with *flat files*, and the storage, extraction and delivery of these collections often involves a DBMS. It therefore falls to the experiments, at least for the short and medium term, to do much of the vertical integration, and to provide many of the 'End-to-end Global Managed Services' required to meet their needs. It is also important to note

---

[23] See Chapter 4, Anatomy of the Grid, Figure 2 and Section 4.4.

that the experiments' code bases already contain hundreds of thousands to millions of lines of code, and users' needs are supported by powerful frameworks [23] or 'problem solving environments' that assist the user in handling persistency, in loading libraries and setting loading and application parameters consistently, in launching jobs for software development and test, and so on.

Hence Grid services, to be effective, must be able to interface effectively to the existing frameworks, and to generalize their use for work across a heterogeneous ensemble of local, continental and transoceanic networks.

HEP Grid architecture should therefore include the following layers, above the Collective layer shown in Chapter 4:[24]

Physics Reconstruction, Simulation and Analysis Code Layer

Experiments' Software Framework Layer
Modular and Grid-aware: Architecture able to interact effectively with the
lower layers (above)

Grid Applications Layer
(Parameters and algorithms that govern system operations)
Policy and priority metrics
Workflow evaluation metrics
Task-site coupling proximity metrics

Global End-to-End System Services Layer
(Mechanisms and services that govern long-term system operation)
Monitoring and Tracking component performance
Workflow monitoring and evaluation mechanisms
Error recovery and redirection mechanisms
System self-monitoring, evaluation and optimization mechanisms

The Global End-to-End System Services Layer consists of services that monitor and track all the subsystem components over long periods, monitor and in some cases try to optimize or improve system performance, as well as resolve problems or inefficiencies caused by contention for scarce resources. This layer is 'self-aware' to the extent that it is continually checking how well the resource usage is matched to the policies, and attempting to steer the system by redirecting tasks and/or altering priorities as needed, while using adaptive learning methods (such as the Self-Organizing Neural Net described in Section 39.8.5) for optimization.

The Grid Applications Layer refers to the parameters, metrics and in some cases the algorithms used in the End-to-End System Services layer. This allows each experiment to express policies relating to resource usage and other desired system behaviors, and such aspects as how tightly coupled the processing tasks of a given subcommunity are to a given geographical region.

---

[24] Alternatively, the Grid layers given in this list could be considered as part of a 'thick' collective layer.

39.8.3 Grid system software design and development requirements

The issues raised in Section 39.8.1 lead to a number of general architectural characteristics that are highly desirable, if not required, for the services composing an open scalable Grid system of global extent, able to fulfill HEP's data-intensive needs.

The system has to be *dynamic*, with software components designed to cooperate across networks and to communicate state changes throughout the system, end-to-end, in a short time. It must be *modular and loosely coupled*, with *resilient* autonomous and/or semi-autonomous service components that will continue to operate, and take appropriate action (individually and cooperatively) in the case other components fail or are isolated because of network failures. It must be *adaptable and heuristic*, able to add new services and/or reconfigure itself without disrupting the overall operation of the system, to deal with a variety of both normal and abnormal situations (often more complex than point failures) that are not known *a priori* in the early and middle stages of system design and development. It must be *designed to intercommunicate*, using standard protocols and *de facto* mechanisms where possible, so that it can be easily integrated while supporting a variety of legacy systems (adopted at some of the main sites for historical reasons or specific functional reasons). It must support a *high degree of parallelism* for ongoing tasks, so that as the system scales the service components are not overwhelmed by service requests.

A prototype distributed services architecture with these characteristics is described in Section 39.9.

Because of the scale of HEP experiments, they are usually executed as managed projects with milestones and deliverables well specified at each stage. The developing Grid systems, therefore, also must serve and support the development of the reconstruction and simulation software, and as a consequence they must also support the vital studies of on-line filtering algorithms, detector performance and the expected physics discovery potential. These studies begin years in advance of the start-up of the accelerator and the experiment,[25] and continue up to and into the operations phase. As a consequence, the development philosophy must be to deploy *working vertically integrated systems,* that are (to an increasing degree) production-ready, with increasing functionality at each development cycle. This development methodology is distinct, and may be at odds with, a 'horizontal' mode of development (depending on the development schedule) that focuses on basic services in the lower layers of the architecture and works its way up.[26]

In order to mitigate these differences in development methods, some experiments (e.g. CMS) have adopted a procedure of 'sideways migration'. Home-grown tools (scripts and applications; sometimes whole working environments) that provide timely and fully functional support for 'productions' of simulated and reconstructed physics events, currently involving tens of terabytes produced at 10 to 20 institutions are progressively integrated with standard services as they become available and are production-tested. The drive towards standardization is sometimes spurred on by the manpower-intensiveness of the

---

[25] Six to eight years before start-up in the case of the LHC experiments.

[26] Specifically many of the major HEP experiments foresee using the Open Grid Services Architecture (OGSA) described in Chapter 5, as the services are deployed and made production ready. Given the current and upcoming milestones, the strategy employed is likely to consist of two steps: (1) deploying and testing a fully functional one-off integrated system developed by an HEP experiment or Grid project, and then (2) doing a progressive 'sideways migration' towards the OGSA.

home-grown tools and procedures. An example is the evolution from the MOP to a more integrated system employing a wide range of basic Grid services.

39.8.4 HEP Grids and networks

As summarized in the introduction to this chapter, HEP requires high-performance networks, with data volumes for large-scale transfers rising from the 100 GB to the 100 Tbyte range (drawn from 100 terabyte to 100 petabyte data stores) over the next decade. This corresponds to throughput requirements for data flows across national and international networks rising from the 100 Mbps range now to the Gbps range within the next 2 to 3 years, and the 10 Gbps range within 4 to 6 years. These bandwidth estimates correspond to static data flows lasting hours, of which only a few could be supported (presumably with high priority) over the 'baseline' networks currently foreseen.

These requirements make HEP a driver of network needs, and make it strongly dependent on the support and rapid advance of the network infrastructures in the US (Internet2 [24], ESnet [25] and the Regional Networks [26]), Europe (GEANT [27]), Japan (Super-SINET [28]), and across the Atlantic (StarLight [29]; the US-CERN Link Consortium and DataTAG [30]) and the Pacific (GTRN [31]). In some cases HEP has become a very active participant in the development and dissemination of information about state-of-the-art networks. Developments include bbcp ('BaBar Copy') [32] and bbftp ('BaBar ftp') [33], the Caltech-DataTAG 'Grid-TCP' project [34] and the Caltech 'Multi-Gbps TCP' project [35], as well as monitoring systems in the Internet End-to-end Performance Monitoring (IEPM) project at SLAC [36]. The development and deployment of standard working methods aimed at high performance is covered in the Internet2 HENP Working Group [37], the Internet2 End-to-end Initiative [38], and the ICFA Standing Committee on Inter-Regional Connectivity [39].

If one takes into account the time-dimension, and the fact that a reliable distributed system needs to have both task queues (including queues for network transfers) of limited length and a modest number of pending transactions at any one time, then the resulting bandwidth (and throughput) requirements are substantially higher than the baseline needs described above. We may assume, for example, that typical transactions are completed in 10 min or less, in order to avoid the inherently fragile state of the distributed system that would result if hundreds to thousands of requests were left pending for long periods, and to avoid the backlog resulting from tens and then hundreds of such 'data-intensive' requests per day. A 100 GB transaction completed in 10 min corresponds to an average throughput of 1.3 Gbps, while a 1 terabyte transaction in 10 min corresponds to 13 Gbps.[27]

In order to meet these needs in a cost-effective way, in cooperation with the major providers and academic and research networks, some of the HEP sites are pursuing plans

---

[27] It is interesting to speculate, for the long term, that a 100-Tbyte transaction completed in 10 min corresponds to 1.3 Tbps. If we consider this as a requirement 10 years from now, this will be well below the capacity of a fiber, but perhaps (roughly) equal to the I/O capability of the largest disk systems of the time (approximately 5 years after the start of LHC operations). It is not possible to guess whether the bandwidth, routing and switching equipment to do this will be affordable within the next 10 years.

to connect to a key point in their network infrastructures using 'dark fiber'.[28] A leading, nearly complete example is the State of Illinois I-WIRE Project[29] interconnecting StarLight, Argonne National Lab, the University of Chicago and several other university campuses in Illinois and Indiana. Plans are also being developed to put 'last mile fiber' in place between Caltech's Center for Advanced Computing Research (CACR) the carrier hotels in downtown Los Angeles by this Fall,[30] and to use 10 Gbps wavelengths on these fiber strands for HEP applications starting in the Spring of 2003. Fermilab is planning to link to Starlight in Chicago using dark fibers, and a similar plan is being investigated to link the IN2P3 Computing Center in Lyon (CCIN2P3; http://webcc.in2p3.fr) to CERN in Geneva.

Beyond the simple requirement of bandwidth, HEP needs networks that interoperate seamlessly across multiple world regions and administrative domains. Until now the Grid services are (implicitly) assumed to run across networks that are able to provide transparent high performance (as above) as well as secure access. The particle physics-related Grid projects PPDG, GriPhyN/iVDGL, EU DataGrid, DataTAG and others are taking steps towards these goals.[31]

But the complexity of the networks HEP uses means that a high degree of awareness of the network properties, loads, and scheduled data flows will be needed to allow the Grid services to function as planned, and to succeed in scheduling the work (consisting of hundreds to thousands of tasks in progress at any point in time) effectively.

Grid and network operations for HEP will therefore require an Operations Center (or an ensemble of centers) to gather and propagate information on the system status, problems and mitigating actions, assist in troubleshooting, and maintain a repository of guidelines and best practices for Grid use. One example is the iVDGL GOC now under development at Indiana University (http://igoc.iu.edu).

39.8.5 Strategic resource planning: the key role of modeling and simulation

HEP data analysis is and will remain resource-constrained, and so large production teams, small workgroups and individuals all will often need to make strategic decisions on where and how to carry out their work. The decisions will have to take into account their quotas and levels of priority for running at each site, the likely time-delays incurred in running at a given site. Grid users will need to be provided with information (to the degree they are willing to deal with) on the state of the various sites and networks, task queues with estimated times, data flows in progress and planned, problems and estimated time-to-repair if known. They will need to choose whether to run remotely, using centralized large-scale resources, or regionally or even locally on their group's servers or desktops, where they have more control and relatively greater rights to resource usage.

---

[28] 'Dark Fiber' refers to otherwise unused optical fiber that the customer purchases or leases for long periods, and provides for the optical transmission and multiplexing/switching equipment himself.

[29] See http://www.iwire.org.

[30] Shared by the TeraGrid (see http://www.teragrid.org), the Caltech campus and Caltech HEP.

[31] See, for example, the PPDG CS-9 Project Activity on Site Authentication, Authorization and Accounting (AAA) at http://www.ppdg.net, or the DataGrid WP7 Security Coordination Group (SCG).

The hope is that eventually many of these functions will be automated, using adaptive learning algorithms and intelligent software agents, to allow the physicists to concentrate on their own work rather than the internal workings of the Grid systems. But in (at least) the early stages, many of these decisions will have to be manual, and interactive.

Since the basic strategies and guidelines for Grid users (and for the operational decisions to be taken by some of the Grid services in a multiuser environment) have yet to be developed, it is clear that Modeling and Simulation (M&S) will have a key role in the successful development of Grids for H E P. M&S is generally considered an essential step in the design, development and deployment of complex distributed systems in a wide range of fields [40], from space missions to networks, from battlefields to agriculture and from the factory floor to microprocessor design. Yet such simulations, with an appropriately high degree of abstraction and focusing on the key component and distributed system behaviors (so that they can scale to very large and complex systems), have so far not been widely adopted in the HEP community or in the Grid projects.

One such simulation system was developed in the MONARC project [41] (http://monarc. web.cern.ch/MONARC/sim_tool/). This system was applied to regional center operations, to data replication strategies [42], and to the optimization of job scheduling among several Regional Center sites using a Self-Organizing Neural Network (SONN) [43], but it has yet to be applied directly to the problem of designing and testing a wide range of user- and service-strategies for HEP Grids. Such a series of studies, using this or a similar system, will be needed to (1) develop scalable Grid services of sufficient robustness, (2) formulate and then validate the architecture and design of effective Grid and decision-support services, as well as guidelines to be provided to users, and (3) determine the achievable level of automation in handling strategic scheduling, job placement, and resource co-scheduling decisions. Because the MONARC system is based on process-oriented discrete event simulation, it is well adapted to real-time operational support for running Grid systems. The system could be applied, for example, to receive monitoring information within a real operational Grid, and return evaluations of the estimated time to completion corresponding to different job placement and scheduling scenarios.

39.9 A DISTRIBUTED SERVER ARCHITECTURE FOR DYNAMIC HEP GRID SERVICES

A scalable agent-based Dynamic Distributed Server Architecture (DDSA), hosting loosely coupled dynamic services for HEP Grids has been developed at Caltech, that meets the general criteria outlined in Section 39.8.3. These systems are able to gather, disseminate and coordinate configuration, time-dependent state and other information across the Grid as a whole. As discussed in this section, this architecture, and the services implemented within it, provide an effective enabling technology for the construction of workflow- and other forms of global higher level end-to-end Grid system management and optimization services (along the lines described in Section 39.8.2).

A prototype distributed architecture based on JINI [44] has been developed,[32] with services written in Java. This has been applied to the development of a flexible real-time monitoring system for heterogeneous regional centers [45] (described in the next section), and to the optimization of the interconnections among the 'reflectors' making up Caltech's Virtual Room Videoconferencing System (VRVS [46]) for worldwide collaboration.

The prototype design is based on a set of 'Station Servers' (generic network server units) dynamically interconnected (peer-to-peer) to form a distributed framework for hosting different types of services. The use of JINI distributed system support allows each Station Server to easily keep a dynamic list of active Station Servers at any moment in time.

The prototype framework has been based on JINI because it allows cooperating services and applications to discover and to access each other seamlessly, to adapt to a dynamic environment, and to share code and configurations transparently. The system design avoids single points of failure, allows service replication and reactivation, and aims to offer reliable support for large-scale distributed applications in real conditions, in which individual (or multiple) components may fail.

39.9.1 The station server framework

The Station Server framework provides support for three types of distributed computing entities:

Dynamic services are hosted by the framework of networked Station Servers and made available to interested clients. The framework allows each service to locate and access information from anywhere in the entire system, and to interact with other services. The Station Server does the service management and facilitates interservice communication.
Mobile agents are dynamic autonomous services (with internal rules governing their behavior), which can move between Station Servers to perform one or more specified tasks. This transfer is done using a transaction management service, which provides a two phase commit and protects the integrity of the operation. Agents may interact synchronously or asynchronously using the Station Servers' support for 'roaming' and a messages mailbox.
'Smart' proxies are flexible services that are deployed to the interested clients and services and act differently according to the rule base encountered at the destination, which includes a set of local and remote parameters.

These types of components work together and interact by using remote event subscription/notification and synchronous and asynchronous message-based communication. Code mobility is also required to provide this functionality in a scalable and manageable way.

---

[32] The prototype systems described here also include services based on WSDL, SOAP and UDDI (http://www.w3.org). As the Open Grid Services Architecture described in Chapter 5 is developed, we will adopt and integrate OGSA services as they become available and production-ready. For a discussion of JINI and the need for a transition from a Java-based system to an open XML-based system, see for example http://www.fawcette.com/javapro/2002_08/magazine/columns/proshop/default_pf.asp

39.9.2 Key distributed system features of the JINI prototype

The purpose of the JINI architecture is to *federate* groups of software components, according to a reliable distributed object model, into an integrated, loosely coupled dynamic system supporting code mobility. The key service features and mechanisms of JINI that we use are:

Lookup discovery service: Services are found and resolved by a *lookup service*. The lookup service is the central bootstrapping mechanism for the system and provides the major point of contact between the system and its users

Leasing mechanism: Access to many of the services in the JINI system environment is *lease* based. Each lease is negotiated between the user of the service and the provider of the service as part of the service protocol, for a specified time period.

Remote events: The JINI architecture supports distributed *events*. An object may allow other objects to register interest in events (changes of state) in the object and receive a notification of the occurrence of such an event. This enables distributed event-based programs to be written with a variety of reliability and scalability guarantees.

Transactions manager: Reliable distributed object models require transaction support to aid in protecting the integrity of the resource layer. The specified transactions are inherited from the JINI programming model and focus on supporting large numbers of heterogeneous resources, rather than a single large resource (e.g. a database). This service provides a series of operations, either within a single service or spanning multiple services, that can be wrapped in one transaction.

The JavaSpaces service:[33] This service supports an ensemble of active programs, distributed over a set of physically dispersed machines. While each program is able to execute independently of the others, they all communicate with each other by releasing data (a tuple) into tuple spaces containing code as well as data. Programs read, write, and take tuples (entries) from tuple spaces that are of interest to them.[34]

The mailbox service: This service can be used to provide asynchronous communications (based on any type of messages) between distributed services.

39.9.3 Station server operation

The interconnections among the Station Servers, and the mechanisms for service registration and notification that keep the Server framework updated, are shown schematically in Figure 39.13 below. Each Station Server registers itself to be a provider of one or more dynamic services with a set of JINI lookup-servers. As a result it receives the necessary code and parameter data (the yellow dots in the figure), downloaded from a JavaSpace. At the same time the Station Server subscribes as a remote listener, to be notified (through remote events) of state changes in any of the other Station Servers. This allows each Server to keep a dynamically updated list of active Station Servers, through the use of a

---

[33] This communication mechanism was heavily influenced by the concept of a tuple space that was first described in 1982 in a programming language called Linda. See for example http://lindaspaces.com/products/linda_overview.html

[34] A similar implementation from IBM that can be also used in the JINI architecture is named Tspaces (http://www.alphaworks. ibm.com/tech/tspaces). An extended, high performance implementation that also supports replication is provided by gigaspaces. (http://www.gigaspaces.com)

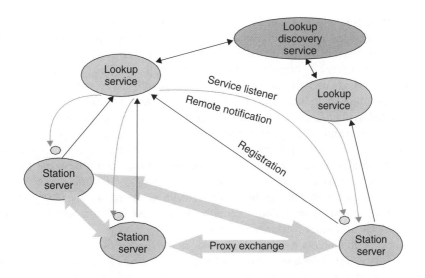

Figure 39.13　Showing the interconnections and mechanisms for registration and notification in the DDSA station server framework.

proxy for each of the other Servers. The JINI lease mechanism is used to inform each unit of changes that occur in other services (or to alert the Servers of other changes, as in the case of network problems).

The use of dynamic remote event subscription allows a service to register to be notified of certain event types, even if there is no provider to do the notification at registration time. The lookup discovery service will then notify the Station Servers when a new provider service, or a new service attribute, becomes available.

In large complex systems such as those foreseen for LHC, the services[35] will be organized and clustered according to a flexible, somewhat-hierarchical structure. Higher-level services that include optimization algorithms are used to provide decision support, or automated decisions, as discussed in the next sections. As the information provider-services are distributed, the algorithms used for decisions also should be distributed: for the sake of efficiency, and to be able to cope with a wide variety of abnormal conditions (e.g. when one or more network links are down).

39.9.4 Possible application to a scalable job scheduling service

As a simple example, we describe how the distributed Station Server framework may be used for job scheduling between Regional Centers.[36] Each Center starts a Station Server, and the Server network is created through the mechanisms described earlier. Each Station

---

[35] Several dynamic services on each station server, managed with a scheduling engine at each site.

[36] As discussed in this section, the example scheduling service shown here is targeted at very large scale heterogeneous systems distributed over many sites. It is meant to be complementary to other local and distributed scheduling systems (such as Condor), which have more sophisticated job queue management features, and to intercommunicate with those systems.

Server registers for, and downloads code and a parameter set for a 'Job Scheduling optimization' service. When a Regional Center (through its Station Server) considers exporting a job to another site, it first sends out a call to (all or a designated subset of) the other Station Servers for 'remote job estimation', requesting that the time to complete the job at each designated site be provided by the Station Server (specifically by a job execution estimation agent housed by the Server) at that site. The answers received within a set time window then are used to decide if and where to export the job for execution.

In order to determine the optimal site to execute the job, a 'thin proxy' is sent to each remote site. The Station Server there does an evaluation, using the characteristics of the job, the site configuration parameters, the present load and other state parameters of the local site. It may also use the historical 'trajectory' of resource usage and other state variables, along with external information (such as the resource usage policies and priorities). The evaluation procedure may also use adaptive learning algorithms, such as a SONN (SONN; introduced in Section 39.8.5).

Having the evaluation done in parallel by each of the Station Servers has a performance advantage. The Server at each site also may have direct access to local monitoring systems that keep track of the available resources and queues, as well as access to local policy rules and possibly to more complex local systems managing job scheduling, based on detailed knowledge of the characteristics and history of each job.

As the remote evaluations are completed, the results are returned to the original site that sent out the request. These results may be as simple as the time to complete the job, or as complex as a set of functions that give the 'cost to complete' expressed in terms of the priority for the job (and the implied level of resource usage). The originating site then makes the final decision based on the information received, as well as its own 'global' evaluation algorithms (such as another SONN).

Once the decision on the (remote) site for execution has been made, the description object for the job is transferred to the remote site using a Transaction Manager and a progress job handle is returned. This provides the mechanisms for the originating site to control the job execution at a remote site and to monitor its progress (through the notification mechanism and/or explicit requests for information). A schematic view of a prototypical Job Scheduling Service using the DDSA is illustrated in Figure 39.14.

39.9.5 An agent-based monitoring system using the DDSA

A prototype agent-based monitoring system MONALISA[37] has been built using the DDSA architecture, based on JINI as well as WSDL and SOAP technologies. The system has been deployed and in its initial implementation is currently monitoring the prototype Regional Centers at Caltech, CERN, Fermilab, Florida, and Bucharest.

The goal of MONALISA is to gather and disseminate real-time and historical monitoring information on the heterogeneous Regional Center site facilities and network links in an HEP Data Grid, and to deliver this information to a set of loosely coupled higher level services. As described in Sections 39.9.3 and 39.9.4, the higher level services may be

---

[37] MONitoring Agents in a large integrated services architecture. http://cil.cern.ch:8080/MONALISA

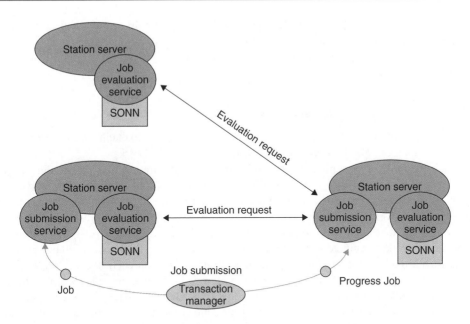

Figure 39.14 Illustration of a basic job scheduling system based on the DDSA architecture.

targeted at managing and optimizing workflow though the Grid, or providing decision support to users interactively. The monitoring system uses autonomous mobile agents and has built-in simple network management protocol (SNMP) support. It is constructed to intercommunicate with any other monitoring system [47], queuing system [48] or procedure, in order to collect parameters describing the characteristics and state of computational nodes, storage management systems, databases, Web servers, and other site components, as well as the performance of network links to and within the site as a function of time.

Some of MONALISA's main features are the following:

- A mechanism to dynamically discover all the 'Farm Units' used by a community.
- Remote event notification for changes in any component in the system.
- A lease mechanism for each registered unit.
- The ability to change the farm/network elements, and the specific parameters being monitored dynamically on the fly, without disturbing system operations.
- The ability to monitor any element through SNMP, and to drill down to get detailed information on a single node, storage or other element.
- Real-time tracking of the network throughput and traffic on each link.
- Active filters (applied through code in the mobile agents in JINI) to process the data and to provide dedicated/customized information to a number of other services or clients.
- Routing of selected (filtered) real-time and/or historical information to each subscribed listener, through a set of predicates with optional time limits.
- Flexible interface to any database supporting Java database connectivity (JDBC) [49].

- When using self-describing WSDL [50] Web services, being able to discover such services via Universal Description, Discovery, and Integration (UDDI) [51], and to automatically generate dynamic proxies to access the available information from each discovered service in a flexible way.

39.9.5.1 Data collection and processing in MONALISA

The data collection in MONALISA is based on dynamically loadable *'Monitoring Modules'* and a set of *'Farm Monitors'*, as illustrated in Figure 39.15.

Each *Farm Monitor* unit is responsible for the configuration and the monitoring of one or several farms. It can dynamically load any monitoring modules from a (set of) Web servers (with http), or a distributed file system. The Farm Monitor then uses the modules to perform monitoring tasks on each node, based on the configuration it receives from a *Regional Center Monitor* unit (not shown) that controls a set of Farm Monitors. The multithreaded engine controlling a dynamic pool of threads shown in Figure 39.16 is used to run the specified monitoring modules concurrently on each node, while limiting the additional load on the system being monitored. Dedicated modules adapted to use parameters collected by other monitoring tools (e.g. Ganglia, MRTG) are controlled by the same engine. The use of multithreading to control the execution of the modules also provides robust operation, since a monitoring task that fails or hangs (because of I/O errors for example) will not disrupt or delay the execution of the other modules. A dedicated control thread is used to stop any threads that encounter errors, and to reschedule the tasks associated with this thread if they have not already been successfully completed. A priority queue is used to handle the monitoring tasks that need to be executed periodically.

A *Monitoring Module* is a dynamically loadable unit that executes a procedure (runs a script or a program, or makes an SNMP request) to monitor a set of values, and to correctly parse the results before reporting them back to the Farm Monitor. Each

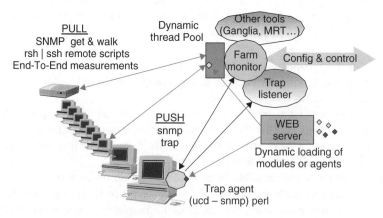

Figure 39.15 Data collection and processing in MONALISA. Farm monitors use dynamically loadable monitoring modules to pull or push information from computational nodes or other elements at one or more sites.

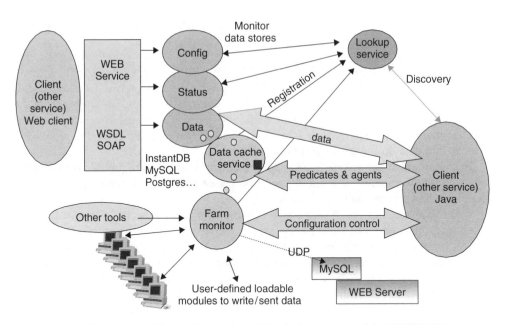

Figure 39.16   Data flow and operation of the farm monitor unit in MONALISA.

Monitoring Module must implement a method that provides the names (identifiers) for the parameters it monitors. Monitoring Modules can be used to pull data once, or with a given frequency. They may also push and install code at a monitored node (the dark dots at the right of the figure), after which they will autonomously push back the monitoring results (using SNMP, user datagram protocol (UDP or TCP) periodically back to the Farm Monitoring module.

Dynamically loading the Monitoring Modules from a relatively small set of sites when they are needed, makes it much easier to keep large monitoring systems updated and able to provide the latest functionality in a timely manner.

39.9.5.2 *Farm monitor unit operation and data handling*

The operation of the Farm Monitor unit and the flow of monitored data are illustrated in Figure 39.16. Each Farm Monitor registers as a JINI service and/or a WSDL service. Clients or other services get the needed system configuration information, and are notified automatically when a change in this information occurs. Clients subscribe as listeners to receive the values of monitored parameters (once or periodically; starting now or in the future; for real-time and/or historical data). The monitoring predicates are based on regular expressions for string selection, including configuration parameters (e.g. system names and parameters), conditions for returning values and time limits specifying when to start and stop monitoring. In addition, predicates may perform elementary functions such as MIN, MAX, average, integral, and so on. The predicate-matching and the client

notification is done in independent threads (one per client IP address) under the control
of the DataCache Service unit (shown in the figure).

The measured values of monitored parameters are currently stored in a relational
database using JDBC (such as InstantDB, MySQL, Postgres, Oracle, etc.) in order to
maintain a historical record of the data. The predicates used to retrieve historical data
are translated into SQL queries, which are then used to select the desired data from the
database and deliver it to the client who made the query. The thread associated with the
query remains active, and as soon as it receives additional data satisfying the predicates,
it sends that data on (as an update) to the client.

The system also allows one to add additional data writers and to provide the collected
values to other programs or tools, through the use of user-defined dynamically loadable
modules (shown at the bottom of the figure).

More complex data processing can be handled by Filter Agents (represented as the
blue square in Figure 39.16). These agents are 'active objects' that may be deployed by
a client or another service to perform a dedicated task using the data collected by a Farm
Monitor unit. Each agent uses a predicate to filter the data it receives, and it may send
the computed values back to a set of registered Farm Monitor units. As an example, a
maximum flow path algorithm[38] can be performed by such an agent. Agents may perform
such tasks without being deployed to a certain service, but in this case the Data Cache
Service unit needs to send all the requested values to the remote sites in which the units
that subscribed to receive these values are located.

The Farm Monitor unit is designed as a service system able to accept values, predicate-
based requests from clients, and Agent Filters, and to manage all of them asynchronously.
The Farm Monitor is then 'published' as a JINI service and a WSDL service at the
same time.

39.9.5.3 MONALISA output example

An example of the monitoring system's output is shown in Figure 39.17. In the upper left,
there is a real-time plot of the farms being monitored, indicating their loads (Green =
low load; Pink = high load; Red = unresponsive node) and the bottleneck bandwidths
(monitored using ping and the round trip time for different packet sizes) on each network
path between them. The upper-right plot shows a list of site components, in a hierarchical
directory-like structure, at the selected site (CERN in this case). One can obtain detailed
information about any site component's configuration and status, drilling down through
several levels in the directory structure if needed.

Various kinds of information can be displayed for each site component or set of components. In this example, the bottom left plot shows the CPU load as a function of time
on a single selected node at the CERN site, with three curves showing the time-averages
over the last N minutes ($N = 5, 10, 15$) using a sliding time window. The bottom center
plot shows the current CPU load for selected nodes in a subcluster of 50 nodes, averaged
over the same three time-intervals. The bottom right plot shows the current bottleneck
bandwidth between CERN and each of the other four sites being monitored.

---

[38] Finding the path with the maximum predicted throughput through a distributed system.

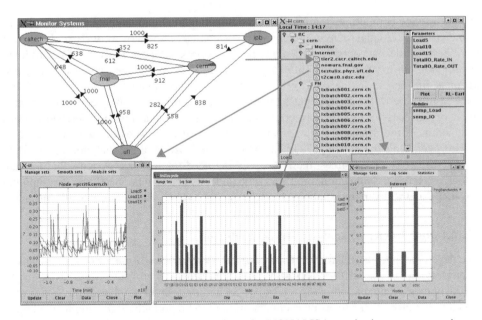

Figure 39.17   Example output (screenshot) from the MONALISA monitoring system, running on farms at (left to right in the upper left plot) Caltech, Fermilab, Florida, and Politechnica University in Bucharest.

The monitored values in the plots are updated frequently in real time, as new measurements are performed and the results come in.

39.10 THE GRID-ENABLED ANALYSIS ENVIRONMENT

Until now, the Grid architecture being developed by the LHC experiments has focused on sets of files and on the relatively well-ordered large-scale production environment. Considerable effort is already being devoted to the preparation of Grid middleware and services (this work being done largely in the context of the PPDG, GriPhyN, EU DataGrid and LCG projects already described). However, in early 2002, the problem of how processed object collections, processing and data-handling resources, and ultimately physics results may be obtained efficiently by global physics collaborations had yet to be tackled head on. Developing Grid-based tools to aid in solving this problem over the next few years, and hence beginning to understand new concepts and the foundations of the analysis solution, was deemed to be essential if the LHC experiments were to be ready for the start of LHC running.

We have described how the prevailing view of the LHC experiments' computing and software models is well developed, being based on the use of the Grid to leverage and exploit a set of computing resources that are distributed around the globe at the collaborating institutes. Analysis environment prototypes based on modern software tools, chosen from both inside and outside HEP have been developed. These tools have been aimed at providing an excellent capability to perform all the standard data analysis tasks, and assumed full access to the data, very significant local computing resources, and a full local installation of the experiment's software. These prototypes have been, and continue to be very successful; a large number of physicists use them to produce detailed physics simulations of the detector, and attempt to analyze large quantities of simulated data.

However, with the advent of Grid computing, the size of the collaborations, and the expected scarcity of resources, there is a pressing need for software systems that manage resources, reduce duplication of effort, and aid physicists who need data, computing resources, and software installations, but who cannot have all they require locally installed.

39.10.1 Requirements: analysis versus production

The development of an interactive *Grid-enabled Analysis Environment* (GAE) for physicists working on the LHC experiments was thus proposed in 2002. In contrast to the production environment, which is typically operated by a small number of physicists dedicated to the task, the analysis environment needed to be portable, lightweight (yet highly functional), and make use of existing and future experimental analysis tools as plug-in components. It needed to consist of tools and utilities that exposed the Grid system functions, parameters and behavior at selectable levels of detail and complexity. It is believed that only by exposing this complexity can an intelligent user learn what is reasonable (and efficient for getting work done) in the highly constrained global system foreseen. The use of Web services to expose the Grid in this way will allow the physicist to interactively request a collection of analysis objects to monitor the process of preparation and production of the collection and to provide 'hints' or control parameters for the individual processes. The GAE will provide various types of feedback to the physicist, such as time of completion of a task, evaluation of the task complexity, diagnostics generated at the different stages of processing, real-time maps of the global system, and so on.

39.10.2 Access to object collections

A key challenge in the development of a GAE for HENP is to develop suitable services for object collection identification, creation and selection. In order to complete an analysis, a physicist needs access to data collections. Data structures and collections not available locally need to be identified. Those collections identified as remote or nonexistent need to be produced and transferred to an accessible location. Collections that already exist need to be obtained using an optimized strategy. Identified computing and storage resources need to be matched with the desired data collections using available network resources.

It is intended to use Web-based portals and Web services to achieve this functionality, with features that include collection browsing within the Grid, enhanced job submission and interactive response times. Scheduling of object collection analysis activities, to make the most efficient use of resources, is especially challenging given the current *ad hoc* systems. Without Grid-integrated tools, scheduling cannot advance beyond the most basic scenarios (involving local resources or resources at one remote site). Accordingly, a set of tools to support the creation of more complex and efficient schedules and resource allocation plans is required. In the long term, the Grid projects intend to deliver advanced schedulers that can efficiently map many jobs to distributed resources. Currently, this work is in the early research phase, and the first tools will focus on the comparatively simpler scheduling of large batch production jobs. Experience with using new scheduling tools should also help the ongoing development efforts for automatic schedulers.

39.10.3 Components of the GAE

The GAE should consist of tools and utilities that expose the Grid system functions, parameters and behavior at selectable levels of detail and complexity. At the request submission level, an end user might interact with the Grid to request a collection of analysis objects. At the progress monitoring level, an end user would monitor the process of preparing and producing this collection. Finally at the control level, the end user would provide 'hints' or control parameters to the Grid for the production process itself. Within this interaction framework, the Grid would provide feedback on whether the request is 'reasonable', for example, by estimating the time to complete a given task, showing the ongoing progress toward completion, and displaying key diagnostic information as required. This complexity must be visible to the users so that they can learn what is reasonable in the highly constrained global Grid system.

By using Web services, it is ensured that the GAE is comparatively easy to deploy, regardless of the platform used by the end user. It is important to preserve the full functionality of the GAE regardless of the end user's platform. The desire is for OS-neutrality, so allowing access from as wide a range of devices as possible. The GAE Web services are integrated with other services such as agent-based monitoring services (as discussed in Section 39.9) and the Virtual Data Catalog Service.

39.10.4 Clarens

At Caltech, we have been developing a key component of the GAE toolset. The Clarens component software aims to build a wide-area network client/server system for remote access to a variety of data and analysis services. It does this by facilitating the integration of several existing HEP analysis and production tools, in a plug-in architecture that features an extremely small client footprint, and a modular server-side implementation.

One example of a Clarens service provides analysis of events stored in a CMS Objectivity database. This service has been extended to support analysis of events stored in RDBMSs such as Oracle and SQLServer. Another Clarens service provides remote access to Globus functionality for non-Globus clients, and includes file transfer, replica catalog access and job scheduling.

Communication between the Clarens client and server is conducted via the lightweight XML-RPC[39] remote procedure call mechanism. This was chosen both for its simplicity,

---

[39] See http://www.xmlrpc.com/

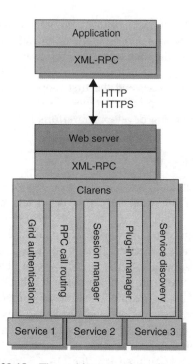

Figure 39.18   The architecture of the Clarens system.

good degree of standardization, and wide support by almost all programming languages. Communication using SOAP is also available.

The modular design of the Clarens server allows functionality to be added to a running server without taking it off-line by way of drop-in components written in Python and/or C++ (see Figure 39.18). The multiprocess model of the underlying Web server (Apache) allows Clarens to handle large numbers of clients as well as long-running client requests. The server processes are protected from other faulty or malicious requests made by other clients since each process runs in its own address space.

Several Clarens features are of particular note:

- The Python command line interface,
- A C++ command line client as an extension of the ROOT analysis environment,
- A Python GUI client in the SciGraphica analysis environment, and
- The transparent GUI and command line access to remote ROOT files, via a download service.

39.10.5 Caigee

The CMS Analysis – an Interactive Grid-Enabled Environment (CAIGEE) proposal to NSF's ITR program,[40] directly addresses the development of a GAE. The proposal

---

[40] Submitted by Caltech together with UCSD, UC Davis, and UC Riverside in February 2002, and now funded for the first year. A proposal to complete the funding is planned for 2003.

describes the development of an interactive GAE for physicists working on the CMS experiment. The environment will be lightweight yet highly functional, and will make use of existing and future CMS analysis tools as plug-in components.

The CAIGEE architecture is based on a traditional client-server scheme, with one or more intercommunicating servers as shown in Figure 39.19. A small set of clients is logically associated with each server, the association being based primarily on geographic location. The architecture is 'tiered', in the sense that a server can delegate the execution of one or more of its advertised services to another server in the Grid, which logically would be at the same or a higher level in the tiered hierarchy. In this way, a client request can be brokered to a server that is better equipped to deal with it than the client's local server, if necessary. In practice, CAIGEE will initially deploy Web services running on Tier2 regional centers at several US CMS institutions. Each of these servers will be the first point of interaction with the 'Front End' clients (local physicist end users) at US CMS sites.

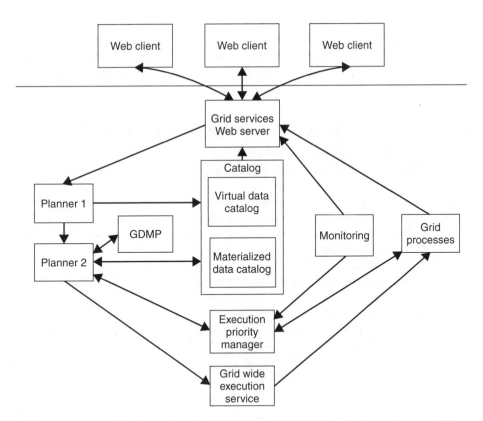

Figure 39.19 The CAIGEE architecture for a Grid-enabled HEP analysis environment. The use of Web services with thin clients makes the architecture amenable to deployment on a wide range of client devices.

The servers will offer a set of Web-based services. This architecture allows the dynamic addition of, or improvement to, services. Also, software clients will always be able to correctly use the services they are configured for. This is a very important feature of CAIGEE from a usability standpoint; it can be contrasted with static protocols between partner software clients, which would make any update or improvement in a large distributed system hard or impossible.

Grid-based data analysis requires information and coordination of hundreds to thousands of computers at each of several Grid locations. Any of these computers may be off-line for maintenance or repairs. There may also be differences in the computational, storage and memory capabilities of each computer in the Grid. At any point in time, the Grid may be performing analysis activities for tens to hundreds of users while simultaneously doing production runs, all with differing priorities. If a production run is proceeding at high priority, there may not be enough CPUs available for the user's data analysis activities at a certain location.

Because of these complexities, CAIGEE will have facilities that allow the user to pose 'what if' scenarios that will guide the user in his/her use of the Grid. An example is the estimation of the time to complete the user's analysis task at each Grid location. CAIGEE monitoring widgets, with selectable levels of detail, will give the user a 'bird's eye view' of the global Grid system, and show the salient features of the prevailing resources in use and scheduled for the future. CAIGEE will also be interfaced to the MONALISA monitoring system (described in Section 39.9.5) for more sophisticated decisions, and decision support.

39.11 CONCLUSION: RELEVANCE OF MEETING THESE CHALLENGES FOR FUTURE NETWORKS AND SOCIETY

The HEP (orHENP, for high-energy and nuclear physics) problems are the most data-intensive known. Hundreds to thousands of scientist-developers around the world continually develop software to better select candidate physics signals, better calibrate the detector and better reconstruct the quantities of interest (energies and decay vertices of particles such as electrons, photons and muons, as well as jets of particles from quarks and gluons). The globally distributed ensemble of facilities, while large by any standard, is less than the physicists require to do their work in an unbridled way. There is thus a need, and a drive to solve the problem of managing global resources in an optimal way, in order to maximize the potential of the major experiments for breakthrough discoveries.

In order to meet these technical goals, priorities have to be set, the system has to managed and monitored globally end-to-end, and a new mode of 'human–Grid' interactions has to be developed and deployed so that the physicists, as well as the Grid system itself, can learn to operate optimally to maximize the workflow through the system. Developing an effective set of trade-offs between high levels of resource utilization, rapid turnaround time, and matching resource usage profiles to the policy of each scientific collaboration over the long term presents new challenges (new in scale and complexity) for distributed systems.

A new scalable Grid agent-based monitoring architecture, a Grid-enabled Data Analysis Environment, and new optimization algorithms coupled to Grid simulations are all under development in the HEP community.

Successful construction of network and Grid systems able to serve the global HEP and other scientific communities with data-intensive needs could have wide-ranging effects: on research, industrial and commercial operations. The key is intelligent, resilient, self-aware, and self-forming systems able to support a large volume of robust terabyte and larger transactions, able to adapt to a changing workload, and capable of matching the use of distributed resources to policies. These systems could provide a strong foundation for managing the large-scale data-intensive operations processes of the largest research organizations, as well as the distributed business processes of multinational corporations in the future.

It is also conceivable that the development of the new generation of systems of this kind could lead to new modes of interaction between people and 'persistent information' in their daily lives. Learning to provide and efficiently manage and absorb this information in a persistent, collaborative environment could have a profound transformational effect on society.

ACKNOWLEDGEMENTS

A great number of people contributed to the work and ideas presented in this chapter. Prominent among these are Richard Mount (SLAC), Paul Messina (Caltech), Laura Perini (INFN/Milan), Paolo Capiluppi (INFN/Bologna), Krzysytof Sliwa (Tufts), Luciano Barone (INFN/Bologna), Paul Avery (Florida), Miron Livny (Wisconsin), Ian Foster (Argonne and Chicago), Carl Kesselman (USC/ISI), Olivier Martin (CERN/IT), Larry Price (Argonne), Ruth Pordes (Fermilab), Lothar Bauerdick (Fermilab), Vicky White (DOE), Alex Szalay (Johns Hopkins), Tom de Fanti (UIC), Fabrizio Gagliardi (CERN/IT), Rob Gardner (Chicago), Les Robertson (CERN/IT), Vincenzo Innocente (CERN/CMS), David Stickland (Princeton), Lucas Taylor (Northeastern), and the members of our Caltech team working on Grids and related issues: Eric Aslakson, Philippe Galvez, Koen Holtman, Saima Iqbal, Iosif Legrand, Sylvain Ravot, Suresh Singh, Edwin Soedarmadji, and Conrad Steenberg. This work has been supported in part by DOE Grants DE-FG03-92-ER40701 and DE-FC03-99ER25410 and by NSF Grants 8002–48195, PHY-0122557 and ACI-96-19020.

REFERENCES

1. *The Compact Muon Solenoid Technical Proposal*, CERN/LHCC 94-38 (1994) and CERN LHCC-P1, http://cmsdoc.cern.ch/.
2. *The ATLAS Technical Proposal*, CERN/LHCC 94-43 (1994) and CERN LHCC-P2, http://atlasinfo.cern.ch/ATLAS/TP/NEW/HTML/tp9new/tp9.html. Also the ALICE experiment at http://www.cern.ch/ALICE and the LHCb experiment at http://lhcb-public.web.cern.ch/lhcb-public/.
3. www.cern.ch/LHC.

4. The BaBar Experiment at SLAC, http://www-public.slac.stanford.edu/babar/.
5. The D0 Experiment at Fermilab, http://www-d0.fnal.gov/.
6. The CDF Experiment at Fermilab, http://www-cdf.fnal.gov/.
7. The Relativistic Heavy Ion Collider at BNL, http://www.bnl.gov/RHIC/.
8. The Grid Physics Network, http://www.griphyn.org, and Section 5.2.
9. The Particle Physics Data Grid, http://www.ppdg.net and Section 5.1.
10. The International Virtual Data Grid Laboratory, http://www.ivdgl.org and Section 5.2.
11. http://eu-datagrid.web.cern.ch/eu-datagrid/.
12. http://www.datatag.org.
13. The LHC Computing Grid Project (LCG), http://lhcgrid.web.cern.ch/LHCgrid/.
14. Report of the Steering Group of the LHC Computing Review, CERN/LHCC 2001–004, http://lhcb-comp.web.cern.ch/lhcb-comp/Reviews/LHCComputing2000/Report_final.pdf.
15. Kunze, M. *et al.* (2002) Report of the LHC Computing Grid Project RTAG6, *Regional Center Category and Service Definition*, June, 2002, http://www.fzk.de/hik/orga/ges/RTAG6.final.doc.
16. The LHC Computer Technology Tracking Teams, http://wwwinfo.cern.ch/di/ttt.html.
17. RD45 – A Persistent Object Manager for HEP, http://wwwinfo.cern.ch/asd/cernlib/rd45/.
18. Holtman, K. On behalf of the CMS collaboration, *CMS Data Grid System Overview and Requirements*, CMS Note 2001/037.
19. http://pcbunn.cacr.caltech.edu/pubs/GIOD_Full_Report.pdf.
20. The MONARC Collaboration. *MONARC: Models of Networked Analysis at Regional Centers for LHC Experiments*, Phase 2, Report CERN/LCB-2000-001, http://www.cern.ch/MONARC/docs/phase2report/Phase2Report.pdf.
21. Legand, I. *et al. The MONARC Distributed Computing Simulation Environment*, http://www.cern.ch/MONARC/sim_tool.
22. http://www.skyserver.sdss.org.
23. For example the CMS Coherent Object Reconstruction for CMS Analysis (COBRA) that includes the CARF framework, http://cmsdoc.cern.ch/swdev/snapshot/COBRA/ReferenceManual/html/COBRA.html or the ATLAS Athena framework, http://cmsdoc.cern.ch/swdev/snapshot/COBRA/ReferenceManual/html/COBRA.html.
24. http://www.internet2.edu.
25. http://www.es.net.
26. http://www.calren2.net and http://www.mren.org.
27. http://www.dante.net/geant/about-geant.html.
28. The new academic and research network in Japan that started in January 2002, http://www.japan-telecom.co.jp/PRdept/NEWSLETTER_Eng/nl13/update2.html.
29. http://www.startap.net/starlight.
30. See DataTAG Work Package 2 on High Performance Networking, http://icfamon.dl.ac.uk/DataTAG-WP2/.
31. Proposal for a Global terabit Research Network, http://www.indiana.edu/~gtrn/.
32. http://www.slac.stanford.edu/~abh/CHEP2001/7-018.pdf.
33. http://doc.in2p3.fr/bbftp/ and http://www.slac.stanford.edu/comp/net/bandwidth-tests/predict/html/bbftp.html.
34. Ravot, S. and Martin-Flatin, J. P. (2003) TCP congestion control in long-distance networks. Submitted to *InfoComm 2003*, 2003..
35. http://netlab.caltech.edu/FAST/bg.htm.
36. http://www-iepm.slac.stanford.edu/.
37. http://www.internet2.edu/henp.
38. http://www.internet2.edu/e2e.
39. http://icfa-scic.web.cern.ch/ICFA-SCIC/.
40. See for example the society for model and simulation international at http://www.scs.org, and the many conferences listed on their calendar.

41. Legrand, I. C. (2001) Multithreaded discrete event simulation of distributed computing systems. *Computer Physics Communications*, **140**, 274, http://clegrand.home.cern.ch/clegrand/MONARC/CHEP2k/sim_chep.pdf.
42. Le Grand, I. Simulation studies in data replication strategies. *CHEP2001*, Beijing, 2001, http://clegrand.home.cern.ch/clegrand/CHEP01/chep01-10-048.pdf and http://www.cern.ch/MONARC/sim_tool/Publish/CMS.
43. Newman, H. B. and Legrand, I. C. *A Self-Organizing Neural Network for Job Scheduling in Distributed Systems*, CMS Note 2001/009, http://clegrand.home.cern.ch/clegrand/note01_009.pdf.
44. http://www.sun.com/jini, A number of interesting example applications may be found at http://wwws.sun.com/software/jini/news/success.html.
45. http://cil.cern.ch:8080/MONALISA/.
46. http://www.vrvs.org.
47. MONALISA is currently interfaced to Ganglia (http://ganglia.sourceforge.net/) and MRTG (http://mrtg.hdl.com/mrtg.html). It will be interfaced to MDS2 (http://www.globus.org) and Hawkeye (http://www.cs.wisc.edu/condor/hawkeye/) in the near future.
48. Such as PBS (http://www.openpbs.org/) and LSF (http://www.platform.com/products/wm/LSF/index.asp). These interfaces will be implemented in the near future.
49. Java Database Connectivity, http://java.sun.com/products/jdbc/
50. Web Services Description Language, http://www.w3.org/TR/wsdl
51. Universal Description, Discovery and Integration, A 'Meta-service' for Discovering Web Services, http://www.uddi.org/.

The new biology and the Grid

Kim Baldridge and Philip E. Bourne

University of California, San Diego, California, United States

40.1 INTRODUCTION

Computational biology is undergoing a revolution from a traditionally compute-intensive science conducted by individuals and small research groups to a high-throughput, data-driven science conducted by teams working in both academia and industry. It is this *new biology* as a data-driven science in the era of Grid Computing that is the subject of this chapter. This chapter is written from the perspective of bioinformatics specialists who seek to fully capitalize on the promise of the Grid and who are working with computer scientists and technologists developing biological applications for the Grid.

To understand what has been developed and what is proposed for utilizing the Grid in the new biology era, it is useful to review the 'first wave' of computational biology application models. In the next section, we describe the first wave of computational models used for computational biology and computational chemistry to date.

40.1.1 The first wave: compute-driven biology applications

The first computational models for biology and chemistry were developed for the classical von Neumann machine model, that is, for sequential, scalar processors. With the

Grid Computing – Making the Global Infrastructure a Reality. Edited by F. Berman, A. Hey and G. Fox
© 2003 John Wiley & Sons, Ltd   ISBN: 0-470-85319-0

emergence of parallel computing, biological applications were developed that could take advantage of multiple processor architectures with distributed or shared memory and locally located disk space to execute a collection of tasks. Applications that compute molecular structure or electronic interactions of a protein fragment are examples of programs developed to take advantage of emerging computational technologies.

As distributed memory parallel architectures became more prevalent, computational biologists became familiar with message passing library toolkits, first with Parallel Virtual Machine (PVM) and more recently with Message Passing Interface (MPI). This enabled biologists to take advantage of distributed computational models as a target for executing applications whose structure is that of a pipelined set of stages, each dependent on the completion of a previous stage. In *pipelined applications*, the computation involved for each stage can be relatively independent from the others. For example, one computer may perform molecular computations and immediately stream results to another computer for visualization and analysis of the data generated. Another application scenario is that of a computer used to collect data from an instrument (say a tilt series from an electron microscope), which is then transferred to a supercomputer with a large shared memory to perform a volumetric reconstruction, which is then rendered on yet a different high-performance graphic engine. The distribution of the application pipeline is driven by the number and the type of different tasks to be performed, the available architectures that can support each task, and the I/O requirements between tasks.

While the need to support these applications continues to be very important in computational biology, an emerging challenge is to support a new generation of applications that analyze and/or process immense amounts of input/output data. In such applications, the computation on each of a large number of data points can be relatively small, and the 'results' of an application are provided by the analysis and often visualization of the input/output data. For such applications, the challenge to infrastructure developers is to provide a software environment that promotes application performance and can leverage large numbers of computational resources for simultaneous data analysis and processing. In this chapter, we consider these new applications that are forming the next wave of computational biology.

40.1.2 The next wave: data-driven applications

The next wave of computational biology is characterized by high-throughput, high technology, data-driven applications. The focus on genomics, exemplified by the human genome project, will engender new science impacting a wide spectrum of areas from crop production to personalized medicine. And this is just the beginning. The amount of raw DNA sequence being deposited in the public databases doubles every 6 to 8 months. Bioinformatics and Computational Biology have become a prime focus of academic and industrial research. The core of this research is the analysis and synthesis of immense amounts of data resulting in a new generation of applications that require information technology as a vehicle for the next generation of advances.

Bioinformatics grew out of the human genome project in the early 1990s. The requests for proposals for the physical and genetic mapping of specific chromosomes called for developments in informatics and computer science, not just for data management but for innovations in algorithms and application of those algorithms to synergistically improve the rate and accuracy of the genetic mapping. A new generation of scientists was born, whose demand still significantly outweighs their supply, and who have been brought up on commodity hardware architectures and fast turnaround. This is a generation that contributed significantly to the fast adoption of the Web by biologists and who want instant gratification, a generation that makes a strong distinction between wall clock and CPU time. It makes no difference if an application runs 10 times as fast on a high-performance architecture (minimizing execution time) if you have to wait 10 times as long for a result by sitting in a long queue (maximizing turnaround time). In data-driven biology, turnaround time is important in part because of sampling: a partial result is generally useful while the full result is being generated. We will see specific examples of this subsequently, for now let us better grasp the scientific field we wish the Grid to support.

The complexity of new biology applications reflects exponential growth rates at different levels of biological complexity. This is illustrated in Figure 40.1 that highlights representative activities at different levels of biological complexity. While bioinformatics is currently focusing on the molecular level, this is just the beginning. Molecules form complexes that are located in different parts of the cell. Cells differentiate into different types forming organs like the brain and liver. Increasingly complex biological systems

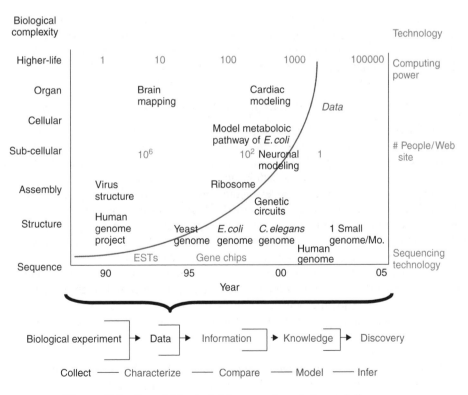

Figure 40.1 From biological data comes knowledge and discovery.

generate increasingly large and complex biological data sets. If we do not solve the problems of processing the data at the level of the molecule, we will not solve problems of higher order biological complexity.

Technology has catalyzed the development of the new biology as shown on the right vertical axis of Figure 40.1. To date, Moore's Law has at least allowed data processing to keep approximate pace with the rate of data produced. Moreover, the cost of disks, as well as the communication access revolution brought about by the Web, has enabled the science to flourish. Today, it costs approximately 1% of what it did 10 to 15 years ago to sequence one DNA base pair. With the current focus on genomics, data rates are anticipated to far outweigh Moore's Law in the near future making Grid and cluster technologies more critical for the new biology to flourish.

Now and in the near future, a critical class of new biology applications will involve large-scale data production, data analysis and synthesis, and access through the Web and/or advanced visualization tools to the processed data from high-performance databases ideally federated with other types of data. In the next section, we illustrate in more detail two new biology applications that fit this profile.

40.2 BIOINFORMATICS GRID APPLICATIONS TODAY

The two applications in this section require large-scale data analysis and management, wide access through Web portals, and visualization. In the sections below, we describe CEPAR (Combinatorial Extension in PARallel), a computational biology application, and CHEMPORT, a computational chemistry framework.

40.2.1 Example 1: CEPAR and CEPort – 3D protein structure comparison

The human genome and the less advertised but very important 800 other genomes that have been mapped, encode genes. Those genes are the blueprints for the proteins that are synthesized by reading the genes. It is the proteins that are considered the building blocks of life. Proteins control all cellular processes and define us as a species and as individuals. A step on the way to understanding protein function is protein structure – the 3D arrangement that recognizes other proteins, drugs, and so on. The growth in the number and complexity of protein structures has undergone the same revolution as shown in Figure 40.1, and can be observed in the evolution of the Protein Data Bank (PDB; http://www.pdb.org), the international repository for protein structure data.

A key element to understanding the relationship between biological structure and function is to characterize all known protein structures. From such a characterization comes the ability to be able to infer the function of the protein once the structure has been determined, since similar structure implies similar function. High-throughput structure determination is now happening in what is known as structure genomics – a follow-on to the human genome project in which one objective is to determine all protein structures encoded by the genome of an organism. While a typical protein consists of 300 of one of 20 different amino acids – a total of 20^{300} possibilities – more than all the atoms in the universe – nature has performed her own reduction, both in the number of sequences and in the number of protein structures as defined by discrete folds. The number of unique protein folds is currently estimated at between 1000 and 10 000. These folds need to be characterized and all new structures tested to see whether they conform to an existing fold or represent a new fold. In short, characterization of how all proteins fold requires that they be compared in 3D to each other in a pairwise fashion.

With approximately 30 000 protein chains currently available in the PDB, and with each pair taking 30 s to compare on a typical desktop processor using any one of several algorithms, we have a (30 000*30 000/2)*30 s size problem to compute all pairwise comparisons, that is, a total of 428 CPU years on one processor. Using a combination of data reduction (a pre-filtering step that permits one structure to represent a number of similar structures), data organization optimization, and efficient scheduling, this computation was performed on 1000 processors of the 1.7 Teraflop IBM Blue Horizon in a matter of days using our Combinatorial Extension (CE) algorithm for pairwise structure comparison [1]. The result is a database of comparisons that is used by a worldwide community of users 5 to 10 000 times per month and has led to a number of interesting discoveries cited in over 80 research papers. The resulting database is maintained by the San Diego Supercomputer Center (SDSC) and is available at http://cl.sdsc.edu/ce.html [2]. The procedure to compute and update this database as new structures become available is equally amenable to Grid and cluster architectures, and a Web portal to permit users to submit their own structures for comparison has been established.

In the next section, we describe the optimization utilized to diminish execution time and increase applicability of the CE application to distributed and Grid resources. The result is a new version of the CE algorithm we refer to as CEPAR. CEPAR distributes each 3D comparison of two protein chains to a separate processor for analysis. Since each pairwise comparison represents an independent calculation, this is an embarrassingly parallel problem.

40.2.1.1 Optimizations of CEPAR

The optimization of CEPAR involves structuring CE as an efficient and scalable master/worker algorithm. While initially implemented on 1024 processors of Blue Horizon, the algorithm and optimization undertaken can execute equally well on a Grid platform. The addition of resources available on demand through the Grid is an important next step for problems requiring computational and data integration resources of this magnitude. We have employed algorithmic and optimization strategies based on numerical studies on CEPAR that have made a major impact on performance and scalability. To illustrate what can be done in distributed environments, we discuss them here. The intent is to familiarize the reader with one approach to optimizing a bioinformatics application for the Grid.

Using a trial version of the algorithm without optimization (Figure 40.2), performance bottlenecks were identified. The algorithm was then redesigned and implemented with the following optimizations:

1. The assignment packets (chunks of data to be worked on) are buffered in advance.
2. The master processor algorithm prioritizes incoming messages from workers since such messages influence the course of further calculations.

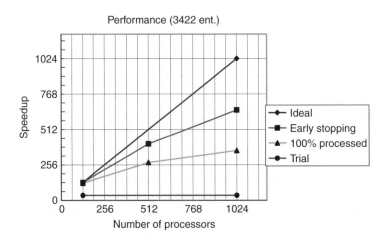

Figure 40.2 Scalability of CEPAR running on a sample database of 3422 data points (protein chains). The circles show the performance of the trial version of the code. The triangles show the improved performance after improvements 1, 2, and 4 were added to the trial version. The squares show the performance based on timing obtained with an early stopping criterion (improvement 3). The diamonds provide an illustration of the ideal scaling.

3. Workers processing a data stream that no longer poses any interest (based on a result from another worker) are halted. We call this *early stopping*.
4. Standard single-processor optimization techniques are applied to the master processor.

With these optimizations, the scalability of CEPAR was significantly improved, as can be seen from Figure 40.2.

The optimizations significantly improved the performance of the code. The MPI implementation on the master processor is straightforward, but it was essential to use buffered sends (or another means such as asynchronous sends) in order to avoid communication channel congestion. In summary, with 1024 processors the CEPAR algorithm outperforms CE (no parallel optimization) by 30 to 1 and scales well. It is anticipated that this scaling would continue even on a larger number of processors.

One final point concerns the end-process load imbalance. That is, a large number of processors can remain idle while the final few do their job. We chose to handle this by breaking the runs involving a large number of processors down into two separate runs. The first run does most of the work and exits when the early stopping criterion is met. Then the second run completes the task for the outliers using a small number of processors, thus freeing these processors for other users. Ease of use of the software is maintained through an automatic two-step job processing utility.

CEPAR has been developed to support our current research efforts on PDB structure similarity analysis on the Grid. CEPAR software uses MPI, which is a universal standard for interprocessor communication. Therefore, it is suitable for running in any parallel environment that has an implementation of MPI, including PC clusters or Grids. There is no dependence on the particular structural alignment algorithm or on the specific application.

The CEPAR design provides a framework that can be applied to other problems facing computational biologists today where large numbers of data points need to be processed in an embarrassingly parallel way. Pairwise sequence comparison as described subsequently is an example. Researchers and programmers working on parallel software for these problems might find useful the information on the bottlenecks and optimization techniques used to overcome them, as well as the general approach of using numerical studies to aid algorithm design briefly reported here and given in more detail in Reference [1]. But what of the naive user wishing to take advantage of high-performance Grid computing?

40.2.1.2 CEPAR portals

One feature of CEPAR is the ability to allow users worldwide provide their own structures for comparison and alignment against the existing database of structures. This service currently runs on a Sun Enterprise server as part of the CE Website (http://cl.sdsc.edu/ce.html) outlined above. Each computation takes on an average three hours of CPU time for a single user request. On occasion, this service must be turned off as the number of requests for structure comparisons far outweighs what can be processed on a Sun Enterprise server. To overcome this shortage of compute resources, a Grid portal has been established to handle this situation (https://gridport.npaci.edu/CE/) using SDSC's GridPort technology [3]. The portal allows this computation to be done using additional resources when available. Initial target compute resources for the portal are the IBM Blue Horizon, a 64-node Sun Enterprise server and a Linux PC cluster of 64 nodes.

The GridPort Toolkit [3] is composed of a collection of modules that are used to provide portal services running on a Web server and template Web pages needed to implement a Web portal. The function of GridPort is simply to act as a Web frontend to Globus services [4], which provide a virtualization layer for distributed resources. The only requirements for adding a new high-performance computing (HPC) resource to the portal are that the CE program is recompiled on the new architecture, and that Globus services are running on it. Together, these technologies allowed the development of a portal with the following capabilities:

- Secure and encrypted access for each user to his/her high-performance computing (HPC) accounts, allowing submission, monitoring, and deletion of jobs and file management;
- Separation of client application (CE) and Web portal services onto separate servers;
- A single, common point of access to multiple heterogeneous compute resources;
- Availability of real-time status information on each compute machine;
- Easily adaptable (e.g. addition of newly available compute resources, modification of user interfaces etc.).

40.2.1.3 Work in progress

While the CE portal is operational, much work remains to be done. A high priority is the implementation of a distributed file system for the databases, user input files, jobs in progress, and results. A single shared, persistent file space is a key component of the distributed abstract machine model on which GridPort was built. At present, files must be explicitly transferred from the server to the compute machine and back again; while this process is invisible to the user, from the point of view of portal development and administration, it is not the most elegant solution to the problem. Furthermore, the present system requires that the all-against-all database must be stored locally on the file system of each compute machine. This means that database updates must be carried out individually on each machine.

These problems could be solved by placing all user files, along with the databases, in a shared file system that is available to the Web server and all HPC machines. Adding Storage Resource Broker (SRB) [5] capability to the portal would achieve this. Work is presently ongoing on automatically creating an SRB collection for each registered GridPort user; once this is complete, SRB will be added to the CE portal.

Another feature that could be added to the portal is the automatic selection of compute machine. Once 'real-world' data on CPU allocation and turnaround time becomes available, it should be possible to write scripts that inspect the queue status on each compute machine and allocate each new CE search to the machine expected to produce results in the shortest time.

Note that the current job status monitoring system could also be improved. Work is underway to add an event daemon to the GridPort system, such that compute machines could notify the portal directly when, for example, searches are scheduled, start and finish. This would alleviate the reliance of the portal on intermittent inspection of the queue of each HPC machine and provide near-instantaneous status updates. Such a system would also allow the portal to be regularly updated with other information, such as warnings when compute machines are about to go down for scheduled maintenance, broadcast messages from HPC system administrators and so on.

40.2.2 Example 2: Chemport – a quantum mechanical biomedical framework

The successes of highly efficient, composite software for molecular structure and dynamics prediction has driven the proliferation of computational tools and the development of first-generation cheminformatics for data storage, analysis, mining, management, and presentations. However, these first-generation cheminformatics tools do not meet the needs of today's researchers. Massive volumes of data are now routinely being created that span the molecular scale, both experimentally and computationally, which are available for access for an expanding scope of research. What is required to continue progress is the integration of individual 'pieces' of the methodologies involved and the facilitation of the computations in the most efficient manner possible.

Towards meeting these goals, applications and technology specialists have made considerable progress towards solving some of the problems associated with integrating the algorithms to span the molecular scale computationally and through the data, as well as providing infrastructure to remove the complexity of logging on to a HPC system in order to submit jobs, retrieve results, and supply 'hooks' into other codes. In this section, we give an example of a framework that serves as a working environment for researchers, which demonstrates new uses of the Grid for computational chemistry and biochemistry studies.

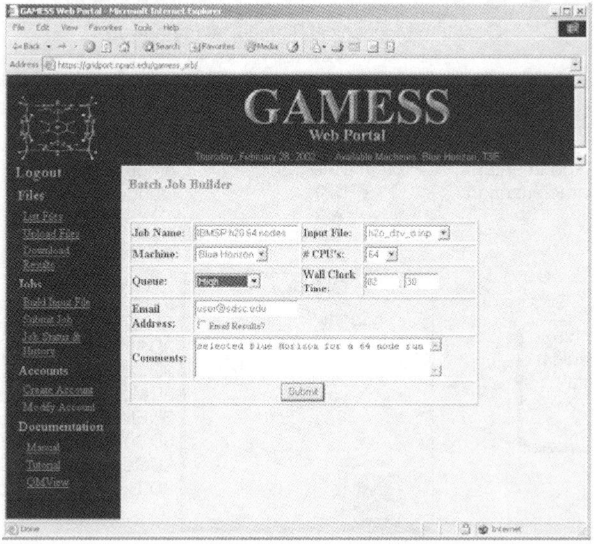

Figure 40.3   The job submission page from the SDSC GAMESS portal.

Using GridPort technologies [3] as described for CEPAR, our efforts began with the creation of a portal for carrying out chemistry computations for understanding various details of structure and property for molecular systems – the General Atomic Molecular Electronic Structure Systems (GAMESS) [6] quantum chemistry portal (http://gridport. npaci.edu/gamess). The GAMESS software has been deployed on a variety of computational platforms, including both distributed and shared memory platforms. The job submission page from the GAMESS portal is shown in Figure 40.3. The portal uses Grid technologies such as the SDSC's GridPort toolkit [3], the SDSC SRB [5] and Globus [7] to assemble and monitor jobs, as well as store the results. One goal in the creation of a new architecture is to improve the user experience by streamlining job creation and management.

Related molecular sequence, structure, and property software have been created using similar frameworks, including the AMBER [8] classical molecular dynamics portal, the EULER [9] genetic sequencing program, and the Adaptive Poisson-Boltzmann Solver (APBS) [10] program for calculating electrostatic potential surfaces around biomolecules. Each type of molecular computational software provides a level of understanding of molecular structure that can be used for a larger scale understanding of the function. What is needed next are strategies to link the molecular scale technologies through the data and/or through novel new algorithmic strategies. Both involve additional Grid technologies.

Development of portal infrastructure has enabled considerable progress towards the integration across scale from molecules to cells, linking the wealth of ligand-based data present in the PDB, and detailed molecular scale quantum chemical structure and

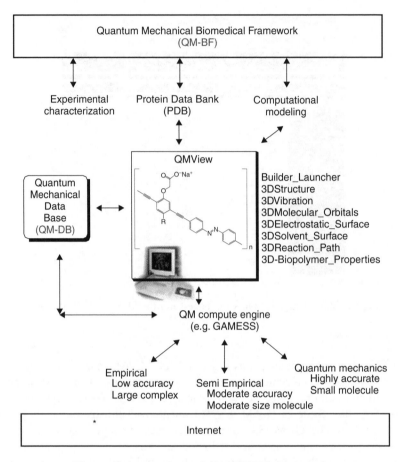

Figure 40.4   Topology of the QM-PDB framework.

property data. As such, accurate quantum mechanical data that has been hitherto under-utilized will be made accessible to the nonexpert for integrated molecule to cell stud-ies, including visualization and analysis, to aid in the understanding of more detailed molecular recognition and interaction studies than is currently available or sufficiently reliable. The resulting QM-PDB framework (Figure 40.4) integrates robust computational quantum chemistry software (e.g. GAMESS) with associated visualization and analysis toolkits, QMView, [11] and associated prototype Quantum Mechanical (QM) database facility, together with the PDB. Educational tools and models are also integrated into the framework.

With the creation of Grid-based toolkits and associated environment spaces, researchers can begin to ask more complex questions in a variety of contexts over a broader range of scales, using seamless transparent computing access. As more realistic molecular com-putations are enabled, extending well into the nanosecond and even microsecond range at a faster turnaround time, and as problems that simply could not fit within the physical constraints of earlier generations of supercomputers become feasible, the ability to integrate methodologies becomes more critical. Such important developments have made possible the transition from the structural characterization of biological systems alone, to the computational analysis of functions in biomolecular systems via classical and quantum mechanical simulations.

40.2.2.1 Work in progress: next generation portals

The next generation of Chemport will require the ability to build complex and *dynamically reconfigurable* workflows. At present, GridPort only facilitates the creation of Web-based portals, limiting its potential use. The next step is to develop an architecture that integrates the myriad of tools (GridPort, XML, Simple Object Access Protocol (SOAP) [12]) into a unified system. Ideally, this architecture will provide users with an interface that shields them from the complexity of available high-performance resources and allows them to concentrate on solving their scientific problem. The overarching goal is to enable scientific programmers to integrate their software into the workflow system in a simple, effective, and uniform manner.

The workflow management system that allows portal construction must be fully integrated with emerging Grid standards (e.g. the Globus Toolkit) and dynamic reconfigurability. By defining XML schema to describe both resources and application codes, and interfaces, using emerging Grid standards (such as Web services, SOAP [13], Open Grid Services Architecture), and building user friendly interfaces like science portals, the next generation portal will include a 'pluggable' event-driven model in which Grid-enabled services can be composed to form more elaborate pipelines of information processing, simulation, and visual analysis.

This approach is currently being adopted at SDSC and involves development of a layered next generation portals architecture, consisting of an *Interface Layer*, a *Middleware Layer*, and a *Resource Layer*. The layers shield the developer from the complexity of the underlying system and provide convenient mechanisms of abstraction of functionality, interface, hardware, and software. The layered approach also helps with the logical design of a complex system by grouping together the related components, while separating them into manageable parts with interfaces to join the layers together. The separation of the Resource Layer from the Middleware Layer provides the ability to design a workflow system that is not bound to a specific application or HPC resource.

A schematic of the architecture is shown in Figure 40.5. Communication between the layers is accomplished by sending SOAP messages or by using Grid technologies like Globus to communicate using the Globus Resource Allocation Manager (GRAM) [7] protocol. The goal is to render a software environment with the ability to support HPC applications from any scientific domain with little configuration effort by developers and users. Definitions of HPC applications will be constructed in XML according to XML DTDs and XML Schemas defined as part of the Application Programming Interface (API). Proper design of these definitions is critical to ensuring usability of the system across a wide spectrum of computational science domains.

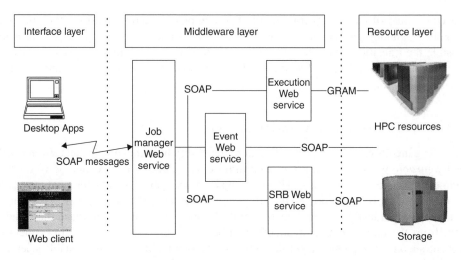

Figure 40.5 The three-layered architecture and the communication through SOAP and GRAM protocols over the Internet.

40.3 THE CHALLENGES OF THE GRID FROM THE PERSPECTIVE OF BIOINFORMATICS RESEARCHERS

Disciplinary researchers care most about results, not the infrastructure used to achieve those results. Computational scientists care about optimizing their codes for the underlying compute environment. Technologists seek to improve the potential capacity and capability of their technologies. All these approaches focus on different goals that can conflict (e.g. the conflicts inherent in optimizing both application performance and resource utilization).

Conflicting goals are evident in an environment in which the technology changes rapidly and new platforms provide greater potential and capability, but in which successful codes that have been stabilized for older platforms must be revamped to run on new platforms yet another time. The challenge to Grid researchers, from the perspective of bioinformaticists is:

- To provide a usable and accessible computational and data management environment which can enhance application performance for bioinformaticists requiring high-end resources above and beyond what is available to them locally.
- To provide sufficient support services to sustain the environment and present a stable configuration of the resources to users.
- To ensure that the science performed on the Grid constitutes the next generation of advances and not just proof-of-concept computations.
- To accept feedback from bioinformaticians that is used in the design and implementation of the current environment and to improve the next generation of Grid infrastructure.

40.3.1 A future data-driven application – the encyclopedia of life

Given the data-driven nature of the new biology, we anticipate the Grid being an enormous asset to large-scale data processing and analysis. The Encyclopedia of Life (EOL) project, being developed by biologists and technologists at SDSC and UCSD, is a good example of the challenges introduced in the previous section.

As stated earlier, while much attention has been given to the complete draft of the human genome, there are 800 genomes of other species – many of them pathogenic bacteria that have significant impact on human health – that have been completed. The majority of the proteins encoded in those genomes are uncharacterized functionally and have no structures available. The EOL Project seeks to characterize these proteins for a worldwide audience of researchers, in short to create a one-stop shop for information on proteins. As such there are two components to EOL: first, putative assignment of function and structure based on comparative analysis with proteins for which this experimental information is known, and second, integration with other sources of data on proteins. The overall topology of the EOL system is given in Figure 40.6.

Public genomic data are processed by a pipeline of applications and the final data stored in a highly normalized *Data Warehouse*, from which different instances (*DataMarts*) are derived for different user groups and made accessible to users and applications. The Grid is a prime facilitator of the comparative analysis needed for putative assignment. The pipeline consists of a series of applications well known in bioinformatics and outlined in Figure 40.7.

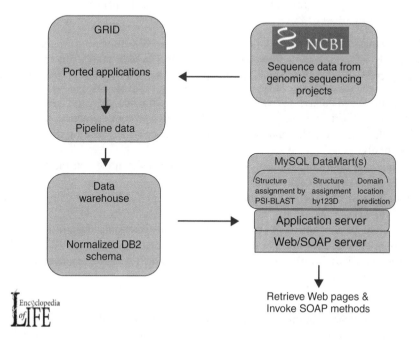

Figure 40.6   Topology of the EOL system.

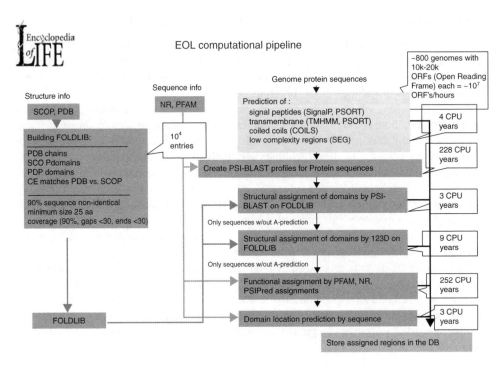

Figure 40.7   The EOL computational pipeline.

The first pass on all genomes is estimated to take 500 processor years, much of this consumed by standard applications like PSI-BLAST which could run in parallel – one gene per processor. The procedure (protocol) just for the PSI-BLAST step is as follows:

1. Distribute the database to which every target sequence is to be compared to each node on the Grid. We use the nonredundant (NR) protein sequence database and the PFAM databases which contain all unique sequences and those organized by families respectively.
2. Schedule and distribute each of 10^7 target sequences to nodes on the Grid.
3. Run PSI-BLAST on each node.
4. Retain the PSI-BLAST profiles in secondary storage for future lookup as needed (estimated 50 TB).
5. Pass output of the run (sequence alignments and functional description) to a central relational database for comparative access (estimated 5 TB).
6. Repeat steps 1 to 5 with a new version of NR and PFAM as they become available.

Other steps on the right-hand side of this pipeline represent similar, albeit less CPU consuming, steps for the most part. Steps 2 and 5 can be run asynchronously and so scheduling overhead is limited. Moreover, data value is accumulative – it is not necessary to await the final result before useful science can be done. Once two genomes have been completed, comparative proteomics can begin.

Interestingly, the major challenge in using the Grid for EOL is not technical at this stage, but sociological (how should policies for sharing and accounting of data and computational resources be formulated?) and logistical (where can we find a sufficient number of resources to support this 'continuous' computation?).

In EOL, data integration (the second component) is a major challenge as well. Database federation is needed in the new biology but is not yet used widely by the bioinformatics community: today either existing data sources are absorbed into a single system and indexed, or sources are extensively hyperlinked. The first approach does not scale, nor does it resolve issues of alternative nomenclatures (semantics remain unresolved). The second approach does not permit meaningful query across data resources. The EOL approach to this problem is to use community accepted ontologies in which they exist, or to define such ontologies that can then be mapped to APIs using either CORBA-based IDLs or Enterprise Java Beans (EJB) and SOAP protocols [13]. A major goal of the Integrated Biosciences (IBS) and Computational Sciences (CS) programs at the San Diego Supercomputer Center is to integrate the approximately 15 different proteomics resources that exist (see http://biology.sdsc.edu/portals.html) as the first step to a more global integration.

40.4 SUMMARY

The Grid offers great promise for many applications in bioinformatics that will potentially lead to improvements in health care and the quality of life. We have focused here on problems at the 'simple' end of biological complexity – simple and complex biological molecules. More complex biological systems – cells, organs, complete organisms with their associated very large datasets – will continue to challenge Grid hardware and software architects. It is the view of these authors that collaboration between the biology and technology communities is critical to enable advances in biology. Technology is the vehicle rather than the destination; biological researchers must be able to achieve new results in their disciplines.

Note that for many bioinformaticists, research decisions are made not just on technical issues, but on resource issues. For example, if funding support is available, researchers must consider how much science can get done by acquiring a relatively small but dedicated set of processors rather than accessing a much larger set of processors that must be shared with others. The challenge for Grid researchers is to enable disciplinary researchers to easily use a hybrid of local dedicated resources and national and international Grids in a way that enhances the science and maximizes the potential for new disciplinary advances.

If this is not challenge enough, consider, as does Richard Dawkins [12], that the new biology represents the 'Son of Moore's Law.' DNA is easily represented in a digital form that shows similar characteristics to Moore's Law. As such, by 2050 we will be able to sequence the genome of a human for about US $160 in hours, and our genetic profile will be a normal part of our patient record with drugs designed specifically for our profile. We will have sequenced a very large number of species and have a tree of life constructed from comparative genomics. From this, we will be able to interpolate and define the missing link in our ancestry. Further, we will be able to follow the ancestry from birds back to the dinosaurs and construct from the assumed genomes (the blueprints) the phenotype (the physical characteristics) of what species are missing in our world today, but existed millions of years ago. The Grid has a lot to do.

ACKNOWLEDGMENTS

The optimization of CE and the associated portal is the work of Dr Ilya Shindyalov, Dimitri Pekurovsky, Gareth Stockwell, and Jerry Greenberg. We would also like to acknowledge the efforts of Jerry Greenberg, Maytal Dahan, and Steve Mock for the technology portal infrastructures for GAMESS and AMBER; Jerry Greenberg and Nathan Baker for APBS, and Jerry Greenberg and Pavel Pevzner for EULER. The topology of the EOL system is the work of Dr Greg Quinn. The EOL pipeline and its implementation is the work of Drs Ilya Shindyalov, Coleman Mosley, and Wilfred Li. This work is funded by the National Partnership for Advanced Computational Infrastructure (NPACI) grant number NSF ACI 9619020, NIGMS 1P01GM063208-01A1, and the NIH-NBCR (RR08605–06), as well as an NSF-DBI award between the two authors (DBI-0078296).

REFERENCES

1. Pekurovsky, D., Shindyalov, I. N. and Bourne, P. E. (2002) High Throughput Biological Data Processing on Massively Parallel Computers. A Case Study of Pairwise Structure Comparison and Alignment Using the Combinatorial Extension (CE) Algorithm. *Bioinformatics*, Submitted.
2. Shindyalov, I. N. and Bourne, P. E. (2001) CE: A resource to compute and review 3-D protein structure alignments. *Nucleic Acids Research*, **29**(1), 228, 229.
3. Thomas, M., Mock, S. and Boisseau, J. (2000) Development of web toolkits for computational science portals: The NPACI hotpage. *Proc. of the Ninth IEEE International Symposium on High Performance Distributed Computing*, 2000.
4. Foster, I. and Kesselman, C. (1997) Globus: A metacomputing infrastructure toolkit. *International Journal of Supercomputer Applications*, **11**, 115–128.
5. Chaitanya Baru, R. M., Rajasekar, A. and Wan, M. (1998) *Proc. CASCON '98 Conference*, Toronto, Canada, 1998.
6. Schmidt, M. W. *et al.* (1993) The general atomic and molecular electronic structure system. *Journal of Computational Chemistry*, **14**, 1347.
7. Foster, I. and Kesselman, C. (1998) *IPPS/SPDP '98 Heterogeneous Workshop*, 1998, pp. 4–18.
8. http://gridport.npaci.edu/amber.
9. Pevzner, P., Tang, H. and Waterman, M. S. (2001) An Eulerian approach to DNA fragment assembly. *Proceedings of the National Academy of Sciences (USA)*, **98**, 9748–9753, http://gridport.npaci.edu/euler_srb.
10. Baker, N. A., Sept, D., Holst, M. J. and McCammon., J. A. (2001) The adaptive multilevel finite element solution of the Poisson-Boltzmann equation on massively parallel computers. *IBM Journal of Research and Development*, **45**, 427–438, http://gridport.npaci.edu/APBS.
11. Baldridge, K. K. and Greenberg, J. P. (1995) QMView: A computational 3D visualization tool at the interface between molecules and man. *Journal of Molecular Graphics*, **13**, 63–66.
12. Dawkins, R. (2002) in Brockman, J. (ed.) *Son of Moore's Law in The Next Fifty Years* New York: Random House.
13. Seely, S. (2002) *SOAP: Cross Platform Web Service Development Using XML*. Upper Saddle River, NJ: Prentice Hall.

eDiamond: a Grid-enabled federated database of annotated mammograms

Michael Brady,[1] David Gavaghan,[2] Andrew Simpson,[3] Miguel Mulet Parada,[3] and Ralph Highnam[3]

[1]*Oxford University, Oxford, United Kingdom,* [2]*Computing Laboratory, Oxford, United Kingdom,* [3]*Oxford Centre for Innovation, Oxford, United Kingdom*

41.1 INTRODUCTION

This chapter introduces a project named *eDiamond*, which aims to develop a Grid-enabled federated database of annotated mammograms, built at a number of sites (initially in the United Kingdom), and which ensures database consistency and reliable image processing. A key feature of *eDiamond* is that images are 'standardised' prior to storage. Section 41.3 describes what this means, and why it is a fundamental requirement for numerous grid applications, particularly in medical image analysis, and especially in mammography. The *eDiamond* database will be developed with two particular applications in mind: teaching and supporting diagnosis. There are several other applications for such a database, as Section 41.4 discusses, which are the subject of related projects. The remainder of

Grid Computing – Making the Global Infrastructure a Reality. Edited by F. Berman, A. Hey and G. Fox
© 2003 John Wiley & Sons, Ltd ISBN: 0-470-85319-0

this section discusses the ways in which information technology (IT) is impacting on the provision of health care – a subject that in Europe is called *Healthcare Informatics*. Section 41.2 outlines some of the issues concerning medical images, and then Section 41.3 describes mammography as an important special case. Section 41.4 is concerned with medical image databases, as a prelude to the description in Section 41.5 of the *eDiamond* e-Science project. Section 41.6 relates the *eDiamond* project to a number of other efforts currently under way, most notably the US *NDMA* project. Finally, we draw some conclusions in Section 41.7.

All Western societies are confronting similar problems in providing effective healthcare at an affordable cost, particularly as the baby boomer generation nears retirement, as the cost of litigation spirals, and as there is a continuing surge of developments in often expensive pharmaceuticals and medical technologies. Interestingly, IT is now regarded as the key to meeting this challenge, unlike the situation as little as a decade ago when IT was regarded as a part of the problem. Of course, some of the reasons for this change in attitude to IT are generic, rather than being specific to healthcare:

- The massive and continuing increase in the power of affordable computing, and the consequent widespread use of PCs in the home, so that much of the population now regard computers and the Internet as aspects of modern living that are equally indispensable as owning a car or a telephone;
- The miniaturisation of electronics, which have made computing devices ubiquitous, in phones and personal organisers;
- The rapid deployment of high-bandwidth communications, key for transmitting large images and other patient data between centres quickly;
- The development of the global network, increasingly transitioning from the Internet to the Grid; and
- The design of methodologies that enable large, robust software systems to be developed, maintained and updated.

In addition, there are a number of factors that contribute to the changed attitude to IT which are specific to healthcare:

- The increasing number of implementations of hospital information systems, including electronic medical records;
- The rapid uptake of Picture Archiving and Communication Systems (PACS) which enable images and signals to be communicated and accessed at high bandwidth around a hospital, enabling clinicians to store images and signals in databases and then to view them at whichever networked workstation that is most appropriate;
- Growing evidence that advanced decision support systems can have a dramatic impact on the consistency and quality of care;
- Novel imaging and signalling systems (see Section 41.2), which provide new ways to see inside the body, and to monitor disease processes non-invasively;
- Miniaturisation of mechatronic systems, which enable minimally invasive surgery, and which in turn benefits the patient by reducing recovery time and the risk of complications, at the same time massively driving down costs for the health service provider;

- Digitisation of information, which means that the sites at which signals, images and other patient data are generated, analysed, and stored need not be the same, as increasingly they are not;[1] and, by no means least;
- The increased familiarity with, and utilisation of, PCs by clinicians. As little as five years ago, few consultant physicians would use a PC in their normal workflow, now almost all do.

Governments have recognised these benefits and have launched a succession of initiatives, for example, the UK Government's widely publicised commitment to electronic delivery of healthcare by 2008, and its National Cancer Plan, in which IT features strongly.

However, these technological developments have also highlighted a number of major challenges. First, the increasing range of imaging modalities allied to fear of litigation,[2] mean that clinicians are drowning in data. We return to this point in Section 41.2. Second, in some areas of medicine – most notably mammography – there are far fewer skilled clinicians than there is a need for. As we point out in Section 41.3, this offers an opportunity for the Grid to contribute significantly to developing teleradiology in order to allow the geographic separation of the skilled clinician from his/her less-skilled colleague and that clinician's patient whilst improving diagnostic capability.

41.2 MEDICAL IMAGES

Röntgen's discovery of X rays in the last decade of the Nineteenth Century was the first of a continuing stream of technologies that enabled clinicians to see inside the body, without first opening the body up. Since bones are calcium-rich, and since calcium attenuates X rays about 26 times more strongly than soft tissues, X-radiographs were quickly used to reveal the skeleton, in particular, to show fractures. X rays are normally used in transmission mode – the two-dimensional spatial distribution is recorded for a given (known) source flux. A variety of reconstruction techniques, for example, based on the Radon transform, have been developed to combine a series of two-dimensional projection images taken from different directions (normally on a circular orbit) to form a three-dimensional 'tomographic' volume. Computed Tomography (CT) is nowadays one of the tools most widely used in medicine. Of course, X rays are intrinsically ionising radiation, so in many applications the energy has to be very carefully controlled, kept as low as possible, and passed through the body for as short a time as possible, with the inevitable result that the signal-to-noise (SNR) of the image/volume is greatly reduced. X rays of the appropriate energies were used increasingly from the 1930s to reveal the properties of soft tissues, and from the 1960s onwards to discover small, non-palpable tumours for which the prognosis is very good. This is most highly developed for mammography, to

---

[1] This technological change, together with the spread of PACS systems, has provoked turf battles between different groups of medical specialists as to who 'owns' the patient at which stage of diagnosis and treatment. The emergence of the Grid will further this restructuring.

[2] It is estimated that fully 12% of malpractice suits filed in the USA concern mammography, with radiologists overwhelmingly heading the 'league table' of clinical specialties that are sued.

which we return in the next section; but it remains the case that X rays are inappropriate for distinguishing many important classes of soft tissues, for example, white and grey matter in the brain.

The most exquisite images of soft tissue are currently produced using magnetic resonance imaging (MRI), see Westbrook and Kaut [1] for a good introduction to MRI. However, to date, no pulse sequence is capable of distinguishing cancerous tissue from normal tissue, except when using a contrast agent such as the paramagnetic chelate of Gadolinium, gadopentetate dimeglumine, abbreviated as DTPA. In contrast-enhanced MRI to detect breast cancer, the patient lies on her front with the breasts pendulous in a special radio frequency (RF) receiver coil; one or more image volumes are taken prior to bolus injection of DTPA and then image volumes are taken as fast as possible, for up to ten minutes. In a typical clinical setting, this generates 12 image volumes, each comprising 24 slice images, each 256×256 pixels, a total of 18 MB per patient per visit. This is not large by medical imaging standards, certainly it is small compared to mammography. Contrast-enhanced MRI is important for detecting cancer because it highlights the neoangeogenesis, a tangled mass of millions of micron-thick leaky blood vessels, grown by a tumour to feed its growth. This is essentially physiological – functional – rather than anatomical – information [2]. Nuclear medicine modalities such as positron-emission tomography (PET) and single photon emission computed tomography (SPECT) currently have the highest sensitivity and specificity for cancer, though PET remains relatively scarce, because of the associated capital and recurrent costs, not least of which involve a cyclotron to produce the necessary quantities of radio-pharmaceuticals.

Finally, in this very brief tour (see [3, 4] for more details about medical imaging), ultrasound image analysis has seen major developments over the past decade, with Doppler, second harmonic, contrast agents, three-dimensional probes, and so on; but image quality, particularly for cancer, remains sufficiently poor to offset its price advantages.

Generally, medical images are large and depict anatomical and pathophysiological information of staggering variety both within a single image and across a population of images. Worse, it is usually the case that clinically significant information is quite subtle. For example, Figure 41.1 shows a particularly straightforward example of a mammogram.

Microcalcifications, the small white spots shown in Figure 41.1, are deposits of calcium or magnesium salts that are smaller than 1 mm. Clusters of microcalcifications are often the earliest sign of non-palpable breast cancer, though it must be stressed that benign clusters are often found, and that many small white dots do not correspond to microcalcifications (see Highnam and Brady [5] for an introduction to the physics of mammography and to microcalcifications). In order to retain the microcalcifications that a skilled radiologist can detect, it is usual to digitise mammograms to a resolution of 50 to 100μ. It has been found that the densities in a mammogram need to be digitised to a resolution of 14 to 16 bits, yielding 2 bytes per pixel. An A4-sized mammogram digitised at the appropriate resolution gives an image that is typically 4000×4000 pixels, that is 32 MB. Generally, two views – craniocaudal (CC, head to toe) and mediolateral oblique (MLO, shoulder to opposite hip) – are taken of each of the breasts, giving 128 MB per patient per visit, approximately an order of magnitude greater than that from a contrast-enhanced MRI

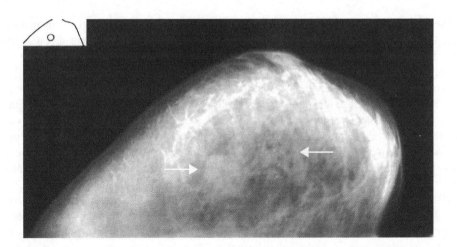

Figure 41.1 A patient aged 61 years presented with a breast lump. Mammography reveals a 2 cm tumour and extensive microcalcifications, as indicated by the arrows. Diagnostically, this is straightforward.

examination. Note that the subtlety of clinical signs means that in practice only loss-less image compression can be used.

Medical images have poor SNR, relative to good quality charge-coupled device (CCD) images (the latter is nowadays less than 1% noise, a factor of 5 to 10 better than most medical images). It is important to realise that there are distortions of many kinds in medical images. As well as high frequency noise (that is rarely Gaussian), there are degrading effects, such as the 'bias field', a low-frequency distortion due to imperfections in the MRI receiver coil. Such a degradation of an image may appear subtle, and may be discounted by the (expert) human eye; but it can distort massively the results of automatic tissue classification and segmentation algorithms, and give wildly erroneous results for algorithms attempting quantitative analysis of an image.

Over the past fifteen years there has been substantial effort aimed at medical image analysis – the interested reader is referred to journals such as IEEE Transactions on Medical Imaging or Medical Image Analysis, as well as conference proceedings such as MICCAI (Medical Image Computation and Computer-Assisted Intervention). There has been particular effort expended upon image segmentation to detect regions-of-interest: shape analysis, motion analysis, and non-rigid registration of data, for example, from different patients. To be deployed in clinical practice, an algorithm has to work 24/7 with extremely high sensitivity and specificity. This is a tough specification to achieve even for images of relatively simple shapes and in cases for which the lighting and camera-subject pose can be controlled; it is doubly difficult for medical images, for which none of these simplifying considerations apply. There is, in fact, a significant difference between image analysis that uses medical images to illustrate the performance of an algorithm, and medical image analysis, in which application-specific information is embedded in algorithms in order to meet the demanding performance specifications.

We noted in the previous section that clinicians often find themselves drowning in data. One potential solution is data fusion – the integration of diverse data sets in a single cohesive framework – which provides the clinician with information rather than data. For example, as we noted above, PET, and SPECT can help identify the microvasculature grown by a tumour. However, the spatial resolution of PET is currently relatively poor (e.g. 3 to 8 mm voxels), too poor to be the basis for planning (say) radiotherapy. On the other hand, CT has excellent spatial resolution; but it does not show soft tissues such as grey matter, white matter, or a brain tumour. Data fusion relates information in the CT with that in the PET image, so that the clinician not only knows *that* there is a tumour but *where* it is. Examples of data fusion can be found by visiting the Website: http://www.mirada-solutions.com

PACS systems have encouraged the adoption of standards in file format, particularly the DICOM standards – digital communication in medicine. In principle, apart from the raw image data, DICOM specifies the patient identity, the time and place at which the image was taken, gives certain technical information (e.g. pulse sequence, acquisition time), specifies out the region imaged, and gives information such as the number of slices, and so on.

Such is the variety of imaging types and the rate of progress in the field that DICOM is currently an often frustrating, emerging set of standards.

41.3 MAMMOGRAPHY

41.3.1 Breast cancer facts

Breast cancer is a major problem for public health in the Western world, where it is the most common cancer among women. In the European Community, for example, breast cancer represents 19% of cancer deaths and fully 24% of all cancer cases. It is diagnosed in a total of 348 000 cases annually in the United States and the European Community and kills almost 115 000 annually. Approximately 1 in 8 of women will develop breast cancer during the course of their lives, and 1 in 28 will die of the disease. According to the World Health Organization, there were 900 000 new cases worldwide in 1997. Such grim statistics are now being replicated in eastern countries as diets and environment become more like their western counterparts.

During the past sixty years, female death rates in the United States from breast cancer stayed remarkably constant while those from almost all other causes declined. The sole exception is lung cancer death rates, which increased sharply from 5 to 26 per 100 000. It is interesting to compare the figures for breast cancer with those from cervical cancer, for which mortality rates declined by 70% after the cervical smear gained widespread acceptance.

The earlier a tumour is detected the better the prognosis. A tumour that is detected when its size is just 0.5 cm has a favourable prognosis in about 99% of cases, since it is highly unlikely to have metastasized. Few women can detect a tumour by palpation (breast self-examination) when it is smaller than 1 cm, by which time (on average) the tumour will have been in the breast for up to 6 to 8 years. The five-year survival rate for localized breast cancer is 97%; this drops to 77% if the cancer has spread by the time of diagnosis and to 22% if distant metastases are found (Journal of the National Cancer Institute).

This is the clear rationale for screening, which is currently based entirely on X ray mammography (though see below). The United Kingdom was the first country to develop a national screening programme, though several other countries have established such programmes: Sweden, Finland, The Netherlands, Australia, and Ireland; France, Germany and Japan are now following suit. The first national screening programme was the UK Breast Screening Programme (BSP), which began in 1987. Currently, the BSP invites women between the ages of 50 and 64 for breast screening every three years. If a mammogram displays any suspicious signs, the woman is invited back to an assessment clinic where other views and other imaging modalities are utilized. Currently, 1.3 million women are screened annually in the United Kingdom. There are 92 screening centres with 230 radiologists, each radiologist reading on average 5000 cases per year, but some read up to 20 000.

The restriction of the BSP to women aged 50 and above stems from fact that the breasts of pre-menopausal women, particularly younger women, are composed primarily of milk-bearing tissue that is calcium-rich; this milk-bearing tissue involutes to fat during the menopause – and fat is transparent to X rays. So, while a mammogram of a young woman appears like a white-out, the first signs of tumours can often be spotted in those of post-menopause women. In essence, the BSP defines the menopause to be substantially complete by age 50!

The UK programme resulted from the Government's acceptance of the report of the committee chaired by Sir Patrick Forrest. The report was quite bullish about the effects of a screening programme:

> by the year 2000 the screening programme is expected to prevent about 25% of deaths from breast cancer in the population of women invited for screening ... On average each of the women in whom breast cancer is prevented will live about 20 years more. Thus by the year 2000 the screening programme is expected to result in about 25 000 extra years of life gained annually in the UK.

To date, the BSP has screened more than eleven million women and has detected over 65 000 cancers. Research published in the BMJ in September 2000 demonstrated that the National Health Service (NHS) Breast Screening Programme is saving at least 300 lives per year. The figure is set to rise to 1250 by 2010. More precisely, Moss (British Medical Journal 16/9/2000), demonstrated that the NHS breast screening program, begun in 1987, resulted in substantial reductions in mortality from breast cancer by 1998. In 1998, mortality was reduced by an average of 14.9% in those aged 50 to 54 and 75 to 79, which would be attributed to treatment improvements. In the age groups also affected by screening (55 to 69), the reduction in mortality was 21.3%. Hence, the estimated direct contribution from screening was 6.4%.

Recent studies suggest that the rate of interval at which cancers appear between successive screening rounds is turning out to be considerably larger than predicted in the Forrest Report. Increasingly, there are calls for mammograms to be taken every two years and for both a CC and MLO image to be taken of each breast.

Currently, some 26 million women are screened in the United States annually (approximately 55 million worldwide). In the United States there are 10 000 mammography-accredited units. Of these, 39% are community and/or public hospitals, 26% are private radiology practices, and 13% are private hospitals. Though there are 10 000 mammography centres, there are only 2500 mammography specific radiologists – there is a worldwide shortage of radiologists and radiologic technologists (the term in the United Kingdom is radiographers). Huge numbers of mammograms are still read by non-specialists, contravening recommended practice, nevertheless continuing with average throughput rates between 5 and 100 per hour. Whereas expert radiologists have cancer detection rates of 76 to 84%, generalists have rates that vary from between 8 to 98% (with varying numbers of false-positives). The number of cancers that are deemed to be visible in retrospect, that is, when the outcome is known, approaches 70% (American Journal of Roentgenology 1993). Staff shortages in mammography seem to stem from the perception that it is 'boring but risky': as we noted earlier, 12% of all malpractice lawsuits in the United States are against radiologists, with the failure to diagnose breast cancer becoming one of the leading reasons for malpractice litigation (AJR 1997 and Clark 1992). The shortage of radiologists is driving the development of specialist centres and technologies (computers) that aspire to replicate their skills. Screening environments are ideally suited to computers, as they are repetitive and require objective measurements.

As we have noted, screening has already produced encouraging results. However, there is much room for improvement. For example, it is estimated that a staggering 25% of cancers are missed at screening. It has been demonstrated empirically that double reading greatly improves screening results; but this is too expensive and in any case there are too few screening radiologists. Indeed, recall rates drop by 15% when using 2 views of each breast (British Medical Journal, 1999). Double reading of screening mammograms has been shown to half the number of cancers missed. However, a study at Yale of board certified, radiologists showed that they disagreed 25% of the times about whether a biopsy was warranted and 19% of the time in assigning patients to 1 of 5 diagnostic categories. Recently, it has been demonstrated that single screening plus the use of computer-aided diagnosis (CAD) tools – image analysis algorithms that aim to detect microcalcifications and small tumours – also greatly improve screening effectiveness, perhaps by as much as 20%.

Post-screening, the patient may be assessed by other modalities such as palpation, ultrasound and increasingly, by MRI. 5 to 10% of those screened have these extended 'work-up'. Post work-up, around 5% of patients have a biopsy. In light of the number of tumours that are missed at screening (which reflects the complexity of diagnosing the disease from a mammogram), it is not surprising that clinicians err on the side of caution and order a large number of biopsies. In the United States, for example, there are over one million biopsies performed each year: a staggering 80% of these reveal benign (non-cancerous) disease.

It has been reported that between screenings 22% of previously taken mammograms are unavailable or are difficult to find, mostly because of the fact that they have been misfiled in large film archives – lost films are a daily headache for radiologists around the world, 50% were obtained only after major effort, Bassett *et al.* (American Journal of Roentgenology, 1997).

41.3.2 Mammographic images and standard mammogram form (SMF)

Figure 41.2 is a schematic of the formation of a (film-screen) mammogram. A collimated beam of X rays passes through the breast and is compressed (typically to a force of 14 N) between two Lucite plates. The X-ray photons that emerge from the lower plate pass through the film before being converted to light photons, which then expose the film, which is subsequently scanned (i.e. converted to electrons) at a resolution (typically) of 50 μ. In the case of full-field digital mammography, the X-ray photons are converted directly to electrons by an amorphous silicon sensor that replaces the film screen. As Figure 41.2 also shows, a part of the X-ray flux passes in a straight line through the breast, losing a proportion of less energetic photons en route as they are attenuated by the tissue that is encountered. The remaining X-ray photon flux is scattered and arrives at the sensor surface from many directions (which are, in practice, reduced by an anti-scatter grid, which has the side-effect of approximately doubling the exposure of the breast). Full details of the physics of image acquisition, including many of the distorting effects, and the way in which image analysis algorithms can be developed to undo these distortions, are presented in Highnam and Brady [5].

For the purposes of this article, it suffices to note that though radiologic technologists are well trained, the control over image formation is intrinsically weak. This is illustrated in Figure 41.3, which shows the *same* breast imaged with two different exposure times. The images appear very different. There are many parameters **p** that affect the appearance of a mammogram, including: tube voltage, film type, exposure time, and placement of an automatic exposure control. If these were to vary freely for the same compressed breast, there would be huge variation in image brightness and contrast. Of course, it would be ethically unacceptable to perform that experiment on a living breast: the accumulated radiation dose would be far too high. However, it *is* possible to develop a mathematical model of the formation of a mammogram, for example, the Highnam-Brady physics model. With such a model in hand, the variation in image appearance can be *simulated*. This is the basis of the teaching system *VirtualMammo* developed

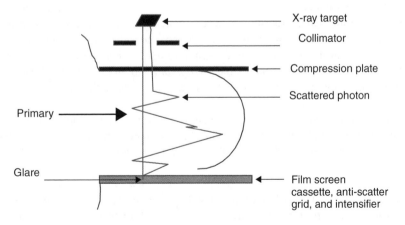

Figure 41.2 Schematic of the formation of a mammogram.

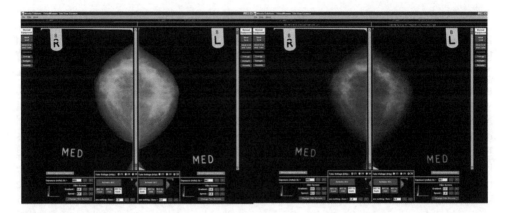

Figure 41.3   Both sets of images are of the same pair of breasts, but the left pair is scanned with a shorter exposure time than the right pair – an event that can easily happen in mammography. Image processing algorithms that search for 'bright spots' will be unable to deal with such changes.

by Mirada Solutions Limited in association with the American Society of Radiologic Technologists (ASRT).

The relatively weak control on image formation, coupled with the huge change in image appearance, at which Figure 41.3 can only hint, severely limits the usefulness of the (huge) databases that are being constructed – images submitted to the database may tell more about the competence of the technologists who took the image, or the state of the equipment on which the image was formed, than about the patient anatomy/physiology, which is the reason for constructing the database in the first place! It is precisely this problem that the *eDiamond* project aims to address.

In the course of developing an algorithm to estimate, and correct for, the scattered radiation shown in Figure 41.2, Highnam and Brady [5] made an unexpected discovery: it is possible to estimate, accurately, the amount of non-fat tissue in each pixel column of the mammogram. More precisely, first note that the X-ray attenuation coefficients of normal, healthy tissue and cancerous tissue are very nearly equal, but are quite different from that of fat. Fat is clinically uninteresting, so normal healthy and cancerous tissues are collectively referred to as 'interesting': Highnam and Brady's method estimates – in millimetres – the amount of interesting tissue in each pixel column, as is illustrated in Figure 41.4.

The critical point to note is that the interesting tissue representation refers only to (projected) anatomical structures – the algorithm has estimated and eliminated the particular parameters $\mathbf{p}(\mathbf{I})$ that were used to form this image $\mathbf{I}$. In short, the image can be regarded as *standardised*. Images in standardised form can be included in a database without the confounding effect of the (mostly irrelevant – see below) image formation parameters. This greatly increases the utility of that database. Note also that the interesting tissue representation is *quantitative*: measurements are in millimetres, not in arbitrary contrast units that have no absolute meaning.

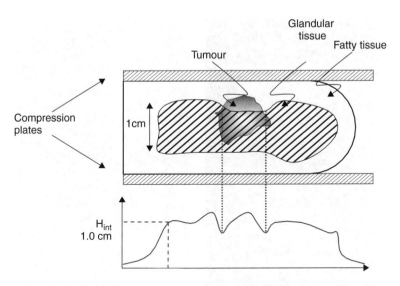

Figure 41.4   The 'interesting tissue' representation developed by Highnam and Brady [5]. Tumours and normal glandular tissue together form the 'interesting' tissue class. The algorithm estimates, for each column the amount of interesting tissue. If, for example, the separation between the Lucite plates is 6.5 cm, the amount of interesting tissue at a location (x, y) might be 4.75 cm, implying 1.75 cm of fat.

Figures 41.5 and 41.6 show two different *depictions* of the interesting tissue representation: one as a surface, one as a standardised image. The information content of these two depictions is precisely the same – whether one chooses to work with a surface depiction, which is useful in some cases, or with an image depiction, which is useful in others, depends only upon the particular application and the user's preference. *The information that is recorded in the database is the same in both cases, and it is freed of the confounding effects of weakly controlled image formation.*

It should now be clear why the first *eDiamond* database has been based on mammography: (1) there are compelling social and healthcare reasons for choosing breast cancer and (2) the interesting tissue representation provides a standardisation that is currently almost unique in medical imaging. Larry Clarke, Chief of Biomedical Imaging at the National Cancer Institute in Washington DC recently said: 'I believe standardisation is crucial to biomedical image processing.'

41.4 MEDICAL DATABASES

Medical image databases represent both huge challenges and huge opportunities. The challenges are many. First, as we have noted earlier, medical images tend to be large, variable across populations, contain subtle clinical signs, have a requirement for loss-less compression, require extremely fast access, have variable quality, and have privacy as

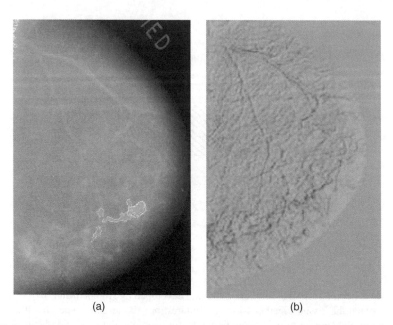

(a)   (b)

Figure 41.5   Depicting the interesting tissue representation as a surface. In this case, the amounts of interesting tissue shown in Figure 41.4 are regarded as heights (b), encouraging analyses of the surface using, for example, local differential geometry to estimate the characteristic 'slopes' of tumours. This depiction is called the h_{int} *image*.

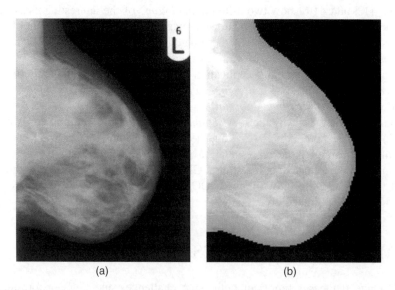

(a)   (b)

Figure 41.6   Depicting the interesting tissue representation as an image. In this case a standard set of imaging parameters $\mathbf{p(S)}$ are chosen and a fresh image is formed (b). Note that this is fundamentally different from applying an enhancement algorithm such as histogram equalisation, whose result would be unpredictable. This depiction is called the *standard mammogram form SMF*.

a major concern. More precisely, medical images often involve 3D+t image sequences taken with multiple imaging protocols. Imaging information aside, metadata concerning personal and clinical information is usually needed: age, gender, prior episodes, general health, incidence of the disease in close relatives, disease status, and so on. Medical records, including images, are subject to strict ethical, legal and clinical protocols that govern what information can be used for what purposes. More prosaically, there are practical problems of multiple, ill-defined data formats, particularly for images. All of this means that the data is difficult to manage, and is often stored in a form in which it is only useful to individuals who know, independently, what it is, how it was acquired, from whom, and with what consent. In addition to these challenges are the facts that inter- and even intra-variability amongst clinicians is often 30 to 35%, and that the ground truth of diagnosis is hard to come by.

These issues are becoming a major issue in view of the rapid growth of on-line storage of medical records, particularly images, and particularly in the United States. (In the United States, each individual owns their records, as is natural in a society that enables individuals to seek second and third opinions, and to approach medical specialists directly.) In the United Kingdom, by way of contrast, the NHS decrees that medical records are owned by the individual's clinician (primary care or specialist), and there are constraints on changing clinicians and on seeking second opinions. This situation is in a state of flux as healthcare provision is being increasingly privatised, complicating even further the development of a data archive that is designed to have lasting value.

As we noted in the previous section, acquiring any kind of medical image involves setting a large number of parameters, which as often as not, reflect the individual preferences of the clinician and whose effects are confounded in the appearance of the image. The practical impact is that it is usually the case even at large medical centres that the number of images that are gathered over any given time interval – say, a year – is insufficient to avoid statistical biases such as clinician preference. In practice, this may be one of the most important contributions of the Grid, for example, to facilitate epidemiological studies using numbers of images that have sufficient statistical power to overcome biases. In a federated database, for example, one might develop a 'leave-one-out' protocol to test each contributing institution against the pooled set of all other institutions. A variant to the leave-one-out methodology has, in principle, the potential for automatic monitoring of other variations between institutions such as quality control. This is currently being studied in a complementary European project *Mammogrid,* which also involves Oxford University and Mirada Solutions.

Section 41.3 also pointed out that in the United States there are currently far fewer mammography specialist radiologists (2500) than there are mammogram readers (25 000) or mammography machines (15 000). This has re-awakened the idea of teleradiology: shipping at least problematical images to centres of expertise, an idea that demands the bandwidth of the Grid. In a related vein, clinicians and technologists alike in the United States are subject to strict continuing education (CE) requirements. Mirada Solutions' *VirtualMammo* is the first of what is sure to be a stream of computer-based teaching and CE credit systems for which answers can be submitted over the net and marked remotely. The ways in which a large database of the type that *eDiamond* envisages can be used to further teaching – for technologists, medical physicists, and radiologists alike – will be a major part of the scientific development of the *eDiamond* project.

Finally, there are a whole range of diagnostic uses for a huge database such as *eDiamond* is building. These range from training classification algorithms on instances for which the 'ground truth' (i.e. diagnosis, confirmed or not by pathology) is known, through to data mining applications. To this latter end, Mirada Solutions has developed a data mining package called *FindOneLikeIt* in which the clinician identifies a region of interest in a mammogram and the data mining system trawls the database to find the 10 image fragments – together with associated metadata including diagnoses. The current system has been developed for a static database comprising 20 000 images; the challenge for *eDiamond* is to further develop both the algorithm and the database search when the database is growing rapidly (so that the statistical weight associated with a particular feature is changing rapidly).

41.5 eDIAMOND

41.5.1 Introduction

The *eDiamond* project is designed to deliver the following major benefits, initially for mammography in the United Kingdom, but eventually wider both geographically and for other image modalities and diseases. The project aims to use the archive to evaluate innovative software based on the SMF standardisation process to compute the quality of each mammogram as it is sent to the archive. This will ensure that any centre feeding into the archive will be automatically checked for quality. A result of the *eDiamond project* will be an understanding of automated methods – based on standardisation – for assessing quality control. Second, the *eDiamond* archive will provide a huge teaching and training resource. Thirdly, radiologists faced with difficult cases will be able to use the archive as a resource for comparing difficult cases with previous cases, both benign and malignant. This has the potential to reduce the number of biopsies for benign disease thus reducing trauma and cost to the NHS. Fourth, it is intended that *eDiamond* will be able to provide a huge resource for epidemiological studies such as those relating to hormone replacement therapy (HRT) [6], to breast cancer risk, and to parenchymal patterns in mammograms [7]. Finally, as regards computer-aided detection CADe, *eDiamond* contributes to the development of a technology that appears able to increase detection rates, and thus save lives.

The project offers the following stakeholder benefits:

- Patients will benefit from secure storage of films, better and faster patient record access, better opinions, and lowering of the likelihood of requiring a biopsy. They will begin to 'own' their mammography medical records.
- Radiologists will benefit from computer assistance, massively reduced storage space requirements, instant access to mammograms without loss of mammograms, improved early diagnosis (because of improved image quality), and greater all-round efficiency leading to a reduction in waiting time for assessment. Radiologists will also benefit by

applying data mining technology to the database to seek out images that are 'similar' to the one under consideration, and for which the diagnosis is already known.

- Administrative staff, although their initial workload will increase, will subsequently benefit from the significantly faster image archiving and retrieval.
- Hospital Trust managers will benefit from the overall reduced cost of providing a better quality service.
- Hospital IT Managers will benefit from the greatly reduced burden on their already over-stretched resources. The national consolidation in storage implied by *eDiamond* promises to reduce drastically their costs through reduced equipment and staffing requirements and support contracts.
- Researchers will benefit, as the *eDiamond* database, together with associated computing tools, will provide an unparalleled resource for epidemiological studies based on images.
- The Government will benefit, as it is intended that *eDiamond* will be the basis of an improved service at greatly reduced cost. Furthermore, *eDiamond* will be a pilot and proving ground for other image-based databases and other areas where secure on-line data storage is important.

41.5.2 e-Science challenges

This project requires that several generic e-Science challenges be addressed both by leveraging existing Grid technology and by developing novel middleware solutions. Key issues include the development of each of the following:

- Ontologies and metadata for the description of demographic data, the physics underpinning the imaging process, key features within images, and relevant clinical information.
- Large, federated databases both of metadata and images.
- Data compression and transfer.
- Effective ways of combining Grid-enabled databases of information that must be protected and which will be based in hospitals that are firewall-protected.
- Very rapid data mining techniques.
- A secure Grid infrastructure for use within a clinical environment.

41.5.3 Objectives

It is currently planned to construct a large federated database of annotated mammograms at St George's and Guy's Hospitals in London, the John Radcliffe Hospital in Oxford, and the Breast Screening Centres in Edinburgh and Glasgow. All mammograms entering the database will be SMF standardised prior to storage to ensure database consistency and reliable image processing. Applications for teaching, aiding detection, and aiding diagnosis will be developed.

There are three main objectives to the initial phase of the project: the development of the Grid technology infrastructure to support federated databases of huge images (and related information) within a secure environment; the design and construction of the Grid-connected workstation and database of standardised images; and the development, testing and validation of the system on a set of important applications. We consider each of these in turn.

41.5.3.1 Development of the Grid infrastructure

There are a number of aspects to the development of a Grid infrastructure for *eDiamond*. The first such aspect is security: ensuring secure file transfer, and tackling the security issues involved in having patient records stored on-line, allowing access to authorised persons but also, potentially, the patients themselves at some time in the future are all key issues.

The design and implementation of Grid-enabled federated databases for both the metadata and the images is another such aspect, as is the issue of data transfer: typically, each image is 30 Mb, or 120 Mb for a set of 4 images, which is the usual number for a complete case. Issues here revolve around (loss-less) data compression and very rapid and secure file transfer.

Data mining issues revolve around speed of access: we aim for *eDiamond* to return the ten most similar cases within 8 to 10 s. Teaching tools that test students by production of 'random' cases will need to work at the speed of current breast screening, which means the next case will need to be displayed within seconds after the 'next' button is hit.

41.5.3.2 Database construction and data mining

Ontologies are being developed for description of patient and demographic data, together with descriptions of the image parameters and of features within images.

Database design is tailored to the needs of rapid search and retrieval of images, and the database is being built within the DB2 framework. The needs of future epidemiological studies and information related to high risk such as family history and known breast cancer causing mutations are also being incorporated.

Some prototype data mining tools have already been developed within the context of a single database. Again, speed of access is crucial: the database architecture is being determined according to the frequency of requests for particular image types.

41.5.3.3 System testing and validation

Each of the clinical partners is being provided with a specialised breast imaging worksta-tion to allow access to the Grid-enabled system. This will allow validation of all aspects of the development process, with the following aspects being focused upon: teaching and education, which will include the ability to undertake random testing from the database; epidemiological studies, which will involve the analysis of breast patterns using various quantification techniques; and diagnosis, which will use the database to study the benefits of computer-aided diagnosis.

41.5.4 Project structure

The overall structure of the *eDiamond* architecture is illustrated in Figure 41.7.

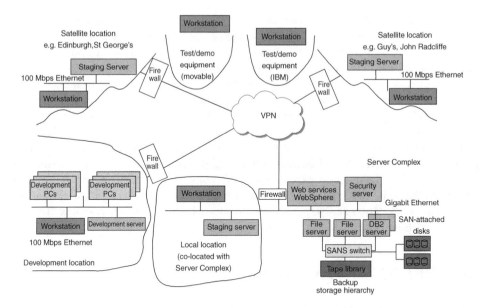

Figure 41.7 The proposed architecture for the eDiamond system. Individual hospitals and development centres are inter-connected by a VPN and protected by a firewall equipped with suitable gatekeeper software. The lower part of the figure, on the right, shows the file server complex for eDiamond..

The system consists of a number of different types of location: the server complex, the satellite locations, the development locations; and the local workstation locations. We consider each in turn.

In the *eDiamond* project, IBM Hursley will provide technical consultancy in areas such as understanding user scenarios, architecture and design, the use of products such as WAS and DB2, security and privacy (including encryption and watermarking); and optimisation for performance. In addition, IBM will provide selected software by means of the IBM Scholars Program, including IBM DB2 Universal Database, IBM DiscoveryLink, IBM WebSphere Application Server (WAS), MQ Series Technology Release and Java and XML tools on Linux and other platforms. Finally, IBM will provide access to relevant IBM Research conferences and papers relevant to this project. A sister project being conducted with Dr Steve Heisig (IBM Watson Lab, New York) looking at the generic problem of Workload Management in a Grid environment is the subject of a separate (submitted) application to the DTI from OeSC. This project will use the *eDiamond* project as its primary application test bed.

The server complex will be relatively conventional, and consist of: a high-speed switched network, a connection to the Wide Area Network (WAN) through which the remote locations connect to the server complex, a firewall, two file servers with disks attached via a Storage Area Network, two database servers with the database contents stored on SAN-attached disks, a small automatic tape library, a security server and a Web Services/WebSphere Application Server machine. In this case, the satellite locations are hospital and/or teaching locations where data is created for loading onto the system, and also where authorised users are able to access the system. Each will have a standard system user workstation and a 'staging server' connected to each other and to the external link via a 100 Mbps Ethernet network. Consideration should be given to making this internal network also be Gigabit Ethernet to maximise the performance of image loading. The staging server is included in the configuration to allow for the pre-loading of images and to avoid delays that might occur if all images had to be retrieved over the WAN. The workstation locations comprise a workstation and staging server configuration, co-located with the server complex, that is, on the same high-speed network, so as to explore the performance characteristics of accessing image files when the network performance is as good as possible. In order to explore the performance characteristics of using a workstation on the same high-speed network as the server complex, it is proposed that there should be a workstation and staging server co-located with the server complex. These are identical to those deployed at the satellite locations.

41.6 RELATED PROJECTS

The *eDiamond* project (in most cases, deliberate) overlaps with several other grid projects, particularly in mammography, and more generally in medical image analysis. First, the project has strong links to the US *NDMA* project, which is exploring the use of Grid technology to enable a database of directly digitised (as opposed to film-screen) mammograms. IBM is also the main industrial partner for the US *NDMA* project, and has provided a Shared University Research (SUR) grant to create the *NDMA* Grid under the leadership of the University of Pennsylvania. Now in Phase II of deployment, the project connects hospitals in Pennsylvania, Chicago, North Carolina, and Toronto. The architecture of the *NDMA* Grid leverages the strengths of the IBM' eServer clusters – running AIX and Linux – with open protocols from Globus. The data volumes will exceed 5 petabytes per year, with network traffic at 28 terabytes per day. In addition, privacy mandates that all image transmissions and information concerning patient data be encrypted across secure public networks.

Teams from the University of Pennsylvania and IBM have worked together to implement a prototype fast access, very large capacity DB2 Universal Database to serve as the secure, highly available index to the digitised X-ray data. Operation of this system is enhanced by DB2 parallel technology that is capable of providing multi-gigabyte performance. This technology also enables scalable performance on large databases by breaking the processing into separate execution components that can be run concurrently on multiple processors. The *eDiamond* project will be able to learn to draw from and utilise the considerable body of practical experience gained in the development of the *NDMA* Grid. It is expected that *eDiamond* and *NDMA* will collaborate increasingly closely. A critical difference between *eDiamond* and the *NDMA* project will be that *eDiamond* will use standardisation techniques prior to image storage in the database.

Second, there is a complementary European project *Mammogrid*, which also involves Oxford University and Mirada Solutions, together with CERN, Cambridge University, the University of Western England, and Udine; but which, as noted in the previous section, will concentrate on three different applications: the use of a federated database for quality control, for training and testing a system to detect microcalcification clusters, and to initiate work on using the grid to support epidemiological studies.

Third, we are just beginning work on a project entitled 'Grid-Enabled Knowledge Services: Collaborative Problem Solving Environments in Medical Informatics'. This is a programme of work between the UK Inter-disciplinary Research Consortium (IRC) entitled MIAS[3] (from Medical Images and Signals to Clinical Information) and one of the other computer-centric IRCs, the Advanced Knowledge Technologies (AKT). The domain of application is collaborative medical problem solving using knowledge services provided via the e-Science Grid infrastructure. In particular, the initial focus will be Triple Assessment in symptomatic focal breast disease. The domain has been chosen because it contains a number of characteristics that make it especially valuable as a common focus for the IRC e-Science research agenda and because it is typical of medical multi-disciplinary team-based evidential reasoning about images, signals, and patient data. Triple assessment is a loosely regulated cooperative decision-making process that involves radiologists, oncologists, pathologists, and breast surgeons. The aim is to determine, often rapidly, the most appropriate management course for a patient: chemotherapy, neo-adjuvant chemotherapy, exploratory biopsy, surgery – at varying levels of severity – or continued surveillance with no treatment.

The vision of the Grid that is taking shape in the UK e-Science community consists of a tier of services ranging across the computational fabric, information and data management, and the use of information in particular problem solving contexts and other knowledge intensive tasks. This project regards the e-Science infrastructure as a set of services that are provided by particular institutions for consumption by others, and, as such, it adopts a service-oriented view of the Grid. Moreover, this view is based upon the notion of various entities providing services to one another under various forms of contract and provides one of the main research themes being investigated – agent-oriented delivery of knowledge services on the Grid.

The project aims to extend the research ambitions of the AKT and MIAS IRCs to include researching the provision of AKT and Medical Image and Signal (MIS) technologies in a Grid services context. Although this work focuses on medical application the majority of the research has generic applicability to many e-Science areas. The project aims at the use of the Grid to solve a pressing – and typical – medical problem rather than seeking primarily to develop the Grid architecture and software base. However, it seeks to provide information enrichment and knowledge services that run in a Grid environment, and so demonstrate applications at the information and knowledge tiers.

The project will complement other projects under development within the Oxford e-Science Centre, and also within the UK e-Science community more generally. Within Oxford, several projects involve the building and interrogation of large databases, whilst others involve methods of interrogating video images, and large-scale data compression for visualisation. Within the wider UK community, Grid-enabling database technologies is a fundamental component of several of the EPSRC Pilot Projects, and is anticipated to be one of the United Kingdom's primary contributions to the proposed Open Grid

---

[3] MIAS is directed by Michael Brady and includes the Universities of Oxford, Manchester, King's College, London, University College, London, and Imperial College of Science, Technology and Medicine, London. It is supported by EPSRC and MRC.

Services Architecture (OGSA). In particular, the MIAS IRC is also developing a *Dynamic Brain Atlas* in which a set of features are determined by a clinician to be relevant to this patient (age, sex, medical history) and a patient-specific atlas is created from those images in a database of brain images that match the features (see References 8 and 9 for more details).

IBM is involved fully in this work, and is participating actively in the development of Data Access and Integration (DAI) for relational and XML databases (the OGSA – DAI project) that can be used in conjunction with future releases of Globus, including OGSA features. Additionally, IBM's strategic database support, DB2, provides XML and Web services support within the product, and it is anticipated that this support will evolve to incorporate any significant developments in the evolution of OGSA.

IBM is also actively involved in the European DataGrid Project, which will provide middleware to enable next-generation exploitation of petabytes datasets of the order of petabytes. Some of the technologies being developed in that project are relevant to this project; overlap between these projects will help drive the integration of the range of different Grid technologies available today.

41.7 CONCLUSIONS

eDiamond is an initial exploration of developing a grid-enabled, federated database to support clinical practice throughout the United Kingdom. One of its key innovations is that images will be standardised, removing as much as possible of the irrelevant image formation information that is confounded in the appearance of the image. Of course, *eDiamond* is only a start – for medical image databases in general, and even for the construction of a federated database of breast cancer images, which will, in the future, need to incorporate MRI, PET, ultrasound and other image types. This in turn will necessitate non-rigid registration of datasets and, perhaps, the development of an appropriate coordinate frame for the breast to complement those such as the Tallairach atlas for the brain. Finally, clinical diagnosis of breast cancer relies as much on *non*-image data: individual records about a patient's medical history, diet, lifestyle, and the incidence of breast cancer in the patient's family, workplace, and region. There has been excellent work done on the use of such non-image data, but it remains a significant challenge to integrate image and non-image data, as is done effortlessly by the skilled clinician. Plainly, there is much to do.

ACKNOWLEDGEMENTS

This project would not have happened without the vision, drive and support of Tony Hey, Anne Trefethen, and Ray Browne. Paul Jeffreys of the Oxford eScience Centre continues to be a strong supporter, and we acknowledge a debt to the drive of Roberto Amendolia, who leads the *Mammogrid* project. Numerous other colleagues in the UK's IRCs, particularly, David Hawkes, Derek Hill, Nigel Shadbolt, and Chris Taylor have had a major influence on the project. Finally, the project could not have taken place without the support of the Engineering and Physical Sciences Research Council and the Medical Research Council. JMB acknowledges many fruitful conversations with Paul Taylor, John Fox, and Andrew Todd-Pokropek of UCL.

REFERENCES

1. Westbrook, C. and Kaut, C. (1998) *MRI in Practice*. Cambridge, MA: Blackwell Science.
2. Armitage, P. A., Behrenbruch C. P., Brady, M. and Moore, N. (2002) Extracting and visualizing physiological parameters using dynamic contrast-enhanced magnetic resonance imaging of the breast. *IEEE Transactions on Medical Imaging*.
3. Webb, S., (ed.) (1988) *The Physics of Medical Imaging*. Medical Science Series, Bristol, Philadelphia: Institute of Physics Publishing.
4. Michael Fitzpatrick, J., and Sonka, M. (eds.) *Handbook of Medical Imaging, Volume 2: Medical Image Processing and Analysis*. SBN 0-8194-3622-4, Monograph Vol. PM80, Bellingham, WA: SPIE Press, p. 1250.
5. Highnam, R. and Brady, M. (1999) *Mammographic Image Analysis*. Dordrecht, Netherlands: Kluwer.
6. Marias, K., Highnam, R., Brady, M., Parbhoo, S. and Seifalian, A. (2002) Assessing the role of quantitative analysis of mammograms in describing breast density changes in women using HRT, To appear in *Proc. Int. Workshop on Digital Mammography*. Springer-Verlag.
7. Marias, K., Petroudi, S., English, R., Adams, R. and Brady, M. (2003) Subjective and computer-based characterisation of mammographic patterns, To appear in *Proc. Int. Workshop on Digital Mammography*. Berlin: Springer-Verlag.
8. Hartkens, T., Hill, D. L. G., Hajnal, J. V., Rueckert, D., Smith, S. M. and McKleish, K. (2002) Dynamic brain atlas, *McGraw Hill Year Book of Science and Technology*. New York: McGraw-Hill; in press.
9. Hill, D. L. G., Hajnal, J. V., Rueckert, D., Smith, S. M., Hartkens, T. and McLeish K. (2002) Springer Verlag, Berlin, 2002 A dynamic brain atlas, *Proc. MICCAI 2002*. Japan Springer Lecture Notes in Computer Science, Berlin: Springer-Verlag.

42

Combinatorial chemistry and the Grid

**Jeremy G. Frey, Mark Bradley, Jonathan W. Essex,
Michael B. Hursthouse, Susan M. Lewis, Michael M. Luck,
Luc Moreau, David C. De Roure, Mike Surridge, and Alan H. Welsh**

University of Southampton, Southampton, United Kingdom

42.1 INTRODUCTION

In line with the usual chemistry seminar speaker who cannot resist changing the advertised title of a talk as the first, action of the talk, we will first, if not actually extend the title, indicate the vast scope of combinatorial chemistry. 'Combinatorial Chemistry' includes not only the synthesis of new molecules and materials, but also the associated purification, formulation, 'parallel experiments' and 'high-throughput screening' covering all areas of chemical discovery. This chapter will demonstrate the potential relationship of all these areas with the Grid.

In fact, observed from a distance all these aspects of combinatorial chemistry may look rather similar, all of them involve applying the same or very similar processes in parallel to a range of different materials. The three aspects often occur in conjunction with each other, for example, the generation of a library of compounds, which are then screened for some specific feature to find the most promising drug or material. However, there are

Grid Computing – Making the Global Infrastructure a Reality. Edited by F. Berman, A. Hey and G. Fox
© 2003 John Wiley & Sons, Ltd ISBN: 0-470-85319-0

many differences in detail and the approaches of the researchers involved in the work and these will have consequences in the way the researchers will use (or be persuaded of the utility of?) the Grid.

42.2 WHAT IS COMBINATORIAL CHEMISTRY?

Combinatorial chemistry often consists of methods of parallel synthesis that enable a large number of combinations of molecular units to be assembled rapidly. The first applications were on the production of materials for the semiconductor industry by IBM back in the 1970s but the area has come into prominence over the last 5 to 10 years because of its application to lead optimisation in the pharmaceutical industry. One early application in this area was the assembly of different combinations of the amino acids to give small peptide sequences. The collection produced is often referred to as a library. The synthetic techniques and methods have now broadened to include many different molecular motifs to generate a wide variety of molecular systems and materials.

42.3 'SPLIT & MIX' APPROACH TO COMBINATORIAL CHEMISTRY

The procedure is illustrated with three different chemical units (represented in Figure 42.1 by a circle, square, and triangle). These units have two reactive areas so that they can be coupled one to another forming, for example, a chain. The molecules are usually 'grown' out from a solid support, typically a polymer bead that is used to 'carry' the results of the reactions through the system. This makes it easy to separate the product from the reactants (not linked to the bead).

At each stage the reactions are carried out in parallel. After the first stage we have three different types of bead, each with only one of the different units on them.

The results of these reactions are then combined together – not something a chemist would usually do having gone to great effort to make separate pure compounds – but each bead only has one type of compound on it, so it is not so hard to separate them if required. The mixture of beads is then split into three containers, and the same reactions as in the first stage are carried out again. This results in beads that now have every combination of two of the construction units. After n synthetic stages, 3^n different compounds have been generated (Figure 42.2) for only $3 \times n$ reactions, thus giving a significant increase in synthetic efficiency.

Other parallel approaches can produce thin films made up of ranges of compositions of two or three different materials. This method reflects the very early use of the combinatorial approach in the production of materials used in the electronics industry (see Figure 42.3).

In methods now being applied to molecular and materials synthesis, computer control of the synthetic process can ensure that the synthetic sequence is reproducible and recorded. The synthesis history can be recorded along with the molecule, for example, by being coded into the beads, to use the method described above, for example, by using an

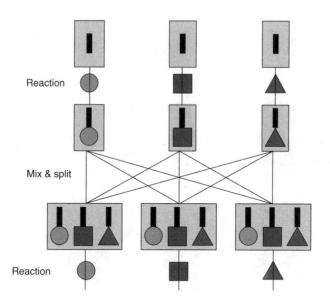

Figure 42.1 The split and mix approach to combinatorial synthesis. The black bar represents the microscopic bead and linker used to anchor the growing molecules. In this example, three molecular units, represented by the circle, the square and the triangle that can be linked in any order are used in the synthesis. In the first step these units are coupled to the bead, the reaction products separated and then mixed up together and split back to three separate reaction vessels. The next coupling stage (essentially the same chemistry as the first step) is then undertaken. The figure shows the outcome of repeating this process a number of times.

RF tag or even using a set of fluorescent molecular tags added in parallel with each synthetic step – identifying the tag is much easier than making the measurements needed to determine the structure of the molecules attached to a given bead. In cases in which materials are formed on a substrate surface or in reaction vessels arranged on a regular Grid, the synthetic sequence is known (i.e. it can be controlled and recorded) simply from the physical location of the selected molecule (i.e. where on a 2D plate it is located, or the particular well selected) [1].

In conjunction with parallel synthesis comes parallel screening of, for example, potential drug molecules. Each of the members of the library is tested against a target and those with the best response are selected for further study. When a significant response is found, then the structure of that particular molecule (i.e. the exact sequence XYZ or YXZ for example) is then determined and used as the basis for further investigation to produce a potential drug molecule.

It will be apparent that in the split and mix approach a library containing 10 000 or 100 000 or more different compounds can be readily generated. In the combinatorial synthesis of thin film materials, if control over the composition can be achieved, then the number of distinct 'patches' deposited could easily form a Grid of 100×100 members. A simple measurement on such a Grid could be the electrical resistance of each patch,

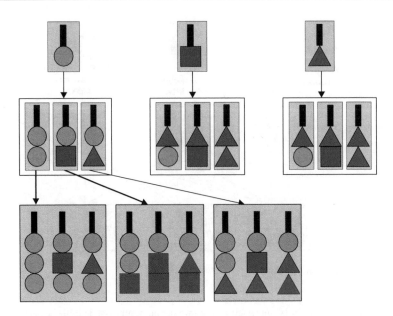

Figure 42.2 A partial enumeration of the different species produced after three parallel synthetic steps of a split & mix combinatorial synthesis. The same representation of the molecular units as in Figure 42.1 is used here. If each parallel synthetic step involves more units (e.g. for peptide synthesis, it could be a selection of all the naturally occurring amino acids) and the process is continued through more stages, a library containing a very large number of different chemical species can be readily generated. In this bead-based example, each microscopic bead would still have only one type of molecule attached.

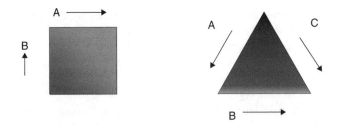

Figure 42.3 A representation of thin films produced by depositing variable proportions of two or three different elements or compounds (A, B & C) using controlled vapour deposition sources. The composition of the film will vary across the target area; in the figure the blending of different colours represents this variation. Control of the vapour deposition means that the proportions of the materials deposited at each point can be predicted simply by knowing the position on the plate. Thus, tying the stochiometry (composition) of the material to the measured properties – measured by a parallel or high throughput serial system – is readily achieved.

already a substantial amount of information, but nothing compared to the amount of the data and information to be handled if the Infrared or Raman vibrational spectrum (each spectrum is an xy plot), or the X-ray crystallographic information, and so on is recorded for each area (see Figure 42.4). In the most efficient application of the parallel screening measurements of such a variable composition thin film, the measurements are all made in parallel and the processing of the information becomes an image processing computation.

Almost all the large chemical and pharmaceutical companies are involved in combinatorial chemistry. There are also many companies specifically dedicated to using combinatorial chemistry to generate lead compounds or materials or to optimise catalysts or process conditions. The business case behind this is the lower cost of generating a large number of compounds to test. The competition is from the highly selective synthesis driven by careful reasoning. In the latter case, chemical understanding is used to predict which species should be made and then only these are produced. This is very effective when the understanding is good but much less so when we do not fully understand the processes involved. Clearly a combination of the two approaches, which one may characterise as 'directed combinatorial chemistry' is possible. In our project, we suggest that the greater use of *statistical experimental design* techniques can make a significant impact on the parallel synthesis experimentation.

The general community is reasonably aware of the huge advances made in understanding the genetic code, genes and associated proteins. They have some comprehension of

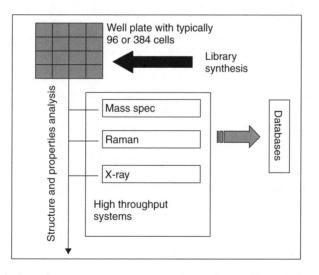

Figure 42.4 High throughput measurements are made on the combinatorial library, often while held in the same well plates used in the robotic driven synthesis. The original plates had 96 wells, now 384 is common with 1556 also being used. Very large quantities of data can be generated in this manner and will be held in associated databases. Electronic laboratory notebook systems correlate the resulting data libraries with the conditions and synthesis. Holding all this information distributed on the Grid ensures that the virtual record of all the data and metadata is available to any authorised users without geographical restriction.

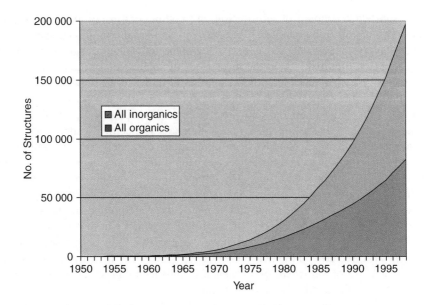

Figure 42.5 The number of X-ray crystal structures of small molecules in the Cambridge Crystallographic Data Centre database (which is one of the main examples of this type of structural databases) as a function of the year. Inorganic and organic represents two major ways in which chemists classify molecules. The rapid increase in numbers started before the high throughput techniques were available. The numbers can be expected to show an even more rapid rise in the near future. This will soon influence the way in which these types of databases are held, maintained and distributed, something with which the gene databases have already had to contend.

the incredibly rapid growth in the quantities of data on genetic sequences and thus by implication some knowledge of new proteins. The size and growth rates of the genetic databases are already almost legendary. In contrast, in the more mature subject of chemistry, the growth in the numbers of what we may call nonprotein, more typical 'small' molecules (not that they have to be that small) and materials has not had the same general impact. Nonetheless the issue is dramatic, in some ways more so, as much more detailed information can be obtained and held about these small molecules.

To give an example of the rapid rise in this 'Chemical' information Figure 42.5 shows the rise in the numbers of fully resolved X-ray structures held on the Cambridge Crystallographic Database (CCDC). The impact of combinatorial synthesis and high throughput crystallography has only just started to make an impact and so we expect even faster rise in the next few years.

42.4 CHEMICAL MARKUP LANGUAGE (cML)

In starting to set up the Comb-e-Chem project, we realized that it is essential to develop mechanisms for exchanging information. This is of course a common feature of all the e-Science projects, but the visual aspects of chemistry do lead to some extra difficulties. Many chemists have been attracted to the graphical interfaces available on computers (indeed this is one of the main reasons why many in the Chemistry community used Macs). The drag-and-drop, point-and-shoot techniques are easy and intuitive to use but present much more of a problem to automate than the simple command line program interface. Fortunately, these two streams of ideas are not impossible to integrate, but it does require a fundamental rethink on how to implement the distributed systems while still retaining (or perhaps providing) the usability required by a bench chemist.

One way in which we will ensure that the output of one machine or program can be fed in to the next program in the sequence is to ensure that all the output is wrapped with appropriate XML. In this we have some advantages as chemists, as Chemical Markup Language (cML) was one of the first (if not the first) of the XML systems to be developed (www.xml-cml.org) by Peter Murray-Rust [2].

Figure 42.6 illustrates this for a common situation in which information needs to be passed between a Quantum Mechanical (QM) calculation that has evaluated molecular properties [3] (e.g. in the author's particular laser research the molecular hyperpolarisibility) and a simulation programme to calculate the properties of a bulk system or interface (surface second harmonic generation to compare with experiments). It equally applies to the exchange between equipment and analysis. A typical chemical application would involve, for example, a search of structure databases for the details of small molecules,

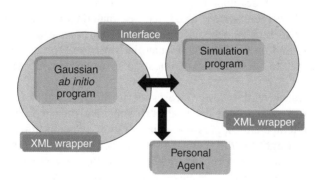

Figure 42.6 Showing the use of XML wrappers to facilitate the interaction between two typical chemistry calculation programs. The program on the left could be calculating a molecular property using an *ab initio* quantum mechanical package. The property could, for example, be the electric field surrounding the molecule, something that has a significant impact on the forces between molecules. The program on the right would be used to simulate a collection of these molecules employing classical mechanics and using the results of the molecular property calculations. The XML (perhaps cML and other schemas) ensures that a transparent, reusable and flexible workflow can be implemented. The resulting workflow system can then be applied to all the elements of a combinatorial library automatically. The problem with this approach is that additional information is frequently required as the sequence of connected programs is traversed. Currently, the expert user adds much of this information ('on the fly') but an Agent may be able to access the required information from other sources on the Grid further improving the automation.

followed by a simulation of the molecular properties of this molecule, then matching these results by further calculations against a protein binding target selected from the protein database and finally visualisation of the resulting matches. Currently, the transfer of data between the programs is accomplished by a combination of macros and Perl scripts each crafted for an individual case with little opportunity for intelligent reuse of scripts. This highlights the use of several large distributed databases and significant cluster computational resources. Proper analysis of this process and the implementation of a workflow will enable much better automation of the whole research process [4].

The example given in Figure 42.6, however, illustrates another issue; more information may be required by the second program than is available as output from the first. Extra knowledge (often experience) needs to be added. The Quantum program provides, for example, a molecular structure, but the simulation program requires a force field (describing the interactions between molecules). This could be simply a choice of one of the standard force fields available in the packages (but a choice nevertheless that must be made) or may be derived from additional calculations from the QM results. This is where the interaction between the 'Agent' and the workflow appears [5, 6] (Figure 42.7).

42.5 STATISTICS & DESIGN OF EXPERIMENTS

Ultimately, the concept of combinatorial chemistry would lead to all the combinations forming a library to be made, or all the variations in conditions being applied to a screen. However, even with the developments in parallel methods the time required to carry out these steps will be prohibitive. Indeed, the raw materials required to accomplish all the synthesis can also rapidly become prohibitive. This is an example in which direction should be imposed on the basic combinatorial structure. The application of modern statistical approaches to 'design of experiments' can make a significant contribution to this process.

Our initial approach to this design process is to the screening of catalysts. In such experiments, the aim is to optimise the catalyst structure and the conditions of the reaction; these may involve temperatures, pressure, concentration, reaction time, solvent – even if only a few 'levels' (high, middle, low) are set for each parameter, this provides a huge parameter space to search even for one molecule and thus a vast space to screen for a library. Thus, despite the speed advantage of the parallel approach and even given the

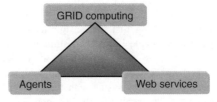

Figure 42.7   The Agent & Web services triangle view of the Grid world. This view encompasses most of the functionality needed for Comb-e-Chem while building on existing industrial based e-Business ideas.

ability to store and process the resulting data, methods of trimming the exponentially large set of experiments is required.

The significant point of this underlying idea is that the interaction of the combinatorial experiments and the data/knowledge on the Grid should take place from the inception of the experiments and not just at the end of the experiment with the results. Furthermore, the interaction of the design and analysis should continue while the experiments are in progress. This links our ideas with some of those from RealityGrid in which the issue of experimental steering of computations is being addressed; in a sense the reverse of our desire for computational steering of experiments. This example shows how the combinatorial approach, perhaps suitably trimmed, can be used for process optimisation as well as for the identification of lead compounds.

42.6 STATISTICAL MODELS

The presence of a large amount of related data such as that obtained from the analysis of a combinatorial library suggests that it would be productive to build simplified statistical models to predict complex properties rapidly. A few extensive detailed calculations on some members of the library will be used to define the statistical approach, building models using, for example, appropriate regression algorithms or neural nets or genetic algorithms, that can then be applied rapidly to the very large datasets.

42.7 THE MULTIMEDIA NATURE OF CHEMISTRY INFORMATION

Chemistry is a multimedia subject – 3D structures are key to our understanding of the way in which molecules interact with each other. The historic presentation of results originally as text and then on a flat sheet of paper is too limiting for current research. 3D projectors are now available; dynamic images and movies are now required to portray adequately the chemist's view of the molecular world. This dramatically changes expectations of what a journal will provide and what is meant by 'publication'; much of this seems to be driven by the available technology–toys for the chemist. While there may be some justification for this view by early adopters, in reality the technology now available is only just beginning to provide for chemists the ability to disseminate the models they previously only held in the 'minds eye'.

Chemistry is becoming an information science [7], but exactly what information should be published? And by whom? The traditional summary of the research with all the important details (but these are not the same for all consumers of the information) will continue to provide a productive means of dissemination of chemical ideas. The databases and journal papers link to reference data provided by the authors and probably held at the journal site or a subject specific authority (see Figure 42.8). Further links back to the original data take you to the author's laboratory records. The extent type of access available to such data will be dependent on the authors as will be the responsibility of archiving these data. There is thus inevitably a growing partnership between the traditional authorities in

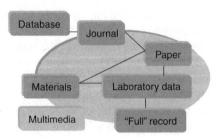

Figure 42.8 Publication @source: e-dissemination rather than simply e-publication of papers on a Web site. The databases and journal papers link to reference data provided by the authors and probably held at the journal site or a subject specific authority. Further links back to the original data take you to the author's laboratory records. The extent and type of access available to such data will be dependent on the authors as will be the responsibility of archiving these data.

publication and the people behind the source of the published information, in the actual publication process.

One of the most frustrating things is reading a paper and finding that the data you would like to use in your own analysis is in a figure so that you have to resort to scanning the image to obtain the numbers. Even if the paper is available as a pdf your problems are not much simpler. In many cases, the numeric data is already provided separately by a link to a database or other similar service (i.e. the crystallographic information provided by the CIF (Crystallographic Information File) data file). In a recent case of the 'publication' of the rice genome, the usual automatic access to this information to subscribers to the journal (i.e. relatively public access) was restricted to some extent by the agreement to place the sequence only on a company controlled Website.

In many cases, if the information required is not of the standard type anticipated by the author then the only way to request the information is to contact the author and hope they can still provide this in a computer readable form (assuming it was ever in this form?). We seek to formalise this process by extending the nature of publication to include these links back to information held in the originating laboratories. In principle, this should lead right back to the original records (spectra, laboratory notebooks as is shown in Figure 42.9). It

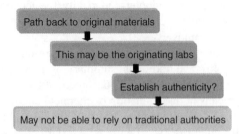

Figure 42.9 What constitutes a trusted authority when publication @ source becomes increasingly important. Will adequate archives be kept? Will versioning be reliably supported? Can access be guaranteed?

may be argued that for publicly funded research we have a responsibility to make all this information available. The immediate impact that many people may imagine on this is that it will make the detection of fraud much easier, but this is in fact a relatively minor issue. The main advantage will be the much greater use and reuse of the original data and the consequent checking of the data and different approaches to the analysis. The scientific process requires that as much as possible of the investigations are repeated and this applies just as much to the analysis as the initial data capture in the experiments.

42.8 THE PERVASIVE GRID AND METADATA

In my view if information (more valuable than just the data) is destined for the Grid, then this should be considered from the inception of the data. The relevant provenance, environment, and so on that traditionally characterise the metadata should not be added as an afterthought, captured separately in a laboratory notebook, but should be part of the data: it is this that differentiates information from mere data. If this background is not there from the beginning, how can we hope to propagate it efficiently over the Grid? This leads to the concept that the computing scaffold should be pervasive as well as forming a background transparent Grid infrastructure. The smart laboratory is another, and a highly challenging, environment in which to deploy pervasive computing [8].

Not only do we need to capture the data, the environment in which the data was generated but also the processes by which it was generated and analysed. In a research environment, the process capture is particularly interesting as the processes, for example, involved in analysing the data is not fully known in advance. Some basic aspects of the workflow will be predefined but it is usual to apply different methods (maybe different models or different computational methods) and the results compared (and compared with theory). We need to improve the tools to facilitate this process to capture and to represent it visually and to ensure that they can be used by a distributed research group/community to discuss the results and analysis. With a proper record of the process having been captured, the chosen route can then be quickly implemented for subsequent data.

The need to capture the experimental environment and the subsequent analysis process followed by integration with the existing knowledge shows the importance of the metadata. However, splitting the information into data and metadata may be counterproductive. The split between these two parts of the information is not well defined and depends on the perceived use of the information by the creator; which may, of course, be quite different from what actually happens. It is all too easy for the loss of the apparently useless metadata to render the data devoid of any real value.

More problematic is the subsequent need for information about the experiment not collected when the experiment was undertaken. This will always be a potential problem, as we cannot foresee all the parameters that may be significant. However, the smart environment will go a long way to help. The information will be automatically recorded and available if needed (and hidden if not to avoid confusion?). What is significant about the way we see the knowledge Grid is that this 'metadata' will always remain accessible from the foreground data even as the information propagates over the Grid. This is another part of the provenance of the information and something for which the current Web is not generally a good example.

Quoting one of the authors (Dave De Roure) ' "Comb-e-Chem" is a *real-time and pervasive semantic Grid*, and as such provides a challenging environment in which to test many of the current ideas of the Human Computer Interface'.

42.9 VIRTUAL DATA

A combinatorial library could be thought of as a 'Library Collection' with the material and information on that material all ideally cross-referenced. In a traditional system, if information is requested from the library collection then it can be provided if it is present in the collection. If the requested item it is not held in the collection, then a search can be made to find it elsewhere. This situation is paralleled with a molecular combinatorial library and applies not only to the physical material but also to the information held on the structure and properties of the molecules or materials in the library.

The power of the Grid-based approach to the handling of the combinatorial data is that we can go further than this 'static' approach. The combination of the laboratory equipment, the resulting information, together with the calculation resources of the Grid allows for a much more interesting system to be created. In the system outlined as an example described in Figure 42.10, an appropriate model can calculate the requested data. These models are themselves validated by comparison with the measured properties of the actual physical members of the library.

Depending on time or resource constraints, different types of models or different levels of implementation of a model can be chosen, ranging from resource hungry high-level QM calculation [9], through extensive simulations, to an empirically based approach; we thus have, in effect, a virtual entry in the database. Ultimately, this process has a close connection with the ideas of virtual screening of combinatorial libraries.

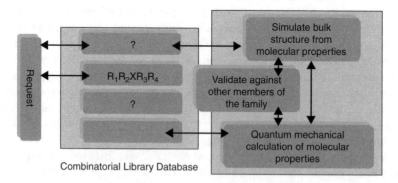

Figure 42.10 An example of virtual data interactions in the context of a combinatorial family. In this example, the combinatorial library is formed from the parallel synthesis of molecules with the form $R_1R_2XR_3R_4$ where X is a common building block and R_1 R_2 R_3 R_4 represent the variety of related groups that are added to this core to give the library. For example, they may be hydrocarbon chains of different length; CH_3-, CH_3CH_2-, $CH_3(CH_2)_n-$.

As in our model, the Grid extends down to the laboratory such that this virtual data idea can be extended to not only calculations but also to new experimental data acquisition or even to automated synthesis. That is, the direction of the synthesis or analysis in the automated laboratory would be controlled via a database request.

42.10 MULTIMEDIA COLLABORATION

A key issue for chemists making use of the Grid will be the support it can provide for distributed collaboration. This includes video, multimedia as well as the traditional need we have for visualisation. We have already demonstrated the need for significant, real-time, video interaction in the area of running a high throughput single crystal X-ray crystallography service. A demonstration Grid-aware system allowing users to interact with the UK Engineering and Physical Sciences Research Council (EPSRC) X-ray crystallography service bases at Southampton has highlighted a number of QoS and security issues that a Grid system must encompass if it is to provide an adequate infrastructure for this type of collaborative interactions. For example, the demands made on a firewall transmitting the video stream are very significant [10].

42.11 A GRID OR INTRA-GRIDS

It is possible that we may be able to enable query access to 'hidden databases' inside companies. In this model, certain types of general queries (e.g. checking structural motifs) could be asked without revealing the full nature of the compounds in the database. This could be useful, as we believe that this hidden information store exceeds considerably the publicly available information. Even with such a 'diode' placed in the dataflow to ensure reasonable isolation of company data, it seems likely that initially there will be intra-Grids in which large multinational companies use the Grid model we are proposing but restrict it to within the company (cf. intranets), so we will not initially have one Grid but many disjoint Grids.

There is also need for information to flow securely out of a company in support of equipment and other items that need to be serviced or run collaboratively with the manufacturers. The idea and indeed implementation of remote equipment diagnostics has been around for many years. In the computer industry, the remote support of PC is not uncommon – the supporting company can remotely control and 'fix' your PC (or router or other similar device). This has also been invaluable for training. Virtual Network Computing (VNC) provides this functionality in an open environment; one of our current projects is to marry this approach with the Web services model to provide more selective and secure interaction of this type. VNC was produced by Olivetti-AT&T labs (AT&T acquired the Olivetti Research Laboratory in 1999) and made open source [11].

Discussions with some of the equipment manufacturers who have used remote diagnostics (usually via a dial-up) indicate that the interaction is insufficient even when the support team may have some physical presence at the site of the equipment and certainly is often inadequate to help the less well-trained users to fix the equipment. What is needed in addition to the direct connection to the machines is a connection with the users; this usually takes place by a concurrent telephone call.

An X-ray crystallography demonstration project based around the UK EPSRC National Crystallography Service (NCS), funded by the DTI e-Science core programme (M.S. Surridge & M.B. Hursthouse) has demonstrated how the use of audio and video over the Grid (despite the limitations of bandwidth, firewalls etc.) adds considerably to the quality of the interaction between users, experts, technicians and equipment. A full account of the security and associated issues uncovered by this demonstrator project will be the subject of a separate paper.

Taking the interactive approach to an experimental service used in the X-ray demonstrator project and linking them with the concerns of remote diagnostics, suggests that extensive remote monitoring of equipment will be possible over the Grid. This should allow pre-emptive maintenance, as frequent monitoring will be feasible. However, the remote diagnostics will often need to be augmented with the person-to-person multimedia links running synchronised with the control of and acquisition from equipment, for instruction and for any cases in which physical intervention is required by the user.

However, this discussion has taken the perspective of the researcher (or perhaps research manager) in the laboratory (or in the 'user' company). It must be married with the view from the other side (or of course as we are using the Grid, the other sides) of the interaction, the manufacturer support service. As already indicated, we believe the multimedia link directly benefits both sides of this interaction, and is already frequently undertaken by using a second separate communication channel (the phone). The use of parallel channels within the Grid is desirable as it allows for more efficient synchronisation of the information between the people and the equipment (or computers).

The Grid model allows for more. In the analysis of the combinatorial chemistry, structural and functional information models and calculations are used to link the data to extract information and knowledge (i.e. it is not just simply a mapping connection of data, though that is a significant activity, but the active development/extraction of new information).

Similarly, the diagnostic activates require comparison with a model that may be a physical model, that is, another copy of the equipment or, as the author suspects, because of the growing frequency of purpose designed individual systems, a physical device together with a computational model of the system. The integration and synchronisation of the computational model (which indeed may be viewed by some as essentially producing an Agent to help with the diagnostics) with the external machines is yet another area where the active properties of the Grid will be important.

The manner in which users, equipment, experts and servicing will be linked over the Grid will depend on which resources are most in demand and which are most limited in supply. The X-ray crystallography demonstrator is built around a very high-end diffractometer – there will not usually be many of these, coupled with experts in the collection and analysis of data from difficult crystal samples – people in even shorter supply. The connection model is thus that of Figure 42.11(a). An alternative model in which the equipment is relatively cheaper and easier to use, but nevertheless generated data that may require the help of an expert to understand fully, is shown in Figure 42.11(b). In both cases the Grid is enabling scarce resources to be shared while ensuring that all

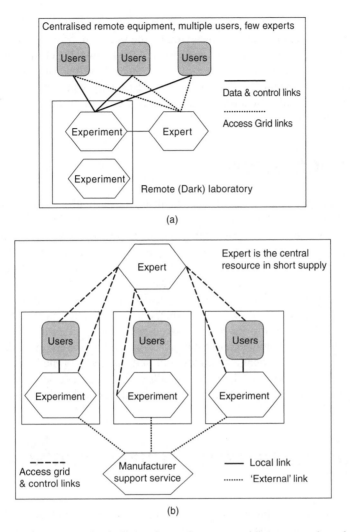

Figure 42.11   (a) shows a connection map for equipment providing a central service; (b) shows the situation for more commonly available equipment needing centralised support.

the 'stakeholders' in an experiment have the possibility of a presence in all stages of the procedures.

42.12 e-SCIENCE AND e-BUSINESS

I hope that this discussion has highlighted some of the specific aspects of chemistry research and, in particular, the way in which the application of combinatorial chemistry ideas together with the needs of collaborative research give rise to demands of the computational infrastructure that can be answered by the Grid. I hope it is clear that chemistry can use and catalyse many of the generic aspects of the knowledge Grid. The requirements of the research community are not that different from those carrying out e-Business. In particular, there is a common need for security (especially where patent-sensitive information is involved), authentication and provenance to ensure that the information can be trusted or at least investigated.

42.13 CONCLUSIONS

Chemistry in general and combinatorial chemistry in particular, will continue to make great demands on computational and network resources both for calculations and for knowledge management. The Grid will make an important impact in both these areas. The pervasive possibilities of the modern computing environment are ideal for extending the idea of a computational Grid down in the laboratory. The ability to automate both the experiments and the data analysis provides new possibilities and requirements for knowledge management. The exponentially growing quantities of data that combinatorial chemistry, in particular, is already delivering demonstrates the need for a Grid-based approach to handling the information generated [12]. The desirability of efficient use of resources (human, computational and equipment) in handling the resulting data is reflected in the need to employ statistical techniques both in the analysis of the large datasets and in the design of the experiments. The Grid will allow the desired close connection between the design, control and analysis of experiments (both physical and computational) to be implemented efficiently.

APPENDIX 1: THE COMB-e-CHEM e-SCIENCE PILOT PROJECT

Comb-e-Chem (www.combechem.org) is an interdisciplinary pilot project involving researchers in chemistry, mathematics and computer science and is funded by the UK Engineering and Physical Science Research Council (www.epsrc.ac.uk) under the Office of Science and Technology e-Science initiative (www.research-councils.ac.uk/escience/).

A major aspect of the crystal structure measurement and modelling involved in this project will be to develop the e-Science techniques to improve our understanding of how molecular structure influences the crystal and material properties (Figure 42.12). The same molecule can crystallise in a number of different forms each with different physical and chemical properties; the same compound can frequently form many different materials – to take a very important example in food – chocolate can crystallise in six different forms but only one of them has the lustre and snap of good quality chocolate – the other forms are cheap and nasty. The same thing can happen with drugs, the wrong formulation can result in none of the drug being absorbed by the body. For one AIDS drug, the appearance of a new less soluble polymorph required $40 M to reformulate.

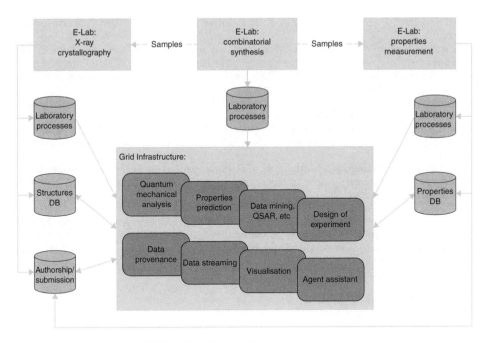

Figure 42.12   The Comb-*e*-Chem project development.

One of us (JGF) as Principle Investigator has the task of coordinating the other researchers who are in Chemistry (www.chemistry.soton.ac.uk) at the University of Southampton: Jeremy Frey, Mike Hursthouse, Jon Essex and Chris Frampton (formally at Roche); Chemistry at the University of Bristol (www.bris.ac.uk/Depts/Chemistry/Bristol_Chemistry.html), Guy Orpen; in Statistics at Southampton (www.maths.soton.ac.uk/stats/), Sue Lewis and Alan Welsh; in Electronics & Computer Science at Southampton, Dave De Roure, Mike Luck, and Luc Moreau; and Mike Surridge at IT-Innovation (www.it-innovation.soton.ac.uk). The project is in its early stages but already we have a very active team of researchers (a list can be found at www.combechem.org). We appreciate the significant help and assistance from our industrial collaborators, in particular, IBM UK at Hursley (www.hursley.ibm.com), The Cambridge Crystallographic Data Centre (CCDC www.ccdc.cam.ac.uk), Astra-Zeneka, and Pfizer.

REFERENCES

1. Scheemeyer, L. F. and van Dover, R. B. (2001) The combinatorial approach to materials chemistry, Chapter 10, in Keinan, E. and Schechter, I. (eds) *Chemistry for the 21st Century*. ISBN 2-527-30235-2, Weinheim, Germany: Wiley-VCH, pp. 151–174.
2. Murray-Rust, P. (1997) Chemical Markup language. *World Wide Web Journal*, 135–147.
3. Crawford, T. D., Wesolowski, S. S., Valeev, E. F., King, R. A., Leininger, M. L. and Schaefer III, H. F. (2001) The past present and future of quantum Chemistry, Chapter 13, in Keinan, E.

and Schechter, I. (eds) *Chemistry for the 21st Century*. ISBN 2-527-30235-2, Weinheim, Germany: Wiley-VCH, pp. 219–246.

4. Leymann, F. and Roller, D. (1997) Workflow-based applications. *IBM Systems Journal*, **36**(1), 102–123.

5. Wooldridge, M. and Jennings, N. R. (1995) Intelligent agents: theory and practice. *The Knowledge Engineering Review*, **10**(2), 115–152.

6. Jennings, N. R. (2001) An agent-based approach for building complex software systems. *Communications of the ACM*, **44**(4), 35–41.

7. Lehn, J. M. 2001 Some reflections on chemistry, Chapter 1, in Keinan, E. and Schechter, I. (eds) *Chemistry for the 21st Century*. ISBN 2-527-30235-2, Weinheim, Germany: Wiley-VCH, pp. 1–7.

8. De Roure, D., Jennings, N., Shadbolt, N. (2001) *Research Agenda for the Semantic Grid: A Future e-Science Infrastructure*, Technical report UKeS-2002-02, Edinburgh: UK National e-Science Centre, January, 2002..

9. Alchemy, Q. and Cohen, M. L. (2001), Chapter 14, in Keinan, E. and Schechter, I. (eds) *Chemistry for the 21st Century*. ISBN 2-527-30235-2., Weinheim, Germany: Wiley-VCH, pp. 247–270.

10. Buckingham Shum, S., De Roure, D., Eisenstadt, M., Shadbolt, N. and Tate, A. (2002) CoAKTinG: collaborative advanced knowledge technologies in the grid. *Proceedings of the Second Workshop on Advanced Collaborative Environments, Eleventh IEEE Int. Symposium on High Performance Distributed Computing (HPDC-11)*, Edinburgh, Scotland, July 24–26, 2002.

11. Richardson, T., Stafford-Fraser, Q., Wood, K. R. and Hopper, A. (1998) Virtual network computing. *IEEE Internet Computing*, **2**(1), 33–38.

12. Maes, P. (1994) Agents that reduce work and information overload. *Communications of the ACM*, **37**(7:31), 40.

43

Education and the enterprise with the Grid

Geoffrey Fox

Indiana University, Bloomington, Indiana, United States

43.1 INTRODUCTION

In this short article, we aim to describe the relevance of Grids in education. As in fact information technology for education builds on that for any organization, we first discuss the implication of Grids and Web services for any organization – we call this an Enterprise to stress the importance of the Enterprise Grids and the different roles of general and specific features in any Grid deployment. The discussion of the importance of Grids for virtual organizations in Chapter 6 already implies its importance in education where our organization involves learners, teachers and other stakeholders such as parents and employers. We describe in Section 43.2, the role of Web services and their hierarchical construction in terms of generic capabilities and applications of increasing specialization. In Section 43.3, we summarize this in terms of a Web service implementation strategy for a hypothetical enterprise. Finally, in Section 43.4, we describe education grids pointing out the differences and similarities to general enterprises. We stress Web service issues, as these require the most substantial enterprise-specific investment for they embody the particular objects and functionalities characteristic of each domain. The Grid provides

Grid Computing – Making the Global Infrastructure a Reality. Edited by F. Berman, A. Hey and G. Fox
© 2003 John Wiley & Sons, Ltd   ISBN: 0-470-85319-0

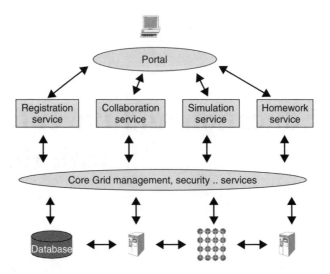

Figure 43.1 Typical Grid (education) enterprise architecture.

the infrastructure on which to build the various Web service implementations. Deploying Grid infrastructure will get easier as commercial support grows and the heroic efforts described in Chapter 5 are packaged properly.

One will of course still have to worry about needed resources – computers, data storage and networks. On these one will install 'core' Grid software infrastructure whose many components and approaches are described in Part B of this book. This is the bottom two layers of Figure 43.1. On top of this, one will need several services – some could be generic like collaboration and others very specific to the enterprise – such as a homework submission service in education. It would be wonderful if there was a clear hierarchy but this will only be approximate with services connected, say, with 'science', 'people' and 'education' not having a clear hierarchical relationship. Rather we will have a complex network of services with an approximate hierarchy; core services at the bottom of Figure 43.1 and portals handling user-facing service ports at the top (Chapter 18). In this chapter we focus on the filling of the portal-core Grid sandwich, which we discuss below for first general enterprises and then education. Although we are not certain as to the details of the 'final' Grid architecture, we are certain that we have a service model and that the interfaces are defined in XML. This we can start to address today without worrying too much about how technology evolves. For this reason, we discuss in most detail how Web services can be deployed in particular domains.

43.2 WEB SERVICE PARADIGM FOR THE ENTERPRISE

We suppose that Web services will be developed for a wide variety of applications ('all of them') and that there will be a corresponding suite of XML schema describing the object and services associated with each application. The net result will be a hierarchical structure of information and services. This has been described earlier in especially Chapters 14 to 19 of this book on the role of data and the Grid. Let us imagine that we are the chief information officer (CIO) of some enterprise and wish to adopt a uniform Grid and Web service enabled view of our information technology environment shown in Figure 43.2. We would of course adopt a service architecture and define this with XML Schema for both our data structures and the functions (Web services) that operate on them. We assume this will eventually be set up hierarchically as sketched in Figure 43.2. Our application would define its schema and, this would be used on top of other standards, for example, those of computing and databases as shown on the top of Figure 43.2. These specific Grid-wide application standards would themselves be built on general Grid, Web and Internet protocols (IP).

Even our application could itself be composite and built up hierarchically internally – suppose our enterprise was a physics department of a university; then the 'application schema' could involve a mixture of those for *physics* (extending a Schema for *science*) *research* and *education*. It could also involve Schema specific to the home university. As we will see later, the *education* schema itself is composite. Notice that this hierarchical information model is projected to the user through application related

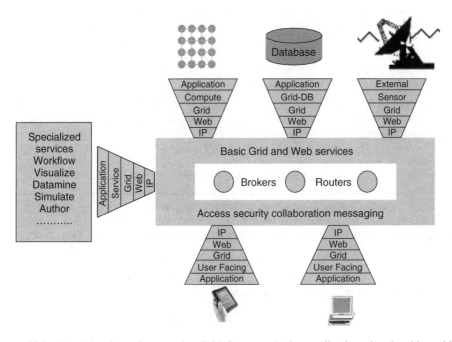

Figure 43.2 Possible view of enterprise Grid for a particular application showing hierarchical information structure with at the top, a parallel computer, database and sensor linked to the Grid. The left part of diagram lists important services with user interface devices at the bottom. The hierarchical interface is shown at top and bottom.

content rendered to clients through user-facing ports on the Web service. As CIO, we would certainly try to ensure that our entire system, respected this single albeit complex representation. Figure 43.2 illustrates that there will be some places we need foreign (external) formats. At the top right, we assume that we have a scientific instrument on our grid and this has some distinct external specification. We imagine that the Grid community has defined some sort of *sensor* schema into which we can add the instrument. We now build a custom conversion web service that maps this device into the common data and service model of our grid. This process allows us to use the same *application* schema for all services and so build an integrated grid.

Another example could be a grid servicing a group of hospitals in which we have devised a single specification of all medical, administrative and patient data. This is the interoperability language of the healthcare grid linking the hospitals together but realistically many hospitals in the chain would have their own (pre-existing) information systems with disparate data representations. In designing our grid, we would represent each hospital's legacy system as an external extension to a base health care schema and then design mapping (Web) services that converted all to the common interoperable representation. This discussion is meant to illustrate that building an enterprise (application) specific grid involves study of the different current representations of related systems and where possible adopting a hierarchical architecture based on more general applications.

The hierarchy of Web services is explored in Tables 43.1 to 43.6. The last three tables describe application of Web services to science, education and research and will be discussed later in Section 43.4. Here we want to describe briefly generic (Tables 43.1 and 43.2), commodity and business services (Table 43.3). We want to make two important points here

- All electronic processes will be implemented as Grid or Web services
- The processes will use objects described by XML defined by Schema agreed by particular organizations. Of course, the Web services are XML described methods (functions) that input and output information specified by the XML application object specifications.

Table 43.1 Some basic Grid technology services

Security services	Authorization, authentication, privacy
Scheduling	Advance reservations, resource co-scheduling
Data services	Data object namespace management, file staging, data stream management, caching (replication)
Database service	Relational, object and XML databases
User services	Trouble tickets, problem resolution
Application management services	Application factories [1], lifetime, tracking, performance analysis,
Autonomy and monitoring service	Keep-alive meta-services. See Reference [2]
Information service	Manage service metadata including service discovery [3]
Composition service	Compose multiple Web services into a single service
Messaging service	Manage linkage of Grid and Web services [4]

Table 43.2 General application-level services

Portal	Customization and aggregation
People collaboration	Access Grid – desktop audio-video
Resource collaboration	Document sharing (WebDAV, Lotus Notes, P2P), news groups, channels, instant messenger, whiteboards, annotation systems. virtual organization technology [5]
Decision-making services	Surveys, consensus, group mediation
Knowledge discovery service	Data mining, indexes (directory based or unstructured), metadata indices, digital library services. semantic Grid
Workflow services	Support flow of information (approval) through some process, secure authentication of this flow. planning and documentation
Universal access	From PDA/phone to disabilities; language translation

Table 43.3 Some commodity and business applications

News & entertainment	The Web
Video-on-demand	Multimedia delivery
Copyright	The issues that troubled Napster done acceptably
Authoring services	Multi-fragment pages, charts, multimedia
Voting, survey service	Political and product review
Advertising service	Marketing as a Web service
e-Commerce	Payment, digital cash, contracting; electronic marketplaces (portals)
Catalogs	As used in on-line sites like Amazon
Human resources; and ERM	Uses privacy, security services; performance, references; employee relationship management (ERM) as a Web service
Enterprise resource planning ERP	Manage internal operations of an enterprise
Customer-relationship management CRM	Business to customer (B2C) as a Web service. Call centers, integration of reseller and customer service Web services.
SFA sales force automation	Manage sales and customer relationship; contacts, training
Supply chain management SCM	Typical Business to business (B2B) Web services; also partner relationship management, collaborative product commerce (CPC) and so on
Health care	Patient and other hospital records, medical instruments, remote monitoring, telemedicine

Note that Web services are combined to form other Web services. All the high-level examples, we discuss here and give in the tables are really composites of many different Web services. In fact, this composition is an active area of research these days [6, 7] and is one service in Table 43.1. Actually deciding on the grain size of Web services will be important in all areas; if the Services are too small, communication overhead between services could be large; if the services are too large, modularity will be decreased and it will be hard to maintain interoperability.

Table 43.4 Science and engineering generic services

Authoring and rendering specialized to science	Storage rendering and authoring of mathematics, scientific whiteboards, n dimensional ($n = 2, 3$) data support, visualization, geographical information systems, virtual worlds
Discipline wide capabilities as network services	Generic mathematics (algebra, statistics, optimization, differential equation solution, image processing)
Sensor services	Support general instruments (time series)
Tenure evaluation	Shared with all scholarly fields; references. Specialization of generic human resources service

Table 43.5 Science and engineering research (e-Science)

Portal shell services	Job control/submission, scheduling, visualization, parameter specification, monitoring
Software development support	Wrapping, application integration, version control, software engineering
Scientific data services	High performance, special formats, virtual data
(Theory) research support services	Scientific notebook/whiteboard, brainstorming, theorem proving
Experiment support	Virtual control rooms (accelerator to satellite), data analysis, virtual instruments, sensors (satellites to field work to wireless to video to medical instruments, multi-instrument federation
Publication	Submission, preservation, review, uses general copyright service
Dissemination and outreach	Virtual seminars, multi-cultural customization, multilevel presentations,

Table 43.1 contains services that have been discussed in detail in Part B of this book, Chapters 6 to 19. These are the services creating the Grid environment from core capabilities such as security [8] and scheduling [9] to those that allow databases to be mounted as a Grid service [10–12]. The services in Table 43.2 have also been largely discussed in the book and consist of core capabilities at the 'application Web service level'. Collaboration is the sharing of Web services as described in References [3, 13], while portals are extensively discussed in Part C of the book, Chapters 20 to 34. Universal access covers the customization of user interactions for different clients coping with physical capabilities of user and nature of network and client device. The same user-facing ports of Web services drive all clients with customization using the universal access service [13]. Workflow builds on the composition service of Table 43.1 but can have additional process and administrative function. Moving from data to information and then knowledge is critical as has been stressed in References [12, 14] and various data mining and metadata tools will be developed to support this. The Semantic Grid is a critical concept [14] capturing the knowledge related services.

Table 43.3 illustrates broad-based application services that are developed to support consumers and business. The Web itself is of course a critical service providing 'web

Table 43.6 Education as a Web service (LMS or learning management system)

Registration	Extends generic human resources service
Student performance	Grading including transcripts
Homework	Submission, answers; needs performance and security services
Quizzes	Set and take – extends voting/survey service
Curriculum (content)	Authoring, prerequisites, completion requirements, standards, extend generic authoring and data management services to get learning content management systems (LCMS)
Assessment	Related to refereeing and reference (tenure) services
Course scheduling	Related to generic event scheduling service in collaboration service
Learning plans	Builds on curriculum and student performance services. Support building of 'degrees' with requirements
Learning	Integrate curriculum, collaboration and knowledge discovery services
Mentoring and teaching	Office hours, (virtual) classrooms
Distance education	Asynchronous and synchronous, integrate curriculum, quiz and so on, services with generic collaboration services

pages' on demand. This is being extended with video-on-demand or high-quality multimedia delivery; given the controversy that music downloading has caused we can expect copyright monitoring to be packaged as a service. Authoring – using Microsoft Word (and of course other packages such as Star Office, Macromedia and Adobe) – is an interesting Web service; implementing this will make it a lot easier to share documents (discussed in Section 43.4) and build composite Web sites consisting of many fragments. We will derive our curriculum preparation service for education by extending this core authoring service. Voting, polling and advertising are commodity capabilities naturally implemented as Web services. The areas of internal enterprise management (ERP), B2B and B2C are being re-implemented as Web services today. Initially this will involve rehosting databases from companies like Oracle, PeopleSoft, SAP and Sybase as Grid services without necessarily much change. However, the new Grid architectures can lead to profound changes as Web services allow richer object structures (XML and not relational tables) and most importantly, interoperability. This will allow tools like security and collaboration to be universally applied and the different Web services to be linked in complex dynamic value chains. The fault tolerance and self-organization (autonomy) of the Grid will lead to more robust powerful environments.

43.3 IMPLEMENTING WEB SERVICES

We have learnt that gradually everything will become a Web service and both objects and functions will be specified in XML. What does this mean for our harried chief information officer or CIO that we introduced in the last section? Clearly the CIO needs to rethink their environment as a Grid of Web services. All data, information and knowledge must be specified in XML and the services built on top of them in Web Services Description Language (WSDL) [15]. The CIO will study the building blocks and related applications as exemplified in Tables 43.1 to 43.3. This will lead each enterprise to define two key specifications – Your Enterprise Internal framework (YEIF) and Your Enterprise External Framework (YEEF). These could be essentially identical to those used in similar enterprises or very different if our CIO has a quite distinct organization. The YEEF is used to interface outside or legacy systems to the enterprise Grid – we gave examples of a physics sensor or a legacy healthcare database when discussing Figure 43.2 above. Internally the enterprise Grid will use the customized XML-based framework YEIF. When you accept bids for new software components, the vendor would be responsible for supporting YEIF. This would be defined by a set of Schemas placed on a (secure) Web resource and always referenced by Universal Resource Identifier (URI). YEIF would inevitably have multiple versions and the support software would need to understand any mappings needed between these. There would be an XML database managing this schema repository which would need to store rich semantic information as discussed in Chapters 17 and 19; the Universal Description, Discovery and Integration (UDDI) effort [16] is trying to define such an enhanced schema storage but much work needs to be done here. Probably software referencing data structures defined by YEIF would not just be written in the programmer's or CIO's favorite programming model – rather the data structures would be generated automatically from the XML specification using technology like Castor [17]. This suggests new programming paradigms in which data structures and method interfaces are defined in XML and control logic in traditional languages. Note that although interfaces are specified in XML, they certainly need not be implemented in this way. For instance, we can use the binding feature of WSDL [15] to indicate that different, perhaps higher-performance protocols are used that preserve the XML specification but have a more efficient implementation than Simple Object Access Protocol (SOAP) [18].

The Web service approach gains interoperability from greater use of standards. Thus, our CIO must be aware of and perhaps involved in the community processes defining Web service–relevant standards for the application areas that are of importance to the Enterprise.

43.4 EDUCATION AS A WEB SERVICE

We will simplify our discussion and only consider education for science and engineering. It will be straightforward to generalize to any curricula area but this is the author's expertise. Further, science and engineering have extensive existing experience on, the use of electronic information, instruments and computer simulations in education. Figure 43.3 extends the generic environment of Figure 43.2 to education. Currently, one uses rather specialized learning (content) management systems as the heart of a sophisticated learning environment. Such systems will be reworked to use generic Web services as much as possible. There will be specialized learning objects but functions like authoring and metadata management will use the generic services of Tables 43.1 to 43.3. Already this field has an excellent XML-based object model through the work of the Instructional Management System (IMS) Global Learning Consortium [19] and Advanced Distributed Learning (ADL) [20] initiatives. These have technical problems – they were not designed for a Grid or even Web Service architecture but rather to the client-server world of yesteryear.

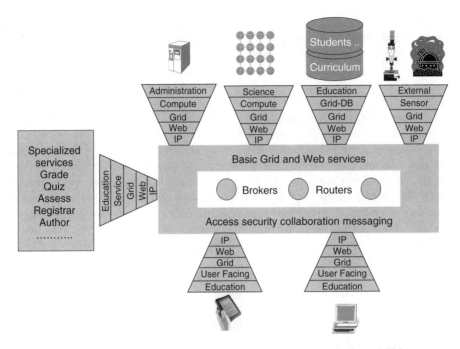

Figure 43.3   A view of Grid in education illustrating typical capabilities.

Further, they are designed to be stand-alone rather than extending existing Service and XML-based data structures. These deficiencies are straightforward to address and these standards give us a clear object model for learning. We currently do not have services defined and these must be added – hopefully these national consortia will recognize this for it will not be difficult if they adopt the Grid architecture.

We assume a similar approach to that described in the last two sections for a typical Enterprise. Education is a very natural and important application of Grid technologies. Although 'Education Grids' are not particularly common, the ideas underlie many of the efforts in distance education such as those of the Department of the Defense with ADL (Advanced Distributed Learning ADL [20]) and the author's own research in this area [21, 22]. The Biology Workbench from NCSA and now SDSC [23] is a particular good early example of an Education and Research Grid for science. There are several other examples developed by NSF's EOT-PACI program [24]. Grids offer support of virtual organizations – and clearly the network of learners, teachers, mentors, parents, and administrators, that is, education form an interesting heterogeneous distributed virtual organization. Education has some special features of relevance to Grids. On the good (easy) side, education does not typically stress performance, as files tend to be of modest size, for even if one uses simulations to illustrate educational issues, these need not be of the highest resolution. Important timescales are illustrated by the 30 ms typical of an audio–video frame. Although this timescale is not in the microsecond range needed by parallel computing, quality of service is critical in education. Learning is hard and poor information delivery such as any distortion of audio packets (which only need some 10 Kb s^{-1} bandwidth) will render the learning environment unacceptable [25]. This is particularly relevant for so-called synchronous learning in which participants are linked in real time in an interactive session – such as a delivery of a class over the Internet with teacher and students in different locations. Although this case is important and should be supported by an education Grid, most technologies in this book are aimed at asynchronous learning. Resources (curriculum – lectures, homework and quizzes) are shared but not accessed simultaneously. Probably in terms of student time, asynchronous learning is nearly always dominant but in many education paradigms, the synchronous case is also essential and a distinctive requirement of an education Grid. One interesting feature of an education Grid is the richness of the (meta) data illustrated by the properties defined by IMS and ADL and by the many special Web services in Table 43.6. Consistent with the lack of emphasis on performance, education does not have individually huge data blocks but rather a myriad (as many students) of very rich XML structures. We can expect XML's natural support of complex objects to be more important in education than some other enterprises.

As mentioned, we will discuss an education Grid for science and engineering fields and adopt the hierarchical model used in Section 43.2. First, we assume that science and engineering will be implemented as Web services and in Table 43.4, give a few simple examples. Note that we will, for brevity, drop engineering in the following text and discuss science even though engineering has essentially identical considerations. Table 43.5 specializes to research, which corresponds to e-Science as discussed in several places in this book – especially Chapters 1, 7, 35 and 36. Table 43.6 lists a set of critical education Web services, which are applicable in many fields. Table 43.4 notes the special importance of mathematics and support of the natural topologies of science – two and three-dimensional spaces are dominant but more general cases must also be supported. Geographical Information Systems (GIS) as a Web service would support both educational curricula on the environment as well as the latest simulations of a new model for earthquake triggering. The general authoring Web service of Table 43.3 would need special extensions for science – in particular, to support mathematical notation as seen in most leading word processing systems today. The network server model of Chapters 24 and 25 (NetSolve and Ninf) is particularly appropriate for some generic science servers. The NEOS optimization resource at Argonne is a nice example of this type of service [26]. This of course developed a long time before Web services and illustrates that Web services are in many cases just following existing best practice. We illustrated the role of sensors in Section 43.2 and 'tenure evaluation' is listed to illustrate how general application Web services (in this case human resource service of Table 43.3) are specialized in particular domains.

Table 43.5 illustrates some of the Web services that are needed by e-Science. We have the suite of computational tools with a portal (controlling user-facing ports) front end described in Part C of the book, Chapters 20 to 34. Unlike education (Table 43.6), we often require the highest performance both in simulation and communication services. Virtual data described in Chapter 16 was developed to support research efforts with multiple data sources and multiple analysis efforts spread around the world – see Chapters 38 and 39. This concept will also be important in distance education in which one builds curricula and mentoring models with geographically distributed resources. For example, a student might take classes with grades from different organizations but these may need to be integrated to form learning plans. The near-term impact of the Grid will perhaps be greater in experimental and phenomenological fields (due to the data deluge of Chapter 36) than theoretical studies. Some of the needed experimental Web services are covered in Part D of the book, Chapters 35 to 43. However, support for theoreticians interacting at a distance has similarities to those needed for education. We mentioned the latter needed excellent quality of service. The same is even more true for 'e-Theory' as the latter must support unstructured interactions at any time – the structure (known schedules) in educational class delivery (and office hours) helps one improve quality of service with careful prelesson testing [25]. Publication services are an interesting area with some possibilities enabled by the Grid discussed in Chapter 36.

Dissemination and outreach for e-Science has close relations to the requirements of an education Grid. A speaker can give his seminar from afar using similar technologies to those needed by distance education. The Web service model has a general and possibility important implication for the ongoing search for better ways to integrate research and education. Thus, in e-Science, we are instructed to build each electronic science application as a network of interlocking, interacting Web services. Typically a 'leading edge' research 'Web service' (say a data analysis or simulation) cannot be easily used in education directly because its complexity often hides the 'essential features' and because it needs substantial resources. Deriving an educational version of a research tour de force is both time-consuming and not easy to 'automate'. The modularity of Web services offers an approach to this. In some cases, we can take a linked set of research Web services and 'just' modify a few of them to get an educational version. This requires thinking carefully through the Web service implementation to isolate complexity (which needs to be simplified for education or outreach) in just a few places.

As already mentioned, IMS and ADL have already defined many of the XML properties needed for education. This is illustrated by the list of available IMS specifications as of August 2002. These are [19]

- accessibility (universal access),
- competency definitions (grades, degrees, 'learning outcomes', skills, knowledge),
- content packaging (defining collections such as courses built of lectures),
- digital repositories (should reflect directly digital libraries and content management),
- enterprise (support people and groups such as organizations),
- learner information package (education record, resume etc. of people),
- metadata including dublin core bibliographical information [27],
- question & test (quizzes), and
- simple sequencing (of lesson components).

These are supported in IMS-compliant Learning management Systems (LMS) to provide the functionalities illustrated in Table 43.6. Essentially all these services are available in familiar systems, which are for the education academic environment Blackboard [28], and WebCT [29]. However, as we discussed, current education standards are not built with a service architecture and so an interoperable Web service Grid with components from different vendors is not possible – ADL [20] however, has tested interoperability within the current model successfully. We suggest that an important next step for the education community is to discuss a service architecture and agree on the needed Web service and Grid interoperable standards. We suggest these should not be built in isolation but rather adopt the hierarchical model described here so that, for instance, learning content management is built on broader commodity standards; further, perhaps aspects of quizzes should build on survey Web services and so on. An important complication of the Web service model is this linkage between the services of different systems with often the more specialized applications 'ahead' of the generic case. We need to develop good ways to cope with this.

Collaboration is a general Grid service of great importance in education. We stress the service model because as described in Reference [13], it is far clearer how to support collaboration for Web services than for general applications. The latter's state is defined by a complex mix of input information from files, user events and other programs. The state of a Web service is uniquely defined by its initial conditions and message-based input information – we ignore the subtle effects of different hosting environments that give different results from the same message-based information. Either the state defining or the user-facing port messages can be replicated (multicast) to give a collaborative Web service. There is no such simple strategy for a general application. Thus, we see significant changes if programs like Microsoft Word are in fact restructured as a Web service.

Currently, collaboration support falls into broad classes of products: instant messenger and other such tools from the major vendors such as AOL, Microsoft and Yahoo; audio- videoconferencing systems such as the Access Grid and Polycom [30] [31]; largely asynchronous peer-to-peer systems such as Groove Networks and JXTA [32, 33]; synchronous shared applications for 'Web conferencing' and virtual seminars and lectures from Centra, Placeware, WebEx, Anabas, Interwise and the public domain Virtual Network Computing (VNC) [34–39]. We can expect the capabilities of these systems to be 'unbundled' and built as Web services. For example, shared display is the most flexible shared application model and it is straightforward to build this as a Web service. Such a Web service would much more easily work with aggregation portals like Jetspeed [40] from Apache; it could link to the universal access Web service to customize the collaboration for different clients. The current clumsy integrations of collaboration systems with LMS would be simplified as we just need to know that LMS is a Web service and capture its input or output messages. We could hope that instant messengers would be integrated as another portlet in such a system; currently, they come from different vendors and can only be easily linked to a distance education session using intermediaries like that from Jabber [41].

We have suggested that education is an important focus area for the Grid. The Grid offers a new framework that can exploit the sophisticated existing Object API's from IMS [19] and ADL [20] to build a Web service environment that can better enable e-Education, which offers learners a much richer environment than available today. We expect education's rich XML structure to lead development of tools for handling distributed metadata of complex structure. We can also expect education to be a natural application for peer-to-peer Grids.

REFERENCES

1. Gannon, D., Ananthakrishnan, R., Krishnan, S., Govindaraju, M., Ramakrishnan, L. and Slominski, A. (2003) Grid web services and application factories, Chapter 9, in Berman, F., Fox, G. and Hey, T. (eds) *Grid Computing: Making the Global Infrastructure a Reality*. Chichester: John Wiley & Sons.
2. Pattnaik, P., Ekanadham, K. and Jann, J. (2003) Autonomic computing and the grid, Chapter 13, in Berman, F., Fox, G. and Hey, T. (eds) *Grid Computing: Making the Global Infrastructure a Reality*. Chichester: John Wiley & Sons.
3. Hoschek, W. (2003) Peer-to-peer grid databases for web service discovery, Chapter 19, in Berman, F., Fox, G. and Hey, T. (eds) *Grid Computing: Making the Global Infrastructure a Reality*. Chichester: John Wiley & Sons.
4. Fox, G. and Pallickara, S. (2003) NaradaBrokering: An event based infrastructure for building scaleable durable peer-to-peer grids, Chapter 22, in Berman, F., Fox, G. and Hey, T. (eds) *Grid Computing: Making the Global Infrastructure a Reality*. Chichester: John Wiley & Sons.
5. Foster, I., Kesselman, C. and Tuecke, S. (2003) Anatomy of the grid, Chapter 6, in Berman, F., Fox, G. and Hey, T. (eds) *Grid Computing: Making the Global Infrastructure a Reality*. Chichester: John Wiley & Sons.
6. Fox, G., Gannon, D. and Thomas, M. (2003) Overview of grid computing environments, Chapter 20, in Berman, F., Fox, G. and Hey, T. (eds) *Grid Computing: Making the Global Infrastructure a Reality*. Chichester: John Wiley & Sons.
7. IBM, Microsoft and BEA, *Business Process Execution Language for Web Services*, or BPEL4WS http://www-3.ibm.com/software/solutions/webservices/pr20020809.html, August 9, 2002.
8. Johnston, B. (2003) Implementing production grids, Chapter 5, in Berman F., Fox, G. and Hey, T. (eds) *Grid Computing: Making the Global Infrastructure a Reality*. Chichester: John Wiley & Sons.
9. Thain, D., Tannenbaum, T. and Livny, M. (2003) Condor and the grid, Chapter 11, in Berman, F., Fox, G. and Hey, T. (eds) *Grid Computing: Making the Global Infrastructure a Reality*. Chichester: John Wiley & Sons.
10. Watson, P. (2003) Databases and the grid, Chapter 14, in Berman, F., Fox, G. and Hey, T. (eds) *Grid Computing: Making the Global Infrastructure a Reality*. Chichester: John Wiley & Sons.
11. Kunszt, P. Z. and Guy, L. P. (2003) The Open Grid Services Architecture and data grids, Chapter 15, in Berman, F., Fox, G. and Hey, T. (eds) *Grid Computing: Making the Global Infrastructure a Reality*. Chichester: John Wiley & Sons.
12. Moore, R. and Baru, C. (2003) Virtualization services for data grids, Chapter 16, in Berman, F., Fox, G. and Hey, T. (eds) *Grid Computing: Making the Global Infrastructure a Reality*. Chichester: John Wiley & Sons.
13. Fox, G. *et al.* (2003) Peer-to-peer grids, Chapter 18, in Berman, F., Fox, G. and Hey, T. (eds) *Grid Computing: Making the Global Infrastructure a Reality*. Chichester: John Wiley & Sons.
14. De Roure, D., Jennings, N. and Shadbolt, N. (2003) The semantic grid: a future e-Science infrastructure, Chapter 17, in Berman, F., Fox, G. and Hey, T. (eds) *Grid Computing: Making the Global Infrastructure a Reality*. Chichester: John Wiley & Sons.
15. Web Services Description Language, http://www.w3.org/TR/wsdl.
16. Universal Description, Discovery and Integration (UDDI) Project, http://www.uddi.org/.
17. Castor open source data binding framework for Java, http://castor.exolab.org/.
18. Simple Object Access Protocol (SOAP), http://www.w3.org/TR/SOAP/.
19. Instructional Management Systems (IMS), http://www.imsproject.org.
20. Advanced Distributed Learning Initiative (ADL), http://www.adlnet.org.
21. Fox, G. C. (2002) From computational science to internetics: integration of science with computer science, in Boisvert, R. F. and Houstis, E. (eds) *Computational Science, Mathematics and Software*. ISBN 1-55753-250-8, West Lafayette, Indiana: Purdue University Press, pp. 217–236, http://grids.ucs.indiana.edu/ptliupages/publications/Internetics2.pdf.

22. Fox, G. Experience with Distance Education 1998–2002, http://grids.ucs.indiana.edu/ptliupages/publications/disted/.
23. Biology Workbench at SDSC (San Diego Supercomputer Center), http://workbench.sdsc.edu/.
24. NSF PACI (Partnership in Advanced Computing Infrastructure), EOT (Education and Outreach) Program, http://www.eot.org.
25. Bernholdt, D. E., Fox, G. C., McCracken, N. J., Markowski, R. and Podgorny, M. (2000) *Reflections on Three Years of Network-Based Distance Education*, Unpublished Report for US Army Corp of Engineers ERDC Vicksburg Miss, July 2000, http://grids.ucs.indiana.edu/ptliupages/publications/disted/erdctraining00.pdf.
26. NEOS Optimization Server from Argonne National Laboratory, http://www-neos.mcs.anl.gov/neos/.
27. The Dublin Core Bibliographic Meta Data, http://dublincore.org/.
28. Blackboard Learning System, http://www.blackboard.com/.
29. WebCT Learning System, http://www.webct.com/.
30. Access Grid Conferencing Environment from Argonne National Laboratory, http://www.accessgrid.org.
31. Polycom Conferencing Environment, http://www.polycom.com.
32. Groove Desktop Collaboration Software, http://www.groove.net/.
33. JXTA Peer to Peer environment from Sun Microsystems http://www.jxta.org.
34. Centra Collaboration Environment, http://www.centra.com.
35. Placeware Collaboration Environment, http://www.placeware.com.
36. WebEx Collaboration Environment, http://www.webex.com.
37. Anabas Collaboration Environment, http://www.anabas.com.
38. Interwise Enterprise Communications Platform, http://www.interwise.com.
39. Virtual Network Computing System (VNC), http://www.uk.research.att.com/vnc.
40. Jetspeed Enterprise Portal from Apache, http://jakarta.apache.org/jetspeed/.
41. Jabber Instant Messenger, http://www.jabber.org/.

Index

INTRODUCTION TO INDEX SECTION

There are three resources here. Firstly there is a traditional index, where we have spelt out essentially all abbreviations to clarify the acronym-soup characterizing the Grid arena. We also note the index entry 'Applications' listing many different fields discussed in the book; the index entry 'Application Classification' pointing to parts of the book discussing categories of applications and their Grid implementation. There are of course many topics under the 'Grid' entry but we highlight 'Grid thought (correctly or incorrectly) of as' which links to parts of the book discussing Grids from different points of view.

The second resource is an 'Indirect Glossary'; rather than give a traditional glossary, we note that many parts of the book summarize key Grid-related topics; we list some of these here.

The final resource is a list of over 70 different Grid projects around the world. This is but a partial summary and we have only listed broad-based activities but even here choice of entries is quite subjective.

We commend the reader to two Web sites; *http://www.grid2002.org* contains up-to-date summary material for the book and will accumulate instructional resources as it gets used in courses; *http://www.gridpp.ac.uk/docs/GAS.html* gives detail on the Grid Acronym Soup.

Grid Computing – Making the Global Infrastructure a Reality. Edited by F. Berman, A. Hey and G. Fox
© 2003 John Wiley & Sons, Ltd   ISBN: 0-470-85319-0

VIEWS OF THE GRID

Here we collect links to parts of the book that describe views of Grid–what it is and what it isn't:

Advanced Networking (Chapter 6)

Artificial Intelligence (Chapter 6)

Architecture as a Protocol architecture (Chapter 6)

Autonomic System (Chapter 13.4)

Collaborative Environment (Chapter 2, Section 2.2; Chapter 18)

Combining powerful resources, federated computing and a security structure (Chapter 38, Section 38.2)

Coordinated resource sharing and problem solving in dynamic multi-institutional virtual organizations (Chapter 6)

Data Grids as Managed Distributed Systems for Global Virtual Organizations (Chapter 39)

Distributed Computing or distributed systems (Chapter 2, Section 2.2; Chapter 10)

Enabling Scalable Virtual Organizations (Chapter 6)

Enabling use of enterprise-wide systems, and someday nationwide systems, that consist of workstations, vector supercomputers, and parallel supercomputers connected by local and wide area networks. Users will be presented the illusion of a single, very powerful computer, rather than a collection of disparate machines. The system will schedule application components on processors, manage data transfer, and provide communication and synchronization in such a manner as to dramatically improve application performance. Further, boundaries between computers will be invisible, as will the location of data and the failure of processors. (Chapter 10)

GCE (Grid Computing Environment) as an integrated set of tools that extends the user's computing environment in order to provide access to Grid Services (Chapter 26)

Grid portal as an effective tool for providing users of computational Grids with simple intuitive interfaces for accessing Grid information and for using Grid resources (Chapter 28)

  as a specialized portal providing an entry point to the Grid to access applications, services, information, and data available within a Grid. (Chapter 26)

  as a community service with a single point of entry to an integrated system providing access to information, data, applications, and services. (Chapter 26)

  as an infrastructure that will provide us with the ability to dynamically link together resources as an ensemble to support the execution of large-scale, resource-intensive, and distributed applications. (Chapter 1)

  makes high-performance computers superfluous (Chapter 6)

  metasystems or metacomputing systems (Chapters 10, 37)

Middleware as the services needed to support a common set of applications in a distributed network environment (Chapter 6)

Next-Generation Internet (Chapter 6)

Peer-to-peer network or peer-to-peer systems (Chapters 10, 18)

INDIRECT GLOSSARY

The following book links provide succinct description of many key Grid-related concepts

List of Grid Projects

Grid Project	Web Link	Topic	Affiliation
ACI Grid	http://www.inria.fr/presse/dossier/gridcomputing/grid2b.en.html	French Grid (INRIA)	France
Advanced Knowledge Technologies (AKT)	http://www.aktors.org/	Management of the knowledge life cycle	UK e-Science IRC (Interdisciplinary Research Collaboration)
Akenti	http://www-itg.lbl.gov/Akenti/homepage.html	Distributed Access Control System	US (Lawrence Berkeley Laboratory)
APGrid	http://www.apgrid.org/	Asia-Pacific Grid	Asia
AstroGrid	http://www.astrogrid.org/	Astronomy	UK
AVO (Astronomical Virtual Observatory)	http://www.euro-vo.org/	Astronomy	Europe
BIRN	http://birn.ncrr.nih.gov/, http://www.nbirn.net/	Biomedical Informatics	US
Budapest Open Access Initiative	http://www.soros.org/openaccess/	Initiative aiming at open access to research articles	International
Butterfly Grid	http://www.butterfly.net/	Online games	Industry
CLIMDB	http://www.fsl.orst.edu/climdb/	Climate Database Project	US
Comb-e-chem	http://www.combechem.org/	Combinatorial Chemistry	UK e-Science Pilot
Condor-G	http://www.cs.wisc.edu/condor/	Grid Resource Management for High throughput computing	UK e-Science Pilot
CrossGrid	http://www.crossgrid.org	Exploitation of Grid concepts for interactive compute and data intensive applications	Europe
DAME	http://www.cs.york.ac.uk/dame/	Distributed Aircraft Maintenance, Engine diagnostics	UK e-Science Pilot

(continued)

(continued)

Grid Project	Web Link	Topic	Affiliation
SRB (Storage Resource Broker)	http://www.npaci.edu/DICE/SRB/	Data Grid technology	US
TeraGrid	http://www.teragrid.org/	Grid linking major NSF Supercomputer Centers	US
UK e-Science	http://www.research-councils.ac.uk/escience/	Particle physics, astronomy, engineering, medicine, bio-informatics	UK e-Science
UK Grid Support Center	http://www.grid-support.ac.uk/	Support UK e-Science Grid	UK e-Science
UNICORE	http://www.unicore.de/	Grid access Portal	Germany
VLAM	http://www.dutchgrid.nl/VLAM-G/AM/doclive-vlam.html	Virtual laboratory for experimental physics	Netherlands

WILEY SERIES ON PARALLEL AND DISTRIBUTED COMPUTING

SERIES EDITOR: Albert Y. Zomaya